LINEAR MODELS IN STATISTICS

THE WILEY BICENTENNIAL–KNOWLEDGE FOR GENERATIONS

Each generation has its unique needs and aspirations. When Charles Wiley first opened his small printing shop in lower Manhattan in 1807, it was a generation of boundless potential searching for an identity. And we were there, helping to define a new American literary tradition. Over half a century later, in the midst of the Second Industrial Revolution, it was a generation focused on building the future. Once again, we were there, supplying the critical scientific, technical, and engineering knowledge that helped frame the world. Throughout the 20th Century, and into the new millennium, nations began to reach out beyond their own borders and a new international community was born. Wiley was there, expanding its operations around the world to enable a global exchange of ideas, opinions, and know-how.

For 200 years, Wiley has been an integral part of each generation's journey, enabling the flow of information and understanding necessary to meet their needs and fulfill their aspirations. Today, bold new technologies are changing the way we live and learn. Wiley will be there, providing you the must-have knowledge you need to imagine new worlds, new possibilities, and new opportunities.

Generations come and go, but you can always count on Wiley to provide you the knowledge you need, when and where you need it!

WILLIAM J. PESCE
PRESIDENT AND CHIEF EXECUTIVE OFFICER

PETER BOOTH WILEY
CHAIRMAN OF THE BOARD

LINEAR MODELS IN STATISTICS

Second Edition

Alvin C. Rencher and G. Bruce Schaalje

Department of Statistics, Brigham Young University, Provo, Utah

WILEY-INTERSCIENCE

A JOHN WILEY & SONS, INC., PUBLICATION

Published by John Wiley & Sons, Inc., Hoboken, New Jersey
Published simultaneously in Canada

Wiley Bicentennial Logo: Richard J. Pacifico

Library of Congress Cataloging-in-Publication Data:
Rencher, Alvin C., 1934-
Linear models in statistics/Alvin C. Rencher, G. Bruce Schaalje. – 2nd ed.
p. cm.
Includes bibliographical references.
ISBN 978-0-471-75498-5 (cloth)
1. Linear models (Statistics) I. Schaalje, G. Bruce. II. Title.
QA276.R425 2007
519.5′35–dc22

2007024268

Printed in the United States of America

10 9 8 7 6 5 4 3 2 1

CONTENTS

PREFACE

In the second edition, we have added chapters on Bayesian inference in linear models (Chapter 11) and linear mixed models (Chapter 17), and have upgraded the material in all other chapters. Our continuing objective has been to introduce the theory of linear models in a clear but rigorous format.

In spite of the availability of highly innovative tools in statistics, the main tool of the applied statistician remains the linear model. The linear model involves the simplest and seemingly most restrictive statistical properties: independence, normality, constancy of variance, and linearity. However, the model and the statistical methods associated with it are surprisingly versatile and robust. More importantly, mastery of the linear model is a prerequisite to work with advanced statistical tools because most advanced tools are generalizations of the linear model. The linear model is thus central to the training of any statistician, applied or theoretical.

This book develops the basic theory of linear models for regression, analysis-of-variance, analysis–of–covariance, and linear mixed models. Chapter 18 briefly introduces logistic regression, generalized linear models, and nonlinear models. Applications are illustrated by examples and problems using real data. This combination of theory and applications will prepare the reader to further explore the literature and to more correctly interpret the output from a linear models computer package.

This introductory linear models book is designed primarily for a one-semester course for advanced undergraduates or MS students. It includes more material than can be covered in one semester so as to give an instructor a choice of topics and to serve as a reference book for researchers who wish to gain a better understanding of regression and analysis-of-variance. The book would also serve well as a text for PhD classes in which the instructor is looking for a one-semester introduction, and it would be a good supplementary text or reference for a more advanced PhD class for which the students need to review the basics on their own.

Our overriding objective in the preparation of this book has been *clarity of exposition*. We hope that students, instructors, researchers, and practitioners will find this linear models text more comfortable than most. In the final stages of development, we asked students for written comments as they read each day's assignment. They made many suggestions that led to improvements in readability of the book. We are grateful to readers who have notified us of errors and other suggestions for improvements of the text, and we will continue to be very grateful to readers who take the time to do so for this second edition.

Another objective of the book is to tie up loose ends. There are many approaches to teaching regression, for example. Some books present estimation of regression coefficients for fixed x's only, other books use random x's, some use centered models, and others define estimated regression coefficients in terms of variances and covariances or in terms of correlations. Theory for linear models has been presented using both an algebraic and a geometric approach. Many books present classical (frequentist) inference for linear models, while increasingly the Bayesian approach is presented. We have tried to cover all these approaches carefully and to show how they relate to each other. We have attempted to do something similar for various approaches to analysis-of-variance. We believe that this will make the book useful as a reference as well as a textbook. An instructor can choose the approach he or she prefers, and a student or researcher has access to other methods as well.

The book includes a large number of theoretical problems and a smaller number of applied problems using real datasets. The problems, along with the extensive set of answers in Appendix A, extend the book in two significant ways: (1) the theoretical problems and answers fill in nearly all gaps in derivations and proofs and also extend the coverage of material in the text, and (2) the applied problems and answers become additional examples illustrating the theory. As instructors, we find that having answers available for the students saves a great deal of class time and enables us to cover more material and cover it better. The answers would be especially useful to a reader who is engaging this material outside the formal classroom setting.

The mathematical prerequisites for this book are multivariable calculus and matrix algebra. The review of matrix algebra in Chapter 2 is intended to be sufficiently complete so that the reader with no previous experience can master matrix manipulation up to the level required in this book. Statistical prerequisites include some exposure to statistical theory, with coverage of topics such as distributions of random variables, expected values, moment generating functions, and an introduction to estimation and testing hypotheses. These topics are briefly reviewed as each is introduced. One or two statistical methods courses would also be helpful, with coverage of topics such as t tests, regression, and analysis-of-variance.

We have made considerable effort to maintain consistency of notation throughout the book. We have also attempted to employ standard notation as far as possible and to avoid exotic characters that cannot be readily reproduced on the chalkboard. With a few exceptions, we have refrained from the use of abbreviations and mnemonic devices. We often find these annoying in a book or journal article.

Equations are numbered sequentially throughout each chapter; for example, (3.29) indicates the twenty-ninth numbered equation in Chapter 3. Tables and figures are also numbered sequentially throughout each chapter in the form "Table 7.4" or "Figure 3.2." On the other hand, examples and theorems are numbered sequentially within a section, for example, Theorems 2.2a and 2.2b.

The solution of most of the problems with real datasets requires the use of the computer. We have not discussed command files or output of any particular program, because there are so many good packages available. Computations for the numerical examples and numerical problems were done with SAS. The datasets and SAS command files for all the numerical examples and problems in the text are available on the Internet; see Appendix B.

The references list is not intended to be an exhaustive survey of the literature. We have provided original references for some of the basic results in linear models and have also referred the reader to many up-to-date texts and reference books useful for further reading. When citing references in the text, we have used the standard format involving the year of publication. For journal articles, the year alone suffices, for example, Fisher (1921). But for a specific reference in a book, we have included a page number or section, as in Hocking (1996, p. 216).

Our selection of topics is intended to prepare the reader for a better understanding of applications and for further reading in topics such as mixed models, generalized linear models, and Bayesian models. Following a brief introduction in Chapter 1, Chapter 2 contains a careful review of all aspects of matrix algebra needed to read the book. Chapters 3, 4, and 5 cover properties of random vectors, matrices, and quadratic forms. Chapters 6, 7, and 8 cover simple and multiple linear regression, including estimation and testing hypotheses and consequences of misspecification of the model. Chapter 9 provides diagnostics for model validation and detection of influential observations. Chapter 10 treats multiple regression with random x's. Chapter 11 covers Bayesian multiple linear regression models along with Bayesian inferences based on those models. Chapter 12 covers the basic theory of analysis-of-variance models, including estimability and testability for the overparameterized model, reparameterization, and the imposition of side conditions. Chapters 13 and 14 cover balanced one-way and two-way analysis-of-variance models using an overparameterized model. Chapter 15 covers unbalanced analysis-of-variance models using a cell means model, including a section on dealing with empty cells in two-way analysis-of-variance. Chapter 16 covers analysis of covariance models. Chapter 17 covers the basic theory of linear mixed models, including residual maximum likelihood estimation of variance components, approximate small-sample inferences for fixed effects, best linear unbiased prediction of random effects, and residual analysis. Chapter 18 introduces additional topics such as nonlinear regression, logistic regression, loglinear models, Poisson regression, and generalized linear models.

In our class for first-year master's-level students, we cover most of the material in Chapters 2–5, 7–8, 10–12, and 17. Many other sequences are possible. For example, a thorough one-semester regression and analysis-of-variance course could cover Chapters 1–10, and 12–15.

Al's introduction to linear models came in classes taught by Dale Richards and Rolf Bargmann. He also learned much from the books by Graybill, Scheffé, and Rao. Al expresses thanks to the following for reading the first edition manuscript and making many valuable suggestions: David Turner, John Walker, Joel Reynolds, and Gale Rex Bryce. Al thanks the following students at Brigham Young University (BYU) who helped with computations, graphics, and typing of the first edition: David Fillmore, Candace Baker, Scott Curtis, Douglas Burton, David Dahl, Brenda Price, Eric Hintze, James Liechty, and Joy Willbur. The students in Al's Linear Models class went through the manuscript carefully and spotted many typographical errors and passages that needed additional clarification.

Bruce's education in linear models came in classes taught by Mel Carter, Del Scott, Doug Martin, Peter Bloomfield, and Francis Giesbrecht, and influential short courses taught by John Nelder and Russ Wolfinger.

We thank Bruce's Linear Models classes of 2006 and 2007 for going through the book and new chapters. They made valuable suggestions for improvement of the text. We thank Paul Martin and James Hattaway for invaluable help with LaTex. The Department of Statistics, Brigham Young University provided financial support and encouragement throughout the project.

Second Edition

For the second edition we added Chapter 11 on Bayesian inference in linear models (including Gibbs sampling) and Chapter 17 on linear mixed models.

We also added a section in Chapter 2 on vector and matrix calculus, adding several new theorems and covering the Lagrange multiplier method. In Chapter 4, we presented a new proof of the conditional distribution of a subvector of a multivariate normal vector. In Chapter 5, we provided proofs of the moment generating function and variance of a quadratic form of a multivariate normal vector. The section on the geometry of least squares was completely rewritten in Chapter 7, and a section on the geometry of least squares in the overparameterized linear model was added to Chapter 12. Chapter 8 was revised to provide more motivation for hypothesis testing and simultaneous inference. A new section was added to Chapter 15 dealing with two-way analysis-of-variance when there are empty cells. This material is not available in any other textbook that we are aware of.

This book would not have been possible without the patience, support, and encouragement of Al's wife LaRue and Bruce's wife Lois. Both have helped and supported us in more ways than they know. This book is dedicated to them.

ALVIN C. RENCHER AND G. BRUCE SCHAALJE

Department of Statistics
Brigham Young University
Provo, Utah

1 Introduction

The scientific method is frequently used as a guided approach to learning. Linear statistical methods are widely used as part of this learning process. In the biological, physical, and social sciences, as well as in business and engineering, linear models are useful in both the planning stages of research and analysis of the resulting data. In Sections 1.1–1.3, we give a brief introduction to simple and multiple linear regression models, and analysis-of-variance (ANOVA) models.

1.1 SIMPLE LINEAR REGRESSION MODEL

In simple linear regression, we attempt to model the relationship between two variables, for example, income and number of years of education, height and weight of people, length and width of envelopes, temperature and output of an industrial process, altitude and boiling point of water, or dose of a drug and response. For a linear relationship, we can use a model of the form

$$y = \beta_0 + \beta_1 x + \varepsilon, \tag{1.1}$$

where y is the *dependent* or *response* variable and x is the *independent* or *predictor* variable. The random variable ε is the error term in the model. In this context, *error* does not mean mistake but is a statistical term representing random fluctuations, measurement errors, or the effect of factors outside of our control.

The linearity of the model in (1.1) is an assumption. We typically add other assumptions about the distribution of the error terms, independence of the observed values of y, and so on. Using observed values of x and y, we estimate β_0 and β_1 and make inferences such as confidence intervals and tests of hypotheses for β_0 and β_1. We may also use the estimated model to forecast or predict the value of y for a particular value of x, in which case a measure of predictive accuracy may also be of interest.

Estimation and inferential procedures for the simple linear regression model are developed and illustrated in Chapter 6.

Linear Models in Statistics, Second Edition, by Alvin C. Rencher and G. Bruce Schaalje

1.2 MULTIPLE LINEAR REGRESSION MODEL

The response y is often influenced by more than one predictor variable. For example, the yield of a crop may depend on the amount of nitrogen, potash, and phosphate fertilizers used. These variables are controlled by the experimenter, but the yield may also depend on uncontrollable variables such as those associated with weather.

A linear model relating the response y to several predictors has the form

$$y = \beta_0 + \beta_1 x_1 + \beta_2 x_2 + \cdots + \beta_k x_k + \varepsilon. \tag{1.2}$$

The parameters $\beta_0, \beta_1, \ldots, \beta_k$ are called *regression coefficients*. As in (1.1), ε provides for random variation in y not explained by the x variables. This random variation may be due partly to other variables that affect y but are not known or not observed.

The model in (1.2) is linear in the β parameters; it is not necessarily linear in the x variables. Thus models such as

$$y = \beta_0 + \beta_1 x_1 + \beta_2 x_1^2 + \beta_3 x_2 + \beta_4 \sin x_2 + \varepsilon$$

are included in the designation *linear model*.

A model provides a theoretical framework for better understanding of a phenomenon of interest. Thus a model is a mathematical construct that we believe may represent the mechanism that generated the observations at hand. The postulated model may be an idealized oversimplification of the complex real-world situation, but in many such cases, empirical models provide useful approximations of the relationships among variables. These relationships may be either associative or causative.

Regression models such as (1.2) are used for various purposes, including the following:

1. *Prediction*. Estimates of the individual parameters $\beta_0, \beta_1, \ldots, \beta_k$ are of less importance for prediction than the overall influence of the x variables on y. However, good estimates are needed to achieve good prediction performance.
2. *Data Description or Explanation*. The scientist or engineer uses the estimated model to summarize or describe the observed data.
3. *Parameter Estimation*. The values of the estimated parameters may have theoretical implications for a postulated model.
4. *Variable Selection or Screening*. The emphasis is on determining the importance of each predictor variable in modeling the variation in y. The predictors that are associated with an important amount of variation in y are retained; those that contribute little are deleted.
5. *Control of Output*. A cause-and-effect relationship between y and the x variables is assumed. The estimated model might then be used to control the

output of a process by varying the inputs. By systematic experimentation, it may be possible to achieve the optimal output.

There is a fundamental difference between purposes 1 and 5. For prediction, we need only assume that the same correlations that prevailed when the data were collected also continue in place when the predictions are to be made. Showing that there is a significant relationship between y and the x variables in (1.2) does not necessarily prove that the relationship is causal. To establish causality in order to control output, the researcher must choose the values of the x variables in the model and use randomization to avoid the effects of other possible variables unaccounted for. In other words, to ascertain the effect of the x variables on y when the x variables are changed, it is necessary to change them.

Estimation and inferential procedures that contribute to the five purposes listed above are discussed in Chapters 7–11.

1.3 ANALYSIS-OF-VARIANCE MODELS

In analysis-of-variance (ANOVA) models, we are interested in comparing several populations or several conditions in a study. Analysis-of-variance models can be expressed as linear models with restrictions on the x values. Typically the x's are 0s or 1s. For example, suppose that a researcher wishes to compare the mean yield for four types of catalyst in an industrial process. If n observations are to be obtained for each catalyst, one model for the $4n$ observations can be expressed as

$$y_{ij} = \mu_i + \varepsilon_{ij}, \quad i = 1, 2, 3, 4, \quad j = 1, 2, \ldots, n, \tag{1.3}$$

where μ_i is the mean corresponding to the ith catalyst. A hypothesis of interest is $H_0 : \mu_1 = \mu_2 = \mu_3 = \mu_4$. The model in (1.3) can be expressed in the alternative form

$$y_{ij} = \mu + \alpha_i + \varepsilon_{ij}, \quad i = 1, 2, 3, 4, \quad j = 1, 2, \ldots, n. \tag{1.4}$$

In this form, α_i is the effect of the ith catalyst, and the hypothesis can be expressed as $H_0 : \alpha_1 = \alpha_2 = \alpha_3 = \alpha_4$.

Suppose that the researcher also wishes to compare the effects of three levels of temperature and that n observations are taken at each of the 12 catalyst–temperature combinations. Then the model can be expressed as

$$\begin{aligned} y_{ijk} &= \mu_{ij} + \varepsilon_{ijk} = \mu + \alpha_i + \beta_j + \gamma_{ij} + \varepsilon_{ijk} \\ & i = 1, 2, 3, 4; \quad j = 1, 2, 3; \quad k = 1, 2, \ldots, n, \end{aligned} \tag{1.5}$$

where μ_{ij} is the mean for the ijth catalyst–temperature combination, α_i is the effect of the ith catalyst, β_j is the effect of the jth level of temperature, and γ_{ij} is the interaction or joint effect of the ith catalyst and jth level of temperature.

In the examples leading to models (1.3)–(1.5), the researcher chooses the type of catalyst or level of temperature and thus applies different *treatments* to the objects or experimental units under study. In other settings, we compare the means of variables measured on natural groupings of units, for example, males and females or various geographic areas.

Analysis-of-variance models can be treated as a special case of regression models, but it is more convenient to analyze them separately. This is done in Chapters 12–15. Related topics, such as analysis-of-covariance and mixed models, are covered in Chapters 16–17.

2 Matrix Algebra

If we write a linear model such as (1.2) for each of n observations in a dataset, the n resulting models can be expressed in a single compact matrix expression. Then the estimation and testing results can be more easily obtained using matrix theory.

In the present chapter, we review the elements of matrix theory needed in the remainder of the book. Proofs that seem instructive are included or called for in the problems. For other proofs, see Graybill (1969), Searle (1982), Harville (1997), Schott (1997), or any general text on matrix theory. We begin with some basic definitions in Section 2.1.

2.1 MATRIX AND VECTOR NOTATION

2.1.1 Matrices, Vectors, and Scalars

A *matrix* is a rectangular or square array of numbers or variables. We use uppercase boldface letters to represent matrices. In this book, all elements of matrices will be real numbers or variables representing real numbers. For example, the height (in inches) and weight (in pounds) for three students are listed in the following matrix:

$$\mathbf{A} = \begin{pmatrix} 65 & 154 \\ 73 & 182 \\ 68 & 167 \end{pmatrix}. \tag{2.1}$$

To represent the elements of $\mathbf{A}$ as variables, we use

$$\mathbf{A} = (a_{ij}) = \begin{pmatrix} a_{11} & a_{12} \\ a_{21} & a_{22} \\ a_{31} & a_{32} \end{pmatrix}. \tag{2.2}$$

The first subscript in a_{ij} indicates the row; the second identifies the column. The notation $\mathbf{A} = (a_{ij})$ represents a matrix by means of a typical element.

The matrix $\mathbf{A}$ in (2.1) or (2.2) has three rows and two columns, and we say that $\mathbf{A}$ is 3×2, or that the *size* of $\mathbf{A}$ is 3×2.

A *vector* is a matrix with a single row or column. Elements in a vector are often identified by a single subscript; for example

$$\mathbf{x} = \begin{pmatrix} x_1 \\ x_2 \\ x_3 \end{pmatrix}.$$

As a convention, we use lowercase boldface letters for column vectors and lowercase boldface letters followed by the prime symbol ($'$) for row vectors; for example

$$\mathbf{x}' = (x_1, x_2, x_3) = (x_1 \; x_2 \; x_3).$$

(Row vectors are regarded as *transposes* of column vectors. The transpose is defined in Section 2.1.3 below). We use either commas or spaces to separate elements of a row vector.

Geometrically, a row or column vector with p elements can be associated with a point in a p-dimensional space. The elements in the vector are the coordinates of the point. Sometimes we are interested in the distance from the origin to the point (vector), the distance between two points (vectors), or the angle between the arrows drawn from the origin to the two points.

In the context of matrices and vectors, a single real number is called a *scalar*. Thus 2.5, -9, and 7.26 are scalars. A variable representing a scalar will be denoted by a lightface letter (usually lowercase), such as c. A scalar is technically distinct from a 1×1 matrix in terms of its uses and properties in matrix algebra. The same notation is often used to represent a scalar and a 1×1 matrix, but the meaning is usually obvious from the context.

2.1.2 Matrix Equality

Two matrices or two vectors are equal if they are of the same size and if the elements in corresponding positions are equal; for example

$$\begin{pmatrix} 3 & -2 & 4 \\ 1 & 3 & 7 \end{pmatrix} = \begin{pmatrix} 3 & -2 & 4 \\ 1 & 3 & 7 \end{pmatrix},$$

but

$$\begin{pmatrix} 5 & 2 & -9 \\ 8 & -4 & 6 \end{pmatrix} \neq \begin{pmatrix} 5 & 3 & -9 \\ 8 & -4 & 6 \end{pmatrix}.$$

2.1.3 Transpose

If we interchange the rows and columns of a matrix $\mathbf{A}$, the resulting matrix is known as the *transpose* of $\mathbf{A}$ and is denoted by $\mathbf{A}'$; for example

$$\mathbf{A} = \begin{pmatrix} 6 & -2 \\ 4 & 7 \\ 1 & 3 \end{pmatrix}, \qquad \mathbf{A}' = \begin{pmatrix} 6 & 4 & 1 \\ -2 & 7 & 3 \end{pmatrix}.$$

Formally, if $\mathbf{A}$ is denoted by $\mathbf{A} = (a_{ij})$, then $\mathbf{A}'$ is defined as

$$\mathbf{A}' = (a_{ij})' = (a_{ji}). \tag{2.3}$$

This notation indicates that the element in the ith row and jth column of $\mathbf{A}$ is found in the jth row and ith column of $\mathbf{A}'$. If the matrix $\mathbf{A}$ is $n \times p$, then $\mathbf{A}'$ is $p \times n$.

If a matrix is transposed twice, the result is the original matrix.

Theorem 2.1. If $\mathbf{A}$ is any matrix, then

$$(\mathbf{A}')' = \mathbf{A}. \tag{2.4}$$

PROOF. By (2.3), $\mathbf{A}' = (a_{ij})' = (a_{ji})$. Then $(\mathbf{A}')' = (a_{ji})' = (a_{ij}) = \mathbf{A}$. □
(The notation □ is used to indicate the end of a theorem proof, corollary proof or example.)

2.1.4 Matrices of Special Form

If the transpose of a matrix $\mathbf{A}$ is the same as the original matrix, that is, if $\mathbf{A}' = \mathbf{A}$ or equivalently $(a_{ji}) = (a_{ij})$, then the matrix $\mathbf{A}$ is said to be *symmetric*. For example

$$\mathbf{A} = \begin{pmatrix} 3 & 2 & 6 \\ 2 & 10 & -7 \\ 6 & -7 & 9 \end{pmatrix}$$

is symmetric. Clearly, all symmetric matrices are square.

$$\mathbf{A} = \begin{pmatrix} a_{11} & a_{12} & \cdots & a_{1p} \\ a_{21} & a_{22} & \cdots & a_{2p} \\ \vdots & \vdots & & \vdots \\ a_{p1} & a_{22} & \cdots & a_{pp} \end{pmatrix}$$

The *diagonal* of a $p \times p$ square matrix $\mathbf{A} = (a_{ij})$ consists of the elements $a_{11}, a_{22}, \ldots, a_{pp}$. If a matrix contains zeros in all off-diagonal positions, it is said to be a *diagonal matrix*; for example, consider the matrix

$$\mathbf{D} = \begin{pmatrix} 8 & 0 & 0 & 0 \\ 0 & -3 & 0 & 0 \\ 0 & 0 & 0 & 0 \\ 0 & 0 & 0 & 4 \end{pmatrix},$$

which can also be denoted as

$$\mathbf{D} = \text{diag}(8, \ -3, 0, \ 4).$$

We also use the notation diag($\mathbf{A}$) to indicate a diagonal matrix with the same diagonal elements as $\mathbf{A}$; for example

$$\mathbf{A} = \begin{pmatrix} 3 & 2 & 6 \\ 2 & 10 & -7 \\ 6 & -7 & 9 \end{pmatrix}, \quad \text{diag}(\mathbf{A}) = \begin{pmatrix} 3 & 0 & 0 \\ 0 & 10 & 0 \\ 0 & 0 & 9 \end{pmatrix}.$$

A diagonal matrix with a 1 in each diagonal position is called an *identity* matrix, and is denoted by $\mathbf{I}$; for example

$$\mathbf{I} = \begin{pmatrix} 1 & 0 & 0 \\ 0 & 1 & 0 \\ 0 & 0 & 1 \end{pmatrix}. \tag{2.5}$$

An *upper triangular matrix* is a square matrix with zeros below the diagonal; for example,

$$\mathbf{T} = \begin{pmatrix} 7 & 2 & 3 & -5 \\ 0 & 0 & -2 & 6 \\ 0 & 0 & 4 & 1 \\ 0 & 0 & 0 & 8 \end{pmatrix}.$$

A *lower triangular matrix* is defined similarly.

A vector of 1s is denoted by $\mathbf{j}$:

$$\mathbf{j} = \begin{pmatrix} 1 \\ 1 \\ \vdots \\ 1 \end{pmatrix}. \tag{2.6}$$

A square matrix of 1s is denoted by $\mathbf{J}$; for example

$$\mathbf{J} = \begin{pmatrix} 1 & 1 & 1 \\ 1 & 1 & 1 \\ 1 & 1 & 1 \end{pmatrix}. \tag{2.7}$$

We denote a vector of zeros by $\mathbf{0}$ and a matrix of zeros by $\mathbf{O}$; for example

$$\mathbf{0} = \begin{pmatrix} 0 \\ 0 \\ 0 \end{pmatrix}, \qquad \mathbf{O} = \begin{pmatrix} 0 & 0 & 0 & 0 \\ 0 & 0 & 0 & 0 \\ 0 & 0 & 0 & 0 \end{pmatrix}. \tag{2.8}$$

2.2 OPERATIONS

We now define sums and products of matrices and vectors and consider some properties of these sums and products.

2.2.1 Sum of Two Matrices or Two Vectors

If two matrices or two vectors are the same size, they are said to be *conformal for addition*. Their *sum* is found by adding corresponding elements. Thus, if $\mathbf{A}$ is $n \times p$ and $\mathbf{B}$ is $n \times p$, then $\mathbf{C} = \mathbf{A} + \mathbf{B}$ is also $n \times p$ and is found as $\mathbf{C} = (c_{ij}) = (a_{ij} + b_{ij})$; for example

$$\begin{pmatrix} 7 & -3 & 4 \\ 2 & 8 & -5 \end{pmatrix} + \begin{pmatrix} 11 & 5 & -6 \\ 3 & 4 & 2 \end{pmatrix} = \begin{pmatrix} 18 & 2 & -2 \\ 5 & 12 & -3 \end{pmatrix}.$$

The *difference* $\mathbf{D} = \mathbf{A} - \mathbf{B}$ between two conformal matrices $\mathbf{A}$ and $\mathbf{B}$ is defined similarly: $\mathbf{D} = (d_{ij}) = (a_{ij} - b_{ij})$.

Two properties of matrix addition are given in the following theorem.

Theorem 2.2a. If $\mathbf{A}$ and $\mathbf{B}$ are both $n \times m$, then

(i) $\mathbf{A} + \mathbf{B} = \mathbf{B} + \mathbf{A}$. (2.9)

(ii) $(\mathbf{A} + \mathbf{B})' = \mathbf{A}' + \mathbf{B}'$. (2.10)

□

2.2.2 Product of a Scalar and a Matrix

Any scalar can be multiplied by any matrix. The product of a scalar and a matrix is defined as the product of each element of the matrix and the scalar:

$$c\mathbf{A} = (ca_{ij}) = \begin{pmatrix} ca_{11} & ca_{12} & \cdots & ca_{1m} \\ ca_{21} & ca_{22} & \cdots & ca_{2m} \\ \vdots & \vdots & & \vdots \\ ca_{n1} & ca_{n2} & \cdots & ca_{nm} \end{pmatrix}. \tag{2.11}$$

Since $ca_{ij} = a_{ij}c$, the product of a scalar and a matrix is commutative:

$$c\mathbf{A} = \mathbf{A}c. \tag{2.12}$$

2.2.3 Product of Two Matrices or Two Vectors

In order for the product $\mathbf{AB}$ to be defined, the number of columns in $\mathbf{A}$ must equal the number of rows in $\mathbf{B}$, in which case $\mathbf{A}$ and $\mathbf{B}$ are said to be *conformal for multiplication*. Then the (ij)th element of the product $\mathbf{C} = \mathbf{AB}$ is defined as

$$c_{ij} = \sum_k a_{ik}b_{kj}, \tag{2.13}$$

which is the sum of products of the elements in the ith row of $\mathbf{A}$ and the elements in the jth column of $\mathbf{B}$. Thus we multiply every row of $\mathbf{A}$ by every column of $\mathbf{B}$. If $\mathbf{A}$ is $n \times m$ and $\mathbf{B}$ is $m \times p$, then $\mathbf{C} = \mathbf{AB}$ is $n \times p$. We illustrate matrix multiplication in the following example.

Example 2.2.3. Let

$$\mathbf{A} = \begin{pmatrix} 2 & 1 & 3 \\ 4 & 6 & 5 \end{pmatrix} \quad \text{and} \quad \mathbf{B} = \begin{pmatrix} 1 & 4 \\ 2 & 6 \\ 3 & 8 \end{pmatrix}.$$

Then

$$\mathbf{AB} = \begin{pmatrix} 2 \cdot 1 + 1 \cdot 2 + 3 \cdot 3 & 2 \cdot 4 + 1 \cdot 6 + 3 \cdot 8 \\ 4 \cdot 1 + 6 \cdot 2 + 5 \cdot 3 & 4 \cdot 4 + 6 \cdot 6 + 5 \cdot 8 \end{pmatrix} = \begin{pmatrix} 13 & 38 \\ 31 & 92 \end{pmatrix},$$

$$\mathbf{BA} = \begin{pmatrix} 18 & 25 & 23 \\ 28 & 38 & 36 \\ 38 & 51 & 49 \end{pmatrix}.$$

□

Note that a 1×1 *matrix* $\mathbf{A}$ can only be multiplied on the right by a $1 \times n$ matrix $\mathbf{B}$ or on the left by an $n \times 1$ matrix $\mathbf{C}$, whereas a *scalar* can be multiplied on the right or left by a matrix of any size.

If $\mathbf{A}$ is $n \times m$ and $\mathbf{B}$ is $m \times p$, where $n \neq p$, then $\mathbf{AB}$ is defined, but $\mathbf{BA}$ is not defined. If $\mathbf{A}$ is $n \times p$ and $\mathbf{B}$ is $p \times n$, then $\mathbf{AB}$ is $n \times n$ and $\mathbf{BA}$ is $p \times p$. In this case, of course, $\mathbf{AB} \neq \mathbf{BA}$, as illustrated in Example 2.2.3. If $\mathbf{A}$ and $\mathbf{B}$ are both $n \times n$, then $\mathbf{AB}$ and $\mathbf{BA}$ are the same size, but, in general

$$\mathbf{AB} \neq \mathbf{BA}. \tag{2.14}$$

[There are a few exceptions to (2.14), for example, two diagonal matrices or a square matrix and an identity.] Thus matrix multiplication is not commutative, and certain familiar manipulations with real numbers cannot be done with matrices. However, matrix multiplication is distributive over addition or subtraction:

$$\mathbf{A}(\mathbf{B} \pm \mathbf{C}) = \mathbf{AB} \pm \mathbf{AC}, \tag{2.15}$$

$$(\mathbf{A} \pm \mathbf{B})\mathbf{C} = \mathbf{AC} \pm \mathbf{BC}. \tag{2.16}$$

Using (2.15) and (2.16), we can expand products such as $(\mathbf{A} - \mathbf{B})(\mathbf{C} - \mathbf{D})$:

$$\begin{aligned} (\mathbf{A} - \mathbf{B})(\mathbf{C} - \mathbf{D}) &= (\mathbf{A} - \mathbf{B})\mathbf{C} - (\mathbf{A} - \mathbf{B})\mathbf{D} && [\text{by } (2.15)] \\ &= \mathbf{AC} - \mathbf{BC} - \mathbf{AD} + \mathbf{BD} && [\text{by } (2.16)]. \end{aligned} \tag{2.17}$$

Multiplication involving vectors follows the same rules as for matrices. Suppose that $\mathbf{A}$ is $n \times p$, $\mathbf{b}$ is $p \times 1$, $\mathbf{c}$ is $p \times 1$, and $\mathbf{d}$ is $n \times 1$. Then $\mathbf{Ab}$ is a column vector of size $n \times 1$, $\mathbf{d}'\mathbf{A}$ is a row vector of size $1 \times p$, $\mathbf{b}'\mathbf{c}$ is a sum of products (1×1), $\mathbf{bc}'$ is a $p \times p$ matrix, and $\mathbf{cd}'$ is a $p \times n$ matrix. Since $\mathbf{b}'\mathbf{c}$ is a 1×1 sum of products, it is equal to $\mathbf{c}'\mathbf{b}$:

$$\begin{aligned} \mathbf{b}'\mathbf{c} &= b_1c_1 + b_2c_2 + \cdots + b_pc_p, \\ \mathbf{c}'\mathbf{b} &= c_1b_1 + c_2b_2 + \cdots + c_pb_p, \\ \mathbf{b}'\mathbf{c} &= \mathbf{c}'\mathbf{b}. \end{aligned} \tag{2.18}$$

The matrix $\mathbf{cd}'$ is given by

$$\mathbf{cd}' = \begin{pmatrix} c_1d_1 & c_1d_2 & \cdots & c_1d_n \\ c_2d_1 & c_2d_2 & \cdots & c_2d_n \\ \vdots & \vdots & & \vdots \\ c_pd_1 & c_pd_2 & \cdots & c_pd_n \end{pmatrix}. \tag{2.19}$$

Similarly

$$\mathbf{b}'\mathbf{b} = b_1^2 + b_2^2 + \cdots + b_p^2, \tag{2.20}$$

$$\mathbf{b}\mathbf{b}' = \begin{pmatrix} b_1^2 & b_1b_2 & \cdots & b_1b_p \\ b_2b_1 & b_2^2 & \cdots & b_2b_p \\ \vdots & \vdots & & \vdots \\ b_pb_1 & b_pb_2 & \cdots & b_p^2 \end{pmatrix}. \tag{2.21}$$

Thus, $\mathbf{b}'\mathbf{b}$ is a sum of squares and $\mathbf{b}\mathbf{b}'$ is a (symmetric) square matrix.

The square root of the sum of squares of the elements of a $p \times 1$ vector $\mathbf{b}$ is the distance from the origin to the point $\mathbf{b}$ and is also referred to as the *length* of $\mathbf{b}$:

$$\text{Length of } \mathbf{b} = \sqrt{\mathbf{b}'\mathbf{b}} = \sqrt{\sum_{i=1}^{p} b_i^2}. \tag{2.22}$$

If $\mathbf{j}$ is an $n \times 1$ vector of 1s as defined in (2.6), then by (2.20) and (2.21), we have

$$\mathbf{j}'\mathbf{j} = n, \quad \mathbf{j}\mathbf{j}' = \begin{pmatrix} 1 & 1 & \cdots & 1 \\ 1 & 1 & \cdots & 1 \\ \vdots & \vdots & & \vdots \\ 1 & 1 & \cdots & 1 \end{pmatrix} = \mathbf{J}, \tag{2.23}$$

where $\mathbf{J}$ is an $n \times n$ square matrix of 1s as illustrated in (2.7). If $\mathbf{a}$ is $n \times 1$ and $\mathbf{A}$ is $n \times p$, then

$$\mathbf{a}'\mathbf{j} = \mathbf{j}'\mathbf{a} = \sum_{i=1}^{n} a_i, \tag{2.24}$$

$$\mathbf{j}'\mathbf{A} = \left(\sum_i a_{i1}, \sum_i a_{i2}, \ldots, \sum_i a_{ip} \right), \qquad \mathbf{A}\mathbf{j} = \begin{pmatrix} \sum_j a_{1j} \\ \sum_j a_{2j} \\ \vdots \\ \sum_j a_{nj} \end{pmatrix}. \tag{2.25}$$

Thus $\mathbf{a}'\mathbf{j}$ is the sum of the elements in $\mathbf{a}$, $\mathbf{j}'\mathbf{A}$ contains the column sums of $\mathbf{A}$, and $\mathbf{A}\mathbf{j}$ contains the row sums of $\mathbf{A}$. Note that in $\mathbf{a}'\mathbf{j}$, the vector $\mathbf{j}$ is $n \times 1$; in $\mathbf{j}'\mathbf{A}$, the vector $\mathbf{j}$ is $n \times 1$; and in $\mathbf{A}\mathbf{j}$, the vector $\mathbf{j}$ is $p \times 1$.

The transpose of the product of two matrices is the product of the transposes in reverse order.

Theorem 2.2b. If **A** is $n \times p$ and **B** is $p \times m$, then

$$(\mathbf{AB})' = \mathbf{B}'\mathbf{A}'. \tag{2.26}$$

PROOF. Let $\mathbf{C} = \mathbf{AB}$. Then by (2.13)

$$\mathbf{C} = (c_{ij}) = \left(\sum_{k=1}^{p} a_{ik} b_{kj} \right).$$

By (2.3), the transpose of $\mathbf{C} = \mathbf{AB}$ becomes

$$\begin{aligned}(\mathbf{AB})' &= \mathbf{C}' = (c_{ij})' = (c_{ji}) \\ &= \left(\sum_{k=1}^{p} a_{jk} b_{ki} \right) = \left(\sum_{k=1}^{p} b_{ki} a_{jk} \right) = \mathbf{B}'\mathbf{A}'.\end{aligned}$$

□

We illustrate the steps in the proof of Theorem 2.2b using a 2×3 matrix **A** and a 3×2 matrix **B**:

$$\mathbf{AB} = \begin{pmatrix} a_{11} & a_{12} & a_{13} \\ a_{21} & a_{22} & a_{23} \end{pmatrix} \begin{pmatrix} b_{11} & b_{12} \\ b_{21} & b_{22} \\ b_{31} & b_{32} \end{pmatrix}$$

$$= \begin{pmatrix} a_{11}b_{11} + a_{12}b_{21} + a_{13}b_{31} & a_{11}b_{12} + a_{12}b_{22} + a_{13}b_{32} \\ a_{21}b_{11} + a_{22}b_{21} + a_{23}b_{31} & a_{21}b_{12} + a_{22}b_{22} + a_{23}b_{32} \end{pmatrix},$$

$$(\mathbf{AB})' = \begin{pmatrix} a_{11}b_{11} + a_{12}b_{21} + a_{13}b_{31} & a_{21}b_{11} + a_{22}b_{21} + a_{23}b_{31} \\ a_{11}b_{12} + a_{12}b_{22} + a_{13}b_{32} & a_{21}b_{12} + a_{22}b_{22} + a_{23}b_{32} \end{pmatrix}$$

$$= \begin{pmatrix} b_{11}a_{11} + b_{21}a_{12} + b_{31}a_{13} & b_{11}a_{21} + b_{21}a_{22} + b_{31}a_{23} \\ b_{12}a_{11} + b_{22}a_{12} + b_{32}a_{13} & b_{12}a_{21} + b_{22}a_{22} + b_{32}a_{23} \end{pmatrix}$$

$$= \begin{pmatrix} b_{11} & b_{21} & b_{31} \\ b_{12} & b_{22} & b_{32} \end{pmatrix} \begin{pmatrix} a_{11} & a_{21} \\ a_{12} & a_{22} \\ a_{13} & a_{23} \end{pmatrix}$$

$$= \mathbf{B}'\mathbf{A}'.$$

The following corollary to Theorem 2.2b gives the transpose of the product of three matrices.

Corollary 1. If $\mathbf{A}, \mathbf{B}$, and $\mathbf{C}$ are conformal so that $\mathbf{ABC}$ is defined, then $(\mathbf{ABC})' = \mathbf{C}'\mathbf{B}'\mathbf{A}'$. □

Suppose that $\mathbf{A}$ is $n \times m$ and $\mathbf{B}$ is $m \times p$. Let $\mathbf{a}_i'$ be the ith *row* of $\mathbf{A}$ and $\mathbf{b}_j$ be the jth *column* of $\mathbf{B}$, so that

$$\mathbf{A} = \begin{pmatrix} \mathbf{a}_1' \\ \mathbf{a}_2' \\ \vdots \\ \mathbf{a}_n' \end{pmatrix}, \qquad \mathbf{B} = (\mathbf{b}_1, \mathbf{b}_2, \ldots, \mathbf{b}_p).$$

Then, by definition, the (ij)th element of $\mathbf{AB}$ is $\mathbf{a}_i'\mathbf{b}_j$:

$$\mathbf{AB} = \begin{pmatrix} \mathbf{a}_1'\mathbf{b}_1 & \mathbf{a}_1'\mathbf{b}_2 & \cdots & \mathbf{a}_1'\mathbf{b}_p \\ \mathbf{a}_2'\mathbf{b}_1 & \mathbf{a}_2'\mathbf{b}_2 & \cdots & \mathbf{a}_2'\mathbf{b}_p \\ \vdots & \vdots & & \vdots \\ \mathbf{a}_n'\mathbf{b}_1 & \mathbf{a}_n'\mathbf{b}_2 & \cdots & \mathbf{a}_n'\mathbf{b}_p \end{pmatrix}.$$

This product can be written in terms of the rows of $\mathbf{A}$:

$$\mathbf{AB} = \begin{pmatrix} \mathbf{a}_1'(\mathbf{b}_1, \mathbf{b}_2, \ldots, \mathbf{b}_p) \\ \mathbf{a}_2'(\mathbf{b}_1, \mathbf{b}_2, \ldots, \mathbf{b}_p) \\ \vdots \\ \mathbf{a}_n'(\mathbf{b}_1, \mathbf{b}_2, \ldots, \mathbf{b}_p) \end{pmatrix} = \begin{pmatrix} \mathbf{a}_1'\mathbf{B} \\ \mathbf{a}_2'\mathbf{B} \\ \vdots \\ \mathbf{a}_n'\mathbf{B} \end{pmatrix} = \begin{pmatrix} \mathbf{a}_1' \\ \mathbf{a}_2' \\ \vdots \\ \mathbf{a}_n' \end{pmatrix}\mathbf{B}. \tag{2.27}$$

The first column of $\mathbf{AB}$ can be expressed in terms of $\mathbf{A}$ as

$$\begin{pmatrix} \mathbf{a}_1'\mathbf{b}_1 \\ \mathbf{a}_2'\mathbf{b}_1 \\ \vdots \\ \mathbf{a}_n'\mathbf{b}_1 \end{pmatrix} = \begin{pmatrix} \mathbf{a}_1' \\ \mathbf{a}_2' \\ \vdots \\ \mathbf{a}_n' \end{pmatrix}\mathbf{b}_1 = \mathbf{A}\mathbf{b}_1.$$

Likewise, the second column is $\mathbf{A}\mathbf{b}_2$, and so on. Thus $\mathbf{AB}$ can be written in terms of the columns of $\mathbf{B}$:

$$\mathbf{AB} = \mathbf{A}(\mathbf{b}_1, \mathbf{b}_2, \ldots, \mathbf{b}_p) = (\mathbf{A}\mathbf{b}_1, \mathbf{A}\mathbf{b}_2, \ldots, \mathbf{A}\mathbf{b}_p). \tag{2.28}$$

Any matrix $\mathbf{A}$ can be multiplied by its transpose to form $\mathbf{A}'\mathbf{A}$ or $\mathbf{A}\mathbf{A}'$. Some properties of these two products are given in the following theorem.

Theorem 2.2c. Let $\mathbf{A}$ be any $n \times p$ matrix. Then $\mathbf{A}'\mathbf{A}$ and $\mathbf{A}\mathbf{A}'$ have the following properties.

(i) $\mathbf{A}'\mathbf{A}$ is $p \times p$ and its elements are products of the *columns* of $\mathbf{A}$.
(ii) $\mathbf{A}\mathbf{A}'$ is $n \times n$ and its elements are products of the *rows* of $\mathbf{A}$.
(iii) Both $\mathbf{A}'\mathbf{A}$ and $\mathbf{A}\mathbf{A}'$ are symmetric.
(iv) If $\mathbf{A}'\mathbf{A} = \mathbf{O}$, then $\mathbf{A} = \mathbf{O}$. □

Let $\mathbf{A}$ be an $n \times n$ matrix and let $\mathbf{D} = \text{diag}(d_1, d_2, \ldots, d_n)$. In the product $\mathbf{DA}$, the ith row of $\mathbf{A}$ is multiplied by d_i, and in $\mathbf{AD}$, the jth column of $\mathbf{A}$ is multiplied by d_j. For example, if $n = 3$, we have

$$\mathbf{DA} = \begin{pmatrix} d_1 & 0 & 0 \\ 0 & d_2 & 0 \\ 0 & 0 & d_3 \end{pmatrix} \begin{pmatrix} a_{11} & a_{12} & a_{13} \\ a_{21} & a_{22} & a_{23} \\ a_{31} & a_{32} & a_{33} \end{pmatrix} = \begin{pmatrix} d_1a_{11} & d_1a_{12} & d_1a_{13} \\ d_2a_{21} & d_2a_{22} & d_2a_{23} \\ d_3a_{31} & d_3a_{32} & d_3a_{33} \end{pmatrix}, \tag{2.29}$$

$$\mathbf{AD} = \begin{pmatrix} a_{11} & a_{12} & a_{13} \\ a_{21} & a_{22} & a_{23} \\ a_{31} & a_{32} & a_{33} \end{pmatrix} \begin{pmatrix} d_1 & 0 & 0 \\ 0 & d_2 & 0 \\ 0 & 0 & d_3 \end{pmatrix} = \begin{pmatrix} d_1a_{11} & d_2a_{12} & d_3a_{13} \\ d_1a_{21} & d_2a_{22} & d_3a_{23} \\ d_1a_{31} & d_2a_{32} & d_3a_{33} \end{pmatrix}, \tag{2.30}$$

$$\mathbf{DAD} = \begin{pmatrix} d_1^2a_{11} & d_1d_2a_{12} & d_1d_3a_{13} \\ d_2d_1a_{21} & d_2^2a_{22} & d_2d_3a_{23} \\ d_3d_1a_{31} & d_3d_2a_{32} & d_3^2a_{33} \end{pmatrix}. \tag{2.31}$$

Note that $\mathbf{DA} \neq \mathbf{AD}$. However, in the special case where the diagonal matrix is the identity, (2.29) and (2.30) become

$$\mathbf{IA} = \mathbf{AI} = \mathbf{A}. \tag{2.32}$$

If $\mathbf{A}$ is rectangular, (2.32) still holds, but the two identities are of different sizes.

If $\mathbf{A}$ is a symmetric matrix and $\mathbf{y}$ is a vector, the product

$$\mathbf{y}'\mathbf{A}\mathbf{y} = \sum_i a_{ii}y_i^2 + \sum_{i \neq j} a_{ij}y_iy_j \tag{2.33}$$

is called a *quadratic form*. If $\mathbf{x}$ is $n \times 1$, $\mathbf{y}$ is $p \times 1$, and $\mathbf{A}$ is $n \times p$, the product

$$\mathbf{x}'\mathbf{A}\mathbf{y} = \sum_{ij} a_{ij}x_iy_j \tag{2.34}$$

is called a *bilinear form*.

2.2.4 Hadamard Product of Two Matrices or Two Vectors

Sometimes a third type of product, called the *elementwise* or *Hadamard product*, is useful. If two matrices or two vectors are of the same size (conformal for addition), the Hadamard product is found by simply multiplying corresponding elements:

$$(a_{ij}b_{ij}) = \begin{pmatrix} a_{11}b_{11} & a_{12}b_{12} & \cdots & a_{1p}b_{1p} \\ a_{21}b_{21} & a_{22}b_{22} & \cdots & a_{2p}b_{2p} \\ \vdots & \vdots & & \vdots \\ a_{n1}b_{n1} & a_{n2}b_{n2} & \cdots & a_{np}b_{np} \end{pmatrix}.$$

2.3 PARTITIONED MATRICES

It is sometimes convenient to partition a matrix into submatrices. For example, a partitioning of a matrix $\mathbf{A}$ into four (square or rectangular) submatrices of appropriate sizes can be indicated symbolically as follows:

$$\mathbf{A} = \begin{pmatrix} \mathbf{A}_{11} & \mathbf{A}_{12} \\ \mathbf{A}_{21} & \mathbf{A}_{22} \end{pmatrix}.$$

To illustrate, let the 4×5 matrix $\mathbf{A}$ be partitioned as

$$\mathbf{A} = \left(\begin{array}{ccc|cc} 7 & 2 & 5 & 8 & 4 \\ -3 & 4 & 0 & 2 & 7 \\ \hline 9 & 3 & 6 & 5 & -2 \\ 3 & 1 & 2 & 1 & 6 \end{array}\right) = \begin{pmatrix} \mathbf{A}_{11} & \mathbf{A}_{12} \\ \mathbf{A}_{21} & \mathbf{A}_{22} \end{pmatrix},$$

where

$$\mathbf{A}_{11} = \begin{pmatrix} 7 & 2 & 5 \\ -3 & 4 & 0 \end{pmatrix}, \quad \mathbf{A}_{12} = \begin{pmatrix} 8 & 4 \\ 2 & 7 \end{pmatrix},$$

$$\mathbf{A}_{21} = \begin{pmatrix} 9 & 3 & 6 \\ 3 & 1 & 2 \end{pmatrix}, \quad \mathbf{A}_{22} = \begin{pmatrix} 5 & -2 \\ 1 & 6 \end{pmatrix}.$$

If two matrices $\mathbf{A}$ and $\mathbf{B}$ are conformal for multiplication, and if $\mathbf{A}$ and $\mathbf{B}$ are partitioned so that the submatrices are appropriately conformal, then the product $\mathbf{AB}$ can be found using the usual pattern of row by column multiplication with the submatrices as if they were single elements; for example

$$\begin{aligned} \mathbf{AB} &= \begin{pmatrix} \mathbf{A}_{11} & \mathbf{A}_{12} \\ \mathbf{A}_{21} & \mathbf{A}_{22} \end{pmatrix} \begin{pmatrix} \mathbf{B}_{11} & \mathbf{B}_{12} \\ \mathbf{B}_{21} & \mathbf{B}_{22} \end{pmatrix} \\ &= \begin{pmatrix} \mathbf{A}_{11}\mathbf{B}_{11} + \mathbf{A}_{12}\mathbf{B}_{21} & \mathbf{A}_{11}\mathbf{B}_{12} + \mathbf{A}_{12}\mathbf{B}_{22} \\ \mathbf{A}_{21}\mathbf{B}_{11} + \mathbf{A}_{22}\mathbf{B}_{21} & \mathbf{A}_{21}\mathbf{B}_{12} + \mathbf{A}_{22}\mathbf{B}_{22} \end{pmatrix}. \end{aligned} \tag{2.35}$$

If $\mathbf{B}$ is replaced by a vector $\mathbf{b}$ partitioned into two sets of elements, and if $\mathbf{A}$ is correspondingly partitioned into two sets of columns, then (2.35) becomes

$$\mathbf{Ab} = (\mathbf{A}_1, \mathbf{A}_2) \begin{pmatrix} \mathbf{b}_1 \\ \mathbf{b}_2 \end{pmatrix} = \mathbf{A}_1\mathbf{b}_1 + \mathbf{A}_2\mathbf{b}_2, \tag{2.36}$$

where the number of columns of $\mathbf{A}_1$ is equal to the number of elements of $\mathbf{b}_1$, and $\mathbf{A}_2$ and $\mathbf{b}_2$ are similarly conformal. Note that the partitioning in $\mathbf{A} = (\mathbf{A}_1, \mathbf{A}_2)$ is indicated by a comma.

The partitioned multiplication in (2.36) can be extended to individual columns of $\mathbf{A}$ and individual elements of $\mathbf{b}$:

$$\mathbf{Ab} = (\mathbf{a}_1, \mathbf{a}_2, \ldots, \mathbf{a}_p) \begin{pmatrix} b_1 \\ b_2 \\ \vdots \\ b_p \end{pmatrix} = b_1\mathbf{a}_1 + b_2\mathbf{a}_2 + \cdots + b_p\mathbf{a}_p. \tag{2.37}$$

Thus **Ab** is expressible as a linear combination of the columns of **A**, in which the coefficients are elements of **b**. We illustrate (2.37) in the following example.

Example 2.3. Let

$$\mathbf{A} = \begin{pmatrix} 6 & -2 & 3 \\ 2 & 1 & 0 \\ 4 & 3 & 2 \end{pmatrix}, \qquad \mathbf{b} = \begin{pmatrix} 4 \\ 2 \\ -1 \end{pmatrix}.$$

Then

$$\mathbf{Ab} = \begin{pmatrix} 17 \\ 10 \\ 20 \end{pmatrix}.$$

Using a linear combination of columns of **A** as in (2.37), we obtain

$$\begin{aligned} \mathbf{Ab} &= b_1\mathbf{a}_1 + b_2\mathbf{a}_2 + b_2\mathbf{a}_3 \\ &= 4\begin{pmatrix} 6 \\ 2 \\ 4 \end{pmatrix} + 2\begin{pmatrix} -2 \\ 1 \\ 3 \end{pmatrix} - \begin{pmatrix} 3 \\ 0 \\ 2 \end{pmatrix} \\ &= \begin{pmatrix} 24 \\ 8 \\ 16 \end{pmatrix} + \begin{pmatrix} -4 \\ 2 \\ 6 \end{pmatrix} - \begin{pmatrix} 3 \\ 0 \\ 2 \end{pmatrix} = \begin{pmatrix} 17 \\ 10 \\ 20 \end{pmatrix}. \end{aligned}$$

□

By (2.28) and (2.37), the columns of the product **AB** are linear combinations of the columns of **A**. The coefficients for the jth column of **AB** are the elements of the jth column of **B**.

The product of a row vector and a matrix, $\mathbf{a}'\mathbf{B}$, can be expressed as a linear combination of the rows of **B**, in which the coefficients are elements of $\mathbf{a}'$:

$$\mathbf{a}'\mathbf{B} = (a_1, a_2, \ldots, a_n)\begin{pmatrix} \mathbf{b}'_1 \\ \mathbf{b}'_2 \\ \vdots \\ \mathbf{b}'_n \end{pmatrix} = a_1\mathbf{b}'_1 + a_2\mathbf{b}'_2 + \cdots + a_n\mathbf{b}'_n. \tag{2.38}$$

By (2.27) and (2.38), the rows of the matrix product **AB** are linear combinations of the rows of **B**. The coefficients for the ith row of **AB** are the elements of the ith row of **A**.

Finally, we note that if a matrix $\mathbf{A}$ is partitioned as $\mathbf{A} = (\mathbf{A}_1, \mathbf{A}_2)$, then

$$\mathbf{A}' = (\mathbf{A}_1, \mathbf{A}_2)' = \begin{pmatrix} \mathbf{A}_1' \\ \mathbf{A}_2' \end{pmatrix}. \tag{2.39}$$

2.4 RANK

Before defining the rank of a matrix, we first introduce the notion of linear independence and dependence. A set of vectors $\mathbf{a}_1, \mathbf{a}_2, \ldots, \mathbf{a}_n$ is said to be *linearly dependent* if scalars $c_1, c_2, \ldots, c_n$ (not all zero) can be found such that

$$c_1\mathbf{a}_1 + c_2\mathbf{a}_2 + \cdots + c_n\mathbf{a}_n = \mathbf{0}. \tag{2.40}$$

If no coefficients $c_1, c_2, \ldots, c_n$ can be found that satisfy (2.40), the set of vectors $\mathbf{a}_1, \mathbf{a}_2, \ldots, \mathbf{a}_n$ is said to be *linearly independent*. By (2.37) this can be restated as follows. The columns of $\mathbf{A}$ are linearly independent if $\mathbf{Ac} = \mathbf{0}$ implies $\mathbf{c} = \mathbf{0}$. (If a set of vectors includes $\mathbf{0}$, the set is linearly dependent.) If (2.40) holds, then at least one of the vectors $\mathbf{a}_i$ can be expressed as a linear combination of the other vectors in the set. Among linearly independent vectors there is no redundancy of this type.

The *rank* of any square or rectangular matrix $\mathbf{A}$ is defined as

$$\begin{aligned} \text{rank}(\mathbf{A}) &= \text{number of linearly independent columns of } \mathbf{A} \\ &= \text{number of linearly independent rows of } \mathbf{A}. \end{aligned}$$

It can be shown that the number of linearly independent columns of any matrix is always equal to the number of linearly independent rows.

If a matrix $\mathbf{A}$ has a single nonzero element, with all other elements equal to 0, then $\text{rank}(\mathbf{A}) = 1$. The vector $\mathbf{0}$ and the matrix $\mathbf{O}$ have rank 0.

Suppose that a rectangular matrix $\mathbf{A}$ is $n \times p$ of rank p, where $p < n$. (We typically shorten this statement to "$\mathbf{A}$ is $n \times p$ of rank $p < n$.") Then $\mathbf{A}$ has maximum possible rank and is said to be of *full rank*. In general, the maximum possible rank of an $n \times p$ matrix $\mathbf{A}$ is $\min(n, p)$. Thus, in a rectangular matrix, the rows or columns (or both) are linearly dependent. We illustrate this in the following example.

Example 2.4a. The rank of

$$\mathbf{A} = \begin{pmatrix} 1 & -2 & 3 \\ 5 & 2 & 4 \end{pmatrix}$$

is 2 because the two rows are linearly independent (neither row is a multiple of the other). Hence, by the definition of rank, the number of linearly independent columns is also 2. Therefore, the columns are linearly dependent, and by (2.40) there exist constants c_1, c_2, and c_3 such that

$$c_1\begin{pmatrix}1\\5\end{pmatrix}+c_2\begin{pmatrix}-2\\2\end{pmatrix}+c_3\begin{pmatrix}3\\4\end{pmatrix}=\begin{pmatrix}0\\0\end{pmatrix}. \tag{2.41}$$

By (2.37), we can write (2.41) in the form

$$\begin{pmatrix}1 & -2 & 3\\5 & 2 & 4\end{pmatrix}\begin{pmatrix}c_1\\c_2\\c_3\end{pmatrix}=\begin{pmatrix}0\\0\end{pmatrix} \quad \text{or} \quad \mathbf{Ac}=\mathbf{0}. \tag{2.42}$$

The solution to (2.42) is given by any multiple of $\mathbf{c}=(14,\ -11,\ -12)'$. In this case, the product $\mathbf{Ac}$ is equal to $\mathbf{0}$, even though $\mathbf{A}\neq\mathbf{O}$ and $\mathbf{c}\neq\mathbf{0}$. This is possible because of the linear dependence of the column vectors of $\mathbf{A}$. □

We can extend (2.42) to products of matrices. It is possible to find $\mathbf{A}\neq\mathbf{O}$ and $\mathbf{B}\neq\mathbf{O}$ such that

$$\mathbf{AB}=\mathbf{O}; \tag{2.43}$$

for example

$$\begin{pmatrix}1 & 2\\2 & 4\end{pmatrix}\begin{pmatrix}2 & 6\\-1 & -3\end{pmatrix}=\begin{pmatrix}0 & 0\\0 & 0\end{pmatrix}.$$

We can also exploit the linear dependence of rows or columns of a matrix to create expressions such as $\mathbf{AB}=\mathbf{CB}$, where $\mathbf{A}\neq\mathbf{C}$. Thus in a matrix equation, we cannot, in general, cancel a matrix from both sides of the equation. There are two exceptions to this rule: (1) if $\mathbf{B}$ is a full-rank square matrix, then $\mathbf{AB}=\mathbf{CB}$ implies $\mathbf{A}=\mathbf{C}$; (2) the other special case occurs when the expression holds for all possible values of the matrix common to both sides of the equation; for example

$$\text{if } \mathbf{Ax}=\mathbf{Bx} \text{ for all possible values of } \mathbf{x}, \tag{2.44}$$

then $\mathbf{A}=\mathbf{B}$. To see this, let $\mathbf{x}=(1,0,\ldots,0)'$. Then, by (2.37) the first column of $\mathbf{A}$ equals the first column of $\mathbf{B}$. Now let $\mathbf{x}=(0,1,0,\ldots,0)'$, and the second column of $\mathbf{A}$ equals the second column of $\mathbf{B}$. Continuing in this fashion, we obtain $\mathbf{A}=\mathbf{B}$.

Example 2.4b. We illustrate the existence of matrices $\mathbf{A}$, $\mathbf{B}$, and $\mathbf{C}$ such that $\mathbf{AB} = \mathbf{CB}$, where $\mathbf{A} \neq \mathbf{C}$. Let

$$\mathbf{A} = \begin{pmatrix} 1 & 3 & 2 \\ 2 & 0 & -1 \end{pmatrix}, \quad \mathbf{B} = \begin{pmatrix} 1 & 2 \\ 0 & 1 \\ 1 & 0 \end{pmatrix}, \quad \mathbf{C} = \begin{pmatrix} 2 & 1 & 1 \\ 5 & -6 & -4 \end{pmatrix}.$$

Then

$$\mathbf{AB} = \mathbf{CB} = \begin{pmatrix} 3 & 5 \\ 1 & 4 \end{pmatrix}.$$

□

The following theorem gives a general case and two special cases for the rank of a product of two matrices.

Theorem 2.4

(i) If the matrices $\mathbf{A}$ and $\mathbf{B}$ are conformal for multiplication, then $\text{rank}(\mathbf{AB}) \leq \text{rank}(\mathbf{A})$ and $\text{rank}(\mathbf{AB}) \leq \text{rank}(\mathbf{B})$.

(ii) Multiplication by a full–rank square matrix does not change the rank; that is, if $\mathbf{B}$ and $\mathbf{C}$ are full–rank square matrices, $\text{rank}(\mathbf{AB}) = \text{rank}(\mathbf{CA}) = \text{rank}(\mathbf{A})$.

(iii) For any matrix $\mathbf{A}$, $\text{rank}(\mathbf{A}'\mathbf{A}) = \text{rank}(\mathbf{AA}') = \text{rank}(\mathbf{A}') = \text{rank}(\mathbf{A})$.

PROOF

(i) All the columns of $\mathbf{AB}$ are linear combinations of the columns of $\mathbf{A}$ (see a comment following Example 2.3). Consequently, the number of linearly independent columns of $\mathbf{AB}$ is less than or equal to the number of linearly independent columns of $\mathbf{A}$, and $\text{rank}(\mathbf{AB}) \leq \text{rank}(\mathbf{A})$. Similarly, all the rows of $\mathbf{AB}$ are linear combinations of the rows of $\mathbf{B}$ [see a comment following (2.38)], and therefore $\text{rank}(\mathbf{AB}) \leq \text{rank}(\mathbf{B})$.

(ii) This will be proved later.

(iii) This will also be proved later.

□

2.5 INVERSE

A full-rank square matrix is said to be *nonsingular*. A nonsingular matrix $\mathbf{A}$ has a unique *inverse*, denoted by $\mathbf{A}^{-1}$, with the property that

$$\mathbf{AA}^{-1} = \mathbf{A}^{-1}\mathbf{A} = \mathbf{I}. \tag{2.45}$$

If $\mathbf{A}$ is square and less than full rank, then it does not have an inverse and is said to be *singular*. Note that full-rank rectangular matrices do not have inverses as in (2.45). From the definition in (2.45), it is clear that $\mathbf{A}$ is the inverse of $\mathbf{A}^{-1}$:

$$(\mathbf{A}^{-1})^{-1} = \mathbf{A}. \tag{2.46}$$

Example 2.5. Let

$$\mathbf{A} = \begin{pmatrix} 4 & 7 \\ 2 & 6 \end{pmatrix}.$$

Then

$$\mathbf{A}^{-1} = \begin{pmatrix} .6 & -.7 \\ -.2 & .4 \end{pmatrix}$$

and

$$\begin{pmatrix} 4 & 7 \\ 2 & 6 \end{pmatrix}\begin{pmatrix} .6 & -.7 \\ -.2 & .4 \end{pmatrix} = \begin{pmatrix} .6 & -.7 \\ -.2 & .4 \end{pmatrix}\begin{pmatrix} 4 & 7 \\ 2 & 6 \end{pmatrix} = \begin{pmatrix} 1 & 0 \\ 0 & 1 \end{pmatrix}.$$ □

We can now prove Theorem 2.4(ii).

PROOF. If $\mathbf{B}$ is a full-rank square (nonsingular) matrix, there exists a matrix $\mathbf{B}^{-1}$ such that $\mathbf{BB}^{-1} = \mathbf{I}$. Then, by Theorem 2.4(i), we have

$$\text{rank}(\mathbf{A}) = \text{rank}(\mathbf{ABB}^{-1}) \leq \text{rank}(\mathbf{AB}) \leq \text{rank}(\mathbf{A}).$$

Thus both inequalities become equalities, and $\text{rank}(\mathbf{A}) = \text{rank}(\mathbf{AB})$. Similarly, $\text{rank}(\mathbf{A}) = \text{rank}(\mathbf{CA})$ for $\mathbf{C}$ nonsingular. □

In applications, inverses are typically found by computer. Many calculators also compute inverses. Algorithms for hand calculation of inverses of small matrices can be found in texts on matrix algebra.

If $\mathbf{B}$ is nonsingular and $\mathbf{AB} = \mathbf{CB}$, then we can multiply on the right by $\mathbf{B}^{-1}$ to obtain $\mathbf{A} = \mathbf{C}$. (If $\mathbf{B}$ is singular or rectangular, we can't cancel it from both sides of $\mathbf{AB} = \mathbf{CB}$; see Example 2.4b and the paragraph preceding the example.) Similarly, if $\mathbf{A}$ is nonsingular, the system of equations $\mathbf{Ax} = \mathbf{c}$ has the unique solution

$$\mathbf{x} = \mathbf{A}^{-1}\mathbf{c}, \tag{2.47}$$

since we can multiply on the left by $\mathbf{A}^{-1}$ to obtain

$$\mathbf{A}^{-1}\mathbf{A}\mathbf{x} = \mathbf{A}^{-1}\mathbf{c}$$
$$\mathbf{I}\mathbf{x} = \mathbf{A}^{-1}\mathbf{c}.$$

Two properties of inverses are given in the next two theorems.

Theorem 2.5a. If $\mathbf{A}$ is nonsingular, then $\mathbf{A}'$ is nonsingular and its inverse can be found as

$$(\mathbf{A}')^{-1} = (\mathbf{A}^{-1})'. \tag{2.48}$$

□

Theorem 2.5b. If $\mathbf{A}$ and $\mathbf{B}$ are nonsingular matrices of the same size, then $\mathbf{AB}$ is nonsingular and

$$(\mathbf{AB})^{-1} = \mathbf{B}^{-1}\mathbf{A}^{-1}. \tag{2.49}$$

□

We now give the inverses of some special matrices. If $\mathbf{A}$ is symmetric and nonsingular and is partitioned as

$$\mathbf{A} = \begin{pmatrix} \mathbf{A}_{11} & \mathbf{A}_{12} \\ \mathbf{A}_{21} & \mathbf{A}_{22} \end{pmatrix},$$

and if $\mathbf{B} = \mathbf{A}_{22} - \mathbf{A}_{21}\mathbf{A}_{11}^{-1}\mathbf{A}_{12}$, then, provided $\mathbf{A}_{11}^{-1}$ and $\mathbf{B}^{-1}$ exist, the inverse of $\mathbf{A}$ is given by

$$\mathbf{A}^{-1} = \begin{pmatrix} \mathbf{A}_{11}^{-1} + \mathbf{A}_{11}^{-1}\mathbf{A}_{12}\mathbf{B}^{-1}\mathbf{A}_{21}\mathbf{A}_{11}^{-1} & -\mathbf{A}_{11}^{-1}\mathbf{A}_{12}\mathbf{B}^{-1} \\ -\mathbf{B}^{-1}\mathbf{A}_{21}\mathbf{A}_{11}^{-1} & \mathbf{B}^{-1} \end{pmatrix}. \tag{2.50}$$

As a special case of (2.50), consider the symmetric nonsingular matrix

$$\mathbf{A} = \begin{pmatrix} \mathbf{A}_{11} & \mathbf{a}_{12} \\ \mathbf{a}'_{12} & a_{22} \end{pmatrix},$$

in which $\mathbf{A}_{11}$ is square, a_{22} is a 1×1 matrix, and $\mathbf{a}_{12}$ is a vector. Then if $\mathbf{A}_{11}^{-1}$ exists, $\mathbf{A}^{-1}$ can be expressed as

$$\mathbf{A}^{-1} = \frac{1}{b}\begin{pmatrix} b\mathbf{A}_{11}^{-1} + \mathbf{A}_{11}^{-1}\mathbf{a}_{12}\mathbf{a}'_{12}\mathbf{A}_{11}^{-1} & -\mathbf{A}_{11}^{-1}\mathbf{a}_{12} \\ -\mathbf{a}'_{12}\mathbf{A}_{11}^{-1} & 1 \end{pmatrix}, \tag{2.51}$$

where $b = a_{22} - \mathbf{a}'_{12}\mathbf{A}_{11}^{-1}\mathbf{a}_{12}$. As another special case of (2.50), we have

$$\begin{pmatrix} \mathbf{A}_{11} & \mathbf{O} \\ \mathbf{O} & \mathbf{A}_{22} \end{pmatrix}^{-1} = \begin{pmatrix} \mathbf{A}_{11}^{-1} & \mathbf{O} \\ \mathbf{O} & \mathbf{A}_{22}^{-1} \end{pmatrix}. \tag{2.52}$$

If a square matrix of the form $\mathbf{B} + \mathbf{cc}'$ is nonsingular, where $\mathbf{c}$ is a vector and $\mathbf{B}$ is a nonsingular matrix, then

$$(\mathbf{B} + \mathbf{cc}')^{-1} = \mathbf{B}^{-1} - \frac{\mathbf{B}^{-1}\mathbf{cc}'\mathbf{B}^{-1}}{1 + \mathbf{c}'\mathbf{B}^{-1}\mathbf{c}}. \tag{2.53}$$

In more generality, if $\mathbf{A}$, $\mathbf{B}$, and $\mathbf{A} + \mathbf{PBQ}$ are nonsingular, then

$$(\mathbf{A} + \mathbf{PBQ})^{-1} = \mathbf{A}^{-1} - \mathbf{A}^{-1}\mathbf{PB}(\mathbf{B} + \mathbf{BQA}^{-1}\mathbf{PB})^{-1}\mathbf{BQA}^{-1}. \tag{2.54}$$

Both (2.53) and (2.54) can be easily verified (Problems 2.33 and 2.34).

2.6 POSITIVE DEFINITE MATRICES

Quadratic forms were introduced in (2.33). For example, the quadratic form $3y_1^2 + y_2^2 + 2y_3^2 + 4y_1y_2 + 5y_1y_3 - 6y_2y_3$ can be expressed as

$$3y_1^2 + y_2^2 + 2y_3^2 + 4y_1y_2 + 5y_1y_3 - 6y_2y_3 = \mathbf{y}'\mathbf{Ay},$$

where

$$\mathbf{y} = \begin{pmatrix} y_1 \\ y_2 \\ y_3 \end{pmatrix}, \qquad \mathbf{A} = \begin{pmatrix} 3 & 4 & 5 \\ 0 & 1 & -6 \\ 0 & 0 & 2 \end{pmatrix}.$$

However, the same quadratic form can also be expressed in terms of the symmetric matrix

$$\frac{1}{2}(\mathbf{A} + \mathbf{A}') = \begin{pmatrix} 3 & 2 & \frac{5}{2} \\ 2 & 1 & -3 \\ \frac{5}{2} & -3 & 2 \end{pmatrix}.$$

In general, any quadratic form $\mathbf{y}'\mathbf{A}\mathbf{y}$ can be expressed as

$$\mathbf{y}'\mathbf{A}\mathbf{y} = \mathbf{y}'\left(\frac{\mathbf{A} + \mathbf{A}'}{2}\right)\mathbf{y}, \tag{2.55}$$

and thus the matrix of a quadratic form can always be chosen to be symmetric (and thereby unique).

The sums of squares we will encounter in regression (Chapters 6–11) and analysis–of–variance (Chapters 12–15) can be expressed in the form $\mathbf{y}'\mathbf{A}\mathbf{y}$, where $\mathbf{y}$ is an observation vector. Such quadratic forms remain positive (or at least nonnegative) for all possible values of $\mathbf{y}$. We now consider quadratic forms of this type.

If the symmetric matrix $\mathbf{A}$ has the property $\mathbf{y}'\mathbf{A}\mathbf{y} > 0$ for all possible $\mathbf{y}$ except $\mathbf{y} = \mathbf{0}$, then the quadratic form $\mathbf{y}'\mathbf{A}\mathbf{y}$ is said to be *positive definite*, and $\mathbf{A}$ is said to be a *positive definite* matrix. Similarly, if $\mathbf{y}'\mathbf{A}\mathbf{y} \geq 0$ for all $\mathbf{y}$ and there is at least one $\mathbf{y} \neq \mathbf{0}$ such that $\mathbf{y}'\mathbf{A}\mathbf{y} = 0$, then $\mathbf{y}'\mathbf{A}\mathbf{y}$ and $\mathbf{A}$ are said to be *positive semidefinite*. Both types of matrices are illustrated in the following example.

Example 2.6. To illustrate a positive definite matrix, consider

$$\mathbf{A} = \begin{pmatrix} 2 & -1 \\ -1 & 3 \end{pmatrix}$$

and the associated quadratic form

$$\mathbf{y}'\mathbf{A}\mathbf{y} = 2y_1^2 - 2y_1y_2 + 3y_2^2 = 2(y_1 - \tfrac{1}{2}y_2)^2 + \tfrac{5}{2}y_2^2,$$

which is clearly positive as long as y_1 and y_2 are not both zero.

To illustrate a positive semidefinite matrix, consider

$$(2y_1 - y_2)^2 + (3y_1 - y_3)^2 + (3y_2 - 2y_3)^2,$$

which can be expressed as $\mathbf{y}'\mathbf{A}\mathbf{y}$, with

$$\mathbf{A} = \begin{pmatrix} 13 & -2 & -3 \\ -2 & 10 & -6 \\ -3 & -6 & 5 \end{pmatrix}.$$

If $2y_1 = y_2, 3y_1 = y_3$, and $3y_2 = 2y_3$, then $(2y_1 - y_2)^2 + (3y_1 - y_3)^2 + (3y_2 - 2y_3)^2 = 0$. Thus $\mathbf{y}'\mathbf{A}\mathbf{y} = 0$ for any multiple of $\mathbf{y} = (1, 2, 3)'$. Otherwise $\mathbf{y}'\mathbf{A}\mathbf{y} > 0$ (except for $\mathbf{y} = \mathbf{0}$). □

In the matrices in Example 2.6, the diagonal elements are positive. For positive definite matrices, this is true in general.

Theorem 2.6a

(i) If $\mathbf{A}$ is positive definite, then all its diagonal elements a_{ii} are positive.
(ii) If $\mathbf{A}$ is positive semidefinite, then all $a_{ii} \geq 0$.

PROOF

(i) Let $\mathbf{y}' = (0, \ldots, 0, 1, 0, \ldots, 0)$ with a 1 in the ith position and 0's elsewhere. Then $\mathbf{y}'\mathbf{A}\mathbf{y} = a_{ii} > 0$.
(ii) Let $\mathbf{y}' = (0, \ldots, 0, 1, 0, \ldots, 0)$ with a 1 in the ith position and 0's elsewhere. Then $\mathbf{y}'\mathbf{A}\mathbf{y} = a_{ii} \geq 0$. □

Some additional properties of positive definite and positive semidefinite matrices are given in the following theorems.

Theorem 2.6b. Let $\mathbf{P}$ be a nonsingular matrix.

(i) If $\mathbf{A}$ is positive definite, then $\mathbf{P}'\mathbf{A}\mathbf{P}$ is positive definite.
(ii) If $\mathbf{A}$ is positive semidefinite, then $\mathbf{P}'\mathbf{A}\mathbf{P}$ is positive semidefinite.

PROOF

(i) To show that $\mathbf{y}'\mathbf{P}'\mathbf{A}\mathbf{P}\mathbf{y} > 0$ for $\mathbf{y} \neq \mathbf{0}$, note that $\mathbf{y}'(\mathbf{P}'\mathbf{A}\mathbf{P})\mathbf{y} = (\mathbf{P}\mathbf{y})'\mathbf{A}(\mathbf{P}\mathbf{y})$. Since $\mathbf{A}$ is positive definite, $(\mathbf{P}\mathbf{y})'\mathbf{A}(\mathbf{P}\mathbf{y}) > 0$ provided that $\mathbf{P}\mathbf{y} \neq \mathbf{0}$. By (2.47), $\mathbf{P}\mathbf{y} = \mathbf{0}$ only if $\mathbf{y} = \mathbf{0}$, since $\mathbf{P}^{-1}\mathbf{P}\mathbf{y} = \mathbf{P}^{-1}\mathbf{0} = \mathbf{0}$. Thus $\mathbf{y}'\mathbf{P}'\mathbf{A}\mathbf{P}\mathbf{y} > 0$ if $\mathbf{y} \neq \mathbf{0}$.
(ii) See problem 2.36. □

Corollary 1. Let $\mathbf{A}$ be a $p \times p$ positive definite matrix and let $\mathbf{B}$ be a $k \times p$ matrix of rank $k \leq p$. Then $\mathbf{B}\mathbf{A}\mathbf{B}'$ is positive definite. □

Corollary 2. Let $\mathbf{A}$ be a $p \times p$ positive definite matrix and let $\mathbf{B}$ be a $k \times p$ matrix. If $k > p$ or if $\text{rank}(\mathbf{B}) = r$, where $r < k$ and $r < p$, then $\mathbf{B}\mathbf{A}\mathbf{B}'$ is positive semidefinite. □

Theorem 2.6c. A symmetric matrix $\mathbf{A}$ is positive definite if and only if there exists a nonsingular matrix $\mathbf{P}$ such that $\mathbf{A} = \mathbf{P}'\mathbf{P}$.

PROOF. We prove the "if" part only. Suppose $\mathbf{A} = \mathbf{P}'\mathbf{P}$ for nonsingular $\mathbf{P}$. Then

$$\mathbf{y}'\mathbf{A}\mathbf{y} = \mathbf{y}'\mathbf{P}'\mathbf{P}\mathbf{y} = (\mathbf{P}\mathbf{y})'(\mathbf{P}\mathbf{y}).$$

This is a sum of squares [see (2.20)] and is positive unless $\mathbf{P}\mathbf{y} = \mathbf{0}$. By (2.47), $\mathbf{P}\mathbf{y} = \mathbf{0}$ only if $\mathbf{y} = \mathbf{0}$. □

Corollary 1. A positive definite matrix is nonsingular. □

One method of factoring a positive definite matrix $\mathbf{A}$ into a product $\mathbf{P}'\mathbf{P}$ as in Theorem 2.6c is provided by the Cholesky decomposition (Seber and Lee 2003, pp. 335–337), by which $\mathbf{A}$ can be factored uniquely into $\mathbf{A} = \mathbf{T}'\mathbf{T}$, where $\mathbf{T}$ is a nonsingular upper triangular matrix.

For any square or rectangular matrix $\mathbf{B}$, the matrix $\mathbf{B}'\mathbf{B}$ is positive definite or positive semidefinite.

Theorem 2.6d. Let $\mathbf{B}$ be an $n \times p$ matrix.

(i) If rank($\mathbf{B}$) $= p$, then $\mathbf{B}'\mathbf{B}$ is positive definite.
(ii) If rank($\mathbf{B}$) $< p$, then $\mathbf{B}'\mathbf{B}$ is positive semidefinite.

PROOF

(i) To show that $\mathbf{y}'\mathbf{B}'\mathbf{B}\mathbf{y} > 0$ for $\mathbf{y} \neq \mathbf{0}$, we note that

$$\mathbf{y}'\mathbf{B}'\mathbf{B}\mathbf{y} = (\mathbf{B}\mathbf{y})'(\mathbf{B}\mathbf{y}),$$

which is a sum of squares and is thereby positive unless $\mathbf{B}\mathbf{y} = \mathbf{0}$. By (2.37), we can express $\mathbf{B}\mathbf{y}$ in the form

$$\mathbf{B}\mathbf{y} = y_1\mathbf{b}_1 + y_2\mathbf{b}_2 + \cdots + y_p\mathbf{b}_p.$$

This linear combination is not $\mathbf{0}$ (for any $\mathbf{y} \neq \mathbf{0}$) because rank($\mathbf{B}$) $= p$, and the columns of $\mathbf{B}$ are therefore linearly independent [see (2.40)].

(ii) If rank($\mathbf{B}$) $< p$, then we can find $\mathbf{y} \neq \mathbf{0}$ such that

$$\mathbf{B}\mathbf{y} = y_1\mathbf{b}_1 + y_2\mathbf{b}_2 + \cdots + y_p\mathbf{b}_p = \mathbf{0}$$

since the columns of $\mathbf{B}$ are linearly dependent [see (2.40)]. Hence $\mathbf{y}'\mathbf{B}'\mathbf{B}\mathbf{y} \geq 0$.

□

Note that if $\mathbf{B}$ is a square matrix, the matrix $\mathbf{BB} = \mathbf{B}^2$ is not necessarily positive semidefinite. For example, let

$$\mathbf{B} = \begin{pmatrix} 1 & -2 \\ 1 & -2 \end{pmatrix}.$$

Then

$$\mathbf{B}^2 = \begin{pmatrix} -1 & 2 \\ -1 & 2 \end{pmatrix}, \quad \mathbf{B}'\mathbf{B} = \begin{pmatrix} 2 & -4 \\ -4 & 8 \end{pmatrix}.$$

In this case, $\mathbf{B}^2$ is not positive semidefinite, but $\mathbf{B}'\mathbf{B}$ is positive semidefinite, since $\mathbf{y}'\mathbf{B}'\mathbf{B}\mathbf{y} = 2(y_1 - 2y_2)^2$.

Two additional properties of positive definite matrices are given in the following theorems.

Theorem 2.6e. If $\mathbf{A}$ is positive definite, then $\mathbf{A}^{-1}$ is positive definite.

PROOF. By Theorem 2.6c, $\mathbf{A} = \mathbf{P}'\mathbf{P}$, where $\mathbf{P}$ is nonsingular. By Theorems 2.5a and 2.5b, $\mathbf{A}^{-1} = (\mathbf{P}'\mathbf{P})^{-1} = \mathbf{P}^{-1}(\mathbf{P}')^{-1} = \mathbf{P}^{-1}(\mathbf{P}^{-1})'$, which is positive definite by Theorem 2.6c. □

Theorem 2.6f. If $\mathbf{A}$ is positive definite and is partitioned in the form

$$\mathbf{A} = \begin{pmatrix} \mathbf{A}_{11} & \mathbf{A}_{12} \\ \mathbf{A}_{21} & \mathbf{A}_{22} \end{pmatrix},$$

where $\mathbf{A}_{11}$ and $\mathbf{A}_{22}$ are square, then $\mathbf{A}_{11}$ and $\mathbf{A}_{22}$ are positive definite.

PROOF. We can write $\mathbf{A}_{11}$, for example, as $\mathbf{A}_{11} = (\mathbf{I}, \mathbf{O})\mathbf{A}\begin{pmatrix} \mathbf{I} \\ \mathbf{O} \end{pmatrix}$, where $\mathbf{I}$ is the same size as $\mathbf{A}_{11}$. Then by Corollary 1 to Theorem 2.6b, $\mathbf{A}_{11}$ is positive definite. □

2.7 SYSTEMS OF EQUATIONS

The system of n (linear) equations in p unknowns

$$\begin{aligned} a_{11}x_1 + a_{12}x_2 + \cdots + a_{1p}x_p &= c_1 \\ a_{21}x_1 + a_{22}x_2 + \cdots + a_{2p}x_p &= c_2 \\ &\vdots \\ a_{n1}x_1 + a_{n2}x_2 + \cdots + a_{np}x_p &= c_n \end{aligned} \tag{2.56}$$

can be written in matrix form as

$$\mathbf{Ax} = \mathbf{c}, \tag{2.57}$$

where $\mathbf{A}$ is $n \times p$, $\mathbf{x}$ is $p \times 1$, and $\mathbf{c}$ is $n \times 1$. Note that if $n \neq p$, $\mathbf{x}$ and $\mathbf{c}$ are of different sizes. If $n = p$ and $\mathbf{A}$ is nonsingular, then by (2.47), there exists a unique solution vector $\mathbf{x}$ obtained as $\mathbf{x} = \mathbf{A}^{-1}\mathbf{c}$. If $n > p$, so that $\mathbf{A}$ has more rows than columns, then $\mathbf{Ax} = \mathbf{c}$ typically has no solution. If $n < p$, so that $\mathbf{A}$ has fewer rows than columns, then $\mathbf{Ax} = \mathbf{c}$ typically has an infinite number of solutions.

If the system of equations $\mathbf{Ax} = \mathbf{c}$ has one or more solution vectors, it is said to be *consistent*. If the system has no solution, it is said to be *inconsistent*.

To illustrate the structure of a consistent system of equations $\mathbf{Ax} = \mathbf{c}$, suppose that $\mathbf{A}$ is $p \times p$ of rank $r < p$. Then the rows of $\mathbf{A}$ are linearly dependent, and there exists some $\mathbf{b}$ such that [see (2.38)]

$$\mathbf{b}'\mathbf{A} = b_1\mathbf{a}_1' + b_2\mathbf{a}_2' + \cdots + b_p\mathbf{a}_p' = \mathbf{0}'.$$

Then we must also have $\mathbf{b}'\mathbf{c} = b_1c_1 + b_2c_2 + \cdots + b_pc_p = 0$, since multiplication of $\mathbf{Ax} = \mathbf{c}$ by $\mathbf{b}'$ gives $\mathbf{b}'\mathbf{Ax} = \mathbf{b}'\mathbf{c}$, or $\mathbf{0}'\mathbf{x} = \mathbf{b}'\mathbf{c}$. Otherwise, if $\mathbf{b}'\mathbf{c} \neq 0$, there is no $\mathbf{x}$ such that $\mathbf{Ax} = \mathbf{c}$. Hence, in order for $\mathbf{Ax} = \mathbf{c}$ to be consistent, the same linear relationships, if any, that exist among the rows of $\mathbf{A}$ must exist among the elements (rows) of $\mathbf{c}$. This is formalized by comparing the rank of $\mathbf{A}$ with the rank of the *augmented matrix* $(\mathbf{A}, \mathbf{c})$. The notation $(\mathbf{A}, \mathbf{c})$ indicates that $\mathbf{c}$ has been appended to $\mathbf{A}$ as an additional column.

Theorem 2.7 The system of equations $\mathbf{Ax} = \mathbf{c}$ has at least one solution vector $\mathbf{x}$ if and only if $\text{rank}(\mathbf{A}) = \text{rank}(\mathbf{A}, \mathbf{c})$.

PROOF. Suppose that $\text{rank}(\mathbf{A}) = \text{rank}(\mathbf{A}, \mathbf{c})$, so that appending $\mathbf{c}$ does not change the rank. Then $\mathbf{c}$ is a linear combination of the columns of $\mathbf{A}$; that is, there exists some $\mathbf{x}$ such that

$$x_1\mathbf{a}_1 + x_2\mathbf{a}_2 + \cdots + x_p\mathbf{a}_p = \mathbf{c},$$

which, by (2.37), can be written as $\mathbf{Ax} = \mathbf{c}$. Thus $\mathbf{x}$ is a solution.

Conversely, suppose that there exists a solution vector $\mathbf{x}$ such that $\mathbf{Ax} = \mathbf{c}$. In general, $\text{rank}\ (\mathbf{A}) \leq \text{rank}(\mathbf{A}, \mathbf{c})$ (Harville 1997, p. 41). But since there exists an $\mathbf{x}$ such that $\mathbf{Ax} = \mathbf{c}$, we have

$$\begin{aligned} \text{rank}(\mathbf{A}, \mathbf{c}) &= \text{rank}(\mathbf{A}, \mathbf{Ax}) = \text{rank}[\mathbf{A}(\mathbf{I}, \mathbf{x})] \\ &\leq \text{rank}(\mathbf{A}) \qquad \text{[by Theorem 2.4(i)]}. \end{aligned}$$

Hence

$$\text{rank}(\mathbf{A}) \leq \text{rank}(\mathbf{A}, \mathbf{c}) \leq \text{rank}(\mathbf{A}),$$

and we have $\text{rank}(\mathbf{A}) = \text{rank}(\mathbf{A}, \mathbf{c})$. □

A consistent system of equations can be solved by the usual methods given in elementary algebra courses for eliminating variables, such as adding a multiple of one equation to another or solving for a variable and substituting into another equation. In the process, one or more variables may end up as arbitrary constants, thus generating an infinite number of solutions. A method of solution involving generalized inverses is given in Section 2.8.2. Some illustrations of systems of equations and their solutions are given in the following examples.

Example 2.7a. Consider the system of equations

$$\begin{aligned} x_1 + 2x_2 &= 4 \\ x_1 - x_2 &= 1 \\ x_1 + x_2 &= 3 \end{aligned}$$

or

$$\begin{pmatrix} 1 & 2 \\ 1 & -1 \\ 1 & 1 \end{pmatrix} \begin{pmatrix} x_1 \\ x_2 \end{pmatrix} = \begin{pmatrix} 4 \\ 1 \\ 3 \end{pmatrix}.$$

The augmented matrix is

$$(\mathbf{A}, \mathbf{c}) = \begin{pmatrix} 1 & 2 & 4 \\ 1 & -1 & 1 \\ 1 & 1 & 3 \end{pmatrix},$$

which has rank = 2 because the third column is equal to twice the first column plus the second:

$$2\begin{pmatrix} 1 \\ 1 \\ 1 \end{pmatrix} + \begin{pmatrix} 2 \\ -1 \\ 1 \end{pmatrix} = \begin{pmatrix} 4 \\ 1 \\ 3 \end{pmatrix}.$$

Since $\text{rank}(\mathbf{A}) = \text{rank}(\mathbf{A}, \mathbf{c}) = 2$, there is at least one solution. If we add twice the first equation to the second, the result is a multiple of the third equation. Thus the third equation is redundant, and the first two can readily be solved to obtain the unique solution $\mathbf{x} = (2, 1)'$.

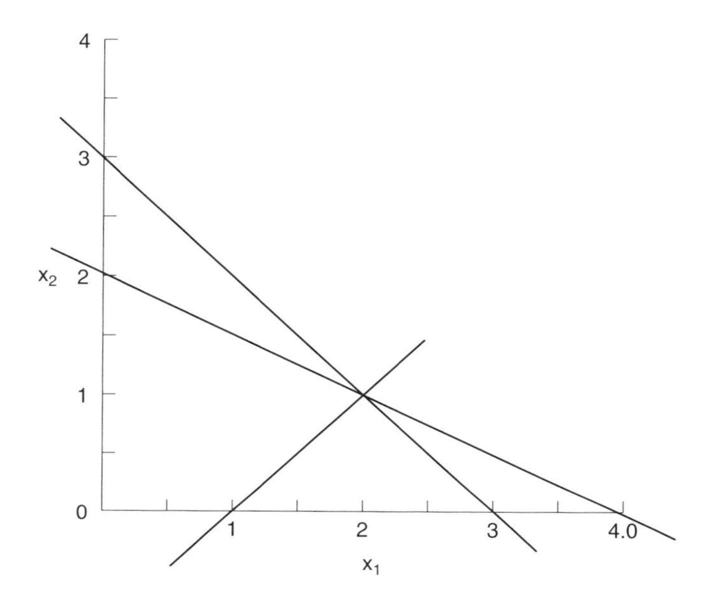

Figure 2.1 Three lines representing the three equations in Example 2.7a.

The three lines representing the three equations are plotted in Figure 2.1. Notice that the three lines intersect at the point (2, 1), which is the unique solution of the three equations. □

Example 2.7b. If we change the 3 to 2 in the third equation in Example 2.7, the augmented matrix becomes

$$(\mathbf{A}, \mathbf{c}) = \begin{pmatrix} 1 & 2 & 4 \\ 1 & -1 & 1 \\ 1 & 1 & 2 \end{pmatrix},$$

which has rank = 3, since no linear combination of columns is $\mathbf{0}$. [Alternatively, $|(\mathbf{A}, \mathbf{c})| \neq 0$, and $(\mathbf{A}, \mathbf{c})$ is nonsingular; see Theorem 2.9(iii)] Hence rank $(\mathbf{A}, \mathbf{c}) = 3 \neq \text{rank}(\mathbf{A}) = 2$, and the system is inconsistent.

The three lines representing the three equations are plotted in Figure 2.2, in which we see that the three lines do not have a common point of intersection. [For the "best" approximate solution, one approach is to use least squares; that is, we find the values of x_1 and x_2 that minimize $(x_1 + 2x_2 - 4)^2 + (x_1 - x_2 - 1)^2 + (x_1 + x_2 - 2)^2$.] □

Example 2.7c. Consider the system

$$\begin{aligned} x_1 + x_2 + x_3 &= 1 \\ 2x_1 + x_2 + 3x_3 &= 5 \\ 3x_1 + 2x_2 + 4x_3 &= 6. \end{aligned}$$

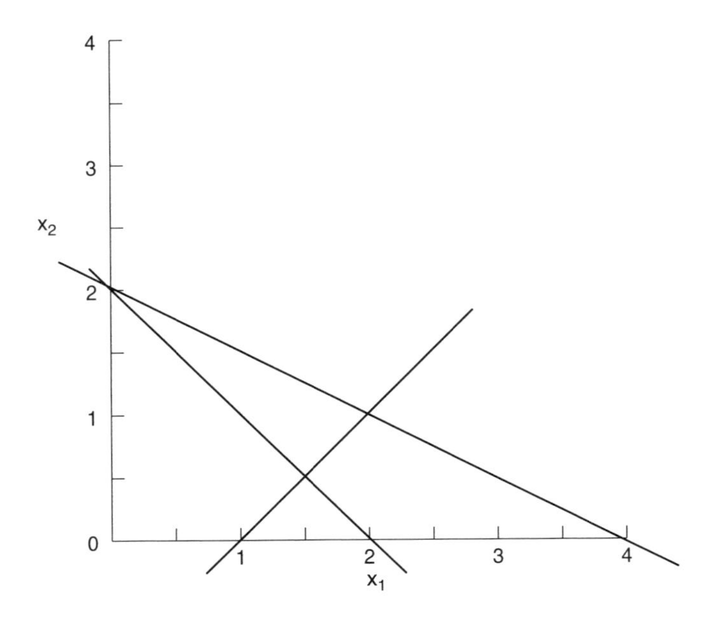

Figure 2.2 Three lines representing the three equations in Example 2.7b.

The third equation is the sum of the first two, but the second is not a multiple of the first. Thus, rank($\mathbf{A}, \mathbf{c}$) = rank($\mathbf{A}$) = 2, and the system is consistent.

By solving the first two equations for x_1 and x_2 in terms of x_3, we obtain

$$x_1 = -2x_3 + 4$$
$$x_2 = x_3 - 3.$$

The solution vector can be expressed as

$$\mathbf{x} = \begin{pmatrix} -2x_3 + 4 \\ x_3 - 3 \\ x_3 \end{pmatrix} = x_3 \begin{pmatrix} -2 \\ 1 \\ 1 \end{pmatrix} + \begin{pmatrix} 4 \\ -3 \\ 0 \end{pmatrix},$$

where x_3 is an arbitrary constant. Geometrically, $\mathbf{x}$ is the line representing the intersection of the two planes corresponding to the first two equations. □

2.8 GENERALIZED INVERSE

We now consider generalized inverses of those matrices that do not have inverses in the usual sense [see (2.45)]. A solution of a consistent system of equations $\mathbf{Ax} = \mathbf{c}$ can be expressed in terms of a generalized inverse of $\mathbf{A}$.

2.8.1 Definition and Properties

A *generalized inverse* of an $n \times p$ matrix $\mathbf{A}$ is any matrix $\mathbf{A}^-$ that satisfies

$$\mathbf{AA}^-\mathbf{A} = \mathbf{A}. \tag{2.58}$$

A generalized inverse is not unique except when $\mathbf{A}$ is nonsingular, in which case $\mathbf{A}^- = \mathbf{A}^{-1}$. A generalized inverse is also called a *conditional inverse*.

Every matrix, whether square or rectangular, has a generalized inverse. This holds even for vectors. For example, let

$$\mathbf{x} = \begin{pmatrix} 1 \\ 2 \\ 3 \\ 4 \end{pmatrix}.$$

Then $\mathbf{x}_1^- = (1, 0, 0, 0)$ is a generalized inverse of $\mathbf{x}$ satisfying (2.58). Other examples are $\mathbf{x}_2^- = (0, \frac{1}{2}, 0, 0)$, $\mathbf{x}_3^- = (0, 0, \frac{1}{3}, 0)$, and $\mathbf{x}_4^- = (0, 0, 0, \frac{1}{4})$. For each $\mathbf{x}_i^-$, we have

$$\mathbf{xx}_i^-\mathbf{x} = \mathbf{x}1 = \mathbf{x}, \quad i = 1, 2, 3, 4.$$

In this illustration, $\mathbf{x}$ is a column vector and $\mathbf{x}_i^-$ is a row vector. This pattern is generalized in the following theorem.

Theorem 2.8a. If $\mathbf{A}$ is $n \times p$, any generalized inverse $\mathbf{A}^-$ is $p \times n$. □

In the following example we give two illustrations of generalized inverses of a singular matrix.

Example 2.8.1. Let

$$\mathbf{A} = \begin{pmatrix} 2 & 2 & 3 \\ 1 & 0 & 1 \\ 3 & 2 & 4 \end{pmatrix}. \tag{2.59}$$

The third row of $\mathbf{A}$ is the sum of the first two rows, and the second row is not a multiple of the first; hence $\mathbf{A}$ has rank 2. Let

$$\mathbf{A}_1^- = \begin{pmatrix} 0 & 1 & 0 \\ \frac{1}{2} & -1 & 0 \\ 0 & 0 & 0 \end{pmatrix}, \quad \mathbf{A}_2^- = \begin{pmatrix} 0 & 1 & 0 \\ 0 & -\frac{3}{2} & \frac{1}{2} \\ 0 & 0 & 0 \end{pmatrix}. \tag{2.60}$$

It is easily verified that $\mathbf{AA}_1^-\mathbf{A} = \mathbf{A}$ and $\mathbf{AA}_2^-\mathbf{A} = \mathbf{A}$. □

The methods used to obtain $\mathbf{A}_1^-$ and $\mathbf{A}_2^-$ in (2.60) are described in Theorem 2.8b and the five-step algorithm following the theorem.

Theorem 2.8b. Suppose $\mathbf{A}$ is $n \times p$ of rank r and that $\mathbf{A}$ is partitioned as

$$\mathbf{A} = \begin{pmatrix} \mathbf{A}_{11} & \mathbf{A}_{12} \\ \mathbf{A}_{21} & \mathbf{A}_{22} \end{pmatrix},$$

where $\mathbf{A}_{11}$ is $r \times r$ of rank r. Then a generalized inverse of $\mathbf{A}$ is given by

$$\mathbf{A}^- = \begin{pmatrix} \mathbf{A}_{11}^{-1} & \mathbf{O} \\ \mathbf{O} & \mathbf{O} \end{pmatrix},$$

where the three $\mathbf{O}$ matrices are of appropriate sizes so that $\mathbf{A}^-$ is $p \times n$.

PROOF. By multiplication of partitioned matrices, as in (2.35), we obtain

$$\mathbf{A}\mathbf{A}^-\mathbf{A} = \begin{pmatrix} \mathbf{I} & \mathbf{O} \\ \mathbf{A}_{21}\mathbf{A}_{11}^{-1} & \mathbf{O} \end{pmatrix}\mathbf{A} = \begin{pmatrix} \mathbf{A}_{11} & \mathbf{A}_{12} \\ \mathbf{A}_{21} & \mathbf{A}_{21}\mathbf{A}_{11}^{-1}\mathbf{A}_{12} \end{pmatrix}.$$

To show that $\mathbf{A}_{21}\mathbf{A}_{11}^{-1}\mathbf{A}_{12} = \mathbf{A}_{22}$, multiply $\mathbf{A}$ by

$$\mathbf{B} = \begin{pmatrix} \mathbf{I} & \mathbf{O} \\ -\mathbf{A}_{21}\mathbf{A}_{11}^{-1} & \mathbf{I} \end{pmatrix},$$

where $\mathbf{O}$ and $\mathbf{I}$ are of appropriate sizes, to obtain

$$\mathbf{B}\mathbf{A} = \begin{pmatrix} \mathbf{A}_{11} & \mathbf{A}_{12} \\ \mathbf{O} & \mathbf{A}_{22} - \mathbf{A}_{21}\mathbf{A}_{11}^{-1}\mathbf{A}_{12} \end{pmatrix}.$$

The matrix $\mathbf{B}$ is nonsingular, and the rank of $\mathbf{BA}$ is therefore $r = \text{rank}(\mathbf{A})$ [see Theorem 2.4(ii)]. In $\mathbf{BA}$, the submatrix $\begin{pmatrix} \mathbf{A}_{11} \\ \mathbf{O} \end{pmatrix}$ is of rank r, and the columns headed by $\mathbf{A}_{12}$ are therefore linear combinations of the columns headed by $\mathbf{A}_{11}$. By a comment following Example 2.3, this relationship can be expressed as

$$\begin{pmatrix} \mathbf{A}_{12} \\ \mathbf{A}_{22} - \mathbf{A}_{21}\mathbf{A}_{11}^{-1}\mathbf{A}_{12} \end{pmatrix} = \begin{pmatrix} \mathbf{A}_{11} \\ \mathbf{O} \end{pmatrix}\mathbf{Q} \qquad (2.61)$$

for some matrix $\mathbf{Q}$. By (2.27), the right side of (2.61) becomes

$$\begin{pmatrix} \mathbf{A}_{11} \\ \mathbf{O} \end{pmatrix} \mathbf{Q} = \begin{pmatrix} \mathbf{A}_{11}\mathbf{Q} \\ \mathbf{O}\mathbf{Q} \end{pmatrix} = \begin{pmatrix} \mathbf{A}_{11}\mathbf{Q} \\ \mathbf{O} \end{pmatrix}.$$

Thus $\mathbf{A}_{22} - \mathbf{A}_{21}\mathbf{A}_{11}^{-1}\mathbf{A}_{12} = \mathbf{O}$, or

$$\mathbf{A}_{22} = \mathbf{A}_{21}\mathbf{A}_{11}^{-1}\mathbf{A}_{12}.$$

□

Corollary 1. Suppose that $\mathbf{A}$ is $n \times p$ of rank r and that $\mathbf{A}$ is partitioned as in Theorem 2.8b, where $\mathbf{A}_{22}$ is $r \times r$ of rank r. Then a generalized inverse of $\mathbf{A}$ is given by

$$\mathbf{A}^{-} = \begin{pmatrix} \mathbf{O} & \mathbf{O} \\ \mathbf{O} & \mathbf{A}_{22}^{-1} \end{pmatrix},$$

where the three $\mathbf{O}$ matrices are of appropriate sizes so that $\mathbf{A}^{-}$ is $p \times n$. □

The nonsingular submatrix need not be in the $\mathbf{A}_{11}$ or $\mathbf{A}_{22}$ position, as in Theorem 2.8b or its corollary. Theorem 2.8b can be extended to the following algorithm for finding a conditional inverse $\mathbf{A}^{-}$ for any $n \times p$ matrix $\mathbf{A}$ of rank r (Searle 1982, p. 218):

1. Find any nonsingular $r \times r$ submatrix $\mathbf{C}$. It is not necessary that the elements of $\mathbf{C}$ occupy adjacent rows and columns in $\mathbf{A}$.
2. Find $\mathbf{C}^{-1}$ and $(\mathbf{C}^{-1})'$.
3. Replace the elements of $\mathbf{C}$ by the elements of $(\mathbf{C}^{-1})'$.
4. Replace all other elements in $\mathbf{A}$ by zeros.
5. Transpose the resulting matrix.

Some properties of generalized inverses are given in the following theorem, which is the theoretical basis for many of the results in Chapter 11.

Theorem 2.8c. Let $\mathbf{A}$ be $n \times p$ of rank r, let $\mathbf{A}^{-}$ be any generalized inverse of $\mathbf{A}$, and let $(\mathbf{A}'\mathbf{A})^{-}$ be any generalized inverse of $\mathbf{A}'\mathbf{A}$. Then

(i) $\text{rank}(\mathbf{A}^{-}\mathbf{A}) = \text{rank}(\mathbf{A}\mathbf{A}^{-}) = \text{rank}(\mathbf{A}) = r$.
(ii) $(\mathbf{A}^{-})'$ is a generalized inverse of $\mathbf{A}'$; that is, $(\mathbf{A}')^{-} = (\mathbf{A}^{-})'$.
(iii) $\mathbf{A} = \mathbf{A}(\mathbf{A}'\mathbf{A})^{-}\mathbf{A}'\mathbf{A}$ and $\mathbf{A}' = \mathbf{A}'\mathbf{A}(\mathbf{A}'\mathbf{A})^{-}\mathbf{A}'$.
(iv) $(\mathbf{A}'\mathbf{A})^{-}\mathbf{A}'$ is a generalized inverse of $\mathbf{A}$; that is, $\mathbf{A}^{-} = (\mathbf{A}'\mathbf{A})^{-}\mathbf{A}'$.

(v) $\mathbf{A}(\mathbf{A}'\mathbf{A})^{-}\mathbf{A}'$ is symmetric, has rank $= r$, and is invariant to the choice of $(\mathbf{A}'\mathbf{A})^{-}$; that is, $\mathbf{A}(\mathbf{A}'\mathbf{A})^{-}\mathbf{A}'$ remains the same, no matter what value of $(\mathbf{A}'\mathbf{A})^{-}$ is used. □

A generalized inverse of a symmetric matrix is not necessarily symmetric. However, it is also true that a symmetric generalized inverse can always be found for a symmetric matrix; see Problem 2.46. In this book, we will assume that generalized inverses of symmetric matrices are symmetric.

2.8.2 Generalized Inverses and Systems of Equations

Generalized inverses can be used to find solutions to a system of equations.

Theorem 2.8d. If the system of equations $\mathbf{Ax} = \mathbf{c}$ is consistent and if $\mathbf{A}^{-}$ is any generalized inverse for $\mathbf{A}$, then $\mathbf{x} = \mathbf{A}^{-}\mathbf{c}$ is a solution.

PROOF. Since $\mathbf{AA}^{-}\mathbf{A} = \mathbf{A}$, we have

$$\mathbf{AA}^{-}\mathbf{Ax} = \mathbf{Ax}.$$

Substituting $\mathbf{Ax} = \mathbf{c}$ on both sides, we obtain

$$\mathbf{AA}^{-}\mathbf{c} = \mathbf{c}.$$

Writing this in the form $\mathbf{A}(\mathbf{A}^{-}\mathbf{c}) = \mathbf{c}$, we see that $\mathbf{A}^{-}\mathbf{c}$ is a solution to $\mathbf{Ax} = \mathbf{c}$. □

Different choices of $\mathbf{A}^{-}$ will result in different solutions for $\mathbf{Ax} = \mathbf{c}$.

Theorem 2.8e. If the system of equations $\mathbf{Ax} = \mathbf{c}$ is consistent, then all possible solutions can be obtained in the following two ways:

(i) Use a specific $\mathbf{A}^{-}$ in $\mathbf{x} = \mathbf{A}^{-}\mathbf{c} + (\mathbf{I} - \mathbf{A}^{-}\mathbf{A})\mathbf{h}$, and use all possible values of the arbitrary vector $\mathbf{h}$.

(ii) Use all possible values of $\mathbf{A}^{-}$ in $\mathbf{x} = \mathbf{A}^{-}\mathbf{c}$ if $\mathbf{c} \neq \mathbf{0}$.

PROOF. See Searle (1982, p. 238). □

A necessary and sufficient condition for the system of equations $\mathbf{Ax} = \mathbf{c}$ to be consistent can be given in terms of a generalized inverse of $\mathbf{A}$ (Graybill 1976, p. 36).

Theorem 2.8f. The system of equations $\mathbf{Ax} = \mathbf{c}$ has a solution if and only if for any generalized inverse $\mathbf{A}^-$ of $\mathbf{A}$

$$\mathbf{AA}^-\mathbf{c} = \mathbf{c}.$$

Proof. Suppose that $\mathbf{Ax} = \mathbf{c}$ is consistent. Then, by Theorem 2.8d, $\mathbf{x} = \mathbf{A}^-\mathbf{c}$ is a solution. Multiply $\mathbf{c} = \mathbf{Ax}$ by $\mathbf{AA}^-$ to obtain

$$\mathbf{AA}^-\mathbf{c} = \mathbf{AA}^-\mathbf{Ax} = \mathbf{Ax} = \mathbf{c}.$$

Conversely, suppose $\mathbf{AA}^-\mathbf{c} = \mathbf{c}$. Multiply $\mathbf{x} = \mathbf{A}^-\mathbf{c}$ by $\mathbf{A}$ to obtain

$$\mathbf{Ax} = \mathbf{AA}^-\mathbf{c} = \mathbf{c}.$$

Hence, a solution exists, namely, $\mathbf{x} = \mathbf{A}^-\mathbf{c}$. □

Theorem 2.8f provides an alternative to Theorem 2.7a for determining whether a system of equations is consistent.

2.9 DETERMINANTS

The *determinant* of an $n \times n$ matrix $\mathbf{A}$ is a scalar function of $\mathbf{A}$ defined as the sum of all $n!$ possible products of n elements such that

1. each product contains one element from every row and every column of $\mathbf{A}$.
2. the factors in each product are written so that the column subscripts appear in order of magnitude and each product is then preceded by a plus or minus sign according to whether the number of inversions in the row subscripts is even or odd. (An *inversion* occurs whenever a larger number precedes a smaller one.)

The determinant of $\mathbf{A}$ is denoted by $|\mathbf{A}|$ or $\det(\mathbf{A})$. The preceding definition is not very useful in evaluating determinants, except in the case of 2×2 or 3×3 matrices. For larger matrices, determinants are typically found by computer. Some calculators also evaluate determinants.

The determinants of some special square matrices are given in the following theorem.

Theorem 2.9a.

(i) If $\mathbf{D} = \text{diag}(d_1, d_2, \ldots, d_n)$, $\quad |\mathbf{D}| = \prod_{i=1}^{n} d_i$.

(ii) The determinant of a triangular matrix is the product of the diagonal elements.

(iii) If $\mathbf{A}$ is singular, $|\mathbf{A}| = 0$.

(iv) If $\mathbf{A}$ is nonsingular, $|\mathbf{A}| \neq 0$.

(v) If $\mathbf{A}$ is positive definite, $|\mathbf{A}| > 0$.

(vi) $|\mathbf{A}'| = |\mathbf{A}|$.

(vii) If $\mathbf{A}$ is nonsingular, $|\mathbf{A}^{-1}| = \dfrac{1}{|\mathbf{A}|}$.

□

Example 2.9a. We illustrate each of the properties in Theorem 2.9a.

(i) diagonal: $\begin{vmatrix} 2 & 0 \\ 0 & 3 \end{vmatrix} = (2)(3) - (0)(0) = (2)(3).$

(ii) triangular: $\begin{vmatrix} 2 & 1 \\ 0 & 3 \end{vmatrix} = (2)(3) - (0)(1) = (2)(3).$

(iii) singular: $\begin{vmatrix} 1 & 2 \\ 3 & 6 \end{vmatrix} = (1)(6) - (3)(2) = 0,$

nonsingular: $\begin{vmatrix} 1 & 2 \\ 3 & 4 \end{vmatrix} = (1)(4) - (3)(2) = -2.$

(iv) positive definite: $\begin{vmatrix} 3 & -2 \\ -2 & 4 \end{vmatrix} = (3)(4) - (-2)(-2) = 8 > 0.$

(v) transpose: $\begin{vmatrix} 3 & -7 \\ 2 & 1 \end{vmatrix} = (3)(1) - (2)(-7) = 17,$

$\begin{vmatrix} 3 & 2 \\ -7 & 1 \end{vmatrix} = (3)(1) - (-7)(2) = 17.$

(vi) inverse:

$$\begin{pmatrix} 3 & 2 \\ 1 & 4 \end{pmatrix}^{-1} = \begin{pmatrix} .4 & -.2 \\ -.1 & .3 \end{pmatrix}, \quad \begin{vmatrix} 3 & 2 \\ 1 & 4 \end{vmatrix} = 10, \quad \begin{vmatrix} .4 & -.2 \\ -.1 & .3 \end{vmatrix} = .1.$$

□

As a special case of (62), suppose that all diagonal elements are equal, say, $\mathbf{D} = \text{diag}(c, c, \ldots, c) = c\mathbf{I}$. Then

$$|\mathbf{D}| = |c\mathbf{I}| = \prod_{i=1}^{n} c = c^n. \qquad (2.68)$$

By extension, if an $n \times n$ matrix is multiplied by a scalar, the determinant becomes

$$|c\mathbf{A}| = c^n|\mathbf{A}|. \tag{2.69}$$

The determinant of certain partitioned matrices is given in the following theorem.

Theorem 2.9b. If the square matrix $\mathbf{A}$ is partitioned as

$$\mathbf{A} = \begin{pmatrix} \mathbf{A}_{11} & \mathbf{A}_{12} \\ \mathbf{A}_{21} & \mathbf{A}_{22} \end{pmatrix}, \tag{2.70}$$

and if $\mathbf{A}_{11}$ and $\mathbf{A}_{22}$ are square and nonsingular (but not necessarily the same size), then

$$|\mathbf{A}| = |\mathbf{A}_{11}|\,|\mathbf{A}_{22} - \mathbf{A}_{21}\mathbf{A}_{11}^{-1}\mathbf{A}_{12}| \tag{2.71}$$

$$= |\mathbf{A}_{22}|\,|\mathbf{A}_{11} - \mathbf{A}_{12}\mathbf{A}_{22}^{-1}\mathbf{A}_{21}|. \tag{2.72}$$

□

Note the analogy of (2.71) and (2.72) to the case of the determinant of a 2×2 matrix:

$$\begin{aligned}
\begin{vmatrix} a_{11} & a_{12} \\ a_{21} & a_{22} \end{vmatrix} &= a_{11}a_{22} - a_{21}a_{12} \\
&= a_{11}\left(a_{22} - \frac{a_{21}a_{12}}{a_{11}}\right) \\
&= a_{22}\left(a_{11} - \frac{a_{12}a_{21}}{a_{22}}\right).
\end{aligned}$$

Corollary 1. Suppose

$$\mathbf{A} = \begin{pmatrix} \mathbf{A}_{11} & \mathbf{O} \\ \mathbf{A}_{21} & \mathbf{A}_{22} \end{pmatrix} \quad \text{or} \quad \mathbf{A} = \begin{pmatrix} \mathbf{A}_{11} & \mathbf{A}_{12} \\ \mathbf{O} & \mathbf{A}_{22} \end{pmatrix},$$

where $\mathbf{A}_{11}$ and $\mathbf{A}_{22}$ are square (but not necessarily the same size). Then in either case

$$|\mathbf{A}| = |\mathbf{A}_{11}|\,|\mathbf{A}_{22}|. \tag{2.73}$$

□

Corollary 2. Let

$$\mathbf{A} = \begin{pmatrix} \mathbf{A}_{11} & \mathbf{O} \\ \mathbf{O} & \mathbf{A}_{22} \end{pmatrix},$$

where $\mathbf{A}_{11}$ and $\mathbf{A}_{22}$ are square (but not necessarily the same size). Then

$$|\mathbf{A}| = |\mathbf{A}_{11}|\,|\mathbf{A}_{22}|. \tag{2.74}$$

□

Corollary 3. If $\mathbf{A}$ has the form $\mathbf{A} = \begin{pmatrix} \mathbf{A}_{11} & \mathbf{a}_{12} \\ \mathbf{a}'_{12} & a_{22} \end{pmatrix}$, where $\mathbf{A}_{11}$ is a nonsingular matrix, $\mathbf{a}_{12}$ is a vector, and a_{22} is a 1×1 matrix, then

$$|\mathbf{A}| = \begin{vmatrix} \mathbf{A}_{11} & \mathbf{a}_{12} \\ \mathbf{a}'_{12} & a_{22} \end{vmatrix} = |\mathbf{A}_{11}|(a_{22} - \mathbf{a}'_{12}\mathbf{A}_{11}^{-1}\mathbf{a}_{12}). \tag{2.75}$$

□

Corollary 4. If $\mathbf{A}$ has the form $\mathbf{A} = \begin{pmatrix} \mathbf{B} & \mathbf{c} \\ -\mathbf{c}' & 1 \end{pmatrix}$, where $\mathbf{c}$ is a vector and $\mathbf{B}$ is a nonsingular matrix, then

$$|\mathbf{B} + \mathbf{cc}'| = |\mathbf{B}|(1 + \mathbf{c}'\mathbf{B}^{-1}\mathbf{c}). \tag{2.76}$$

□

The determinant of the product of two square matrices is given in the following theorem.

Theorem 2.9c. If $\mathbf{A}$ and $\mathbf{B}$ are square and the same size, then the determinant of the product is the product of the determinants:

$$|\mathbf{AB}| = |\mathbf{A}|\,|\mathbf{B}|. \tag{2.77}$$

□

Corollary 1

$$|\mathbf{AB}| = |\mathbf{BA}|. \tag{2.78}$$

□

Corollary 2

$$|\mathbf{A}^2| = |\mathbf{A}|^2. \tag{2.79}$$

□

Example 2.9b. To illustrate Theorem 2.9c, let

$$\mathbf{A} = \begin{pmatrix} 1 & 2 \\ 3 & 4 \end{pmatrix} \quad \text{and} \quad \mathbf{B} = \begin{pmatrix} 3 & -2 \\ 1 & 2 \end{pmatrix}.$$

Then

$$\mathbf{AB} = \begin{pmatrix} 5 & 2 \\ 13 & 2 \end{pmatrix}, \quad |\mathbf{AB}| = -16,$$

$$|\mathbf{A}| = -2, \quad |\mathbf{B}| = 8, \quad |\mathbf{A}|\,|\mathbf{B}| = -16.$$

□

2.10 ORTHOGONAL VECTORS AND MATRICES

Two $n \times 1$ vectors **b** and **b** are said to be *orthogonal* if

$$\mathbf{a}'\mathbf{b} = a_1b_1 + a_2b_2 + \cdots + a_nb_n = 0. \tag{2.80}$$

Note that the term *orthogonal* applies to *two* vectors, not to a single vector.

Geometrically, two orthogonal vectors are perpendicular to each other. This is illustrated in Figure 2.3 for the vectors $\mathbf{x}_1 = (4, 2)'$ and $\mathbf{x}_2 = (-1, 2)'$. Note that $\mathbf{x}_1'\mathbf{x}_2 = (4)(-1) + (2)(2) = 0$.

To show that two orthogonal vectors are perpendicular, let θ be the angle between vectors **a** and **b** in Figure 2.4. The vector from the terminal point of **a** to the terminal point of **b** can be represented as $\mathbf{c} = \mathbf{b} - \mathbf{a}$. The law of cosines for the relationship of

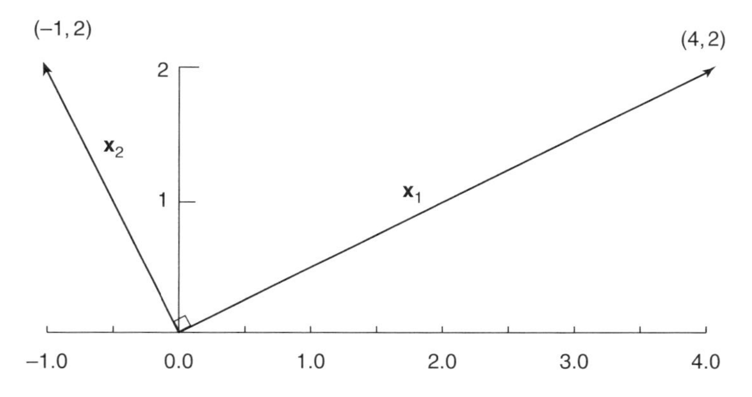

Figure 2.3 Two orthogonal (perpendicular) vectors.

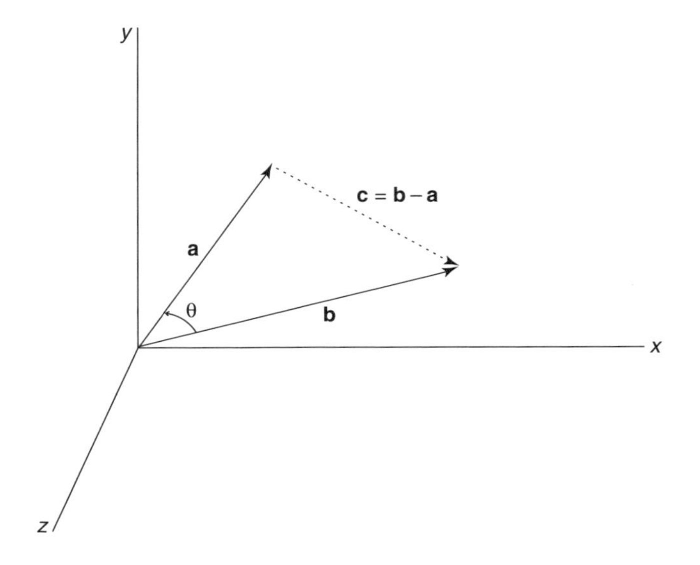

Figure 2.4 Vectors **a** and **b** in 3-space.

θ to the sides of the triangle can be stated in vector form as

$$\begin{aligned}\cos\theta &= \frac{\mathbf{a}'\mathbf{a} + \mathbf{b}'\mathbf{b} - (\mathbf{b}-\mathbf{a})'(\mathbf{b}-\mathbf{a})}{2\sqrt{(\mathbf{a}'\mathbf{a})(\mathbf{b}'\mathbf{b})}} \\ &= \frac{\mathbf{a}'\mathbf{a} + \mathbf{b}'\mathbf{b} - (\mathbf{b}'\mathbf{b} + \mathbf{a}'\mathbf{a} - 2\mathbf{a}'\mathbf{b})}{2\sqrt{(\mathbf{a}'\mathbf{a})(\mathbf{b}'\mathbf{b})}} \\ &= \frac{\mathbf{a}'\mathbf{b}}{\sqrt{(\mathbf{a}'\mathbf{a})(\mathbf{b}'\mathbf{b})}}. \end{aligned} \tag{2.81}$$

When $\theta = 90^\circ$, $\mathbf{a}'\mathbf{b} = 0$ since $\cos(90^\circ) = 0$. Thus **a** and **b** are *perpendicular* when $\mathbf{a}'\mathbf{b} = 0$.

If $\mathbf{a}'\mathbf{a} = 1$, the vector **a** is said to be *normalized*. A vector **b** can be normalized by dividing by its length, $\sqrt{\mathbf{b}'\mathbf{b}}$. Thus

$$\mathbf{c} = \frac{\mathbf{b}}{\sqrt{\mathbf{b}'\mathbf{b}}} \tag{2.82}$$

is normalized so that $\mathbf{c}'\mathbf{c} = 1$.

A set of $p \times 1$ vectors $\mathbf{c}_1, \mathbf{c}_2, \ldots, \mathbf{c}_p$ that are normalized ($\mathbf{c}_i'\mathbf{c}_i = 1$ for all i) and mutually orthogonal ($\mathbf{c}_i'\mathbf{c}_j = 0$ for all $i \neq j$) is said to be an *orthonormal set* of vectors. If the $p \times p$ matrix $\mathbf{C} = (\mathbf{c}_1, \mathbf{c}_2, \ldots, \mathbf{c}_p)$ has orthonormal columns, $\mathbf{C}$ is called an *orthogonal* matrix. Since the elements of $\mathbf{C}'\mathbf{C}$ are products of columns of

C [see Theorem 2.2c(i)], an orthogonal matrix **C** has the property

$$\mathbf{C}'\mathbf{C} = \mathbf{I}. \tag{2.83}$$

It can be shown that an orthogonal matrix **C** also satisfies

$$\mathbf{C}\mathbf{C}' = \mathbf{I}. \tag{2.84}$$

Thus an orthogonal matrix **C** has orthonormal rows as well as orthonormal columns. It is also clear from (2.83) and (2.84) that $\mathbf{C}' = \mathbf{C}^{-1}$ if **C** is orthogonal.

Example 2.10. To illustrate an orthogonal matrix, we start with

$$\mathbf{A} = \begin{pmatrix} 1 & 1 & 1 \\ 1 & -2 & 0 \\ 1 & 1 & -1 \end{pmatrix},$$

whose columns are mutually orthogonal but not orthonormal. To normalize the three columns, we divide by their respective lengths, $\sqrt{3}$, $\sqrt{6}$, and $\sqrt{2}$, to obtain the matrix

$$\mathbf{C} = \begin{pmatrix} 1/\sqrt{3} & 1/\sqrt{6} & 1/\sqrt{2} \\ 1/\sqrt{3} & -2/\sqrt{6} & 0 \\ 1/\sqrt{3} & 1/\sqrt{6} & -1/\sqrt{2} \end{pmatrix},$$

whose columns are orthonormal. Note that the rows of **C** are also orthonormal, so that **C** satisfies (2.84) as well as (2.83). □

Multiplication of a vector by an orthogonal matrix has the effect of rotating axes; that is, if a point **x** is transformed to $\mathbf{z} = \mathbf{C}\mathbf{x}$, where **C** is orthogonal, then the distance from the origin to **z** is the same as the distance to **x**:

$$\mathbf{z}'\mathbf{z} = (\mathbf{C}\mathbf{x})'(\mathbf{C}\mathbf{x}) = \mathbf{x}'\mathbf{C}'\mathbf{C}\mathbf{x} = \mathbf{x}'\mathbf{I}\mathbf{x} = \mathbf{x}'\mathbf{x}. \tag{2.85}$$

Hence, the transformation from **x** to **z** is a rotation.

Some properties of orthogonal matrices are given in the following theorem.

Theorem 2.10. If the $p \times p$ matrix **C** is orthogonal and if **A** is any $p \times p$ matrix, then

(i) $|\mathbf{C}| = +1$ or -1.

(ii) $|\mathbf{C}'\mathbf{A}\mathbf{C}| = |\mathbf{A}|$.
(iii) $-1 \leq c_{ij} \leq 1$, where c_{ij} is any element of $\mathbf{C}$.

2.11 TRACE

The *trace* of an $n \times n$ matrix $\mathbf{A} = (a_{ij})$ is a scalar function defined as the sum of the diagonal elements of $\mathbf{A}$; that is, $\text{tr}(\mathbf{A}) = \sum_{i=1}^{n} a_{ii}$. For example, suppose

$$\mathbf{A} = \begin{pmatrix} 8 & 4 & 2 \\ 2 & -3 & 6 \\ 3 & 5 & 9 \end{pmatrix}.$$

Then

$$\text{tr}(\mathbf{A}) = 8 - 3 + 9 = 14.$$

Some properties of the trace are given in the following theorem.

Theorem 2.11

(i) If $\mathbf{A}$ and $\mathbf{B}$ are $n \times n$, then

$$\text{tr}(\mathbf{A} \pm \mathbf{B}) = \text{tr}(\mathbf{A}) \pm \text{tr}(\mathbf{B}). \tag{2.86}$$

(ii) If $\mathbf{A}$ is $n \times p$ and $\mathbf{B}$ is $p \times n$, then

$$\text{tr}(\mathbf{AB}) = \text{tr}(\mathbf{BA}). \tag{2.87}$$

Note that in (2.87) n can be less than, equal to, or greater than p.

(iii) If $\mathbf{A}$ is $n \times p$, then

$$\text{tr}(\mathbf{A}'\mathbf{A}) = \sum_{i=1}^{p} \mathbf{a}_i'\mathbf{a}_i, \tag{2.88}$$

where $\mathbf{a}_i$ is the ith *column* of $\mathbf{A}$.

(iv) If $\mathbf{A}$ is $n \times p$, then

$$\text{tr}(\mathbf{A}\mathbf{A}') = \sum_{i=1}^{n} \mathbf{a}_i'\mathbf{a}_i, \tag{2.89}$$

where $\mathbf{a}_i'$ is the ith *row* of $\mathbf{A}$.

(v) If $\mathbf{A} = (a_{ij})$ is an $n \times p$ matrix with representative element a_{ij}, then

$$\text{tr}(\mathbf{A}'\mathbf{A}) = \text{tr}(\mathbf{A}\mathbf{A}') = \sum_{i=1}^{n} \sum_{j=1}^{p} a_{ij}^2. \tag{2.90}$$

(vi) If $\mathbf{A}$ is any $n \times n$ matrix and $\mathbf{P}$ is any $n \times n$ nonsingular matrix, then

$$\text{tr}(\mathbf{P}^{-1}\mathbf{A}\mathbf{P}) = \text{tr}(\mathbf{A}). \tag{2.91}$$

(vii) If $\mathbf{A}$ is any $n \times n$ matrix and $\mathbf{C}$ is any $n \times n$ orthogonal matrix, then

$$\text{tr}(\mathbf{C}'\mathbf{A}\mathbf{C}) = \text{tr}(\mathbf{A}). \tag{2.92}$$

(viii) If $\mathbf{A}$ is $n \times p$ of rank r and $\mathbf{A}^-$ is a generalized inverse of $\mathbf{A}$, then

$$\text{tr}(\mathbf{A}^-\mathbf{A}) = \text{tr}(\mathbf{A}\mathbf{A}^-) = r. \tag{2.93}$$

Proof. We prove parts (ii), (iii), and (vi).

(ii) By (2.13), the ith diagonal element of $\mathbf{E} = \mathbf{AB}$ is $e_{ii} = \sum_k a_{ik}b_{ki}$. Then

$$\text{tr}(\mathbf{AB}) = \text{tr}(\mathbf{E}) = \sum_i e_{ii} = \sum_i \sum_k a_{ik}b_{ki}.$$

Similarly, the ith diagonal element of $\mathbf{F} = \mathbf{BA}$ is $f_{ii} = \sum_k b_{ik}a_{ki}$, and

$$\begin{aligned}\text{tr}(\mathbf{BA}) = \text{tr}(\mathbf{F}) = \sum_i f_{ii} &= \sum_i \sum_k b_{ik}a_{ki} \\ &= \sum_k \sum_i a_{ki}b_{ik} = \text{tr}(\mathbf{E}) = \text{tr}(\mathbf{AB}).\end{aligned}$$

(iii) By Theorem 2.2c(i), $\mathbf{A}'\mathbf{A}$ is obtained as products of columns of $\mathbf{A}$. If $\mathbf{a}_i$ is the ith column of $\mathbf{A}$, then the ith diagonal element of $\mathbf{A}'\mathbf{A}$ is $\mathbf{a}_i'\mathbf{a}_i$.

(vi) By (2.87) we obtain

$$\text{tr}(\mathbf{P}^{-1}\mathbf{A}\mathbf{P}) = \text{tr}(\mathbf{A}\mathbf{P}\mathbf{P}^{-1}) = \text{tr}(\mathbf{A}).$$

□

Example 2.11. We illustrate parts (ii) and (viii) of Theorem 2.11.

(ii) Let

$$\mathbf{A} = \begin{pmatrix} 1 & 3 \\ 2 & -1 \\ 4 & 6 \end{pmatrix} \quad \text{and} \quad \mathbf{B} = \begin{pmatrix} 3 & -2 & 1 \\ 2 & 4 & 5 \end{pmatrix}.$$

Then

$$\mathbf{AB} = \begin{pmatrix} 9 & 10 & 16 \\ 4 & -8 & -3 \\ 24 & 16 & 34 \end{pmatrix}, \quad \mathbf{BA} = \begin{pmatrix} 3 & 17 \\ 30 & 32 \end{pmatrix},$$

$$\text{tr}(\mathbf{AB}) = 9 - 8 + 34 = 35, \quad \text{tr}(\mathbf{BA}) = 3 + 32 = 35.$$

(viii) Using $\mathbf{A}$ in (2.59) and $\mathbf{A}_1^-$ in (2.60), we obtain

$$\mathbf{A}^-\mathbf{A} = \begin{pmatrix} 1 & 0 & 1 \\ 0 & 1 & \frac{1}{2} \\ 0 & 0 & 0 \end{pmatrix}, \quad \mathbf{AA}^- = \begin{pmatrix} 1 & 0 & 0 \\ 0 & 1 & 0 \\ 1 & 1 & 0 \end{pmatrix},$$

$$\text{tr}(\mathbf{A}^-\mathbf{A}) = 1 + 1 + 0 = 2 = \text{rank}(\mathbf{A}),$$
$$\text{tr}(\mathbf{AA}^-) = 1 + 1 + 0 = 2 = \text{rank}(\mathbf{A}).$$

□

2.12 EIGENVALUES AND EIGENVECTORS

2.12.1 Definition

For every square matrix $\mathbf{A}$, a scalar λ and a nonzero vector $\mathbf{x}$ can be found such that

$$\mathbf{Ax} = \lambda\mathbf{x}, \tag{2.94}$$

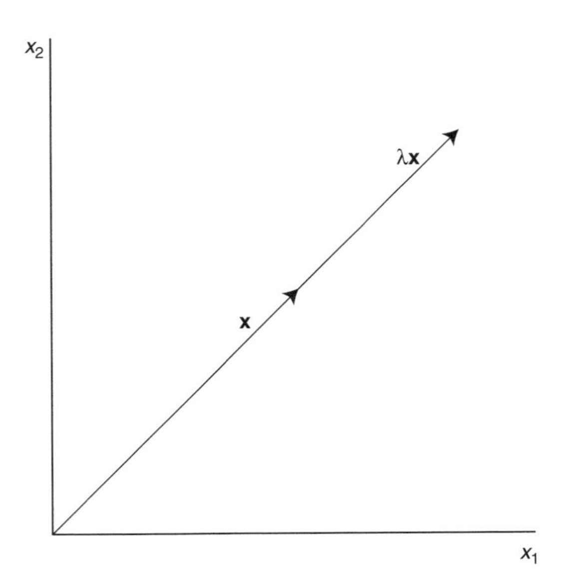

Figure 2.5 An eigenvector $\mathbf{x}$ is transformed to $\lambda\mathbf{x}$.

where λ is an *eigenvalue* of $\mathbf{A}$ and $\mathbf{x}$ is an *eigenvector*. (These terms are sometimes referred to as characteristic root and characteristic vector, respectively.) Note that in (2.94), the vector $\mathbf{x}$ is transformed by $\mathbf{A}$ onto a multiple of itself, so that the point $\mathbf{Ax}$ is on the line passing through $\mathbf{x}$ and the origin. This is illustrated in Figure 2.5.

To find λ and $\mathbf{x}$ for a matrix $\mathbf{A}$, we write (2.94) as

$$(\mathbf{A} - \lambda\mathbf{I})\mathbf{x} = \mathbf{0}. \tag{2.95}$$

By (2.37), $(\mathbf{A} - \lambda\mathbf{I})\mathbf{x}$ is a linear combination of the columns of $\mathbf{A} - \lambda\mathbf{I}$, and by (2.40) and (2.95), these columns are linearly dependent. Thus the square matrix $(\mathbf{A} - \lambda\mathbf{I})$ is singular, and by Theorem 2.9a(iii), we can solve for λ using

$$|\mathbf{A} - \lambda\mathbf{I}| = 0, \tag{2.96}$$

which is known as the *characteristic equation*.

If $\mathbf{A}$ is $n \times n$, the characteristic equation (2.96) will have n roots; that is, $\mathbf{A}$ will have n eigenvalues $\lambda_1, \lambda_2, \ldots, \lambda_n$. The λ's will not necessarily all be distinct, or all nonzero, or even all real. (However, the eigenvalues of a symmetric matrix are real; see Theorem 2.12c.) After finding $\lambda_1, \lambda_2, \ldots, \lambda_n$ using (2.96), the accompanying eigenvectors $\mathbf{x}_1, \mathbf{x}_2, \ldots, \mathbf{x}_n$ can be found using (2.95).

If an eigenvalue is 0, the corresponding eigenvector is not $\mathbf{0}$. To see this, note that if $\lambda = 0$, then $(\mathbf{A} - \lambda\mathbf{I})\mathbf{x} = \mathbf{0}$ becomes $\mathbf{Ax} = \mathbf{0}$, which has solutions for $\mathbf{x}$ because $\mathbf{A}$ is singular, and the columns are therefore linearly dependent. [The matrix $\mathbf{A}$ is singular because it has a zero eigenvalue; see (63) and (2.107).]

If we multiply both sides of (2.95) by a scalar k, we obtain

$$k(\mathbf{A} - \lambda\mathbf{I})\mathbf{x} = k\mathbf{0} = \mathbf{0},$$

which can be rewritten as

$$(\mathbf{A} - \lambda\mathbf{I})k\mathbf{x} = \mathbf{0} \qquad \text{[by (2.12)]}.$$

Thus if $\mathbf{x}$ is an eigenvector of $\mathbf{A}$, $k\mathbf{x}$ is also an eigenvector. Eigenvectors are therefore unique only up to multiplication by a scalar. (There are many solution vectors $\mathbf{x}$ because $\mathbf{A} - \lambda\mathbf{I}$ is singular; see Section 2.8) Hence, the length of $\mathbf{x}$ is arbitrary, but its direction from the origin is unique; that is, the relative values of (ratios of) the elements of $\mathbf{x} = (x_1, x_2, \ldots, x_n)'$ are unique. Typically, an eigenvector $\mathbf{x}$ is scaled to normalized form as in (2.82), $\mathbf{x}'\mathbf{x} = 1$.

Example 2.12.1. To illustrate eigenvalues and eigenvectors, consider the matrix

$$\mathbf{A} = \begin{pmatrix} 1 & 2 \\ -1 & 4 \end{pmatrix}.$$

By (2.96), the characteristic equation is

$$|\mathbf{A} - \lambda\mathbf{I}| = \begin{vmatrix} 1-\lambda & 2 \\ -1 & 4-\lambda \end{vmatrix} = (1-\lambda)(4-\lambda) + 2 = 0,$$

which becomes

$$\lambda^2 - 5\lambda + 6 = (\lambda - 3)(\lambda - 2) = 0,$$

with roots $\lambda_1 = 3$ and $\lambda_2 = 2$.

To find the eigenvector $\mathbf{x}_1$ corresponding to $\lambda_1 = 3$, we use (2.95)

$$(\mathbf{A} - \lambda_1\mathbf{I})\mathbf{x}_1 = \mathbf{0},$$

$$\begin{pmatrix} 1-3 & 2 \\ -1 & 4-3 \end{pmatrix}\begin{pmatrix} x_1 \\ x_2 \end{pmatrix} = \begin{pmatrix} 0 \\ 0 \end{pmatrix},$$

which can be written as

$$\begin{aligned} -2x_1 + 2x_2 &= 0 \\ -x_1 + x_2 &= 0. \end{aligned}$$

The second equation is a multiple of the first, and either equation yields $x_1 = x_2$. The solution vector can be written with $x_1 = x_2 = c$ as an arbitrary constant:

$$\mathbf{x}_1 = \begin{pmatrix} x_1 \\ x_2 \end{pmatrix} = \begin{pmatrix} x_1 \\ x_1 \end{pmatrix} = x_1\begin{pmatrix} 1 \\ 1 \end{pmatrix} = c\begin{pmatrix} 1 \\ 1 \end{pmatrix}.$$

If c is set equal to $1/\sqrt{2}$ to normalize the eigenvector, we obtain

$$\mathbf{x}_1 = \begin{pmatrix} 1/\sqrt{2} \\ 1/\sqrt{2} \end{pmatrix}.$$

Similarly, corresponding to $\lambda_2 = 2$, we obtain

$$\mathbf{x}_2 = \begin{pmatrix} 2/\sqrt{5} \\ 1/\sqrt{5} \end{pmatrix}.$$

□

2.12.2 Functions of a Matrix

If λ is an eigenvalue of $\mathbf{A}$ with corresponding eigenvector $\mathbf{x}$, then for certain functions $g(\mathbf{A})$, an eigenvalue is given by $g(\lambda)$ and $\mathbf{x}$ is the corresponding eigenvector of $g(\mathbf{A})$ as well as of $\mathbf{A}$. We illustrate some of these cases:

1. If λ is an eigenvalue of $\mathbf{A}$, then $c\lambda$ is an eigenvalue of $c\mathbf{A}$, where c is an arbitrary constant such that $c \neq 0$. This is easily demonstrated by multiplying the defining relationship $\mathbf{Ax} = \lambda\mathbf{x}$ by c:

$$c\mathbf{Ax} = c\lambda\mathbf{x}. \tag{2.97}$$

   Note that $\mathbf{x}$ is an eigenvector of $\mathbf{A}$ corresponding to λ, and $\mathbf{x}$ is also an eigenvector of $c\mathbf{A}$ corresponding to $c\lambda$.

2. If λ is an eigenvalue of the $\mathbf{A}$ and $\mathbf{x}$ is the corresponding eigenvector of $\mathbf{A}$, then $c\lambda + k$ is an eigenvalue of the matrix $c\mathbf{A} + k\mathbf{I}$ and $\mathbf{x}$ is an eigenvector of $c\mathbf{A} + k\mathbf{I}$, where c and k are scalars. To show this, we add $k\mathbf{x}$ to (2.97):

$$c\mathbf{Ax} + k\mathbf{x} = c\lambda\mathbf{x} + k\mathbf{x},$$
$$(c\mathbf{A} + k\mathbf{I})\mathbf{x} = (c\lambda + k)\mathbf{x}. \tag{2.98}$$

   Thus $c\lambda + k$ is an eigenvalue of $c\mathbf{A} + k\mathbf{I}$ and $\mathbf{x}$ is the corresponding eigenvector of $c\mathbf{A} + k\mathbf{I}$. Note that (2.98) does not extend to $\mathbf{A} + \mathbf{B}$ for arbitrary $n \times n$ matrices $\mathbf{A}$ and $\mathbf{B}$; that is, $\mathbf{A} + \mathbf{B}$ does not have $\lambda_A + \lambda_B$ for an eigenvalue, where λ_A is an eigenvalue of $\mathbf{A}$ and λ_B is an eigenvalue of $\mathbf{B}$.

3. If λ is an eigenvalue of $\mathbf{A}$, then λ^2 is an eigenvalue of $\mathbf{A}^2$. This can be demonstrated by multiplying the defining relationship $\mathbf{Ax} = \lambda\mathbf{x}$ by $\mathbf{A}$:

$$\mathbf{A}(\mathbf{Ax}) = \mathbf{A}(\lambda\mathbf{x}),$$
$$\mathbf{A}^2\mathbf{x} = \lambda\mathbf{Ax} = \lambda(\lambda\mathbf{x}) = \lambda^2\mathbf{x}. \tag{2.99}$$

   Thus λ^2 is an eigenvalue of $\mathbf{A}^2$, and $\mathbf{x}$ is the corresponding eigenvector of $\mathbf{A}^2$. This can be extended to any power of $\mathbf{A}$:

$$\mathbf{A}^k\mathbf{x} = \lambda^k\mathbf{x}; \tag{2.100}$$

   that is, λ^k is an eigenvalue of $\mathbf{A}^k$, and $\mathbf{x}$ is the corresponding eigenvector.

4. If λ is an eigenvalue of the nonsingular matrix $\mathbf{A}$, then $1/\lambda$ is an eigenvalue of $\mathbf{A}^{-1}$. To demonstrate this, we multiply $\mathbf{Ax} = \lambda\mathbf{x}$ by $\mathbf{A}^{-1}$ to obtain

$$\begin{aligned}\mathbf{A}^{-1}\mathbf{Ax} &= \mathbf{A}^{-1}\lambda\mathbf{x},\\ \mathbf{x} &= \lambda\mathbf{A}^{-1}\mathbf{x},\\ \mathbf{A}^{-1}\mathbf{x} &= \frac{1}{\lambda}\mathbf{x}.\end{aligned} \tag{2.101}$$

   Thus $1/\lambda$ is an eigenvalue of $\mathbf{A}^{-1}$, and $\mathbf{x}$ is an eigenvector of both $\mathbf{A}$ and $\mathbf{A}^{-1}$.

5. The results in (2.97) and (2.100) can be used to obtain eigenvalues and eigenvectors of a polynomial in $\mathbf{A}$. For example, if λ is an eigenvalue of $\mathbf{A}$, then

$$\begin{aligned}(\mathbf{A}^3 + 4\mathbf{A}^2 - 3\mathbf{A} + 5\mathbf{I})\mathbf{x} &= \mathbf{A}^3\mathbf{x} + 4\mathbf{A}^2\mathbf{x} - 3\mathbf{Ax} + 5\mathbf{x}\\ &= \lambda^3\mathbf{x} + 4\lambda^2\mathbf{x} - 3\lambda\mathbf{x} + 5\mathbf{x}\\ &= (\lambda^3 + 4\lambda^2 - 3\lambda + 5)\mathbf{x}.\end{aligned}$$

   Thus $\lambda^3 + 4\lambda^2 - 3\lambda + 5$ is an eigenvalue of $\mathbf{A}^3 + 4\mathbf{A}^2 - 3\mathbf{A} + 5\mathbf{I}$, and $\mathbf{x}$ is the corresponding eigenvector.

For certain matrices, property 5 can be extended to an infinite series. For example, if λ is an eigenvalue of $\mathbf{A}$, then, by (2.98), $1 - \lambda$ is an eigenvalue of $\mathbf{I} - \mathbf{A}$. If $\mathbf{I} - \mathbf{A}$ is nonsingular, then, by (2.101), $1/(1 - \lambda)$ is an eigenvalue of $(\mathbf{I} - \mathbf{A})^{-1}$. If $-1 < \lambda < 1$, then $1/(1 - \lambda)$ can be represented by the series

$$\frac{1}{1 - \lambda} = 1 + \lambda + \lambda^2 + \lambda^3 + \cdots.$$

Correspondingly, if all eigenvalues of $\mathbf{A}$ satisfy $-1 < \lambda < 1$, then

$$(\mathbf{I} - \mathbf{A})^{-1} = \mathbf{I} + \mathbf{A} + \mathbf{A}^2 + \mathbf{A}^3 + \cdots. \tag{2.102}$$

2.12.3 Products

It was noted in a comment following (2.98) that the eigenvalues of $\mathbf{A} + \mathbf{B}$ are not of the form $\lambda_A + \lambda_B$, where λ_A is an eigenvalue of $\mathbf{A}$ and λ_B is an eigenvalue of $\mathbf{B}$. Similarly, the eigenvalues of $\mathbf{AB}$ are not products of the form $\lambda_A\lambda_B$. However, the eigenvalues of $\mathbf{AB}$ are the same as those of $\mathbf{BA}$.

Theorem 2.12a. If $\mathbf{A}$ and $\mathbf{B}$ are $n \times n$ or if $\mathbf{A}$ is $n \times p$ and $\mathbf{B}$ is $p \times n$, then the (nonzero) eigenvalues of $\mathbf{AB}$ are the same as those of $\mathbf{BA}$. If $\mathbf{x}$ is an eigenvector of $\mathbf{AB}$, then $\mathbf{Bx}$ is an eigenvector of $\mathbf{BA}$. □

Two additional results involving eigenvalues of products are given in the following theorem.

Theorem 2.12b. Let $\mathbf{A}$ be any $n \times n$ matrix.

(i) If $\mathbf{P}$ is any $n \times n$ nonsingular matrix, then $\mathbf{A}$ and $\mathbf{P}^{-1}\mathbf{AP}$ have the same eigenvalues.

(ii) If $\mathbf{C}$ is any $n \times n$ orthogonal matrix, then $\mathbf{A}$ and $\mathbf{C}'\mathbf{AC}$ have the same eigenvalues.

□

2.12.4 Symmetric Matrices

Two properties of the eigenvalues and eigenvectors of a symmetric matrix are given in the following theorem.

Theorem 2.12c. Let $\mathbf{A}$ be an $n \times n$ symmetric matrix.

(i) The eigenvalues $\lambda_1, \lambda_2, \ldots, \lambda_n$ of $\mathbf{A}$ are real.

(ii) The eigenvectors $\mathbf{x}_1, \mathbf{x}_2, \ldots, \mathbf{x}_k$ of $\mathbf{A}$ corresponding to distinct eigenvalues $\lambda_1, \lambda_2, \ldots, \lambda_k$ are mutually orthogonal; the eigenvectors $\mathbf{x}_{k+1}, \mathbf{x}_{k+2}, \ldots, \mathbf{x}_n$ corresponding to the nondistinct eigenvalues can be chosen to be mutually orthogonal to each other and to the other eigenvectors; that is, $\mathbf{x}_i'\mathbf{x}_j = 0$ for $i \neq j$.

□

If the eigenvectors of a symmetric matrix $\mathbf{A}$ are normalized and placed as columns of a matrix $\mathbf{C}$, then by Theorem 2.12c(ii), $\mathbf{C}$ is an orthogonal matrix. This orthogonal matrix can be used to express $\mathbf{A}$ in terms of its eigenvalues and eigenvectors.

Theorem 2.12d. If $\mathbf{A}$ is an $n \times n$ symmetric matrix with eigenvalues $\lambda_1, \lambda_2, \ldots, \lambda_n$ and normalized eigenvectors $\mathbf{x}_1, \mathbf{x}_2, \ldots, \mathbf{x}_n$, then $\mathbf{A}$ can be expressed as

$$\mathbf{A} = \mathbf{CDC}' \tag{2.103}$$

$$= \sum_{i=1}^{n} \lambda_i \mathbf{x}_i \mathbf{x}_i', \tag{2.104}$$

where $\mathbf{D} = \text{diag}(\lambda_1, \lambda_2, \ldots, \lambda_n)$ and $\mathbf{C}$ is the orthogonal matrix $\mathbf{C} = (\mathbf{x}_1, \mathbf{x}_2, \ldots, \mathbf{x}_n)$. The result in either (2.103) or (2.104) is often called the *spectral decomposition* of $\mathbf{A}$.

Proof. By Theorem 2.12c(ii), $\mathbf{C}$ is orthogonal. Then by (2.84), $\mathbf{I} = \mathbf{CC}'$, and multiplication by $\mathbf{A}$ gives

$$\mathbf{A} = \mathbf{ACC}'.$$

We now substitute $\mathbf{C} = (\mathbf{x}_1, \mathbf{x}_2, \ldots, \mathbf{x}_n)$ to obtain

$$\begin{aligned}
\mathbf{A} &= \mathbf{A}(\mathbf{x}_1, \mathbf{x}_2, \ldots, \mathbf{x}_n)\mathbf{C}' \\
&= (\mathbf{A}\mathbf{x}_1, \mathbf{A}\mathbf{x}_2, \ldots, \mathbf{A}\mathbf{x}_n)\mathbf{C}' \qquad \text{[by (2.28)]} \\
&= (\lambda_1\mathbf{x}_1, \lambda_2\mathbf{x}_2, \ldots, \lambda_n\mathbf{x}_n)\mathbf{C}' \qquad \text{[by (2.94)]} \\
&= \mathbf{CDC}', \qquad (2.105)
\end{aligned}$$

since multiplication on the right by $\mathbf{D} = \text{diag}(\lambda_1, \lambda_2, \ldots, \lambda_n)$ multiplies columns of $\mathbf{C}$ by elements of $\mathbf{D}$ [see (2.30)]. Now writing $\mathbf{C}'$ in the form

$$\mathbf{C}' = (\mathbf{x}_1, \mathbf{x}_2, \ldots, \mathbf{x}_n)' = \begin{pmatrix} \mathbf{x}_1' \\ \mathbf{x}_2' \\ \vdots \\ \mathbf{x}_n' \end{pmatrix} \qquad \text{[by (2.39)]},$$

(2.105) becomes

$$\begin{aligned}
\mathbf{A} &= (\lambda_1\mathbf{x}_1, \lambda_2\mathbf{x}_2, \ldots, \lambda_n\mathbf{x}_n)\begin{pmatrix} \mathbf{x}_1' \\ \mathbf{x}_2' \\ \vdots \\ \mathbf{x}_n' \end{pmatrix} \\
&= \lambda_1\mathbf{x}_1\mathbf{x}_1' + \lambda_2\mathbf{x}_2\mathbf{x}_2' + \cdots + \lambda_n\mathbf{x}_n\mathbf{x}_n'.
\end{aligned}$$

□

Corollary 1. If $\mathbf{A}$ is symmetric and $\mathbf{C}$ and $\mathbf{D}$ are defined as in Theorem 2.12d, then $\mathbf{C}$ *diagonalizes* $\mathbf{A}$:

$$\mathbf{C}'\mathbf{A}\mathbf{C} = \mathbf{D}. \qquad (2.106)$$

□

We can express the determinant and trace of a square matrix $\mathbf{A}$ in terms of its eigenvalues.

Theorem 2.12e. If $\mathbf{A}$ is any $n \times n$ matrix with eigenvalues $\lambda_1, \lambda_2, \ldots, \lambda_n$, then

(i)

$$|\mathbf{A}| = \prod_{i=1}^{n} \lambda_i. \qquad (2.107)$$

(ii)

$$\text{tr}(\mathbf{A}) = \sum_{i=1}^{n} \lambda_i. \qquad (2.108)$$

□

We have included Theorem 2.12e here because it is easy to prove for a symmetric matrix $\mathbf{A}$ using Theorem 2.12d (see Problem 2.72). However, the theorem is true for any square matrix (Searle 1982, p. 278).

Example 2.12.4. To illustrate Theorem 2.12e, consider the matrix **A** in Example 2.12.1

$$\mathbf{A} = \begin{pmatrix} 1 & 2 \\ -1 & 4 \end{pmatrix},$$

which has eigenvalues $\lambda_1 = 3$ and $\lambda_2 = 2$. The product $\lambda_1 \lambda_2 = 6$ is the same as $|\mathbf{A}| = 4 - (-1)(2) = 6$. The sum $\lambda_1 + \lambda_2 = 3 + 2 = 5$ is the same as $\text{tr}(\mathbf{A}) = 1 + 4 = 5$. □

2.12.5 Positive Definite and Semidefinite Matrices

The eigenvalues $\lambda_1, \lambda_2, \ldots, \lambda_n$ of positive definite and positive semidefinite matrices (Section 2.6) are positive and nonnegative, respectively.

Theorem 2.12f. Let **A** be $n \times n$ with eigenvalues $\lambda_1, \lambda_2, \ldots, \lambda_n$.

(i) If **A** is positive definite, then $\lambda_i > 0$ for $i = 1, 2, \ldots, n$.
(ii) If **A** is positive semidefinite, then $\lambda_i \geq 0$ for $i = 1, 2, \ldots, n$. The number of eigenvalues λ_i for which $\lambda_i > 0$ is the rank of **A**.

PROOF.

(i) For any λ_i, we have $\mathbf{A}\mathbf{x}_i = \lambda_i \mathbf{x}_i$. Multiplying by $\mathbf{x}_i'$, we obtain

$$\mathbf{x}_i'\mathbf{A}\mathbf{x}_i = \lambda_i \mathbf{x}_i'\mathbf{x}_i,$$

$$\lambda_i = \frac{\mathbf{x}_i'\mathbf{A}\mathbf{x}_i}{\mathbf{x}_i'\mathbf{x}_i} > 0.$$

In the second expression, $\mathbf{x}_i'\mathbf{A}\mathbf{x}_i$ is positive because **A** is positive definite, and $\mathbf{x}_i'\mathbf{x}_i$ is positive because $\mathbf{x}_i \neq \mathbf{0}$. □

If a matrix **A** is positive definite, we can find a *square root matrix* $\mathbf{A}^{1/2}$ as follows. Since the eigenvalues of **A** are positive, we can substitute the square roots $\sqrt{\lambda_i}$ for λ_i in the spectral decomposition of **A** in (2.103), to obtain

$$\mathbf{A}^{1/2} = \mathbf{C}\mathbf{D}^{1/2}\mathbf{C}', \tag{2.109}$$

where $\mathbf{D}^{1/2} = \text{diag}(\sqrt{\lambda_1}, \sqrt{\lambda_2}, \ldots, \sqrt{\lambda_n})$. The matrix $\mathbf{A}^{1/2}$ is symmetric and has the property

$$\mathbf{A}^{1/2}\mathbf{A}^{1/2} = (\mathbf{A}^{1/2})^2 = \mathbf{A}. \tag{2.110}$$

2.13 IDEMPOTENT MATRICES

A square matrix $\mathbf{A}$ is said to be *idempotent* if $\mathbf{A}^2 = \mathbf{A}$. Most idempotent matrices in this book are symmetric. Many of the sums of squares in regression (Chapters 6–11) and analysis of variance (Chapters 12–15) can be expressed as quadratic forms $\mathbf{y}'\mathbf{A}\mathbf{y}$. The idempotence of $\mathbf{A}$ or of a product involving $\mathbf{A}$ will be used to establish that $\mathbf{y}'\mathbf{A}\mathbf{y}$ (or a multiple of $\mathbf{y}'\mathbf{A}\mathbf{y}$) has a chi-square distribution.

An example of an idempotent matrix is the identity matrix $\mathbf{I}$.

Theorem 2.13a. The only nonsingular idempotent matrix is the identity matrix $\mathbf{I}$.

PROOF. If $\mathbf{A}$ is idempotent and nonsingular, then $\mathbf{A}^2 = \mathbf{A}$ and the inverse $\mathbf{A}^{-1}$ exists. If we multiply $\mathbf{A}^2 = \mathbf{A}$ by $\mathbf{A}^{-1}$, we obtain

$$\mathbf{A}^{-1}\mathbf{A}^2 = \mathbf{A}^{-1}\mathbf{A},$$
$$\mathbf{A} = \mathbf{I}.$$

□

Many of the matrices of quadratic forms we will encounter in later chapters are singular idempotent matrices. We now give some properties of such matrices.

Theorem 2.13b. If $\mathbf{A}$ is singular, symmetric, and idempotent, then $\mathbf{A}$ is positive semidefinite.

PROOF. Since $\mathbf{A} = \mathbf{A}'$ and $\mathbf{A} = \mathbf{A}^2$, we have

$$\mathbf{A} = \mathbf{A}^2 = \mathbf{A}\mathbf{A} = \mathbf{A}'\mathbf{A},$$

which is positive semidefinite by Theorem 2.6d(ii). □

If a is a real number such that $a^2 = a$, then a is either 0 or 1. The analogous property for matrices is that if $\mathbf{A}^2 = \mathbf{A}$, then the eigenvalues of $\mathbf{A}$ are 0s and 1s.

Theorem 2.13c. If $\mathbf{A}$ is an $n \times n$ symmetric idempotent matrix of rank r, then $\mathbf{A}$ has r eigenvalues equal to 1 and $n - r$ eigenvalues equal to 0.

PROOF. By (2.99), if $\mathbf{A}\mathbf{x} = \lambda\mathbf{x}$, then $\mathbf{A}^2\mathbf{x} = \lambda^2\mathbf{x}$. Since $\mathbf{A}^2 = \mathbf{A}$, we have $\mathbf{A}^2\mathbf{x} = \mathbf{A}\mathbf{x} = \lambda\mathbf{x}$. Equating the right sides of $\mathbf{A}^2\mathbf{x} = \lambda^2\mathbf{x}$ and $\mathbf{A}^2\mathbf{x} = \lambda\mathbf{x}$, we have

$$\lambda\mathbf{x} = \lambda^2\mathbf{x} \quad \text{or} \quad (\lambda - \lambda^2)\mathbf{x} = \mathbf{0}.$$

But $\mathbf{x} \neq \mathbf{0}$, and therefore $\lambda - \lambda^2 = 0$, from which, λ is either 0 or 1.

By Theorem 2.13b, $\mathbf{A}$ is positive semidefinite, and therefore by Theorem 2.12f(ii), the number of nonzero eigenvalues is equal to rank($\mathbf{A}$). Thus r eigenvalues of $\mathbf{A}$ are equal to 1 and the remaining $n - r$ eigenvalues are equal to 0. □

We can use Theorems 2.12e and 2.13c to find the rank of a symmetric idempotent matrix.

Theorem 2.13d. If $\mathbf{A}$ is symmetric and idempotent of rank r, then $\text{rank}(\mathbf{A}) = \text{tr}(\mathbf{A}) = r$.

PROOF. By Theorem 2.12e(ii), $\text{tr}(\mathbf{A}) = \sum_{i=1}^{n} \lambda_i$, and by Theorem 2.13c, $\sum_{i=1}^{n} \lambda_i = r$. □

Some additional properties of idempotent matrices are given in the following four theorems.

Theorem 2.13e. If $\mathbf{A}$ is an $n \times n$ idempotent matrix, $\mathbf{P}$ is an $n \times n$ nonsingular matrix, and $\mathbf{C}$ is an $n \times n$ orthogonal matrix, then

(i) $\mathbf{I} - \mathbf{A}$ is idempotent.
(ii) $\mathbf{A}(\mathbf{I} - \mathbf{A}) = \mathbf{O}$ and $(\mathbf{I} - \mathbf{A})\mathbf{A} = \mathbf{O}$.
(iii) $\mathbf{P}^{-1}\mathbf{A}\mathbf{P}$ is idempotent.
(iv) $\mathbf{C}'\mathbf{A}\mathbf{C}$ is idempotent. (If $\mathbf{A}$ is symmetric, $\mathbf{C}'\mathbf{A}\mathbf{C}$ is a symmetric idempotent matrix.) □

Theorem 2.13f. Let $\mathbf{A}$ be $n \times p$ of rank r, let $\mathbf{A}^-$ be any generalized inverse of $\mathbf{A}$, and let $(\mathbf{A}'\mathbf{A})^-$ be any generalized inverse of $\mathbf{A}'\mathbf{A}$. Then $\mathbf{A}^-\mathbf{A}, \mathbf{A}\mathbf{A}^-$, and $\mathbf{A}(\mathbf{A}'\mathbf{A})^-\mathbf{A}'$ are all idempotent. □

Theorem 2.13g. Suppose that the $n \times n$ symmetric matrix $\mathbf{A}$ can be written as $\mathbf{A} = \sum_{i=1}^{k} \mathbf{A}_i$ for some k, where each $\mathbf{A}_i$ is an $n \times n$ symmetric matrix. Then any two of the following conditions implies the third condition.

(i) $\mathbf{A}$ is idempotent.
(ii) Each of $\mathbf{A}_1, \mathbf{A}_2, \ldots, \mathbf{A}_k$ is idempotent.
(iii) $\mathbf{A}_i\mathbf{A}_j = \mathbf{O}$ for $i \neq j$.

□

Theorem 2.13h. If $\mathbf{I} = \sum_{i=1}^{k} \mathbf{A}_i$, where each $n \times n$ matrix $\mathbf{A}_i$ is symmetric of rank r_i, and if $n = \sum_{i=1}^{k} r_i$, then both of the following are true:

(i) Each of $\mathbf{A}_1, \mathbf{A}_2, \ldots, \mathbf{A}_k$ is idempotent.
(ii) $\mathbf{A}_i\mathbf{A}_j = \mathbf{O}$ for $i \neq j$.

□

2.14 VECTOR AND MATRIX CALCULUS

2.14.1 Derivatives of Functions of Vectors and Matrices

Let $u = f(\mathbf{x})$ be a function of the variables $x_1, x_2, \ldots, x_p$ in $\mathbf{x} = (x_1, x_2, \ldots, x_p)'$, and let $\partial u/\partial x_1, \partial u/\partial x_2, \ldots, \partial u/\partial x_p$ be the partial derivatives. We define $\partial u/\partial \mathbf{x}$ as

$$\frac{\partial u}{\partial \mathbf{x}} = \begin{pmatrix} \frac{\partial u}{\partial x_1} \\ \frac{\partial u}{\partial x_2} \\ \vdots \\ \frac{\partial u}{\partial x_p} \end{pmatrix}. \tag{2.111}$$

Two specific functions of interest are $u = \mathbf{a}'\mathbf{x}$ and $u = \mathbf{x}'\mathbf{A}\mathbf{x}$. Their derivatives with respect to $\mathbf{x}$ are given in the following two theorems.

Theorem 2.14a. Let $u = \mathbf{a}'\mathbf{x} = \mathbf{x}'\mathbf{a}$, where $\mathbf{a}' = (a_1, a_2, \ldots, a_p)$ is a vector of constants. Then

$$\frac{\partial u}{\partial \mathbf{x}} = \frac{\partial(\mathbf{a}'\mathbf{x})}{\partial \mathbf{x}} = \frac{\partial(\mathbf{x}'\mathbf{a})}{\partial \mathbf{x}} = \mathbf{a}. \tag{2.112}$$

PROOF

$$\frac{\partial u}{\partial x_i} = \frac{\partial(a_1x_1 + a_2x_2 + \cdots + a_px_p)}{\partial x_i} = a_i.$$

Thus by (2.111) we obtain

$$\frac{\partial u}{\partial \mathbf{x}} = \begin{pmatrix} a_1 \\ a_2 \\ \vdots \\ a_p \end{pmatrix} = \mathbf{a}.$$

□

Theorem 2.14b. Let $u = \mathbf{x}'\mathbf{A}\mathbf{x}$, where $\mathbf{A}$ is a symmetric matrix of constants. Then

$$\frac{\partial u}{\partial \mathbf{x}} = \frac{\partial(\mathbf{x}'\mathbf{A}\mathbf{x})}{\partial \mathbf{x}} = 2\mathbf{A}\mathbf{x}. \tag{2.113}$$

PROOF. We demonstrate that (2.113) holds for the special case in which $\mathbf{A}$ is 3×3. The illustration could be generalized to a symmetric $\mathbf{A}$ of any size. Let

$$\mathbf{x} = \begin{pmatrix} x_1 \\ x_2 \\ x_3 \end{pmatrix} \quad \text{and} \quad \mathbf{A} = \begin{pmatrix} a_{11} & a_{12} & a_{13} \\ a_{12} & a_{22} & a_{23} \\ a_{13} & a_{23} & a_{33} \end{pmatrix} = \begin{pmatrix} \mathbf{a}_1' \\ \mathbf{a}_2' \\ \mathbf{a}_3' \end{pmatrix}.$$

Then $\mathbf{x}'\mathbf{A}\mathbf{x} = x_1^2 a_{11} + 2x_1x_2a_{12} + 2x_1x_3a_{13} + x_2^2 a_{22} + 2x_2x_3a_{23} + x_3^2 a_{33}$, and we have

$$\frac{\partial(\mathbf{x}'\mathbf{A}\mathbf{x})}{\partial x_1} = 2x_1a_{11} + 2x_2a_{12} + 2x_3a_{13} = 2\mathbf{a}_1'\mathbf{x}$$

$$\frac{\partial(\mathbf{x}'\mathbf{A}\mathbf{x})}{\partial x_2} = 2x_1a_{12} + 2x_2a_{22} + 2x_3a_{23} = 2\mathbf{a}_2'\mathbf{x}$$

$$\frac{\partial(\mathbf{x}'\mathbf{A}\mathbf{x})}{\partial x_3} = 2x_1a_{13} + 2x_2a_{23} + 2x_3a_{33} = 2\mathbf{a}_3'\mathbf{x}.$$

Thus by (2.11), (2.27), and (2.111), we obtain

$$\frac{\partial(\mathbf{x}'\mathbf{A}\mathbf{x})}{\partial \mathbf{x}} = \begin{pmatrix} \dfrac{\partial(\mathbf{x}'\mathbf{A}\mathbf{x})}{\partial x_1} \\ \dfrac{\partial(\mathbf{x}'\mathbf{A}\mathbf{x})}{\partial x_2} \\ \dfrac{\partial(\mathbf{x}'\mathbf{A}\mathbf{x})}{\partial x_3} \end{pmatrix} = 2\begin{pmatrix} \mathbf{a}_1'\mathbf{x} \\ \mathbf{a}_2'\mathbf{x} \\ \mathbf{a}_3'\mathbf{x} \end{pmatrix} = 2\mathbf{A}\mathbf{x}.$$

□

Now let $u = f(\mathbf{X})$ be a function of the variables $x_{11}, x_{12}, \ldots, x_{pp}$ in the $p \times p$ matrix $\mathbf{X}$, and let $(\partial u/\partial x_{11}), (\partial u/\partial x_{12}), \ldots, (\partial u/\partial x_{pp})$ be the partial derivatives. Similarly to (2.111), we define $\partial u/\partial \mathbf{X}$ as

$$\frac{\partial u}{\partial \mathbf{X}} = \begin{pmatrix} \dfrac{\partial u}{\partial x_{11}} & \cdots & \dfrac{\partial u}{\partial x_{1p}} \\ \vdots & & \vdots \\ \dfrac{\partial u}{\partial x_{p1}} & \cdots & \dfrac{\partial u}{\partial x_{pp}} \end{pmatrix}. \tag{2.114}$$

Two functions of interest of this type are $u = \text{tr}(\mathbf{XA})$ and $u = \ln|\mathbf{X}|$ for a positive definite matrix $\mathbf{X}$.

Theorem 2.14c. Let $u = \text{tr}(\mathbf{XA})$, where $\mathbf{X}$ is a $p \times p$ positive definite matrix and $\mathbf{A}$ is a $p \times p$ matrix of constants. Then

$$\frac{\partial u}{\partial \mathbf{X}} = \frac{\partial[\text{tr}(\mathbf{XA})]}{\partial \mathbf{X}} = \mathbf{A} + \mathbf{A}' - \text{diag}\,\mathbf{A}. \tag{2.115}$$

PROOF. Note that $\text{tr}(\mathbf{XA}) = \sum_{i=1}^{p} \sum_{j=1}^{p} x_{ij} a_{ji}$ [see the proof of Theorem 2.11(ii)]. Since $x_{ij} = x_{ji}$, $[\partial \text{tr}(\mathbf{XA})]/\partial x_{ij} = a_{ji} + a_{ij}$ if $i \neq j$, and $[\partial \text{tr}(\mathbf{XA})]/\partial x_{ii} = a_{ii}$. The result follows. □

Theorem 2.14d. Let $u = \ln |\mathbf{X}|$ where $\mathbf{X}$ is a $p \times p$ positive definite matrix. Then

$$\frac{\partial \ln |\mathbf{X}|}{\partial \mathbf{X}} = 2\mathbf{X}^{-1} - \text{diag}(\mathbf{X}^{-1}). \qquad (2.116)$$

PROOF. See Harville (1997, p. 306). See Problem 2.83 for a demonstration that this theorem holds for 2×2 matrices. □

2.14.2 Derivatives Involving Inverse Matrices and Determinants

Let $\mathbf{A}$ be an $n \times n$ nonsingular matrix with elements a_{ij} that are functions of a scalar x. We define $\partial \mathbf{A}/\partial x$ as the $n \times n$ matrix with elements $\partial a_{ij}/\partial x$. The related derivative $\partial \mathbf{A}^{-1}/\partial x$ is often of interest. If $\mathbf{A}$ is positive definite, the derivative $(\partial/\partial x) \log |\mathbf{A}|$ is also often of interest.

Theorem 2.14e. Let $\mathbf{A}$ be nonsingular of order n with derivative $\partial \mathbf{A}/\partial x$. Then

$$\frac{\partial \mathbf{A}^{-1}}{\partial x} = -\mathbf{A}^{-1} \frac{\partial \mathbf{A}}{\partial x} \mathbf{A}^{-1} \qquad (2.117)$$

PROOF. Because $\mathbf{A}$ is nonsingular, we have

$$\mathbf{A}^{-1}\mathbf{A} = \mathbf{I}.$$

Thus

$$\frac{\partial \mathbf{A}^{-1}}{\partial x} \mathbf{A} + \mathbf{A}^{-1} \frac{\partial \mathbf{A}}{\partial x} = \mathbf{O}.$$

Hence

$$\frac{\partial \mathbf{A}^{-1}}{\partial x} \mathbf{A} = -\mathbf{A}^{-1} \frac{\partial \mathbf{A}}{\partial x},$$

and so

$$\frac{\partial \mathbf{A}^{-1}}{\partial x} = -\mathbf{A}^{-1} \frac{\partial \mathbf{A}}{\partial x} \mathbf{A}^{-1}.$$

□

Theorem 2.14f. Let $\mathbf{A}$ be an $n \times n$ positive define matrix. Then

$$\frac{\partial \log |\mathbf{A}|}{\partial x} = \text{tr}\left(\mathbf{A}^{-1} \frac{\partial \mathbf{A}}{\partial x} \right). \qquad (2.118)$$

PROOF. Since $\mathbf{A}$ is positive definite, its spectral decomposition (Theorem 2.12d) can be written as $\mathbf{CDC}'$, where $\mathbf{C}$ is an orthogonal matrix and $\mathbf{D}$ is a diagonal matrix of positive eigenvalues, λ_i. Using Theorem 2.12e, we obtain

$$\begin{aligned}\frac{\partial \log|\mathbf{A}|}{\partial x} &= \frac{\partial \log \prod_{i=1}^n \lambda_i}{\partial x} \\ &= \frac{\partial \sum_{i=1}^n \log \lambda_i}{\partial x} \\ &= \sum_{i=1}^n \frac{1}{\lambda_i}\frac{\partial \lambda_i}{\partial x} \\ &= \mathrm{tr}\left(\mathbf{D}^{-1}\frac{\partial \mathbf{D}}{\partial x}\right).\end{aligned}$$

Now

$$\begin{aligned}\mathbf{A}^{-1}\frac{\partial \mathbf{A}}{\partial x} &= \mathbf{CD}^{-1}\mathbf{C}'\frac{\partial \mathbf{CDC}'}{\partial x} \\ &= \mathbf{CD}^{-1}\mathbf{C}'\left[\mathbf{C}\frac{\partial \mathbf{DC}'}{\partial x} + \frac{\partial \mathbf{C}}{\partial x}\mathbf{DC}'\right] \\ &= \mathbf{CD}^{-1}\mathbf{C}'\left[\mathbf{C}\frac{\partial \mathbf{D}}{\partial x}\mathbf{C}' + \mathbf{CD}\frac{\partial \mathbf{C}'}{\partial x} + \frac{\partial \mathbf{C}}{\partial x}\mathbf{DC}'\right] \\ &= \mathbf{CD}^{-1}\frac{\partial \mathbf{D}}{\partial x}\mathbf{C}' + \mathbf{C}\frac{\partial \mathbf{C}'}{\partial x} + \mathbf{CD}^{-1}\mathbf{C}'\frac{\partial \mathbf{C}}{\partial x}\mathbf{DC}'.\end{aligned}$$

Using Theorem 2.11(i) and (ii), we have

$$\mathrm{tr}\left(\mathbf{A}^{-1}\frac{\partial \mathbf{A}}{\partial x}\right) = \mathrm{tr}\left(\mathbf{D}^{-1}\frac{\partial \mathbf{D}}{\partial x} + \mathbf{C}\frac{\partial \mathbf{C}'}{\partial x} + \mathbf{C}'\frac{\partial \mathbf{C}}{\partial x}\right).$$

Since $\mathbf{C}$ is orthogonal, $\mathbf{C}'\mathbf{C} = \mathbf{I}$ which implies that

$$\frac{\partial \mathbf{C}'\mathbf{C}}{\partial x} = \mathbf{C}'\frac{\partial \mathbf{C}}{\partial x} + \frac{\partial \mathbf{C}'}{\partial x}\mathbf{C} = \mathbf{O}$$

and

$$\mathrm{tr}\left(\mathbf{C}'\frac{\partial \mathbf{C}}{\partial x} + \frac{\partial \mathbf{C}'\mathbf{C}}{\partial x}\right) = \mathrm{tr}\left(\mathbf{C}'\frac{\partial \mathbf{C}}{\partial x} + \mathbf{C}\frac{\partial \mathbf{C}'}{\partial x}\right) = 0.$$

Thus $\mathrm{tr}[\mathbf{A}^{-1}(\partial \mathbf{A}/\partial x)] = \mathrm{tr}[\mathbf{D}^{-1}(\partial \mathbf{D}/\partial x)]$ and the result follows. □

2.14.3 Maximization or Minimization of a Function of a Vector

Consider a function $u = f(\mathbf{x})$ of the p variables in $\mathbf{x}$. In many cases we can find a maximum or minimum of u by solving the system of p equations

$$\frac{\partial u}{\partial \mathbf{x}} = \mathbf{0}. \qquad (2.119)$$

Occasionally the situation requires the maximization or minimization of the function u, subject to q constraints on $\mathbf{x}$. We denote the constraints as $h_1(\mathbf{x}) = 0,\ h_2(\mathbf{x}) = 0, \ldots, h_q(\mathbf{x}) = 0$ or, more succinctly, $\mathbf{h}(\mathbf{x}) = \mathbf{0}$. Maximization or minimization of u subject to $\mathbf{h}(\mathbf{x}) = \mathbf{0}$ can often be carried out by the method of Lagrange multipliers. We denote a vector of q unknown constants (the *Lagrange multipliers*) by $\boldsymbol{\lambda}$ and let $\mathbf{y}' = (\mathbf{x}', \boldsymbol{\lambda}')$. We then let $v = u + \boldsymbol{\lambda}'\mathbf{h}(\mathbf{x})$. The maximum or minimum of u subject to $\mathbf{h}(\mathbf{x}) = \mathbf{0}$ is obtained by solving the equations

$$\frac{\partial v}{\partial \mathbf{y}} = \mathbf{0}$$

or, equivalently

$$\frac{\partial u}{\partial \mathbf{x}} + \frac{\partial \mathbf{h}}{\partial \mathbf{x}}\boldsymbol{\lambda} = \mathbf{0} \quad \text{and} \quad \mathbf{h}(\mathbf{x}) = \mathbf{0}, \qquad (2.120)$$

where

$$\frac{\partial \mathbf{h}}{\partial \mathbf{x}} = \begin{pmatrix} \dfrac{\partial h_1}{\partial x_1} & \cdots & \dfrac{\partial h_q}{\partial x_1} \\ \vdots & & \vdots \\ \dfrac{\partial h_1}{\partial x_p} & \cdots & \dfrac{\partial h_q}{\partial x_p} \end{pmatrix}.$$

PROBLEMS

2.1 Prove Theorem 2.2a.

2.2 Let $\mathbf{A} = \begin{pmatrix} 7 & -3 & 2 \\ 4 & 9 & 5 \end{pmatrix}$.

(**a**) Find $\mathbf{A}'$.

(**b**) Verify that $(\mathbf{A}')' = \mathbf{A}$, thus illustrating Theorem 2.1.

(**c**) Find $\mathbf{A}'\mathbf{A}$ and $\mathbf{A}\mathbf{A}'$.

2.3 Let $\mathbf{A} = \begin{pmatrix} 2 & 4 \\ -1 & 3 \end{pmatrix}$ and $\mathbf{B} = \begin{pmatrix} 1 & 3 \\ 2 & -1 \end{pmatrix}$.

(**a**) Find **AB** and **BA**.

(**b**) Find $|\mathbf{A}|$, $|\mathbf{B}|$, and $|\mathbf{AB}|$, and verify that Theorem 2.9c holds in this case.

(**c**) Find $|\mathbf{BA}|$ and compare to $|\mathbf{AB}|$.

(**d**) Find $(\mathbf{AB})'$ and compare to $\mathbf{B}'\mathbf{A}'$.

(**e**) Find tr(**AB**) and compare to tr(**BA**).

(**f**) Find the eigenvalues of **AB** and of **BA**, thus illustrating Theorem 2.12a.

2.4 Let $\mathbf{A} = \begin{pmatrix} 1 & 3 & -4 \\ 5 & -7 & 2 \end{pmatrix}$ and $\mathbf{B} = \begin{pmatrix} 3 & -2 & 5 \\ 6 & 9 & 7 \end{pmatrix}$.

(**a**) Find $\mathbf{A} + \mathbf{B}$ and $\mathbf{A} - \mathbf{B}$.

(**b**) Find $\mathbf{A}'$ and $\mathbf{B}'$.

(**c**) Find $(\mathbf{A} + \mathbf{B})'$ and $\mathbf{A}' + \mathbf{B}'$, thus illustrating Theorem 2.2a(ii).

2.5 Verify the distributive law in (2.15), $\mathbf{A}(\mathbf{B} + \mathbf{C}) = \mathbf{AB} + \mathbf{AC}$.

2.6 Let $\mathbf{A} = \begin{pmatrix} 8 & 3 & 7 \\ -2 & 5 & -3 \end{pmatrix}$, $\mathbf{B} = \begin{pmatrix} -2 & 5 \\ 3 & 7 \\ 6 & -4 \end{pmatrix}$, $\mathbf{C} = \begin{pmatrix} 1 & 2 \\ -3 & 1 \\ 2 & 4 \end{pmatrix}$.

(**a**) Find **AB** and **BA**.

(**b**) Find $\mathbf{B} + \mathbf{C}$, **AC**, and $\mathbf{A}(\mathbf{B} + \mathbf{C})$. Compare $\mathbf{A}(\mathbf{B} + \mathbf{C})$ with $\mathbf{AB} + \mathbf{AC}$, thus illustrating (2.15).

(**c**) Compare $(\mathbf{AB})'$ with $\mathbf{B}'\mathbf{A}'$, thus illustrating Theorem 2.2b.

(**d**) Compare tr(**AB**) with tr(**BA**) and confirm that (2.87) holds in this case.

(**e**) Let $\mathbf{a}_1'$ and $\mathbf{a}_2'$ be the two rows of **A**. Find $\begin{pmatrix} \mathbf{a}_1'\mathbf{B} \\ \mathbf{a}_2'\mathbf{B} \end{pmatrix}$ and compare with **AB** in part (a), thus illustrating (2.27).

(**f**) Let $\mathbf{b}_1$ and $\mathbf{b}_2$ be the two columns of **B**. Find $(\mathbf{Ab}_1, \mathbf{Ab}_2)$ and compare with **AB** in part (a), thus illustrating (2.28).

2.7 Let $\mathbf{A} = \begin{pmatrix} 3 & 2 & 1 \\ 6 & 4 & 2 \\ 12 & 8 & 4 \end{pmatrix}$, $\mathbf{B} = \begin{pmatrix} 1 & -1 & 2 \\ -1 & 1 & -2 \\ -1 & 1 & -2 \end{pmatrix}$.

(**a**) Show that $\mathbf{AB} = \mathbf{O}$.

(**b**) Find a vector **x** such that $\mathbf{Ax} = \mathbf{0}$.

(**c**) What is the rank of **A** and the rank of **B**?

2.8 If **j** is a vector of 1s, as defined in (2.6), show that

(**a**) $\mathbf{j}'\mathbf{a} = \mathbf{a}'\mathbf{j} = \sum_i a_i$, as in (2.24).

(**b**) **Aj** is a column vector whose elements are the row sums of **A**, as in (2.25).

(**c**) $\mathbf{j}'\mathbf{A}$ is a row vector whose elements are the column sums of **A**, as in (2.25).

2.9 Prove Corollary 1 to Theorem 2.2b; that is, assuming that **A**, **B**, and **C** are conformal, show that $(\mathbf{ABC})' = \mathbf{C}'\mathbf{B}'\mathbf{A}'$.

2.10 Prove Theorem 2.2c.

2.11 Use matrix **A** in Problem 2.6 and let

$$\mathbf{D}_1 = \begin{pmatrix} 3 & 0 \\ 0 & -2 \end{pmatrix}, \qquad \mathbf{D}_2 = \begin{pmatrix} 5 & 0 & 0 \\ 0 & 3 & 0 \\ 0 & 0 & 6 \end{pmatrix}.$$

Find $\mathbf{D}_1\mathbf{A}$ and $\mathbf{AD}_2$, thus illustrating (2.29) and (2.30).

2.12 Let $\mathbf{A} = \begin{pmatrix} 1 & 2 & 3 \\ 4 & 5 & 6 \\ 7 & 8 & 9 \end{pmatrix}, \qquad \mathbf{D} = \begin{pmatrix} a & 0 & 0 \\ 0 & b & 0 \\ 0 & 0 & c \end{pmatrix}.$

Find **DA**, **AD**, and **DAD**.

2.13 For $\mathbf{y}' = (y_1, y_2, y_3)$ and the symmetric matrix

$$\mathbf{A} = \begin{pmatrix} a_{11} & a_{12} & a_{13} \\ a_{12} & a_{22} & a_{23} \\ a_{13} & a_{23} & a_{33} \end{pmatrix},$$

express $\mathbf{y}'\mathbf{Ay}$ in the form given in (2.33).

2.14 Let $\mathbf{A} = \begin{pmatrix} 5 & -1 & 3 \\ -1 & 1 & 2 \\ 3 & 2 & 7 \end{pmatrix}, \quad \mathbf{B} = \begin{pmatrix} 6 & -2 & 3 \\ 7 & 1 & 0 \\ 2. & -3 & 5 \end{pmatrix}, \quad \mathbf{C} = \begin{pmatrix} 2 & -3 \\ -1 & 4 \\ 3 & 1 \end{pmatrix},$

$$\mathbf{x} = \begin{pmatrix} 3 \\ -1 \\ 2 \end{pmatrix}, \quad \mathbf{y} = \begin{pmatrix} 3 \\ 2 \\ 4 \end{pmatrix}, \quad \mathbf{z} = \begin{pmatrix} 2 \\ 5 \end{pmatrix}.$$

Find the following:

(a) $\mathbf{Bx}$
(b) $\mathbf{y}'\mathbf{B}$
(c) $\mathbf{x}'\mathbf{Ax}$
(d) $\mathbf{x}'\mathbf{Cz}$
(e) $\mathbf{x}'\mathbf{x}$
(f) $\mathbf{x}'\mathbf{y}$
(g) $\mathbf{xx}'$
(h) $\mathbf{xy}'$
(i) $\mathbf{B}'\mathbf{B}$
(j) $\mathbf{yz}'$
(k) $\mathbf{zy}'$
(l) $\sqrt{\mathbf{y}'\mathbf{y}}$
(m) $\mathbf{C}'\mathbf{C}$

2.15 Use **x**, **y**, **A**, and **B** as defined in Problem 2.14.

(a) Find $\mathbf{x} + \mathbf{y}$ and $\mathbf{x} - \mathbf{y}$.

(b) Find tr(**A**), tr(**B**), **A** + **B**, and tr(**A** + **B**).
(c) Find **AB** and **BA**.
(d) Find tr(**AB**) and tr(**BA**).
(e) Find |**AB**| and |**BA**|.
(f) Find $(\mathbf{AB})'$ and $\mathbf{B}'\mathbf{A}'$.

2.16 Using **B** and **x** in Problem 2.14, find **Bx** as a linear combination of the columns of **B**, as in (2.37), and compare with **Bx** as found in Problem 2.14(a).

2.17 Let $\mathbf{A} = \begin{pmatrix} 2 & 5 \\ 1 & 3 \end{pmatrix}$, $\mathbf{B} = \begin{pmatrix} 1 & -6 & 2 \\ 5 & 0 & 3 \end{pmatrix}$, $\mathbf{I} = \begin{pmatrix} 1 & 0 \\ 0 & 1 \end{pmatrix}$.

(a) Show that $(\mathbf{AB})' = \mathbf{B}'\mathbf{A}'$ as in (2.26).
(b) Show that $\mathbf{AI} = \mathbf{A}$ and that $\mathbf{IB} = \mathbf{B}$.
(c) Find |**A**|.
(d) Find $\mathbf{A}^{-1}$.
(e) Find $(\mathbf{A}^{-1})^{-1}$ and compare with **A**, thus verifying (2.46).
(f) Find $(\mathbf{A}')^{-1}$ and verify that it is equal to $(\mathbf{A}^{-1})'$ as in Theorem 2.5a.

2.18 Let **A** and **B** be defined and partitioned as follows:

$$\mathbf{A} = \left(\begin{array}{cc|c} 2 & 1 & 2 \\ 3 & 2 & 0 \\ \hline 1 & 0 & 1 \end{array}\right), \qquad \mathbf{B} = \left(\begin{array}{ccc|c} 1 & 1 & 1 & 0 \\ 2 & 1 & 1 & 2 \\ \hline 2 & 3 & 1 & 2 \end{array}\right).$$

(a) Find **AB** as in (2.35), using the indicated partitioning.
(b) Check by finding **AB** in the usual way, ignoring the partitioning.

2.19 Partition the matrices **A** and **B** in Problem 2.18 as follows:

$$\mathbf{A} = \left(\begin{array}{c|cc} 2 & 1 & 2 \\ 3 & 2 & 0 \\ 1 & 0 & 1 \end{array}\right) = (\mathbf{a}_1, \mathbf{A}_2),$$

$$\mathbf{B} = \left(\begin{array}{cccc} 1 & 1 & 1 & 0 \\ \hline 2 & 1 & 1 & 2 \\ 2 & 3 & 1 & 2 \end{array}\right) = \begin{pmatrix} \mathbf{b}_1' \\ \mathbf{B}_2 \end{pmatrix}.$$

Repeat parts (a) and (b) of Problem 2.18. Note that in this case, (2.35) becomes $\mathbf{AB} = \mathbf{a}_1\mathbf{b}_1' + \mathbf{A}_2\mathbf{B}_2$.

2.20 Let $\mathbf{A} = \begin{pmatrix} 5 & -2 & 3 \\ 7 & 3 & 1 \end{pmatrix}$, $\quad \mathbf{b} = \begin{pmatrix} 2 \\ 4 \\ -3 \end{pmatrix}$.

Find $\mathbf{Ab}$ as a linear combination of the columns of $\mathbf{A}$ as in (2.37) and check the result by finding $\mathbf{Ab}$ in the usual way.

2.21 Show that each column of the product $\mathbf{AB}$ can be expressed as a linear combination of the columns of $\mathbf{A}$, with coefficients arising from the corresponding column of $\mathbf{B}$, as noted following Example 2.3.

2.22 Let $\mathbf{A} = \begin{pmatrix} 3 & 0 & 2 \\ 1 & -1 & 1 \\ 2 & 1 & 0 \end{pmatrix}$, $\quad \mathbf{B} = \begin{pmatrix} -2 & -1 \\ 3 & 1 \\ 1 & -1 \end{pmatrix}$.

Express the columns of $\mathbf{AB}$ as linear combinations of the columns of $\mathbf{A}$.

2.23 Show that if a set of vectors includes $\mathbf{0}$, the set is linearly dependent, as noted following (2.40).

2.24 Suppose that $\mathbf{A}$ and $\mathbf{B}$ are $n \times n$ and that $\mathbf{AB} = \mathbf{O}$ as in (2.43). Show that $\mathbf{A}$ and $\mathbf{B}$ are both singular or one of them is $\mathbf{O}$.

2.25 Let $\mathbf{A} = \begin{pmatrix} 1 & 3 & 2 \\ 2 & 0 & -1 \end{pmatrix}$, $\quad \mathbf{B} = \begin{pmatrix} 1 & 2 \\ 0 & 1 \\ 1 & 0 \end{pmatrix}$, $\quad \mathbf{C} = \begin{pmatrix} 2 & 1 & 1 \\ 5 & -6 & -4 \end{pmatrix}$.

Find $\mathbf{AB}$ and $\mathbf{CB}$. Are they equal? What are the ranks of $\mathbf{A}$, $\mathbf{B}$, and $\mathbf{C}$?

2.26 Let $\mathbf{A} = \begin{pmatrix} 3 & 1 & 2 \\ 1 & 0 & -1 \end{pmatrix}$, $\quad \mathbf{B} = \begin{pmatrix} 2 & 1 \\ 0 & 2 \\ 1 & 0 \end{pmatrix}$.

(**a**) Find a matrix $\mathbf{C}$ such that $\mathbf{AB} = \mathbf{CB}$. Is $\mathbf{C}$ unique?

(**b**) Find a vector $\mathbf{x}$ such that $\mathbf{Ax} = \mathbf{0}$. Can you do this for $\mathbf{B}$?

2.27 Let $\mathbf{A} = \begin{pmatrix} 3 & 1 & 2 \\ 4 & -2 & 3 \\ 1 & 0 & -1 \end{pmatrix}$, $\quad \mathbf{x} = \begin{pmatrix} 5 \\ 2 \\ 3 \end{pmatrix}$.

(**a**) Find a matrix $\mathbf{B} \neq \mathbf{A}$ such that $\mathbf{Ax} = \mathbf{Bx}$. Why is this possible? Can $\mathbf{A}$ and $\mathbf{B}$ be nonsingular? Can $\mathbf{A} - \mathbf{B}$ be nonsingular?

(**b**) Find a matrix $\mathbf{C} \neq \mathbf{O}$ such that $\mathbf{Cx} = \mathbf{0}$. Can $\mathbf{C}$ be nonsingular?

2.28 Prove Theorem 2.5a.

2.29 Prove Theorem 2.5b.

2.30 Use the matrix $\mathbf{A}$ in Problem 2.17, and let $\mathbf{B} = \begin{pmatrix} 4 & -2 \\ 3 & 1 \end{pmatrix}$. Find $\mathbf{AB}$, $\mathbf{B}^{-1}$, and $(\mathbf{AB})^{-1}$. Verify that Theorem 2.5b holds in this case.

2.31 Show that the partitioned matrix $\mathbf{A} = \begin{pmatrix} \mathbf{A}_{11} & \mathbf{A}_{12} \\ \mathbf{A}_{21} & \mathbf{A}_{22} \end{pmatrix}$ has the inverse indicated in (2.50).

2.32 Show that the partitioned matrix $\mathbf{A} = \begin{pmatrix} \mathbf{A}_{11} & \mathbf{a}_{12} \\ \mathbf{a}'_{12} & a_{22} \end{pmatrix}$ has the inverse given in (2.51).

2.33 Show that $\mathbf{B} + \mathbf{cc}'$ has the inverse indicated in (2.53).

2.34 Show that $\mathbf{A} + \mathbf{PBQ}$ has the inverse indicated in (2.54).

2.35 Show that $\mathbf{y}'\mathbf{A}\mathbf{y} = \mathbf{y}'\left[\frac{1}{2}(\mathbf{A} + \mathbf{A}')\right]\mathbf{y}$ as in (2.55).

2.36 Prove Theorem 2.6b(ii).

2.37 Prove Corollaries 1 and 2 of Theorem 2.6b.

2.38 Prove the "only if" part of Theorem 2.6c.

2.39 Prove Corollary 1 to Theorem 2.6c.

2.40 Compare the rank of the augmented matrix with the rank of the coefficient matrix for each of the following systems of equations. Find solutions where they exist.

(**a**)
$$\begin{aligned} x_1 + 2x_2 + 3x_3 &= 6 \\ x_1 - x_2 &= 2 \\ x_1 - x_3 &= -1 \end{aligned}$$

(**b**)
$$\begin{aligned} x_1 - x_2 + 2x_3 &= 2 \\ x_1 - x_2 - x_3 &= -1 \\ 2x_1 - 2x_2 + x_3 &= 2 \end{aligned}$$

(**c**)
$$\begin{aligned} x_1 + x_2 + x_3 + x_4 &= 8 \\ x_1 - x_2 - x_3 - x_4 &= 6 \\ 3x_1 + x_2 + x_3 + x_4 &= 22 \end{aligned}$$

2.41 Prove Theorem 2.8a.

2.42 For the matrices $\mathbf{A}, \mathbf{A}_1^-$, and $\mathbf{A}_2^-$ in (2.59) and (2.60), show that $\mathbf{A}\mathbf{A}_1^-\mathbf{A} = \mathbf{A}$ and $\mathbf{A}\mathbf{A}_2^-\mathbf{A} = \mathbf{A}$.

2.43 Show that $\mathbf{A}_1^-$ in (2.60) can be obtained using Theorem 2.8b.

2.44 Show that $\mathbf{A}_2^-$ in (2.60) can be obtained using the five-step algorithm following Theorem 2.8b.

2.45 Prove Theorem 2.8c.

2.46 Show that if $\mathbf{A}$ is symmetric, there exists a symmetric generalized inverse for $\mathbf{A}$, as noted following Theorem 2.8c.

2.47 Let $\mathbf{A} = \begin{pmatrix} 4 & 2 & 2 \\ 2 & 2 & 0 \\ 2 & 0 & 2 \end{pmatrix}$.

(**a**) Find a symmetric generalized inverse for $\mathbf{A}$.

(b) Find a nonsymmetric generalized inverse for $\mathbf{A}$.

2.48 (a) Show that if $\mathbf{A}$ is nonsingular, then $\mathbf{A}^- = \mathbf{A}^{-1}$.

(b) Show that if $\mathbf{A}$ is $n \times p$ of rank $p < n$, then $\mathbf{A}^-$ is a "left inverse" of $\mathbf{A}$, that is, $\mathbf{A}^-\mathbf{A} = \mathbf{I}$.

2.49 Prove Theorem 2.9a parts (iv) and (vi).

2.50 Use $\mathbf{A} = \begin{pmatrix} 2 & 5 \\ 1 & 3 \end{pmatrix}$ from Problem 2.17 to illustrate (64), (2.66), and (2.67) in Theorem 2.9a.

2.51 (a) Multiply $\mathbf{A}$ in Problem 2.50 by 10 and verify that (2.69) holds in this case.
(b) Verify that (2.69) holds in general.

2.52 Prove Corollaries 1, 2, 3, and 4 of Theorem 2.9b.

2.53 Prove Corollaries 1 and 2 of Theorem 2.9c.

2.54 Use $\mathbf{A}$ in Problem 2.50 and let $\mathbf{B} = \begin{pmatrix} 4 & -2 \\ 3 & 1 \end{pmatrix}$.

(a) Find $|\mathbf{A}|, |\mathbf{B}|, \mathbf{AB}$, and $|\mathbf{AB}|$ and illustrate (2.77).
(b) Find $|\mathbf{A}|^2$ and $|\mathbf{A}^2|$ and illustrate (2.79).

2.55 Use Theorem 2.9c and Corollary 1 of Theorem 2.9b to prove Theorem 2.9b.

2.56 Show that if $\mathbf{C}'\mathbf{C} = \mathbf{I}$, then $\mathbf{CC}' = \mathbf{I}$ as in (2.84).

2.57 The columns of the following matrix are mutually orthogonal:

$$\mathbf{A} = \begin{pmatrix} 1 & -1 & 1 \\ -1 & 0 & 2 \\ 1 & 1 & 1 \end{pmatrix}.$$

(a) Normalize the columns of $\mathbf{A}$ by dividing each column by its length; denote the resulting matrix by $\mathbf{C}$.
(b) Show that $\mathbf{C}'\mathbf{C} = \mathbf{CC}' = \mathbf{I}$.

2.58 Prove Theorem 2.10a.

2.59 Prove Theorem 2.11 parts (i), (iv), (v), and (vii).

2.60 Use matrix $\mathbf{B}$ in Problem 2.26 to illustrate Theorem 2.11 parts (iii) and (iv).

2.61 Use matrix $\mathbf{A}$ in Problem 2.26 to illustrate Theorem 2.11(v), that is, $\text{tr}(\mathbf{A}'\mathbf{A}) = \text{tr}(\mathbf{AA}') = \sum_{ij} a_{ij}^2$.

2.62 Show that $\text{tr}(\mathbf{A}^-\mathbf{A}) = \text{tr}(\mathbf{AA}^-) = r = \text{rank}(\mathbf{A})$, as in (2.93).

2.63 Use $\mathbf{A}$ in (2.59) and $\mathbf{A}_2^-$ in (2.60) to illustrate Theorem 2.11(viii), that is, $\text{tr}(\mathbf{A}^-\mathbf{A}) = \text{tr}(\mathbf{AA}^-) = r = \text{rank}(\mathbf{A})$.

2.64 Obtain $\mathbf{x}_2 = (2/\sqrt{5}, 1/\sqrt{5})'$ in Example 2.12.1.

2.65 For $k = 3$, show that $\mathbf{A}^k\mathbf{x} = \lambda^k\mathbf{x}$ as in (2.100).

2.66 Show that $\lim_{k\to\infty} \mathbf{A}^k = \mathbf{O}$ in (2.102) if $\mathbf{A}$ is symmetric and if all eigenvalues of $\mathbf{A}$ satisfy $-1 < \lambda < 1$.

2.67 Prove Theorem 2.12a.

2.68 Prove Theorem 2.12b.

2.69 Prove Theorem 2.12c(ii) for the case where the eigenvalues $\lambda_1, \lambda_2, \ldots, \lambda_n$ are distinct.

2.70 Prove Corollary 1 to Theorem 2.12d.

2.71 Let $\mathbf{A} = \begin{pmatrix} 3 & 1 & 1 \\ 1 & 0 & 2 \\ 1 & 2 & 0 \end{pmatrix}$.

(**a**) The eigenvalues of $\mathbf{A}$ are $1, 4, -2$. Find the normalized eigenvectors and use them as columns in an orthogonal matrix $\mathbf{C}$.

(**b**) Show that $\mathbf{A} = \mathbf{CDC}'$, as in (2.103), where $\mathbf{D} = \text{diag}(1, 4, -2)$.

(**c**) Show that $\mathbf{C}'\mathbf{AC} = \mathbf{D}$ as in (2.106).

2.72 Prove Theorem 2.12e for a symmetric matrix $\mathbf{A}$.

2.73 Let $\mathbf{A} = \begin{pmatrix} 1 & 1 & -2 \\ -1 & 2 & 1 \\ 0 & 1 & -1 \end{pmatrix}$.

(**a**) Find the eigenvalues and associated normalized eigenvectors.

(**b**) Find $\text{tr}(\mathbf{A})$ and $|\mathbf{A}|$ and verify that $\text{tr}(\mathbf{A}) = \sum_{i=1}^{3} \lambda_i$ and $|\mathbf{A}| = \prod_{i=1}^{3} \lambda_i$, as in Theorem 2.12e.

2.74 Prove Theorem 2.12f(ii).

2.75 Let $\mathbf{A} = \begin{pmatrix} 1 & 0 & -1 \\ 0 & 1 & -1 \\ -1 & -1 & 3 \end{pmatrix}$.

(**a**) Show that $|\mathbf{A}| > 0$.

(**b**) Find the eigenvalues of $\mathbf{A}$. Are they all positive?

2.76 Let $\mathbf{A}^{1/2}$ be defined as in (2.109).

(a) Show that $\mathbf{A}^{1/2}$ is symmetric.
(b) Show that $(\mathbf{A}^{1/2})^2 = \mathbf{A}$ as in (2.110).

2.77 For the positive definite matrix $\mathbf{A} = \begin{pmatrix} 2 & -1 \\ -1 & 2 \end{pmatrix}$, calculate the eigenvalues and eigenvectors and find the square root matrix $\mathbf{A}^{1/2}$ as in (2.109). Check by showing $(\mathbf{A}^{1/2})^2 = \mathbf{A}$.

2.78 Prove Theorem 2.13e.

2.79 Prove Theorem 2.13f.

2.80 Let $\mathbf{A} = \begin{pmatrix} \frac{2}{3} & 0 & \frac{\sqrt{2}}{3} \\ 0 & 1 & 0 \\ \frac{\sqrt{2}}{3} & 0 & \frac{1}{3} \end{pmatrix}$.

(a) Find the rank of $\mathbf{A}$.
(b) Show that $\mathbf{A}$ is idempotent.
(c) Show that $\mathbf{I} - \mathbf{A}$ is dempotent.
(d) Show that $\mathbf{A}(\mathbf{I} - \mathbf{A}) = \mathbf{O}$.
(e) Find $\text{tr}(\mathbf{A})$.
(f) Find the eigenvalues of $\mathbf{A}$.

2.81 Consider a $p \times p$ matrix $\mathbf{A}$ with eigenvalues $\lambda_1, \lambda_2, \ldots, \lambda_p$. Show that $[\text{tr}(\mathbf{A})]^2 = \text{tr}(\mathbf{A}^2) + 2\sum\sum_{i \neq j} \lambda_i \lambda_j$.

2.82 Consider a nonsingular $n \times n$ matrix $\mathbf{A}$ whose elements are functions of the scalar x. Also consider the full-rank $p \times n$ matrix $\mathbf{B}$. Let $\mathbf{H} = \mathbf{B}'(\mathbf{B}\mathbf{A}\mathbf{B}')^{-1}\mathbf{B}$. Show that

$$\frac{\partial \mathbf{H}}{\partial \mathbf{x}} = -\mathbf{H}\frac{\partial \mathbf{A}}{\partial \mathbf{x}}\mathbf{H}.$$

2.83 Show that

$$\frac{\partial \ln |\mathbf{X}|}{\partial \mathbf{X}} = 2\mathbf{X}^{-1} - \text{diag}\,\mathbf{X}^{-1}$$

for a 2×2 positive definite matrix $\mathbf{X}$.

2.84 Let $u = \mathbf{x}'\mathbf{A}\mathbf{x}$ where $\mathbf{x}$ is a 3×1 vector and $\mathbf{A} = \begin{pmatrix} 1 & 0 & 0 \\ 0 & 2 & 0 \\ 0 & 0 & 3 \end{pmatrix}$. Use the Lagrange multiplier method to find the vector $\mathbf{x}$ that minimizes u subject to the constraints $x_1 + x_2 = 2$, and $x_2 + x_3 = 3$.

3 Random Vectors and Matrices

3.1 INTODUCTION

As we work with linear models, it is often convenient to express the observed data (or data that will be observed) in the form of a vector or matrix. A *random vector* or *random matrix* is a vector or matrix whose elements are random variables. Informally, a *random variable* is defined as a variable whose value depends on the outcome of a chance experiment. (Formally, a random variable is a function defined for each element of a sample space.)

In terms of experimental structure, we can distinguish two kinds of random vectors:

1. A vector containing a measurement on each of n different individuals or experimental units. In this case, where the same variable is observed on each of n units selected at random, the n random variables $y_1, y_2, \ldots, y_n$ in the vector are typically uncorrelated and have the same variance.
2. A vector consisting of p different measurements on one individual or experimental unit. The p random variables thus obtained are typically correlated and have different variances.

To illustrate the first type of random vector, consider the multiple regression model

$$y_i = \beta_0 + \beta_1 x_{i1} + \beta_2 x_{i2} + \cdots + \beta_k x_{ik} + \varepsilon_i, \quad i = 1, 2, \ldots, n,$$

as given in (1.2). In Chapters 7–9, we treat the x variables as constants, in which case we have two random vectors:

$$\mathrm{y} = \begin{pmatrix} y_1 \\ y_2 \\ \vdots \\ y_n \end{pmatrix} \quad \text{and} \quad \boldsymbol{\varepsilon} = \begin{pmatrix} \varepsilon_1 \\ \varepsilon_2 \\ \vdots \\ \varepsilon_n \end{pmatrix}. \tag{3.1}$$

Linear Models in Statistics, Second Edition, by Alvin C. Rencher and G. Bruce Schaalje

The y_i values are observable, but the ε_i's are not observable unless the β's are known.

To illustrate the second type of random vector, consider regression of y on several random x variables (this regression case is discussed in Chapter 10). For the ith individual in the sample, we observe the $k+1$ random variables $y_i, x_{i1}, x_{i2}, \ldots, x_{ik}$, which constitute the random vector $(y_i, x_{i1}, \ldots, x_{ik})'$. In some cases, the $k+1$ variables $y_i, x_{i1}, \ldots, x_{ik}$ are all measured using the same units or scale of measurement, but typically the scales differ.

3.2 MEANS, VARIANCES, COVARIANCES, AND CORRELATIONS

In this section, we review some properties of univariate and bivariate random variables. We begin with a univariate random variable y. We do not distinguish notationally between the random variable y and an observed value of y. In many texts, an uppercase letter is used for the random variable and the corresponding lowercase letter represents a realization of the random variable, as in the expression $P(Y \leq y)$. This practice is convenient in a univariate context but would be confusing in the present text where we use uppercase letters for matrices and lowercase letters for vectors.

If $f(y)$ is the *density* of the random variable y, the *mean* or *expected value* of y is defined as

$$\mu = E(y) = \int_{-\infty}^{\infty} yf(y)\,dy. \tag{3.2}$$

This is the population mean. Later (beginning in Chapter 5), we also use the sample mean of y, obtained from a random sample of n observed values of y.

The expected value of a function of y such as y^2 can be found directly without first finding the density of y^2. In general, for a function $u(y)$, we have

$$E[u(y)] = \int_{-\infty}^{\infty} u(y)f(y)\,dy. \tag{3.3}$$

For a constant a and functions $u(y)$ and $v(y)$, it follows from (3.3) that

$$E(ay) = aE(y), \tag{3.4}$$

$$E[u(y) + v(y)] = E[u(y)] + E[v(y)]. \tag{3.5}$$

The *variance* of a random variable y is defined as

$$\sigma^2 = \mathrm{var}(y) = E(y - \mu)^2, \tag{3.6}$$

This is the population variance. Later (beginning in Chapter 5), we also use the sample variance of y, obtained from a random sample of n observed values of y. The square root of the variance is known as the *standard deviation*:

$$\sigma = \sqrt{\text{var}(y)} = \sqrt{E(y - \mu)^2}. \tag{3.7}$$

Using (3.4) and (3.5), we can express the variance of y in the form

$$\sigma^2 = \text{var}(y) = E(y^2) - \mu^2. \tag{3.8}$$

If a is a constant, we can use (3.4) and (3.6) to show that

$$\text{var}(ay) = a^2\text{var}(y) = a^2\sigma^2. \tag{3.9}$$

For any two variables y_i and y_j in the random vector $\mathbf{y}$ in (3.1), we define the *covariance* as

$$\sigma_{ij} = \text{cov}(y_i, y_j) = E[(y_i - \mu_i)(y_j - \mu_j)], \tag{3.10}$$

where $\mu_i = E(y_i)$ and $\mu_j = E(y_j)$. Using (3.4) and (3.5), we can express σ_{ij} in the form

$$\sigma_{ij} = \text{cov}(y_i, y_j) = E(y_i y_j) - \mu_i \mu_j. \tag{3.11}$$

Two random variables y_i and y_j are said to be *independent* if their joint density factors into the product of their marginal densities

$$f(y_i, y_j) = f_i(y_i) f_j(y_j), \tag{3.12}$$

where the marginal density $f_i(y_i)$ is defined as

$$f_i(y_i) = \int_{-\infty}^{\infty} f(y_i, y_j) dy_j. \tag{3.13}$$

From the definition of independence in (3.12), we obtain the following properties:

1. $E(y_i, y_j) = E(y_i)E(y_j)$ if y_i and y_j are independent. (3.14)

2. $\sigma_{ij} = \text{cov}(y_i, y_j) = 0$ if y_i and y_j are independent. (3.15)

The second property follows from the first.

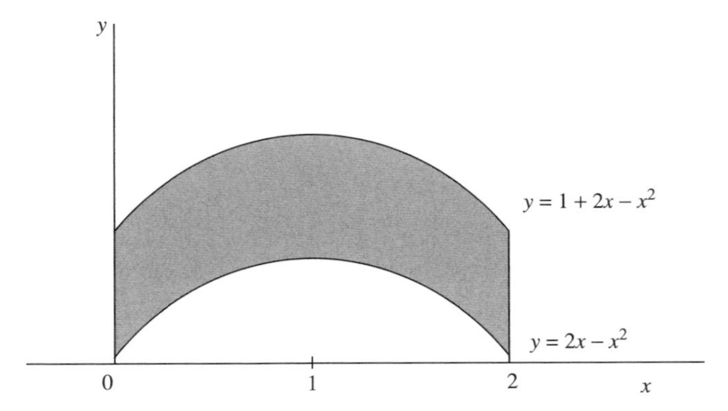

Figure 3.1 Region for $f(x, y)$ in Example 3.2.

In the first type of random vector defined in Section 3.1, the variables $y_1, y_2, \ldots, y_n$ would typically be independent if obtained from a random sample, and we would thus have $\sigma_{ij} = 0$ for all $i \neq j$. However, for the variables in the second type of random vector, we would typically have $\sigma_{ij} \neq 0$ for at least some values of i and j.

The converse of the property in (3.15) is not true; that is, $\sigma_{ij} = 0$ does not imply independence. This is illustrated in the following example.

Example 3.2. Suppose that the bivariate random variable (x, y) is distributed uniformly over the region $0 \leq x \leq 2$, $2x - x^2 \leq y \leq 1 + 2x - x^2$; see Figure 3.1.

The area of the region is given by

$$\text{Area} = \int_0^2 \int_{2x-x^2}^{1+2x-x^2} dy\,dx = 2.$$

Hence, for a uniform distribution over the region, we set

$$f(x, y) = \tfrac{1}{2}, \quad 0 \leq x \leq 2, \quad 2x - x^2 \leq y \leq 1 + 2x - x^2,$$

so that $\int\int f(x, y)dx\,dy = 1$.

To find σ_{xy} using (3.11), we need $E(xy)$, $E(x)$, and $E(y)$. The first of these is given by

$$\begin{aligned} E(xy) &= \int_0^2 \int_{2x-x^2}^{1+2x-x^2} xy\left(\tfrac{1}{2}\right) dy\,dx \\ &= \int_0^2 \frac{x}{4}(1 + 4x - 2x^2)dx = \frac{7}{6}. \end{aligned}$$

To find $E(x)$ and $E(y)$, we first find the marginal distributions of x and y. For $f_1(x)$, we have, by (3.13),

$$f_1(x) = \int_{2x-x^2}^{1+2x-x^2} \tfrac{1}{2}\,dy = \tfrac{1}{2}, \quad 0 \le x \le 2.$$

For $f_2(y)$, we obtain different results for $0 \le y \le 1$ and $1 \le y \le 2$:

$$f_2(y) = \int_0^{1-\sqrt{1-y}} \tfrac{1}{2}\,dx + \int_{1+\sqrt{1-y}}^{2} \tfrac{1}{2}\,dx = 1 - \sqrt{1-y}, \quad 0 \le y \le 1, \tag{3.16}$$

$$f_2(y) = \int_{1-\sqrt{2-y}}^{1+\sqrt{2-y}} \tfrac{1}{2}\,dx = \sqrt{2-y}, \quad 1 \le y \le 2. \tag{3.17}$$

Then

$$E(x) = \int_0^2 x\left(\tfrac{1}{2}\right)dx = 1,$$

$$E(y) = \int_0^1 y(1 - \sqrt{1-y})dy + \int_1^2 y\sqrt{2-y} \quad dy = \tfrac{7}{6}.$$

Now by (3.11), we obtain

$$\begin{aligned}\sigma_{xy} &= E(xy) - E(x)E(y)\\ &= \tfrac{7}{6} - (1)\left(\tfrac{7}{6}\right) = 0.\end{aligned}$$

However, x and y are clearly dependent since the range of y for each x depends on the value of x.

As a further indication of the dependence of y on x, we examine $E(y|x)$, the expected value of y for a given value of x, which is found as

$$E(y|x) = \int y f(y|x) dy.$$

The conditional density $f(y|x)$ is defined as

$$f(y|x) = \frac{f(x, y)}{f_1(x)}, \tag{3.18}$$

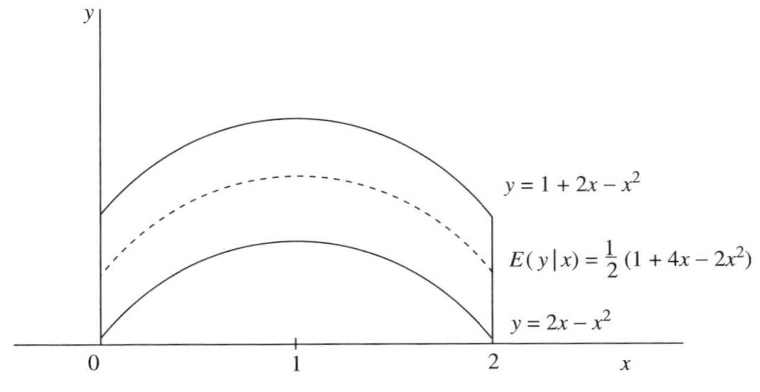

Figure 3.2 $E(y|x)$ in Example 3.2.

which becomes

$$f(y|x) = \frac{\frac{1}{2}}{\frac{1}{2}} = 1, \quad 2x - x^2 \leq y \leq 1 + 2x - x^2.$$

Thus

$$\begin{aligned} E(y|x) &= \int_{2x-x^2}^{1+2x-x^2} y(1)dy \\ &= \tfrac{1}{2}(1 + 4x - 2x^2). \end{aligned}$$

Since $E(y|x)$ depends on x, the two variables are dependent. Note that $E(y|x) = \frac{1}{2}(1 + 4x - 2x^2)$ is the average of the two curves $y = 2x - x^2$ and $y = 1 + 2x - x^2$. This is illustrated in Figure 3.2. □

In Example 3.2 we have two dependent random variables x and y for which $\sigma_{xy} = 0$. In cases such as this, σ_{xy} is not a good measure of relationship. However, if x and y have a bivariate normal distribution (see Section 4.2), then $\sigma_{xy} = 0$ implies independence of x and y (see Corollary 1 to Theorem 4.4c). In the bivariate normal case, $E(y|x)$ is a linear function of x (see Theorem 4.4d), and curves such as $E(y|x) = \frac{1}{2}(1 + 4x - 2x^2)$ do not occur.

The covariance σ_{ij} as defined in (3.10) depends on the scale of measurement of both y_i and y_j. To standardize σ_{ij}, we divide it by (the product of) the standard deviations of y_i and y_j to obtain the *correlation*:

$$\rho_{ij} = \text{corr}(y_i, y_j) = \frac{\sigma_{ij}}{\sigma_i \sigma_j}. \tag{3.19}$$

3.3 MEAN VECTORS AND COVARIANCE MATRICES FOR RANDOM VECTORS

3.3.1 Mean Vectors

The expected value of a $p \times 1$ random vector $\mathbf{y}$ is defined as the vector of expected values of the p random variables $y_1, y_2, \ldots, y_p$ in $\mathbf{y}$:

$$E(\mathbf{y}) = E\begin{pmatrix} y_1 \\ y_2 \\ \vdots \\ y_p \end{pmatrix} = \begin{pmatrix} E(y_1) \\ E(y_2) \\ \vdots \\ E(y_p) \end{pmatrix} = \begin{pmatrix} \mu_1 \\ \mu_2 \\ \vdots \\ \mu_p \end{pmatrix} = \boldsymbol{\mu}, \qquad (3.20)$$

where $E(y_i) = \mu_i$ is obtained as $E(y_i) = \int y_i f_i(y_i)\, dy_i$, using $f_i(y_i)$, the marginal density of y_i.

If $\mathbf{x}$ and $\mathbf{y}$ are $p \times 1$ random vectors, it follows from (3.20) and (3.5) that the expected value of their sum is the sum of their expected values:

$$E(\mathbf{x} + \mathbf{y}) = E(\mathbf{x}) + E(\mathbf{y}). \qquad (3.21)$$

3.3.2 Covariance Matrix

The variances $\sigma_1^2, \sigma_2^2, \ldots, \sigma_p^2$ of $y_1, y_2, \ldots, y_p$ and the covariances σ_{ij} for all $i \neq j$ can be conveniently displayed in the *covariance matrix*, which is denoted by $\boldsymbol{\Sigma}$, the uppercase version of σ_{ij}:

$$\boldsymbol{\Sigma} = \mathrm{cov}(\mathbf{y}) = \begin{pmatrix} \sigma_{11} & \sigma_{12} & \cdots & \sigma_{1p} \\ \sigma_{21} & \sigma_{22} & \cdots & \sigma_{2p} \\ \vdots & \vdots & & \vdots \\ \sigma_{p1} & \sigma_{p2} & \cdots & \sigma_{pp} \end{pmatrix}. \qquad (3.22)$$

The ith row of $\boldsymbol{\Sigma}$ contains the variance of y_i and the covariance of y_i with each of the other y variables. To be consistent with the notation σ_{ij}, we have used $\sigma_{ii} = \sigma_i^2$, $i = 1, 2, \ldots, p$, for the variances. The variances are on the diagonal of $\boldsymbol{\Sigma}$, and the covariances occupy off-diagonal positions. There is a distinction in the font used for $\boldsymbol{\Sigma}$ as the covariance matrix and $\sum$ as the summation symbol. Note also the distinction in meaning between the notation $\mathrm{cov}(\mathbf{y}) = \boldsymbol{\Sigma}$ and $\mathrm{cov}(y_i, y_j) = \sigma_{ij}$.

The covariance matrix $\boldsymbol{\Sigma}$ is symmetric because $\sigma_{ij} = \sigma_{ji}$ [see (3.10)]. In many applications, $\boldsymbol{\Sigma}$ is assumed to be positive definite. This will ordinarily hold if the y variables are continuous random variables and if there are no linear relationships among them. (If there are linear relationships among the y variables, $\boldsymbol{\Sigma}$ will be positive semidefinite.)

By analogy with (3.20), we define the expected value of a random matrix $\mathbf{Z}$ as the matrix of expected values:

$$E(\mathbf{Z}) = E\begin{pmatrix} z_{11} & z_{12} & \cdots & z_{1p} \\ z_{21} & z_{22} & \cdots & z_{2p} \\ \vdots & \vdots & & \vdots \\ z_{n1} & z_{n2} & \cdots & z_{np} \end{pmatrix} = \begin{pmatrix} E(z_{11}) & E(z_{12}) & \cdots & E(z_{1p}) \\ E(z_{21}) & E(z_{22}) & \cdots & E(z_{2p}) \\ \vdots & \vdots & & \vdots \\ E(z_{n1}) & E(z_{n2}) & \cdots & E(z_{np}) \end{pmatrix}. \tag{3.23}$$

We can express $\boldsymbol{\Sigma}$ as the expected value of a random matrix. By (2.21), the (ij)th element of the matrix $(\mathbf{y}-\boldsymbol{\mu})(\mathbf{y}-\boldsymbol{\mu})'$ is $(y_i-\mu_i)(y_j-\mu_j)$. Thus, by (3.10) and (3.23), the (ij)th element of $E\,[(\mathbf{y}-\boldsymbol{\mu})\,(\mathbf{y}-\boldsymbol{\mu})']$ is $E[\,(y_i-\mu_i)\,(y_j-\mu_j)] = \sigma_{ij}$. Hence

$$E[(\mathbf{y}-\boldsymbol{\mu})(\mathbf{y}-\boldsymbol{\mu}')] = \begin{pmatrix} \sigma_{11} & \sigma_{12} & \cdots & \sigma_{1p} \\ \sigma_{21} & \sigma_{22} & \cdots & \sigma_{2p} \\ \vdots & \vdots & & \vdots \\ \sigma_{p1} & \sigma_{p2} & \cdots & \sigma_{pp} \end{pmatrix} = \boldsymbol{\Sigma}. \tag{3.24}$$

We illustrate (3.24) for $p = 3$:

$$\begin{aligned}
\boldsymbol{\Sigma} &= E[(\mathbf{y}-\boldsymbol{\mu})(\mathbf{y}-\boldsymbol{\mu})'] \\
&= E\left[\begin{pmatrix} y_1-\mu_1 \\ y_2-\mu_2 \\ y_3-\mu_3 \end{pmatrix}(y_1-\mu_1,\, y_2-\mu_2,\, y_3-y_3)\right] \\
&= E\begin{bmatrix} (y_1-\mu_1)^2 & (y_1-\mu_1)(y_2-\mu_2) & (y_1-\mu_1)(y_3-\mu_3) \\ (y_2-\mu_2)(y_1-\mu_1) & (y_2-\mu_2)^2 & (y_2-\mu_2)(y_3-\mu_3) \\ (y_3-\mu_3)(y_1-\mu_1) & (y_3-\mu_3)(y_2-\mu_2) & (y_3-\mu_3)^2 \end{bmatrix} \\
&= \begin{bmatrix} E(y_1-\mu_1)^2 & E[(y_1-\mu_1)(y_2-\mu_2)] & E[(y_1-\mu_1)(y_3-\mu_3)] \\ E[(y_2-\mu_2)(y_1-\mu_1)] & E(y_2-\mu_2)^2 & E[(y_2-\mu_2)(y_3-\mu_3)] \\ E[(y_3-\mu_3)(y_1-\mu_1)] & E[(y_3-\mu_3)(y_2-\mu_2)] & E(y_3-\mu_3)^2 \end{bmatrix} \\
&= \begin{pmatrix} \sigma_1^2 & \sigma_{12} & \sigma_{13} \\ \sigma_{21} & \sigma_2^2 & \sigma_{23} \\ \sigma_{31} & \sigma_{32} & \sigma_3^2 \end{pmatrix}.
\end{aligned}$$

We can write (3.24) in the form

$$\boldsymbol{\Sigma} = E[(\mathbf{y}-\boldsymbol{\mu})(\mathbf{y}-\boldsymbol{\mu})'] = E(\mathbf{y}\mathbf{y}') - \boldsymbol{\mu}\boldsymbol{\mu}', \tag{3.25}$$

which is analogous to (3.8) and (3.11).

3.3.3 Generalized Variance

A measure of overall variability in the population of $\mathbf{y}$ variables can be defined as the determinant of $\boldsymbol{\Sigma}$:

$$\text{Generalized variance} = |\boldsymbol{\Sigma}|. \tag{3.26}$$

If $|\boldsymbol{\Sigma}|$ is small, the $\mathbf{y}$ variables are concentrated closer to $\boldsymbol{\mu}$ than if $|\boldsymbol{\Sigma}|$ is large. A small value of $|\boldsymbol{\Sigma}|$ may also indicate that the variables $y_1, y_2, \ldots, y_p$ in $\mathbf{y}$ are highly intercorrelated, in which case the $\mathbf{y}$ variables tend to occupy a subspace of the p dimensions [this corresponds to one or more small eigenvalues; see Rencher (1998, Section 2.1.3)].

3.3.4 Standardized Distance

To obtain a useful measure of distance between $\mathbf{y}$ and $\boldsymbol{\mu}$, we need to account for the variances and covariances of the y_i variables in $\mathbf{y}$. By analogy with the univariate standardized variable $(y - \mu)/\sigma$, which has mean 0 and variance 1, the *standardized distance* is defined as

$$\text{Standardized distance} = (\mathbf{y} - \boldsymbol{\mu})'\boldsymbol{\Sigma}^{-1}(\mathbf{y} - \boldsymbol{\mu}). \tag{3.27}$$

The use of $\boldsymbol{\Sigma}^{-1}$ standardizes the (transformed) y_i variables so that they have means equal to 0 and variances equal to 1 and are also uncorrelated (see Problem 3.11). A distance such as (3.27) is often called a *Mahalanobis distance* (Mahalanobis 1936).

3.4 CORRELATION MATRICES

By analogy with $\boldsymbol{\Sigma}$ in (3.22), the *correlation matrix* is defined as

$$\mathbf{P}_\rho = (\rho_{ij}) = \begin{pmatrix} 1 & \rho_{12} & \cdots & \rho_{1p} \\ \rho_{21} & 1 & \cdots & \rho_{2p} \\ \vdots & \vdots & & \vdots \\ \rho_{p1} & \rho_{p2} & \cdots & 1 \end{pmatrix}, \tag{3.28}$$

where $\rho_{ij} = \sigma_{ij}/\sigma_i\sigma_j$ is the correlation of y_i and y_j defined in (3.19). The second row of $\mathbf{P}_\rho$, for example, contains the correlation of y_2 with each of the other y variables. We use the subscript ρ in $\mathbf{P}_\rho$ to emphasize that $\mathbf{P}$ is the uppercase version of ρ.

If we define

$$\mathbf{D}_\sigma = [\text{diag}(\boldsymbol{\Sigma})]^{1/2} = \text{diag}(\sigma_1, \sigma_2, \ldots, \sigma_p), \tag{3.29}$$

then by (2.31), we can obtain $\mathbf{P}_\rho$ from $\boldsymbol{\Sigma}$ and vice versa:

$$\mathbf{P}_\rho = \mathbf{D}_\sigma^{-1}\boldsymbol{\Sigma}\mathbf{D}_\sigma^{-1}, \tag{3.30}$$

$$\boldsymbol{\Sigma} = \mathbf{D}_\sigma\mathbf{P}_\rho\mathbf{D}_\sigma. \tag{3.31}$$

3.5 MEAN VECTORS AND COVARIANCE MATRICES FOR PARTITIONED RANDOM VECTORS

Suppose that the random vector $\mathbf{v}$ is partitioned into two subsets of variables, which we denote by $\mathbf{y}$ and $\mathbf{x}$:

$$\mathbf{v} = \begin{pmatrix} \mathbf{y} \\ \mathbf{x} \end{pmatrix} = \begin{pmatrix} y_1 \\ \vdots \\ y_p \\ x_1 \\ \vdots \\ x_q \end{pmatrix}.$$

Thus there are $p + q$ random variables in $\mathbf{v}$.

The mean vector and covariance matrix for $\mathbf{v}$ partitioned as above can be expressed in the following form

$$\boldsymbol{\mu} = E(\mathbf{v}) = E\begin{pmatrix} \mathbf{y} \\ \mathbf{x} \end{pmatrix} = \begin{pmatrix} E(\mathbf{y}) \\ E(\mathbf{x}) \end{pmatrix} = \begin{pmatrix} \boldsymbol{\mu}_y \\ \boldsymbol{\mu}_x \end{pmatrix}, \tag{3.32}$$

$$\boldsymbol{\Sigma} = \text{cov}(\mathbf{v}) = \text{cov}\begin{pmatrix} \mathbf{y} \\ \mathbf{x} \end{pmatrix} = \begin{pmatrix} \boldsymbol{\Sigma}_{yy} & \boldsymbol{\Sigma}_{yx} \\ \boldsymbol{\Sigma}_{xy} & \boldsymbol{\Sigma}_{xx} \end{pmatrix}, \tag{3.33}$$

where $\boldsymbol{\Sigma}_{xy} = \boldsymbol{\Sigma}'_{yx}$. In (3.32), the submatrix $\boldsymbol{\mu}_y = [E(y_1), E(y_2), \ldots, E(y_p)]'$ contains the means of $y_1, y_2, \ldots, y_p$. Similarly $\boldsymbol{\mu}_x$ contains the means of the x variables. In (3.33), the submatrix $\boldsymbol{\Sigma}_{yy} = \text{cov}(\mathbf{y})$ is a $p \times p$ covariance matrix for $\mathbf{y}$ containing the variances of $y_1, y_2, \ldots, y_p$ on the diagonal and the covariance of each y_i with each y_j $(i \neq j)$ off the diagonal:

$$\boldsymbol{\Sigma}_{yy} = \begin{pmatrix} \sigma_{y_1}^2 & \sigma_{y_1 y_2} & \cdots & \sigma_{y_1 y_p} \\ \sigma_{y_2 y_1} & \sigma_{y_2}^2 & \cdots & \sigma_{y_2 y_p} \\ \vdots & \vdots & & \vdots \\ \sigma_{y_p y_1} & \sigma_{y_p y_2} & \cdots & \sigma_{y_p}^2 \end{pmatrix}.$$

Similarly, $\boldsymbol{\Sigma}_{xx} = \text{cov}(\mathbf{x})$ is the $q \times q$ covariance matrix of $x_1, x_2, \ldots, x_q$. The matrix $\boldsymbol{\Sigma}_{yx}$ in (3.33) is $p \times q$ and contains the covariance of each y_i with each x_j:

$$\boldsymbol{\Sigma}_{yx} = \begin{pmatrix} \sigma_{y_1 x_1} & \sigma_{y_1 x_2} & \cdots & \sigma_{y_1 x_q} \\ \sigma_{y_2 x_1} & \sigma_{y_2 x_2} & \cdots & \sigma_{y_2 x_q} \\ \vdots & \vdots & & \vdots \\ \sigma_{y_p x_1} & \sigma_{y_p x_2} & \cdots & \sigma_{y_p x_q} \end{pmatrix}.$$

Thus $\boldsymbol{\Sigma}_{yx}$ is rectangular unless $p = q$. The covariance matrix $\boldsymbol{\Sigma}_{yx}$ is also denoted by cov($\mathbf{y}$, $\mathbf{x}$) and can be defined as

$$\boldsymbol{\Sigma}_{yx} = \text{cov}(\mathbf{y}, \mathbf{x}) = E[(\mathbf{y} - \boldsymbol{\mu}_y)(\mathbf{x} - \boldsymbol{\mu}_x)']. \tag{3.34}$$

Note the difference in meaning between $\text{cov}\begin{pmatrix} \mathbf{y} \\ \mathbf{x} \end{pmatrix}$ in (3.33) and $\text{cov}(\mathbf{y}, \mathbf{x}) = \boldsymbol{\Sigma}_{yx}$ in (3.34). We have now used the notation cov in three ways: (1) cov(y_i, y_j), (2) cov($\mathbf{y}$), and (3) cov($\mathbf{y}$, $\mathbf{x}$). The first of these is a scalar, the second is a symmetric (usually positive definite) matrix, and the third is a rectangular matrix.

3.6 LINEAR FUNCTIONS OF RANDOM VECTORS

We often use linear combinations of the variables $y_1, y_2, \ldots, y_p$ from a random vector $\mathbf{y}$. Let $\mathbf{a} = (a_1, a_2, \ldots, a_p)'$ be a vector of constants. Then, by an expression preceding (2.18), the linear combination using the a terms as coefficients can be written as

$$z = a_1 y_1 + a_2 y_2 + \cdots + a_p y_p = \mathbf{a}'\mathbf{y}. \tag{3.35}$$

We consider the means, variances, and covariances of such linear combinations in Sections 3.6.1 and 3.6.2.

3.6.1 Means

Since $\mathbf{y}$ is a random vector, the linear combination $z = \mathbf{a}'\mathbf{y}$ is a (univariate) random variable. The mean of $\mathbf{a}'\mathbf{y}$ is given the following theorem.

Theorem 3.6a. If $\mathbf{a}$ is a $p \times 1$ vector of constants and $\mathbf{y}$ is a $p \times 1$ random vector with mean vector $\boldsymbol{\mu}$, then

$$\mu_z = E(\mathbf{a}'\mathbf{y}) = \mathbf{a}'E(\mathbf{y}) = \mathbf{a}'\boldsymbol{\mu}. \tag{3.36}$$

PROOF. Using (3.4), (3.5), and (3.35), we obtain

$$\begin{aligned} E(\mathbf{a}'\mathbf{y}) &= E(a_1y_1 + a_2y_2 + \cdots + a_py_p) \\ &= E(a_1y_1) + E(a_2y_2) + \cdots + E(a_py_p) \\ &= a_1E(y_1) + a_2E(y_2) + \cdots + a_pE(y_p) \\ &= (a_1, a_2, \ldots, a_p)\begin{pmatrix} E(y_1) \\ E(y_2) \\ \vdots \\ E(y_p) \end{pmatrix} \\ &= \mathbf{a}'E(\mathbf{y}) = \mathbf{a}'\boldsymbol{\mu}. \end{aligned}$$

□

Suppose that we have several linear combinations of $\mathbf{y}$ with constant coefficients:

$$\begin{aligned} z_1 &= a_{11}y_1 + a_{12}y_2 + \ldots + a_{1p}y_p = \mathbf{a}_1'\mathbf{y} \\ z_2 &= a_{21}y_1 + a_{22}y_2 + \ldots + a_{2p}y_p = \mathbf{a}_2'\mathbf{y} \\ &\vdots \qquad\qquad\qquad\qquad\qquad\qquad \vdots \\ z_k &= a_{k1}y_1 + a_{k2}y_2 + \ldots + a_{kp}y_p = \mathbf{a}_k'\mathbf{y}, \end{aligned}$$

where $\mathbf{a}_i' = (a_{i1}, a_{i2}, \ldots, a_{ip})$ and $\mathbf{y} = (y_1, y_2, \ldots, y_p)'$. These k linear functions can be written in the form

$$\mathbf{z} = \mathbf{A}\mathbf{y}, \tag{3.37}$$

where

$$\mathbf{z} = \begin{pmatrix} z_1 \\ z_2 \\ \vdots \\ z_k \end{pmatrix}, \quad \mathbf{A} = \begin{pmatrix} \mathbf{a}_1' \\ \mathbf{a}_2' \\ \vdots \\ \mathbf{a}_k' \end{pmatrix} = \begin{pmatrix} a_{11} & a_{12} & \cdots & a_{1p} \\ a_{21} & a_{22} & \cdots & a_{2p} \\ \vdots & \vdots & & \vdots \\ a_{k1} & a_{k2} & \cdots & a_{kp} \end{pmatrix}.$$

It is possible to have $k > p$, but we typically have $k \leq p$ with the rows of $\mathbf{A}$ linearly independent, so that $\mathbf{A}$ is full-rank. Since $\mathbf{y}$ is a random vector, each $z_i = \mathbf{a}_i'\mathbf{y}$ is a random variable and $\mathbf{z} = (z_1, z_2, \ldots, z_k)'$ is a random vector. The expected value of $\mathbf{z} = \mathbf{Ay}$ is given in the following theorem, as well as some extensions.

Theorem 3.6b. Suppose that $\mathbf{y}$ is a random vector, $\mathbf{X}$ is a random matrix, $\mathbf{a}$ and $\mathbf{b}$ are vectors of constants, and $\mathbf{A}$ and $\mathbf{B}$ are matrices of constants. Then, assuming the matrices and vectors in each product are conformal, we have the following expected values:

$$\text{(i)} \quad E(\mathbf{Ay}) = \mathbf{A}E(\mathbf{y}). \tag{3.38}$$

$$\text{(ii)} \quad E(\mathbf{a}'\mathbf{Xb}) = \mathbf{a}'E(\mathbf{X})\mathbf{b}. \tag{3.39}$$

$$\text{(iii)} \quad E(\mathbf{AXB}) = \mathbf{A}E(\mathbf{X})\mathbf{B}. \tag{3.40}$$

PROOF. These results follow from Theorem 3.6A (see Problem 3.14). □

Corollary 1. If $\mathbf{A}$ is a $k \times p$ matrix of constants, $\mathbf{b}$ is a $k \times 1$ vector of constants, and $\mathbf{y}$ is a $p \times 1$ random vector, then

$$E(\mathbf{Ay} + \mathbf{b}) = \mathbf{A}E(\mathbf{y}) + \mathbf{b}. \tag{3.41}$$

□

3.6.2 Variances and Covariances

The variance of the random variable $z = \mathbf{a}'\mathbf{y}$ is given in the following theorem.

Theorem 3.6c. If $\mathbf{a}$ is a $p \times 1$ vector of constants and $\mathbf{y}$ is a $p \times 1$ random vector with covariance matrix $\boldsymbol{\Sigma}$, then the variance of $z = \mathbf{a}'\mathbf{y}$ is given by

$$\sigma_z^2 = \text{var}(\mathbf{a}'\mathbf{y}) = \mathbf{a}'\boldsymbol{\Sigma}\mathbf{a}. \tag{3.42}$$

PROOF. By (3.6) and Theorem 3.6a, we obtain

$$\begin{aligned} \text{var}(\mathbf{a}'\mathbf{y}) &= E(\mathbf{a}'\mathbf{y} - \mathbf{a}'\boldsymbol{\mu})^2 = E[\mathbf{a}'(\mathbf{y} - \boldsymbol{\mu})]^2 \\ &= E[\mathbf{a}'(\mathbf{y} - \boldsymbol{\mu})\mathbf{a}'(\mathbf{y} - \boldsymbol{\mu})] \\ &= E[\mathbf{a}'(\mathbf{y} - \boldsymbol{\mu})(\mathbf{y} - \boldsymbol{\mu})'\mathbf{a}] \qquad \text{[by (2.18)]} \\ &= \mathbf{a}'E[(\mathbf{y} - \boldsymbol{\mu})(\mathbf{y} - \boldsymbol{\mu})']\mathbf{a} \qquad \text{[by Theorem 3.6b(ii)]} \\ &= \mathbf{a}'\boldsymbol{\Sigma}\mathbf{a} \qquad \text{[by(3.24)].} \end{aligned}$$

□

We illustrate 3.42 for $p = 3$:

$$\begin{aligned} \text{var}(\mathbf{a}'\mathbf{y}) &= \text{var}(a_1 y_1 + a_2 y_2 + a_3 y_3) = \mathbf{a}'\boldsymbol{\Sigma}\mathbf{a} \\ &= a_1^2\sigma_1^2 + a_2^2\sigma_2^2 + a_3^2\sigma_3^2 + 2a_1a_2\sigma_{12} + 2a_1a_3\sigma_{13} + 2a_2a_3\sigma_{23}. \end{aligned}$$

Thus, $\text{var}(\mathbf{a}'\mathbf{y}) = \mathbf{a}'\boldsymbol{\Sigma}\mathbf{a}$ involves all the variances and covariances of y_1, y_2, and y_3.

The covariance of two linear combinations is given in the following corollary to Theorem 3.6c.

Corollary 1. If $\mathbf{a}$ and $\mathbf{b}$ are $p \times 1$ vectors of constants, then

$$\text{cov}(\mathbf{a}'\mathbf{y}, \mathbf{b}'\mathbf{y}) = \mathbf{a}'\boldsymbol{\Sigma}\mathbf{b}. \tag{3.43}$$

□

Each variable z_i in the random vector $\mathbf{z} = (z_1, z_2, \ldots, z_k)' = \mathbf{Ay}$ in (3.37) has a variance, and each pair z_i and z_j $(i \neq j)$ has a covariance. These variances and covariances are found in the covariance matrix for $\mathbf{z}$, which is given in the following theorem, along with $\text{cov}(\mathbf{z}, \mathbf{w})$, where $\mathbf{w} = \mathbf{By}$ is another set of linear functions.

Theorem 3.6d. Let $\mathbf{z} = \mathbf{Ay}$ and $\mathbf{w} = \mathbf{By}$, where $\mathbf{A}$ is a $k \times p$ matrix of constants, $\mathbf{B}$ is an $m \times p$ matrix of constants, and $\mathbf{y}$ is a $p \times 1$ random vector with covariance matrix $\boldsymbol{\Sigma}$. Then

(i) $$\text{cov}(\mathbf{z}) = \text{cov}(\mathbf{Ay}) = \mathbf{A}\boldsymbol{\Sigma}\mathbf{A}', \tag{3.44}$$

(ii) $$\text{cov}(\mathbf{z}, \mathbf{w}) = \text{cov}(\mathbf{Ay}, \mathbf{By}) = \mathbf{A}\boldsymbol{\Sigma}\mathbf{B}'. \tag{3.45}$$

□

Typically, $k \leq p$, and the $k \times p$ matrix $\mathbf{A}$ is full rank, in which case, by Corollary 1 to 2.6b, $\mathbf{A}\boldsymbol{\Sigma}\mathbf{A}'$ is positive definite (assuming $\boldsymbol{\Sigma}$ to be positive definite). If $k > p$, then by Corollary 2 to Theorem 2.6b, $\mathbf{A}\boldsymbol{\Sigma}\mathbf{A}'$ is positive semidefinite. In this case, $\mathbf{A}\boldsymbol{\Sigma}\mathbf{A}'$ is still a covariance matrix, but it cannot be used in either the numerator or denominator of the multivariate normal density given in (4.9) in Chapter 4.

Note that $\text{cov}(\mathbf{z}, \mathbf{w}) = \mathbf{A\Sigma B}'$ is a $k \times m$ rectangular matrix containing the covariance of each z_i with each w_j, that is, $\text{cov}(\mathbf{z}, \mathbf{w})$ contains $\text{cov}(z_i, w_j)$, $i = 1, 2, \ldots, k$, $j = 1, 2, \ldots, m$. These km covariances can also be found individually by (3.43).

Corollary 1. If $\mathbf{b}$ is a $k \times 1$ vector of constants, then

$$\text{cov}(\mathbf{Ay} + \mathbf{b}) = \mathbf{A\Sigma A}'. \tag{3.46}$$

□

The covariance matrix of linear functions of two different random vectors is given in the following theorem.

Theorem 3.6e. Let $\mathbf{y}$ be a $p \times 1$ random vector and $\mathbf{x}$ be a $q \times 1$ random vector such that $\text{cov}(\mathbf{y}, \mathbf{x}) = \mathbf{\Sigma}_{yx}$. Let $\mathbf{A}$ be a $k \times p$ matrix of constants and $\mathbf{B}$ be an $h \times q$ matrix of constants. Then

$$\text{cov}(\mathbf{Ay}, \mathbf{Bx}) = \mathbf{A\Sigma}_{yx}\mathbf{B}'. \tag{3.47}$$

PROOF. Let

$$\mathbf{v} = \begin{pmatrix} \mathbf{y} \\ \mathbf{x} \end{pmatrix} \quad \text{and} \quad \mathbf{C} = \begin{pmatrix} \mathbf{A} & \mathbf{O} \\ \mathbf{O} & \mathbf{B} \end{pmatrix}.$$

Use Theorem 3.6d(i) to obtain $\text{cov}(\mathbf{Cv})$. The result follows. □

PROBLEMS

3.1 Show that $E(ay) = aE(y)$ as in (3.4).

3.2 Show that $E(y - \mu)^2 = E(y^2) - \mu^2$ as in (3.8).

3.3 Show that $\text{var}(ay) = a^2\sigma^2$ as in (3.9).

3.4 Show that $\text{cov}(y_i, y_j) = E(y_i y_j) - \mu_i\mu_j$ as in (3.11).

3.5 Show that if y_i and y_j are independent, then $E(y_i y_j) = E(y_i)E(y_j)$ as in (3.14).

3.6 Show that if y_i and y_j are independent, then $\sigma_{ij} = 0$ as in (3.15).

3.7 Establish the following results in Example 3.2:

(a) Show that $f_2(y) = 1 - \sqrt{1 - y}$ for $0 \leq y \leq 1$ and $f_2(y) = \sqrt{2 - y}$ for $1 \leq y \leq 2$.

(b) Show that $E(y) = \frac{7}{6}$ and $E(xy) = \frac{7}{6}$.

(c) Show that $E(y|x) = \frac{1}{2}(1 + 4x - 2x^2)$.

3.8 Suppose the bivariate random variable (x, y) is uniformly distributed over the region bounded below by $y = x - 1$ for $1 \leq x \leq 2$ and by $y = 3 - x$ for $2 \leq x \leq 3$ and bounded above by $y = x$ for $1 \leq x \leq 2$ and by $y = 4 - x$ for $2 \leq x \leq 3$.

(a) Show that the area of this region is 2, so that $f(x, y) = \frac{1}{2}$.

(b) Find $f_1(x)$, $f_2(y)$, $E(x)$, $E(y)$, $E(xy)$, and σ_{xy}, as was done in Example 3.2. Are x and y independent?

(c) Find $f(y|x)$ and $E(y|x)$.

3.9 Show that $E(\mathbf{x} + \mathbf{y}) = E(\mathbf{x}) + E(\mathbf{y})$ as in (3.21).

3.10 Show that $E[(\mathbf{y} - \boldsymbol{\mu})(\mathbf{y} - \boldsymbol{\mu})'] = E(\mathbf{y}\mathbf{y}') - \boldsymbol{\mu}\boldsymbol{\mu}'$ as in (3.25).

3.11 Show that the standardized distance transforms the variables so that they are uncorrelated and have means equal to 0 and variances equal to 1 as noted following (3.27).

3.12 Illustrate $\mathbf{P}_\rho = \mathbf{D}_\sigma^{-1}\boldsymbol{\Sigma}\mathbf{D}_\sigma^{-1}$ in (3.30) for $p = 3$.

3.13 Using (3.24), show that

$$\text{cov}(\mathbf{v}) = \text{cov}\begin{pmatrix} \mathbf{y} \\ \mathbf{x} \end{pmatrix} = \begin{pmatrix} \boldsymbol{\Sigma}_{yy} & \boldsymbol{\Sigma}_{yx} \\ \boldsymbol{\Sigma}_{xy} & \boldsymbol{\Sigma}_{xx} \end{pmatrix}$$

as in (3.33).

3.14 Prove Theorem 3.6b.

3.15 Prove Corollary 1 to Theorem 3.6b.

3.16 Prove Corollary 1 to Theorem 3.6c.

3.17 Prove Theorem 3.6d.

3.18 Prove Corollary 1 to Theorem 3.6d.

3.19 Consider four $k \times 1$ random vectors $\mathbf{y}$, $\mathbf{x}$, $\mathbf{v}$, and $\mathbf{w}$, and four $h \times k$ constant matrices $\mathbf{A}$, $\mathbf{B}$, $\mathbf{C}$, and $\mathbf{D}$. Find $\text{cov}(\mathbf{Ay} + \mathbf{Bx}, \mathbf{Cv} + \mathbf{Dw})$.

3.20 Let $\mathbf{y} = (y_1, y_2, y_3)'$ be a random vector with mean vector and covariance matrix

$$\boldsymbol{\mu} = \begin{pmatrix} 1 \\ -1 \\ 3 \end{pmatrix}, \quad \boldsymbol{\Sigma} = \begin{pmatrix} 1 & 1 & 0 \\ 1 & 2 & 3 \\ 0 & 3 & 10 \end{pmatrix}.$$

(a) Let $z = 2y_1 - 3y_2 + y_3$. Find $E(z)$ and $\text{var}(z)$.

(b) Let $z_1 = y_1 + y_2 + y_3$ and $z_2 = 3y_1 + y_2 - 2y_3$. Find $E(\mathbf{z})$ and $\text{cov}(\mathbf{z})$, where $\mathbf{z} = (z_1, z_2)'$.

3.21 Let $\mathbf{y}$ be a random vector with mean vector and covariance matrix $\boldsymbol{\mu}$ and $\boldsymbol{\Sigma}$ as given in Problem 3.20, and define $\mathbf{w} = (w_1, w_2, w_3)'$ as follows:

$$\begin{aligned} w_1 &= 2y_1 - y_2 + y_3 \\ w_2 &= y_1 + 2y_2 - 3y_3 \\ w_3 &= y_1 + y_2 + 2y_3. \end{aligned}$$

(a) Find $E(\mathbf{w})$ and $\text{cov}(\mathbf{w})$.

(b) Using $\mathbf{z}$ as defined in Problem 3.20b, find $\text{cov}(\mathbf{z}, \mathbf{w})$.

4 Multivariate Normal Distribution

In order to make inferences, we often assume that the random vector of interest has a multivariate normal distribution. Before developing the multivariate normal density function and its properties, we first review the univariate normal distribution.

4.1 UNIVARIATE NORMAL DENSITY FUNCTION

We begin with a standard normal random variable z with mean 0 and variance 1. We then transform z to a random variable y with arbitrary mean μ and variance σ^2, and we find the density of y from that of z. In Section 4.2, we will follow an analogous procedure to obtain the density of a multivariate normal random vector.

The standard normal density is given by

$$g(z) = \frac{1}{\sqrt{2\pi}} e^{-z^2/2}, \qquad -\infty < z < \infty, \tag{4.1}$$

for which $E(z) = 0$ and $\text{var}(z) = 1$. When z has the density (4.1), we say that z is distributed as $N(0, 1)$, or simply that z is $N(0,1)$.

To obtain a normal random variable y with arbitrary mean μ and variance σ^2, we use the transformation $z = (y - \mu)/\sigma$ or $y = \sigma z + \mu$, so that $E(y) = \mu$ and $\text{var}(y) = \sigma^2$. We now find the density $f(y)$ from $g(z)$ in (4.1). For a continuous increasing function (such as $y = \sigma z + \mu$) or for a continuous decreasing function, the change-of-variable technique for a definite integral gives

$$f(y) = g(z)\left|\frac{dz}{dy}\right|, \tag{4.2}$$

where $|dz/dy|$ is the absolute value of dz/dy (Hogg and Craig 1995, p. 169). To use (4.2) to find the density of y, it is clear that both z and dz/dy on the right side must be expressed in terms of y.

Let us apply (4.2) to $y = \sigma z + \mu$. The density $g(z)$ is given in (4.1), and for $z = (y - \mu)/\sigma$, we have $|dz/dy| = 1/\sigma$. Thus

$$f(y) = g(z)\left|\frac{dz}{dy}\right| = g\left(\frac{y-\mu}{\sigma}\right)\frac{1}{\sigma}$$

$$= \frac{1}{\sqrt{2\pi}\sigma} e^{-(y-\mu)^2/2\sigma^2}, \tag{4.3}$$

which is the normal density with $E(y) = \mu$ and $\text{var}(y) = \sigma^2$. When y has the density (4.3), we say that y is distributed as $N(\mu, \sigma^2)$ or simply that y is $N(\mu, \sigma^2)$.

In Section 4.2, we use a multivariate extension of this technique to find the multivariate normal density function.

4.2 MULTIVARIATE NORMAL DENSITY FUNCTION

We begin with independent standard normal random variables $z_1, z_2, \ldots, z_p$, with $\mu_i = 0$ and $\sigma_i^2 = 1$ for all i and $\sigma_{ij} = 0$ for $i \neq j$, and we then transform the z_i's to multivariate normal variables $y_1, y_2, \ldots, y_p$, with arbitrary means, variances, and covariances. We thus start with a random vector $\mathbf{z} = (z_1, z_2, \ldots, z_p)'$, where $E(\mathbf{z}) = \mathbf{0}$, $\text{cov}(\mathbf{z}) = \mathbf{I}$, and each z_i has the $N(0,1)$ density in (4.1). We wish to transform $\mathbf{z}$ to a multivariate normal random vector $\mathbf{y} = (y_1, y_2 \ldots, y_p)'$ with $E(\mathbf{y}) = \boldsymbol{\mu}$ and $\text{cov}(\mathbf{y}) = \boldsymbol{\Sigma}$, where $\boldsymbol{\mu}$ is any $p \times 1$ vector and $\boldsymbol{\Sigma}$ is any $p \times p$ positive definite matrix.

By (4.1) and an extension of (3.12), we have

$$g(z_1, z_2, \ldots, z_p) = g(\mathbf{z}) = g_1(z_1)g_2(z_2)\cdots g_p(z_p)$$

$$= \frac{1}{\sqrt{2\pi}} e^{-z_1^2/2} \frac{1}{\sqrt{2\pi}} e^{-z_2^2/2} \cdots \frac{1}{\sqrt{2\pi}} e^{-z_p^2/2}$$

$$= \frac{1}{(\sqrt{2\pi})^p} e^{-\sum_i z_i^2/2}$$

$$= \frac{1}{(\sqrt{2\pi})^p} e^{-\mathbf{z}'\mathbf{z}/2} \quad \text{[by (2.20)]}. \tag{4.4}$$

If $\mathbf{z}$ has the density (4.4), we say that $\mathbf{z}$ has a multivariate normal density with mean vector $\mathbf{0}$ and covariance matrix $\mathbf{I}$ or simply that $\mathbf{z}$ is distributed as $N_p(\mathbf{0}, \mathbf{I})$, where p is the dimension of the distribution and corresponds to the number of variables in $\mathbf{y}$.

To transform $\mathbf{z}$ to $\mathbf{y}$ with arbitrary mean vector $E(\mathbf{y}) = \boldsymbol{\mu}$ and arbitrary (positive definite) covariance matrix $\text{cov}(\mathbf{y}) = \boldsymbol{\Sigma}$, we define the transformation

$$\mathbf{y} = \boldsymbol{\Sigma}^{1/2}\mathbf{z} + \boldsymbol{\mu}, \tag{4.5}$$

where $\boldsymbol{\Sigma}^{1/2}$ is the (symmetric) square root matrix defined in (2.109). By (3.41) and (3.46), we obtain

$$E(\mathbf{y}) = E(\boldsymbol{\Sigma}^{1/2}\mathbf{z} + \boldsymbol{\mu}) = \boldsymbol{\Sigma}^{1/2}E(\mathbf{z}) + \boldsymbol{\mu} = \boldsymbol{\Sigma}^{1/2}\mathbf{0} + \boldsymbol{\mu} = \boldsymbol{\mu},$$

$$\text{cov}(\mathbf{y}) = \text{cov}(\boldsymbol{\Sigma}^{1/2}\mathbf{z} + \boldsymbol{\mu}) = \boldsymbol{\Sigma}^{1/2}\text{cov}(\mathbf{z})(\boldsymbol{\Sigma}^{1/2})' = \boldsymbol{\Sigma}^{1/2}\mathbf{I}\boldsymbol{\Sigma}^{1/2} = \boldsymbol{\Sigma}.$$

Note the analogy of (4.5) to $y = \sigma z + \mu$ in Section 4.1.

Let us now find the density of $\mathbf{y} = \boldsymbol{\Sigma}^{1/2}\mathbf{z} + \boldsymbol{\mu}$ from the density of $\mathbf{z}$ in (4.4). By (4.2), the density of $y = \sigma z + \mu$ is $f(y) = g(z)|dz/dy = g(z)|1/\sigma|$. The analogous expression for the multivariate linear transformation $\mathbf{y} = \boldsymbol{\Sigma}^{1/2}\mathbf{z} + \boldsymbol{\mu}$ is

$$f(\mathbf{y}) = g(\mathbf{z})\text{abs}(|\boldsymbol{\Sigma}^{-1/2}|), \tag{4.6}$$

where $\boldsymbol{\Sigma}^{-1/2}$ is defined as $(\boldsymbol{\Sigma}^{1/2})^{-1}$ and $\text{abs}(|\boldsymbol{\Sigma}^{-1/2}|)$ represents the absolute value of the determinant of $\boldsymbol{\Sigma}^{-1/2}$, which parallels the absolute value expression $|dz/dy| = |1/\sigma|$ in the univariate case. (The determinant $|\boldsymbol{\Sigma}^{-1/2}|$ is the *Jacobian* of the transformation; see any advanced calculus text.) Since $\boldsymbol{\Sigma}^{-1/2}$ is positive definite, we can dispense with the absolute value and write (4.6) as

$$f(\mathbf{y}) = g(\mathbf{z})|\boldsymbol{\Sigma}^{-1/2}| \tag{4.7}$$

$$= g(\mathbf{z})|\boldsymbol{\Sigma}|^{-1/2} \qquad \text{[by (2.67)]}. \tag{4.8}$$

In order to express $\mathbf{z}$ in terms of $\mathbf{y}$, we use (4.5) to obtain $\mathbf{z} = \boldsymbol{\Sigma}^{-1/2}(\mathbf{y} - \boldsymbol{\mu})$. Then using (4.4) and (4.8), we can write the density of $\mathbf{y}$ as

$$\begin{aligned} f(\mathbf{y}) = g(\mathbf{z})|\boldsymbol{\Sigma}|^{-1/2} &= \frac{1}{(\sqrt{2\pi})^p|\boldsymbol{\Sigma}|^{1/2}}\, e^{-\mathbf{z}'\mathbf{z}/2} \\ &= \frac{1}{(\sqrt{2\pi})^p|\boldsymbol{\Sigma}|^{1/2}}\, e^{-[\boldsymbol{\Sigma}^{-1/2}(\mathbf{y}-\boldsymbol{\mu})]'[\boldsymbol{\Sigma}^{-1/2}(\mathbf{y}-\boldsymbol{\mu})]/2} \\ &= \frac{1}{(\sqrt{2\pi})^p|\boldsymbol{\Sigma}|^{1/2}}\, e^{-(\mathbf{y}-\boldsymbol{\mu})'(\boldsymbol{\Sigma}^{1/2}\boldsymbol{\Sigma}^{1/2})^{-1}(\mathbf{y}-\boldsymbol{\mu})/2} \\ &= \frac{1}{(\sqrt{2\pi})^p|\boldsymbol{\Sigma}|^{1/2}}\, e^{-(\mathbf{y}-\boldsymbol{\mu})'\boldsymbol{\Sigma}^{-1}(\mathbf{y}-\boldsymbol{\mu})/2}, \end{aligned} \tag{4.9}$$

which is the multivariate normal density function with mean vector $\boldsymbol{\mu}$ and covariance matrix $\boldsymbol{\Sigma}$. When $\mathbf{y}$ has the density (4.9), we say that $\mathbf{y}$ is distributed as $N_p(\boldsymbol{\mu}, \boldsymbol{\Sigma})$ or simply that $\mathbf{y}$ is $N_p(\boldsymbol{\mu}, \boldsymbol{\Sigma})$. The subscript p is the dimension of the p-variate normal distribution and indicates the number of variables, that is, $\mathbf{y}$ is $p \times 1$, $\boldsymbol{\mu}$ is $p \times 1$, and $\boldsymbol{\Sigma}$ is $p \times p$.

A comparison of (4.9) and (4.3) shows the standardized distance $(\mathbf{y} - \boldsymbol{\mu})' \boldsymbol{\Sigma}^{-1}(\mathbf{y} - \boldsymbol{\mu})$ in place of $(y - \mu)^2/\sigma^2$ in the exponent and the square root of the generalized variance $|\boldsymbol{\Sigma}|$ in place of the square root of σ^2 in the denominator. [For standardized distance, see (3.27), and for generalized variance, see (3.26).] These distance and variance functions serve analogous purposes in the densities (4.9) and (4.3). In (4.9), $f(\mathbf{y})$ decreases as the distance from $\mathbf{y}$ to $\boldsymbol{\mu}$ increases, and a small value of $|\boldsymbol{\Sigma}|$ indicates that the $\mathbf{y}'$s are concentrated closer to $\boldsymbol{\mu}$ than is the case when $|\boldsymbol{\Sigma}|$ is large. A small value of $|\boldsymbol{\Sigma}|$ may also indicate a high degree of multicollinearity among the variables. High *multicollinearity* indicates that the variables are highly intercorrelated, in which case the $\mathbf{y}'$s tend to occupy a subspace of the p dimensions.

4.3 MOMENT GENERATING FUNCTIONS

We now review moment generating functions, which can be used to obtain some of the properties of multivariate normal random variables. We begin with the univariate case.

The *moment generating function* for a univariate random variable y is defined as

$$M_y(t) = E(e^{ty}), \tag{4.10}$$

provided $E(e^{ty})$ exists for every real number t in the neighborhood $-h < t < h$ for some positive number h. For the univariate normal $N(\mu, \sigma^2)$, the moment generating function of y is given by

$$M_y(t) = e^{t\mu + t^2\sigma^2/2}. \tag{4.11}$$

Moment generating functions characterize a distribution in some important ways that prove very useful (see the two properties at the end of this section). As their name implies, moment generating functions can also be used to generate moments. We now demonstrate this. For a continuous random variable y, the moment generating function can be written as $M_y(t) = E(e^{ty}) = \int_{-\infty}^{\infty} e^{ty} f(y)\, dy$. Then, provided we can interchange the order of integration and differentiation,we have

$$\frac{dM_y(t)}{dt} = M_y'(t) = \int_{-\infty}^{\infty} y e^{ty} f(y)\, dy. \tag{4.12}$$

Setting $t = 0$ gives the first moment or mean:

$$M_y'(0) = \int_{-\infty}^{\infty} y f(y)\, dy = E(y). \tag{4.13}$$

Similarly, the kth moment can be obtained using the kth derivative evaluated at 0:

$$M_y^{(k)}(0) = E(y^k). \tag{4.14}$$

The second moment, $E(y^2)$, can be used to find the variance [see (3.8)].

For a random vector $\mathbf{y}$, the moment generating function is defined as

$$M_{\mathbf{y}}(\mathbf{t}) = E(e^{t_1 y_1 + t_2 y_2 + \cdots + t_p y_p}) = E(e^{\mathbf{t}'\mathbf{y}}). \tag{4.15}$$

By analogy with (4.13), we have

$$\frac{\partial M_{\mathbf{y}}(\mathbf{0})}{\partial \mathbf{t}} = E(\mathbf{y}), \tag{4.16}$$

where the notation $\partial M_{\mathbf{y}}(\mathbf{0})/\partial \mathbf{t}$ indicates that $\partial M_{\mathbf{y}}(\mathbf{t})/\partial \mathbf{t}$ is evaluated at $\mathbf{t} = \mathbf{0}$. Similarly, $\partial^2 M_{\mathbf{y}}(\mathbf{t})/\partial t_r \partial t_s$ evaluated at $t_r = t_s = 0$ gives $E(y_r y_s)$:

$$\frac{\partial^2 M_{\mathbf{y}}(\mathbf{0})}{\partial t_r \partial t_s} = E(y_r y_s). \tag{4.17}$$

For a multivariate normal random vector $\mathbf{y}$, the moment generating function is given in the following theorem.

Theorem 4.3. If $\mathbf{y}$ is distributed as $N_p(\boldsymbol{\mu}, \boldsymbol{\Sigma})$, its moment generating function is given by

$$M_{\mathbf{y}}(\mathbf{t}) = e^{\mathbf{t}'\boldsymbol{\mu} + \mathbf{t}'\boldsymbol{\Sigma}\mathbf{t}/2}. \tag{4.18}$$

Proof. By (4.15) and (4.9), the moment generating function is

$$M_{\mathbf{y}}(\mathbf{t}) = \int_{-\infty}^{\infty} \cdots \int_{-\infty}^{\infty} k e^{\mathbf{t}'\mathbf{y} - (\mathbf{y}-\boldsymbol{\mu})'\boldsymbol{\Sigma}^{-1}(\mathbf{y}-\boldsymbol{\mu})/2} \, d\mathbf{y}, \tag{4.19}$$

where $k = 1/(\sqrt{2\pi})^p |\boldsymbol{\Sigma}|^{1/2}$ and $d\mathbf{y} = dy_1\, dy_2 \cdots dy_p$. By rewriting the exponent, we obtain

$$M_{\mathbf{y}}(\mathbf{t}) = \int_{-\infty}^{\infty} \cdots \int_{-\infty}^{\infty} k\, e^{\mathbf{t}'\boldsymbol{\mu} + \mathbf{t}'\boldsymbol{\Sigma}\mathbf{t}/2 - (\mathbf{y}-\boldsymbol{\mu}-\boldsymbol{\Sigma}\mathbf{t})'\boldsymbol{\Sigma}^{-1}(\mathbf{y}-\boldsymbol{\mu}-\boldsymbol{\Sigma}\mathbf{t})/2} \, d\mathbf{y} \tag{4.20}$$

$$= e^{\mathbf{t}'\boldsymbol{\mu} + \mathbf{t}'\boldsymbol{\Sigma}\mathbf{t}/2} \int_{-\infty}^{\infty} \cdots \int_{-\infty}^{\infty} k\, e^{-[\mathbf{y}-(\boldsymbol{\mu}+\boldsymbol{\Sigma}\mathbf{t})]'\boldsymbol{\Sigma}^{-1}[\mathbf{y}-(\boldsymbol{\mu}+\boldsymbol{\Sigma}\mathbf{t})]/2} \, d\mathbf{y} \tag{4.21}$$

$$= e^{\mathbf{t}'\boldsymbol{\mu} + \mathbf{t}'\boldsymbol{\Sigma}\mathbf{t}/2}.$$

The multiple integral in (4.21) is equal to 1 because the multivariate normal density in (4.9) integrates to 1 for any mean vector, including $\boldsymbol{\mu} + \boldsymbol{\Sigma}\mathbf{t}$. □

Corollary 1. The moment generating function for $\mathbf{y} - \boldsymbol{\mu}$ is

$$M_{\mathbf{y}-\boldsymbol{\mu}}(\mathbf{t}) = e^{\mathbf{t}'\boldsymbol{\Sigma}\mathbf{t}/2}. \tag{4.22}$$

□

We now list two important properties of moment generating functions.

1. If two random vectors have the same moment generating function, they have the same density.
2. Two random vectors are independent if and only if their joint moment generating function factors into the product of their two separate moment generating functions; that is, if $\mathbf{y}' = (\mathbf{y}_1', \mathbf{y}_2')$ and $\mathbf{t}' = (\mathbf{t}_1', \mathbf{t}_2')$, then $\mathbf{y}_1$ and $\mathbf{y}_2$ are independent if and only if

$$M_{\mathbf{y}}(\mathbf{t}) = M_{\mathbf{y}_1}(\mathbf{t}_1)M_{\mathbf{y}_2}(\mathbf{t}_2). \tag{4.23}$$

4.4 PROPERTIES OF THE MULTIVARIATE NORMAL DISTRIBUTION

We first consider the distribution of linear functions of multivariate normal random variables.

Theorem 4.4a. Let the $p \times 1$ random vector $\mathbf{y}$ be $N_p(\boldsymbol{\mu}, \boldsymbol{\Sigma})$, let $\mathbf{a}$ be any $p \times 1$ vector of constants, and let $\mathbf{A}$ be any $k \times p$ matrix of constants with rank $k \leq p$. Then

(i) $z = \mathbf{a}'\mathbf{y}$ is $N(\mathbf{a}'\boldsymbol{\mu}, \mathbf{a}'\boldsymbol{\Sigma}\mathbf{a})$
(ii) $\mathbf{z} = \mathbf{A}\mathbf{y}$ is $N_k(\mathbf{A}\boldsymbol{\mu}, \mathbf{A}\boldsymbol{\Sigma}\mathbf{A}')$.

PROOF

(i) The moment generating function for $z = \mathbf{a}'\mathbf{y}$ is given by

$$\begin{aligned} M_z(t) &= E(e^{tz}) = E(e^{t\mathbf{a}'\mathbf{y}}) = E(e^{(t\mathbf{a})'\mathbf{y}}) \\ &= e^{(t\mathbf{a})'\boldsymbol{\mu} + (t\mathbf{a})'\boldsymbol{\Sigma}(t\mathbf{a})/2} \qquad \text{[by (4.18)]} \\ &= e^{(\mathbf{a}'\boldsymbol{\mu})t + (\mathbf{a}'\boldsymbol{\Sigma}\mathbf{a})t^2/2}. \end{aligned} \tag{4.24}$$

On comparing (4.24) with (4.11), it is clear that $z = \mathbf{a}'\mathbf{y}$ is univariate normal with mean $\mathbf{a}'\boldsymbol{\mu}$ and variance $\mathbf{a}'\boldsymbol{\Sigma}\mathbf{a}$.

(ii) The moment generating function for $\mathbf{z} = \mathbf{Ay}$ is given by

$$M_{\mathbf{z}}(\mathbf{t}) = E(e^{\mathbf{t}'\mathbf{z}}) = E(e^{\mathbf{t}'\mathbf{Ay}}),$$

which becomes

$$M_{\mathbf{z}}(\mathbf{t}) = e^{\mathbf{t}'(\mathbf{A}\boldsymbol{\mu})+\mathbf{t}'(\mathbf{A}\boldsymbol{\Sigma}\mathbf{A}')\mathbf{t}/2} \tag{4.25}$$

(see Problem 4.7). By Corollary 1 to Theorem 2.6b, the covariance matrix $\mathbf{A}\boldsymbol{\Sigma}\mathbf{A}'$ is positive definite. Thus, by (4.18) and (4.25), the $k \times 1$ random vector $\mathbf{z} = \mathbf{Ay}$ is distributed as the k-variate normal $N_k(\mathbf{A}\boldsymbol{\mu}, \mathbf{A}\boldsymbol{\Sigma}\mathbf{A}')$. □

Corollary 1. If $\mathbf{b}$ is any $k \times 1$ vector of constants, then

$$\mathbf{z} = \mathbf{Ay} + \mathbf{b} \quad \text{is} \quad N_k(\mathbf{A}\boldsymbol{\mu} + \mathbf{b}, \mathbf{A}\boldsymbol{\Sigma}\mathbf{A}').$$ □

The marginal distributions of multivariate normal variables are also normal, as shown in the following theorem.

Theorem 4.4b. If $\mathbf{y}$ is $N_p(\boldsymbol{\mu}, \boldsymbol{\Sigma})$, then any $r \times 1$ subvector of $\mathbf{y}$ has an r-variate normal distribution with the same means, variances, and covariances as in the original p-variate normal distribution.

PROOF. Without loss of generality, let $\mathbf{y}$ be partitioned as $\mathbf{y}' = (\mathbf{y}_1', \mathbf{y}_2')$, where $\mathbf{y}_1$ is the $r \times 1$ subvector of interest. Let $\boldsymbol{\mu}$ and $\boldsymbol{\Sigma}$ be partitioned accordingly:

$$\mathbf{y} = \begin{pmatrix} \mathbf{y}_1 \\ \mathbf{y}_2 \end{pmatrix}, \quad \boldsymbol{\mu} = \begin{pmatrix} \boldsymbol{\mu}_1 \\ \boldsymbol{\mu}_2 \end{pmatrix}, \quad \boldsymbol{\Sigma} = \begin{pmatrix} \boldsymbol{\Sigma}_{11} & \boldsymbol{\Sigma}_{12} \\ \boldsymbol{\Sigma}_{21} & \boldsymbol{\Sigma}_{22} \end{pmatrix}.$$

Define $\mathbf{A} = (\mathbf{I}_r, \mathbf{O})$, where $\mathbf{I}_r$ is an $r \times r$ identity matrix and $\mathbf{O}$ is an $r \times (p - r)$ matrix of 0s. Then $\mathbf{Ay} = \mathbf{y}_1$, and by Theorem 4.4a (ii), $\mathbf{y}_1$ is distributed as $N_r(\boldsymbol{\mu}_1, \boldsymbol{\Sigma}_{11})$. □

Corollary 1. If $\mathbf{y}$ is $N_p(\boldsymbol{\mu}, \boldsymbol{\Sigma})$, then any individual variable y_i in $\mathbf{y}$ is distributed as $N(\mu_i, \sigma_{ii})$. □

For the next two theorems, we use the notation of Section 3.5, in which the random vector $\mathbf{v}$ is partitioned into two subvectors denoted by $\mathbf{y}$ and $\mathbf{x}$, where $\mathbf{y}$ is $p \times 1$ and $\mathbf{x}$

is $q \times 1$, with a corresponding partitioning of $\boldsymbol{\mu}$ and $\boldsymbol{\Sigma}$ [see (3.32) and (3.33)]:

$$\mathbf{v} = \begin{pmatrix} \mathbf{y} \\ \mathbf{x} \end{pmatrix}, \quad \boldsymbol{\mu} = E\begin{pmatrix} \mathbf{y} \\ \mathbf{x} \end{pmatrix} = \begin{pmatrix} \boldsymbol{\mu}_y \\ \boldsymbol{\mu}_x \end{pmatrix}, \quad \boldsymbol{\Sigma} = \text{cov}\begin{pmatrix} \mathbf{y} \\ \mathbf{x} \end{pmatrix} = \begin{pmatrix} \boldsymbol{\Sigma}_{yy} & \boldsymbol{\Sigma}_{yx} \\ \boldsymbol{\Sigma}_{xy} & \boldsymbol{\Sigma}_{xx} \end{pmatrix}.$$

By (3.15), if two random variables y_i and y_j are independent, then $\sigma_{ij} = 0$. The converse of this is not true, as illustrated in Example 3.2. By extension, if two random vectors $\mathbf{y}$ and $\mathbf{x}$ are independent (i.e., each y_i is independent of each x_j), then $\boldsymbol{\Sigma}_{yx} = \mathbf{O}$ (the covariance of each y_i with each x_j is 0). The converse is not true in general, but it is true for multivariate normal random vectors.

Theorem 4.4c. If $\mathbf{v} = \begin{pmatrix} \mathbf{y} \\ \mathbf{x} \end{pmatrix}$ is $N_{p+q}(\boldsymbol{\mu}, \boldsymbol{\Sigma})$, then $\mathbf{y}$ and $\mathbf{x}$ are independent if $\boldsymbol{\Sigma}_{yx} = \mathbf{O}$.

PROOF. Suppose $\boldsymbol{\Sigma}_{yx} = \mathbf{O}$. Then

$$\boldsymbol{\Sigma} = \begin{pmatrix} \boldsymbol{\Sigma}_{yy} & \mathbf{O} \\ \mathbf{O} & \boldsymbol{\Sigma}_{xx} \end{pmatrix},$$

and the exponent of the moment generating function in (4.18) becomes

$$\begin{aligned} \mathbf{t}'\boldsymbol{\mu} + \tfrac{1}{2}\mathbf{t}'\boldsymbol{\Sigma}\mathbf{t} &= (\mathbf{t}'_y, \mathbf{t}'_x)\begin{pmatrix} \boldsymbol{\mu}_y \\ \boldsymbol{\mu}_x \end{pmatrix} + \tfrac{1}{2}(\mathbf{t}'_y, \mathbf{t}'_x)\begin{pmatrix} \boldsymbol{\Sigma}_{yy} & \mathbf{O} \\ \mathbf{O} & \boldsymbol{\Sigma}_{xx} \end{pmatrix}\begin{pmatrix} \mathbf{t}_y \\ \mathbf{t}_x \end{pmatrix} \\ &= \mathbf{t}'_y\boldsymbol{\mu}_y + \mathbf{t}'_x\boldsymbol{\mu}_x + \tfrac{1}{2}\mathbf{t}'_y\boldsymbol{\Sigma}_{yy}\mathbf{t}_y + \tfrac{1}{2}\mathbf{t}'_x\boldsymbol{\Sigma}_{xx}\mathbf{t}_x. \end{aligned}$$

The moment generating function can then be written as

$$M_{\mathbf{v}}(\mathbf{t}) = e^{\mathbf{t}'_y\boldsymbol{\mu}_y + \mathbf{t}'_y\boldsymbol{\Sigma}_{yy}\mathbf{t}_y/2} e^{\mathbf{t}'_x\boldsymbol{\mu}_x + \mathbf{t}'_x\boldsymbol{\Sigma}_{xx}\mathbf{t}_x/2},$$

which is the product of the moment generating functions of $\mathbf{y}$ and $\mathbf{x}$. Hence, by (4.23), $\mathbf{y}$ and $\mathbf{x}$ are independent. □

Corollary 1. If $\mathbf{y}$ is $N_p(\boldsymbol{\mu}, \boldsymbol{\Sigma})$, then any two individual variables y_i and y_j are independent if $\sigma_{ij} = 0$.

Corollary 2. If $\mathbf{y}$ is $N_p(\boldsymbol{\mu}, \boldsymbol{\Sigma})$ and if $\text{cov}(\mathbf{Ay}, \mathbf{By}) = \mathbf{A}\boldsymbol{\Sigma}\mathbf{B}' = \mathbf{O}$, then $\mathbf{Ay}$ and $\mathbf{By}$ are independent. □

The relationship between subvectors $\mathbf{y}$ and $\mathbf{x}$ when they are not independent ($\boldsymbol{\Sigma}_{yx} \neq \mathbf{O}$) is given in the following theorem.

Theorem 4.4d. If $\mathbf{y}$ and $\mathbf{x}$ are jointly multivariate normal with $\boldsymbol{\Sigma}_{yx} \neq \mathbf{O}$, then the conditional distribution of $\mathbf{y}$ given $\mathbf{x}$, $f(\mathbf{y}|x)$, is multivariate normal with mean vector and covariance matrix

$$E(\mathbf{y}|\mathbf{x}) = \boldsymbol{\mu}_y + \boldsymbol{\Sigma}_{yx}\boldsymbol{\Sigma}_{xx}^{-1}(\mathbf{x} - \boldsymbol{\mu}_x), \tag{4.26}$$

$$\text{cov}(\mathbf{y}|\mathbf{x}) = \boldsymbol{\Sigma}_{yy} - \boldsymbol{\Sigma}_{yx}\boldsymbol{\Sigma}_{xx}^{-1}\boldsymbol{\Sigma}_{xy}. \tag{4.27}$$

Proof. By an extension of (3.18), the conditional density of $\mathbf{y}$ given $\mathbf{x}$ is

$$f(\mathbf{y}|\mathbf{x}) = \frac{g(\mathbf{y}, \mathbf{x})}{h(\mathbf{x})}, \tag{4.28}$$

where $g(\mathbf{y}, \mathbf{x})$ is the joint density of $\mathbf{y}$ and $\mathbf{x}$, and $h(\mathbf{x})$ is the marginal density of $\mathbf{x}$. The proof can be carried out by directly evaluating the ratio on the right hand side of (4.28), using results (2.50) and (2.71) (see Problem 4.13). For variety, we use an alternative approach that avoids working explicitly with $g(\mathbf{y}, \mathbf{x})$ and $h(\mathbf{x})$ and the resulting partitioned matrix formulas.

Consider the function

$$\begin{pmatrix} \mathbf{w} \\ \mathbf{u} \end{pmatrix} = \mathbf{A}\left[\begin{pmatrix} \mathbf{y} \\ \mathbf{x} \end{pmatrix} - \begin{pmatrix} \boldsymbol{\mu}_y \\ \boldsymbol{\mu}_x \end{pmatrix}\right], \tag{4.29}$$

where

$$\mathbf{A} = \begin{pmatrix} \mathbf{A}_1 \\ \mathbf{A}_2 \end{pmatrix} = \begin{pmatrix} \mathbf{I} & -\boldsymbol{\Sigma}_{yx}\boldsymbol{\Sigma}_{xx}^{-1} \\ \mathbf{O} & \mathbf{I} \end{pmatrix}.$$

To be conformal, the identity matrix in $\mathbf{A}_1$ is $p \times p$ while the identity in $\mathbf{A}_2$ is $q \times q$. Simplifying and rearranging (4.29), we obtain $\mathbf{w} = \mathbf{y} - [\boldsymbol{\mu}_y + \boldsymbol{\Sigma}_{yx}\boldsymbol{\Sigma}_{xx}^{-1}(\mathbf{x} - \boldsymbol{\mu}_x)]$ and $\mathbf{u} = \mathbf{x} - \boldsymbol{\mu}_x$. Using the multivariate change-of-variable technique [referred to in (4.6], the joint density of $(\mathbf{w}, \mathbf{u})$ is

$$p(\mathbf{w}, \mathbf{u}) = g(\mathbf{y}, \mathbf{x})|\mathbf{A}^{-1}| = g(\mathbf{y}, \mathbf{x})$$

[employing Theorem 2.9a (ii) and (vi)]. Similarly, the marginal density of $\mathbf{u}$ is

$$q(\mathbf{u}) = h(\mathbf{x})|\mathbf{I}^{-1}| = h(\mathbf{x}).$$

Using (3.45), it also turns out that

$$\text{cov}(\mathbf{w}, \mathbf{u}) = \mathbf{A}_1\boldsymbol{\Sigma}\mathbf{A}_2 = \boldsymbol{\Sigma}_{yx} - \boldsymbol{\Sigma}_{yx}\boldsymbol{\Sigma}_{xx}^{-1}\boldsymbol{\Sigma}_{xx} = \mathbf{O} \tag{4.30}$$

(see Problem 4.14). Thus, by Theorem 4.4c, $\mathbf{w}$ is independent of $\mathbf{u}$. Hence

$$p(\mathbf{w}, \mathbf{u}) = r(\mathbf{w})q(\mathbf{u}),$$

where $r(\mathbf{w})$ is the density of $\mathbf{w}$. Since $p(\mathbf{w}, \mathbf{u}) = g(\mathbf{y}, \mathbf{x})$ and $q(\mathbf{u}) = h(\mathbf{x})$, we also have

$$g(\mathbf{y}, \mathbf{x}) = r(\mathbf{w})h(\mathbf{x}),$$

and by (4.28),

$$r(\mathbf{w}) = \frac{g(\mathbf{y}, \mathbf{x})}{h(\mathbf{x})} = f(\mathbf{y}|\mathbf{x}).$$

Hence we obtain $f(\mathbf{y}|\mathbf{x})$ simply by finding $r(\mathbf{w})$. By Corollary 1 to Theorem 4.4a, $r(\mathbf{w})$ is the multivariate normal density with

$$\boldsymbol{\mu}_w = \mathbf{A}_1 \left[\begin{pmatrix} \boldsymbol{\mu}_y \\ \boldsymbol{\mu}_x \end{pmatrix} - \begin{pmatrix} \boldsymbol{\mu}_y \\ \boldsymbol{\mu}_x \end{pmatrix} \right] = \mathbf{0}, \tag{4.31}$$

$$\begin{aligned} \boldsymbol{\Sigma}_{ww} &= \mathbf{A}_1 \boldsymbol{\Sigma} \mathbf{A}_1' \\ &= (\mathbf{I}, \ -\boldsymbol{\Sigma}_{yx}\boldsymbol{\Sigma}_{xx}^{-1}) \begin{pmatrix} \boldsymbol{\Sigma}_{yy} & \boldsymbol{\Sigma}_{yx} \\ \boldsymbol{\Sigma}_{xy} & \boldsymbol{\Sigma}_{xx} \end{pmatrix} \begin{pmatrix} \mathbf{I} \\ -\boldsymbol{\Sigma}_{xx}^{-1}\boldsymbol{\Sigma}_{xy} \end{pmatrix} \\ &= \boldsymbol{\Sigma}_{yy} - \boldsymbol{\Sigma}_{yx}\boldsymbol{\Sigma}_{xx}^{-1}\boldsymbol{\Sigma}_{xy}. \end{aligned} \tag{4.32}$$

Thus $r(\mathbf{w}) = r(\mathbf{y} - [\boldsymbol{\mu}_y + \boldsymbol{\Sigma}_{yx}\boldsymbol{\Sigma}_{xx}^{-1}(\mathbf{x} - \boldsymbol{\mu}_x)])$ is of the form $N_p(\mathbf{0}, \boldsymbol{\Sigma}_{yy} - \boldsymbol{\Sigma}_{yx}\boldsymbol{\Sigma}_{xx}^{-1}\boldsymbol{\Sigma}_{xy})$. Equivalently, $\mathbf{y}|\mathbf{x}$ is $N_p[\boldsymbol{\mu}_y + \boldsymbol{\Sigma}_{yx}\boldsymbol{\Sigma}_{xx}^{-1}(\mathbf{x} - \boldsymbol{\mu}_x), \ \boldsymbol{\Sigma}_{yy} - \boldsymbol{\Sigma}_{yx}\boldsymbol{\Sigma}_{xx}^{-1}\boldsymbol{\Sigma}_{xy}]$. □

Since $E(\mathbf{y}|\mathbf{x}) = \boldsymbol{\mu}_y + \boldsymbol{\Sigma}_{yx}\boldsymbol{\Sigma}_{xx}^{-1}(\mathbf{x} - \boldsymbol{\mu}_x)$ in (4.26) is a linear function of $\mathbf{x}$, any pair of variables y_i and y_j in a multivariate normal vector exhibits a linear trend $E(y_i|y_j) = \mu_i + (\sigma_{ij}/\sigma_{jj})(y_j - \mu_j)$. Thus the covariance σ_{ij} is related to the slope of the line representing the trend, and σ_{ij} is a useful measure of relationship between two normal variables. In the case of nonnormal variables that exhibit a curved trend, σ_{ij} may give a very misleading indication of the relationship, as illustrated in Example 3.2.

The conditional covariance matrix $\text{cov}(\mathbf{y}|\mathbf{x}) = \boldsymbol{\Sigma}_{yy} - \boldsymbol{\Sigma}_{yx}\boldsymbol{\Sigma}_{xx}^{-1}\boldsymbol{\Sigma}_{xy}$ in (4.27) does not involve $\mathbf{x}$. For some nonnormal distributions, on the other hand, $\text{cov}(\mathbf{y}|\mathbf{x})$ is a function of $\mathbf{x}$.

If there is only one y, so that $\mathbf{v}$ is partitioned in the form $\mathbf{v} = (y, x_1, x_2, \ldots, x_q) = (y, \mathbf{x}')$, then $\boldsymbol{\mu}$ and $\boldsymbol{\Sigma}$ have the form

$$\boldsymbol{\mu} = \begin{pmatrix} \mu_y \\ \boldsymbol{\mu}_x \end{pmatrix}, \quad \boldsymbol{\Sigma} = \begin{pmatrix} \sigma_y^2 & \boldsymbol{\sigma}_{yx}' \\ \boldsymbol{\sigma}_{yx} & \boldsymbol{\Sigma}_{xx} \end{pmatrix},$$

where μ_y and σ_y^2 are the mean and variance of y, $\boldsymbol{\sigma}_{yx}' = (\sigma_{y1}, \sigma_{y2}, \ldots, \sigma_{yq})$ contains the covariances $\sigma_{yi} = \text{cov}(y, x_i)$, and $\boldsymbol{\Sigma}_{xx}$ contains the variances and covariances of the x variables. The conditional distribution is given in the following corollary to Theorem 4.4d.

Corollary 1. If $\mathbf{v} = (y, x_1, x_2, \ldots, x_q) = (y, \mathbf{x}')$, with

$$\boldsymbol{\mu} = \begin{pmatrix} \mu_y \\ \boldsymbol{\mu}_x \end{pmatrix}, \quad \boldsymbol{\Sigma} = \begin{pmatrix} \sigma_y^2 & \boldsymbol{\sigma}'_{yx} \\ \boldsymbol{\sigma}_{yx} & \boldsymbol{\Sigma}_{xx} \end{pmatrix},$$

then $y|\mathbf{x}$ is normal with

$$E(y|\mathbf{x}) = \mu_y + \boldsymbol{\sigma}'_{yx}\boldsymbol{\Sigma}_{xx}^{-1}(\mathbf{x} - \boldsymbol{\mu}_x), \tag{4.33}$$

$$\text{var}(y|\mathbf{x}) = \sigma_y^2 - \boldsymbol{\sigma}'_{yx}\boldsymbol{\Sigma}_{xx}^{-1}\boldsymbol{\sigma}_{yx}. \tag{4.34}$$

□

In (4.34), $\boldsymbol{\sigma}'_{yx}\boldsymbol{\Sigma}_{xx}^{-1}\boldsymbol{\sigma}_{yx} \geq 0$ because $\boldsymbol{\Sigma}_{xx}^{-1}$ is positive definite. Therefore

$$\text{var}(y|\mathbf{x}) \leq \text{var}(y). \tag{4.35}$$

Example 4.4a. To illustrate Theorems 4.4a–c, suppose that $\mathbf{y}$ is $N_3(\boldsymbol{\mu}, \boldsymbol{\Sigma})$, where

$$\boldsymbol{\mu} = \begin{pmatrix} 3 \\ 1 \\ 2 \end{pmatrix}, \quad \boldsymbol{\Sigma} = \begin{pmatrix} 4 & 0 & 2 \\ 0 & 1 & -1 \\ 2 & -1 & 3 \end{pmatrix}.$$

For $z = y_1 - 2y_2 + y_3 = (1, -2, 1)\mathbf{y} = \mathbf{a}'\mathbf{y}$, we have $\mathbf{a}'\boldsymbol{\mu} = 3$ and $\mathbf{a}'\boldsymbol{\Sigma}\mathbf{a} = 19$. Hence by Theorem 4.4a(i), z is $N(3, 19)$.

The linear functions

$$z_1 = y_1 - y_2 + y_3, \quad z_2 = 3y_1 + y_2 - 2y_3$$

can be written as

$$\mathbf{z} = \begin{pmatrix} z_1 \\ z_2 \end{pmatrix} = \begin{pmatrix} 1 & -1 & 1 \\ 3 & 1 & -2 \end{pmatrix}\begin{pmatrix} y_1 \\ y_2 \\ y_3 \end{pmatrix} = \mathbf{Ay}.$$

Then by Theorem 3.6b(i) and Theorem 3.6d(i), we obtain

$$\mathbf{A}\boldsymbol{\mu} = \begin{pmatrix} 4 \\ 6 \end{pmatrix}, \quad \mathbf{A}\boldsymbol{\Sigma}\mathbf{A}' = \begin{pmatrix} 14 & 4 \\ 4 & 29 \end{pmatrix},$$

and by Theorem 4.4a(ii), we have

$$\mathbf{z} \text{ is } N_2\left[\begin{pmatrix} 4 \\ 6 \end{pmatrix}, \begin{pmatrix} 14 & 4 \\ 4 & 29 \end{pmatrix}\right].$$

To illustrate the marginal distributions in Theorem 4.4b, note that y_1 is $N(3, 4)$, y_3 is

$$N(2, 3), \begin{pmatrix} y_1 \\ y_2 \end{pmatrix} \text{ is } N_2\left[\begin{pmatrix} 3 \\ 1 \end{pmatrix}, \begin{pmatrix} 4 & 0 \\ 0 & 1 \end{pmatrix}\right], \text{ and } \begin{pmatrix} y_1 \\ y_3 \end{pmatrix} \text{ is } N_2\left[\begin{pmatrix} 3 \\ 2 \end{pmatrix}, \begin{pmatrix} 4 & 2 \\ 2 & 3 \end{pmatrix}\right].$$

To illustrate Theorem 4.4c, we note that $\sigma_{12} = 0$, and therefore y_1 and y_2 are independent. □

Example 4.4b. To illustrate Theorem 4.4d, let the random vector $\mathbf{v}$ be $N_4(\boldsymbol{\mu}, \boldsymbol{\Sigma})$, where

$$\boldsymbol{\mu} = \begin{pmatrix} 2 \\ 5 \\ -2 \\ 1 \end{pmatrix}, \quad \boldsymbol{\Sigma} = \begin{pmatrix} 9 & 0 & 3 & 3 \\ 0 & 1 & -1 & 2 \\ 3 & -1 & 6 & -3 \\ 3 & 2 & -3 & 7 \end{pmatrix}.$$

If $\mathbf{v}$ is partitioned as $\mathbf{v} = (y_1, y_2, x_1, x_2)'$, then $\boldsymbol{\mu}_y = \begin{pmatrix} 2 \\ 5 \end{pmatrix}$, $\boldsymbol{\mu}_x = \begin{pmatrix} -2 \\ 1 \end{pmatrix}$, $\boldsymbol{\Sigma}_{yy} = \begin{pmatrix} 9 & 0 \\ 0 & 1 \end{pmatrix}$, $\boldsymbol{\Sigma}_{yx} = \begin{pmatrix} 3 & 3 \\ -1 & 2 \end{pmatrix}$, and $\boldsymbol{\Sigma}_{xx} = \begin{pmatrix} 6 & -3 \\ -3 & 7 \end{pmatrix}$. By (4.26), we obtain

$$\begin{aligned} E(\mathbf{y}|\mathbf{x}) &= \boldsymbol{\mu}_y + \boldsymbol{\Sigma}_{yx}\boldsymbol{\Sigma}_{xx}^{-1}(\mathbf{x} - \boldsymbol{\mu}_x) \\ &= \begin{pmatrix} 2 \\ 5 \end{pmatrix} + \begin{pmatrix} 3 & 3 \\ -1 & 2 \end{pmatrix}\begin{pmatrix} 6 & -3 \\ -3 & 7 \end{pmatrix}^{-1}\begin{pmatrix} x_1 + 2 \\ x_2 - 1 \end{pmatrix} \\ &= \begin{pmatrix} 2 \\ 5 \end{pmatrix} + \frac{1}{33}\begin{pmatrix} 30 & 27 \\ -1 & 9 \end{pmatrix}\begin{pmatrix} x_1 + 2 \\ x_2 - 1 \end{pmatrix} \\ &= \begin{pmatrix} 3 + \dfrac{10}{11}x_1 + \dfrac{9}{11}x_2 \\ \dfrac{14}{3} - \dfrac{1}{33}x_1 + \dfrac{3}{11}x_2 \end{pmatrix}. \end{aligned}$$

By (4.27), we have

$$\begin{aligned} \text{cov}(\mathbf{y}|\mathbf{x}) &= \boldsymbol{\Sigma}_{yy} - \boldsymbol{\Sigma}_{yx}\boldsymbol{\Sigma}_{xx}^{-1}\boldsymbol{\Sigma}_{xy} \\ &= \begin{pmatrix} 9 & 0 \\ 0 & 1 \end{pmatrix} - \begin{pmatrix} 3 & 3 \\ -1 & 2 \end{pmatrix}\begin{pmatrix} 6 & -3 \\ -3 & 7 \end{pmatrix}^{-1}\begin{pmatrix} 3 & -1 \\ 3 & 2 \end{pmatrix} \\ &= \begin{pmatrix} 9 & 0 \\ 0 & 1 \end{pmatrix} - \tfrac{1}{33}\begin{pmatrix} 171 & 24 \\ 24 & 19 \end{pmatrix} \\ &= \tfrac{1}{33}\begin{pmatrix} 126 & -24 \\ -24 & 14 \end{pmatrix}. \end{aligned}$$

Thus

$$\mathbf{y}|\mathbf{x} \text{ is } N_2\left[\begin{pmatrix} 3 + \frac{10}{11}x_1 + \frac{9}{11}x_2 \\ \frac{14}{3} - \frac{1}{33}x_1 + \frac{3}{11}x_2 \end{pmatrix}, \frac{1}{33}\begin{pmatrix} 126 & -24 \\ -24 & 14 \end{pmatrix}\right].$$

□

Example 4.4c. To illustrate Corollary 1 to Theorem 4.4d, let $\mathbf{v}$ be $N_4(\boldsymbol{\mu}, \boldsymbol{\Sigma})$, where $\boldsymbol{\mu}$ and $\boldsymbol{\Sigma}$ are as given in Example 4.4b. If $\mathbf{v}$ is partitioned as $\mathbf{v} = (y, x_1, x_2, x_3)'$, then $\boldsymbol{\mu}$ and $\boldsymbol{\Sigma}$ are partitioned as follows:

$$\boldsymbol{\mu} = \begin{pmatrix} \mu_y \\ \boldsymbol{\mu}_x \end{pmatrix} = \begin{pmatrix} 2 \\ 5 \\ -2 \\ 1 \end{pmatrix},$$

$$\boldsymbol{\Sigma} = \begin{pmatrix} \sigma_y^2 & \boldsymbol{\sigma}_{yx}' \\ \boldsymbol{\sigma}_{yx} & \boldsymbol{\Sigma}_{xx} \end{pmatrix} = \begin{pmatrix} 9 & 0 & 3 & 3 \\ 0 & 1 & -1 & 2 \\ 3 & -1 & 6 & -3 \\ 3 & 2 & -3 & 7 \end{pmatrix}.$$

By (4.33), we have

$$\begin{aligned} E(y|x_1, x_2, x_3) &= \mu_y + \boldsymbol{\sigma}_{yx}'\boldsymbol{\Sigma}_{xx}^{-1}(\mathbf{x} - \boldsymbol{\mu}_x) \\ &= 2 + (0, 3, 3)\begin{pmatrix} 1 & -1 & 2 \\ -1 & 6 & -3 \\ 2 & -3 & 7 \end{pmatrix}^{-1}\begin{pmatrix} x_1 - 5 \\ x_2 + 2 \\ x_3 + 1 \end{pmatrix} \\ &= \tfrac{95}{7} - \tfrac{12}{7}x_1 + \tfrac{6}{7}x_2 + \tfrac{9}{7}x_3. \end{aligned}$$

By (4.34), we obtain

$$\begin{aligned} \operatorname{var}(y|x_1, x_2, x_3) &= \sigma_y^2 - \boldsymbol{\sigma}_{yx}'\boldsymbol{\Sigma}_{xx}^{-1}\boldsymbol{\sigma}_{yx} \\ &= 9 - (0, 3, 3)\begin{pmatrix} 1 & -1 & 2 \\ -1 & 6 & -3 \\ 2 & -3 & 7 \end{pmatrix}^{-1}\begin{pmatrix} 0 \\ 3 \\ 3 \end{pmatrix} \\ &= 9 - \tfrac{45}{7} = \tfrac{18}{7}. \end{aligned}$$

Hence $y|x_1, x_2, x_3$ is $N(\frac{95}{7} - \frac{12}{7}x_1 + \frac{6}{7}x_2 + \frac{9}{7}x_3, \frac{18}{7})$. Note that $\operatorname{var}(y|x_1, x_2, x_3) = \frac{18}{7}$ is less than $\operatorname{var}(y) = 9$, which illustrates (4.35). □

4.5 PARTIAL CORRELATION

We now define the partial correlation of y_i and y_j adjusted for a subset of other y variables. For convenience, we use the notation of Theorems 4.4c and 4.4d. The subset of y's containing y_i and y_j is denoted by $\mathbf{y}$, and the other subset of y's is denoted by $\mathbf{x}$.

Let $\mathbf{v}$ be $N_{p+q}(\boldsymbol{\mu}, \boldsymbol{\Sigma})$ and let $\mathbf{v}$, $\boldsymbol{\mu}$, and $\boldsymbol{\Sigma}$ be partitioned as in Theorem 4.4c and 4.4d:

$$\mathbf{v} = \begin{pmatrix} \mathbf{y} \\ \mathbf{x} \end{pmatrix}, \quad \boldsymbol{\mu} = \begin{pmatrix} \boldsymbol{\mu}_y \\ \boldsymbol{\mu}_x \end{pmatrix}, \quad \boldsymbol{\Sigma} = \begin{pmatrix} \boldsymbol{\Sigma}_{yy} & \boldsymbol{\Sigma}_{yx} \\ \boldsymbol{\Sigma}_{xy} & \boldsymbol{\Sigma}_{xx} \end{pmatrix}.$$

The covariance of y_i and y_j in the conditional distribution of $\mathbf{y}$ given $\mathbf{x}$ will be denoted by $\sigma_{ij\cdot rs\ldots q}$, where y_i and y_j are two of the variables in $\mathbf{y}$ and $y_r, y_s, \ldots, y_q$ are all the variables in $\mathbf{x}$. Thus $\sigma_{ij\cdot rs\ldots q}$ is the (ij)th element of $\text{cov}(\mathbf{y}|\mathbf{x}) = \boldsymbol{\Sigma}_{yy} - \boldsymbol{\Sigma}_{yx}\boldsymbol{\Sigma}_{xx}^{-1}\boldsymbol{\Sigma}_{xy}$. For example, $\sigma_{13\cdot 567}$ represents the covariance between y_1 and y_3 in the conditional distribution of y_1, y_2, y_3, y_4 given y_5, y_6, and y_7 [in this case $\mathbf{x} = (y_5, y_6, y_7)'$]. Similarly, $\sigma_{22\cdot 567}$ represents the variance of y_2 in the conditional distribution of y_1, y_2, y_3, y_4 given y_5, y_6, y_7.

We now define the *partial correlation coefficient* $\rho_{ij\cdot rs\ldots q}$ to be the correlation between y_i and y_j in the conditional distribution of $\mathbf{y}$ given $\mathbf{x}$, where $\mathbf{x} = (y_r, y_s, \ldots, y_q)'$. From the usual definition of a correlation [see (3.19)], we can obtain $\rho_{ij\cdot rs\ldots q}$ from $\sigma_{ij\cdot rs\ldots q}$:

$$\rho_{ij\cdot rs\ldots q} = \frac{\sigma_{ij\cdot rs\ldots q}}{\sqrt{\sigma_{ii\cdot rs\ldots q}\sigma_{jj\cdot rs\ldots q}}}. \tag{4.36}$$

This is the population partial correlation. The sample partial correlation $r_{ij\cdot rs}\cdots q$ is discussed in Section 10.7, including a formulation that does not require normality.

The matrix of partial correlations, $\mathbf{P}_{y\cdot x} = (\rho_{ij\cdot rs\ldots q})$ can be found by (3.30) and (4.27) as

$$\mathbf{P}_{y\cdot x} = \mathbf{D}_{y\cdot x}^{-1}\boldsymbol{\Sigma}_{y\cdot x}\mathbf{D}_{y\cdot x}^{-1}, \tag{4.37}$$

where $\boldsymbol{\Sigma}_{y\cdot x} = \text{cov}(\mathbf{y}|\mathbf{x}) = \boldsymbol{\Sigma}_{yy} - \boldsymbol{\Sigma}_{yx}\boldsymbol{\Sigma}_{xx}^{-1}\boldsymbol{\Sigma}_{xy}$ and $\mathbf{D}_{y\cdot x} = [\text{diag}(\boldsymbol{\Sigma}_{y\cdot x})]^{1/2}$.

Unless $\mathbf{y}$ and $\mathbf{x}$ are independent ($\boldsymbol{\Sigma}_{yx} = \mathbf{O}$), the partial correlation $\rho_{ij\cdot rs\ldots q}$ is different from the usual correlation $\rho_{ij} = \sigma_{ij}/\sqrt{\sigma_{ii}\sigma_{jj}}$. In fact, $\rho_{ij\cdot rs\ldots q}$ and ρ_{ij} can be of opposite signs (for an illustration, see Problem 4.16 g, h). To show this, we express $\sigma_{ij\cdot rs\ldots q}$ in terms of σ_{ij}. We first write $\boldsymbol{\Sigma}_{yx}$ in terms of its rows

$$\boldsymbol{\Sigma}_{yx} = \text{cov}(\mathbf{y}, \mathbf{x}) = \begin{pmatrix} \sigma_{y_1x_1} & \sigma_{y_1x_2} & \cdots & \sigma_{y_1x_q} \\ \sigma_{y_2x_1} & \sigma_{y_2x_2} & \cdots & \sigma_{y_2x_q} \\ \vdots & \vdots & & \vdots \\ \sigma_{y_px_1} & \sigma_{y_px_2} & \cdots & \sigma_{y_px_q} \end{pmatrix} = \begin{pmatrix} \boldsymbol{\sigma}'_{1x} \\ \boldsymbol{\sigma}'_{2x} \\ \vdots \\ \boldsymbol{\sigma}'_{px} \end{pmatrix}, \tag{4.38}$$

where $\boldsymbol{\sigma}'_{ix} = (\sigma_{y_i x_1}, \sigma_{y_i x_2}, \ldots, \sigma_{y_i x_q})$. Then $\sigma_{ij \cdot rs \ldots q}$, the (ij)th element of $\boldsymbol{\Sigma}_{yy} - \boldsymbol{\Sigma}_{yx}\boldsymbol{\Sigma}_{xx}^{-1}\boldsymbol{\Sigma}_{xy}$, can be written as

$$\sigma_{ij \cdot rs \ldots q} = \sigma_{ij} - \boldsymbol{\sigma}'_{ix}\boldsymbol{\Sigma}_{xx}^{-1}\boldsymbol{\sigma}_{jx}. \tag{4.39}$$

Suppose that σ_{ij} is positive. Then $\sigma_{ij \cdot rs \ldots q}$ is negative if $\boldsymbol{\sigma}'_{ix}\boldsymbol{\Sigma}_{xx}^{-1}\boldsymbol{\sigma}_{jx} > \sigma_{ij}$. Note also that since $\boldsymbol{\Sigma}_{xx}^{-1}$ is positive definite, (4.39) shows that

$$\sigma_{ii \cdot rs \ldots q} = \sigma_{ii} - \boldsymbol{\sigma}'_{ix}\boldsymbol{\Sigma}_{xx}^{-1}\boldsymbol{\sigma}_{ix} \leq \sigma_{ii}.$$

Example 4.5. We compare ρ_{12} and $\rho_{12 \cdot 34}$ using $\boldsymbol{\mu}$ and $\boldsymbol{\Sigma}$ in Example 4.4b. From $\boldsymbol{\Sigma}$, we obtain

$$\rho_{12} = \frac{\sigma_{12}}{\sqrt{\sigma_{11}\sigma_{22}}} = \frac{0}{\sqrt{(9)(1)}} = 0.$$

From $\text{cov}(\mathbf{y}|\mathbf{x}) = \frac{1}{33}\begin{pmatrix} 126 & -24 \\ -24 & 14 \end{pmatrix}$ in Example 4.4b, we obtain

$$\begin{aligned} \rho_{12 \cdot 34} &= \frac{\sigma_{12 \cdot 34}}{\sqrt{\sigma_{11 \cdot 34}\sigma_{22 \cdot 34}}} = \frac{-24/33}{\sqrt{(126/33)(14/33)}} = \frac{-24}{\sqrt{(36)(49)}} \\ &= \frac{-4}{7} = -.571. \end{aligned}$$

□

PROBLEMS

4.1 Show that $E(z) = 0$ and $\text{var}(z) = 1$ when z has the standard normal density (4.1).

4.2 Obtain (4.8) from (4.7); that is, show that $|\boldsymbol{\Sigma}^{-1/2}| = |\boldsymbol{\Sigma}|^{-1/2}$.

4.3 Show that $\partial M_{\mathbf{y}}(\mathbf{0})/\partial \mathbf{t} = E(\mathbf{y})$ as in (4.16).

4.4 Show that $\partial^2 M_{\mathbf{y}}(\mathbf{0})/\partial t_r \partial t_s = E(y_r y_s)$ as in (4.17).

4.5 Show that the exponent in (4.19) can be expressed as in (4.20); that is, show that $\mathbf{t}'\mathbf{y} - (\mathbf{y} - \boldsymbol{\mu})'\boldsymbol{\Sigma}^{-1}(\mathbf{y} - \boldsymbol{\mu})/2 = \mathbf{t}'\boldsymbol{\mu} + \mathbf{t}'\boldsymbol{\Sigma}\mathbf{t}/2 - (\mathbf{y} - \boldsymbol{\mu} - \boldsymbol{\Sigma}\mathbf{t})'\boldsymbol{\Sigma}^{-1}$ $(\mathbf{y} - \boldsymbol{\mu} - \boldsymbol{\Sigma}\mathbf{t})/2$.

4.6 Prove Corollary 1 to Theorem 4.3.

4.7 Show that $E(e^{\mathbf{t}'\mathbf{A}\mathbf{y}}) = e^{\mathbf{t}'(\mathbf{A}\boldsymbol{\mu}) + \mathbf{t}'(\mathbf{A}\boldsymbol{\Sigma}\mathbf{A}')\mathbf{t}/2}$ as in (4.25).

4.8 Consider a random variable with moment generating function $M(t)$. Show that the second derivative of $\ln[M(t)]$ evaluated at $t = 0$ is the variance of the random variable.

4.9 Assuming that $\mathbf{y}$ is $N_p(\boldsymbol{\mu}, \sigma^2\mathbf{I})$ and $\mathbf{C}$ is an orthogonal matrix, show that $\mathbf{Cy}$ is $N_p(\mathbf{C}\boldsymbol{\mu}, \sigma^2\mathbf{I})$.

4.10 Prove Corollary 1 to Theorem 4.4a.

4.11 Let $\mathbf{A} = (\mathbf{I}_r, \mathbf{O})$, as defined in the proof of Theorem 4.4b. Show that $\mathbf{Ay} = \mathbf{y}_1$, $\mathbf{A}\boldsymbol{\mu} = \boldsymbol{\mu}_1$, and $\mathbf{A\Sigma A}' = \boldsymbol{\Sigma}_{11}$.

4.12 Prove Corollary 2 to Theorem 4.4c.

4.13 Prove Theorem 4.4d by direct evaluation of (4.28).

4.14 Given $\mathbf{w} = \mathbf{y} - \mathbf{Bx}$, show that $\text{cov}(\mathbf{w}, \mathbf{x}) = \boldsymbol{\Sigma}_{yx} - \mathbf{B}\boldsymbol{\Sigma}_{xx}$, as in (4.30).

4.15 Show that $E(\mathbf{y} - \boldsymbol{\Sigma}_{yx}\boldsymbol{\Sigma}_{xx}^{-1}\mathbf{x}) = \boldsymbol{\mu}_y - \boldsymbol{\Sigma}_{yx}\boldsymbol{\Sigma}_{xx}^{-1}\boldsymbol{\mu}_x$ as in (4.31) and that $\text{cov}(\mathbf{y} - \boldsymbol{\Sigma}_{yx}\boldsymbol{\Sigma}_{xx}^{-1}\mathbf{x}) = \boldsymbol{\Sigma}_{yy} - \boldsymbol{\Sigma}_{yx}\boldsymbol{\Sigma}_{xx}^{-1}\boldsymbol{\Sigma}_{xy}$ as in (4.32).

4.16 Suppose that $\mathbf{y}$ is $N_4(\boldsymbol{\mu}, \boldsymbol{\Sigma})$, where

$$\boldsymbol{\mu} = \begin{pmatrix} 1 \\ 2 \\ 3 \\ -2 \end{pmatrix}, \quad \boldsymbol{\Sigma} = \begin{pmatrix} 4 & 2 & -1 & 2 \\ 2 & 6 & 3 & -2 \\ -1 & 3 & 5 & -4 \\ 2 & -2 & -4 & 4 \end{pmatrix}.$$

Find the following.

(a) The joint marginal distribution of y_1 and y_3
(b) The marginal distribution of y_2
(c) The distribution of $z = y_1 + 2y_2 - y_3 + 3y_4$
(d) The joint distribution of $z_1 = y_1 + y_2 - y_3 - y_4$ and $z_2 = -3y_1 + y_2 + 2y_3 - 2y_4$
(e) $f(y_1, y_2 | y_3, y_4)$
(f) $f(y_1, y_3 | y_2, y_4)$
(g) ρ_{13}
(h) $\rho_{13\cdot 24}$
(i) $f(y_1 | y_2, y_3, y_4)$

4.17 Let $\mathbf{y}$ be distributed as $N_3(\boldsymbol{\mu}, \boldsymbol{\Sigma})$, where

$$\boldsymbol{\mu} = \begin{pmatrix} 2 \\ -1 \\ 3 \end{pmatrix}, \quad \boldsymbol{\Sigma} = \begin{pmatrix} 4 & 1 & 0 \\ 1 & 2 & 1 \\ 0 & 1 & 3 \end{pmatrix}.$$

Find the following.

(a) The distribution of $z = 4y_1 - 6y_2 + y_3$
(b) The distribution of $\mathbf{z} = \begin{pmatrix} y_1 - y_2 + y_3 \\ 2y_1 + y_2 - y_3 \end{pmatrix}$
(c) $f(y_2 | y_1, y_3)$
(d) $f(y_1, y_2 | y_3)$
(e) ρ_{12} and $\rho_{12\cdot 3}$

4.18 If $\mathbf{y}$ is $N_3(\boldsymbol{\mu}, \boldsymbol{\Sigma})$, where

$$\boldsymbol{\Sigma} = \begin{pmatrix} 2 & 0 & -1 \\ 0 & 4 & 0 \\ -1 & 0 & 3 \end{pmatrix},$$

which variables are independent? (See Corollary 1 to Theorem 4.4a)

4.19 If $\mathbf{y}$ is $N_4(\boldsymbol{\mu}, \boldsymbol{\Sigma})$, where

$$\boldsymbol{\Sigma} = \begin{pmatrix} 1 & 0 & 0 & 0 \\ 0 & 2 & 0 & 0 \\ 0 & 0 & 3 & -4 \\ 0 & 0 & -4 & 6 \end{pmatrix},$$

which variables are independent?

4.20 Show that $\sigma_{ij \cdot rs} \cdots q = \sigma_{ij} - \boldsymbol{\sigma}'_{ix} \boldsymbol{\Sigma}_{xx}^{-1} \boldsymbol{\sigma}_{jx}$ as in (4.39).

5 Distribution of Quadratic Forms in y

5.1 SUMS OF SQUARES

In Chapters 3 and 4, we discussed some properties of linear functions of the random vector $\mathbf{y}$. We now consider quadratic forms in $\mathbf{y}$. We will find it useful in later chapters to express a sum of squares encountered in regression or analysis of variance as a quadratic form $\mathbf{y}'\mathbf{A}\mathbf{y}$, where $\mathbf{y}$ is a random vector and $\mathbf{A}$ is a symmetric matrix of constants [see (2.33)]. In this format, we will be able to show that certain sums of squares have chi-square distributions and are independent, thereby leading to F tests.

Example 5.1. We express some simple sums of squares as quadratic forms in $\mathbf{y}$. Let $y_1, y_2, \ldots, y_n$ be a random sample from a population with mean μ and variance σ^2. In the following identity, the total sum of squares $\sum_{i=1}^{n} y_i^2$ is partitioned into a sum of squares about the sample mean $\bar{y} = \sum_{i=1}^{n} y_i/n$ and a sum of squares due to the mean:

$$\begin{aligned} \sum_{i=1}^{n} y_i^2 &= \left(\sum_{i=1}^{n} y_i^2 - n\bar{y}^{+2} \right) + n\bar{y}^2 \\ &= \sum_{i=1}^{n} (y_i - \bar{y})^2 + n\bar{y}^2. \end{aligned} \tag{5.1}$$

Using (2.20), we can express $\sum_{i=1}^{n} y_i^2$ as a quadratic form

$$\sum_{i=1}^{n} y_i^2 = \mathbf{y}'\mathbf{y} = \mathbf{y}'\mathbf{I}\mathbf{y},$$

where $\mathbf{y}' = (y_1, y_2, \ldots, y_n)$. Using $\mathbf{j} = (1, 1, \ldots, 1)'$ as defined in (2.6), we can

---

write $\bar{y}$ as

$$\bar{y} = \frac{1}{n}\sum_{i=1}^{n} y_i = \frac{1}{n}\mathbf{j}'\mathbf{y}$$

[see (2.24)]. Then $n\bar{y}^2$ becomes

$$\begin{aligned} n\bar{y}^2 &= n\left(\frac{1}{n}\mathbf{j}'\mathbf{y}\right)^2 = n\left(\frac{1}{n}\mathbf{j}'\mathbf{y}\right)\left(\frac{1}{n}\mathbf{j}'\mathbf{y}\right) \\ &= n\left(\frac{1}{n}\right)^2 \mathbf{y}'\mathbf{j}\mathbf{j}'\mathbf{y} \quad \text{[by (2.18)]} \\ &= n\left(\frac{1}{n}\right)^2 \mathbf{y}'\mathbf{J}\mathbf{y} \quad \text{[by (2.23)]} \\ &= \mathbf{y}'\left(\frac{1}{n}\mathbf{J}\right)\mathbf{y}. \end{aligned}$$

We can now write $\sum_{i=1}^{n}(y_i - \bar{y})^2$ as

$$\begin{aligned} \sum_{i=1}^{n}(y_i - \bar{y})^2 &= \sum_{i=1}^{n} y_i^2 - n\bar{y}^2 = \mathbf{y}'\mathbf{I}\mathbf{y} - \mathbf{y}'\left(\frac{1}{n}\mathbf{J}\right)\mathbf{y} \\ &= \mathbf{y}'\left(\mathbf{I} - \frac{1}{n}\mathbf{J}\right)\mathbf{y}. \end{aligned} \tag{5.2}$$

Hence (5.1) can be written in terms of quadratic forms as

$$\mathbf{y}'\mathbf{I}\mathbf{y} = \mathbf{y}'\left(\mathbf{I} - \frac{1}{n}\mathbf{J}\right)\mathbf{y} + \mathbf{y}'\left(\frac{1}{n}\mathbf{J}\right)\mathbf{y}. \tag{5.3}$$

□

The matrices of the three quadratic forms in (5.3) have the following properties:

1. $\mathbf{I} = \left(\mathbf{I} - \frac{1}{n}\mathbf{J}\right) + \frac{1}{n}\mathbf{J}$.
2. $\mathbf{I}, \mathbf{I} - \frac{1}{n}\mathbf{J}$, and $\frac{1}{n}\mathbf{J}$ are idempotent.
3. $\left(\mathbf{I} - \frac{1}{n}\mathbf{J}\right)\left(\frac{1}{n}\mathbf{J}\right) = \mathbf{O}$.

Using theorems given later in this chapter (and assuming normality of the y_i's), these three properties lead to the conclusion that $\sum_{i=1}^{n}(y_i - \bar{y})^2/\sigma^2$ and $n\bar{y}^2/\sigma^2$ have chi-square distributions and are independent.

5.2 MEAN AND VARIANCE OF QUADRATIC FORMS

We first consider the mean of a quadratic form $\mathbf{y}'\mathbf{A}\mathbf{y}$.

Theorem 5.2a. If $\mathbf{y}$ is a random vector with mean $\boldsymbol{\mu}$ and covariance matrix $\boldsymbol{\Sigma}$ and if $\mathbf{A}$ is a symmetric matrix of constants, then

$$E(\mathbf{y}'\mathbf{A}\mathbf{y}) = \mathrm{tr}(\mathbf{A}\boldsymbol{\Sigma}) + \boldsymbol{\mu}'\mathbf{A}\boldsymbol{\mu}. \tag{5.4}$$

Proof. By (3.25), $\Sigma = E(\mathbf{y}\mathbf{y})' - \boldsymbol{\mu}\boldsymbol{\mu}'$, which can be written as

$$E(\mathbf{y}\mathbf{y}') = \Sigma + \boldsymbol{\mu}\boldsymbol{\mu}'. \tag{5.5}$$

Since $\mathbf{y}'\mathbf{A}\mathbf{y}$ is a scalar, it is equal to its trace. We thus have

$$\begin{aligned}
E(\mathbf{y}'\mathbf{A}\mathbf{y}) &= E[\mathrm{tr}(\mathbf{y}'\mathbf{A}\mathbf{y})] \\
&= E[\mathrm{tr}(\mathbf{A}\mathbf{y}\mathbf{y}')] && \text{[by (2.87)]} \\
&= \mathrm{tr}[E(\mathbf{A}\mathbf{y}\mathbf{y}')] && \text{[by (3.5)]} \\
&= \mathrm{tr}[\mathbf{A}E(\mathbf{y}\mathbf{y}')] && \text{[by (3.40)]} \\
&= \mathrm{tr}[\mathbf{A}(\boldsymbol{\Sigma} + \boldsymbol{\mu}\boldsymbol{\mu}')] && \text{[by (5.5)]} \\
&= \mathrm{tr}[\mathbf{A}\boldsymbol{\Sigma} + \boldsymbol{A}\boldsymbol{\mu}\boldsymbol{\mu}'] && \text{[by (2.15)]} \\
&= \mathrm{tr}(\mathbf{A}\boldsymbol{\Sigma}) + \mathrm{tr}(\boldsymbol{\mu}'\mathbf{A}\boldsymbol{\mu}) && \text{[by (2.86)]} \\
&= \mathrm{tr}(\mathbf{A}\boldsymbol{\Sigma}) + \boldsymbol{\mu}'\mathbf{A}\boldsymbol{\mu}
\end{aligned}$$

Note that since $\mathbf{y}'\mathbf{A}\mathbf{y}$ is not a linear function of $\mathbf{y}$, $E(\mathbf{y}'\mathbf{A}\mathbf{y}) \neq E(\mathbf{y}')\mathbf{A}E(\mathbf{y})$. □

Example 5.2a. To illustrate Theorem 5.2a, consider the sample variance

$$s^2 = \frac{\sum_{i=1}^{n}(y_i - \bar{y})^2}{n-1}. \tag{5.6}$$

By (5.2), the numerator of (5.6) can be written as

$$\sum_{i=1}^{n}(y_i - \bar{y})^2 = \mathbf{y}'\left(\mathbf{I} - \frac{1}{n}\mathbf{J}\right)\mathbf{y},$$

where $\mathbf{y} = (y_1, y_2, \ldots, y_n)'$. If the y's are assumed to be independently distributed with mean μ and variance σ^2, then $E(\mathbf{y}) = (\mu, \mu, \ldots, \mu)' = \mu\mathbf{j}$ and $\text{cov}(\mathbf{y}) = \sigma^2\mathbf{I}$. Thus for use in (5.4) we have $\mathbf{A} = \mathbf{I} - (1/n)\mathbf{J}$, $\boldsymbol{\Sigma} = \sigma^2\mathbf{I}$, and $\boldsymbol{\mu} = \mu\mathbf{j}$; hence

$$
\begin{aligned}
E\left[\sum_{i=1}^{n}(y_i - \bar{y})^2\right] &= \text{tr}\left[\left(\mathbf{I} - \frac{1}{n}\mathbf{J}\right)(\sigma^2\mathbf{I})\right] + \mu\mathbf{j}'\left(\mathbf{I} - \frac{1}{n}\mathbf{J}\right)\mu\mathbf{j} \\
&= \sigma^2 \text{tr}\left(\mathbf{I} - \frac{1}{n}\mathbf{J}\right) + \mu^2(\mathbf{j}'\mathbf{j} - \mathbf{j}'\mathbf{j}\mathbf{j}'\mathbf{j}) \qquad \text{[by (2.23)]} \\
&= \sigma^2\left(n - \frac{n}{n}\right) + \mu^2\left(n - \frac{1}{n}n^2\right) \qquad \text{[by (2.23)]} \\
&= \sigma^2(n-1) + 0.
\end{aligned}
$$

Therefore

$$
E(s^2) = \frac{E\left[\sum_{i=1}^{n}(y_i - \bar{y})^2\right]}{n-1} = \frac{(n-1)\sigma^2}{n-1} = \sigma^2. \tag{5.7}
$$

□

Note that normality of the y's is not assumed in Theorem 5.2a. However, normality is assumed in obtaining the moment generating function of $\mathbf{y}'\mathbf{A}\mathbf{y}$ and $\text{var}(\mathbf{y}'\mathbf{A}\mathbf{y})$ in the following theorems.

Theorem 5.2b. If $\mathbf{y}$ is $N_p(\boldsymbol{\mu}, \boldsymbol{\Sigma})$, then the moment generating function of $\mathbf{y}'\mathbf{A}\mathbf{y}$ is

$$
M_{\mathbf{y}'\mathbf{A}\mathbf{y}}(t) = |\mathbf{I} - 2t\mathbf{A}\boldsymbol{\Sigma}|^{-1/2} e^{-\boldsymbol{\mu}'[\mathbf{I} - (\mathbf{I} - 2t\mathbf{A}\boldsymbol{\Sigma})^{-1}]\boldsymbol{\Sigma}^{-1}\boldsymbol{\mu}/2} \tag{5.8}
$$

PROOF. By the multivariate analog of (3.3), we obtain

$$
\begin{aligned}
M_{\mathbf{y}'\mathbf{A}\mathbf{y}}(t) = E(e^{t\mathbf{y}'\mathbf{A}\mathbf{y}}) &= \int_{-\infty}^{\infty} \cdots \int_{-\infty}^{\infty} e^{t\mathbf{y}'\mathbf{A}\mathbf{y}} k_1 e^{-(\mathbf{y}-\boldsymbol{\mu})'\boldsymbol{\Sigma}^{-1}(\mathbf{y}-\boldsymbol{\mu})/2} d\mathbf{y} \\
&= k_1 \int_{-\infty}^{\infty} \cdots \int_{-\infty}^{\infty} e^{-[\mathbf{y}'(\mathbf{I} - 2t\mathbf{A}\boldsymbol{\Sigma})\boldsymbol{\Sigma}^{-1}\mathbf{y} - 2\boldsymbol{\mu}'\boldsymbol{\Sigma}^{-1}\mathbf{y} + \boldsymbol{\mu}'\boldsymbol{\Sigma}^{-1}\boldsymbol{\mu}]/2} d\mathbf{y},
\end{aligned}
$$

where $k_1 = 1/[(\sqrt{2\pi})^p |\Sigma|^{1/2}]$ and $d\mathbf{y} = dy_1\, dy_2 \ldots dy_p$. For t sufficiently close to 0, $\mathbf{I} - 2t\mathbf{A}\boldsymbol{\Sigma}$ is nonsingular. Letting $\boldsymbol{\theta}' = \boldsymbol{\mu}'(\mathbf{I} - 2t\mathbf{A}\boldsymbol{\Sigma})^{-1}$ and $\mathbf{V}^{-1} = (\mathbf{I} - 2t\mathbf{A}\boldsymbol{\Sigma})\boldsymbol{\Sigma}^{-1}$, we obtain

$$
M_{\mathbf{y}'\mathbf{A}\mathbf{y}}(t) = k_1 k_2 \int_{-\infty}^{\infty} \cdots \int_{-\infty}^{\infty} k_3 e^{-(\mathbf{y}-\boldsymbol{\theta})'\mathbf{V}^{-1}(\mathbf{y}-\boldsymbol{\theta})/2} d\mathbf{y}
$$

(Problem 5.4), where $k_2 = (\sqrt{(2\pi)}^p |\mathbf{V}|^{1/2} e^{-[\boldsymbol{\mu}'\boldsymbol{\Sigma}^{-1}\boldsymbol{\mu} - \boldsymbol{\theta}'\mathbf{V}^{-1}\boldsymbol{\theta}]/2}$ and $k_3 = 1/[(\sqrt{(2\pi)}^p |\mathbf{V}|^{1/2}]$. The multiple integral is equal to 1 since the multivariate normal density integrates to 1. Thus $M_{\mathbf{y}'\mathbf{A}\mathbf{y}}(t) = k_1 k_2$. Substituting and simplifying, we obtain (5.8) (see Problem 5.5). □

Theorem 5.2c. If $\mathbf{y}$ is $N_p(\boldsymbol{\mu}, \boldsymbol{\Sigma})$, then

$$\text{var}(\mathbf{y}'\mathbf{A}\mathbf{y}) = 2\text{tr}[(\mathbf{A}\boldsymbol{\Sigma})^2] + 4\boldsymbol{\mu}'\mathbf{A}\boldsymbol{\Sigma}\mathbf{A}\boldsymbol{\mu}. \tag{5.9}$$

PROOF. The variance of a random variable can be obtained by evaluating the second derivative of the natural logarithm of its moment generating function at $t = 0$ (see hint to Problem 5.14). Let $\mathbf{C} = \mathbf{I} - 2t\mathbf{A}\boldsymbol{\Sigma}$. Then, from (5.8)

$$k(t) = \ln [M_{\mathbf{y}'\mathbf{A}\mathbf{y}}(t)] = -\frac{1}{2}\ln |\mathbf{C}| - \frac{1}{2}\boldsymbol{\mu}'(\mathbf{I} - \mathbf{C}^{-1})\boldsymbol{\Sigma}^{-1}\boldsymbol{\mu}.$$

Using (2.117), we differentiate $k(t)$ twice to obtain

$$k''(t) = \frac{1}{2}\frac{1}{|\mathbf{C}|^2}\left[\frac{d|\mathbf{C}|}{dt}\right]^2 - \frac{1}{2}\frac{1}{|\mathbf{C}|}\frac{d^2|\mathbf{C}|}{dt^2} - \frac{1}{2}\boldsymbol{\mu}'\mathbf{C}^{-1}\frac{d^2\mathbf{C}}{dt^2}\mathbf{C}^{-1}\boldsymbol{\Sigma}^{-1}\boldsymbol{\mu}$$
$$+ \boldsymbol{\mu}\left[\mathbf{C}^{-1}\frac{d\mathbf{C}}{dt}\right]^2\mathbf{C}^{-1}\boldsymbol{\Sigma}^{-1}\boldsymbol{\mu}$$

(Problem 5.6). A useful expression for $|\mathbf{C}|$ can be found using (2.97) and (2.107). Thus, if the eigenvalues of $\mathbf{A}\boldsymbol{\Sigma}$ are λ_i, $i = 1, \ldots, p$, we obtain

$$\begin{aligned}|\mathbf{C}| &= \prod_{i=1}^{p}(1 - 2t\lambda_i) \\ &= 1 - 2t\sum_i \lambda_i + 4t^2\sum_{i \neq j}\lambda_i\lambda_j - \cdots + (-1)^p 2^p t^p \lambda_1\lambda_2\cdots\lambda_p.\end{aligned}$$

Then $(d|\mathbf{C}|/dt) = -2\Sigma_i\lambda_i + 8t\Sigma_{i \neq j}\lambda_i\lambda_j +$ higher-order terms in t, and $(d^2|\mathbf{C}|/dt^2) = 8\Sigma_{i \neq j}\lambda_i\lambda_j +$ higher-order terms in t. Evaluating these expressions at $t = 0$, we obtain $|\mathbf{C}| = 1$, $(d|\mathbf{C}|/dt)|_{t=0} = -2\Sigma_i\lambda_i = -2\,\text{tr}(\mathbf{A}\boldsymbol{\Sigma})$ and $(d^2|\mathbf{C}|/dt^2)|_{t=0} = 8\Sigma_{i \neq j}\lambda_i\lambda_j$. For $t = 0$ it is also true that $\mathbf{C} = \mathbf{I}$, $\mathbf{C}^{-1} = \mathbf{I}$, $(d\mathbf{C}/dt)|_{t=0} = 2\mathbf{A}\boldsymbol{\Sigma}$ and $(d^2\mathbf{C}/dt)|_{t=0} = \mathbf{O}$. Thus

$$\begin{aligned}k''(0) &= 2[\text{tr}(\mathbf{A}\boldsymbol{\Sigma})]^2 - 4\sum_{i \neq j}\lambda_i\lambda_j + 0 + 4\boldsymbol{\mu}'\mathbf{A}\boldsymbol{\Sigma}\mathbf{A}\boldsymbol{\mu} \\ &= 2\left\{[\text{tr}(\mathbf{A}\boldsymbol{\Sigma})]^2 - 2\sum_{i \neq j}\lambda_i\lambda_j\right\} + 4\boldsymbol{\mu}'\mathbf{A}\boldsymbol{\Sigma}\mathbf{A}\boldsymbol{\mu}.\end{aligned}$$

By Problem 2.81, this can be written as

$$2\,\mathrm{tr}[(\mathbf{A\Sigma})^2] + 4\boldsymbol{\mu}'\mathbf{A\Sigma A}\boldsymbol{\mu}.$$

□

We now consider $\mathrm{cov}(\mathbf{y}, \mathbf{y}'\mathbf{Ay})$. To clarify the meaning of the expression $\mathrm{cov}(\mathbf{y}, \mathbf{y}'\mathbf{Ay})$, we denote $\mathbf{y}'\mathbf{Ay}$ by the scalar random variable v. Then $\mathrm{cov}(\mathbf{y}, v)$ is a column vector containing the covariance of each y_i and v:

$$\mathrm{cov}(\mathbf{y}, v) = E\{[\mathbf{y} - E(\mathbf{y})][v - E(v)]\} = \begin{pmatrix} \sigma_{y_1 v} \\ \sigma_{y_2 v} \\ \vdots \\ \sigma_{y_p v} \end{pmatrix}. \tag{5.10}$$

[On the other hand, $\mathrm{cov}(v, \mathbf{y})$ would be a row vector.] An expression for $\mathrm{cov}(\mathbf{y}, \mathbf{y}'\mathbf{Ay})$ is given in the next theorem.

Theorem 5.2d. If $\mathbf{y}$ is $N_p(\boldsymbol{\mu}, \boldsymbol{\Sigma})$, then

$$\mathrm{cov}(\mathbf{y}, \mathbf{y}'\mathbf{Ay}) = 2\boldsymbol{\Sigma}\mathbf{A}\boldsymbol{\mu}. \tag{5.11}$$

Proof. By the definition in (5.10), we have

$$\mathrm{cov}(\mathbf{y}, \mathbf{y}'\mathbf{Ay}) = E\{[\mathbf{y} - E(\mathbf{y})][\mathbf{y}'\mathbf{Ay} - E(\mathbf{y}'\mathbf{Ay})]\}.$$

By Theorem 5.2a, this becomes

$$\mathrm{cov}(\mathbf{y}, \mathbf{y}'\mathbf{Ay}) = E\{(\mathbf{y} - \boldsymbol{\mu})[\mathbf{y}'\mathbf{Ay} - \mathrm{tr}(\mathbf{A\Sigma}) - \boldsymbol{\mu}'\mathbf{A}\boldsymbol{\mu}]\}.$$

Rewriting $\mathbf{y}'\mathbf{Ay} - \boldsymbol{\mu}'\mathbf{A}\boldsymbol{\mu}$ in terms of $\mathbf{y} - \boldsymbol{\mu}$ (see Problem 5.7), we obtain

$$\begin{aligned} \mathrm{cov}(\mathbf{y}, \mathbf{y}'\mathbf{Ay}) &= E\{(\mathbf{y} - \boldsymbol{\mu})[(\mathbf{y} - \boldsymbol{\mu})'\mathbf{A}(\mathbf{y} - \boldsymbol{\mu}) + 2(\mathbf{y} - \boldsymbol{\mu})'\mathbf{A}\boldsymbol{\mu} - \mathrm{tr}(\mathbf{A\Sigma})]\} \\ &= E[(\mathbf{y} - \boldsymbol{\mu})(\mathbf{y} - \boldsymbol{\mu})'\mathbf{A}(\mathbf{y} - \boldsymbol{\mu})] + 2E[(\mathbf{y} - \boldsymbol{\mu})(\mathbf{y} - \boldsymbol{\mu})'\mathbf{A}\boldsymbol{\mu}] \\ &\quad - E[(\mathbf{y} - \boldsymbol{\mu})\mathrm{tr}(\mathbf{A\Sigma})] \\ &= \mathbf{0} + 2\boldsymbol{\Sigma}\mathbf{A}\boldsymbol{\mu} - \mathbf{0}. \end{aligned} \tag{5.12}$$

The first term on the right side is $\mathbf{0}$ because all third central moments of the multivariate normal are zero. The results for the other two terms do not depend on normality (see Problem 5.7). □

Corollary 1. Let $\mathbf{B}$ be a $k \times p$ matrix of constants. Then

$$\text{cov}(\mathbf{By}, \mathbf{y}'\mathbf{Ay}) = 2\mathbf{B\Sigma A}\boldsymbol{\mu}. \tag{5.13}$$

□

For the partitioned random vector $\mathbf{v} = \begin{pmatrix} \mathbf{y} \\ \mathbf{x} \end{pmatrix}$, the bilinear form $\mathbf{x}'\mathbf{Ay}$ was introduced in (2.34). The expected value of $\mathbf{x}'\mathbf{Ay}$ is given in the following theorem.

Theorem 5.2e. Let $\mathbf{v} = \begin{pmatrix} \mathbf{y} \\ \mathbf{x} \end{pmatrix}$ be a partitioned random vector with mean vector and covariance matrix given by (3.32) and (3.33)

$$E\begin{pmatrix} \mathbf{y} \\ \mathbf{x} \end{pmatrix} = \begin{pmatrix} \boldsymbol{\mu}_y \\ \boldsymbol{\mu}_x \end{pmatrix} \quad \text{and} \quad \text{cov}\begin{pmatrix} \mathbf{y} \\ \mathbf{x} \end{pmatrix} = \begin{pmatrix} \boldsymbol{\Sigma}_{yy} & \boldsymbol{\Sigma}_{yx} \\ \boldsymbol{\Sigma}_{xy} & \boldsymbol{\Sigma}_{xx} \end{pmatrix},$$

where $\mathbf{y}$ is $p \times 1$, $\mathbf{x}$ is $q \times 1$, and $\boldsymbol{\Sigma}_{yx}$ is $p \times q$. Let $\mathbf{A}$ be a $q \times p$ matrix of constants. Then

$$E(\mathbf{x}'\mathbf{Ay}) = \text{tr}(\mathbf{A}\boldsymbol{\Sigma}_{yx}) + \boldsymbol{\mu}_x'\mathbf{A}\boldsymbol{\mu}_y. \tag{5.14}$$

Proof. The proof is similar to that of Theorem 5.2a; see Problem 5.10. □

Example 5.2b. To estimate the population covariance $\sigma_{xy} = E[(x - \mu_x)(y - \mu_y)]$ in (3.10), we use the sample covariance

$$s_{xy} = \frac{\sum_{i=1}^{n} (x_i - \bar{x})(y_i - \bar{y})}{n - 1}, \tag{5.15}$$

where $(x_1, y_1), (x_2, y_2), \ldots, (x_n, y_n)$ is a bivariate random sample from a population with means μ_x and μ_y, variances σ_x^2 and σ_y^2, and covariance σ_{xy}. We can write (5.15) in the form

$$s_{xy} = \frac{\sum_{i=1}^{n} x_i y_i - n\bar{x}\bar{y}}{n - 1} = \frac{\mathbf{x}'[\mathbf{I} - (1/n)\mathbf{J}]\mathbf{y}}{n - 1}, \tag{5.16}$$

where $\mathbf{x} = (x_1, x_2, \ldots, x_n)'$ and $\mathbf{y} = (y_1, y_2, \ldots, y_n)'$. Since (x_i, y_i) is independent of (x_j, y_j) for $i \neq j$, the random vector $\mathbf{v} = \begin{pmatrix} \mathbf{y} \\ \mathbf{x} \end{pmatrix}$ has mean vector and covariance matrix

$$E\begin{pmatrix} \mathbf{y} \\ \mathbf{x} \end{pmatrix} = \begin{pmatrix} \boldsymbol{\mu}_y \\ \boldsymbol{\mu}_x \end{pmatrix} = \begin{pmatrix} \mu_y\mathbf{j} \\ \mu_x\mathbf{j} \end{pmatrix},$$

$$\text{cov}\begin{pmatrix} \mathbf{y} \\ \mathbf{x} \end{pmatrix} = \begin{pmatrix} \boldsymbol{\Sigma}_{yy} & \boldsymbol{\Sigma}_{yx} \\ \boldsymbol{\Sigma}_{xy} & \boldsymbol{\Sigma}_{xx} \end{pmatrix} = \begin{pmatrix} \sigma_y^2\mathbf{I} & \sigma_{xy}\mathbf{I} \\ \sigma_{xy}\mathbf{I} & \sigma_x^2\mathbf{I} \end{pmatrix},$$

where each $\mathbf{I}$ is $n \times n$. Thus for use in (5.14), we have $\mathbf{A} = \mathbf{I} - (1/n)\mathbf{J}$, $\boldsymbol{\Sigma}_{yx} = \sigma_{xy}\mathbf{I}$, $\boldsymbol{\mu}_x = \mu_x\mathbf{j}$, and $\boldsymbol{\mu}_y = \mu_y\mathbf{j}$. Hence

$$
\begin{aligned}
E\left[\mathbf{x}'\left(\mathbf{I} - \frac{1}{n}\mathbf{J}\right)\mathbf{y}\right] &= \text{tr}\left[\left(\mathbf{I} - \frac{1}{n}\mathbf{J}\right)\sigma_{xy}\mathbf{I}\right] + \mu_x\mathbf{j}'\left(\mathbf{I} - \frac{1}{n}\mathbf{J}\right)\mu_y\mathbf{j} \\
&= \sigma_{xy}\text{tr}\left(\mathbf{I} - \frac{1}{n}\mathbf{J}\right) + \mu_x\mu_y\left(\mathbf{j}'\mathbf{j} - \frac{1}{n}\mathbf{j}'\mathbf{j}\mathbf{j}'\mathbf{j}\right) \\
&= \sigma_{xy}(n-1) + 0.
\end{aligned}
$$

Therefore

$$
E(s_{xy}) = \frac{E\left[\sum_{i=1}^n (x_i - \bar{x})(y_i - \bar{y})\right]}{n-1} = \frac{(n-1)\sigma_{xy}}{n-1} = \sigma_{xy}. \tag{5.17}
$$

□

5.3 NONCENTRAL CHI-SQUARE DISTRIBUTION

Before discussing the noncentral chi-square distribution, we first review the central chi-square distribution. Let $z_1, z_2, \ldots, z_n$ be a random sample from the standard normal distribution $N(0, 1)$. Since the z's are independent (by definition of random sample) and each z_i is $N(0, 1)$, the random vector $\mathbf{z} = (z_1, z_2, \ldots, z_n)'$ is distributed as $N_n(\mathbf{0}, \mathbf{I})$. By definition

$$
\sum_{i=1}^n z_i^2 = \mathbf{z}'\mathbf{z} \text{ is } \chi^2(n); \tag{5.18}
$$

that is, the sum of squares of n independent standard normal random variables is distributed as a (central) chi-square random variable with n degrees of freedom.

The mean, variance, and moment generating function of a chi-square random variable are given in the following theorem.

Theorem 5.3a. If u is distributed as $\chi^2(n)$, then

$$
E(u) = n, \tag{5.19}
$$

$$
\text{var}(u) = 2n, \tag{5.20}
$$

$$
M_u(t) = \frac{1}{(1-2t)^{n/2}}. \tag{5.21}
$$

Proof. Since u is the quadratic form $\mathbf{z}'\mathbf{I}\mathbf{z}$, $E(u)$, var(u), and $M_u(t)$ can be obtained by applying Theorems 5.2a, 5.2c, and 5.2b, respectively. □

Now suppose that $y_1, y_2, \ldots, y_n$ are independently distributed as $N(\mu_i, 1)$, so that $\mathbf{y}$ is $N_n(\boldsymbol{\mu}, \mathbf{I})$, where $\boldsymbol{\mu} = (\mu_1, \mu_2, \ldots, \mu_n)'$. In this case, $\Sigma_{i=1}^n y_i^2 = \mathbf{y}'\mathbf{y}$ does not have a chi-square distribution, but $\Sigma(y_i - \mu_i)^2 = (\mathbf{y} - \boldsymbol{\mu})'(\mathbf{y} - \boldsymbol{\mu})$ is $\chi^2(n)$ since $y_i - \mu_i$ is distributed as $N(0,1)$.

The density of $v = \Sigma_{i=1}^n y_i^2 = \mathbf{y}'\mathbf{y}$, where the y's are independently distributed as $N(\mu_i, 1)$, is called the *noncentral chi-square distribution* and is denoted by $\chi^2(n, \lambda)$. The *noncentrality parameter* λ is defined as

$$\lambda = \frac{1}{2}\sum_{i=1}^{n} \mu_i^2 = \frac{1}{2}\boldsymbol{\mu}'\boldsymbol{\mu}. \tag{5.22}$$

Note that λ is not an eigenvalue here and that the mean of $v = \Sigma_{i=1}^n y_i^2$ is greater than the mean of $u = \Sigma_{i=1}^n (y_i - \mu_i)^2$:

$$E\left[\sum_{i=1}^{n} (y_i - \mu_i)^2\right] = \sum_{i=1}^{n} E(y_i - \mu_i)^2 = \sum_{i=1}^{n} \text{var}(y_i) = \sum_{i=1}^{n} 1 = n,$$

$$E\left(\sum_{i=1}^{n} y_i^2\right) = \sum_{i=1}^{n} E(y_i^2) \underset{(5.4)}{=} \sum_{i=1}^{n} (\sigma_i^2 + \mu_i^2) = \sum_{i=1}^{n} (1 + \mu_i^2)$$

$$= n + \sum_{i=1}^{n} \mu_i^2 = n + 2\lambda,$$

where λ is as defined in (5.22). The densities of u and v are illustrated in Figure 5.1.

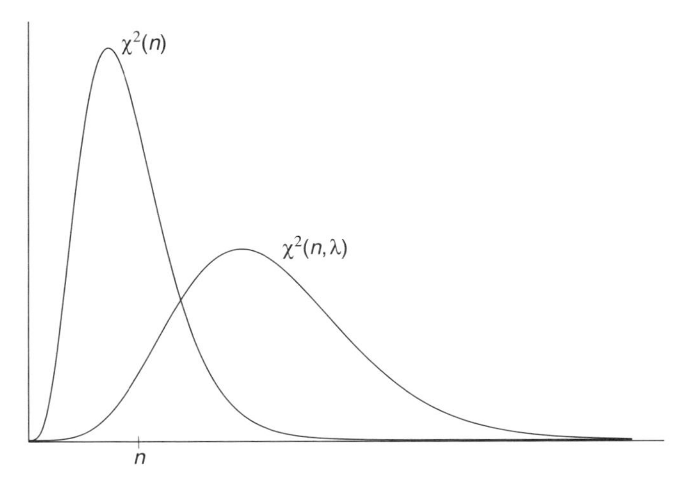

Figure 5.1 Central and noncentral chi-square densities.

The mean, variance, and moment generating function of a noncentral chi-square random variable are given in the following theorem.

Theorem 5.3b. If v is distributed as $\chi^2(n, \lambda)$, then

$$E(v) = n + 2\lambda, \tag{5.23}$$

$$\text{var}(v) = 2n + 8\lambda, \tag{5.24}$$

$$M_v(t) = \frac{1}{(1-2t)^{n/2}} e^{-\lambda[1-1/(1-2t)]}. \tag{5.25}$$

PROOF. For $E(v)$ and var(v), see Problems 5.13 and 5.14. For $M_v(t)$, use Theorem 5.2b. □

Corollary 1. If $\lambda = 0$ (which corresponds to $\mu_i = 0$ for all i), then $E(v)$, var(v), and $M_v(t)$ in Theorem 5.3b reduce to $E(u)$, var(u), $M_u(t)$ for the central chi-square distribution in Theorem 5.3a. Thus

$$\chi^2(n, 0) = \chi^2(n). \tag{5.26}$$

□

The chi-square distribution has an additive property, as shown in the following theorem.

Theorem 5.3c. If $v_1, v_2, \ldots, v_k$ are independently distributed as $\chi^2(n_i, \lambda_i)$, then

$$\sum_{i=1}^{k} v_i \text{ is distributed as } \chi^2\left(\sum_{i=1}^{k} n_i, \sum_{i=1}^{k} \lambda_i\right). \tag{5.27}$$

□

Corollary 1. If $u_1, u_2, \ldots, u_k$ are independently distributed as $\chi^2(n_i)$, then

$$\sum_{i=1}^{k} u_i \text{ is distributed as } \chi^2\left(\sum_{i=1}^{k} n_i\right).$$

□

5.4 NONCENTRAL F AND t DISTRIBUTIONS

5.4.1 Noncentral F Distribution

Before defining the noncentral F distribution, we first review the central F. If u is $\chi^2(p)$, v is $\chi^2(q)$, and u and v are independent, then by definition

$$w = \frac{u/p}{v/q} \text{ is distributed as } F(p, q), \tag{5.28}$$

the (central) F distribution with p and q degress of freedom. The mean and variance of w are given by

$$E(w) = \frac{q}{q-2}, \quad \text{var}(w) = \frac{2q^2(p+q-2)}{p(q-1)^2(q-4)}. \tag{5.29}$$

Now suppose that u is distributed as a noncentral chi-square random variable, $\chi^2(p, \lambda)$, while v remains central chi-square random variable, $\chi^2(q)$, with u and v independent. Then

$$z = \frac{u/p}{v/q} \text{ is distributed as } F(p, q, \lambda), \tag{5.30}$$

the *noncentral F distribution* with noncentrality parameter λ, where λ is the same noncentrality parameter as in the distribution of u (noncentral chi-square distribution). The mean of z is

$$E(z) = \frac{q}{q-2}\left(1 + \frac{2\lambda}{p}\right), \tag{5.31}$$

which is, course, greater than $E(w)$ in (5.29).

When an F statistic is used to test a hypothesis H_0, the distribution will typically be central if the (null) hypothesis is true and noncentral if the hypothesis is false. Thus the noncentral F distribution can often be used to evaluate the power of an F test. The *power* of a test is the probability of rejecting H_0 for a given value of λ. If F_α is the upper α percentage point of the central F distribution, then the power, $P(p, q, \alpha, \lambda)$, can be defined as

$$P(p, q, \alpha, \lambda) = \text{Prob}\ (z \geq F_\alpha), \tag{5.32}$$

where z is the noncentral F random variable defined in (5.30). Ghosh (1973) showed that $P(p, q, \alpha, \lambda)$ increases if q or α or λ increases, and $P(p, q, \alpha, \lambda)$ decreases if p increases. The power is illustrated in Figure 5.2.

The power as defined in (5.32) can be evaluated from tables (Tiku 1967) or directly from distribution functions available in many software packages. For example, in SAS, the noncentral F-distribution function PROBF can be used to find the power in (5.32) as follows:

$$P(p, q, \alpha, \lambda) = 1 - \text{PROBF}(F_\alpha, p, q, \lambda).$$

A probability calculator for the F and other distributions is available free of charge from NCSS (download at www.ncss.com).

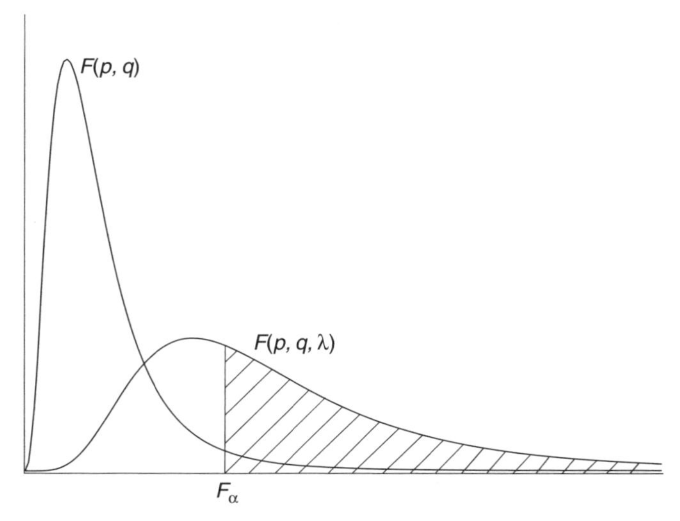

Figure 5.2 Central F, noncentral F, and power of the F test (shaded area).

5.4.2 Noncentral t Distribution

We first review the central t distribution. If z is $N(0,1)$, u is $\chi^2(p)$, and z and u are independent, then by definition

$$t = \frac{z}{\sqrt{u/p}} \text{ is distributed as } t(p), \tag{5.33}$$

the (central) t distribution with p degrees of freedom.

Now suppose that y is $N(\mu, 1)$, u is $\chi^2(p)$, and y and u are independent. Then

$$t = \frac{y}{\sqrt{u/p}} \text{ is distributed as } t(p, \mu), \tag{5.34}$$

the noncentral t distribution with p degrees of freedom and noncentrality parameter μ. If y is $N(\mu, \sigma^2)$, then

$$t = \frac{y/\sigma}{\sqrt{u/p}} \text{ is distributed as } t(p, \mu/\sigma),$$

since by (3.4), (3.9), and Theorem 4.4a(i), y/σ is distributed as $N(\mu/\sigma, 1)$.

5.5 DISTRIBUTION OF QUADRATIC FORMS

It was noted following Theorem 5.3a that if $\mathbf{y}$ is $N_n(\boldsymbol{\mu}, \mathbf{I})$, then $(\mathbf{y}-\boldsymbol{\mu})'(\mathbf{y}-\boldsymbol{\mu})$ is $\chi^2(n)$. If $\mathbf{y}$ is $N_n(\boldsymbol{\mu}, \boldsymbol{\Sigma})$, we can extend this to

$$(\mathbf{y}-\boldsymbol{\mu})'\boldsymbol{\Sigma}^{-1}(\mathbf{y}-\boldsymbol{\mu}) \text{ is } \chi^2(n). \tag{5.35}$$

To show this, we write $(\mathbf{y}-\boldsymbol{\mu})'\boldsymbol{\Sigma}^{-1}(\mathbf{y}-\boldsymbol{\mu})$ in the form

$$\begin{aligned}(\mathbf{y}-\boldsymbol{\mu})'\boldsymbol{\Sigma}^{-1}(\mathbf{y}-\boldsymbol{\mu}) &= (\mathbf{y}-\boldsymbol{\mu})'\boldsymbol{\Sigma}^{-1/2}\boldsymbol{\Sigma}^{-1/2}(\mathbf{y}-\boldsymbol{\mu}) \\ &= \left[\boldsymbol{\Sigma}^{-1/2}(\mathbf{y}-\boldsymbol{\mu})\right]'\left[\boldsymbol{\Sigma}^{-1/2}(\mathbf{y}-\boldsymbol{\mu})\right] \\ &= \mathbf{z}'\mathbf{z},\end{aligned}$$

where $\mathbf{z} = \boldsymbol{\Sigma}^{-1/2}(\mathbf{y}-\boldsymbol{\mu})$ and $\boldsymbol{\Sigma}^{-1/2} = (\boldsymbol{\Sigma}^{1/2})^{-1}$, with $\boldsymbol{\Sigma}^{1/2}$ given by (2.109). The vector $\mathbf{z}$ is distributed as $N_n(\mathbf{0}, \mathbf{I})$ (see Problem 5.17); therefore, $\mathbf{z}'\mathbf{z}$ is $\chi^2(n)$ by definition [see (5.18)]. Note the analogy of $(\mathbf{y}-\boldsymbol{\mu})'\boldsymbol{\Sigma}^{-1}(\mathbf{y}-\boldsymbol{\mu})$ to the univariate random variable $(y-\mu)^2/\sigma^2$, which is distributed as $\chi^2(1)$ if y is $N(\mu, \sigma^2)$.

In the following theorem, we consider the distribution of quadratic forms in general. In the proof we follow Searle (1971, p. 57). For alternative proofs, see Graybill (1976, pp. 134–136) and Hocking (1996, p. 51).

Theorem 5.5. Let $\mathbf{y}$ be distributed as $N_p(\boldsymbol{\mu}, \boldsymbol{\Sigma})$, let $\mathbf{A}$ be a symmetric matrix of constants of rank r, and let $\lambda = \frac{1}{2}\boldsymbol{\mu}'\mathbf{A}\boldsymbol{\mu}$. Then $\mathbf{y}'\mathbf{A}\mathbf{y}$ is $\chi^2(r, \lambda)$, if and only if $\mathbf{A}\boldsymbol{\Sigma}$ is idempotent.

Proof. By Theorem 5.2b the moment generating function of $\mathbf{y}'\mathbf{A}\mathbf{y}$ is

$$M_{\mathbf{y}'\mathbf{A}\mathbf{y}}(t) = |\mathbf{I} - 2t\mathbf{A}\boldsymbol{\Sigma}|^{-1/2} e^{-(1/2)\boldsymbol{\mu}'[\mathbf{I}-(\mathbf{I}-2t\mathbf{A}\boldsymbol{\Sigma})^{-1}]\boldsymbol{\Sigma}^{-1}\boldsymbol{\mu}}.$$

By (2.98), the eigenvalues of $\mathbf{I} - 2t\mathbf{A}\boldsymbol{\Sigma}$ are $1 - 2t\lambda_i$, $i = 1, 2, \ldots, p$, where λ_i is an eigenvalue of $\mathbf{A}\boldsymbol{\Sigma}$. By (2.107), $|\mathbf{I} - 2t\mathbf{A}\boldsymbol{\Sigma}| = \prod_{i=1}^{p}(1 - 2t\lambda_i)$. By (2.102), $(\mathbf{I} - 2t\mathbf{A}\boldsymbol{\Sigma})^{-1} = \mathbf{I} + \sum_{k=1}^{\infty}(2t)^k(\mathbf{A}\boldsymbol{\Sigma})^k$, provided $-1 < 2t\lambda_i < 1$ for all i. Thus $M_{\mathbf{y}'\mathbf{A}\mathbf{y}}(t)$ can be written as

$$M_{\mathbf{y}'\mathbf{A}\mathbf{y}}(t) = \left(\prod_{i=1}^{p}(1 - 2t\lambda_i)^{-1/2}\right) e^{-(1/2)\boldsymbol{\mu}'\left[-\sum_{k=1}^{\infty}(2t)^k(\mathbf{A}\boldsymbol{\Sigma})^k\right]\boldsymbol{\Sigma}^{-1}\boldsymbol{\mu}}.$$

Suppose that $\mathbf{A\Sigma}$ is idempotent of rank r (the rank of $\mathbf{A}$); then r of the λ_i's are equal to 1, $p - r$ of the λ_i's are equal to 0, and $(\mathbf{A\Sigma})^k = \mathbf{A\Sigma}$. Therefore,

$$M_{\mathbf{y}'\mathbf{Ay}}(t) = \left(\prod_{i=1}^{r} (1 - 2t)^{-1/2} \right) e^{-(1/2)\boldsymbol{\mu}'\left[-\sum_{k=1}^{\infty} (2t)^k \right] \mathbf{A\Sigma\Sigma}^{-1}\boldsymbol{\mu}}$$

$$= (1 - 2t)^{-r/2} e^{-1/2\boldsymbol{\mu}'[1-(1-2t)-1]\mathbf{A}\boldsymbol{\mu}},$$

provided $-1 < 2t < 1$ or $-\frac{1}{2} < t < \frac{1}{2}$, which is compatible with the requirement that the moment generating function exists for t in a neighborhood of 0. Thus

$$M_{\mathbf{y}'\mathbf{Ay}}(t) = \frac{1}{(1 - 2t)^{r/2}} e^{-(1/2)\boldsymbol{\mu}'\mathbf{A}\boldsymbol{\mu}[1-1/(1-2t)]},$$

which by (5.25) is the moment generating function of a noncentral chi-square random variable with degrees of freedom $r = \text{rank}(\mathbf{A})$ and noncentrality parameter $\lambda = \frac{1}{2}\boldsymbol{\mu}'\mathbf{A}\boldsymbol{\mu}$.

For a proof of the converse, namely, if $\mathbf{y}'\mathbf{Ay}$ is $\chi^2(r, \lambda)$, then $\mathbf{A\Sigma}$ is idempotent; see Driscoll (1999). □

Some corollaries of interest are the following (for additional corollaries, see Problem 5.20).

Corollary 1. If $\mathbf{y}$ is $N_p(\mathbf{0}, \mathbf{I})$, then $\mathbf{y}'\mathbf{Ay}$ is $\chi^2(r)$ if and only if $\mathbf{A}$ is idempotent of rank r. □

Corollary 2. If $\mathbf{y}$ is $N_p(\boldsymbol{\mu}, \sigma^2\mathbf{I})$, then $\mathbf{y}'\mathbf{Ay}/\sigma^2$ is $\chi^2(r, \boldsymbol{\mu}'\mathbf{A}\boldsymbol{\mu}/2\sigma^2)$ if and only if $\mathbf{A}$ is idempotent of rank r. □

Example 5. To illustrate Corollary 2 to Theorem 5.5, consider the distribution of $(n-1)s^2/\sigma^2 = \sum_{i=1}^{n} (y_i - \bar{y})^2/\sigma^2$, where $\mathbf{y} = (y_1, y_2, \ldots, y_n)'$ is distributed as $N_n(\mu\mathbf{j}, \sigma^2\mathbf{I})$ as in Examples 5.1 and 5.2 In (5.2) we have $\sum_{i=1}^{n} (y_i - \bar{y})^2 = \mathbf{y}'[\mathbf{I} - (1/n)\mathbf{J}]\mathbf{y}$. The matrix $\mathbf{I} - (1/n)\mathbf{J}$ is shown to be idempotent in Problem 5.2. Then by Theorem 2.13d, rank $[\mathbf{I} - (1/n)\mathbf{J}] = \text{tr}[\mathbf{I} - (1/n)\mathbf{J}] = n - 1$. We next find λ, which is given by

$$\lambda = \frac{\boldsymbol{\mu}'\mathbf{A}\boldsymbol{\mu}}{2\sigma^2} = \frac{\mu\mathbf{j}'(\mathbf{I} - \frac{1}{n}\mathbf{J})\mu\mathbf{j}}{2\sigma^2} = \frac{\mu^2(\mathbf{j}'\mathbf{j} - \frac{1}{n}\mathbf{j}'\mathbf{J}\mathbf{j})}{2\sigma^2}$$

$$= \frac{\mu^2(n - \frac{1}{n}\mathbf{j}'\mathbf{j}\mathbf{j}'\mathbf{j})}{2\sigma^2} = \frac{\mu^2[n - \frac{1}{n}(n)(n)]}{2\sigma^2} = 0.$$

Therefore, $\mathbf{y}'[\mathbf{I} - (1/n)\mathbf{J}]\mathbf{y}/\sigma^2$ is $\chi^2(n-1)$. □

5.6 INDEPENDENCE OF LINEAR FORMS AND QUADRATIC FORMS

In this section, we discuss the independence of (1) a linear form and a quadratic form, (2) two quadratic forms, and (3) several quadratic forms.

For an example of (1), consider $\bar{y}$ and s^2 in a simple random sample or $\hat{\beta}$ and s^2 in a regression setting. To illustrate (2), consider the sum of squares due to regression and the sum of squares due to error. An example of (3) is given by the sums of squares due to main effects and interaction in a balanced two-way analysis of variance.

We begin with the independence of a linear form and a quadratic form.

Theorem 5.6a. Suppose that $\mathbf{B}$ is a $k \times p$ matrix of constants, $\mathbf{A}$ is a $p \times p$ symmetric matrix of constants, and $\mathbf{y}$ is distributed as $N_p(\boldsymbol{\mu}, \boldsymbol{\Sigma})$. Then $\mathbf{By}$ and $\mathbf{y}'\mathbf{Ay}$ are independent if and only if $\mathbf{B\Sigma A} = \mathbf{O}$.

PROOF. Suppose $\mathbf{B\Sigma A} = \mathbf{O}$. We prove that $\mathbf{By}$ and $\mathbf{y}'\mathbf{Ay}$ are independent for the special case in which $\mathbf{A}$ is symmetric and idempotent. For a general proof, see Searle (1971, p. 59).

Assuming that $\mathbf{A}$ is symmetric and idempotent, $\mathbf{y}'\mathbf{Ay}$ can be written as

$$\mathbf{y}'\mathbf{Ay} = \mathbf{y}'\mathbf{A}'\mathbf{Ay} = (\mathbf{Ay})'\mathbf{Ay}.$$

If $\mathbf{B\Sigma A} = \mathbf{O}$, we have by (3.45)

$$\mathbf{B\Sigma A} = \mathrm{cov}(\mathbf{By}, \mathbf{Ay}) = \mathbf{O}.$$

Hence, by Corollary 2 to Theorem 4.4c, $\mathbf{By}$ and $\mathbf{Ay}$ are independent, and therefore $\mathbf{By}$ and the function $(\mathbf{Ay})'\mathbf{Ay}$ are also independent (Seber 1977, pp. 17, 33–34).

We now establish the converse, namely, if $\mathbf{By}$ and $\mathbf{y}'\mathbf{Ay}$ are independent, then $\mathbf{B\Sigma A} = \mathbf{O}$. By Corollary 1 to Theorem 5.2d, $\mathrm{cov}(\mathbf{By}, \mathbf{y}'\mathbf{Ay}) = \mathbf{0}$ becomes

$$2\mathbf{B\Sigma A}\boldsymbol{\mu} = \mathbf{0}.$$

Since this holds for all possible $\boldsymbol{\mu}$, we have $\mathbf{B\Sigma A} = \mathbf{O}$ [see (2.44)]. □

Note that $\mathbf{B\Sigma A} = \mathbf{O}$ does not imply $\mathbf{A\Sigma B} = \mathbf{O}$. In fact, the product $\mathbf{A\Sigma B}$ will not be defined unless $\mathbf{B}$ has p rows.

Corollary 1. If $\mathbf{y}$ is $N_p(\boldsymbol{\mu}, \sigma^2\mathbf{I})$, then $\mathbf{By}$ and $\mathbf{y}'\mathbf{Ay}$ are independent if and only if $\mathbf{BA} = \mathbf{O}$. □

Example 5.6a. To illustrate Corollary 1, consider $s^2 = \sum_{i=1}^{n} (y_i - \bar{y})^2/(n-1)$ and $\bar{y} = \sum_{i=1}^{n} y_i/n$, where $\mathbf{y} = (y_1, y_2, \ldots, y_n)'$ is $N_n(\mu\mathbf{j}, \sigma^2\mathbf{I})$. As in Example 5.1, $\bar{y}$ and s^2 can be written as $\bar{y} = (1/n)\mathbf{j}'\mathbf{y}$ and $s^2 = \mathbf{y}'[\mathbf{I} - (1/n)\mathbf{J}]\mathbf{y}/(n-1)$. By Corollary 1, $\bar{y}$ is independent of s^2 since $(1/n)\mathbf{j}'[\mathbf{I} - (1/n)\mathbf{J}] = \mathbf{0}'$. □

We now consider the independence of two quadratic forms.

Theorem 5.6b. Let $\mathbf{A}$ and $\mathbf{B}$ be symmetric matrices of constants. If $\mathbf{y}$ is $N_p(\boldsymbol{\mu}, \boldsymbol{\Sigma})$, then $\mathbf{y}'\mathbf{A}\mathbf{y}$ and $\mathbf{y}'\mathbf{B}\mathbf{y}$ are independent if and only if $\mathbf{A}\boldsymbol{\Sigma}\mathbf{B} = \mathbf{O}$.

PROOF. Suppose $\mathbf{A}\boldsymbol{\Sigma}\mathbf{B} = \mathbf{O}$. We prove that $\mathbf{y}'\mathbf{A}\mathbf{y}$ and $\mathbf{y}'\mathbf{B}\mathbf{y}$ are independent for the special case in which $\mathbf{A}$ and $\mathbf{B}$ are symmetric and idempotent. For a general proof, see Searle (1971, pp. 59–60) or Hocking (1996, p. 52).

Assuming that $\mathbf{A}$ and $\mathbf{B}$ are symmetric and idempotent, $\mathbf{y}'\mathbf{A}\mathbf{y}$ and $\mathbf{y}'\mathbf{B}\mathbf{y}$ can be written as $\mathbf{y}'\mathbf{A}\mathbf{y} = \mathbf{y}'\mathbf{A}'\mathbf{A}\mathbf{y} = (\mathbf{A}\mathbf{y})'\mathbf{A}\mathbf{y}$ and $\mathbf{y}'\mathbf{B}\mathbf{y} = \mathbf{y}'\mathbf{B}'\mathbf{B}\mathbf{y} = (\mathbf{B}\mathbf{y})'\mathbf{B}\mathbf{y}$. If $\mathbf{A}\boldsymbol{\Sigma}\mathbf{B} = \mathbf{O}$, we have [see (3.45)]

$$\mathbf{A}\boldsymbol{\Sigma}\mathbf{B} = \mathrm{cov}(\mathbf{A}\mathbf{y}, \mathbf{B}\mathbf{y}) = \mathbf{O}.$$

Hence, by Corollary 2 to Theorem 4.4c, $\mathbf{A}\mathbf{y}$ and $\mathbf{B}\mathbf{y}$ are independent. It follows that the functions $(\mathbf{A}\mathbf{y})'(\mathbf{A}\mathbf{y}) = \mathbf{y}'\mathbf{A}\mathbf{y}$ and $(\mathbf{B}\mathbf{y})'(\mathbf{B}\mathbf{y}) = \mathbf{y}'\mathbf{B}\mathbf{y}$ are independent (Seber 1977, pp. 17, 33–34). □

Note that $\mathbf{A}\boldsymbol{\Sigma}\mathbf{B} = \mathbf{O}$ is equivalent to $\mathbf{B}\boldsymbol{\Sigma}\mathbf{A} = \mathbf{O}$ since transposing both sides of $\mathbf{A}\boldsymbol{\Sigma}\mathbf{B} = \mathbf{O}$ gives $\mathbf{B}\boldsymbol{\Sigma}\mathbf{A} = \mathbf{O}$ ($\mathbf{A}$ and $\mathbf{B}$ are symmetric).

Corollary 1. If $\mathbf{y}$ is $N_p(\boldsymbol{\mu}, \sigma^2\mathbf{I})$, then $\mathbf{y}'\mathbf{A}\mathbf{y}$ and $\mathbf{y}'\mathbf{B}\mathbf{y}$ are independent if and only if $\mathbf{A}\mathbf{B} = \mathbf{O}$ (or, equivalently, $\mathbf{B}\mathbf{A} = \mathbf{O}$). □

Example 5.6b. To illustrate Corollary 1, consider the partitioning in (5.1), $\sum_{i=1}^{n} y_i^2 = \sum_{i=1}^{n} (y_i - \bar{y})^2 + n\bar{y}^2$, which was expressed in (5.3) as

$$\mathbf{y}'\mathbf{y} = \mathbf{y}'(\mathbf{I} - (1/n)\mathbf{J})\mathbf{y} + \mathbf{y}'((1/n)\mathbf{J})\mathbf{y}.$$

If $\mathbf{y}$ is $N_n(\mu\mathbf{j}, \sigma^2\mathbf{I})$, then by Corollary 1, $\mathbf{y}'[\mathbf{I} - (1/n)\mathbf{J}]\mathbf{y}$ and $\mathbf{y}'[(1/n)\mathbf{J}]\mathbf{y}$ are independent if and only if $[\mathbf{I} - (1/n)\mathbf{J}][(1/n)\mathbf{J}] = \mathbf{O}$, which is shown in Problem 5.2. □

The distribution and independence of several quadratic forms are considered in the following theorem.

Theorem 5.6c. Let $\mathbf{y}$ be $N_n(\boldsymbol{\mu}, \sigma^2\mathbf{I})$, let $\mathbf{A}_i$ be symmetric of rank r_i for $i = 1, 2, \ldots, k$, and let $\mathbf{y}'\mathbf{A}\mathbf{y} = \sum_{i=1}^k \mathbf{y}'\mathbf{A}_i\mathbf{y}$, where $\mathbf{A} = \sum_{i=1}^k \mathbf{A}_i$ is symmetric of rank r. Then

(i) $\mathbf{y}'\mathbf{A}_i\mathbf{y}/\sigma^2$ is $\chi^2(r_i, \boldsymbol{\mu}'\mathbf{A}_i\boldsymbol{\mu}/2\sigma^2)$, $i = 1, 2, \ldots, k$.
(ii) $\mathbf{y}'\mathbf{A}_i\mathbf{y}$ and $\mathbf{y}'\mathbf{A}_j\mathbf{y}$ are independent for all $i \neq j$.
(iii) $\mathbf{y}'\mathbf{A}\mathbf{y}/\sigma^2$ is $\chi^2(r, \boldsymbol{\mu}'\mathbf{A}\boldsymbol{\mu}/2\sigma^2)$.

These results are obtained if and only if any two of the following three statements are true:

(a) Each $\mathbf{A}_i$ is idempotent.
(b) $\mathbf{A}_i\mathbf{A}_j = \mathbf{O}$ for all $i \neq j$.
(c) $\mathbf{A} = \sum_{i=1}^k \mathbf{A}_i$ is idempotent.

Or if and only if (c) and (d) are true, where (d) is the following statement:

(d) $r = \sum_{i=1}^k r_i$.

PROOF. See Searle (1971, pp. 61–64). □

Note that by Theorem 2.13g, any two of (a), (b), or (c) implies the third.

Theorem 5.6c pertains to partitioning a sum of squares into several component sums of squares. The following corollary treats the special case where $\mathbf{A} = \mathbf{I}$; that is, the case of partitioning the total sum of squares $\mathbf{y}'\mathbf{y}$ into several sums of squares.

Corollary 1. Let $\mathbf{y}$ be $N_n(\boldsymbol{\mu}, \sigma^2\mathbf{I})$, let $\mathbf{A}_i$ be symmetric of rank r_i for $i = 1, 2, \ldots, k$, and let $\mathbf{y}'\mathbf{y} = \sum_{i=1}^k \mathbf{y}'\mathbf{A}_i\mathbf{y}$. Then (i) each $\mathbf{y}'\mathbf{A}_i\mathbf{y}/\sigma^2$ is $\chi^2(r_i, \boldsymbol{\mu}'\mathbf{A}_i\boldsymbol{\mu}/2\sigma^2)$ and (ii) the $\mathbf{y}'\mathbf{A}_i\mathbf{y}$ terms are mutually independent if and only if any *one* of the following statements holds:

(a) Each $\mathbf{A}_i$ is idempotent.
(b) $\mathbf{A}_i\mathbf{A}_j = \mathbf{O}$ for all $i \neq j$.
(c) $n = \sum_{i=1}^k r_i$. □

Note that by Theorem 2.13h, condition (c) implies the other two conditions. Cochran (1934) first proved a version of Corollary 1 to Theorem 5.6c.

PROBLEMS

5.1 Show that $\sum_{i=1}^n (y_i - \bar{y})^2 = \sum_{i=1}^n y_i^2 - n\bar{y}^2$ as in (5.1).

5.2 Show that $(1/n)\mathbf{J}$ is idempotent, $\mathbf{I}-(1/n)\mathbf{J}$ is idempotent, and $[\mathbf{I}-(1/n)\mathbf{J}][(1/n)\mathbf{J}]=\mathbf{O}$, as noted in Section 5.1.

5.3 Obtain $\text{var}(s^2)$ in the following two ways, where s^2 is defined in (5.6) as $s^2=\sum_{i=1}^n(y_i-\bar{y})^2/(n-1)$ and we assume that $\mathbf{y}=(y_1, y_2, \ldots, y_n)'$ is $N_n(\mu\mathbf{j}, \sigma^2\mathbf{I})$.

(a) Write s^2 as $s^2=\mathbf{y}'[\mathbf{I}-(1/n)\mathbf{J}]\mathbf{y}/(n-1)$ and use Theorem 5.2b.

(b) The function $u=(n-1)s^2/\sigma^2$ is distributed as $\chi^2(n-1)$, and therefore $\text{var}(u)=2(n-1)$. Then $\text{var}(s^2)=\text{var}[\sigma^2u/(n-1)]$.

5.4 Show that

$$|\boldsymbol{\Sigma}|^{-(1/2)}|\mathbf{V}|^{(1/2)}\mathbf{e}^{-(\boldsymbol{\mu}'\boldsymbol{\Sigma}^{-1}\boldsymbol{\mu}-\boldsymbol{\theta}'\mathbf{V}^{-1}\boldsymbol{\theta})/2}$$
$$=|\mathbf{I}-2t\mathbf{A}\boldsymbol{\Sigma}|e^{-(1/2)\boldsymbol{\mu}'}[\mathbf{I}-(\mathbf{I}-2t\mathbf{A}\boldsymbol{\Sigma})^{-1}]\boldsymbol{\Sigma}^{-1}\boldsymbol{\mu}/2$$

as in the proof of Theorem 5.2b, where $\boldsymbol{\theta}'=\boldsymbol{\mu}'(\mathbf{I}-2t\mathbf{A}\boldsymbol{\Sigma})^{-1}$ and $\mathbf{V}^{-1}=(\mathbf{I}-2t\mathbf{A}\boldsymbol{\Sigma})\boldsymbol{\Sigma}^{-1}$.

5.5 Show that

$$e^{-[\mathbf{y}'(\mathbf{I}-2t\mathbf{A}\boldsymbol{\Sigma})\boldsymbol{\Sigma}^{-1}\mathbf{y}-2\boldsymbol{\mu}'\boldsymbol{\Sigma}^{-1}\mathbf{y}+\boldsymbol{\mu}'\boldsymbol{\Sigma}^{-1}\boldsymbol{\mu}]/2}=e^{-[\boldsymbol{\mu}'\boldsymbol{\Sigma}^{-1}\boldsymbol{\mu}-\boldsymbol{\theta}'\mathbf{V}^{-1}\boldsymbol{\theta}]/2}e^{-(\mathbf{y}-\boldsymbol{\theta})'\mathbf{V}^{-1}(\mathbf{y}-\boldsymbol{\theta})/2}$$

as in the proof of Theorem 5.2b, where $\boldsymbol{\theta}'=\boldsymbol{\mu}'(\mathbf{I}-2t\mathbf{A}\boldsymbol{\Sigma})^{-1}$ and $\mathbf{V}^{-1}=(\mathbf{I}-2t\mathbf{A}\boldsymbol{\Sigma})\boldsymbol{\Sigma}^{-1}$.

5.6 Let $k(t)=-\frac{1}{2}\ln|\mathbf{C}|-\frac{1}{2}\boldsymbol{\mu}'(\mathbf{I}-\mathbf{C}^{-1})\boldsymbol{\Sigma}^{-1}\boldsymbol{\mu}$ as in the proof of Theorem 5.2c, where $\mathbf{C}$ is a nonsingular matrix. Derive $k''(t)$.

5.7 Show that $\mathbf{y}'\mathbf{A}\mathbf{y}-\boldsymbol{\mu}'\mathbf{A}\boldsymbol{\mu}=(\mathbf{y}-\boldsymbol{\mu})'\mathbf{A}(\mathbf{y}-\boldsymbol{\mu})+2(\mathbf{y}-\boldsymbol{\mu})'\mathbf{A}\boldsymbol{\mu}$ as in (5.12).

5.8 Obtain the three terms $\mathbf{0}, 2\boldsymbol{\Sigma}\mathbf{A}\boldsymbol{\mu}$, and $\mathbf{0}$ in the proof of Theorem 5.2d.

5.9 Prove Corollary 1 to Theorem 5.2d.

5.10 Prove Theorem 5.2e.

5.11 **(a)** Show that $\sum_{i=1}^n(x_i-\bar{x})(y_i-\bar{y})$ in (5.15) is equal to $\sum_{i=1}^n x_iy_i-n\bar{x}\bar{y}$ in (5.16).

(b) Show that $\sum_{i=1}^n x_iy_i-n\bar{x}\bar{y}=\mathbf{x}'[\mathbf{I}-(1/n)\mathbf{J}]\mathbf{y}$, as in (5.16) in Example 5.2.

5.12 Prove Theorem 5.3a.

5.13 If $v=\chi^2(n, \lambda)$, use Theorem 5.2c to show that $\text{var}(v)=2n+8\lambda$ as in (5.24).

5.14 If v is $\chi^2(n, \lambda)$, use the moment generating function in (5.25) to find $\text{E}(v)$ and $\text{var}(v)$. [*Hint*: Use $\ln[M_v(t)]$; then $d\ln[M_v(0)]/dt=E(v)$ and $d^2\ln[M_v(0)]/dt=\text{var}(v)$ (see Problem 4.8). The notation $d\ln[M_v(0)]/dt$

indicates that $d \ln[M_y(t)]/dt$ is evaluated at $t = 0$; the notation $d^2 \ln[M_y(0)]/dt^2$ is defined similarly.]

5.15 Prove Theorem 5.3c.

5.16 **(a)** Show that if $t = z/\sqrt{u/p}$ is $t(p)$ as in (5.33), then t^2 is $F(1, p)$.
(b) Show that if $t = y/\sqrt{u/p}$ is $t(p, \mu)$ as in (5.34), then t^2 is $F(1, p, \frac{1}{2}\mu^2)$.

5.17 Show that $\mathbf{\Sigma}^{-1/2}(\mathbf{y} - \boldsymbol{\mu})$ is $N_n(\mathbf{0}, \mathbf{I})$, as used in the illustration at the beginning of Section 5.5.

5.18 **(a)** Prove Corollary 1 of Theorem 5.5a.
(b) Prove Corollary 2 of Theorem 5.5a.

5.19 If $\mathbf{y}$ is $N_n(\boldsymbol{\mu}, \mathbf{\Sigma})$, verify that $(\mathbf{y} - \boldsymbol{\mu})'\mathbf{\Sigma}^{-1}(\mathbf{y} - \boldsymbol{\mu})$ is $\chi^2(n)$, as in (5.25), by using Theorem 5.5a. What is the distribution of $\mathbf{y}'\mathbf{\Sigma}^{-1}\mathbf{y}$?

5.20 Prove the following additional corollaries to Theorem 5.5a:

(a) If $\mathbf{y}$ is $N_p(\mathbf{0}, \mathbf{\Sigma})$, then $\mathbf{y}'\mathbf{A}\mathbf{y}$ is $\chi^2(r)$ if and only if $\mathbf{A\Sigma}$ is idempotent of rank r.
(b) If $\mathbf{y}$ is $N_p(\boldsymbol{\mu}, \sigma^2\mathbf{I})$, then $\mathbf{y}'\mathbf{y}/\sigma^2$ is $\chi^2(p, \boldsymbol{\mu}'\boldsymbol{\mu}/2\sigma^2)$.
(c) If $\mathbf{y}$ is $N_p(\boldsymbol{\mu}, \mathbf{I})$, then $\mathbf{y}'\mathbf{A}\mathbf{y}$ is $\chi^2(r, \frac{1}{2}\boldsymbol{\mu}'\mathbf{A}\boldsymbol{\mu})$ if and only if $\mathbf{A}$ is idempotent of rank r.
(d) If $\mathbf{y}$ is $N_p(\boldsymbol{\mu}, \sigma^2\mathbf{\Sigma})$, then $\mathbf{y}'\mathbf{A}\mathbf{y}/\sigma^2$ is $\chi^2(r, \boldsymbol{\mu}'\mathbf{A}\boldsymbol{\mu}/2\sigma^2)$ if and only if $\mathbf{A\Sigma}$ is idempotent of rank r.
(e) If $\mathbf{y}$ is $N_p(\boldsymbol{\mu}, \sigma^2\mathbf{\Sigma})$, then $\mathbf{y}'\mathbf{\Sigma}^{-1}\mathbf{y}/\sigma^2$ is $\chi^2(p, \boldsymbol{\mu}'\mathbf{\Sigma}^{-1}\boldsymbol{\mu}/2\sigma^2)$.

5.21 Prove Corollary 1 of Theorem 5.6a.

5.22 Show that $\mathbf{j}'[\mathbf{I} - (1/n)\mathbf{J}] = \mathbf{0}'$, as in Example 5.6a.

5.23 Prove Corollary 1 of Theorem 5.6b.

5.24 Suppose that $y_1, y_2, \ldots, y_n$ is a random sample from $N(\mu, \sigma^2)$ so that $\mathbf{y} = (y_1, y_2, \ldots, y_n)'$ is $N_n(\mu\mathbf{j}, \sigma^2\mathbf{I})$. It was shown in Example 5.5 that $(n-1)s^2/\sigma^2 = \sum_{i=1}^n (y_i - \bar{y})^2/\sigma^2$ is $\chi^2(n-1)$. In Example 5.6a, it was demonstrated that $\bar{y}$ and $s^2 = \sum_{i=1}^n (y_i - \bar{y})^2/(n-1)$ are independent.

(a) Show that $\bar{y}$ is $N(\mu, \sigma^2/n)$.
(b) Show that $t = (\bar{y} - \mu)/(s/\sqrt{n})$ is distributed as $t\ (n-1)$.
(c) Given $\mu_0 \neq \mu$, show that $t = (\bar{y} - \mu_0)/(s/\sqrt{n})$ is distributed as $t(n-1, \delta)$. Find δ.

5.25 Suppose that $\mathbf{y}$ is $N_n(\mu\mathbf{j}, \sigma^2\mathbf{I})$. Find the distribution of

$$u = \frac{n\bar{y}^2}{\sum_{i=1}^n (y_i - \bar{y})^2/(n-1)}.$$

(This statistic could be used to test H_0: $\mu = 0$.)

5.26 Suppose that $\mathbf{y}$ is $N_n(\boldsymbol{\mu}, \boldsymbol{\Sigma})$, where $\boldsymbol{\mu} = \mu\mathbf{j}$ and

$$\boldsymbol{\Sigma} = \sigma^2 \begin{pmatrix} 1 & \rho & \cdots & \rho \\ \rho & 1 & \cdots & \rho \\ \vdots & \vdots & & \vdots \\ \rho & \rho & \cdots & 1 \end{pmatrix}.$$

Thus $E(y_i) = \mu$ for all i, $\text{var}(y_i) = \sigma^2$ for all i, and $\text{cov}(y_i, y_j) = \sigma^2\rho$ for all $i \neq j$; that is, the y's are equicorrelated.

(a) Show that $\boldsymbol{\Sigma}$ can be written in the form $\boldsymbol{\Sigma} = \sigma^2[(1-\rho)\mathbf{I} + \rho\mathbf{J}]$.
(b) Show that $\sum_{i=1}^n (y_i - \bar{y})^2/[\sigma^2(1-\rho)]$ is $\chi^2(n-1)$.

5.27 Suppose that $\mathbf{y}$ is $N_3(\boldsymbol{\mu}, \boldsymbol{\Sigma})$, where

$$\boldsymbol{\mu} = \begin{pmatrix} 2 \\ -1 \\ 3 \end{pmatrix}, \quad \boldsymbol{\Sigma} = \begin{pmatrix} 4 & 1 & 0 \\ 1 & 2 & 1 \\ 0 & 1 & 3 \end{pmatrix}.$$

Let

$$\mathbf{A} = \begin{pmatrix} 1 & -3 & -8 \\ -3 & 2 & -6 \\ -8 & -6 & 3 \end{pmatrix}.$$

(a) Find $E(\mathbf{y}'\mathbf{A}\mathbf{y})$.
(b) Find $\text{var}(\mathbf{y}'\mathbf{A}\mathbf{y})$.
(c) Does $\mathbf{y}'\mathbf{A}\mathbf{y}$ have a chi-square distribution?
(d) If $\boldsymbol{\Sigma} = \sigma^2\mathbf{I}$, does $\mathbf{y}'\mathbf{A}\mathbf{y}/\sigma^2$ have a chi-square distribution?

5.28 Assuming that $\mathbf{y}$ is $N_3(\boldsymbol{\mu}, \boldsymbol{\Sigma})$, where

$$\boldsymbol{\mu} = \begin{pmatrix} 3 \\ -2 \\ 1 \end{pmatrix}, \quad \boldsymbol{\Sigma} = \begin{pmatrix} 2 & 0 & 0 \\ 0 & 4 & 0 \\ 0 & 0 & 3 \end{pmatrix},$$

find a symmetric matrix $\mathbf{A}$ such that $\mathbf{y}'\mathbf{A}\mathbf{y}$ is $\chi^2(3, \frac{1}{2}\boldsymbol{\mu}'\mathbf{A}\boldsymbol{\mu})$. What is $\lambda = \frac{1}{2}\boldsymbol{\mu}'\mathbf{A}\boldsymbol{\mu}$?

5.29 Assuming that $\mathbf{y}$ is $N_4(\boldsymbol{\mu}, \boldsymbol{\Sigma})$, where

$$\boldsymbol{\mu} = \begin{pmatrix} 3 \\ -2 \\ 1 \\ 4 \end{pmatrix}, \quad \boldsymbol{\Sigma} = \begin{pmatrix} 1 & 0 & 0 & 0 \\ 0 & 2 & 0 & 0 \\ 0 & 0 & 3 & -4 \\ 0 & 0 & -4 & 6 \end{pmatrix},$$

find a matrix $\mathbf{A}$ such that $\mathbf{y}'\mathbf{A}\mathbf{y}$ is $\chi^2(4, \frac{1}{2}\boldsymbol{\mu}'\mathbf{A}\boldsymbol{\mu})$. What is $\lambda = \frac{1}{2}\boldsymbol{\mu}'\mathbf{A}\boldsymbol{\mu}$?

5.30 Suppose that $\mathbf{y}$ is $N_3(\boldsymbol{\mu}, \sigma^2\mathbf{I})$ and let

$$\boldsymbol{\mu} = \begin{pmatrix} 3 \\ -2 \\ 1 \end{pmatrix}, \quad \mathbf{A} = \frac{1}{3}\begin{pmatrix} 2 & -1 & -1 \\ -1 & 2 & -1 \\ -1 & -1 & 2 \end{pmatrix}, \quad \mathbf{B} = \begin{pmatrix} 1 & 1 & 1 \\ 1 & 0 & -1 \end{pmatrix}.$$

(a) What is the distribution of $\mathbf{y}'\mathbf{A}\mathbf{y}/\sigma^2$?

(b) Are $\mathbf{y}'\mathbf{A}\mathbf{y}$ and $\mathbf{B}\mathbf{y}$ independent?

(c) Are $\mathbf{y}'\mathbf{A}\mathbf{y}$ and $y_1 + y_2 + y_3$ independent?

5.31 Suppose that $\mathbf{y}$ is $N_3(\boldsymbol{\mu}, \sigma^2\mathbf{I})$, where $\boldsymbol{\mu} = (1, 2, 3)'$, and let

$$\mathbf{B} = \tfrac{1}{3}\begin{pmatrix} 1 & 1 & 1 \\ 1 & 1 & 1 \\ 1 & 1 & 1 \end{pmatrix}.$$

(a) What is the distribution of $\mathbf{y}'\mathbf{B}\mathbf{y}/\sigma^2$?

(b) Is $\mathbf{y}'\mathbf{B}\mathbf{y}$ independent of $\mathbf{y}'\mathbf{A}\mathbf{y}$, where $\mathbf{A}$ is as defined in Problem 5.30?

5.32 Suppose that $\mathbf{y}$ is $N_n(\boldsymbol{\mu}, \sigma^2\mathbf{I})$ and that $\mathbf{X}$ is an $n \times p$ matrix of constants with rank $p < n$.

(a) Show that $\mathbf{H} = \mathbf{X}(\mathbf{X}'\mathbf{X})^{-1}\mathbf{X}'$ and $\mathbf{I} - \mathbf{H} = \mathbf{I} - \mathbf{X}(\mathbf{X}'\mathbf{X})^{-1}\mathbf{X}'$ are idempotent, and find the rank of each.

(b) Assuming $\boldsymbol{\mu}$ is a linear combination of the columns of $\mathbf{X}$, that is $\boldsymbol{\mu} = \mathbf{X}\mathbf{b}$ for some $\mathbf{b}$ [see (2.37)], find $E(\mathbf{y}'\mathbf{H}\mathbf{y})$ and $E[\mathbf{y}'(\mathbf{I} - \mathbf{H})\mathbf{y}]$, where $\mathbf{H}$ is as defined in part (a) .

(c) Find the distributions of $\mathbf{y}'\mathbf{H}\mathbf{y}/\sigma^2$ and $\mathbf{y}'(\mathbf{I} - \mathbf{H})\mathbf{y}/\sigma^2$.

(d) Show that $\mathbf{y}'\mathbf{H}\mathbf{y}$ and $\mathbf{y}'(\mathbf{I} - \mathbf{H})\mathbf{y}$ are independent.

(e) Find the distribution of

$$\frac{\mathbf{y}'\mathbf{H}\mathbf{y}/p}{\mathbf{y}'(\mathbf{I} - \mathbf{H})\mathbf{y}/(n - p)}.$$

6 Simple Linear Regression

6.1 THE MODEL

By (1.1), the *simple linear regression* model for n observations can be written as

$$y_i = \beta_0 + \beta_1 x_i + \varepsilon_i, \quad i = 1, 2, \ldots, n. \tag{6.1}$$

The designation *simple* indicates that there is only one x to predict the response y, and *linear* means that the model (6.1) is linear in β_0 and β_1. [Actually, it is the assumption $E(y_i) = \beta_0 + \beta_1 x_i$ that is linear; see assumption 1 below.] For example, a model such as $y_i = \beta_0 + \beta_1 x_i^2 + \varepsilon_i$ is linear in β_0 and β_1, whereas the model $y_i = \beta_0 + e^{\beta_1 x_i} + \varepsilon_i$ is not linear.

In this chapter, we assume that y_i and ε_i are random variables and that the values of x_i are known constants, which means that the same values of $x_1, x_2, \ldots, x_n$ would be used in repeated sampling. The case in which the x variables are random variables is treated in Chapter 10.

To complete the model in (6.1), we make the following additional assumptions:

1. $E(\varepsilon_i) = 0$ for all $i = 1, 2, \ldots, n$, or, equivalently, $E(y_i) = \beta_0 + \beta_1 x_i$.
2. $\text{var}(\varepsilon_i) = \sigma^2$ for all $i = 1, 2, \ldots, n$, or, equivalently, $\text{var}(y_i) = \sigma^2$.
3. $\text{cov}(\varepsilon_i, \varepsilon_j) = 0$ for all $i \neq j$, or, equivalently, $\text{cov}(y_i, y_j) = 0$.

Assumption 1 states that the model (6.1) is correct, implying that y_i depends only on x_i and that all other variation in y_i is random. Assumption 2 asserts that the variance of ε or y does not depend on the values of x_i. (Assumption 2 is also known as the assumption of *homoscedasticity*, *homogeneous variance* or *constant variance*.) Under assumption 3, the ε variables (or the y variables) are uncorrelated with each other. In Section 6.3, we will add a normality assumption, and the y (or the ε) variables will thereby be independent as well as uncorrelated. Each assumption has been stated in terms of the ε's or the y's. For example, if $\text{var}(\varepsilon_i) = \sigma^2$, then $\text{var}(y_i) = E[y_i - E(y_i)]^2 = E(y_i - \beta_0 - \beta_1 x_i)^2 = E(\varepsilon_i^2) = \sigma^2$.

Linear Models in Statistics, Second Edition, by Alvin C. Rencher and G. Bruce Schaalje

Any of these assumptions may fail to hold with real data. A plot of the data will often reveal departures from assumptions 1 and 2 (and to a lesser extent assumption 3). Techniques for checking on the assumptions are discussed in Chapter 9.

6.2 ESTIMATION OF β_0, β_1, AND σ^2

Using a random sample of n observations $y_1, y_2, \ldots, y_n$ and the accompanying fixed values $x_1, x_2, \ldots, x_n$, we can estimate the parameters β_0, β_1, and σ^2. To obtain the estimates $\hat{\beta}_0$ and $\hat{\beta}_1$, we use the method of least squares, which does not require any distributional assumptions (for maximum likelihood estimators based on normality, see Section 7.6.2).

In the *least-squares* approach, we seek estimators $\hat{\beta}_0$ and $\hat{\beta}_1$ that minimize the sum of squares of the deviations $y_i - \hat{y}_i$ of the n observed y_i's from their predicted values $\hat{y}_i = \hat{\beta}_0 + \hat{\beta}_1 x_i$:

$$\hat{\boldsymbol{\varepsilon}}'\hat{\boldsymbol{\varepsilon}} = \sum_{i=1}^{n} \hat{\varepsilon}_i^2 = \sum_{i=1}^{n} (y_i - \hat{y}_i)^2 = \sum_{i=1}^{n} (y_i - \hat{\beta}_0 - \hat{\beta}_1 x_i)^2. \tag{6.2}$$

Note that the predicted value $\hat{y}_i$ estimates $E(y_i)$, not y_i; that is, $\hat{\beta}_0 + \hat{\beta}_1 x_i$ estimates $\beta_0 + \beta_1 x_i$, not $\beta_0 + \beta_1 x_i + \varepsilon_i$. A better notation would be $\widehat{E(y_i)}$, but $\hat{y}_i$ is commonly used.

To find the values of $\hat{\beta}_0$ and $\hat{\beta}_1$ that minimize $\hat{\boldsymbol{\varepsilon}}'\hat{\boldsymbol{\varepsilon}}$ in (6.2), we differentiate with respect to $\hat{\beta}_0$ and $\hat{\beta}_1$ and set the results equal to 0:

$$\frac{\partial \hat{\boldsymbol{\varepsilon}}'\hat{\boldsymbol{\varepsilon}}}{\partial \hat{\beta}_0} = -2\sum_{i=1}^{n} (y_i - \hat{\beta}_0 - \hat{\beta}_1 x_i) = 0, \tag{6.3}$$

$$\frac{\partial \hat{\boldsymbol{\varepsilon}}'\hat{\boldsymbol{\varepsilon}}}{\partial \hat{\beta}_1} = -2\sum_{i=1}^{n} (y_i - \hat{\beta}_0 - \hat{\beta}_1 x_i)x_i = 0. \tag{6.4}$$

The solution to (6.3) and (6.4) is given by

$$\hat{\beta}_1 = \frac{\sum_{i=1}^{n} x_i y_i - n\bar{x}\bar{y}}{\sum_{i=1}^{n} x_i^2 - n\bar{x}^2} = \frac{\sum_{i=1}^{n} (x_i - \bar{x})(y_i - \bar{y})}{\sum_{i=1}^{n} (x_i - \bar{x})^2}, \tag{6.5}$$

$$\hat{\beta}_0 = \bar{y} - \hat{\beta}_1 \bar{x}. \tag{6.6}$$

To verify that $\hat{\beta}_0$ and $\hat{\beta}_1$ in (6.5) and (6.6) minimize $\hat{\boldsymbol{\varepsilon}}'\hat{\boldsymbol{\varepsilon}}$ in (6.2), we can examine the second derivatives or simply observe that $\hat{\boldsymbol{\varepsilon}}'\hat{\boldsymbol{\varepsilon}}$ has no maximum and therefore the first derivatives yield a minimum. For an algebraic proof that $\hat{\beta}_0$ and $\hat{\beta}_1$ minimize (6.2), see (7.10) in Section 7.3.1.

Example 6.2. Students in a statistics class (taught by one of the authors) claimed that doing the homework had not helped prepare them for the midterm exam. The exam score y and homework score x (averaged up to the time of the midterm) for the 18 students in the class were as follows:

y	x	y	x	y	x
95	96	72	89	35	0
80	77	66	47	50	30
0	0	98	90	72	59
0	0	90	93	55	77
79	78	0	18	75	74
77	64	95	86	66	67

Using (6.5) and (6.6), we obtain

$$\begin{aligned}\hat{\beta}_1 &= \frac{\sum_{i=1}^n x_i y_i - n\bar{x}\bar{y}}{\sum_{i=1}^n x_i^2 - n\bar{x}^2} \\ &= \frac{81{,}195 - 18(58.056)(61.389)}{80{,}199 - 18(58.056)^2} = .8726, \\ \hat{\beta}_0 &= \bar{y} - \hat{\beta}_1\bar{x} = 61.389 - .8726(58.056) = 10.73.\end{aligned}$$

The prediction equation is thus given by

$$\hat{y} = 10.73 + .8726x.$$

This equation and the 18 points are plotted in Figure 6.1. It is readily apparent in the plot that the slope $\hat{\beta}_1$ is the rate of change of $\hat{y}$ as x varies and that the intercept $\hat{\beta}_0$ is the value of $\hat{y}$ at $x = 0$.

The apparent linear trend in Figure 6.1 does not establish cause and effect between homework and test results (for inferences that can be drawn, see Section 6.3). The assumption $\text{var}(\varepsilon_i) = \sigma^2$ (constant variance) for all $i = 1, 2, \ldots, 18$ appears to be reasonable. □

Note that the three assumptions in Section 6.1 were not used in deriving the least-squares estimators $\hat{\beta}_0$ and $\hat{\beta}_1$ in (6.5) and (6.6). It is not necessary that $\hat{y}_i = \hat{\beta}_0 + \hat{\beta}_1 x_i$ be based on $E(y_i) = \beta_0 + \beta_1 x_i$; that is, $\hat{y}_i = \hat{\beta}_0 + \hat{\beta}_1 x_i$ can be fit to a set of data for which $E(y_i) \neq \beta_0 + \beta_1 x_i$. This is illustrated in Figure 6.2, where a straight line has been fitted to curved data.

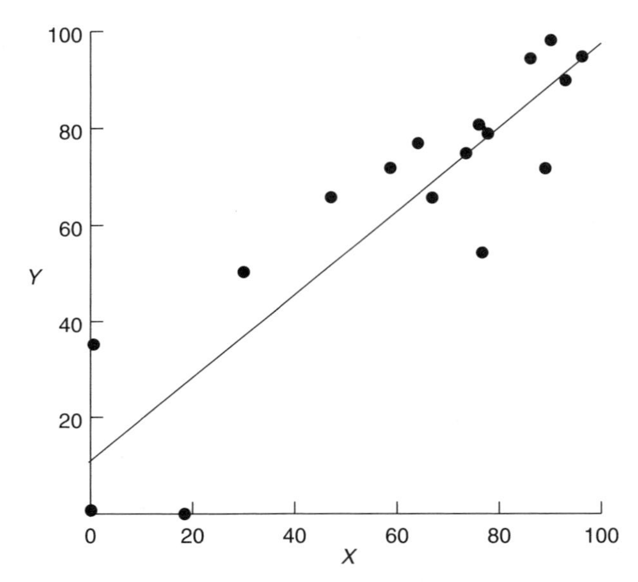

Figure 6.1 Regression line and data for homework and test scores.

However, if the three assumptions in Section 6.1 hold, then the least-squares estimators $\hat{\beta}_0$ and $\hat{\beta}_1$ are unbiased and have minimum variance among all linear unbiased estimators (for the minimum variance property, see Theorem 7.3d in Section 7.3.2; note that $\hat{\beta}_0$ and $\hat{\beta}_1$ are linear functions of $y_1, y_2, \ldots, y_n$). Using the three

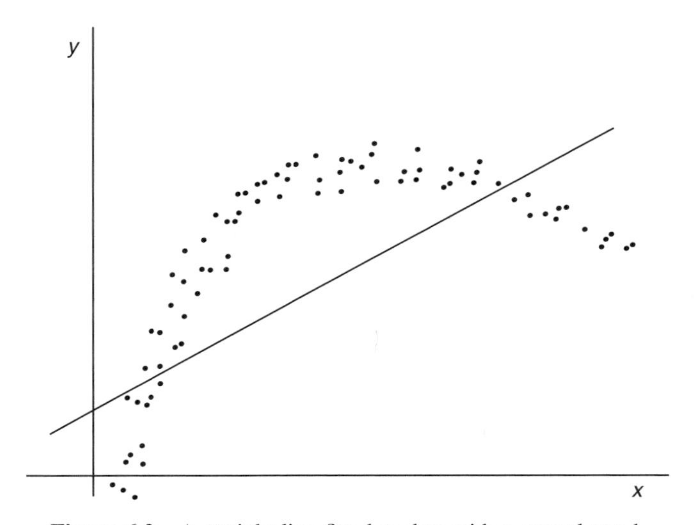

Figure 6.2 A straight line fitted to data with a curved trend.

assumptions, we obtain the following means and variances of $\hat{\beta}_0$ and $\hat{\beta}_1$:

$$E(\hat{\beta}_1) = \beta_1 \tag{6.7}$$

$$E(\hat{\beta}_0) = \beta_0 \tag{6.8}$$

$$\text{var}(\hat{\beta}_1) = \frac{\sigma^2}{\sum_{i=1}^{n}(x_i - \bar{x})^2} \tag{6.9}$$

$$\text{var}(\hat{\beta}_0) = \sigma^2\left[\frac{1}{n} + \frac{\bar{x}^2}{\sum_{i=1}^{n}(x_i - \bar{x})^2}\right]. \tag{6.10}$$

Note that in discussing $E(\hat{\beta}_1)$ and $\text{var}(\hat{\beta}_1)$, for example, we are considering random variation of $\hat{\beta}_1$ from sample to sample. It is assumed that the n values x_1, $x_2, \ldots, x_n$ would remain the same in future samples so that $\text{var}(\hat{\beta}_1)$ and $\text{var}(\hat{\beta}_0)$ are constant.

In (6.9), we see that $\text{var}(\hat{\beta}_1)$ is minimized when $\sum_{i=1}^{n}(x_i - \bar{x})^2$ is maximized. If the x_i values have the range $a \leq x_i \leq b$, then $\sum_{i=1}^{n}(x_i - \bar{x})^2$ is maximized if half the x's are selected equal to a and half equal to b (assuming that n is even; see Problem 6.4). In (6.10), it is clear that $\text{var}(\hat{\beta}_0)$ is minimized when $\bar{x} = 0$.

The method of least squares does not yield an estimator of $\text{var}(y_i) = \sigma^2$; minimization of $\hat{\boldsymbol{\varepsilon}}'\hat{\boldsymbol{\varepsilon}}$ yields only $\hat{\beta}_0$ and $\hat{\beta}_1$. To estimate σ^2, we use the definition in (3.6), $\sigma^2 = E[y_i - E(y_i)]^2$. By assumption 2 in Section 6.1, σ^2 is the same for each y_i, $i = 1, 2, \ldots, n$. Using $\hat{y}_i$ as an estimator of $E(y_i)$, we estimate σ^2 by an average from the sample, that is

$$s^2 = \frac{\sum_{i=1}^{n}(y_i - \hat{y}_i)^2}{n-2} = \frac{\sum_i (y_i - \hat{\beta}_0 - \hat{\beta}_1 x_i)^2}{n-2} = \frac{\text{SSE}}{n-2}, \tag{6.11}$$

where $\hat{\beta}_0$ and $\hat{\beta}_1$ are given by (6.5) and (6.6) and $\text{SSE} = \sum_i (y_i - \hat{y}_i)^2$. The deviation $\hat{\varepsilon}_i = y_i - \hat{y}_i$ is often called the *residual* of y_i, and SSE is called the *residual sum of squares* or *error sum of squares*. With $n-2$ in the denominator, s^2 is an unbiased estimator of σ^2:

$$E(s^2) = \frac{E(\text{SSE})}{n-2} = \frac{(n-2)\sigma^2}{n-2} = \sigma^2. \tag{6.12}$$

Intuitively, we divide by $n-2$ in (6.11) instead of $n-1$ as in $s^2 = \sum_i (y_i - \bar{y})^2/(n-1)$ in (5.6), because $\hat{y}_i = \hat{\beta}_0 + \hat{\beta}_1 x_i$ has two estimated parameters and should thereby be a better estimator of $E(y_i)$ than $\bar{y}$. Thus we expect $\text{SSE} = \sum_i (y_i - \hat{y}_i)^2$ to be less than $\sum_i (y_i - \bar{y})^2$. In fact, using (6.5) and (6.6), we can write the numerator of (6.11) in the form

$$\text{SSE} = \sum_{i=1}^{n} (y_i - \hat{y}_i)^2 = \sum_{i=1}^{n} (y_i - \bar{y})^2 - \frac{\left[\sum_{i=1}^{n} (x_i - \bar{x})(y_i - \bar{y})\right]^2}{\sum_{i=1}^{n} (x_i - \bar{x})^2}, \tag{6.13}$$

which shows that $\sum_i (y_i - \hat{y}_i)^2$ is indeed smaller than $\sum_i (y_i - \bar{y})^2$.

6.3 HYPOTHESIS TEST AND CONFIDENCE INTERVAL FOR β_1

Typically, hypotheses about β_1 are of more interest than hypotheses about β_0, since our first priority is to determine whether there is a linear relationship between y and x. (See Problem 6.9 for a test and confidence interval for β_0.) In this section, we consider the hypothesis H_0: $\beta_1 = 0$, which states that there is no linear relationship between y and x in the model $y_i = \beta_0 + \beta_1 x_i + \varepsilon_i$. The hypothesis H_0:$\beta_1 = c$ (for $c \neq 0$) is of less interest.

In order to obtain a test for H_0: $\beta_1 = 0$, we assume that y_i is $N(\beta_0 + \beta_1 x_i, \sigma^2)$. Then $\hat{\beta}_1$ and s^2 have the following properties (these are special cases of results established in Theorem 7.6b in Section 7.6.3):

1. $\hat{\beta}_1$ is $N\left[\beta_1, \sigma^2 / \sum_i (x_i - \bar{x})^2\right]$.
2. $(n-2)s^2/\sigma^2$ is $\chi^2(n-2)$.
3. $\hat{\beta}_1$ and s^2 are independent.

From these three properties it follows by (5.29) that

$$t = \frac{\hat{\beta}_1}{s \Big/ \sqrt{\sum_i (x_i - \bar{x})^2}} \tag{6.14}$$

is distributed as $t(n-2, \delta)$, the noncentral t with noncentrality parameter δ. By a comment following (5.29), δ is given by $\delta = E(\hat{\beta}_1)/\sqrt{\text{var}(\hat{\beta}_1)} = \beta_1/[\sigma/\sqrt{\sum_i (x_i - \bar{x})^2}]$. If $\beta_1 = 0$, then by (5.28), t is distributed as $t(n-2)$. For a two-sided alternative hypothesis $H_1: \beta_1 \neq 0$, we reject $H_0: \beta_1 = 0$ if $|t| \geq t_{\alpha/2,\, n-2}$, where $t_{\alpha/2,\, n-2}$ is the upper $\alpha/2$ percentage point of the central t distribution and α is the desired significance level of the test (probability of rejecting H_0 when it is true). Alternatively, we reject H_0 if $p \leq \alpha$, where p is the p value. For a two-sided test, the p value is defined as twice the probability that $t(n-2)$ exceeds the absolute value of the observed t.

A $100(1-\alpha)\%$ confidence interval for β_1 is given by

$$\hat{\beta}_1 \pm t_{\alpha/2,\, n-2} \frac{s}{\sqrt{\sum_{i=1}^{n}(x_i-\bar{x})^2}}. \tag{6.15}$$

Confidence intervals are defined and discussed further in Section 8.6. A confidence interval for $E(y)$ and a prediction interval for y are also given in Section 8.6.

Example 6.3. We test the hypothesis H_0: $\beta_1 = 0$ for the grades data in Example 6.2. By (6.14), the t statistic is

$$t = \frac{\hat{\beta}_1}{s/\sqrt{\sum_{i=1}^{n}(x_i-\bar{x})^2}} = \frac{.8726}{(13.8547)/(139.753)} = 8.8025.$$

Since $t = 8.8025 > t_{.025,\, 16} = 2.120$, we reject H_0: $\beta_1 = 0$ at the $\alpha = .05$ level of significance. Alternatively, the p value is 1.571×10^{-7}, which is less than .05.

A 95% confidence interval for β_1 is given by (6.15) as

$$\hat{\beta}_1 \pm t_{.025,\, 16} \frac{s}{\sqrt{\sum_{i=1}^{n}(x_i-\bar{x})^2}}$$

$$.8726 \pm 2.120(.09914)$$

$$.8726 \pm .2102$$

$$(.6624,\ 1.0828).$$

6.4 COEFFICIENT OF DETERMINATION

The *coefficient of determination* r^2 is defined as

$$r^2 = \frac{\text{SSR}}{\text{SST}} = \frac{\sum_{i=1}^{n}(\hat{y}_i-\bar{y})^2}{\sum_{i=1}^{n}(y_i-\bar{y})^2}, \tag{6.16}$$

where $\text{SSR} = \sum_i (\hat{y}_i-\bar{y})^2$ is the regression sum of squares and $\text{SST} = \sum_i (y_i-\bar{y})^2$ is the total sum of squares. The total sum of squares can be partitioned into SST = SSR + SSE, that is,

$$\sum_{i=1}^{n}(y_i-\bar{y})^2 = \sum_{i=1}^{n}(\hat{y}_i-\bar{y})^2 + \sum_{i=1}^{n}(y_i-\hat{y}_i)^2. \tag{6.17}$$

Thus r^2 in (6.16) gives the proportion of variation in y that is explained by the model or, equivalently, accounted for by regression on x.

We have labeled (6.16) as r^2 because it is the same as the square of the *sample correlation coefficient* r between y and x

$$r = \frac{s_{xy}}{\sqrt{s_x^2 s_y^2}} = \frac{\sum_{i=1}^n (x_i - \bar{x})(y_i - \bar{y})}{\sqrt{\left[\sum_{i=1}^n (x_i - \bar{x})^2\right]\left[\sum_{i=1}^n (y_i - \bar{y})^2\right]}}, \tag{6.18}$$

where s_{xy} is given by 5.15 (see Problem 6.11). When x is a random variable, r estimates the population correlation in (3.19). The coefficient of determination r^2 is discussed further in Sections 7.7, 10.4, and 10.5.

Example 6.4. For the grades data of Example 6.2, we have

$$r^2 = \frac{\text{SSR}}{\text{SST}} = \frac{14,873.0}{17,944.3} = .8288.$$

The correlation between homework score and exam score is $r = \sqrt{.8288} = .910$.

The t statistic in (6.14) can be expressed in terms of r as follows:

$$t = \frac{\hat{\beta}_1}{s\Big/\sqrt{\sum_i (x_i - \bar{x})^2}} \tag{6.19}$$

$$= \frac{\sqrt{n-2}\, r}{\sqrt{1-r^2}}. \tag{6.20}$$

If H_0: $\beta_1 = 0$ is true, then, as noted following (6.14), the statistic in (6.19) is distributed as $t(n-2)$ under the assumption that the x_i's are fixed and the y_i's are independently distributed as $N(\beta_0 + \beta_1 x_i, \sigma^2)$. If x is a random variable such that x and y have a bivariate normal distribution, then $t = \sqrt{n-2}\, r/\sqrt{1-r^2}$ in (6.20) also has the $t(n-2)$ distribution provided that $H_0 : \rho = 0$ is true, where ρ is the population correlation coefficient defined in (3.19) (see Theorem 10.5). However, (6.19) and (6.20) have different distributions if $H_0 : \beta_1 = 0$ and $H_0 : \rho = 0$ are false (see Section 10.4). If $\beta_1 \neq 0$, then (6.19) has a noncentral t distribution, but if $\rho \neq 0$, (6.20) does not have a noncentral t distribution.

PROBLEMS

6.1 Obtain the least-squares solutions (6.5) and (6.6) from (6.3) and (6.4).

6.2 **(a)** Show that $E(\hat{\beta}_1) = \beta_1$ as in (6.7).

(b) Show that $E(\hat{\beta}_0) = \beta_0$ as in (6.8).

6.3 **(a)** Show that $\text{var}(\hat{\beta}_1) = \sigma^2 / \sum_{i=1}^n (x_i - \bar{x})^2$ as in (6.9).

(b) Show that $\text{var}(\hat{\beta}_0) = \sigma^2 \left[1/n + \bar{x}^2 / \sum_{i=1}^n (x_i - \bar{x})^2\right]$ as in (6.10).

6.4 Suppose that n is even and the n values of x_i can be selected anywhere in the interval from a to b. Show that $\text{var}(\hat{\beta}_1)$ is a minimum if $n/2$ values of x_i are equal to a and $n/2$ values are equal to b.

6.5 Show that $\text{SSE} = \sum_{i=1}^n (y_i - \hat{y}_i)^2$ in (6.11) can be expressed in the form given in (6.13).

6.6 Show that $E(s^2) = \sigma^2$ as in (6.12).

6.7 Show that $t = \hat{\beta}_1 / [s/\sqrt{\sum_i (x_i - \bar{x})^2}]$ in (6.14) is distributed as $t(n-2,\ \delta)$, where $\delta = \beta_1 / [\sigma/\sqrt{\sum_i (x_i - \bar{x})^2}]$.

6.8 Obtain a test for $H_0 : \beta_1 = c$ versus $H_1 : \beta_1 \neq c$.

6.9 **(a)** Obtain a test for $H_0 : \beta_0 = a$ versus $H_1 : \beta_0 \neq a$.

(b) Obtain a confidence interval for β_0.

6.10 Show that $\sum_{i=1}^n (y_i - \bar{y})^2 = \sum_{i=1}^n (\hat{y}_i - \bar{y})^2 + \sum_{i=1}^n (y_i - \hat{y}_i)^2$ as in (6.17).

6.11 Show that r^2 in (6.16) is the square of the correlation

$$r = \frac{\sum_{i=1}^n (x_i - \bar{x})(y_i - \bar{y})}{\sqrt{\left[\sum_{i=1}^n (x_i - \bar{x})^2\right]\left[\sum_{i=1}^n (y_i - \bar{y})^2\right]}}$$

as given by (6.18).

TABLE 6.1 Eruptions of Old Faithful Geyser, August 1–4, 1978[a]

y	x	y	x	y	x	y	x
78	4.4	80	4.3	76	4.5	75	4.0
74	3.9	56	1.7	82	3.9	73	3.7
68	4.0	80	3.9	84	4.3	67	3.7
76	4.0	69	3.7	53	2.3	68	4.3
80	3.5	57	3.1	86	3.8	86	3.6
84	4.1	90	4.0	51	1.9	72	3.8
50	2.3	42	1.8	85	4.6	75	3.8
93	4.7	91	4.1	45	1.8	75	3.8
55	1.7	51	1.8	88	4.7	66	2.5
76	4.9	79	3.2	51	1.8	84	4.5
58	1.7	53	1.9	80	4.6	70	4.1
74	4.6	82	4.6	49	1.9	79	3.7
75	3.4	51	2.0	82	3.5	60	3.8
—	—	—	—	—	—	86	3.4

[a]Where x = duration, y = interval (both in minutes).

6.12 Show that $r = \cos\theta$, where θ is the angle between the vectors $\mathbf{x} - \bar{x}\mathbf{j}$ and $\mathbf{y} - \bar{y}\mathbf{j}$, where $\mathbf{x} - \bar{x}\mathbf{j} = (x_1 - \bar{x}, x_2 - \bar{x}, \ldots, x_n - \bar{x})'$ and $\mathbf{y} - \bar{y}\mathbf{j} = (y_1 - \bar{y}, y_2 - \bar{y}, \ldots, y_n - \bar{y})'$.

6.13 Show that $t = \hat{\beta}_1/[s/\sqrt{\sum_{i=1}^n (x_i - \bar{x})^2}]$ in (6.19) is equal to $t = \sqrt{n-2}\, r/\sqrt{1-r^2}$ in (6.20).

6.14 Table 6.1 (Weisberg 1985, p. 231) gives the data on daytime eruptions of Old Faithful Geyser in Yellowstone National Park during August 1–4, 1978. The variables are $x =$ duration of an eruption and $y =$ interval to the next eruption. Can x be used to successfully predict y using a simple linear model $y_i = \beta_0 + \beta_1 x_i + \varepsilon_i$?

(a) Find $\hat{\beta}_0$ and $\hat{\beta}_1$.

(b) Test $H_0 : \beta_1 = 0$ using (6.14).

(c) Find a confidence interval for β_1.

(d) Find r^2 using (6.16).

7 Multiple Regression: Estimation

7.1 INTRODUCTION

In *multiple regression*, we attempt to predict a *dependent* or *response* variable y on the basis of an assumed linear relationship with several *independent* or *predictor* variables $x_1, x_1, \ldots, x_k$. In addition to constructing a model for prediction, we may wish to assess the extent of the relationship between y and the x variables. For this purpose, we use the multiple correlation coefficient R (Section 7.7).

In this chapter, y is a continuous random variable and the x variables are fixed constants (either discrete or continuous) that are controlled by the experimenter. The case in which the x variables are random variables is covered in Chapter 10. In analysis-of-variance (Chapters 12–15), the x variables are fixed and discrete.

Useful applied expositions of multiple regression for the fixed-x case can be found in Morrison (1983), Myers (1990), Montgomery and Peck (1992), Graybill and Iyer (1994), Mendenhall and Sincich (1996), Ryan (1997), Draper and Smith (1998), and Kutner et al. (2005). Theoretical treatments are given by Searle (1971), Graybill (1976), Guttman (1982), Kshirsagar (1983), Myers and Milton (1991), Jørgensen (1993), Wang and Chow (1994), Christensen (1996), Seber and Lee (2003), and Hocking (1976, 1985, 2003).

7.2 THE MODEL

The multiple linear regression model, as introduced in Section 1.2, can be expressed as

$$y = \beta_0 + \beta_1 x_1 + \beta_2 x_2 + \cdots + \beta_k x_k + \varepsilon. \tag{7.1}$$

We discuss estimation of the β parameters when the model is linear in the β's. An example of a model that is linear in the β's but not the x's is the second-order

Linear Models in Statistics, Second Edition, by Alvin C. Rencher and G. Bruce Schaalje

response surface model

$$y = \beta_0 + \beta_1 x_1 + \beta_2 x_2 + \beta_3 x_1^2 + \beta_4 x_2^2 + \beta_5 x_1 x_2 + \varepsilon. \qquad (7.2)$$

To estimate the β's in (7.1), we will use a sample of n observations on y and the associated x variables. The model for the ith observation is

$$y_i = \beta_0 + \beta_1 x_{i1} + \beta_2 x_{i2} + \cdots + \beta_k x_{ik} + \varepsilon_i, \quad i = 1, 2, \ldots, n. \qquad (7.3)$$

The assumptions for ε_i or y_i are essentially the same as those for simple linear regression in Section 6.1:

1. $E(\varepsilon_i) = 0$ for $i = 1, 2, \ldots, n$, or, equivalently, $E(y_i) = \beta_0 + \beta_1 x_{i1} + \beta_2 x_{i2} + \cdots + \beta_k x_{ik}$.
2. $\text{var}(\varepsilon_i) = \sigma^2$ for $i = 1, 2, \ldots, n$, or, equivalently, $\text{var}(y_i) = \sigma^2$.
3. $\text{cov}(\varepsilon_i, \varepsilon_j) = 0$ for all $i \neq j$, or, equivalently, $\text{cov}(y_i, y_j) = 0$.

Assumption 1 states that the model is correct, in other words that all relevant x's are included and the model is indeed linear. Assumption 2 asserts that the variance of y is constant and therefore does not depend on the x's. Assumption 3 states that the y's are uncorrelated with each other, which usually holds in a random sample (the observations would typically be correlated in a time series or when repeated measurements are made on a single plant or animal). Later we will add a normality assumption (Section 7.6), under which the y variable will be independent as well as uncorrelated.

When all three assumptions hold, the least-squares estimators of the β's have some good properties (Section 7.3.2). If one or more assumptions do not hold, the estimators may be poor. Under the normality assumption (Section 7.6), the maximum likelihood estimators have excellent properties.

Any of the three assumptions may fail to hold with real data. Several procedures have been devised for checking the assumptions. These diagnostic techniques are discussed in Chapter 9.

Writing (7.3) for each of the n observations, we have

$$\begin{aligned} y_1 &= \beta_0 + \beta_1 x_{11} + \beta_2 x_{12} + \cdots + \beta_k x_{1k} + \varepsilon_1 \\ y_2 &= \beta_0 + \beta_1 x_{21} + \beta_2 x_{22} + \cdots + \beta_k x_{2k} + \varepsilon_2 \\ &\vdots \\ y_n &= \beta_0 + \beta_1 x_{n1} + \beta_2 x_{n2} + \cdots + \beta_k x_{nk} + \varepsilon_n. \end{aligned}$$

These n equations can be written in matrix form as

$$\begin{pmatrix} y_1 \\ y_2 \\ \vdots \\ y_n \end{pmatrix} = \begin{pmatrix} 1 & x_{11} & x_{12} & \dots & x_{1k} \\ 1 & x_{21} & x_{22} & \dots & x_{2k} \\ \vdots & \vdots & \vdots & & \vdots \\ 1 & x_{n1} & x_{n2} & \dots & x_{nk} \end{pmatrix} \begin{pmatrix} \beta_0 \\ \beta_1 \\ \vdots \\ \beta_k \end{pmatrix} + \begin{pmatrix} \varepsilon_1 \\ \varepsilon_2 \\ \vdots \\ \varepsilon_n \end{pmatrix}$$

or

$$\mathbf{y} = \mathbf{X}\boldsymbol{\beta} + \boldsymbol{\varepsilon}. \tag{7.4}$$

The preceding three assumptions on ε_i or y_i can be expressed in terms of the model in (7.4):

1. $E(\boldsymbol{\varepsilon}) = \mathbf{0}$ or $E(\mathbf{y}) = \mathbf{X}\boldsymbol{\beta}$.
2. $\text{cov}(\boldsymbol{\varepsilon}) = \sigma^2\mathbf{I}$ or $\text{cov}(\mathbf{y}) = \sigma^2\mathbf{I}$.

Note that the assumption $\text{cov}(\boldsymbol{\varepsilon}) = \sigma^2\mathbf{I}$ includes both the previous assumptions $\text{var}(\varepsilon_i) = \sigma^2$ and $\text{cov}(\varepsilon_i, \varepsilon_j) = 0$.

The matrix $\mathbf{X}$ in (7.4) is $n \times (k+1)$. In this chapter we assume that $n > k+1$ and rank $(\mathbf{X}) = k+1$. If $n < k+1$ or if there is a linear relationship among the x's, for example, $x_5 = \sum_{j=1}^{4} x_j/4$, then $\mathbf{X}$ will not have full column rank. If the values of the x_{ij}'s are planned (chosen by the researcher), then the $\mathbf{X}$ matrix essentially contains the experimental design and is sometimes called the *design matrix*.

The β parameters in (7.1) or (7.4) are called *regression coefficients*. To emphasize their collective effect, they are sometimes referred to as *partial regression coefficients*. The word *partial* carries both a mathematical and a statistical meaning. Mathematically, the partial derivative of $E(y) = \beta_0 + \beta_1 x_1 + \beta_2 x_2 + \cdots + \beta_k x_k$ with respect to x_1, for example, is β_1. Thus β_1 indicates the change in $E(y)$ with a unit increase in x_1 when $x_2, x_3, \ldots, x_k$ are held constant. Statistically, β_1 shows the effect of x_1 on $E(y)$ in the presence of the other x's. This effect would typically be different from the effect of x_1 on $E(y)$ if the other x's were not present in the model. Thus, for example, β_0 and β_1 in

$$y = \beta_0 + \beta_1 x_1 + \beta_2 x_2 + \varepsilon$$

will usually be different from β_0^* and β_1^* in

$$y = \beta_0^* + \beta_1^* x_1 + \varepsilon^*.$$

[If x_1 and x_2 are orthogonal, that is, if $\mathbf{x}_1'\mathbf{x}_2 = 0$ or if $(\mathbf{x}_1 - \bar{x}_1\mathbf{j})'(\mathbf{x}_2 - \bar{x}_2\mathbf{j}) = 0$, where $\mathbf{x}_1$ and $\mathbf{x}_2$ are columns in the $\mathbf{X}$ matrix, then $\beta_0 = \beta_0^*$ and $\beta_1 = \beta_1^*$; see Corollary 1 to Theorem 7.9a and Theorem 7.10]. The change in parameters when an x is deleted from the model is illustrated (with estimates) in the following example.

TABLE 7.1 Data for Example 7.2

Observation Number	y	x_1	x_2
1	2	0	2
2	3	2	6
3	2	2	7
4	7	2	5
5	6	4	9
6	8	4	8
7	10	4	7
8	7	6	10
9	8	6	11
10	12	6	9
11	11	8	15
12	14	8	13

Example 7.2. [See Freund and Minton (1979, pp. 36–39)]. Consider the (contrived) data in Table 7.1.

Using (6.5) and (6.6) from Section 6.2 and (7.6) in Section 7.3 (see Example 7.3.1), we obtain prediction equations for y regressed on x_1 alone, on x_2 alone, and on both x_1 and x_2:

$$\hat{y} = 1.86 + 1.30x_1,$$
$$\hat{y} = .86 + .78x_2,$$
$$\hat{y} = 5.37 + 3.01x_1 - 1.29x_2.$$

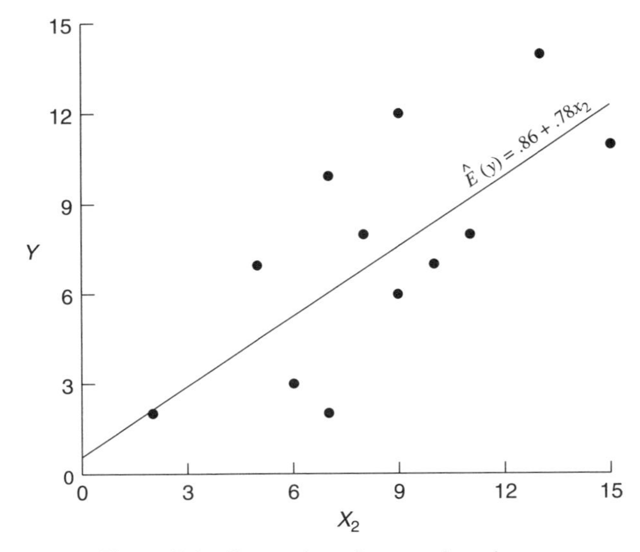

Figure 7.1 Regression of y on x_2 ignoring x_1.

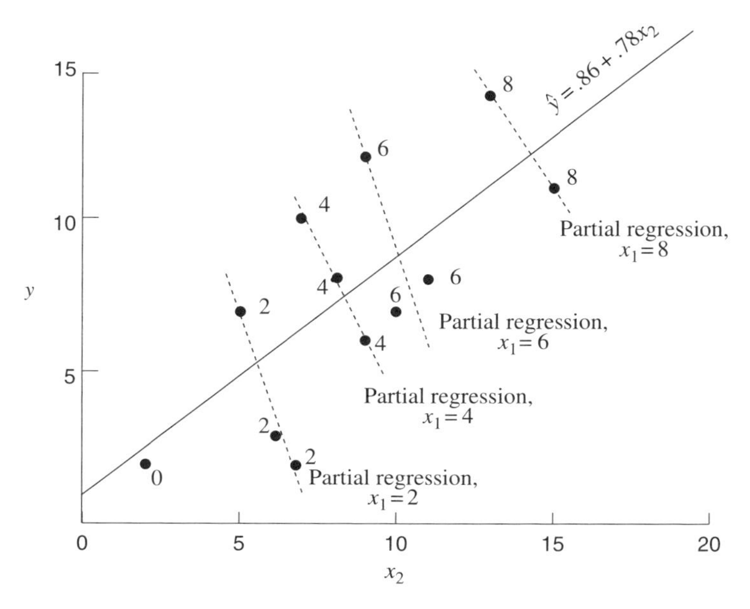

Figure 7.2 Regression of y on x_2 showing the value of x_1 at each point and partial regressions of y on x_2.

As expected, the coefficients change from either of the reduced models to the full model. Note the sign change as the coefficient of x_2 changes from .78 to -1.29.

The values of y and x_2 are plotted in Figure 7.1 along with the prediction equation $\hat{y} = .86 + .78x_2$. The linear trend is clearly evident.

In Figure 7.2 we have the same plot as in Figure 7.1, except that each point is labeled with the value of x_1. Examining values of y and x_2 for a fixed value of x_1 (2, 4, 6, or 8) shows a negative slope for the relationship. These negative relationships are shown as partial regressions of y on x_2 for each value of x_1. The partial regression coefficient $\hat{\beta}_2 = -1.29$ reflects the negative slopes of these four partial regressions.

Further insight into the meaning of the partial regression coefficients is given in Section 7.10. □

7.3 ESTIMATION OF β AND σ^2

7.3.1 Least-Squares Estimator for β

In this section, we discuss the *least-squares approach* to estimation of the β's in the fixed-x model (7.1) or (7.4). No distributional assumptions on y are required to obtain the estimators.

For the parameters $\beta_0, \beta_1, \ldots, \beta_k$, we seek estimators that minimize the sum of squares of deviations of the n observed y's from their predicted values $\hat{y}$. By extension of (6.2), we seek $\hat{\beta}_0, \hat{\beta}_1, \ldots, \hat{\beta}_k$ that minimize

$$\begin{aligned}\sum_{i=1}^{n} \hat{\varepsilon}_i^2 &= \sum_{i=1}^{n} (y_i - \hat{y}_i)^2 \\ &= \sum_{i=1}^{n} (y_i - \hat{\beta}_0 - \hat{\beta}_1 x_{i1} - \hat{\beta}_2 x_{i2} - \cdots - \hat{\beta}_k x_{ik})^2. \end{aligned} \qquad (7.5)$$

Note that the predicted value $\hat{y}_i = \hat{\beta}_0 + \hat{\beta}_1 x_{i1} + \cdots + \hat{\beta}_k x_{ik}$ estimates $E(y_i)$, not y_i. A better notation would be $\widehat{E(y_i)}$, but $\hat{y}_i$ is commonly used.

To obtain the least-squares estimators, it is not necessary that the prediction equation $\hat{y}_i = \hat{\beta}_0 + \hat{\beta}_1 x_{i1} + \cdots + \hat{\beta}_k x_{ik}$ be based on $E(y_i)$. It is only necessary to postulate an empirical model that is linear in the $\hat{\beta}$'s, and the least-squares method will find the "best" fit to this model. This was illustrated in Figure 6.2.

To find the values of $\hat{\beta}_0, \hat{\beta}_1, \ldots, \hat{\beta}_k$ that minimize (7.5), we could differentiate $\sum_i \hat{\varepsilon}_i^2$ with respect to each $\hat{\beta}_j$ and set the results equal to zero to yield $k + 1$ equations that can be solved simultaneously for the $\hat{\beta}_j$'s. However, the procedure can be carried out in more compact form with matrix notation. The result is given in the following theorem.

Theorem 7.3a. If $\mathbf{y} = \mathbf{X}\boldsymbol{\beta} + \boldsymbol{\varepsilon}$, where $\mathbf{X}$ is $n \times (k + 1)$ of rank $k + 1 < n$, then the value of $\hat{\boldsymbol{\beta}} = (\hat{\beta}_0, \hat{\beta}_1, \ldots, \hat{\beta}_k)'$ that minimizes (7.5) is

$$\hat{\boldsymbol{\beta}} = (\mathbf{X}'\mathbf{X})^{-1}\mathbf{X}'\mathbf{y}. \qquad (7.6)$$

PROOF. Using (2.20) and (2.27), we can write (7.5) as

$$\hat{\boldsymbol{\varepsilon}}'\hat{\boldsymbol{\varepsilon}} = \sum_{i=1}^{n} (y_i - \mathbf{x}_i'\hat{\boldsymbol{\beta}})^2 = (\mathbf{y} - \mathbf{X}\hat{\boldsymbol{\beta}})'(\mathbf{y} - \mathbf{X}\hat{\boldsymbol{\beta}}), \qquad (7.7)$$

where $\mathbf{x}_i' = (1, x_{i1}, \ldots, x_{ik})$ is the ith row of $\mathbf{X}$. When the product $(\mathbf{y} - \mathbf{X}\hat{\boldsymbol{\beta}})'(\mathbf{y} - \mathbf{X}\hat{\boldsymbol{\beta}})$ in (7.7) is expanded as in (2.17), two of the resulting four terms can be combined to yield

$$\hat{\boldsymbol{\varepsilon}}'\hat{\boldsymbol{\varepsilon}} = \mathbf{y}'\mathbf{y} - 2\mathbf{y}'\mathbf{X}\hat{\boldsymbol{\beta}} + \hat{\boldsymbol{\beta}}'\mathbf{X}'\mathbf{X}\hat{\boldsymbol{\beta}}.$$

We can find the value of $\hat{\boldsymbol{\beta}}$ that minimizes $\hat{\boldsymbol{\varepsilon}}'\hat{\boldsymbol{\varepsilon}}$ by differentiating $\hat{\boldsymbol{\varepsilon}}'\hat{\boldsymbol{\varepsilon}}$ with respect to $\hat{\boldsymbol{\beta}}$ [using (2.112) and (2.113)] and setting the result equal to zero:

$$\frac{\partial \hat{\boldsymbol{\varepsilon}}'\hat{\boldsymbol{\varepsilon}}}{\partial \hat{\boldsymbol{\beta}}} = \mathbf{0} - 2\mathbf{X}'\mathbf{y} + 2\mathbf{X}'\mathbf{X}\hat{\boldsymbol{\beta}} = \mathbf{0},$$

This gives the *normal equations*

$$\mathbf{X}'\mathbf{X}\hat{\boldsymbol{\beta}} = \mathbf{X}'\mathbf{y}. \qquad (7.8)$$

By Theorems 2.4(iii) and 2.6d(i) and Corollary 1 of Theorem 2.6c, if $\mathbf{X}$ is full-rank, $\mathbf{X}'\mathbf{X}$ is nonsingular, and the solution to (7.8) is given by (7.6). □

Since $\hat{\boldsymbol{\beta}}$ in (7.6) minimizes the sum of squares in (7.5), $\hat{\boldsymbol{\beta}}$ is called the *least-squares estimator*. Note that each $\hat{\beta}_j$ in $\hat{\boldsymbol{\beta}}$ is a linear function of $\mathbf{y}$; that is, $\hat{\beta}_j = \mathbf{a}_j'\mathbf{y}$, where $\mathbf{a}_j'$ is the jth row of $(\mathbf{X}'\mathbf{X})^{-1}\mathbf{X}'$. This usage of the word *linear* in *linear estimator* is different from that in *linear model*, which indicates that the model is linear in the β's.

We now show that $\hat{\boldsymbol{\beta}} = (\mathbf{X}'\mathbf{X})^{-1}\mathbf{X}'\mathbf{y}$ minimizes $\hat{\boldsymbol{\varepsilon}}'\hat{\boldsymbol{\varepsilon}}$. Let $\mathbf{b}$ be an alternative estimator that may do better than $\hat{\boldsymbol{\beta}}$ so that $\hat{\boldsymbol{\varepsilon}}'\hat{\boldsymbol{\varepsilon}}$ is

$$\hat{\boldsymbol{\varepsilon}}'\hat{\boldsymbol{\varepsilon}} = (\mathbf{y} - \mathbf{X}\mathbf{b})'(\mathbf{y} - \mathbf{X}\mathbf{b}).$$

Now adding and subtracting $\mathbf{X}\hat{\boldsymbol{\beta}}$, we obtain

$$= (\mathbf{y} - \mathbf{X}\hat{\boldsymbol{\beta}} + \mathbf{X}\hat{\boldsymbol{\beta}} - \mathbf{X}\mathbf{b})'(\mathbf{y} - \mathbf{X}\hat{\boldsymbol{\beta}} + \mathbf{X}\hat{\boldsymbol{\beta}} - \mathbf{X}\mathbf{b}) \tag{7.9}$$

$$= (\mathbf{y} - \mathbf{X}\hat{\boldsymbol{\beta}})'(\mathbf{y} - \mathbf{X}\hat{\boldsymbol{\beta}}) + (\hat{\boldsymbol{\beta}} - \mathbf{b})'\mathbf{X}'\mathbf{X}(\hat{\boldsymbol{\beta}} - \mathbf{b}) + 2(\hat{\boldsymbol{\beta}} - \mathbf{b})'(\mathbf{X}'\mathbf{y} - \mathbf{X}'\mathbf{X}\hat{\boldsymbol{\beta}}). \tag{7.10}$$

The third term on the right side of (7.10) vanishes because of the normal equations $\mathbf{X}'\mathbf{y} = \mathbf{X}'\mathbf{X}\hat{\boldsymbol{\beta}}$ in (7.8). The second term is a positive definite quadratic form (assuming that $\mathbf{X}$ is full-rank; see Theorem 2.6d), and $\hat{\boldsymbol{\varepsilon}}'\hat{\boldsymbol{\varepsilon}}$ is therefore minimized when $\mathbf{b} = \hat{\boldsymbol{\beta}}$.

To examine the structure of $\mathbf{X}'\mathbf{X}$ and $\mathbf{X}'\mathbf{y}$, note that by Theorem 2.2c(i), the $(k+1) \times (k+1)$ matrix $\mathbf{X}'\mathbf{X}$ can be obtained as products of columns of $\mathbf{X}$; similarly, $\mathbf{X}'\mathbf{y}$ contains products of columns of $\mathbf{X}$ and $\mathbf{y}$:

$$\mathbf{X}'\mathbf{X} = \begin{pmatrix} n & \sum_i x_{i1} & \sum_i x_{i2} & \cdots & \sum_i x_{ik} \\ \sum_i x_{i1} & \sum_i x_{i1}^2 & \sum_i x_{i1}x_{i2} & \cdots & \sum_i x_{i1}x_{ik} \\ \vdots & \vdots & \vdots & & \vdots \\ \sum x_{ik} & \sum_i x_{i1}x_{ik} & \sum_i x_{i2}x_{ik} & \cdots & \sum_i x_{ik}^2 \end{pmatrix},$$

$$\mathbf{X}'\mathbf{y} = \begin{pmatrix} \sum_i y_i \\ \sum_i x_{i1}y_i \\ \vdots \\ \sum_i x_{ik}y_i \end{pmatrix}.$$

If $\hat{\boldsymbol{\beta}} = (\mathbf{X}'\mathbf{X})^{-1}\mathbf{X}'\mathbf{y}$ as in (7.6), then

$$\hat{\boldsymbol{\varepsilon}} = \mathbf{y} - \mathbf{X}\hat{\boldsymbol{\beta}} = \mathbf{y} - \hat{\mathbf{y}} \tag{7.11}$$

is the vector of *residuals*, $\hat{\varepsilon}_1 = y_1 - \hat{y}_1, \hat{\varepsilon}_2 = y_2 - \hat{y}_2, \ldots, \hat{\varepsilon}_n = y_n - \hat{y}_n$. The residual vector $\hat{\boldsymbol{\varepsilon}}$ estimates $\boldsymbol{\varepsilon}$ in the model $\mathbf{y} = \mathbf{X}\boldsymbol{\beta} + \boldsymbol{\varepsilon}$ and can be used to check the validity of the model and attendant assumptions; see Chapter 9.

Example 7.3.1a. We use the data in Table 7.1 to illustrate computation of $\hat{\boldsymbol{\beta}}$ using (7.6).

$$\mathbf{y} = \begin{pmatrix} 2 \\ 3 \\ 2 \\ 7 \\ 6 \\ 8 \\ 10 \\ 7 \\ 8 \\ 12 \\ 11 \\ 14 \end{pmatrix}, \quad \mathbf{X} = \begin{pmatrix} 1 & 0 & 2 \\ 1 & 2 & 6 \\ 1 & 2 & 7 \\ 1 & 2 & 5 \\ 1 & 4 & 9 \\ 1 & 4 & 8 \\ 1 & 4 & 7 \\ 1 & 6 & 10 \\ 1 & 6 & 11 \\ 1 & 6 & 9 \\ 1 & 8 & 15 \\ 1 & 8 & 13 \end{pmatrix}, \quad \mathbf{X}'\mathbf{X} = \begin{pmatrix} 12 & 52 & 102 \\ 52 & 395 & 536 \\ 102 & 536 & 1004 \end{pmatrix},$$

$$\mathbf{X}'\mathbf{y} = \begin{pmatrix} 90 \\ 482 \\ 872 \end{pmatrix}, \quad (\mathbf{X}'\mathbf{X})^{-1} = \begin{pmatrix} .97476 & .24290 & -.22871 \\ .24290 & .16207 & -.11120 \\ -.22871 & -.11120 & .08360 \end{pmatrix},$$

$$\hat{\boldsymbol{\beta}} = (\mathbf{X}'\mathbf{X})^{-1}\mathbf{X}'\mathbf{y} = \begin{pmatrix} 5.3754 \\ 3.0118 \\ -1.2855 \end{pmatrix}.$$

□

Example 7.3.1b. Simple linear regression from Chapter 6 can also be expressed in matrix terms:

$$\mathbf{y} = \begin{pmatrix} y_1 \\ y_2 \\ \vdots \\ y_n \end{pmatrix}, \quad \mathbf{X} = \begin{pmatrix} 1 & x_1 \\ 1 & x_2 \\ \vdots & \vdots \\ 1 & x_n \end{pmatrix}, \quad \boldsymbol{\beta} = \begin{pmatrix} \beta_0 \\ \beta_1 \end{pmatrix},$$

$$\mathbf{X}'\mathbf{X} = \begin{pmatrix} n & \sum_i x_i \\ \sum_i x_i & \sum_i x_i^2 \end{pmatrix}, \quad \mathbf{X}'\mathbf{y} = \begin{pmatrix} \sum_i y_i \\ \sum_i x_i y_i \end{pmatrix},$$

$$(\mathbf{X}'\mathbf{X})^{-1} = \frac{1}{n\sum_i x_i^2 - (\sum_i x_i)^2} \begin{pmatrix} \sum_i x_i^2 & -\sum_i x_i \\ -\sum_i x_i & n \end{pmatrix}.$$

Then $\hat{\beta}_0$ and $\hat{\beta}_1$ can be obtained using (7.6), $\hat{\boldsymbol{\beta}} = (\mathbf{X}'\mathbf{X})^{-1}\mathbf{X}'\mathbf{y}$:

$$\hat{\boldsymbol{\beta}} = \begin{pmatrix} \hat{\beta}_0 \\ \hat{\beta}_1 \end{pmatrix} = \frac{1}{n\sum_i x_i^2 - (\sum_i x_i)^2} \begin{pmatrix} (\sum_i x_i^2)(\sum y_i) - (\sum_i x_i)(\sum_i x_i y_i) \\ -(\sum_i x_i)(\sum y_i) + n\sum_i x_i y_i \end{pmatrix}. \quad (7.12)$$

The estimators $\hat{\beta}_0$ and $\hat{\beta}_1$ in (7.11) are the same as those in (6.5) and (6.6). □

7.3.2 Properties of the Least-Squares Estimator $\hat{\boldsymbol{\beta}}$

The least-squares estimator $\hat{\boldsymbol{\beta}} = (\mathbf{X}'\mathbf{X})^{-1}\mathbf{X}'\mathbf{y}$ in Theorem 7.3a was obtained without using the assumptions $E(\mathbf{y}) = \mathbf{X}\boldsymbol{\beta}$ and $\text{cov}(\mathbf{y}) = \sigma^2\mathbf{I}$ given in Section 7.2. We merely postulated a model $\mathbf{y} = \mathbf{X}\boldsymbol{\beta} + \boldsymbol{\varepsilon}$ as in (7.4) and fitted it. If $E(\mathbf{y}) \neq \mathbf{X}\boldsymbol{\beta}$, the model $\mathbf{y} = \mathbf{X}\boldsymbol{\beta} + \boldsymbol{\varepsilon}$ could still be fitted to the data, in which case, $\hat{\boldsymbol{\beta}}$ may have poor properties. If $\text{cov}(\mathbf{y}) \neq \sigma^2\mathbf{I}$, there may be additional adverse effects on the estimator $\hat{\boldsymbol{\beta}}$. However, if $E(\mathbf{y}) = \mathbf{X}\boldsymbol{\beta}$ and $\text{cov}(\mathbf{y}) = \sigma^2\mathbf{I}$ hold, $\hat{\boldsymbol{\beta}}$ has some good properties, as noted in the four theorems in this section. Note that $\hat{\boldsymbol{\beta}}$ is a random vector (from sample to sample). We discuss its mean vector and covariance matrix in this section (with no distributional assumptions on $\mathbf{y}$) and its distribution (assuming that the y variables are normal) in Section 7.6.3. In the following theorems, we assume that $\mathbf{X}$ is fixed (remains constant in repeated sampling) and full rank.

Theorem 7.3b. If $E(\mathbf{y}) = \mathbf{X}\boldsymbol{\beta}$, then $\hat{\boldsymbol{\beta}}$ is an unbiased estimator for $\boldsymbol{\beta}$.

PROOF

$$\begin{aligned} E(\hat{\boldsymbol{\beta}}) &= E[(\mathbf{X}'\mathbf{X})^{-1}\mathbf{X}'\mathbf{y}] \\ &= (\mathbf{X}'\mathbf{X})^{-1}\mathbf{X}'E(\mathbf{y}) \qquad \text{[by (3.38)]} \\ &= (\mathbf{X}'\mathbf{X})^{-1}\mathbf{X}'\mathbf{X}\boldsymbol{\beta} \\ &= \boldsymbol{\beta}. \end{aligned} \quad (7.13)$$

□

Theorem 7.3c. If $\text{cov}(\mathbf{y}) = \sigma^2\mathbf{I}$, the covariance matrix for $\hat{\boldsymbol{\beta}}$ is given by $\sigma^2(\mathbf{X}'\mathbf{X})^{-1}$.

PROOF

$$\begin{aligned} \text{cov}(\hat{\boldsymbol{\beta}}) &= \text{cov}[(\mathbf{X}'\mathbf{X})^{-1}\mathbf{X}'\mathbf{y}] \\ &= (\mathbf{X}'\mathbf{X})^{-1}\mathbf{X}'\text{cov}(\mathbf{y})[(\mathbf{X}'\mathbf{X})^{-1}\mathbf{X}']' \qquad \text{[by (3.44)]} \\ &= (\mathbf{X}'\mathbf{X})^{-1}\mathbf{X}'(\sigma^2\mathbf{I})\mathbf{X}(\mathbf{X}'\mathbf{X})^{-1} \\ &= \sigma^2(\mathbf{X}'\mathbf{X})^{-1}\mathbf{X}'\mathbf{X}(\mathbf{X}'\mathbf{X})^{-1} \\ &= \sigma^2(\mathbf{X}'\mathbf{X})^{-1}. \end{aligned} \quad (7.14)$$

□

Example 7.3.2a. Using the matrix $(\mathbf{X}'\mathbf{X})^{-1}$ for simple linear regression given in Example 7.3.1, we obtain

$$\text{cov}(\hat{\boldsymbol{\beta}}) = \text{cov}\begin{pmatrix} \hat{\beta}_0 \\ \hat{\beta}_1 \end{pmatrix} = \begin{pmatrix} \text{var}(\hat{\beta}_0) & \text{cov}(\hat{\beta}_0, \hat{\beta}_1) \\ \text{cov}(\hat{\beta}_0, \hat{\beta}_1) & \text{var}(\hat{\beta}_1) \end{pmatrix} = \sigma^2(\mathbf{X}'\mathbf{X})^{-1}$$

$$= \frac{\sigma^2}{n\sum_i x_i^2 - (\sum_i x_i)^2} \begin{pmatrix} \sum_i x_i^2 & -\sum_i x_i \\ -\sum_i x_i & n \end{pmatrix} \tag{7.15}$$

$$= \frac{\sigma^2}{\sum_i (x_i - \bar{x})^2} \begin{pmatrix} \sum_i x_i^2/n & -\bar{x} \\ -\bar{x} & 1 \end{pmatrix}. \tag{7.16}$$

Thus

$$\text{var}(\hat{\beta}_0) = \frac{\sigma^2 \sum_i x_i^2/n}{\sum_i (x_i - \bar{x})^2}, \quad \text{var}(\hat{\beta}_1) = \frac{\sigma^2}{\sum_i (x_i - \bar{x})^2},$$

$$\text{cov}(\hat{\beta}_0, \hat{\beta}_1) = \frac{-\sigma^2 \bar{x}}{\sum_i (x_i - \bar{x})^2}.$$

We found $\text{var}(\hat{\beta}_0)$ and $\text{var}(\hat{\beta}_1)$ in Section 6.2 but did not obtain $\text{cov}(\hat{\beta}_0, \hat{\beta}_1)$. Note that if $\bar{x} > 0$, then $\text{cov}(\hat{\beta}_0, \hat{\beta}_1)$ is negative and the estimated slope and intercept are negatively correlated. In this case, if the estimate of the slope increases from one sample to another, the estimate of the intercept tends to decrease (assuming the x's stay the same). □

Example 7.3.2b. For the data in Table 7.1, $(\mathbf{X}'\mathbf{X})^{-1}$ is as given in Example 7.3.1. Thus, $\text{cov}(\hat{\boldsymbol{\beta}})$ is given by

$$\text{cov}(\hat{\boldsymbol{\beta}}) = \sigma^2(\mathbf{X}'\mathbf{X})^{-1} = \sigma^2 \begin{pmatrix} .975 & .243 & -.229 \\ .243 & .162 & -.111 \\ -.229 & -.111 & .084 \end{pmatrix}.$$

The negative value of $\text{cov}(\hat{\beta}_1, \hat{\beta}_2) = -.111$ indicates that in repeated sampling (using the same 12 values of x_1 and x_2), $\hat{\beta}_1$ and $\hat{\beta}_2$ would tend to move in opposite directions; that is, an increase in one would be accompanied by a decrease in the other. □

In addition to $E(\hat{\boldsymbol{\beta}}) = \boldsymbol{\beta}$ and $\text{cov}(\hat{\boldsymbol{\beta}}) = \sigma^2(\mathbf{X}'\mathbf{X})^{-1}$, a third important property of $\hat{\boldsymbol{\beta}}$ is that under the standard assumptions, the variance of each $\hat{\beta}_j$ is minimum (see the following theorem).

Theorem 7.3d (Gauss–Markov Theorem). If $E(\mathbf{y}) = \mathbf{X}\boldsymbol{\beta}$ and $\text{cov}(\mathbf{y}) = \sigma^2\mathbf{I}$, the least-squares estimators $\hat{\beta}_j, j = 0, 1, \ldots, k$, have minimum variance among all linear unbiased estimators.

PROOF. We consider a linear estimator $\mathbf{Ay}$ of $\boldsymbol{\beta}$ and seek the matrix $\mathbf{A}$ for which $\mathbf{Ay}$ is a minimum variance unbiased estimator of $\boldsymbol{\beta}$. In order for $\mathbf{Ay}$ to be an unbiased estimator of $\boldsymbol{\beta}$, we must have $E(\mathbf{Ay}) = \boldsymbol{\beta}$. Using the assumption $E(\mathbf{y}) = \mathbf{X}\boldsymbol{\beta}$, this can be expressed as

$$E(\mathbf{Ay}) = \mathbf{A}E(\mathbf{y}) = \mathbf{AX}\boldsymbol{\beta} = \boldsymbol{\beta},$$

which gives the unbiasedness condition

$$\mathbf{AX} = \mathbf{I}$$

since the relationship $\mathbf{AX}\boldsymbol{\beta} = \boldsymbol{\beta}$ must hold for any possible value of $\boldsymbol{\beta}$ [see (2.44)].

The covariance matrix for the estimator $\mathbf{Ay}$ is given by

$$\text{cov}(\mathbf{Ay}) = \mathbf{A}(\sigma^2\mathbf{I})\mathbf{A}' = \sigma^2\mathbf{AA}'.$$

The variances of the $\hat{\beta}_j$'s are on the diagonal of $\sigma^2\mathbf{AA}'$, and we therefore need to choose $\mathbf{A}$ (subject to $\mathbf{AX} = \mathbf{I}$) so that the diagonal elements of $\mathbf{AA}'$ are minimized. To relate $\mathbf{Ay}$ to $\hat{\boldsymbol{\beta}} = (\mathbf{X}'\mathbf{X})^{-1}\mathbf{X}'\mathbf{y}$, we add and subtract $(\mathbf{X}'\mathbf{X})^{-1}\mathbf{X}'$ to obtain

$$\mathbf{AA}' = [\mathbf{A} - (\mathbf{X}'\mathbf{X})^{-1}\mathbf{X}' + (\mathbf{X}'\mathbf{X})^{-1}\mathbf{X}'][\mathbf{A} - (\mathbf{X}'\mathbf{X})^{-1}\mathbf{X}' + (\mathbf{X}'\mathbf{X})^{-1}\mathbf{X}']'.$$

Expanding this in terms of $\mathbf{A} - (\mathbf{X}'\mathbf{X})^{-1}\mathbf{X}'$ and $(\mathbf{X}'\mathbf{X})^{-1}\mathbf{X}'$, we obtain four terms, two of which vanish because of the restriction $\mathbf{AX} = \mathbf{I}$. The result is

$$\mathbf{AA}' = [\mathbf{A} - (\mathbf{X}'\mathbf{X})^{-1}\mathbf{X}'][\mathbf{A} - (\mathbf{X}'\mathbf{X})^{-1}\mathbf{X}']' + (\mathbf{X}'\mathbf{X})^{-1}. \tag{7.17}$$

The matrix $[\mathbf{A} - (\mathbf{X}'\mathbf{X})^{-1}\mathbf{X}'][\mathbf{A} - (\mathbf{X}'\mathbf{X})^{-1}\mathbf{X}']'$ on the right side of (7.17) is positive semidefinite (see Theorem 2.6d), and, by Theorem 2.6a (ii), the diagonal elements are greater than or equal to zero. These diagonal elements can be made equal to zero by choosing $\mathbf{A} = (\mathbf{X}'\mathbf{X})^{-1}\mathbf{X}'$. (This value of $\mathbf{A}$ also satisfies the unbiasedness condition $\mathbf{AX} = \mathbf{I}$.) The resulting minimum variance estimator of $\boldsymbol{\beta}$ is

$$\mathbf{Ay} = (\mathbf{X}'\mathbf{X})^{-1}\mathbf{X}'\mathbf{y},$$

which is equal to the least–squares estimator $\hat{\boldsymbol{\beta}}$. □

The Gauss–Markov theorem is sometimes stated as follows. If $E(\mathbf{y}) = \mathbf{X}\boldsymbol{\beta}$ and $\text{cov}(\mathbf{y}) = \sigma^2\mathbf{I}$, the least-squares estimators $\hat{\beta}_0, \hat{\beta}_1, \ldots, \hat{\beta}_k$ are *best linear unbiased estimators* (BLUE). In this expression, *best* means minimum variance and *linear* indicates that the estimators are linear functions of $\mathbf{y}$.

The remarkable feature of the Gauss–Markov theorem is its distributional generality. The result holds for any distribution of $\mathbf{y}$; normality is not required. The only assumptions used in the proof are $E(\mathbf{y}) = \mathbf{X}\boldsymbol{\beta}$ and $\text{cov}(\mathbf{y}) = \sigma^2\mathbf{I}$. If these assumptions do not hold, $\hat{\boldsymbol{\beta}}$ may be biased or each $\hat{\beta}_j$ may have a larger variance than that of some other estimator.

The Gauss–Markov theorem is easily extended to a linear combination of the $\hat{\beta}$'s, as follows.

Corollary 1. If $E(\mathbf{y}) = \mathbf{X}\boldsymbol{\beta}$ and $\text{cov}(\mathbf{y}) = \sigma^2\mathbf{I}$, the best linear unbiased estimator of $\mathbf{a}'\boldsymbol{\beta}$ is $\mathbf{a}'\hat{\boldsymbol{\beta}}$, where $\hat{\boldsymbol{\beta}}$ is the least–squares estimator $\hat{\boldsymbol{\beta}} = (\mathbf{X}'\mathbf{X})^{-1}\mathbf{X}'\mathbf{y}$.

PROOF. See Problem 7.7. □

Note that Theorem 7.3d is concerned with the form of the estimator $\hat{\boldsymbol{\beta}}$ for a given $\mathbf{X}$ matrix. Once $\mathbf{X}$ is chosen, the variances of the $\hat{\beta}_j$'s are minimized by $\hat{\boldsymbol{\beta}} = (\mathbf{X}'\mathbf{X})^{-1}\mathbf{X}'\mathbf{y}$. However, in Theorem 7.3c, we have $\text{cov}(\hat{\boldsymbol{\beta}}) = \sigma^2(\mathbf{X}'\mathbf{X})^{-1}$ and therefore $\text{var}(\hat{\beta}_j)$ and $\text{cov}(\hat{\beta}_i, \hat{\beta}_j)$ depend on the values of the x_j's. Thus the configuration of $\mathbf{X}'\mathbf{X}$ is important in estimation of the β_j's (this was illustrated in Problem 6.4).

In both estimation and testing, there are advantages to choosing the x's (or the centered x's) to be orthogonal so that $\mathbf{X}'\mathbf{X}$ is diagonal. These advantages include minimizing the variances of the $\hat{\beta}_j$'s and maximizing the power of tests about the β_j's (Chapter 8). For clarification, we note that orthogonality is necessary but not sufficient for minimizing variances and maximizing power. For example, if there are two x's, with values to be selected in a rectangular space, the points could be evenly placed on a grid, which would be an orthogonal pattern. However, the optimal orthogonal pattern would be to place one-fourth of the points at each corner of the rectangle.

A fourth property of $\hat{\boldsymbol{\beta}}$ is as follows. The predicted value $\hat{y} = \hat{\beta}_0 + \hat{\beta}_1 x_1 + \cdots + \hat{\beta}_k x_k = \hat{\boldsymbol{\beta}}'\mathbf{x}$ is invariant to simple linear changes of scale on the x's, where $\mathbf{x} = (1, x_1, x_2, \ldots, x_k)'$. Let the rescaled variables be denoted by $z_j = c_j x_j$, $j = 1, 2, \ldots, k$, where the c_j terms are constants. Thus $\mathbf{x}$ is transformed to $\mathbf{z} = (1, c_1 x_1, \ldots, c_k x_k)'$. The following theorem shows that $\hat{y}$ based on $\mathbf{z}$ is the same as $\hat{y}$ based on $\mathbf{x}$.

Theorem 7.3e. If $\mathbf{x} = (1, x_1, \ldots, x_k)'$ and $\mathbf{z} = (1, c_1 x_1, \ldots, c_k x_k)'$, then $\hat{y} = \hat{\boldsymbol{\beta}}'\mathbf{x} = \hat{\boldsymbol{\beta}}_z'\mathbf{z}$, where $\hat{\boldsymbol{\beta}}_z$ is the least squares estimator from the regression of y on $\mathbf{z}$.

PROOF. From (2.29), we can rewrite $\mathbf{z}$ as $\mathbf{z} = \mathbf{D}\mathbf{x}$, where $\mathbf{D} = \text{diag}(1, c_1, c_2, \ldots, c_k)$. Then, the $\mathbf{X}$ matrix is transformed to $\mathbf{Z} = \mathbf{X}\mathbf{D}$ [see (2.28)]. We substitute $\mathbf{Z} = \mathbf{X}\mathbf{D}$ in the least-squares estimator $\hat{\boldsymbol{\beta}}_z = (\mathbf{Z}'\mathbf{Z})^{-1}\mathbf{Z}'\mathbf{y}$ to obtain

$$\begin{aligned} \hat{\boldsymbol{\beta}}_z &= (\mathbf{Z}'\mathbf{Z})^{-1}\mathbf{Z}'\mathbf{y} = [(\mathbf{X}\mathbf{D})'(\mathbf{X}\mathbf{D})]^{-1}(\mathbf{X}\mathbf{D})'\mathbf{y} \\ &= \mathbf{D}^{-1}(\mathbf{X}'\mathbf{X})^{-1}\mathbf{X}'\mathbf{y} \qquad \text{[by (2.49)]} \\ &= \mathbf{D}^{-1}\hat{\boldsymbol{\beta}}, \end{aligned} \tag{7.18}$$

where $\hat{\boldsymbol{\beta}}$ is the usual estimator for y regressed on the x's. Then

$$\hat{\boldsymbol{\beta}}_z'\mathbf{z} = (\mathbf{D}^{-1}\hat{\boldsymbol{\beta}})'\mathbf{D}\mathbf{x} = \hat{\boldsymbol{\beta}}'\mathbf{x}.$$

□

In the following corollary to Theorem 7.3e, the invariance of $\hat{y}$ is extended to any full-rank linear transformation of the x variables.

Corollary 1. The predicted value $\hat{y}$ is invariant to a full-rank linear transformation on the x's.

Proof. We can express a full-rank linear transformation of the x's as

$$\mathbf{Z} = \mathbf{XK} = (\mathbf{j}, \mathbf{X}_1)\begin{pmatrix} 1 & \mathbf{0}' \\ \mathbf{0} & \mathbf{K}_1 \end{pmatrix} = (\mathbf{j} + \mathbf{X}_1\mathbf{0}, \mathbf{j}\mathbf{0}' + \mathbf{X}_1\mathbf{K}_1) = (\mathbf{j}, \mathbf{X}_1\mathbf{K}_1),$$

where $\mathbf{K}_1$ is nonsingular and

$$\mathbf{X}_1 = \begin{pmatrix} x_{11} & x_{12} & \dots & x_{1k} \\ x_{21} & x_{22} & \dots & x_{2k} \\ \vdots & \vdots & & \vdots \\ x_{n1} & x_{n2} & \dots & x_{nk} \end{pmatrix}. \tag{7.19}$$

We partition $\mathbf{X}$ and $\mathbf{K}$ in this way so as to transform only the x's in $\mathbf{X}_1$, leaving the first column of $\mathbf{X}$ unaffected. Now $\hat{\boldsymbol{\beta}}_z$ becomes

$$\hat{\boldsymbol{\beta}}_z = (\mathbf{Z}'\mathbf{Z})^{-1}\mathbf{Z}'\mathbf{y} = \mathbf{K}^{-1}\hat{\boldsymbol{\beta}}, \tag{7.20}$$

and we have

$$\hat{y} = \hat{\boldsymbol{\beta}}_z'\mathbf{z} = \hat{\boldsymbol{\beta}}'\mathbf{x}, \tag{7.21}$$

where $\mathbf{z} = \mathbf{K}'\mathbf{x}$. □

In addition to $\hat{y}$, the sample variance s^2 (Section 7.3.3) is also invariant to changes of scale on the x variable (see Problem 7.10). The following are invariant to changes of scale on y as well as on the x's (but not to a joint linear transformation on y and the x's): t statistics (Section 8.5), F statistics (Chapter 8), and R^2 (Sections 7.7 and 10.3).

7.3.3 An Estimator for σ^2

The method of least squares does not yield a function of the y and x values in the sample that we can minimize to obtain an estimator of σ^2. However, we can devise an unbiased estimator for σ^2 based on the least-squares estimator $\hat{\boldsymbol{\beta}}$. By assumption 2 following (7.3), σ^2 is the same for each y_i, $i = 1, 2, \dots, n$. By (3.6), σ^2 is defined by $\sigma^2 = E[y_i - E(y_i)]^2$, and by assumption 1, we obtain

$$E(y_i) = \beta_0 + \beta_i x_{i1} + \beta_2 x_{i2} + \dots + \beta_k x_{ik} = \mathbf{x}_i'\boldsymbol{\beta},$$

where $\mathbf{x}_i'$ is the ith row of $\mathbf{X}$. Thus σ^2 becomes

$$\sigma^2 = E[y_i - \mathbf{x}_i'\boldsymbol{\beta}]^2.$$

We estimate σ^2 by a corresponding average from the sample

$$s^2 = \frac{1}{n-k-1}\sum_{i=1}^{n}(y_i - \mathbf{x}_i'\hat{\boldsymbol{\beta}})^2, \tag{7.22}$$

where n is the sample size and k is the number of x's. Note that, by the corollary to Theorem 7.3d, $\mathbf{x}_i'\hat{\boldsymbol{\beta}}$ is the BLUE of $\mathbf{x}_i'\boldsymbol{\beta}$.

Using (7.7), we can write (7.22) as

$$s^2 = \frac{1}{n-k-1}(\mathbf{y} - \mathbf{X}\hat{\boldsymbol{\beta}})'(\mathbf{y} - \mathbf{X}\hat{\boldsymbol{\beta}}) \tag{7.23}$$

$$= \frac{\mathbf{y}'\mathbf{y} - \hat{\boldsymbol{\beta}}'\mathbf{X}'\mathbf{y}}{n-k-1} = \frac{\text{SSE}}{n-k-1}, \tag{7.24}$$

where $\text{SSE} = (\mathbf{y} - \mathbf{X}\hat{\boldsymbol{\beta}})'(\mathbf{y} - \mathbf{X}\hat{\boldsymbol{\beta}}) = \mathbf{y}'\mathbf{y} - \hat{\boldsymbol{\beta}}'\mathbf{X}'\mathbf{y}$. With the denominator $n-k-1$, s^2 is an unbiased estimator of σ^2, as shown below.

Theorem 7.3f. If s^2 is defined by (7.22), (7.23), or (7.24) and if $E(\mathbf{y}) = \mathbf{X}\boldsymbol{\beta}$ and $\text{cov}(\mathbf{y}) = \sigma^2\mathbf{I}$, then

$$E(s^2) = \sigma^2. \tag{7.25}$$

PROOF. Using (7.24) and (7.6), we write SSE as a quadratic form:

$$\begin{aligned} \text{SSE} &= \mathbf{y}'\mathbf{y} - \hat{\boldsymbol{\beta}}'\mathbf{X}'\mathbf{y} = \mathbf{y}'\mathbf{y} - \mathbf{y}'\mathbf{X}(\mathbf{X}'\mathbf{X})^{-1}\mathbf{X}'\mathbf{y} \\ &= \mathbf{y}'\left[\mathbf{I} - \mathbf{X}(\mathbf{X}'\mathbf{X})^{-1}\mathbf{X}'\right]\mathbf{y}. \end{aligned} \tag{7.26}$$

By Theorem 5.2a, we have

$$\begin{aligned} E(\text{SSE}) &= \text{tr}\left\{\left[\mathbf{I} - \mathbf{X}(\mathbf{X}'\mathbf{X})^{-1}\mathbf{X}'\right]\sigma^2\mathbf{I}\right\} \\ &\quad + E(\mathbf{y}')\left[\mathbf{I} - \mathbf{X}(\mathbf{X}'\mathbf{X})^{-1}\mathbf{X}'\right]E(\mathbf{y}) \\ &= \sigma^2\text{tr}\left[\mathbf{I} - \mathbf{X}(\mathbf{X}'\mathbf{X})^{-1}\mathbf{X}'\right] \\ &\quad + \boldsymbol{\beta}'\mathbf{X}'\left[\mathbf{I} - \mathbf{X}(\mathbf{X}'\mathbf{X})^{-1}\mathbf{X}'\right]\mathbf{X}\boldsymbol{\beta} \\ &= \sigma^2\left\{n - \text{tr}\left[\mathbf{X}(\mathbf{X}'\mathbf{X})^{-1}\mathbf{X}'\right]\right\} \\ &\quad + \boldsymbol{\beta}'\mathbf{X}'\mathbf{X}\boldsymbol{\beta} - \boldsymbol{\beta}'\mathbf{X}'\mathbf{X}(\mathbf{X}'\mathbf{X})^{-1}\mathbf{X}'\mathbf{X}\boldsymbol{\beta} \\ &= \sigma^2\left\{n - \text{tr}[\mathbf{X}'\mathbf{X}(\mathbf{X}'\mathbf{X})^{-1}]\right\} \\ &\quad + \boldsymbol{\beta}'\mathbf{X}'\mathbf{X}\boldsymbol{\beta} - \boldsymbol{\beta}'\mathbf{X}'\mathbf{X}\boldsymbol{\beta} \qquad \text{[by (2.87)]}. \end{aligned}$$

Since $\mathbf{X}'\mathbf{X}$ is $(k+1) \times (k+1)$, this becomes

$$E(\text{SSE}) = \sigma^2[n - \text{tr}(\mathbf{I}_{k+1})] = \sigma^2(n - k - 1).$$

□

Corollary 1. An unbiased estimator of $\text{cov}(\hat{\boldsymbol{\beta}})$ in (7.14) is given by

$$\widehat{\text{cov}}(\hat{\boldsymbol{\beta}}) = s^2(\mathbf{X}'\mathbf{X})^{-1}. \tag{7.27}$$

□

Note the correspondence between $n - (k+1)$ and $\mathbf{y}'\mathbf{y} - \hat{\boldsymbol{\beta}}'\mathbf{X}'\mathbf{y}$; there are n terms in $\mathbf{y}'\mathbf{y}$ and $k+1$ terms in $\hat{\boldsymbol{\beta}}'\mathbf{X}'\mathbf{y} = \hat{\boldsymbol{\beta}}'\mathbf{X}'\mathbf{X}\hat{\boldsymbol{\beta}}$ [see (7.8)]. A corresponding property of the sample is that each additional x (and $\hat{\beta}$) in the model reduces SSE (see Problem 7.13).

Since SSE is a quadratic function of $\mathbf{y}$, it is not a best *linear* unbiased estimator. The optimality property of s^2 is given in the following theorem.

Theorem 7.3g. If $E(\boldsymbol{\varepsilon}) = \mathbf{0}$, $\text{cov}(\boldsymbol{\varepsilon}) = \sigma^2\mathbf{I}$, and $E(\varepsilon_i^4) = 3\sigma^4$ for the linear model $\mathbf{y} = \mathbf{X}\boldsymbol{\beta} + \boldsymbol{\varepsilon}$, then s^2 in (7.23) or (7.24) is the best (minimum variance) *quadratic* unbiased estimator of σ^2.

Proof. See Graybill (1954), Graybill and Wortham (1956), or Wang and Chow (1994, pp. 161–163). □

Example 7.3.3. For the data in Table 7.1, we have

$$\begin{aligned}
\text{SSE} &= \mathbf{y}'\mathbf{y} - \hat{\boldsymbol{\beta}}'\mathbf{X}\mathbf{y} \\
&= 840 - (5.3754,\ 3.0118,\ -1.2855)\begin{pmatrix} 90 \\ 482 \\ 872 \end{pmatrix} \\
&= 840 - 814.541 = 25.459, \\
s^2 &= \frac{\text{SSE}}{n-k-1} = \frac{25.459}{12-2-1} = 2.829.
\end{aligned}$$

□

7.4 GEOMETRY OF LEAST SQUARES

In Sections 7.1–7.3 we presented the multiple linear regression model as the matrix equation $\mathbf{y} = \mathbf{X}\boldsymbol{\beta} + \boldsymbol{\varepsilon}$ in (7.4). We defined the principle of least-squares estimation in terms of deviations from the model [see (7.7)], and then used matrix calculus and matrix algebra to derive the estimators of $\boldsymbol{\beta}$ in (7.6) and of σ^2 in (7.23) and (7.24). We now present an alternate but equivalent derivation of these estimators based completely on geometric ideas.

It is important to clarify first what the geometric approach to least squares is *not*. In two dimensions, we illustrated the principle of least squares by creating a two-dimensional scatter plot (Fig. 6.1) of the n points $(x_1, y_1), (x_2, y_2), \ldots, (x_n, y_n)$. We then visualized the least-squares regression line as the best-fitting straight line to the data. This approach can be generalized to present the least-squares estimate in multiple linear regression on the basis of the best-fitting hyperplane in $(k+1)$-dimensional space to the n points $(x_{11}, x_{12}, \ldots, x_{1k}, y_1), (x_{21}, x_{22}, \ldots, x_{2k}, y_2), \ldots, (x_{n1}, x_{n2}, \ldots, x_{nk}, y_n)$. Although this approach is somewhat useful in visualizing multiple linear regression, the geometric approach to least-squares estimation in multiple linear regression does *not* involve this high-dimensional generalization.

The geometric approach to be discussed below is appealing because of its mathematical elegance. For example, the estimator is derived without the use of matrix calculus. Also, the geometric approach provides deeper insight into statistical inference. Several advanced statistical methods including kernel smoothing (Eubank and Eubank 1999), Fourier analysis (Bloomfield 2000), and wavelet analysis (Ogden 1997) can be understood as generalizations of this geometric approach. The geometric approach to linear models was first proposed by Fisher (Mahalanobis 1964). Christensen (1996) and Jammalamadaka and Sengupta (2003) discuss the linear statistical model almost completely from the geometric perspective.

7.4.1 Parameter Space, Data Space, and Prediction Space

The geometric approach to least squares begins with two high-dimensional spaces, a $(k+1)$-dimensional space and an n-dimensional space. The unknown parameter vector $\boldsymbol{\beta}$ can be viewed as a single point in $(k+1)$-dimensional space, with axes corresponding to the $k+1$ regression coefficients $\beta_0, \beta_1, \beta_0, \ldots, \beta_k$. Hence we call this space the *parameter space* (Fig. 7.3). Similarly, the data vector $\mathbf{y}$ can be viewed as a

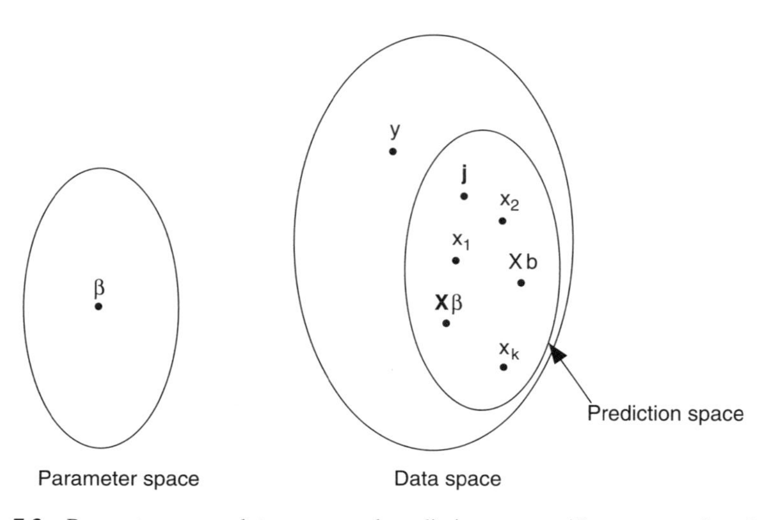

Figure 7.3 Parameter space, data space, and prediction space with representative elements.

single point in n-dimensional space with axes corresponding to the n observations. We call this space the *data space*.

The $\mathbf{X}$ matrix of the multiple regression model (7.4) can be written as a partitioned matrix in terms of its $k+1$ columns as

$$\mathbf{X} = (\mathbf{j}, \mathbf{x}_1, \mathbf{x}_2, \mathbf{x}_3, \ldots, \mathbf{x}_k).$$

The columns of $\mathbf{X}$, including $\mathbf{j}$, are all n-dimensional vectors and are therefore points in the data space. Note that because we assumed that $\mathbf{X}$ is of rank $k+1$, these vectors are linearly independent. The set of all possible linear combinations of the columns of $\mathbf{X}$ (Section 2.3) constitutes a subset of the data space. Elements of this subset can be written as

$$\mathbf{Xb} = b_0\mathbf{j} + b_1\mathbf{x}_1 + b_2\mathbf{x}_2 + \cdots + b_k\mathbf{x}_k, \tag{7.28}$$

where $\mathbf{b}$ is any $k+1$ vector, that is, any vector in the parameter space. This subset actually has the status of a *subspace* because it is closed under addition and scalar multiplication (Harville 1997, pp. 28–29). This subset is said to be the subspace generated or *spanned* by the columns of $\mathbf{X}$, and we will call this subspace the *prediction space*. The columns of $\mathbf{X}$ constitute a *basis set* for the prediction space.

7.4.2 Geometric Interpretation of the Multiple Linear Regression Model

The multiple linear regression model (7.4) states that $\mathbf{y}$ is equal to a vector in the prediction space, $E(\mathbf{y}) = \mathbf{X}\boldsymbol{\beta}$, plus a vector of random errors, $\boldsymbol{\varepsilon}$ (Fig. 7.4). The

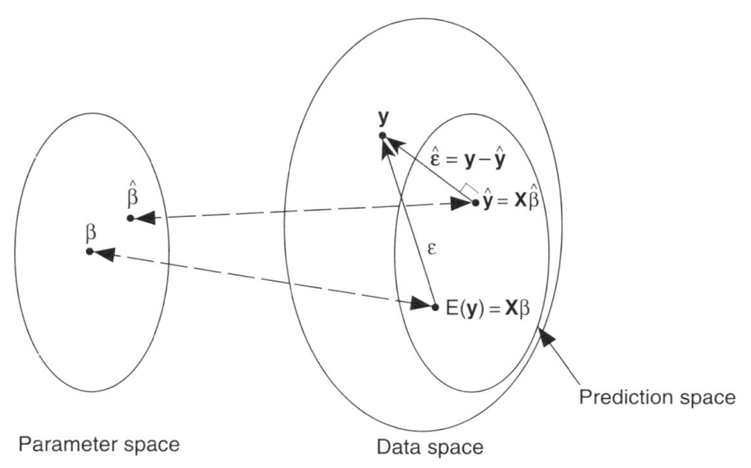

Figure 7.4 Geometric relationships of vectors associated with the multiple linear regression model.

problem is that neither $\boldsymbol{\beta}$ nor $\boldsymbol{\varepsilon}$ is known. However, the data vector $\mathbf{y}$, which is not in the prediction space, is known. And it is known that $E(\mathbf{y})$ is in the prediction space.

Multiple linear regression can be understood geometrically as the process of finding a sensible estimate of $E(\mathbf{y})$ in the prediction space and then determining the vector in the parameter space that is associated with this estimate (Fig. 7.4). The estimate of $E(\mathbf{y})$ is denoted as $\hat{\mathbf{y}}$, and the associated vector in the parameter space is denoted as $\hat{\boldsymbol{\beta}}$.

A reasonable geometric idea is to estimate $E(\mathbf{y})$ using the point in the prediction space that is closest to $\mathbf{y}$. It turns out that $\hat{\mathbf{y}}$, the closest point in the prediction space to $\mathbf{y}$, can be found by noting that the difference vector $\hat{\boldsymbol{\varepsilon}} = \mathbf{y} - \hat{\mathbf{y}}$ must be orthogonal (perpendicular) to the prediction space (Harville 1997, p. 170). Furthermore, because the prediction space is spanned by the columns of $\mathbf{X}$, the point $\hat{\mathbf{y}}$ must be such that $\hat{\boldsymbol{\varepsilon}}$ is orthogonal to the columns of $\mathbf{X}$. Using an extension of (2.80), we therefore seek $\hat{\mathbf{y}}$ such that

$$\mathbf{X}'\hat{\boldsymbol{\varepsilon}} = \mathbf{0}$$

or

$$\mathbf{X}'(\mathbf{y} - \hat{\mathbf{y}}) = \mathbf{X}'(\mathbf{y} - \mathbf{X}\hat{\boldsymbol{\beta}}) = \mathbf{X}'\mathbf{y} - \mathbf{X}'\mathbf{X}\hat{\boldsymbol{\beta}} = \mathbf{0}, \tag{7.29}$$

which implies that

$$\mathbf{X}'\mathbf{X}\hat{\boldsymbol{\beta}} = \mathbf{X}'\mathbf{y}.$$

Thus, using purely geometric ideas, we obtain the normal equations (7.8) and consequently the usual least-squares estimator $\hat{\boldsymbol{\beta}}$ in (7.6). We can then calculate $\hat{\mathbf{y}}$ as $\mathbf{X}\hat{\boldsymbol{\beta}} = \mathbf{X}(\mathbf{X}'\mathbf{X})^{-1}\mathbf{X}'\mathbf{y} = \mathbf{H}\mathbf{y}$. Also, $\hat{\boldsymbol{\varepsilon}} = \mathbf{y} - \mathbf{X}\hat{\boldsymbol{\beta}} = (\mathbf{I} - \mathbf{H})\mathbf{y}$ can be taken as an estimate of $\boldsymbol{\varepsilon}$. Since $\hat{\boldsymbol{\varepsilon}}$ is a vector in $(n - k - 1)$-dimensional space, it seems reasonable to estimate σ^2 as the squared length (2.22) of $\hat{\boldsymbol{\varepsilon}}$ divided by $n - k - 1$. In other words, a sensible estimator of σ^2 is $s^2 = \mathbf{y}'(\mathbf{I} - \mathbf{H})\mathbf{y}/(n - k - 1)$, which is equal to (7.25).

7.5 THE MODEL IN CENTERED FORM

The model in (7.3) for each y_i can be written in terms of centered x variables as

$$\begin{aligned} y_i &= \beta_0 + \beta_1 x_{i1} + \beta_2 x_{i2} + \cdots + \beta_k x_{ik} + \varepsilon_i \\ &= \alpha + \beta_1(x_{i1} - \bar{x}_1) + \beta_2(x_{i2} - \bar{x}_2) + \cdots + \beta_k(x_{ik} - \bar{x}_k) + \varepsilon_i, \end{aligned} \tag{7.30}$$

$i = 1, 2, \ldots, n$, where

$$\alpha = \beta_0 + \beta_1\bar{x}_1 + \beta_2\bar{x}_2 + \ldots + \beta_k\bar{x}_k \tag{7.31}$$

and $\bar{x}_j = \sum_{i=1}^n x_{ij}/n$, $j = 1, 2, \ldots, k$. The centered form of the model is useful in expressing certain hypothesis tests (Section 8.1), in a search for influential observations (Section 9.2), and in providing other insights.

In matrix form, the centered model (7.30) for $y_1, y_2, \ldots, y_n$ becomes

$$\mathbf{y} = (\mathbf{j}, \mathbf{X}_c)\begin{pmatrix} \alpha \\ \boldsymbol{\beta}_1 \end{pmatrix} + \boldsymbol{\varepsilon}, \tag{7.32}$$

where $\boldsymbol{\beta}_1 = (\beta_1, \beta_2, \ldots, \beta_k)'$,

$$\mathbf{X}_c = \left(\mathbf{I} - \frac{1}{n}\mathbf{J}\right)\mathbf{X}_1 = \begin{pmatrix} x_{11} - \bar{x}_1 & x_{12} - \bar{x}_2 & \ldots & x_{1k} - \bar{x}_k \\ x_{21} - \bar{x}_1 & x_{22} - \bar{x}_2 & \ldots & x_{2k} - \bar{x}_k \\ \vdots & \vdots & & \vdots \\ x_{n1} - \bar{x}_1 & x_{n2} - \bar{x}_2 & \ldots & x_{nk} - \bar{x}_k \end{pmatrix}, \tag{7.33}$$

and $\mathbf{X}_1$ is as given in (7.19). The matrix $\mathbf{I} - (1/n)\mathbf{J}$ is sometimes called the *centering matrix*.

As in (7.8), the normal equations for the model in (7.32) are

$$(\mathbf{j}, \mathbf{X}_c)'(\mathbf{j}, \mathbf{X}_c)\begin{pmatrix} \hat{\alpha} \\ \hat{\boldsymbol{\beta}}_1 \end{pmatrix} = (\mathbf{j}, \mathbf{X}_c)'\mathbf{y}. \tag{7.34}$$

By (2.35) and (2.39), the product $(\mathbf{j}, \mathbf{X}_c)'(\mathbf{j}, \mathbf{X}_c)$ on the left side of (7.34) becomes

$$\begin{aligned} (\mathbf{j}, \mathbf{X}_c)'(\mathbf{j}, \mathbf{X}_c) &= \begin{pmatrix} \mathbf{j}' \\ \mathbf{X}_c' \end{pmatrix}(\mathbf{j}, \mathbf{X}_c) = \begin{pmatrix} \mathbf{j}'\mathbf{j} & \mathbf{j}'\mathbf{X}_c \\ \mathbf{X}_c'\mathbf{j} & \mathbf{X}_c'\mathbf{X}_c \end{pmatrix} \\ &= \begin{pmatrix} n & \mathbf{0}' \\ \mathbf{0} & \mathbf{X}_c'\mathbf{X}_c \end{pmatrix}, \end{aligned} \tag{7.35}$$

where $\mathbf{j}'\mathbf{X}_c = \mathbf{0}'$ because the columns of $\mathbf{X}_c$ sum to zero (Problem 7.16). The right side of (7.34) can be written as

$$(\mathbf{j}, \mathbf{X}_c)'\mathbf{y} = \begin{pmatrix} \mathbf{j}' \\ \mathbf{X}_c' \end{pmatrix}\mathbf{y} = \begin{pmatrix} n\bar{y} \\ \mathbf{X}_c'\mathbf{y} \end{pmatrix}.$$

The least-squares estimators are then given by

$$\begin{aligned} \begin{pmatrix} \hat{\alpha} \\ \hat{\boldsymbol{\beta}}_1 \end{pmatrix} &= [(\mathbf{j}, \mathbf{X}_c)'(\mathbf{j}, \mathbf{X}_c)]^{-1}(\mathbf{j}, \mathbf{X}_c)'\mathbf{y} = \begin{pmatrix} n & \mathbf{0}' \\ \mathbf{0} & \mathbf{X}_c'\mathbf{X}_c \end{pmatrix}^{-1}\begin{pmatrix} n\bar{y} \\ \mathbf{X}_c'\mathbf{y} \end{pmatrix} \\ &= \begin{pmatrix} 1/n & \mathbf{0}' \\ \mathbf{0} & (\mathbf{X}_c'\mathbf{X}_c)^{-1} \end{pmatrix}\begin{pmatrix} n\bar{y} \\ \mathbf{X}_c'\mathbf{y} \end{pmatrix} = \begin{pmatrix} \bar{y} \\ (\mathbf{X}_c'\mathbf{X}_c)^{-1}\mathbf{X}_c'\mathbf{y} \end{pmatrix}, \end{aligned}$$

or

$$\hat{\alpha} = \bar{y}, \tag{7.36}$$

$$\hat{\boldsymbol{\beta}}_1 = (\mathbf{X}_c'\mathbf{X}_c)^{-1}\mathbf{X}_c'\mathbf{y}. \tag{7.37}$$

These estimators are the same as the usual least-squares estimators $\hat{\boldsymbol{\beta}} = (\mathbf{X}'\mathbf{X})^{-1}\mathbf{X}'\mathbf{y}$ in (7.6), with the adjustment

$$\hat{\beta}_0 = \hat{\alpha} - \hat{\beta}_1\bar{x}_1 - \hat{\beta}_2\bar{x} - \cdots - \hat{\beta}_k\bar{x}_k = \bar{y} - \hat{\boldsymbol{\beta}}_1'\bar{\mathbf{x}} \tag{7.38}$$

obtained from an estimator of α in (7.31) (see Problem 7.17).

When we express $\hat{\mathbf{y}}$ in centered form

$$\hat{\mathbf{y}} = \hat{\alpha} + \hat{\beta}_1(x_1 - \bar{x}_1) + \cdots + \hat{\beta}_k(x_k - \bar{x}_k),$$

it is clear that the fitted regression plane passes through the point $(\bar{x}_1, \bar{x}_2, \ldots, \bar{x}_k, \bar{y})$.

Adapting the expression for SSE (7.24) to the centered model with centered $\hat{y}$'s, we obtain

$$\text{SSE} = \sum_{i=1}^{n}(y_i - \bar{y})^2 - \hat{\boldsymbol{\beta}}_1'\mathbf{X}_c'\mathbf{y}, \tag{7.39}$$

which turns out to be equal to $\text{SSE} = \mathbf{y}'\mathbf{y} - \hat{\boldsymbol{\beta}}'\mathbf{X}'\mathbf{y}$ (see Problem 7.19).

We can use (7.36)–(7.38) to express $\hat{\boldsymbol{\beta}}_1$ and $\hat{\boldsymbol{\beta}}_0$ in terms of sample variances and covariances, which will be useful in comparing these estimators with those for the random-x case in Chapter 10. We first define a sample covariance matrix for the x variables and a vector of sample covariances between y and the x's

$$\mathbf{S}_{xx} = \begin{pmatrix} s_1^2 & s_{12} & \cdots & s_{1k} \\ s_{21} & s_2^2 & \cdots & s_{2k} \\ \vdots & \vdots & & \vdots \\ s_{k1} & s_{k2} & \cdots & s_k^2 \end{pmatrix}, \qquad \mathbf{s}_{yx} = \begin{pmatrix} s_{y1} \\ s_{y2} \\ \vdots \\ s_{yk} \end{pmatrix}, \tag{7.40}$$

where, s_i^2, s_{ij}, and s_{yi} are analogous to s^2 and s_{xy} defined in (5.6) and (5.15); for example

$$s_2^2 = \frac{\sum_{i=1}^{n}(x_{i2} - \bar{x}_2)^2}{n-1}, \tag{7.41}$$

$$s_{12} = \frac{\sum_{i=1}^{n}(x_{i1} - \bar{x}_1)(x_{i2} - \bar{x}_2)}{n-1}, \tag{7.42}$$

$$s_{y2} = \frac{\sum_{i=1}^{n}(x_{i2} - \bar{x}_2)(y_i - \bar{y})}{n-1}, \tag{7.43}$$

with $\bar{x}_2 = \sum_{i=1}^{n} x_{i2}/n$. However, since the x's are fixed, these sample variances and covariances do not estimate population variances and covariances. If the x's were random variables, as in Chapter 10, the s_i^2, s_{ij}, and s_{yi} values would estimate population parameters.

To express $\hat{\boldsymbol{\beta}}_1$ and $\hat{\boldsymbol{\beta}}_0$ in terms of $\mathbf{S}_{xx}$ and $\mathbf{s}_{yx}$, we first write $\mathbf{S}_{xx}$ and $\mathbf{s}_{yx}$ in terms of the centered matrix $\mathbf{X}_c$:

$$\mathbf{S}_{xx} = \frac{\mathbf{X}_c'\mathbf{X}_c}{n-1}, \tag{7.44}$$

$$\mathbf{s}_{yx} = \frac{\mathbf{X}_c'\mathbf{y}}{n-1}. \tag{7.45}$$

Note that $\mathbf{X}_c'\mathbf{y}$ in (7.45) contains terms of the form $\sum_{i=1}^n (x_{ij} - \bar{x}_j)y_i$ rather than $\sum_{i=1}^n (x_{ij} - \bar{x}_j)(y_i - \bar{y})$ as in (7.43). It can readily be shown that $\sum_i (x_{ij} - \bar{x}_j)(y_i - \bar{y}) = \sum_i (x_{ij} - \bar{x}_j)y_i$ (see Problem 6.2).

From (7.37), (7.44), and (7.45), we have

$$\hat{\boldsymbol{\beta}}_1 = (n-1)(\mathbf{X}_c'\mathbf{X}_c)^{-1}\frac{\mathbf{X}_c'\mathbf{y}}{n-1} = \left(\frac{\mathbf{X}_c'\mathbf{X}_c}{n-1}\right)^{-1}\frac{\mathbf{X}_c'\mathbf{y}}{n-1} = \mathbf{S}_{xx}^{-1}\mathbf{s}_{yx}, \tag{7.46}$$

and from (7.38) and (7.46), we obtain

$$\hat{\beta}_0 = \hat{\alpha} - \hat{\boldsymbol{\beta}}_1'\bar{\mathbf{x}} = \bar{y} - \mathbf{s}_{yx}'\mathbf{S}_{xx}^{-1}\bar{\mathbf{x}}. \tag{7.47}$$

Example 7.5. For the data in Table 7.1, we calculate $\hat{\boldsymbol{\beta}}_1$ and $\hat{\beta}_0$ using (7.46) and (7.47).

$$\hat{\boldsymbol{\beta}}_1 = \mathbf{S}_{xx}^{-1}\mathbf{s}_{yx} = \begin{pmatrix} 6.4242 & 8.5455 \\ 8.5455 & 12.4545 \end{pmatrix}^{-1}\begin{pmatrix} 8.3636 \\ 9.7273 \end{pmatrix}$$

$$= \begin{pmatrix} 3.0118 \\ -1.2855 \end{pmatrix},$$

$$\hat{\beta}_0 = \bar{y} - \mathbf{s}_{yx}'\mathbf{S}_{xx}^{-1}\bar{\mathbf{x}}$$

$$= 7.5000 - (3.0118, \; -1.2855)\begin{pmatrix} 4.3333 \\ 8.5000 \end{pmatrix}$$

$$= 7.500 - 2.1246 = 5.3754.$$

These values are the same as those obtained in Example 7.3.1a. □

7.6 NORMAL MODEL

7.6.1 Assumptions

Thus far we have made no normality assumptions about the random variables $y_1, y_2, \ldots, y_n$. To the assumptions in Section 7.2, we now add that

$$\mathbf{y} \text{ is } N_n(\mathbf{X}\boldsymbol{\beta}, \sigma^2\mathbf{I}) \quad \text{or} \quad \boldsymbol{\varepsilon} \text{ is } N_n(\mathbf{0}, \sigma^2\mathbf{I}).$$

Under normality, $\sigma_{ij} = 0$ implies that the y (or ε) variables are independent, as well as uncorrelated.

7.6.2 Maximum Likelihood Estimators for $\boldsymbol{\beta}$ and σ^2

With the normality assumption, we can obtain maximum likelihood estimators. The likelihood function is the joint density of the y's, which we denote by $L(\boldsymbol{\beta}, \sigma^2)$. We seek values of the unknown $\boldsymbol{\beta}$ and σ^2 that maximize $L(\boldsymbol{\beta}, \sigma^2)$ for the given y and x values in the sample.

In the case of the normal density function, it is possible to find maximum likelihood estimators $\hat{\boldsymbol{\beta}}$ and $\hat{\sigma}^2$ by differentiation. Because the normal density involves a product and an exponential, it is simpler to work with $\ln L(\boldsymbol{\beta}, \sigma^2)$, which achieves its maximum for the same values of $\boldsymbol{\beta}$ and σ^2 as does $L(\boldsymbol{\beta}, \sigma^2)$.

The maximum likelihood estimators for $\boldsymbol{\beta}$ and σ^2 are given in the following theorem.

Theorem 7.6a. If $\mathbf{y}$ is $N_n(\mathbf{X}\boldsymbol{\beta}, \sigma^2\mathbf{I})$, where $\mathbf{X}$ is $n \times (k+1)$ of rank $k+1 < n$, the maximum likelihood estimators of $\boldsymbol{\beta}$ and σ^2 are

$$\hat{\boldsymbol{\beta}} = (\mathbf{X}'\mathbf{X})^{-1}\mathbf{X}'\mathbf{y}, \tag{7.48}$$

$$\hat{\sigma}^2 = \frac{1}{n}(\mathbf{y} - \mathbf{X}\hat{\boldsymbol{\beta}})'(\mathbf{y} - \mathbf{X}\hat{\boldsymbol{\beta}}). \tag{7.49}$$

PROOF. We sketch the proof. For the remaining steps, see Problem 7.21. The likelihood function (joint density of $y_1, y_2, \ldots, y_n$) is given by the multivariate normal density (4.9)

$$\begin{aligned} L(\boldsymbol{\beta}, \sigma^2) = f(\mathbf{y}; \boldsymbol{\beta}, \sigma^2) &= \frac{1}{(2\pi)^{n/2}|\sigma^2\mathbf{I}|^{1/2}} e^{-(\mathbf{y}-\mathbf{X}\boldsymbol{\beta})'(\sigma^2\mathbf{I})^{-1}(\mathbf{y}-\mathbf{X}\boldsymbol{\beta})/2} \\ &= \frac{1}{(2\pi\sigma^2)^{n/2}} e^{-(\mathbf{y}-\mathbf{X}\boldsymbol{\beta})'(\mathbf{y}-\mathbf{X}\boldsymbol{\beta})/2\sigma^2}. \end{aligned} \tag{7.50}$$

[Since the y_i's are independent, $L(\boldsymbol{\beta}, \sigma^2)$ can also be obtained as $\prod_{i=1}^{n} f(y_i; \mathbf{x}_i'\boldsymbol{\beta}, \sigma^2)$.] Then $\ln L(\boldsymbol{\beta}, \sigma^2)$ becomes

$$\ln L(\boldsymbol{\beta}, \sigma^2) = -\frac{n}{2}\ln(2\pi) - \frac{n}{2}\ln\sigma^2 - \frac{1}{2\sigma^2}(\mathbf{y} - \mathbf{X}\boldsymbol{\beta})'(\mathbf{y} - \mathbf{X}\boldsymbol{\beta}). \tag{7.51}$$

Taking the partial derivatives of $\ln L(\boldsymbol{\beta}, \sigma^2)$ with respect to $\boldsymbol{\beta}$ and σ^2 and setting the results equal to zero will produce (7.48) and (7.49). To verify that $\hat{\boldsymbol{\beta}}$ maximizes (7.50) or (7.51), see (7.10). □

The maximum likelihood estimator $\hat{\boldsymbol{\beta}}$ in (7.48) is the same as the least-squares estimator $\hat{\boldsymbol{\beta}}$ in Theorem 7.3a. The estimator $\hat{\sigma}^2$ in (7.49) is biased since the denominator is n rather than $n - k - 1$. We often use the unbiased estimator s^2 given in (7.23) or (7.24).

7.6.3 Properties of $\hat{\boldsymbol{\beta}}$ and $\hat{\sigma}^2$

We now consider some properties of $\hat{\boldsymbol{\beta}}$ and $\hat{\sigma}^2$ (or s^2) under the normal model. The distributions of $\hat{\boldsymbol{\beta}}$ and $\hat{\sigma}^2$ are given in the following theorem.

Theorem 7.6b. Suppose that $\mathbf{y}$ is $N_n(\mathbf{X}\boldsymbol{\beta}, \sigma^2\mathbf{I})$, where $\mathbf{X}$ is $n \times (k+1)$ of rank $k + 1 < n$ and $\boldsymbol{\beta} = (\beta_0, \beta_1, \ldots, \beta_k)'$. Then the maximum likelihood estimators $\hat{\boldsymbol{\beta}}$ and $\hat{\sigma}^2$ given in Theorem 7.6a have the following distributional properties:

(i) $\hat{\boldsymbol{\beta}}$ is $N_{k+1}[\boldsymbol{\beta}, \sigma^2(\mathbf{X}'\mathbf{X})^{-1}]$.
(ii) $n\hat{\sigma}^2/\sigma^2$ is $\chi^2(n-k-1)$, or equivalently, $(n-k-1)s^2/\sigma^2$ is $\chi^2(n-k-1)$.
(iii) $\hat{\boldsymbol{\beta}}$ and $\hat{\sigma}^2$ (or s^2) are independent.

PROOF

(i) Since $\hat{\boldsymbol{\beta}} = (\mathbf{X}'\mathbf{X})^{-1}\mathbf{X}'\mathbf{y}$ is a linear function of $\mathbf{y}$ of the form $\hat{\boldsymbol{\beta}} = \mathbf{A}\mathbf{y}$, where $\mathbf{A} = (\mathbf{X}'\mathbf{X})^{-1}\mathbf{X}'$ is a constant matrix, then by Theorem 4.4a(ii), $\hat{\boldsymbol{\beta}}$ is $N_{k+1}[\boldsymbol{\beta}, \sigma^2(\mathbf{X}'\mathbf{X})^{-1}]$.
(ii) The result follows from Corollary 2 to Theorem 5.5.
(iii) The result follows from Corollary 1 to Theorem 5.6a.

□

Another property of $\hat{\boldsymbol{\beta}}$ and $\hat{\sigma}^2$ under normality is that they are sufficient statistics. Intuitively, a statistic is sufficient for a parameter if the statistic summarizes all the information in the sample about the parameter. Sufficiency of $\hat{\boldsymbol{\beta}}$ and $\hat{\sigma}^2$ can be established by the Neyman factorization theorem [see Hogg and Craig (1995, p. 318) or Graybill (1976, pp. 69–70)], which states that $\hat{\boldsymbol{\beta}}$ and $\hat{\sigma}^2$ are jointly sufficient for $\boldsymbol{\beta}$ and σ^2 if the density $f(\mathbf{y}; \boldsymbol{\beta}, \sigma^2)$ can be factored as $f(\mathbf{y}; \boldsymbol{\beta}, \sigma^2) = g(\hat{\boldsymbol{\beta}}, \hat{\sigma}^2, \boldsymbol{\beta}, \sigma^2)h(\mathbf{y})$, where $h(\mathbf{y})$ does not depend on $\boldsymbol{\beta}$ or σ^2. The following theorem shows that $\hat{\boldsymbol{\beta}}$ and $\hat{\sigma}^2$ satisfy this criterion.

Theorem 7.6c. If $\mathbf{y}$ is $N_n(\mathbf{X}\boldsymbol{\beta}, \sigma^2\mathbf{I})$, then $\hat{\boldsymbol{\beta}}$ and $\hat{\sigma}^2$ are jointly sufficient for $\boldsymbol{\beta}$ and σ^2.

PROOF. The density $f(\mathbf{y}; \boldsymbol{\beta}, \sigma^2)$ is given in (7.50). In the exponent, we add and subtract $\mathbf{X}\hat{\boldsymbol{\beta}}$ to obtain

$$
\begin{aligned}
(\mathbf{y} - \mathbf{X}\boldsymbol{\beta})'(\mathbf{y} - \mathbf{X}\boldsymbol{\beta}) &= (\mathbf{y} - \mathbf{X}\hat{\boldsymbol{\beta}} + \mathbf{X}\hat{\boldsymbol{\beta}} - \mathbf{X}\boldsymbol{\beta})'(\mathbf{y} - \mathbf{X}\hat{\boldsymbol{\beta}} + \mathbf{X}\hat{\boldsymbol{\beta}} - \mathbf{X}\boldsymbol{\beta}) \\
&= [(\mathbf{y} - \mathbf{X}\hat{\boldsymbol{\beta}}) + \mathbf{X}(\hat{\boldsymbol{\beta}} - \boldsymbol{\beta})]'[(\mathbf{y} - \mathbf{X}\hat{\boldsymbol{\beta}}) + \mathbf{X}(\hat{\boldsymbol{\beta}} - \boldsymbol{\beta})].
\end{aligned}
$$

Expanding this in terms of $\mathbf{y} - \mathbf{X}\hat{\boldsymbol{\beta}}$ and $\mathbf{X}(\hat{\boldsymbol{\beta}} - \boldsymbol{\beta})$, we obtain four terms, two of which vanish because of the normal equations $\mathbf{X}'\mathbf{X}\hat{\boldsymbol{\beta}} = \mathbf{X}'\mathbf{y}$. The result is

$$\begin{aligned}(\mathbf{y} - \mathbf{X}\boldsymbol{\beta})'(\mathbf{y} - \mathbf{X}\boldsymbol{\beta}) &= (\mathbf{y} - \mathbf{X}\hat{\boldsymbol{\beta}})'(\mathbf{y} - \mathbf{X}\hat{\boldsymbol{\beta}}) + (\hat{\boldsymbol{\beta}} - \boldsymbol{\beta})'\mathbf{X}'\mathbf{X}(\hat{\boldsymbol{\beta}} - \boldsymbol{\beta}) \\ &= n\hat{\sigma}^2 + (\hat{\boldsymbol{\beta}} - \boldsymbol{\beta})'\mathbf{X}'\mathbf{X}(\hat{\boldsymbol{\beta}} - \boldsymbol{\beta}).\end{aligned} \tag{7.52}$$

We can now write the density (7.50) as

$$f(\mathbf{y}; \boldsymbol{\beta}, \sigma^2) = \frac{1}{(2\pi\sigma^2)^{n/2}} e^{-[n\hat{\sigma}^2 + (\hat{\boldsymbol{\beta}} - \boldsymbol{\beta})'\mathbf{X}'\mathbf{X}(\hat{\boldsymbol{\beta}} - \boldsymbol{\beta})]/2\sigma^2},$$

which is of the form

$$f(\mathbf{y}; \boldsymbol{\beta}, \sigma^2) = g(\hat{\boldsymbol{\beta}}, \hat{\sigma}^2, \boldsymbol{\beta}, \sigma^2)h(\mathbf{y}),$$

where $h(\mathbf{y}) = 1$. Therefore, by the Neyman factorization theorem, $\hat{\boldsymbol{\beta}}$ and $\hat{\sigma}^2$ are jointly sufficient for $\boldsymbol{\beta}$ and σ^2. □

Note that $\hat{\boldsymbol{\beta}}$ and $\hat{\sigma}^2$ are jointly sufficient for $\boldsymbol{\beta}$ and σ^2, not independently sufficient; that is, $f(\mathbf{y}; \boldsymbol{\beta}, \sigma^2)$ does not factor into the form $g_1(\hat{\boldsymbol{\beta}}, \boldsymbol{\beta})g_2(\hat{\sigma}^2, \sigma^2)h(\mathbf{y})$. Also note that because $s^2 = n\hat{\sigma}^2/(n - k - 1)$, the proof to Theorem 7.6c can be easily modified to show that $\hat{\boldsymbol{\beta}}$ and s^2 are also jointly sufficient for $\boldsymbol{\beta}$ and σ^2.

Since $\hat{\boldsymbol{\beta}}$ and s^2 are sufficient, no other estimators can improve on the information they extract from the sample to estimate $\boldsymbol{\beta}$ and σ^2. Thus, it is not surprising that $\hat{\boldsymbol{\beta}}$ and s^2 are minimum variance unbiased estimators (each $\hat{\beta}_j$ in $\hat{\boldsymbol{\beta}}$ has minimum variance). This result is given in the following theorem.

Theorem 7.6d. If $\mathbf{y}$ is $N_n(\mathbf{X}\boldsymbol{\beta}, \sigma^2\mathbf{I})$, then $\hat{\boldsymbol{\beta}}$ and s^2 have minimum variance among all unbiased estimators.

PROOF. See Graybill (1976, p. 176) or Christensen (1996, pp. 25–27). □

In Theorem 7.3d, the elements of $\hat{\boldsymbol{\beta}}$ were shown to have minimum variance among all *linear unbiased* estimators. With the normality assumption added in Theorem 7.6d, the elements of $\hat{\boldsymbol{\beta}}$ have minimum variance among all *unbiased* estimators. Similarly, by Theorem 7.3g, s^2 has minimum variance among all *quadratic unbiased* estimators. With the added normality assumption in Theorem 7.6d, s^2 has minimum variance among all *unbiased* estimators.

The following corollary to Theorem 7.6d is analogous to Corollary 1 of Theorem 7.3d.

Corollary 1. If $\mathbf{y}$ is $N_n(\mathbf{X}\boldsymbol{\beta}, \sigma^2\mathbf{I})$, then the minimum variance unbiased estimator of $\mathbf{a}'\boldsymbol{\beta}$ is $\mathbf{a}'\hat{\boldsymbol{\beta}}$, where $\hat{\boldsymbol{\beta}}$ is the maximum likelihood estimator given in (7.48). □

7.7 R^2 IN FIXED-x REGRESSION

In (7.39), we have SSE $= \sum_{i=1}^{n}(y_i - \bar{y})^2 - \hat{\boldsymbol{\beta}}_1'\mathbf{X}_c'\mathbf{y}$. Thus the corrected total sum of squares SST $= \sum_i (y_i - \bar{y})^2$ can be partitioned as

$$\sum_{i=1}^{n}(y_i - \bar{y})^2 = \hat{\boldsymbol{\beta}}_1'\mathbf{X}_c'\mathbf{y} + \text{SSE}, \tag{7.53}$$

$$\text{SST} = \text{SSR} + \text{SSE},$$

where SSR $= \hat{\boldsymbol{\beta}}_1'\mathbf{X}_c'\mathbf{y}$ is the *regression sum of squares*. From (7.37), we obtain $\mathbf{X}_c'\mathbf{y} = \mathbf{X}_c'\mathbf{X}_c\hat{\boldsymbol{\beta}}_1$, and multiplying this by $\hat{\boldsymbol{\beta}}_1'$ gives $\hat{\boldsymbol{\beta}}_1'\mathbf{X}_c'\mathbf{y} = \hat{\boldsymbol{\beta}}_1'\mathbf{X}_c'\mathbf{X}_c\hat{\boldsymbol{\beta}}_1$. Then SSR $= \hat{\boldsymbol{\beta}}_1'\mathbf{X}_c'\mathbf{y}$ can be written as

$$\text{SSR} = \hat{\boldsymbol{\beta}}_1'\mathbf{X}_c'\mathbf{X}_c\hat{\boldsymbol{\beta}}_1 = (\mathbf{X}_c\hat{\boldsymbol{\beta}}_1)'(\mathbf{X}_c\hat{\boldsymbol{\beta}}_1). \tag{7.54}$$

In this form, it is clear that SSR is due to $\boldsymbol{\beta}_1 = (\beta_1, \beta_2, \ldots, \beta_k)'$.

The proportion of the total sum of squares due to regression is

$$R^2 = \frac{\hat{\boldsymbol{\beta}}_1'\mathbf{X}_c'\mathbf{X}_c\hat{\boldsymbol{\beta}}_1}{\sum_{i=1}^{n}(y_i - \bar{y})^2} = \frac{\text{SSR}}{\text{SST}}, \tag{7.55}$$

which is known as the *coefficient of determination* or the *squared multiple correlation*. The ratio in (7.55) is a measure of model fit and provides an indication of how well the x's predict y.

The partitioning in (7.53) can be rewritten as the identity

$$\begin{aligned}\sum_{i=1}^{n}(y_i - \bar{y})^2 = \mathbf{y}'\mathbf{y} - n\bar{y}^2 &= (\hat{\boldsymbol{\beta}}'\mathbf{X}'\mathbf{y} - n\bar{y}^2) + (\mathbf{y}'\mathbf{y} - \hat{\boldsymbol{\beta}}'\mathbf{X}'\mathbf{y}) \\ &= \text{SSR} + \text{SSE},\end{aligned}$$

which leads to an alternative expression for R^2:

$$R^2 = \frac{\hat{\boldsymbol{\beta}}'\mathbf{X}'\mathbf{y} - n\bar{y}^2}{\mathbf{y}'\mathbf{y} - n\bar{y}^2}. \tag{7.56}$$

The positive square root R obtained from (7.55) or (7.56) is called the *multiple correlation coefficient*. If the x variables were random, R would estimate a population multiple correlation (see Section (10.4)).

We list some properties of R^2 and R:

1. The range of R^2 is $0 \le R^2 \le 1$. If all the $\hat{\beta}_j$'s were zero, except for $\hat{\beta}_0$, R^2 would be 0. (This event has probability 0 for continuous data.) If all the y values fell on the fitted surface, that is, if $y_i = \hat{y}_i$, $i = 1, 2, \ldots, n$, then R^2 would be 1.

2. $R = r_{y\hat{y}}$; that is, the multiple correlation is equal to the simple correlation [see (6.18)] between the observed y_i's and the fitted $\hat{y}_i$'s.
3. Adding a variable x to the model increases (cannot decrease) the value of R^2.
4. If $\beta_1 = \beta_2 = \cdots = \beta_k = 0$, then

$$E(R^2) = \frac{k}{n-1}. \tag{7.57}$$

   Note that the $\hat{\beta}_j$'s will not be 0 when the β_j's are 0.
5. R^2 cannot be partitioned into k components, each of which is uniquely attributable to an x_j, unless the x's are mutually orthogonal, that is, $\sum_{i=1}^{n}(x_{ij} - \bar{x}_j)(x_{im} - \bar{x}_m) = 0$ for $j \neq m$.
6. R^2 is invariant to full-rank linear transformations on the x's and to a scale change on y (but not invariant to a joint linear transformation including y and the x's).

In properties 3 and 4 we see that if k is a relatively large fraction of n, it is possible to have a large value of R^2 that is not meaningful. In this case, x's that do not contribute to predicting y may appear to do so in a particular example, and the estimated regression equation may not be a useful estimator of the population model. To correct for this tendency, an adjusted R^2, denoted by R_a^2, was proposed by Ezekiel (1930). To obtain R_a^2, we first subtract $k/(n-1)$ in (7.57) from R^2 in order to correct for the bias when $\beta_1 = \beta_2 = \ldots = \beta_k = 0$. This correction, however, would make R_a^2 too small when the β's are large, so a further modification is made so that $R_a^2 = 1$ when $R^2 = 1$. Thus R_a^2 is defined as

$$R_a^2 = \frac{(R^2 - \frac{k}{n-1})(n-1)}{n-k-1} = \frac{(n-1)R^2 - k}{n-k-1}. \tag{7.58}$$

Example 7.7. For the data in Table 7.1 in Example 7.2, we obtain R^2 by (7.56) and R_a^2 by (7.58). The values of $\hat{\boldsymbol{\beta}}'\mathbf{X}'\mathbf{y}$ and $\mathbf{y}'\mathbf{y}$ are given in Example 7.3.3.

$$\begin{aligned} R^2 &= \frac{\hat{\boldsymbol{\beta}}'\mathbf{X}'\mathbf{y} - n\bar{y}^2}{\mathbf{y}'\mathbf{y} - n\bar{y}^2} = \frac{814.5410 - 12(7.5)^2}{840 - 12(7.5)^2} \\ &= \frac{139.5410}{165.0000} = .8457, \\ R_a^2 &= \frac{(n-1)R^2 - k}{n-k-1} = \frac{(11)(.8457) - 2}{9} = .8114. \end{aligned}$$

□

Using (7.44) and (7.46), we can express R^2 in (7.55) in terms of sample variances and covariances:

$$R^2 = \frac{\hat{\boldsymbol{\beta}}_1'\mathbf{X}_c'\mathbf{X}_c\hat{\boldsymbol{\beta}}_1}{\sum_{i=1}^{n}(y_i - \bar{y})^2} = \frac{\mathbf{s}_{yx}'\mathbf{S}_{xx}^{-1}(n-1)\mathbf{S}_{xx}\mathbf{S}_{xx}^{-1}\mathbf{s}_{yx}}{\sum_{i=1}^{n}(y_i - \bar{y})^2} = \frac{\mathbf{s}_{yx}'\mathbf{S}_{xx}^{-1}\mathbf{s}_{yx}}{s_y^2}. \tag{7.59}$$

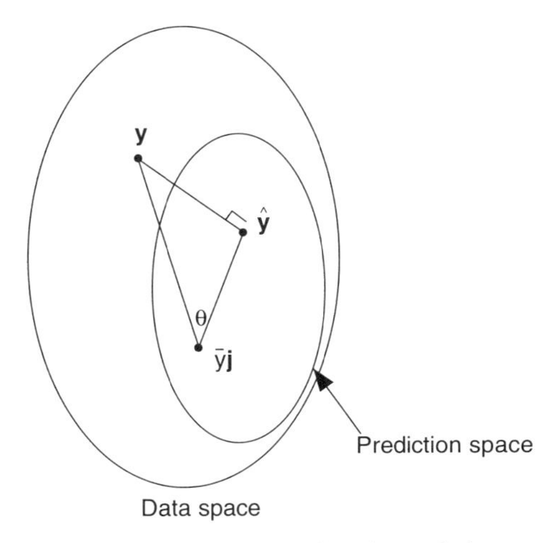

Figure 7.5 Multiple correlation R as cosine of θ, the angle between $\mathbf{y} - \bar{y}\mathbf{j}$ and $\hat{\mathbf{y}} - \bar{y}\mathbf{j}$.

This form of R^2 will facilitate a comparison with R^2 for the random-x case in Section (10.4) [see (10.34)].

Geometrically, R is the cosine of the angle θ between $\mathbf{y}$ and $\hat{\mathbf{y}}$ corrected for their means. The mean of $\hat{y}_1, \hat{y}_2, \ldots, \hat{y}_n$ is $\bar{y}$, the same as the mean of $y_1, y_2, \ldots, y_n$ (see Problem 7.30). Thus the centered forms of $\mathbf{y}$ and $\hat{\mathbf{y}}$ are $\mathbf{y} - \bar{y}\mathbf{j}$ and $\hat{\mathbf{y}} - \bar{y}\mathbf{j}$. The angle between them is illustrated in Figure 7.5. (Note that $\bar{y}\mathbf{j}$ is in the estimation space since it is a multiple of the first column of $\mathbf{X}$.)

To show that $\cos\theta$ is equal to the square root of R^2 as given by (7.56), we use (2.81) for the cosine of the angle between two vectors:

$$\cos\theta = \frac{(\mathbf{y} - \bar{y}\mathbf{j})'(\hat{\mathbf{y}} - \bar{y}\mathbf{j})}{\sqrt{[(\mathbf{y} - \bar{y}\mathbf{j})'(\mathbf{y} - \bar{y}\mathbf{j})][(\hat{\mathbf{y}} - \bar{y}\mathbf{j})'(\hat{\mathbf{y}} - \bar{y}\mathbf{j})]}}. \tag{7.60}$$

To simplify (7.60), we use the identity $\mathbf{y} - \bar{y}\mathbf{j} = (\hat{\mathbf{y}} - \bar{y}\mathbf{j}) + (\mathbf{y} - \hat{\mathbf{y}})$, which can also be seen geometrically in Figure 7.5. The vectors $\hat{\mathbf{y}} - \bar{y}\mathbf{j}$ and $\mathbf{y} - \hat{\mathbf{y}}$ on the right side of this identity are orthogonal since $\hat{\mathbf{y}} - \bar{y}\mathbf{j}$ is in the prediction space. Thus the numerator of (7.60) can be written as

$$\begin{aligned}(\mathbf{y} - \bar{y}\mathbf{j})'(\hat{\mathbf{y}} - \bar{y}\mathbf{j}) &= [(\hat{\mathbf{y}} - \bar{y}\mathbf{j}) + (\mathbf{y} - \hat{\mathbf{y}})]'(\hat{\mathbf{y}} - \bar{y}\mathbf{j}) \\ &= (\hat{\mathbf{y}} - \bar{y}\mathbf{j})'(\hat{\mathbf{y}} - \bar{y}\mathbf{j}) + (\mathbf{y} - \hat{\mathbf{y}})'(\hat{\mathbf{y}} - \bar{y}\mathbf{j}) \\ &= (\hat{\mathbf{y}} - \bar{y}\mathbf{j})'(\hat{\mathbf{y}} - \bar{y}\mathbf{j}) + 0.\end{aligned}$$

Then (7.60) becomes

$$\cos\theta = \frac{\sqrt{(\hat{\mathbf{y}} - \bar{y}\mathbf{j})'(\hat{\mathbf{y}} - \bar{y}\mathbf{j})}}{\sqrt{(\mathbf{y} - \bar{y}\mathbf{j})'(\mathbf{y} - \bar{y}\mathbf{j})}} = R, \tag{7.61}$$

which is easily shown to be the square root of R^2 as given by (7.56). This is equivalent to property 2 following (7.56): $R = r_{y\hat{y}}$.

We can write (7.61) in the form

$$R^2 = \frac{\sum_{i=1}^n (\hat{y}_i - \bar{y})^2}{\sum_{i=1}^n (y_i - \bar{y})^2} = \frac{\text{SSR}}{\text{SST}},$$

in which $\text{SSR} = \sum_{i=1}^n (\hat{y}_i - \bar{y})^2$ is a sum of squares for the $\hat{y}_i$'s. Then the partitioning $\text{SST} = \text{SSR} + \text{SSE}$ below (7.53) can be written as

$$\sum_{i=1}^n (y_i - \bar{y})^2 = \sum_{i=1}^n (\hat{y}_i - \bar{y})^2 + \sum_{i=1}^n (y_i - \hat{y}_i)^2,$$

which is analogous to (6.17) for simple linear regression.

7.8 GENERALIZED LEAST SQUARES: $\text{cov}(\mathbf{Y}) = \sigma^2\mathbf{V}$

We now consider models in which the y variables are correlated or have differing variances, so that $\text{cov}(\mathbf{y}) \neq \sigma^2\mathbf{I}$. In simple linear regression, larger values of x_i may lead to larger values of $\text{var}(y_i)$. In either simple or multiple regression, if $y_1, y_2, \ldots, y_n$ occur at sequential points in time, they are typically correlated. For cases such as these, in which the assumption $\text{cov}(\mathbf{y}) = \sigma^2\mathbf{I}$ is no longer appropriate, we use the model

$$\mathbf{y} = \mathbf{X}\boldsymbol{\beta} + \boldsymbol{\varepsilon}, \quad E(\mathbf{y}) = \mathbf{X}\boldsymbol{\beta}, \quad \text{cov}(\mathbf{y}) = \Sigma = \sigma^2\mathbf{V}, \tag{7.62}$$

where $\mathbf{X}$ is full-rank and $\mathbf{V}$ is a known positive definite matrix. The usage $\Sigma = \sigma^2\mathbf{V}$ permits estimation of σ^2 in some convenient contexts (see Examples 7.8.1 and 7.8.2). The $n \times n$ matrix $\mathbf{V}$ has n diagonal elements and $\binom{n}{2}$ elements above (or below) the diagonal. If $\mathbf{V}$ were unknown, these $\binom{n}{2} + n$ distinct elements could not be estimated from a sample of n observations. In certain applications, a simpler structure for $\mathbf{V}$ is assumed that permits estimation. Such structures are illustrated in Examples 7.8.1 and 7.8.2.

7.8.1 Estimation of $\boldsymbol{\beta}$ and σ^2 when $\text{cov}(\mathbf{y}) = \sigma^2\mathbf{V}$

In the following theorem we give estimators of $\boldsymbol{\beta}$ and σ^2 for the model in (7.62).

Theorem 7.8a. Let $\mathbf{y} = \mathbf{X}\boldsymbol{\beta} + \boldsymbol{\varepsilon}$, let $E(\mathbf{y}) = \mathbf{X}\boldsymbol{\beta}$, and let $\text{cov}(\mathbf{y}) = \text{cov}(\boldsymbol{\varepsilon}) = \sigma^2\mathbf{V}$, where $\mathbf{X}$ is a full-rank matrix and $\mathbf{V}$ is a known positive definite matrix. For this model, we obtain the following results:

(i) The best linear unbiased estimator (BLUE) of $\boldsymbol{\beta}$ is

$$\hat{\boldsymbol{\beta}} = (\mathbf{X}'\mathbf{V}^{-1}\mathbf{X})^{-1}\mathbf{X}'\mathbf{V}^{-1}\mathbf{y}. \tag{7.63}$$

(ii) The covariance matrix for $\hat{\boldsymbol{\beta}}$ is

$$\text{cov}(\hat{\boldsymbol{\beta}}) = \sigma^2(\mathbf{X}'\mathbf{V}^{-1}\mathbf{X})^{-1}. \tag{7.64}$$

(iii) An unbiased estimator of σ^2 is

$$s^2 = \frac{(\mathbf{y} - \mathbf{X}\hat{\boldsymbol{\beta}})'\mathbf{V}^{-1}(\mathbf{y} - \mathbf{X}\hat{\boldsymbol{\beta}})}{n-k-1} \tag{7.65}$$

$$= \frac{\mathbf{y}'[\mathbf{V}^{-1} - \mathbf{V}^{-1}\mathbf{X}(\mathbf{X}'\mathbf{V}^{-1}\mathbf{X})^{-1}\mathbf{X}'\mathbf{V}^{-1}]\mathbf{y}}{n-k-1}, \tag{7.66}$$

where $\hat{\boldsymbol{\beta}}$ is as given by (7.63).

PROOF. We prove part (i). For parts (ii) and (iii), see Problems (7.32) and (7.33).

1. Since $\mathbf{V}$ is positive definite, there exists an $n \times n$ nonsingular matrix $\mathbf{P}$ such that $\mathbf{V} = \mathbf{P}\mathbf{P}'$ (see Theorem 2.6c). Multiplying $\mathbf{y} = \mathbf{X}\boldsymbol{\beta} + \boldsymbol{\varepsilon}$ by $\mathbf{P}^{-1}$, we obtain $\mathbf{P}^{-1}\mathbf{y} = \mathbf{P}^{-1}\mathbf{X}\boldsymbol{\beta} + \mathbf{P}^{-1}\boldsymbol{\varepsilon}$, for which $E(\mathbf{P}^{-1}\boldsymbol{\varepsilon}) = \mathbf{P}^{-1}E(\boldsymbol{\varepsilon}) = \mathbf{P}^{-1}\mathbf{0} = \mathbf{0}$ and

$$\begin{aligned} \text{cov}(\mathbf{P}^{-1}\boldsymbol{\varepsilon}) &= \mathbf{P}^{-1}\text{cov}(\boldsymbol{\varepsilon})(\mathbf{P}^{-1})' \qquad \text{[by (3.44)]} \\ &= \mathbf{P}^{-1}\sigma^2\mathbf{V}(\mathbf{P}^{-1})' = \sigma^2\mathbf{P}^{-1}\mathbf{P}\mathbf{P}'(\mathbf{P}')^{-1} = \sigma^2\mathbf{I}. \end{aligned}$$

Thus the assumptions for Theorem 7.3d are satisfied for the model $\mathbf{P}^{-1}\mathbf{y} = \mathbf{P}^{-1}\mathbf{X}\boldsymbol{\beta} + \mathbf{P}^{-1}\boldsymbol{\varepsilon}$, and the least-squares estimator $\hat{\boldsymbol{\beta}} = [(\mathbf{P}^{-1}\mathbf{X})'(\mathbf{P}^{-1}\mathbf{X})]^{-1}(\mathbf{P}^{-1}\mathbf{X})'\mathbf{P}^{-1}\mathbf{y}$ is BLUE. Using Theorems 2.2b and 2.5b, this can be written as

$$\begin{aligned} \hat{\boldsymbol{\beta}} &= [\mathbf{X}'(\mathbf{P}^{-1})'\mathbf{P}^{-1}\mathbf{X}]^{-1}\mathbf{X}'(\mathbf{P}^{-1})'\mathbf{P}^{-1}\mathbf{y} \\ &= [\mathbf{X}'(\mathbf{P}')^{-1}\mathbf{P}^{-1}\mathbf{X}]^{-1}\mathbf{X}'(\mathbf{P}')^{-1}\mathbf{P}^{-1}\mathbf{y} \qquad \text{[by (2.48)]} \\ &= [\mathbf{X}'(\mathbf{P}\mathbf{P}')^{-1}\mathbf{X}]^{-1}\mathbf{X}'(\mathbf{P}\mathbf{P}')^{-1}\mathbf{y} \qquad \text{[by (2.49)]} \\ &= (\mathbf{X}'\mathbf{V}^{-1}\mathbf{X})^{-1}\mathbf{X}'\mathbf{V}^{-1}\mathbf{y}. \end{aligned}$$

□

Note that since $\mathbf{X}$ is full-rank, $\mathbf{X}'\mathbf{V}^{-1}\mathbf{X}$ is positive definite (see Theorem 2.6b). The estimator $\hat{\boldsymbol{\beta}} = (\mathbf{X}'\mathbf{V}^{-1}\mathbf{X})^{-1}\mathbf{X}'\mathbf{V}^{-1}\mathbf{y}$ is usually called the *generalized least-squares* estimator. The same estimator is obtained under a normality assumption.

Theorem 7.8b. If $\mathbf{y}$ is $N_n(\mathbf{X}\boldsymbol{\beta}, \sigma^2\mathbf{V})$, where $\mathbf{X}$ is full-rank and $\mathbf{V}$ is a known positive definite matrix, where $\mathbf{X}$ is $n \times (k+1)$ of rank $k+1$, then the maximum likelihood estimators for $\boldsymbol{\beta}$ and σ^2 are

$$\hat{\boldsymbol{\beta}} = (\mathbf{X}'\mathbf{V}^{-1}\mathbf{X})^{-1}\mathbf{X}'\mathbf{V}^{-1}\mathbf{y},$$

$$\hat{\sigma}^2 = \frac{1}{n}(\mathbf{y} - \mathbf{X}\hat{\boldsymbol{\beta}})'\mathbf{V}^{-1}(\mathbf{y} - \mathbf{X}\hat{\boldsymbol{\beta}}).$$

Proof. The likelihood function is

$$L(\boldsymbol{\beta}, \sigma^2) = \frac{1}{(2\pi)^{n/2}|\sigma^2\mathbf{V}|^{1/2}} e^{-(\mathbf{y}-\mathbf{X}\boldsymbol{\beta})'(\sigma^2\mathbf{V})^{-1}(\mathbf{y}-\mathbf{X}\boldsymbol{\beta})/2}.$$

By (2.69), $|\sigma^2\mathbf{V}| = (\sigma^2)^n|\mathbf{V}|$. Hence

$$L(\boldsymbol{\beta}, \sigma^2) = \frac{1}{(2\pi\sigma^2)^{n/2}|\mathbf{V}|^{1/2}} e^{-(\mathbf{y}-\mathbf{X}\boldsymbol{\beta})'\mathbf{V}^{-1}(\mathbf{y}-\mathbf{X}\boldsymbol{\beta})/2\sigma^2}.$$

The results can be obtained by differentiation of $\ln L(\boldsymbol{\beta}, \sigma^2)$ with respect to $\boldsymbol{\beta}$ and with respect to σ^2. □

We illustrate an application of generalized least squares.

Example 7.8.1. Consider the centered model in (7.32)

$$\mathbf{y} = (\mathbf{j}, \mathbf{X}_c)\begin{pmatrix} \alpha \\ \boldsymbol{\beta}_1 \end{pmatrix} + \boldsymbol{\varepsilon},$$

with covariance pattern

$$\Sigma = \sigma^2[(1-\rho)\mathbf{I} + \rho\mathbf{J}] = \sigma^2\mathbf{V} \tag{7.67}$$

$$= \sigma^2\begin{pmatrix} 1 & \rho & \cdots & \rho \\ \rho & 1 & \cdots & \rho \\ \vdots & \vdots & & \vdots \\ \rho & \rho & \cdots & 1 \end{pmatrix},$$

in which all variables have the same variance σ^2 and all pairs of variables have the same correlation ρ. This covariance pattern was introduced in Problem 5.26 and is assumed for certain repeated measures and intraclass correlation designs. See (3.19) for a definition of ρ.

By (7.63), we have

$$\hat{\boldsymbol{\beta}} = \begin{pmatrix} \hat{\alpha} \\ \hat{\boldsymbol{\beta}}_1 \end{pmatrix} = (\mathbf{X}'\mathbf{V}^{-1}\mathbf{X})^{-1}\mathbf{X}'\mathbf{V}^{-1}\mathbf{y}.$$

For the centered model with $\mathbf{X} = (\mathbf{j}, \mathbf{X}_c)$, the matrix $\mathbf{X}'\mathbf{V}^{-1}\mathbf{X}$ becomes

$$\begin{aligned} \mathbf{X}'\mathbf{V}^{-1}\mathbf{X} &= \begin{pmatrix} \mathbf{j}' \\ \mathbf{X}_c' \end{pmatrix} \mathbf{V}^{-1}(\mathbf{j}, \mathbf{X}_c) \\ &= \begin{pmatrix} \mathbf{j}'\mathbf{V}^{-1}\mathbf{j} & \mathbf{j}'\mathbf{V}^{-1}\mathbf{X}_c \\ \mathbf{X}_c'\mathbf{V}^{-1}\mathbf{j} & \mathbf{X}_c'\mathbf{V}^{-1}\mathbf{X}_c \end{pmatrix}. \end{aligned}$$

The inverse of the $n \times n$ matrix $\mathbf{V} = (1 - \rho)\mathbf{I} + \rho\mathbf{J}$ in (7.67) is given by

$$\mathbf{V}^{-1} = a(\mathbf{I} - b\rho\mathbf{J}), \tag{7.68}$$

where $a = 1/(1 - \rho)$ and $b = 1/[1 + (n - 1)\rho]$. Using $\mathbf{V}^{-1}$ in (7.68), $\mathbf{X}'\mathbf{V}^{-1}\mathbf{X}$ becomes

$$\mathbf{X}'\mathbf{V}^{-1}\mathbf{X} = \begin{pmatrix} bn & \mathbf{0}' \\ \mathbf{0} & a\mathbf{X}_c'\mathbf{X}_c \end{pmatrix}. \tag{7.69}$$

Similarly

$$\mathbf{X}'\mathbf{V}^{-1}\mathbf{y} = \begin{pmatrix} bn\bar{y} \\ a\mathbf{X}_c'\mathbf{y} \end{pmatrix}. \tag{7.70}$$

We therefore have

$$\begin{pmatrix} \hat{\alpha} \\ \hat{\boldsymbol{\beta}}_1 \end{pmatrix} = (\mathbf{X}'\mathbf{V}^{-1}\mathbf{X})^{-1}\mathbf{X}'\mathbf{V}^{-1}\mathbf{y} = \begin{pmatrix} \bar{y} \\ (\mathbf{X}_c'\mathbf{X}_c)^{-1}\mathbf{X}_c'\mathbf{y} \end{pmatrix},$$

which is the same as (7.36) and (7.37). Thus the usual least-squares estimators are BLUE for a covariance structure with equal variances and equal correlations. □

7.8.2 Misspecification of the Error Structure

Suppose that the model is $\mathbf{y} = \mathbf{X}\boldsymbol{\beta} + \boldsymbol{\varepsilon}$ with $\text{cov}(\mathbf{y}) = \sigma^2\mathbf{V}$, as in (7.62), and we mistakenly (or deliberately) use the ordinary least-squares estimator $\hat{\boldsymbol{\beta}}^* = (\mathbf{X}'\mathbf{X})^{-1}\mathbf{X}'\mathbf{y}$ in (7.6), which we denote here by $\hat{\boldsymbol{\beta}}^*$ to distinguish it from the BLUE estimator $\hat{\boldsymbol{\beta}} = (\mathbf{X}'\mathbf{V}^{-1}\mathbf{X})^{-1}\mathbf{X}'\mathbf{V}^{-1}\mathbf{y}$ in (7.63). Then the mean vector and covariance matrix for $\hat{\boldsymbol{\beta}}^*$ are

$$E(\hat{\boldsymbol{\beta}}^*) = \boldsymbol{\beta}, \tag{7.71}$$

$$\text{cov}(\hat{\boldsymbol{\beta}}^*) = \sigma^2(\mathbf{X}'\mathbf{X})^{-1}\mathbf{X}'\mathbf{V}\mathbf{X}(\mathbf{X}'\mathbf{X})^{-1}. \tag{7.72}$$

Thus the ordinary least-squares estimators are unbiased, but the covariance matrix differs from (7.64). Because of Theorem 7.8a(i), the variances of the $\hat{\beta}_j^*$'s in (7.72) cannot be smaller than the variances in $\text{cov}(\hat{\boldsymbol{\beta}}) = \sigma^2(\mathbf{X}'\mathbf{V}^{-1}\mathbf{X})^{-1}$ in (7.64). This is illustrated in the following example.

Example 7.8.2. Suppose that we have a simple linear regression model $y_i = \beta_0 + \beta_1 x_i + \varepsilon_i$, where $\text{var}(y_i) = \sigma^2 x_i$ and $\text{cov}(y_i, y_j) = 0$ for $i \neq j$. Thus

$$\text{cov}(\mathbf{y}) = \sigma^2\mathbf{V} = \sigma^2 \begin{pmatrix} x_1 & 0 & \dots & 0 \\ 0 & x_2 & \dots & 0 \\ \vdots & \vdots & & \vdots \\ 0 & 0 & \dots & x_n \end{pmatrix}.$$

This is an example of *weighted least squares*, which typically refers to the case where $\mathbf{V}$ is diagonal with functions of the x's on the diagonal. In this case

$$\mathbf{X} = \begin{pmatrix} 1 & x_1 \\ 1 & x_2 \\ \vdots & \vdots \\ 1 & x_n \end{pmatrix},$$

and by (7.63), we have

$$\hat{\boldsymbol{\beta}} = \begin{pmatrix} \hat{\beta}_0 \\ \hat{\beta}_1 \end{pmatrix} = (\mathbf{X}'\mathbf{V}^{-1}\mathbf{X})^{-1}\mathbf{X}'\mathbf{V}^{-1}\mathbf{y}$$

$$= \frac{1}{\left(\sum_{i=1}^n x_i\right)\left(\sum_{i=1}^n \frac{1}{x_i}\right) - n^2} \begin{pmatrix} \left(\sum_{i=1}^n x_i\right)\left(\sum_{i=1}^n \frac{y_i}{x_i}\right) - n\sum_{i=1}^n y_i \\ \left(\sum_{i=1}^n y_i\right)\left(\sum_{i=1}^n \frac{1}{x_i}\right) - n\sum_{i=1}^n \frac{y_i}{x_i} \end{pmatrix}. \tag{7.73}$$

The covariance matrix for $\hat{\boldsymbol{\beta}}$ is given by (7.64):

$$\text{cov}(\hat{\boldsymbol{\beta}}) = \sigma^2(\mathbf{X}'\mathbf{V}^{-1}\mathbf{X})^{-1}$$

$$= \frac{\sigma^2}{\sum_i x_i \sum_i \frac{1}{x_i} - n^2} \begin{pmatrix} \sum_i x_i & -n \\ -n & \sum_i \frac{1}{x_i} \end{pmatrix}. \tag{7.74}$$

If we use the ordinary least-squares estimator $\hat{\boldsymbol{\beta}}^* = (\mathbf{X}'\mathbf{X})^{-1}\mathbf{X}'\mathbf{y}$ as given in (6.5) and (6.6) or in (7.12) in Example 7.3.1b, then $\text{cov}(\hat{\boldsymbol{\beta}}^*)$ is given by (7.72); that is,

$$\begin{aligned}
\text{cov}(\hat{\boldsymbol{\beta}}^*) &= \sigma^2(\mathbf{X}'\mathbf{X})^{-1}\mathbf{X}'\mathbf{V}\mathbf{X}(\mathbf{X}'\mathbf{X})^{-1} \\
&= \sigma^2 \begin{pmatrix} n & \sum_i x_i \\ \sum_i x_i & \sum_i x_i^2 \end{pmatrix}^{-1} \begin{pmatrix} \sum_i x_i & \sum_i x_i^2 \\ \sum_i x_i^2 & \sum_i x_i^3 \end{pmatrix} \begin{pmatrix} n & \sum_i x_i \\ \sum_i x_i & \sum_i x_i^2 \end{pmatrix}^{-1} \\
&= \sigma^2 c \begin{pmatrix} \sum_i x_i^3 (\sum_i x_i)^2 - \sum_i x_i (\sum_i x_i^2)^2 & n(\sum_i x_i^2)^2 - n\sum_i x_i \sum_i x_i^3 \\ n(\sum_i x_i^2)^2 - n\sum_i x_i \sum_i x_i^3 & n^2 \sum_i x_i^3 - 2n \sum_i x_i \sum_i x_i^2 + (\sum_i x_i)^3 \end{pmatrix},
\end{aligned} \tag{7.75}$$

where $c = 1/\left[n\sum_i x_i^2 - \left(\sum_i x_i\right)^2\right]^2$. The variance of the estimator $\hat{\beta}_1^*$ is given by the lower right diagonal element of (7.75):

$$\text{var}(\hat{\beta}_1^*) = \sigma^2 \frac{n^2 \sum_i x_i^3 - 2n\sum_i x_i \sum_i x_i^2 + (\sum_i x_i)^3}{\left[n\sum_i x_i^2 - \left(\sum_i x_i\right)^2\right]^2}, \tag{7.76}$$

and the variance of the estimator $\hat{\beta}_1$ is given by the corresponding element of (7.74):

$$\text{var}(\hat{\beta}_1) = \sigma^2 \frac{\sum_i (1/x_i)}{\sum_i x_i \sum_i (1/x_i) - n^2}. \tag{7.77}$$

Consider the following seven values of x: 1, 2, 3, 4, 5, 6, 7. Using (7.76), we obtain $\text{var}(\hat{\beta}_1^*) = .1429\sigma^2$, and from (7.77), we have $\text{var}(\hat{\beta}_1) = .1099\sigma^2$. Thus for these values of x, the use of ordinary least squares yields a slope estimator with a larger variance, as expected. □

Further consequences of using a wrong model are discussed in the next section.

7.9 MODEL MISSPECIFICATION

In Section 7.8.2, we discussed some consequences of misspecification of cov(**y**). We now consider consequences of misspecification of $E(\mathbf{y})$. As a framework for discussion, let the model $\mathbf{y} = \mathbf{X}\boldsymbol{\beta} + \boldsymbol{\varepsilon}$ be partitioned as

$$\begin{aligned}
\mathbf{y} = \mathbf{X}\boldsymbol{\beta} + \boldsymbol{\varepsilon} &= (\mathbf{X}_1, \mathbf{X}_2)\begin{pmatrix} \boldsymbol{\beta}_1 \\ \boldsymbol{\beta}_2 \end{pmatrix} + \boldsymbol{\varepsilon} \\
&= \mathbf{X}_1\boldsymbol{\beta}_1 + \mathbf{X}_2\boldsymbol{\beta}_2 + \boldsymbol{\varepsilon}.
\end{aligned} \tag{7.78}$$

If we leave out $\mathbf{X}_2\boldsymbol{\beta}_2$ when it should be included (i.e., when $\boldsymbol{\beta}_2 \neq \mathbf{0}$), we are *underfitting*. If we include $\mathbf{X}_2\boldsymbol{\beta}_2$ when it should be excluded (i.e., when $\boldsymbol{\beta}_2 = \mathbf{0}$), we are *overfitting*. We discuss the effect of underfitting or overfitting on the bias and the variance of the $\hat{\beta}_j$, $\hat{y}$, and s^2 values.

We first consider estimation of $\boldsymbol{\beta}_1$ when underfitting. We write the *reduced* model as

$$\mathbf{y} = \mathbf{X}_1\boldsymbol{\beta}_1^* + \boldsymbol{\varepsilon}^*, \tag{7.79}$$

using $\boldsymbol{\beta}_1^*$ to emphasize that these parameters (and their estimates $\hat{\boldsymbol{\beta}}_1^*$) will be different from $\boldsymbol{\beta}_1$ (and $\hat{\boldsymbol{\beta}}_1$) in the *full* model (7.78) (unless the x's are orthogonal; see Corollary 1 to Theorem 7.9a and Theorem 7.10). This was illustrated in Example 7.2. In the following theorem, we discuss the bias in the estimator $\hat{\boldsymbol{\beta}}_1^*$ obtained from (7.79) and give the covariance matrix for $\hat{\boldsymbol{\beta}}_1^*$.

Theorem 7.9a. If we fit the model $\mathbf{y} = \mathbf{X}_1\boldsymbol{\beta}_1^* + \boldsymbol{\varepsilon}^*$ when the correct model is $\mathbf{y} = \mathbf{X}_1\boldsymbol{\beta}_1 + \mathbf{X}_2\boldsymbol{\beta}_2 + \boldsymbol{\varepsilon}$ with $\text{cov}(\mathbf{y}) = \sigma^2\mathbf{I}$, then the mean vector and covariance matrix for the least-squares estimator $\hat{\boldsymbol{\beta}}_1^* = (\mathbf{X}_1'\mathbf{X}_1)^{-1}\mathbf{X}_1'\mathbf{y}$ are as follows:

$$\text{(i) } E(\hat{\boldsymbol{\beta}}_1^*) = \boldsymbol{\beta}_1 + \mathbf{A}\boldsymbol{\beta}_2, \text{ where } \mathbf{A} = (\mathbf{X}_1'\mathbf{X}_1)^{-1}\mathbf{X}_1'\mathbf{X}_2, \tag{7.80}$$

$$\text{(ii) } \text{cov}(\hat{\boldsymbol{\beta}}_1^*) = \sigma^2(\mathbf{X}_1'\mathbf{X}_1)^{-1}. \tag{7.81}$$

PROOF

$$\begin{aligned}
\text{(i) } E(\hat{\boldsymbol{\beta}}_1^*) &= E[(\mathbf{X}_1'\mathbf{X}_1)^{-1}\mathbf{X}_1'\mathbf{y}] = (\mathbf{X}_1'\mathbf{X}_1)^{-1}\mathbf{X}_1'E(\mathbf{y}) \\
&= (\mathbf{X}_1'\mathbf{X}_1)^{-1}\mathbf{X}_1'(\mathbf{X}_1\boldsymbol{\beta}_1 + \mathbf{X}_2\boldsymbol{\beta}_2) \\
&= \boldsymbol{\beta}_1 + (\mathbf{X}_1'\mathbf{X}_1)^{-1}\mathbf{X}_1'\mathbf{X}_2\boldsymbol{\beta}_2. \\
\text{(ii) } \text{cov}(\hat{\boldsymbol{\beta}}_1^*) &= \text{cov}[(\mathbf{X}_1'\mathbf{X}_1)^{-1}\mathbf{X}_1'\mathbf{y}] \\
&= (\mathbf{X}_1'\mathbf{X}_1)^{-1}\mathbf{X}_1'(\sigma^2\mathbf{I})\mathbf{X}_1(\mathbf{X}_1'\mathbf{X}_1)^{-1} \qquad \text{[by (3.44)]} \\
&= \sigma^2(\mathbf{X}_1'\mathbf{X}_1)^{-1}.
\end{aligned}$$

□

Thus, when underfitting, $\hat{\boldsymbol{\beta}}_1^*$ is biased by an amount that depends on the values of the x's in both $\mathbf{X}_1$ and $\mathbf{X}_2$. The matrix $\mathbf{A} = (\mathbf{X}_1'\mathbf{X}_1)^{-1}\mathbf{X}_1'\mathbf{X}_2$ in (7.81) is called the *alias* matrix.

Corollary 1. If $\mathbf{X}_1'\mathbf{X}_2 = \mathbf{O}$, that is, if the columns of $\mathbf{X}_1$ are orthogonal to the columns of $\mathbf{X}_2$, then $\hat{\boldsymbol{\beta}}_1^*$ is unbiased: $E(\hat{\boldsymbol{\beta}}_1^*) = \boldsymbol{\beta}_1$. □

In the next three theorems, we discuss the effect of underfitting or overfitting on $\hat{y}$, s^2, and the variances of the $\hat{\beta}_j$'s. In some of the proofs we follow Hocking (1996, pp. 245–247).

Let $\mathbf{x}_0 = (1, x_{01}, x_{02}, \ldots, x_{0k})'$ be a particular value of $\mathbf{x}$ for which we desire to estimate $E(y_0) = \mathbf{x}_0'\boldsymbol{\beta}$. If we partition $\mathbf{x}_0'$ into $(\mathbf{x}_{01}', \mathbf{x}_{02}')$ corresponding to the partitioning $\mathbf{X} = (\mathbf{X}_1, \mathbf{X}_2)$ and $\boldsymbol{\beta}' = (\boldsymbol{\beta}_1', \boldsymbol{\beta}_2')$, then we can use either $\hat{y}_0 = \mathbf{x}_0'\hat{\boldsymbol{\beta}}$ or $\hat{y}_{01} = \mathbf{x}_{01}'\hat{\boldsymbol{\beta}}_1^*$ to estimate $\mathbf{x}_0'\boldsymbol{\beta}$. In the following theorem, we consider the mean of $\hat{y}_{01}$.

Theorem 7.9b. Let $\hat{y}_{01} = \mathbf{x}_{01}'\hat{\boldsymbol{\beta}}_1^*$, where $\hat{\boldsymbol{\beta}}_1^* 1 = (\mathbf{X}_1'\mathbf{X}_1)^{-1}\mathbf{X}_1'\mathbf{y}$. Then, if $\boldsymbol{\beta}_2 \neq \mathbf{0}$, we obtain

$$E(\mathbf{x}_{01}'\hat{\boldsymbol{\beta}}_1^*) = \mathbf{x}_{01}'(\boldsymbol{\beta}_1 + \mathbf{A}\boldsymbol{\beta}_2), \tag{7.82}$$

$$= \mathbf{x}_0'\boldsymbol{\beta} - (\mathbf{x}_{02} - \mathbf{A}'\mathbf{x}_{01})'\boldsymbol{\beta}_2 \neq \mathbf{x}_0'\boldsymbol{\beta}. \tag{7.83}$$

PROOF. See Problem 7.43. □

In Theorem 7.9b, we see that, when underfitting, $\mathbf{x}_{01}'\hat{\boldsymbol{\beta}}_1^*$ is biased for estimating $\mathbf{x}_0'\boldsymbol{\beta}$. [When overfitting, $\mathbf{x}_0'\hat{\boldsymbol{\beta}}$ is unbiased since $E(\mathbf{x}_0'\hat{\boldsymbol{\beta}}) = \mathbf{x}_0'\boldsymbol{\beta} = \mathbf{x}_{01}'\boldsymbol{\beta}_1 + \mathbf{x}_{02}'\boldsymbol{\beta}_2$, which is equal to $\mathbf{x}_{01}'\boldsymbol{\beta}_1$ if $\boldsymbol{\beta}_2 = \mathbf{0}$.]

In the next theorem, we compare the variances of $\hat{\beta}_j^*$ and $\hat{\beta}_j$, where $\hat{\beta}_j^*$ is from $\hat{\boldsymbol{\beta}}_1^*$ and $\hat{\beta}_j$ is from $\hat{\boldsymbol{\beta}}_1$. We also compare the variances of $\mathbf{x}_{01}'\hat{\boldsymbol{\beta}}_1^*$ and $\mathbf{x}_0'\hat{\boldsymbol{\beta}}$.

Theorem 7.9c. Let $\hat{\boldsymbol{\beta}} = (\mathbf{X}'\mathbf{X})^{-1}\mathbf{X}'\mathbf{y}$ from the full model be partitioned as $\hat{\boldsymbol{\beta}} = \begin{pmatrix} \hat{\boldsymbol{\beta}}_1 \\ \hat{\boldsymbol{\beta}}_2 \end{pmatrix}$, and let $\hat{\boldsymbol{\beta}}_1^* = (\mathbf{X}_1'\mathbf{X}_1)^{-1}\mathbf{X}_1'\mathbf{y}$ be the estimator from the reduced model. Then

(i) $\text{cov}(\hat{\boldsymbol{\beta}}_1) - \text{cov}(\hat{\boldsymbol{\beta}}_1^*) = \sigma^2\mathbf{A}\mathbf{B}^{-1}\mathbf{A}'$, which is a positive definite matrix, where $\mathbf{A} = (\mathbf{X}_1'\mathbf{X}_1)^{-1}\mathbf{X}_1'\mathbf{X}_2$ and $\mathbf{B} = \mathbf{X}_2'\mathbf{X}_2 - \mathbf{X}_2'\mathbf{X}_1\mathbf{A}$. Thus $\text{var}(\hat{\beta}_j) > \text{var}(\hat{\beta}_j^*)$.

(ii) $\text{var}(\mathbf{x}_0'\hat{\boldsymbol{\beta}}) \geq \text{var}(\mathbf{x}_{01}'\hat{\boldsymbol{\beta}}_1^*)$.

PROOF

(i) Using $\mathbf{X}'\mathbf{X}$ partitioned to conform to $\mathbf{X} = (\mathbf{X}_1, \mathbf{X}_2)$, we have

$$\text{cov}(\hat{\boldsymbol{\beta}}) = \text{cov}\begin{pmatrix} \hat{\boldsymbol{\beta}}_1 \\ \hat{\boldsymbol{\beta}}_2 \end{pmatrix} = \sigma^2(\mathbf{X}'\mathbf{X})^{-1} = \sigma^2\begin{pmatrix} \mathbf{X}_1'\mathbf{X}_1 & \mathbf{X}_1'\mathbf{X}_2 \\ \mathbf{X}_2'\mathbf{X}_1 & \mathbf{X}_2'\mathbf{X}_2 \end{pmatrix}^{-1}$$

$$= \sigma^2\begin{pmatrix} \mathbf{G}_{11} & \mathbf{G}_{12} \\ \mathbf{G}_{21} & \mathbf{G}_{22} \end{pmatrix}^{-1} = \sigma^2\begin{pmatrix} \mathbf{G}^{11} & \mathbf{G}^{12} \\ \mathbf{G}^{21} & \mathbf{G}^{22} \end{pmatrix},$$

where $\mathbf{G}_{ij} = \mathbf{X}_i'\mathbf{X}_j$ and $\mathbf{G}^{ij}$ is the corresponding block of the partitioned inverse matrix $(\mathbf{X}'\mathbf{X})^{-1}$. Thus $\text{cov}(\hat{\boldsymbol{\beta}}_1) = \sigma^2\mathbf{G}^{11}$. By (2.50), $\mathbf{G}^{11} = \mathbf{G}_{11}^{-1} + \mathbf{G}_{11}^{-1}\mathbf{G}_{12}\mathbf{B}^{-1}\mathbf{G}_{21}\mathbf{G}_{11}^{-1}$, where $\mathbf{B} = \mathbf{G}_{22} - \mathbf{G}_{21}\mathbf{G}_{11}^{-1}\mathbf{G}_{12}$. By (7.81), $\text{cov}(\hat{\boldsymbol{\beta}}_1^*) = \sigma^2(\mathbf{X}_1'\mathbf{X}_1)^{-1} = \sigma^2\mathbf{G}_{11}^{-1}$. Hence

$$\begin{aligned} \text{cov}(\hat{\boldsymbol{\beta}}_1) - \text{cov}(\hat{\boldsymbol{\beta}}_1^*) &= \sigma^2(\mathbf{G}^{11} - \mathbf{G}_{11}^{-1}) \\ &= \sigma^2(\mathbf{G}_{11}^{-1} + \mathbf{G}_{11}^{-1}\mathbf{G}_{12}\mathbf{B}^{-1}\mathbf{G}_{21}\mathbf{G}_{11}^{-1} - \mathbf{G}_{11}^{-1}) \\ &= \sigma^2\mathbf{A}\mathbf{B}^{-1}\mathbf{A}'. \end{aligned}$$

(ii)
$$\begin{aligned}\text{var}(\mathbf{x}_0'\hat{\boldsymbol{\beta}}) &= \sigma^2\mathbf{x}_0'(\mathbf{X}'\mathbf{X})^{-1}\mathbf{x}_0\\ &= \sigma^2(\mathbf{x}_{01}', \mathbf{x}_{02}')\begin{pmatrix}\mathbf{G}^{11} & \mathbf{G}^{12}\\ \mathbf{G}^{21} & \mathbf{G}^{22}\end{pmatrix}\begin{pmatrix}\mathbf{x}_{01}\\ \mathbf{x}_{02}\end{pmatrix}\\ &= \sigma^2(\mathbf{x}_{01}'\mathbf{G}^{11}\mathbf{x}_{01} + \mathbf{x}_{01}'\mathbf{G}^{12}\mathbf{x}_{02} + \mathbf{x}_{02}'\mathbf{G}^{21}\mathbf{x}_{01} + \mathbf{x}_{02}'\mathbf{G}^{22}\mathbf{x}_{02}).\end{aligned}$$

Using (2.50), it can be shown that

$$\text{var}(\mathbf{x}_0'\hat{\boldsymbol{\beta}}) - \text{var}(\mathbf{x}_{01}'\hat{\boldsymbol{\beta}}_1^*) = \sigma^2(\mathbf{x}_{02} - \mathbf{A}'\mathbf{x}_{01})'\mathbf{G}^{22}(\mathbf{x}_{02} - \mathbf{A}'\mathbf{x}_{01}) \geq 0$$

because $\mathbf{G}^{22}$ is positive definite. □

By Theorem 7.9c(i), $\text{var}(\hat{\beta}_j)$ in the full model is greater than $\text{var}(\hat{\beta}_j^*)$ in the reduced model. Thus underfitting reduces the variance of the $\hat{\beta}_j$'s but introduces bias. On the other hand, overfitting increases the variance of the $\hat{\beta}_j$'s. In Theorem 7.9c (ii), $\text{var}(\hat{y}_0)$ based on the full model is greater than $\text{var}(\hat{y}_{01})$ based on the reduced model. Again, underfitting reduces the variance of the estimate of $E(y_0)$ but introduces bias. Overfitting increases the variance of the estimate of $E(y_0)$.

We now consider s^2 for the full model and for the reduced model. For the full model $\mathbf{y} = \mathbf{X}\boldsymbol{\beta} + \boldsymbol{\varepsilon} = \mathbf{X}_1\boldsymbol{\beta}_1 + \mathbf{X}_2\boldsymbol{\beta}_2 + \boldsymbol{\varepsilon}$, the sample variance s^2 is given by (7.23) as

$$s^2 = \frac{(\mathbf{y} - \mathbf{X}\hat{\boldsymbol{\beta}})'(\mathbf{y} - \mathbf{X}\hat{\boldsymbol{\beta}})}{n - k - 1}.$$

In Theorem 7.3f, we have $E(s^2) = \sigma^2$. The expected value of s^2 for the reduced model is given in the following theorem.

Theorem 7.9d. If $\mathbf{y} = \mathbf{X}\boldsymbol{\beta} + \boldsymbol{\varepsilon}$ is the correct model, then for the reduced model $\mathbf{y} = \mathbf{X}_1\boldsymbol{\beta}_1^* + \boldsymbol{\varepsilon}^*$ (underfitting), where $\mathbf{X}_1$ is $n \times (p+1)$ with $p < k$, the variance estimator

$$s_1^2 = \frac{(\mathbf{y} - \mathbf{X}_1\hat{\boldsymbol{\beta}}_1^*)'(\mathbf{y} - \mathbf{X}_1\hat{\boldsymbol{\beta}}_1^*)}{n - p - 1} \tag{7.84}$$

has expected value

$$E(s_1^2) = \sigma^2 + \frac{\boldsymbol{\beta}_2'\mathbf{X}_2'[\mathbf{I} - \mathbf{X}_1(\mathbf{X}_1'\mathbf{X}_1)^{-1}\mathbf{X}_1']\mathbf{X}_2\boldsymbol{\beta}_2}{n - p - 1}. \tag{7.85}$$

Proof. We write the numerator of (7.84) as

$$\begin{aligned}\text{SSE}_1 &= \mathbf{y}'\mathbf{y} - \hat{\boldsymbol{\beta}}_1^{*\prime}\mathbf{X}_1'\mathbf{y} = \mathbf{y}'\mathbf{y} - \mathbf{y}'\mathbf{X}_1(\mathbf{X}_1'\mathbf{X}_1)^{-1}\mathbf{X}_1'\mathbf{y}\\ &= \mathbf{y}'[\mathbf{I} - \mathbf{X}_1(\mathbf{X}_1'\mathbf{X}_1)^{-1}\mathbf{X}_1']\mathbf{y}.\end{aligned}$$

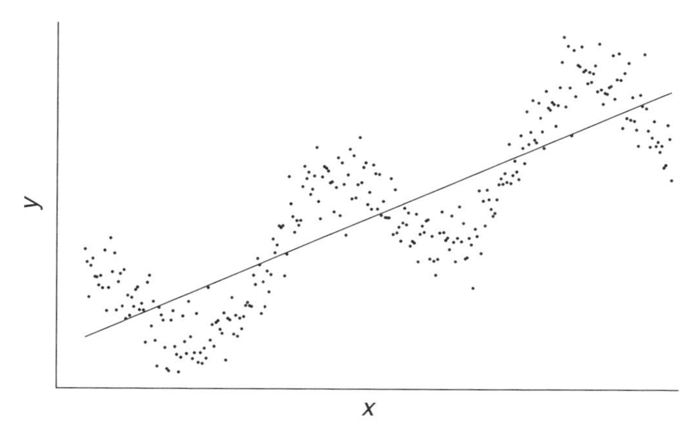

Figure 7.6 Straight-line fit to a curved pattern of points.

Since $E(\mathbf{y}) = \mathbf{X}\boldsymbol{\beta}$ by assumption, we have, by Theorem 5.2a,

$$E(\text{SSE}_1) = \text{tr}\{[\mathbf{I} - \mathbf{X}_1(\mathbf{X}_1'\mathbf{X}_1)^{-1}\mathbf{X}_1']\sigma^2\mathbf{I}\} + \boldsymbol{\beta}'\mathbf{X}'[\mathbf{I} - \mathbf{X}_1(\mathbf{X}_1'\mathbf{X}_1)^{-1}\mathbf{X}_1']\mathbf{X}\boldsymbol{\beta}$$
$$= (n - p - 1)\sigma^2 + \boldsymbol{\beta}_2'\mathbf{X}_2'[\mathbf{I} - \mathbf{X}_1(\mathbf{X}_1'\mathbf{X}_1)^{-1}\mathbf{X}_1']\mathbf{X}_2\boldsymbol{\beta}_2$$

(see Problem 7.45). □

Since the quadratic form in (7.85) is positive semidefinite, s^2 is biased upward when underfitting (see Fig. 7.6). We can also examine (7.85) from the perspective of overfitting, in which case $\boldsymbol{\beta}_2 = \mathbf{0}$ and s^2 is unbiased.

To summarize the results in this section, underfitting leads to biased $\hat{\beta}_j$'s, biased $\hat{y}$'s, and biased s^2. Overfitting increases the variances of the $\hat{\beta}_j$'s and of the $\hat{y}$'s. We are thus compelled to seek an appropriate balance between a biased model and one with large variances. This is the task of the model builder and serves as motivation for seeking an optimum subset of x's.

Example 7.9a. Suppose that the model $y_i = \beta_0^* + \beta_1^* x_i + \varepsilon_i^*$ has been fitted when the true model is $y_i = \beta_0 + \beta_1 x_i + \beta_2 x_i^2 + \varepsilon_i$. (This situation is similar to that illustrated in Figure 6.2.) In this case, $\hat{\beta}_0^*$, $\hat{\beta}_1^*$, and s_1^2 would be biased by an amount dependent on the choice of the x_i's [see (7.80) and (7.86)]. The error term $\hat{\varepsilon}_i^*$ in the misspecified model $y_i = \beta_0^* + \beta_1^* x_i + \varepsilon_i^*$ does not have a mean of 0:

$$\begin{aligned} E(\varepsilon_i^*) &= E(y_i - \beta_0^* - \beta_1^* x_i) \\ &= E(y_i) - \beta_0^* - \beta_1^* x_i \\ &= \beta_0 + \beta_1 x_i + \beta_2 x_i^2 - \beta_0^* - \beta_1^* x_i \\ &= \beta_0 - \beta_0^* + (\beta_1 - \beta_1^*) x_i + \beta_2 x_i^2. \end{aligned}$$

□

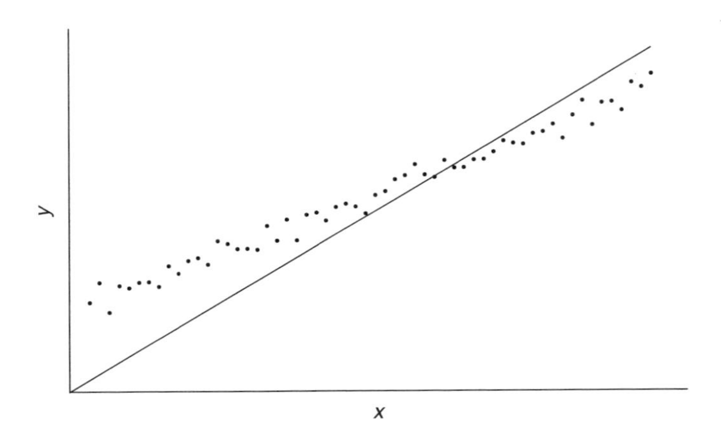

Figure 7.7 No-intercept model fit to data from an intercept model.

Example 7.9b. Suppose that the true model is $y_i = \beta_0 + \beta_1 x_i + \varepsilon_i$ and we fit the model $y_i = \beta_1^* x_i + \varepsilon_i^*$, as illustrated in Figure 7.7.

For the model $y_i = \beta_1^* x_i + \varepsilon_i^*$, the least-squares estimator is

$$\hat{\beta}_1^* = \frac{\sum_{i=1}^n x_i y_i}{\sum_{i=1}^n x_i^2} \tag{7.86}$$

(see Problem 7.46). Then, under the full model $y_i = \beta_0 + \beta_1 x_i + \varepsilon_i$, we have

$$\begin{aligned} E(\hat{\beta}_1^*) &= \frac{1}{\sum_i x_i^2} \sum_i x_i E(y_i) \\ &= \frac{1}{\sum_i x_i^2} \sum_i x_i(\beta_0 + \beta_1 x_i) \\ &= \frac{1}{\sum_i x_i^2} \left(\beta_0 \sum_i x_i + \beta_1 \sum_i x_i^2 \right) \\ &= \beta_0 \frac{\sum_i x_i}{\sum_i x_i^2} + \beta_1. \end{aligned} \tag{7.87}$$

Thus $\hat{\beta}_1^*$ is biased by an amount that depends on β_0 and the values of the x's. □

7.10 ORTHOGONALIZATION

In Section 7.9, we discussed estimation of $\boldsymbol{\beta}_1^*$ in the model $\mathbf{y} = \mathbf{X}_1 \boldsymbol{\beta}_1^* + \boldsymbol{\varepsilon}^*$ when the true model is $\mathbf{y} = \mathbf{X}_1 \boldsymbol{\beta}_1 + \mathbf{X}_2 \boldsymbol{\beta}_2 + \boldsymbol{\varepsilon}$. By Theorem 7.9a, $E(\hat{\boldsymbol{\beta}}_1^*) = \boldsymbol{\beta}_1 + (\mathbf{X}_1'\mathbf{X}_1)^{-1}\mathbf{X}_1'\mathbf{X}_2\boldsymbol{\beta}_2$, so that estimation of $\boldsymbol{\beta}_1$ is affected by the presence of $\mathbf{X}_2$, unless $\mathbf{X}_1'\mathbf{X}_2 = \mathbf{O}$, in which case, $E(\hat{\boldsymbol{\beta}}_1^*) = \boldsymbol{\beta}_1$. In the following theorem, we show that if $\mathbf{X}_1'\mathbf{X}_2 = \mathbf{O}$, the estimators of $\boldsymbol{\beta}_1^*$ and $\boldsymbol{\beta}_1$ not only have the same expected value, but are exactly the same.

Theorem 7.10. If $\mathbf{X}_1'\mathbf{X}_2 = \mathbf{O}$, then the estimator of $\boldsymbol{\beta}_1$ in the full model $\mathbf{y} = \mathbf{X}_1\boldsymbol{\beta}_1 + \mathbf{X}_2\boldsymbol{\beta}_2 + \boldsymbol{\varepsilon}$ is the same as the estimator of $\boldsymbol{\beta}_1^*$ in the reduced model $\mathbf{y} = \mathbf{X}_1\boldsymbol{\beta}_1^* + \boldsymbol{\varepsilon}^*$.

Proof. The least-squares estimator of $\boldsymbol{\beta}_1^*$ is $\hat{\boldsymbol{\beta}}_1^* = (\mathbf{X}_1'\mathbf{X}_1)^{-1}\mathbf{X}_1'\mathbf{y}$. For the estimator of $\boldsymbol{\beta}_1$ in the full model, we partition $\hat{\boldsymbol{\beta}} = (\mathbf{X}'\mathbf{X})^{-1}\mathbf{X}'\mathbf{y}$ to obtain

$$\begin{pmatrix} \hat{\boldsymbol{\beta}}_1 \\ \hat{\boldsymbol{\beta}}_2 \end{pmatrix} = \begin{pmatrix} \mathbf{X}_1'\mathbf{X}_1 & \mathbf{X}_1'\mathbf{X}_2 \\ \mathbf{X}_2'\mathbf{X}_1 & \mathbf{X}_2'\mathbf{X}_2 \end{pmatrix}^{-1} \begin{pmatrix} \mathbf{X}_1'\mathbf{y} \\ \mathbf{X}_2'\mathbf{y} \end{pmatrix}.$$

Using the notation in the proof of Theorem 7.9c, this becomes

$$\begin{aligned} \begin{pmatrix} \hat{\boldsymbol{\beta}}_1 \\ \hat{\boldsymbol{\beta}}_2 \end{pmatrix} &= \begin{pmatrix} \mathbf{G}_{11} & \mathbf{G}_{12} \\ \mathbf{G}_{21} & \mathbf{G}_{22} \end{pmatrix}^{-1} \begin{pmatrix} \mathbf{X}_1'\mathbf{y} \\ \mathbf{X}_2'\mathbf{y} \end{pmatrix} \\ &= \begin{pmatrix} \mathbf{G}^{11} & \mathbf{G}^{12} \\ \mathbf{G}^{21} & \mathbf{G}^{22} \end{pmatrix} \begin{pmatrix} \mathbf{X}_1'\mathbf{y} \\ \mathbf{X}_2'\mathbf{y} \end{pmatrix}. \end{aligned}$$

By (2.50), we obtain

$$\begin{aligned} \hat{\boldsymbol{\beta}}_1 &= \mathbf{G}^{11}\mathbf{X}_1'\mathbf{y} + \mathbf{G}^{12}\mathbf{X}_2'\mathbf{y} \\ &= (\mathbf{G}_{11}^{-1} + \mathbf{G}_{11}^{-1}\mathbf{G}_{12}\mathbf{B}^{-1}\mathbf{G}_{21}\mathbf{G}_{11}^{-1})\mathbf{X}_1'\mathbf{y} - \mathbf{G}_{11}^{-1}\mathbf{G}_{12}\mathbf{B}^{-1}\mathbf{X}_2'\mathbf{y}, \end{aligned}$$

where $\mathbf{B} = \mathbf{G}_{22} - \mathbf{G}_{21}\mathbf{G}_{11}^{-1}\mathbf{G}_{12}$. If $\mathbf{G}_{12} = \mathbf{X}_1'\mathbf{X}_2 = \mathbf{O}$, then $\hat{\boldsymbol{\beta}}_1$ reduces to

$$\hat{\boldsymbol{\beta}}_1 = \mathbf{G}_{11}^{-1}\mathbf{X}_1'\mathbf{y} = (\mathbf{X}_1'\mathbf{X}_1)^{-1}\mathbf{X}_1'\mathbf{y},$$

which is the same as $\hat{\boldsymbol{\beta}}_1^*$. □

Note that Theorem 7.10 will also hold if $\mathbf{X}_1$ and $\mathbf{X}_2$ are "essentially orthogonal," that is, if the centered columns of $\mathbf{X}_1$ are orthogonal to the centered columns of $\mathbf{X}_2$.

In Theorem 7.9a, we discussed estimation of $\boldsymbol{\beta}_1^*$ in the presence of $\boldsymbol{\beta}_2$ when $\mathbf{X}_1'\mathbf{X}_2 \neq \mathbf{O}$. We now consider a process of orthogonalization to give additional insights into the meaning of partial regression coefficients.

In Example 7.2, we illustrated the change in the estimate of a regression coefficient when another x was added to the model. We now use the same data to further examine this change.The prediction equation obtained in Example 7.2 was

$$\hat{y} = 5.3754 + 3.0118x_1 - 1.2855x_2, \tag{7.88}$$

and the negative partial regressions of y on x_2 were shown in Figure 7.2. By means of orthogonalization, we can give additional meaning to the term $-1.2855x_2$. In order to add x_2 to the prediction equation containing only x_1, we need to determine how much variation in y is due to x_2 after the effect of x_1 has been accounted for, and we must also correct for the relationship between x_1 and x_2. Our approach is to consider the relationship between the residual variation after regressing y on x_1 and the residual variation after regressing x_2 on x_1. We follow a three-step process.

1. Regress y on x_1, and calculate residuals [see (7.11)]. The prediction equation is

$$\hat{y} = 1.8585 + 1.3019x_1, \tag{7.89}$$

and the residuals $y_i - \hat{y}_i(x_1)$ are given in Table 7.2, where $\hat{y}_i(x_1)$ indicates that $\hat{y}$ is based on a regression of y on x_1 as in (7.89).

2. Regress x_2 on x_1 and calculate residuals. The prediction equation is

$$\hat{x}_2 = 2.7358 + 1.3302x_1, \tag{7.90}$$

and the residuals $x_{2i} - \hat{x}_{2i}(x_1)$ are given in Table 7.2, where $\hat{x}_{2i}(x_1)$ indicates that x_2 has been regressed on x_1 as in (7.90).

3. Now regress $y - \hat{y}(x_1)$ on $x_2 - \hat{x}_2(x_1)$, which gives

$$\widehat{y - \hat{y}} = -1.2855(x_2 - \hat{x}_2). \tag{7.91}$$

There is no intercept in (7.91) because both sets of residuals have a mean of 0.

TABLE 7.2 Data from Table 7.1 and Residuals

y	x_1	x_2	$y - \hat{y}(x_1)$	$x_2 - \hat{x}_2(x_1)$
2	0	2	0.1415	−0.7358
3	2	6	−1.4623	0.6038
2	2	7	−2.4623	1.6038
7	2	5	2.5377	−0.3962
6	4	9	−1.0660	0.9434
8	4	8	0.9340	−0.0566
10	4	7	2.9340	−1.0566
7	6	10	−2.6698	−0.7170
8	6	11	−1.6698	0.2830
12	6	9	2.3302	−1.7170
11	8	15	−1.2736	1.6226
14	8	13	1.7264	−0.3774

In (7.91), we obtain a clearer insight into the meaning of the partial regression coefficient -1.2855 in (7.88). We are using the "unexplained" portion of x_2 (after x_1 is accounted for) to predict the "unexplained" portion of y (after x_1 is accounted for).

Since $x_2 - \hat{x}_2(x_1)$ is orthogonal to x_1 [see Section 7.4.2, in particular (7.29)], fitting $y - \hat{y}(x_1)$ to $x_2 - \hat{x}_2(x_1)$ yields the same coefficient, -1.2855, as when fitting y to x_1 and x_2 together. Thus -1.2855 represents the additional effect of x_2 beyond the effect of x_1 and also after taking into account the overlap between x_1 and x_2 in their effect on y. The orthogonality of x_1 and $x_2 - \hat{x}_2(x_1)$ makes this simplified breakdown of effects possible.

We can substitute $\hat{y}(x_1)$ and $\hat{x}_2(x_1)$ in (7.91) to obtain

$$\widehat{y - \hat{y}} = \hat{y}(x_1, x_2) - \hat{y}(x_1) = -1.2855[x_2 - \hat{x}_2(x_1)],$$

or

$$\hat{y} - (1.8585 + 1.3019x_1) = -1.2855[x_2 - (2.7358 + 1.3302x_1)], \tag{7.92}$$

which reduces to

$$\hat{y} = 5.3754 + 3.0118x_1 - 1.2855x_2, \tag{7.93}$$

the same as (7.88). If we regress y (rather than $y - \hat{y}$) on $x_2 - \hat{x}_2(x_1)$, we will still obtain $-1.2855x_2$, but we will not have $5.3754 + 3.0118x_1$.

The correlation between the residuals $y - \hat{y}(x_1)$ and $x_2 - \hat{x}_2(x_1)$ is the same as the (sample) partial correlation of y and x_2 with x_1 held fixed:

$$r_{y2\cdot 1} = r_{y-\hat{y},\, x_2-\hat{x}_2}. \tag{7.94}$$

This is discussed further in Section 10.8.

We now consider the general case with full model

$$\mathbf{y} = \mathbf{X}_1\boldsymbol{\beta}_1 + \mathbf{X}_2\boldsymbol{\beta}_2 + \boldsymbol{\varepsilon}$$

and reduced model

$$\mathbf{y} = \mathbf{X}_1\boldsymbol{\beta}_1^* + \boldsymbol{\varepsilon}^*.$$

We use an orthogonalization approach to obtain an estimator of $\boldsymbol{\beta}_2$, following the same three steps as in the illustration with x_1 and x_2 above:

1. Regress $\mathbf{y}$ on $\mathbf{X}_1$ and calculate residuals $\mathbf{y} - \hat{\mathbf{y}}(\mathbf{X}_1)$, where $\hat{\mathbf{y}}(\mathbf{X}_1) = \mathbf{X}_1\hat{\boldsymbol{\beta}}_1^* = \mathbf{X}_1(\mathbf{X}_1'\mathbf{X}_1)^{-1}\mathbf{X}_1'\mathbf{y}$ [see (7.11)].
2. Regress the columns of $\mathbf{X}_2$ on $\mathbf{X}_1$ and obtain residuals $\mathbf{X}_{2\cdot 1} = \mathbf{X}_2 - \hat{\mathbf{X}}_2(\mathbf{X}_1)$. If $\mathbf{X}_2$ is written in terms of its columns as $\mathbf{X}_2 = (\mathbf{x}_{21}, \ldots, \mathbf{x}_{2j}, \ldots, \mathbf{x}_{2p})$, then the

regression coefficient vector for $\mathbf{x}_{2j}$ on $\mathbf{X}_1$ is $\mathbf{b}_j = (\mathbf{X}_1'\mathbf{X}_1)^{-1}\mathbf{X}_1'\mathbf{x}_{2j}$, and $\hat{\mathbf{x}}_{2j} = \mathbf{X}_1\mathbf{b}_j = \mathbf{X}_1(\mathbf{X}_1'\mathbf{X}_1)^{-1}\mathbf{X}_1'\mathbf{x}_{2j}$. For all columns of $\mathbf{X}_2$, this becomes $\hat{\mathbf{X}}_2(\mathbf{X}_1) = \mathbf{X}_1(\mathbf{X}_1'\mathbf{X}_1)^{-1}\mathbf{X}_1'\mathbf{X}_2 = \mathbf{X}_1\mathbf{A}$, where $\mathbf{A} = (\mathbf{X}_1'\mathbf{X}_1)^{-1}\mathbf{X}_1'\mathbf{X}_2$ is the alias matrix defined in (7.80). Note that $\mathbf{X}_{2\cdot 1} = \mathbf{X}_2 - \hat{\mathbf{X}}_2(\mathbf{X}_1)$ is orthogonal to $\mathbf{X}_1$:

$$\mathbf{X}_1'\mathbf{X}_{2\cdot 1} = \mathbf{O}. \tag{7.95}$$

Using the alias matrix $\mathbf{A}$, the residual matrix can be expressed as

$$\mathbf{X}_{2\cdot 1} = \mathbf{X}_2 - \hat{\mathbf{X}}_2(\mathbf{X}_1) \tag{7.96}$$

$$= \mathbf{X}_2 - \mathbf{X}_1(\mathbf{X}_1'\mathbf{X}_1)^{-1}\mathbf{X}_1'\mathbf{X}_2 = \mathbf{X}_2 - \mathbf{X}_1\mathbf{A}. \tag{7.97}$$

3. Regress $\mathbf{y} - \hat{\mathbf{y}}(\mathbf{X}_1)$ on $\mathbf{X}_{2\cdot 1} = \mathbf{X}_2 - \hat{\mathbf{X}}_2(\mathbf{X}_1)$. Since $\mathbf{X}_{2\cdot 1}$ is orthogonal to $\mathbf{X}_1$, we obtain the same $\hat{\boldsymbol{\beta}}_2$ as in the full model $\hat{\mathbf{y}} = \mathbf{X}_1\hat{\boldsymbol{\beta}}_1 + \mathbf{X}_2\hat{\boldsymbol{\beta}}_2$. Adapting the notation of (7.91) and (7.92), this can be expressed as

$$\hat{\mathbf{y}}(\mathbf{X}_1, \mathbf{X}_2) - \hat{\mathbf{y}}(\mathbf{X}_1) = \mathbf{X}_{2\cdot 1}\hat{\boldsymbol{\beta}}_2. \tag{7.98}$$

If we substitute $\hat{\mathbf{y}}(\mathbf{X}_1) = \mathbf{X}_1\hat{\boldsymbol{\beta}}_1^*$ and $\mathbf{X}_{2\cdot 1} = \mathbf{X}_2 - \mathbf{X}_1\mathbf{A}$ into (7.98) and use $\hat{\boldsymbol{\beta}}_1^* = \hat{\boldsymbol{\beta}}_1 + \mathbf{A}\hat{\boldsymbol{\beta}}_2$ from (7.80), we obtain

$$\begin{aligned}
\hat{\mathbf{y}}(\mathbf{X}_1, \mathbf{X}_2) &= \mathbf{X}_1\hat{\boldsymbol{\beta}}_1^* + (\mathbf{X}_2 - \mathbf{X}_1\mathbf{A})\hat{\boldsymbol{\beta}}_2 \\
&= \mathbf{X}_1(\hat{\boldsymbol{\beta}}_1 + \mathbf{A}\hat{\boldsymbol{\beta}}_2) + (\mathbf{X}_2 - \mathbf{X}_1\mathbf{A})\hat{\boldsymbol{\beta}}_2 \\
&= \mathbf{X}_1\hat{\boldsymbol{\beta}}_1 + \mathbf{X}_2\hat{\boldsymbol{\beta}}_2
\end{aligned}$$

which is analogous to (7.93). This confirms that the orthogonality of $\mathbf{X}_1$ and $\mathbf{X}_{2\cdot 1}$ leads to the estimator $\hat{\boldsymbol{\beta}}_2$ in (7.98). For a formal proof, see Problem 7.50.

PROBLEMS

7.1 Show that $\sum_{i=1}^{n}(y_i - \mathbf{x}_i'\hat{\boldsymbol{\beta}})^2 = (\mathbf{y} - \mathbf{X}\hat{\boldsymbol{\beta}})'(\mathbf{y} - \mathbf{X}\hat{\boldsymbol{\beta}})$, thus verifying (7.7).

7.2 Show that (7.10) follows from (7.9). Why is $\mathbf{X}'\mathbf{X}$ positive definite, as noted below (7.10)?

7.3 Show that $\hat{\beta}_0$ and $\hat{\beta}_1$ in (7.12) in Example 7.3.1 are the same as in (6.5) and (6.6).

7.4 Obtain $\text{cov}(\hat{\boldsymbol{\beta}})$ in (7.16) from (7.15).

7.5 Show that $\text{var}(\hat{\beta}_0) = \sigma^2(\sum_i x_i^2/n)/\sum_i (x_i - \bar{x})^2$ in (7.16) in Example 7.3.2a is the same as $\text{var}(\hat{\beta}_0)$ in (6.10).

7.6 Show that $\mathbf{AA}'$ can be expressed as $\mathbf{AA}' = [\mathbf{A} - (\mathbf{X}'\mathbf{X})^{-1}\mathbf{X}'][\mathbf{A} - (\mathbf{X}'\mathbf{X})^{-1}\mathbf{X}']' + (\mathbf{X}'\mathbf{X})^{-1}$ as in (7.17) in Theorem 7.3d.

7.7 Prove Corollary 1 to Theorem 7.3d in the following two ways:

(**a**) Use an approach similar to the proof of Theorem 7.3d.

(**b**) Use the method of Lagrange multipliers (Section 2.14.3).

7.8 Show that if the x's are rescaled as $z_j = c_j x_j$, $j = 1, 2, \ldots, k$, then $\hat{\boldsymbol{\beta}}_z = \mathbf{D}^{-1}\hat{\boldsymbol{\beta}}$, as in (7.18) in the proof of the Theorem 7.3e.

7.9 Verify (7.20) and (7.21) in the proof of Corollary 1 to Theorem 7.3e.

7.10 Show that s^2 is invariant to changes of scale on the x's, as noted following Corollary 1 to Theorem 7.3e.

7.11 Show that $(\mathbf{y} - \mathbf{X}\hat{\boldsymbol{\beta}})'(\mathbf{y} - \mathbf{X}\hat{\boldsymbol{\beta}}) = \mathbf{y}'\mathbf{y} - \hat{\boldsymbol{\beta}}'\mathbf{X}'\mathbf{y}$ as in (7.24).

7.12 Show that $E(\text{SSE}) = \sigma^2(n - k - 1)$, as in Theorem 7.3f, using the following approach. Show that $\text{SSE} = \mathbf{y}'\mathbf{y} - \hat{\boldsymbol{\beta}}'\mathbf{X}'\mathbf{X}\hat{\boldsymbol{\beta}}$. Show that $E(\mathbf{y}'\mathbf{y}) = n\sigma^2 + \boldsymbol{\beta}'\mathbf{X}'\mathbf{X}\boldsymbol{\beta}$ and that $E(\hat{\boldsymbol{\beta}}'\mathbf{X}'\mathbf{X}\hat{\boldsymbol{\beta}}) = (k + 1)\sigma^2 + \boldsymbol{\beta}'\mathbf{X}'\mathbf{X}\boldsymbol{\beta}$.

7.13 Prove that an additional x reduces SSE, as noted following Theorem 7.3f.

7.14 Show that the noncentered model preceding (7.30) can be written in the centered form in (7.30), with α defined as in (7.31).

7.15 Show that $\mathbf{X}_c = [\mathbf{I} - (1/n)\mathbf{J}]\mathbf{X}_1$ as in (7.33), where $\mathbf{X}_1$ is as given in (7.19).

7.16 Show that $\mathbf{j}'\mathbf{X}_c = \mathbf{0}'$, as in (7.35), where $\mathbf{X}_c$ is the centered $\mathbf{X}$ matrix defined in (7.33).

7.17 Show that the estimators $\hat{\alpha} = \bar{y}$ and $\hat{\boldsymbol{\beta}}_1 = (\mathbf{X}_c'\mathbf{X}_c)^{-1}\mathbf{X}_c'\mathbf{y}$ in (7.36) and (7.37) are the same as $\hat{\boldsymbol{\beta}} = (\mathbf{X}'\mathbf{X})^{-1}\mathbf{X}'\mathbf{y}$ in (7.6). Use the following two methods:

(**a**) Work with the normal equations in both cases.

(**b**) Use the inverse of $\mathbf{X}'\mathbf{X}$ in partitioned form: $(\mathbf{X}'\mathbf{X})^{-1} = [(\mathbf{j}, \mathbf{X}_1)'(\mathbf{j}, \mathbf{X}_1)]^{-1}$.

7.18 Show that the fitted regression plane $\hat{y} = \hat{\alpha} + \hat{\beta}_1(x_1 - \bar{x}_1) + \cdots + \hat{\beta}_k(x_k - \bar{x}_k)$ passes through the point $(\bar{x}_1, \bar{x}_2, \ldots, \bar{x}_k, \bar{y})$, as noted below (7.38).

7.19 Show that $\text{SSE} = \sum_i (y_i - \bar{y})^2 - \hat{\boldsymbol{\beta}}_1'\mathbf{X}_c'\mathbf{y}$ in (7.39) is the same as $\text{SSE} = \mathbf{y}'\mathbf{y} - \hat{\boldsymbol{\beta}}'\mathbf{X}'\mathbf{y}$ in (7.24).

7.20 (**a**) Show that $\mathbf{S}_{xx} = \mathbf{X}_c'\mathbf{X}_c/(n-1)$ as in (7.44).

(**b**) Show that $\mathbf{s}_{yx} = \mathbf{X}_c'\mathbf{y}/(n-1)$ as in (7.45).

7.21 (**a**) Show that if $\mathbf{y}$ is $N_n(\mathbf{X}\boldsymbol{\beta}, \sigma^2\mathbf{I})$, the likelihood function is

$$L(\boldsymbol{\beta}, \sigma^2) = \frac{1}{(2\pi\sigma^2)^{n/2}} e^{-(\mathbf{y}-\mathbf{X}\boldsymbol{\beta})'(\mathbf{y}-\mathbf{X}\boldsymbol{\beta})/2\sigma^2},$$

as in (7.50) in the proof of Theorem 7.6a.

(**b**) Differentiate $\ln L(\boldsymbol{\beta}, \sigma^2)$ in (7.51) with respect to $\boldsymbol{\beta}$ to obtain $\hat{\boldsymbol{\beta}} = (\mathbf{X}'\mathbf{X})^{-1}\mathbf{X}'\mathbf{y}$ in (7.48).

(**c**) Differentiate $\ln L(\boldsymbol{\beta}, \sigma^2)$ with respect to σ^2 to obtain $\hat{\sigma}^2 = (\mathbf{y} - \mathbf{X}\hat{\boldsymbol{\beta}})'(\mathbf{y} - \mathbf{X}\hat{\boldsymbol{\beta}})/n$ as in (7.49).

7.22 Prove parts (ii) and (iii) of Theorem 7.6b.

7.23 Show that $(\mathbf{y} - \mathbf{X}\boldsymbol{\beta})'(\mathbf{y} - \mathbf{X}\boldsymbol{\beta}) = (\mathbf{y} - \mathbf{X}\hat{\boldsymbol{\beta}})'(\mathbf{y} - \mathbf{X}\hat{\boldsymbol{\beta}}) + (\hat{\boldsymbol{\beta}} - \boldsymbol{\beta})'\mathbf{X}'\mathbf{X}(\hat{\boldsymbol{\beta}} - \boldsymbol{\beta})$ as in (7.52) in the proof of Theorem 7.6c.

7.24 Explain why $f(\mathbf{y}; \boldsymbol{\beta}, \sigma^2)$ does not factor into $g_1(\hat{\boldsymbol{\beta}}, \boldsymbol{\beta})g_2(\hat{\sigma}^2, \sigma^2)h(\mathbf{y})$, as noted following Theorem 7.6c.

7.25 Verify the equivalence of (7.55) and (7.56); that is, show that $\hat{\boldsymbol{\beta}}'\mathbf{X}'\mathbf{y} - n\bar{y}^2 = \hat{\boldsymbol{\beta}}_1'\mathbf{X}_c'\mathbf{X}_c\hat{\boldsymbol{\beta}}_1$.

7.26 Verify the comments in property 1 in Section 7.7, namely, that if $\hat{\beta}_1 = \hat{\beta}_2 = \cdots = \hat{\beta}_k = 0$, then $R^2 = 0$, and if $y_i = \hat{y}_i$, $i = 1, 2, \ldots, n$, then $R^2 = 1$.

7.27 Show that adding an x to the model increases (cannot decrease) the value of R^2, as in property 3 in Section 7.7.

7.28 (**a**) Verify that R^2 is invariant to full-rank linear transformations on the x's as in property 6 in Section 7.7.

(**b**) Show that R^2 is invariant to a scale change $\mathbf{z} = c\mathbf{y}$ on $\mathbf{y}$.

7.29 (**a**) Show that R^2 in (7.55) can be written in the form $R^2 = 1 - \text{SSE}/\sum_i (y_i - \bar{y})^2$.

(**b**) Replace SSE and $\sum_i (y_i - \bar{y})^2$ in part (a) by variance estimators $\text{SSE}/(n-k-1)$ and $\sum_i (y_i - \bar{y})^2/(n-1)$ and show that the result is the same as R_a^2 in (7.56).

7.30 Show that $\sum_{i=1}^n \hat{y}_i/n = \sum_{i=1}^n y_i/n$, as noted following (7.59) in Section 7.7.

7.31 Show that $\cos\theta = R$ as in (7.61), where R^2 is as given by (7.56).

7.32 (**a**) Show that $E(\hat{\boldsymbol{\beta}}) = \boldsymbol{\beta}$, where $\hat{\boldsymbol{\beta}} = (\mathbf{X}'\mathbf{V}^{-1}\mathbf{X})^{-1}\mathbf{X}'\mathbf{V}^{-1}\mathbf{y}$ as in (7.63).

(**b**) Show that $\text{cov}(\hat{\boldsymbol{\beta}}) = \sigma^2(\mathbf{X}'\mathbf{V}^{-1}\mathbf{X})^{-1}$ as in (7.64).

7.33 (**a**) Show that the two forms of s^2 in (7.65) and (7.66) are equal.

(**b**) Show that $E(s^2) = \sigma^2$, where s^2 is as given by (7.66).

7.34 Complete the steps in the proof of Theorem 7.8b.

7.35 Show that for $\mathbf{V} = (1-\rho)\mathbf{I} + \rho\mathbf{J}$ in (7.67), the inverse is given by $\mathbf{V}^{-1} = a(\mathbf{I} - b\rho\mathbf{J})$ as in (7.68), where $a = 1/(1-\rho)$ and $b = 1/[1+(n-1)\rho]$.

7.36 (**a**) Show that $\mathbf{X}'\mathbf{V}^{-1}\mathbf{X} = \begin{pmatrix} bn & \mathbf{0}' \\ \mathbf{0} & a\mathbf{X}_c'\mathbf{X}_c \end{pmatrix}$ as in (7.69).

(**b**) Show that $\mathbf{X}'\mathbf{V}^{-1}\mathbf{y} = \begin{pmatrix} bn\bar{y} \\ a\mathbf{X}_c'\mathbf{y} \end{pmatrix}$ as in (7.70).

7.37 Show that $\text{cov}(\hat{\boldsymbol{\beta}}^*) = \sigma^2(\mathbf{X}'\mathbf{X})^{-1}\mathbf{X}'\mathbf{V}\mathbf{X}(\mathbf{X}'\mathbf{X})^{-1}$ as in (7.72), where $\hat{\boldsymbol{\beta}}^* = (\mathbf{X}'\mathbf{X})^{-1}\mathbf{X}'\mathbf{y}$ and $\text{cov}(\mathbf{y}) = \sigma^2\mathbf{V}$.

7.38 (**a**) Show that the weighted least-squares estimator $\hat{\boldsymbol{\beta}} = (\hat{\beta}_0, \hat{\beta}_1)'$ for the model $y_i = \beta_0 + \beta_1 x_i + \varepsilon_i$ with $\text{var}(y_i) = \sigma^2 x_i$ has the form given in (7.73).

(**b**) Verify the expression for $\text{cov}(\hat{\boldsymbol{\beta}})$ in (7.74).

7.39 Obtain the expression for $\text{cov}(\hat{\boldsymbol{\beta}}^*)$ in (7.75).

7.40 As an alternative derivation of $\text{var}(\hat{\beta}_1^*)$ in (7.76), use the following two steps to find $\text{var}(\hat{\beta}_1^*)$ using $\hat{\beta}_1^* = \sum_i (x_i - \bar{x})y_i / \sum_i (x_i - \bar{x})^2$ from the answer to Problem 6.2:

(**a**) Using $\text{var}(y_i) = \sigma^2 x_i$, show that $\text{var}(\hat{\beta}_1^*) = \sigma^2 \sum_i (x_i - \bar{x})^2 x_i / \left[\sum_i (x_i - \bar{x})^2\right]^2$.

(**b**) Show that this expression for $\text{var}(\hat{\beta}_1^*)$ is equal to that in (7.76).

7.41 Using $x = 2, 3, 5, 7, 8, 10$, compare $\text{var}(\hat{\beta}_1^*)$ in (7.76) with $\text{var}(\hat{\beta}_1)$ in (7.77).

7.42 Provide an alternative proof of $\text{cov}(\hat{\boldsymbol{\beta}}_1^*) = \sigma^2(\mathbf{X}_1'\mathbf{X}_1)^{-1}$ in (7.81) using the definition in (3.24), $\text{cov}(\hat{\boldsymbol{\beta}}_1^*) = E\{[\hat{\boldsymbol{\beta}}_1^* - E(\hat{\boldsymbol{\beta}}_1^*)][\hat{\boldsymbol{\beta}}_1^* - E(\hat{\boldsymbol{\beta}}_1^*)]'\}$.

7.43 Prove Theorem 7.9b.

7.44 Provide the missing steps in the proof of Theorem 7.9c(ii).

7.45 Show that $\mathbf{x}_{01}\hat{\boldsymbol{\beta}}_1^*$ is biased for estimating $\mathbf{x}_{01}\boldsymbol{\beta}_1$ if $\boldsymbol{\beta}_2 \neq \mathbf{0}$ and $\mathbf{X}_1'\mathbf{X}_2 \neq \mathbf{O}$.

7.46 Show that $\text{var}(\mathbf{x}_{01}\hat{\boldsymbol{\beta}}_1) \geq \text{var}(\mathbf{x}_{01}\hat{\boldsymbol{\beta}}_1^*)$.

7.47 Complete the steps in the proof of Theorem 7.9d.

7.48 Show that for the no-intercept model $y_i = \beta_1^* x_i + \varepsilon_i^*$, the least-squares estimator is $\hat{\beta}_1^* = \sum_i x_i y_i / \sum_i x_i^2$ as in (7.86).

7.49 Obtain $E(\hat{\beta}_1^*) = \beta_0 \sum_i x_i / \sum_i x_i^2 + \beta_1$ in (7.87) using (7.80), $E(\hat{\boldsymbol{\beta}}_1^*) = \boldsymbol{\beta}_1 + \mathbf{A}\boldsymbol{\beta}_2$.

7.50 Suppose that we use the model $y_i = \beta_0^* + \beta_1^* x_i + \varepsilon_i^*$ when the true model is $y_i = \beta_0 + \beta_1 x_i + \beta_2 x_i^2 + \beta_3 x_i^3 + \varepsilon_i$.

(**a**) Using (7.80), find $E(\hat{\beta}_0^*)$ and $E(\hat{\beta}_1^*)$ if observations are taken at $x = -3, -2, -1, 0, 1, 2, 3$.

(**b**) Using (7.85), find $E(s_1^2)$ for the same values of x.

7.51 Show that $\mathbf{X}_{2\cdot 1} = \mathbf{X}_2 - \hat{\mathbf{X}}_2(\mathbf{X}_1)$ is orthogonal to $\mathbf{X}_1$, that is, $\mathbf{X}_1'\mathbf{X}_{2\cdot 1} = \mathbf{O}$, as in (7.95).

7.52 Show that $\hat{\boldsymbol{\beta}}_2$ in (7.98) is the same as in the full fitted model $\hat{\mathbf{y}} = \mathbf{X}_1\hat{\boldsymbol{\beta}}_1 + \mathbf{X}_2\hat{\boldsymbol{\beta}}_2$.

7.53 When gasoline is pumped into the tank of a car, vapors are vented into the atmosphere. An experiment was conducted to determine whether y, the amount of vapor, can be predicted using the following four variables based on initial conditions of the tank and the dispensed gasoline:

$$x_1 = \text{tank temperature } (^\circ\text{F})$$
$$x_2 = \text{gasoline temperature } (^\circ\text{F})$$
$$x_3 = \text{vapor pressure in tank (psi)}$$
$$x_4 = \text{vapor pressure of gasoline (psi)}$$

The data are given in Table 7.3 (Weisberg 1985, p. 138).

(**a**) Find $\hat{\boldsymbol{\beta}}$ and s^2.

(**b**) Find an estimate of $\text{cov}(\hat{\boldsymbol{\beta}})$.

(**c**) Find $\hat{\boldsymbol{\beta}}_1$ and $\hat{\beta}_0$ using $\mathbf{S}_{xx}$ and $\mathbf{s}_{yx}$ as in (7.46) and (7.47).

(**d**) Find R^2 and R_a^2.

7.54 In an effort to obtain maximum yield in a chemical reaction, the values of the following variables were chosen by the experimenter:

$$x_1 = \text{temperature } (^\circ\text{C})$$
$$x_2 = \text{concentration of a reagent } (\%)$$
$$x_3 = \text{time of reaction (hours)}$$

Two different response variables were observed:

$$y_1 = \text{percent of unchanged starting material}$$
$$y_2 = \text{percent converted to the desired product}$$

TABLE 7.3 Gas Vapor Data

y	x_1	x_2	x_3	x_4	y	x_1	x_2	x_3	x_4
29	33	53	3.32	3.42	40	90	64	7.32	6.70
24	31	36	3.10	3.26	46	90	60	7.32	7.20
26	33	51	3.18	3.18	55	92	92	7.45	7.45
22	37	51	3.39	3.08	52	91	92	7.27	7.26
27	36	54	3.20	3.41	29	61	62	3.91	4.08
21	35	35	3.03	3.03	22	59	42	3.75	3.45
33	59	56	4.78	4.57	31	88	65	6.48	5.80
34	60	60	4.72	4.72	45	91	89	6.70	6.60
32	59	60	4.60	4.41	37	63	62	4.30	4.30
34	60	60	4.53	4.53	37	60	61	4.02	4.10
20	34	35	2.90	2.95	33	60	62	4.02	3.89
36	60	59	4.40	4.36	27	59	62	3.98	4.02
34	60	62	4.31	4.42	34	59	62	4.39	4.53
23	60	36	4.27	3.94	19	37	35	2.75	2.64
24	62	38	4.41	3.49	16	35	35	2.59	2.59
32	62	61	4.39	4.39	22	37	37	2.73	2.59

The data are listed in Table 7.4 (Box and Youle 1955, Andrews and Herzberg 1985, p. 188). Carry out the following for y_1:

(a) Find $\hat{\boldsymbol{\beta}}$ and s^2.

(b) Find an estimate of $\text{cov}(\hat{\boldsymbol{\beta}})$.

TABLE 7.4 Chemical Reaction Data

y_1	y_2	x_1	x_2	x_3
41.5	45.9	162	23	3
33.8	53.3	162	23	8
27.7	57.5	162	30	5
21.7	58.8	162	30	8
19.9	60.6	172	25	5
15.0	58.0	172	25	8
12.2	58.6	172	30	5
4.3	52.4	172	30	8
19.3	56.9	167	27.5	6.5
6.4	55.4	177	27.5	6.5
37.6	46.9	157	27.5	6.5
18.0	57.3	167	32.5	6.5
26.3	55.0	167	22.5	6.5
9.9	58.9	167	27.5	9.5
25.0	50.3	167	27.5	3.5
14.1	61.1	177	20	6.5
15.2	62.9	177	20	6.5
15.9	60.0	160	34	7.5
19.6	60.6	160	34	7.5

TABLE 7.5 Land Rent Data

y	x_1	x_2	x_3	y	x_1	x_2	x_3
18.38	15.50	17.25	.24	8.50	9.00	8.89	.08
20.00	22.29	18.51	.20	36.50	20.64	23.81	.24
11.50	12.36	11.13	.12	60.00	81.40	4.54	.05
25.00	31.84	5.54	.12	16.25	18.92	29.62	.72
52.50	83.90	5.44	.04	50.00	50.32	21.36	.19
82.50	72.25	20.37	.05	11.50	21.33	1.53	.10
25.00	27.14	31.20	.27	35.00	46.85	5.42	.08
30.67	40.41	4.29	.10	75.00	65.94	22.10	.09
12.00	12.42	8.69	.41	31.56	38.68	14.55	.17
61.25	69.42	6.63	.04	48.50	51.19	7.59	.13
60.00	48.46	27.40	.12	77.50	59.42	49.86	.13
57.50	69.00	31.23	.08	21.67	24.64	11.46	.21
31.00	26.09	28.50	.21	19.75	26.94	2.48	.10
60.00	62.83	29.98	.17	56.00	46.20	31.62	.26
72.50	77.06	13.59	.05	25.00	26.86	53.73	.43
60.33	58.83	45.46	.16	40.00	20.00	40.18	.56
49.75	59.48	35.90	.32	56.67	62.52	15.89	.05

(**c**) Find R^2 and R_a^2.

(**d**) In order to find the maximum yield for y_1, a second-order model is of interest. Find $\hat{\boldsymbol{\beta}}$ and s^2 for the model $y_1 = \beta_0 + \beta_1 x_1 + \beta_2 x_2 + \beta_3 x_3 + \beta_4 x_1^2 + \beta_5 x_2^2 + \beta_6 x_3^2 + \beta_7 x_1 x_2 + \beta_8 x_1 x_3 + \beta_9 x_2 x_3 + \varepsilon$.

(**e**) Find R^2 and R_a^2 for the second-order model.

7.55 The following variables were recorded for several counties in Minnesota in 1977:

$$y = \text{average rent paid per acre of land with alfalfa}$$
$$x_1 = \text{average rent paid per acre for all land}$$
$$x_2 = \text{average number of dairy cows per square mile}$$
$$x_3 = \text{proportion of farmland in pasture}$$

The data for 34 counties are given in Table 7.5 (Weisberg 1985, p. 162). Can rent for alfalfa land be predicted from the other three variables?

(**a**) Find $\hat{\boldsymbol{\beta}}$ and s^2.

(**b**) Find $\hat{\boldsymbol{\beta}}_1$ and $\hat{\beta}_0$ using $\mathbf{S}_{xx}$ and $\mathbf{s}_{yx}$ as in (7.46) and (7.47).

(**c**) Find R^2 and R_a^2.

8 Multiple Regression: Tests of Hypotheses and Confidence Intervals

In this chapter we consider hypothesis tests and confidence intervals for the parameters $\beta_0, \beta_1, \ldots, \beta_k$ in $\boldsymbol{\beta}$ in the model $\mathbf{y} = \mathbf{X}\boldsymbol{\beta} + \boldsymbol{\varepsilon}$. We also provide a confidence interval for $\sigma^2 = \text{var}(y_i)$. We will assume throughout the chapter that $\mathbf{y}$ is $N_n(\mathbf{X}\boldsymbol{\beta}, \sigma^2\mathbf{I})$, where $\mathbf{X}$ is $n \times (k+1)$ of rank $k + 1 < n$.

8.1 TEST OF OVERALL REGRESSION

We noted in Section 7.9 that the problems associated with both overfitting and underfitting motivate us to seek an optimal model. Hypothesis testing is a formal tool for, among other things, choosing between a reduced model and an associated full model. The hypothesis H_0, expresses the reduced model in terms of values of a subset of the β_j's in $\boldsymbol{\beta}$. The alternative hypothesis, H_1, is associated with the full model.

To illustrate this tool we begin with a common test, the test of the overall regression hypothesis that none of the x variables predict y. This hypothesis (leading to the reduced model) can be expressed as $H_0 : \boldsymbol{\beta}_1 = \mathbf{0}$, where $\boldsymbol{\beta}_1 = (\beta_1, \beta_2, \ldots, \beta_k)'$. Note that we wish to test $H_0 : \boldsymbol{\beta}_1 = \mathbf{0}$, not $H_0 : \boldsymbol{\beta} = \mathbf{0}$, where

$$\boldsymbol{\beta} = \begin{pmatrix} \beta_0 \\ \boldsymbol{\beta}_1 \end{pmatrix}.$$

Since β_0 is usually not zero, we would rarely be interested in including $\beta_0 = 0$ in the hypothesis. Rejection of H_0: $\boldsymbol{\beta} = \mathbf{0}$ might be due solely to β_0, and we would not learn whether the x variables predict y. For a test of $H_0 : \boldsymbol{\beta} = \mathbf{0}$, see Problem 8.6.

We proceed by proposing a test statistic that is distributed as a central F if H_0 is true and as a noncentral F otherwise. Our approach to obtaining a test statistic is somewhat

Linear Models in Statistics, Second Edition, by Alvin C. Rencher and G. Bruce Schaalje

simplified if we use the centered model (7.32)

$$\mathbf{y} = (\mathbf{j}, \mathbf{X}_c)\begin{pmatrix} \alpha \\ \boldsymbol{\beta}_1 \end{pmatrix} + \boldsymbol{\varepsilon},$$

where $\mathbf{X}_c = [\mathbf{I} - (1/n)\mathbf{J}]\mathbf{X}_1$ is the centered matrix [see (7.33)] and $\mathbf{X}_1$ contains all the columns of $\mathbf{X}$ except the first [see (7.19)]. The corrected total sum of squares $\text{SST} = \sum_{i=1}^n (y_i - \bar{y})^2$ can be partitioned as

$$\sum_{i=1}^n (y_i - \bar{y})^2 = \hat{\boldsymbol{\beta}}_1'\mathbf{X}_c'\mathbf{y} + \left[\sum_{i=1}^n (y_i - \bar{y})^2 - \hat{\boldsymbol{\beta}}_1'\mathbf{X}_c'\mathbf{y}\right] \quad \text{[by (7.53)]}$$

$$= \hat{\boldsymbol{\beta}}_1'\mathbf{X}_c'\mathbf{X}_c\hat{\boldsymbol{\beta}}_1 + \text{SSE} = \text{SSR} + \text{SSE} \quad \text{[by (7.54)]}, \tag{8.1}$$

where SSE is as given in (7.39). The regression sum of squares $\text{SSR} = \hat{\boldsymbol{\beta}}_1'\mathbf{X}_c'\mathbf{X}_c\hat{\boldsymbol{\beta}}_1$ is clearly due to $\boldsymbol{\beta}_1$.

In order to construct an F test, we first express the sums of squares in (8.1) as quadratic forms in $\mathbf{y}$ so that we can use theorems from Chapter 5 to show that SSR and SSE have chi-square distributions and are independent. Using $\sum_i (y_i - \bar{y})^2 = \mathbf{y}'[\mathbf{I} - (1/n)\mathbf{J}]\mathbf{y}$ in (5.2), $\hat{\boldsymbol{\beta}}_1 = (\mathbf{X}_c'\mathbf{X}_c)^{-1}\mathbf{X}_c'\mathbf{y}$ in (7.37), and $\text{SSE} = \sum_{i=1}^n (y_i - \bar{y})^2 - \hat{\boldsymbol{\beta}}_1'\mathbf{X}_c'\mathbf{y}$ in (7.39), we can write (8.1) as

$$\mathbf{y}'\left(\mathbf{I} - \frac{1}{n}\mathbf{J}\right)\mathbf{y} = \text{SSR} + \text{SSE}$$

$$= \mathbf{y}'\mathbf{X}_c(\mathbf{X}_c'\mathbf{X}_c)^{-1}\mathbf{X}_c'\mathbf{y} + \mathbf{y}'\left(\mathbf{I} - \frac{1}{n}\mathbf{J}\right)\mathbf{y} - \mathbf{y}'\mathbf{X}_c(\mathbf{X}_c'\mathbf{X}_c)^{-1}\mathbf{X}_c'\mathbf{y}$$

$$= \mathbf{y}'\mathbf{H}_c\mathbf{y} + \mathbf{y}'\left(\mathbf{I} - \frac{1}{n}\mathbf{J} - \mathbf{H}_c\right)\mathbf{y}, \tag{8.2}$$

where $\mathbf{H}_c = \mathbf{X}_c(\mathbf{X}_c'\mathbf{X}_c)^{-1}\mathbf{X}_c'$.

In the following theorem we establish some properties of the three matrices of the quadratic forms in (8.2).

Theorem 8.1a. The matrices $\mathbf{I} - (1/n)\,\mathbf{J}$, $\mathbf{H}_c = \mathbf{X}_c(\mathbf{X}_c'\mathbf{X}_c)^{-1}\mathbf{X}_c'$, and $\mathbf{I} - (1/n)\,\mathbf{J} - \mathbf{H}_c$ have the following properties:

(i) $\mathbf{H}_c[\mathbf{I} - (1/n)\,\mathbf{J}] = \mathbf{H}_c$. (8.3)

(ii) $\mathbf{H}_c$ is idempotent of rank k.

(iii) $\mathbf{I} - (1/n)\,\mathbf{J} - \mathbf{H}_c$ is idempotent of rank $n - k - 1$.

(iv) $\mathbf{H}_c[\mathbf{I} - (1/n)\,\mathbf{J} - \mathbf{H}_c] = \mathbf{O}$. (8.4)

Proof. Part (i) follows from $\mathbf{X}_c'\mathbf{j} = \mathbf{0}$, which was established in Problem 7.16. Part (ii) can be shown by direct multiplication. Parts (iii) and (iv) follow from (i) and (ii). □

The distributions of SSR/σ^2 and SSE/σ^2 are given in the following theorem.

Theorem 8.1b. If $\mathbf{y}$ is $N_n(\mathbf{X}\boldsymbol{\beta}, \sigma^2\mathbf{I})$, then $\mathrm{SSR}/\sigma^2 = \hat{\boldsymbol{\beta}}_1'\mathbf{X}_c'\mathbf{X}_c\hat{\boldsymbol{\beta}}_1/\sigma^2$ and $\mathrm{SSE}/\sigma^2 = \left[\sum_{i=1}^n (y_i - \bar{y})^2 - \hat{\boldsymbol{\beta}}_1'\mathbf{X}_c'\mathbf{X}_c\hat{\boldsymbol{\beta}}_1\right]/\sigma^2$ have the following distributions:

(i) SSR/σ^2 is $\chi^2(k, \lambda_1)$, where $\lambda_1 = \boldsymbol{\mu}'\mathbf{A}\boldsymbol{\mu}/2\sigma^2 = \boldsymbol{\beta}_1'\mathbf{X}_c'\mathbf{X}_c\boldsymbol{\beta}_1/2\sigma^2$.
(ii) SSE/σ^2 is $\chi^2(n-k-1)$.

Proof. These results follow from (8.2), Theorem 8.1a(ii) and (iii), and Corollary 2 to Theorem 5.5. □

The independence of SSR and SSE is demonstrated in the following theorem.

Theorem 8.1c. If $\mathbf{y}$ is $N_n(\mathbf{X}\boldsymbol{\beta}, \sigma^2\mathbf{I})$, then SSR and SSE are independent, where SSR and SSE are defined in (8.1) and (8.2).

Proof. This follows from Theorem 8.1a(iv) and Corollary 1 to Theorem 5.6b. □

We can now establish an F test for H_0: $\boldsymbol{\beta}_1 = \mathbf{0}$ versus H_1: $\boldsymbol{\beta}_1 \neq \mathbf{0}$.

Theorem 8.1d. If $\mathbf{y}$ is $N_n(\mathbf{X}\boldsymbol{\beta}, \sigma^2\mathbf{I})$, the distribution of

$$F = \frac{\mathrm{SSR}/(k\sigma^2)}{\mathrm{SSE}/[(n-k-1)\sigma^2]} = \frac{\mathrm{SSR}/k}{\mathrm{SSE}/(n-k-1)} \tag{8.5}$$

is as follows:

(i) If $H_0: \boldsymbol{\beta}_1 = \mathbf{0}$ is false, then

$$F \text{ is distributed as } F(k, n-k-1, \lambda_1),$$

where $\lambda_1 = \boldsymbol{\beta}_1'\mathbf{X}_c'\mathbf{X}_c\boldsymbol{\beta}_1/2\sigma^2$.
(ii) If $H_0: \boldsymbol{\beta}_1 = \mathbf{0}$ is true, then $\lambda_1 = 0$ and

$$F \text{ is distributed as } F(k, n-k-1).$$

Proof

(i) This result follows from (5.30) and Theorems 8.1b and 8.1c.
(ii) This result follows from (5.28) and Theorems 8.1b and 8.1c. □

Note that $\lambda_1 = 0$ if and only if $\boldsymbol{\beta}_1 = \mathbf{0}$, since $\mathbf{X}_c'\mathbf{X}_c$ is positive definite (see Corollary 1 to Theorem 2.6b).

TABLE 8.1 ANOVA Table for the F Test of $H_0 : \boldsymbol{\beta}_1 = \mathbf{0}$

Source of Variation	df	Sum of Squares	Mean Square	Expected Mean Square
Due to $\boldsymbol{\beta}_1$	k	$\text{SSR} = \hat{\boldsymbol{\beta}}_1'\mathbf{X}_c'\mathbf{y} = \hat{\boldsymbol{\beta}}'\mathbf{X}'\mathbf{y} - n\bar{y}^2$	SSR/k	$\sigma^2 + \frac{1}{k}\boldsymbol{\beta}_1'\mathbf{X}_c'\mathbf{X}_c\boldsymbol{\beta}_1$
Error	$n-k-1$	$\text{SSE} = \sum_i (y_i - \bar{y})^2 - \hat{\boldsymbol{\beta}}_1'\mathbf{X}_c'\mathbf{y} = \mathbf{y}'\mathbf{y} - \hat{\boldsymbol{\beta}}'\mathbf{X}'\mathbf{y}$	$\text{SSE}/(n-k-1)$	σ^2
Total	$n-1$	$\text{SST} = \sum_i (y_i - \bar{y})^2$		

The test for $H_0 : \boldsymbol{\beta}_1 = \mathbf{0}$ is carried out as follows. Reject H_0 if $F \geq F_{\alpha,k,n-k-1}$, where $F_{\alpha,k,n-k-1}$ is the upper α percentage point of the (central) F distribution. Alternatively, a p value can be used to carry out the test. A p value is the tail area of the central F distribution beyond the calculated F value, that is, the probability of exceeding the calculated F value, assuming $H_0 : \boldsymbol{\beta}_1 = \mathbf{0}$ to be true. A p value less than α is equivalent to $F > F_{\alpha,k,n-k-1}$.

The analysis-of-variance (ANOVA) table (Table 8.1) summarizes the results and calculations leading to the overall F test. Mean squares are sums of squares divided by the degrees of freedom of the associated chi-square (χ^2) distributions.

The entries in the column for expected mean squares in Table 8.1 are simply $E(\text{SSR}/k)$ and $E[\text{SSE}/(n-k-1)]$. The first of these can be established by Theorem 5.2a or by (5.20). The second was established by Theorem 7.3f.

If $H_0 : \boldsymbol{\beta}_1 = \mathbf{0}$ is true, both of the expected mean squares in Table 8.1 are equal to σ^2, and we expect F to be near 1. If $\boldsymbol{\beta}_1 \neq \mathbf{0}$, then $E(\text{SSR}/k) > \sigma^2$ since $\mathbf{X}_c'\mathbf{X}_c$ is positive definite, and we expect F to exceed 1. We therefore reject H_0 for large values of F.

The test of $H_0 : \boldsymbol{\beta}_1 = \mathbf{0}$ in Table 8.1 has been developed using the centered model (7.32). We can also express SSR and SSE in terms of the noncentered model $\mathbf{y} = \mathbf{X}\boldsymbol{\beta} + \boldsymbol{\varepsilon}$ in (7.4):

$$\text{SSR} = \hat{\boldsymbol{\beta}}'\mathbf{X}'\mathbf{y} - n\bar{y}^2, \quad \text{SSE} = \mathbf{y}'\mathbf{y} - \hat{\boldsymbol{\beta}}'\mathbf{X}'\mathbf{y}. \tag{8.6}$$

These are the same as SSR and SSE in (8.1) [see (7.24), (7.39), (7.54), and Problems 7.19, 7.25].

Example 8.1. Using the data in Table 7.1, we illustrate the test of $H_0 : \boldsymbol{\beta}_1 = \mathbf{0}$ where, in this case, $\boldsymbol{\beta}_1 = (\beta_1, \beta_2)'$. In Example 7.3.1(a), we found $\mathbf{X}'\mathbf{y} = (90, 482, 872)'$ and $\hat{\boldsymbol{\beta}} = (5.3754, \ 3.0118, \ -1.2855)'$. The quantities $\mathbf{y}'\mathbf{y}$, $\hat{\boldsymbol{\beta}}'\mathbf{X}'\mathbf{y}$, and $n\bar{y}^2$ are given by

$$\mathbf{y}'\mathbf{y} = \sum_{i=1}^{12} y_i^2 = 2^2 + 3^2 + \cdots + 14^2 = 840,$$

$$\hat{\boldsymbol{\beta}}'\mathbf{X}'\mathbf{y} = (5.3754, 3.0118, -1.2855)\begin{pmatrix} 90 \\ 482 \\ 872 \end{pmatrix} = 814.5410,$$

TABLE 8.2 ANOVA for Overall Regression Test for Data in Table 7.1

Source	df	SS	MS	F
Due to β_1	2	139.5410	69.7705	24.665
Error	9	25.4590	2.8288	
Total	11	165.0000		

$$n\bar{y}^2 = n\left(\frac{\sum_i y_i}{n}\right)^2 = 12\left(\frac{90}{12}\right)^2 = 675.$$

Thus, by (8.6), we obtain

$$\text{SSR} = \hat{\boldsymbol{\beta}}'\mathbf{X}'\mathbf{y} - n\bar{y}^2 = 139.5410,$$

$$\text{SSE} = \mathbf{y}'\mathbf{y} - \hat{\boldsymbol{\beta}}'\mathbf{X}'\mathbf{y} = 25.4590,$$

$$\sum_{i=1}^{n}(y_i - \bar{y})^2 = \mathbf{y}'\mathbf{y} - n\bar{y}^2 = 165.$$

The F test is given in Table 8.2. Since $24.665 > F_{.05,2,9} = 4.26$, we reject $H_0 : \boldsymbol{\beta}_1 = \mathbf{0}$ and conclude that at least one of β_1 or β_2 is not zero. The p value is .000223. □

8.2 TEST ON A SUBSET OF THE β'S

In more generality, suppose that we wish to test the hypothesis that a subset of the x's is not useful in predicting y. A simple example is $H_0 : \beta_j = 0$ for a single β_j. If H_0 is rejected, we would retain $\beta_j x_j$ in the model. As another illustration, consider the model in (7.2)

$$y = \beta_0 + \beta_1 x_1 + \beta_2 x_2 + \beta_3 x_1^2 + \beta_4 x_2^2 + \beta_5 x_1 x_2 + \varepsilon,$$

for which we may wish to test the hypothesis $H_0 : \beta_3 = \beta_4 = \beta_5 = 0$. If H_0 is rejected, we would choose the full second-order model over the reduced first-order model.

Without loss of generality, we assume that the β's to be tested have been arranged last in $\boldsymbol{\beta}$, with a corresponding arrangement of the columns of $\mathbf{X}$. Then $\boldsymbol{\beta}$ and $\mathbf{X}$ can be partitioned accordingly, and by (7.78), the model for all n observations becomes

$$\begin{aligned}\mathbf{y} = \mathbf{X}\boldsymbol{\beta} + \boldsymbol{\varepsilon} &= (\mathbf{X}_1, \mathbf{X}_2)\begin{pmatrix}\boldsymbol{\beta}_1 \\ \boldsymbol{\beta}_2\end{pmatrix} + \boldsymbol{\varepsilon} \\ &= \mathbf{X}_1\boldsymbol{\beta}_1 + \mathbf{X}_2\boldsymbol{\beta}_2 + \boldsymbol{\varepsilon}, \end{aligned} \tag{8.7}$$

where $\boldsymbol{\beta}_2$ contains the β's to be tested. The intercept β_0 would ordinarily be included in $\boldsymbol{\beta}_1$.

The hypothesis of interest is $H_0: \boldsymbol{\beta}_2 = \mathbf{0}$. If we designate the number of parameters in $\boldsymbol{\beta}_2$ by h, then $\mathbf{X}_2$ is $n \times h$, $\boldsymbol{\beta}_1$ is $(k-h+1) \times 1$, and $\mathbf{X}_1$ is $n \times (k-h+1)$. Thus $\boldsymbol{\beta}_1 = (\beta_0, \beta_1, \cdots, \beta_{k-h})'$ and $\boldsymbol{\beta}_2 = (\beta_{k-h+1}, \cdots, \beta_k)'$. In terms of the illustration at the beginning of this section, we would have $\boldsymbol{\beta}_1 = (\beta_0, \beta_1, \beta_2)'$ and $\boldsymbol{\beta}_2 = (\beta_3, \beta_4, \beta_5)'$. Note that $\boldsymbol{\beta}_1$ in (8.7) is different from $\boldsymbol{\beta}_1$ in Section 8.1, in which $\boldsymbol{\beta}$ was partitioned as $\boldsymbol{\beta} = \begin{pmatrix} \beta_0 \\ \boldsymbol{\beta}_1 \end{pmatrix}$ and $\boldsymbol{\beta}_1$ constituted all of $\boldsymbol{\beta}$ except β_0.

To test $H_0: \boldsymbol{\beta}_2 = \mathbf{0}$ versus $H_1: \boldsymbol{\beta}_2 \neq \mathbf{0}$, we use a full–reduced-model approach. The full model is given by (8.7). Under $H_0: \boldsymbol{\beta}_2 = \mathbf{0}$, the reduced model becomes

$$\mathbf{y} = \mathbf{X}_1\boldsymbol{\beta}_1^* + \boldsymbol{\varepsilon}^*. \tag{8.8}$$

We use the notation $\boldsymbol{\beta}_1^*$ and $\boldsymbol{\varepsilon}^*$ as in Section 7.9, because in the reduced model, $\boldsymbol{\beta}_1^*$ and $\boldsymbol{\varepsilon}^*$ will typically be different from $\boldsymbol{\beta}_1$ and $\boldsymbol{\varepsilon}$ in the full model (unless $\mathbf{X}_1$ and $\mathbf{X}_2$ are orthogonal; see Theorem 7.9a and its corollary). The estimator of $\boldsymbol{\beta}_1^*$ in the reduced model (8.8) is $\hat{\boldsymbol{\beta}}_1^* = (\mathbf{X}_1'\mathbf{X}_1)^{-1}\mathbf{X}_1'\mathbf{y}$, which is, in general, not the same as the first $k-h+1$ elements of $\hat{\boldsymbol{\beta}} = (\mathbf{X}'\mathbf{X})^{-1}\mathbf{X}'\mathbf{y}$ from the full model (8.7) (unless $\mathbf{X}_1$ and $\mathbf{X}_2$ are orthogonal; see Theorem 7.10).

In order to compare the fit of the full model (8.7) to the fit of the reduced model (8.8), we add and subtract $\hat{\boldsymbol{\beta}}'\mathbf{X}'\mathbf{y}$ and $\hat{\boldsymbol{\beta}}_1^{*\prime}\mathbf{X}_1'\mathbf{y}$ to the total corrected sum of squares $\sum_{i=1}^n (y_i - \bar{y})^2 = \mathbf{y}'\mathbf{y} - n\bar{y}^2$ so as to obtain the partitioning

$$\mathbf{y}'\mathbf{y} - n\bar{y}^2 = (\mathbf{y}'\mathbf{y} - \hat{\boldsymbol{\beta}}'\mathbf{X}\mathbf{y}) + (\hat{\boldsymbol{\beta}}'\mathbf{X}'\mathbf{y} - \hat{\boldsymbol{\beta}}_1^{*\prime}\mathbf{X}_1'\mathbf{y}) + (\hat{\boldsymbol{\beta}}_1^{*\prime}\mathbf{X}_1\mathbf{y} - n\bar{y}^2) \tag{8.9}$$

or

$$\text{SST} = \text{SSE} + \text{SS}(\boldsymbol{\beta}_2|\boldsymbol{\beta}_1) + \text{SSR(reduced)}, \tag{8.10}$$

where $\text{SS}(\boldsymbol{\beta}_2|\boldsymbol{\beta}_1) = \hat{\boldsymbol{\beta}}'\mathbf{X}'\mathbf{y} - \hat{\boldsymbol{\beta}}_1^{*\prime}\mathbf{X}_1'\mathbf{y}$ is the "extra" regression sum of squares due to $\boldsymbol{\beta}_2$ after adjusting for $\boldsymbol{\beta}_1$. Note that $\text{SS}(\boldsymbol{\beta}_2|\boldsymbol{\beta}_1)$ can also be expressed as

$$\begin{aligned} \text{SS}(\boldsymbol{\beta}_2|\boldsymbol{\beta}_1) &= \hat{\boldsymbol{\beta}}'\mathbf{X}'\mathbf{y} - n\bar{y}^2 - (\hat{\boldsymbol{\beta}}_1^{*\prime}\mathbf{X}_1'\mathbf{y} - n\bar{y}^2) \\ &= \text{SSR(full)} - \text{SSR(reduced)}, \end{aligned}$$

which is the difference between the overall regression sum of squares for the full model and the overall regression sum of squares for the reduced model [see (8.6)].

If $H_0: \boldsymbol{\beta}_2 = \mathbf{0}$ is true, we would expect $\text{SS}(\boldsymbol{\beta}_2|\boldsymbol{\beta}_1)$ to be small so that SST in (8.10) is composed mostly of SSR(reduced) and SSE. If $\boldsymbol{\beta}_2 \neq \mathbf{0}$, we expect $\text{SS}(\boldsymbol{\beta}_2|\boldsymbol{\beta}_1)$ to be larger and account for more of SST. Thus we are testing $H_0: \boldsymbol{\beta}_2 = \mathbf{0}$ in the full model in which there are no restrictions on $\boldsymbol{\beta}_1$. We are not *ignoring* $\boldsymbol{\beta}_1$ (assuming $\boldsymbol{\beta}_1 = \mathbf{0}$) but are testing $H_0: \boldsymbol{\beta}_2 = \mathbf{0}$ *in the presence of* $\boldsymbol{\beta}_1$, that is, above and beyond whatever $\boldsymbol{\beta}_1$ contributes to SST.

To develop a test statistic based on $\text{SS}(\boldsymbol{\beta}_2|\boldsymbol{\beta}_1)$, we first write (8.9) in terms of quadratic forms in $\mathbf{y}$. Using $\hat{\boldsymbol{\beta}} = (\mathbf{X}'\mathbf{X})^{-1}\mathbf{X}'\mathbf{y}$ and $\hat{\boldsymbol{\beta}}_1^* = (\mathbf{X}_1'\mathbf{X}_1)^{-1}\mathbf{X}_1'\mathbf{y}$ and (5.2), (8.9) becomes

$$\begin{aligned}
\mathbf{y}'\left(\mathbf{I} - \frac{1}{n}\mathbf{J}\right)\mathbf{y} &= \mathbf{y}'\mathbf{y} - \mathbf{y}'\mathbf{X}(\mathbf{X}'\mathbf{X})^{-1}\mathbf{X}'\mathbf{y} + \mathbf{y}'\mathbf{X}(\mathbf{X}'\mathbf{X})^{-1}\mathbf{X}'\mathbf{y} \\
&\quad - \mathbf{y}'\mathbf{X}_1(\mathbf{X}_1'\mathbf{X}_1)^{-1}\mathbf{X}_1'\mathbf{y} + \mathbf{y}'\mathbf{X}_1(\mathbf{X}_1'\mathbf{X}_1)^{-1}\mathbf{X}_1'\mathbf{y} - \mathbf{y}'\frac{1}{n}\mathbf{J}\mathbf{y} \\
&= \mathbf{y}'\left[\mathbf{I} - \mathbf{X}(\mathbf{X}'\mathbf{X})^{-1}\mathbf{X}'\right]\mathbf{y} + \mathbf{y}'[\mathbf{X}(\mathbf{X}'\mathbf{X})^{-1}\mathbf{X}' - \mathbf{X}_1(\mathbf{X}_1'\mathbf{X}_1)^{-1}\mathbf{X}_1']\mathbf{y} \\
&\quad + \mathbf{y}'\left[\mathbf{X}_1(\mathbf{X}_1'\mathbf{X}_1)^{-1}\mathbf{X}_1' - \frac{1}{n}\mathbf{J}\right]\mathbf{y} \qquad (8.11) \\
&= \mathbf{y}'(\mathbf{I} - \mathbf{H})\mathbf{y} + \mathbf{y}'(\mathbf{H} - \mathbf{H}_1)\mathbf{y} + \mathbf{y}'\left(\mathbf{H}_1 - \frac{1}{n}\mathbf{J}\right)\mathbf{y}, \qquad (8.12)
\end{aligned}$$

where $\mathbf{H} = \mathbf{X}(\mathbf{X}'\mathbf{X})^{-1}\mathbf{X}'$ and $\mathbf{H}_1 = \mathbf{X}_1(\mathbf{X}_1'\mathbf{X}_1)^{-1}\mathbf{X}_1'$. The matrix $\mathbf{I} - \mathbf{H}$ was shown to be idempotent in Problem 5.32a, with rank $n - k - 1$, where $k + 1$ is the rank of $\mathbf{X}$ ($k + 1$ is also the number of elements in $\boldsymbol{\beta}$). The matrix $\mathbf{H} - \mathbf{H}_1$ is shown to be idempotent in the following theorem.

Theorem 8.2a. The matrix $\mathbf{H} - \mathbf{H}_1 = \mathbf{X}(\mathbf{X}'\mathbf{X})^{-1}\mathbf{X}' - \mathbf{X}_1(\mathbf{X}_1'\mathbf{X}_1)^{-1}\mathbf{X}_1'$ is idempotent with rank h, where h is the number of elements in $\boldsymbol{\beta}_2$.

Proof. Premultiplying $\mathbf{X}$ by $\mathbf{H}$, we obtain

$$\mathbf{H}\mathbf{X} = \mathbf{X}(\mathbf{X}'\mathbf{X})^{-1}\mathbf{X}'\mathbf{X} = \mathbf{X}$$

or

$$\mathbf{X} = \left[\mathbf{X}(\mathbf{X}'\mathbf{X})^{-1}\mathbf{X}'\right]\mathbf{X}. \qquad (8.13)$$

Partitioning $\mathbf{X}$ on the left side of (8.13) and the last $\mathbf{X}$ on the right side, we obtain [by an extension of (2.28)]

$$\begin{aligned}
(\mathbf{X}_1, \mathbf{X}_2) &= \left[\mathbf{X}(\mathbf{X}'\mathbf{X})^{-1}\mathbf{X}'\right](\mathbf{X}_1, \mathbf{X}_2) \\
&= \left[\mathbf{X}(\mathbf{X}'\mathbf{X})^{-1}\mathbf{X}'\mathbf{X}_1, \mathbf{X}(\mathbf{X}'\mathbf{X})^{-1}\mathbf{X}'\mathbf{X}_2\right].
\end{aligned}$$

Thus

$$\begin{aligned}
\mathbf{X}_1 &= \mathbf{X}(\mathbf{X}'\mathbf{X})^{-1}\mathbf{X}'\mathbf{X}_1, \\
\mathbf{X}_2 &= \mathbf{X}(\mathbf{X}'\mathbf{X})^{-1}\mathbf{X}'\mathbf{X}_2.
\end{aligned} \qquad (8.14)$$

Simplifying $\mathbf{HH}_1$ and $\mathbf{H}_1\mathbf{H}$ by (8.14) and its transpose, we obtain

$$\mathbf{HH}_1 = \mathbf{H}_1 \quad \text{and} \quad \mathbf{H}_1\mathbf{H} = \mathbf{H}_1. \tag{8.15}$$

The matrices $\mathbf{H}$ and $\mathbf{H}_1$ are idempotent (see Problem 5.32). Thus

$$\begin{aligned}(\mathbf{H} - \mathbf{H}_1)^2 &= \mathbf{H}^2 - \mathbf{HH}_1 - \mathbf{H}_1\mathbf{H} + \mathbf{H}_1^2 \\ &= \mathbf{H} - \mathbf{H}_1 - \mathbf{H}_1 + \mathbf{H}_1 \\ &= \mathbf{H} - \mathbf{H}_1,\end{aligned}$$

and $\mathbf{H} - \mathbf{H}_1$ is idempotent. For the rank of $\mathbf{H} - \mathbf{H}_1$, we have (by Theorem 2.13d)

$$\begin{aligned}\text{rank}(\mathbf{H} - \mathbf{H}_1) &= \text{tr}(\mathbf{H} - \mathbf{H}_1) \\ &= \text{tr}(\mathbf{H}) - \text{tr}(\mathbf{H}_1) && \text{[by (2.86)]} \\ &= \text{tr}\left[\mathbf{X}(\mathbf{X}'\mathbf{X})^{-1}\mathbf{X}'\right] - \text{tr}\left[\mathbf{X}_1(\mathbf{X}_1'\mathbf{X}_1)^{-1}\mathbf{X}_1'\right] \\ &= \text{tr}\left[\mathbf{X}'\mathbf{X}(\mathbf{X}'\mathbf{X})^{-1}\right] - \text{tr}\left[\mathbf{X}_1'\mathbf{X}_1(\mathbf{X}_1'\mathbf{X}_1)^{-1}\right] && \text{[by (2.87)]} \\ &= \text{tr}(\mathbf{I}_{k+1}) - \text{tr}(\mathbf{I}_{k-h+1}) = k + 1 - (k - h + 1) = h.\end{aligned}$$

□

We now find the distributions of $\mathbf{y}'(\mathbf{I} - \mathbf{H})\mathbf{y}$ and $\mathbf{y}'(\mathbf{H} - \mathbf{H}_1)\mathbf{y}$ in (8.12) and show that they are independent.

Theorem 8.2b. If $\mathbf{y}$ is $N_n(\mathbf{X}\boldsymbol{\beta}, \sigma^2\mathbf{I})$ and $\mathbf{H}$ and $\mathbf{H}_1$ are as defined in (8.11) and (8.12), then

(i) $\mathbf{y}'(\mathbf{I} - \mathbf{H})\mathbf{y}/\sigma^2$ is $\chi^2(n - k - 1)$.
(ii) $\mathbf{y}'(\mathbf{H} - \mathbf{H}_1)\mathbf{y}/\sigma^2$ is $\chi^2(h, \lambda_1)$, $\lambda_1 = \boldsymbol{\beta}_2'\left[\mathbf{X}_2'\mathbf{X}_2 - \mathbf{X}_2'\mathbf{X}_1(\mathbf{X}_1'\mathbf{X}_1)^{-1}\mathbf{X}_1'\mathbf{X}_2\right]\boldsymbol{\beta}_2/2\sigma^2$.
(iii) $\mathbf{y}'(\mathbf{I} - \mathbf{H})\mathbf{y}$ and $\mathbf{y}'(\mathbf{H} - \mathbf{H}_1)\mathbf{y}$ are independent.

PROOF. Adding $\mathbf{y}'(1/n)\mathbf{J}\mathbf{y}$ to both sides of (8.12), we obtain the decomposition $\mathbf{y}'\mathbf{y} = \mathbf{y}'(\mathbf{I} - \mathbf{H})\mathbf{y} + \mathbf{y}'(\mathbf{H} - \mathbf{H}_1)\mathbf{y} + \mathbf{y}'\mathbf{H}_1\mathbf{y}$. The matrices $\mathbf{I} - \mathbf{H}$, $\mathbf{H} - \mathbf{H}_1$, and $\mathbf{H}_1$ were shown to be idempotent in Problem 5.32 and Theorem 8.2a. Hence by Corollary 1 to Theorem 5.6c, all parts of the theorem follow. See Problem 8.9 for the derivation of λ_1. □

If $\lambda_1 = 0$ in Theorem 8.2b(ii), then $\mathbf{y}'(\mathbf{H} - \mathbf{H}_1)\mathbf{y}/\sigma^2$ has the central chi-square distribution $\chi^2(h)$. Since $\mathbf{X}_2'\mathbf{X}_2 - \mathbf{X}_2'\mathbf{X}_1(\mathbf{X}_1'\mathbf{X}_1)^{-1}\mathbf{X}_1'\mathbf{X}_2$ is positive definite (see Problem 8), $\lambda_1 = 0$ if and only if $\boldsymbol{\beta}_2 = \mathbf{0}$.

An F test for $H_0: \boldsymbol{\beta}_2 = \mathbf{0}$ versus $H_1: \boldsymbol{\beta}_2 \neq \mathbf{0}$ is given in the following theorem.

Theorem 8.2c. Let $\mathbf{y}$ be $N_n(\mathbf{X}\boldsymbol{\beta}, \sigma^2\mathbf{I})$ and define an F statistic as follows:

$$F = \frac{\mathbf{y}'(\mathbf{H} - \mathbf{H}_1)\mathbf{y}/h}{\mathbf{y}'(\mathbf{I} - \mathbf{H})\mathbf{y}/(n-k-1)} = \frac{\text{SS}(\boldsymbol{\beta}_2|\boldsymbol{\beta}_1)/h}{\text{SSE}/(n-k-1)} \tag{8.16}$$

$$= \frac{(\hat{\boldsymbol{\beta}}'\mathbf{X}'\mathbf{y} - \hat{\boldsymbol{\beta}}_1^{*\prime}\mathbf{X}_1'\mathbf{y})/h}{(\mathbf{y}'\mathbf{y} - \hat{\boldsymbol{\beta}}'\mathbf{X}'\mathbf{y})/(n-k-1)}, \tag{8.17}$$

where $\hat{\boldsymbol{\beta}} = (\mathbf{X}'\mathbf{X})^{-1}\mathbf{X}'\mathbf{y}$ is from the full model $\mathbf{y} = \mathbf{X}\boldsymbol{\beta} + \boldsymbol{\varepsilon}$ and $\hat{\boldsymbol{\beta}}_1^* = (\mathbf{X}_1'\mathbf{X}_1)^{-1}\mathbf{X}_1'\mathbf{y}$ is from the reduced model $\mathbf{y} = \mathbf{X}_1\boldsymbol{\beta}_1^* + \boldsymbol{\varepsilon}^*$. The distribution of F in (8.17) is as follows:

(i) If $H_0: \boldsymbol{\beta}_2 = \mathbf{0}$ is false, then

$$F \text{ is distributed as } F(h, n-k-1, \lambda_1),$$

where $\lambda_1 = \boldsymbol{\beta}_2'\left[\mathbf{X}_2'\mathbf{X}_2 - \mathbf{X}_2'\mathbf{X}_1(\mathbf{X}_1'\mathbf{X}_1)^{-1}\mathbf{X}_1'\mathbf{X}_2\right]\boldsymbol{\beta}_2/2\sigma^2$.

(ii) If $H_0: \boldsymbol{\beta}_2 = \mathbf{0}$ is true, then $\lambda_1 = 0$ and

$$F \text{ is distributed as } F(h, n-k-1).$$

PROOF

(i) This result follows from (5.30) and Theorem 8.2b.
(ii) This result follows from (5.28) and Theorem 8.2b. □

The test for $H_0: \boldsymbol{\beta}_2 = \mathbf{0}$ is carried out as follows: Reject H_0 if $F \geq F_{\alpha,h,n-k-1}$, where $F_{\alpha,h,n-k-1}$ is the upper α percentage point of the (central) F distribution. Alternatively, we reject H_0 if $p < \alpha$, where p is the p value. Since $\mathbf{X}_2'\mathbf{X}_2 - \mathbf{X}_2'\mathbf{X}_1(\mathbf{X}_1'\mathbf{X}_1)^{-1}\mathbf{X}_1'\mathbf{X}_2$ is positive definite (see Problem 8.10), $\lambda_1 > 0$ if $H_0: \boldsymbol{\beta}_2 = \mathbf{0}$ is false. This justifies rejection of H_0 for large values of F.

Results and calculations leading to this F test are summarized in the ANOVA table (Table 8.3), where $\boldsymbol{\beta}_1$ is $(k-h+1) \times 1$, $\boldsymbol{\beta}_2$ is $h \times 1$, $\mathbf{X}_1$ is $n \times (k-h+1)$, and $\mathbf{X}_2$ is $n \times h$.

The entries in the column for expected mean squares are $E[\text{SS}(\boldsymbol{\beta}_2|\boldsymbol{\beta}_1)/h]$ and $E[\text{SSE}/(n-k-1)]$. For $E[\text{SS}(\boldsymbol{\beta}_2|\boldsymbol{\beta}_1)/h]$, see Problem 8.11. Note that if H_0 is true, both expected mean squares (Table 8.3) are equal to σ^2, and if H_0 is false, $E[\text{SS}(\boldsymbol{\beta}_2|\boldsymbol{\beta}_1)/h] > E[\text{SSE}/(n-k-1)]$ since $\mathbf{X}_2'\mathbf{X}_2 - \mathbf{X}_2'\mathbf{X}_1(\mathbf{X}_1'\mathbf{X}_1)^{-1}\mathbf{X}_1'\mathbf{X}_2$ is positive definite. This inequality provides another justification for rejecting H_0 for large values of F.

TABLE 8.3 ANOVA Table for F-Test of $H_0 : \boldsymbol{\beta}_2 = \mathbf{0}$

Source of Variation	df	Sum of Squares	Mean Square	Expected Mean Square
Due to $\boldsymbol{\beta}_2$ adjusted for $\boldsymbol{\beta}_1$	h	$\text{SS}(\beta_2\|\beta_1) = \hat{\boldsymbol{\beta}}'\mathbf{X}'\mathbf{y} - \hat{\boldsymbol{\beta}}_1^{*\prime}\mathbf{X}_1'\mathbf{y}$	$\text{SS}(\boldsymbol{\beta}_2\|\beta_1)/h$	$\sigma^2 + \frac{1}{h}\boldsymbol{\beta}_2'[\mathbf{X}_2'\mathbf{X}_2 - \mathbf{X}_2'\mathbf{X}_1(\mathbf{X}_1'\mathbf{X}_1)^{-1}\mathbf{X}_1'\mathbf{X}_2]\boldsymbol{\beta}_2$
Error	$n-k-1$	$\text{SSE} = \mathbf{y}'\mathbf{y} - \hat{\boldsymbol{\beta}}'\mathbf{X}'\mathbf{y}$	$\text{SSE}/(n-k-1)$	σ^2
Total	$n-1$	$\text{SST} = \mathbf{y}'\mathbf{y} - n\bar{y}^2$		

Example 8.2a. Consider the dependent variable y_2 in the chemical reaction data in Table 7.4 (see Problem 7.52 for a description of the variables). In order to check the usefulness of second-order terms in predicting y_2, we use as a full model, $y_2 = \beta_0 + \beta_1 x_1 + \beta_2 x_2 + \beta_3 x_3 + \beta_4 x_1^2 + \beta_5 x_2^2 + \beta_6 x_3^2 + \beta_7 x_1 x_2 + \beta_8 x_1 x_3 + \beta_9 x_2 x_3 + \varepsilon$, and test $H_0: \beta_4 = \beta_5 = \ldots = \beta_9 = 0$. For the full model, we obtain $\hat{\boldsymbol{\beta}}'\mathbf{X}'\mathbf{y} - n\bar{y}^2 = 339.7888$, and for the reduced model $y_2 = \beta_0^* + \beta_1^* x_1 + \beta_2^* x_2 + \beta_3^* x_3 + \varepsilon*$, we have $\hat{\boldsymbol{\beta}}_1^{*\prime}\mathbf{X}_1'\mathbf{y} - n\bar{y}^2 = 151.0022$. The difference is $\hat{\boldsymbol{\beta}}'\mathbf{X}'\mathbf{y} - \hat{\boldsymbol{\beta}}_1^{*\prime}\mathbf{X}_1'\mathbf{y} = 188.7866$. The error sum of squares is SSE $= 60.6755$, and the F statistic is given by (8.16) or Table 8.3 as

$$F = \frac{188.7866/6}{60.6755/9} = \frac{31.4644}{6.7417} = 4.6671,$$

which has a p value of .0198. Thus the second-order terms are useful in prediction of y_2. In fact, the overall F in (8.5) for the reduced model is 3.027 with $p = .0623$, so that x_1, x_2, and x_3 are inadequate for predicting y_2. The overall F for the full model is 5.600 with $p = .0086$. □

In the following theorem, we express $\mathrm{SS}(\boldsymbol{\beta}_2|\boldsymbol{\beta}_1)$ as a quadratic form in $\hat{\boldsymbol{\beta}}_2$ that corresponds to λ_1 in Theorem 8.2b(ii).

Theorem 8.2d. If the model is partitioned as in (8.7), then $\mathrm{SS}(\boldsymbol{\beta}_2|\boldsymbol{\beta}_1) = \hat{\boldsymbol{\beta}}'\mathbf{X}'\mathbf{y} - \hat{\boldsymbol{\beta}}_1^{*\prime}\mathbf{X}_1'\mathbf{y}$ can be written as

$$\mathrm{SS}(\boldsymbol{\beta}_2|\boldsymbol{\beta}_1) = \hat{\boldsymbol{\beta}}_2'\left[\mathbf{X}_2'\mathbf{X}_2 - \mathbf{X}_2'\mathbf{X}_1(\mathbf{X}_1'\mathbf{X}_1)^{-1}\mathbf{X}_1'\mathbf{X}_2\right]\hat{\boldsymbol{\beta}}_2, \tag{8.18}$$

where $\hat{\boldsymbol{\beta}}_2$ is from a partitioning of $\hat{\boldsymbol{\beta}}$ in the full model:

$$\hat{\boldsymbol{\beta}} = \begin{pmatrix} \hat{\boldsymbol{\beta}}_1 \\ \hat{\boldsymbol{\beta}}_2 \end{pmatrix} = (\mathbf{X}'\mathbf{X})^{-1}\mathbf{X}'\mathbf{y}. \tag{8.19}$$

Proof. We can write $\mathbf{X}\hat{\boldsymbol{\beta}}$ in terms of $\hat{\boldsymbol{\beta}}_1$ and $\hat{\boldsymbol{\beta}}_2$ as $\mathbf{X}\hat{\boldsymbol{\beta}} = (\mathbf{X}_1, \mathbf{X}_2)\begin{pmatrix} \hat{\boldsymbol{\beta}}_1 \\ \hat{\boldsymbol{\beta}}_2 \end{pmatrix} = \mathbf{X}_1\hat{\boldsymbol{\beta}}_1 + \mathbf{X}_2\hat{\boldsymbol{\beta}}_2$. To write $\hat{\boldsymbol{\beta}}_1^*$ in terms of $\hat{\boldsymbol{\beta}}_1$ and $\hat{\boldsymbol{\beta}}_2$, we note that by (7.80), $E(\hat{\boldsymbol{\beta}}_1^*) = \boldsymbol{\beta}_1 + \mathbf{A}\boldsymbol{\beta}_2$, where $\mathbf{A} = (\mathbf{X}_1'\mathbf{X}_1)^{-1}\mathbf{X}_1'\mathbf{X}_2$ is the alias matrix defined in Theorem 7.9a. This can be estimated by $\hat{\boldsymbol{\beta}}_1^* = \hat{\boldsymbol{\beta}}_1 + \mathbf{A}\hat{\boldsymbol{\beta}}_2$, where $\hat{\boldsymbol{\beta}}_1$ and $\hat{\boldsymbol{\beta}}_2$ are from the full model, as in (8.19). Then $\mathrm{SS}(\boldsymbol{\beta}_2|\boldsymbol{\beta}_1)$ in (8.10) or Table 8.3 can be written as

$$\begin{aligned}
\mathrm{SS}(\boldsymbol{\beta}_2|\boldsymbol{\beta}_1) &= \hat{\boldsymbol{\beta}}'\mathbf{X}'\mathbf{y} - \hat{\boldsymbol{\beta}}_1^{*\prime}\mathbf{X}_1'\mathbf{y} \\
&= \hat{\boldsymbol{\beta}}'\mathbf{X}'\mathbf{X}\hat{\boldsymbol{\beta}} - \hat{\boldsymbol{\beta}}_1^{*\prime}\mathbf{X}_1'\mathbf{X}_1\hat{\boldsymbol{\beta}}^* \quad \text{[by (7.8)]} \\
&= (\hat{\boldsymbol{\beta}}_1'\mathbf{X}_1' + \hat{\boldsymbol{\beta}}_2'\mathbf{X}_2')(\mathbf{X}_1\hat{\boldsymbol{\beta}}_1 + \mathbf{X}_2\hat{\boldsymbol{\beta}}_2) - (\hat{\boldsymbol{\beta}}_1' + \hat{\boldsymbol{\beta}}_2'\mathbf{A}')\mathbf{X}_1'\mathbf{X}_1(\hat{\boldsymbol{\beta}}_1 + \mathbf{A}\hat{\boldsymbol{\beta}}_2).
\end{aligned}$$

Multiplying this out and substituting $(\mathbf{X}_1'\mathbf{X}_1)^{-1}\mathbf{X}_1'\mathbf{X}_2$ for $\mathbf{A}$, we obtain (8.18). □

In (8.18), it is clear that $\text{SS}(\boldsymbol{\beta}_2|\boldsymbol{\beta}_1)$ is due to $\boldsymbol{\beta}_2$. We also see in (8.18) a direct correspondence between $\text{SS}(\boldsymbol{\beta}_2|\boldsymbol{\beta}_1)$ and the noncentrality parameter λ_1 in Theorem 8.2b (ii) or the expected mean square in Table 8.3.

Example 8.2b. The full–reduced-model test of $H_0: \boldsymbol{\beta}_2 = \mathbf{0}$ in Table 8.3 can be used to test for significance of a single $\hat{\beta}_j$. To illustrate, suppose that we wish to test $H_0: \beta_k = 0$, where $\boldsymbol{\beta}$ is partitioned as

$$\boldsymbol{\beta} = \begin{pmatrix} \beta_0 \\ \beta_1 \\ \vdots \\ \beta_{k-1} \\ \hline \beta_k \end{pmatrix} = \begin{pmatrix} \boldsymbol{\beta}_1 \\ \beta_k \end{pmatrix}.$$

Then $\mathbf{X}$ is partitioned as $\mathbf{X} = (\mathbf{X}_1, \mathbf{x}_k)$, where $\mathbf{x}_k$ is the last column of $\mathbf{X}$ and $\mathbf{X}_1$ contains all columns except $\mathbf{x}_k$. The reduced model is $\mathbf{y} = \mathbf{X}_1\boldsymbol{\beta}_1^* + \boldsymbol{\varepsilon}^*$, and $\boldsymbol{\beta}_1^*$ is estimated as $\hat{\boldsymbol{\beta}}_1^* = (\mathbf{X}_1'\mathbf{X}_1)^{-1}\mathbf{X}_1'\mathbf{y}$. In this case, $h = 1$, and the F statistic in (8.17) becomes

$$F = \frac{\hat{\boldsymbol{\beta}}'\mathbf{X}'\mathbf{y} - \hat{\boldsymbol{\beta}}_1^{*\prime}\mathbf{X}_1'\mathbf{y}}{(\mathbf{y}'\mathbf{y} - \hat{\boldsymbol{\beta}}'\mathbf{X}'\mathbf{y})/(n-k-1)}, \tag{8.20}$$

which is distributed as $F(1, n-k-1)$ if $H_0: \beta_k = 0$ is true. □

Example 8.2c. The test in Section 8.1 for overall regression can be obtained as a full–reduced-model test. In this case, the partitioning of $\mathbf{X}$ and of $\boldsymbol{\beta}$ is $\mathbf{X} = (\mathbf{j}, \mathbf{X}_1)$ and

$$\boldsymbol{\beta} = \begin{pmatrix} \beta_0 \\ \hline \beta_1 \\ \vdots \\ \beta_k \end{pmatrix} = \begin{pmatrix} \beta_0 \\ \boldsymbol{\beta}_1 \end{pmatrix}.$$

The reduced model is $\mathbf{y} = \beta_0^*\mathbf{j} + \boldsymbol{\varepsilon}^*$, for which we have

$$\hat{\beta}_0^* = \bar{y} \quad \text{and} \quad \text{SS}(\beta_0^*) = n\bar{y}^2 \tag{8.21}$$

(see Problem 8.13). Then $\text{SS}(\boldsymbol{\beta}_1|\beta_0) = \hat{\boldsymbol{\beta}}'\mathbf{X}'\mathbf{y} - n\bar{y}^2$, which is the same as (8.6). □

8.3 *F* TEST IN TERMS OF R^2

The F statistics in Sections 8.1 and 8.2 can be expressed in terms of R^2 as defined in (7.56).

Theorem 8.3. The F statistics in (8.5) and (8.17) for testing $H_0 : \boldsymbol{\beta}_1 = \mathbf{0}$ and $H_0 : \boldsymbol{\beta}_2 = \mathbf{0}$, respectively, can be written in terms of R^2 as

$$F = \frac{(\hat{\boldsymbol{\beta}}'\mathbf{X}'\mathbf{y} - n\bar{y}^2)/k}{(\mathbf{y}'\mathbf{y} - \hat{\boldsymbol{\beta}}'\mathbf{X}'\mathbf{y})/(n-k-1)} \tag{8.22}$$

$$= \frac{R^2/k}{(1-R^2)/(n-k-1)} \tag{8.23}$$

and

$$F = \frac{(\hat{\boldsymbol{\beta}}'\mathbf{X}'\mathbf{y} - \hat{\boldsymbol{\beta}}_1^{*\prime}\mathbf{X}_1'\mathbf{y})/h}{(\mathbf{y}'\mathbf{y} - \hat{\boldsymbol{\beta}}'\mathbf{X}'\mathbf{y})/(n-k-1)} \tag{8.24}$$

$$= \frac{(R^2 - R_r^2)/h}{(1-R^2)/(n-k-1)}, \tag{8.25}$$

where R^2 for the full model is given in (7.56) as $R^2 = (\hat{\boldsymbol{\beta}}'\mathbf{X}'\mathbf{y} - n\bar{y}^2)/(\mathbf{y}'\mathbf{y} - n\bar{y}^2)$ and R_r^2 for the reduced model $\mathbf{y} = \mathbf{X}_1\boldsymbol{\beta}_1^* + \boldsymbol{\varepsilon}$ in (8.8) is similarly defined as

$$R_r^2 = \frac{\hat{\boldsymbol{\beta}}_1^{*\prime}\mathbf{X}_1'\mathbf{y} - n\bar{y}^2}{\mathbf{y}'\mathbf{y} - n\bar{y}^2}. \tag{8.26}$$

Proof. Adding and subtracting $n\bar{y}^2$ in the denominator of (8.22) gives

$$F = \frac{(\hat{\boldsymbol{\beta}}'\mathbf{X}'\mathbf{y} - n\bar{y}^2)/k}{[\mathbf{y}'\mathbf{y} - n\bar{y}^2 - (\hat{\boldsymbol{\beta}}'\mathbf{X}'\mathbf{y} - n\bar{y}^2)]/(n-k-1)}.$$

Dividing numerator and denominator by $\mathbf{y}'\mathbf{y} - n\bar{y}^2$ yields (8.23). For (8.25), see Problem 8.15. □

In (8.25), we see that the F test for $H_0 : \boldsymbol{\beta}_2 = \mathbf{0}$ is equivalent to a test for significant reduction in R^2. Note also that since $F \geq 0$ in (8.25), we have $R^2 \geq R_r^2$, which is an additional confirmation of property 3 in Section 7.7, namely, that adding an x to the model increases R^2.

Example 8.3. For the dependent variable y_2 in the chemical reaction data in Table 7.4, a full model with nine x's and a reduced model with three x's were considered in Example 8.2a. The values of R^2 for the full model and reduced model are .8485 and .3771, respectively. To test the significance of the increase in R^2

from .3771 to .8485, we use (8.25)

$$F = \frac{(R^2 - R_r^2)/h}{(1 - R^2)/(n - k - 1)} = \frac{(.8485 - .3771)/6}{(1 - .8485)/9}$$
$$= \frac{.07857}{.01683} = 4.6671,$$

which is the same as the value obtained for F in Example 8.2a. □

8.4 THE GENERAL LINEAR HYPOTHESIS TESTS FOR $H_0: \mathbf{C}\boldsymbol{\beta} = \mathbf{0}$ AND $H_0: \mathbf{C}\boldsymbol{\beta} = \mathbf{t}$

We discuss a test for $H_0 : \mathbf{C}\boldsymbol{\beta} = \mathbf{0}$ in Section 8.4.1 and a test for $H_0 : \mathbf{C}\boldsymbol{\beta} = \mathbf{t}$ in Section 8.4.2.

8.4.1 The Test for $H_0 : \mathbf{C}\boldsymbol{\beta} = \mathbf{0}$

The hypothesis $H_0 : \mathbf{C}\boldsymbol{\beta} = \mathbf{0}$, where $\mathbf{C}$ is a known $q \times (k + 1)$ coefficient matrix of rank $q \leq k + 1$, is known as the *general linear hypothesis*. The alternative hypothesis is H_1: $\mathbf{C}\boldsymbol{\beta} \neq \mathbf{0}$. The formulation $H_0 : \mathbf{C}\boldsymbol{\beta} = \mathbf{0}$ includes as special cases the hypotheses in Sections 8.1 and 8.2. The hypothesis $H_0 : \boldsymbol{\beta}_1 = \mathbf{0}$ in Section 8.1 can be expressed in the form $H_0 : \mathbf{C}\boldsymbol{\beta} = \mathbf{0}$ as follows

$$H_0 : \mathbf{C}\boldsymbol{\beta} = (\mathbf{0}, \mathbf{I}_k)\begin{pmatrix} \beta_0 \\ \boldsymbol{\beta}_1 \end{pmatrix} = \boldsymbol{\beta}_1 = \mathbf{0} \quad \text{[by (2.36)]},$$

where $\mathbf{0}$ is a $k \times 1$ vector. Similarly, the hypothesis $H_0 : \boldsymbol{\beta}_2 = \mathbf{0}$ in Section 8.2 can be expressed in the form $H_0 : \mathbf{C}\boldsymbol{\beta} = \mathbf{0}$:

$$H_0 : \mathbf{C}\boldsymbol{\beta} = (\mathbf{O}, \mathbf{I}_h)\begin{pmatrix} \boldsymbol{\beta}_1 \\ \boldsymbol{\beta}_2 \end{pmatrix} = \boldsymbol{\beta}_2 = \mathbf{0},$$

where the matrix $\mathbf{O}$ is $h \times (k - h + 1)$ and the vector $\mathbf{0}$ is $h \times 1$.

The formulation $H_0 : \mathbf{C}\boldsymbol{\beta} = \mathbf{0}$ also allows for more general hypotheses such as

$$H_0 : 2\beta_1 - \beta_2 = \beta_2 - 2\beta_3 + 3\beta_4 = \beta_1 - \beta_4 = 0,$$

which can be expressed as follows:

$$H_0 : \begin{pmatrix} 0 & 2 & -1 & 0 & 0 \\ 0 & 0 & 1 & -2 & 3 \\ 0 & 1 & 0 & 0 & -1 \end{pmatrix}\begin{pmatrix} \beta_0 \\ \beta_1 \\ \beta_2 \\ \beta_3 \\ \beta_4 \end{pmatrix} = \begin{pmatrix} 0 \\ 0 \\ 0 \end{pmatrix}.$$

As another illustration, the hypothesis $H_0 : \beta_1 = \beta_2 = \beta_3 = \beta_4$ can be expressed in terms of three differences, $H_0 : \beta_1 - \beta_2 = \beta_2 - \beta_3 = \beta_3 - \beta_4 = 0$, or, equivalently, as $H_0 : \mathbf{C}\boldsymbol{\beta} = \mathbf{0}$:

$$H_0 : \begin{pmatrix} 0 & 1 & -1 & 0 & 0 \\ 0 & 0 & 1 & -1 & 0 \\ 0 & 0 & 0 & 1 & -1 \end{pmatrix} \begin{pmatrix} \beta_0 \\ \beta_1 \\ \beta_2 \\ \beta_3 \\ \beta_4 \end{pmatrix} = \begin{pmatrix} 0 \\ 0 \\ 0 \end{pmatrix}.$$

In the following theorem, we give the sums of squares used in the test of $H_0 : \mathbf{C}\boldsymbol{\beta} = \mathbf{0}$ versus H_1: $\mathbf{C}\boldsymbol{\beta} \neq \mathbf{0}$, along with the properties of these sums of squares. We denote the sum of squares due to $\mathbf{C}\boldsymbol{\beta}$ (due to the hypothesis) as SSH.

Theorem 8.4a. If $\mathbf{y}$ is distributed $N_n(\mathbf{X}\boldsymbol{\beta}, \sigma^2\mathbf{I})$ and $\mathbf{C}$ is $q \times (k+1)$ of rank $q \leq k+1$, then

(i) $\mathbf{C}\hat{\boldsymbol{\beta}}$ is $N_q[\mathbf{C}\boldsymbol{\beta}, \sigma^2\mathbf{C}(\mathbf{X}'\mathbf{X})^{-1}\mathbf{C}']$.
(ii) $\text{SSH}/\sigma^2 = (\mathbf{C}\hat{\boldsymbol{\beta}})'[\mathbf{C}(\mathbf{X}'\mathbf{X})^{-1}\mathbf{C}']^{-1}\mathbf{C}\hat{\boldsymbol{\beta}}/\sigma^2$ is $\chi^2(q, \lambda)$, where $\lambda = (\mathbf{C}\boldsymbol{\beta})'[\mathbf{C}(\mathbf{X}'\mathbf{X})^{-1}\mathbf{C}']^{-1}\mathbf{C}\boldsymbol{\beta}/2\sigma^2$.
(iii) $\text{SSE}/\sigma^2 = \mathbf{y}'[\mathbf{I} - \mathbf{X}(\mathbf{X}'\mathbf{X})^{-1}\mathbf{X}']\mathbf{y}/\sigma^2$ is $\chi^2(n-k-1)$.
(iv) SSH and SSE are independent.

PROOF

(i) By Theorem 7.6b (i), $\hat{\boldsymbol{\beta}}$ is $N_{k+1}[\boldsymbol{\beta}, \sigma^2(\mathbf{X}'\mathbf{X})^{-1}]$. The result then follows by Theorem 4.4a (ii).
(ii) Since $\text{cov}(\mathbf{C}\hat{\boldsymbol{\beta}}) = \sigma^2\mathbf{C}(\mathbf{X}'\mathbf{X})^{-1}\mathbf{C}'$ and $\sigma^2[\mathbf{C}(\mathbf{X}'\mathbf{X})^{-1}\mathbf{C}]^{-1}\mathbf{C}(\mathbf{X}'\mathbf{X})^{-1}\mathbf{C}'/\sigma^2 = \mathbf{I}$, the result follows by Theorem 5.5.
(iii) This was established in Theorem 8.1b(ii).
(iv) Since $\hat{\boldsymbol{\beta}}$ and SSE are independent [see Theorem 7.6b(iii)], $\text{SSH} = \hat{\boldsymbol{\beta}}\mathbf{C}'[\mathbf{C}(\mathbf{X}'\mathbf{X})^{-1}\mathbf{C}']\mathbf{C}\hat{\boldsymbol{\beta}}$ and SSE are also independent (Seber 1977, pp. 17, 33–34). For a more formal proof, see Problem 8.16. □

The F test for $H_0 : \mathbf{C}\boldsymbol{\beta} = \mathbf{0}$ versus $H_1 : \mathbf{C}\boldsymbol{\beta} \neq \mathbf{0}$ is given in the following theorem.

Theorem 8.4b. Let $\mathbf{y}$ be $N_n(\mathbf{X}\boldsymbol{\beta}, \sigma^2\mathbf{I})$ and define the statistic

$$F = \frac{\text{SSH}/q}{\text{SSE}/(n-k-1)} = \frac{(\mathbf{C}\hat{\boldsymbol{\beta}})'[\mathbf{C}(\mathbf{X}'\mathbf{X})^{-1}\mathbf{C}']^{-1}\mathbf{C}\hat{\boldsymbol{\beta}}/q}{\text{SSE}/(n-k-1)}, \tag{8.27}$$

where $\mathbf{C}$ is $q \times (k+1)$ of rank $q \leq k+1$ and $\hat{\boldsymbol{\beta}} = (\mathbf{X}'\mathbf{X})^{-1}\mathbf{X}'\mathbf{y}$. The distribution of F in (8.27) is as follows:

(i) If $H_0: \mathbf{C}\boldsymbol{\beta} = \mathbf{0}$ is false, then

$$F \text{ is distributed as } F(q, n-k-1, \lambda),$$

where $\lambda = (\mathbf{C}\boldsymbol{\beta})'\left[\mathbf{C}(\mathbf{X}'\mathbf{X})^{-1}\mathbf{C}'\right]^{-1}\mathbf{C}\boldsymbol{\beta}/2\sigma^2$.

(ii) If $H_0: \mathbf{C}\boldsymbol{\beta} = \mathbf{0}$ is true, then

$$F \text{ is distributed as } F(q, n-k-1).$$

Proof

(i) This result follows from (5.30) and Theorem 8.4a.
(ii) This result follows from (5.28) and Theorem 8.4a. □

The F test for $H_0: \mathbf{C}\boldsymbol{\beta} = \mathbf{0}$ in Theorem 8.4b is usually called the *general linear hypothesis test*. The degrees of freedom q is the number of linear combinations in $\mathbf{C}\beta$. The test for $H_0: \mathbf{C}\boldsymbol{\beta} = \mathbf{0}$ is carried out as follows. Reject H_0 if $F \geq F_{\alpha,q,n-k-1}$, where F is as given in (8.27) and $F_{\alpha,q,n-k-1}$ is the upper α percentage point of the (central) F distribution. Alternatively, we can reject H_0 if $p \leq \alpha$ where p is the p value for F. [The p value is the probability that $F(q, n-k-1)$ exceeds the observed F value.] Since $\mathbf{C}(\mathbf{X}'\mathbf{X})^{-1}\mathbf{C}'$ is positive definite (see Problem 8.17), $\lambda > 0$ if H_0 is false, where $\lambda = (\mathbf{C}\boldsymbol{\beta})'[\mathbf{C}(\mathbf{X}'\mathbf{X})^{-1}\mathbf{C}']^{-1}\mathbf{C}\boldsymbol{\beta}/2\sigma^2$. Hence we reject $H_0: \mathbf{C}\boldsymbol{\beta} = \mathbf{0}$ for large values of F.

In Theorems 8.4a and 8.4b, SSH could be written as $(\mathbf{C}\hat{\boldsymbol{\beta}} - \mathbf{0})'[\mathbf{C}(\mathbf{X}'\mathbf{X})^{-1}\mathbf{C}']^{-1}$ $(\mathbf{C}\hat{\boldsymbol{\beta}} - \mathbf{0})$, which is the squared distance between $\mathbf{C}\hat{\boldsymbol{\beta}}$ and the hypothesized value of $\mathbf{C}\boldsymbol{\beta}$. The distance is standardized by the covariance matrix of $\mathbf{C}\hat{\boldsymbol{\beta}}$. Intuitively, if H_0 is true, $\mathbf{C}\hat{\boldsymbol{\beta}}$ tends to be close to $\mathbf{0}$ so that the numerator of F in (8.27) is small. On the other hand, if $\mathbf{C}\boldsymbol{\beta}$ is very different from $\mathbf{0}$, the numerator of F tends to be large.

The expected mean squares for the F test are given by

$$E\left(\frac{\text{SSH}}{q}\right) = \sigma^2 + \frac{1}{q}(\mathbf{C}\boldsymbol{\beta})'\left[\mathbf{C}(\mathbf{X}'\mathbf{X})^{-1}\mathbf{C}'\right]^{-1}\mathbf{C}\boldsymbol{\beta}, \tag{8.28}$$

$$E\left(\frac{\text{SSE}}{n-k-1}\right) = \sigma^2.$$

These expected mean squares provide additional motivation for rejecting H_0 for large values of F. If H_0 is true, both expected mean squares are equal to σ^2; if H_0 is false, $E(\text{SSH}/q) > E[\text{SSE}/(n-q-1)]$.

The F statistic in (8.27) is invariant to full-rank linear transformations on the x's or on y.

Theorem 8.4c. Let $\mathbf{z} = c\mathbf{y}$ and $\mathbf{W} = \mathbf{XK}$, where $\mathbf{K}$ is nonsingular (see Corollary 1 to Theorem 7.3e for the form of $\mathbf{K}$). The F statistic in (8.27) is unchanged by these transformations.

PROOF. See Problem 8.18. □

In the first paragraph of this section, it was noted that the hypothesis $H_0 : \boldsymbol{\beta}_2 = \mathbf{0}$ can be expressed in the form $H_0 : \mathbf{C}\boldsymbol{\beta} = \mathbf{0}$. Since we used a full–reduced-model approach to develop the test for $H_0 : \boldsymbol{\beta}_2 = \mathbf{0}$, we expect that the general linear hypothesis test is also a full–reduced-model test. This is confirmed in the following theorem.

Theorem 8.4d. The F test in Theorem 8.4b for the general linear hypothesis $H_0 : \mathbf{C}\boldsymbol{\beta} = \mathbf{0}$ is a full–reduced-model test.

PROOF. The reduced model under H_0 is

$$\mathbf{y} = \mathbf{X}\boldsymbol{\beta} + \boldsymbol{\varepsilon} \text{ subject to } \mathbf{C}\boldsymbol{\beta} = \mathbf{0}. \tag{8.29}$$

Using Lagrange multipliers (Section 2.14.3), it can be shown (see Problem 8.19) that the estimator for $\boldsymbol{\beta}$ in this reduced model is

$$\hat{\boldsymbol{\beta}}_c = \hat{\boldsymbol{\beta}} - (\mathbf{X}'\mathbf{X})^{-1}\mathbf{C}'[\mathbf{C}(\mathbf{X}'\mathbf{X})^{-1}\mathbf{C}']^{-1}\mathbf{C}\hat{\boldsymbol{\beta}}, \tag{8.30}$$

where $\hat{\boldsymbol{\beta}} = (\mathbf{X}'\mathbf{X})^{-1}\mathbf{X}'\mathbf{y}$ is estimated from the full model unrestricted by the hypothesis and the subscript c in $\hat{\boldsymbol{\beta}}_c$ indicates that $\boldsymbol{\beta}$ is estimated subject to the constraint $\mathbf{C}\boldsymbol{\beta} = \mathbf{0}$. In (8.29), the $\mathbf{X}$ matrix for the reduced model is unchanged from the full model, and the regression sum of squares for the reduced model is therefore $\hat{\boldsymbol{\beta}}_c'\mathbf{X}'\mathbf{y}$ (for a more formal justification of $\hat{\boldsymbol{\beta}}_c'\mathbf{X}'\mathbf{y}$, see Problem 8.20). Hence, the regression sum of squares due to the hypothesis is

$$\text{SSH} = \hat{\boldsymbol{\beta}}'\mathbf{X}'\mathbf{y} - \hat{\boldsymbol{\beta}}_c'\mathbf{X}'\mathbf{y}. \tag{8.31}$$

By substituting $\hat{\boldsymbol{\beta}}_c$ [as given by (8.30)] into (8.31), we obtain

$$\text{SSH} = (\mathbf{C}\hat{\boldsymbol{\beta}})'[\mathbf{C}(\mathbf{X}'\mathbf{X})^{-1}\mathbf{C}']^{-1}\mathbf{C}\hat{\boldsymbol{\beta}} \tag{8.32}$$

(see Problem 8.21), thus establishing that the F test in Theorem 8.4b for $H_0 : \mathbf{C}\boldsymbol{\beta} = \mathbf{0}$, is a full–reduced-model test. □

Example 8.4.1a. In many cases, the hypothesis can be incorporated directly into the model to obtain the reduced model. Suppose that the full model is

$$y_i = \beta_0 + \beta_1 x_{i1} + \beta_2 x_{i2} + \beta_3 x_{i3} + \varepsilon_i$$

and the hypothesis is $H_0 : \beta_1 = 2\beta_2$. Then the reduced model becomes

$$\begin{aligned} y_i &= \beta_0 + 2\beta_2 x_{i1} + \beta_2 x_{i2} + \beta_3 x_{i3} + \varepsilon_i \\ &= \beta_{c0} + \beta_{c2}(2x_{i1} + x_{i2}) + \beta_{c3} x_{i3} + \varepsilon_i, \end{aligned}$$

where β_{ci} indicates a parameter subject to the constraint $\beta_1 = 2\beta_2$. The full model and reduced model could be fit, and the difference $\text{SS}(\boldsymbol{\beta}_2|\boldsymbol{\beta}_1) = \hat{\boldsymbol{\beta}}'\mathbf{X}'\mathbf{y} - \hat{\boldsymbol{\beta}}^{*\prime}\mathbf{X}_1'\mathbf{y}$ would be the same as SSH in (8.32). □

If $\mathbf{C}\boldsymbol{\beta} \neq \mathbf{0}$, the estimator $\hat{\boldsymbol{\beta}}_c$ in (8.30) is a biased estimator of $\boldsymbol{\beta}$, but the variances of the $\hat{\beta}_{cj}$'s in $\hat{\boldsymbol{\beta}}_c$ are reduced, as shown in the following theorem.

Theorem 8.4e. The mean vector and covariance matrix of $\hat{\boldsymbol{\beta}}_c$ in (8.30) are as follows:

$$\text{(i)}\quad E(\hat{\boldsymbol{\beta}}_c) = \boldsymbol{\beta} - (\mathbf{X}'\mathbf{X})^{-1}\mathbf{C}'\left[\mathbf{C}(\mathbf{X}'\mathbf{X})^{-1}\mathbf{C}'\right]^{-1}\mathbf{C}\boldsymbol{\beta}. \tag{8.33}$$

$$\text{(ii)}\quad \text{cov}(\hat{\boldsymbol{\beta}}_c) = \sigma^2(\mathbf{X}'\mathbf{X})^{-1} - \sigma^2(\mathbf{X}'\mathbf{X})^{-1}\mathbf{C}'\left[\mathbf{C}(\mathbf{X}'\mathbf{X})^{-1}\mathbf{C}'\right]^{-1}\mathbf{C}(\mathbf{X}'\mathbf{X})^{-1}. \tag{8.34}$$

PROOF. See Problem 8.22. □

Since the second matrix on the right side of (8.34) is positive semidefinite, the diagonal elements of $\text{cov}(\hat{\boldsymbol{\beta}}_c)$ are less than those of $\text{cov}(\hat{\boldsymbol{\beta}}) = \sigma^2(\mathbf{X}'\mathbf{X})^{-1}$; that is, $\text{var}(\hat{\beta}_{cj}) \leq \text{var}(\hat{\beta}_j)$ for $j = 0, 1, 2, \cdots, k$, where $\hat{\beta}_{cj}$ is the jth diagonal element of $\text{cov}(\hat{\boldsymbol{\beta}}_c)$ in (8.34). This is analogous to the inequality $\text{var}(\hat{\beta}_j^*) < \text{var}(\hat{\beta}_j)$ in Theorem 7.9c, where $\hat{\beta}_j^*$ is from the reduced model.

Example 8.4.1b. Consider the dependent variable y_1 in the chemical reaction data in Table 7.4. For the model $y_1 = \beta_0 + \beta_1 x_1 + \beta_2 x_2 + \beta_3 x_3 + \varepsilon$, we test $H_0 : 2\beta_1 = 2\beta_2 = \beta_3$ using (8.27) in Theorem 8.4b. To express H_0 in the form $\mathbf{C}\boldsymbol{\beta} = \mathbf{0}$, the matrix $\mathbf{C}$ becomes

$$\mathbf{C} = \begin{pmatrix} 0 & 1 & -1 & 0 \\ 0 & 0 & 2 & -1 \end{pmatrix},$$

and we obtain

$$\mathbf{C}\hat{\boldsymbol{\beta}} = \begin{pmatrix} -.1214 \\ -.6118 \end{pmatrix},$$

$$\mathbf{C}(\mathbf{X}'\mathbf{X})^{-1}\mathbf{C}' = \begin{pmatrix} .003366 & -.006943 \\ -.006943 & .044974 \end{pmatrix},$$

$$F = \frac{\begin{pmatrix} -.1214 \\ -.6118 \end{pmatrix}' \begin{pmatrix} .003366 & -.006943 \\ -.006943 & .044974 \end{pmatrix}^{-1} \begin{pmatrix} -.1214 \\ -.6118 \end{pmatrix} \Big/ 2}{5.3449}$$

$$= \frac{28.62301/2}{5.3449} = 2.6776,$$

which has $p = .101$. □

8.4.2 The Test for $H_0 : \mathbf{C}\boldsymbol{\beta} = \mathbf{t}$

The test for $H_0 : \mathbf{C}\boldsymbol{\beta} = \mathbf{t}$ is a straightforward extension of the test for $H_0 : \mathbf{C}\boldsymbol{\beta} = \mathbf{0}$. With the additional flexibility provided by $\boldsymbol{t}$, we can test hypotheses such as $H_0 : \beta_2 = \beta_1 + 5$. We assume that the system of equations $\mathbf{C}\boldsymbol{\beta} = \mathbf{t}$ is consistent, that is, that $\text{rank}(\mathbf{C}) = \text{rank}(\mathbf{C}, \mathbf{t})$ (see Theorem 2.7). The requisite sums of squares and their properties are given in the following theorem, which is analogous to Theorem 8.4a.

Theorem 8.4f. If $\mathbf{y}$ is $N_n(\mathbf{X}\boldsymbol{\beta}, \sigma^2\mathbf{I})$ and $\mathbf{C}$ is $q \times (k+1)$ of rank $q \leq k+1$, then

(i) $\mathbf{C}\hat{\boldsymbol{\beta}} - \mathbf{t}$ is $N_q[\mathbf{C}\boldsymbol{\beta} - \mathbf{t}, \sigma^2\mathbf{C}(\mathbf{X}'\mathbf{X})^{-1}\mathbf{C}']$.
(ii) $\text{SSH}/\sigma^2 = (\mathbf{C}\hat{\boldsymbol{\beta}} - \mathbf{t})'\left[\mathbf{C}(\mathbf{X}'\mathbf{X})^{-1}\mathbf{C}'\right]^{-1}(\mathbf{C}\hat{\boldsymbol{\beta}} - \mathbf{t})/\sigma^2$ is $\chi^2(q, \lambda)$ where $\lambda = (\mathbf{C}\boldsymbol{\beta} - \mathbf{t})'\,[\mathbf{C}(\mathbf{X}'\mathbf{X})^{-1}\mathbf{C}']^{-1}(\mathbf{C}\boldsymbol{\beta} - \mathbf{t})/2\sigma^2$.
(iii) $\text{SSE}/\sigma^2 = \mathbf{y}'[\mathbf{I} - \mathbf{X}(\mathbf{X}'\mathbf{X})^{-1}\mathbf{X}']\mathbf{y}/\sigma^2$ is $\chi^2(n-k-1)$.
(iv) SSH and SSE are independent.

PROOF

(i) By Theorem 7.6b (i), $\hat{\boldsymbol{\beta}}$ is $N_{k+1}[\boldsymbol{\beta}, \sigma^2(\mathbf{X}'\mathbf{X})^{-1}]$. The result follows by Corollary 1 to Theorem 4.4a.
(ii) By part (i), $\text{cov}(\mathbf{C}\hat{\boldsymbol{\beta}} - \mathbf{t}) = \sigma^2\mathbf{C}(\mathbf{X}'\mathbf{X})^{-1}\mathbf{C}'$. The result follows as in the proof of Theorem 8.4a (ii).
(iii) See Theorem 8.1b (ii).
(iv) Since $\hat{\boldsymbol{\beta}}$ and SSE are independent [see Theorem 7.6b (iii)], SSH and SSE are independent [see Seber (1977, pp. 17, 33–34)]. For a more formal proof, see Problem 8.23. □

An F test for $H_0 : \mathbf{C}\boldsymbol{\beta} = \mathbf{t}$ versus $H_1 : \mathbf{C}\boldsymbol{\beta} \neq \mathbf{t}$ is given in the following theorem, which is analogous to Theorem 8.4b.

Theorem 8.4g. Let $\mathbf{y}$ be $N_n(\mathbf{X}\boldsymbol{\beta}, \sigma^2\mathbf{I})$ and define an F statistic as follows:

$$F = \frac{\text{SSH}/q}{\text{SSE}/(n-k-1)} = \frac{(\mathbf{C}\hat{\boldsymbol{\beta}} - \mathbf{t})'\left[\mathbf{C}(\mathbf{X}'\mathbf{X})^{-1}\mathbf{C}'\right]^{-1}(\mathbf{C}\hat{\boldsymbol{\beta}} - \mathbf{t})/q}{\text{SSE}/(n-k-1)}, \tag{8.35}$$

where $\hat{\boldsymbol{\beta}} = (\mathbf{X}'\mathbf{X})^{-1}\mathbf{X}'\mathbf{y}$. The distribution of F in (8.35) is as follows:

(i) If $H_0 : \mathbf{C}\boldsymbol{\beta} = \mathbf{t}$ is false, then

$$F \text{ is distributed as } F(q, n-k-1, \lambda),$$

where $\lambda = (\mathbf{C}\boldsymbol{\beta} - \mathbf{t})'[\mathbf{C}(\mathbf{X}'\mathbf{X})^{-1}\mathbf{C}']^{-1}(\mathbf{C}\boldsymbol{\beta} - \mathbf{t})/2\sigma^2$.

(ii) If $H_0: \mathbf{C}\boldsymbol{\beta} = \mathbf{t}$ is true, then $\lambda = 0$ and

$$F \text{ is distributed as } F(q, n - k - 1).$$

PROOF

(i) This result follows from (5.28) and Theorem 8.4f.
(ii) This result follows from (5.30) and Theorem 8.4f. □

The test for $H_0: \mathbf{C}\boldsymbol{\beta} = \mathbf{t}$ is carried out as follows. Reject H_0 if $F \geq F_{\alpha,q,n-k-1}$, where $F_{\alpha,q,\,n-k-1}$ is the upper α percentage point of the central F distribution. Alternatively, we can reject H_0 if $p \leq \alpha$, where p is the p value for F.

The expected mean squares for the F test are given by

$$E\left(\frac{\text{SSH}}{q}\right) = \sigma^2 + \frac{1}{q}(\mathbf{C}\boldsymbol{\beta} - \mathbf{t})'\left[\mathbf{C}(\mathbf{X}'\mathbf{X})^{-1}\mathbf{C}'\right]^{-1}(\mathbf{C}\boldsymbol{\beta} - \mathbf{t}),$$
$$E\left(\frac{\text{SSE}}{n-k-1}\right) = \sigma^2. \tag{8.36}$$

By extension of Theorem 8.4d, the F test for $H_0: \mathbf{C}\boldsymbol{\beta} = \mathbf{t}$ in Theorem 8.4g is a full–reduced-model test (see Problem 8.24 for a partial result).

8.5 TESTS ON β_j AND $\mathbf{a}'\boldsymbol{\beta}$

We consider tests for a single β_j or a single linear combination $\mathbf{a}'\boldsymbol{\beta}$ in Section 8.5.1 and tests for several β_j's or several $\mathbf{a}_i'\boldsymbol{\beta}$'s in Section 8.5.2.

8.5.1 Testing One β_j or One $\mathbf{a}'\boldsymbol{\beta}$

Tests for an individual β_j can be obtained using either the full–reduced-model approach in Section 8.2 or the general linear hypothesis approach in Section 8.4 The test statistic for $H_0: \beta_k = 0$ using a full–reduced–model is given in (8.20) as

$$F = \frac{\hat{\boldsymbol{\beta}}'\mathbf{X}'\mathbf{y} - \hat{\boldsymbol{\beta}}_1^{*\prime}\mathbf{X}_1'\mathbf{y}}{\text{SSE}/(n-k-1)}, \tag{8.37}$$

which is distributed as $F(1, n-k-1)$ if H_0 is true. In this case, β_k is the last β, so that $\boldsymbol{\beta}$ is partitioned as $\boldsymbol{\beta} = \begin{pmatrix} \boldsymbol{\beta}_1 \\ \beta_k \end{pmatrix}$ and $\mathbf{X}$ is partitioned as $\mathbf{X} = (\mathbf{X}_1, \mathbf{x}_k)$, where $\mathbf{x}_k$ is the last column of $\mathbf{X}$. Then $\mathbf{X}_1$ in the reduced model $\mathbf{y} = \mathbf{X}_1\boldsymbol{\beta}_1^* + \boldsymbol{\varepsilon}$ contains all the columns of $\mathbf{X}$ except the last.

To test $H_0 : \beta_j = 0$ by means of the general linear hypothesis test of $H_0 : \mathbf{C}\boldsymbol{\beta} = \mathbf{0}$ (Section 8.4.1), we first consider a test of $H_0 : \mathbf{a}'\boldsymbol{\beta} = 0$ for a single linear combination, for example, $\mathbf{a}'\boldsymbol{\beta} = (0, 2, -2, 3, 1)\boldsymbol{\beta}$. Using $\mathbf{a}'$ in place of the matrix $\mathbf{C}$ in $\mathbf{C}\boldsymbol{\beta} = \mathbf{0}$, we have $q = 1$, and (8.27) becomes

$$F = \frac{(\mathbf{a}'\hat{\boldsymbol{\beta}})'\left[\mathbf{a}'(\mathbf{X}'\mathbf{X})^{-1}\mathbf{a}\right]^{-1}\mathbf{a}'\hat{\boldsymbol{\beta}}}{\text{SSE}/(n-k-1)} = \frac{(\mathbf{a}'\hat{\boldsymbol{\beta}})^2}{s^2\mathbf{a}'(\mathbf{X}'\mathbf{X})^{-1}\mathbf{a}}, \tag{8.38}$$

where $s^2 = \text{SSE}/(n-k-1)$. The F statistic in (8.38) is distributed as $F(1, n-k-1)$ if $H_0 : \mathbf{a}'\boldsymbol{\beta} = 0$ is true.

To test $H_0 : \beta_j = 0$ using (8.38), we define $\mathbf{a}' = (0, \ldots, 0, 1, 0, \ldots, 0)$, where the 1 is in the jth position. This gives

$$F = \frac{\hat{\beta}_j^2}{s^2 g_{jj}}, \tag{8.39}$$

where g_{jj} is the jth diagonal element of $(\mathbf{X}'\mathbf{X})^{-1}$. If $H_0 : \beta_j = 0$ is true, F in (8.39) is distributed as $F(1, n-k-1)$. We reject $H_0 : \beta_j = 0$ if $F \geq F_{\alpha,1,n-k-1}$ or, equivalently, if $p \leq \alpha$, where p is the p value for F.

By Theorem 8.4d (see also Problem 8.25), the F statistics in (8.37) and (8.39) are the same (for $j = k$). This confirms that (8.39) tests $H_0 : \beta_j = 0$ adjusted for the other β's.

Since the F statistic in (8.39) has 1 and $n-k-1$ degrees of freedom, we can equivalently use the t statistic

$$t_j = \frac{\hat{\beta}_j}{s\sqrt{g_{jj}}} \tag{8.40}$$

to test the effect of β_j above and beyond the other β's (see Problem 5.16). We reject $H_0 : \beta_j = 0$ if $|t_j| \geq t_{\alpha/2,n-k-1}$ or, equivalently, if $p \leq \alpha$, where p is the p value. For a two-tailed t test such as this one, the p value is twice the probability that $t(n-k-1)$ exceeds the absolute value of the observed t.

For $j = 1$, (8.40) becomes $t = \hat{\beta}_1/s\sqrt{g_{11}}$, which is not the same as $t = \hat{\beta}_1/\left[s/\sqrt{\sum_i (x_i - \bar{x})^2}\right]$ in (6.14). Unless the x's are orthogonal, $g_{11}^{-1} \neq \sum_i (x_{1i} - \bar{x}_1)^2$.

8.5.2 Testing Several β_j's or $\mathbf{a}_i'\boldsymbol{\beta}$'s

We sometimes want to carry out several separate tests rather than a single joint test of the hypotheses. For example, the test in (8.40) might be carried out separately for each $\boldsymbol{\beta}_i$, $i = 1, \ldots, k$ rather than the joint test of $H_0 : \boldsymbol{\beta}_1 = 0$ in (8.5). Similarly, we might want to carry out separate tests for several (say, d) $\mathbf{a}_i\boldsymbol{\beta}$'s using (8.38) rather than the joint test of $H_0 : \mathbf{C}\boldsymbol{\beta} = 0$ using (8.27), where

$$\mathbf{C} = \begin{pmatrix} \mathbf{a}_1 \\ \mathbf{a}_2 \\ \vdots \\ \mathbf{a}_d \end{pmatrix}.$$

In such situations there are two different α levels, the overall or *familywise* α level (α_f) and the α level for each test or *comparisonwise* α level (α_c). In some cases researchers desire to control α_c when doing several tests (Saville 1990), and so no changes are needed in the testing procedure. In other cases, the desire is to control α_f. In yet other cases, especially those involving thousands of separate tests (e.g., microarray data), it makes sense to control other quantities such as the false discovery rate (Benjamini and Hochberg 1995, Benjamini and Yekutieli 2001). This will not be discussed further here. We consider two ways to control α_f when several tests are made.

The first of these methods is the Bonferroni approach (Bonferroni 1936), which reduces α_c for each test, so that α_f is less than the desired level of α^*. As an example, suppose that we carry out the k tests of H_{0j}: $\beta_j = 0, j = 1, 2, \ldots, k$. Let E_j be the event that the jth test rejects H_{0j} when it is true, where $P(E_j) = \alpha_c$. The overall α_f can be defined as

$$\begin{aligned} \alpha_f &= P(\text{reject at least one } H_{0j} \text{ when all } H_{0j} \text{ are true}) \\ &= P(E_1 \text{ or } E_2 \ldots \text{ or } E_k). \end{aligned}$$

Expressing this more formally and applying the Bonferroni inequality, we obtain

$$\begin{aligned} \alpha_f &= P(E_1 \cup E_2 \cup \cdots \cup E_k) \\ &\leq \sum_{j=1}^{k} P(E_j) = \sum_{j=1}^{k} \alpha_c = k\alpha_c. \end{aligned} \tag{8.41}$$

We can thus ensure that α_f is less than or equal to the desired α^* by simply setting $\alpha_c = \alpha^*/k$. Since α_f in (8.41) is at most α^*, the Bonferroni procedure is a conservative approach.

To test $H_{0j} : \beta_j = 0, j = 1, 2, \ldots, k$, with $\alpha_f \leq \alpha^*$, we use (8.40)

$$t_j = \frac{\hat{\beta}_j}{s\sqrt{g_{jj}}}, \tag{8.42}$$

and reject H_{0j} if $|t_j| \geq t_{\alpha^*/2k, n-k-1}$. Bonferroni critical values $t_{\alpha^*/2k,\nu}$ are available in Bailey (1977). See also Rencher (2002, pp. 562–565). The critical values $t_{\alpha^*/2k,\nu}$ can also be found using many software packages. Alternatively, we can carry out the test by the use of p values and reject H_{0j} if $p \leq \alpha^*/k$.

More generally, to test $H_{0i}: \mathbf{a}_i'\boldsymbol{\beta} = 0$ for $i = 1, 2, \ldots, d$ with $\alpha_f \leq \alpha^*$, we use (8.38)

$$F_i = \frac{(\mathbf{a}_i'\hat{\boldsymbol{\beta}})'\left[\mathbf{a}_i'(\mathbf{X}'\mathbf{X})^{-1}\mathbf{a}_i\right]^{-1}\mathbf{a}_i'\hat{\boldsymbol{\beta}}}{s^2} \tag{8.43}$$

and reject H_{0i} if $F_i \geq F_{\alpha^*/d, 1, n-k-1}$. The critical values $F_{\alpha^*/d}$ are available in many software packages. To use p values, reject H_{0i} if $p \leq \alpha^*/d$.

The above Bonferroni procedures do not require independence of the $\hat{\beta}_j$'s; they are valid for any covariance structure on the $\hat{\beta}_j$'s. However, the logic of the Bonferroni procedure for testing H_{0i}: $\mathbf{a}_i'\boldsymbol{\beta} = 0$ for $i = 1, 2, \ldots, d$ requires that the coefficient vectors $\mathbf{a}_1, \mathbf{a}_2, \ldots, \mathbf{a}_d$ be specified before seeing the data. If we wish to choose values of $\mathbf{a}_i$ after looking at the data, we must use the Scheffé procedure described below. Modifications of the Bonferroni approach have been proposed that are less conservative but still control α_f. For examples of these modified procedures, see Holm (1979), Shaffer (1986), Simes (1986), Holland and Copenhaver (1987), Hochberg (1988), Hommel (1988), Rom (1990), and Rencher (1995, Section 3.4.4). Comparisons of these procedures have been made by Holland (1991) and Broadbent (1993).

A second approach to controlling α_f due to Scheffé (1953; 1959, p. 68) yields simultaneous tests of $H_0: \mathbf{a}'\boldsymbol{\beta} = 0$ for all possible values of $\mathbf{a}$ including those chosen after looking at the data. We could also test $H_0: \mathbf{a}'\boldsymbol{\beta} = t$ for arbitrary t. For any given $\mathbf{a}$, the hypothesis $H_0: \mathbf{a}'\boldsymbol{\beta} = 0$ is tested as usual by (8.38)

$$\begin{aligned} F &= \frac{(\mathbf{a}'\hat{\boldsymbol{\beta}})'\left[\mathbf{a}'(\mathbf{X}'\mathbf{X})^{-1}\mathbf{a}\right]^{-1}\mathbf{a}'\hat{\boldsymbol{\beta}}}{s^2} \\ &= \frac{(\mathbf{a}'\hat{\boldsymbol{\beta}})^2}{s^2\mathbf{a}'(\mathbf{X}'\mathbf{X})^{-1}\mathbf{a}}, \end{aligned} \tag{8.44}$$

but the test proceeds by finding a critical value large enough to hold for all possible $\mathbf{a}$. Accordingly, we now find the distribution of $\max_{\mathbf{a}} F$.

Theorem 8.5

(i) The maximum value of F in (8.44) is given by

$$\max_{\mathbf{a}} \frac{(\mathbf{a}'\hat{\boldsymbol{\beta}})^2}{s^2\mathbf{a}'(\mathbf{X}'\mathbf{X})^{-1}\mathbf{a}} = \frac{\hat{\boldsymbol{\beta}}'\mathbf{X}'\mathbf{X}\hat{\boldsymbol{\beta}}}{s^2}. \tag{8.45}$$

(ii) If $\mathbf{y}$ is $N_n(\mathbf{X}\boldsymbol{\beta}, \sigma^2\mathbf{I})$, then $\hat{\boldsymbol{\beta}}'\mathbf{X}'\mathbf{X}\hat{\boldsymbol{\beta}}/(k+1)s^2$ is distributed as $F(k+1, n-k-1)$. Thus

$$\max_{\mathbf{a}} \frac{(\mathbf{a}'\hat{\boldsymbol{\beta}})^2}{s^2\mathbf{a}'(\mathbf{X}'\mathbf{X})^{-1}\mathbf{a}(k+1)}$$

is distributed as $F(k+1, n-k-1)$.

PROOF

(i) Using the quotient rule, chain rule, and Section 2.14.1, we differentiate $(\mathbf{a}'\hat{\boldsymbol{\beta}})^2/\mathbf{a}'(\mathbf{X}'\mathbf{X})^{-1}\mathbf{a}$ with respect to $\mathbf{a}$ and set the result equal to $\mathbf{0}$:

$$\frac{\partial}{\partial \mathbf{a}} \frac{(\mathbf{a}'\hat{\boldsymbol{\beta}})^2}{\mathbf{a}'(\mathbf{X}'\mathbf{X})^{-1}\mathbf{a}} = \frac{[\mathbf{a}'(\mathbf{X}'\mathbf{X})^{-1}\mathbf{a}]2(\mathbf{a}'\hat{\boldsymbol{\beta}})\hat{\boldsymbol{\beta}} - (\mathbf{a}'\hat{\boldsymbol{\beta}})^2 2(\mathbf{X}'\mathbf{X})^{-1}\mathbf{a}}{[\mathbf{a}'(\mathbf{X}'\mathbf{X})^{-1}\mathbf{a}]^2} = \mathbf{0}.$$

Multiplying by $[\mathbf{a}'(\mathbf{X}'\mathbf{X})^{-1}\mathbf{a}]^2/2\mathbf{a}'\hat{\boldsymbol{\beta}}$ and treating 1×1 matrices as scalars, we obtain

$$[\mathbf{a}'(\mathbf{X}'\mathbf{X})^{-1}\mathbf{a}]\,\hat{\boldsymbol{\beta}} - \mathbf{a}'\hat{\boldsymbol{\beta}}(\mathbf{X}'\mathbf{X})^{-1}\mathbf{a} = \mathbf{0},$$

$$\mathbf{a} = \frac{\mathbf{a}'(\mathbf{X}'\mathbf{X})^{-1}\mathbf{a}}{\mathbf{a}'\hat{\boldsymbol{\beta}}}\mathbf{X}'\mathbf{X}\hat{\boldsymbol{\beta}} = c\mathbf{X}'\mathbf{X}\hat{\boldsymbol{\beta}},$$

where $c = \mathbf{a}'(\mathbf{X}'\mathbf{X})^{-1}\mathbf{a}/\mathbf{a}'\hat{\boldsymbol{\beta}}$. Substituting $\mathbf{a} = c\mathbf{X}'\mathbf{X}\hat{\boldsymbol{\beta}}$ into (8.44) gives

$$\max_{\mathbf{a}} \frac{(\mathbf{a}'\hat{\boldsymbol{\beta}})^2}{s^2\mathbf{a}'(\mathbf{X}'\mathbf{X})^{-1}\mathbf{a}} = \frac{(c\hat{\boldsymbol{\beta}}'\mathbf{X}'\mathbf{X}\hat{\boldsymbol{\beta}})^2}{s^2 c\hat{\boldsymbol{\beta}}'\mathbf{X}'\mathbf{X}(\mathbf{X}'\mathbf{X})^{-1}c\mathbf{X}'\mathbf{X}\hat{\boldsymbol{\beta}}} = \frac{c^2(\hat{\boldsymbol{\beta}}'\mathbf{X}'\mathbf{X}\hat{\boldsymbol{\beta}})^2}{s^2c^2\hat{\boldsymbol{\beta}}'\mathbf{X}'\mathbf{X}\hat{\boldsymbol{\beta}}} = \frac{\hat{\boldsymbol{\beta}}'\mathbf{X}'\mathbf{X}\hat{\boldsymbol{\beta}}}{s^2}.$$

(ii) Using $\mathbf{C} = \mathbf{I}_{k+1}$ in (8.27), we have, by Theorem 8.4b (ii), that

$$F = \frac{\hat{\boldsymbol{\beta}}'\mathbf{X}'\mathbf{X}\hat{\boldsymbol{\beta}}}{(k+1)s^2} \text{ is distributed as } F(k+1, n-k-1).$$

□

By Theorem 8.5(ii), we have

$$P\left[\max_{\mathbf{a}} \frac{(\mathbf{a}'\hat{\boldsymbol{\beta}})^2}{s^2\mathbf{a}'(\mathbf{X}'\mathbf{X})^{-1}\mathbf{a}(k+1)} \geq F_{\alpha^*,k+1,n-k-1}\right] = \alpha^*,$$

$$P\left[\max_{\mathbf{a}} \frac{(\mathbf{a}'\hat{\boldsymbol{\beta}})^2}{s^2\mathbf{a}'(\mathbf{X}'\mathbf{X})^{-1}\mathbf{a}} \geq (k+1)F_{\alpha^*,k+1,n-k-1}\right] = \alpha^*.$$

Thus, to test $H_0 : \mathbf{a}'\boldsymbol{\beta} = 0$ for any and all $\mathbf{a}$ (including values of $\mathbf{a}$ chosen after seeing the data) with $\alpha_f \leq \alpha^*$, we calculate F in (8.44) and reject H_0 if $F \geq (k+1)F_{\alpha^*,k+1,n-k-1}$.

To test for individual β_j's using using Scheffé's procedure, we set $\mathbf{a}' = (0, \ldots, 0, 1, 0, \ldots, 0)$ with a 1 in the jth position. Then F in (8.44) reduces to $F = \hat{\beta}_j^2/s^2 g_{jj}$ in (8.39), and the square root is $t_j = \hat{\beta}_j/s\sqrt{g_{jj}}$ in (8.42). By Theorem 8.5, we reject $H_0 : \mathbf{a}'\boldsymbol{\beta} = \beta_j = 0$ if $|t_j| \geq \sqrt{(k+1)F_{\alpha^*,k+1,n-k-1}}$.

For practical purposes $[k \leq (n-3)]$, we have

$$t_{\alpha^*/2k,\,n-k-1} < \sqrt{(k+1)F_{\alpha^*,k+1,n-k-1}},$$

and thus the Bonferroni tests for individual β_j's in (8.42) are usually more powerful than the Scheffé tests. On the other hand, for a large number of linear combinations $\mathbf{a}'\boldsymbol{\beta}$, the Scheffé test is better since $(k+1)F_{\alpha^*, k+1, n-k-1}$ is constant, while the critical value $F_{\alpha^*/d,1,n-k-1}$ for Bonferroni tests in (8.43) increases with the number of tests d and eventually exceeds the critical value for Scheffé tests.

It has been assumed that the tests in this section for $H_0 : \beta_j = 0$ are carried out without regard to whether the overall hypothesis $H_0 : \boldsymbol{\beta}_1 = \mathbf{0}$ is rejected. However, if the test statistics $t_j = \hat{\beta}_j / s\sqrt{g_{jj}}, j = 1, 2, \ldots, k$, in (8.42) are calculated only if $H_0 : \boldsymbol{\beta}_1 = \mathbf{0}$ is rejected using F in (8.5), then clearly α_f is reduced and the conservative critical values $t_{\alpha^*/2k, n-k-1}$ and $\sqrt{(k+1)F_{\alpha^*,k+1,n-k-1}}$ become even more conservative. Using this protected testing principle (Hocking 1996, p. 106), we can even use the critical value $t_{\alpha^*/2,n-k-1}$ for all k tests and α_f will still be close to α^*. [For illustrations of this familywise error rate structure, see Hummel and sligo (1971) and Rencher and Scott (1990).] A similar statement can be made for testing the overall hypothesis $H_0 : \mathbf{C}\boldsymbol{\beta} = \mathbf{0}$ followed by t tests or F tests of $H_0 : \mathbf{c}_i'\boldsymbol{\beta} = 0$ using the rows of $\mathbf{C}$.

Example 8.5.2. We test $H_{01} : \beta_1 = 0$ and $H_{02} : \beta_2 = 0$ for the data in Table 7.1. Using (8.42) and the results in Examples 7.3.1(a), 7.33 and 8.1, we have

$$t_1 = \frac{\hat{\beta}_1}{s\sqrt{g_{11}}} = \frac{3.0118}{\sqrt{2.8288}\sqrt{.16207}} = \frac{3.0118}{.67709} = 4.448,$$

$$t_2 = \frac{\hat{\beta}_2}{s\sqrt{g_{22}}} = \frac{-1.2855}{\sqrt{2.8288}\sqrt{.08360}} = \frac{-1.2855}{0.48629} = -2.643.$$

Using α=.05 for each test, we reject both H_{01} and H_{02} because $t_{.025,9} = 2.262$. The (two-sided) p values are .00160 and .0268, respectively. If we use $\alpha = .05/2 = .025$ for a Bonferroni test, we would not reject H_{02} since $p = .0268 > .025$. However, using the protected testing principle, we would reject H_{02} because the overall regression hypothesis $H_0 : \boldsymbol{\beta}_1 = \mathbf{0}$ was rejected in Example 8.1. □

8.6 CONFIDENCE INTERVALS AND PREDICTION INTERVALS

In this section we consider a confidence region for $\boldsymbol{\beta}$, confidence intervals for β_j, $\mathbf{a}'\boldsymbol{\beta}$, $E(y)$, and σ^2, and prediction intervals for future observations. We assume throughout Section 8.6 that $\mathbf{y}$ is $N_n(\mathbf{X}\boldsymbol{\beta}, \sigma^2\mathbf{I})$.

8.6.1 Confidence Region for $\boldsymbol{\beta}$

If $\mathbf{C}$ is equal to $\mathbf{I}$ and $\mathbf{t}$ is equal to $\boldsymbol{\beta}$ in (8.35), q becomes $k+1$, we obtain a central F distribution, and we can make the probability statement

$$P[(\hat{\boldsymbol{\beta}} - \boldsymbol{\beta})'\mathbf{X}'\mathbf{X}(\hat{\boldsymbol{\beta}} - \boldsymbol{\beta})/(k+1)s^2 \leq F_{\alpha,k+1,n-k-1}] = 1 - \alpha,$$

where $s^2 = \text{SSE}/(n-k-1)$. From this statement, a $100(1-\alpha)\%$ joint confidence region for $\beta_0, \beta_1, \ldots, \beta_k$ in $\boldsymbol{\beta}$ is defined to consist of all vectors $\boldsymbol{\beta}$ that satisfy

$$(\hat{\boldsymbol{\beta}} - \boldsymbol{\beta})'\mathbf{X}'\mathbf{X}(\hat{\boldsymbol{\beta}} - \boldsymbol{\beta}) \leq (k+1)s^2 F_{\alpha,k+1,n-k-1}. \tag{8.46}$$

For $k = 1$, this region can be plotted as an ellipse in two dimensions. For $k > 1$, the ellipsoidal region in (8.46) is unwieldy to interpret and report, and we therefore consider intervals for the individual β_j's.

8.6.2 Confidence Interval for β_j

If $\beta_j \neq 0$, we can subtract β_j in (8.40) so that $t_j = (\hat{\beta}_j - \beta_j)/s\sqrt{g_{jj}}$ has the central t distribution, where g_{jj} is the jth diagonal element of $(\mathbf{X}'\mathbf{X})^{-1}$. Then

$$P\left[-t_{\alpha/2,n-k-1} \leq \frac{\hat{\beta}_j - \beta_j}{s\sqrt{g_{jj}}} \leq t_{\alpha/2,n-k-1}\right] = 1 - \alpha.$$

Solving the inequality for β_j gives

$$P(\hat{\beta}_j - t_{\alpha/2,n-k-1}s\sqrt{g_{jj}} \leq \beta_j \leq \hat{\beta}_j + t_{\alpha/2,n-k-1}s\sqrt{g_{jj}}) = 1 - \alpha.$$

Before taking the sample, the probability that the random interval will contain β_j is $1 - \alpha$. *After* taking the sample, the $100(1-\alpha)\%$ confidence interval for β_j

$$\hat{\beta}_j \pm t_{\alpha/2,\, n-k-1}s\sqrt{g_{jj}} \tag{8.47}$$

is no longer random, and thus we say that we are $100(1-\alpha)\%$ *confident* that the interval contains β_j.

Note that the confidence coefficient $1 - \alpha$ holds only for a single confidence interval for one of the β_j's. For confidence intervals for all $k+1$ of the β's that hold simultaneously with overall confidence coefficient $1 - \alpha$, see Section 8.6.7.

Example 8.6.2. We compute a 95% confidence interval for each β_j using y_2 in the chemical reaction data in Table 7.4 (see Example 8.2a). The matrix $(\mathbf{X}'\mathbf{X})^{-1}$ (see the answer to Problem 7.52) and the estimate $\hat{\boldsymbol{\beta}}$ have the following values:

$$(\mathbf{X}'\mathbf{X})^{-1} = \begin{pmatrix} 65.37550 & -0.33885 & -0.31252 & -0.02041 \\ -0.33885 & 0.00184 & 0.00127 & -0.00043 \\ -0.31252 & 0.00127 & 0.00408 & -0.00176 \\ -0.02041 & -0.00043 & -0.00176 & 0.02161 \end{pmatrix},$$

$$\hat{\boldsymbol{\beta}} = \begin{pmatrix} -26.0353 \\ 0.4046 \\ 0.2930 \\ 1.0338 \end{pmatrix}.$$

For β_1, we obtain by (8.47),

$$\begin{aligned}
&\hat{\beta}_1 \pm t_{.025,15} s\sqrt{g_{11}} \\
&.4046 \pm (2.1314)(4.0781)\sqrt{.00184} \\
&.4046 \pm .3723, \\
&(.0322, .7769).
\end{aligned}$$

For the other β_j's, we have

$$\begin{aligned}
\beta_0:&\quad -26.0353 \pm 70.2812 \\
&\quad (-96.3165, 44.2459), \\
\beta_2:&\quad .2930 \pm .5551 \\
&\quad (-.2621, .8481), \\
\beta_3:&\quad 1.0338 \pm 1.27777 \\
&\quad (-.2439, 2.3115).
\end{aligned}$$

The confidence coefficient .95 holds for only one of the four confidence intervals. For more than one interval, see Example 8.6.7. □

8.6.3 Confidence Interval for $\mathbf{a}'\boldsymbol{\beta}$

If $\mathbf{a}'\boldsymbol{\beta} \neq 0$, we can subtract $\mathbf{a}'\boldsymbol{\beta}$ from $\mathbf{a}'\hat{\boldsymbol{\beta}}$ in (8.44) to obtain

$$F = \frac{(\mathbf{a}'\hat{\boldsymbol{\beta}} - \mathbf{a}'\boldsymbol{\beta})^2}{s^2\mathbf{a}'(\mathbf{X}'\mathbf{X})^{-1}\mathbf{a}},$$

which is distributed as $F(1, n-k-1)$. Then by Problem 5.16,

$$t = \frac{\mathbf{a}'\hat{\boldsymbol{\beta}} - \mathbf{a}'\boldsymbol{\beta}}{s\sqrt{\mathbf{a}'(\mathbf{X}'\mathbf{X})^{-1}\mathbf{a}}} \tag{8.48}$$

is distributed as $t(n-k-1)$, and a $100(1-\alpha)\%$ confidence interval for a single value of $\mathbf{a}'\boldsymbol{\beta}$ is given by

$$\mathbf{a}'\hat{\boldsymbol{\beta}} \pm t_{\alpha/2,n-k-1} s\sqrt{\mathbf{a}'(\mathbf{X}'\mathbf{X})^{-1}\mathbf{a}}. \tag{8.49}$$

8.6.4 Confidence Interval for $E(y)$

Let $\mathbf{x}_0 = (1, x_{01}, x_{02}, \ldots, x_{0k})'$ denote a particular choice of $\mathbf{x} = (1, x_1, x_2, \ldots, x_k)'$. Note that $\mathbf{x}_0$ need not be one of the $\mathbf{x}$'s in the sample; that is, $\mathbf{x}_0'$ need not be a row of $\mathbf{X}$. If $\mathbf{x}_0$ is very far outside the area covered by the sample however, the prediction may be poor. Let y_0 be an observation corresponding to $\mathbf{x}_0$. Then

$$y_0 = \mathbf{x}_0'\boldsymbol{\beta} + \varepsilon,$$

and [assuming that the model is correct so that $E(\varepsilon) = 0$]

$$E(y_0) = \mathbf{x}_0'\boldsymbol{\beta}. \tag{8.50}$$

We wish to find a confidence interval for $E(y_0)$, that is, for the mean of the distribution of y-values corresponding to $\boldsymbol{x}_0$.

By Corollary 1 to Theorem 7.6d, the minimum variance unbiased estimator of $E(y_0)$ is given by

$$\widehat{E(y_0)} = \mathbf{x}_0'\hat{\boldsymbol{\beta}}. \tag{8.51}$$

Since (8.50) and (8.51) are of the form $\mathbf{a}'\boldsymbol{\beta}$ and $\mathbf{a}'\hat{\boldsymbol{\beta}}$, respectively, we obtain a $100(1-\alpha)\%$ confidence interval for $E(y_0) = \mathbf{x}_0'\boldsymbol{\beta}$ from (8.49):

$$\mathbf{x}_0'\hat{\boldsymbol{\beta}} \pm t_{\alpha/2,n-k-1} s\sqrt{\mathbf{x}_0'(\mathbf{X}'\mathbf{X})^{-1}\mathbf{x}_0}. \tag{8.52}$$

The confidence coefficient $1-\alpha$ for the interval in (8.52) holds only for a single choice of the vector $\mathbf{x}_0$. For intervals covering several values of $\mathbf{x}_0$ or all possible values of $\mathbf{x}_0$, see Section 8.6.7.

We can express the confidence interval in (8.52) in terms of the centered model in Section 7.5, $y_i = \alpha + \boldsymbol{\beta}_1'(\mathbf{x}_{01} - \bar{\mathbf{x}}_1) + \varepsilon_i$, where $\mathbf{x}_{01} = (x_{01}, x_{02}, \ldots, x_{0k})'$ and $\bar{\mathbf{x}}_1 = (\bar{x}_1, \bar{x}_2, \ldots, \bar{x}_k)'$. [We use the notation $\mathbf{x}_{01}$ to distinguish this vector from $\mathbf{x}_0 = (1, x_{01}, x_{02}, \ldots, x_{0k})'$ above.] For the centered model, (8.50), (8.51), and (8.52) become

$$E(y_0) = \alpha + \boldsymbol{\beta}_1'(\mathbf{x}_{01} - \bar{\mathbf{x}}_1), \tag{8.53}$$

$$\widehat{E(y_0)} = \bar{y} + \hat{\boldsymbol{\beta}}_1'(\mathbf{x}_{01} - \bar{\mathbf{x}}_1), \tag{8.54}$$

$$\bar{y} + \hat{\boldsymbol{\beta}}_1'(\mathbf{x}_{01} - \bar{\mathbf{x}}_1) \pm t_{\alpha/2,n-k-1} s\sqrt{\frac{1}{n} + (\mathbf{x}_{01} - \bar{\mathbf{x}}_1)'(\mathbf{X}_c'\mathbf{X}_c)^{-1}(\mathbf{x}_{01} - \bar{\mathbf{x}}_1)}. \tag{8.55}$$

Note that in the form shown in (8.55), it is clear that if $\mathbf{x}_{01}$ is close to $\bar{\mathbf{x}}_1$ the interval is narrower; in fact, it is narrowest for $\mathbf{x}_{01} = \bar{\mathbf{x}}$. The width of the interval increases as the distance of $\mathbf{x}_{01}$ from $\bar{\mathbf{x}}_1$ increases.

For the special case of simple linear regression, (8.50), (8.51), and (8.55) reduce to

$$E(y_0) = \beta_0 + \beta_1 x_0, \tag{8.56}$$

$$\widehat{E(y_0)} = \hat{\beta}_0 + \hat{\beta}_1 x_0, \tag{8.57}$$

$$\hat{\beta}_0 + \hat{\beta}_1 x_0 \pm t_{\alpha/2,n-2} s\sqrt{\frac{1}{n} + \frac{(x_0 - \bar{x})^2}{\sum_{i=1}^n (x_i - \bar{x})^2}}, \tag{8.58}$$

where s is given by (6.11). The width of the interval in (8.58) depends on how far x_0 is from $\bar{x}$.

Example 8.6.4. For the grades data in Example 6.2, we find a 95% confidence interval for $E(y_0)$, where $x_0 = 80$. Using (8.58), we obtain

$$\begin{gathered}\hat{\beta}_0 + \hat{\beta}_1(80) \pm t_{.025,16}s\sqrt{\frac{1}{18} + \frac{(80 - 58.056)^2}{19530.944}},\\ 80.5386 \pm 2.1199(13.8547)(.2832),\\ 80.5386 \pm 8.3183,\\ (72.2204, 88.8569).\end{gathered}$$

□

8.6.5 Prediction Interval for a Future Observation

A "confidence interval" for a future observation y_0 corresponding to $\mathbf{x}_0$ is called a *prediction interval.* We speak of a prediction interval rather than a confidence interval because y_0 is an individual observation and is thereby a random variable rather than a parameter. To be $100(1-\alpha)\%$ confident that the interval contains y_0, the prediction interval will clearly have to be wider than a confidence interval for the parameter $E(y_0)$.

Since $y_0 = \mathbf{x}_0'\beta + \varepsilon_0$, we predict y_0 by $\hat{y}_0 = \mathbf{x}_0'\hat{\boldsymbol{\beta}}$, which is also the estimator of $E(y_0) = \mathbf{x}_0'\boldsymbol{\beta}$. The random variables y_0 and $\hat{y}_0$ are independent because y_0 is a future observation to be obtained independently of the n observations used to compute $\hat{y}_0 = \mathbf{x}_0'\hat{\boldsymbol{\beta}}$. Hence the variance of $y_0 - \hat{y}_0$ is

$$\text{var}(y_0 - \hat{y}_0) = \text{var}(y_0 - \mathbf{x}_0'\hat{\boldsymbol{\beta}}) = \text{var}(\mathbf{x}_0'\boldsymbol{\beta} + \varepsilon_0 - \mathbf{x}_0'\hat{\boldsymbol{\beta}}).$$

Since $\mathbf{x}_0'\boldsymbol{\beta}$ is a constant, this becomes

$$\begin{aligned}\text{var}(y_0 - \hat{y}_0) &= \text{var}(\varepsilon_0) + \text{var}(\mathbf{x}_0'\hat{\boldsymbol{\beta}}) = \sigma^2 + \sigma^2\mathbf{x}_0'(\mathbf{X}'\mathbf{X})^{-1}\mathbf{x}_0\\ &= \sigma^2\left[1 + \mathbf{x}_0'(\mathbf{X}'\mathbf{X})^{-1}\mathbf{x}_0\right],\end{aligned} \tag{8.59}$$

which is estimated by $s^2[1 + \mathbf{x}_0'(\mathbf{X}'\mathbf{X})^{-1}\mathbf{x}_0]$. It can be shown that $E(y_0 - \hat{y}_0) = 0$ and that s^2 is independent of both y_0 and $\hat{y}_0 = \mathbf{x}_0'\hat{\boldsymbol{\beta}}$. Therefore, the t statistic

$$t = \frac{y_0 - \hat{y}_0 - 0}{s\sqrt{1 + \mathbf{x}_0'(\mathbf{X}'\mathbf{X})^{-1}\mathbf{x}_0}} \tag{8.60}$$

is distributed as $t(n - k - 1)$, and

$$P = \left[-t_{\alpha/2,n-k-1} \leq \frac{y_0 - \hat{y}_0}{s\sqrt{1 + \mathbf{x}_0'(\mathbf{X}'\mathbf{X})^{-1}\mathbf{x}_0}} \leq t_{\alpha/2,n-k-1}\right] = 1 - \alpha.$$

The inequality can be solved for y_0 to obtain the $100(1-\alpha)\%$ prediction interval

$$\hat{y}_0 - t_{\alpha/2,\,n-k-1} s\sqrt{1 + \mathbf{x}_0'(\mathbf{X}'\mathbf{X})^{-1}\mathbf{x}_0} \leq y_0 \leq \hat{y}_0 + t_{\alpha/2,\,n-k-1} s\sqrt{1 + \mathbf{x}_0'(\mathbf{X}'\mathbf{X})^{-1}\mathbf{x}_0}$$

or, using $\hat{y}_0 = \mathbf{x}_0'\hat{\boldsymbol{\beta}}$, we have

$$\mathbf{x}_0'\hat{\boldsymbol{\beta}} \pm t_{\alpha/2,n-k-1} s\sqrt{1 + \mathbf{x}_0'(\mathbf{X}'\mathbf{X})^{-1}\mathbf{x}_0}. \tag{8.61}$$

Note that the confidence coefficient $1-\alpha$ for the prediction interval in (8.61) holds for only one value of $\mathbf{x}_0$.

In $1 + \mathbf{x}_0'(\mathbf{X}'\mathbf{X})^{-1}\mathbf{x}_0$, the second term, $\mathbf{x}_0'(\mathbf{X}'\mathbf{X})^{-1}\mathbf{x}_0$, is typically much smaller than 1 (provided k is much smaller than n) because the variance of $\hat{y}_0 = \mathbf{x}_0'\hat{\boldsymbol{\beta}}$ is much less than the variance of y_0. [To illustrate, if $\mathbf{X}'\mathbf{X}$ were diagonal and $\mathbf{x}_0$ were in the area covered by the rows of $\mathbf{X}$, then $\mathbf{x}_0'(\mathbf{X}'\mathbf{X})^{-1}\mathbf{x}_0$ would be a sum with $k+1$ terms, each of the form $x_{0j}^2/\sum_{i=1}^n x_{ij}^2$, which is of the order of $1/n$.] Thus prediction intervals for y_0 are generally much wider than confidence intervals for $E(y_0) = \mathbf{x}_0'\boldsymbol{\beta}$.

In terms of the centered model in Section 7.5, the $100(1-\alpha)\%$ prediction interval in (8.61) becomes

$$\bar{y} + \hat{\boldsymbol{\beta}}_1'(\mathbf{x}_{01} - \bar{\mathbf{x}}_1) \pm t_{\alpha/2,n-k-1} s\sqrt{1 + \frac{1}{n} + (\mathbf{x}_{01} - \bar{\mathbf{x}}_1)'(\mathbf{X}_c'\mathbf{X}_c)^{-1}(\mathbf{x}_{01} - \bar{\mathbf{x}}_1)}. \tag{8.62}$$

For the case of simple linear regression, (8.61) and (8.62) reduce to

$$\hat{\beta}_0 + \hat{\beta}_1 x_0 \pm t_{\alpha/2,\,n-2} s\sqrt{1 + \frac{1}{n} + \frac{(x_0 - \bar{x})^2}{\sum_{i=1}^n (x_i - \bar{x})^2}}, \tag{8.63}$$

where s is given by (6.11). In (8.63), it is clear that the second and third terms within the square root are much smaller than 1 unless x_0 is far removed from the interval bounded by the smallest and largest x's.

For a prediction interval for the mean of q future observations, see Problem 8.30.

Example 8.6.5. Using the data from Example 6.2, we find a 95% prediction interval for y_0 when $x_0 = 80$. Using (8.63), we obtain

$$\hat{\beta}_0 + \hat{\beta}_1(80) \pm t_{.025,16} s\sqrt{1 + \frac{1}{18} + \frac{(80 - 58.056)^2}{19530.944}},$$

$$80.5386 \pm 2.1199(13.8547)(1.0393),$$

$$80.5386 \pm 30.5258,$$

$$(50.0128,\ 111.0644).$$

Note that the prediction interval for y_0 here is much wider than the confidence interval for $E(y_0)$ in Example 8.6.4. □

8.6.6 Confidence Interval for σ^2

By Theorem 7.6b(ii), $(n-k-1)s^2/\sigma^2$ is $\chi^2(n-k-1)$. Therefore

$$P\left[\chi^2_{1-\alpha/2,\,n-k-1} \leq \frac{(n-k-1)s^2}{\sigma^2} \leq \chi^2_{\alpha/2,\,n-k-1}\right] = 1-\alpha, \tag{8.64}$$

where $\chi^2_{\alpha/2,\,n-k-1}$ is the upper $\alpha/2$ percentage point of the chi-square distribution and $\chi^2_{1-\alpha/2,\,n-k-1}$ is the lower $\alpha/2$ percentage point. Solving the inequality for σ^2 yields the $100(1-\alpha)\%$ confidence interval

$$\frac{(n-k-1)s^2}{\chi^2_{\alpha/2,\,n-k-1}} \leq \sigma^2 \leq \frac{(n-k-1)s^2}{\chi^2_{1-\alpha/2,\,n-k-1}}. \tag{8.65}$$

A $100(1-\alpha)\%$ confidence interval for σ is given by

$$\sqrt{\frac{(n-k-1)s^2}{\chi^2_{\alpha/2,n-k-1}}} \leq \sigma \leq \sqrt{\frac{(n-k-1)s^2}{\chi^2_{1-\alpha/2,\,n-k-1}}}. \tag{8.66}$$

8.6.7 Simultaneous Intervals

By analogy to the discussion of testing several hypotheses (Section 8.5.2), when several intervals are computed, two confidence coefficients can be considered: familywise confidence $(1-\alpha_f)$ and individual confidence $(1-\alpha_c)$. Familywise confidence of $1-\alpha_f$ means that we are $100(1-\alpha_f)\%$ confident that every interval contains its respective parameter.

In some cases, our goal is simply to control $1-\alpha_c$ for each one of several confidence or prediction intervals so that no changes are needed to expressions (8.47), (8.49), (8.52), and (8.61). In other cases the desire is to control $1-\alpha_f$. To do so, both the Bonferroni and Scheffé methods can be adapted to the situation of multiple intervals. In yet other cases we may want to control other properties of multiple intervals (Benjamini and Yekutieli 2005).

The Bonferroni procedure increases the width of each individual interval so that $1-\alpha_f$ for the set of intervals is greater than or equal to the desired value $1-\alpha^*$. As an example suppose that it is desired to calculate the k confidence intervals for $\beta_1, \ldots, \beta_k$. Let E_j be the event that the jth interval includes β_j, and E_j^c be the complement of that event. Then by definition

$$\begin{aligned} 1-\alpha_f &= P(E_1 \cap E_2 \cap \ldots \cap E_k) \\ &= 1 - P(E_1^c \cup E_2^c \cup \ldots \cup E_k^c). \end{aligned}$$

Assuming that $P(E_j^c) = \alpha_c$ for $j = 1, \ldots, k$, the Bonferroni inequality now implies that

$$1 - \alpha_f \geq 1 - k\alpha_c.$$

Hence we can ensure that $1 - \alpha_f$ is greater than or equal to the desired $1 - \alpha^*$ by setting $1 - \alpha_c = 1 - \alpha^*/k$ for the individual intervals.

Using this approach, Bonferroni confidence intervals for $\beta_1, \beta_2, \ldots, \beta_k$ are given by

$$\hat{\beta}_j \pm t_{\alpha^*/2k,\, n-k-1} s\sqrt{g_{jj}}, \quad j = 1, 2, \ldots, k, \tag{8.67}$$

where g_{jj} is the jth element of $(\mathbf{X}'\mathbf{X})^{-1}$. Bonferroni t values $t_{\alpha^*/2k}$ are available in Bailey (1977) and can also be obtained in many software programs. For example, a probability calculator for the t, the F, and other distributions is available free from NCSS (download at www.ncss.com).

Similarly for d linear functions $\mathbf{a}_1'\boldsymbol{\beta}, \mathbf{a}_2'\boldsymbol{\beta}, \ldots, \mathbf{a}_d'\boldsymbol{\beta}$ (chosen before seeing the data), Bonferroni confidence intervals are given by

$$\mathbf{a}_i'\hat{\boldsymbol{\beta}} \pm t_{\alpha^*/2d,\, n-k-1} s\sqrt{\mathbf{a}_i'(\mathbf{X}'\mathbf{X})^{-1}\mathbf{a}_i}, \quad i = 1, 2, \ldots, d. \tag{8.68}$$

These intervals hold simultaneously with familywise confidence of at least $1 - \alpha^*$.

Bonferroni confidence intervals for $E(y_0) = \mathbf{x}_0'\boldsymbol{\beta}$ for a few values of $\mathbf{x}_0$, say, $\mathbf{x}_{01}, \mathbf{x}_{02}, \ldots, \mathbf{x}_{0d}$ are given by

$$\mathbf{x}_{0i}'\hat{\boldsymbol{\beta}} \pm t_{\alpha^*/2d,\, n-k-1} s\sqrt{\mathbf{x}_{0i}'(\mathbf{X}'\mathbf{X})^{-1}\mathbf{x}_{0i}}, \quad i = 1, 2, \ldots, d. \tag{8.69}$$

[Note that $\mathbf{x}_{01}$ here differs from $\mathbf{x}_{01}$ in (8.53)–(8.55).]

For simultaneous prediction of d new observations $y_{01}, y_{02}, \ldots, y_{0d}$ at d values of $\mathbf{x}_0$, say, $\mathbf{x}_{01}, \mathbf{x}_{02}, \ldots, \mathbf{x}_{0d}$, we can use the Bonferroni prediction intervals

$$\mathbf{x}_{0i}'\hat{\boldsymbol{\beta}} \pm t_{\alpha^*/2d, n-k-1} s\sqrt{1 + \mathbf{x}_{0i}'(\mathbf{X}'\mathbf{X})^{-1}\mathbf{x}_{0i}} \quad i = 1, 2, \ldots, d \tag{8.70}$$

[see (8.61) and (8.69)].

Simultaneous Scheffé confidence intervals for all possible linear functions $\mathbf{a}'\boldsymbol{\beta}$ (including those chosen after seeing the data) can be based on the distribution of $\max_{\mathbf{a}} F$ [Theorem 8.5(ii)]. Thus a conservative confidence interval for any and all $\mathbf{a}'\boldsymbol{\beta}$ is

$$\mathbf{a}'\hat{\boldsymbol{\beta}} \pm s\sqrt{(k+1)F_{\alpha^*,\, k+1,\, n-k-1}\mathbf{a}'(\mathbf{X}'\mathbf{X})^{-1}\mathbf{a}}. \tag{8.71}$$

The (potentially infinite number of) intervals in (8.71) have an overall confidence coefficient of at least $1 - \alpha^*$. For a few linear functions, the intervals in (8.68) will be narrower, but for a large number of linear functions, the intervals in (8.71) will be narrower. A comparison of $t_{\alpha^*/2d, n-k-1}$ and $\sqrt{(k+1)F_{\alpha^*,\, k+1,\, n-k-1}}$ will show which is preferred in a given case.

For confidence limits for $E(y_0) = \mathbf{x}_0'\boldsymbol{\beta}$ for all possible values of $\mathbf{x}_0$, we use (8.71):

$$\mathbf{x}_0'\hat{\boldsymbol{\beta}} \pm s\sqrt{(k+1)F_{\alpha^*, k+1, n-k-1}\mathbf{x}_0'(\mathbf{X}'\mathbf{X})^{-1}\mathbf{x}_0}. \tag{8.72}$$

These intervals hold simultaneously with a confidence coefficient of $1 - \alpha^*$. Thus, (8.72) becomes a confidence region that can be applied to the entire regression surface for all values of $\mathbf{x}_0$. The intervals in (8.71) and (8.72) are due to Scheffé (1953; 1959, p. 68) and Working and Hotelling (1929).

Scheffé-type prediction intervals for $y_{01}, y_{02}, \ldots, y_{0d}$ are given by

$$\mathbf{x}_{0i}'\hat{\boldsymbol{\beta}} \pm s\sqrt{dF_{\alpha^*, d, h-k-1}[1 + \mathbf{x}_{0i}'(\mathbf{X}'\mathbf{X})^{-1}\mathbf{x}_{0i}]} \quad i = 1, 2, \ldots, d \tag{8.73}$$

(see Problem 8.32). These d prediction intervals hold simultaneously with overall confidence coefficient at least $1 - \alpha^*$, but note that $dF_{\alpha^*, d, n-k-1}$ is not constant. It depends on the number of predictions.

Example 8.6.7. We compute 95% Bonferroni confidence limits for β_1, β_2, and β_3, using y_2 in the chemical reaction data in Table 7.4; see Example 8.6.2 for $(\mathbf{X}'\mathbf{X})^{-1}$ and $\hat{\boldsymbol{\beta}}$. By (8.67), we have

$$\hat{\beta}_1 \pm t_{.025/3, 15}s\sqrt{g_{11}}$$
$$.4056 \pm (2.6937)(4.0781)\sqrt{.00184}$$
$$.4056 \pm .4706$$
$$(-.0660, .8751),$$

$$\beta_2: \quad .2930 \pm .7016$$
$$(-.4086, .9946),$$
$$\beta_3: \quad 1.0338 \pm 1.6147$$
$$(-.5809, 2.6485).$$

These three intervals hold simultaneously with confidence coefficient at least .95. □

8.7 LIKELIHOOD RATIO TESTS

The tests in Sections 8.1, 8.2, and 8.4 were derived using informal methods based on finding sums of squares that have chi-square distributions and are independent. These same tests can be obtained more formally by the likelihood ratio approach. Likelihood ratio tests have some good properties and sometimes have optimal properties.

We describe the likelihood ratio method in the simple context of testing $H_0: \boldsymbol{\beta} = \mathbf{0}$ versus $H_1: \boldsymbol{\beta} \neq \mathbf{0}$. The likelihood function $L(\boldsymbol{\beta}, \sigma^2)$ was defined in Section 7.6.2 as the joint density of the y's. For a random sample $\mathbf{y} = (y_1, y_2, \ldots, y_n)'$ with density $N_n(\mathbf{X}\boldsymbol{\beta}, \sigma^2\mathbf{I})$, the likelihood function is given by (7.50) as

$$L(\boldsymbol{\beta},\sigma^2) = \frac{1}{(2\pi\sigma^2)^{n/2}} e^{-(\mathbf{y}-\mathbf{X}\boldsymbol{\beta})'(\mathbf{y}-\mathbf{X}\boldsymbol{\beta})/2\sigma^2}. \tag{8.74}$$

The likelihood ratio method compares the maximum value of $L(\boldsymbol{\beta}, \sigma^2)$ restricted by $H_0 : \boldsymbol{\beta} = \mathbf{0}$ to the maximum value of $L(\boldsymbol{\beta}, \sigma^2)$ under H_1: $\boldsymbol{\beta}_1 \neq \mathbf{0}$, which is essentially unrestricted. We denote the maximum value of $L(\boldsymbol{\beta},\sigma^2)$ restricted by $\boldsymbol{\beta} = \mathbf{0}$ as $\max_{H_0} L(\boldsymbol{\beta}, \sigma^2)$ and the unrestricted maximum as $\max_{H_1} L(\boldsymbol{\beta}, \sigma^2)$. If $\boldsymbol{\beta}$ is equal (or close) to $\mathbf{0}$, then $\max_{H_0} L(\boldsymbol{\beta}, \sigma^2)$ should be close to $\max_{H_1} L(\boldsymbol{\beta}, \sigma^2)$. If $\max_{H_0} L(\boldsymbol{\beta}, \sigma^2)$ is not close to $\max_{H_1} L(\boldsymbol{\beta}, \sigma^2)$, we would conclude that $y = (y_1, y_2, \ldots, y_n)'$ apparently did not come from $N_n(\mathbf{X}\boldsymbol{\beta}, \sigma^2\mathbf{I})$ with $\boldsymbol{\beta} = \mathbf{0}$.

In this illustration, we can find $\max_{H_0} L(\boldsymbol{\beta}, \sigma^2)$ by setting $\boldsymbol{\beta} = \mathbf{0}$ and then estimating σ^2 as the value that maximizes $L(\mathbf{0}, \sigma^2)$. Under $H_1 : \boldsymbol{\beta} \neq \mathbf{0}$, both $\boldsymbol{\beta}$ and σ^2 are estimated without restriction as the values that maximize $L(\boldsymbol{\beta}, \sigma^2)$. [In designating the unrestricted maximum as $\max_{H_1} L(\boldsymbol{\beta}, \sigma^2)$, we are ignoring the restriction in H_1 that $\boldsymbol{\beta} \neq \mathbf{0}$.]

It is customary to describe the likelihood ratio method in terms of maximizing L subject to ω, the set of all values of $\boldsymbol{\beta}$ and σ^2 satisfying H_0, and subject to Ω, the set of all values of β and σ^2 without restrictions (other than natural restrictions such as $\sigma^2 > 0$). However, to simplify notation in cases such as this in which H_1 includes all values of $\boldsymbol{\beta}$ except $\mathbf{0}$, we refer to maximizing L under H_0 and H_1.

We compare the restricted maximum under H_0 with the unrestricted maximum under H_1 by the *likelihood ratio*

$$\begin{aligned} \text{LR} &= \frac{\max_{H_0} L(\boldsymbol{\beta}, \sigma^2)}{\max_{H_1} L(\boldsymbol{\beta}, \sigma^2)} \\ &= \frac{\max L(\mathbf{0}, \sigma^2)}{\max L(\boldsymbol{\beta}, \sigma^2)}. \end{aligned} \tag{8.75}$$

It is clear that $0 \leq \text{LR} \leq 1$, because the maximum of L restricted to $\boldsymbol{\beta} = \mathbf{0}$ cannot exceed the unrestricted maximum. Smaller values of **LR** would favor H_1, and larger values would favor H_0. We thus reject H_0 if $\text{LR} \leq c$, where c is chosen so that $P(\text{LR} \leq c) = \alpha$ if H_0 is true.

Wald (1943) showed that, under H_0

$$-2 \ln \text{LR} \text{ is approximately } \chi^2(\nu)$$

for large n, where ν is the number of parameters estimated under H_1 minus the number estimated under H_0. In the case of $H_0 : \boldsymbol{\beta} = \mathbf{0}$ versus $H_1 : \boldsymbol{\beta} \neq \mathbf{0}$, we have $\nu = k + 2 - 1 = k + 1$ because $\boldsymbol{\beta}$ and σ^2 are estimated under H_1 while only σ^2 is estimated under H_0. In some cases, the χ^2 approximation is not needed because LR turns out to be a function of a familiar test statistic, such as t or F, whose exact distribution is available.

We now obtain the likelihood ratio test for $H_0 : \boldsymbol{\beta} = \mathbf{0}$. The resulting likelihood ratio is a function of the F statistic obtained in Problem 8.6 by partitioning the total sum of squares.

Theorem 8.7a. If $\mathbf{y}$ is $N_n(\mathbf{X}\boldsymbol{\beta}, \sigma^2\mathbf{I})$, the likelihood ratio test for $H_0 : \boldsymbol{\beta} = \mathbf{0}$ can be based on

$$F = \frac{\hat{\boldsymbol{\beta}}'\mathbf{X}'\mathbf{y}/(k+1)}{(\mathbf{y}'\mathbf{y} - \hat{\boldsymbol{\beta}}'\mathbf{X}'\mathbf{y})/(n-k-1)}.$$

We reject H_0 if $F > F_{\alpha, k+1, n-k-1}$.

Proof. To find $\max_{H_1} L(\boldsymbol{\beta}, \sigma^2) = \max L(\boldsymbol{\beta}, \sigma^2)$, we use the maximum likelihood estimators $\hat{\boldsymbol{\beta}} = (\mathbf{X}'\mathbf{X})^{-1}\mathbf{X}'\mathbf{y}$ and $\hat{\sigma}^2 = (\mathbf{y} - \mathbf{X}\hat{\boldsymbol{\beta}})'(\mathbf{y} - \mathbf{X}\hat{\boldsymbol{\beta}})/n$ from Theorem 7.6a. Substituting these in (8.74), we obtain

$$\begin{aligned}
\max_{H_1} L(\boldsymbol{\beta}, \sigma^2) &= \max L(\boldsymbol{\beta}, \sigma^2) = L(\hat{\boldsymbol{\beta}}, \hat{\sigma}^2) \\
&= \frac{1}{(2\pi\hat{\sigma}^2)^{n/2}} e^{-(\mathbf{y}-\mathbf{X}\hat{\boldsymbol{\beta}})'(\mathbf{y}-\mathbf{X}\hat{\boldsymbol{\beta}})/2\hat{\sigma}^2} \\
&= \frac{n^{n/2}e^{-n/2}}{(2\pi)^{n/2}\left[(\mathbf{y} - \mathbf{X}\hat{\boldsymbol{\beta}})'(\mathbf{y} - \mathbf{X}\hat{\boldsymbol{\beta}})\right]^{n/2}}. \qquad (8.76)
\end{aligned}$$

To find $\max_{H_0} L(\boldsymbol{\beta}, \sigma^2) = \max L(\mathbf{0}, \sigma^2)$, we solve $\partial \ln L(\mathbf{0}, \sigma^2)/\partial\sigma^2 = 0$ to obtain

$$\hat{\sigma}_0^2 = \frac{\mathbf{y}'\mathbf{y}}{n}. \qquad (8.77)$$

Then

$$\begin{aligned}
\max_{H_0} L(\boldsymbol{\beta}, \sigma^2) &= \max L(\mathbf{0}, \sigma^2) = L(\mathbf{0}, \hat{\sigma}_0^2) \\
&= \frac{1}{(2\pi\hat{\sigma}_0^2)^{n/2}} e^{-\mathbf{y}'\mathbf{y}/2\hat{\sigma}_0^2} \\
&= \frac{n^{n/2}e^{-n/2}}{(2\pi)^{n/2}(\mathbf{y}'\mathbf{y})^{n/2}}. \qquad (8.78)
\end{aligned}$$

Substituting (8.76) and (8.78) into (8.75), we obtain

$$\begin{aligned} \mathrm{LR} = \frac{\max_{H_0} L(\boldsymbol{\beta}, \sigma^2)}{\max_{H_1} L(\boldsymbol{\beta}, \sigma^2)} &= \left[\frac{(\mathbf{y} - \mathbf{X}\hat{\boldsymbol{\beta}})'(\mathbf{y} - \mathbf{X}\hat{\boldsymbol{\beta}})}{\mathbf{y}'\mathbf{y}} \right]^{n/2} \\ &= \left[\frac{1}{1 + (k+1)F/(n-k-1)} \right]^{n/2}, \end{aligned} \tag{8.79}$$

where

$$F = \frac{\hat{\boldsymbol{\beta}}'\mathbf{X}'\mathbf{y}/(k+1)}{(\mathbf{y}'\mathbf{y} - \hat{\boldsymbol{\beta}}'\mathbf{X}'\mathbf{y})/(n-k-1)}.$$

Thus, rejecting $H_0 : \boldsymbol{\beta} = \mathbf{0}$ for a small value of LR is equivalent to rejecting H_0 for a large value of F. □

We now show that the F test in Theorem 8.4b for the general linear hypothesis $H_0 : \mathbf{C}\boldsymbol{\beta} = \mathbf{0}$ is a likelihood ratio test.

Theorem 8.7b. If $\mathbf{y}$ is $N_n(\mathbf{X}\boldsymbol{\beta}, \sigma^2\mathbf{I})$, then the F test for $H_0 : \mathbf{C}\boldsymbol{\beta} = \mathbf{0}$ in Theorem 8.4b is equivalent to the likelihood ratio test.

Proof. Under $H_1 : \mathbf{C}\boldsymbol{\beta} \neq \mathbf{0}$, which is essentially unrestricted, $\max_{H_1} L(\boldsymbol{\beta}, \sigma^2)$ is given by (8.76). To find $\max_{H_0} L(\boldsymbol{\beta}, \sigma^2) = \max L(\boldsymbol{\beta}, \sigma^2)$ subject to $\mathbf{C}\boldsymbol{\beta} = \mathbf{0}$, we use the method of Lagrange multipliers (Section 2.14.3) and work with $L(\boldsymbol{\beta}, \sigma^2)$ to simplify the differentiation:

$$\begin{aligned} v &= \ln L(\boldsymbol{\beta}, \sigma^2) + \boldsymbol{\lambda}'(\mathbf{C}\boldsymbol{\beta} - \mathbf{0}) \\ &= -\frac{n}{2}\ln(2\pi) - \frac{n}{2}\ln\sigma^2 - \frac{(\mathbf{y} - \mathbf{X}\boldsymbol{\beta})'(\mathbf{y} - \mathbf{X}\boldsymbol{\beta})}{2\sigma^2} + \boldsymbol{\lambda}'\mathbf{C}\boldsymbol{\beta}. \end{aligned}$$

Expanding $(\mathbf{y} - \mathbf{X}\boldsymbol{\beta})'(\mathbf{y} - \mathbf{X}\boldsymbol{\beta})$ and differentiating with respect to $\boldsymbol{\beta}, \boldsymbol{\lambda},$ and σ^2, we obtain

$$\frac{\partial v}{\partial \boldsymbol{\beta}} = (2\mathbf{X}'\mathbf{y} - 2\mathbf{X}'\mathbf{X}\boldsymbol{\beta})/2\sigma^2 + \mathbf{C}'\boldsymbol{\lambda} = \mathbf{0}, \tag{8.80}$$

$$\frac{\partial v}{\partial \boldsymbol{\lambda}} = \mathbf{C}\boldsymbol{\beta} = \mathbf{0}, \tag{8.81}$$

$$\frac{\partial v}{\partial \sigma^2} = -\frac{n}{2\sigma^2} + \frac{1}{2(\sigma^2)^2}(\mathbf{y} - \mathbf{X}\boldsymbol{\beta})'(\mathbf{y} - \mathbf{X}\boldsymbol{\beta}) = 0. \tag{8.82}$$

Eliminating $\boldsymbol{\lambda}$ and solving for $\boldsymbol{\beta}$ and σ^2 gives

$$\hat{\boldsymbol{\beta}}_0 = \hat{\boldsymbol{\beta}} - (\mathbf{X}'\mathbf{X})^{-1}\mathbf{C}'[\mathbf{C}(\mathbf{X}'\mathbf{X})^{-1}\mathbf{C}']^{-1}\mathbf{C}\hat{\boldsymbol{\beta}}, \tag{8.83}$$

$$\hat{\sigma}_0^2 = \frac{1}{n}(\mathbf{y} - \mathbf{X}\hat{\boldsymbol{\beta}}_0)'(\mathbf{y} - \mathbf{X}\hat{\boldsymbol{\beta}}_0) \tag{8.84}$$

$$= \hat{\sigma}^2 + \frac{1}{n}(\mathbf{C}\hat{\boldsymbol{\beta}})'[\mathbf{C}(\mathbf{X}'\mathbf{X})^{-1}\mathbf{C}']^{-1}\mathbf{C}\hat{\boldsymbol{\beta}} \tag{8.85}$$

(Problems 8.35 and 8.36), where $\hat{\sigma}^2 = (\mathbf{y} - \mathbf{X}\hat{\boldsymbol{\beta}})'(\mathbf{y} - \mathbf{X}\hat{\boldsymbol{\beta}})/n$ and $\hat{\boldsymbol{\beta}} = (\mathbf{X}'\mathbf{X})^{-1}\mathbf{X}'\mathbf{y}$ are the maximum likelihood estimates from Theorem 7.6a. Thus

$$\begin{aligned}\max_{H_0} L(\boldsymbol{\beta}, \sigma^2) &= L(\hat{\boldsymbol{\beta}}_0, \hat{\sigma}_0^2) \\ &= \frac{1}{(2\pi)^{n/2}(\hat{\sigma}_0^2)^{n/2}} e^{-(\mathbf{y}-\mathbf{X}\hat{\boldsymbol{\beta}}_0)'(\mathbf{y}-\mathbf{X}\hat{\boldsymbol{\beta}}_0)/2\hat{\sigma}_0^2} \\ &= \frac{n^{n/2}e^{-n/2}}{(2\pi)^{n/2}\left\{\text{SSE} + (\mathbf{C}\hat{\boldsymbol{\beta}})'\left[\mathbf{C}(\mathbf{X}'\mathbf{X})^{-1}\mathbf{C}'\right]^{-1}\mathbf{C}\hat{\boldsymbol{\beta}}\right\}^{n/2}},\end{aligned}$$

and

$$\begin{aligned}\text{LR} &= \frac{\max_{H_0} L(\boldsymbol{\beta}, \sigma^2)}{\max_{H_1} L(\boldsymbol{\beta}, \sigma^2)} \\ &= \left[\frac{\text{SSE}}{\text{SSE} + (\mathbf{C}\hat{\boldsymbol{\beta}})'[\mathbf{C}(\mathbf{X}'\mathbf{X})^{-1}\mathbf{C}']^{-1}\mathbf{C}\hat{\boldsymbol{\beta}}}\right]^{n/2} \\ &= \left[\frac{1}{1 + \text{SSH}/\text{SSE}}\right]^{n/2} = \left[\frac{1}{1 + qF/(n-k-1)}\right]^{n/2},\end{aligned}$$

where $\text{SSH} = (\mathbf{C}\hat{\boldsymbol{\beta}})'[\mathbf{C}(\mathbf{X}'\mathbf{X})^{-1}\mathbf{C}']^{-1}\mathbf{C}\hat{\boldsymbol{\beta}}$, $\text{SSE} = (\mathbf{y} - \mathbf{X}\hat{\boldsymbol{\beta}})'(\mathbf{y} - \mathbf{X}\hat{\boldsymbol{\beta}})$, and F is given in (8.27). □

PROBLEMS

8.1 Show that $\text{SSR} = \hat{\boldsymbol{\beta}}_1'\mathbf{X}_c'\mathbf{X}_c\hat{\boldsymbol{\beta}}_1$ in (8.1) becomes $\mathbf{y}'\mathbf{X}_c(\mathbf{X}_c'\mathbf{X}_c)^{-1}\mathbf{X}_c'\mathbf{y}$ as in (8.2).

8.2 (**a**) Show that $\mathbf{H}_c[\mathbf{I} - (1/n)\mathbf{J}] = \mathbf{H}_c$, as in (8.3) in Theorem 8.1a(i), where $\mathbf{H}_c = \mathbf{X}_c(\mathbf{X}_c'\mathbf{X}_c)^{-1}\mathbf{X}_c'$.

(**b**) Prove Theorem 8.1a(ii).

(**c**) Prove Theorem 8.1a(iii).

(**d**) Prove Theorem 8.1a(iv).

8.3 Show that $\lambda_1 = \boldsymbol{\beta}_1' \mathbf{X}_c \mathbf{X}_c \boldsymbol{\beta}_1 / 2\sigma^2$ as in Theorem 8.1b(i).

8.4 Prove Theorem 8.1b(ii).

8.5 Show that $E(\text{SSR}/k) = \sigma^2 + (1/k)\boldsymbol{\beta}_1' \mathbf{X}_c' \mathbf{X}_c \boldsymbol{\beta}_1$, as in the expected mean square column of Table 8.1. Employ the following two approaches:

(a) Use Theorem 5.2a.

(b) Use the noncentrality parameter in (5.19).

8.6 Develop a test for $H_0 : \boldsymbol{\beta} = \mathbf{0}$ in the model $\mathbf{y} = \mathbf{X}\boldsymbol{\beta} + \boldsymbol{\varepsilon}$, where $\mathbf{y}$ is $N_n(\mathbf{X}\boldsymbol{\beta}, \sigma^2 \mathbf{I})$. (It was noted at the beginning of Section 8.1 that this hypothesis is of little practical interest because it includes $\beta_0 = 0$.) Use the partitioning $\mathbf{y}'\mathbf{y} = (\mathbf{y}'\mathbf{y} - \hat{\boldsymbol{\beta}}'\mathbf{X}'\mathbf{y}) + \hat{\boldsymbol{\beta}}'\mathbf{X}'\mathbf{y}$, and proceed as follows:

(a) Show that $\hat{\boldsymbol{\beta}}'\mathbf{X}'\mathbf{y} = \mathbf{y}'\mathbf{X}(\mathbf{X}'\mathbf{X})^{-1}\mathbf{X}'\mathbf{y}$ and $\mathbf{y}'\mathbf{y} - \hat{\boldsymbol{\beta}}'\mathbf{X}'\mathbf{y} = \mathbf{y}'[\mathbf{I} - \mathbf{X}(\mathbf{X}'\mathbf{X})^{-1} \mathbf{X}']\mathbf{y}$.

(b) Let $\mathbf{H} = \mathbf{X}(\mathbf{X}'\mathbf{X})^{-1}\mathbf{X}'$. Show that $\mathbf{H}$ and $\mathbf{I} - \mathbf{H}$ are idempotent of rank $k+1$ and $n-k-1$, respectively.

(c) Show that $\mathbf{y}'\mathbf{H}\mathbf{y}/\sigma^2$ is $\chi^2(k+1, \lambda_1)$, where $\lambda_1 = \boldsymbol{\beta}'\mathbf{X}'\mathbf{X}\boldsymbol{\beta}/2\sigma^2$, and that $\mathbf{y}'(\mathbf{I} - \mathbf{H})\mathbf{y}/\sigma^2$ is $\chi^2(n-k-1)$.

(d) Show that $\mathbf{y}'\mathbf{H}\mathbf{y}$ and $\mathbf{y}'(\mathbf{I} - \mathbf{H})\mathbf{y}$ are independent.

(e) Show that

$$\frac{\hat{\boldsymbol{\beta}}'\mathbf{X}'\mathbf{y}}{(k+1)s^2} = \frac{\mathbf{y}'\mathbf{H}\mathbf{y}/(k+1)}{\mathbf{y}'(\mathbf{I} - \mathbf{H})\mathbf{y}/(n-k-1)}$$

is distributed as $F(k+1, n-k-1, \lambda_1)$.

8.7 Show that $\mathbf{H}\mathbf{H}_1 = \mathbf{H}_1$ and $\mathbf{H}_1\mathbf{H} = \mathbf{H}_1$, as in (8.15), where $\mathbf{H}$ and $\mathbf{H}_1$ are as defined in (8.11) and (8.12).

8.8 Show that conditions (a) and (b) of Corollary 1 to Theorem 5.6c are satisfied for the sum of quadratic forms in (8.12), as noted in the proof of Theorem 8.2b.

8.9 Show that $\lambda_1 = \boldsymbol{\beta}_2'[\mathbf{X}_2'\mathbf{X}_2 - \mathbf{X}_2'\mathbf{X}_1(\mathbf{X}_1'\mathbf{X}_1)^{-1}\mathbf{X}_1'\mathbf{X}_2]\boldsymbol{\beta}_2/2\sigma^2$ as in Theorem 8.2b(ii).

8.10 Show that $\mathbf{X}_2'\mathbf{X}_2 - \mathbf{X}_2'\mathbf{X}_1(\mathbf{X}_1'\mathbf{X}_1)^{-1}\mathbf{X}_1'\mathbf{X}_2$ is positive definite, as noted below Theorem 8.2b.

8.11 Show that $E[\text{SS}(\boldsymbol{\beta}_2|\boldsymbol{\beta}_1)/h] = \sigma^2 + \boldsymbol{\beta}_2'[\mathbf{X}_2'\mathbf{X}_2 - \mathbf{X}_2'\mathbf{X}_1(\mathbf{X}_1'\mathbf{X}_1)^{-1}\mathbf{X}_1'\mathbf{X}_2]\boldsymbol{\beta}_2/h$ as in Table 8.3.

8.12 Find the expected mean square corresponding to the numerator of the F statistic in (8.20) in Example 8.2b.

8.13 Show that $\hat{\beta}_0^* = \bar{y}$ and $\text{SS}(\beta_0^*) = n\bar{y}^2$, as in (8.21) in Example 8.2c.

8.14 In the proof of Theorem 8.2d, show that $(\hat{\boldsymbol{\beta}}_1'\mathbf{X}_1' + \hat{\boldsymbol{\beta}}_2'\mathbf{X}_2')(\mathbf{X}_1\hat{\boldsymbol{\beta}}_1 + \mathbf{X}_2\hat{\boldsymbol{\beta}}_2) - (\hat{\boldsymbol{\beta}}_1' + \hat{\boldsymbol{\beta}}_2'\mathbf{A}')\mathbf{X}_1'\mathbf{X}_1(\hat{\boldsymbol{\beta}}_1 + \mathbf{A}\hat{\boldsymbol{\beta}}_2) = \hat{\boldsymbol{\beta}}_2'[\mathbf{X}_2'\mathbf{X}_2 - \mathbf{X}_2'\mathbf{X}_1(\mathbf{X}_1'\mathbf{X}_1)^{-1}\mathbf{X}_1'\mathbf{X}_2]\hat{\boldsymbol{\beta}}_2$.

8.15 Express the test for $H_0 : \boldsymbol{\beta}_2 = \mathbf{0}$ in terms of R^2, as in (8.25) in Theorem 8.3.

8.16 Prove Theorem 8.4a(iv).

8.17 Show that $\mathbf{C}(\mathbf{X}'\mathbf{X})^{-1}\mathbf{C}'$ is positive definite, as noted following Theorem 8.4b.

8.18 Prove Theorem 8.4c.

8.19 Show that in the model $\mathbf{y} = \mathbf{X}\boldsymbol{\beta} + \boldsymbol{\varepsilon}$ subject to $\mathbf{C}\boldsymbol{\beta} = \mathbf{0}$ in (8.29), the estimator of $\boldsymbol{\beta}$ is $\hat{\boldsymbol{\beta}}_c = \hat{\boldsymbol{\beta}} - (\mathbf{X}'\mathbf{X})^{-1}\mathbf{C}'[\mathbf{C}(\mathbf{X}'\mathbf{X})^{-1}\mathbf{C}']^{-1}\mathbf{C}\hat{\boldsymbol{\beta}}$ as in (8.30), where $\hat{\boldsymbol{\beta}} = (\mathbf{X}'\mathbf{X})^{-1}\mathbf{X}'\mathbf{y}$. Use a Lagrange multiplier $\boldsymbol{\lambda}$ and minimize $u = (\mathbf{y} - \mathbf{X}\boldsymbol{\beta})'(\mathbf{y} - \mathbf{X}\boldsymbol{\beta}) + \boldsymbol{\lambda}'(\mathbf{C}\boldsymbol{\beta} - \mathbf{0})$ with respect to $\boldsymbol{\beta}$ and $\boldsymbol{\lambda}$ as follows:

(**a**) Differentiate u with respect to $\boldsymbol{\lambda}$ and set the result equal to $\mathbf{0}$ to obtain $\mathbf{C}\hat{\boldsymbol{\beta}}_c = \mathbf{0}$.

(**b**) Differentiate u with respect to $\boldsymbol{\beta}$ and set the result equal to $\mathbf{0}$ to obtain

$$\hat{\boldsymbol{\beta}}_c = \hat{\boldsymbol{\beta}} - \tfrac{1}{2}(\mathbf{X}'\mathbf{X})^{-1}\mathbf{C}'\boldsymbol{\lambda}, \tag{1}$$

where $\hat{\boldsymbol{\beta}} = (\mathbf{X}'\mathbf{X})^{-1}\mathbf{X}'\mathbf{y}$.

(**c**) Multiply (1) in part (b) by $\mathbf{C}$, use $\mathbf{C}\hat{\boldsymbol{\beta}}_c = \mathbf{0}$ from part (a), solve for $\boldsymbol{\lambda}$, and substitute back into (1).

8.20 Show that $\hat{\boldsymbol{\beta}}_c'\mathbf{X}'\mathbf{X}\hat{\boldsymbol{\beta}}_c = \hat{\boldsymbol{\beta}}_c'\mathbf{X}'\mathbf{y}$, thus demonstrating directly that the sum of squares due to the reduced model is $\hat{\boldsymbol{\beta}}_c'\mathbf{X}'\mathbf{y}$ and that (8.31) holds.

8.21 Show that for the general linear hypothesis $H_0 : \mathbf{C}\boldsymbol{\beta} = \mathbf{0}$ in Theorem 8.4d, we have $\hat{\boldsymbol{\beta}}'\mathbf{X}'\mathbf{y} - \hat{\boldsymbol{\beta}}_c'\mathbf{X}'\mathbf{y} = (\mathbf{C}\hat{\boldsymbol{\beta}})'[\mathbf{C}(\mathbf{X}'\mathbf{X})^{-1}\mathbf{C}']^{-1}\mathbf{C}\hat{\boldsymbol{\beta}}$ as in (8.32), where $\hat{\boldsymbol{\beta}}_c$ is as given in (8.30).

8.22 Prove Theorem 8.4e.

8.23 Prove Theorem 8.4f(iv) by expressing SSH and SSE as quadratic forms in the same normally distributed random vector.

8.24 Show that the estimator for $\boldsymbol{\beta}$ in the reduced model $\mathbf{y} = \mathbf{X}\boldsymbol{\beta} + \boldsymbol{\varepsilon}$ subject to $\mathbf{C}\boldsymbol{\beta} = \mathbf{t}$ is given by $\hat{\boldsymbol{\beta}}_c = \hat{\boldsymbol{\beta}} - (\mathbf{X}'\mathbf{X})^{-1}\mathbf{C}'[\mathbf{C}(\mathbf{X}'\mathbf{X})^{-1}\mathbf{C}']^{-1}(\mathbf{C}\hat{\boldsymbol{\beta}} - \mathbf{t})$, where $\hat{\boldsymbol{\beta}} = (\mathbf{X}'\mathbf{X})^{-1}\mathbf{X}'\mathbf{y}$.

8.25 Show that $\hat{\boldsymbol{\beta}}'\mathbf{X}'\mathbf{y} - \hat{\boldsymbol{\beta}}_1^{*\prime}\mathbf{X}_1'\mathbf{y}$ in (8.37) is equal to $\hat{\beta}_k^2/g_{kk}$ in (8.39) (for $j = k$), as noted below (8.39).

8.26 Obtain the confidence interval for $\mathbf{a}'\boldsymbol{\beta}$ in (8.49) from the t statistic in (8.48).

8.27 Show that the confidence interval for $\mathbf{x}_0'\boldsymbol{\beta}$ in (8.52) is the same as that for the centered model in (8.55).

8.28 Show that the confidence interval for $\beta_0 + \beta_1 x_0$ in (8.58) follows from (8.55).

8.29 Show that $t = (y_0 - \hat{y}_0)/s\sqrt{1 + \mathbf{x}_0'(\mathbf{X}'\mathbf{X})^{-1}\mathbf{x}_0}$ in (8.60) is distributed as $t(n - k - 1)$.

8.30 (**a**) Given that $\bar{y}_0 = \sum_i^q = y_{0i}/q$ is the mean of q future observations at $\mathbf{x}_0$, show that a $100(1 - \alpha)\%$ prediction interval for $\bar{y}_0$ is given by $\mathbf{x}_0'\hat{\boldsymbol{\beta}} \pm t_{\alpha/2,\, n-k-1}s\sqrt{1/q + \mathbf{x}_0'(\mathbf{X}'\mathbf{X})^{-1}\mathbf{x}_0}$.

(**b**) Show that for simple linear regression, the prediction interval for $\bar{y}_0$ in part (a) reduces to $\hat{\beta}_0 + \hat{\beta}_1 x_0 \pm t_{\alpha/2,\, n-2}s\sqrt{1/q + 1/n + (x_0 - \bar{x})^2/\sum_{i=1}^n (x_i - \bar{x})^2}$.

8.31 Obtain the confidence interval for σ^2 in (8.65) from the probability statement in (8.64).

8.32 Show that the Scheffé prediction intervals for d future observations are given by (8.73).

8.33 Verify (8.76)–(8.79) in the proof of Theorem 8.7a.

8.34 Verify (8.80), $\partial v/\partial\boldsymbol{\beta} = (2\mathbf{X}'\mathbf{y} - 2\mathbf{X}'\mathbf{X}\boldsymbol{\beta})/2\sigma^2 + \mathbf{C}'\boldsymbol{\lambda}$.

8.35 Show that the solution to (8.80)–(8.82) is given by $\hat{\boldsymbol{\beta}}_0$ and $\hat{\sigma}_0^2$ in (8.83) and (8.84).

8.36 Show that $(\mathbf{y} - \mathbf{X}\hat{\boldsymbol{\beta}}_0)'(\mathbf{y} - \mathbf{X}\hat{\boldsymbol{\beta}}_0) = n\hat{\sigma}^2 + (\mathbf{C}\hat{\boldsymbol{\beta}})'[\mathbf{C}(\mathbf{X}'\mathbf{X})^{-1}\mathbf{C}']^{-1}\mathbf{C}\hat{\boldsymbol{\beta}}$ as in (8.85).

8.37 Use the gas vapor data in Table 7.3.

(**a**) Test the overall regression hypothesis $H_0 : \boldsymbol{\beta}_1 = \mathbf{0}$ using (8.5) [or (8.22)] and (8.23).

(**b**) Test $H_0 : \beta_1 = \beta_3 = 0$, that is, that x_1 and x_3 do not significantly contribute above and beyond x_2 and x_4.

(**c**) Test $H_0 : \beta_j = 0$ for $j = 1, 2, 3, 4$ using t_j in (8.40). Use $t_{.05/2}$ for each test and also use a Bonferroni approach based on $t_{.05/8}$ (or compare the p value to $.05/4$).

(**d**) Using general linear hypothesis tests, test $H_0 : \beta_1 = \beta_2 = 12\beta_3 = 12\beta_4$, $H_{01} : \beta_1 = \beta_2, H_{02} : \beta_2 = 12\beta_3, H_{03} : \beta_3 = \beta_4$, and $H_{04} : \beta_1 = \beta_2$ and $\beta_3 = \beta_4$.

(**e**) Find confidence intervals for $\beta_1, \beta_2, \beta_3$ and β_4 using both (8.47) and (8.67).

8.38 Use the land rent data in Table 7.5.

(**a**) Test the overall regression hypothesis $H_0 : \boldsymbol{\beta}_1 = \mathbf{0}$ using (8.5) [or (8.22)] and (8.23).

(**b**) Test $H_0 : \beta_j = 0$ for $j = 1, 2, 3$ using t_j in (8.40). Use $t_{.05/2}$ for each test and also use a Bonferroni approach based on $t_{.05/6}$ (or compare the p value to $.05/3$).

(**c**) Find confidence intervals for $\beta_1, \beta_2, \beta_3$ using both (8.47) and (8.67).

(**d**) Using (8.52), find a 95% confidence interval for $E(y_0) = \mathbf{x}_0'\boldsymbol{\beta}$, where $\mathbf{x}_0' = (1,15,30,.5)$.

(e) Using (8.61), find a 95% prediction interval for $y_0 = \mathbf{x}_0'\boldsymbol{\beta} + \varepsilon$, where $\mathbf{x}_0' = (1,15,30,.5)$.

8.39 Use y_2 in the chemical reaction data in Table 7.4.

(a) Using (8.52), find a 95% confidence interval for $E(y_0) = \mathbf{x}_0'\boldsymbol{\beta}$, where $\mathbf{x}_0' = (1, 165, 32,5)$.

(b) Using (8.61), find a 95% prediction interval for $y_0 = \mathbf{x}_0'\boldsymbol{\beta} + \varepsilon$, where $\mathbf{x}_0' = (1,165,32,5)$.

(c) Test $H_0{:}2\beta_1 = 2\beta_2 = \beta_3$ using (8.27). (This was done for y_1 in Example 8.4.b.)

8.40 Use y_1 in the chemical reaction data in Table 7.4. The full model with second-order terms and the reduced model with only linear terms were fit in Problem 7.52.

(a) Test $H_0: \beta_4 = \beta_5 = \cdots = \beta_9 = 0$, that is, that the second-order terms are not useful in predicting y_1. (This was done for y_2 in Example 8.2a.)

(b) Test the significance of the increase in R^2 from the reduced model to the full model. (This was done for y^2 in Example 8.3. See Problem 7.52 for values of R^2.)

(c) Find a 95% confidence interval for each of β_0, β_1, β_2, β_3 using (8.47).

(d) Find Bonferroni confidence intervals for β_1, β_2, β_3 using (8.67).

(e) Using (8.52), find a 95% confidence interval for $E(y_0) = \mathbf{x}_0'\boldsymbol{\beta}$, where $\mathbf{x}_0' = (1,165,32,5)$.

(f) Using (8.61), find a 95%, prediction interval for $y_0 = \mathbf{x}_0'\boldsymbol{\beta} + \varepsilon$, where $\mathbf{x}_0' = (1,165,32,5)$.

9 Multiple Regression: Model Validation and Diagnostics

In Sections 7.8.2 and 7.9 we discussed some consequences of misspecification of the model. In this chapter we consider various approaches to checking the model and the attendant assumptions for adequacy and validity. Some properties of the residuals [see (7.11)] and the hat matrix are developed in Sections 9.1 and 9.2. We discuss outliers, the influence of individual observations, and leverage in Sections 9.3 and 9.4.

For additional reading, see Snee (1977), Cook (1977), Belsley et al. (1980), Draper and Smith (1981, Chapter 6), Cook and Weisberg (1982), Beckman and Cook (1983), Weisberg (1985, Chapters 5, 6), Chatterjee and Hadi (1988), Myers (1990, Chapters 5–8), Sen and Srivastava (1990, Chapter 8), Montgomery and Peck (1992, pp. 67–113, 159–192), Jørgensen (1993, Chapter 5), Graybill and Iyer (1994, Chapter 5), Hocking (1996, Chapter 9), Christensen (1996, Chapter 13), Ryan (1997, Chapters 2, 5), Fox (1997, Chapters 11–13) and Kutner et al. (2005, Chapter 10).

9.1 RESIDUALS

The usual model is given by (7.4) as $\mathbf{y} = \mathbf{X}\boldsymbol{\beta} + \boldsymbol{\varepsilon}$ with assumptions $E(\boldsymbol{\varepsilon}) = \mathbf{0}$ and $\text{cov}(\boldsymbol{\varepsilon}) = \sigma^2\mathbf{I}$, where $\mathbf{y}$ is $n \times 1$, $\mathbf{X}$ is $n \times (k+1)$ of rank $k + 1 < n$, and β is $(k+1) \times 1$. The error vector $\boldsymbol{\varepsilon}$ is unobservable unless $\boldsymbol{\beta}$ is known. To estimate $\boldsymbol{\varepsilon}$ for a given sample, we use the residual vector

$$\hat{\boldsymbol{\varepsilon}} = \mathbf{y} - \mathbf{X}\hat{\beta} = \mathbf{y} - \hat{\mathbf{y}} \tag{9.1}$$

as defined in (7.11). The n residuals in (9.1), $\hat{\varepsilon}_1, \hat{\varepsilon}_2, \ldots, \hat{\varepsilon}_n$, are used in various plots and procedures for checking on the validity or adequacy of the model.

We first consider some properties of the residual vector $\hat{\varepsilon}$. Using the least-squares estimator $\hat{\boldsymbol{\beta}} = (\mathbf{X}'\mathbf{X})^{-1}\mathbf{X}'\mathbf{y}$ in (7.6), the vector of predicted values $\hat{\mathbf{y}} = \mathbf{X}\hat{\boldsymbol{\beta}}$ can be written as

$$\begin{aligned}\hat{\mathbf{y}} &= \mathbf{X}\hat{\boldsymbol{\beta}} = \mathbf{X}(\mathbf{X}'\mathbf{X})^{-1}\mathbf{X}'\mathbf{y} \\ &= \mathbf{H}\mathbf{y},\end{aligned} \tag{9.2}$$

where $\mathbf{H} = \mathbf{X}(\mathbf{X}'\mathbf{X})^{-1}\mathbf{X}'$ (see Section 8.2). The $n \times n$ matrix $\mathbf{H}$ is called the *hat matrix* because it transforms $\mathbf{y}$ to $\hat{\mathbf{y}}$. We also refer to $\mathbf{H}$ as a *projection matrix* for essentially the same reason; geometrically it projects $\mathbf{y}$ (perpendicularly) onto $\hat{\mathbf{y}}$ (see Fig. 7.4). The hat matrix $\mathbf{H}$ is symmetric and idempotent (see Problem 5.32a).

Multiplying $\mathbf{X}$ by $\mathbf{H}$, we obtain

$$\mathbf{H}\mathbf{X} = \mathbf{X}(\mathbf{X}'\mathbf{X})^{-1}\mathbf{X}'\mathbf{X} = \mathbf{X}. \tag{9.3}$$

Writing $\mathbf{X}$ in terms of its columns and using (2.28), we can write (9.3) as

$$\mathbf{H}\mathbf{X} = \mathbf{H}(\mathbf{j}, \mathbf{x}_1, \ldots \mathbf{x}_k) = (\mathbf{H}\mathbf{j}, \mathbf{H}\mathbf{x}_1, \ldots, \mathbf{H}\mathbf{x}_k),$$

so that

$$\mathbf{j} = \mathbf{H}\mathbf{j}, \quad \mathbf{x}_i = \mathbf{H}\mathbf{x}_i, \quad i = 1, 2, \ldots, k. \tag{9.4}$$

Using (9.2), the residual vector $\hat{\boldsymbol{\varepsilon}}$ (9.1) can be expressed in terms of $\mathbf{H}$:

$$\begin{aligned}\hat{\boldsymbol{\varepsilon}} &= \mathbf{y} - \hat{\mathbf{y}} = \mathbf{y} - \mathbf{H}\mathbf{y} \\ &= (\mathbf{I} - \mathbf{H})\mathbf{y}.\end{aligned} \tag{9.5}$$

We can rewrite (9.5) to express the residual vector $\hat{\boldsymbol{\varepsilon}}$ in terms of $\boldsymbol{\varepsilon}$:

$$\begin{aligned}\hat{\boldsymbol{\varepsilon}} &= (\mathbf{I} - \mathbf{H})\mathbf{y} = (\mathbf{I} - \mathbf{H})(\mathbf{X}\boldsymbol{\beta} + \boldsymbol{\varepsilon}) \\ &= (\mathbf{X}\boldsymbol{\beta} - \mathbf{H}\mathbf{X}\boldsymbol{\beta}) + (\mathbf{I} - \mathbf{H})\boldsymbol{\varepsilon} \\ &= (\mathbf{X}\boldsymbol{\beta} - \mathbf{X}\boldsymbol{\beta}) + (\mathbf{I} - \mathbf{H})\boldsymbol{\varepsilon} \qquad \text{[by (9.3)]} \\ &= (\mathbf{I} - \mathbf{H})\boldsymbol{\varepsilon}.\end{aligned} \tag{9.6}$$

In terms of the elements h_{ij} of $\mathbf{H}$, we have $\hat{\varepsilon}_i = \varepsilon_i - \sum_{j=1}^{n} h_{ij}\varepsilon_j$, $i = 1, 2, \ldots, n$. Thus, if the h_{ij}'s are small (in absolute value), $\hat{\boldsymbol{\varepsilon}}$ is close to $\boldsymbol{\varepsilon}$.

The following are some of the properties of $\hat{\boldsymbol{\varepsilon}}$ (see Problem 9.1). For the first four, we assume that $E(\mathbf{y}) = \mathbf{X}\boldsymbol{\beta}$ and $\text{cov}(\mathbf{y}) = \sigma^2\mathbf{I}$:

$$E(\hat{\boldsymbol{\varepsilon}}) = \mathbf{0} \tag{9.7}$$

$$\text{cov}(\hat{\boldsymbol{\varepsilon}}) = \sigma^2[\mathbf{I} - \mathbf{X}(\mathbf{X}'\mathbf{X})^{-1}\mathbf{X}'] = \sigma^2(\mathbf{I} - \mathbf{H}) \tag{9.8}$$

$$\text{cov}(\hat{\boldsymbol{\varepsilon}}, \mathbf{y}) = \sigma^2[\mathbf{I} - \mathbf{X}(\mathbf{X}'\mathbf{X})^{-1}\mathbf{X}'] = \sigma^2(\mathbf{I} - \mathbf{H}) \tag{9.9}$$

$$\text{cov}(\hat{\boldsymbol{\varepsilon}}, \hat{\mathbf{y}}) = \mathbf{O} \tag{9.10}$$

$$\bar{\hat{\varepsilon}} = \sum_{i=1}^{n} \hat{\varepsilon}_i/n = \hat{\boldsymbol{\varepsilon}}'\mathbf{j}/n = 0 \tag{9.11}$$

$$\hat{\boldsymbol{\varepsilon}}'\mathbf{y} = \text{SSE} = \mathbf{y}'[\mathbf{I} - \mathbf{X}(\mathbf{X}'\mathbf{X})^{-1}\mathbf{X}']\mathbf{y} = \mathbf{y}'(\mathbf{I} - \mathbf{H})\mathbf{y} \tag{9.12}$$

$$\hat{\boldsymbol{\varepsilon}}'\hat{\mathbf{y}} = 0 \tag{9.13}$$

$$\hat{\boldsymbol{\varepsilon}}'\mathbf{X} = \mathbf{0}' \tag{9.14}$$

In (9.7), the residual vector $\hat{\boldsymbol{\varepsilon}}$ has the same mean as the error term $\boldsymbol{\varepsilon}$, but in (9.8) $\text{cov}(\hat{\boldsymbol{\varepsilon}}) = \sigma^2(\mathbf{I} - \mathbf{H})$ differs from the assumption $\text{cov}(\boldsymbol{\varepsilon}) = \sigma^2\mathbf{I}$. Thus the residuals $\hat{\varepsilon}_1, \hat{\varepsilon}_2, \ldots, \hat{\varepsilon}_n$ are not independent. However, in many cases, especially if n is large, the h_{ij}'s tend to be small (for $i \neq j$), and the dependence shown in $\sigma^2(\mathbf{I} - \mathbf{H})$ does not unduly affect plots and other techniques for model validation. Each $\hat{\varepsilon}_i$ is seen to be correlated with each y_j in (9.9), but in (9.10) the $\hat{\varepsilon}_i$'s are uncorrelated with the $\hat{y}_j$'s.

Some sample properties of the residuals are given in (9.11)–(9.14). The sample mean of the residuals is zero, as shown in (9.11). By (9.12), it can be seen that $\hat{\varepsilon}$ and y are correlated in the sample since $\hat{\boldsymbol{\varepsilon}}'\mathbf{y}$ is the numerator of

$$r_{\hat{\varepsilon}y} = \frac{\hat{\boldsymbol{\varepsilon}}'(\mathbf{y} - \bar{y}\mathbf{j})}{\sqrt{(\hat{\boldsymbol{\varepsilon}}'\hat{\boldsymbol{\varepsilon}})(\mathbf{y} - \bar{y}\mathbf{j})'(\mathbf{y} - \bar{y}\mathbf{j})}} = \frac{\hat{\boldsymbol{\varepsilon}}'\mathbf{y}}{\sqrt{(\hat{\boldsymbol{\varepsilon}}'\hat{\boldsymbol{\varepsilon}})(\mathbf{y} - \bar{y}\mathbf{j})'(\mathbf{y} - \bar{y}\mathbf{j})}}.$$

However, $\hat{\boldsymbol{\varepsilon}}$ and $\hat{\mathbf{y}}$ are orthogonal by (9.13), and therefore

$$r_{\hat{\varepsilon}\hat{y}} = 0. \tag{9.15}$$

Similarly, by (9.14), $\hat{\boldsymbol{\varepsilon}}$ is orthogonal to each column of $\mathbf{X}$ and

$$r_{\hat{\varepsilon}x_i} = 0, \quad i = 1, 2, \ldots, k. \tag{9.16}$$

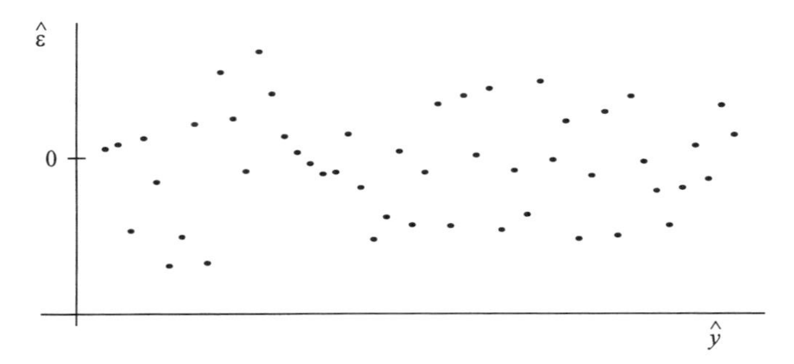

Figure 9.1 Ideal residual plot when model is correct.

If the model and attendant assumptions are correct, then by (9.15), a plot of the residuals versus predicted values, $(\hat{\varepsilon}_1, \hat{y}_1), (\hat{\varepsilon}_2, \hat{y}_2), \ldots, (\hat{\varepsilon}_n, \hat{y}_n)$, should show no systematic pattern. Likewise, by (9.16), the k plots of the residuals versus each of $x_1, x_2, \ldots, x_k$ should show only random variation. These plots are therefore useful for checking the model. A typical plot of this type is shown in Figure 9.1. It may also be useful to plot the residuals on normal probability paper and to plot residuals in time sequence (Christensen 1996, Section 13.2).

If the model is incorrect, various plots involving residuals may show departures from the fitted model such as outliers, curvature, or nonconstant variance. The plots may also suggest remedial measures to improve the fit of the model. For example, the residuals could be plotted versus any of the x_i's, and a simple curved pattern might suggest the addition of x_i^2 to the model. We will consider various approaches for detecting outliers in Section 9.3 and for finding influential observations in Section 9.4. Before doing so, we discuss some properties of the hat matrix in Section 9.2.

9.2 THE HAT MATRIX

It was noted following (9.2) that the hat matrix $\mathbf{H} = \mathbf{X}(\mathbf{X}'\mathbf{X})^{-1}\mathbf{X}'$ is symmetric and idempotent. We now present some additional properties of this matrix. These properties will be useful in the discussion of outliers and influential observations in Sections 9.3 and 9.4.

For the centered model

$$\mathbf{y} = \alpha\mathbf{j} + \mathbf{X}_c\boldsymbol{\beta}_1 + \boldsymbol{\varepsilon} \tag{9.17}$$

in (7.32), $\hat{\mathbf{y}}$ becomes

$$\hat{\mathbf{y}} = \hat{\alpha}\mathbf{j} + \mathbf{X}_c\hat{\boldsymbol{\beta}}_1, \tag{9.18}$$

and the hat matrix is $\mathbf{H}_c = \mathbf{X}_c(\mathbf{X}_c'\mathbf{X}_c)^{-1}\mathbf{X}_c'$, where

$$\mathbf{X}_c = \left(\mathbf{I} - \frac{1}{n}\mathbf{J}\right)\mathbf{X}_1 = \begin{pmatrix} x_{11} - \bar{x}_1 & x_{12} - \bar{x}_2 & \cdots & x_{1k} - \bar{x}_k \\ x_{21} - \bar{x}_1 & x_{22} - \bar{x}_2 & \cdots & x_{2k} - \bar{x}_k \\ \vdots & \vdots & & \vdots \\ x_{n1} - \bar{x}_1 & x_{n2} - \bar{x}_2 & \cdots & x_{nk} - \bar{x}_k \end{pmatrix}.$$

By (7.36) and (7.37), we can write (9.18) as

$$\begin{aligned} \hat{\mathbf{y}} &= \bar{y}\mathbf{j} + \mathbf{X}_c(\mathbf{X}_c'\mathbf{X}_c)^{-1}\mathbf{X}_c'\mathbf{y} = \left(\frac{1}{n}\mathbf{j}'\mathbf{y}\right)\mathbf{j} + \mathbf{H}_c\mathbf{y} \\ &= \left(\frac{1}{n}\mathbf{J} + \mathbf{H}_c\right)\mathbf{y}. \end{aligned} \tag{9.19}$$

Comparing (9.19) and (9.2), we have

$$\mathbf{H} = \frac{1}{n}\mathbf{J} + \mathbf{H}_c = \frac{1}{n}\mathbf{J} + \mathbf{X}_c(\mathbf{X}_c'\mathbf{X}_c)^{-1}\mathbf{X}_c'. \tag{9.20}$$

We now examine some properties of the elements h_{ij} of $\mathbf{H}$.

Theorem 9.2. If $\mathbf{X}$ is $n \times (k+1)$ of rank $k+1 < n$, and if the first column of $\mathbf{X}$ is $\mathbf{j}$, then the elements h_{ij} of $\mathbf{H} = \mathbf{X}(\mathbf{X}'\mathbf{X})^{-1}\mathbf{X}'$ have the following properties:

(i) $(1/n) \leq h_{ii} \leq 1$ for $i = 1, 2, \ldots, n$.
(ii) $-.5 \leq h_{ij} \leq .5$ for all $j \neq i$.
(iii) $h_{ii} = (1/n) + (\mathbf{x}_{1i} - \bar{\mathbf{x}}_1)'(\mathbf{X}_c'\mathbf{X}_c)^{-1}(\mathbf{x}_{1i} - \bar{\mathbf{x}}_1)$, where $\mathbf{x}_{1i}' = (x_{i1}, x_{i2}, \ldots, x_{ik})$, $\bar{\mathbf{x}}_1' = (\bar{x}_1, \bar{x}_2, \ldots, \bar{x}_k)$, and $(\mathbf{x}_{1i} - \bar{\mathbf{x}}_1)'$ is the ith row of the centered matrix $\mathbf{X}_c$.
(iv) $\text{tr}(\mathbf{H}) = \sum_{i=1}^{n} h_{ii} = k + 1$.

PROOF

(i) The lower bound follows from (9.20), since $\mathbf{X}_c'\mathbf{X}_c$ is positive definite. Since $\mathbf{H}$ is symmetric and idempotent, we use the relationship $\mathbf{H} = \mathbf{H}^2$ to find an upper bound on h_{ii}. Let $\mathbf{h}_i'$ be the ith row of $\mathbf{H}$. Then

$$\begin{aligned} h_{ii} &= \mathbf{h}_i'\mathbf{h}_i = (h_{i1}, h_{i2}, \ldots, h_{in}) \begin{pmatrix} h_{i1} \\ h_{i2} \\ \vdots \\ h_{in} \end{pmatrix} = \sum_{j=1}^{n} h_{ij}^2 \\ &= h_{ii}^2 + \sum_{j \neq i} h_{ij}^2. \end{aligned} \tag{9.21}$$

Dividing both sides of (9.21) by h_{ii} [which is positive since $h_{ii} \geq (1/n)$], we obtain

$$1 = h_{ii} + \frac{\sum_{j \neq i} h_{ij}^2}{h_{ii}}, \tag{9.22}$$

which implies $h_{ii} \leq 1$.

(ii) (Chatterjee and Hadi 1988, p. 18.) We can write (9.21) in the form

$$h_{ii} = h_{ii}^2 + h_{ij}^2 + \sum_{r \neq i,j} h_{ir}^2$$

or

$$h_{ii} - h_{ii}^2 = h_{ij}^2 + \sum_{r \neq i,j} h_{ir}^2.$$

Thus, $h_{ij}^2 \leq h_{ii} - h_{ii}^2$, and since the maximum value of $h_{ii} - h_{ii}^2$ is $\frac{1}{4}$, we have $h_{ij}^2 \leq \frac{1}{4}$ for $j \neq i$.

(iii) This follows from (9.20); see Problem 9.2b.

(iv) See Problem 9.2c. □

By Theorem 9.2(iv), we see that as n increases, the values of h_{ii} will tend to decrease.

The function $(\mathbf{x}_{1i} - \bar{\mathbf{x}}_1)'(\mathbf{X}_c'\mathbf{X}_c)^{-1}(\mathbf{x}_{1i} - \bar{\mathbf{x}}_1)$ in Theorem 9.2(iii) is a standardized distance. The standardized distance (Mahalanobis distance) defined in (3.27) is for a population covariance matrix. The matrix $\mathbf{X}_c'\mathbf{X}_c$ is proportional to a sample covariance matrix [see (7.44)]. Thus, $(\mathbf{x}_{1i} - \bar{\mathbf{x}}_1)'(\mathbf{X}_c'\mathbf{X}_c)^{-1}(\mathbf{x}_{1i} - \bar{\mathbf{x}}_1)$ is an estimated standardized distance and provides a good measure of the relative distance of each $\mathbf{x}_{1i}$ from the center of the points as represented by $\bar{\mathbf{x}}_1$.

9.3 OUTLIERS

In some cases, the model appears to be correct for most of the data, but one residual is much larger (in absolute value) than the others. Such an outlier may be due to an error in recording or may be from another population or may simply be an unusual observation from the assumed distribution. For example, if the errors ε_i are distributed as $N(0, \sigma^2)$, a value of ε_i greater than 3σ or less than -3σ would occur with frequency .0027.

If no explanation for an apparent outlier can be found, the dataset could be analyzed both with and without the outlying observation. If the results differ sufficiently to affect the conclusions, then both analyses could be maintained until additional data become available. Another alternative is to discard the outlier, even though no explanation has been found. A third possibility is to use *robust* methods that accommodate the outlying observation (Huber 1973, Andrews 1974, Hampel 1974, Welsch 1975, Devlin et al. 1975, Mosteller and Turkey 1977, Birch 1980, Krasker and Welsch 1982).

One approach to checking for outliers is to plot the residuals $\hat{\varepsilon}_i$ versus $\hat{y}_i$ or versus i, the observation number. In our examination of residuals, we need to keep in mind that by (9.8), the variance of the residuals is not constant:

$$\text{var}(\hat{\varepsilon}_i) = \sigma^2(1 - h_{ii}). \tag{9.23}$$

By Theorem 9.2(i), $h_{ii} \leq 1$; hence, $\text{var}(\hat{\varepsilon}_i)$ will be small if h_{ii} is near 1. By Theorem 9.2(iii), h_{ii} will be large if $\mathbf{x}_{1i}$ is far from $\bar{\mathbf{x}}_1$, where $\mathbf{x}_{1i} = (x_{i1}, x_{i2}, \ldots, x_{ik})'$ and $\bar{\mathbf{x}}_1 = (\bar{x}_1, \bar{x}_2, \ldots, \bar{x}_k)'$. By (9.23), such observations will tend to have small residuals, which seems unfortunate because the model is less likely to hold far from $\bar{\mathbf{x}}_1$. A small residual at a point where $\mathbf{x}_{1i}$ is far from $\bar{\mathbf{x}}_1$ may result because the fitted model will tend to pass close to a point isolated from the bulk of the points, with a resulting poorer fit to the bulk of the data. This may mask an inadequacy of the true model in the region of $\mathbf{x}_{1i}$.

An additional verification that large values of h_{ii} are accompanied by small residuals is provided by the following inequality (see Problem 9.4):

$$\frac{1}{n} \leq h_{ii} + \frac{\hat{\varepsilon}_i^2}{\hat{\boldsymbol{\varepsilon}}'\hat{\boldsymbol{\varepsilon}}} \leq 1. \tag{9.24}$$

For the reasons implicit in (9.23) and (9.24), it is desirable to scale the residuals so that they have the same variance. There are two common (and related) methods of scaling.

For the first method of scaling, we use $\text{var}(\hat{\varepsilon}_i) = \sigma^2(1 - h_{ii})$ in (9.23) to obtain the standardized residuals $\hat{\varepsilon}_i/\sigma\sqrt{1 - h_{ii}}$, which have mean 0 and variance 1. Replacing σ by s yields the *studentized residual*

$$r_i = \frac{\hat{\varepsilon}_i}{s\sqrt{1 - h_{ii}}}, \tag{9.25}$$

where $s^2 = \text{SSE}/(n - k - 1)$ is as defined in (7.24). The use of r_i in place of $\hat{\varepsilon}_i$ eliminates the location effect (due to h_{ii}) on the size of residuals, as discussed following (9.23).

A second method of scaling the residuals uses an estimate of σ that excludes the ith observation

$$t_i = \frac{\hat{\varepsilon}_i}{s_{(i)}\sqrt{1 - h_{ii}}}, \tag{9.26}$$

where $s_{(i)}$ is the standard error computed with the $n - 1$ observations remaining after omitting $(y_i, \mathbf{x}_i') = (y_{i1}, x_{i1}, \ldots, x_{ik})$, in which y_i is the ith element of $\mathbf{y}$ and $\mathbf{x}_i'$ is the ith row of $\mathbf{X}$. If the ith observation is an outlier, it will more likely show up as such with the standardization in (9.26), which is called the *externally studentized residual* or the *studentized deleted residual* or *R student*.

Another option is to examine the *deleted residuals*. The ith deleted residual, $\varepsilon_{(i)}$, is computed with $\hat{\boldsymbol{\beta}}_{(i)}$ on the basis of $n-1$ observations with $(y_i, \mathbf{x}_i')$ deleted:

$$\hat{\varepsilon}_{(i)} = y_i - \hat{y}_{(i)} = y_i - \mathbf{x}_i'\hat{\boldsymbol{\beta}}_{(i)}. \tag{9.27}$$

By definition

$$\hat{\boldsymbol{\beta}}_{(i)} = (\mathbf{X}_{(i)}'\mathbf{X}_{(i)})^{-1}\mathbf{X}_{(i)}'\mathbf{y}_{(i)}, \tag{9.28}$$

where $\mathbf{X}_{(i)}$ is the $(n-1)\times(k+1)$ matrix obtained by deleting $\mathbf{x}_i' = (1, x_{i1}, \ldots, x_{ik})$, the ith row of $\mathbf{X}$, and $\mathbf{y}_{(i)}$ is the corresponding $(n-1)\times 1$ $\mathbf{y}$ vector after deleting y_i. The deleted vector $\hat{\boldsymbol{\beta}}_{(i)}$ can also be found without actually deleting $(y_i, \mathbf{x}_i')$ since

$$\hat{\boldsymbol{\beta}}_{(i)} = \hat{\boldsymbol{\beta}} - \frac{\hat{\varepsilon}_i}{1-h_{ii}}(\mathbf{X}'\mathbf{X})^{-1}\mathbf{x}_i \tag{9.29}$$

(see Problem 9.5).

The deleted residual $\hat{\varepsilon}_{(i)} = y_i - \mathbf{x}_i'\hat{\boldsymbol{\beta}}_{(i)}$ in (9.27) can be expressed in terms of $\hat{\varepsilon}_i$ and h_{ii} as

$$\hat{\varepsilon}_{(i)} = \frac{\hat{\varepsilon}_i}{1-h_{ii}} \tag{9.30}$$

(see Problem 9.6). Thus the n deleted residuals can be obtained without computing n regressions. The scaled residual t_i in (9.26) can be expressed in terms of $\hat{\varepsilon}_{(i)}$ in (9.30) as

$$t_i = \frac{\hat{\varepsilon}_{(i)}}{\sqrt{\widehat{\text{var}}(\varepsilon_{(i)})}} \tag{9.31}$$

(see Problem 9.7).

The deleted sample variance $s_{(i)}^2$ used in (9.26) is defined as $s_{(i)}^2 = \text{SSE}_{(i)}/(n-k-2)$, where $\text{SSE}_{(i)} = \mathbf{y}_{(i)}'\mathbf{y}_{(i)} - \hat{\boldsymbol{\beta}}_{(i)}'\mathbf{X}_{(i)}'\mathbf{y}_{(i)}$. This can be found without excluding the ith observation as

$$s_{(i)}^2 = \frac{\text{SSE}_{(i)}}{n-k-2} = \frac{\text{SSE} - \hat{\varepsilon}_i^2/(1-h_{ii})}{n-k-2} \tag{9.32}$$

(see Problem 9.8).

Another option for outlier detection is to plot the ordinary residuals $\hat{\varepsilon}_i = y_i - \mathbf{x}_i'\hat{\boldsymbol{\beta}}$ against the deleted residuals $\hat{\varepsilon}_{(i)}$ in (9.27) or (9.30). If the fit does not change substantially when the ith observation is deleted in computation of $\hat{\boldsymbol{\beta}}$, the plotted points should approximately follow a straight line with a slope of 1. Any points that are relatively far from this line are potential outliers.

If an outlier is from a distribution with a different mean, the model can be expressed as $E(y_i) = \mathbf{x}_i'\boldsymbol{\beta} + \theta$, where $\mathbf{x}_i'$ is the ith row of $\mathbf{X}$. This is called the *mean-shift outlier model*. The distribution of t_i in (9.26) or (9.31) is $t(n-k-1)$, and t_i can therefore be used in a test of the hypothesis $H_0 : \theta = 0$. Since n tests will be made, a Bonferroni adjustment to the critical values can be used, or we can simply focus on the largest t_i values.

The n deleted residuals in (9.30) can be used for model validation or selection by defining the *prediction sum of squares* (PRESS):

$$\text{PRESS} = \sum_{i=1}^{n} \hat{\varepsilon}_{(i)}^2 = \sum_{i=1}^{n} \left(\frac{\hat{\varepsilon}_i}{1 - h_{ii}} \right)^2. \tag{9.33}$$

Thus, a residual $\hat{\varepsilon}_i$ that corresponds to a large value of h_{ii} contributes more to PRESS. For a given dataset, PRESS may be a better measure than SSE of how well the model will predict future observations. To use PRESS to compare alternative models when the objective is prediction, preference would be shown to models with small values of PRESS.

9.4 INFLUENTIAL OBSERVATIONS AND LEVERAGE

In Section 9.3, we emphasized a search for outliers that did not fit the model. In this section, we consider the effect that deletion of an observation $(y_i, \mathbf{x}_i')$ has on the estimates $\hat{\boldsymbol{\beta}}$ and $\mathbf{X}\hat{\boldsymbol{\beta}}$. An observation that makes a major difference on these estimates is called an *influential observation*. A point $(y_i, \mathbf{x}_i')$ is potentially influential if it is an outlier in the y direction or if it is unusually far removed from the center of the x's.

We illustrate influential observations for the case of one x in Figure 9.2. Points 1 and 3 are extreme in the x direction; points 2 and 3 would likely appear as outliers in the y direction. Even though point 1 is extreme in x, it will not unduly influence the slope or intercept. Point 3 will have a dramatic influence on the slope and intercept since the regression line would pass near point 3. Point 2 is also influential, but much less so than point 3.

Thus, influential points are likely to be found in areas where little or no other data were collected. Such points may be fitted very well, sometimes to the detriment of the fit to the other data.

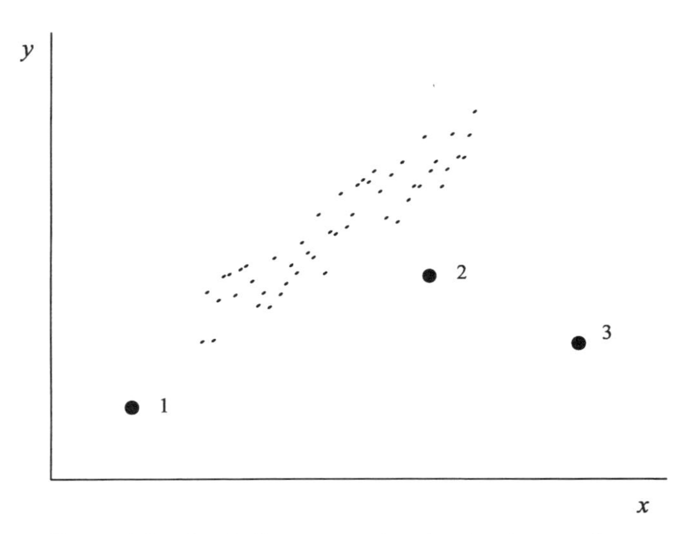

Figure 9.2 Simple linear regression showing three outliers.

To investigate the influence of each observation, we begin with $\hat{\mathbf{y}} = \mathbf{Hy}$ in (9.2), the elements of which are

$$\hat{y}_i = \sum_{j=1}^{n} h_{ij}y_j = h_{ii}y_i + \sum_{j \neq i} h_{ij}y_i. \tag{9.34}$$

By (9.22), if h_{ii} is large (close to 1), then the h'_{ij}s, $j \neq i$, are all small, and y_i contributes much more than the other y's to $\hat{y}_i$. Hence, h_{ii} is called the *leverage* of y_i. Points with high leverage have high potential for influencing regression results. In general, if an observation $(y_i, \mathbf{x}'_i)$ has a value of h_{ii} near 1, then the estimated regression equation will be close to y_i; that is, $\hat{y}_i - y_i$ will be small.

By Theorem 9.2(iv), the average value of the h_{ii}'s is $(k+1)/n$. Hoaglin and Welsch (1978) suggest that a point with $h_{ii} > 2(k+1)/n$ is a high leverage point. Alternatively, we can simply examine any observation whose value of h_{ii} is unusually large relative to the other values of h_{ii}.

In terms of fitting the model to the bulk of the data, high leverage points can be either good or bad, as illustrated by points 1 and 3 in Figure 9.2. Point 1 may reduce the variance of $\hat{\beta}_0$ and $\hat{\beta}_1$. On the other hand, point 3 will drastically alter the fitted model. If point 3 is not the result of a recording error, then the researcher must choose between two competing fitted models. Typically, the model that fits the bulk of the data might be preferred until additional points can be observed in other areas.

To formalize the influence of a point $(y_i, \mathbf{x}'_i)$, we consider the effect of its deletion on $\boldsymbol{\beta}$ and $\hat{\mathbf{y}} = \mathbf{X}\hat{\boldsymbol{\beta}}$. The estimate of $\boldsymbol{\beta}$ obtained by deleting the ith observation $(y_i, \mathbf{x}'_i)$ is defined in (9.28) as $\hat{\boldsymbol{\beta}}_{(i)} = (\mathbf{X}'_{(i)}\mathbf{X}_{(i)})^{-1}\mathbf{X}'_{(i)}\mathbf{y}_{(i)}$. We can compare $\hat{\boldsymbol{\beta}}_{(i)}$ to $\hat{\boldsymbol{\beta}}$ by means of *Cook's distance*, defined as

$$D_i = \frac{(\hat{\boldsymbol{\beta}}_{(i)} - \hat{\boldsymbol{\beta}})'\mathbf{X}'\mathbf{X}(\hat{\boldsymbol{\beta}}_{(i)} - \hat{\boldsymbol{\beta}})}{(k+1)s^2}. \tag{9.35}$$

This can be rewritten as

$$\begin{aligned} D_i &= \frac{(\mathbf{X}\hat{\boldsymbol{\beta}}_{(i)} - \mathbf{X}\hat{\boldsymbol{\beta}})'(\mathbf{X}\hat{\boldsymbol{\beta}}_{(i)} - \mathbf{X}\hat{\boldsymbol{\beta}})}{(k+1)s^2} \\ &= \frac{(\hat{\mathbf{y}}_{(i)} - \hat{\mathbf{y}})'(\hat{\mathbf{y}}_{(i)} - \hat{\mathbf{y}})}{(k+1)s^2}, \end{aligned} \tag{9.36}$$

in which D_i is proportional to the ordinary Euclidean distance between $\hat{\mathbf{y}}_{(i)}$ and $\hat{\mathbf{y}}$. Thus if D_i is large, the observation $(y_i, \mathbf{x}_i')$ has substantial influence on both $\hat{\boldsymbol{\beta}}$ and $\hat{\mathbf{y}}$. A more computationally convenient form of D_i is given by

$$D_i = \frac{r_i^2}{k+1}\left(\frac{h_{ii}}{1-h_{ii}}\right) \tag{9.37}$$

TABLE 9.1 Residuals and Influence Measures for the Chemical Data with Dependent Variable y_1

Observation	y_i	$\hat{y}_i$	$\hat{\varepsilon}_i$	h_{ii}	r_i	t_i	D_i
1	41.5	42.19	−0.688	0.430	−0.394	−0.383	0.029
2	33.8	31.00	2.798	0.310	1.457	1.520	0.239
3	27.7	27.74	−0.042	0.155	−0.020	−0.019	0.000
4	21.7	21.03	0.670	0.139	0.313	0.303	0.004
5	19.9	19.40	0.495	0.129	0.230	0.222	0.002
6	15.0	12.69	2.307	0.140	1.076	1.082	0.047
7	12.2	12.28	−0.082	0.228	−0.040	−0.039	0.000
8	4.3	5.57	−1.270	0.186	−0.609	−0.596	0.021
9	19.3	20.22	−0.917	0.053	−0.408	−0.396	0.002
10	6.4	4.76	1.642	0.233	0.811	0.801	0.050
11	37.6	35.68	1.923	0.240	0.954	0.951	0.072
12	18.0	13.09	4.906	0.164	2.320	2.800	0.264
13	26.3	27.34	−1.040	0.146	−0.487	−0.474	0.010
14	9.9	13.51	−3.605	0.245	−1.795	−1.956	0.261
15	25.0	26.93	−1.929	0.250	−0.964	−0.961	0.077
16	14.1	15.44	−1.342	0.258	−0.674	−0.661	0.039
17	15.2	15.44	−0.242	0.258	−0.121	−0.117	0.001
18	15.9	19.54	−3.642	0.217	−1.780	−1.937	0.220
19	19.6	19.54	0.058	0.217	0.028	0.027	0.000

(see Problem 9.9). Muller and Mok (1997) discuss the distribution of D_i and provide a table of critical values.

Example 9.4. We illustrate several diagnostic tools for the chemical reaction data of Table 7.4 using y_1. In Table 9.1, we give $\hat{\varepsilon}_i$, h_{ii}, and some functions of these from Sections 9.3 and 9.4.

The guideline for h_{ii} in Section 9.4 is $2(k+1)/n = 2(4)/19 = .421$. The only value of h_{ii} that exceeds .421 is the first, $h_{11} = .430$. Thus the first observation has potential for influencing the model fit, but this influence does not appear in $t_1 = -.383$ and $D_1 = .029$. Other relatively large values of h_{ii} are seen for observations 2, 11, 14, 15, 16, and 17. Of these only observation 14 has a very large (absolute) value of t_i. Observation 12 has large values of $\hat{\varepsilon}_i$, r_i, t_i and D_i and is a potentially influential outlier.

The value of PRESS as defined in (9.33) is PRESS = 130.76, which can be compared to SSE = 80.17. □

PROBLEMS

9.1 Verify the following properties of the residual vector $\hat{\boldsymbol{\varepsilon}}$ as given in (9.7)–(9.14):

(**a**) $E(\hat{\boldsymbol{\varepsilon}}) = \mathbf{0}$

(**b**) $\text{cov}(\hat{\boldsymbol{\varepsilon}}) = \sigma^2(\mathbf{I} - \mathbf{H})$

(**c**) $\text{cov}(\hat{\boldsymbol{\varepsilon}}, \mathbf{y}) = \sigma^2(\mathbf{I} - \mathbf{H})$

(**d**) $\text{cov}(\hat{\boldsymbol{\varepsilon}}, \hat{\mathbf{y}}) = \mathbf{O}$

(**e**) $\bar{\hat{\boldsymbol{\varepsilon}}} = \sum_{i=1}^{n} \hat{\varepsilon}_i / n = 0$

(**f**) $\hat{\boldsymbol{\varepsilon}}'\mathbf{y} = \mathbf{y}'(\mathbf{I} - \mathbf{H})\mathbf{y}$

(**g**) $\hat{\boldsymbol{\varepsilon}}'\hat{\mathbf{y}} = 0$

(**h**) $\hat{\boldsymbol{\varepsilon}}'\mathbf{X} = \mathbf{0}'$

9.2 (**a**) In the proof of Theorem 9.2(ii), verify that the maximum value of $h_{ii} - h_{ii}^2$ is $\frac{1}{4}$.

(**b**) Prove Theorem 9.2(iii).

(**c**) Prove Theorem 9.2(iv).

9.3 Show that an alternative expression for h_{ii} in Theorem 9.2(iii) is the following:

$$h_{ii} = \frac{1}{n} + (\mathbf{x}_{1i} - \bar{\mathbf{x}}_1)'(\mathbf{x}_{1i} - \bar{\mathbf{x}}_1) \sum_{r=1}^{k} \frac{1}{\lambda_r} \cos^2 \theta_{ir},$$

where θ_{ir} is the angle between $\mathbf{x}_{1i} - \bar{\mathbf{x}}_1$ and $\mathbf{a}_r$, the rth eigenvector of $\mathbf{X}_c'\mathbf{X}_c$ (Cook and Weisberg 1982, p. 13). Thus h_{ii} is large if $(\mathbf{x}_{1i} - \bar{\mathbf{x}}_1)'(\mathbf{x}_{1i} - \bar{\mathbf{x}}_1)$ is large or if θ_{ir} is small for some r.

9.4 Show that $\frac{1}{n} \leq h_{ii} + \hat{\varepsilon}_i^2/\hat{\boldsymbol{\varepsilon}}'\hat{\boldsymbol{\varepsilon}} \leq 1$ as in (9.24). The following steps are suggested:

(**a**) Let $\mathbf{H}^*$ be the hat matrix corresponding to the augmented matrix $(\mathbf{X}, \mathbf{y})$. Then

$$\mathbf{H}^* = (\mathbf{X},\ \mathbf{y})[(\mathbf{X}, \mathbf{y})'(\mathbf{X}, \mathbf{y})]^{-1}(\mathbf{X}, \mathbf{y})'$$
$$= (\mathbf{X}, \mathbf{y})\begin{pmatrix} \mathbf{X}'\mathbf{X} & \mathbf{X}'\mathbf{y} \\ \mathbf{y}'\mathbf{X} & \mathbf{y}'\mathbf{y} \end{pmatrix}^{-1}\begin{pmatrix} \mathbf{X}' \\ \mathbf{y}' \end{pmatrix}.$$

Use the inverse of a partitioned matrix in (2.50) with $\mathbf{A}_{11} = \mathbf{X}'\mathbf{X}$, $\mathbf{a}_{12} = \mathbf{X}'\mathbf{y}$, and $a_{22} = \mathbf{y}'\mathbf{y}$ to obtain

$$\mathbf{H}^* = \mathbf{H} + \frac{1}{b}[\mathbf{X}(\mathbf{X}'\mathbf{X})^{-1}\mathbf{X}'\mathbf{y}\mathbf{y}'\mathbf{X}(\mathbf{X}'\mathbf{X})^{-1}\mathbf{X}' - \mathbf{y}\mathbf{y}'\mathbf{X}(\mathbf{X}'\mathbf{X})^{-1}\mathbf{X}'$$
$$- \mathbf{X}(\mathbf{X}'\mathbf{X})^{-1}\mathbf{X}'\mathbf{y}\mathbf{y}' + \mathbf{y}\mathbf{y}']$$
$$= \mathbf{H} + \frac{1}{b}[\mathbf{H}\mathbf{y}\mathbf{y}'\mathbf{H} - \mathbf{y}\mathbf{y}'\mathbf{H} - \mathbf{H}\mathbf{y}\mathbf{y}' + \mathbf{y}\mathbf{y}'],$$

where $b = \mathbf{y}'\mathbf{y} - \mathbf{y}'\mathbf{X}(\mathbf{X}'\mathbf{X})^{-1}\mathbf{X}'\mathbf{y}$.

(**b**) Show that the above expression factors into

$$\mathbf{H}^* = \mathbf{H} + \frac{(\mathbf{I} - \mathbf{H})\mathbf{y}\mathbf{y}'(\mathbf{I} - \mathbf{H})}{\mathbf{y}'(\mathbf{I} - \mathbf{H})\mathbf{y}} = \mathbf{H} + \frac{\hat{\boldsymbol{\varepsilon}}\hat{\boldsymbol{\varepsilon}}'}{\hat{\boldsymbol{\varepsilon}}'\hat{\boldsymbol{\varepsilon}}},$$

which gives $h_{ii}^* = h_{ii} + \hat{\varepsilon}_i^2/\hat{\boldsymbol{\varepsilon}}'\hat{\boldsymbol{\varepsilon}}$.

(**c**) The proof is easily completed by noting that $\mathbf{H}^*$ is a hat matrix and therefore $(1/n) \leq h_{ii}^* \leq 1$ by Theorem 9.2(i).

9.5 Show that $\hat{\boldsymbol{\beta}}_{(i)} = \hat{\boldsymbol{\beta}} - \hat{\varepsilon}_i(\mathbf{X}'\mathbf{X})^{-1}\mathbf{x}_i/(1 - h_{ii})$ as in (9.29). The following steps are suggested:

(**a**) Show that $\mathbf{X}'\mathbf{X} = \mathbf{X}_{(i)}'\mathbf{X}_{(i)} + \mathbf{x}_i\mathbf{x}_i'$ and that $\mathbf{X}'\mathbf{y} = \mathbf{X}_{(i)}'\mathbf{y}_{(i)} + \mathbf{x}_i y_i$.

(**b**) Show that $(\mathbf{X}'\mathbf{X})^{-1}\mathbf{X}_{(i)}'\mathbf{y}_{(i)} = \hat{\boldsymbol{\beta}} - (\mathbf{X}'\mathbf{X})^{-1}\mathbf{x}_i y_i$.

(**c**) Using the following adaptation of (2.53)

$$(\mathbf{B} - \mathbf{c}\mathbf{c}')^{-1} = \mathbf{B}^{-1} + \frac{\mathbf{B}^{-1}\mathbf{c}\mathbf{c}'\mathbf{B}^{-1}}{1 - \mathbf{c}'\mathbf{B}^{-1}\mathbf{c}}.$$

show that

$$\hat{\boldsymbol{\beta}}_{(i)} = \left[(\mathbf{X}'\mathbf{X})^{-1} + \frac{(\mathbf{X}'\mathbf{X})^{-1}\mathbf{x}_i\mathbf{x}_i'(\mathbf{X}'\mathbf{X})^{-1}}{1 - h_{ii}}\right]\mathbf{X}_{(i)}'\mathbf{y}_{(i)}.$$

(d) Using the result of parts (b) and (c), show that

$$\hat{\boldsymbol{\beta}}_{(i)} = \hat{\boldsymbol{\beta}} - \frac{\hat{\varepsilon}_i}{1 - h_{ii}} (\mathbf{X}'\mathbf{X})^{-1}\mathbf{x}_i.$$

9.6 Show that $\hat{\varepsilon}_{(i)} = \hat{\varepsilon}_i/(1 - h_{ii})$ as in (9.30).

9.7 Show that $t_i = \hat{\varepsilon}_{(i)}/\sqrt{\widehat{\text{var}}(\hat{\varepsilon}_{(i)})}$ in (9.31) is the same as $t_i = \hat{\varepsilon}_i/s_{(i)}\sqrt{1 - h_{ii}}$ in (9.26). The following steps are suggested:

(a) Using $\hat{\varepsilon}_{(i)} = \hat{\varepsilon}_i/(1 - h_{ii})$ in (9.30), show that $\text{var}(\hat{\varepsilon}_{(i)}) = \sigma^2/(1 - h_{ii})$.

(b) If $\text{var}(\hat{\varepsilon}_{(i)})$ in part (a) is estimated by $\widehat{\text{var}}(\hat{\varepsilon}_{(i)}) = s^2_{(i)}/(1 - h_{ii})$, show that $\hat{\varepsilon}_{(i)}/\sqrt{\widehat{\text{var}}(\varepsilon_{(i)})} = \hat{\varepsilon}_i/s_{(i)}\sqrt{1 - h_{ii}}$.

9.8 Show that $\text{SSE}_{(i)} = \mathbf{y}'_{(i)}\mathbf{y}_{(i)} - \mathbf{y}'_{(i)}\mathbf{X}_{(i)}\hat{\boldsymbol{\beta}}_{(i)}$ can be written in the form

$$\text{SSE}_{(i)} = \text{SSE} - \hat{\varepsilon}_i^2/(1 - h_{ii})$$

as in (9.32). One way to do this is as follows:

(a) Show that $\mathbf{y}'_{(i)}\mathbf{y}_{(i)} = \mathbf{y}'\mathbf{y} - y_i^2$.

(b) Using Problem 9.5a,d, we have

$$\mathbf{y}'_{(i)}\mathbf{X}_{(i)}\hat{\boldsymbol{\beta}}_{(i)} = (\mathbf{y}'\mathbf{X} - y_i\mathbf{x}'_i)\left[\hat{\boldsymbol{\beta}} - \frac{\hat{\varepsilon}_i}{1 - h_{ii}}(\mathbf{X}'\mathbf{X})^{-1}\mathbf{x}_i\right].$$

Show that this can be written as

$$\mathbf{y}'_{(i)}\mathbf{X}_{(i)}\hat{\boldsymbol{\beta}}_{(i)} = \mathbf{y}'\mathbf{X}\hat{\boldsymbol{\beta}} - y_i^2 + \frac{\hat{\varepsilon}_i^2}{1 - h_{ii}}.$$

(c) Show that

$$\text{SSE}_{(i)} = \text{SSE} - \hat{\varepsilon}_i^2/(1 - h_{ii}).$$

9.9 Show that $D_i = r_i^2 h_{ii}/(k + 1)(1 - h_{ii})$ in (9.37) is the same as D_i in (9.35). This may be done by substituting (9.29) into (9.35).

9.10 For the gas vapor data in Table 7.3, compute the diagnostic measures $\hat{y}_i$, $\hat{\varepsilon}_i$, h_{ii}, r_i, t_i, and D_i. Display these in a table similar to Table 9.1. Are there outliers or potentially influential observations? Calculate PRESS and compare to SSE.

9.11 For the land rent data in Table 7.5, compute the diagnostic measures $\hat{y}_i$, $\hat{\varepsilon}_i$, h_{ii}, r_i, t_i, and D_i. Display these in a table similar to Table 9.1. Are there outliers or potentially influential observations? Calculate PRESS and compare to SSE.

9.12 For the chemical reaction data of Table 7.4 with dependent variable y_2, compute the diagnostic measures $\hat{y}_i$, $\hat{\varepsilon}_i$, h_{ii}, r_i, t_i, and D_i. Display these in a table similar to Table 9.1. Are there outliers or potentially influential observations? Calculate PRESS and compare to SSE.

10 Multiple Regression: Random x's

Throughout Chapters 7–9 we assumed that the x variables were fixed; that is, that they remain constant in repeated sampling. However, in many regression applications, they are random variables. In this chapter we obtain estimators and test statistics for a regression model with random x variables. Many of these estimators and test statistics are the same as those for fixed x's, but their properties are somewhat different.

In the random-x case, $k + 1$ variables $y, x_1, x_2, \ldots, x_k$ are measured on each of the n subjects or experimental units in the sample. These n observation vectors yield the data

$$\begin{array}{ccccc} y_1 & x_{11} & x_{12} & \cdots & x_{1k} \\ y_2 & x_{21} & x_{22} & \cdots & x_{2k} \\ \vdots & \vdots & \vdots & & \vdots \\ y_n & x_{n1} & x_{n2} & \cdots & x_{nk}. \end{array} \tag{10.1}$$

The rows of this array are random vectors of the second type described in Section 3.1. The variables $y, x_1, x_2, \ldots, x_k$ in a row are typically correlated and have different variances; that is, for the random vector $(y, x_1, \ldots, x_k) = (y, \mathbf{x}')$, we have

$$\text{cov}\begin{pmatrix} y \\ x_1 \\ \vdots \\ x_k \end{pmatrix} = \text{cov}\begin{pmatrix} y \\ \mathbf{x} \end{pmatrix} = \boldsymbol{\Sigma},$$

where $\boldsymbol{\Sigma}$ is not a diagonal matrix. The vectors themselves [rows of the array in (10.1)] are ordinarily mutually independent (uncorrelated) if they arise from a random sample.

In Sections 10.1–10.5 we assume that y and the x variables have a multivariate normal distribution. Many of the results in Sections 10.6–10.8 do not require a normality assumption.

10.1 MULTIVARIATE NORMAL REGRESSION MODEL

The estimation and testing results in Sections 10.1–10.5 are based on the assumption that $(y, x_1, \ldots, x_k) = (y, \mathbf{x}')$ is distributed as $N_{k+1}(\boldsymbol{\mu}, \boldsymbol{\Sigma})$ with

$$\boldsymbol{\mu} = \begin{pmatrix} \mu_y \\ \hline \mu_1 \\ \vdots \\ \mu_k \end{pmatrix} = \begin{pmatrix} \mu_y \\ \boldsymbol{\mu}_x \end{pmatrix}$$

$$\boldsymbol{\Sigma} = \left(\begin{array}{c|ccc} \sigma_{yy} & \sigma_{y1} & \cdots & \sigma_{yk} \\ \hline \sigma_{1y} & \sigma_{11} & \cdots & \sigma_{1k} \\ \vdots & \vdots & & \vdots \\ \sigma_{ky} & \sigma_{k1} & \cdots & \sigma_{kk} \end{array}\right) = \begin{pmatrix} \sigma_{yy} & \boldsymbol{\sigma}'_{yx} \\ \boldsymbol{\sigma}_{yx} & \boldsymbol{\Sigma}_{xx} \end{pmatrix}, \tag{10.3}$$

where $\boldsymbol{\mu}_x$ is the mean vector for the x's, $\boldsymbol{\sigma}_{yx}$ is the vector of covariances between y and the x's, and $\boldsymbol{\Sigma}_{xx}$ is the covariance matrix for the x's.

From Corollary 1 to Theorem 4.4d, we have

$$E(y|\mathbf{x}) = \mu_y + \boldsymbol{\sigma}'_{yx}\boldsymbol{\Sigma}_{xx}^{-1}(\mathbf{x} - \boldsymbol{\mu}_x) \tag{10.4}$$

$$= \beta_0 + \boldsymbol{\beta}'_1\mathbf{x}, \tag{10.5}$$

where

$$\beta_0 = \mu_y - \boldsymbol{\sigma}'_{yx}\boldsymbol{\Sigma}_{xx}^{-1}\boldsymbol{\mu}_x, \tag{10.6}$$

$$\boldsymbol{\beta}_1 = \boldsymbol{\Sigma}_{xx}^{-1}\boldsymbol{\sigma}_{yx}. \tag{10.7}$$

From Corollary 1 to Theorem 4.4d, we also obtain

$$\mathrm{var}(y|\mathbf{x}) = \sigma_{yy} - \boldsymbol{\sigma}'_{yx}\boldsymbol{\Sigma}_{xx}^{-1}\boldsymbol{\sigma}_{yx} = \sigma^2. \tag{10.8}$$

The mean, $E(y|\mathbf{x}) = \mu_y + \boldsymbol{\sigma}'_{yx}\boldsymbol{\Sigma}_{xx}^{-1}(\mathbf{x} - \boldsymbol{\mu}_x)$, is a linear function of $\mathbf{x}$, but the variance, $\sigma^2 = \sigma_{yy} - \boldsymbol{\sigma}'_{yx}\boldsymbol{\Sigma}_{xx}^{-1}\boldsymbol{\sigma}_{yx}$, is not a function $\mathbf{x}$. Thus under the multivariate normal assumption, (10.4) and (10.8) provide a linear model with constant variance, which is analogous to the fixed-x case. Note, however, that $E(y|\mathbf{x}) = \beta_0 + \boldsymbol{\beta}_1'\mathbf{x}$ in (10.5) does not allow for curvature such as $E(y) = \beta_0 + \beta_1 x + \beta_2 x^2$. Thus $E(y|\mathbf{x}) = \beta_0 + \boldsymbol{\beta}_1'\mathbf{x}$ represents a model that is linear in the x's as well as the β's. This differs from the linear model in the fixed-x case, which requires only linearity in the β's.

10.2 ESTIMATION AND TESTING IN MULTIVARIATE NORMAL REGRESSION

Before obtaining estimators of β_0, $\boldsymbol{\beta}_1$, and σ^2 in (10.6)–(10.8), we must first estimate $\boldsymbol{\mu}$ and $\boldsymbol{\Sigma}$. Maximum likelihood estimators of $\boldsymbol{\mu}$ and $\boldsymbol{\Sigma}$ are given in the following theorem.

Theorme 10.2a. If $(y_1, \mathbf{x}_1'), (y_2, \mathbf{x}_2'), \ldots, (y_n, \mathbf{x}_n')$ [rows of the array in (10.1)] is a random sample from $N_{k+1}(\boldsymbol{\mu}, \boldsymbol{\Sigma})$, with $\boldsymbol{\mu}$ and $\boldsymbol{\Sigma}$ as given in (10.2) and (10.3), the maximum likelihood estimators are

$$\hat{\boldsymbol{\mu}} = \begin{pmatrix} \hat{\mu}_y \\ \hat{\boldsymbol{\mu}}_x \end{pmatrix} = \begin{pmatrix} \bar{y} \\ \bar{\mathbf{x}} \end{pmatrix}, \tag{10.9}$$

$$\hat{\boldsymbol{\Sigma}} = \frac{n-1}{n}\mathbf{S} = \frac{n-1}{n}\begin{pmatrix} s_{yy} & \mathbf{s}_{yx}' \\ \mathbf{s}_{yx} & \mathbf{S}_{xx} \end{pmatrix}, \tag{10.10}$$

where the partitioning of $\hat{\boldsymbol{\mu}}$ and $\mathbf{S}$ is analogous to the partitioning of $\boldsymbol{\mu}$ and $\boldsymbol{\Sigma}$ in (10.2) and (10.3). The elements of the sample covariance matrix $\mathbf{S}$ are defined in (7.40) and in (10.14).

PROOF. Denote $(y_i, \mathbf{x}_i')$ by $\mathbf{v}_i'$, $i = 1, 2, \ldots, n$. As noted below (10.1), $\mathbf{v}_1, \mathbf{v}_2, \ldots, \mathbf{v}_n$ are independent because they arise from a random sample. The likelihood function (joint density) is therefore given by the product

$$\begin{aligned} L(\boldsymbol{\mu}, \boldsymbol{\Sigma}) &= \prod_{i=1}^{n} f(\mathbf{v}_i; \boldsymbol{\mu}, \boldsymbol{\Sigma}) \\ &= \prod_{i=1}^{n} \frac{1}{(\sqrt{2\pi})^{k+1}|\boldsymbol{\Sigma}|^{1/2}} e^{-(\mathbf{v}_i-\boldsymbol{\mu})'\boldsymbol{\Sigma}^{-1}(\mathbf{v}_i-\boldsymbol{\mu})/2} \\ &= \frac{1}{(\sqrt{2\pi})^{n(k+1)}|\boldsymbol{\Sigma}|^{n/2}} e^{-\sum_{i=1}^{n}(\mathbf{v}_i-\boldsymbol{\mu})'\boldsymbol{\Sigma}^{-1}(\mathbf{v}_i-\boldsymbol{\mu})/2}. \end{aligned} \tag{10.11}$$

Note that $L(\boldsymbol{\mu}, \boldsymbol{\Sigma}) = \prod_{i=1}^{n} f(\mathbf{v}_i; \boldsymbol{\mu}, \boldsymbol{\Sigma})$ is a product of n multivariate normal densities, each involving $k+1$ random variables. Thus there are $n(k+1)$ random variables as compared to the likelihood $L(\boldsymbol{\beta}, \sigma^2)$ in (7.50) that involves n random variables $y_1, y_2, \ldots, y_n$ [the x's are fixed in (7.50)].

To find the maximum likelihood estimator for $\boldsymbol{\mu}$, we expand and sum the exponent in (10.11) and then take the logarithm to obtain

$$\ln L(\boldsymbol{\mu}, \boldsymbol{\Sigma}) = -n(k+1)\ln\sqrt{2\pi} - \frac{n}{2}\ln|\boldsymbol{\Sigma}| - \frac{1}{2}\sum_i \mathbf{v}_i'\boldsymbol{\Sigma}^{-1}\mathbf{v}_i + \boldsymbol{\mu}'\boldsymbol{\Sigma}^{-1}\sum_i \mathbf{v}_i - \frac{n}{2}\boldsymbol{\mu}'\boldsymbol{\Sigma}^{-1}\boldsymbol{\mu}. \tag{10.12}$$

Differentiating (10.12) with respect to $\boldsymbol{\mu}$ using (2.112) and (2.113) and setting the result equal to $\mathbf{0}$, we obtain

$$\frac{\partial \ln L(\boldsymbol{\mu}, \boldsymbol{\Sigma})}{\partial \boldsymbol{\mu}} = -\mathbf{0} - \mathbf{0} - \mathbf{0} + \boldsymbol{\Sigma}^{-1}\sum_i \mathbf{v}_i - \frac{2n}{2}\boldsymbol{\Sigma}^{-1}\boldsymbol{\mu} = \mathbf{0},$$

which gives

$$\hat{\boldsymbol{\mu}} = \frac{1}{n}\sum_{i=1}^{n} \mathbf{v}_i = \bar{\mathbf{v}} = \begin{pmatrix} \bar{y} \\ \bar{\mathbf{x}} \end{pmatrix},$$

where $\bar{\mathbf{x}} = (\bar{x}_1, \bar{x}_2, \ldots, \bar{x}_k)'$ is the vector of sample means of the x's. To find the maximum likelihood estimator of $\boldsymbol{\Sigma}$, we rewrite the exponent of (10.11) and then take the logarithm to obtain

$$\begin{aligned}
\ln L(\boldsymbol{\mu}, \boldsymbol{\Sigma}^{-1}) &= -n(k+1)\ln\sqrt{2\pi} + \frac{n}{2}\ln|\boldsymbol{\Sigma}^{-1}| - \frac{1}{2}\sum_i (\mathbf{v}_i - \bar{\mathbf{v}})'\boldsymbol{\Sigma}^{-1}(\mathbf{v}_i - \bar{\mathbf{v}}) \\
&\quad - \frac{n}{2}(\bar{\mathbf{v}} - \boldsymbol{\mu})'\boldsymbol{\Sigma}^{-1}(\bar{\mathbf{v}} - \boldsymbol{\mu}) \\
&= -n(k+1)\ln\sqrt{2\pi} + \frac{n}{2}\ln|\boldsymbol{\Sigma}^{-1}| - \frac{1}{2}\mathrm{tr}\left[\boldsymbol{\Sigma}^{-1}\sum_i (\mathbf{v}_i - \bar{\mathbf{v}})(\mathbf{v}_i - \bar{\mathbf{v}})'\right] \\
&\quad - \frac{n}{2}\mathrm{tr}[\boldsymbol{\Sigma}^{-1}(\bar{\mathbf{v}} - \boldsymbol{\mu})(\bar{\mathbf{v}} - \boldsymbol{\mu})'].
\end{aligned}$$

Differentiating this with respect to $\boldsymbol{\Sigma}^{-1}$ using (2.115) and (2.116), and setting the result equal to $\mathbf{0}$, we obtain

$$\begin{aligned}
\frac{\partial \ln L(\boldsymbol{\mu}, \boldsymbol{\Sigma}^{-1})}{\partial \boldsymbol{\Sigma}^{-1}} &= n\boldsymbol{\Sigma} - \frac{n}{2}\mathrm{diag}(\boldsymbol{\Sigma}) - \sum_i (\mathbf{v}_i - \bar{\mathbf{v}})(\mathbf{v}_i - \bar{\mathbf{v}})' + \frac{1}{2}\mathrm{diag}\left[\sum_i (\mathbf{v}_i - \bar{\mathbf{v}})(\mathbf{v}_i - \bar{\mathbf{v}})\right] \\
&\quad - n(\bar{\mathbf{v}} - \boldsymbol{\mu})(\bar{\mathbf{v}} - \boldsymbol{\mu})' + \frac{n}{2}\mathrm{diag}[(\bar{\mathbf{v}} - \boldsymbol{\mu})(\bar{\mathbf{v}} - \boldsymbol{\mu})'] = \mathbf{0}.
\end{aligned}$$

Since $\hat{\boldsymbol{\mu}} = \bar{\mathbf{v}}$, the last two terms disappear and we obtain

$$\hat{\boldsymbol{\Sigma}} = \frac{1}{n}\sum_{i=1}^{n}(\mathbf{v}_i - \bar{\mathbf{v}})(\mathbf{v}_i - \bar{\mathbf{v}})' = \frac{n-1}{n}\mathbf{S}. \tag{10.13}$$

See Problem 10.1 for verification that $\sum_i (\mathbf{v}_i - \bar{\mathbf{v}})(\mathbf{v}_i - \bar{\mathbf{v}})' = (n-1)\mathbf{S}$. □

In partitioned form, the sample covariance matrix $\mathbf{S}$ can be written as in (10.10)

$$\mathbf{S} = \begin{pmatrix} s_{yy} & \mathbf{s}'_{yx} \\ \mathbf{s}_{yx} & \mathbf{S}_{xx} \end{pmatrix} = \left(\begin{array}{c|ccc} s_{yy} & s_{y1} & \cdots & s_{yk} \\ \hline s_{1y} & s_{11} & \cdots & s_{1k} \\ \vdots & \vdots & & \vdots \\ s_{ky} & s_{k1} & \cdots & s_{kk} \end{array}\right), \tag{10.14}$$

where $\mathbf{s}_{yx}$ is the vector of sample covariances between y and the x's and $\mathbf{S}_{xx}$ is the sample covariance matrix for the x's. For example

$$s_{y1} = \frac{\sum_{i=1}^{n}(y_i - \bar{y})(x_{i1} - \bar{x}_1)}{n-1},$$

$$s_{11} = \frac{\sum_{i=1}^{n}(x_{i1} - \bar{x}_1)^2}{n-1},$$

$$s_{12} = \frac{\sum_{i=1}^{n}(x_{i1} - \bar{x}_1)(x_{i2} - \bar{x}_2)}{(n-1)}$$

[see (7.41)–(7.43)]. By (5.7), $E(s_{yy}) = \sigma_{yy}$ and $E(s_{jj}) = \sigma_{jj}$. By (5.17), $E(s_{yj}) = \sigma_{yj}$ and $E(s_{ij}) = \sigma_{ij}$. Thus $E(\mathbf{S}) = \boldsymbol{\Sigma}$, where $\boldsymbol{\Sigma}$ is given in (10.3). The maximum likelihood estimator $\hat{\boldsymbol{\Sigma}} = (n-1)\mathbf{S}/n$ is therefore biased.

In order to find maximum likelihood estimators of $\boldsymbol{\beta}_0$, $\boldsymbol{\beta}_1$, and σ^2 we first note the *invariance property* of maximum likelihood estimators.

Theorem 10.2b. The maximum likelihood estimator of a function of one or more parameters is the same function of the corresponding estimators; that is, if $\hat{\boldsymbol{\theta}}$ is the maximum likelihood estimator of the vector or matrix of parameters $\boldsymbol{\theta}$, then $g(\hat{\boldsymbol{\theta}})$ is the maximum likelihood estimator of $g(\boldsymbol{\theta})$.

PROOF. See Hogg and Craig (1995, p. 265). □

Example 10.2. We illustrate the use of the invariance property in Theorem 10.2b by showing that the sample correlation matrix $\mathbf{R}$ is the maximum likelihood estimator of the population correlation matrix $\mathbf{P}_\rho$ when sampling from the multivariate normal distribution. By (3.30), the relationship between $\mathbf{P}_\rho$ and $\boldsymbol{\Sigma}$ is given by $\mathbf{P}_\rho = \mathbf{D}_\sigma^{-1}\boldsymbol{\Sigma}\mathbf{D}_\sigma^{-1}$, where $\mathbf{D}_\sigma = [\text{diag}(\boldsymbol{\Sigma})]^{1/2}$, so that

$$\mathbf{D}_\sigma^{-1} = \text{diag}\left(\frac{1}{\sqrt{\sigma_{11}}}, \frac{1}{\sqrt{\sigma_{22}}}, \ldots, \frac{1}{\sqrt{\sigma_{pp}}}\right).$$

The maximum likelihood estimator of $1/\sqrt{\sigma_{jj}}$ is $1/\sqrt{\hat{\sigma}_{jj}}$, where $\hat{\sigma}_{jj} = (1/n)\Sigma_{i=1}^n (y_{ij} - \bar{y}_j)^2$. Thus $\hat{\mathbf{D}}_\sigma^{-1} = \text{diag}(1/\sqrt{\hat{\sigma}_{11}}, 1/\sqrt{\hat{\sigma}_{22}}, \ldots, 1/\sqrt{\hat{\sigma}_{pp}}$, and we obtain

$$\begin{aligned}
\hat{\mathbf{P}}_\rho = \hat{\mathbf{D}}_\sigma^{-1}\hat{\boldsymbol{\Sigma}}\hat{\mathbf{D}}_\sigma^{-1} &= \left(\frac{\hat{\sigma}_{jk}}{\sqrt{\hat{\sigma}_{jj}}\sqrt{\hat{\sigma}_{kk}}}\right) \\
&= \left(\frac{\sum_i (y_{ij} - \bar{y}_j)(y_{ik} - \bar{y}_k)/n}{\sqrt{\sum_i (y_{ij} - \bar{y}_j)^2/n}\sqrt{\sum_i (y_{ik} - \bar{y}_k)^2/n}}\right) \\
&= \left(\frac{\sum_i (y_{ij} - \bar{y}_j)(y_{ik} - \bar{y}_k)}{\sqrt{\sum_i (y_{ij} - \bar{y}_j)^2}\sqrt{\sum_i (y_{ik} - \bar{y}_k)^2}}\right) \\
&= (r_{jk}) = \mathbf{R}.
\end{aligned}$$

□

Maximum likelihood estimators of β_0, $\boldsymbol{\beta}_1$, and σ^2 are now given in the following theorem.

Theorem 10.2c. If $(y_1, \mathbf{x}_1'), (y_2, \mathbf{x}_2'), \ldots, (y_n, \mathbf{x}_n')$, is a random sample from $N_{k+1}(\boldsymbol{\mu}, \boldsymbol{\Sigma})$, where $\boldsymbol{\mu}$ and $\boldsymbol{\Sigma}$ are given by (10.2) and (10.3), the maximum likelihood estimators for β_0, $\boldsymbol{\beta}_1$, and σ^2 in (10.6)–(10.8) are as follows:

$$\hat{\beta}_0 = \bar{y} - \mathbf{s}_{yx}'\mathbf{S}_{xx}^{-1}\bar{\mathbf{x}}, \tag{10.15}$$

$$\hat{\boldsymbol{\beta}}_1 = \mathbf{S}_{xx}^{-1}\mathbf{s}_{yx}, \tag{10.16}$$

$$\hat{\sigma}^2 = \frac{n-1}{n}s^2 \quad \text{where} \quad s^2 = s_{yy} - \mathbf{s}_{yx}'\mathbf{S}_{xx}^{-1}\mathbf{s}_{yx}. \tag{10.17}$$

The estimator s^2 is a bias-corrected estimator of σ^2.

Proof. By the invariance property of maximum likelihood estimators (Theorem 10.2b), we insert (10.9) and (10.10) into (10.6), (10.7), and (10.8) to obtain the desired results (using the unbiased estimator $\mathbf{S}$ in place of $\hat{\boldsymbol{\Sigma}}$). □

The estimators $\hat{\beta}_0$, $\hat{\boldsymbol{\beta}}_1$, and s^2 have a minimum variance property analogous to that of the corresponding estimators for the case of normal y's and fixed x's in Theorem 7.6d. It can be shown that $\hat{\boldsymbol{\mu}}$ and $\mathbf{S}$ in (10.9) and (10.10) are jointly sufficient for $\boldsymbol{\mu}$ and $\boldsymbol{\Sigma}$ (see Problem 10.2). Then, with some additional properties that can be demonstrated, it follows that $\hat{\beta}_0$, $\hat{\boldsymbol{\beta}}_1$, and s^2 are minimum variance unbiased estimators for β_0, $\boldsymbol{\beta}_1$, and σ^2 (Graybill 1976, p. 380).

The maximum likelihood estimators $\hat{\beta}_0$ and $\hat{\boldsymbol{\beta}}_1$ in (10.15) and (10.16) are the same algebraic functions of the observations as the least-squares estimators given in (7.47) and (7.46) for the fixed-x case. The estimators in (10.15) and (10.16) are also identical to the maximum likelihood estimators for normal y's and fixed x's in Section 7.6.2 (see Problem 7.17). However, even though the estimators in the random-x case and fixed-x case are the same, their distributions differ. When y and the x's are multivariate normal, $\hat{\boldsymbol{\beta}}_1$ does not have a multivariate normal distribution as it does in the fixed-x case with normal y's [Theorem 7.6b(i)]. For large n, the distribution is similar to the multivariate normal, but for small n, the distribution has heavier tails than the multivariate normal.

In spite of the nonnormality of $\hat{\boldsymbol{\beta}}_1$ in the random-x model, the F tests and t tests and associated confidence regions and intervals of Chapter 8 (fixed-x model) are still appropriate. To see this, note that since the conditional distribution of y for a given value of $\mathbf{x}$ is normal (Corollary 1 to Theorem 4.4d), the conditional distribution of the vector of observations $\mathbf{y} = (y_1, y_2, \ldots, y_n)'$ for a given value of the $\mathbf{X}$ matrix is multivariate normal. Therefore, a test statistic such as (8.35) is distributed conditionally as an F for the given value of $\mathbf{X}$ when H_0 is true. However, the central F distribution depends only on degrees of freedom; it does not depend on $\mathbf{X}$. Thus under H_0, the statistic has (unconditionally) an F distribution for all values of $\mathbf{X}$, and so tests can be carried out exactly as in the fixed-x case.

The main difference is that when H_0 is false, the noncentrality parameter is a function of $\mathbf{X}$, which is random. Hence the noncentral F distribution does not apply to the random-x case. This only affects such things as power calculations.

Confidence intervals for the β_j's in Section 8.6.2 and for linear functions of the β_j's in Section 8.6.3 are based on the central t distribution [e.g., see (8.48)]. Thus they also remain valid for the random-x case. However, the expected width of the interval differs in the two cases (random x's and fixed x's) because of randomness in $\mathbf{X}$.

In Section 10.5, we obtain the F test for $H_0 : \boldsymbol{\beta}_1 = \mathbf{0}$ using the likelihood ratio approach.

10.3 STANDARDIZED REGRESSION COEFFICENTS

We now show that the regression coefficient vector $\hat{\boldsymbol{\beta}}_1$ in (10.16) can be expressed in terms of sample correlations. By analogy to (10.14), the sample correlation matrix can be written in partitioned form as

$$\mathbf{R} = \begin{pmatrix} 1 & \mathbf{r}'_{yx} \\ \mathbf{r}_{yx} & \mathbf{R}_{xx} \end{pmatrix} = \left(\begin{array}{c|cccc} 1 & r_{y1} & r_{y2} & \cdots & r_{yk} \\ \hline r_{1y} & 1 & r_{12} & \cdots & r_{1k} \\ r_{2y} & r_{21} & 1 & \cdots & r_{2k} \\ \vdots & \vdots & \vdots & & \vdots \\ r_{ky} & r_{k1} & r_{k2} & \cdots & 1 \end{array}\right), \tag{10.18}$$

where $\mathbf{r}_{yx}$ is the vector of correlations between y and the x's and $\mathbf{R}_{xx}$ is the correlation matrix for the x's. For example

$$r_{y2} = \frac{s_{y2}}{\sqrt{s_y^2 s_2^2}} = \frac{\sum_{i=1}^n (y_i - \bar{y})(x_{i2} - \bar{x}_2)}{\sqrt{\sum_{i=1}^n (y_i - \bar{y})^2 \sum_{i=1}^n (x_{i2} - \bar{x}_2)^2}},$$

$$r_{12} = \frac{s_{12}}{\sqrt{s_1^2 s_2^2}} = \frac{\sum_{i=1}^n (x_{i1} - \bar{x}_1)(x_{i2} - \bar{x}_2)}{\sqrt{\sum_{i=1}^n (x_{i1} - \bar{x}_1)^2 \sum_{i=1}^n (x_{i2} - \bar{x}_2)^2}}.$$

By analogy to (3.31), $\mathbf{R}$ can be converted to $\mathbf{S}$ by

$$\mathbf{S} = \mathbf{DRD},$$

where $\mathbf{D} = [\text{diag}(\mathbf{S})]^{1/2}$, which can be written in partitioned form as

$$\mathbf{D} = \left(\begin{array}{c|cccc} s_y & 0 & 0 & \cdots & 0 \\ \hline 0 & \sqrt{s_{11}} & 0 & \cdots & 0 \\ 0 & 0 & \sqrt{s_{22}} & \cdots & 0 \\ \vdots & \vdots & \vdots & & \vdots \\ 0 & 0 & 0 & \cdots & \sqrt{s_{kk}} \end{array}\right) = \begin{pmatrix} s_y & \mathbf{0}' \\ \mathbf{0} & \mathbf{D}_x \end{pmatrix}.$$

Using the partitioned form of $\mathbf{S}$ in (10.14), $\mathbf{S} = \mathbf{DRD}$ can be written as

$$\mathbf{S} = \begin{pmatrix} s_{yy} & \mathbf{s}'_{yx} \\ \mathbf{s}_{yx} & \mathbf{S}_{xx} \end{pmatrix} = \begin{pmatrix} s_y^2 & s_y \mathbf{r}'_{yx} \mathbf{D}_x \\ s_y \mathbf{D}_x \mathbf{r}_{yx} & \mathbf{D}_x \mathbf{R}_{xx} \mathbf{D}_x \end{pmatrix}, \tag{10.19}$$

so that

$$\mathbf{S}_{xx} = \mathbf{D}_x \mathbf{R}_{xx} \mathbf{D}_x, \tag{10.20}$$

$$\mathbf{s}_{yx} = s_y \mathbf{D}_x \mathbf{r}_{yx}, \tag{10.21}$$

where $\mathbf{D}_x = \text{diag}(s_1, s_2, \ldots, s_k)$ and $s_y = \sqrt{s_y^2} = \sqrt{s_{yy}}$ is the sample standard deviation of y. When (10.20) and (10.21) are substituted into (10.16), we obtain an expression for $\hat{\boldsymbol{\beta}}_1$ in terms of correlations:

$$\hat{\boldsymbol{\beta}}_1 = s_y \mathbf{D}_x^{-1} \mathbf{R}_{xx}^{-1} \mathbf{r}_{yx}. \tag{10.22}$$

The regression coefficients $\hat{\beta}_1, \hat{\beta}_2, \ldots, \hat{\beta}_k$ in $\hat{\boldsymbol{\beta}}_1$ can be standardized so as to show the effect of standardized x values (sometimes called *z scores*). We illustrate this for $k = 2$. The model in centered form [see (7.30) and an expression following (7.38)] is

$$\hat{y}_i = \bar{y} + \hat{\beta}_1(x_{i1} - \bar{x}_1) + \hat{\beta}_2(x_{i2} - \bar{x}_2).$$

This can be expressed in terms of standardized variables as

$$\frac{\hat{y}_i - \bar{y}}{s_y} = \frac{s_1}{s_y}\hat{\beta}_1\left(\frac{x_{i1} - \bar{x}_1}{s_1}\right) + \frac{s_2}{s_y}\hat{\beta}_2\left(\frac{x_{i2} - \bar{x}_2}{s_2}\right), \tag{10.23}$$

where $s_j = \sqrt{s_{jj}}$ is the standard deviation of x_j. We thus define the standardized coefficients as

$$\hat{\beta}_j^* = \frac{s_j}{s_y}\hat{\beta}_j.$$

These coefficients are often referred to as *beta weights* or *beta coefficients*. Since they are used with standardized variables $(x_{ij} - \bar{x}_j)/s_j$ in (10.23), the $\hat{\beta}_j^*$'s can be readily compared to each other, whereas the $\hat{\beta}_j$'s cannot be so compared. [Division by s_y in (10.23) is customary but not necessary; the relative values of $s_1\hat{\beta}_1$ and $s_2\hat{\beta}_2$ are the same as those of $s_1\hat{\beta}_1/s_y$ and $s_2\hat{\beta}_2/s_y$.]

The beta weights can be expressed in vector form as

$$\hat{\boldsymbol{\beta}}_1^* = \frac{1}{s_y}\mathbf{D}_x\hat{\boldsymbol{\beta}}_1.$$

Using (10.22), this can be written as

$$\hat{\boldsymbol{\beta}}_1^* = \mathbf{R}_{xx}^{-1}\mathbf{r}_{yx}. \tag{10.24}$$

Note that $\hat{\boldsymbol{\beta}}_1^*$ in (10.24) is not the same as $\hat{\boldsymbol{\beta}}_1^*$ from the reduced model in (8.8). Note also the analogy of $\hat{\boldsymbol{\beta}}_1^* = \mathbf{R}_{xx}^{-1}\mathbf{r}_{yx}$ in (10.24) to $\hat{\boldsymbol{\beta}}_1 = \mathbf{S}_{xx}^{-1}\mathbf{s}_{yx}$ in (10.16). In effect, $\mathbf{R}_{xx}$ and $\mathbf{r}_{xy}$ are the covariance matrix and covariance vector for standardized variables.

Replacing $\mathbf{S}_{xx}^{-1}$ and $\mathbf{s}_{yx}$ by $\mathbf{R}_{xx}^{-1}$ and $\mathbf{r}_{yx}$ leads to regression coefficients for standardized variables.

Example 10.3. The following six hematology variables were measured on 51 workers (Royston 1983):

$y =$ lymphocyte count
$x_1 =$ hemoglobin concentration
$x_2 =$ packed-cell volume
$x_3 =$ white blood cell count ($\times .01$)
$x_4 =$ neutrophil count
$x_5 =$ serum lead concentration

The data are given in Table 10.1.

For $\bar{y}, \bar{\mathbf{x}}, \mathbf{S}_{xx}$ and $\mathbf{s}_{yx}$, we have

$$\bar{y} = 22.902, \quad \bar{\mathbf{x}}' = (15.108, 45.196, 53.824, 25.529, 21.039),$$

$$\mathbf{S}_{xx} = \begin{pmatrix} 0.691 & 1.494 & 3.255 & 0.422 & -0.268 \\ 1.494 & 5.401 & 10.155 & 1.374 & 1.292 \\ 3.255 & 10.155 & 200.668 & 64.655 & 4.067 \\ 0.422 & 1.374 & 64.655 & 56.374 & 0.579 \\ -0.268 & 1.292 & 4.067 & 0.579 & 18.078 \end{pmatrix},$$

$$\mathbf{s}_{yx} = \begin{pmatrix} 1.535 \\ 4.880 \\ 106.202 \\ 3.753 \\ 3.064 \end{pmatrix}.$$

By (10.15) to (10.17), we obtain

$$\hat{\boldsymbol{\beta}}_1 = \mathbf{S}_{xx}^{-1}\mathbf{s}_{yx} = \begin{pmatrix} -0.491 \\ -0.316 \\ 0.837 \\ -0.882 \\ 0.025 \end{pmatrix},$$

$$\hat{\beta}_0 = \bar{y} - \mathbf{s}_{yx}'\mathbf{S}_{xx}^{-1}\bar{\mathbf{x}} = 22.902 - 1.355 = 21.547,$$

$$s^2 = s_{yy} - \mathbf{s}_{yx}'\mathbf{S}_{xx}^{-1}\mathbf{s}_{yx} = 90.2902 - 83.3542 = 6.9360.$$

TABLE 10.1 Hematology Data

Observation Number	y	x_1	x_2	x_3	x_4	x_5
1	14	13.4	39	41	25	17
2	15	14.6	46	50	30	20
3	19	13.5	42	45	21	18
4	23	15.0	46	46	16	18
5	17	14.6	44	51	31	19
6	20	14.0	44	49	24	19
7	21	16.4	49	43	17	18
8	16	14.8	44	44	26	29
9	27	15.2	46	41	13	27
10	34	15.5	48	84	42	36
11	26	15.2	47	56	27	22
12	28	16.9	50	51	17	23
13	24	14.8	44	47	20	23
14	26	16.2	45	56	25	19
15	23	14.7	43	40	13	17
16	9	14.7	42	34	22	13
17	18	16.5	45	54	32	17
18	28	15.4	45	69	36	24
19	17	15.1	45	46	29	17
20	14	14.2	46	42	25	28
21	8	15.9	46	52	34	16
22	25	16.0	47	47	14	18
23	37	17.4	50	86	39	17
24	20	14.3	43	55	31	19
25	15	14.8	44	42	24	29
26	9	14.9	43	43	32	17
27	16	15.5	45	52	30	20
28	18	14.5	43	39	18	25
29	17	14.4	45	60	37	23
30	23	14.6	44	47	21	27
31	43	15.3	45	79	23	23
32	17	14.9	45	34	15	24
33	23	15.8	47	60	32	21
34	31	14.4	44	77	39	23
35	11	14.7	46	37	23	23
36	25	14.8	43	52	19	22
37	30	15.4	45	60	25	18
38	32	16.2	50	81	38	18
39	17	15.0	45	49	26	24
40	22	15.1	47	60	33	16
41	20	16.0	46	46	22	22
42	20	15.3	48	55	23	23

(Continued)

TABLE 10.1 *Continued*

Observation Number	y	x_1	x_2	x_3	x_4	x_5
43	20	14.5	41	62	36	21
44	26	14.2	41	49	20	20
45	40	15.0	45	72	25	25
46	22	14.2	46	58	31	22
47	61	14.9	45	84	17	17
48	12	16.2	48	31	15	18
49	20	14.5	45	40	18	20
50	35	16.4	49	69	22	24
51	38	14.7	44	78	34	16

The correlations are given by

$$\mathbf{R}_{xx} = \begin{pmatrix} 1.000 & 0.774 & 0.277 & 0.068 & -0.076 \\ 0.774 & 1.000 & 0.308 & 0.079 & 0.131 \\ 0.277 & 0.308 & 1.000 & 0.608 & 0.068 \\ 0.068 & 0.079 & 0.608 & 1.000 & 0.018 \\ -0.076 & 0.131 & 0.068 & 0.018 & 1.000 \end{pmatrix}, \quad \mathbf{r}_{yx} = \begin{pmatrix} 0.194 \\ 0.221 \\ 0.789 \\ 0.053 \\ 0.076 \end{pmatrix}.$$

By (10.24), the standardized coefficient vector is given by

$$\hat{\boldsymbol{\beta}}_1^* = \mathbf{R}_{xx}^{-1}\mathbf{r}_{yx} = \begin{pmatrix} -0.043 \\ -0.077 \\ 1.248 \\ -0.697 \\ 0.011 \end{pmatrix}.$$

□

10.4 R^2 IN MULTIVARIATE NORMAL REGRESSION

In the case of fixed x's, we defined R^2 as the proportion of variation in y due to regression [see (7.55)]. In the case of random x's, we obtain R as an estimate of a population multiple correlation between y and the x's. Then R^2 is the square of this sample multiple correlation.

The *population multiple correlation coefficient* $\rho_{y|\mathbf{x}}$ is defined as the correlation between y and the linear function $w = \mu_y + \boldsymbol{\sigma}'_{yx}\boldsymbol{\Sigma}_{xx}^{-1}(\mathbf{x} - \boldsymbol{\mu}_x)$:

$$\rho_{y|\mathbf{x}} = \text{corr}(y, w) = \frac{\sigma_{yw}}{\sigma_y \sigma_w}. \tag{10.25}$$

(We use the subscript $y|\mathbf{x}$ to distinguish $\rho_{y|\mathbf{x}}$ from ρ, the correlation between y and x in the bivariate normal case; see Sections 3.2, 6.4, and 10.5). By (10.4), w is equal to $E(y|\mathbf{x})$, which is the population analogue of $\hat{y} = \hat{\beta}_0 + \hat{\boldsymbol{\beta}}_1'\mathbf{x}_1$, the sample predicted value of y. As $\mathbf{x}$ varies randomly, the *population predicted value* $w = \mu_y + \boldsymbol{\sigma}_{yx}'\boldsymbol{\Sigma}_{xx}^{-1}(\mathbf{x} - \boldsymbol{\mu}_x)$ becomes a random variable.

It is easily established that $\text{cov}(y, w)$ and $\text{var}(w)$ have the same value:

$$\text{cov}(y, w) = \text{var}(w) = \boldsymbol{\sigma}_{yx}'\boldsymbol{\Sigma}_{xx}^{-1}\boldsymbol{\sigma}_{yx}. \tag{10.26}$$

Then the population multiple correlation $\rho_{y|\mathbf{x}}$ in (10.25) becomes

$$\rho_{y|\mathbf{x}} = \frac{\text{cov}(y, w)}{\sqrt{\text{var}(y)\text{var}(w)}} = \sqrt{\frac{\boldsymbol{\sigma}_{yx}'\boldsymbol{\Sigma}_{xx}^{-1}\boldsymbol{\sigma}_{yx}}{\boldsymbol{\sigma}_{yy}}},$$

and the *population coefficient of determination* or *population squared multiple correlation* $\rho_{y|\mathbf{x}}^2$ is given by

$$\rho_{y|\mathbf{x}}^2 = \frac{\boldsymbol{\sigma}_{yx}'\boldsymbol{\Sigma}_{xx}^{-1}\boldsymbol{\sigma}_{yx}}{\boldsymbol{\sigma}_{yy}}. \tag{10.27}$$

We now list some properties of $\rho_{y|\mathbf{x}}$ and $\rho_{y|\mathbf{x}}^2$.

1. $\rho_{y|\mathbf{x}}$ is the maximum correlation between y and any linear function of $\mathbf{x}$:

$$\rho_{y|\mathbf{x}} = \max_{\alpha} \rho_{y,\, \boldsymbol{\alpha}'\mathbf{x}}. \tag{10.28}$$

   This is an alternative definition of $\rho_{y|\mathbf{x}}$ that is not based on the multivariate normal distribution as is the definition in (10.25).

2. $\rho_{y|\mathbf{x}}^2$ can be expressed in terms of determinants:

$$\rho_{y|\mathbf{x}}^2 = 1 - \frac{|\boldsymbol{\Sigma}|}{\sigma_{yy}|\boldsymbol{\Sigma}_{xx}|}, \tag{10.29}$$

   where $\boldsymbol{\Sigma}$ and $\boldsymbol{\Sigma}_{xx}$ are defined in (10.3).

3. $\rho_{y|\mathbf{x}}^2$ is invariant to linear transformations on y or on the x's; that is, if $u = ay$ and $\mathbf{v} = \mathbf{B}\mathbf{x}$, where $\mathbf{B}$ is nonsingular, then

$$\rho_{u|\mathbf{v}}^2 = \rho_{y|\mathbf{x}}^2. \tag{10.30}$$

   (Note that $\mathbf{v}$ here is not the same as $\mathbf{v}_i$ used in the proof of Theorem 10.2a.)

4. Using $\text{var}(w) = \boldsymbol{\sigma}'_{yx}\boldsymbol{\Sigma}_{xx}^{-1}\boldsymbol{\sigma}_{yx}$ in (10.26), $\rho^2_{y|\mathbf{x}}$ in (10.27) can be written in the form

$$\rho^2_{y|\mathbf{x}} = \frac{\text{var}(w)}{\text{var}(y)}. \qquad (10.31)$$

Since $w = \mu_y + \boldsymbol{\sigma}'_{yx}\boldsymbol{\Sigma}_{xx}^{-1}(\mathbf{x} - \boldsymbol{\mu}_x)$ is the population regression equation, $\rho^2_{y|\mathbf{x}}$ in (10.31) represents the proportion of the variance of y that can be attributed to the regression relationship with the variables in $\mathbf{x}$. In this sense, $\rho^2_{y|\mathbf{x}}$ is analogous to R^2 in the fixed-x case in (7.55).

5. By (10.8) and (10.27), $\text{var}(y|\mathbf{x})$ can be expressed in terms of $\rho^2_{y|\mathbf{x}}$:

$$\begin{aligned} \text{var}(y|\mathbf{x}) &= \sigma_{yy} - \boldsymbol{\sigma}'_{yx}\boldsymbol{\Sigma}_{xx}^{-1}\boldsymbol{\sigma}_{yx} = \sigma_{yy} - \sigma_{yy}\rho^2_{y|\mathbf{x}} \\ &= \sigma_{yy}(1 - \rho^2_{y|\mathbf{x}}). \end{aligned} \qquad (10.32)$$

6. If we consider $y - w$ as a residual or error term, then $y - w$ is uncorrelated with the x's

$$\text{cov}(y - w, \mathbf{x}) = \mathbf{0}' \qquad (10.33)$$

(see Problem 10.8).

We can obtain a maximum likelihood estimator for $\rho^2_{y|\mathbf{x}}$ by substituting estimators from (10.14) for the parameters in (10.27):

$$R^2 = \frac{\mathbf{s}'_{yx}\mathbf{S}_{xx}^{-1}\mathbf{s}_{yx}}{s_{yy}} \qquad (10.34)$$

We use the notation R^2 rather than $\hat{\rho}^2_{y|\mathbf{x}}$ because (10.34) is recognized as having the same form as R^2 for the fixed-x case in (7.59). We refer to R^2 as the *sample coefficient of determination* or as the *sample squared multiple correlation*. The square root of R^2

$$R = \sqrt{\frac{\mathbf{s}'_{yx}\mathbf{S}_{xx}^{-1}\mathbf{s}_{yx}}{s_{yy}}} \qquad (10.35)$$

is the *sample multiple correlation coefficient*.

We now list several properties of R and R^2, some of which are analogous to properties of $\rho^2_{y|\mathbf{x}}$ above.

1. R is equal to the correlation between y and $\hat{y} = \hat{\beta}_0 + \hat{\beta}_1 x_1 + \cdots + \hat{\beta}_k x_k = \hat{\beta}_0 + \hat{\boldsymbol{\beta}}'_1\mathbf{x}$:

$$R = r_{y\hat{y}}. \qquad (10.36)$$

2. R is equal to the maximum correlation between y and any linear combination of the x's, $\mathbf{a}'\mathbf{x}$:

$$R = \max_{\mathbf{a}} r_{y,\mathbf{a}'\mathbf{x}}. \qquad (10.37)$$

3. R^2 can be expressed in terms of correlations:

$$R^2 = \mathbf{r}'_{yx}\mathbf{R}_{xx}^{-1}\mathbf{r}_{yx}, \tag{10.38}$$

where $\mathbf{r}_{yx}$ and $\mathbf{R}_{xx}$ are from the sample correlation matrix $\mathbf{R}$ partitioned as in (10.18).

4. R^2 can be obtained from $\mathbf{R}^{-1}$:

$$R^2 = 1 - \frac{1}{r^{yy}}, \tag{10.39}$$

where r^{yy} is the first diagonal element of $\mathbf{R}^{-1}$. Using the other diagonal elements of $\mathbf{R}^{-1}$, this relationship can be extended to give the multiple correlation of any x_j with the other x's and y. Thus from $\mathbf{R}^{-1}$ we obtain multiple correlations, as opposed to the simple correlations in $\mathbf{R}$.

5. R^2 can be expressed in terms of determinants:

$$R^2 = 1 - \frac{|\mathbf{S}|}{s_{yy}|\mathbf{S}_{xx}|} \tag{10.40}$$

$$= 1 - \frac{|\mathbf{R}|}{|\mathbf{R}_{xx}|}, \tag{10.41}$$

where $\mathbf{S}_{xx}$ and $\mathbf{R}_{xx}$ are defined in (10.14) and (10.18).

6. From (10.24) and (10.38), we can express R^2 in terms of beta weights:

$$R^2 = \mathbf{r}'_{yx}\hat{\boldsymbol{\beta}}_1^*, \tag{10.42}$$

where $\hat{\boldsymbol{\beta}}_1^* = \mathbf{R}_{xx}^{-1}\mathbf{r}_{yx}$. This equation does *not* imply that R^2 is the sum of squared partial correlations (Section 10.8).

7. If $\rho^2_{y|\mathbf{x}} = 0$, the expected value of R^2 is given by

$$E(R^2) = \frac{k}{n-1}. \tag{10.43}$$

Thus R^2 is biased when $\rho^2_{y|\mathbf{x}}$ is 0 [this is analogous to (7.57)].

8. $R^2 \geq \max_j r^2_{yj}$, where r_{yj} is an element of $\mathbf{r}'_{yx} = (r_{y1}, r_{y2}, \ldots, r_{yk})$.
9. R^2 is invariant to full rank linear transformations on y or on the x's.

Example 10.4. For the hematology data in Table 10.1, $\mathbf{S}_{xx}, \mathbf{s}_{yx}, \mathbf{R}_{xx}$, and $\mathbf{r}_{yx}$ were obtained in Example 10.3. Using either (10.34) or (10.38), we obtain

$$R^2 = .9232.$$

□

10.5 TESTS AND CONFIDENCE INTERVALS FOR R^2

Note that by (10.27), $\rho^2_{y|\mathbf{x}} = 0$ becomes

$$\rho^2_{y|\mathbf{x}} = \frac{\boldsymbol{\sigma}'_{yx}\boldsymbol{\Sigma}^{-1}_{xx}\boldsymbol{\sigma}_{yx}}{\sigma_{yy}} = 0,$$

which leads to $\boldsymbol{\sigma}_{yx} = \mathbf{0}$ since $\boldsymbol{\Sigma}_{xx}$ is positive definite. Then by (10.7), $\boldsymbol{\beta}_1 = \boldsymbol{\Sigma}^{-1}_{xx}\boldsymbol{\sigma}_{yx} = \mathbf{0}$, and $H_0: \rho^2_{y|\mathbf{x}} = 0$ is equivalent to $H_0: \boldsymbol{\beta}_1 = \mathbf{0}$.

The F statistic for fixed x's is given in (8.5), (8.22), and (8.23) as

$$\begin{aligned} F &= \frac{(\hat{\boldsymbol{\beta}}'\mathbf{X}'\mathbf{y} - n\bar{y}^2)/k}{(\mathbf{y}'\mathbf{y} - \hat{\boldsymbol{\beta}}'\mathbf{X}'\mathbf{y})/(n-k-1)} \\ &= \frac{R^2/k}{(1-R^2)/(n-k-1)}. \end{aligned} \tag{10.44}$$

The test statistic in (10.44) can be obtained by the likelihood ratio approach in the case of random x's (Anderson 1984, pp. 140–142):

Theorem 10.5. If $(y_1, \mathbf{x}'_1), (y_2, \mathbf{x}'_2), \ldots, (y_n, \mathbf{x}'_n)$ is a random sample from $N_{k+1}(\boldsymbol{\mu}, \boldsymbol{\Sigma})$, where $\boldsymbol{\mu}$ and $\boldsymbol{\Sigma}$ are given by (10.2) and (10.3), the likelihood ratio test for $H_0: \boldsymbol{\beta}_1 = \mathbf{0}$ or equivalently $H_0: \rho^2_{y|\mathbf{x}} = 0$ can be based on F in (10.44). We reject H_0 if $F \geq F_{\alpha,k,n-k-1}$.

PROOF. Using the notation $\mathbf{v}'_i = (y_i, \mathbf{x}'_i)$, as in the proof of Theorem 10.2a, the likelihood function $L(\boldsymbol{\mu}, \boldsymbol{\Sigma}) = \prod_{i=1}^n f(\mathbf{v}_i; \boldsymbol{\mu}, \boldsymbol{\Sigma})$ is given by (10.11), and the likelihood ratio is

$$\mathrm{LR} = \frac{\max_{H_0} L(\boldsymbol{\mu}, \boldsymbol{\Sigma})}{\max_{H_1} L(\boldsymbol{\mu}, \boldsymbol{\Sigma})}.$$

Under H_1, the parameters $\boldsymbol{\mu}$ and $\boldsymbol{\Sigma}$ are essentially unrestricted, and we have

$$\max_{H_1} L(\boldsymbol{\mu}, \boldsymbol{\Sigma}) = \max L(\boldsymbol{\mu}, \boldsymbol{\Sigma}) = L(\hat{\boldsymbol{\mu}}, \hat{\boldsymbol{\Sigma}}),$$

where $\hat{\boldsymbol{\mu}}$ and $\hat{\boldsymbol{\Sigma}}$ are the maximum likelihood estimators in (10.9) and (10.10). Since $(\mathbf{v}_i - \boldsymbol{\mu})'\boldsymbol{\Sigma}^{-1}(\mathbf{v}_i - \boldsymbol{\mu})$ is a scalar, the exponent of $L(\boldsymbol{\mu}, \boldsymbol{\Sigma})$ in (10.11) can be written as

$$\frac{\sum_{i=1}^{n} \mathrm{tr}\left[(\mathbf{v}_i - \boldsymbol{\mu})'\boldsymbol{\Sigma}^{-1}(\mathbf{v}_i - \boldsymbol{\mu})\right]}{2} = \frac{\sum_{i=1}^{n} \mathrm{tr}\left[\boldsymbol{\Sigma}^{-1}(\mathbf{v}_i - \boldsymbol{\mu})(\mathbf{v}_i - \boldsymbol{\mu})'\right]}{2}$$

$$= \frac{\mathrm{tr}\left[\boldsymbol{\Sigma}^{-1} \sum_{i=1}^{n} (\mathbf{v}_i - \boldsymbol{\mu})(\mathbf{v}_i - \boldsymbol{\mu})'\right]}{2}.$$

Then substitution of $\hat{\boldsymbol{\mu}}$ and $\hat{\boldsymbol{\Sigma}}$ for $\boldsymbol{\mu}$ and $\boldsymbol{\Sigma}$ in $L(\boldsymbol{\mu}, \boldsymbol{\Sigma})$ gives

$$\max_{H_1} L(\boldsymbol{\mu}, \boldsymbol{\Sigma}) = L(\hat{\boldsymbol{\mu}}, \hat{\boldsymbol{\Sigma}}) = \frac{1}{(\sqrt{2\pi})^{n(k+1)}|\hat{\boldsymbol{\Sigma}}|^{n/2}} e^{-\mathrm{tr}(\hat{\boldsymbol{\Sigma}}^{-1} n\hat{\boldsymbol{\Sigma}}/2)}$$

$$= \frac{e^{-n(k+1)/2}}{(\sqrt{2\pi})^{n(k+1)}|\hat{\boldsymbol{\Sigma}}|^{n/2}}.$$

Under $H_0 : \rho^2_{y|\mathbf{x}} = 0$, we have $\boldsymbol{\sigma}_{yx} = \mathbf{0}$, and $\boldsymbol{\Sigma}$ in (10.3) becomes

$$\boldsymbol{\Sigma}_0 = \begin{pmatrix} \sigma_{yy} & \mathbf{0}' \\ \mathbf{0} & \boldsymbol{\Sigma}_{xx} \end{pmatrix}, \tag{10.45}$$

whose maximum likelihood estimator is

$$\hat{\boldsymbol{\Sigma}}_0 = \begin{pmatrix} \hat{\sigma}_{yy} & \mathbf{0}' \\ \mathbf{0} & \hat{\boldsymbol{\Sigma}}_{xx} \end{pmatrix}. \tag{10.46}$$

Using $\hat{\boldsymbol{\Sigma}}_0$ in (10.46) and $\hat{\boldsymbol{\mu}} = \bar{\mathbf{v}}$ in (10.9), we have

$$\max_{H_0} L(\boldsymbol{\mu}, \boldsymbol{\Sigma}) = L(\hat{\boldsymbol{\mu}}, \hat{\boldsymbol{\Sigma}}_0) = \frac{1}{(\sqrt{2\pi})^{n(k+1)}|\hat{\boldsymbol{\Sigma}}_0|^{n/2}} e^{-\mathrm{tr}(\hat{\boldsymbol{\Sigma}}_0^{-1} n\hat{\boldsymbol{\Sigma}}_0/2)}.$$

By (2.74), this becomes

$$L(\hat{\boldsymbol{\mu}}, \hat{\boldsymbol{\Sigma}}_0) = \frac{e^{-n(k+1)/2}}{(\sqrt{2\pi})^{n(k+1)}\hat{\sigma}_{yy}^{n/2}|\hat{\boldsymbol{\Sigma}}_{xx}|^{n/2}}. \tag{10.47}$$

Thus

$$\mathrm{LR} = \frac{|\hat{\boldsymbol{\Sigma}}|^{n/2}}{\hat{\sigma}_{yy}^{n/2}|\hat{\boldsymbol{\Sigma}}_{xx}|^{n/2}}. \tag{10.48}$$

Substituting $\hat{\boldsymbol{\Sigma}} = (n-1)\mathbf{S}/n$ and using (10.40), we obtain

$$\text{LR} = (1 - R^2)^{n/2}. \tag{10.49}$$

We reject H_0 for $(1 - R^2)^{n/2} \leq c$, which is equivalent to

$$F = \frac{R^2/k}{(1 - R^2)/(n - k - 1)} \geq F_{\alpha, k, n-k-1},$$

since $R^2/(1 - R^2)$ is a monotone increasing function of R^2 and F is distributed as $F(k, n - k - 1)$ when H_0 is true (Anderson 1984, pp. 138–139). □

When $k = 1$, F in (10.44) reduces to $F = (n-2)r^2/(1 - r^2)$. Then, by Problem 5.16

$$t = \frac{\sqrt{n-2}r}{\sqrt{1 - r^2}}$$

[see (6.20)] has a t distribution with $n - 2$ degrees of freedom (df) when (y, x) has a bivariate normal distribution with $\rho = 0$.

If (y, x) is bivariate normal and $\rho \neq 0$, then $\text{var}(r) = (1 - \rho^2)^2/n$ and the function

$$u = \frac{\sqrt{n}(r - \rho)}{1 - \rho^2} \tag{10.50}$$

is approximately standard normal for large n. However, the distribution of u approaches normality very slowly as n increases (Kendall and Stuart 1969, p. 236). Its use is questionable for $n < 500$.

Fisher (1921) found a function of r that approaches normality much faster than does (10.50) and can thereby be used with much smaller n than that required for (10.50). In addition, the variance is almost independent of ρ. Fisher's function is

$$z = \frac{1}{2}\ln\frac{1 + r}{1 - r} = \tanh^{-1} r, \tag{10.51}$$

where $\tanh^{-1} r$ is the inverse hyperbolic tangent of r. The approximate mean and variance of z are

$$E(z) \cong \frac{1}{2}\ln\frac{1 + \rho}{1 - \rho} = \tanh^{-1} \rho, \tag{10.52}$$

$$\text{var}(z) \cong \frac{1}{n - 3}. \tag{10.53}$$

We can use Fisher's z transformation in (10.51) to test hypotheses such as $H_0: \rho = \rho_0$ or $H_0: \rho_1 = \rho_2$. To test $H_0: \rho = \rho_0$ vs. $H_1: \rho \neq \rho_0$, we calculate

$$v = \frac{z - \tanh^{-1}\rho_0}{\sqrt{1/(n-3)}}, \tag{10.54}$$

which is approximately distributed as the standard normal $N(0, 1)$. We reject H_0 if $|v| \geq z_{\alpha/2}$, where $z = \tanh^{-1} r$ and $z_{\alpha/2}$ is the upper $\alpha/2$ percentage point of the standard normal distribution. To test $H_0: \rho_1 = \rho_2$ vs. $H_1: \rho_1 \neq \rho_2$ for two independent samples of sizes n_1 and n_2 yielding sample correlations r_1 and r_2, we calculate

$$v = \frac{z_1 - z_2}{\sqrt{1/(n_1-3) + 1/(n_2-3)}} \tag{10.55}$$

and reject H_0 if $|v| \geq z_{\alpha/2}$, where $z_1 = \tanh^{-1} r_1$ and $z_2 = \tanh^{-1} r_2$. To test $H_0: \rho_1 = \cdots = \rho_q$ for $q > 2$, see Problem 10.18.

To obtain a confidence interval for ρ, we note that since z in (10.51) is approximately normal, we can write

$$P\left(-z_{\alpha/2} \leq \frac{z - \tanh^{-1}\rho}{1/\sqrt{n-3}} \leq z_{\alpha/2}\right) \cong 1 - \alpha. \tag{10.56}$$

Solving the inequality for ρ, we obtain the approximate $100(1-\alpha)\%$ confidence interval

$$\tanh\left(z - \frac{z_{\alpha/2}}{\sqrt{n-3}}\right) \leq \rho \leq \tanh\left(z + \frac{z_{\alpha/2}}{\sqrt{n-3}}\right). \tag{10.57}$$

A confidence interval for $\rho^2_{y|x}$ was given by Helland (1987).

Example 10.5a. For the hematology data in Table 10.1, we obtained R^2 in Example 10.4. The overall F test of $H_0: \boldsymbol{\beta}_1 = \mathbf{0}$ or $H_0: \rho^2_{y|\mathbf{x}} = 0$ is carried out using F in (10.44):

$$\begin{aligned} F &= \frac{R^2/k}{(1-R^2)/(n-k-1)} \\ &= \frac{.9232/5}{(1-.9232)/45} = 108.158. \end{aligned}$$

The p value is less than 10^{-16}. □

Example 10.5b. To illustrate Fisher's z transformation in (10.51) and its use to compare two independent correlations in (10.55), we divide the hematology data in Table 10.1 into two subsamples of sizes $n_1 = 26$ and $n_2 = 25$ (the first 26 observations and the last 25 observations). For the correlation between y and x_1 in each of the two subsamples, we obtain $r_1 = .4994$ and $r_2 = .0424$. The z transformation in (10.51) for each of these two values is given by

$$z_1 = \tanh^{-1} r_1 = .5485,$$
$$z_2 = \tanh^{-1} r_2 = .0425.$$

To test $H_0 : \rho_1 = \rho_2$, we use the approximate test statistic (10.55) to obtain

$$v = \frac{.5485 - .0425}{\sqrt{1/(26-3) + 1/(25-3)}} = 1.6969.$$

Since $1.6969 < z_{.025} = 1.96$, we do not reject H_0.

To obtain approximate 95% confidence limits for ρ_1, we use (10.57):

$$\text{Lower limit for } \rho_1: \quad \tanh\left(.5485 - \frac{1.96}{\sqrt{23}}\right) = .1389,$$

$$\text{Upper limit for } \rho_1: \quad \tanh\left(.5485 + \frac{1.96}{\sqrt{23}}\right) = .7430.$$

For ρ_2, the limits are given by

$$\text{Lower limit for } \rho_2: \quad \tanh\left(.0425 - \frac{1.96}{\sqrt{22}}\right) = -.3587,$$

$$\text{Upper limit for } \rho_2: \quad \tanh\left(.0425 + \frac{1.96}{\sqrt{22}}\right) = .4303.$$

□

10.6 EFFECT OF EACH VARIABLE ON R^2

The contribution of a variable x_j to the multiple correlation R will, in general, be different from its bivariate correlation with y; that is, the increase in R^2 when x_j is added is not equal to $r^2_{yx_j}$. This increase in R_2 can be either more or less than $r^2_{yx_j}$. It seems clear that relationships with other variables can render a variable partially redundant and thereby reduce the contribution of x_j to R^2, but it is not intuitively apparent how the contribution of x_j to R^2 can exceed $r^2_{yx_j}$. The latter phenomenon has been illustrated numerically by Flury (1989) and Hamilton (1987).

In this section, we provide a breakdown of the factors that determine how much each variable adds to R^2 and show how the increase in R^2 can exceed $r^2_{yx_j}$ (Rencher 1993). We first introduce some notation. The variable of interest is denoted by z, which can be one of the x's or a new variable added to the x's. We make the following additional notational definitions:

R^2_{yw} = squared multiple correlation between y and $\mathbf{w} = (x_1, x_2, \ldots, x_k, z)'$.

R^2_{yx} = squared multiple correlation between y and $\mathbf{x} = (x_1, x_2, \ldots, x_k)'$.

$R^2_{zx} = \mathbf{s}'_{zx}\mathbf{S}^{-1}_{xx}\mathbf{s}_{zx}/s^2_z$ = squared multiple correlation between z and $\mathbf{x}$.

r_{yz} = simple correlation between y and z.

$\mathbf{r}_{yx} = (r_{yx_1}, r_{yx_2}, \ldots, r_{yx_k})'$ = vector of correlations between y and $\mathbf{x}$.

$\mathbf{r}_{zx} = (r_{zx_1}, r_{zx_2}, \ldots, r_{zx_k})'$ = vector of correlations between z and $\mathbf{x}$.

$\hat{\boldsymbol{\beta}}^*_{zx} = \mathbf{R}^{-1}_{xx}\mathbf{r}_{zx}$ is the vector of standardized regression coefficients (beta weights) of z regressed on $\mathbf{x}$ [see (10.24)].

The effect of z on R^2 is formulated in the following theorem.

Theorem 10.6. The increase in R^2 due to z can be expressed as

$$R^2_{yw} - R^2_{yx} = \frac{(\hat{r}_{yz} - r_{yz})^2}{1 - R^2_{zx}}, \tag{10.58}$$

where $\hat{r}_{yz} = \hat{\boldsymbol{\beta}}^{*\prime}_{zx}\mathbf{r}_{yx}$ is a "predicted" value of r_{yz} based on the relationship of z to the x's.

PROOF. See Problem 10.19. □

Since the right side of (10.58) is positive, R^2 cannot decrease with an additional variable, which is a verification of property 3 in Section 7.7. If z is orthogonal to $\mathbf{x}$ (i.e., if $\mathbf{r}_{zx} = \mathbf{0}$), then $\hat{\boldsymbol{\beta}}^*_{zx} = \mathbf{0}$, which implies that $\hat{r}_{yz} = 0$ and $R^2_{zx} = 0$. In this case, (10.58) can be written as $R^2_{yw} = R^2_{yx} + r^2_{yz}$, which verifies property 5 of Section 7.7.

It is clear in Theorem 10.6 that the contribution of z to R^2 can either be less than or greater than r^2_{yz}. If $\hat{r}_{yz}$ is close to r_{yz}, the contribution of z is less than r^2_{yz}. There are three ways in which the contribution of z can exceed r^2_{yz}: (1) $\hat{r}_{yz}$ is substantially larger in absolute value than r_{yz}, (2) $\hat{r}_{yz}$ and r_{yz} are of opposite signs, and (3) R^2_{zx} is large.

In many cases, the researcher may find it helpful to know why a variable contributed more than expected or less than expected. For example, admission to a university or professional school may be based on previous grades and the score on a standardized national test. An applicant for admission to a university with limited enrollment would submit high school grades and a national test score. These might be entered into a regression equation to obtain a predicted value of first-year grade-point average at the university. It is typically found that the standardized test increases R^2 only slightly above that based on high school grades alone. This small increase in R^2 would be disappointing to admissions officials who had hoped that the national test score might be a more useful predictor than high school grades. The designers of such standardized tests may find it beneficial to know precisely why the test makes such an unexpectedly small contribution relative to high school grades.

In Theorem 10.6, we have available the specific information needed by the designer of the standardized test. To illustrate the use of (10.58), let y be the grade-point average for the first year at the university, let z be the score on the standardized test, and let $x_1, x_2, \ldots, x_k$ be high school grades in key subject areas. By (10.58), the increase in R^2 due to z is $(\hat{r}_{yz} - r_{yz})^2/(1 - R^2_{zx})$, in which we see that z adds little to R^2 if $\hat{r}_{yz}$ is close to r_{yz}. We could examine the coefficients in $\hat{r}_{yz} = \hat{\boldsymbol{\beta}}^{*\prime}_{zx}\mathbf{r}_{yx}$ to determine which of the r_{yx_j}'s in $\mathbf{r}_{yx}$ have the most effect. This information could be used in redesigning the questions so as to reduce these particular r_{yx_j}'s. It may also be possible to increase the contribution of z to R^2_{yw} by increasing R^2_{zx} (thereby reducing $1 - R^2_{zx}$). This might be done by designing the questions in the standardized test so that the test score z is more correlated with high school grades, $x_1, x_2, \ldots, x_q$.

Theil and Chung (1988) proposed a measure of the relative importance of a variable in multiple regression based on information theory.

Example 10.6. For the hematology data in Table 10.1, the overall R^2_{yw} was found in Example 10.4 to be .92318. From Theorem 10.6, the increase in R^2 due to a variable z has the breakdown $R^2_{yw} - R^2_{yx} = (\hat{r}_{yz} - r_{yz})^2/(1 - R^2_{zx})$, where z represents any one of $x_1, x_2, \ldots, x_5$, and x represents the other four variables. The values of $\hat{r}_{yz}$, r_{yz}, R^2_{zx}, $R^2_{yw} - R^2_{yx}$, and F are given below for each variable in turn as z:

z	$\hat{r}_{yz}$	r_{yz}	R^2_{zx}	$R^2_{yw} - R^2_{yx}$	F	p value
x_1	.2101	.1943	.6332	.00068	0.4	.53
x_2	.2486	.2210	.6426	.00213	1.25	.26
x_3	.0932	.7890	.4423	.86820	508.6	0
x_4	.4822	.0526	.3837	.29945	175.4	0
x_5	.0659	.0758	.0979	.00011	0.064	.81

The F value is from the partial F test in (8.25), (8.37), or (8.39) for the significance of the increase in R^2 due to each variable.

An interesting variable here is x_4, whose value of r_{yz} is .0526, the smallest among the five variables. Despite this small individual correlation with y, x_4 contributes much more to R^2_{yw} than do all other variables except x_3 because $\hat{r}_{yz}$ is much greater for x_4 than for the other variables. This illustrates how the contribution of a variable can be augmented in the presence of other variables as reflected in $\hat{r}_{yz}$.

The difference between the two major contributors x_3 and x_4 may be very revealing to the researcher. The contribution of x_3 to R^2_{yw} is due mostly to its own correlation with y, whereas virtually all the effect of x_4 comes from its association with the other variables as reflected in $\hat{r}_{yz}$. □

10.7 PREDICTION FOR MULTIVARIATE NORMAL OR NONNORMAL DATA

In this section, we consider an approach to modeling and estimation in the random-x case that is somewhat reminiscent of least squares in the fixed-x case. Suppose that $(y, \mathbf{x}') = (y, x_1, x_2, \ldots, x_k)$ is not necessarily assumed to be multivariate normal and we wish to find a function $t(\mathbf{x})$ for predicting y. In order to find a predicted value $t(\mathbf{x})$ that is expected to be "close" to y, we will choose the function $t(\mathbf{x})$ that minimizes the mean squared error $E[y - t(\mathbf{x})]^2$, where the expectation is in the joint distribution of $y, x_1, \ldots, x_k$. This function is given in the following theorem.

Theorem 10.7. For the random vector $(y, \mathbf{x}')$, the function $t(\mathbf{x})$ that minimizes the mean squared error $E[y - t(\mathbf{x})]^2$ is given by $E(y|\mathbf{x})$.

PROOF. For notational simplicity, we use $k = 1$. By (4.28), the joint density $g(y, x)$ can be written as $g(y, x) = f(y|x)h(x)$. Then

$$
\begin{aligned}
E[y - t(x)]^2 &= \int\int [y - t(x)]^2 g(y, x)\, dy\, dx \\
&= \int\int [y - t(x)]^2 f(y|x)h(x)\, dy\, dx \\
&= \int h(x)\left\{\int [y - t(x)]^2 f(y|x)\, dy\right\} dx.
\end{aligned}
$$

To find the function $t(x)$ that minimizes $E(y - t)^2$, we differentiate with respect to t and set the result equal to 0 [for a more general proof not involving differentiation, see Graybill (1976, pp. 432–434) or Christensen (1996, p. 119)]. Assuming that we can interchange integration and differentiation, we obtain

$$
\frac{\partial E[y - t(x)]^2}{\partial t} = \int h(x)\left\{\int 2(-1)[y - t(x)]\, f(y|x) dy\right\} dx = 0,
$$

which gives

$$
2\int h(x)\left[\int y f(y|x) dy - \int t(x) f(y|x) dy\right] dx = 0,
$$

$$
2\int h(x)[E(y|x) - t(x)] dx = 0.
$$

The left side is 0 if

$$
t(x) = E(y|x). \qquad \square
$$

In the case of the multivariate normal, the prediction function $E(y|\mathbf{x})$ is a linear function of $\mathbf{x}$ [see (10.4) and (10.5)]. However, in general, $E(y|\mathbf{x})$ is not linear. For an illustration of a nonlinear $E(y|x)$, see Example 3.2, in which we have $E(y|x) = \frac{1}{2}(1 + 4x - 2x^2)$.

If we restrict $t(\mathbf{x})$ to *linear* functions of $\mathbf{x}$, then the optimal result is the same linear function as in the multivariate normal case [see (10.6) and (10.7)].

Theorem 10.7b. The linear function $t(\mathbf{x})$ that minimizes $E[y - t(\mathbf{x})]^2$ is given by $t(\mathbf{x}) = \beta_0 + \boldsymbol{\beta}_1'\mathbf{x}$, where

$$\beta_0 = \mu_y - \boldsymbol{\sigma}_{yx}'\boldsymbol{\Sigma}_{xx}^{-1}\boldsymbol{\mu}_x, \quad (10.59)$$

$$\boldsymbol{\beta}_1 = \boldsymbol{\Sigma}_{xx}^{-1}\boldsymbol{\sigma}_{yx}. \quad (10.60)$$

PROOF. See Problem 10.21. □

We can find estimators $\hat{\beta}_0$ and $\hat{\boldsymbol{\beta}}_1$ for β_0 and $\boldsymbol{\beta}_1$ in (10.59) and (10.60) by minimizing the sample mean squared error, $\sum_{i=1}^n (y_i - \hat{\beta}_0 - \hat{\boldsymbol{\beta}}_1'\mathbf{x}_i)^2/n$. The results are given in the following theorem.

Theorem 10.7c. If $(y_1, \mathbf{x}_1'), (y_2, \mathbf{x}_2'), \ldots, (y_n, \mathbf{x}_n')$ is a random sample with mean vector and covariance matrix

$$\hat{\boldsymbol{\mu}} = \begin{pmatrix} \bar{y} \\ \bar{\mathbf{x}} \end{pmatrix}, \quad \mathbf{S} = \begin{pmatrix} s_{yy} & \mathbf{s}_{yx}' \\ \mathbf{s}_{yx} & \mathbf{S}_{xx} \end{pmatrix},$$

then the estimators $\hat{\beta}_0$ and $\hat{\boldsymbol{\beta}}_1$ that minimize $\sum_{i=1}^n (y_i - \hat{\beta}_0 - \hat{\boldsymbol{\beta}}_1'\mathbf{x}_i)^2/n$ are given by

$$\hat{\beta}_0 = \bar{y} - \mathbf{s}_{yx}'\mathbf{S}_{xx}^{-1}\bar{\mathbf{x}}, \quad (10.61)$$

$$\hat{\boldsymbol{\beta}}_1 = \mathbf{S}_{xx}^{-1}\mathbf{s}_{yx}. \quad (10.62)$$

PROOF. See Problem 10.22. □

The estimators $\hat{\beta}_0$ and $\hat{\boldsymbol{\beta}}_1$ in (10.61) and (10.62) are the same as the maximum likelihood estimators in the normal case [see (10.15) and (10.16)].

10.8 SAMPLE PARTIAL CORRELATIONS

Partial correlations were introduced in Sections 4.5 and 7.10. Assuming multivariate normality, the population partial correlation $\rho_{ij\cdot rs\cdots q}$ is the correlation between y_i and y_j in the conditional distribution of $\mathbf{y}$ given $\mathbf{x}$, where y_i and y_j are in $\mathbf{y}$ and the subscripts $r, s, \ldots, q$ represent all the variables in $\mathbf{x}$. By (4.36), we obtain

$$\rho_{ij \cdot rs \ldots q} = \frac{\sigma_{ij \cdot rs \ldots q}}{\sqrt{\sigma_{ii \cdot rs \ldots q} \sigma_{jj \cdot rs \ldots q}}}, \tag{10.63}$$

where $\sigma_{ij \cdot rs \cdots q,}$ is the (ij) element of $\boldsymbol{\Sigma}_{y \cdot x} = \text{cov}(\mathbf{y}|\mathbf{x})$. For normal populations, $\boldsymbol{\Sigma}_{y \cdot x}$ is given by (4.27) as $\boldsymbol{\Sigma}_{y \cdot x} = \boldsymbol{\Sigma}_{yy} - \boldsymbol{\Sigma}_{yx}\boldsymbol{\Sigma}_{xx}^{-1}\boldsymbol{\Sigma}_{xy}$, where $\boldsymbol{\Sigma}_{yy}, \boldsymbol{\Sigma}_{yx}, \boldsymbol{\Sigma}_{xx}$, and $\boldsymbol{\Sigma}_{yx}$ are from the partitioned covariance matrix

$$\text{cov}\begin{pmatrix} \mathbf{y} \\ \mathbf{x} \end{pmatrix} = \boldsymbol{\Sigma} = \begin{pmatrix} \boldsymbol{\Sigma}_{yy} & \boldsymbol{\Sigma}_{yx} \\ \boldsymbol{\Sigma}_{xy} & \boldsymbol{\Sigma}_{xx} \end{pmatrix}$$

[see (3.33)]. The matrix of (population) partial correlations $\rho_{ij \cdot rs \ldots q}$ can be found by (4.37):

$$\mathbf{P}_{y \cdot x} = \mathbf{D}_{y \cdot x}^{-1} \boldsymbol{\Sigma}_{y \cdot x} \mathbf{D}_{y \cdot x}^{-1} = \mathbf{D}_{y \cdot x}^{-1} (\boldsymbol{\Sigma}_{yy} - \boldsymbol{\Sigma}_{yx}\boldsymbol{\Sigma}_{xx}^{-1}\boldsymbol{\Sigma}_{xy}) \mathbf{D}_{y \cdot x}^{-1}, \tag{10.64}$$

where $\mathbf{D}_{y \cdot x} = [\text{diag}(\boldsymbol{\Sigma}_{y \cdot x})]^{1/2}$.

To obtain a maximum likelihood estimator $\mathbf{R}_{y \cdot x} = (r_{ij \cdot rs \ldots q})$ of $\mathbf{P}_{y \cdot x} = (\rho_{ij \cdot rs \ldots q})$ in (10.64), we use the invariance property of maximum likelihood estimators (Theorem 10.2b) to obtain

$$\mathbf{R}_{y \cdot x} = \mathbf{D}_s^{-1} (\mathbf{S}_{yy} - \mathbf{S}_{yx}\mathbf{S}_{xx}^{-1}\mathbf{S}_{xy}) \mathbf{D}_s^{-1}, \tag{10.65}$$

where

$$\mathbf{D}_s = [\text{diag}(\mathbf{S}_{yy} - \mathbf{S}_{yx}\mathbf{S}_{xx}^{-1}\mathbf{S}_{xy})]^{1/2}.$$

The matrices $\mathbf{S}_{yy}, \mathbf{S}_{yx}, \mathbf{S}_{xx}$, and $\mathbf{S}_{xy}$ are from the sample covariance matrix partitioned by analogy to $\boldsymbol{\Sigma}$ above

$$\mathbf{S} = \begin{pmatrix} \mathbf{S}_{yy} & \mathbf{S}_{yx} \\ \mathbf{S}_{xy} & \mathbf{S}_{xx} \end{pmatrix},$$

where

$$\mathbf{S}_{yy} = \begin{pmatrix} s_{y_1}^2 & s_{y_1 y_2} & \cdots & s_{y_1 y_p} \\ s_{y_2 y_1} & s_{y_2}^2 & \cdots & s_{y_2 y_p} \\ \vdots & \vdots & & \vdots \\ s_{y_p y_1} & s_{y_p y_2} & \cdots & s_{y_p}^2 \end{pmatrix} \quad \text{and}$$

$$\mathbf{S}_{yx} = \begin{pmatrix} s_{y_1 x_1} & s_{y_1 x_2} & \cdots & s_{y_1 x_q} \\ s_{y_2 x_1} & s_{y_2 x_2} & \cdots & s_{y_2 x_q} \\ \vdots & \vdots & & \vdots \\ s_{y_p x_1} & s_{y_p x_2} & \cdots & s_{y_p x_q} \end{pmatrix}$$

are estimators of $\boldsymbol{\Sigma}_{yy}$ and $\boldsymbol{\Sigma}_{yx}$. Thus the maximum likelihood estimator of $\rho_{ij\cdot rs\ldots q}$ in (10.63) is $r_{ij\cdot rs\ldots q}$, the (ij) th element of $\mathbf{R}_{y\cdot x}$ in (10.65).

We now consider two other expressions for partial correlation and show that they are equivalent to $r_{ij\cdot rs\ldots q}$ in (10.65). To simplify exposition, we illustrate with $r_{12\cdot 3}$. The sample partial correlation of y_1 and y_2 with y_3 held fixed is usually given as

$$r_{12\cdot 3} = \frac{r_{12} - r_{13}r_{23}}{\sqrt{(1 - r_{13}^2)(1 - r_{23}^2)}}, \tag{10.66}$$

where r_{12}, r_{13}, and r_{23} are the ordinary correlations between y_1 and y_2, y_1 and y_3, and y_2 and y_3, respectively. In the following theorem, we relate $r_{12\cdot 3}$ to two previous definitions of partial correlation.

Theorem 10.8a. The expression for $r_{12\cdot 3}$ in (10.66) is equivalent to an element of $\mathbf{R}_{y\cdot x}$ in (10.65) and is also equal to $r_{y_1-\hat{y}_1,y_2-\hat{y}_2}$ from (7.94), where $y_1 - \hat{y}_1$ and $y_2 - \hat{y}_2$ are residuals from regression of y_1 on y_3 and y_2 on y_3.

Proof. We first consider $r_{y_1-\hat{y}_1,y_2-\hat{y}_2}$, which is not a maximum likelihood estimator and can therefore be used when the data are not normal. We obtain $\hat{y}_1$ and $\hat{y}_1$ by regressing y_1 on y_3 and y_2 on y_3. Using the notation in Section 7.10, we indicate the predicted value of y_1 based on regression of y_1 on y_3 as $\hat{y}_1(y_3)$. With a similar definition of $\hat{y}_2(y_3)$, the residuals can be expressed as

$$u_1 = y_1 - \hat{y}_1(y_3) = y_1 - (\hat{\beta}_{01} + \hat{\beta}_{11}y_3),$$
$$u_2 = y_2 - \hat{y}_2(y_3) = y_2 - (\hat{\beta}_{02} + \hat{\beta}_{12}y_3),$$

where, by (6.5), $\hat{\beta}_{11}$ and $\hat{\beta}_{12}$ are the usual least-squares estimators

$$\hat{\beta}_{11} = \frac{\sum_{i=1}^n (y_{1i} - \bar{y}_1)(y_{3i} - \bar{y}_3)}{\sum_{i=1}^n (y_{3i} - \bar{y}_3)^2}, \tag{10.67}$$

$$\hat{\beta}_{12} = \frac{\sum_{i=1}^n (y_{2i} - \bar{y}_2)(y_{3i} - \bar{y}_3)}{\sum_{i=1}^n (y_{3i} - \bar{y}_3)^2}. \tag{10.68}$$

Then the sample correlation between $u_1 = y_1 - \hat{y}_1(y_3)$ and $u_2 = y_2 - \hat{y}_2(y_3)$ [see (7.94)] is

$$\begin{aligned} r_{u_1u_2} &= r_{y_1-\hat{y}_1,y_2-\hat{y}_2} \\ &= \frac{\widehat{\text{cov}}(u_1, u_2)}{\sqrt{\widehat{\text{var}}(u_1)\widehat{\text{var}}(u_2)}}. \end{aligned} \tag{10.69}$$

Since the sample mean of the residuals u_1 and u_2 is 0 [see (9.11)], $r_{u_1u_2}$ can be written as

$$
\begin{aligned}
r_{u_1u_2} &= \frac{\sum_{i=1}^{n} u_{1i}u_{2i}}{\sqrt{\sum_{i=1}^{n} u_{1i}^2 \sum_{i=1}^{n} u_{2i}^2}} \\
&= \frac{\sum_{i=1}^{n} (y_{1i} - \hat{y}_{1i})(y_{2i} - \hat{y}_{2i})}{\sqrt{\sum_{i=1}^{n} (y_{1i} - \hat{y}_{1i})^2 \sum_{i=1}^{n} (y_{2i} - \hat{y}_{2i})^2}}. \qquad (10.70)
\end{aligned}
$$

We now show that $r_{u_1u_2}$ in (10.70) can be expressed as an element of $\mathbf{R}_{y\cdot x}$ in (10.65). Note that in this illustration, $\mathbf{R}_{y\cdot x}$ is 2×2. The numerator of (10.70) can be written as

$$
\begin{aligned}
\sum_{i=1}^{n} u_{1i}u_{2i} &= \sum_{i=1}^{n} (y_{1i} - \hat{y}_{1i})(y_{2i} - \hat{y}_{2i}) \\
&= \sum_{i=1}^{n} (y_{1i} - \hat{\beta}_{01} - \hat{\beta}_{11}y_{3i})(y_{2i} - \hat{\beta}_{02} - \hat{\beta}_{12}y_{3i}).
\end{aligned}
$$

Using $\hat{\beta}_{01} = \bar{y}_1 - \hat{\beta}_{11}\bar{y}_3$ and $\hat{\beta}_{02} = \bar{y}_2 - \hat{\beta}_{12}\bar{y}_3$, we obtain

$$
\begin{aligned}
\sum_{i=1}^{n} u_{1i}u_{2i} &= \sum_{i=1}^{n} [y_{1i} - \bar{y}_1 - \hat{\beta}_{11}(y_{3i} - \bar{y}_3)][y_{2i} - \bar{y}_2 - \hat{\beta}_{12}(y_{3i} - \bar{y}_3)] \\
&= \sum_{i} (y_{1i} - \bar{y}_1)(y_{2i} - \bar{y}_2) - \hat{\beta}_{11}\hat{\beta}_{12} \sum_{i} (y_{3i} - \bar{y}_3)^2. \qquad (10.71)
\end{aligned}
$$

The other two terms in (10.71) sum to zero. Using (10.67) and (10.68), the second term on the right side of (10.71) can be written as

$$
\hat{\beta}_{11}\hat{\beta}_{12} \sum_{i} (y_{3i} - \bar{y}_3)^2 = \frac{\left[\sum_{i=1}^{n} (y_{1i} - \bar{y}_1)(y_{3i} - \bar{y}_3)\right]\left[\sum_{i=1}^{n} (y_{2i} - \bar{y}_2)(y_{3i} - \bar{y}_3)\right]}{\sum_{i=1}^{n} (y_{3i} - \bar{y}_3)^2}. \qquad (10.72)
$$

If we divide (10.71) by $n - 1$, divide numerator and denominator of (10.72) by $n - 1$, and substitute (10.72) into (10.71), we obtain

$$
\widehat{\text{cov}}(u_1, u_2) = \widehat{\text{cov}}(y_1 - \hat{y}_1, y_2 - \hat{y}_2) = s_{12} - \frac{s_{13}s_{23}}{s_{33}}.
$$

This is the element in the first row and second column of $\mathbf{S}_{yy} - \mathbf{S}_{yx}\mathbf{S}_{xx}^{-1}\mathbf{S}_{xy}$ in (10.65), where $\mathbf{S}_{yy} = \begin{pmatrix} s_{11} & s_{12} \\ s_{21} & s_{22} \end{pmatrix}$, $\mathbf{S}_{yx} = \mathbf{s}_{yx} = \begin{pmatrix} s_{13} \\ s_{23} \end{pmatrix}$, $\mathbf{S}_{xx} = s_{33}$, and $\mathbf{S}_{xy} = \mathbf{s}'_{yx}$. In this case, the 2×2 matrix $\mathbf{S}_{yy} - \mathbf{S}_{yx}\mathbf{S}_{xx}^{-1}\mathbf{S}_{xy}$ is given by

$$\begin{aligned}\mathbf{S}_{yy} - \mathbf{S}_{yx}\mathbf{S}_{xx}^{-1}\mathbf{S}_{xy} &= \begin{pmatrix} s_{11} & s_{12} \\ s_{21} & s_{22} \end{pmatrix} - \frac{1}{s_{33}} \begin{pmatrix} s_{13} \\ s_{23} \end{pmatrix} (s_{13}, s_{23}) \\ &= \begin{pmatrix} s_{11} & s_{12} \\ s_{21} & s_{22} \end{pmatrix} - \frac{1}{s_{33}} \begin{pmatrix} s_{13}^2 & s_{13}s_{23} \\ s_{23}s_{13} & s_{23}^2 \end{pmatrix}.\end{aligned}$$

Thus $r_{u_1u_2}$, as based on residuals in (10.69), is equivalent to the maximum likelihood estimator in (10.65).

We now use (10.71) to convert $r_{u_1u_2}$ in (10.69) into the familiar formula for $r_{12\cdot3}$ given in (10.66). By (10.70), we obtain

$$r_{u_1u_2} = \frac{\sum_i u_{1i}u_{2i}}{\sqrt{\sum_i u_{1i}^2 \sum_i u_{2i}^2}}. \tag{10.73}$$

By an extension of (10.71), we further obtain

$$\sum_{i=1}^{n} u_{1i}^2 = \sum_i (y_{1i} - \bar{y}_1)^2 - \hat{\beta}_{11}^2 \sum_i (y_{3i} - \bar{y}_3)^2, \tag{10.74}$$

$$\sum_{i=1}^{n} u_{2i}^2 = \sum_i (y_{2i} - \bar{y}_2)^2 - \hat{\beta}_{12}^2 \sum_i (y_{3i} - \bar{y}_3)^2. \tag{10.75}$$

Then (10.73) becomes

$$r_{u_1u_2} = \frac{\sum_i (y_{1i} - \bar{y}_1)(y_{2i} - \bar{y}_2) - \hat{\beta}_{11}\hat{\beta}_{12}\sum_i (y_{3i} - \bar{y}_3)^2}{\sqrt{\left[\sum_i (y_{1i} - \bar{y}_1)^2 - \hat{\beta}_{11}^2 \sum_i (y_{3i} - \bar{y}_3)^2\right]\left[\sum_i (y_{2i} - \bar{y}_2)^2 - \hat{\beta}_{12}^2 \sum_i (y_{3i} - \bar{y}_3)^2\right]}}. \tag{10.76}$$

We now substitute for $\hat{\beta}_{11}$ and $\hat{\beta}_{12}$ as defined in (10.67) and (10.68) and divide numerator and denominator by $\sqrt{\sum_i (y_{1i} - \bar{y}_1)^2 \sum_i (y_{2i} - \bar{y}_2)^2}$ to obtain

$$r_{u_1u_2} = r_{12\cdot3} = \frac{r_{12} - r_{13}r_{23}}{\sqrt{(1 - r_{13}^2)(1 - r_{23}^2)}}. \tag{10.77}$$

Thus $r_{u_1u_2}$ based on residuals as in (10.69) is equivalent to the usual formulation $r_{12\cdot3}$ in (10.66). □

For the general case $r_{ij\cdot rs\ldots q}$, where i and j are subscripts pertaining to $\mathbf{y}$ and $r, s, \ldots, q$ are all the subscripts associated with $\mathbf{x}$, we define a residual vector $\mathbf{y}_i - \hat{\mathbf{y}}_i(\mathbf{x})$, where $\hat{\mathbf{y}}_i(\mathbf{x})$ is the vector of predicted values from the regression of $\mathbf{y}$ on $\mathbf{x}$. [Note that i is used differently in $r_{ij\cdot rs\ldots q}$ and $\mathbf{y}_i - \hat{\mathbf{y}}_i(\mathbf{x})$.] In Theorem 10.8a, $r_{12\cdot 3}$ was found to be equal to $r_{y_1-\hat{y}_1, y_2-\hat{y}_2}$, the ordinary correlation of the two residuals, and to be equivalent to the partial correlation defined as an element of $\mathbf{R}_{y\cdot x}$ in (10.65). In the following theorem, this is extended to the vectors $\mathbf{y}$ and $\mathbf{x}$.

Theorem 10.8b. The sample covariance matrix of the residual vector $\mathbf{y}_i - \hat{\mathbf{y}}_i(\mathbf{x})$ is equivalent to $\mathbf{S}_{yy} - \mathbf{S}_{yx}\mathbf{S}_{xx}^{-1}\mathbf{S}_{xy}$ in (10.65), that is, $\mathbf{S}_{\mathbf{y}-\hat{\mathbf{y}}} = \mathbf{S}_{yy} - \mathbf{S}_{yx}\mathbf{S}_{xx}^{-1}\mathbf{S}_{xy}$.

PROOF. The sample predicted value $\hat{\mathbf{y}}_i(\mathbf{x})$ is an estimator of $E(\mathbf{y}|\mathbf{x}_i) = \boldsymbol{\mu}_y + \boldsymbol{\Sigma}_{yx}\boldsymbol{\Sigma}_{xx}^{-1}(\mathbf{x}_i - \boldsymbol{\mu}_x)$ given in (4.26). For $\hat{\mathbf{y}}_i(\mathbf{x})$, we use the maximum likelihood estimator of $E(\mathbf{y}|\mathbf{x}_i)$:

$$\hat{\mathbf{y}}_i(\mathbf{x}) = \bar{\mathbf{y}} + \mathbf{S}_{yx}\mathbf{S}_{xx}^{-1}(\mathbf{x}_i - \bar{\mathbf{x}}). \tag{10.78}$$

[The same result can be obtained without reference to normality; see Rencher (1998, p. 304).]

Since the sample mean of $\mathbf{y}_i - \hat{\mathbf{y}}_i(\mathbf{x})$ is $\mathbf{0}$ (see Problem 10.26), the sample covariance matrix of $\mathbf{y}_i - \hat{\mathbf{y}}_i(\mathbf{x})$ is defined as

$$\mathbf{S}_{\mathbf{y}-\hat{\mathbf{y}}} = \frac{1}{n-1}\sum_{i=1}^{n}[\mathbf{y}_i - \hat{\mathbf{y}}_i(\mathbf{x})][\mathbf{y}_i - \hat{\mathbf{y}}_i(\mathbf{x})]' \tag{10.79}$$

(see Problem 10.1).We first note that by extension of (10.13), we have $\mathbf{S}_{yy} = \sum_i (\mathbf{y}_i - \bar{\mathbf{y}})(\mathbf{y}_i - \bar{\mathbf{y}})'/(n-1)$, $\mathbf{S}_{yx} = \sum_i (\mathbf{y}_i - \bar{\mathbf{y}})(\mathbf{x}_i - \bar{\mathbf{x}})'/(n-1)$, and $\mathbf{S}_{xx} = \sum_i (\mathbf{x}_i - \bar{\mathbf{x}})(\mathbf{x}_i - \bar{\mathbf{x}})'/(n-1)$ (see Problem 10.27). Using these expressions, after substituting (10.78) in (10.79), we obtain

$$\begin{aligned}
\mathbf{S}_{\mathbf{y}-\hat{\mathbf{y}}} &= \frac{1}{n-1}\sum_{i=1}^{n}[\mathbf{y}_i - \bar{\mathbf{y}} - \mathbf{S}_{yx}\mathbf{S}_{xx}^{-1}(\mathbf{x}_i - \bar{\mathbf{x}})][\mathbf{y}_i - \bar{\mathbf{y}} - \mathbf{S}_{yx}\mathbf{S}_{xx}^{-1}(\mathbf{x}_i - \bar{\mathbf{x}})]' \\
&= \frac{1}{n-1}\left[\sum_{i=1}^{n}(\mathbf{y}_i - \bar{\mathbf{y}})(\mathbf{y}_i - \bar{\mathbf{y}})' - \sum_{i=1}^{n}(\mathbf{y}_i - \bar{\mathbf{y}})(\mathbf{x}_i - \bar{\mathbf{x}})'\mathbf{S}_{xx}^{-1}\mathbf{S}_{xy}\right. \\
&\quad \left. -\mathbf{S}_{yx}\mathbf{S}_{xx}^{-1}\sum_{i=1}^{n}(\mathbf{x}_i - \bar{\mathbf{x}})(\mathbf{y}_i - \bar{\mathbf{y}})' + \mathbf{S}_{yx}\mathbf{S}_{xx}^{-1}\sum_{i=1}^{n}(\mathbf{x}_i - \bar{\mathbf{x}})(\mathbf{x}_i - \bar{\mathbf{x}})'\mathbf{S}_{xx}^{-1}\mathbf{S}_{xy}\right] \\
&= \mathbf{S}_{yy} - \mathbf{S}_{yx}\mathbf{S}_{xx}^{-1}\mathbf{S}_{xy} - \mathbf{S}_{yx}\mathbf{S}_{xx}^{-1}\mathbf{S}_{xy} + \mathbf{S}_{yx}\mathbf{S}_{xx}^{-1}\mathbf{S}_{xx}\mathbf{S}_{xx}^{-1}\mathbf{S}_{xy} \\
&= \mathbf{S}_{yy} - \mathbf{S}_{yx}\mathbf{S}_{xx}^{-1}\mathbf{S}_{xy}.
\end{aligned}$$

Thus the covariance matrix of residuals gives the same result as the maximum likelihood estimator of conditional covariances and correlations in (10.65). □

Example 10.8. We illustrate some partial correlations for the hematology data in Table 10.1. To find $r_{y1\cdot 2345}$, for example, we use (10.65), $\mathbf{R}_{y\cdot x} = \mathbf{D}_s^{-1}(\mathbf{S}_{yy} - \mathbf{S}_{yx}\mathbf{S}_{xx}^{-1}\mathbf{S}_{xy})\mathbf{D}_s^{-1}$. In this case, $\mathbf{y} = (y, x_1)'$ and $\mathbf{x} = (x_2, x_3, x_4, x_5)'$. The matrix $\mathbf{S}$ is therefore partitioned as

$$\mathbf{S} = \left(\begin{array}{cc|cccc} 90.290 & 1.535 & 4.880 & 106.202 & 3.753 & 3.064 \\ 1.535 & 0.691 & 1.494 & 3.255 & 0.422 & -0.268 \\ \hline 4.880 & 1.494 & 5.401 & 10.155 & 1.374 & 1.292 \\ 106.202 & 3.255 & 10.155 & 200.668 & 64.655 & 4.067 \\ 3.753 & 0.422 & 1.374 & 64.655 & 56.374 & 0.579 \\ 3.064 & -0.268 & 1.292 & 4.067 & 0.579 & 18.078 \end{array}\right)$$

$$= \begin{pmatrix} \mathbf{S}_{yy} & \mathbf{S}_{yx} \\ \mathbf{S}_{xy} & \mathbf{S}_{xx} \end{pmatrix}.$$

The matrix $\mathbf{D}_s = [\text{diag}(\mathbf{S}_{yy} - \mathbf{S}_{yx}\mathbf{S}_{xx}^{-1}\mathbf{S}_{xy})]^{1/2}$ is given by

$$\mathbf{D}_s = \begin{pmatrix} 2.645 & 0 \\ 0 & .503 \end{pmatrix},$$

and we have

$$\mathbf{R}_{y\cdot x} = \begin{pmatrix} 1.0000 & -0.0934 \\ -0.0934 & 1.000 \end{pmatrix}.$$

Thus, $r_{y1\cdot 2345} = -.0934$. On the other hand, $r_{y1} = .1934$.

To find $r_{y2\cdot 1345}$, we have $\mathbf{y} = (y, x_2)'$ and $\mathbf{x} = (x_1, x_3, x_4, x_5)'$. Thus

$$\mathbf{S}_{yy} = \begin{pmatrix} 90.290 & 4.880 \\ 4.880 & 5.401 \end{pmatrix},$$

and there are corresponding matrices for $\mathbf{S}_{yx}$, $\mathbf{S}_{xy}$, and $\mathbf{S}_{xx}$. The diagonal matrix $\mathbf{D}_s$ is given by $\mathbf{D}_s = \text{diag}(2.670, 1.389)$, and we have

$$\mathbf{R}_{y\cdot x} = \begin{pmatrix} 1.000 & -0.164 \\ -0.164 & 1.000 \end{pmatrix}.$$

Thus, $r_{y2\cdot 1345} = -.164$, which can be compared with $r_{y2} = .221$.

To find $r_{y3\cdot45}$, we have $\mathbf{y} = (y, x_1, x_2, x_3)'$ and $\mathbf{x} = (x_4, x_5)'$. Then, for example, we obtain

$$\mathbf{S}_{yy} = \begin{pmatrix} 90.290 & 1.535 & 4.880 & 106.202 \\ 1.535 & 0.691 & 1.494 & 3.255 \\ 4.880 & 1.494 & 5.401 & 10.155 \\ 106.202 & 3.255 & 10.155 & 200.668 \end{pmatrix}.$$

The diagonal matrix $\mathbf{D}_s$ is given by

$$\mathbf{D}_s = \text{diag}(9.462, .827, 2.297, 11.219),$$

and we have

$$\mathbf{R}_{y\cdot x} = \begin{pmatrix} 1.000 & 0.198 & 0.210 & 0.954 \\ 0.198 & 1.000 & 0.792 & 0.304 \\ 0.210 & 0.792 & 1.000 & 0.324 \\ 0.954 & 0.304 & 0.324 & 1.000 \end{pmatrix}.$$

Thus, for example, $r_{y1\cdot45} = .198, r_{y3\cdot45} = .954, r_{12\cdot45} = .792,$ and $r_{23\cdot45} = .324$. In this case, $\mathbf{R}_{y\cdot x}$ is little changed from $\mathbf{R}_{yy}$:

$$\mathbf{R}_{yy} = \begin{pmatrix} 1.000 & 0.194 & 0.221 & 0.789 \\ 0.194 & 1.000 & 0.774 & 0.277 \\ 0.221 & 0.774 & 1.000 & 0.308 \\ 0.789 & 0.277 & 0.308 & 1.000 \end{pmatrix}.$$

□

PROBLEMS

10.1 Show that $\mathbf{S}$ in (10.14) can be found as $\mathbf{S} = \sum_{i=1}^{n} (\mathbf{v}_i - \bar{\mathbf{v}})(\mathbf{v}_i - \bar{\mathbf{v}})'/(n-1)$ as in (10.13).

10.2 Show that $\hat{\boldsymbol{\mu}}$ and $\mathbf{S}$ in (10.9) and (10.10) are jointly sufficient for $\boldsymbol{\mu}$ and $\boldsymbol{\Sigma}$, as noted following Theorem 10.2c.

10.3 Show that $\mathbf{S} = \mathbf{DRD}$ gives the partitioned result in (10.19).

10.4 Show that $\text{cov}(y, w) = \boldsymbol{\sigma}'_{yx}\boldsymbol{\Sigma}_{xx}^{-1}\boldsymbol{\sigma}_{yx}$ and $\text{var}(w) = \boldsymbol{\sigma}'_{yx}\boldsymbol{\Sigma}_{xx}^{-1}\boldsymbol{\sigma}_{yx}$ as in (10.26), where $w = \mu_y + \boldsymbol{\sigma}'_{yx}\boldsymbol{\Sigma}_{xx}^{-1}(\mathbf{x} - \boldsymbol{\mu}_x)$.

10.5 Show that $\rho^2_{y|\mathbf{x}}$ in (10.27) is the maximum squared correlation between y and any linear function of $\mathbf{x}$, as in (10.28).

10.6 Show that $\rho^2_{y|\mathbf{x}}$ can be expressed as $\rho^2_{y|\mathbf{x}} = 1 - |\boldsymbol{\Sigma}|/(\sigma_{yy}|\boldsymbol{\Sigma}_{xx}|)$ as in (10.29).

10.7 Show that $\rho^2_{y|\mathbf{x}}$ is invariant to linear transformations $u = ay$ and $\mathbf{v} = \mathbf{Bx}$, where $\mathbf{B}$ is nonsingular, as in (10.30).

10.8 Show that $\text{cov}(y - w, \mathbf{x}) = \mathbf{0}'$ as in (10.33).

10.9 Verify that $R^2 = r_{y\hat{y}}^2$, as in (10.36), using the following two definitions of $r_{y\hat{y}}^2$:

(a) $r_{y\hat{y}}^2 = \left[\sum_{i=1}^n (y_i - \bar{y}_i)(\hat{y}_i - \bar{\hat{y}})\right]^2 / \left[\sum_{i=1}^n (y_i - \bar{y})^2 \sum_{i=1}^n (\hat{y}_i - \bar{\hat{y}})^2\right]$

(b) $r_{y\hat{y}} = s_{y\hat{y}}/(s_y s_{\hat{y}})$

10.10 Show that $R^2 = \max_{\mathbf{a}} r_{y,\mathbf{a}'\mathbf{x}}^2$ as in (10.37).

10.11 Show that $R^2 = \mathbf{r}_{yx}'\mathbf{R}_{xx}^{-1}\mathbf{r}_{yx}$ as in (10.38) .

10.12 Show that $R^2 = 1 - 1/r^{yy}$ as in (10.39), where r^{yy} is the upper left-hand diagonal element of $\mathbf{R}^{-1}$, with $\mathbf{R}$ partitioned as in (10.18).

10.13 Verify that R^2 can be expressed in terms of determinants as in (10.40) and (10.41).

10.14 Show that R^2 is invariant to full-rank linear transformations on y or the x's, as in property 9 in Section 10.4.

10.15 Show that $\hat{\boldsymbol{\Sigma}}_0$ in (10.46) is the maximum likelihood estimator of $\boldsymbol{\Sigma}_0$ in (10.45) and that $\max_{H_0} L(\boldsymbol{\mu}, \boldsymbol{\Sigma})$ is given by (10.47).

10.16 Show that LR in (10.48) is equal to $\text{LR} = (1 - R^2)^{n/2}$ in (10.49).

10.17 Obtain the confidence interval in (10.57) from the inequality in (10.56).

10.18 Suppose that we have three independent samples of bivariate normal data. The three sample correlations are r_1, r_2, and r_3 based, respectively, on sample sizes n_1, n_2, and n_3.

(a) Find the covariance matrix $\mathbf{V}$ of $\mathbf{z} = (z_1 \quad z_2 \quad z_3)'$ where $z_i = \frac{1}{2}\ln[(1 + r_i)/(1 - r_i)]$.

(b) Let $\boldsymbol{\mu}_z' = (\tanh^{-1}\rho_1, \tanh^{-1}\rho_2, \tanh^{-1}\rho_3)$, and let

$$\mathbf{C} = \begin{pmatrix} 1 & -1 & 0 \\ 1 & 0 & -1 \end{pmatrix}.$$

Find the distribution of $[\mathbf{C}(\mathbf{z} - \boldsymbol{\mu}_z)]' \, [\mathbf{C}\mathbf{V}\mathbf{C}']^{-1}[\mathbf{C}(\mathbf{z} - \boldsymbol{\mu}_z)]$.

(c) Using (b), propose a test for $H_0 : \rho_1 = \rho_2 = \rho_3$ or equivalently $H_0 : \mathbf{C}\boldsymbol{\mu}_z = \mathbf{0}$.

10.19 Prove Theorem 10.6.

10.20 Show that if z were orthogonal to the x's, (10.58) could be written in the form $R_{yw}^2 = R_{yx}^2 + r_{yz}^2$, as noted following Theorem 10.6.

10.21 Prove Theorem 10.7b.

10.22 Prove Theorem 10.7c.

10.23 Show that $\sum_{i=1}^{n} u_{1i}u_{2i} = \sum_{i=1}^{n} (y_{1i} - \bar{y}_1)(y_{2i} - \bar{y}_2) - \hat{\beta}_{11}\hat{\beta}_{12} \sum_{i=1}^{n} (y_{3i} - \bar{y}_3)^2$ as in (10.71).

10.24 Show that $\sum_{i=1}^{n} u_{1i}^2 = \sum_{i=1}^{n} (y_{1i} - \bar{y}_1)^2 - \hat{\beta}_{11}^2 \sum_{i=1}^{n} (y_{3i} - \bar{y}_3)^2$ as in (10.74).

10.25 Obtain $r_{12\cdot 3}$ in (10.77) from $r_{u_1 u_2}$ in (10.76).

10.26 Show that $\Sigma_{i=1}^{n} [\mathbf{y}_i - \hat{\mathbf{y}}_i(\mathbf{x})] = \mathbf{0}$, as noted following (10.78).

10.27 Show that $\mathbf{S}_{yy} = \sum_i (\mathbf{y}_i - \bar{\mathbf{y}})(\mathbf{y}_i - \bar{\mathbf{y}})'/(n-1)$, $\mathbf{S}_{yx} = \sum_i (\mathbf{y}_i - \bar{\mathbf{y}})(\mathbf{x}_i - \bar{\mathbf{x}})'/(n-1)$, and $\mathbf{S}_{xx} = \sum_i (\mathbf{x}_i - \bar{\mathbf{x}})(\mathbf{x}_i - \bar{\mathbf{x}})'/(n-1)$, as noted following (10.79).

10.28 In an experiment with rats, the concentration of a particular drug in the liver was of interest. For 19 rats the following variables were observed:

$$y = \text{percentage of the dose in the liver}$$
$$x_1 = \text{body weight}$$
$$x_2 = \text{liver weight}$$
$$x_3 = \text{relative dose}$$

The data are given in Table 10.2 (Weisberg 1985, p. 122).

(a) Find $\mathbf{S}_{xx}$, $\mathbf{s}_{yx}$, $\hat{\boldsymbol{\beta}}_1$, $\hat{\beta}_0$, and s^2.

(b) Find $\mathbf{R}_{xx}$, $\mathbf{r}_{yx}$, and $\hat{\boldsymbol{\beta}}_1^*$.

(c) Find R^2.

(d) Test $H_0 : \boldsymbol{\beta}_1 = \mathbf{0}$.

10.29 Use the hematology data in Table 10.1 as divided into two subsamples of sizes 26 and 25 in Example 10.5b (the first 26 observations and the last 25 observations). For each pair of variables below, find r_1 and r_2 for the two subsamples, find z_1 and z_2 as in (10.51), test $H_0 : \rho_1 = \rho_2$ as in (10.55), and find confidence limits for ρ_1 and ρ_2 as in (10.57).

(a) y and x_2

(b) y and x_3

TABLE 10.2 Rat Data

y	x_1	x_2	x_3	y	x_1	x_2	x_3
.42	176	6.5	0.88	.27	158	6.9	.80
.25	176	9.5	0.88	.36	148	7.3	.74
.56	190	9.0	1.00	.21	149	5.2	.75
.23	176	8.9	0.88	.28	163	8.4	.81
.23	200	7.2	1.00	.34	170	7.2	.85
.32	167	8.9	0.83	.28	186	6.8	.94
.37	188	8.0	0.94	.30	164	7.3	.73
.41	195	10.0	0.98	.37	181	9.0	.90
.33	176	8.0	0.88	.46	149	6.4	.75
.38	165	7.9	0.84				

(**c**) y and x_4

(**d**) y and x_5

10.30 For the rat data in Table 10.2, check the effect of each variable on R^2 as in Section 10.6.

10.31 Using the rat data in Table 10.2.

(**a**) Find $r_{y1\cdot 23}$ and compare to r_{y1}.

(**b**) Find $r_{y2\cdot 13}$

(**c**) Find $\mathbf{R}_{y\cdot x}$, where $\mathbf{y} = (y, x_1, x_2)'$ and $\mathbf{x} = x_3$, in order to obtain $r_{y1\cdot 3}, r_{y2\cdot 3}$, and $r_{12\cdot 3}$.

11 Multiple Regression: Bayesian Inference

We now consider Bayesian estimation and prediction for the multiple linear regression model in which the x variables are fixed constants as in Chapters 7–9. The Bayesian statistical paradigm is conceptually simple and general because inferences involve only probability calculations as opposed to maximization of a function like the log likelihood. On the other hand, the probability calculations usually entail complicated or even intractable integrals. The Bayesian approach has become popular more recently because of the development of computer-intensive approximations to these integrals (Evans and Swartz 2000) and user-friendly programs to carry out the computations (Gilks et al. 1998). We discuss both analytical and computer-intensive approaches to the Bayesian multiple regression model.

Throughout Chapters 7 and 8 we assumed that the parameters $\boldsymbol{\beta}$ and σ^2 were unknown fixed constants. We couldn't really do otherwise because to this point (at least implicitly) we have only allowed probability distributions to represent variability due to such things as random sampling or imprecision of measurement instruments. The Bayesian approach additionally allows probability distributions to represent conjectural uncertainty. Thus $\boldsymbol{\beta}$ and σ^2 can be treated as if they are random variables because we are uncertain about their values. The technical property that allows one to treat parameters as random variables is *exchangeability* of the observational units in the study (Lindley and Smith 1972).

11.1 ELEMENTS OF BAYESIAN STATISTICAL INFERENCE

In Bayesian statistics, uncertainty about the value of a parameter is expressed using the tools of probability theory (e.g., a density function—see Section 3.2). Density functions of parameters like $\boldsymbol{\beta}$ and σ^2 reflect the current credibility of possible values for these parameters. The goal of the Bayesian approach is to use data to update the uncertainty distributions for parameters, and then draw sensible conclusions using these updated distributions.

The Bayesian approach can be used in any inference situation. However, it seems especially natural in the following type of problem. Consider an industrial process in which it is desired to estimate β_0 and β_1 for the straight-line relationship in (6.1) between a response y and a predictor x for a particular batch of product. Suppose that it is known from experience that β_0 and β_1 vary randomly from batch to batch. Bayesian inference allows historical (or prior) knowledge of the distributions of β_0 and β_1 among batches to be expressed in probabilistic form, and then to be combined with (x, y) data from a specific batch in order to give improved estimates of β_0 and β_1 for that specific batch.

Bayesian inference is based on two general equations. In these equations as presented below, $\boldsymbol{\theta}$ is a vector of m continuous parameters, $\mathbf{y}$ is a vector of n continuous observations, and f, g, h, k, p, q, r and t are probability density functions.

We begin with the definition of the conditional density of $\boldsymbol{\theta}$ given $\mathbf{y}$ [see (3.18)]

$$g(\boldsymbol{\theta} \mid \mathbf{y}) = \frac{k(\mathbf{y}, \boldsymbol{\theta})}{h(\mathbf{y})}, \tag{11.1}$$

where $k(\mathbf{y}, \boldsymbol{\theta})$ is the joint density of $y_1, y_2, \ldots, y_n$ and $\theta_1, \theta_2, \ldots, \theta_m$. Using the definition of the conditional density $f(\mathbf{y} \mid \boldsymbol{\theta})$, we can write $k(\mathbf{y}, \boldsymbol{\theta}) = f(\mathbf{y} \mid \boldsymbol{\theta}) p(\boldsymbol{\theta})$, and (11.1) becomes

$$g(\boldsymbol{\theta} \mid \mathbf{y}) = \frac{f(\mathbf{y} \mid \boldsymbol{\theta}) p(\boldsymbol{\theta})}{h(\mathbf{y})}, \tag{11.2}$$

an expression that is commonly referred to as *Bayes' theorem*. By an extension of (3.13), the marginal density $h(\mathbf{y})$ can be obtained by integrating $\boldsymbol{\theta}$ out of $k(\mathbf{y}, \boldsymbol{\theta}) = f(\mathbf{y} \mid \boldsymbol{\theta}) p(\boldsymbol{\theta})$ so that (11.2) becomes

$$\begin{aligned} g(\boldsymbol{\theta} \mid \mathbf{y}) &= \frac{f(\mathbf{y} \mid \boldsymbol{\theta}) p(\boldsymbol{\theta})}{\int_{-\infty}^{\infty} \cdots \int_{-\infty}^{\infty} f(\mathbf{y} \mid \boldsymbol{\theta}) p(\boldsymbol{\theta}) d\boldsymbol{\theta}} \\ &= c f(\mathbf{y} \mid \boldsymbol{\theta}) p(\boldsymbol{\theta}), \end{aligned} \tag{11.3}$$

where $d\boldsymbol{\theta} = d\theta_1 \cdots d\theta_m$. In this expression, $p(\boldsymbol{\theta})$ is known as the *prior density* of $\boldsymbol{\theta}$, and $g(\boldsymbol{\theta} \mid \mathbf{y})$ is called the *posterior density* of $\boldsymbol{\theta}$. The definite integral in the denominator of (11.3) is often replaced by a constant (c) because after integration, it no longer involves the random vector $\boldsymbol{\theta}$. This definite integral is often very complicated, but can sometimes be obtained by noting that c is a *normalizing constant*, that is, a value such that the posterior density integrates to 1. Rearranging this expression and reinterpreting the joint density function $f(\mathbf{y} \mid \boldsymbol{\theta})$ of the data as the likelihood function $L(\boldsymbol{\theta} \mid \mathbf{y})$ (see Section 7.6.2), we obtain

$$g(\boldsymbol{\theta} \mid \mathbf{y}) = c p(\boldsymbol{\theta}) L(\boldsymbol{\theta} \mid \mathbf{y}). \tag{11.4}$$

Thus (11.2), the first general equation of Bayesian inference, merely states that the posterior density of $\boldsymbol{\theta}$ given the data (representing the updated uncertainty in $\boldsymbol{\theta}$) is proportional to the prior density of $\boldsymbol{\theta}$ times the likelihood function. Point and interval estimates of the parameters are taken as mathematical features of this joint posterior density or associated marginal posterior densities of individual parameters θ_i. For example, the mode or mean of the marginal posterior density of a parameter may be used as a point estimate of the parameter. A central or highest density interval (Gelman et al. 2004, pp. 38–39) over which the marginal posterior density of a parameter integrates to $1 - \omega$ may be taken as a $100(1 - \omega)\%$ interval estimate of the parameter.

For the second general equation of Bayesian inference, we consider a future observation y_0. In the Bayesian approach, y_0 is not independent of $\mathbf{y}$ as was assumed in Section 8.6.5 because its density depends on $\boldsymbol{\theta}$, a random vector whose current uncertainty depends on $\mathbf{y}$. Since y_0, $\mathbf{y}$ and $\boldsymbol{\theta}$ are jointly distributed, the posterior predictive density of y_0 given $\mathbf{y}$ is obtained by integrating $\boldsymbol{\theta}$ out of the joint conditional density of y_0 and $\boldsymbol{\theta}$ given $\mathbf{y}$:

$$
\begin{aligned}
r(y_0|\mathbf{y}) &= \int_{-\infty}^{\infty} \cdots \int_{-\infty}^{\infty} t(y_0, \boldsymbol{\theta}|\mathbf{y}) d\boldsymbol{\theta} \\
&= \int_{-\infty}^{\infty} \cdots \int_{-\infty}^{\infty} q(y_0|\boldsymbol{\theta}, \mathbf{y}) g(\boldsymbol{\theta}|\mathbf{y}) d\boldsymbol{\theta} \quad \text{[by (4.28)]}
\end{aligned}
$$

where $q(y_0|\boldsymbol{\theta}, \mathbf{y})$ is the conditional density function of the sampling distribution for a future observation y_0. Since y_0 is dependent on $\mathbf{y}$ only through $\boldsymbol{\theta}$, $q(y_0|\boldsymbol{\theta}, \mathbf{y})$ simplifies, and we have

$$
r(y_0|\mathbf{y}) = \int_{-\infty}^{\infty} \cdots \int_{-\infty}^{\infty} q(y_0|\boldsymbol{\theta}) g(\boldsymbol{\theta}|\mathbf{y}) \, d\boldsymbol{\theta}. \tag{11.5}
$$

Equation (11.5) expresses the intuitive idea that uncertainty associated with the predicted value of a future observation has two components: sampling variability and uncertainty in the parameters. As before, point and interval predictions can be taken as mathematical features (such as the mean, mode, or specified integral) of this posterior predictive density.

11.2 A BAYESIAN MULTIPLE LINEAR REGRESSION MODEL

Bayesian multiple regression models are similar to the classical multiple regression model (see Section 7.6.1) except that they include specifications of the prior distributions for the parameters. Prior specification is an important part of the art and practice of Bayesian modeling, but since the focus of this text is the basic theory of linear models, we discuss only one set of prior specifications—one that is chosen for its mathematical convenience rather than actual prior information.

11.2.1 A Bayesian Multiple Regression Model with a Conjugate Prior

Although not necessary, it is often convenient to parameterize Bayesian models using *precision* (τ) rather than variance (σ^2), where

$$\tau = \frac{1}{\sigma^2}.$$

Using this parameterization, as an example of a Bayesian linear regression model, let

$$\mathbf{y}|\boldsymbol{\beta}, \tau \text{ be } N_n\left(X\boldsymbol{\beta}, \frac{1}{\tau}\mathbf{I}\right),$$

$$\boldsymbol{\beta}|\tau \text{ be } N_{k+1}\left(\boldsymbol{\phi}, \frac{1}{\tau}\mathbf{V}\right),$$

$$\tau \text{ be gamma}(\alpha, \delta).$$

The second and third distributions here are prior distributions, and we assume that $\boldsymbol{\phi}$, $\mathbf{V}$, α, and δ (the parameters of the prior distributions), are known. Although we will not do so here, this model could be extended by specifying *hyperprior* distributions for $\boldsymbol{\phi}$, $\mathbf{V}$, α, and δ (Lindley and Smith 1972).

As in previous chapters, the number of predictor variables is denoted by k (so that the rank of $\mathbf{X}$ is $k+1$) and the number of observations by n. The prior density function for $\boldsymbol{\beta}|\tau$ is, using (4.9)

$$p_1(\boldsymbol{\beta}|\tau) = \frac{1}{(2\pi)^{(k+1)/2}|\tau^{-1}\mathbf{V}|^{\frac{1}{2}}} e^{-\tau(\boldsymbol{\beta}-\boldsymbol{\phi})'\mathbf{V}^{-1}(\boldsymbol{\beta}-\boldsymbol{\phi})/2}. \tag{11.6}$$

The prior density function for τ is the gamma density (Gelman et al. 2004, pp. 574–575)

$$p_2(\tau) = \frac{\delta^\alpha}{\Gamma(\alpha)} \tau^{\alpha-1} e^{-\delta\tau}, \tag{11.7}$$

where $\alpha > 0, \delta > 0$, and by definition

$$\Gamma(\alpha) = \int_0^\infty x^{\alpha-1} e^{-x} dx \tag{11.8}$$

(see any advanced calculus text). For the gamma density in (11.7),

$$E(\tau) = \frac{\alpha}{\delta} \quad \text{and} \quad \text{var}(\tau) = \frac{\alpha}{\delta^2}.$$

These prior distributions could be formulated with small enough variances that the prior knowledge strongly influences posterior distributions of the parameters in the model. If so, they are called *informative priors*. On the other hand, both of these priors could be formulated with large variances so that they have very little effect on the posterior distributions. If so, they are called *diffuse priors*. The priors would be diffuse if, for example, $\mathbf{V}$ in (11.6) were a diagonal matrix with very large diagonal elements, and if δ in (11.7) were very close to zero.

The prior specifications in (11.6) and (11.7) are flexible and reasonable, and they also have nice mathematical properties, as will be shown in Theorem 11.2a. Other specifications for the prior distributions could be used. However, even the minor modification of proposing a prior distribution for $\boldsymbol{\beta}$ that is not conditional on τ makes the model far less mathematically tractable.

The joint prior for $\boldsymbol{\beta}$ and τ in our model is called a *conjugate prior* because its use results in a posterior distribution of the same form as the prior. We prove this in the following theorem.

Theorem 11.2a. Consider the Bayesian multiple regression model in which $\mathbf{y}|\boldsymbol{\beta}, \tau$ is $N_n(\mathbf{X}\boldsymbol{\beta}, \tau^{-1}\mathbf{I})$, $\boldsymbol{\beta}|\tau$ is $N_{k+1}(\boldsymbol{\phi}, \tau^{-1}\mathbf{V})$, and τ is gamma(α, δ). The joint prior distribution is conjugate, that is, $g(\boldsymbol{\beta}, \tau|\mathbf{y})$ is of the same form as $p(\boldsymbol{\beta}, \tau)$.

PROOF. Combining (11.6) and (11.7), the joint prior density is

$$\begin{aligned} p(\boldsymbol{\beta}, \tau) &= p_1(\boldsymbol{\beta}|\tau)p_2(\tau) \\ &= c_1 \tau^{(k+1/)2} e^{-\tau(\boldsymbol{\beta}-\boldsymbol{\phi})'\mathbf{V}^{-1}(\boldsymbol{\beta}-\boldsymbol{\phi})/2} \tau^{\alpha-1} e^{-\delta\tau} \\ &= c_1 \tau^{(\alpha_*+k+1)/2} e^{-\tau[(\boldsymbol{\beta}-\boldsymbol{\phi})'\mathbf{V}^{-1}(\boldsymbol{\beta}-\boldsymbol{\phi})+\delta_*]/2}, \end{aligned} \tag{11.9}$$

where $\alpha_* = 2\alpha - 2$, $\delta_* = 2\delta$ and all the factors not involving random variables are collected into the normalizing constant c_1. Using (11.4), the joint posterior density is then

$$\begin{aligned} g(\boldsymbol{\beta}, \tau|\mathbf{y}) &= cp(\boldsymbol{\beta}, \tau)L(\boldsymbol{\beta}, \tau\,|\,\mathbf{y}) \\ &= c_2 \tau^{(\alpha_*+k+1)/2} e^{-\tau[(\boldsymbol{\beta}-\boldsymbol{\phi})'\mathbf{V}^{-1}(\boldsymbol{\beta}-\boldsymbol{\phi})+\delta_*]/2} \tau^{n/2} e^{-\tau(\mathbf{y}-\mathbf{X}\boldsymbol{\beta})'(\mathbf{y}-\mathbf{X}\boldsymbol{\beta})/2} \\ &= c_2 \tau^{(\alpha_{**}+k+1)/2} e^{-\tau[(\boldsymbol{\beta}-\boldsymbol{\phi})'\mathbf{V}^{-1}(\boldsymbol{\beta}-\boldsymbol{\phi})+(\mathbf{y}-\mathbf{X}\boldsymbol{\beta})'(\mathbf{y}-\mathbf{X}\boldsymbol{\beta})+\delta_*]/2}, \end{aligned}$$

where $\alpha_{**} = 2\alpha - 2 + n$, and all the factors not involving random variables are collected into the normalizing constant c_2. By expanding and completing the square in the exponent (Problem 11.1), we obtain

$$g(\boldsymbol{\beta}, \tau \,|\, \mathbf{y}) = c_2 \tau^{(\alpha_{**}+k+1)/2} e^{-\tau[(\boldsymbol{\beta}-\boldsymbol{\phi}_*)'\mathbf{V}_*^{-1}(\boldsymbol{\beta}-\boldsymbol{\phi}_*)+\delta_{**}]/2}, \tag{11.10}$$

where $\mathbf{V}_* = (\mathbf{V}^{-1} + \mathbf{X}'\mathbf{X})^{-1}\boldsymbol{\phi}_* = \mathbf{V}_*(\mathbf{V}^{-1}\boldsymbol{\phi} + \mathbf{X}'\mathbf{y})$, and $\delta_{**} = -\boldsymbol{\phi}_*'\mathbf{V}_*^{-1}\boldsymbol{\phi}_* + \boldsymbol{\phi}'\mathbf{V}^{-1}\boldsymbol{\phi} + \mathbf{y}'\mathbf{y} + \delta_*$. Hence the joint posterior density has exactly the same form as the joint prior density in (11.9). □

It might seem odd to include terms like $\mathbf{X}'\mathbf{y}$ and $\mathbf{y}'\mathbf{y}$ in the "*constants*" of a probability distribution, while considering parameters like $\boldsymbol{\beta}$ and τ to be random, but this is completely characteristic of Bayesian inference. In this sense, inference in a Bayesian linear model is opposite to inference in the classical linear model.

11.2.2 Marginal Posterior Density of $\boldsymbol{\beta}$

In order to carry out inferences for $\boldsymbol{\beta}$, the marginal posterior density of $\boldsymbol{\beta}$ [see (3.13)] must be obtained by integrating τ out of the joint posterior density in (11.10). The following theorem gives the form of this marginal distribution.

Theorem 11.2b. Consider the Bayesian multiple regression model in which $\mathbf{y}|\boldsymbol{\beta}, \tau$ is $N_n(\mathbf{X}\boldsymbol{\beta}, \tau^{-1}\mathbf{I})$, $\boldsymbol{\beta}|\tau$ is $N_{k+1}(\boldsymbol{\phi}, \tau^{-1}\mathbf{V})$, and τ is gamma(α, δ). The marginal posterior distribution $u(\boldsymbol{\beta}|\mathbf{y})$ is a multivariate t distribution with parameters $(n + 2\alpha, \boldsymbol{\phi}_*, \mathbf{W}_*)$, where

$$\boldsymbol{\phi}_* = (\mathbf{V}^{-1} + \mathbf{X}'\mathbf{X})^{-1}(\mathbf{V}^{-1}\boldsymbol{\phi} + \mathbf{X}'\mathbf{y}) \tag{11.11}$$

and

$$\mathbf{W}_* = \left[\frac{(\mathbf{y} - \mathbf{X}\boldsymbol{\phi})'(\mathbf{I} + \mathbf{X}\mathbf{V}\mathbf{X}')^{-1}(\mathbf{y} - \mathbf{X}\boldsymbol{\phi}) + 2\delta}{n + 2\alpha}\right](\mathbf{V}^{-1} + \mathbf{X}'\mathbf{X})^{-1}. \tag{11.12}$$

PROOF. The marginal distribution of $\boldsymbol{\beta}|\mathbf{y}$ is obtained by integration as

$$u(\boldsymbol{\beta}|\mathbf{y}) = \int_0^\infty g(\boldsymbol{\beta}, \tau|\mathbf{y})d\tau.$$

By (11.10), this becomes

$$u(\boldsymbol{\beta}|\mathbf{y}) = c_2 \int_0^\infty \tau^{(\alpha_{**}+k+1)/2} e^{-\tau[(\boldsymbol{\beta}-\boldsymbol{\phi}_*)'\mathbf{V}_*^{-1}(\boldsymbol{\beta}-\boldsymbol{\phi}_*)+\delta_{**}]/2} d\tau.$$

Using (11.8) together with integration by substitution, the integral in this expression can be solved (Problem 11.2) to give the posterior distribution of $\boldsymbol{\beta}|\mathbf{y}$ as

$$u(\boldsymbol{\beta}|\mathbf{y}) = c_2 \Gamma\left(\frac{\alpha_{**}+2+k+1}{2}\right)\left[\frac{(\boldsymbol{\beta}-\boldsymbol{\phi}_*)'\mathbf{V}_*^{-1}(\boldsymbol{\beta}-\boldsymbol{\phi}_*)+\delta_{**}}{2}\right]^{-(\alpha_{**}+2+k+1)/2}$$
$$= c_3[(\boldsymbol{\beta}-\boldsymbol{\phi}_*)'\mathbf{V}_*^{-1}(\boldsymbol{\beta}-\boldsymbol{\phi}_*)-\boldsymbol{\phi}_*'\mathbf{V}_*^{-1}\boldsymbol{\phi}_*+\boldsymbol{\phi}'\mathbf{V}^{-1}\boldsymbol{\phi}+\mathbf{y}'\mathbf{y}+\delta_*]^{-(\alpha_{**}+2+k+1)/2}.$$

To show that this is the multivariate t density, several algebraic steps are required as outlined in Problems 11.3a–c and 11.4. See also Seber and Lee (2003, pp. 100–110). After these steps, the preceding expression becomes

$$u(\boldsymbol{\beta}|\mathbf{y}) = c_3[(\boldsymbol{\beta}-\boldsymbol{\phi}_*)'\mathbf{V}_*^{-1}(\boldsymbol{\beta}-\boldsymbol{\phi}_*)+(\mathbf{y}-\mathbf{X}\boldsymbol{\phi})'(\mathbf{I}+\mathbf{X}\mathbf{V}\mathbf{X}')^{-1}(\mathbf{y}-\mathbf{X}\boldsymbol{\phi})$$
$$+2\delta]^{-(\alpha_{**}+2+k+1)/2}.$$

Dividing the expression in square brackets by $(\mathbf{y}-\mathbf{X}\boldsymbol{\phi})'(\mathbf{I}+\mathbf{X}\mathbf{V}\mathbf{X}')^{-1}(\mathbf{y}-\mathbf{X}\boldsymbol{\phi})+2\delta$, modifying the normalizing constant accordingly, and replacing α_{**} by $2\alpha-2+n$, we obtain

$$u(\boldsymbol{\beta}|\mathbf{y}) = c_4\left[1+\frac{(\boldsymbol{\beta}-\boldsymbol{\phi}_*)'\mathbf{V}_*^{-1}(\boldsymbol{\beta}-\boldsymbol{\phi}_*)/(n+2\alpha)}{[(\mathbf{y}-\mathbf{X}\boldsymbol{\phi})'(\mathbf{I}+\mathbf{X}\mathbf{V}\mathbf{X}')^{-1}(\mathbf{y}-\mathbf{X}\boldsymbol{\phi})+2\delta]/(n+2\alpha)}\right]^{-(n+2\alpha+k+1)/2}$$
$$= c_4\left(\frac{1+(\boldsymbol{\beta}-\boldsymbol{\phi}_*)'\mathbf{W}_*^{-1}(\boldsymbol{\beta}-\boldsymbol{\phi}_*)}{n+2\alpha}\right)^{-(n+2\alpha+k+1)/2}, \qquad (11.13)$$

where $\mathbf{W}_*$ is as given in (11.12). The expression in (11.13) can now be recognized as the density function of the multivariate t distribution (Gelman et al. 2004, pp. 576–577; Rencher 1998, p. 56) with parameters $(n+2\alpha, \boldsymbol{\phi}_*, \mathbf{W}_*)$. Note that $\boldsymbol{\phi}_*$ is the mean vector and $[(n+2\alpha)/(n+2\alpha-2)]\mathbf{W}_*$ is the covariance matrix of $\boldsymbol{\beta}|\mathbf{y}$. □

As a historical note, the reasoning in this section is closely related to the work of W. S. Gosset or "Student" (Pearson et al. 1990, pp. 49–53, 72–73) on the small-sample distribution of

$$t = \frac{\bar{y}}{s}.$$

Gosset used Bayesian reasoning ("inverse probability") with a uniform prior distribution ("equal distribution of ignorance") to show through a combination of proof, conjecture, and simulation that the posterior density of t is related to what we now call Student's t distribution with $n-1$ degrees of freedom.

11.2.3 Marginal Posterior Densities of τ and σ^2

Inferences regarding τ and σ^2 require knowledge of the marginal posterior distribution of $\tau|\mathbf{y}$. We derive the posterior density of $\tau|\mathbf{y}$ in the following theorem.

Theorem 11.2c. Consider the Bayesian multiple regression model in which $\mathbf{y}|\boldsymbol{\beta},\tau$ is $N_n(\mathbf{X}\boldsymbol{\beta}, \tau^{-1}\mathbf{I})$, $\boldsymbol{\beta}|\tau$ is $N_{k+1}(\boldsymbol{\phi}, \tau^{-1}\mathbf{V})$, and τ is gamma(α, δ). The marginal posterior distribution $v(\tau|\mathbf{y})$ is a gamma distribution with parameters $\alpha + n/2$ and $(-\boldsymbol{\phi}_*'\mathbf{V}_*^{-1}\boldsymbol{\phi}_* + \boldsymbol{\phi}'\mathbf{V}^{-1}\boldsymbol{\phi} + \mathbf{y}'\mathbf{y} + 2\delta)/2$, where $\mathbf{V}_* = (\mathbf{V}^{-1} + \mathbf{X}'\mathbf{X})^{-1}$ and $\boldsymbol{\phi}_* = \mathbf{V}_*(\mathbf{V}^{-1}\boldsymbol{\phi} + \mathbf{X}'\mathbf{y})$.

Proof. The marginal distribution of $\tau|\mathbf{y}$ is obtained by integration as

$$
\begin{aligned}
\mathbf{v}(\tau|\mathbf{y}) &= \int_{-\infty}^{\infty} \cdots \int_{-\infty}^{\infty} g(\boldsymbol{\beta}, \tau|\mathbf{y}) d\boldsymbol{\beta} \\
&= c_2 \int_{-\infty}^{\infty} \cdots \int_{-\infty}^{\infty} \tau^{(\alpha_{**}+k+1)/2} e^{-\tau[(\boldsymbol{\beta}-\boldsymbol{\phi}_*)'\mathbf{V}_*-1(\boldsymbol{\beta}-\boldsymbol{\phi}_*)+\delta_{**}]/2} d\boldsymbol{\beta} \\
&= c_2 \tau^{(\alpha_{**}+k+1)/2} e^{-\tau\delta_{**}/2} \int_{-\infty}^{\infty} \cdots \int_{-\infty}^{\infty} e^{-\tau[(\boldsymbol{\beta}-\boldsymbol{\phi}_*)'\mathbf{V}_*^{-1}(\boldsymbol{\beta}-\boldsymbol{\phi}_*)]/2} d\boldsymbol{\beta}
\end{aligned}
$$

where all the factors not involving random variables are collected into the normalizing constant c_2 as in (11.10). Since the integral in the preceding expression is proportional to the integral of a joint multivariate normal density, we obtain

$$
\begin{aligned}
\mathbf{v}(\tau|\mathbf{y}) &= c_2 \tau^{(\alpha_{**}+k+1)/2} e^{-(\delta_{**}/2)\tau} (2\pi)^{(k+1)/2} |\mathbf{V}_*|^{1/2} \tau^{-(k+1)/2} \\
&= c_5 \tau^{(\alpha_{**}+k+1)/2-(k+1)/2} e^{-(\delta_{**}/2)\tau} \\
&= c_5 \tau^{(\alpha+n)/2-1} e^{-[(-\boldsymbol{\phi}_*'\mathbf{V}_*^{-1}\boldsymbol{\phi}_*+\boldsymbol{\phi}'\mathbf{V}^{-1}\boldsymbol{\phi}+\mathbf{y}'\mathbf{y}+2\delta)/2]\tau},
\end{aligned}
\tag{11.14}
$$

which is the density function of the specified gamma distribution. □

The marginal posterior density of σ^2 can now be obtained by the univariate change-of-variable technique (4.2) as

$$
\mathbf{w}(\sigma^2|\mathbf{y}) = c_6(\sigma^2)^{-(\alpha+n)/2-1} e^{-[(-\boldsymbol{\phi}_*'\mathbf{V}_*^{-1}\boldsymbol{\phi}_*+\boldsymbol{\phi}'\mathbf{V}^{-1}\boldsymbol{\phi}+\mathbf{y}'\mathbf{y}+2\delta)/2]/\sigma^2} \tag{11.15}
$$

which is the density function of the inverse gamma distribution with parameters $\alpha + n/2$ and $(-\boldsymbol{\phi}_*'\mathbf{V}_*^{-1}\boldsymbol{\phi}_* + \boldsymbol{\phi}'\mathbf{V}^{-1}\boldsymbol{\phi} + \mathbf{y}'\mathbf{y} + 2\delta)/2$ (Gelman et al. 2004, pp. 574–575).

11.3 INFERENCE IN BAYESIAN MULTIPLE LINEAR REGRESSION

11.3.1 Bayesian Point and Interval Estimates of Regression Coefficients

A sensible Bayesian point estimator of $\boldsymbol{\beta}$ is the mean of the marginal posterior density in (11.13)

$$\boldsymbol{\phi}_* = (\mathbf{V}^{-1} + \mathbf{X}'\mathbf{X})^{-1}(\mathbf{V}^{-1}\boldsymbol{\phi} + \mathbf{X}'\mathbf{y}), \tag{11.16}$$

and a sensible $100(1 - \omega)\%$ Bayesian confidence region for $\boldsymbol{\beta}$ is the highest-density region $\boldsymbol{\Omega}$ such that

$$c_4 \int_{\Omega} \cdots \int \left[\frac{1 + (\boldsymbol{\beta} - \boldsymbol{\phi}_*)'\mathbf{W}_*^{-1}(\boldsymbol{\beta} - \boldsymbol{\phi}_*)}{n + 2\alpha} \right]^{-(n+2\alpha+k+1)/2} d\boldsymbol{\beta} = 1 - \omega. \tag{11.17}$$

A convenient property of the multivariate t distribution is that linear functions of the random vector follow the (univariate) t distribution. Thus, given $\mathbf{y}$,

$$\frac{\mathbf{a}'\boldsymbol{\beta} - \mathbf{a}'\boldsymbol{\phi}_*}{\mathbf{a}'\mathbf{W}_*\mathbf{a}} \quad \text{is} \quad t(\mathrm{n} + 2\alpha)$$

and, as an important special case,

$$\frac{\beta_i - \phi_{*i}}{w_{*ii}} \quad \text{is} \quad \mathrm{t}(n + 2\alpha), \tag{11.18}$$

where ϕ_{*i} is the ith element of $\boldsymbol{\phi}_*$ and w_{*ii} is the ith diagonal element of $\mathbf{W}_*$. Thus a Bayesian point estimate of β_i is ϕ_{*i} and a $100(1 - \omega)\%$ Bayesian confidence interval for β_i is

$$\phi_{*i} \pm t_{\omega/2,n+2\alpha} w_{*ii}. \tag{11.19}$$

One very appealing aspect of Bayesian inference is that intervals like (11.19) have a natural interpretation. Instead of the careful classical interpretation of a confidence interval in terms of hypothetical repeated sampling, one can simply and correctly say that the probability is $1 - \omega$ that β_i is in (11.19).

An interesting final note on Bayesian estimation of $\boldsymbol{\beta}$ is that the Bayesian estimator $\boldsymbol{\phi}_*$ in (11.16) can be obtained as the generalized least-squares estimator of $\boldsymbol{\beta}$ in (7.63). To see this, consider adding the prior information to the data as if it constituted a set of additional observations. The idea is to augment $\mathbf{y}$ with $\boldsymbol{\phi}$, and to consider the mean vector and covariance matrix of the augmented vector $\begin{pmatrix} \mathbf{y} \\ \boldsymbol{\phi} \end{pmatrix}$ to be, respectively

$$\begin{pmatrix} \mathbf{X} \\ \mathbf{I}_{k+1} \end{pmatrix} \boldsymbol{\beta} \quad \text{and} \quad \frac{1}{\tau} \begin{pmatrix} \mathbf{I} & \mathbf{O} \\ \mathbf{O} & \mathbf{V} \end{pmatrix}.$$

Generalized least squares estimation expressed in terms of these partitioned matrices then gives $\boldsymbol{\phi}_*$ in (11.16) as an estimate of $\boldsymbol{\beta}$ (Problem 11.6). The implication of this is that prior information on the regression coefficients can be incorporated into a multiple linear regression model by the intuitive informal process of "adding" observations.

11.3.2 Hypothesis Tests for Regression Coefficients in Bayesian Inference

Classical hypothesis testing is not a natural part of Bayesian inference (Gelman et al. 2004, p. 162). Nonetheless, if the question addressed by a classical hypothesis test is whether the data support the conclusion (i.e., alternative hypothesis) that β_i is greater than β_{i0}, a sensible approach is to use the posterior distribution (in this case the t distribution with $n + 2\alpha$ degrees of freedom) to compute the probability

$$P\left(t(n + 2\alpha) > \frac{\beta_{i0} - \phi_{*i}}{w_{*ii}}\right).$$

The larger this probability is, the more credible is the hypothesis that $\beta_i > \beta_{i0}$.

If, alternatively, classical hypothesis testing is used to select a model from a set of candidate models, the corresponding Bayesian approach is to compute an information statistic for each model in question. For example, Schwarz (1978) proposed the Bayesian Information Criterion (BIC) for multiple linear regression models, and Spiegelhalter et al. (2002) proposed the Deviance Information Criterion (DIC) for more general Bayesian models. The model with the lowest value of the information criterion is selected. Model selection in Bayesian analysis is an area of current research.

11.3.3 Special Cases of Inference in Bayesian Multiple Regression Models

Two special cases of inference in this Bayesian linear model are of particular interest. First, consider the use of a diffuse prior. Let $\boldsymbol{\phi} = \mathbf{0}$, let $\mathbf{V}$ be a diagonal matrix with all diagonal elements equal to a large constant (say, 10^6), and let α and δ both be equal to a small constant (say, 10^{-6}). In this case, $\mathbf{V}^{-1}$ is close to $\mathbf{O}$, and so $\boldsymbol{\phi}_*$, the Bayesian point estimate of $\boldsymbol{\beta}$ in (11.16), is approximately equal to

$$(\mathbf{X}'\mathbf{X})^{-1}\mathbf{X}'\mathbf{y},$$

the classical least-squares estimate. Also, since $(\mathbf{I} + \mathbf{X}\mathbf{V}\mathbf{X}')^{-1} = \mathbf{I} - \mathbf{X}(\mathbf{X}'\mathbf{X} + \mathbf{V}^{-1})^{-1}\mathbf{X}'$ (see Problem 11.3a), the covariance matrix $\mathbf{W}_*$ approaches

$$\mathbf{W}_* = \frac{\mathbf{y}'[\mathbf{I} - \mathbf{X}(\mathbf{X}'\mathbf{X})^{-1}\mathbf{X}']\mathbf{y}}{n}(\mathbf{X}'\mathbf{X})^{-1}$$

$$= \frac{n-1}{n}s^2(\mathbf{X}'\mathbf{X})^{-1} \qquad \text{[by (7.26)]}.$$

Thus, in the case of diffuse priors, the Bayesian confidence region (11.17) reduces to a region similar to (8.46), and Bayesian confidence intervals for the regression coefficients in (11.19) are similar to classical confidence intervals in (8.47); the only differences are the multiplicative factor $(n-1)/n$ and the use of the t distribution with n degrees of freedom rather than $n-k-1$ degrees of freedom. If a Bayesian multiple linear regression model with independent uniformly distributed priors for $\boldsymbol{\beta}$ and $\ln(\tau^{-1})$ is considered, Bayesian confidence intervals for the regression coefficients are exactly equal to classical confidence intervals (Problem 11.5). One result of this is that simple Bayesian interpretations can be validly applied to confidence intervals for the classical linear model. In fact, most inferences for the classical linear model can be stated in terms of properties of posterior distributions.

The second special case of inference in this Bayesian linear model is the case in which $\boldsymbol{\phi} = \mathbf{0}$ and $\mathbf{V}$ is a diagonal matrix with a constant on the diagonal. Thus $\mathbf{V} = a\mathbf{I}$, where a is a positive number, and the Bayesian estimator of $\boldsymbol{\beta}$ in (11.16) becomes

$$\left(\mathbf{X}'\mathbf{X} + \frac{1}{a}\mathbf{I}\right)^{-1}\mathbf{X}'\mathbf{y}.$$

For the centered model (Section 7.5) this estimator is also known as the "ridge estimator" (Hoerl and Kennard 1970). It was originally proposed as a method for dealing with collinearity, the situation in which the columns of the $\mathbf{X}$ matrix have near-linear dependence so that $\mathbf{X}'\mathbf{X}$ is nearly singular. However, the estimator may also be understood as a "shrinkage estimator" in which prior information causes the estimates of the coefficients to be shrunken toward zero (Seber and Lee 2003, pp. 321–322). The use of a Bayesian linear model with *hyperpriors* (prior distributions for the parameters of the prior distributions) leads to a reasonable choice of value for a in terms of variances of the prior and hyperprior distributions (Lindley and Smith 1972).

11.3.4 Bayesian Point and Interval Estimation of σ^2

A possible Bayesian point estimator of σ^2 is the mean of the marginal inverse gamma density in (11.15)

$$\frac{(-\boldsymbol{\phi}_*'\mathbf{V}_*^{-1}\boldsymbol{\phi}_* + \boldsymbol{\phi}'\mathbf{V}^{-1}\boldsymbol{\phi} + \mathbf{y}'\mathbf{y} + 2\delta)/2}{\alpha + n/2 - 1}$$

and a $100(1-\omega)\%$ Bayesian confidence interval for σ^2 is given by the $1-\omega/2$ and $\omega/2$ quantiles of the appropriate inverse gamma distribution.

As a special case, note that if α and δ are both close to 0, $\boldsymbol{\phi} = \mathbf{0}$, and $\mathbf{V}$ is a diagonal matrix with all diagonal elements equal to a large constant so that $\mathbf{V}^{-1}$ is close to $\mathbf{O}$, then the Bayesian point estimator of σ^2 is approximately

$$\begin{aligned}\frac{(\mathbf{y}'\mathbf{y}-\boldsymbol{\phi}_*'\mathbf{V}_*^{-1}\boldsymbol{\phi}_*)/2}{n/2-1} &= \frac{\mathbf{y}'\mathbf{y}-\mathbf{y}'\mathbf{X}(\mathbf{X}'\mathbf{X})^{-1}\mathbf{X}'\mathbf{y}}{n-2} \\ &= \frac{\mathbf{y}'[\mathbf{I}-\mathbf{X}(\mathbf{X}'\mathbf{X})^{-1}\mathbf{X}']\mathbf{y}}{n-2} \\ &= \frac{n-k-1}{n-2}s^2,\end{aligned}$$

and the centered Bayesian confidence limits are the $1-\omega/2$ quantile and the $\omega/2$ quantile of the inverse gamma distribution with parameters $n/2$ and $\mathbf{y}'[\mathbf{I}-\mathbf{X}(\mathbf{X}'\mathbf{X})^{-1}\mathbf{X}']\mathbf{y}/2$.

11.4 BAYESIAN INFERENCE THROUGH MARKOV CHAIN MONTE CARLO SIMULATION

The inability to derive a closed-form marginal posterior distribution for a parameter is extremely common in Bayesian inference (Gilks et al. 1998, p. 3). For example, if the Bayesian multiple regression model of Section 11.2.1 had involved a prior distribution for $\boldsymbol{\beta}$ that was *not* conditional on τ, closed-form marginal distributions for the parameters could not have been derived (Lindley and Smith 1972). In actual practice, the exception in Bayesian inference is to be able to derive closed-form marginal posterior distributions. However, this difficulty turns out to be only a minor hindrance when modern computing resources are available.

If it were possible, an ideal solution would be to draw a large number of samples from the joint posterior distribution. Then marginal means, marginal highest density intervals, and other properties of the posterior distribution could be approximated using sample statistics. Furthermore, functions of the sampled values could be calculated in order to approximate marginal posterior distributions of these functions. The big question, of course, is how it would be possible to draw samples from a distribution for which a familiar closed-form joint density function is not available.

A general approach for accomplishing this is referred to as *Markov Chain Monte Carlo* (MCMC) simulation (Gilks et al. 1998). A *Markov Chain* is a special sequence of random variables (Ross 2006, p. 185). Probability laws for general sequences of random variables are specified in terms of the conditional distribution of the current value in the sequence, given all past values. A Markov Chain is a simple sequence in which the conditional distribution of the current value is completely specified, given only the most recent value.

Markov Chain Monte Carlo simulation in Bayesian inference is based on sequences of alternating random draws from conditional posterior distributions of each of the parameters in the model given the most recent values of the other parameters. This process generates a Markov Chain for each parameter. Moreover, the unconditional distribution for each parameter converges to the marginal posterior distribution of the parameter, and the unconditional joint distribution of the vector of parameters for any complete iteration of MCMC converges to the joint posterior distribution. Thus after discarding a number of initial draws (the "burn-in"), draws may be considered to constitute sequences of samples from marginal posterior distributions of the parameters. The samples are not independent, but the nonindependence can be ignored if the number of draws is sufficiently large. Plots of sample values can be examined to determine whether a sufficiently large number of draws has been obtained (Gilks et al. 1998).

When the prior distributions are conjugate, closed-form density functions of the conditional posterior distributions of the parameters are available regardless of whether closed-form marginal posterior distributions can be derived. In the case of conjugate priors, a simple form of MCMC called "Gibbs sampling" (Gilks et al. 1998, Casella and George 1992) can be used by which draws are made successively from each of the conditional distributions of the parameters, given the current draws for the other parameters.

We now illustrate this procedure. Consider again the Bayesian multiple regression model in which $\mathbf{y}|\boldsymbol{\beta}, \tau$ is $N_n(\mathbf{X}\boldsymbol{\beta}, \tau^{-1}\mathbf{I})$, $\boldsymbol{\beta}|\tau$ is $N_{k+1}(\boldsymbol{\phi}, \tau^{-1}\mathbf{V})$, and τ is gamma(α, δ). The joint posterior density function is given in (11.10). The conditional posterior density (or "full conditional") of $\boldsymbol{\beta}|\tau, \mathbf{y}$ can be obtained by picking the terms out of (11.10) that involve $\boldsymbol{\beta}$, and considering everything else to be part of the normalizing constant. Thus, the conditional density of $\boldsymbol{\beta}|\tau, \mathbf{y}$ is

$$\varphi(\boldsymbol{\beta}|\tau, \mathbf{y}) = c_6 e^{-\tau(\boldsymbol{\beta}-\boldsymbol{\phi}_*)'\mathbf{V}_*^{-1}(\boldsymbol{\beta}-\boldsymbol{\phi}_*)/2}.$$

Clearly $\boldsymbol{\beta}|\tau, \mathbf{y}$ is $N_{k+1}(\boldsymbol{\phi}_*, \tau^{-1}\mathbf{V}_*)$. Similarly, the conditional posterior density for $\tau|\boldsymbol{\beta}, \mathbf{y}$ is

$$\psi(\tau|\boldsymbol{\beta}, \mathbf{y}) = c_7 \tau^{[(\alpha_{**}+k+3)/2]-1} e^{-\tau[(\boldsymbol{\beta}-\boldsymbol{\phi}_*)'\mathbf{V}_*^{-1}(\boldsymbol{\beta}-\boldsymbol{\phi}_*)+\delta_{**}]/2}$$

so that $\tau|\boldsymbol{\beta}, \mathbf{y}$ can be seen to be gamma $\{(\alpha_{**} + k + 3)/2, [(\boldsymbol{\beta} - \boldsymbol{\phi}_*)'\mathbf{V}_*^{-1}(\boldsymbol{\beta} - \boldsymbol{\phi}_*) + \delta_{**}]/2\}$.

Gibbs sampling for this model proceeds as follows:

- Specify a starting value τ_0 [possibly $1/s^2$ from (7.23)].
- For $i = 1$ to M: draw $\boldsymbol{\beta}_i$ from $N_{k+1}(\boldsymbol{\phi}_*, \tau_{i-1}^{-1}\mathbf{V}_*)$, draw τ_i from gamma $\{(\alpha_{**} + k + 3)/2, [(\boldsymbol{\beta}_i - \boldsymbol{\phi}_*)'V_*^{-1}(\boldsymbol{\beta}_i - \boldsymbol{\phi}_*) + \delta_{**}]/2\}$.
- Discard the first Q draws (as burn-in), and consider the last $M - Q$ draws $(\boldsymbol{\beta}_i, \tau_i)$ to be draws from the joint posterior distribution. For this model, using the starting value of $1/s^2$, Q would usually be very small (say, 0), and M would be large (say, 10,000).

Bayesian inferences for all parameters of the model could now be carried out using sample statistics of this empirical joint posterior distribution. For example, a Bayesian point estimate of τ could be calculated as the sample mean or median of the draws of τ from the joint posterior distribution. If we calculate (or "monitor") $1/\tau$ on each iteration, a Bayesian point estimate of $\sigma^2 = 1/\tau$ could be calculated as the mean or

TABLE 11.1 Body Fat Data

y	x_1	x_2
11.9	19.5	29.1
22.8	24.7	28.2
18.7	30.7	37.0
20.1	29.8	31.1
12.9	19.1	30.9
21.7	25.6	23.7
27.1	31.4	27.6
25.4	27.9	30.6
21.3	22.1	23.2
19.3	25.5	24.8
25.4	31.1	30.0
27.2	30.4	28.3
11.7	18.7	23.0
17.8	19.7	28.6
12.8	14.6	21.3
23.9	29.5	30.1
22.6	27.7	25.7
25.4	30.2	24.6
14.8	22.7	27.1
21.1	25.2	27.5

median of $1/\tau$. A 95% Bayesian interval estimate of σ^2 could be computed as the central 95% interval of the sample distribution of σ^2. Other inferences could similarly be drawn on the basis of sample draws from the joint posterior distribution.

Example 11.4. Table 11.1 contains body fat data for a sample of 20 females aged 25–34 (Kutner et al. 2005, p. 256). The response variable was body fat (y), and two predictor variables were triceps skinfold thickness (x_1) and midarm circumference (x_2). The data were analyzed using the Bayesian multiple regression model of Section 11.2.1 with diffuse priors in which $\boldsymbol{\phi}' = (0, 0, 0)$, $\mathbf{V} = 10^6\mathbf{I}_3$, $\alpha = 0.0001$, and $\delta = 0.0001$. Density functions of the marginal posterior distributions of β_0, β_1, and β_2 from (11.13) as well as the marginal posterior density of σ^2 from (11.15) are graphed in Figure 11.1. Superimposed on these (and almost indistinguishable from them) are smooth estimates (Silverman 1999) of the same posterior densities based on Gibbs sampling with $Q = 0$ and $M = 10{,}000$. □

11.5 POSTERIOR PREDICTIVE INFERENCE

As a final aspect of Bayesian inference for the multiple regression model, we consider Bayesian prediction of the value of the response variable for a future individual. If we again use the Bayesian multiple regression model of Section 11.2.1 in which $\mathbf{y}|\boldsymbol{\beta}, \tau$ is $N_n(\mathbf{X}\boldsymbol{\beta}, \tau^{-1}\mathbf{I})$, $\boldsymbol{\beta}|\tau$ is $N_{k+1}(\boldsymbol{\phi}, \tau^{-1}\mathbf{V})$, and τ is gamma(α, δ), the posterior predictive density for a future observation y_0 with predictor variables $\mathbf{x}_0$ can be

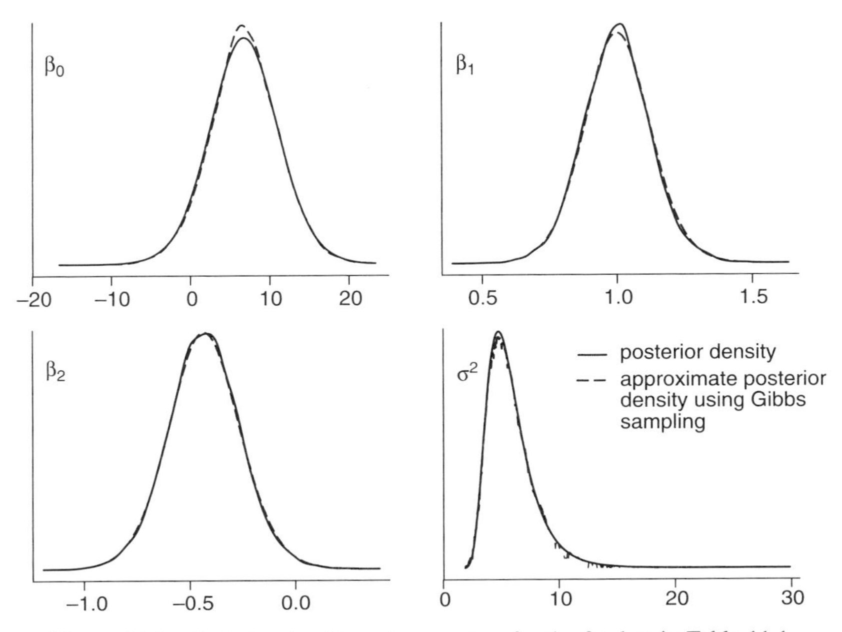

Figure 11.1 Posterior densities of parameters for the fat data in Table 11.1.

expressed using (11.5) as

$$
\begin{aligned}
r(y_0|\mathbf{y}) &= \int_0^{\infty}\int_{-\infty}^{\infty}\cdots\int_{-\infty}^{\infty} q(y_0|\boldsymbol{\beta},\tau)g(\boldsymbol{\beta},\tau|y)d\boldsymbol{\beta}\,d\tau \\
&= c\int_0^{\infty}\int_{-\infty}^{\infty}\cdots\int_{-\infty}^{\infty} \tau^{1/2}e^{-\tau(y_0-\mathbf{x}_0'\boldsymbol{\beta})^2/2}\tau^{(\alpha_{**}+k+1)/2} \\
&\quad\times e^{-\tau[(\boldsymbol{\beta}-\boldsymbol{\phi}_*)'\mathbf{V}_*^{-1}(\boldsymbol{\beta}-\boldsymbol{\phi}_*)+\delta_{**}]/2}d\boldsymbol{\beta}\,d\tau \\
&= c\int_{-\infty}^{\infty}\cdots\int_{-\infty}^{\infty} [(\boldsymbol{\beta}-\boldsymbol{\phi}_*)'\mathbf{V}_*^{-1}(\boldsymbol{\beta}-\boldsymbol{\phi}_*) \\
&\quad + (y_0-\mathbf{x}_0'\boldsymbol{\beta})^2+\delta_{**}]^{-(\alpha_{**}+k+4)/2}d\boldsymbol{\beta}.
\end{aligned}
$$

Further analytical progress with this integral is difficult. Nonetheless, Gibbs sampling as in Section 11.4 can be easily extended to simulate the posterior predictive distribution of y_0 as follows:

- Specify a starting value τ_0 [possibly $1/s^2$ from (7.23)].
- For $i = 1$ to M: draw $\boldsymbol{\beta}_i$ from $N_{k+1}(\boldsymbol{\phi}_*, \tau_{i-1}^{-1}\mathbf{V}_*)$, draw τ_i from gamma$\{(\alpha_{**}+k+3)/2, [(\boldsymbol{\beta}_i-\boldsymbol{\phi}_*)'\mathbf{V}_*^{-1}(\boldsymbol{\beta}_i-\boldsymbol{\phi}_*)+\delta_{**}]/2\}$, draw y_{0i} from $N(x_0'\boldsymbol{\beta}_i, \tau_i^{-1})$.

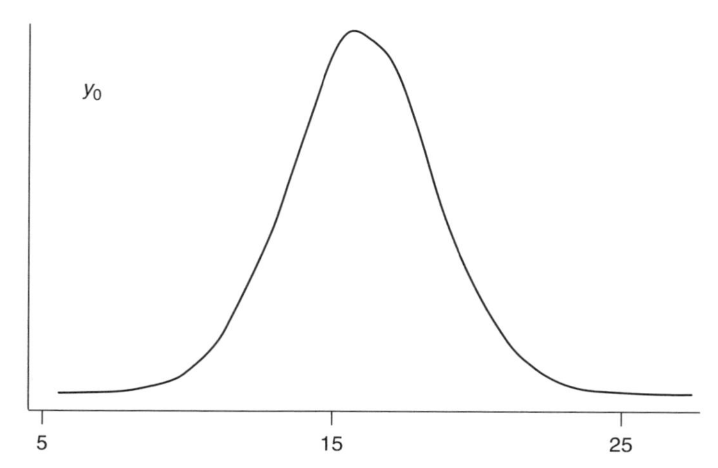

Figure 11.2 Approximate posterior predictive density using Gibbs sampling for a future observation y_0 with $\mathbf{x}_0' = (1, 20, 25)$ for the fat data in Table 11.1.

- Discard the first Q draws (as burn-in), and consider the last $M - Q$ draws of y_{0i} to be draws from the posterior predictive distribution.

Example 11.5. Example 11.4(continued). Consider a new individual with $x_1 = 20$ and $x_2 = 25$. Thus $\mathbf{x}_0' = (1,\ 20,\ 25)$. Figure 11.2 gives a smooth estimate of the posterior predictive density of y_0 based on Gibbs sampling with $Q = 0$ and $M = 10{,}000$. □

The approximate Bayesian 95% prediction interval derived from this density is (11.83, 20.15), which may be compared to the 95% prediction interval (10.46, 21.57) for the same future individual using the non-Bayesian approach (8.62).

This chapter gives a small taste of the calculations associated with the modern Bayesian multiple regression model. With very little additional work, many aspects of the model can be modified and customized, especially if the MCMC approach is used. Versatility is one of the great advantages of the Bayesian approach.

PROBLEMS

11.1 As in Theorem 11.2a, show that $(\boldsymbol{\beta} - \boldsymbol{\phi})'\mathbf{V}^{-1}(\boldsymbol{\beta} - \boldsymbol{\phi}) + (\mathbf{y} - \mathbf{X}\boldsymbol{\beta})'(\mathbf{y} - \mathbf{X}\boldsymbol{\beta}) + \delta_* = (\boldsymbol{\beta} - \boldsymbol{\phi}_*)'\mathbf{V}_*^{-1}(\boldsymbol{\beta} - \boldsymbol{\phi}_*) + \delta_{**}$, where $\mathbf{V}_* = (\mathbf{V}^{-1} + \mathbf{X}'\mathbf{X})^{-1}$, $\boldsymbol{\phi}_* = \mathbf{V}_*(\mathbf{V}^{-1}\boldsymbol{\phi} + \mathbf{X}'\mathbf{y})$, and $\delta_{**} = -\boldsymbol{\phi}_*'\mathbf{V}_*^{-1}\boldsymbol{\phi}_* + \boldsymbol{\phi}'\mathbf{V}^{-1}\boldsymbol{\phi} + \mathbf{y}'\mathbf{y} + \delta_*$.

11.2 As used in the proof to Theorem 11.2b, show that

$$\int_0^{\infty} t^a e^{-bt} dt = b^{-(a+1)}\Gamma(a+1).$$

11.3 (**a**) Show that $(\mathbf{I}+\mathbf{XVX}')^{-1}=\mathbf{I}-\mathbf{X}(\mathbf{X}'\mathbf{X}+\mathbf{V}^{-1})^{-1}\mathbf{X}'$.

(**b**) Show that $(\mathbf{I}+\mathbf{XVX}')^{-1}\mathbf{X}=\mathbf{X}(\mathbf{X}'\mathbf{X}+\mathbf{V}^{-1})^{-1}\mathbf{V}^{-1}$.

(**c**) Show that $\mathbf{V}^{-1}-\mathbf{V}^{-1}(\mathbf{X}'\mathbf{X}+\mathbf{V}^{-1})^{-1}\mathbf{V}^{-1}=\mathbf{X}'(\mathbf{I}+\mathbf{XVX}')^{-1}\mathbf{X}$.

11.4 As in the proof to Theorem 11.2b, show that $\mathbf{y}'\mathbf{y}+\boldsymbol{\phi}'\mathbf{V}^{-1}\boldsymbol{\phi}-\boldsymbol{\phi}_*'\mathbf{V}_*^{-1}\boldsymbol{\phi}_*$ $=(\mathbf{y}-\mathbf{X}\boldsymbol{\phi})'(\mathbf{I}+\mathbf{XVX}')^{-1}(\mathbf{y}-\mathbf{X}\boldsymbol{\phi})$, where $\mathbf{V}_*=(\mathbf{X}'\mathbf{X}+\mathbf{V}^{-1})^{-1}$ and $\boldsymbol{\phi}_*=\mathbf{V}_*(\mathbf{X}'\mathbf{y}+\mathbf{V}^{-1}\boldsymbol{\phi})$.

11.5 Consider the Bayesian multiple linear regression model in which $\mathbf{y}|\boldsymbol{\beta},\tau$ is $N_n(\mathbf{X}\boldsymbol{\beta},\tau^{-1}\mathbf{I})$, $\boldsymbol{\beta}$ is uniform (R^{k+1}) [i.e., uniform over $(k+1)$ -dimensional space], and $\ln(\tau^{-1})$ is uniform $(-\infty,\infty)$. Show that the marginal posterior distribution of $\boldsymbol{\beta}|\mathbf{y}$ is the multivariate t distribution with parameters $[n-k-1,\hat{\boldsymbol{\beta}},s^2(\mathbf{X}'\mathbf{X})^{-1}]$, where $\hat{\boldsymbol{\beta}}$ and s^2 are defined in the usual way [see (7.6) and (7.23)]. These prior distributions are called *improper priors* because uniform distributions must be defined for bounded sets of values. Nonetheless, the sets can be very large, and so we can proceed as if they were unbounded.

11.6 Consider the augmented data vector $\begin{pmatrix} y \\ \boldsymbol{\phi} \end{pmatrix}$ with mean vector $\begin{pmatrix} \mathbf{X} \\ \mathbf{I}_{k+1} \end{pmatrix}\boldsymbol{\beta}$ and covariance matrix

$$\begin{pmatrix} \frac{1}{\tau}\mathbf{I} & \mathbf{O} \\ \mathbf{O} & \frac{1}{\tau}\mathbf{V} \end{pmatrix}.$$

Show that the generalized least-squares estimator of $\boldsymbol{\beta}$ is the Bayesian estimator in (11.16), $(\mathbf{V}^{-1}+\mathbf{X}'\mathbf{X})^{-1}(\mathbf{V}^{-1}\boldsymbol{\phi}+\mathbf{X}'\mathbf{y})$.

11.7 Given that τ is gamma(α,δ) as in (11.7), find $E(\tau)$ and var(τ).

11.8 Use the Bayesian multiple regression model in which $\mathbf{y}|\boldsymbol{\beta},\tau$ is $N_n(\mathbf{X}\boldsymbol{\beta},\tau^{-1}\mathbf{I})$, $\boldsymbol{\beta}|\tau$ is $N_{k+1}(\boldsymbol{\phi},\tau^{-1}\mathbf{V})$, and τ is gamma(α,δ). Derive the marginal posterior density function for $\sigma^2|\mathbf{y}$, where $\sigma^2=1/\tau$.

11.9 Consider the Bayesian simple linear regression model in which $y_i|\beta_0,\beta_1$, is $N(\beta_0+\beta_1x_i,1/\tau)$ for $i=1,\ldots,n$, $\beta_0|\tau$ is $N(a,\sigma_0^2/\tau)$, $\beta_1|\tau$ is $N(b,\sigma_1^2/\tau)$, cov$(\beta_0,\beta_1|\tau)=\sigma_{12}$, and τ is gamma(α,δ).

(**a**) Find the marginal posterior density of $\beta_1|\mathbf{y}$. (Do not simplify the results.)

(**b**) Find Bayesian point and interval estimates of β_1.

11.10 Consider the Bayesian multiple regression model in which $\mathbf{y}|\boldsymbol{\beta},\tau$ is $N_n(\mathbf{X}\boldsymbol{\beta},\tau^{-1}\mathbf{I})$, $\boldsymbol{\beta}$ is $N_{k+1}(\boldsymbol{\phi},\mathbf{V})$, and τ is gamma(α,δ). Note that this is similar to the model of Section 11.2 except that the prior distribution of $\boldsymbol{\beta}$ is *not conditional* on τ.

(**a**) Find the joint posterior density of $\boldsymbol{\beta},\tau|\mathbf{y}$ up to a normalizing constant.

(**b**) Find the conditional posterior density of $\boldsymbol{\beta}|\tau, \mathbf{y}$ up to a normalizing constant.

(**c**) Find the conditional posterior density of $\tau|\boldsymbol{\beta}, \mathbf{y}$ up to a normalizing constant.

(**d**) Develop a Gibbs sampling procedure for estimating the marginal posterior distributions of $\boldsymbol{\beta}|\mathbf{y}$ and $(1/\tau)|\mathbf{y}$.

11.11 Use the land rent data in Table 7.5.

(**a**) Find 95% Bayesian confidence intervals for β_1, β_2, and β_3 using (11.19) in connection with the model in which $\mathbf{y}|\boldsymbol{\beta}, \tau$ is $N_n(\mathbf{X}\boldsymbol{\beta}, \tau^{-1}\mathbf{I})$, $\boldsymbol{\beta}|\tau$ is $N_{k+1}(\boldsymbol{\phi}, \tau^{-1}\mathbf{V})$, and τ is gamma(α, δ), where $\boldsymbol{\phi} = \mathbf{0}, \mathbf{V} = 100\mathbf{I}$, $\alpha = .0001$, and $\delta = .0001$.

(**b**) Repeat part (a), but use Gibbs sampling to approximate the confidence intervals.

(**c**) Use Gibbs sampling to obtain a 95% Bayesian posterior prediction interval for a future individual with $\mathbf{x}_0' = (1, 15, 30, .5)$.

(**d**) Repeat part (b), but use the model in which

$$\begin{aligned}\mathbf{y}|\boldsymbol{\beta}, \tau &\text{ is } N_n(\mathbf{X}\boldsymbol{\beta}, \tau^{-1}\mathbf{I}),\\ \boldsymbol{\beta} &\text{ is } N_{k+1}(\boldsymbol{\phi}, \mathbf{V}),\\ \tau &\text{ is gamma}(\alpha, \delta)\end{aligned} \tag{11.20}$$

where $\boldsymbol{\phi} = \mathbf{0}$, $\mathbf{V} = 100\mathbf{I}$, $\alpha = 0.0001$, and $\delta = 0.0001$.

11.12 As in Section 11.5, show that

$$\int_0^\infty \int_{-\infty}^\infty \cdots \int_{-\infty}^\infty \tau^{1/2} e^{-\tau(y_0 - \mathbf{x}_0'\boldsymbol{\beta})^2/2} \tau^{(\alpha_{**}+k+1)/2} e^{-\tau[(\boldsymbol{\beta}-\boldsymbol{\phi}_*)'\mathbf{V}_*^{-1}(\boldsymbol{\beta}-\boldsymbol{\phi}_*)+\delta_{**}]/2} d\boldsymbol{\beta}\, d\tau$$

$$= c \int_{-\infty}^\infty \cdots \int_{-\infty}^\infty [(\boldsymbol{\beta} - \boldsymbol{\phi}_*)'\mathbf{V}_*^{-1}(\boldsymbol{\beta} - \boldsymbol{\phi}_*) + (y_0 - \mathbf{x}_0'\boldsymbol{\beta})^2 + \delta_{**}]^{(-\alpha_{**}+k+4)/2} d\boldsymbol{\beta}.$$

12 Analysis-of-Variance Models

In many experimental situations, a researcher applies several treatments or treatment combinations to randomly selected experimental units and then wishes to compare the treatment means for some response y. In analysis-of-variance (ANOVA), we use linear models to facilitate a comparison of these means. The model is often expressed with more parameters than can be estimated, which results in an $\mathbf{X}$ matrix that is not of full rank. We consider procedures for estimation and testing hypotheses for such models.

The results are illustrated using balanced models, in which we have an equal number of observations in each cell or treatment combination. Unbalanced models are treated in more detail in Chapter 15.

12.1 NON-FULL-RANK MODELS

In Section 12.1.1 we illustrate a simple one-way model, and in Section 12.1.2 we illustrate a two-way model without interaction.

12.1.1 One-Way Model

Suppose that a researcher has developed two chemical additives for increasing the mileage of gasoline. To formulate the model, we might start with the notion that without additives, a gallon yields an average of μ miles. Then if chemical 1 is added, the mileage is expected to increase by τ_1 miles per gallon, and if chemical 2 is added, the mileage would increase by τ_2 miles per gallon.

The model could be expressed as

$$y_1 = \mu + \tau_1 + \varepsilon_1, \quad y_2 = \mu + \tau_2 + \varepsilon_2,$$

where y_1 is the miles per gallon from a tank of gasoline containing chemical 1 and ε_1 is a random error term. The variables y_2 and ε_2 are defined similarly. The researcher

---

would like to estimate the parameters μ, τ_1, and τ_2 and test hypotheses such as $H_0 : \tau_1 = \tau_2$.

To make reasonable estimates, the researcher needs to observe the mileage per gallon for more than one tank of gasoline for each chemical. Suppose that the experiment consists of filling the tanks of six identical cars with gas, then adding chemical 1 to three tanks and chemical 2 to the other three tanks. We can write a model for each of the six observations as follows:

$$
\begin{aligned}
y_{11} &= \mu + \tau_1 + \varepsilon_{11}, \quad y_{12} = \mu + \tau_1 + \varepsilon_{12}, \quad y_{13} = \mu + \tau_1 + \varepsilon_{13}, \\
y_{21} &= \mu + \tau_2 + \varepsilon_{21}, \quad y_{22} = \mu + \tau_2 + \varepsilon_{22}, \quad y_{23} = \mu + \tau_2 + \varepsilon_{23},
\end{aligned} \tag{12.1}
$$

or

$$
y_{ij} = \mu + \tau_i + \varepsilon_{ij}, \quad i = 1, 2, \quad j = 1, 2, 3 \tag{12.2}
$$

where y_{ij} is the observed miles per gallon of the jth car that contains the ith chemical in its tank and ε_{ij} is the associated random error. The six equations in (12.1) can be written in matrix form as

$$
\begin{pmatrix} y_{11} \\ y_{12} \\ y_{13} \\ y_{21} \\ y_{22} \\ y_{23} \end{pmatrix} = \begin{pmatrix} 1 & 1 & 0 \\ 1 & 1 & 0 \\ 1 & 1 & 0 \\ 1 & 0 & 1 \\ 1 & 0 & 1 \\ 1 & 0 & 1 \end{pmatrix} \begin{pmatrix} \mu \\ \tau_1 \\ \tau_2 \end{pmatrix} + \begin{pmatrix} \varepsilon_{11} \\ \varepsilon_{12} \\ \varepsilon_{13} \\ \varepsilon_{21} \\ \varepsilon_{22} \\ \varepsilon_{23} \end{pmatrix} \tag{12.3}
$$

or

$$
\mathbf{y} = \mathbf{X}\boldsymbol{\beta} + \boldsymbol{\varepsilon}.
$$

In (12.3), $\mathbf{X}$ is a 6×3 matrix whose rank is 2 since the first column is the sum of the second and third columns, which are linearly independent. Since $\mathbf{X}$ is not of full rank, the theorems of Chapters 7 and 8 cannot be used directly for estimating $\boldsymbol{\beta} = (\mu, \tau_1, \tau_2)'$ and testing hypotheses. Thus, for example, the parameters μ, τ_1, and τ_2 cannot be estimated by $\hat{\boldsymbol{\beta}} = (\mathbf{X}'\mathbf{X})^{-1}\mathbf{X}'\mathbf{y}$ in (7.6), because $(\mathbf{X}'\mathbf{X})^{-1}$ does not exist.

To further explore the reasons for the failure of (12.3) to be a full-rank model, let us reconsider the meaning of the parameters. The parameter μ was introduced as the mean before adding chemicals, and τ_1 and τ_2 represented the increase due to chemicals 1 and 2, respectively. However, the model $y_{ij} = \mu + \tau_i + \varepsilon_{ij}$ in (12.2) cannot uniquely support this characterization. For example, if $\mu = 15$, $\tau_1 = 1$, and $\tau_2 = 3$, the model becomes

$$
\begin{aligned}
y_{1j} &= 15 + 1 + \varepsilon_{1j} = 16 + \varepsilon_{1j}, \quad j = 1, 2, 3, \\
y_{2j} &= 15 + 3 + \varepsilon_{2j} = 18 + \varepsilon_{2j}, \quad j = 1, 2, 3.
\end{aligned} \tag{12.4}
$$

However, from $y_{1j} = 16 + \varepsilon_{1j}$ and $y_{2j} = 18 + \varepsilon_{2j}$, we cannot determine that $\mu = 15$, $\tau_1 = 1$, and $\tau_2 = 3$, because the model can also be written as

$$\begin{aligned} y_{1j} &= 10 + 6 + \varepsilon_{1j}, \quad j = 1, 2, 3, \\ y_{2j} &= 10 + 8 + \varepsilon_{2j}, \quad j = 1, 2, 3, \end{aligned}$$

or alternatively as

$$\begin{aligned} y_{1j} &= 25 - 9 + \varepsilon_{1j}, \quad j = 1, 2, 3, \\ y_{2j} &= 25 - 7 + \varepsilon_{2j}, \quad j = 1, 2, 3, \end{aligned}$$

or in infinitely many other ways.

Thus in (12.1) or (12.2), μ, τ_1, and τ_2 are not *unique* and therefore cannot be estimated. With three parameters and $\text{rank}(\mathbf{X}) = 2$, the model is said to be *overparameterized*. Note that increasing the number of observations (replications) for each of the two additives will not change the rank of $\mathbf{X}$.

There are various ways—each with its own advantages and disadvantages—to remedy this lack of uniqueness of the parameters in the overparameterized model. Three such approaches are (1) redefine the model using a smaller number of new parameters that are unique, (2) use the overparameterized model but place constraints on the parameters so that they become unique, and (3) in the overparameterized model, work with linear combinations of the parameters that are unique and can be unambiguously estimated. We briefly illustrate these three techniques.

1. To reduce the number of parameters, consider the illustration in (12.4):

$$y_{1j} = 16 + \varepsilon_{1j} \quad \text{and} \quad y_{2j} = 18 + \varepsilon_{2j}.$$

The values 16 and 18 are the means after the two treatments have been applied. In general, these means could be labeled μ_1 and μ_2 and the model could be written as

$$y_{1j} = \mu_1 + \varepsilon_{1j} \quad \text{and} \quad y_{2j} = \mu_2 + \varepsilon_{2j}.$$

The means μ_1 and μ_2 are unique and can be estimated. The redefined model for all six observations in (12.1) or (12.2) takes the form

$$\begin{pmatrix} y_{11} \\ y_{12} \\ y_{13} \\ y_{21} \\ y_{22} \\ y_{23} \end{pmatrix} = \begin{pmatrix} 1 & 0 \\ 1 & 0 \\ 1 & 0 \\ 0 & 1 \\ 0 & 1 \\ 0 & 1 \end{pmatrix} \begin{pmatrix} \mu_1 \\ \mu_2 \end{pmatrix} + \begin{pmatrix} \varepsilon_{11} \\ \varepsilon_{12} \\ \varepsilon_{13} \\ \varepsilon_{21} \\ \varepsilon_{22} \\ \varepsilon_{23} \end{pmatrix},$$

which we write as

$$\mathbf{y} = \mathbf{W}\boldsymbol{\mu} + \boldsymbol{\varepsilon}.$$

The matrix $\mathbf{W}$ is full-rank, and we can use (7.6) to estimate $\boldsymbol{\mu}$ as

$$\hat{\boldsymbol{\mu}} = \begin{pmatrix} \hat{\mu}_1 \\ \hat{\mu}_2 \end{pmatrix} = (\mathbf{W}'\mathbf{W})^{-1}\mathbf{W}'\mathbf{y}.$$

This solution is called *reparameterization*.

2. An alternative to reducing the number of parameters is to incorporate constraints on the parameters μ, τ_1, and τ_2. We denote the constrained parameters as μ^*, τ_1^*, and τ_2^*. In (12.1) or (12.2), the constraint $\tau_1^* + \tau_2^* = 0$ has the specific effect of defining μ^* to be the new mean after the treatments are applied and τ_1^* and τ_2^* to be deviations from this mean. With this constraint, $y_{1j} = 16 + \varepsilon_{1j}$ and $y_{2j} = 18 + \varepsilon_{2j}$ in (12.4) can be written only as

$$y_{1j} = 17 - 1 + \varepsilon_{1j}, \quad y_{2j} = 17 + 1 + \varepsilon_{2j}.$$

This model is now unique because there is no other way to express it so that $\tau_1^* + \tau_2^* = 0$. Such constraints are often called *side conditions*. The model $y_{ij} = \mu^* + \tau_i^* + \varepsilon_{ij}$ subject to $\tau_1^* + \tau_2^* = 0$ can be expressed in a full-rank format by using $\tau_2^* = -\tau_1^*$ to obtain $y_{1j} = \mu^* + \tau_1^* + \varepsilon_{1j}$ and $y_{2j} = \mu^* - \tau_1^* + \varepsilon_{ij}$. The six observations can then be written in matrix form as

$$\begin{pmatrix} y_{11} \\ y_{12} \\ y_{13} \\ y_{21} \\ y_{22} \\ y_{23} \end{pmatrix} = \begin{pmatrix} 1 & 1 \\ 1 & 1 \\ 1 & 1 \\ 1 & -1 \\ 1 & -1 \\ 1 & -1 \end{pmatrix} \begin{pmatrix} \mu^* \\ \tau_1^* \end{pmatrix} + \begin{pmatrix} \varepsilon_{11} \\ \varepsilon_{12} \\ \varepsilon_{13} \\ \varepsilon_{21} \\ \varepsilon_{22} \\ \varepsilon_{23} \end{pmatrix}$$

or

$$\mathbf{y} = \mathbf{X}^*\boldsymbol{\mu}^* + \boldsymbol{\varepsilon}.$$

The matrix $\mathbf{X}^*$ is full-rank, and the parameters μ^* and τ_1^* can be estimated. It must be kept in mind, however, that specific constraints impose specific definitions on the parameters.

3. As we examine the parameters in the model illustrated in (12.4), we see some linear combinations that are unique. For example, $\tau_1 - \tau_2 = -2$, $\mu + \tau_1 = 16$, and $\mu + \tau_2 = 18$ remain the same for all alternative values of μ, τ_1, and τ_2. Such unique linear combinations can be estimated.

In the following example, we illustrate these three approaches to parameter definition in a simple two-way model without interaction.

12.1.2 Two-Way Model

Suppose that a researcher wants to measure the effect of two different vitamins and two different methods of administering the vitamins on the weight gain of chicks. This leads to a two-way model. Let α_1 and α_2 be the effects of the two vitamins, and let β_1 and β_2 be the effects of the two methods of administration. If the researcher assumes that these effects are additive (no interaction; see the last paragraph in this example for some comments on interaction), the model can be written as

$$\begin{aligned} y_{11} &= \mu + \alpha_1 + \beta_1 + \varepsilon_{11}, \quad & y_{12} &= \mu + \alpha_1 + \beta_2 + \varepsilon_{12}, \\ y_{21} &= \mu + \alpha_2 + \beta_1 + \varepsilon_{21}, \quad & y_{22} &= \mu + \alpha_2 + \beta_2 + \varepsilon_{22}, \end{aligned}$$

or as

$$y_{ij} = \mu + \alpha_i + \beta_j + \varepsilon_{ij}, \quad i = 1, 2,\ j = 1, 2, \tag{12.5}$$

where y_{ij} is the weight gain of the ijth chick and ε_{ij} is a random error. (To simplify exposition, we show only one replication for each vitamin–method combination.)

In matrix form, the model can be expressed as

$$\begin{pmatrix} y_{11} \\ y_{12} \\ y_{21} \\ y_{22} \end{pmatrix} = \begin{pmatrix} 1 & 1 & 0 & 1 & 0 \\ 1 & 1 & 0 & 0 & 1 \\ 1 & 0 & 1 & 1 & 0 \\ 1 & 0 & 1 & 0 & 1 \end{pmatrix} \begin{pmatrix} \mu \\ \alpha_1 \\ \alpha_2 \\ \beta_1 \\ \beta_2 \end{pmatrix} + \begin{pmatrix} \varepsilon_{11} \\ \varepsilon_{12} \\ \varepsilon_{21} \\ \varepsilon_{22} \end{pmatrix} \tag{12.6}$$

or

$$\mathbf{y} = \mathbf{X}\boldsymbol{\beta} + \boldsymbol{\varepsilon}.$$

In the $\mathbf{X}$ matrix, the third column is equal to the first column minus the second column, and the fifth column is equal to the first column minus the fourth column. Thus $\text{rank}(\mathbf{X}) = 3$, and the 5×5 matrix $\mathbf{X}'\mathbf{X}$ does not have an inverse. Many of the theorems of Chapters 7 and 8 are therefore not applicable. Note that if there were replications leading to additional rows in the $\mathbf{X}$ matrix, the rank of $\mathbf{X}$ would still be 3.

Since $\text{rank}(\mathbf{X}) = 3$, there are only three possible unique parameters unless side conditions are imposed on the five parameters. There are many ways to reparameterize in order to reduce to three parameters in the model. For example, consider the parameters γ_1, γ_2, and γ_3 defined as

$$\gamma_1 = \mu + \alpha_1 + \beta_1, \quad \gamma_2 = \alpha_2 - \alpha_1, \quad \gamma_3 = \beta_2 - \beta_1.$$

The model can be written in terms of the γ terms as

$$\begin{aligned}
y_{11} &= (\mu + \alpha_1 + \beta_1) + \varepsilon_{11} = \gamma_1 + \varepsilon_{11} \\
y_{12} &= (\mu + \alpha_1 + \beta_1) + (\beta_2 - \beta_1) + \varepsilon_{12} = \gamma_1 + \gamma_3 + \varepsilon_{12} \\
y_{21} &= (\mu + \alpha_1 + \beta_1) + (\alpha_2 - \alpha_1) + \varepsilon_{21} = \gamma_1 + \gamma_2 + \varepsilon_{21} \\
y_{22} &= (\mu + \alpha_1 + \beta_1) + (\alpha_2 - \alpha_1) + (\beta_2 - \beta_1) + \varepsilon_{22} = \gamma_1 + \gamma_2 + \gamma_3 + \varepsilon_{22}.
\end{aligned}$$

In matrix form, this becomes

$$\begin{pmatrix} y_{11} \\ y_{12} \\ y_{21} \\ y_{22} \end{pmatrix} = \begin{pmatrix} 1 & 0 & 0 \\ 1 & 0 & 1 \\ 1 & 1 & 0 \\ 1 & 1 & 1 \end{pmatrix} \begin{pmatrix} \gamma_1 \\ \gamma_2 \\ \gamma_3 \end{pmatrix} + \begin{pmatrix} \varepsilon_{11} \\ \varepsilon_{12} \\ \varepsilon_{21} \\ \varepsilon_{22} \end{pmatrix}$$

or

$$\mathbf{y} = \mathbf{Z}\boldsymbol{\gamma} + \boldsymbol{\varepsilon}. \tag{12.7}$$

The rank of $\mathbf{Z}$ is clearly 3, and we have a full-rank model for which $\boldsymbol{\gamma}$ can be estimated by $\hat{\boldsymbol{\gamma}} = (\mathbf{Z}'\mathbf{Z})^{-1}\mathbf{Z}'\mathbf{y}$. This provides estimates of $\gamma_2 = \alpha_2 - \alpha_1$ and $\gamma_3 = \beta_2 - \beta_1$, which are typically of interest to the researcher.

In Section 12.2.2, we will discuss methods for showing that linear functions such as $\mu + \alpha_1 + \beta_1$, $\alpha_2 - \alpha_1$, and $\beta_2 - \beta_1$ are unique and estimable, even though μ, α_1, α_2, β_1, β_2 are not unique and not estimable.

We now consider side conditions on the parameters. Since rank($\mathbf{X}$) = 3 and there are five parameters, we need two (linearly independent) side conditions. If these two constraints are appropriately chosen, the five parameters become unique and thereby estimable. We denote the constrained parameters by μ^*, α_i^*, and β_j^* and consider the side conditions $\alpha_1^* + \alpha_2^* = 0$ and $\beta_1^* + \beta_2^* = 0$. These lead to unique definitions of α_i^* and β_j^* as deviations from means. To show this, we start by writing the model as

$$\begin{aligned}
y_{11} &= \mu_{11} + \varepsilon_{11}, \quad & y_{12} &= \mu_{12} + \varepsilon_{12}, \\
y_{21} &= \mu_{21} + \varepsilon_{21}, \quad & y_{22} &= \mu_{22} + \varepsilon_{22},
\end{aligned} \tag{12.8}$$

where $\mu_{ij} = E(y_{ij})$ is the mean weight gain with vitamin i and method j. The means are displayed in Table 12.1, and the parameters α_1^*, α_2^*, β_1^*, β_2^* are defined as row (α) and column (β) effects.

The means in Table 12.1 are defined as follows:

$$\bar{\mu}_{i.} = \frac{\mu_{i1} + \mu_{i2}}{2}, \quad \bar{\mu}_{.j} = \frac{\mu_{1j} + \mu_{2j}}{2}, \quad \bar{\mu}_{..} = \frac{\mu_{11} + \mu_{12} + \mu_{21} + \mu_{22}}{4}.$$

TABLE 12.1 Means and Effects for the Model in (12.8)

	Columns (β)			
Rows (α)	1	2	Row Means	Row Effects
Row 1	μ_{11}	μ_{12}	$\bar{\mu}_{1.}$	$\alpha_1^* = \bar{\mu}_{1.} - \bar{\mu}_{..}$
Row 2	μ_{21}	μ_{22}	$\bar{\mu}_{2.}$	$\alpha_2^* = \bar{\mu}_{2.} - \bar{\mu}_{..}$
Column means	$\bar{\mu}_{.1}$	$\bar{\mu}_{.2}$	$\bar{\mu}_{..}$	—
Column effects	$\beta_1^* = \bar{\mu}_{.1} - \bar{\mu}_{..}$	$\beta_2^* = \bar{\mu}_{.2} - \bar{\mu}_{..}$	—	—

The first row effect, $\alpha_1^* = \bar{\mu}_{1.} - \bar{\mu}_{..}$, is the deviation of the mean for vitamin 1 from the overall mean (after treatments) and is unique. The parameters α_2^*, β_1^*, and β_2^* are likewise uniquely defined. From the definitions in Table 12.1, we obtain

$$\begin{aligned} \alpha_1^* + \alpha_2^* &= \bar{\mu}_{1.} - \bar{\mu}_{..} + \bar{\mu}_{2.} - \bar{\mu}_{..} = \bar{\mu}_{1.} + \bar{\mu}_{2.} - 2\bar{\mu}_{..} \\ &= 2\bar{\mu}_{..} - 2\bar{\mu}_{..} = 0, \end{aligned} \tag{12.9}$$

and similarly, $\beta_1^* + \beta_2^* = 0$. Thus with the side conditions $\alpha_1^* + \alpha_2^* = 0$ and $\beta_1^* + \beta_2^* = 0$, the redefined parameters are both unique and interpretable.

In (12.5), it is assumed that the effects of vitamin and method are additive. To make this notion more precise, we write the model (12.5) in terms of $\mu^* = \mu_{..}$, $\alpha_i^* = \bar{\mu}_{i.} - \bar{\mu}_{..}$, and $\beta_j^* = \bar{\mu}_{.j} - \bar{\mu}_{..}$:

$$\begin{aligned} \mu_{ij} &= \bar{\mu}_{..} + (\bar{\mu}_{i.} - \bar{\mu}_{..}) + (\bar{\mu}_{.j} - \bar{\mu}_{..}) + (\mu_{ij} - \bar{\mu}_{i.} - \bar{\mu}_{.j} + \bar{\mu}_{..}) \\ &= \mu^* + \alpha_i^* + \beta_j^*. \end{aligned}$$

The term $\mu_{ij} - \bar{\mu}_{i.} - \bar{\mu}_{.j} + \bar{\mu}_{..}$, which is required to balance the equation, is associated with the interaction between vitamins and methods. In order for α_i^* and β_j^* to be additive effects, the interaction $\mu_{ij} - \bar{\mu}_{i.} - \bar{\mu}_{.j} + \bar{\mu}_{..}$ must be zero. Interaction will be treated in Chapter 14.

12.2 ESTIMATION

In this section, we consider various aspects of estimation of $\boldsymbol{\beta}$ in the non-full-rank model $\mathbf{y} = \mathbf{X}\boldsymbol{\beta} + \boldsymbol{\varepsilon}$. We do not reparameterize or impose side conditions. These two approaches to estimation are discussed in Sections 12.5 and 12.6, respectively. Normality of $\mathbf{y}$ is not assumed in the present section.

12.2.1 Estimation of β

Consider the model

$$\mathbf{y} = \mathbf{X}\boldsymbol{\beta} + \boldsymbol{\varepsilon},$$

where $E(\mathbf{y}) = \mathbf{X}\boldsymbol{\beta}$, $\mathrm{cov}(\mathbf{y}) = \sigma^2\mathbf{I}$, and $\mathbf{X}$ is $n \times p$ of rank $k < p \leq n$. [We will say "$\mathbf{X}$ is $n \times p$ of rank $k < p \leq n$" to indicate that $\mathbf{X}$ is not of full rank; that is, $\mathrm{rank}(\mathbf{X}) < p$ and $\mathrm{rank}(\mathbf{X}) < n$. In some cases, we have $k < n < p$.] In this non-full-rank model, the p parameters in $\boldsymbol{\beta}$ are not unique. We now ascertain whether $\boldsymbol{\beta}$ can be estimated.

Using least-squares, we seek a value of $\hat{\boldsymbol{\beta}}$ that minimizes

$$\hat{\boldsymbol{\varepsilon}}'\hat{\boldsymbol{\varepsilon}} = (\mathbf{y} - \mathbf{X}\hat{\boldsymbol{\beta}})'(\mathbf{y} - \mathbf{X}\hat{\boldsymbol{\beta}}).$$

We can expand $\hat{\boldsymbol{\varepsilon}}'\hat{\boldsymbol{\varepsilon}}$ to obtain

$$\hat{\boldsymbol{\varepsilon}}'\hat{\boldsymbol{\varepsilon}} = \mathbf{y}'\mathbf{y} - 2\hat{\boldsymbol{\beta}}'\mathbf{X}'\mathbf{y} + \hat{\boldsymbol{\beta}}'\mathbf{X}'\mathbf{X}\hat{\boldsymbol{\beta}}, \tag{12.10}$$

which can be differentiated with respect to $\hat{\boldsymbol{\beta}}$ and set equal to $\mathbf{0}$ to produce the familiar normal equations

$$\mathbf{X}'\mathbf{X}\hat{\boldsymbol{\beta}} = \mathbf{X}'\mathbf{y}. \tag{12.11}$$

Since $\mathbf{X}$ is not full rank, $\mathbf{X}'\mathbf{X}$ has no inverse, and (12.11) does not have a unique solution. However, $\mathbf{X}'\mathbf{X}\hat{\boldsymbol{\beta}} = \mathbf{X}'\mathbf{y}$ has (an infinite number of) solutions:

Theorem 12.2a. If $\mathbf{X}$ is $n \times p$ of rank $k < p \leq n$, the system of equations $\mathbf{X}'\mathbf{X}\hat{\boldsymbol{\beta}} = \mathbf{X}'\mathbf{y}$ is consistent.

PROOF. By Theorem 2.8f, the system is consistent if and only if

$$\mathbf{X}'\mathbf{X}(\mathbf{X}'\mathbf{X})^{-}\mathbf{X}'\mathbf{y} = \mathbf{X}'\mathbf{y}, \tag{12.12}$$

where $(\mathbf{X}'\mathbf{X})^{-}$ is any generalized inverse of $\mathbf{X}'\mathbf{X}$. By Theorem 2.8c(iii), $\mathbf{X}'\mathbf{X}(\mathbf{X}'\mathbf{X})^{-}\mathbf{X}' = \mathbf{X}'$, and (12.12) therefore holds. (An alternative proof is suggested in Problem 12.3.) □

Since the normal equations $\mathbf{X}'\mathbf{X}\hat{\boldsymbol{\beta}} = \mathbf{X}'\mathbf{y}$ are consistent, a solution is given by Theorem 2.8d as

$$\hat{\boldsymbol{\beta}} = (\mathbf{X}'\mathbf{X})^{-}\mathbf{X}'\mathbf{y}, \tag{12.13}$$

where $(\mathbf{X}'\mathbf{X})^-$ is any generalized inverse of $\mathbf{X}'\mathbf{X}$. For a particular generalized inverse $(\mathbf{X}'\mathbf{X})^-$, the expected value of $\hat{\boldsymbol{\beta}}$ is

$$\begin{aligned} E(\hat{\boldsymbol{\beta}}) &= (\mathbf{X}'\mathbf{X})^-\mathbf{X}'E(\mathbf{y}) \\ &= (\mathbf{X}'\mathbf{X})^-\mathbf{X}'\mathbf{X}\boldsymbol{\beta}. \end{aligned} \tag{12.14}$$

Thus, $\hat{\boldsymbol{\beta}}$ is an unbiased estimator of $(\mathbf{X}'\mathbf{X})^-\mathbf{X}'\mathbf{X}\boldsymbol{\beta}$. Since $(\mathbf{X}'\mathbf{X})^-\mathbf{X}'\mathbf{X} \neq \mathbf{I}$, $\hat{\boldsymbol{\beta}}$ is not an unbiased estimator of $\boldsymbol{\beta}$. The expression $(\mathbf{X}'\mathbf{X})^-\mathbf{X}'\mathbf{X}\boldsymbol{\beta}$ is not invariant to the choice of $(\mathbf{X}'\mathbf{X})^-$; that is, $E(\hat{\boldsymbol{\beta}})$ is different for each choice of $(\mathbf{X}'\mathbf{X})^-$. [An implication in (12.14) is that having selected a value of $(\mathbf{X}'\mathbf{X})^-$, we would use that same value of $(\mathbf{X}'\mathbf{X})^-$ in repeated sampling.]

Thus, $\hat{\boldsymbol{\beta}}$ in (12.13) does not estimate $\boldsymbol{\beta}$. Next, we inquire as to whether there are any linear functions of $\mathbf{y}$ that are unbiased estimators for the elements of $\boldsymbol{\beta}$; that is, whether there exists a $p \times n$ matrix $\mathbf{A}$ such that $E(\mathbf{Ay}) = \boldsymbol{\beta}$. If so, then

$$\boldsymbol{\beta} = E(\mathbf{Ay}) = E[\mathbf{A}(\mathbf{X}\boldsymbol{\beta} + \boldsymbol{\varepsilon})] = E(\mathbf{AX}\boldsymbol{\beta}) + \mathbf{A}E(\boldsymbol{\varepsilon}) = \mathbf{AX}\boldsymbol{\beta}.$$

Since this must hold for all $\boldsymbol{\beta}$, we have $\mathbf{AX} = \mathbf{I}_p$ [see (2.44)]. But by Theorem 2.4(i), $\text{rank}(\mathbf{AX}) < p$ since the rank of $\mathbf{X}$ is less than p. Hence $\mathbf{AX}$ cannot be equal to I_p, and there are no linear functions of the observations that yield unbiased estimators for the elements of $\boldsymbol{\beta}$.

Example 12.2.1. Consider the model $y_{ij} = \mu + \tau_i + \varepsilon_{ij}$; $i = 1, 2$; $j = 1, 2, 3$ in (12.2). The matrix $\mathbf{X}$ and the vector $\boldsymbol{\beta}$ are given in (12.3) as

$$\mathbf{X} = \begin{pmatrix} 1 & 1 & 0 \\ 1 & 1 & 0 \\ 1 & 1 & 0 \\ 1 & 0 & 1 \\ 1 & 0 & 1 \\ 1 & 0 & 1 \end{pmatrix}, \quad \boldsymbol{\beta} = \begin{pmatrix} \mu \\ \tau_1 \\ \tau_2 \end{pmatrix}.$$

By Theorem 2.2c(i), we obtain

$$\mathbf{X}'\mathbf{X} = \begin{pmatrix} 6 & 3 & 3 \\ 3 & 3 & 0 \\ 3 & 0 & 3 \end{pmatrix}.$$

By Corollary 1 to Theorem 2.8b, a generalized inverse of $\mathbf{X}'\mathbf{X}$ is given by

$$(\mathbf{X}'\mathbf{X})^- = \begin{pmatrix} 0 & 0 & 0 \\ 0 & \frac{1}{3} & 0 \\ 0 & 0 & \frac{1}{3} \end{pmatrix}.$$

The vector $\mathbf{X}'\mathbf{y}$ is given by

$$\mathbf{X}'\mathbf{y} = \begin{pmatrix} 1 & 1 & 1 & 1 & 1 & 1 \\ 1 & 1 & 1 & 0 & 0 & 0 \\ 0 & 0 & 0 & 1 & 1 & 1 \end{pmatrix} \begin{pmatrix} y_{11} \\ y_{12} \\ y_{13} \\ y_{21} \\ y_{22} \\ y_{23} \end{pmatrix} = \begin{pmatrix} y_{..} \\ y_{1.} \\ y_{2.} \end{pmatrix},$$

where $y_{..} = \sum_{i=1}^{2} \sum_{j=1}^{3} y_{ij}$ and $y_{i.} = \sum_{j=1}^{3} y_{ij}$. Then

$$\hat{\boldsymbol{\beta}} = (\mathbf{X}'\mathbf{X})^{-}\mathbf{X}'\mathbf{y} = \begin{pmatrix} 0 & 0 & 0 \\ 0 & \frac{1}{3} & 0 \\ 0 & 0 & \frac{1}{3} \end{pmatrix} \begin{pmatrix} y_{..} \\ y_{1.} \\ y_{2.} \end{pmatrix} = \begin{pmatrix} 0 \\ \bar{y}_{1.} \\ \bar{y}_{2.} \end{pmatrix},$$

where $\bar{y}_{i.} = \sum_{j=1}^{3} y_{ij}/3 = y_{i.}/3$.

To find $E(\hat{\boldsymbol{\beta}})$, we need $E(\bar{y}_{i.})$. Since $E(\boldsymbol{\varepsilon}) = \mathbf{0}$, we have $E(\varepsilon_{ij}) = 0$. Then

$$\begin{aligned} E(\bar{y}_{i.}) &= E\left(\sum_{j=1}^{3} y_{ij}/3\right) = \tfrac{1}{3}\textstyle\sum_{j=1}^{3} E(y_{ij}) \\ &= \tfrac{1}{3}\textstyle\sum_{j=1}^{3} E(\mu + \tau_i + \varepsilon_{ij}) = \tfrac{1}{3}(3\mu + 3\tau_i + 0) \\ &= \mu + \tau_i. \end{aligned}$$

Thus

$$E(\hat{\boldsymbol{\beta}}) = \begin{pmatrix} 0 \\ \mu + \tau_1 \\ \mu + \tau_2 \end{pmatrix}.$$

The same result is obtained in (12.14):

$$\begin{aligned} E(\hat{\boldsymbol{\beta}}) &= (\mathbf{X}'\mathbf{X})^{-}\mathbf{X}'\mathbf{X}\hat{\boldsymbol{\beta}} \\ &= \begin{pmatrix} 0 & 0 & 0 \\ 0 & \frac{1}{3} & 0 \\ 0 & 0 & \frac{1}{3} \end{pmatrix} \begin{pmatrix} 6 & 3 & 3 \\ 3 & 3 & 0 \\ 3 & 0 & 3 \end{pmatrix} \begin{pmatrix} \mu \\ \tau_1 \\ \tau_2 \end{pmatrix} \\ &= \begin{pmatrix} 0 \\ \mu + \tau_1 \\ \mu + \tau_2 \end{pmatrix}. \end{aligned}$$

□

12.2.2 Estimable Functions of β

Having established that we cannot estimate $\boldsymbol{\beta}$, we next inquire as to whether we can estimate any linear combination of the $\boldsymbol{\beta}$'s, say, $\boldsymbol{\lambda}'\boldsymbol{\beta}$. For example, in Section 12.1.1, we considered the model $y_{ij} = \mu + \tau_i + \varepsilon_{ij}$, $i = 1, 2$, and found that μ, τ_1, and τ_2 in $\boldsymbol{\beta} = (\mu, \tau_1, \tau_2)'$ are not unique but that the linear function $\tau_1 - \tau_2 = (0, 1, -1)\boldsymbol{\beta}$ is unique. In order to show that functions such as $\tau_1 - \tau_2$ can be estimated, we first give a definition of an estimable function $\boldsymbol{\lambda}'\boldsymbol{\beta}$.

A linear function of parameters $\boldsymbol{\lambda}'\boldsymbol{\beta}$ is said to be *estimable* if there exists a linear combination of the observations with an expected value equal to $\boldsymbol{\lambda}'\boldsymbol{\beta}$; that is, $\boldsymbol{\lambda}'\boldsymbol{\beta}$ is estimable if there exists a vector $\mathbf{a}$ such that $E(\mathbf{a}'\mathbf{y}) = \boldsymbol{\lambda}'\boldsymbol{\beta}$.

In the following theorem we consider three methods for determining whether a particular linear function $\boldsymbol{\lambda}'\boldsymbol{\beta}$ is estimable.

Theorem 12.2b. In the model $\mathbf{y} = \mathbf{X}\boldsymbol{\beta} + \boldsymbol{\varepsilon}$, where $E(\mathbf{y}) = \mathbf{X}\boldsymbol{\beta}$ and $\mathbf{X}$ is $n \times p$ of rank $k < p \leq n$, the linear function $\boldsymbol{\lambda}'\boldsymbol{\beta}$ is estimable if and only if any one of the following equivalent conditions holds:

(i) $\boldsymbol{\lambda}'$ is a linear combination of the rows of $\mathbf{X}$; that is, there exists a vector $\mathbf{a}$ such that

$$\mathbf{a}'\mathbf{X} = \boldsymbol{\lambda}'. \tag{12.15}$$

(ii) $\boldsymbol{\lambda}'$ is a linear combination of the rows of $\mathbf{X}'\mathbf{X}$ or $\boldsymbol{\lambda}$ is a linear combination of the columns of $\mathbf{X}'\mathbf{X}$, that is, there exists a vector $\mathbf{r}$ such that

$$\mathbf{r}'\mathbf{X}'\mathbf{X} = \boldsymbol{\lambda}' \quad \text{or} \quad \mathbf{X}'\mathbf{X}\mathbf{r} = \boldsymbol{\lambda}. \tag{12.16}$$

(iii) $\boldsymbol{\lambda}$ or $\boldsymbol{\lambda}'$ is such that

$$\mathbf{X}'\mathbf{X}(\mathbf{X}'\mathbf{X})^{-}\boldsymbol{\lambda} = \boldsymbol{\lambda} \quad \text{or} \quad \boldsymbol{\lambda}'(\mathbf{X}'\mathbf{X})^{-}\mathbf{X}'\mathbf{X} = \boldsymbol{\lambda}', \tag{12.17}$$

where $(\mathbf{X}'\mathbf{X})^{-}$ is any (symmetric) generalized inverse of $\mathbf{X}'\mathbf{X}$.

PROOF. For (ii) and (iii), we prove the "if" part. For (i), we prove both "if" and "only if."

(i) If there exists a vector $\mathbf{a}$ such that $\boldsymbol{\lambda}' = \mathbf{a}'\mathbf{X}$, then, using this vector $\mathbf{a}$, we have

$$E(\mathbf{a}'\mathbf{y}) = \mathbf{a}'E(\mathbf{y}) = \mathbf{a}'\mathbf{X}\boldsymbol{\beta} = \boldsymbol{\lambda}'\boldsymbol{\beta}.$$

Conversely, if $\boldsymbol{\lambda}'\boldsymbol{\beta}$ is estimable, then there exists a vector $\mathbf{a}$ such that $E(\mathbf{a}'\mathbf{y}) = \boldsymbol{\lambda}'\boldsymbol{\beta}$. Thus $\mathbf{a}'\mathbf{X}\boldsymbol{\beta} = \boldsymbol{\lambda}'\boldsymbol{\beta}$, which implies, among other things, that $\mathbf{a}'\mathbf{X} = \boldsymbol{\lambda}'$.

(ii) If there exists a solution $\mathbf{r}$ for $\mathbf{X}'\mathbf{X}\mathbf{r} = \boldsymbol{\lambda}$, then, by defining $\mathbf{a} = \mathbf{X}\mathbf{r}$, we obtain

$$\begin{aligned} E(\mathbf{a}'\mathbf{y}) &= E(\mathbf{r}'\mathbf{X}'\mathbf{y}) = \mathbf{r}'\mathbf{X}'E(\mathbf{y}) \\ &= \mathbf{r}'\mathbf{X}'\mathbf{X}\boldsymbol{\beta} = \boldsymbol{\lambda}'\boldsymbol{\beta}. \end{aligned}$$

(iii) If $\mathbf{X}'\mathbf{X}(\mathbf{X}'\mathbf{X})^{-}\boldsymbol{\lambda} = \boldsymbol{\lambda}$, then $(\mathbf{X}'\mathbf{X})^{-}\boldsymbol{\lambda}$ is a solution to $\mathbf{X}'\mathbf{X}\mathbf{r} = \boldsymbol{\lambda}$ in part(ii). (For proof of the converse, see Problem 12.4.) □

We illustrate the use of Theorem 12.2b in the following example.

Example 12.2.2a. For the model $y_{ij} = \mu + \tau_i + \varepsilon_{ij}; \quad i = 1, 2; \quad j = 1, 2, 3$ in Example 12.2.1, the matrix $\mathbf{X}$ and the vector $\boldsymbol{\beta}$ are given as

$$\mathbf{X} = \begin{pmatrix} 1 & 1 & 0 \\ 1 & 1 & 0 \\ 1 & 1 & 0 \\ 1 & 0 & 1 \\ 1 & 0 & 1 \\ 1 & 0 & 1 \end{pmatrix}, \quad \boldsymbol{\beta} = \begin{pmatrix} \mu \\ \tau_1 \\ \tau_2 \end{pmatrix}.$$

We noted in Section 12.1.1 that $\tau_1 - \tau_2$ is unique. We now show that $\tau_1 - \tau_2 = (0, 1, -1)\boldsymbol{\beta} = \boldsymbol{\lambda}'\boldsymbol{\beta}$ is estimable, using all three conditions of Theorem 12.2b.

(i) To find a vector $\mathbf{a}$ such that $\mathbf{a}'\mathbf{X} = \boldsymbol{\lambda}' = (0, 1, -1)$, consider $\mathbf{a}' = (0, 0, 1, -1, 0, 0)$, which gives

$$\begin{aligned} \mathbf{a}'\mathbf{X} &= (0, 0, 1, -1, 0, 0)\mathbf{X} = (1, 1, 0) - (1, 0, 1) \\ &= (0, 1, -1) = \boldsymbol{\lambda}'. \end{aligned}$$

There are many other choices for $\mathbf{a}$, of course, that will yield $\mathbf{a}'\mathbf{X} = \boldsymbol{\lambda}'$, for example $\mathbf{a}' = (1, 0, 0, 0, 0, -1)$ or $\mathbf{a}' = (2, -1, 0, 0, 1, -2)$. Note that we can likewise obtain $\boldsymbol{\lambda}'\boldsymbol{\beta}$ from $E(\mathbf{y})$:

$$\begin{aligned} \boldsymbol{\lambda}'\boldsymbol{\beta} &= \mathbf{a}'\mathbf{X}\boldsymbol{\beta} = \mathbf{a}'E(\mathbf{y}) = (0, 0, 1, -1, 0, 0)E(\mathbf{y}) \\ &= (0, 0, 1, \ -1, 0, 0) \begin{pmatrix} E(y_{11}) \\ E(y_{12}) \\ E(y_{13}) \\ E(y_{21}) \\ E(y_{22}) \\ E(y_{23}) \end{pmatrix} \\ &= E(y_{13}) - E(y_{21}) = \mu + \tau_1 - (\mu + \tau_2) = \tau_1 - \tau_2. \end{aligned}$$

(ii) The matrix $\mathbf{X}'\mathbf{X}$ is given in Example 12.2.1 as

$$\mathbf{X}'\mathbf{X} = \begin{pmatrix} 6 & 3 & 3 \\ 3 & 3 & 0 \\ 3 & 0 & 3 \end{pmatrix}.$$

To find a vector $\mathbf{r}$ such that $\mathbf{X}'\mathbf{X}\mathbf{r} = \boldsymbol{\lambda} = (0, 1, -1)'$, consider $\mathbf{r} = (0, \frac{1}{3}, -\frac{1}{3})'$, which gives

$$\mathbf{X}'\mathbf{X}\mathbf{r} = \begin{pmatrix} 6 & 3 & 3 \\ 3 & 3 & 0 \\ 3 & 0 & 3 \end{pmatrix} \begin{pmatrix} 0 \\ \frac{1}{3} \\ -\frac{1}{3} \end{pmatrix} = \begin{pmatrix} 0 \\ 1 \\ -1 \end{pmatrix} = \boldsymbol{\lambda}.$$

There are other possible values of $\mathbf{r}$, of course, such as $\mathbf{r} = (-\frac{1}{3}, \frac{2}{3}, 0)'$.

(iii) Using the generalized inverse $(\mathbf{X}'\mathbf{X})^- = \text{diag}(0, \frac{1}{3}, \frac{1}{3})$ given in Example 12.2.1, the product $\mathbf{X}'\mathbf{X}(\mathbf{X}'\mathbf{X})^-$ becomes

$$\mathbf{X}'\mathbf{X}(\mathbf{X}'\mathbf{X})^- = \begin{pmatrix} 0 & 1 & 1 \\ 0 & 1 & 0 \\ 0 & 0 & 1 \end{pmatrix}.$$

Then, for $\boldsymbol{\lambda} = (0, 1, -1)'$, we see that $\mathbf{X}'\mathbf{X}(\mathbf{X}'\mathbf{X})^- \boldsymbol{\lambda} = \boldsymbol{\lambda}$ in (12.17) holds:

$$\begin{pmatrix} 0 & 1 & 1 \\ 0 & 1 & 0 \\ 0 & 0 & 1 \end{pmatrix} \begin{pmatrix} 0 \\ 1 \\ -1 \end{pmatrix} = \begin{pmatrix} 0 \\ 1 \\ -1 \end{pmatrix}.$$ □

A set of functions $\boldsymbol{\lambda}_1'\boldsymbol{\beta}, \boldsymbol{\lambda}_2'\boldsymbol{\beta}, \ldots, \boldsymbol{\lambda}_m'\boldsymbol{\beta}$ is said to be linearly independent if the coefficient vectors $\boldsymbol{\lambda}_1, \boldsymbol{\lambda}_2, \ldots, \boldsymbol{\lambda}_m$ are linearly independent [see (2.40)]. The number of linearly independent estimable functions is given in the next theorem.

Theorem 12.2c. In the non-full-rank model $\mathbf{y} = \mathbf{X}\boldsymbol{\beta} + \boldsymbol{\varepsilon}$, the number of linearly independent estimable functions of $\boldsymbol{\beta}$ is the rank of $\mathbf{X}$.

PROOF. See Graybill (1976, pp. 485–486). □

From Theorem 12.2b(i), we see that $\mathbf{x}_i'\boldsymbol{\beta}$ is estimable for $i = 1, 2, \ldots, n$, where $\mathbf{x}_i'$ is the ith row of $\mathbf{X}$. Thus every row (element) of $\mathbf{X}\boldsymbol{\beta}$ is estimable, and $\mathbf{X}\boldsymbol{\beta}$ itself can be said to be estimable. Likewise, from Theorem 12.2b(ii), every row (element) of $\mathbf{X}'\mathbf{X}\boldsymbol{\beta}$ is estimable, and $\mathbf{X}'\mathbf{X}\boldsymbol{\beta}$ is therefore estimable. Conversely, all estimable functions can be obtained from $\mathbf{X}\boldsymbol{\beta}$ or $\mathbf{X}'\mathbf{X}\boldsymbol{\beta}$:

Thus we can examine linear combinations of the rows of $\mathbf{X}$ or of $\mathbf{X}'\mathbf{X}$ to see what functions of the parameters are estimable. In the following example, we illustrate the use of linear combinations of the rows of **X** to obtain a set of estimable functions of the parameters.

Example 12.2.2b. Consider the model in (12.6) in Section 12.1.2 with

$$\mathbf{X} = \begin{pmatrix} 1 & 1 & 0 & 1 & 0 \\ 1 & 1 & 0 & 0 & 1 \\ 1 & 0 & 1 & 1 & 0 \\ 1 & 0 & 1 & 0 & 1 \end{pmatrix}, \quad \boldsymbol{\beta} = \begin{pmatrix} \mu \\ \alpha_1 \\ \alpha_2 \\ \beta_1 \\ \beta_2 \end{pmatrix}.$$

To examine what is estimable, we take linear combinations $\mathbf{a}'\mathbf{X}$ of the rows of **X** to obtain three linearly independent rows. For example, if we subtract the first row of **X** from the third row and multiply by $\boldsymbol{\beta}$, we obtain $(0\ -1\ 1\ 0\ 0)\boldsymbol{\beta} = -\alpha_1 + \alpha_2$, which involves only the α's. Subtracting the first row of **X** from the third row can be expressed as $\mathbf{a}'\mathbf{X} = (-1\ 0\ 1\ 0)\mathbf{X} = -\mathbf{x}_1' + \mathbf{x}_3'$, where $\mathbf{x}_1'$ and $\mathbf{x}_3'$ are the first and third rows of **X**.

Subtracting the first row from each succeeding row in **X** gives

$$\begin{pmatrix} 1 & 1 & 0 & 1 & 0 \\ 0 & 0 & 0 & -1 & 1 \\ 0 & -1 & 1 & 0 & 0 \\ 0 & -1 & 1 & -1 & 1 \end{pmatrix}.$$

Subtracting the second and third rows from the fourth row of this matrix yields

$$\begin{pmatrix} 1 & 1 & 0 & 1 & 0 \\ 0 & 0 & 0 & -1 & 1 \\ 0 & -1 & 1 & 0 & 0 \\ 0 & 0 & 0 & 0 & 0 \end{pmatrix}.$$

Multiplying the first three rows by $\boldsymbol{\beta}$, we obtain the three linearly independent estimable functions

$$\boldsymbol{\lambda}_1'\boldsymbol{\beta} = \mu + \alpha_1 + \beta_1, \quad \boldsymbol{\lambda}_2'\boldsymbol{\beta} = \beta_2 - \beta_1, \quad \boldsymbol{\lambda}_3'\boldsymbol{\beta} = \alpha_2 - \alpha_1.$$

These functions are identical to the functions γ_1, γ_2, and γ_3 used in Section 12.1.2 to reparameterize to a full-rank model. Thus, in that example, linearly independent estimable functions of the parameters were used as the new parameters.

In Example 12.2.2.b, the two estimable functions $\beta_2 - \beta_1$ and $\alpha_2 - \alpha_1$ are such that the coefficients of the $\boldsymbol{\beta}$'s or of the α's sum to zero. A linear combination of this type is called a *contrast*.

12.3 ESTIMATORS

12.3.1 Estimators of $\boldsymbol{\lambda}'\boldsymbol{\beta}$

From Theorem 12.2b(i) and (ii) we have the estimators $\mathbf{a}'\mathbf{y}$ and $\mathbf{r}'\mathbf{X}'\mathbf{y}$ for $\boldsymbol{\lambda}'\boldsymbol{\beta}$, where $\mathbf{a}'$ and $\mathbf{r}'$ satisfy $\boldsymbol{\lambda}' = \mathbf{a}'\mathbf{X}$ and $\boldsymbol{\lambda}' = \mathbf{r}'\mathbf{X}'\mathbf{X}$, respectively. A third estimator of $\boldsymbol{\lambda}'\boldsymbol{\beta}$ is $\boldsymbol{\lambda}'\hat{\boldsymbol{\beta}}$, where $\hat{\boldsymbol{\beta}}$ is a solution of $\mathbf{X}'\mathbf{X}\hat{\boldsymbol{\beta}} = \mathbf{X}'\mathbf{y}$. In the following theorem, we discuss some properties of $\mathbf{r}'\mathbf{X}'\mathbf{y}$ and $\boldsymbol{\lambda}'\hat{\boldsymbol{\beta}}$. We do not discuss the estimator $\mathbf{a}'\mathbf{y}$ because it is not guaranteed to have minimum variance (see Theorem 12.3d).

Theorem 12.3a. Let $\boldsymbol{\lambda}'\boldsymbol{\beta}$ be an estimable function of $\boldsymbol{\beta}$ in the model $\mathbf{y} = \mathbf{X}\boldsymbol{\beta} + \boldsymbol{\varepsilon}$, where $E(\mathbf{y}) = \mathbf{X}\boldsymbol{\beta}$ and $\mathbf{X}$ is $n \times p$ of rank $k < p \leq n$. Let $\hat{\boldsymbol{\beta}}$ be any solution to the normal equations $\mathbf{X}'\mathbf{X}\hat{\boldsymbol{\beta}} = \mathbf{X}'\mathbf{y}$, and let $\mathbf{r}$ be any solution to $\mathbf{X}'\mathbf{X}\mathbf{r} = \boldsymbol{\lambda}$. Then the two estimators $\boldsymbol{\lambda}'\hat{\boldsymbol{\beta}}$ and $\mathbf{r}'\mathbf{X}'\mathbf{y}$ have the following properties:

(i) $E(\boldsymbol{\lambda}'\hat{\boldsymbol{\beta}}) = E(\mathbf{r}'\mathbf{X}'\mathbf{y}) = \boldsymbol{\lambda}'\boldsymbol{\beta}$.
(ii) $\boldsymbol{\lambda}'\hat{\boldsymbol{\beta}}$ is equal to $\mathbf{r}'\mathbf{X}'\mathbf{y}$ for any $\hat{\boldsymbol{\beta}}$ or any $\mathbf{r}$.
(iii) $\boldsymbol{\lambda}'\hat{\boldsymbol{\beta}}$ and $\mathbf{r}'\mathbf{X}'\mathbf{y}$ are invariant to the choice of $\hat{\boldsymbol{\beta}}$ or $\mathbf{r}$.

PROOF

(i) By (12.14)

$$E(\boldsymbol{\lambda}'\hat{\boldsymbol{\beta}}) = \boldsymbol{\lambda}'E(\hat{\boldsymbol{\beta}}) = \boldsymbol{\lambda}'(\mathbf{X}'\mathbf{X})^{-}\mathbf{X}'\mathbf{X}\boldsymbol{\beta}.$$

By Theorem 12.2b(iii), $\boldsymbol{\lambda}'(\mathbf{X}'\mathbf{X})^{-}\mathbf{X}'\mathbf{X} = \boldsymbol{\lambda}'$, and $E(\boldsymbol{\lambda}'\hat{\boldsymbol{\beta}})$ becomes

$$E(\boldsymbol{\lambda}'\hat{\boldsymbol{\beta}}) = \boldsymbol{\lambda}'\boldsymbol{\beta}.$$

By Theorem 12.2b(ii)

$$E(\mathbf{r}'\mathbf{X}'\mathbf{y}) = \mathbf{r}'\mathbf{X}'E(\mathbf{y}) = \mathbf{r}'\mathbf{X}'\mathbf{X}\boldsymbol{\beta} = \boldsymbol{\lambda}'\boldsymbol{\beta}.$$

(ii) By Theorem 12.2b(ii), if $\boldsymbol{\lambda}'\boldsymbol{\beta}$ is estimable, $\boldsymbol{\lambda}' = \mathbf{r}'\mathbf{X}'\mathbf{X}$ for some $\mathbf{r}$. Multiplying the normal equations $\mathbf{X}'\mathbf{X}\hat{\boldsymbol{\beta}} = \mathbf{X}'\mathbf{y}$ by $\mathbf{r}'$ gives

$$\mathbf{r}'\mathbf{X}'\mathbf{X}\hat{\boldsymbol{\beta}} = \mathbf{r}'\mathbf{X}'\mathbf{y}.$$

Since $\mathbf{r}'\mathbf{X}'\mathbf{X} = \boldsymbol{\lambda}'$, we have

$$\boldsymbol{\lambda}'\hat{\boldsymbol{\beta}} = \mathbf{r}'\mathbf{X}'\mathbf{y}.$$

(iii) To show that $\mathbf{r}'\mathbf{X}'\mathbf{y}$ is invariant to the choice of $\mathbf{r}$, let $\mathbf{r}_1$ and $\mathbf{r}_2$ be such that $\mathbf{X}'\mathbf{X}\mathbf{r}_1 = \mathbf{X}'\mathbf{X}\mathbf{r}_2 = \boldsymbol{\lambda}$. Then

$$\mathbf{r}_1'\mathbf{X}'\mathbf{X}\hat{\boldsymbol{\beta}} = \mathbf{r}_1'\mathbf{X}'\mathbf{y} \quad \text{and} \quad \mathbf{r}_2'\mathbf{X}'\mathbf{X}\hat{\boldsymbol{\beta}} = \mathbf{r}_2'\mathbf{X}'\mathbf{y}.$$

Since $\mathbf{r}_1'\mathbf{X}'\mathbf{X} = \mathbf{r}_2'\mathbf{X}'\mathbf{X}$, we have $\mathbf{r}_1'\mathbf{X}'\mathbf{y} = \mathbf{r}_2'\mathbf{X}'\mathbf{y}$. It is clear that each is equal to $\boldsymbol{\lambda}'\hat{\boldsymbol{\beta}}$. (For a direct proof that $\boldsymbol{\lambda}'\hat{\boldsymbol{\beta}}$ is invariant to the choice of $\hat{\boldsymbol{\beta}}$, see Problem 12.6.) □

We illustrate the estimators $\mathbf{r}'\mathbf{X}'\mathbf{y}$ and $\boldsymbol{\lambda}\hat{\boldsymbol{\beta}}$ in the following example.

Example 12.3.1. The linear function $\boldsymbol{\lambda}'\boldsymbol{\beta} = \tau_1 - \tau_2$ was shown to be estimable in Example 12.2.2a. To estimate $\tau_1 - \tau_2$ with $\mathbf{r}'\mathbf{X}'\mathbf{y}$, we use $\mathbf{r}' = (0, \frac{1}{3}, -\frac{1}{3})$ from Example 12.2.2a to obtain

$$\mathbf{r}'\mathbf{X}'\mathbf{y} = \left(0, \tfrac{1}{3}, -\tfrac{1}{3}\right)\begin{pmatrix} 1 & 1 & 1 & 1 & 1 & 1 \\ 1 & 1 & 1 & 0 & 0 & 0 \\ 0 & 0 & 0 & 1 & 1 & 1 \end{pmatrix}\begin{pmatrix} y_{11} \\ y_{12} \\ y_{13} \\ y_{21} \\ y_{22} \\ y_{23} \end{pmatrix}$$

$$= \left(0, \tfrac{1}{3}, -\tfrac{1}{3}\right)\begin{pmatrix} y_{..} \\ y_{1.} \\ y_{2.} \end{pmatrix} = \frac{y_{1.}}{3} - \frac{y_{2.}}{3} = \bar{y}_{1.} - \bar{y}_{2.},$$

where $y_{..} = \sum_{i=1}^{2}\sum_{j=1}^{3} y_{ij}$, $y_{i.} = \sum_{j=1}^{3} y_{ij}$, and $\bar{y}_{i.} = y_{i.}/3 = \sum_{j=1}^{3} y_{ij}/3$.

To obtain the same result using $\boldsymbol{\lambda}'\hat{\boldsymbol{\beta}}$, we first find a solution to the normal equations $\mathbf{X}'\mathbf{X}\hat{\boldsymbol{\beta}} = \mathbf{X}'\mathbf{y}$

$$\begin{pmatrix} 6 & 3 & 3 \\ 3 & 3 & 0 \\ 3 & 0 & 3 \end{pmatrix}\begin{pmatrix} \hat{\mu} \\ \hat{\tau}_1 \\ \hat{\tau}_2 \end{pmatrix} = \begin{pmatrix} y_{..} \\ y_{1.} \\ y_{2.} \end{pmatrix}$$

or

$$\begin{aligned} 6\hat{\mu} + 3\hat{\tau}_1 + 3\hat{\tau}_2 &= y_{..} \\ 3\hat{\mu} + 3\hat{\tau}_1 &= y_{1.} \\ 3\hat{\mu} + 3\hat{\tau}_2 &= y_{2.}. \end{aligned}$$

The first equation is redundant since it is the sum of the second and third equations. We can take $\hat{\mu}$ to be an arbitrary constant and obtain

$$\hat{\tau}_1 = \tfrac{1}{3}y_{1.} - \hat{\mu} = \bar{y}_{1.} - \hat{\mu}, \quad \hat{\tau}_2 = \tfrac{1}{3}y_{2.} - \hat{\mu} = \bar{y}_{2.} - \hat{\mu}.$$

Thus

$$\hat{\boldsymbol{\beta}} = \begin{pmatrix} \hat{\mu} \\ \hat{\tau}_1 \\ \hat{\tau}_2 \end{pmatrix} = \begin{pmatrix} 0 \\ \bar{y}_{1.} \\ \bar{y}_{2.} \end{pmatrix} + \hat{\mu}\begin{pmatrix} 1 \\ -1 \\ -1 \end{pmatrix}.$$

To estimate $\tau_1 - \tau_2 = (0, 1, -1)\boldsymbol{\beta} = \boldsymbol{\lambda}'\boldsymbol{\beta}$, we can set $\hat{\mu} = 0$ to obtain $\hat{\boldsymbol{\beta}} = (0, \bar{y}_{1.}, \bar{y}_{2.})'$ and $\boldsymbol{\lambda}'\hat{\boldsymbol{\beta}} = \bar{y}_{1.} - \bar{y}_{2.}$. If we leave $\hat{\mu}$ arbitrary, we likewise obtain

$$\begin{aligned}\boldsymbol{\lambda}'\hat{\boldsymbol{\beta}} &= (0, 1, -1)\begin{pmatrix} \hat{\mu} \\ \bar{y}_{1.} - \hat{\mu} \\ \bar{y}_{2.} - \hat{\mu} \end{pmatrix} \\ &= \bar{y}_{1.} - \hat{\mu} - (\bar{y}_{2.} - \hat{\mu}) = \bar{y}_{1.} - \bar{y}_{2.}. \end{aligned}$$

□

Since $\hat{\boldsymbol{\beta}} = (\mathbf{X}'\mathbf{X})^{-}\mathbf{X}'\mathbf{y}$ is not unique for the non-full-rank model $\mathbf{y} = \mathbf{X}\boldsymbol{\beta} + \boldsymbol{\varepsilon}$ with $\text{cov}(\mathbf{y}) = \sigma^2\mathbf{I}$, it does not have a unique covariance matrix. However, for a particular (symmetric) generalized inverse $(\mathbf{X}'\mathbf{X})^{-}$, we can use Theorem 3.6d(i) to obtain the following covariance matrix:

$$\begin{aligned}\text{cov}(\hat{\boldsymbol{\beta}}) &= \text{cov}[(\mathbf{X}'\mathbf{X})^{-}\mathbf{X}'\mathbf{y}] \\ &= (\mathbf{X}'\mathbf{X})^{-}\mathbf{X}'(\sigma^2\mathbf{I})\mathbf{X}[(\mathbf{X}'\mathbf{X})^{-}]' \\ &= \sigma^2(\mathbf{X}'\mathbf{X})^{-}\mathbf{X}'\mathbf{X}(\mathbf{X}'\mathbf{X})^{-}. \end{aligned} \tag{12.18}$$

The expression in (12.18) is not invariant to the choice of $(\mathbf{X}'\mathbf{X})^{-}$.

The variance of $\boldsymbol{\lambda}'\hat{\boldsymbol{\beta}}$ or of $\mathbf{r}'\mathbf{X}'\mathbf{y}$ is given in the following theorem.

Theorem 12.3b. Let $\boldsymbol{\lambda}'\boldsymbol{\beta}$ be an estimable function in the model $\mathbf{y} = \mathbf{X}\boldsymbol{\beta} + \boldsymbol{\varepsilon}$, where $\mathbf{X}$ is $n \times p$ of rank $k < p \leq n$ and $\text{cov}(\mathbf{y}) = \sigma^2\mathbf{I}$. Let $\mathbf{r}$ be any solution to $\mathbf{X}'\mathbf{X}\mathbf{r} = \boldsymbol{\lambda}$, and let $\hat{\boldsymbol{\beta}}$ be any solution to $\mathbf{X}'\mathbf{X}\hat{\boldsymbol{\beta}} = \mathbf{X}'\mathbf{y}$. Then the variance of $\boldsymbol{\lambda}'\hat{\boldsymbol{\beta}}$ or $\mathbf{r}'\mathbf{X}'\mathbf{y}$ has the following properties:

(i) $\text{var}(\mathbf{r}'\mathbf{X}'\mathbf{y}) = \sigma^2\mathbf{r}'\mathbf{X}'\mathbf{X}\mathbf{r} = \sigma^2\mathbf{r}'\boldsymbol{\lambda}$.
(ii) $\text{var}(\boldsymbol{\lambda}'\hat{\boldsymbol{\beta}}) = \sigma^2\boldsymbol{\lambda}'(\mathbf{X}'\mathbf{X})^{-}\boldsymbol{\lambda}$.
(iii) $\text{var}(\boldsymbol{\lambda}'\hat{\boldsymbol{\beta}})$ is unique, that is, invariant to the choice of $\mathbf{r}$ or $(\mathbf{X}'\mathbf{X})^{-}$.

PROOF

(i)

$$\begin{aligned}\text{var}(\mathbf{r}'\mathbf{X}'\mathbf{y}) &= \mathbf{r}'\mathbf{X}'\text{cov}(\mathbf{y})\mathbf{X}\mathbf{r} \qquad \text{[by (3.42)]} \\ &= \mathbf{r}'\mathbf{X}'(\sigma^2\mathbf{I})\mathbf{X}\mathbf{r} = \sigma^2\mathbf{r}'\mathbf{X}'\mathbf{X}\mathbf{r} \\ &= \sigma^2\mathbf{r}'\boldsymbol{\lambda}. \qquad \text{[by (12.16)]}. \end{aligned}$$

(ii)

$$\begin{aligned}\text{var}(\boldsymbol{\lambda}'\hat{\boldsymbol{\beta}}) &= \boldsymbol{\lambda}'\text{cov}(\hat{\boldsymbol{\beta}})\boldsymbol{\lambda} \\ &= \sigma^2\boldsymbol{\lambda}'(\mathbf{X}'\mathbf{X})^{-}\mathbf{X}'\mathbf{X}(\mathbf{X}'\mathbf{X})^{-}\boldsymbol{\lambda} \qquad \text{[by (12.18)]}. \end{aligned}$$

By (12.17), $\boldsymbol{\lambda}'(\mathbf{X}'\mathbf{X})^{-}\mathbf{X}'\mathbf{X} = \boldsymbol{\lambda}'$, and therefore

$$\text{var}(\boldsymbol{\lambda}'\hat{\boldsymbol{\beta}}) = \sigma^2 \boldsymbol{\lambda}'(\mathbf{X}'\mathbf{X})^{-}\boldsymbol{\lambda}.$$

(iii) To show that $\mathbf{r}'\boldsymbol{\lambda}$ is invariant to $\mathbf{r}$, let $\mathbf{r}_1$ and $\mathbf{r}_2$ be such that $\mathbf{X}'\mathbf{X}\mathbf{r}_1 = \boldsymbol{\lambda}$ and $\mathbf{X}'\mathbf{X}\mathbf{r}_2 = \boldsymbol{\lambda}$. Multiplying these two equations by $\mathbf{r}_2'$ and $\mathbf{r}_1'$, we obtain

$$\mathbf{r}_2'\mathbf{X}'\mathbf{X}\mathbf{r}_1 = \mathbf{r}_2'\boldsymbol{\lambda} \quad \text{and} \quad \mathbf{r}_1'\mathbf{X}'\mathbf{X}\mathbf{r}_2 = \mathbf{r}_1'\boldsymbol{\lambda}.$$

The left sides of these two equations are equal since they are scalars and are transposes of each other. Therefore the right sides are also equal:

$$\mathbf{r}_2'\boldsymbol{\lambda} = \mathbf{r}_1'\boldsymbol{\lambda}.$$

To show that $\boldsymbol{\lambda}'(\mathbf{X}'\mathbf{X})^{-}\boldsymbol{\lambda}$ is invariant to the choice of $\mathbf{X}'\mathbf{X}^{-}$, let $\mathbf{G}_1$ and $\mathbf{G}_2$ be two generalized inverses of $\mathbf{X}'\mathbf{X}$. Then by Theorem 2.8c(v), we have

$$\mathbf{X}\mathbf{G}_1\mathbf{X}' = \mathbf{X}\mathbf{G}_2\mathbf{X}'.$$

Multiplying both sides by $\mathbf{a}$ such that $\mathbf{a}'\mathbf{X} = \boldsymbol{\lambda}'$ [see Theorem 12.2b(i)], we obtain

$$\mathbf{a}'\mathbf{X}\mathbf{G}_1\mathbf{X}'\mathbf{a} = \mathbf{a}'\mathbf{X}\mathbf{G}_2\mathbf{X}'\mathbf{a},$$
$$\boldsymbol{\lambda}'\mathbf{G}_1\boldsymbol{\lambda} = \boldsymbol{\lambda}'\mathbf{G}_2\boldsymbol{\lambda}. \qquad \square$$

The covariance of the estimators of two estimable functions is given in the following theorem.

Theorem 12.3c. If $\boldsymbol{\lambda}_1'\boldsymbol{\beta}$ and $\boldsymbol{\lambda}_2'\boldsymbol{\beta}$ are two estimable functions in the model $\mathbf{y} = \mathbf{X}\boldsymbol{\beta} + \boldsymbol{\varepsilon}$, where $\mathbf{X}$ is $n \times p$ of rank $k < p \leq n$ and $\text{cov}(\mathbf{y}) = \sigma^2\mathbf{I}$, the covariance of their estimators is given by

$$\text{cov}(\boldsymbol{\lambda}_1'\hat{\boldsymbol{\beta}}, \boldsymbol{\lambda}_2'\hat{\boldsymbol{\beta}}) = \sigma^2\mathbf{r}_1'\boldsymbol{\lambda}_2 = \sigma^2\boldsymbol{\lambda}_1'\mathbf{r}_2 = \sigma^2\boldsymbol{\lambda}_1'(\mathbf{X}'\mathbf{X})^{-}\boldsymbol{\lambda}_2,$$

where $\mathbf{X}'\mathbf{X}\mathbf{r}_1 = \boldsymbol{\lambda}_1$ and $\mathbf{X}'\mathbf{X}\mathbf{r}_2 = \boldsymbol{\lambda}_2$.

PROOF. See Problem 12.12. $\square$

The estimators $\boldsymbol{\lambda}'\hat{\boldsymbol{\beta}}$ and $\mathbf{r}'\mathbf{X}'\mathbf{y}$ have an optimality property analogous to that in Corollary 1 to Theorem 7.3d.

Theorem 12.3d. If $\boldsymbol{\lambda}'\boldsymbol{\beta}$ is an estimable function in the model $\mathbf{y} = \mathbf{X}\boldsymbol{\beta} + \boldsymbol{\varepsilon}$, where $\mathbf{X}$ is $n \times p$ of rank $k < p \leq n$, then the estimators $\boldsymbol{\lambda}'\hat{\boldsymbol{\beta}}$ and $\mathbf{r}'\mathbf{X}'\mathbf{y}$ are BLUE.

PROOF. Let a linear estimator of $\boldsymbol{\lambda}'\boldsymbol{\beta}$ be denoted by $\mathbf{a}'\mathbf{y}$, where without loss of generality $\mathbf{a}'\mathbf{y} = \mathbf{r}'\mathbf{X}'\mathbf{y} + \mathbf{c}'\mathbf{y}$, that is, $\mathbf{a}' = \mathbf{r}'\mathbf{X}' + \mathbf{c}'$, where $\mathbf{r}'$ is a solution to $\boldsymbol{\lambda}' = \mathbf{r}'\mathbf{X}'\mathbf{X}$. For unbiasedness we must have

$$\boldsymbol{\lambda}'\boldsymbol{\beta} = E(\mathbf{a}'\mathbf{y}) = \mathbf{a}'\mathbf{X}\boldsymbol{\beta} = \mathbf{r}'\mathbf{X}'\mathbf{X}\boldsymbol{\beta} + \mathbf{c}'\mathbf{X}\boldsymbol{\beta} = (\mathbf{r}'\mathbf{X}'\mathbf{X} + \mathbf{c}'\mathbf{X})\boldsymbol{\beta}.$$

This must hold for all $\boldsymbol{\beta}$, and we therefore have

$$\boldsymbol{\lambda}' = \mathbf{r}'\mathbf{X}'\mathbf{X} + \mathbf{c}'\mathbf{X}.$$

Since $\boldsymbol{\lambda}' = \mathbf{r}'\mathbf{X}'\mathbf{X}$, it follows that $\mathbf{c}'\mathbf{X} = \mathbf{0}'$. Using (3.42) and $\mathbf{c}'\mathbf{X} = \mathbf{0}'$, we obtain

$$\begin{aligned} \text{var}(\mathbf{a}'\mathbf{y}) &= \mathbf{a}'\text{cov}(\mathbf{y})\mathbf{a} = \mathbf{a}'\sigma^2\mathbf{I}\mathbf{a} = \sigma^2\mathbf{a}'\mathbf{a} \\ &= \sigma^2(\mathbf{r}'\mathbf{X}' + \mathbf{c}')(\mathbf{X}\mathbf{r} + \mathbf{c}) \\ &= \sigma^2(\mathbf{r}'\mathbf{X}'\mathbf{X}\mathbf{r} + \mathbf{r}'\mathbf{X}'\mathbf{c} + \mathbf{c}'\mathbf{X}\mathbf{r} + \mathbf{c}'\mathbf{c}) \\ &= \sigma^2(\mathbf{r}'\mathbf{X}'\mathbf{X}\mathbf{r} + \mathbf{c}'\mathbf{c}). \end{aligned}$$

Therefore, to minimize $\text{var}(\mathbf{a}'\mathbf{y})$, we must minimize $\mathbf{c}'\mathbf{c} = \sum_i c_i^2$. This is a minimum when $\mathbf{c} = \mathbf{0}$, which is compatible with $\mathbf{c}'\mathbf{X} = \mathbf{0}'$. Hence $\mathbf{a}'$ is equal to $\mathbf{r}'\mathbf{X}'$, and the BLUE for the estimable function $\boldsymbol{\lambda}'\boldsymbol{\beta}$ is $\mathbf{a}'\mathbf{y} = \mathbf{r}'\mathbf{X}'\mathbf{y}$. □

12.3.2 Estimation of σ^2

By analogy with (7.23), we define

$$\text{SSE} = (\mathbf{y} - \mathbf{X}\hat{\boldsymbol{\beta}})'(\mathbf{y} - \mathbf{X}\hat{\boldsymbol{\beta}}), \tag{12.19}$$

where $\hat{\boldsymbol{\beta}}$ is any solution to the normal equations $\mathbf{X}'\mathbf{X}\hat{\boldsymbol{\beta}} = \mathbf{X}'\mathbf{y}$. Two alternative expressions for SSE are

$$\text{SSE} = \mathbf{y}'\mathbf{y} - \hat{\boldsymbol{\beta}}'\mathbf{X}'\mathbf{y}, \tag{12.20}$$

$$\text{SSE} = \mathbf{y}'[\mathbf{I} - \mathbf{X}(\mathbf{X}'\mathbf{X})^{-}\mathbf{X}']\mathbf{y}. \tag{12.21}$$

For an estimator of σ^2, we define

$$s^2 = \frac{\text{SSE}}{n-k}, \tag{12.22}$$

where n is the number of rows of $\mathbf{X}$ and $k = \text{rank}(\mathbf{X})$.

Two properties of s^2 are given in the following theorem.

Theorem 12.3e. For s^2 defined in (12.22) for the non-full-rank model, we have the following properties:

(i) $E(s^2) = \sigma^2$.

(ii) s^2 is invariant to the choice of $\hat{\boldsymbol{\beta}}$ or to the choice of generalized inverse $(\mathbf{X}'\mathbf{X})^-$.

PROOF

(i) Using (12.21), we have $E(\text{SSE}) = E\{\mathbf{y}'[\mathbf{I} - \mathbf{X}(\mathbf{X}'\mathbf{X})^-\mathbf{X}']\mathbf{y}\}$. By Theorem 5.2a, this becomes

$$E(\text{SSE}) = \text{tr}\{[\mathbf{I} - \mathbf{X}(\mathbf{X}'\mathbf{X})^-\mathbf{X}'](\sigma^2\mathbf{I})\} + \boldsymbol{\beta}'\mathbf{X}'[\mathbf{I} - \mathbf{X}(\mathbf{X}'\mathbf{X})^-\mathbf{X}']\mathbf{X}\boldsymbol{\beta}.$$

It can readily be shown that the second term on the right side vanishes. For the first term, we have, by Theorem 2.11(i), (ii), and (viii)

$$\begin{aligned}\sigma^2\text{tr}[\mathbf{I} - \mathbf{X}(\mathbf{X}'\mathbf{X})^-\mathbf{X}'] &= \sigma^2\{\text{tr}(\mathbf{I}) - \mathbf{tr}[\mathbf{X}'\mathbf{X}(\mathbf{X}'\mathbf{X})^-]\} \\ &= (n-k)\sigma^2,\end{aligned}$$

where $k = \text{rank}(\mathbf{X}'\mathbf{X}) = \text{rank}(\mathbf{X})$.

(ii) Since $\mathbf{X}\boldsymbol{\beta}$ is estimable, $\mathbf{X}\hat{\boldsymbol{\beta}}$ is invariant to $\hat{\boldsymbol{\beta}}$ [see Theorem 12.3a(iii)], and therefore $\text{SSE} = (\mathbf{y} - \mathbf{X}\hat{\boldsymbol{\beta}})'(\mathbf{y} - \mathbf{X}\hat{\boldsymbol{\beta}})$ in (12.19) is invariant. To show that SSE in (12.21) is invariant to choice of $(\mathbf{X}'\mathbf{X})^-$, we note that $\mathbf{X}(\mathbf{X}'\mathbf{X})^-\mathbf{X}'$ is invariant by Theorem 2.8c(v). □

12.3.3 Normal Model

For the non-full-rank model $\mathbf{y} = \mathbf{X}\boldsymbol{\beta} + \boldsymbol{\varepsilon}$, we now assume that

$$\mathbf{y} \text{ is } N_n(\mathbf{X}\boldsymbol{\beta}, \sigma^2\mathbf{I}) \quad \text{or} \quad \boldsymbol{\varepsilon} \text{ is } N_n(\mathbf{0}, \sigma^2\mathbf{I}).$$

With the normality assumption we can obtain maximum likelihood estimators.

Theorem 12.3f. If $\mathbf{y}$ is $N_n(\mathbf{X}\boldsymbol{\beta}, \sigma^2\mathbf{I})$, where $\mathbf{X}$ is $n \times p$ of rank $k < p \leq n$, then the maximum likelihood estimators for $\boldsymbol{\beta}$ and σ^2 are given by

$$\hat{\boldsymbol{\beta}} = (\mathbf{X}'\mathbf{X})^- \mathbf{X}'\mathbf{y}, \tag{12.23}$$

$$\hat{\sigma}^2 = \frac{1}{n}(\mathbf{y} - \mathbf{X}\hat{\boldsymbol{\beta}})'(\mathbf{y} - \mathbf{X}\hat{\boldsymbol{\beta}}). \tag{12.24}$$

PROOF. For the non-full-rank model, the likelihood function $L(\boldsymbol{\beta}, \sigma^2)$ and its logarithm $\ln L(\boldsymbol{\beta}, \sigma^2)$ can be written in the same form as those for the full-rank model in (7.50) and (7.51):

$$L(\boldsymbol{\beta}, \sigma^2) = \frac{1}{(2\pi\sigma^2)^{n/2}} e^{-(\mathbf{y}-\mathbf{X}\boldsymbol{\beta})'(\mathbf{y}-\mathbf{X}\boldsymbol{\beta})/2\sigma^2}, \tag{12.25}$$

$$\ln L(\boldsymbol{\beta}, \sigma^2) = -\frac{n}{2}\ln(2\pi) - \frac{n}{2}\ln\sigma^2 - \frac{1}{2\sigma^2}(\mathbf{y} - \mathbf{X}\boldsymbol{\beta})'(\mathbf{y} - \mathbf{X}\boldsymbol{\beta}). \tag{12.26}$$

Differentiation of $\ln L(\boldsymbol{\beta}, \sigma^2)$ with respect to $\boldsymbol{\beta}$ and σ^2 and setting the results equal to zero gives

$$\mathbf{X}'\mathbf{X}\hat{\boldsymbol{\beta}} = \mathbf{X}'\mathbf{y}, \tag{12.27}$$

$$\hat{\sigma}^2 = \frac{1}{n}(\mathbf{y} - \mathbf{X}\hat{\boldsymbol{\beta}})'(\mathbf{y} - \mathbf{X}\hat{\boldsymbol{\beta}}), \tag{12.28}$$

where $\hat{\boldsymbol{\beta}}$ in (12.28) is any solution to (12.27). If $(\mathbf{X}'\mathbf{X})^-$ is any generalized inverse of $\mathbf{X}'\mathbf{X}$, a solution to (12.27) is given by

$$\hat{\boldsymbol{\beta}} = (\mathbf{X}'\mathbf{X})^- \mathbf{X}'\mathbf{y}. \tag{12.29}$$

□

The form of the maximum likelihood estimator $\hat{\boldsymbol{\beta}}$ in (12.29) is the same as that of the least-squares estimator in (12.13). The estimator $\hat{\sigma}^2$ is biased. We often use the unbiased estimator s^2 given in (12.22).

The mean vector and covariance matrix for $\hat{\boldsymbol{\beta}}$ are given in (12.14) and (12.18) as $E(\hat{\boldsymbol{\beta}}) = (\mathbf{X}'\mathbf{X})^- \mathbf{X}'\mathbf{X}\boldsymbol{\beta}$ and $\text{cov}(\hat{\boldsymbol{\beta}}) = \sigma^2(\mathbf{X}'\mathbf{X})^- \mathbf{X}'\mathbf{X}(\mathbf{X}'\mathbf{X})^-$. In the next theorem, we give some additional properties of $\hat{\boldsymbol{\beta}}$ and s^2. Note that some of these follow because $\hat{\boldsymbol{\beta}} = (\mathbf{X}'\mathbf{X})^- \mathbf{X}'\mathbf{y}$ is a linear function of the observations.

Theorem 12.3g. If $\mathbf{y}$ is $N_n(\mathbf{X}\boldsymbol{\beta}, \sigma^2\mathbf{I})$, where $\mathbf{X}$ is $n \times p$ of rank $k < p \leq n$, then the maximum likelihood estimators $\hat{\boldsymbol{\beta}}$ and s^2 (corrected for bias) have the following properties:

(i) $\hat{\boldsymbol{\beta}}$ is $N_p[(\mathbf{X}'\mathbf{X})^-\mathbf{X}'\mathbf{X}\boldsymbol{\beta},\ \sigma^2(\mathbf{X}'\mathbf{X})^-\mathbf{X}'\mathbf{X}(\mathbf{X}'\mathbf{X})^-]$.
(ii) $(n-k)s^2/\sigma^2$ is $\chi^2(n-k)$.
(iii) $\hat{\boldsymbol{\beta}}$ and s^2 are independent.

PROOF. Adapting the proof of Theorem 7.6b for the non-full-rank case yields the desired results. □

The expected value, covariance matrix, and distribution of $\hat{\boldsymbol{\beta}}$ in Theorem 12.3g are valid only for a particular value of $(\mathbf{X}'\mathbf{X})^-$, whereas, s^2 is invariant to the choice of $\hat{\boldsymbol{\beta}}$ or $(\mathbf{X}'\mathbf{X})^-$ [see Theorem 12.3e(ii)].

The following theorem is an adaptation of Corollary 1 to Theorem 7.6d.

Theorem 12.3h. If $\mathbf{y}$ is $N_n(\mathbf{X}\boldsymbol{\beta}, \sigma^2\mathbf{I})$, where $\mathbf{X}$ is $n \times p$ of rank $k < p \leq n$, and if $\boldsymbol{\lambda}'\boldsymbol{\beta}$ is an estimable function, then $\boldsymbol{\lambda}'\hat{\boldsymbol{\beta}}$ has minimum variance among all unbiased estimators. □

In Theorem 12.3d, the estimator $\boldsymbol{\lambda}'\hat{\boldsymbol{\beta}}$ was shown to have minimum variance among all *linear unbiased* estimators. With the normality assumption added in Theorem 12.3g, $\boldsymbol{\lambda}\hat{\boldsymbol{\beta}}$ has minimum variance among all *unbiased* estimators.

12.4 GEOMETRY OF LEAST-SQUARES IN THE OVERPARAMETERIZED MODEL

The geometric approach to least-squares in the overparameterized model is similar to that for the full-rank model (Section 7.4), but there are crucial differences. The approach involves two spaces, a p-dimensional parameter space and an n-dimensional data space. The unknown parameter vector $\boldsymbol{\beta}$ is an element of the parameter space with axes corresponding to the coefficients, and the known data vector $\mathbf{y}$ is an element of the data space with axes corresponding to the observations (Fig. 12.1).

The $n \times p$ partitioned $\mathbf{X}$ matrix of the overparameterized linear model (Section 12.2.1) is

$$\mathbf{X} = (\mathbf{x}_1, \mathbf{x}_2, \ldots, \mathbf{x}_p).$$

The columns of $\mathbf{X}$ are vectors in the data space, but since $\text{rank}(\mathbf{X}) = k < p$, the set of vectors is not linearly independent. Nonetheless, the set of all possible linear combinations of these column vectors constitutes the prediction space. The distinctive

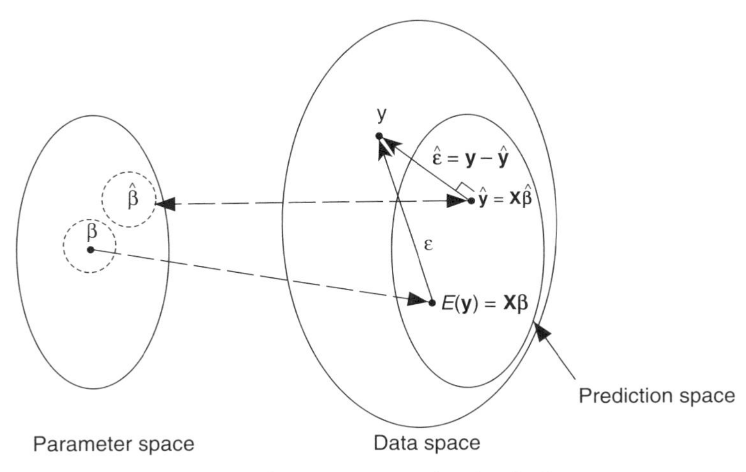

Figure 12.1 A geometric view of least-squares estimation in the overparameterized model.

geometric characteristic of the overparameterized model is that the prediction space is of dimension $k < p$ while the parameter space is of dimension p. Thus the product $\mathbf{Xu}$, where $\mathbf{u}$ is any vector in the parameter space, defines a *many-to-one* relationship between the parameter space and the prediction space (Fig. 12.1). An infinite number of vectors in the parameter space correspond to any particular vector in the prediction space.

As was the case for the full-rank linear model, the overparameterized linear model states that $\mathbf{y}$ is equal to a vector in the prediction space, $E(y) = \mathbf{X}\beta$, plus a vector of random errors $\boldsymbol{\varepsilon}$. Neither $\boldsymbol{\beta}$ nor $\boldsymbol{\varepsilon}$ is known. Geometrically, least-squares estimation for the overparametrized model is the process of finding a sensible guess of $E(\mathbf{y})$ in the prediction space and then determining the *subset* of the parameter space that is associated with this guess (Fig. 12.1).

As in the full-rank model, a reasonable geometric idea is to estimate $E(\mathbf{y})$ using $\hat{\mathbf{y}}$, the unique point in the prediction space that is closest to $\mathbf{y}$. This implies that the difference vector $\hat{\boldsymbol{\varepsilon}} = \mathbf{y} - \hat{\mathbf{y}}$ must be orthogonal to the prediction space, and thus we seek $\hat{\mathbf{y}}$ such that

$$\mathbf{X}'\hat{\boldsymbol{\varepsilon}} = \mathbf{0},$$

which leads to the normal equations

$$\mathbf{X}'\mathbf{X}\hat{\boldsymbol{\beta}} = \mathbf{X}'\mathbf{y}.$$

However, these equations do not have a single solution since $\mathbf{X}'\mathbf{X}$ is not full-rank. Using Theorem 2.8e(ii), all possible solutions to this system of equations are given by $\hat{\boldsymbol{\beta}} = (\mathbf{X}'\mathbf{X})^{-}\mathbf{X}'\mathbf{y}$ using all possible values of $(\mathbf{X}'\mathbf{X})^{-}$. These solutions constitute an infinite subset of the parameter space (Fig. 12.1), but this subset is not a subspace.

Since the solutions are infinite in number, none of the $\hat{\boldsymbol{\beta}}$ values themselves have any meaning. Nonetheless, $\hat{\mathbf{y}} = \mathbf{X}\hat{\boldsymbol{\beta}}$ is unique [see Theorem 2.8c(v)], and therefore, to be unambiguous, all further inferences must be restricted to linear functions of $\mathbf{X}\hat{\boldsymbol{\beta}}$ rather than of $\hat{\boldsymbol{\beta}}$.

Also note that the n rows of $\mathbf{X}$ generate a k-dimensional subspace of p-dimensional space. The matrix products of the row vectors in this space with $\boldsymbol{\beta}$ constitute the set of all possible estimable functions. The matrix products of the row vectors in this space with any $\hat{\boldsymbol{\beta}}$ (these products are invariant to the choice of a generalized inverse) constitute the unambiguous set of corresponding estimates of these functions.

Finally, $\hat{\boldsymbol{\varepsilon}} = \mathbf{y} - \mathbf{X}\hat{\boldsymbol{\beta}} = (\mathbf{I} - \mathbf{H})\mathbf{y}$ can be taken as an unambiguous predictor of $\boldsymbol{\varepsilon}$. Since $\hat{\boldsymbol{\varepsilon}}$ is now a vector in $(n-k)$-dimensional space, it seems reasonable to estimate σ^2 as the squared length (2.22) of $\hat{\boldsymbol{\varepsilon}}$ divided by $n-k$. In other words, a sensible estimator of σ^2 is $s^2 = \mathbf{y}'(\mathbf{I} - \mathbf{H})\mathbf{y}/(n-k)$, which is equal to (12.22).

12.5 REPARAMETERIZATION

Reparameterization was defined and illustrated in Section 12.1.1. We now formalize and extend this approach to obtaining a model based on estimable parameters.

In reparameterization, we transform the non-full-rank model $\mathbf{y} = \mathbf{X}\boldsymbol{\beta} + \boldsymbol{\varepsilon}$, where $\mathbf{X}$ is $n \times p$ of rank $k < p \leq n$, to the full-rank model $\mathbf{y} = \mathbf{Z}\boldsymbol{\gamma} + \boldsymbol{\varepsilon}$, where $\mathbf{Z}$ is $n \times k$ of rank k and $\boldsymbol{\gamma} = \mathbf{U}\boldsymbol{\beta}$ is a set of k linearly independent estimable functions of $\boldsymbol{\beta}$. Thus $\mathbf{Z}\boldsymbol{\gamma} = \mathbf{X}\boldsymbol{\beta}$, and we can write

$$\mathbf{Z}\boldsymbol{\gamma} = \mathbf{Z}\mathbf{U}\boldsymbol{\beta} = \mathbf{X}\boldsymbol{\beta}, \tag{12.30}$$

where $\mathbf{X} = \mathbf{Z}\mathbf{U}$. Since $\mathbf{U}$ is $k \times p$ of rank $k < p$, the matrix $\mathbf{U}\mathbf{U}'$ is nonsingular by Theorem 2.4(iii), and we can multiply $\mathbf{Z}\mathbf{U} = \mathbf{X}$ by $\mathbf{U}'$ to solve for $\mathbf{Z}$ in terms of $\mathbf{X}$ and $\mathbf{U}$:

$$\mathbf{Z}\mathbf{U}\mathbf{U}' = \mathbf{X}\mathbf{U}'$$

$$\mathbf{Z} = \mathbf{X}\mathbf{U}'(\mathbf{U}\mathbf{U}')^{-1}. \tag{12.31}$$

To establish that $\mathbf{Z}$ is full-rank, note that rank($\mathbf{Z}$) $\geq$ rank($\mathbf{Z}\mathbf{U}$) $=$ rank($\mathbf{X}$) $= k$ by Theorem 2.4(i). However, $\mathbf{Z}$ cannot have rank greater than k since $\mathbf{Z}$ has k columns. Thus rank($\mathbf{Z}$) $= k$, and the model $\mathbf{y} = \mathbf{Z}\boldsymbol{\gamma} + \boldsymbol{\varepsilon}$ is a full-rank model. We can therefore use the theorems of Chapters 7 and 8; for example, the normal equations $\mathbf{Z}'\mathbf{Z}\hat{\boldsymbol{\gamma}} = \mathbf{Z}'\mathbf{y}$ have the unique solution $\hat{\boldsymbol{\gamma}} = (\mathbf{Z}'\mathbf{Z})^{-1}\mathbf{Z}'\mathbf{y}$.

In the reparameterized full-rank model $\mathbf{y} = \mathbf{Z}\boldsymbol{\gamma} + \boldsymbol{\varepsilon}$, the unbiased estimator of σ^2 is given by

$$s^2 = \frac{1}{n-k}(\mathbf{y} - \mathbf{Z}\hat{\boldsymbol{\gamma}})'(\mathbf{y} - \mathbf{Z}\hat{\boldsymbol{\gamma}}). \tag{12.32}$$

Since $\mathbf{Z}\boldsymbol{\gamma} = \mathbf{X}\boldsymbol{\beta}$, the estimators $\mathbf{Z}\hat{\boldsymbol{\gamma}}$ and $\mathbf{X}\hat{\boldsymbol{\beta}}$ are also equal

$$\mathbf{Z}\hat{\boldsymbol{\gamma}} = \mathbf{X}\hat{\boldsymbol{\beta}},$$

and SSE in (12.19) and SSE in (12.32) are the same:

$$(\mathbf{y} - \mathbf{X}\hat{\boldsymbol{\beta}})'(\mathbf{y} - \mathbf{X}\hat{\boldsymbol{\beta}}) = (\mathbf{y} - \mathbf{Z}\hat{\boldsymbol{\gamma}})'(\mathbf{y} - \mathbf{Z}\hat{\boldsymbol{\gamma}}). \tag{12.33}$$

The set $\mathbf{U}\boldsymbol{\beta} = \boldsymbol{\gamma}$ is only one possible set of linearly independent estimable functions. Let $\mathbf{V}\boldsymbol{\beta} = \boldsymbol{\delta}$ be another set of linearly independent estimable functions. Then there exists a matrix $\mathbf{W}$ such that $\mathbf{y} = \mathbf{W}\boldsymbol{\delta} + \boldsymbol{\varepsilon}$. Now an estimable function $\boldsymbol{\lambda}'\boldsymbol{\beta}$ can be expressed as a function of $\boldsymbol{\gamma}$ or of $\boldsymbol{\delta}$:

$$\boldsymbol{\lambda}'\boldsymbol{\beta} = \mathbf{b}'\boldsymbol{\gamma} = \mathbf{c}'\boldsymbol{\delta}. \tag{12.34}$$

Hence

$$\widehat{\boldsymbol{\lambda}'\boldsymbol{\beta}} = \mathbf{b}'\hat{\boldsymbol{\gamma}} = \mathbf{c}'\hat{\boldsymbol{\delta}},$$

and either reparameterization gives the same estimator of $\boldsymbol{\lambda}'\boldsymbol{\beta}$.

Example 12.5. We illustrate a reparameterization for the model $y_{ij} = \mu + \tau_i + \varepsilon_{ij}, \quad i = 1, 2, \quad j = 1, 2$. In matrix form, the model can be written as

$$\mathbf{y} = \mathbf{X}\boldsymbol{\beta} + \boldsymbol{\varepsilon} = \begin{pmatrix} 1 & 1 & 0 \\ 1 & 1 & 0 \\ 1 & 0 & 1 \\ 1 & 0 & 1 \end{pmatrix} \begin{pmatrix} \mu \\ \tau_1 \\ \tau_2 \end{pmatrix} + \begin{pmatrix} \varepsilon_{11} \\ \varepsilon_{12} \\ \varepsilon_{21} \\ \varepsilon_{22} \end{pmatrix}.$$

Since $\mathbf{X}$ has rank 2, there exist two linearly independent estimable functions (see Theorem 12.2c). We can choose these in many ways, one of which is $\mu + \tau_1$ and $\mu + \tau_2$. Thus

$$\boldsymbol{\gamma} = \begin{pmatrix} \gamma_1 \\ \gamma_2 \end{pmatrix} = \begin{pmatrix} \mu + \tau_1 \\ \mu + \tau_2 \end{pmatrix} = \begin{pmatrix} 1 & 1 & 0 \\ 1 & 0 & 1 \end{pmatrix} \begin{pmatrix} \mu \\ \tau_1 \\ \tau_2 \end{pmatrix} = \mathbf{U}\boldsymbol{\beta}.$$

To reparameterize in terms of $\boldsymbol{\gamma}$, we can use

$$\mathbf{Z} = \begin{pmatrix} 1 & 0 \\ 1 & 0 \\ 0 & 1 \\ 0 & 1 \end{pmatrix},$$

so that $\mathbf{Z}\boldsymbol{\alpha} = \mathbf{X}\boldsymbol{\beta}$:

$$\mathbf{Z}\boldsymbol{\gamma} = \begin{pmatrix} 1 & 0 \\ 1 & 0 \\ 0 & 1 \\ 0 & 1 \end{pmatrix} \begin{pmatrix} \gamma_1 \\ \gamma_2 \end{pmatrix} = \begin{pmatrix} \gamma_1 \\ \gamma_1 \\ \gamma_2 \\ \gamma_2 \end{pmatrix} = \begin{pmatrix} \mu + \tau_1 \\ \mu + \tau_1 \\ \mu + \tau_2 \\ \mu + \tau_2 \end{pmatrix}.$$

[The matrix $\mathbf{Z}$ can also be obtained directly using (12.31).] It is easy to verify that $\mathbf{ZU} = \mathbf{X}$.

$$\mathbf{ZU} = \begin{pmatrix} 1 & 0 \\ 1 & 0 \\ 0 & 1 \\ 0 & 1 \end{pmatrix} \begin{pmatrix} 1 & 1 & 0 \\ 1 & 0 & 1 \end{pmatrix} = \begin{pmatrix} 1 & 1 & 0 \\ 1 & 1 & 0 \\ 1 & 0 & 1 \\ 1 & 0 & 1 \end{pmatrix} = \mathbf{X}.$$

□

12.6 SIDE CONDITIONS

The technique of imposing side conditions was introduced and illustrated in Section 12.1 Side conditions provide (linear) constraints that make the parameters unique and individually estimable, but side conditions also impose specific definitions on the parameters. Another use for side conditions is to impose arbitrary constraints on the estimates so as to simplify the normal equations. In this case the estimates have exactly the same status as those based on a particular generalized inverse (12.13), and only estimable functions of $\boldsymbol{\beta}$ can be interpreted.

Let $\mathbf{X}$ be $n \times p$ of rank $k < p \leq n$. Then, by Theorem 12.2b(ii), $\mathbf{X}'\mathbf{X}\boldsymbol{\beta}$ represents a set of p estimable functions of $\boldsymbol{\beta}$. If a side condition were an estimable function of $\boldsymbol{\beta}$, it could be expressed as a linear combination of the rows of $\mathbf{X}'\mathbf{X}\boldsymbol{\beta}$ and would contribute nothing to the rank deficiency in $\mathbf{X}$ or to obtaining a solution vector $\hat{\boldsymbol{\beta}}$ for $\mathbf{X}'\mathbf{X}\hat{\boldsymbol{\beta}} = \mathbf{X}'\mathbf{y}$. Therefore, side conditions must be nonestimable functions of $\boldsymbol{\beta}$.

The matrix $\mathbf{X}$ is $n \times p$ of rank $k < p$. Hence the deficiency in the rank of $\mathbf{X}$ is $p - k$. In order for all the parameters to be unique or to obtain a unique solution vector $\hat{\boldsymbol{\beta}}$, we must define side conditions that make up this deficiency in rank. Accordingly, we define side conditions $\mathbf{T}\boldsymbol{\beta} = \mathbf{0}$ or $\mathbf{T}\hat{\boldsymbol{\beta}} = \mathbf{0}$, where $\mathbf{T}$ is a $(p - k) \times p$ matrix of rank $p - k$ such that $\mathbf{T}\boldsymbol{\beta}$ is a set of nonestimable functions.

In the following theorem, we consider a solution vector $\hat{\boldsymbol{\beta}}$ for both $\mathbf{X}'\mathbf{X}\hat{\boldsymbol{\beta}} = \mathbf{X}'\mathbf{y}$ and $\mathbf{T}\hat{\boldsymbol{\beta}} = \mathbf{0}$.

Theorem 12.6a. If $\mathbf{y} = \mathbf{X}\boldsymbol{\beta} + \boldsymbol{\varepsilon}$, where $\mathbf{X}$ is $n \times p$ of rank $k < p \leq n$, and if $\mathbf{T}$ is a $(p - k) \times p$ matrix of rank $p - k$ such that $\mathbf{T}\boldsymbol{\beta}$ is a set of nonestimable functions, then there is a unique vector $\hat{\boldsymbol{\beta}}$ that satisfies both $\mathbf{X}'\mathbf{X}\hat{\boldsymbol{\beta}} = \mathbf{X}'\mathbf{y}$ and $\mathbf{T}\hat{\boldsymbol{\beta}} = \mathbf{0}$.

PROOF. The two sets of equations

$$\mathbf{y} = \mathbf{X}\boldsymbol{\beta} + \boldsymbol{\varepsilon}$$
$$\mathbf{0} = \mathbf{T}\boldsymbol{\beta} + \mathbf{0}$$

can be combined into

$$\begin{pmatrix} \mathbf{y} \\ \mathbf{0} \end{pmatrix} = \begin{pmatrix} \mathbf{X} \\ \mathbf{T} \end{pmatrix} \boldsymbol{\beta} + \begin{pmatrix} \boldsymbol{\varepsilon} \\ \mathbf{0} \end{pmatrix}. \tag{12.35}$$

Since the rows of $\mathbf{T}$ are linearly independent and are not functions of the rows of $\mathbf{X}$, the matrix $\begin{pmatrix} \mathbf{X} \\ \mathbf{T} \end{pmatrix}$ is $(n + p - k) \times p$ of rank p. Thus $\begin{pmatrix} \mathbf{X} \\ \mathbf{T} \end{pmatrix}' \begin{pmatrix} \mathbf{X} \\ \mathbf{T} \end{pmatrix}$ is $p \times p$ of rank p, and the system of equations

$$\begin{pmatrix} \mathbf{X} \\ \mathbf{T} \end{pmatrix}' \begin{pmatrix} \mathbf{X} \\ \mathbf{T} \end{pmatrix} \hat{\boldsymbol{\beta}} = \begin{pmatrix} \mathbf{X} \\ \mathbf{T} \end{pmatrix}' \begin{pmatrix} \mathbf{y} \\ \mathbf{0} \end{pmatrix} \tag{12.36}$$

has the unique solution

$$\begin{aligned} \hat{\boldsymbol{\beta}} &= \left[\begin{pmatrix} \mathbf{X} \\ \mathbf{T} \end{pmatrix}' \begin{pmatrix} \mathbf{X} \\ \mathbf{T} \end{pmatrix} \right]^{-1} \begin{pmatrix} \mathbf{X} \\ \mathbf{T} \end{pmatrix}' \begin{pmatrix} \mathbf{y} \\ \mathbf{0} \end{pmatrix} \\ &= \left[(\mathbf{X}', \mathbf{T}') \begin{pmatrix} \mathbf{X} \\ \mathbf{T} \end{pmatrix} \right]^{-1} (\mathbf{X}', \mathbf{T}') \begin{pmatrix} \mathbf{y} \\ \mathbf{0} \end{pmatrix} \\ &= (\mathbf{X}'\mathbf{X} + \mathbf{T}'\mathbf{T})^{-1} (\mathbf{X}'\mathbf{y} + \mathbf{T}'\mathbf{0}) \\ &= (\mathbf{X}'\mathbf{X} + \mathbf{T}'\mathbf{T})^{-1} \mathbf{X}'\mathbf{y}. \end{aligned} \tag{12.37}$$

This approach to imposing constraints on the parameters does not work for full-rank models [see (8.30) and Problem 8.19] or for overparameterized models if the constraints involve estimable functions. However if $\mathbf{T}\boldsymbol{\beta}$ is a set of nonestimable functions, the least-squares criterion guarantees that $\mathbf{T}\hat{\boldsymbol{\beta}} = \mathbf{0}$. The solution $\hat{\boldsymbol{\beta}}$ in (12.37) also satisfies the original normal equations $\mathbf{X}'\mathbf{X}\hat{\boldsymbol{\beta}} = \mathbf{X}'\mathbf{y}$, since, by (12.36)

$$\begin{aligned} (\mathbf{X}'\mathbf{X} + \mathbf{T}'\mathbf{T})\hat{\boldsymbol{\beta}} &= \mathbf{X}'\mathbf{y} + \mathbf{T}'\mathbf{0} \\ \mathbf{X}'\mathbf{X}\hat{\boldsymbol{\beta}} + \mathbf{T}'\mathbf{T}\hat{\boldsymbol{\beta}} &= \mathbf{X}'\mathbf{y}. \end{aligned} \tag{12.38}$$

But $\mathbf{T}\hat{\boldsymbol{\beta}} = \mathbf{0}$, and (12.38) reduces to $\mathbf{X}'\mathbf{X}\hat{\boldsymbol{\beta}} = \mathbf{X}'\mathbf{y}$. □

Example 12.6. Consider the model $y_{ij} = \mu + \tau_i + \varepsilon_{ij}$, $i = 1, 2, j = 1, 2$ as in Example 12.5. The function $\tau_1 + \tau_2$ was shown to be nonestimable in Problem 12.5b. The side condition $\tau_1 + \tau_2 = 0$ can be expressed as $(0, 1, 1)\boldsymbol{\beta} = 0$, and $\mathbf{X}'\mathbf{X} + \mathbf{T}'\mathbf{T}$ becomes

$$\begin{pmatrix} 4 & 2 & 2 \\ 2 & 2 & 0 \\ 2 & 0 & 2 \end{pmatrix} + \begin{pmatrix} 0 \\ 1 \\ 1 \end{pmatrix} (0 \quad 1 \quad 1) = \begin{pmatrix} 4 & 2 & 2 \\ 2 & 3 & 1 \\ 2 & 1 & 3 \end{pmatrix}.$$

Then

$$(\mathbf{X}'\mathbf{X} + \mathbf{T}'\mathbf{T})^{-1} = \frac{1}{4} \begin{pmatrix} 2 & -1 & -1 \\ -1 & 2 & 0 \\ -1 & 0 & 2 \end{pmatrix}.$$

With $\mathbf{X}'\mathbf{y} = (y_{..}, y_{1.}, y_{2.})'$, we obtain, by (12.37)

$$\begin{aligned} \hat{\boldsymbol{\beta}} &= (\mathbf{X}'\mathbf{X} + \mathbf{T}'\mathbf{T})^{-1}\mathbf{X}'\mathbf{y} \\ &= \frac{1}{4} \begin{pmatrix} 2y_{..} - y_{1.} - y_{2.} \\ 2y_{1.} - y_{..} \\ 2y_{2.} - y_{..} \end{pmatrix} = \begin{pmatrix} \bar{y}_{..} \\ \bar{y}_{1.} - \bar{y}_{..} \\ \bar{y}_{2.} - \bar{y}_{..} \end{pmatrix}, \end{aligned} \tag{12.39}$$

since $y_{1.} + y_{2.} = y_{..}$.

We now show that $\hat{\boldsymbol{\beta}}$ in (12.39) is also a solution to the normal equations $\mathbf{X}'\mathbf{X}\hat{\boldsymbol{\beta}} = \mathbf{X}'\mathbf{y}$:

$$\begin{pmatrix} 4 & 2 & 2 \\ 2 & 2 & 0 \\ 2 & 0 & 2 \end{pmatrix} \begin{pmatrix} \bar{y}_{..} \\ \bar{y}_{1.} - \bar{y}_{..} \\ \bar{y}_{2.} - \bar{y}_{..} \end{pmatrix} = \begin{pmatrix} y_{..} \\ y_{1.} \\ y_{2.} \end{pmatrix}, \quad \text{or}$$

$$\begin{aligned} 4\bar{y}_{..} + 2(\bar{y}_{1.} - \bar{y}_{..}) + 2(\bar{y}_{2.} - \bar{y}_{..}) &= y_{..} \\ 2\bar{y}_{..} + 2(\bar{y}_{1.} - \bar{y}_{..}) &= y_{1.} \\ 2\bar{y}_{..} + 2(\bar{y}_{2.} - \bar{y}_{..}) &= y_{2.} \end{aligned}$$

These simplify to

$$\begin{aligned} 2\bar{y}_{1.} + 2\bar{y}_{2.} &= y_{..} \\ 2\bar{y}_{1.} &= y_{1.} \\ 2\bar{y}_{2.} &= y_{2.}, \end{aligned}$$

which hold because $\bar{y}_{1.} = y_{1.}/2$, $\bar{y}_{2.} = y_{2.}/2$ and $y_{1.} + y_{2.} = y_{..}$. □

12.7 TESTING HYPOTHESES

We now consider hypotheses about the β's in the model $\mathbf{y} = \mathbf{X}\boldsymbol{\beta} + \boldsymbol{\varepsilon}$, where $\mathbf{X}$ is $n \times p$ of rank $k < p \leq n$. In this section, we assume that $\mathbf{y}$ is $N_n(\mathbf{X}\boldsymbol{\beta}, \sigma^2\mathbf{I})$.

12.7.1 Testable Hypotheses

It can be shown that unless a hypothesis can be expressed in terms of estimable functions, it cannot be tested (Searle 1971, pp. 193–196). This leads to the following definition.

A hypothesis such as $H_0: \beta_1 = \beta_2 = \cdots = \beta_q$ is said to be *testable* if there exists a set of linearly independent estimable functions $\boldsymbol{\lambda}_1'\boldsymbol{\beta}, \boldsymbol{\lambda}_2'\boldsymbol{\beta}, \ldots, \boldsymbol{\lambda}_t'\boldsymbol{\beta}$ such that H_0 is true if and only if $\boldsymbol{\lambda}_1'\boldsymbol{\beta} = \boldsymbol{\lambda}_2'\boldsymbol{\beta} = \cdots = \boldsymbol{\lambda}_t'\boldsymbol{\beta} = 0$.

Sometimes the subset of $\boldsymbol{\beta}$'s whose equality we wish to test is such that every contrast $\sum_i c_i\beta_i$ is estimable ($\sum_i c_i\beta_i$ is a contrast if $\sum_i c_i = 0$). In this case, it is easy to find a set of $q-1$ linearly independent estimable functions that can be set equal to zero to express $\beta_1 = \cdots = \beta_q$. One such set is the following:

$$
\begin{aligned}
\boldsymbol{\lambda}_1'\boldsymbol{\beta} &= (q-1)\beta_1 - (\beta_2 + \beta_3 + \cdots + \beta_q) \\
\boldsymbol{\lambda}_2'\boldsymbol{\beta} &= (q-2)\beta_2 - (\beta_3 + \cdots + \beta_q) \\
&\ \ \vdots \\
\boldsymbol{\lambda}_{q-1}'\boldsymbol{\beta} &= (1)\beta_{q-1} - (\beta_q).
\end{aligned}
$$

These $q-1$ contrasts $\boldsymbol{\lambda}_1'\boldsymbol{\beta}, \ldots, \boldsymbol{\lambda}_{q-1}'\boldsymbol{\beta}$ constitute a set of linearly independent estimable functions such that

$$
\begin{pmatrix} \boldsymbol{\lambda}_1'\boldsymbol{\beta} \\ \vdots \\ \boldsymbol{\lambda}_{q-1}'\boldsymbol{\beta} \end{pmatrix} = \begin{pmatrix} 0 \\ \vdots \\ 0 \end{pmatrix}
$$

if and only if $\beta_1 = \beta_2 = \cdots = \beta_q$.

To illustrate a testable hypothesis, suppose that we have the model $y_{ij} = \mu + \alpha_i + \beta_j + \varepsilon_{ij}$, $i = 1, 2, 3$, $j = 1, 2, 3$, and a hypothesis of interest is H_0: $\alpha_1 = \alpha_2 = \alpha_3$. By taking linear combinations of the rows of $\mathbf{X}\boldsymbol{\beta}$, we can obtain the two linearly independent estimable functions $\alpha_1 - \alpha_2$ and $\alpha_1 + \alpha_2 - 2\alpha_3$. The hypothesis H_0: $\alpha_1 = \alpha_2 = \alpha_3$ is true if and only if $\alpha_1 - \alpha_2$ and $\alpha_1 + \alpha_2 - 2\alpha_3$ are simultaneously equal to zero (see Problem 12.21). Therefore, H_0 is a testable hypothesis and is equivalent to

$$H_0: \begin{pmatrix} \alpha_1 - \alpha_2 \\ \alpha_1 + \alpha_2 - 2\alpha_3 \end{pmatrix} = \begin{pmatrix} 0 \\ 0 \end{pmatrix}. \tag{12.40}$$

We now discuss tests for testable hypotheses. In Section 12.7.2, we describe a procedure that is based on the full-reduced-model methods of Section 8.2. Since (12.40) is of the form $H_0\colon \mathbf{C}\boldsymbol{\beta} = \mathbf{0}$, we could alternatively use a general linear hypothesis test (see Section 8.4.1). This approach is discussed in Section 12.7.3.

12.7.2 Full-Reduced-Model Approach

Suppose that we are interested in testing $H_0\colon \beta_1 = \beta_2 = \cdots = \beta_q$ in the non-full-rank model $\mathbf{y} = \mathbf{X}\boldsymbol{\beta} + \boldsymbol{\varepsilon}$, where $\boldsymbol{\beta}$ is $p \times 1$ and $\mathbf{X}$ is $n \times p$ of rank $k < p \leq n$. If H_0 is testable, we can find a set of linearly independent estimable functions $\boldsymbol{\lambda}_1'\boldsymbol{\beta}, \boldsymbol{\lambda}_2'\boldsymbol{\beta}, \ldots, \boldsymbol{\lambda}_t'\boldsymbol{\beta}$ such that $H_0\colon \beta_1 = \beta_2 = \cdots = \beta_q$ is equivalent to

$$H_0\colon \boldsymbol{\gamma}_1 = \begin{pmatrix} \boldsymbol{\lambda}_1'\boldsymbol{\beta} \\ \boldsymbol{\lambda}_2'\boldsymbol{\beta} \\ \vdots \\ \boldsymbol{\lambda}_t'\boldsymbol{\beta} \end{pmatrix} = \begin{pmatrix} 0 \\ 0 \\ \vdots \\ 0 \end{pmatrix}.$$

It is also possible to find

$$\boldsymbol{\gamma}_2 = \begin{pmatrix} \boldsymbol{\lambda}_{t+1}'\boldsymbol{\beta} \\ \vdots \\ \boldsymbol{\lambda}_k'\boldsymbol{\beta} \end{pmatrix}$$

such that the k functions $\boldsymbol{\lambda}_1'\boldsymbol{\beta}, \ldots, \boldsymbol{\lambda}_t'\boldsymbol{\beta}, \boldsymbol{\lambda}_{t+1}'\boldsymbol{\beta}, \ldots, \boldsymbol{\lambda}_k'\boldsymbol{\beta}$ are linearly independent and estimable, where $k = \text{rank}(\mathbf{X})$. Let

$$\boldsymbol{\gamma} = \begin{pmatrix} \boldsymbol{\gamma}_1 \\ \boldsymbol{\gamma}_2 \end{pmatrix}.$$

We can now reparameterize (see Section 12.5) from the non-full-rank model $\mathbf{y} = \mathbf{X}\boldsymbol{\beta} + \boldsymbol{\varepsilon}$ to the full-rank model

$$\mathbf{y} = \mathbf{Z}\boldsymbol{\gamma} + \boldsymbol{\varepsilon} = \mathbf{Z}_1\boldsymbol{\gamma}_1 + \mathbf{Z}_2\boldsymbol{\gamma}_2 + \boldsymbol{\varepsilon},$$

where $\mathbf{Z} = (\mathbf{Z}_1, \mathbf{Z}_2)$ is partitioned to conform with the number of elements in $\boldsymbol{\gamma}_1$ and $\boldsymbol{\gamma}_2$.

For the hypothesis H_0: $\boldsymbol{\gamma}_1 = \mathbf{0}$, the reduced model is $\mathbf{y} = \mathbf{Z}_2\boldsymbol{\gamma}_2^* + \boldsymbol{\varepsilon}^*$. By Theorem 7.10, the estimate of $\boldsymbol{\gamma}_2^*$ in the reduced model is the same as the estimate of $\boldsymbol{\gamma}_2$ in the full model if the columns of $\mathbf{Z}_2$ are orthogonal to those of $\mathbf{Z}_1$, that is, if $\mathbf{Z}_2'\mathbf{Z}_1 = \mathbf{O}$. For the balanced models we are considering in this chapter, the orthogonality will typically hold (see Section 12.8.3). Accordingly, we refer to $\boldsymbol{\gamma}_2$ and $\hat{\boldsymbol{\gamma}}_2$ rather than to $\boldsymbol{\gamma}_2^*$ and $\hat{\boldsymbol{\gamma}}_2^*$.

Since $\mathbf{y} = \mathbf{Z}\boldsymbol{\gamma} + \boldsymbol{\varepsilon}$ is a full-rank model, the hypothesis H_0: $\boldsymbol{\gamma}_1 = \mathbf{0}$ can be tested as in Section 8.2. The test is outlined in Table 12.2, which is analogous to Table 8.3. Note that the degrees of freedom t for SS$(\boldsymbol{\gamma}_1|\boldsymbol{\gamma}_2)$ is the number of linearly independent estimable functions required to express H_0.

In Table 12.2, the sum of squares $\hat{\boldsymbol{\gamma}}'\mathbf{Z}\mathbf{y}$ is obtained from the full model $\mathbf{y} = \mathbf{Z}\boldsymbol{\gamma} + \boldsymbol{\varepsilon}$. The sum of squares $\hat{\boldsymbol{\gamma}}_2'\mathbf{Z}_2'\mathbf{y}$ is obtained from the reduced model $\mathbf{y} = \mathbf{Z}_2\boldsymbol{\gamma}_2 + \boldsymbol{\varepsilon}$, which assumes the hypothesis is true.

The reparameterization procedure presented above seems straightforward. However, finding the matrix $\mathbf{Z}$ in practice can be time-consuming. Fortunately, this step is actually not necessary.

From (12.20) and (12.33), we obtain

$$\mathbf{y}'\mathbf{y} - \hat{\boldsymbol{\beta}}'\mathbf{X}'\mathbf{y} = \mathbf{y}'\mathbf{y} - \hat{\boldsymbol{\gamma}}'\mathbf{Z}\mathbf{y},$$

which gives

$$\hat{\boldsymbol{\beta}}'\mathbf{X}'\mathbf{y} = \hat{\boldsymbol{\gamma}}'\mathbf{Z}'\mathbf{y}, \tag{12.41}$$

where $\hat{\boldsymbol{\beta}}$ represents any solution to the normal equations $\mathbf{X}'\mathbf{X}\hat{\boldsymbol{\beta}} = \mathbf{X}'\mathbf{y}$. Similarly, corresponding to $\mathbf{y} = \mathbf{Z}\boldsymbol{\gamma}_2^* + \boldsymbol{\varepsilon}^*$, we have a reduced model $\mathbf{y} = \mathbf{X}_2\boldsymbol{\beta}_2^* + \boldsymbol{\varepsilon}^*$ obtained by setting $\beta_1 = \beta_2 = \cdots = \beta_q$. Then

$$\hat{\boldsymbol{\beta}}_2^{*\prime}\mathbf{X}_2'\mathbf{y} = \hat{\boldsymbol{\gamma}}_2^{*\prime}\mathbf{Z}_2'\mathbf{y}, \tag{12.42}$$

where $\hat{\boldsymbol{\beta}}_2^*$ is any solution to the reduced normal equations $\mathbf{X}_2'\mathbf{X}_2\hat{\boldsymbol{\beta}}_2^* = \mathbf{X}_2'\mathbf{y}$. We can often use side conditions to find $\hat{\boldsymbol{\beta}}$ and $\hat{\boldsymbol{\beta}}_2^*$.

We noted above (see also Section 12.8.3) that if $\mathbf{Z}_2'\mathbf{Z}_1 = \mathbf{O}$ holds in a reparameterized full-rank model, then by Theorem 7.10, the estimate of $\boldsymbol{\gamma}_2^*$ in the reduced

TABLE 12.2 ANOVA for Testing H_0: $\boldsymbol{\gamma}_1 = \mathbf{0}$ in Reparameterized Balanced Models

Source of Variation	df	Sum of Squares	F Statistic
Due to $\boldsymbol{\gamma}_1$ adjusted for $\boldsymbol{\gamma}_2$	t	$\text{SS}(\boldsymbol{\gamma}_1\|\boldsymbol{\gamma}_2) = \hat{\boldsymbol{\gamma}}'\mathbf{Z}'\mathbf{y} - \hat{\boldsymbol{\gamma}}_2'\mathbf{Z}_2'\mathbf{y}$	$\dfrac{\text{SS}(\boldsymbol{\gamma}_1\|\boldsymbol{\gamma}_2)/t}{\text{SSE}/(n-k)}$
Error	$n-k$	$\text{SSE} = \mathbf{y}'\mathbf{y} - \hat{\boldsymbol{\gamma}}'\mathbf{Z}'\mathbf{y}$	—
Total	$n-1$	$\text{SST} = \mathbf{y}'\mathbf{y} - n\bar{y}^2$	

TABLE 12.3 ANOVA for Testing H_0: $\boldsymbol{\beta}_1 = \boldsymbol{\beta}_2 = \cdots = \boldsymbol{\beta}_q$ in Balanced Non-Full-Rank Models

Source of Variation	df	Sum of Squares	F Statistic
Due to $\boldsymbol{\beta}_1$ adjusted for $\boldsymbol{\beta}_2$	t	$\text{SS}(\boldsymbol{\beta}_1 \vert \boldsymbol{\beta}_2) = \hat{\boldsymbol{\beta}}'\mathbf{X}'\mathbf{y} - \hat{\boldsymbol{\beta}}_2'\mathbf{X}_2'\mathbf{y}$	$\dfrac{\text{SS}(\boldsymbol{\beta}_1 \vert \boldsymbol{\beta}_2)/t}{\text{SSE}/(n-k)}$
Error	$n-k$	$\text{SSE} = \mathbf{y}'\mathbf{y} - \hat{\boldsymbol{\beta}}'\mathbf{X}'\mathbf{y}$	—
Total	$n-1$	$\text{SST} = \mathbf{y}'\mathbf{y} - n\bar{y}^2$	—

model is the same as the estimate of $\boldsymbol{\gamma}_2$ in the full model. The following is an analogous theorem for the non-full-rank case.

Theorem 12.7a. Consider the partitioned model $\mathbf{y} = \mathbf{X}\boldsymbol{\beta} + \boldsymbol{\varepsilon} = \mathbf{X}_1\boldsymbol{\beta}_1 + \mathbf{X}_2\boldsymbol{\beta}_2 + \boldsymbol{\varepsilon}$, where $\mathbf{X}$ is $n \times p$ of rank $k < p \leq n$. If $\mathbf{X}_2'\mathbf{X}_1 = \mathbf{O}$ (see Section 12.8.3), any estimate of $\boldsymbol{\beta}_2^*$ in the reduced model $\mathbf{y} = \mathbf{X}_2\boldsymbol{\beta}_2^* + \boldsymbol{\varepsilon}^*$ is also an estimate of $\boldsymbol{\beta}_2$ in the full model.

PROOF. There is a generalized inverse of

$$\mathbf{X}'\mathbf{X} = \begin{pmatrix} \mathbf{X}_1'\mathbf{X}_1 & \mathbf{X}_1'\mathbf{X}_2 \\ \mathbf{X}_2'\mathbf{X}_1 & \mathbf{X}_2'\mathbf{X}_2 \end{pmatrix}$$

analogous to the inverse of a nonsingular symmetric partitioned matrix in (2.50) (Harville 1997, pp. 121–122). The proof then parallels that of Theorem 7.10. □

In the balanced non-full-rank models we are considering in this chapter, the orthogonality of $\mathbf{X}_1$ and $\mathbf{X}_2$ will typically hold. (This will be illustrated in Section 12.8.3) Accordingly, we refer to $\boldsymbol{\beta}_2$ and $\hat{\boldsymbol{\beta}}_2$, rather than to $\boldsymbol{\beta}_2^*$ and $\hat{\boldsymbol{\beta}}_2^*$.

The test can be expressed as in Table 12.3, in which $\hat{\boldsymbol{\beta}}'\mathbf{X}'\mathbf{y}$ is obtained from the full model $\mathbf{y} = \mathbf{X}\boldsymbol{\beta} + \boldsymbol{\varepsilon}$ and $\hat{\boldsymbol{\beta}}_2'\mathbf{X}_2'\mathbf{y}$ is obtained from the model $\mathbf{y} = \mathbf{X}_2\boldsymbol{\beta}_2 + \boldsymbol{\varepsilon}$, which has been reduced by the hypothesis $H_0: \beta_1 = \beta_2 = \cdots = \beta_q$. Note that the degrees of freedom t for $\text{SS}(\boldsymbol{\beta}_1|\boldsymbol{\beta}_2)$ is the same as for $\text{SS}(\boldsymbol{\gamma}_1|\boldsymbol{\gamma}_2)$ in Table 12.2, namely, the number of linearly independent estimable functions required to express H_0. Typically, this is given by $t = q - 1$. A set of $q - 1$ linearly independent estimable functions was illustrated at the beginning of Section 12.7.1. The test in Table 12.3 will be illustrated in Section 12.8.2.

12.7.3 General Linear Hypothesis

As illustrated in (12.40), a hypothesis such as $H_0: \alpha_1 = \alpha_2 = \alpha_3$ can be expressed in the form $H_0: \mathbf{C}\boldsymbol{\beta} = \mathbf{0}$. We can test this hypothesis in a manner analogous to that used for the general linear hypothesis test for the full-rank model in Section 8.4.1 The following theorem is an extension of Theorem 8.4a to the non-full-rank case.

Theorem 12.7b. If $\mathbf{y}$ is $N_n(\mathbf{X}\boldsymbol{\beta}, \sigma^2\mathbf{I})$, where $\mathbf{X}$ is $n \times p$ of rank $k < p \leq n$, if $\mathbf{C}$ is $m \times p$ of rank $m \leq k$ such that $C\boldsymbol{\beta}$ is a set of m linearly independent estimable functions, and if $\hat{\boldsymbol{\beta}} = (\mathbf{X}'\mathbf{X})^-\mathbf{X}'\mathbf{y}$, then

(i) $\mathbf{C}(\mathbf{X}'\mathbf{X})^-\mathbf{C}'$ is nonsingular.
(ii) $\mathbf{C}\hat{\boldsymbol{\beta}}$ is $N_m[\mathbf{C}\boldsymbol{\beta}, \sigma^2\mathbf{C}(\mathbf{X}'\mathbf{X})^-\mathbf{C}']$.
(iii) $\text{SSH}/\sigma^2 = (\mathbf{C}\hat{\boldsymbol{\beta}})'[\mathbf{C}(\mathbf{X}'\mathbf{X})^-\mathbf{C}']^{-1}\mathbf{C}\hat{\boldsymbol{\beta}}/\sigma^2$ is $\chi^2(m, \lambda)$, where $\lambda = (\mathbf{C}\boldsymbol{\beta})'[\mathbf{C}(\mathbf{X}'\mathbf{X})^-\mathbf{C}']^{-1}\mathbf{C}\boldsymbol{\beta}/2\sigma^2$.
(iv) $\text{SSE}/\sigma^2 = \mathbf{y}'[\mathbf{I} - \mathbf{X}(\mathbf{X}'\mathbf{X})^-\mathbf{X}']\mathbf{y}/\sigma^2$ is $\chi^2(n - k)$.
(v) SSH and SSE are independent.

PROOF

(i) Since

$$\mathbf{C}\boldsymbol{\beta} = \begin{pmatrix} \mathbf{c}_1'\boldsymbol{\beta} \\ \mathbf{c}_2'\boldsymbol{\beta} \\ \vdots \\ \mathbf{c}_m'\boldsymbol{\beta} \end{pmatrix}$$

is a set of m linearly independent estimable functions, then by Theorem 12.2b(iii) we have $\mathbf{c}_i'(\mathbf{X}'\mathbf{X})^-\mathbf{X}'\mathbf{X} = \mathbf{c}_i'$ for $i = 1, 2, \ldots, m$. Hence

$$\mathbf{C}(\mathbf{X}'\mathbf{X})^-\mathbf{X}'\mathbf{X} = \mathbf{C}. \tag{12.43}$$

Writing (12.43) as the product

$$[\mathbf{C}(\mathbf{X}'\mathbf{X})^-\mathbf{X}']\mathbf{X} = \mathbf{C},$$

we can use Theorem 2.4(i) to obtain the inequalities

$$\text{rank}(\mathbf{C}) \leq \text{rank}[\mathbf{C}(\mathbf{X}'\mathbf{X})^-\mathbf{X}'] \leq \text{rank}(\mathbf{C}).$$

Hence $\text{rank}[\mathbf{C}(\mathbf{X}'\mathbf{X})^-\mathbf{X}'] = \text{rank}(\mathbf{C}) = m$. Now, by Theorem 2.4(iii), which states that $\text{rank}(\mathbf{A}) = \text{rank}(\mathbf{A}\mathbf{A}')$, we can write

$$\begin{aligned}
\text{rank}(\mathbf{C}) &= \text{rank}[\mathbf{C}(\mathbf{X}'\mathbf{X})^-\mathbf{X}'] \\
&= \text{rank}[\mathbf{C}(\mathbf{X}'\mathbf{X})^-\mathbf{X}'][\mathbf{C}(\mathbf{X}'\mathbf{X})^-\mathbf{X}']' \\
&= \text{rank}[\mathbf{C}(\mathbf{X}'\mathbf{X})^-\mathbf{X}'\mathbf{X}(\mathbf{X}'\mathbf{X})^-\mathbf{C}'].
\end{aligned}$$

By (12.43), $\mathbf{C}(\mathbf{X}'\mathbf{X})^{-}\mathbf{X}'\mathbf{X} = \mathbf{C}$, and we have

$$\text{rank}(\mathbf{C}) = \text{rank}[\mathbf{C}(\mathbf{X}'\mathbf{X})^{-}\mathbf{C}'].$$

Thus the $m \times m$ matrix $\mathbf{C}(\mathbf{X}'\mathbf{X})^{-}\mathbf{C}'$ is nonsingular. [Note that we are assuming that $(\mathbf{X}'\mathbf{X})^{-}$ is symmetric. See Problem 2.46 and a comment following Theorem 2.8c(v).]

(ii) By (3.38) and (12.14), we obtain

$$E(\mathbf{C}\hat{\boldsymbol{\beta}}) = \mathbf{C}E(\hat{\boldsymbol{\beta}}) = \mathbf{C}(\mathbf{X}'\mathbf{X})^{-}\mathbf{X}'\mathbf{X}\boldsymbol{\beta}.$$

By (12.43), $\mathbf{C}(\mathbf{X}'\mathbf{X})^{-}\mathbf{X}'\mathbf{X} = \mathbf{C}$, and therefore

$$E(\mathbf{C}\hat{\boldsymbol{\beta}}) = \mathbf{C}\boldsymbol{\beta}. \tag{12.44}$$

By (3.44) and (12.18), we have

$$\text{cov}(\mathbf{C}\hat{\boldsymbol{\beta}}) = \mathbf{C}\,\text{cov}(\hat{\boldsymbol{\beta}})\mathbf{C}' = \sigma^2\mathbf{C}(\mathbf{X}'\mathbf{X})^{-}\mathbf{X}'\mathbf{X}(\mathbf{X}'\mathbf{X})^{-}\mathbf{C}'.$$

By (12.43), this becomes

$$\text{cov}(\mathbf{C}\hat{\boldsymbol{\beta}}) = \sigma^2\mathbf{C}(\mathbf{X}'\mathbf{X})^{-}\mathbf{C}'. \tag{12.45}$$

By Theorem 12.3g(i), $\hat{\boldsymbol{\beta}}$ is $N_p[(\mathbf{X}'\mathbf{X})^{-}\mathbf{X}'\mathbf{X}\boldsymbol{\beta}, \sigma^2(\mathbf{X}'\mathbf{X})^{-}\mathbf{X}'\mathbf{X}(\mathbf{X}'\mathbf{X})^{-}]$ for a particular $(\mathbf{X}'\mathbf{X})^{-}$. Then by (12.44), (12.45), and Theorem 4.4a(ii), we obtain

$$\mathbf{C}\hat{\boldsymbol{\beta}} \text{ is } N_m[\mathbf{C}\boldsymbol{\beta}, \sigma^2\mathbf{C}(\mathbf{X}'\mathbf{X})^{-}\mathbf{C}'].$$

(iii) By part (ii), $\text{cov}(\mathbf{C}\hat{\boldsymbol{\beta}}) = \sigma^2\mathbf{C}(\mathbf{X}'\mathbf{X})^{-}\mathbf{C}'$. Since $\sigma^2[\mathbf{C}(\mathbf{X}'\mathbf{X})^{-}\mathbf{C}']^{-1}\mathbf{C}(\mathbf{X}'\mathbf{X})^{-}\mathbf{C}'/\sigma^2 = \mathbf{I}$, the result follows by Theorem 5.5.

(iv) This was established in Theorem 12.3g(ii).

(v) By Theorem 12.3g(iii), $\hat{\boldsymbol{\beta}}$ and SSE are independent. Hence $\text{SSH} = (\mathbf{C}\hat{\boldsymbol{\beta}})'[\mathbf{C}(\mathbf{X}'\mathbf{X})^{-}\mathbf{C}']^{-1}\mathbf{C}\hat{\boldsymbol{\beta}}$ and SSE are independent [see Seber (1977, pp. 17–18) for a proof that continuous functions of independent random variables and vectors are independent]. For a more formal proof, see Problem 12.22. □

Using the results in Theorem 12.7b, we obtain an F test for $H_0: \mathbf{C}\boldsymbol{\beta} = \mathbf{0}$, as given in the following theorem, which is analogous to Theorem 8.4b.

Theorem 12.7c. Let $\mathbf{y}$ be $N_n(\mathbf{X}\boldsymbol{\beta}, \sigma^2\mathbf{I})$, where $\mathbf{X}$ is $n \times p$ of rank $k < p \leq n$, and let $\mathbf{C}$, $\mathbf{C}\boldsymbol{\beta}$, and $\hat{\boldsymbol{\beta}}$ be defined as in Theorem 12.7b. Then, if $H_0 : \mathbf{C}\boldsymbol{\beta} = \mathbf{0}$ is true, the statistic

$$F = \frac{\text{SSH}/m}{\text{SSE}/(n-k)} = \frac{(\mathbf{C}\hat{\boldsymbol{\beta}})'[\mathbf{C}(\mathbf{X}'\mathbf{X})^{-}\mathbf{C}']^{-1}\mathbf{C}\hat{\boldsymbol{\beta}}/m}{\text{SSE}/(n-k)} \tag{12.46}$$

is distributed as $F(m, n-k)$.

Proof. This follows from (5.28) and Theorem 12.7b. □

12.8 AN ILLUSTRATION OF ESTIMATION AND TESTING

Suppose we have the additive (no-interaction) model

$$y_{ij} = \mu + \alpha_i + \beta_j + \varepsilon_{ij}, \quad i = 1, 2, 3, \quad j = 1, 2,$$

and that the hypotheses of interest are $H_0 : \alpha_1 = \alpha_2 = \alpha_3$ and $H_0 : \beta_1 = \beta_2$. The six observations can be written in the form $\mathbf{y} = \mathbf{X}\boldsymbol{\beta} + \boldsymbol{\varepsilon}$ as

$$\begin{pmatrix} y_{11} \\ y_{12} \\ y_{21} \\ y_{22} \\ y_{31} \\ y_{32} \end{pmatrix} = \begin{pmatrix} 1 & 1 & 0 & 0 & 1 & 0 \\ 1 & 1 & 0 & 0 & 0 & 1 \\ 1 & 0 & 1 & 0 & 1 & 0 \\ 1 & 0 & 1 & 0 & 0 & 1 \\ 1 & 0 & 0 & 1 & 1 & 0 \\ 1 & 0 & 0 & 1 & 0 & 1 \end{pmatrix} \begin{pmatrix} \mu \\ \alpha_1 \\ \alpha_2 \\ \alpha_3 \\ \beta_1 \\ \beta_2 \end{pmatrix} + \begin{pmatrix} \varepsilon_{11} \\ \varepsilon_{12} \\ \varepsilon_{21} \\ \varepsilon_{22} \\ \varepsilon_{31} \\ \varepsilon_{32} \end{pmatrix}. \tag{12.47}$$

The matrix $\mathbf{X}'\mathbf{X}$ is given by

$$\mathbf{X}'\mathbf{X} = \begin{pmatrix} 6 & 2 & 2 & 2 & 3 & 3 \\ 2 & 2 & 0 & 0 & 1 & 1 \\ 2 & 0 & 2 & 0 & 1 & 1 \\ 2 & 0 & 0 & 2 & 1 & 1 \\ 3 & 1 & 1 & 1 & 3 & 0 \\ 3 & 1 & 1 & 1 & 0 & 3 \end{pmatrix}.$$

The rank of both $\mathbf{X}$ and $\mathbf{X}'\mathbf{X}$ is 4.

12.8.1 Estimable Functions

The hypothesis $H_0: \alpha_1 = \alpha_2 = \alpha_3$ can be expressed as $H_0: \alpha_1 - \alpha_2 = 0$ and $\alpha_1 - \alpha_3 = 0$. Thus H_0 is testable if $\alpha_1 - \alpha_2$ and $\alpha_1 - \alpha_3$ are estimable. To check $\alpha_1 - \alpha_2$ for estimability, we write it as

$$\alpha_1 - \alpha_2 = (0, 1, -1, 0, 0, 0)\boldsymbol{\beta} = \boldsymbol{\lambda}_1'\boldsymbol{\beta}$$

and then note that $\boldsymbol{\lambda}_1'$ can be obtained from $\mathbf{X}$ as

$$(1, 0, -1, 0, 0, 0)\mathbf{X} = (0, 1, -1, 0, 0, 0)$$

and from $\mathbf{X}'\mathbf{X}$ as

$$(0, \tfrac{1}{2}, -\tfrac{1}{2}, 0, 0, 0)\mathbf{X}'\mathbf{X} = (0, 1, -1, 0, 0, 0)$$

(see Theorem 12.2b). Alternatively, we can obtain $\alpha_1 - \alpha_2$ as a linear combination of the rows (elements) of $E(\mathbf{y}) = \mathbf{X}\boldsymbol{\beta}$:

$$\begin{aligned} E(y_{11} - y_{21}) &= E(y_{11}) - E(y_{21}) \\ &= \mu + \alpha_1 + \beta_1 - (\mu + \alpha_2 + \beta_1) \\ &= \alpha_1 - \alpha_2. \end{aligned}$$

Similarly, $\alpha_1 - \alpha_3$ can be expressed as

$$\alpha_1 - \alpha_3 = (0, 1, 0, -1, 0, 0)\boldsymbol{\beta} = \boldsymbol{\lambda}_2'\boldsymbol{\beta},$$

and $\boldsymbol{\lambda}_2'$ can be obtained from $\mathbf{X}$ or $\mathbf{X}'\mathbf{X}$:

$$\begin{aligned} (1, 0, 0, 0, -1, 0)\mathbf{X} &= (0, 1, 0, -1, 0, 0), \\ (0, \tfrac{1}{2}, 0, -\tfrac{1}{2}, 0, 0)\mathbf{X}'\mathbf{X} &= (0, 1, 0, -1, 0, 0). \end{aligned}$$

It is also of interest to examine a complete set of linearly independent estimable functions obtained as linear combinations of the rows of $\mathbf{X}$ [see Theorem 12.2b(i) and Example 12.2.2b]. If we subtract the first row from each succeeding row of $\mathbf{X}$, we obtain

$$\begin{pmatrix} 1 & 1 & 0 & 0 & 1 & 0 \\ 0 & 0 & 0 & 0 & -1 & 1 \\ 0 & -1 & 1 & 0 & 0 & 0 \\ 0 & -1 & 1 & 0 & -1 & 1 \\ 0 & -1 & 0 & 1 & 0 & 0 \\ 0 & -1 & 0 & 1 & -1 & 1 \end{pmatrix}.$$

We multiply the second and third rows by -1 and then add them to the fourth row, with similar operations involving the second, fifth, and sixth rows. The result is

$$\begin{pmatrix} 1 & 1 & 0 & 0 & 1 & 0 \\ 0 & 0 & 0 & 0 & 1 & -1 \\ 0 & 1 & -1 & 0 & 0 & 0 \\ 0 & 0 & 0 & 0 & 0 & 0 \\ 0 & 1 & 0 & -1 & 0 & 0 \\ 0 & 0 & 0 & 0 & 0 & 0 \end{pmatrix}.$$

Multiplying this matrix by $\boldsymbol{\beta}$, we obtain a complete set of linearly independent estimable functions: $\mu + \alpha_1 + \beta_1$, $\beta_1 - \beta_2$, $\alpha_1 - \alpha_2$, $\alpha_1 - \alpha_3$. Note that the estimable functions not involving μ are contrasts in the α's or β's.

12.8.2 Testing a Hypothesis

As noted at the beginning of Section 12.8.1, $H_0: \alpha_1 = \alpha_2 = \alpha_3$ is equivalent to $H_0: \alpha_1 - \alpha_2 = \alpha_1 - \alpha_3 = 0$. Since two linearly independent estimable functions of the α's are needed to express $H_0: \alpha_1 = \alpha_2 = \alpha_3$ (see Theorems 12.7b and 12.7c), the sum of squares for testing $H_0: \alpha_1 = \alpha_2 = \alpha_3$ has 2 degrees of freedom. Similarly, $H_0: \beta_1 = \beta_2$ is testable with 1 degree of freedom.

The normal equations $\mathbf{X}'\mathbf{X}\hat{\boldsymbol{\beta}} = \mathbf{X}'\mathbf{y}$ are given by

$$\begin{pmatrix} 6 & 2 & 2 & 2 & 3 & 3 \\ 2 & 2 & 0 & 0 & 1 & 1 \\ 2 & 0 & 2 & 0 & 1 & 1 \\ 2 & 0 & 0 & 2 & 1 & 1 \\ 3 & 1 & 1 & 1 & 3 & 0 \\ 3 & 1 & 1 & 1 & 0 & 3 \end{pmatrix} \begin{pmatrix} \hat{\mu} \\ \hat{\alpha}_1 \\ \hat{\alpha}_2 \\ \hat{\alpha}_3 \\ \hat{\beta}_1 \\ \hat{\beta}_2 \end{pmatrix} = \begin{pmatrix} y_{..} \\ y_{1.} \\ y_{2.} \\ y_{3.} \\ y_{.1} \\ y_{.2} \end{pmatrix}. \tag{12.48}$$

If we impose the side conditions $\hat{\alpha}_1 + \hat{\alpha}_2 + \hat{\alpha}_3 = 0$ and $\hat{\beta}_1 + \hat{\beta}_2 = 0$, we obtain the following solution to the normal equations:

$$\begin{aligned} \hat{\mu} &= \bar{y}_{..}, & \hat{\alpha}_1 &= \bar{y}_{1.} - \bar{y}_{..}, & \hat{\alpha}_2 &= \bar{y}_{2.} - \bar{y}_{..}, \\ \hat{\alpha}_3 &= \bar{y}_{3.} - \bar{y}_{..}, & \hat{\beta}_1 &= \bar{y}_{.1} - \bar{y}_{..}, & \hat{\beta}_2 &= \bar{y}_{.2} - \bar{y}_{..}, \end{aligned} \tag{12.49}$$

where $\bar{y}_{..} = \sum_{ij} y_{ij}/6$, $\bar{y}_{1.} = \sum_j y_{1j}/2$, and so on.

If we impose the side conditions on both the parameters and the estimates, equations (12.49) are unique estimates of unique meaningful parameters. Thus, for example, α_1 becomes $\alpha_1^* = \bar{\mu}_{1.} - \bar{\mu}_{..}$, the expected deviation from the mean due to treatment 1 (see Section 12.1.1), and $\bar{y}_{1.} - \bar{y}_{..}$ is a reasonable estimate. On the other hand, if the side conditions are used only to obtain estimates and are not imposed on the parameters, then α_1 is not unique, and $\bar{y}_{1.} - \bar{y}_{..}$ does not estimate a parameter. In this case, $\hat{\alpha}_1 = \bar{y}_{1.} - \bar{y}_{..}$ can be used only together with other elements in $\hat{\boldsymbol{\beta}}$ [as given by (12.49)] to obtain estimates $\boldsymbol{\lambda}'\hat{\boldsymbol{\beta}}$ of estimable functions $\boldsymbol{\lambda}'\boldsymbol{\beta}$.

We now proceed to obtain the test for $H_0: \alpha_1 = \alpha_2 = \alpha_3$ following the outline in Table 12.3. First, for the full model, we need $\hat{\boldsymbol{\beta}}'\mathbf{X}'\mathbf{y} = \text{SS}(\mu, \alpha_1, \alpha_2, \alpha_3, \beta_1, \beta_2)$, which we denote by $\text{SS}(\mu, \alpha, \beta)$. By (12.48) and (12.49), we obtain

$$\begin{aligned}
\text{SS}(\mu, \alpha, \beta) &= \hat{\boldsymbol{\beta}}'\mathbf{X}'\mathbf{y} = (\hat{\mu}, \hat{\alpha}_1, \ldots, \hat{\beta}_2)\begin{pmatrix} y_{..} \\ y_{1.} \\ \vdots \\ y_{.2} \end{pmatrix} \\
&= \hat{\mu}y_{..} + \hat{\alpha}_1 y_{1.} + \hat{\alpha}_2 y_{2.} + \hat{\alpha}_3 y_{3.} + \hat{\beta}_1 y_{.1} + \hat{\beta}_2 y_{.2} \\
&= \bar{y}_{..}y_{..} + \sum_{i=1}^{3}(\bar{y}_{i.} - \bar{y}_{..})y_{i.} + \sum_{j=1}^{2}(\bar{y}_{.j} - \bar{y}_{..})y_{.j} \\
&= \frac{y_{..}^2}{6} + \sum_{i=1}^{3}\left(\frac{y_{i.}}{2} - \frac{y_{..}}{6}\right)y_{i.} + \sum_{j=1}^{2}\left(\frac{y_{.j}}{3} - \frac{y_{..}}{6}\right)y_{.j} \\
&= \frac{y_{..}^2}{6} + \left(\sum_{i=1}^{3}\frac{y_{i.}^2}{2} - \frac{y_{..}^2}{6}\right) + \left(\sum_{j=1}^{2}\frac{y_{.j}^2}{3} - \frac{y_{..}^2}{6}\right), \qquad (12.50)
\end{aligned}$$

since $\sum_i y_{i.} = y_{..}$ and $\sum_j y_{.j} = y_{..}$. The error sum of squares SSE is given by

$$\mathbf{y}'\mathbf{y} - \hat{\boldsymbol{\beta}}'\mathbf{X}'\mathbf{y} = \sum_{ij} y_{ij}^2 - \frac{y_{..}^2}{6} - \left(\sum_{i=1}^{3}\frac{y_{i.}^2}{2} - \frac{y_{..}^2}{6}\right) - \left(\sum_{j=1}^{2}\frac{y_{.j}^2}{3} - \frac{y_{..}^2}{6}\right).$$

To obtain $\hat{\boldsymbol{\beta}}_2\mathbf{X}_2'\mathbf{y}$ in Table 12.3, we use the reduced model $y_{ij} = \mu + \alpha + \beta_j + \varepsilon_{ij} = \mu + \beta_j + \varepsilon_{ij}$, where $\alpha_1 = \alpha_2 = \alpha_3 = \alpha$ and $\mu + \alpha$ is replaced by μ. The normal equations $\mathbf{X}_2'\mathbf{X}_2\hat{\boldsymbol{\beta}}_2 = \mathbf{X}_2'\mathbf{y}$ for the reduced model are

$$\begin{aligned}
6\hat{\mu} + 3\hat{\beta}_1 + 3\hat{\beta}_2 &= y_{..} \\
3\hat{\mu} + 3\hat{\beta}_1 &= y_{.1} \\
3\hat{\mu} + 3\hat{\beta}_2 &= y_{.2}.
\end{aligned} \qquad (12.51)$$

Using the side condition $\hat{\beta}_1 + \hat{\beta}_2 = 0$, the solution to the reduced normal equations in (12.51) is easily obtained as

$$\hat{\mu} = \bar{y}_{..}, \quad \hat{\beta}_1 = \bar{y}_{.1} - \bar{y}_{..}, \quad \hat{\beta}_2 = \bar{y}_{.2} - \bar{y}_{..}. \qquad (12.52)$$

By (12.51) and (12.52), we have

$$\text{SS}(\mu, \beta) = \hat{\boldsymbol{\beta}}_2'\mathbf{X}_2'\mathbf{y} = \hat{\mu}y_{..} + \hat{\beta}_1 y_{.1} + \hat{\beta}_2 y_{.2} = \frac{y_{..}^2}{6} + \left(\sum_{j=1}^{2}\frac{y_{.j}^2}{3} - \frac{y_{..}^2}{6}\right). \qquad (12.53)$$

TABLE 12.4 ANOVA for Testing $H_0 : \alpha_1 = \alpha_2 = \alpha_3$

Source of Variation	df	Sum of Squares	F Statistic
Due to α adjusted for μ, β	2	$\text{SS}(\alpha \vert \mu, \beta) = \sum_i \frac{y_{i.}^2}{2} - \frac{y_{..}^2}{6}$	$\dfrac{\left(\sum_i \frac{y_{i.}^2}{2} - \frac{y_{..}^2}{6}\right)/2}{\text{SSE}/2}$
Error	2	$\text{SSE} = \sum_{ij} y_{ij}^2 - \hat{\boldsymbol{\beta}}'\mathbf{X}'\mathbf{y}$	—
Total	5	$\text{SST} = \sum_{ij} y_{ij}^2 - y_{..}^2/6$	—

Abbreviating $\text{SS}(\alpha_1, \alpha_2, \alpha_3 | \mu, \beta_1\beta_2)$ as $\text{SS}(\alpha | \mu, \alpha)$, we have

$$\text{SS}(\alpha|\mu, \beta) = \hat{\boldsymbol{\beta}}'\mathbf{X}'\mathbf{y} - \hat{\boldsymbol{\beta}}_2'\mathbf{X}_2'\mathbf{y} = \sum_i \frac{y_{i.}^2}{2} - \frac{y_{..}^2}{6}. \tag{12.54}$$

The test is summarized in Table 12.4. [Note that $\text{SS}(\beta | \mu, \alpha)$ is not included.]

12.8.3 Orthogonality of Columns of X

The estimates of μ, β_1, and β_2 given in (12.52) for the reduced model are the same as those of μ, β_1, and β_2 given in (12.49) for the full model. The sum of squares $\hat{\boldsymbol{\beta}}_2'\mathbf{X}_2'\mathbf{y}$ in (12.53) is clearly a part of $\hat{\boldsymbol{\beta}}'\mathbf{X}'\mathbf{y}$ in (12.50). In fact, (12.54) can be expressed as $\text{SS}(\alpha|\mu, \beta) = \text{SS}(\alpha)$, and (12.50) becomes $\text{SS}(\mu, \alpha, \beta) = \text{SS}(\mu) + \text{SS}(\alpha) + \text{SS}(\beta)$. These simplified results are due to the essential orthogonality in the $\mathbf{X}$ matrix in (12.47) as required by Theorem 12.7a. There are three groups of columns in the $\mathbf{X}$ matrix in (12.47), the first column corresponding to μ, the next three columns corresponding to α_1, α_2, and α_3, and the last two columns corresponding to β_1 and β_2. The columns of $\mathbf{X}$ in (12.47) are orthogonal within each group but not among groups as required by Theorem 12.7a. However, consider the same $\mathbf{X}$ matrix if each column after the first is centered using the mean of the column:

$$(\mathbf{j}, \mathbf{X}_c) = \begin{pmatrix} 1 & \frac{2}{3} & -\frac{1}{3} & -\frac{1}{3} & \frac{1}{2} & -\frac{1}{2} \\ 1 & \frac{2}{3} & -\frac{1}{3} & -\frac{1}{3} & -\frac{1}{2} & \frac{1}{2} \\ 1 & -\frac{1}{3} & \frac{2}{3} & -\frac{1}{3} & \frac{1}{2} & -\frac{1}{2} \\ 1 & -\frac{1}{3} & \frac{2}{3} & -\frac{1}{3} & -\frac{1}{2} & \frac{1}{2} \\ 1 & -\frac{1}{3} & -\frac{1}{3} & \frac{2}{3} & \frac{1}{2} & -\frac{1}{2} \\ 1 & -\frac{1}{3} & -\frac{1}{3} & \frac{2}{3} & -\frac{1}{2} & \frac{1}{2} \end{pmatrix}. \tag{12.55}$$

Now the columns are orthogonal among the groups. For example, each of columns 2, 3, and 4 is orthogonal to each of columns 5 and 6, but columns 2, 3, and 4 are not orthogonal to each other. Note that rank($\mathbf{j}$, $\mathbf{X}_c$) = 4 since the sum of columns 2, 3, and 4 is $\mathbf{0}$ and the sum of columns 5 and 6 is $\mathbf{0}$. Thus rank($\mathbf{j}$, $\mathbf{X}_c$) is the same as the rank of $\mathbf{X}$ in (12.47).

We now illustrate the use of side conditions to obtain an orthogonalization that is full-rank (this was illustrated for a one-way model in Section 12.1.1.). Consider the two-way model with interaction

$$y_{ijk} = \mu + \alpha_i + \beta_j + \gamma_{ij} + \varepsilon_{ijk}, \quad i = 1,2;\ j = 1,2;\ k = 1,2. \tag{12.56}$$

In matrix form, the model is

$$\begin{pmatrix} y_{111} \\ y_{112} \\ y_{121} \\ y_{122} \\ y_{211} \\ y_{212} \\ y_{221} \\ y_{222} \end{pmatrix} = \begin{pmatrix} 1 & 1 & 0 & 1 & 0 & 1 & 0 & 0 & 0 \\ 1 & 1 & 0 & 1 & 0 & 1 & 0 & 0 & 0 \\ 1 & 1 & 0 & 0 & 1 & 0 & 1 & 0 & 0 \\ 1 & 1 & 0 & 0 & 1 & 0 & 1 & 0 & 0 \\ 1 & 0 & 1 & 1 & 0 & 0 & 0 & 1 & 0 \\ 1 & 0 & 1 & 1 & 0 & 0 & 0 & 1 & 0 \\ 1 & 0 & 1 & 0 & 1 & 0 & 0 & 0 & 1 \\ 1 & 0 & 1 & 0 & 1 & 0 & 0 & 0 & 1 \end{pmatrix} \begin{pmatrix} \mu \\ \alpha_1 \\ \alpha_2 \\ \beta_1 \\ \beta_2 \\ \gamma_{11} \\ \gamma_{12} \\ \gamma_{21} \\ \gamma_{22} \end{pmatrix} + \begin{pmatrix} \varepsilon_{111} \\ \varepsilon_{112} \\ \varepsilon_{121} \\ \varepsilon_{122} \\ \varepsilon_{211} \\ \varepsilon_{212} \\ \varepsilon_{221} \\ \varepsilon_{222} \end{pmatrix}. \tag{12.57}$$

Useful side conditions become apparent in the context of the normal equations, which are given by

$$\begin{aligned} 8\hat{\mu} + 4(\hat{\alpha}_1 + \hat{\alpha}_2) + 4(\hat{\beta}_1 + \hat{\beta}_2) + 2(\hat{\gamma}_{11} + \hat{\gamma}_{12} + \hat{\gamma}_{21} + \hat{\gamma}_{22}) &= y_{..} \\ 4\hat{\mu} + 4\hat{\alpha}_i + 2(\hat{\beta}_1 + \hat{\beta}_2) + 2(\hat{\gamma}_{i1} + \hat{\gamma}_{i2}) &= y_{i..}, \quad i = 1,2 \\ 4\hat{\mu} + 2(\hat{\alpha}_1 + \hat{\alpha}_2) + 4\hat{\beta}_j + 2(\hat{\gamma}_{1j} + \hat{\gamma}_{2j}) &= y_{.j.}, \quad j = 1,2 \\ 2\hat{\mu} + 2\hat{\alpha}_i + 2\hat{\beta}_j + 2\hat{\gamma}_{ij} &= y_{ij.}, \quad i = 1,2, \quad j = 1,2 \end{aligned} \tag{12.58}$$

Solution of the equations in (12.58) would be simplified by the following side conditions:

$$\begin{aligned} \hat{\alpha}_1 + \hat{\alpha}_2 &= 0, \quad \hat{\beta}_1 + \hat{\beta}_2 = 0, \\ \hat{\gamma}_{i1} + \hat{\gamma}_{i2} &= 0, \quad i = 1,2, \\ \hat{\gamma}_{1j} + \hat{\gamma}_{2j} &= 0, \quad j = 1,2. \end{aligned} \tag{12.59}$$

In (12.57), the $\mathbf{X}$ matrix is 8×9 of rank 4 since the first five columns are all expressible as linear combinations of the last four columns, which are linearly independent. Thus $\mathbf{X}'\mathbf{X}$ is 9×9 and has a rank deficiency of $9 - 4 = 5$. However, there are six side conditions in (12.59). This apparent discrepancy is resolved by noting that there are only three restrictions among the last four equations in (12.59). We can obtain any one of these four from the other three. To illustrate, we obtain the first equation from the last three. Adding the third and fourth equations gives $\hat{\gamma}_{11} + \hat{\gamma}_{21} + \hat{\gamma}_{12} + \hat{\gamma}_{22} = 0$. Then substitution of the second, $\hat{\gamma}_{21} + \hat{\gamma}_{22} = 0$, reduces this to the first, $\hat{\gamma}_{11} + \hat{\gamma}_{12} = 0$.

We can obtain a full-rank orthogonalization by imposing the side conditions in (12.59) on the parameters and using these relationships to express redundant parameters in terms of the four parameters μ, α_1, β_1, and γ_{11}. (For expositional convenience, we do not use $*$ on the parameters subject to side conditions.) This gives

$$\begin{aligned} &\alpha_2 = -\alpha_1, \quad \beta_2 = -\beta_1, \\ &\gamma_{12} = -\gamma_{11}, \quad \gamma_{21} = -\gamma_{11}, \quad \gamma_{22} = \gamma_{11}. \end{aligned} \tag{12.60}$$

The last of these, for example, is obtained from the side condition $\gamma_{12} + \gamma_{22} = 0$. Thus $\gamma_{22} = -\gamma_{12} = -(-\gamma_{11})$.

Using (12.60), we can express the eight y_{ijk} values in (12.56) in terms of μ, α_1, β_1, and γ_{11}:

$$\begin{aligned} y_{11k} &= \mu + \alpha_1 + \beta_1 + \gamma_{11} + \varepsilon_{11k}, \quad k = 1, 2, \\ y_{12k} &= \mu + \alpha_1 + \beta_2 + \gamma_{12} + \varepsilon_{12k} \\ &= \mu + \alpha_1 - \beta_1 - \gamma_{11} + \varepsilon_{12k}, \quad k = 1, 2, \\ y_{21k} &= \mu + \alpha_2 + \beta_1 + \gamma_{21} + \varepsilon_{21k} \\ &= \mu - \alpha_1 + \beta_1 - \gamma_{11} + \varepsilon_{21k}, \quad k = 1, 2, \\ y_{22k} &= \mu + \alpha_2 + \beta_2 + \gamma_{22} + \varepsilon_{22k} \\ &= \mu - \alpha_1 - \beta_1 + \gamma_{11} + \varepsilon_{22k}, \quad k = 1, 2. \end{aligned}$$

The redefined **X** matrix thus becomes

$$\begin{pmatrix} 1 & 1 & 1 & 1 \\ 1 & 1 & 1 & 1 \\ 1 & 1 & -1 & -1 \\ 1 & 1 & -1 & -1 \\ 1 & -1 & 1 & -1 \\ 1 & -1 & 1 & -1 \\ 1 & -1 & -1 & 1 \\ 1 & -1 & -1 & 1 \end{pmatrix},$$

which is a full-rank matrix with orthogonal columns. The methods of Chapters 7 and 8 can now be used for estimation and testing hypotheses.

PROBLEMS

12.1 Show that $\bar{\mu}_{1.} + \bar{\mu}_{2.} = 2\bar{\mu}_{..}$ as in (12.9).

12.2 Show that $\hat{\boldsymbol{\varepsilon}}'\hat{\boldsymbol{\varepsilon}}$ in (12.10) is minimized by $\hat{\boldsymbol{\beta}}$, the solution to $\mathbf{X}'\mathbf{X}\hat{\boldsymbol{\beta}} = \mathbf{X}'\mathbf{y}$ in (12.11).

12.3 Use Theorem 2.7 to prove Theorem 12.2a.

12.4 **(a)** Give an alternative proof of Theorem 12.2b(iii) based on Theorem 2.8c(iii).

(b) Give a second alternative proof of Theorem 12.2b(iii) based on Theorem 2.8f.

12.5 **(a)** Using all three conditions in Theorem 12.2b, show that $\boldsymbol{\lambda}'\boldsymbol{\beta} = \mu + \tau_2 = (1, 0, 1)\boldsymbol{\beta}$ is estimable (use the model in Example 12.2.2a).

(b) Using all three conditions in Theorem 12.2b, show that $\boldsymbol{\lambda}'\boldsymbol{\beta} = \tau_1 + \tau_2 = (0, 1, 1)\boldsymbol{\beta}$ is not estimable.

12.6 If $\boldsymbol{\lambda}'\boldsymbol{\beta}$ is estimable and $\hat{\boldsymbol{\beta}}_1$ and $\hat{\boldsymbol{\beta}}_2$ are two solutions to the normal equations, show that $\boldsymbol{\lambda}'\hat{\boldsymbol{\beta}}_1 = \boldsymbol{\lambda}'\hat{\boldsymbol{\beta}}_2$ as in Theorem 12.3a(iii).

12.7 Obtain an estimate of $\mu + \tau_2$ using $\mathbf{r}'\mathbf{X}'\mathbf{y}$ and $\lambda'\hat{\boldsymbol{\beta}}$ from the model in Example 12.3.1.

12.8 Consider the model $y_{ij} = \mu + \tau_i + \varepsilon_{ij}$, $i = 1, 2$, $j = 1, 2, 3$:

(a) For $\boldsymbol{\lambda}'\boldsymbol{\beta} = (1, 1, 0)\boldsymbol{\beta} = \mu + \tau_1$, show that

$$\mathbf{r} = c\begin{pmatrix} -1 \\ 1 \\ 1 \end{pmatrix} + \begin{pmatrix} 0 \\ \frac{1}{3} \\ 0 \end{pmatrix},$$

with arbitrary c, represents all solutions to $\mathbf{X}'\mathbf{X}\mathbf{r} = \boldsymbol{\lambda}$.

(b) Obtain the BLUE [best linear unbiased estimator] for $\mu + \tau_1$ using $\mathbf{r}$ obtained in part (a).

(c) Find the BLUE for $\tau_1 - \tau_2$ using the method of parts (a) and (b).

12.9 **(a)** In Example 12.2.2b, we found the estimable functions $\boldsymbol{\lambda}_1'\boldsymbol{\beta} = \mu + \alpha_1 + \beta_1$, $\boldsymbol{\lambda}_2'\boldsymbol{\beta} = \beta_1 - \beta_2$, and $\boldsymbol{\lambda}_3'\boldsymbol{\beta} = \alpha_1 - \alpha_2$. Find the BLUE for each of these using $\mathbf{r}'\mathbf{X}'\mathbf{y}$ in each case.

(b) For each estimator in part (a), show that $E(\mathbf{r}_i'\mathbf{X}'\mathbf{y}) = \boldsymbol{\lambda}_i'\boldsymbol{\beta}$.

12.10 In the model $y_{ij} = \mu + \tau_i + \varepsilon_{ij}$, $i = 1, 2, \ldots, k; j = 1, 2, \ldots, n$, show that $\sum_{i=1}^{k} c_i\tau_i$ is estimable if and only if $\sum_{i=1}^{k} c_i = 0$, as suggested following Example 12.2.2b. Use the following two approaches:

(a) In $\boldsymbol{\lambda}'\boldsymbol{\beta} = \sum_{i=1}^{k} c_i\tau_i$, express $\boldsymbol{\lambda}'$ as a linear combination of the rows of $\mathbf{X}$.

(b) Express $\sum_{i=1}^{k} c_i\tau_i$ as a linear combination of the elements of $E(\mathbf{y}) = \mathbf{X}\boldsymbol{\beta}$.

12.11 In Example 12.3.1, find all solutions $\mathbf{r}$ for $\mathbf{X}'\mathbf{X}\mathbf{r} = \boldsymbol{\lambda}$ and show that all of them give $\mathbf{r}'\mathbf{X}'\mathbf{y} = \bar{y}_{1.} - \bar{y}_{2.}$.

12.12 Show that $\text{cov}(\boldsymbol{\lambda}_1'\hat{\boldsymbol{\beta}}, \boldsymbol{\lambda}_2'\hat{\boldsymbol{\beta}}) = \sigma^2\mathbf{r}_1'\boldsymbol{\lambda}_2 = \sigma^2\boldsymbol{\lambda}_1'\mathbf{r}_2 = \sigma^2\boldsymbol{\lambda}_1'(\mathbf{X}'\mathbf{X})^-\boldsymbol{\lambda}_2$ as in Theorem 12.3c.

12.13 **(a)** Show that $(\mathbf{y} - \mathbf{X}\hat{\boldsymbol{\beta}})'(\mathbf{y} - \mathbf{X}\hat{\boldsymbol{\beta}}) = \mathbf{y}'\mathbf{y} - \hat{\boldsymbol{\beta}}'\mathbf{X}'\mathbf{y}$ as in (12.20).

(b) Show that $\mathbf{y}'\mathbf{y} - \hat{\boldsymbol{\beta}}'\mathbf{X}'\mathbf{y} = \mathbf{y}'[\mathbf{I} - \mathbf{X}(\mathbf{X}'\mathbf{X})^-\mathbf{X}']\mathbf{y}$ as in (12.21).

12.14 Show that $\boldsymbol{\beta}'\mathbf{X}'[\mathbf{I} - \mathbf{X}(\mathbf{X}'\mathbf{X})^-\mathbf{X}']\mathbf{X}\boldsymbol{\beta} = 0$, as in the proof of Theorem 12.3e(i).

12.15 Differentiate $\ln L(\boldsymbol{\beta}, \sigma^2)$ in (12.26) with respect to $\boldsymbol{\beta}$ and σ^2 to obtain (12.27) and (12.28).

12.16 Prove Theorem 12.3g.

12.17 Show that $\boldsymbol{\lambda}'\boldsymbol{\beta} = \mathbf{b}'\boldsymbol{\gamma} = \mathbf{c}'\boldsymbol{\delta}$ as in (12.34).

12.18 Show that the matrix $\mathbf{Z}$ in Example 12.5 can be obtained using (12.31), $\mathbf{Z} = \mathbf{X}\mathbf{U}'(\mathbf{U}\mathbf{U}')^{-1}$.

12.19 Redo Example 12.5 with the parameterization

$$\boldsymbol{\gamma} = \begin{pmatrix} \mu + \tau_1 \\ \tau_1 - \tau_2 \end{pmatrix}.$$

Find $\mathbf{Z}$ and $\mathbf{U}$ by inspection and show that $\mathbf{Z}\mathbf{U} = \mathbf{X}$. Then show that $\mathbf{Z}$ can be obtained as $\mathbf{Z} = \mathbf{X}\mathbf{U}'(\mathbf{U}\mathbf{U}')^{-1}$.

12.20 Show that $\hat{\boldsymbol{\beta}}$ in (12.39) is a solution to the normal equations $\mathbf{X}'\mathbf{X}\hat{\boldsymbol{\beta}} = \mathbf{X}'\mathbf{y}$.

12.21 Show that $\begin{pmatrix} \alpha_1 - \alpha_2 \\ \alpha_1 + \alpha_2 - 2\alpha_3 \end{pmatrix} = \begin{pmatrix} 0 \\ 0 \end{pmatrix}$ in (12.40) implies $\alpha_1 = \alpha_2 = \alpha_3$, as noted preceding (12.40).

12.22 Prove Theorem 12.7b(v).

12.23 Multiply $\mathbf{X}'\mathbf{X}$ in (12.48) by $\hat{\boldsymbol{\beta}}$ to obtain the six normal equations. Show that with the side conditions $\hat{\alpha}_1 + \hat{\alpha}_2 + \hat{\alpha}_3 = 0$ and $\hat{\beta}_1 + \hat{\beta}_2 = 0$, the solution is given by (12.49).

12.24 Obtain the reduced normal equations $\mathbf{X}_2'\mathbf{X}_2\hat{\boldsymbol{\beta}}_2 = \mathbf{X}_2'\mathbf{y}$ in (12.51) by writing $\mathbf{X}_2$ and $\mathbf{X}_2'\mathbf{X}_2$ for the reduced model $y_{ij} = \mu + \beta_j + \varepsilon_{ij}$, $i = 1, 2, 3$, $j = 1, 2$.

12.25 Consider the model $y_{ij} = \mu + \tau_i + \varepsilon_{ij}$, $i = 1, 2, 3$, $j = 1, 2, 3$:

(a) Write $\mathbf{X}$, $\mathbf{X}'\mathbf{X}$, $\mathbf{X}'\mathbf{y}$, and the normal equations.

(**b**) What is the rank of $\mathbf{X}$ or $\mathbf{X}'\mathbf{X}$? Find a set of linearly independent estimable functions.

(**c**) Define an appropriate side condition, and find the resulting solution to the normal equations.

(**d**) Show that $H_0: \tau_1 = \tau_2 = \tau_3$ is testable. Find $\hat{\boldsymbol{\beta}}'\mathbf{X}'\mathbf{y} = \text{SS}(\mu, \tau)$ and $\hat{\boldsymbol{\beta}}_2'\mathbf{X}_2'\mathbf{y} = \text{SS}(\mu)$.

(**e**) Construct an ANOVA table for the test of $H_0: \tau_1 = \tau_2 = \tau_3$.

12.26 Consider the model $y_{ijk} = \mu + \alpha_i + \beta_j + \gamma_{ij} + \varepsilon_{ijk}$, $i = 1, 2$, $j = 1, 2$, $k = 1, 2, 3$.

(**a**) Write $\mathbf{X}'\mathbf{X}$, $\mathbf{X}'\mathbf{y}$, and the normal equations.

(**b**) Find a set of linearly independent estimable functions. Are $\alpha_1 - \alpha_2$ and $\beta_1 - \beta_2$ estimable?

12.27 Consider the model $y_{ijk} = \mu + \alpha_i + \beta_j + \gamma_k + \varepsilon_{ijk}$, $i = 1, 2$, $j = 1, 2$, $k = 1, 2$.

(**a**) Write $\mathbf{X}'\mathbf{X}$, $\mathbf{X}'\mathbf{y}$, and the normal equations.

(**b**) Find a set of linearly independent estimable functions.

(**c**) Define appropriate side conditions, and find the resulting solution to the normal equations.

(**d**) Show that $H_0: \alpha_1 = \alpha_2$ is testable. Find $\hat{\boldsymbol{\beta}}'\mathbf{X}'\mathbf{y} = \text{SS}(\mu, \alpha, \beta, \gamma)$ and $\hat{\boldsymbol{\beta}}_2'\mathbf{X}_2'\mathbf{y} = \text{SS}(\mu, \beta, \gamma)$.

(**e**) Construct an ANOVA table for the test of $H_0: \alpha_1 = \alpha_2$.

12.28 For the model $y_{ijk} = \mu + \alpha_i + \beta_j + \gamma_{ij} + \varepsilon_{ijk}$, $i = 1, 2, j = 1, 2, k = 1, 2$ in (12.56), write $\mathbf{X}'\mathbf{X}$ and obtain the normal equations in (12.58).

13 One-Way Analysis-of-Variance: Balanced Case

The one-way analysis-of-variance (ANOVA) model has been illustrated in Sections 12.1.1, 12.2.2, 12.3.1, 12.5, and 12.6. We now analyze this model more fully. To solve the normal equations in Section 13.3, we use side conditions as well as a generalized inverse approach. For hypothesis tests in Section 13.4, we use both the full–reduced-model approach and the general linear hypothesis. Expected mean squares are obtained in Section 13.5 using both a full–reduced-model approach and a general linear hypothesis approach. In Section 13.6, we discuss contrasts on the means, including orthogonal polynomials. Throughout this chapter, we consider only the balanced model. The unbalanced case is discussed in Chapter 15.

13.1 THE ONE-WAY MODEL

The one-way balanced model can be expressed as follows:

$$y_{ij} = \mu + \alpha_i + \varepsilon_{ij}, \quad i = 1, 2, \ldots, k, \; j = 1, 2, \ldots, n. \tag{13.1}$$

If $\alpha_1, \alpha_2, \ldots, \alpha_k$ represent the effects of k treatments, each of which is applied to n experimental units, then y_{ij} is the response of the jth observation among the n units that receive the ith treatment. For example, in an agricultural experiment, the treatments may be different fertilizers or different amounts of a given fertilizer. On the other hand, in some experimental situations, the k groups may represent samples from k populations whose means we wish to compare, populations that are not created by applying treatments. For example, suppose that we wish to compare the average lifetimes of several brands of batteries or the mean grade-point averages for freshmen, sophomores, juniors, and seniors. Three additional assumptions that form part of the model in (13.1) are

1. $E(\varepsilon_{ij}) = 0$ for all i, j.

Linear Models in Statistics, Second Edition, by Alvin C. Rencher and G. Bruce Schaalje

2. $\text{var}(\varepsilon_{ij}) = \sigma^2$ for all i, j.
3. $\text{cov}(\varepsilon_{ij}, \varepsilon_{rs}) = 0$ for all $(i, j) \neq (r, s)$.
4. We sometimes add the assumption that ε_{ij} is distributed as $N(0, \sigma^2)$.
5. In addition, we often use the constraint (side condition) $\sum_{i=1}^{k} \alpha_i = 0$.

The mean for the ith treatment or population can be denoted by μ_i. Thus $E_{ij} = \mu_i$, and using assumption 1, we have $\mu_i = \mu + \alpha_i$. We can thus write (13.1) in the form

$$y_{ij} = \mu_i + \varepsilon_{ij}, \quad i = 1, 2, \ldots, k, \;\; j = 1, 2, \ldots, n. \tag{13.2}$$

In this form of the model, the hypothesis $H_0: \mu_1 = \mu_2 = \cdots = \mu_k$ is of interest.

In the context of design of experiments, the one-way layout is sometimes called a *completely randomized design*. In this design, the experimental units are assigned at random to the k treatments.

13.2 ESTIMABLE FUNCTIONS

To illustrate the model (13.1) in matrix form, let $k = 3$ and $n = 2$. The resulting six equations, $y_{ij} = \mu + \alpha_i + \varepsilon_{ij}, \; i = 1, 2, 3, \; j = 1, 2$, can be expressed as

$$\begin{pmatrix} y_{11} \\ y_{12} \\ y_{21} \\ y_{22} \\ y_{31} \\ y_{32} \end{pmatrix} = \begin{pmatrix} \mu + \alpha_1 \\ \mu + \alpha_1 \\ \mu + \alpha_2 \\ \mu + \alpha_2 \\ \mu + \alpha_3 \\ \mu + \alpha_3 \end{pmatrix} + \begin{pmatrix} \varepsilon_{11} \\ \varepsilon_{12} \\ \varepsilon_{21} \\ \varepsilon_{22} \\ \varepsilon_{31} \\ \varepsilon_{32} \end{pmatrix}$$

$$= \begin{pmatrix} 1 & 1 & 0 & 0 \\ 1 & 1 & 0 & 0 \\ 1 & 0 & 1 & 0 \\ 1 & 0 & 1 & 0 \\ 1 & 0 & 0 & 1 \\ 1 & 0 & 0 & 1 \end{pmatrix} \begin{pmatrix} \mu \\ \alpha_1 \\ \alpha_2 \\ \alpha_3 \end{pmatrix} + \begin{pmatrix} \varepsilon_{11} \\ \varepsilon_{12} \\ \varepsilon_{21} \\ \varepsilon_{22} \\ \varepsilon_{31} \\ \varepsilon_{32} \end{pmatrix}, \tag{13.3}$$

or

$$\mathbf{y} = \mathbf{X}\boldsymbol{\beta} + \boldsymbol{\varepsilon}.$$

In (13.3), $\mathbf{X}$ is 6×4 and is clearly of rank 3 because the first column is the sum of the other three columns. Thus $\boldsymbol{\beta} = (\mu, \alpha_1, \alpha_2, \alpha_3)'$ is not unique and not estimable; hence the individual parameters $\mu, \alpha_1, \alpha_2, \alpha_3$ cannot be estimated unless they are subject to constraints (side conditions). In general, the $\mathbf{X}$ matrix for the one-way balanced model is $kn \times (k+1)$ of rank k.

We discussed estimable functions $\boldsymbol{\lambda}'\boldsymbol{\beta}$ in Section 12.2.2. It was shown in Problem 12.10 that for the one-way balanced model, contrasts in the α's are estimable. Thus $\sum_i c_i\alpha_i$ is estimable if and only if $\sum_i c_i = 0$. For example, contrasts such as $\alpha_1 - \alpha_2$ and $\alpha_1 - 2\alpha_2 + \alpha_3$ are estimable.

If we impose a side condition on the α_i's and denote the constrained parameters as μ^* and α_i^*, then $\mu^*, \alpha_1^*, \ldots, \alpha_k^*$ are uniquely defined and estimable. Under the usual side condition, $\sum_{i=1}^k \alpha_i^* = 0$, the parameters are defined as $\mu^* = \bar{\mu}_.$ and $\alpha_i^* = \mu_i - \bar{\mu}_.$, where $\bar{\mu}_. = \sum_{i=1}^k \mu_i/k$. To see this, we rewrite (13.1) and (13.2) in the form $E(y_{ij}) = \mu_i = \mu^* + \alpha_i^*$ to obtain

$$\begin{aligned} \bar{\mu}_. &= \sum_{i=1}^k \frac{\mu_i}{k} = \sum_i \frac{\mu^* + \alpha_i^*}{k} \\ &= \mu^* + \sum_i \frac{\alpha_i^*}{k} = \mu^*. \end{aligned} \tag{13.4}$$

Then, from $\mu_i = \mu^* + \alpha_i^*$, we have

$$\alpha_i^* = \mu_i - \mu^* = \mu_i - \bar{\mu}_.. \tag{13.5}$$

13.3 ESTIMATION OF PARAMETERS

13.3.1 Solving the Normal Equations

Extending (13.3) to a general k and n, the one-way model can be written in matrix form as

$$\begin{pmatrix} \mathbf{y}_1 \\ \mathbf{y}_2 \\ \vdots \\ \mathbf{y}_k \end{pmatrix} = \begin{pmatrix} \mathbf{j} & \mathbf{j} & \mathbf{0} & \cdots & \mathbf{0} \\ \mathbf{j} & \mathbf{0} & \mathbf{j} & \cdots & \mathbf{0} \\ \vdots & \vdots & \vdots & & \vdots \\ \mathbf{j} & \mathbf{0} & \mathbf{0} & \cdots & \mathbf{j} \end{pmatrix} \begin{pmatrix} \mu \\ \alpha_1 \\ \alpha_2 \\ \vdots \\ \alpha_k \end{pmatrix} + \begin{pmatrix} \boldsymbol{\varepsilon}_1 \\ \boldsymbol{\varepsilon}_2 \\ \vdots \\ \boldsymbol{\varepsilon}_k \end{pmatrix}, \tag{13.6}$$

or

$$\mathbf{y} = \mathbf{X}\boldsymbol{\beta} + \boldsymbol{\varepsilon},$$

where $\mathbf{j}$ and $\mathbf{0}$ are each of size $n \times 1$, and $\mathbf{y}_i$ and $\boldsymbol{\varepsilon}_i$ are defined as

$$\mathbf{y}_i = \begin{pmatrix} y_{i1} \\ y_{i2} \\ \vdots \\ y_{in} \end{pmatrix}, \quad \boldsymbol{\varepsilon}_i = \begin{pmatrix} \varepsilon_{i1} \\ \varepsilon_{i2} \\ \vdots \\ \varepsilon_{in} \end{pmatrix}.$$

For (13.6), the normal equations $\mathbf{X}'\mathbf{X}\hat{\boldsymbol{\beta}} = \mathbf{X}'\mathbf{y}$ take the form

$$\begin{pmatrix} kn & n & n & \cdots & n \\ n & n & 0 & \cdots & 0 \\ n & 0 & n & \cdots & 0 \\ \vdots & \vdots & \vdots & & \vdots \\ n & 0 & 0 & \cdots & n \end{pmatrix} \begin{pmatrix} \hat{\mu} \\ \hat{\alpha}_1 \\ \hat{\alpha}_2 \\ \vdots \\ \hat{\alpha}_k \end{pmatrix} = \begin{pmatrix} y_{..} \\ y_{1.} \\ y_{2.} \\ \vdots \\ y_{k.} \end{pmatrix}, \qquad (13.7)$$

where $y_{..} = \sum_{ij} y_{ij}$ and $y_{i.} = \sum_j y_{ij}$.

In Section 13.3.1.1, we find a solution of (13.7) using side conditions, and in Section 13.3.1.2 we find another solution using a generalized inverse of $\mathbf{X}'\mathbf{X}$.

13.3.1.1 Side Conditions

The $k + 1$ normal equations in (13.7) can be expressed as

$$\begin{aligned} kn\hat{\mu} + n\hat{\alpha}_1 + n\hat{\alpha}_2 + \cdots + n\hat{\alpha}_k &= y_{..}, \\ n\hat{\mu} + n\hat{\alpha}_i &= y_{i.}, \quad i = 1, 2, \ldots, k. \end{aligned} \qquad (13.8)$$

Using the side condition $\sum_i \hat{\alpha}_i = 0$, the solution to (13.8) is given by

$$\begin{aligned} \hat{\mu} &= \frac{y_{..}}{kn} = \bar{y}_{..}, \\ \hat{\alpha}_i &= \frac{y_{i.}}{n} - \hat{\mu} = \bar{y}_{i.} - \bar{y}_{..}, \quad i = 1, 2, \ldots, k. \end{aligned} \qquad (13.9)$$

In vector form, this solution $\hat{\boldsymbol{\beta}}$ for $\mathbf{X}'\mathbf{X}\hat{\boldsymbol{\beta}} = \mathbf{X}'\mathbf{y}$ is expressed as

$$\hat{\boldsymbol{\beta}} = \begin{pmatrix} \bar{y}_{..} \\ \bar{y}_{1.} - \bar{y}_{..} \\ \vdots \\ \bar{y}_{k.} - \bar{y}_{..} \end{pmatrix}. \qquad (13.10)$$

If the side condition $\sum_i \alpha_i^* = 0$ is imposed on the parameters, then the elements of $\hat{\boldsymbol{\beta}}$ are unique estimators of the (constrained) parameters $\mu^* = \bar{\mu}_.$ and

$\alpha_i^* = \mu_i - \bar{\mu}_., i = 1, 2, \ldots, k$, in (13.4) and (13.5). Otherwise, the estimators in (13.9) or (13.10) are to be used in estimable functions. For example, by Theorem 12.3a(i), the estimator of $\boldsymbol{\lambda}'\boldsymbol{\beta} = \alpha_1 - \alpha_2$ is given by $\boldsymbol{\lambda}'\hat{\boldsymbol{\beta}}$:

$$\boldsymbol{\lambda}'\hat{\boldsymbol{\beta}} = \widehat{\alpha_1 - \alpha_2} = \hat{\alpha}_1 - \hat{\alpha}_2 = \bar{y}_{1.} - \bar{y}_{..} - (\bar{y}_{2.} - \bar{y}_{..}) = \bar{y}_{1.} - \bar{y}_{2.}.$$

By Theorem 12.3d, such estimators are BLUE. If ε_{ij} is $N(0, \sigma^2)$, then, by Theorem 12.3h, the estimators are minimum variance unbiased estimators.

13.3.1.2 Generalized Inverse

By Corollary 1 to Theorem 2.8b, a generalized inverse of $\mathbf{X}'\mathbf{X}$ in (13.7) is given by

$$(\mathbf{X}'\mathbf{X})^- = \begin{pmatrix} 0 & 0 & \cdots & 0 \\ 0 & \frac{1}{n} & \cdots & 0 \\ \vdots & \vdots & & \vdots \\ 0 & 0 & \cdots & \frac{1}{n} \end{pmatrix}. \tag{13.11}$$

Then by (12.13) and (13.7), a solution to the normal equations is obtained as

$$\hat{\boldsymbol{\beta}} = (\mathbf{X}'\mathbf{X})^- \mathbf{X}'\mathbf{y} = \begin{pmatrix} 0 \\ \bar{y}_{1.} \\ \vdots \\ \bar{y}_{k.} \end{pmatrix}. \tag{13.12}$$

The estimators in (13.12) are different from those in (13.10), but they give the same estimates of estimable functions. For example, using $\hat{\boldsymbol{\beta}}$ from (13.12) to estimate $\boldsymbol{\lambda}'\boldsymbol{\beta} = \alpha_1 - \alpha_2$, we have

$$\boldsymbol{\lambda}'\hat{\boldsymbol{\beta}} = \widehat{\alpha_1 - \alpha_2} = \hat{\alpha}_1 - \hat{\alpha}_2 = \bar{y}_{1.} - \bar{y}_{2.},$$

which is the same estimate as that obtained above in Section 13.3.1.1 using $\hat{\boldsymbol{\beta}}$ from (13.10).

13.3.2 An Estimator for σ^2

In assumption 2 for the one-way model in (13.1), we have $\text{var}(\varepsilon_{ij}) = \sigma^2$ for all i, j. To estimate σ^2, we use (12.22)

$$s^2 = \frac{\text{SSE}}{k(n-1)},$$

where SSE is as given by (12.20) or (12.21):

$$\text{SSE} = \mathbf{y}'\mathbf{y} - \hat{\boldsymbol{\beta}}'\mathbf{X}'\mathbf{y} = \mathbf{y}'[\mathbf{I} - \mathbf{X}(\mathbf{X}'\mathbf{X})^{-}\mathbf{X}']\mathbf{y}.$$

The rank of the idempotent matrix $\mathbf{I} - \mathbf{X}(\mathbf{X}'\mathbf{X})^{-}\mathbf{X}'$ is $kn - k$ because $\text{rank}(\mathbf{X}) = k$, $\text{tr}(\mathbf{I}) = kn$, and $\text{tr}[\mathbf{X}(\mathbf{X}'\mathbf{X})^{-}\mathbf{X}'] = k$ (see Theorem 2.13d). Then $s^2 = \text{SSE}/k(n-1)$ is an unbiased estimator of σ^2 [see Theorem 12.3e(i)].

Using $\hat{\boldsymbol{\beta}}$ from (13.12), we can express $\text{SSE} = \mathbf{y}'\mathbf{y} - \hat{\boldsymbol{\beta}}'\mathbf{X}\mathbf{y}$ in the following form:

$$\text{SSE} = \mathbf{y}'\mathbf{y} - \hat{\boldsymbol{\beta}}'\mathbf{X}'\mathbf{y} = \sum_{i=1}^{k}\sum_{j=1}^{n} y_{ij}^2 - \sum_{i=1}^{k} \bar{y}_{i.}y_{i.}$$

$$= \sum_{ij} y_{ij}^2 - \sum_i \frac{y_{i.}^2}{n}. \tag{13.13}$$

It can be shown (see Problem 13.4) that (13.13) can be written as

$$\text{SSE} = \sum_{ij} (y_{ij} - \bar{y}_{i.})^2. \tag{13.14}$$

Thus s^2 is given by either of the two forms

$$s^2 = \frac{\sum_{ij} (y_{ij} - \bar{y}_{i.})^2}{k(n-1)} \tag{13.15}$$

$$= \frac{\sum_{ij} y_{ij}^2 - \sum_i y_{i.}^2/n}{k(n-1)}. \tag{13.16}$$

13.4 TESTING THE HYPOTHESIS $H_0: \mu_1 = \mu_2 = \cdots = \mu_k$

Using the model in (13.2), the hypothesis of equality of means can be expressed as $H_0: \mu_1 = \mu_2 = \cdots = \mu_k$. The alternative hypothesis is that at least two means are unequal. Using $\mu_i = \mu + \alpha_i$ [see (13.1) and (13.2)], the hypothesis can be expressed as $H_0: \alpha_1 = \alpha_2 = \cdots = \alpha_k$, which is testable because it can be written in terms of $k-1$ linearly independent estimable contrasts, for example, $H_0: \alpha_1 - \alpha_2 = \alpha_1 - \alpha_3 = \cdots = \alpha_1 - \alpha_k = 0$ (see the second paragraph in Section 12.7.1). In Section 13.4.1 we develop the test using the full–reduced-model approach, and in Section 13.4.2 we use the general linear hypothesis approach. In the model $\mathbf{y} = \mathbf{X}\boldsymbol{\beta} + \boldsymbol{\varepsilon}$, the vector $\mathbf{y}$ is $kn \times 1$ [see (13.6)]. Throughout Section 13.4, we assume that $\mathbf{y}$ is $N_{kn}(\mathbf{X}\boldsymbol{\beta}, \sigma^2\mathbf{I})$.

13.4.1 Full–Reduced-Model Approach

The hypothesis

$$H_0: \alpha_1 = \alpha_2 = \cdots = \alpha_k \tag{13.17}$$

is equivalent to

$$H_0: \alpha_1^* = \alpha_2^* = \cdots = \alpha_k^*, \tag{13.18}$$

where the α_i^* terms are subject to the side condition $\sum_i \alpha_i^* = 0$. With this constraint, H_0 in (13.18) is also equivalent to

$$H_0: \alpha_1^* = \alpha_2^* = \cdots = \alpha_k^* = 0. \tag{13.19}$$

The full model, $y_{ij} = \mu + \alpha_i + \varepsilon_{ij}$, $i = 1, 2, \ldots, k$, $j = 1, 2, \ldots, n$, is expressed in matrix form $\mathbf{y} = \mathbf{X}\boldsymbol{\beta} + \boldsymbol{\varepsilon}$ in (13.6). If the full model is written in terms of μ^* and α_i^* as $y_{ij} = \mu^* + \alpha_i^* + \varepsilon_{ij}$, then the reduced model under H_0 in (13.19) is $y_{ij} = \mu^* + \varepsilon_{ij}$. In matrix form, this becomes $\mathbf{y} = \mu^*\mathbf{j} + \boldsymbol{\varepsilon}$, where $\mathbf{j}$ is $kn \times 1$. To be consistent with the full model $\mathbf{y} = \mathbf{X}\boldsymbol{\beta} + \boldsymbol{\varepsilon}$, we write the reduced model as

$$\mathbf{y} = \mu\mathbf{j} + \boldsymbol{\varepsilon}. \tag{13.20}$$

For the full model, the sum of squares $\text{SS}(\mu, \alpha) = \hat{\boldsymbol{\beta}}'\mathbf{X}'\mathbf{y}$ is given as part of (13.13) as

$$\text{SS}(\mu, \alpha) = \hat{\boldsymbol{\beta}}'\mathbf{X}'\mathbf{y} = \sum_{i=1}^{k} \frac{y_{i.}^2}{n},$$

where the sum of squares $\text{SS}(\mu, \alpha_1, \ldots, \alpha_k)$ is abbreviated as $\text{SS}(\mu, \alpha)$. For the reduced model in (13.20), the estimator "$\hat{\boldsymbol{\beta}} = (\mathbf{X}'\mathbf{X})^{-1}\mathbf{X}'\mathbf{y}$" and the sum of squares "$\hat{\boldsymbol{\beta}}'\mathbf{X}'\mathbf{y}$" become

$$\hat{\mu} = (\mathbf{j}'\mathbf{j})^{-1}\mathbf{j}'\mathbf{y} = \frac{1}{kn} y_{..} = \bar{y}_{..}, \tag{13.21}$$

$$\text{SS}(\mu) = (\hat{\mu})'\mathbf{j}'\mathbf{y} = \bar{y}_{..}y_{..} = \frac{y_{..}^2}{kn}, \tag{13.22}$$

where $\mathbf{j}$ is $kn \times 1$.

From Table 12.3, the sum of squares for the α's adjusted for μ is given by

$$\begin{aligned} \text{SS}(\alpha|\mu) = \text{SS}(\mu, \alpha) - \text{SS}(\mu) &= \hat{\boldsymbol{\beta}}'\mathbf{X}'\mathbf{y} - \frac{y_{..}^2}{kn} \\ &= \frac{1}{n}\sum_{i=1}^{k} y_{i.}^2 - \frac{y_{..}^2}{kn} \qquad (13.23) \\ &= n\sum_{i=1}^{k} (\bar{y}_{i.} - \bar{y}_{..})^2. \qquad (13.24) \end{aligned}$$

TABLE 13.1 ANOVA for Testing $H_0: \alpha_1 = \alpha_2 = \cdots = \alpha_k$ in the One-Way Model

Source of Variation	df	Sum of Squares	Mean Square	F Statistic
Treatments	$k-1$	$SS(\alpha\|\mu) = \frac{1}{n}\sum_i y_{i.}^2 - \frac{y_{..}^2}{kn}$	$\frac{SS(\alpha\|\mu)}{k-1}$	$\frac{SS(\alpha\|\mu)/(k-1)}{SSE/k(n-1)}$
Error	$k(n-1)$	$SSE = \sum_{ij} y_{ij}^2 - \frac{1}{n}\sum_i y_{i.}^2$	$\frac{SSE}{k(n-1)}$	—
Total	$kn-1$	$SST = \sum_{ij} y_{ij}^2 - \frac{y_{..}^2}{kn}$		

The test is summarized in Table 13.1 using $SS(\alpha|\mu)$ in (13.23) and SSE in (13.13). The chi-square and independence properties of $SS(\alpha|\mu)$ and SSE follow from results established in Section 12.7.2.

To facilitate comparison of (13.23) with the result of the general linear hypothesis approach in Section 13.4.2, we now express $SS(\alpha|\mu)$ as a quadratic form in $\mathbf{y}$. By (12.13), $\hat{\boldsymbol{\beta}} = (\mathbf{X}'\mathbf{X})^-\mathbf{X}'\mathbf{y}$, and therefore $\hat{\boldsymbol{\beta}}'\mathbf{X}'\mathbf{y} = \mathbf{y}'\mathbf{X}(\mathbf{X}'\mathbf{X})^-\mathbf{X}'\mathbf{y}$. Then with (13.21) and (13.22), we can write

$$
\begin{aligned}
SS(\alpha|\mu) &= \hat{\boldsymbol{\beta}}'\mathbf{X}'\mathbf{y} - \frac{y_{..}^2}{kn} \\
&= \mathbf{y}'\mathbf{X}(\mathbf{X}'\mathbf{X})^-\mathbf{X}'\mathbf{y} - \mathbf{y}'\mathbf{j}_{kn}(\mathbf{j}_{kn}'\mathbf{j}_{kn})^{-1}\mathbf{j}_{kn}'\mathbf{y} \\
&= \mathbf{y}'\mathbf{X}(\mathbf{X}'\mathbf{X})^-\mathbf{X}'\mathbf{y} - \mathbf{y}'\left(\frac{\mathbf{j}_{kn}\mathbf{j}_{kn}'}{kn}\right)\mathbf{y} \\
&= \mathbf{y}'\left[\mathbf{X}(\mathbf{X}'\mathbf{X})^-\mathbf{X}' - \frac{1}{kn}\mathbf{J}_{kn}\right]\mathbf{y}.
\end{aligned} \tag{13.25}
$$

Using some results in the answer to Problem 13.3, this can be expressed as

$$
SS(\alpha|\mu) = \mathbf{y}'\left[\frac{1}{n}\begin{pmatrix} \mathbf{J} & \mathbf{O} & \cdots & \mathbf{O} \\ \mathbf{O} & \mathbf{J} & \cdots & \mathbf{O} \\ \vdots & \vdots & & \vdots \\ \mathbf{O} & \mathbf{O} & \cdots & \mathbf{J} \end{pmatrix} - \frac{1}{kn}\begin{pmatrix} \mathbf{J} & \mathbf{J} & \cdots & \mathbf{J} \\ \mathbf{J} & \mathbf{J} & \cdots & \mathbf{J} \\ \vdots & \vdots & & \vdots \\ \mathbf{J} & \mathbf{J} & \cdots & \mathbf{J} \end{pmatrix}\right]\mathbf{y} \tag{13.26}
$$

$$
= \frac{1}{kn}\mathbf{y}'\begin{pmatrix} (k-1)\mathbf{J} & -\mathbf{J} & \cdots & -\mathbf{J} \\ -\mathbf{J} & (k-1)\mathbf{J} & \cdots & -\mathbf{J} \\ \vdots & \vdots & & \vdots \\ -\mathbf{J} & -\mathbf{J} & \cdots & (k-1)\mathbf{J} \end{pmatrix}\mathbf{y},
$$

where each $\mathbf{J}$ in (13.26) and (13.27) is $n \times n$.

TABLE 13.2 Ascorbic Acid (mg/100 g) for Three Packaging Methods

Method	A	B	C
	14.29	20.06	20.04
	19.10	20.64	26.23
	19.09	18.00	22.74
	16.25	19.56	24.04
	15.09	19.47	23.37
	16.61	19.07	25.02
	19.63	18.38	23.27
Totals ($y_{i.}$)	120.06	135.18	164.71
Means ($\bar{y}_i$)	17.15	19.31	23.53

Example 13.4. Three methods of packaging frozen foods were compared by Daniel (1974, p. 196). The response variable was ascorbic acid (mg/100 g). The data are in Table 13.2.

To make the test comparing the means of the three methods, we calculate

$$\frac{y_{..}^2}{kn} = \frac{(419.95)^2}{(3)(7)} = 8298.0001,$$

$$\frac{1}{7}\sum_{i=1}^{3} y_{i.}^2 = \frac{1}{7}\left[(120.06)^2 + (135.18)^2 + (164.71)^2\right]$$

$$= \frac{1}{7}(59,817.4201) = 8545.3457,$$

$$\sum_{i=1}^{3}\sum_{j=1}^{7} y_{ij}^2 = 8600.3127.$$

The sums of squares for treatments, error, and total are then

$$\text{SS}(\alpha|\mu) = \frac{1}{7}\sum_{i=1}^{3} y_{i.}^2 - \frac{y_{..}^2}{21} = 8545.3457 - 8398.0001 = 147.3456,$$

$$\text{SSE} = \sum_{ij} y_{ij}^2 - \frac{1}{7}\sum_{i} y_{i.}^2 = 8600.3127 - 8545.3457 = 54.9670,$$

$$\text{SST} = \sum_{ij} y_{ij}^2 - \frac{y_{..}^2}{21} = 8600.3127 - 8398.0001 = 202.3126.$$

These sums of squares can be used to obtain an F test, as in Table 13.3. The p value for $F = 24.1256$ is 8.07×10^{-6}. Thus we reject $H_0: \mu_1 = \mu_2 = \mu_3$. □

TABLE 13.3 ANOVA for the Ascorbic Acid Data in Table 13.2

Source	df	Sum of Squares	Mean Square	F
Method	2	147.3456	73.6728	24.1256
Error	18	54.9670	3.0537	—
Total	20	202.3126		

13.4.2 General Linear Hypothesis

For simplicity of exposition, we illustrate all results in this section with $k = 4$. In this case, $\boldsymbol{\beta} = (\mu, \alpha_1, \alpha_2, \alpha_3, \alpha_4)'$, and the hypothesis is $H_0 : \alpha_1 = \alpha_2 = \alpha_3 = \alpha_4$. Using three linearly independent estimable contrasts, the hypothesis can be written in the form

$$H_0 : \begin{pmatrix} \alpha_1 - \alpha_2 \\ \alpha_1 - \alpha_3 \\ \alpha_1 - \alpha_4 \end{pmatrix} = \begin{pmatrix} 0 \\ 0 \\ 0 \end{pmatrix},$$

which can be expressed as $H_0 : \mathbf{C}\boldsymbol{\beta} = \mathbf{0}$, where

$$\mathbf{C} = \begin{pmatrix} 0 & 1 & -1 & 0 & 0 \\ 0 & 1 & 0 & -1 & 0 \\ 0 & 1 & 0 & 0 & -1 \end{pmatrix}. \tag{13.28}$$

The matrix $\mathbf{C}$ in (13.28) used to express $H_0 : \alpha_1 = \alpha_2 = \alpha_3 = \alpha_4$ is not unique. Other contrasts could be used in $\mathbf{C}$, for example

$$\mathbf{C}_1 = \begin{pmatrix} 0 & 1 & -1 & 0 & 0 \\ 0 & 0 & 1 & -1 & 0 \\ 0 & 0 & 0 & 1 & -1 \end{pmatrix} \quad \text{or} \quad \mathbf{C}_2 = \begin{pmatrix} 0 & 1 & 1 & -1 & -1 \\ 0 & 1 & -1 & 0 & 0 \\ 0 & 0 & 0 & 1 & -1 \end{pmatrix}.$$

From (12.13) and Theorem 12.7b(iii), we have

$$\begin{aligned} \text{SSH} &= (\mathbf{C}\hat{\boldsymbol{\beta}})'[\mathbf{C}(\mathbf{X}'\mathbf{X})^{-}\mathbf{C}']^{-1}\mathbf{C}\hat{\boldsymbol{\beta}} \\ &= \mathbf{y}'\mathbf{X}(\mathbf{X}'\mathbf{X})^{-}\mathbf{C}'[\mathbf{C}(\mathbf{X}'\mathbf{X})^{-}\mathbf{C}']^{-1}\mathbf{C}(\mathbf{X}'\mathbf{X})^{-}\mathbf{X}'\mathbf{y}. \end{aligned} \tag{13.29}$$

Using $\mathbf{C}$ in (13.28) and $(\mathbf{X}'\mathbf{X})^-$ in (13.11), we obtain

$$\mathbf{C}(\mathbf{X}'\mathbf{X})^-\mathbf{C}' = \frac{1}{n}\begin{pmatrix} 0 & 1 & -1 & 0 & 0 \\ 0 & 1 & 0 & -1 & 0 \\ 0 & 1 & 0 & 0 & -1 \end{pmatrix}\begin{pmatrix} 0 & 0 & 0 & 0 & 0 \\ 0 & 1 & 0 & 0 & 0 \\ 0 & 0 & 1 & 0 & 0 \\ 0 & 0 & 0 & 1 & 0 \\ 0 & 0 & 0 & 0 & 1 \end{pmatrix}\begin{pmatrix} 0 & 0 & 0 \\ 1 & 1 & 1 \\ -1 & 0 & 0 \\ 0 & -1 & 0 \\ 0 & 0 & -1 \end{pmatrix}$$

$$= \frac{1}{n}\begin{pmatrix} 2 & 1 & 1 \\ 1 & 2 & 1 \\ 1 & 1 & 2 \end{pmatrix}. \tag{13.30}$$

To find the inverse of (13.30), we write it in the form

$$\mathbf{C}(\mathbf{X}'\mathbf{X})^-\mathbf{C}' = \frac{1}{n}\left[\begin{pmatrix} 1 & 0 & 0 \\ 0 & 1 & 0 \\ 0 & 0 & 1 \end{pmatrix} + \begin{pmatrix} 1 & 1 & 1 \\ 1 & 1 & 1 \\ 1 & 1 & 1 \end{pmatrix}\right] = \frac{1}{n}(\mathbf{I}_3 + \mathbf{j}_3\mathbf{j}_3').$$

Then by (2.53), the inverse is

$$[\mathbf{C}(\mathbf{X}'\mathbf{X})^-\mathbf{C}']^{-1} = n\left(\mathbf{I}_3 - \frac{\mathbf{I}_3^{-1}\mathbf{j}_3\mathbf{j}_3'\mathbf{I}_3^{-1}}{1 + \mathbf{j}_3'\mathbf{I}_3^{-1}\mathbf{j}_3}\right)$$

$$= n\left(\mathbf{I}_3 - \frac{1}{4}\mathbf{J}_3\right), \tag{13.31}$$

where $\mathbf{J}_3$ is 3×3.

For $\mathbf{C}(\mathbf{X}'\mathbf{X})^-\mathbf{X}'$ in (13.29), we obtain

$$\mathbf{C}(\mathbf{X}'\mathbf{X})^-\mathbf{X}' = \frac{1}{n}\begin{pmatrix} \mathbf{j}_n' & -\mathbf{j}_n' & \mathbf{0}' & \mathbf{0}' \\ \mathbf{j}_n' & \mathbf{0}' & -\mathbf{j}_n' & \mathbf{0}' \\ \mathbf{j}_n' & \mathbf{0}' & \mathbf{0}' & -\mathbf{j}_n' \end{pmatrix} = \frac{1}{n}\mathbf{A}, \tag{13.32}$$

where $\mathbf{j}_n'$ and $\mathbf{0}'$ are $1 \times n$.

Using (13.31) and (13.32), the matrix of the quadratic form for SSH in (13.29) can be expressed as

$$\mathbf{X}(\mathbf{X}'\mathbf{X})^-\mathbf{C}'[\mathbf{C}(\mathbf{X}'\mathbf{X})^-\mathbf{C}']^{-1}\mathbf{C}(\mathbf{X}'\mathbf{X})^-\mathbf{X}' = \frac{1}{n}\mathbf{A}'n\left(\mathbf{I}_3 - \frac{1}{4}\mathbf{J}_3\right)\frac{1}{n}\mathbf{A}$$

$$= \frac{1}{n}\mathbf{A}'\mathbf{I}_3\mathbf{A} - \frac{1}{4n}\mathbf{A}'\mathbf{J}_3\mathbf{A}. \tag{13.33}$$

The first term of (13.33) is given by

$$\frac{1}{n}\mathbf{A}'\mathbf{A} = \frac{1}{n}\begin{pmatrix} \mathbf{j}_n & \mathbf{j}_n & \mathbf{j}_n \\ -\mathbf{j}_n & \mathbf{0} & \mathbf{0} \\ \mathbf{0} & -\mathbf{j}_n & \mathbf{0} \\ \mathbf{0} & \mathbf{0} & -\mathbf{j}_n \end{pmatrix}\begin{pmatrix} \mathbf{j}_n' & -\mathbf{j}_n' & \mathbf{0}' & \mathbf{0}' \\ \mathbf{j}_n' & \mathbf{0}' & -\mathbf{j}_n' & \mathbf{0}' \\ \mathbf{j}_n' & \mathbf{0}' & \mathbf{0}' & -\mathbf{j}_n' \end{pmatrix}$$

$$= \frac{1}{n}\begin{pmatrix} 3\mathbf{J}_n & -\mathbf{J}_n & -\mathbf{J}_n & -\mathbf{J}_n \\ -\mathbf{J}_n & \mathbf{J}_n & \mathbf{O} & \mathbf{O} \\ -\mathbf{J}_n & \mathbf{O} & \mathbf{J}_n & \mathbf{O} \\ -\mathbf{J}_n & \mathbf{O} & \mathbf{O} & \mathbf{J}_n \end{pmatrix}, \tag{13.34}$$

since $\mathbf{j}_n\mathbf{j}_n' = \mathbf{J}_n$ and $\mathbf{j}_n\mathbf{0}' = \mathbf{O}$, where $\mathbf{O}$ is $n \times n$. Similarly (see Problem 13.10), the second term of (13.33) is given by

$$\frac{1}{4n}\mathbf{A}'\mathbf{J}_3\mathbf{A} = \frac{1}{4n}\begin{pmatrix} 9\mathbf{J}_n & -3\mathbf{J}_n & -3\mathbf{J}_n & -3\mathbf{J}_n \\ -3\mathbf{J}_n & \mathbf{J}_n & \mathbf{J}_n & \mathbf{J}_n \\ -3\mathbf{J}_n & \mathbf{J}_n & \mathbf{J}_n & \mathbf{J}_n \\ -3\mathbf{J}_n & \mathbf{J}_n & \mathbf{J}_n & \mathbf{J}_n \end{pmatrix}. \tag{13.35}$$

Then (13.33) becomes

$$\frac{1}{4n}(4\mathbf{A}'\mathbf{A}) - \frac{1}{4n}\mathbf{A}'\mathbf{J}_3\mathbf{A} = \frac{1}{4n}\begin{pmatrix} 12\mathbf{J}_n & -4\mathbf{J}_n & -4\mathbf{J}_n & -4\mathbf{J}_n \\ -4\mathbf{J}_n & 4\mathbf{J}_n & \mathbf{O} & \mathbf{O} \\ -4\mathbf{J}_n & \mathbf{O} & 4\mathbf{J}_n & \mathbf{O} \\ -4\mathbf{J}_n & \mathbf{O} & \mathbf{O} & 4\mathbf{J}_n \end{pmatrix}$$

$$- \frac{1}{4n}\begin{pmatrix} 9\mathbf{J}_n & -3\mathbf{J}_n & -3\mathbf{J}_n & -3\mathbf{J}_n \\ -3\mathbf{J}_n & \mathbf{J}_n & \mathbf{J}_n & \mathbf{J}_n \\ -3\mathbf{J}_n & \mathbf{J}_n & \mathbf{J}_n & \mathbf{J}_n \\ -3\mathbf{J}_n & \mathbf{J}_n & \mathbf{J}_n & \mathbf{J}_n \end{pmatrix}$$

$$= \frac{1}{4n}\begin{pmatrix} 3\mathbf{J}_n & -\mathbf{J}_n & -\mathbf{J}_n & -\mathbf{J}_n \\ -\mathbf{J}_n & 3\mathbf{J}_n & -\mathbf{J}_n & -\mathbf{J}_n \\ -\mathbf{J}_n & -\mathbf{J}_n & 3\mathbf{J}_n & -\mathbf{J}_n \\ -\mathbf{J}_n & -\mathbf{J}_n & -\mathbf{J}_n & 3\mathbf{J}_n \end{pmatrix} = \frac{1}{4n}\mathbf{B}. \tag{13.36}$$

Note that the matrix for SSH in (13.36) is the same as the matrix for $SS(\alpha|\mu)$ in (13.27) with $k = 4$.

For completeness, we now express SSH in (13.29) in terms of the y_{ij}'s. We begin by writing (13 .36) in the form

$$\frac{1}{4n}\mathbf{B} = \frac{1}{4n}\begin{pmatrix} 4\mathbf{J}_n & \mathbf{O} & \mathbf{O} & \mathbf{O} \\ \mathbf{O} & 4\mathbf{J}_n & \mathbf{O} & \mathbf{O} \\ \mathbf{O} & \mathbf{O} & 4\mathbf{J}_n & \mathbf{O} \\ \mathbf{O} & \mathbf{O} & \mathbf{O} & 4\mathbf{J}_n \end{pmatrix} - \frac{1}{4n}\begin{pmatrix} \mathbf{J}_n & \mathbf{J}_n & \mathbf{J}_n & \mathbf{J}_n \\ \mathbf{J}_n & \mathbf{J}_n & \mathbf{J}_n & \mathbf{J}_n \\ \mathbf{J}_n & \mathbf{J}_n & \mathbf{J}_n & \mathbf{J}_n \\ \mathbf{J}_n & \mathbf{J}_n & \mathbf{J}_n & \mathbf{J}_n \end{pmatrix}$$

$$= \frac{1}{n}\begin{pmatrix} \mathbf{J}_n & \mathbf{O} & \mathbf{O} & \mathbf{O} \\ \mathbf{O} & \mathbf{J}_n & \mathbf{O} & \mathbf{O} \\ \mathbf{O} & \mathbf{O} & \mathbf{J}_n & \mathbf{O} \\ \mathbf{O} & \mathbf{O} & \mathbf{O} & \mathbf{J}_n \end{pmatrix} - \frac{1}{4n}\mathbf{J}_{4n}.$$

Using $\mathbf{y}' = (\mathbf{y}_1', \mathbf{y}_2', \mathbf{y}_3', \mathbf{y}_4')$ as defined in (13.6), SSH in (13.29) becomes

$$\begin{aligned} \text{SSH} &= \mathbf{y}'\mathbf{X}(\mathbf{X}'\mathbf{X})^-\mathbf{C}'[\mathbf{C}(\mathbf{X}'\mathbf{X})^-\mathbf{C}']^{-1}\mathbf{C}(\mathbf{X}'\mathbf{X})^-\mathbf{X}'\mathbf{y} \\ &= \mathbf{y}'\left(\frac{1}{4n}\mathbf{B}\right)\mathbf{y} \\ &= \frac{1}{n}(\mathbf{y}_1', \mathbf{y}_2', \mathbf{y}_3', \mathbf{y}_4')\begin{pmatrix} \mathbf{J}_n & \mathbf{O} & \mathbf{O} & \mathbf{O} \\ \mathbf{O} & \mathbf{J}_n & \mathbf{O} & \mathbf{O} \\ \mathbf{O} & \mathbf{O} & \mathbf{J}_n & \mathbf{O} \\ \mathbf{O} & \mathbf{O} & \mathbf{O} & \mathbf{J}_n \end{pmatrix}\begin{pmatrix} \mathbf{y}_1 \\ \mathbf{y}_2 \\ \mathbf{y}_3 \\ \mathbf{y}_4 \end{pmatrix} - \frac{1}{4n}\mathbf{y}'\mathbf{J}_{4n}\mathbf{y} \\ &= \frac{1}{n}\sum_{i=1}^{4}\mathbf{y}_i'\mathbf{J}_n\mathbf{y}_i - \frac{1}{4n}\mathbf{y}'\mathbf{J}_{4n}\mathbf{y} \\ &= \frac{1}{n}\sum_{i=1}^{4}\mathbf{y}_i'\mathbf{j}_n\mathbf{j}_n'\mathbf{y}_i - \frac{1}{4n}\mathbf{y}'\mathbf{j}_{4n}\mathbf{j}_{4n}'\mathbf{y} \\ &= \frac{1}{n}\sum_{i=1}^{4}y_{i.}^2 - \frac{1}{4n}y_{..}^2, \end{aligned}$$

which is the same as $\text{SS}(\alpha|\mu)$ in (13.23).

13.5 EXPECTED MEAN SQUARES

The expected mean squares for a one-way ANOVA are given in Table 13.4. The expected mean squares are defined as $E[\text{SS}(\alpha|\mu)/(k-1)]$ and $E[\text{SSE}/k(n-1)]$. The result is given in terms of parameters α_i^* such that $\sum_i \alpha_i^* = 0$.

TABLE 13.4 Expected Mean Squares for One-Way ANOVA

Source of Variation	df	Sum of Squares	Mean Square	Expected Mean Squares
Treatments	$k-1$	$\text{SS}(\alpha\|\mu)$	$\frac{\text{SS}(\alpha\|\mu)}{k-1}$	$\sigma^2 + \frac{n}{k-1}\sum_{i=1}^k \alpha_i^{*2}$
Error	$k(n-1)$	SSE	$\frac{\text{SSE}}{k(n-1)}$	σ^2
Total	$kn-1$	$\sum_{ij} y_{ij}^2 - \frac{y_{..}^2}{kn}$		

If $H_0: \alpha_1^* = \alpha_2^* = \cdots = \alpha_k^* = 0$ is true, both of the expected mean squares are equal to σ^2, and we expect F to be close to 1. On the other hand, if H_0 is false, $E[\text{SS}(\alpha|\mu)/(k-1)] > E[\text{SSE}/k(n-1)]$, and we expect F to exceed 1. We therefore reject H_0 for large values of F.

The expected mean squares in Table 13.4 can be derived using the model $y_{ij} = \mu^* + \alpha_i^* + \varepsilon_{ij}$ in $E[\text{SS}(\alpha|\mu)]$ and E(SSE) (see Problem 13.11). In Sections 13.5.1 and 13.5.2, we obtain the expected mean squares using matrix methods similar to those in Sections 13.4.1 and 13.4.2.

13.5.1 Full–Reduced-Model Approach

For the error term in Table 13.4, we have

$$E(\text{SSE}) = E\{\mathbf{y}'[\mathbf{I} - \mathbf{X}(\mathbf{X}'\mathbf{X})^-\mathbf{X}']\mathbf{y}\} = k(n-1)\sigma^2, \tag{13.37}$$

which was proved in Theorem 12.3e(i).

Using a full–reduced-model approach the sum of squares for the α's adjusted for μ is given by (13.25) as $\text{SS}(\alpha|\mu) = \mathbf{y}'\mathbf{X}(\mathbf{X}'\mathbf{X})^-\mathbf{X}'\mathbf{y} - \mathbf{y}'[(1/kn)\mathbf{J}_{kn}]\mathbf{y}$. Thus

$$E[\text{SS}(\alpha|\mu)] = E[\mathbf{y}'\mathbf{X}(\mathbf{X}'\mathbf{X})^-\mathbf{X}'\mathbf{y}] - E\left[\mathbf{y}'\left(\frac{1}{kn}\mathbf{J}_{kn}\right)\mathbf{y}\right]. \tag{13.38}$$

Using Theorem 5.2a, the first term on the right side of (13.38) becomes

$$\begin{aligned} E[\mathbf{y}'\mathbf{X}(\mathbf{X}'\mathbf{X})^-\mathbf{X}'\mathbf{y}] &= \text{tr}[\mathbf{X}(\mathbf{X}'\mathbf{X})^-\mathbf{X}'\sigma^2\mathbf{I}] + (\mathbf{X}\boldsymbol{\beta})'\mathbf{X}(\mathbf{X}'\mathbf{X})^-\mathbf{X}'(\mathbf{X}\boldsymbol{\beta}) \\ &= \sigma^2\text{tr}[\mathbf{X}(\mathbf{X}'\mathbf{X})^-\mathbf{X}'] + \boldsymbol{\beta}'\mathbf{X}'\mathbf{X}(\mathbf{X}'\mathbf{X})^-\mathbf{X}'\mathbf{X}\boldsymbol{\beta} \\ &= \sigma^2\text{tr}[\mathbf{X}(\mathbf{X}'\mathbf{X})^-\mathbf{X}'] + \boldsymbol{\beta}'\mathbf{X}'\mathbf{X}\boldsymbol{\beta} \qquad \text{[by (2.58)]}. \end{aligned} \tag{13.39}$$

By Theorem 2.13f, the matrix $\mathbf{X}(\mathbf{X}'\mathbf{X})^-\mathbf{X}'$ is idempotent. Hence, by Theorems 2.13d and 2.8c(v), we obtain

$$\text{tr}[\mathbf{X}(\mathbf{X}'\mathbf{X})^-\mathbf{X}'] = \text{rank}[\mathbf{X}(\mathbf{X}'\mathbf{X})^-\mathbf{X}] = \text{rank}(\mathbf{X}) = k. \tag{13.40}$$

To evaluate the second term on the right side of (13.39), we use $\mathbf{X}'\mathbf{X}$ in (13.7) and use $\boldsymbol{\beta}' = (\mu^*, \alpha_1^*, \ldots, \alpha_k^*)$ subject to $\sum_i \alpha_i^* = 0$. Then

$$\boldsymbol{\beta}'\mathbf{X}'\mathbf{X}\boldsymbol{\beta} = n(\mu^*, \alpha_1^*, \ldots, \alpha_k^*) \begin{pmatrix} k & 1 & 1 & \ldots & 1 \\ 1 & 1 & 0 & \ldots & 0 \\ 1 & 0 & 1 & \ldots & 0 \\ \vdots & \vdots & \vdots & & \vdots \\ 1 & 0 & 0 & \ldots & 1 \end{pmatrix} \begin{pmatrix} \mu^* \\ \alpha_1^* \\ \vdots \\ \alpha_k^* \end{pmatrix}$$

$$= n\left(k\mu^* + \sum_i \alpha_i^*, \mu^* + \alpha_1^*, \ldots, \mu^* + \alpha_k^*\right) \begin{pmatrix} \mu^* \\ \alpha_1^* \\ \vdots \\ \alpha_k^* \end{pmatrix}$$

$$= n\left[k\mu^{*2} + \sum_i (\mu^* + \alpha_i^*)\alpha_i^*\right]$$

$$= n\left(k\mu^{*2} + \mu^* \sum_i \alpha_i^* + \sum_i \alpha_i^{*2}\right)$$

$$= kn\mu^{*2} + n\sum_i \alpha_i^{*2}. \tag{13.41}$$

Hence, using (13.40) and (13.41), $E[\mathbf{y}'\mathbf{X}(\mathbf{X}'\mathbf{X})^-\mathbf{X}'\mathbf{y}]$ in (13.39) becomes

$$E[\mathbf{y}'\mathbf{X}(\mathbf{X}'\mathbf{X})^-\mathbf{X}'\mathbf{y}] = k\sigma^2 + kn\mu^{*2} + n\sum_i \alpha_i^{*2}. \tag{13.42}$$

For the second term on the right side of (13.38), we obtain

$$\begin{aligned} E\left[\mathbf{y}'\left(\frac{1}{kn}\mathbf{J}_{kn}\right)\mathbf{y}\right] &= \sigma^2 \mathrm{tr}\left(\frac{1}{kn}\mathbf{J}_{kn}\right) + \boldsymbol{\beta}'\mathbf{X}'\left(\frac{1}{kn}\mathbf{J}_{kn}\right)\mathbf{X}\boldsymbol{\beta} \\ &= \frac{\sigma^2 kn}{kn} + \frac{1}{kn}\boldsymbol{\beta}'\mathbf{X}'\mathbf{j}_{kn}\mathbf{j}_{kn}'\mathbf{X}\boldsymbol{\beta} \\ &= \sigma^2 + \frac{1}{kn}(\boldsymbol{\beta}'\mathbf{X}'\mathbf{j}_{kn})(\mathbf{j}_{kn}'\mathbf{X}\boldsymbol{\beta}). \end{aligned} \tag{13.43}$$

Using $\mathbf{X}$ as given in (13.6), $\mathbf{j}'_{kn}\mathbf{X}\boldsymbol{\beta}$ becomes

$$\mathbf{j}'_{kn}\mathbf{X}\boldsymbol{\beta} = (\mathbf{j}'_n, \mathbf{j}'_n, \ldots, \mathbf{j}'_n)\begin{pmatrix} \mathbf{j}_n & \mathbf{j}_n & \mathbf{0} & \cdots & \mathbf{0} \\ \mathbf{j}_n & \mathbf{0} & \mathbf{j}_n & \cdots & \mathbf{0} \\ \vdots & \vdots & \vdots & & \vdots \\ \mathbf{j}_n & \mathbf{0} & \mathbf{0} & \cdots & \mathbf{j}_n \end{pmatrix}\begin{pmatrix} \mu^* \\ \alpha_1^* \\ \vdots \\ \alpha_k^* \end{pmatrix}$$

$$= (kn, n, n, \ldots, n)\begin{pmatrix} \mu^* \\ \alpha_1^* \\ \vdots \\ \alpha_k^* \end{pmatrix} \qquad (\text{since } \mathbf{j}'_n\mathbf{j}_n = n)$$

$$= kn\mu^* + n\sum_{i=1}^{k}\alpha_i^* = kn\mu^* \qquad \left(\text{since } \sum_i \alpha_i^* = 0\right).$$

The second term on the right side of (13.43) is then given by

$$\frac{1}{kn}(\boldsymbol{\beta}'\mathbf{X}'\mathbf{j}_{kn})(\mathbf{j}'_{kn}\mathbf{X}\boldsymbol{\beta}) = \frac{1}{kn}(\mathbf{j}'\mathbf{X}\boldsymbol{\beta})^2 = \frac{k^2n^2\mu^{*2}}{kn} = kn\mu^{*2},$$

so that (13.43) becomes

$$E\left[\mathbf{y}'\left(\frac{1}{kn}\mathbf{J}_{kn}\right)\mathbf{y}\right] = \sigma^2 + kn\mu^{*2}. \tag{13.44}$$

Now, using (13.42) and (13.44), $E[\text{SS}(\alpha|\mu)]$ in (13.38) becomes

$$E[\text{SS}(\alpha|\mu)] = k\sigma^2 + kn\mu^{*2} + n\sum_{i=1}^{k}\alpha_i^{*2} - (\sigma^2 + kn\mu^{*2})$$

$$= (k-1)\sigma^2 + n\sum_i \alpha_i^{*2}. \tag{13.45}$$

13.5.2 General Linear Hypothesis

To simplify exposition, we use $k = 4$ to illustrate results in this section, as was done in Section 13.4.2. It was shown in Section 13.4.2 that $\text{SSH} = (\mathbf{C}\hat{\boldsymbol{\beta}})'[\mathbf{C}(\mathbf{X}'\mathbf{X})^-\mathbf{C}']^{-1}\mathbf{C}\hat{\boldsymbol{\beta}}$ is the same as $\text{SS}(\alpha|\mu) = \sum_i y_{i.}^2/n - y_{..}^2/kn$ in (13.23). Note that for $k = 4$, $\mathbf{C}$ is 3×5 [see (13.28)] and $\mathbf{C}(\mathbf{X}'\mathbf{X})^-\mathbf{C}'$ is 3×3 [see (13.30)]. To obtain $E[\text{SS}(\alpha|\mu)]$, we first note that by (12.44), (12.45), and (13.31), $E(\mathbf{C}\hat{\boldsymbol{\beta}}) = \mathbf{C}\boldsymbol{\beta}$, $\text{cov}(\mathbf{C}\hat{\boldsymbol{\beta}}) = \sigma^2\mathbf{C}(\mathbf{X}'\mathbf{X})^-\mathbf{C}'$, and $[\mathbf{C}(\mathbf{X}'\mathbf{X})^-\mathbf{C}']^{-1} = n(\mathbf{I}_3 - \frac{1}{4}\mathbf{J}_3)$.

Then, by Theorem 5.2a, we have

$$\begin{aligned}
E[\mathrm{SS}(\alpha|\mu)] &= E\{(\mathbf{C}\hat{\boldsymbol{\beta}})'[\mathbf{C}(\mathbf{X}'\mathbf{X})^{-}\mathbf{C}']^{-1}\mathbf{C}\hat{\boldsymbol{\beta}}\} \\
&= \mathrm{tr}\{[\mathbf{C}(\mathbf{X}'\mathbf{X})^{-}\mathbf{C}']^{-1}\mathrm{cov}(\mathbf{C}\hat{\boldsymbol{\beta}})\} + [E(\mathbf{C}\hat{\boldsymbol{\beta}})]'[\mathbf{C}(\mathbf{X}'\mathbf{X})^{-}\mathbf{C}']^{-1}E(\mathbf{C}\hat{\boldsymbol{\beta}}) \\
&= \mathrm{tr}\{[\mathbf{C}(\mathbf{X}'\mathbf{X})^{-}\mathbf{C}']^{-1}\sigma^2\mathbf{C}(\mathbf{X}'\mathbf{X})^{-}\mathbf{C}'\} + n(\mathbf{C}\boldsymbol{\beta})'[\mathbf{I}_3 - \tfrac{1}{4}\mathbf{J}_3]\mathbf{C}\boldsymbol{\beta} \\
&= \sigma^2\mathrm{tr}(\mathbf{I}_3) + n\boldsymbol{\beta}'\mathbf{C}'(\mathbf{I}_3 - \tfrac{1}{4}\mathbf{J}_3)\mathbf{C}\boldsymbol{\beta} \\
&= 3\sigma^2 + n\boldsymbol{\beta}'(\mathbf{C}'\mathbf{C} - \tfrac{1}{4}\mathbf{C}'\mathbf{J}_3\mathbf{C})\boldsymbol{\beta}. \qquad (13.46)
\end{aligned}$$

Using $\mathbf{C}$ in (13.28), we obtain

$$\mathbf{C}'\mathbf{C} = \begin{pmatrix} 0 & 0 & 0 & 0 & 0 \\ 0 & 3 & -1 & -1 & -1 \\ 0 & -1 & 1 & 0 & 0 \\ 0 & -1 & 0 & 1 & 0 \\ 0 & -1 & 0 & 0 & 1 \end{pmatrix}, \qquad (13.47)$$

$$\mathbf{C}'\mathbf{J}_3\mathbf{C} = \begin{pmatrix} 0 & 0 & 0 & 0 & 0 \\ 0 & 9 & -3 & -3 & -3 \\ 0 & -3 & 1 & 1 & 1 \\ 0 & -3 & 1 & 1 & 1 \\ 0 & -3 & 1 & 1 & 1 \end{pmatrix}. \qquad (13.48)$$

From (13.47) and (13.48), we have

$$\begin{aligned}
\mathbf{C}'\mathbf{C} - \tfrac{1}{4}\mathbf{C}'\mathbf{J}_3\mathbf{C} &= \tfrac{1}{4}(4\mathbf{C}'\mathbf{C} - \mathbf{C}'\mathbf{J}_3\mathbf{C}) \\
&= \tfrac{1}{4}\begin{pmatrix} 0 & 0 & 0 & 0 & 0 \\ 0 & 3 & -1 & -1 & -1 \\ 0 & -1 & 3 & -1 & -1 \\ 0 & -1 & -1 & 3 & -1 \\ 0 & -1 & -1 & -1 & 3 \end{pmatrix} \\
&= \tfrac{1}{4}\begin{pmatrix} 0 & 0 & 0 & 0 & 0 \\ 0 & 4 & 0 & 0 & 0 \\ 0 & 0 & 4 & 0 & 0 \\ 0 & 0 & 0 & 4 & 0 \\ 0 & 0 & 0 & 0 & 4 \end{pmatrix} - \tfrac{1}{4}\begin{pmatrix} 0 & 0 & 0 & 0 & 0 \\ 0 & 1 & 1 & 1 & 1 \\ 0 & 1 & 1 & 1 & 1 \\ 0 & 1 & 1 & 1 & 1 \\ 0 & 1 & 1 & 1 & 1 \end{pmatrix} \\
&= \begin{pmatrix} 0 & \mathbf{0}' \\ \mathbf{0} & \mathbf{I}_4 \end{pmatrix} - \tfrac{1}{4}\begin{pmatrix} 0 & \mathbf{0}' \\ \mathbf{0} & \mathbf{J}_4 \end{pmatrix}.
\end{aligned}$$

Thus the second term on the right side of (13.46) is given by

$$
\begin{aligned}
& n\boldsymbol{\beta}'(\mathbf{C}'\mathbf{C} - \tfrac{1}{4}\mathbf{C}'\mathbf{J}_3\mathbf{C})\boldsymbol{\beta} \\
&\quad = n\boldsymbol{\beta}'\begin{pmatrix} 0 & \mathbf{0}' \\ \mathbf{0} & \mathbf{I}_4 \end{pmatrix}\boldsymbol{\beta} - \tfrac{1}{4}n\boldsymbol{\beta}'\begin{pmatrix} 0 & \mathbf{0}' \\ \mathbf{0} & \mathbf{J}_4 \end{pmatrix}\boldsymbol{\beta} \\
&\quad = n(\mu^*, \alpha_1^*, \alpha_2^*, \alpha_3^*, \alpha_4^*)\begin{pmatrix} 0 & \mathbf{0}' \\ \mathbf{0} & \mathbf{I}_4 \end{pmatrix}\begin{pmatrix} \mu^* \\ \alpha_1^* \\ \alpha_2^* \\ \alpha_3^* \\ \alpha_4^* \end{pmatrix} \\
&\qquad - \tfrac{1}{4}n(\mu^*, \alpha_1^*, \alpha_2^*, \alpha_3^*, \alpha_4^*)\begin{pmatrix} 0 & \mathbf{0}' \\ \mathbf{0} & \mathbf{J}_4 \end{pmatrix}\begin{pmatrix} \mu^* \\ \alpha_1^* \\ \alpha_2^* \\ \alpha_3^* \\ \alpha_4^* \end{pmatrix} \\
&\quad = n\sum_{i=1}^{4}\alpha_i^{*2} - \tfrac{1}{4}n\left(0, \sum_i \alpha_i^*, \sum_i \alpha_i^*, \sum_i \alpha_i^*, \sum_i \alpha_i^*\right)\begin{pmatrix} \mu^* \\ \alpha_1^* \\ \alpha_2^* \\ \alpha_3^* \\ \alpha_4^* \end{pmatrix} \\
&\quad = n\sum_{i=1}^{4}\alpha_i^{*2}.
\end{aligned}
$$

Hence, (13.46) becomes

$$
E[\mathrm{SS}(\alpha|\mu)] = 3\sigma^2 + n\sum_{i=1}^{4}\alpha_i^{*2}. \tag{13.49}
$$

This result is for the special case $k = 4$. For a general k, (13.49) becomes

$$
E[\mathrm{SS}(\alpha|\mu)] = (k-1)\sigma^2 + n\sum_{i=1}^{k}\alpha_i^{*2}.
$$

For the case in which $\boldsymbol{\beta}' = (\boldsymbol{\mu}, \alpha_1, \ldots, \alpha_k)$ is not subject to $\sum_i \alpha_i = 0$, see Problem 13.14.

13.6 CONTRASTS

We noted in Section 13.2 that a linear combination $\sum_{i=1}^{k} c_i\alpha_i$ in the α's is estimable if and only if $\sum_{i=1}^{k} c_i = 0$. In Section 13.6.1, we develop a test of significance for such contrasts. In Section 13.6.2, we show that if the contrasts are formulated appropriately, the sum of squares for treatments can be partitioned into $k-1$ independent sums of squares for contrasts. In Section 13.6.3, we develop orthogonal polynomial contrasts for the special case in which the treatments have equally spaced quantitative levels.

13.6.1 Hypothesis Test for a Contrast

For the one-way model, a contrast $\sum_i c_i\alpha_i$, where $\sum_i c_i = 0$, is equivalent to $\sum_i c_i\mu_i$ since

$$\sum c_i\mu_i = \sum_i c_i(\mu + \alpha_i) = \mu \sum_i c_i + \sum_i c_i\alpha_i = \sum_i c_i\alpha_i.$$

A hypothesis of interest is

$$H_0: \sum c_i\alpha_i = 0 \quad \text{or} \quad H_0: \sum c_i\mu_i = 0, \tag{13.50}$$

which represents a comparison of means if $\sum_i c_i = 0$. For example, the hypothesis

$$H_0: 3\mu_1 - \mu_2 - \mu_3 - \mu_4 = 0$$

can be written as

$$H_0: \mu_1 = \tfrac{1}{3}(\mu_2 + \mu_3 + \mu_4),$$

which compares μ_1 with the average of μ_2, μ_3, and μ_4.

The hypothesis in (13.50) can be expressed as $H_0: \mathbf{c}'\boldsymbol{\beta} = 0$, where $\mathbf{c}' = (0, c_1, c_2, \ldots, c_k)$ and $\boldsymbol{\beta} = (\mu, \alpha_1, \ldots, \alpha_k)'$. Assuming that $\mathbf{y}$ is $N_{kn}(\mathbf{X}\boldsymbol{\beta}, \sigma^2\mathbf{I})$, H_0 can be tested using Theorem 12.7c. In this case, we have $m = 1$, and the test statistic becomes

$$\begin{aligned} F &= \frac{(\mathbf{c}'\hat{\boldsymbol{\beta}})'[\mathbf{c}'(\mathbf{X}'\mathbf{X})^-\mathbf{c}]^{-1}\mathbf{c}'\hat{\boldsymbol{\beta}}}{\text{SSE}/k(n-1)} \\ &= \frac{(\mathbf{c}'\hat{\boldsymbol{\beta}})^2}{s^2\mathbf{c}'(\mathbf{X}'\mathbf{X})^-\mathbf{c}} \qquad (13.51) \\ &= \frac{\left(\sum_{i=1}^{k} c_i\bar{y}_{i.}\right)^2}{s^2\sum_{i=1}^{k} c_i^2/n}, \qquad (13.52) \end{aligned}$$

where $s^2 = \text{SSE}/k(n-1)$, and $(\mathbf{X}'\mathbf{X})^-$ and $\hat{\boldsymbol{\beta}}$ are as given by (13.11) and (13.12). The sum of squares for the contrast is $(\mathbf{c}'\hat{\boldsymbol{\beta}})^2/\mathbf{c}'(\mathbf{X}'\mathbf{X})^-\mathbf{c}$ or $n(\sum_i c_i\bar{y}_{i.})^2/(\sum_i c_i^2)$.

13.6.2 Orthogonal Contrasts

Two contrasts $\mathbf{c}_i'\hat{\boldsymbol{\beta}}$ and $\mathbf{c}_j'\hat{\boldsymbol{\beta}}$ are said to be *orthogonal* if $\mathbf{c}_i'\mathbf{c}_j = 0$. We now show that if $\mathbf{c}_i'\hat{\boldsymbol{\beta}}$ and $\mathbf{c}_j'\hat{\boldsymbol{\beta}}$ are orthogonal, they are independent. Since we are assuming normality, $\mathbf{c}_i'\hat{\boldsymbol{\beta}}$ and $\mathbf{c}_j'\hat{\boldsymbol{\beta}}$ are independent if

$$\operatorname{cov}(\mathbf{c}_i'\hat{\boldsymbol{\beta}}, \mathbf{c}_j'\hat{\boldsymbol{\beta}}) = 0 \tag{13.53}$$

(see Problem 13.16). By Theorem 12.3c, $\operatorname{cov}(\mathbf{c}_i'\hat{\boldsymbol{\beta}}, \mathbf{c}_j'\hat{\boldsymbol{\beta}}) = \sigma^2\mathbf{c}_i'(\mathbf{X}'\mathbf{X})^-\mathbf{c}_j$. By (13.11), $(\mathbf{X}'\mathbf{X})^- = \operatorname{diag}[0, (1/n), \ldots, (1/n)]$, and therefore

$$\operatorname{cov}(\mathbf{c}_i'\hat{\boldsymbol{\beta}}, \mathbf{c}_j'\hat{\boldsymbol{\beta}}) = \mathbf{c}_i'(\mathbf{X}'\mathbf{X})^-\mathbf{c}_j = 0 \quad \text{if} \quad \mathbf{c}_i'\mathbf{c}_j = 0 \tag{13.54}$$

(assuming that the first element of $\mathbf{c}_i$ is 0 for all i). By an argument similar to that used in the proofs of Corollary 1 to Theorem 5.6b and in Theorem 12.7b(v), the sums of squares $(\mathbf{c}_i'\hat{\boldsymbol{\beta}})^2/\mathbf{c}_i'(\mathbf{X}'\mathbf{X})^-\mathbf{c}_i$ and $(\mathbf{c}_j'\hat{\boldsymbol{\beta}})^2/\mathbf{c}_j'(\mathbf{X}'\mathbf{X})^-\mathbf{c}_j$ are also independent. Thus, if two contrasts are orthogonal, they are independent and their corresponding sums of squares are independent.

We now show that if the rows of $\mathbf{C}$ (Section 13.4.2) are mutually orthogonal contrasts, SSH is the sum of $(\mathbf{c}_i'\hat{\boldsymbol{\beta}})^2/\mathbf{c}_i'(\mathbf{X}'\mathbf{X})^-\mathbf{c}_i$ for all rows of $\mathbf{C}$.

Theorem 13.6a. In the balanced one-way model, if $\mathbf{y}$ is $N_{kn}(\mathbf{X}\boldsymbol{\beta}, \sigma^2\mathbf{I})$ and if $H_0: \alpha_1 = \alpha_2 = \cdots = \alpha_k$ is expressed as $\mathbf{C}\boldsymbol{\beta} = \mathbf{0}$, where the rows of

$$\mathbf{C} = \begin{pmatrix} \mathbf{c}_1' \\ \mathbf{c}_2' \\ \vdots \\ \mathbf{c}_{k-1}' \end{pmatrix}$$

are mutually orthogonal contrasts, then $\mathrm{SSH} = (\mathbf{C}\hat{\boldsymbol{\beta}})'[\mathbf{C}(\mathbf{X}'\mathbf{X})^-\mathbf{C}']^{-1}\mathbf{C}\hat{\boldsymbol{\beta}}$ can be expressed (partitioned) as

$$\mathrm{SSH} = \sum_{i=1}^{k-1} \frac{(\mathbf{c}_i'\hat{\boldsymbol{\beta}})^2}{\mathbf{c}_i'(\mathbf{X}'\mathbf{X})^-\mathbf{c}_i}, \tag{13.55}$$

where the sums of squares $(\mathbf{c}_i'\hat{\boldsymbol{\beta}})^2/\mathbf{c}_i'(\mathbf{X}'\mathbf{X})^-\mathbf{c}_i$, $i = 1, 2, \ldots, k-1$, are independent.

PROOF. By (13.54), $\mathbf{C}(\mathbf{X}'\mathbf{X})^-\mathbf{C}'$ is a diagonal matrix with $\mathbf{c}_i'(\mathbf{X}'\mathbf{X})^-\mathbf{c}_i$, $i = 1, 2, \ldots, k-1$, on the diagonal. Thus, with $(\mathbf{C}\hat{\boldsymbol{\beta}})' = (\mathbf{c}_1'\hat{\boldsymbol{\beta}}, \mathbf{c}_2'\hat{\boldsymbol{\beta}}, \ldots, \mathbf{c}_{k-1}'\hat{\boldsymbol{\beta}})$, (13.55) follows. Since the rows $\mathbf{c}_1', \mathbf{c}_2', \ldots, \mathbf{c}_{k-1}'$ of $\mathbf{C}$ are orthogonal, the independence of the sums of squares for the contrasts follows from (13.53) and (13.54). □

An interesting implication of Theorem 13.6a is that the overall F for treatments (Table 13.1) is the average of the F statistics for each of the orthogonal contrasts:

$$F = \frac{\text{SSH}/(k-1)}{s^2} = \frac{1}{k-1} \sum_{i=1}^{k-1} \frac{(\mathbf{c}_i \hat{\boldsymbol{\beta}})^2}{s^2 \mathbf{c}_i'(\mathbf{X}\mathbf{X})^- \mathbf{c}_i}$$

$$= \frac{1}{k-1} \sum_{i=1}^{k-1} F_i.$$

It is possible that the overall F would lead to rejection of the overall H_0 while some of the F_i's for individual contrasts would not lead to rejection of the corresponding H_0's. Likewise, since one or more of the F_i's will be larger than the overall F, it is possible that an individual H_0 would be rejected, while the overall H_0 is not rejected.

Example 13.6a. We illustrate the use of orthogonal contrasts with the ascorbic acid data of Table 13.2. Consider the orthogonal contrasts $2\mu_1 - \mu_2 - \mu_3$ and $\mu_2 - \mu_3$. By (13.50), these can be expressed as

$$2\mu_1 - \mu_2 - \mu_3 = 2\alpha_1 - \alpha_2 - \alpha_3 = (0, 2, -1, -1)\boldsymbol{\beta} = \mathbf{c}_1'\boldsymbol{\beta},$$

$$\mu_2 - \mu_3 = \alpha_2 - \alpha_3 = (0, 0, 1, -1)\boldsymbol{\beta} = \mathbf{c}_2'\boldsymbol{\beta}.$$

The hypotheses $H_{01} : \mathbf{c}_1'\boldsymbol{\beta} = 0$ and $H_{02} : \mathbf{c}_2'\boldsymbol{\beta} = 0$ compare the first treatment versus the other two and the second treatment versus the third.

The means are given in Table 13.2 as $\bar{y}_{1.} = 17.15$, $\bar{y}_{2.} = 19.31$, and $\bar{y}_{3.} = 23.53$. Then by (13.52), the sums of squares for the two contrasts are

$$\text{SS}_1 = \frac{n(\sum_{i=1}^{3} c_i \bar{y}_{i.})^2}{\sum_{i=1}^{3} c_i^2} = \frac{7[2(17.15) - 19.31 - 23.53]^2}{4 + 1 + 1} = 85.0584,$$

$$\text{SS}_2 = \frac{7(19.31 - 23.53)^2}{1 + 1} = 62.2872.$$

By (13.52), the corresponding F statistics are

$$F_1 = \frac{\text{SS}_1}{s^2} = \frac{85.0584}{3.0537} = 27.85, \qquad F_2 = \frac{\text{SS}_2}{s^2} = \frac{62.2872}{3.0537} = 20.40,$$

where $s^2 = 3.0537$ is from Table 13.3. Both F_1 and F_2 exceed $F_{.05,1,18} = 4.41$. The p values are .0000511and .000267, respectively.

Note that the sums of squares for the two orthogonal contrasts add to the sum of squares for treatments given in Example 13.4; that is, $147.3456 = 85.0584 + 62.2872$, as in (13.55). □

The partitioning of the treatment sum of squares in Theorem 13.6a is always possible. First note that $\text{SSH} = \mathbf{y}'\mathbf{A}\mathbf{y}$ as in (13.29), where $\mathbf{A}$ is idempotent. We now show that any such quadratic form can be partitioned into independent components.

Theorem 13.6b. Let $\mathbf{y}'\mathbf{Ay}$ be a quadratic form, let $\mathbf{A}$ be symmetric and idempotent of rank r, let $N = kn$, and let the $N \times 1$ random vector $\mathbf{y}$ be $N_N(\mathbf{X}\boldsymbol{\beta}, \sigma^2\mathbf{I})$. Then there exist r idempotent matrices $\mathbf{A}_1, \mathbf{A}_2, \ldots, \mathbf{A}_r$ such that $\mathbf{A} = \sum_{i=1}^{r} \mathbf{A}_i$, rank$(\mathbf{A}_i) = 1$ for $i = 1, 2, \ldots, r$, and $\mathbf{A}_i\mathbf{A}_j = \mathbf{O}$ for $i \neq j$. Furthermore, $\mathbf{y}'\mathbf{Ay}$ can be partitioned as

$$\mathbf{y}'\mathbf{Ay} = \sum_{i=1}^{r} \mathbf{y}'\mathbf{A}_i\mathbf{y}, \tag{13.56}$$

where each $\mathbf{y}'\mathbf{A}_i\mathbf{y}$ in (13.56) is $\chi^2(1, \lambda_i)$ and $\mathbf{y}'\mathbf{A}_i\mathbf{y}$ and $\mathbf{y}'\mathbf{A}_j\mathbf{y}$ are independent for $i \neq j$ (note that λ_i is a noncentrality parameter).

Proof. Since $\mathbf{A}$ is $N \times N$ of rank r and is symmetric and idempotent, then by Theorem 2.13c, r of its eigenvalues are equal to 1 and the others are 0. Using the spectral decomposition (2.104), we can express $\mathbf{A}$ in the form

$$\mathbf{A} = \sum_{i=1}^{r} \mathbf{v}_i\mathbf{v}_i' = \sum_{i=1}^{r} \mathbf{A}_i, \tag{13.57}$$

where $\mathbf{v}_1, \mathbf{v}_2, \ldots, \mathbf{v}_r$ are normalized orthogonal eigenvectors corresponding to the nonzero eigenvalues and $\mathbf{A}_i = \mathbf{v}_i\mathbf{v}_i'$. It is easily shown that rank$(\mathbf{A}_i) = 1$, $\mathbf{A}_i\mathbf{A}_j = \mathbf{O}$ for $i \neq j$, and $\mathbf{A}_i$ is symmetric and idempotent (see Problem 13.17). Then by Corollary 2 to Theorem 5.5 and Corollary 1 to Theorem 5.6b, $\mathbf{y}'\mathbf{A}_i\mathbf{y}$ is $\chi^2(1, \lambda_i)$ and $\mathbf{y}'\mathbf{A}_i\mathbf{y}$ and $\mathbf{y}'\mathbf{A}_j\mathbf{y}$ are independent. □

If $\mathbf{y}'\mathbf{Ay}$ in Theorem 13.6b is used to represent SSH, the eigenvectors corresponding to nonzero eigenvalues of $\mathbf{A}$ always define contrasts of the cell means. In other words, the partitioning of $\mathbf{y}'\mathbf{Ay}$ in (13.56) is always in terms of orthogonal contrasts. To see this, note that

$$\text{SST} = \text{SSH} + \text{SSE},$$

which, in the case of the one-way balanced model, implies that

$$\mathbf{y}'\left(\mathbf{I} - \frac{1}{kn}\mathbf{J}\right)\mathbf{y} = \sum_{i=1}^{k} \mathbf{y}'\mathbf{A}_i\mathbf{y} + \mathbf{y}'[\mathbf{I} - \mathbf{X}(\mathbf{X}'\mathbf{X})^{-}\mathbf{X}']\mathbf{y}. \tag{13.58}$$

If we let

$$\mathbf{K} = \frac{1}{n}\begin{pmatrix} \mathbf{J} & \mathbf{0} & \cdots & \mathbf{0} \\ \mathbf{0} & \mathbf{J} & \cdots & \mathbf{0} \\ \vdots & \vdots & \ddots & \vdots \\ \mathbf{0} & \mathbf{0} & \cdots & \mathbf{J} \end{pmatrix}$$

as in (13.26), then (13.58) can be rewritten as

$$\mathbf{y}'\mathbf{y} = \mathbf{y}'\frac{1}{kn}\mathbf{Jy} + \sum_{i=1}^{k} \mathbf{y}'(\mathbf{v}_i\mathbf{v}_i')\mathbf{y} + \mathbf{y}'(\mathbf{I} - \mathbf{K})\mathbf{y}. \tag{13.59}$$

By Theorem 2.13h, each $\mathbf{v}_i$ must be orthogonal to the columns of $(1/n)\mathbf{J}$ and $\mathbf{I} - \mathbf{K}$. Orthogonality to $(1/n)\mathbf{J}$ implies that $\mathbf{v}_i\mathbf{j} = 0$; that is, $\mathbf{v}_i$ defines a contrast in the elements of $\mathbf{y}$. Orthogonality to $\mathbf{I} - \mathbf{K}$ implies that the elements of $\mathbf{v}_i$ corresponding to units associated with a particular treatment are constants. Together these results imply that $\mathbf{v}_i$ defines a contrast of the estimated treatment means.

Example 13.6b. Using a one-way model, we demonstrate that orthogonal contrasts in the treatment means can be expressed in terms of contrasts in the observations and that the coefficients in these contrasts form eigenvectors. For simplicity of exposition, let $k = 4$. The model is then

$$y_{ij} = \mu + \alpha_i + \varepsilon_{ij}, \quad i = 1, 2, 3, 4, \quad j = 1, 2, \ldots, n.$$

The sums of squares in (13.59) can be written in the form

$$\begin{aligned} \mathbf{y}'\mathbf{y} &= \mathrm{SS}(\mu) + \mathrm{SS}(\alpha|\mu) + \mathrm{SSE} \\ &= \frac{y_{..}^2}{kn} + \left(\hat{\boldsymbol{\beta}}'\mathbf{X}'\mathbf{y} - \frac{y_{..}^2}{kn} \right) + (\mathbf{y}'\mathbf{y} - \hat{\boldsymbol{\beta}}'\mathbf{X}'\mathbf{y}). \end{aligned}$$

With $k = 4$, the sum of squares for treatments, $\mathbf{y}'\mathbf{A}\mathbf{y} = \hat{\boldsymbol{\beta}}'\mathbf{X}'\mathbf{y} - y_{..}^2/4n$, has 3 degrees of freedom. Any set of three orthogonal contrasts in the treatment means will serve to illustrate. As an example, consider $\mathbf{c}_1'\boldsymbol{\beta} = (0, 1, -1, 0, 0)\boldsymbol{\beta}$, $\mathbf{c}_2'\boldsymbol{\beta} = (0, 1, 1, -2, 0)\boldsymbol{\beta}$, and $\mathbf{c}_3'\boldsymbol{\beta} = (0, 1, 1, 1, -3)\boldsymbol{\beta}$, where $\boldsymbol{\beta} = (\mu, \alpha_1, \alpha_2, \alpha_3, \alpha_4)'$. Thus, we are comparing the first mean to the second, the first two means to the third, and the first three to the fourth (see a comment at the beginning of Section 13.4 for the equivalence of $H_0: \alpha_1 = \alpha_2 = \alpha_3 = \alpha_4$ and $H_0: \mu_1 = \mu_2 = \mu_3 = \mu_4$). Using the format in (13.55), we can write the three contrasts as

$$\frac{\mathbf{c}_1'\hat{\boldsymbol{\beta}}}{\sqrt{\mathbf{c}_1'(\mathbf{X}'\mathbf{X})^-\mathbf{c}_1}} = \frac{\bar{y}_{1.} - \bar{y}_{2.}}{\sqrt{2/n}}$$

$$\frac{\mathbf{c}_2'\hat{\boldsymbol{\beta}}}{\sqrt{\mathbf{c}_2'(\mathbf{X}'\mathbf{X})^-\mathbf{c}_2}} = \frac{\bar{y}_{1.} + \bar{y}_{2.} - 2\bar{y}_{3.}}{\sqrt{6/n}}$$

$$\frac{\mathbf{c}_3'\hat{\boldsymbol{\beta}}}{\sqrt{\mathbf{c}_3'(\mathbf{X}'\mathbf{X})^-\mathbf{c}_3}} = \frac{\bar{y}_{1.} + \bar{y}_{2.} + \bar{y}_{3.} - 3\bar{y}_{4.}}{\sqrt{12/n}},$$

where $(\mathbf{X}'\mathbf{X})^- = \mathrm{diag}[0, (1/n), \ldots, (1/n)]$ is given in (13.11) and $\hat{\boldsymbol{\beta}} = (0, \bar{y}_{1.}, \ldots, \bar{y}_{4.})'$ is from (13.12).

To write these in the form $\mathbf{v}_1'\mathbf{y}$, $\mathbf{v}_2'\mathbf{y}$, and $\mathbf{v}_3'\mathbf{y}$ [as in (13.59)] we start with the first:

$$\begin{aligned} \frac{\bar{y}_{1.} - \bar{y}_{2.}}{\sqrt{2/n}} &= \frac{1}{\sqrt{2/n}} \left(\frac{\sum_{j=1}^{n} y_{1j}}{n} - \frac{\sum_{j=1}^{n} y_{2j}}{n} \right) \\ &= \frac{1/n}{\sqrt{2/n}} (1, 1, \ldots, 1, -1, -1, \ldots, -1, 0, 0, \ldots, 0)\mathbf{y} \\ &= \mathbf{v}_1'\mathbf{y}, \end{aligned}$$

where the number of 1s is n, the number of -1s is n, and the number of 0s is $2n$. Thus $\mathbf{v}_1' = (1/\sqrt{2n})(\mathbf{j}_n', -\mathbf{j}_n', \mathbf{0}', \mathbf{0}')$, and

$$\mathbf{v}_1'\mathbf{v}_1 = \frac{2n}{2n} = 1.$$

Similarly, $\mathbf{v}_2'$ and $\mathbf{v}_3'$ can be expressed as $\mathbf{v}_2' = (1/\sqrt{6n})(\mathbf{j}_n', \mathbf{j}_n' - 2\mathbf{j}_n', \mathbf{0}')$ and $\mathbf{v}_3' = (1/\sqrt{12n})(\mathbf{j}_n', \mathbf{j}_n', \mathbf{j}_n', -3\mathbf{j}_n')$. We now show that $\mathbf{v}_1$, $\mathbf{v}_2$, and $\mathbf{v}_3$ serve as eigenvectors in the spectral decomposition [see (2.104)] of the matrix $\mathbf{A}$ in $\text{SS}(\alpha|\mu) = \mathbf{y}'\mathbf{A}\mathbf{y}$. Since $\mathbf{A}$ is idempotent of rank 3, it has three nonzero eigenvalues, each equal to 1. Thus the spectral decomposition of $\mathbf{A}$ is

$$\begin{aligned}
\mathbf{A} &= \mathbf{v}_1\mathbf{v}_1' + \mathbf{v}_2\mathbf{v}_2' + \mathbf{v}_3\mathbf{v}_3' \\
&= \frac{1}{2n}\begin{pmatrix} \mathbf{j}_n \\ -\mathbf{j}_n \\ \mathbf{0} \\ \mathbf{0} \end{pmatrix}(\mathbf{j}_n', -\mathbf{j}_n', \mathbf{0}', \mathbf{0}') + \frac{1}{6n}\begin{pmatrix} \mathbf{j}_n \\ \mathbf{j}_n \\ -2\mathbf{j}_n \\ \mathbf{0} \end{pmatrix}(\mathbf{j}_n', \mathbf{j}_n', -2\mathbf{j}_n', \mathbf{0}') \\
&\quad + \frac{1}{12n}\begin{pmatrix} \mathbf{j}_n \\ \mathbf{j}_n \\ \mathbf{j}_n \\ -3\mathbf{j}_n \end{pmatrix}(\mathbf{j}_n', \mathbf{j}_n', \mathbf{j}_n', -3\mathbf{j}_n') \\
&= \frac{1}{2n}\begin{pmatrix} \mathbf{J}_n & -\mathbf{J}_n & \mathbf{O} & \mathbf{O} \\ -\mathbf{J}_n & \mathbf{J}_n & \mathbf{O} & \mathbf{O} \\ \mathbf{O} & \mathbf{O} & \mathbf{O} & \mathbf{O} \\ \mathbf{O} & \mathbf{O} & \mathbf{O} & \mathbf{O} \end{pmatrix} + \frac{1}{6n}\begin{pmatrix} \mathbf{J}_n & \mathbf{J}_n & -2\mathbf{J}_n & \mathbf{O} \\ \mathbf{J}_n & \mathbf{J}_n & -2\mathbf{J}_n & \mathbf{O} \\ -2\mathbf{J}_n & -2\mathbf{J}_n & 4\mathbf{J}_n & \mathbf{O} \\ \mathbf{O} & \mathbf{O} & \mathbf{O} & \mathbf{O} \end{pmatrix} \\
&\quad + \frac{1}{12n}\begin{pmatrix} \mathbf{J}_n & \mathbf{J}_n & \mathbf{J}_n & -3\mathbf{J}_n \\ \mathbf{J}_n & \mathbf{J}_n & \mathbf{J}_n & -3\mathbf{J}_n \\ \mathbf{J}_n & \mathbf{J}_n & \mathbf{J}_n & -3\mathbf{J}_n \\ -3\mathbf{J}_n & -3\mathbf{J}_n & -3\mathbf{J}_n & 9\mathbf{J}_n \end{pmatrix} \\
&= \frac{1}{4n}\begin{pmatrix} 3\mathbf{J}_n & -\mathbf{J}_n & -\mathbf{J}_n & -\mathbf{J}_n \\ -\mathbf{J}_n & 3\mathbf{J}_n & -\mathbf{J}_n & -\mathbf{J}_n \\ -\mathbf{J}_n & -\mathbf{J}_n & 3\mathbf{J}_n & -\mathbf{J}_n \\ -\mathbf{J}_n & -\mathbf{J}_n & -\mathbf{J}_n & 3\mathbf{J}_n \end{pmatrix},
\end{aligned}$$

which is the matrix of the quadratic form for $\text{SS}(\alpha|\mu)$ in (13.27) with $k = 4$.

For $\text{SS}(\mu) = y_{..}^2/4n$, we have

$$\frac{y_{..}^2}{4n} = \mathbf{y}'\left(\frac{\mathbf{j}_{4n}\mathbf{j}_{4n}'}{4n}\right)\mathbf{y} = (\mathbf{v}_0'\mathbf{y})^2,$$

where $\mathbf{v}_0' = \mathbf{j}_{4n}'/2\sqrt{n}$. It is easily shown that $\mathbf{v}_0'\mathbf{v}_0 = 1$ and that $\mathbf{v}_0'\mathbf{v}_1 = 0$. It is also clear that $\mathbf{v}_0$ is an eigenvector of $\mathbf{j}_{4n}\mathbf{j}_{4n}'/4n$, because $\mathbf{j}_{4n}\mathbf{j}_{4n}'/4n$ has one eigenvalue equal to 1 and the others equal to 0, so that $\mathbf{j}_{4n}\mathbf{j}_{4n}'/4n$ is already in the form of a spectral decomposition with $\mathbf{j}_{4n}/2\sqrt{n}$ as the eigenvector corresponding to the eigenvalue 1 (see Problem 13.18b). □

13.6.3 Orthogonal Polynomial Contrasts

Suppose the treatments in a one-way analysis of variance have equally spaced quantitative levels, for example, 5, 10, 15, and 20lb of fertilizer per plot of ground. The researcher may then wish to investigate how the response varies with the level of fertilizer. We can check for a linear trend, a quadratic trend, or a cubic trend by fitting a third-order polynomial regression model

$$y_{ij} = \beta_0 + \beta_1 x_i + \beta_2 x_i^2 + \beta_3 x_i^3 + \varepsilon_{ij}, \tag{13.60}$$

$$i = 1, 2, 3, 4, \quad j = 1, 2, \ldots, n,$$

where $x_1 = 5$, $x_2 = 10$, $x_3 = 15$, and $x_4 = 20$. We now show that tests on the β's in (13.60) can be carried out using orthogonal contrasts on the means $\bar{y}_{i.}$ that are estimates of μ_i in the ANOVA model

$$y_{ij} = \mu + \alpha_i + \varepsilon_{ij} = \mu_i + \varepsilon_{ij}, \quad i = 1, 2, 3, 4, \quad j = 1, 2, \ldots, n. \tag{13.61}$$

The sum of squares for the full–reduced-model test of $H_0: \beta_3 = 0$ is

$$\hat{\boldsymbol{\beta}}'\mathbf{X}'\mathbf{y} - \hat{\boldsymbol{\beta}}_1^{*\prime}\mathbf{X}_1'\mathbf{y}, \tag{13.62}$$

where $\hat{\boldsymbol{\beta}}$ is from the full model in (13.60) and $\hat{\boldsymbol{\beta}}_1^*$ is from the reduced model with $\beta_3 = 0$ [see (8.9), (8.20), and Table 8.3]. The $\mathbf{X}$ matrix is of the form

$$\mathbf{X} = \begin{pmatrix} 1 & x_1 & x_1^2 & x_1^3 \\ \vdots & \vdots & \vdots & \vdots \\ 1 & x_1 & x_1^2 & x_1^3 \\ 1 & x_2 & x_2^2 & x_2^3 \\ \vdots & \vdots & \vdots & \vdots \\ 1 & x_2 & x_2^2 & x_2^3 \\ 1 & x_3 & x_3^2 & x_3^3 \\ \vdots & \vdots & \vdots & \vdots \\ 1 & x_3 & x_3^2 & x_3^3 \\ 1 & x_4 & x_4^2 & x_4^3 \\ \vdots & \vdots & \vdots & \vdots \\ 1 & x_4 & x_4^2 & x_4^3 \end{pmatrix}. \tag{13.63}$$

For testing $H_0 : \beta_3 = 0$, we can use (8.37)

$$F = \frac{\hat{\boldsymbol{\beta}}'\mathbf{X}'\mathbf{y} - \hat{\boldsymbol{\beta}}_1^{*\prime}\mathbf{X}_1'\mathbf{y}}{s^2},$$

or (8.39)

$$F = \frac{\hat{\beta}_3^2}{s^2 g_{33}}, \tag{13.64}$$

where $\mathbf{X}_1$ consists of the first three columns of $\mathbf{X}$ in (13.63), $s^2 = \text{SSE}/(n-3-1)$, and g_{33} is the last diagonal element of $(\mathbf{X}'\mathbf{X})^{-1}$. We now carry out this full–reduced-model test using contrasts.

Since the columns of $\mathbf{X}$ are not orthogonal, the sums of squares for the β's analogous to $\hat{\beta}_3^2/g_{33}$ in (13.64) are not independent. Thus, the interpretation in terms of the degree of curvature for $E(y_{ij})$ is more difficult. We therefore orthogonalize the columns of $\mathbf{X}$ so that the sums of squares become independent.

To simplify computations, we first transform $x_1 = 5, x_2 = 10, x_3 = 15$, and $x_4 = 20$ by dividing by 5, the common distance between them. The x's then become $x_1 = 1, x_2 = 2, x_3 = 3$, and $x_4 = 4$. The transformed $4n \times 4$ matrix $\mathbf{X}$ in (13.63) is given by

$$\mathbf{X} = \begin{pmatrix} 1 & 1 & 1^2 & 1^3 \\ \vdots & \vdots & \vdots & \vdots \\ 1 & 1 & 1^2 & 1^3 \\ 1 & 2 & 2^2 & 2^3 \\ \vdots & \vdots & \vdots & \vdots \\ 1 & 2 & 2^2 & 2^3 \\ 1 & 3 & 3^2 & 3^3 \\ \vdots & \vdots & \vdots & \vdots \\ 1 & 3 & 3^2 & 3^3 \\ 1 & 4 & 4^2 & 4^3 \\ \vdots & \vdots & \vdots & \vdots \\ 1 & 4 & 4^2 & 4^3 \end{pmatrix} = (\mathbf{j}, \mathbf{x}_1, \mathbf{x}_2, \mathbf{x}_3),$$

where $\mathbf{j}$ is $4n \times 1$. Note that by Theorem 8.4c, the resulting F statistics such as (13.64) will be unaffected by this transformation.

To obtain orthogonal columns, we use the orthogonalization procedure in Section 7.10 based on regressing columns of $\mathbf{X}$ on other columns and taking residuals. We begin by orthogonalizing $\mathbf{x}_1$. Denoting the first column by $\mathbf{x}_0$, we use

(7.97) to obtain

$$\begin{aligned}\mathbf{x}_{1\cdot 0} &= \mathbf{x}_1 - \mathbf{x}_0(\mathbf{x}_0'\mathbf{x}_0)^{-1}\mathbf{x}_0'\mathbf{x}_1 \\ &= \mathbf{x}_1 - \mathbf{j}(\mathbf{j}'\mathbf{j})^{-1}\mathbf{j}'\mathbf{x}_1 = \mathbf{x}_1 - \mathbf{j}(4n)^{-1}n\sum_{i=1}^{4} x_i \\ &= \mathbf{x}_1 - \bar{x}\mathbf{j}. \end{aligned} \tag{13.65}$$

The residual vector $\mathbf{x}_{1\cdot 0}$ is orthogonal to $\mathbf{x}_0 = \mathbf{j}$:

$$\mathbf{j}'\mathbf{x}_{1\cdot 0} = \mathbf{j}'(\mathbf{x}_1 - \bar{x}\mathbf{j}) = \mathbf{j}'\mathbf{x}_1 - \bar{x}\mathbf{j}'\mathbf{j} = 4n\bar{x} - 4n\bar{x} = 0. \tag{13.66}$$

We apply this procedure successively to the other two columns of $\mathbf{X}$. To transform the third column, $\mathbf{x}_2$, so that it is orthogonal to the first two columns, we use (7.97) to obtain

$$\mathbf{x}_{2\cdot 01} = \mathbf{x}_2 - \mathbf{Z}_1(\mathbf{Z}_1'\mathbf{Z}_1)^{-1}\mathbf{Z}_1'\mathbf{x}_2, \tag{13.67}$$

where $\mathbf{Z}_1 = (\mathbf{j}, \mathbf{x}_{1\cdot 0})$. We use the notation $\mathbf{Z}_1$ instead of $\mathbf{X}_1$ because $\mathbf{x}_{1\cdot 0}$, the second column of $\mathbf{Z}_1$, is different from $\mathbf{x}_1$, the second column of $\mathbf{X}_1$. The matrix $\mathbf{Z}_1'\mathbf{Z}_1$ is given by

$$\begin{aligned}\mathbf{Z}_1'\mathbf{Z}_1 &= \begin{pmatrix} \mathbf{j}' \\ \mathbf{x}_{1\cdot 0}' \end{pmatrix}(\mathbf{j}, \mathbf{x}_{1\cdot 0}) \\ &= \begin{pmatrix} \mathbf{j}'\mathbf{j} & 0 \\ 0 & \mathbf{x}_{1\cdot 0}'\mathbf{x}_{1\cdot 0} \end{pmatrix} \qquad [\text{by } (13.66)], \end{aligned}$$

and (13.67) becomes

$$\begin{aligned}\mathbf{x}_{2\cdot 01} &= \mathbf{x}_2 - \mathbf{Z}_1(\mathbf{Z}_1'\mathbf{Z}_1)^{-1}\mathbf{Z}_1'\mathbf{x}_2 \\ &= \mathbf{x}_2 - (\mathbf{j}, \mathbf{x}_{1\cdot 0})\begin{pmatrix} \mathbf{j}'\mathbf{j} & 0 \\ 0 & \mathbf{x}_{1\cdot 0}'\mathbf{x}_{1\cdot 0} \end{pmatrix}^{-1}\begin{pmatrix} \mathbf{j}' \\ \mathbf{x}_{1\cdot 0}' \end{pmatrix}\mathbf{x}_2 \\ &= \mathbf{x}_2 - \frac{\mathbf{j}'\mathbf{x}_2}{\mathbf{j}'\mathbf{j}}\mathbf{j} - \frac{\mathbf{x}_{1\cdot 0}'\mathbf{x}_2}{\mathbf{x}_{1\cdot 0}'\mathbf{x}_{1\cdot 0}}\mathbf{x}_{1\cdot 0}. \end{aligned} \tag{13.68}$$

The residual vector $\mathbf{x}_{2\cdot 01}$ is orthogonal to $\mathbf{x}_0 = \mathbf{j}$ and to $\mathbf{x}_{1\cdot 0}$:

$$\mathbf{j}'\mathbf{x}_{2\cdot 01} = 0, \quad \mathbf{x}_{1\cdot 0}'\mathbf{x}_{2\cdot 01} = 0. \tag{13.69}$$

The fourth column of $\mathbf{Z}$ becomes

$$\mathbf{x}_{3\cdot 012} = \mathbf{x}_3 - \frac{\mathbf{j}'\mathbf{x}_3}{\mathbf{j}'\mathbf{j}}\mathbf{j} - \frac{\mathbf{x}_{1\cdot 0}'\mathbf{x}_3}{\mathbf{x}_{1\cdot 0}'\mathbf{x}_{1\cdot 0}}\mathbf{x}_{1\cdot 0} - \frac{\mathbf{x}_{2\cdot 01}'\mathbf{x}_3}{\mathbf{x}_{2\cdot 01}'\mathbf{x}_{2\cdot 01}}\mathbf{x}_{2\cdot 01}, \tag{13.70}$$

which is orthogonal to the first three columns, $\mathbf{j}$, $\mathbf{x}_{1\cdot 0}$, and $\mathbf{x}_{2\cdot 01}$.

We have thus transformed $\mathbf{y} = \mathbf{X}\boldsymbol{\beta} + \boldsymbol{\varepsilon}$ to

$$\mathbf{y} = \mathbf{Z}\boldsymbol{\theta} + \boldsymbol{\varepsilon}, \tag{13.71}$$

where the columns of $\mathbf{Z}$ are mutually orthogonal and the elements of $\boldsymbol{\theta}$ are functions of the β's. The columns of $\mathbf{Z}$ are given in (13.65), (13.68), and (13.70):

$$\mathbf{z}_0 = \mathbf{j}, \quad \mathbf{z}_1 = \mathbf{x}_{1\cdot 0}, \quad \mathbf{z}_2 = \mathbf{x}_{2\cdot 01}, \quad \mathbf{z}_3 = \mathbf{x}_{3\cdot 012}.$$

We now evaluate $\mathbf{z}_1$, $\mathbf{z}_2$, and $\mathbf{z}_3$ for our illustration, in which $x_1 = 1$, $x_2 = 2$, $x_3 = 3$, and $x_4 = 4$. By (13.65), we obtain

$$\begin{aligned}\mathbf{z}_1 &= \mathbf{x}_{1\cdot 0} = \mathbf{x}_1 - \bar{x}\mathbf{j} = \mathbf{x}_1 - 2.5\mathbf{j} \\ &= (-1.5, \ldots, -1.5, -.5, \ldots, -.5, .5, \ldots, .5, 1.5, \ldots, 1.5)',\end{aligned}$$

which we multiply by 2 so as to obtain integer values:

$$\mathbf{z}_1 = \mathbf{x}_{1\cdot 0} = (-3, \ldots, -3, -1, \ldots, -1, 1, \ldots, 1, 3, \ldots, 3)'. \tag{13.72}$$

Note that multiplying by 2 preserves the orthogonality and does not affect the F values.

To obtain $\mathbf{z}_2$, by (13.68), we first compute

$$\frac{\mathbf{j}'\mathbf{x}_2}{\mathbf{j}'\mathbf{j}} = \frac{n\sum_{i=1}^{4} x_i^2}{4n} = \frac{\sum_{i=1}^{4} i^2}{4} = \frac{30}{4} = 7.5,$$

$$\frac{\mathbf{x}_{1\cdot 0}'\mathbf{x}_2}{\mathbf{x}_{1\cdot 0}'\mathbf{x}_{1\cdot 0}} = \frac{n[-3(1^2) - 1(2^2) + 1(3^2) + 3(4^2)]}{n[(-3)^2 + (-1)^2 + 1^2 + 3^2]} = \frac{50}{20} = 2.5.$$

Then, by (13.68), we obtain

$$\begin{aligned}
\mathbf{z}_2 &= \mathbf{x}_2 - \frac{\mathbf{j}'\mathbf{x}_2}{\mathbf{j}'\mathbf{j}}\mathbf{j} - \frac{\mathbf{x}'_{1\cdot 0}\mathbf{x}_2}{\mathbf{x}'_{1\cdot 0}\mathbf{x}_{1\cdot 0}}\mathbf{x}_{1\cdot 0} \\
&= \mathbf{x}_2 - 7.5\mathbf{j} - 2.5\mathbf{x}_{1\cdot 0} \\
&= \begin{pmatrix} 1^2 \\ \vdots \\ 1^2 \\ 2^2 \\ \vdots \\ 2^2 \\ 3^2 \\ \vdots \\ 3^2 \\ 4^2 \\ \vdots \\ 4^2 \end{pmatrix} - 7.5 \begin{pmatrix} 1 \\ \vdots \\ 1 \\ 1 \\ \vdots \\ 1 \\ 1 \\ \vdots \\ 1 \\ 1 \\ \vdots \\ 1 \end{pmatrix} - 2.5 \begin{pmatrix} -3 \\ \vdots \\ -3 \\ -1 \\ \vdots \\ -1 \\ 1 \\ \vdots \\ 1 \\ 3 \\ \vdots \\ 3 \end{pmatrix} = \begin{pmatrix} 1 \\ \vdots \\ 1 \\ -1 \\ \vdots \\ -1 \\ -1 \\ \vdots \\ -1 \\ 1 \\ \vdots \\ 1 \end{pmatrix}.
\end{aligned} \tag{13.73}$$

Similarly, using (13.70), we obtain

$$\mathbf{z}_3 = (-1, \ldots, -1, 3, \ldots, 3, -3, \ldots, -3, 1, \ldots, 1)'. \tag{13.74}$$

Thus $\mathbf{Z}$ is given by

$$\mathbf{Z} = \begin{pmatrix} 1 & -3 & 1 & -1 \\ \vdots & \vdots & \vdots & \vdots \\ 1 & -3 & 1 & -1 \\ 1 & -1 & -1 & 3 \\ \vdots & \vdots & \vdots & \vdots \\ 1 & -1 & -1 & 3 \\ 1 & 1 & -1 & -3 \\ \vdots & \vdots & \vdots & \vdots \\ 1 & 1 & -1 & -3 \\ 1 & 3 & 1 & 1 \\ \vdots & \vdots & \vdots & \vdots \\ 1 & 3 & 1 & 1 \end{pmatrix}.$$

Since

$$\mathbf{X}\boldsymbol{\beta} = \mathbf{Z}\boldsymbol{\theta},$$

we can find the θ's in terms of the β's or the β's in terms of the θ's. For our illustration, these relationships are given by (see Problem 13.24)

$$\begin{aligned} &\beta_0 = \theta_0 - 5\theta_1 + 5\theta_2 - 35\theta_3, \quad \beta_1 = 2\theta_1 - 5\theta_2 + \frac{16.7}{.3}\theta_3, \\ &\beta_2 = \theta_2 - 25\theta_3, \quad \beta_3 = \frac{\theta_3}{.3}. \end{aligned} \tag{13.75}$$

Since the columns of $\mathbf{Z} = (\mathbf{j}, \mathbf{z}_1, \mathbf{z}_2, \mathbf{z}_3)$ are orthogonal ($\mathbf{z}_i'\mathbf{z}_j = 0$ for all $i \neq j$), we have $\mathbf{Z}'\mathbf{Z} = \text{diag}(\mathbf{j}'\mathbf{j}, \mathbf{z}_1'\mathbf{z}_1, \mathbf{z}_2'\mathbf{z}_2, \mathbf{z}_3'\mathbf{z}_3)$. Thus

$$\hat{\boldsymbol{\theta}} = (\mathbf{Z}'\mathbf{Z})^{-1}\mathbf{Z}'\mathbf{y} = \begin{pmatrix} \mathbf{j}'\mathbf{y}/\mathbf{j}'\mathbf{j} \\ \mathbf{z}_1'\mathbf{y}/\mathbf{z}_1'\mathbf{z}_1 \\ \mathbf{z}_2'\mathbf{y}/\mathbf{z}_2'\mathbf{z}_2 \\ \mathbf{z}_3'\mathbf{y}/\mathbf{z}_3'\mathbf{z}_3 \end{pmatrix}. \tag{13.76}$$

The regression sum of squares (uncorrected for θ_0) is

$$\text{SS}(\boldsymbol{\theta}) = \hat{\boldsymbol{\theta}}'\mathbf{Z}'\mathbf{y} = \sum_{i=0}^{3} \frac{(\mathbf{z}_i'\mathbf{y})^2}{\mathbf{z}_i'\mathbf{z}_i}, \tag{13.77}$$

where $\mathbf{z}_0 = \mathbf{j}$. By an argument similar to that following (13.54), the sums of squares on the right side of (13.77) are independent.

Since the sums of squares $\text{SS}(\theta_i) = (\mathbf{z}_i'\mathbf{y})^2/\mathbf{z}_i'\mathbf{z}_i$, $i = 1, 2, 3$, are independent, each $\text{SS}(\theta_i)$ tests the significance of $\hat{\theta}_i$ by itself (regressing y on $\mathbf{z}_i$ alone) as well as in the presence of the other $\hat{\theta}_i$'s; that is, for a general k, we have

$$\begin{aligned} \text{SS}(\theta_i|\theta_0, \ldots, \theta_{i-1}, \theta_{i+1}, \ldots, \theta_k) &= \text{SS}(\theta_0, \ldots, \theta_k) - \text{SS}(\theta_0, \ldots, \theta_{i-1}, \theta_{i+1}, \ldots, \theta_k) \\ &= \sum_{j=0}^{k} \frac{(\mathbf{z}_j'\mathbf{y})^2}{\mathbf{z}_j'\mathbf{z}_j} - \sum_{j \neq i} \frac{(\mathbf{z}_j'\mathbf{y})^2}{\mathbf{z}_j'\mathbf{z}_j} \\ &= \frac{(\mathbf{z}_i'\mathbf{y})^2}{\mathbf{z}_i'\mathbf{z}_i} = \text{SS}(\theta_i). \end{aligned}$$

In terms of the $\hat{\beta}_i$'s, it can be shown that each $\text{SS}(\theta_i)$ tests the significance of $\hat{\beta}_i$ in the presence of $\hat{\beta}_0, \hat{\beta}_1, \ldots, \hat{\beta}_{i-1}$. For example, for β_k (the last β), the sum of squares can be written as

$$\mathrm{SS}(\theta_k) = \frac{(\mathbf{z}_k'\mathbf{y})^2}{\mathbf{z}_k'\mathbf{z}_k} = \hat{\boldsymbol{\beta}}'\mathbf{X}'\mathbf{y} - \hat{\boldsymbol{\beta}}_1^{*\prime}\mathbf{X}_1'\mathbf{y} \tag{13.78}$$

(see Problem 13.26), where $\hat{\boldsymbol{\beta}}$ is from the full model $\mathbf{y} = \mathbf{X}\boldsymbol{\beta} + \boldsymbol{\varepsilon}$ and $\hat{\boldsymbol{\beta}}_1^*$ is from the reduced model $\mathbf{y} = \mathbf{X}_1\boldsymbol{\beta}_1^* + \boldsymbol{\varepsilon}$, in which $\boldsymbol{\beta}_1$ contains all the β's except β_k and $\mathbf{X}_1$ consists of all columns of $\mathbf{X}$ except the last.

The sum of squares $\mathrm{SS}(\theta_i) = (\mathbf{z}_i'\mathbf{y})^2/\mathbf{z}_i'\mathbf{z}_i$ is equivalent to a sum of squares for a contrast on the means $\bar{y}_{1.}, \bar{y}_{2.}, \ldots, \bar{y}_{k.}$ as in (13.52). For example

$$\begin{aligned}
\mathbf{z}_1'\mathbf{y} &= -3y_{11} - 3y_{12} - \cdots - 3y_{1n} - y_{21} - \cdots - y_{2n} \\
&\quad + y_{31} + \cdots + y_{3n} + 3y_{41} + \cdots + 3y_{4n} \\
&= -3\sum_{j=1}^{n} y_{1j} - \sum_{j=1}^{n} y_{2j} + \sum_{j=1}^{n} y_{3j} + 3\sum_{j=1}^{n} y_{4j} \\
&= -3y_{1.} - y_{2.} + y_{3.} + 3y_{4.} \\
&= n(-3\bar{y}_{1.} - \bar{y}_{2.} + \bar{y}_{3.} + 3\bar{y}_{4.}) \\
&= n\sum_{i=1}^{4} c_i\bar{y}_{i.},
\end{aligned}$$

where $c_1 = -3$, $c_2 = -1$, $c_3 = 1$, and $c_4 = 3$. Similarly

$$\begin{aligned}
\mathbf{z}_1'\mathbf{z}_1 &= n(-3)^2 + n(-1)^2 + n(1)^2 + n(3)^2 \\
&= n[(-3)^2 + (-1)^2 + 1^2 + 3^2] \\
&= n\sum_{i=1}^{4} c_i^2.
\end{aligned}$$

Then

$$\frac{(\mathbf{z}_1'\mathbf{y})^2}{\mathbf{z}_1'\mathbf{z}_1} = \frac{(n\sum_{i=1}^{4} c_i\bar{y}_{i.})^2}{n\sum_{i=1}^{4} c_i^2} = \frac{n(\sum_{i=1}^{4} c_i\bar{y}_{i.})^2}{\sum_{i=1}^{4} c_i^2},$$

which is the sum of squares for the contrast in (13.52). Note that the coefficients $-3, -1, 1,$ and 3 correspond to a linear trend.

Likewise, $\mathbf{z}_2'\mathbf{y}$ becomes

$$\mathbf{z}_2'\mathbf{y} = n(\bar{y}_{1.} - \bar{y}_{2.} - \bar{y}_{3.} + \bar{y}_{4.}),$$

whose coefficients show a quadratic trend, and $\mathbf{z}_3'\mathbf{y}$ can be written as

$$\mathbf{z}_3'\mathbf{y} = n(-\bar{y}_{1.} + 3\bar{y}_{2.} - 3\bar{y}_{3.} + \bar{y}_{4.})$$

with coefficients that exhibit a cubic pattern.

These contrasts in the $\bar{y}_{i.}$'s have a meaningful interpretation in terms of the shape of the response curve. For example, suppose that the $\bar{y}_{i.}$'s fall on a straight line. Then, for some b_0 and b_1, we have

$$\bar{y}_{i.} = b_0 + b_1 x_i = b_0 + b_1 i, \quad i = 1, 2, 3, 4,$$

since $x_i = i$. In this case, the linear contrast is nonzero and the quadratic and cubic contrasts are zero:

$$\begin{aligned} -3\bar{y}_{1.} - \bar{y}_{2.} + \bar{y}_{3.} + 3\bar{y}_{4.} &= \\ -3(b_0 + b_1) - (b_0 + 2b_1) + b_0 + 3b_1 + 3(b_0 + 4b_1) &= 10b_1, \\ b_0 + b_1 - (b_0 + 2b_1) - (b_0 + 3b_1) + (b_0 + 4b_1) &= 0, \\ -(b_0 + b_1) + 3(b_0 + 2b_1) - 3(b_0 + 3b_1) + (b_0 + 4b_1) &= 0. \end{aligned}$$

This demonstration could be simplified by choosing the linear trend $\bar{y}_{1.} = 1$, $\bar{y}_{2.} = 2$, $\bar{y}_{3.} = 3$, and $\bar{y}_{4.} = 4$.

Similarly, if the $\bar{y}_{i.}$'s follow a quadratic trend, say

$$\bar{y}_{1.} = 1, \quad \bar{y}_{2.} = 2, \quad \bar{y}_{3.} = 2, \quad \bar{y}_{4.} = 1,$$

then the linear and cubic contrasts are zero.

In many cases it is not necessary to find the orthogonal polynomial coefficients by the orthogonalization process illustrated in this section. Tables of orthogonal polynomials are available [see, e.g., Rencher (2002, p. 587) or Guttman (1982, pp. 349–354)]. We give a brief illustration of some orthogonal polynomial coefficients in Table 13.5, including those we found above for $k = 4$.

TABLE 13.5 Orthogonal Polynomial Coefficients for k = 3, 4, 5

	$k=3$			$k=4$				$k=5$				
Linear	−1	0	1	−3	−1	1	3	−2	−1	0	1	2
Quadratic	1	−2	1	1	−1	−1	1	2	−1	−2	−1	2
Cubic				−1	3	−3	1	−1	2	0	−2	1
Quartic								1	−4	6	−4	1

In Table 13.5, we can see some relationships among the coefficients for each value of k. For example, if $k = 3$ and the three means $\bar{y}_{1.}, \bar{y}_{2.}, \bar{y}_{3.}$ have a linear relationship, then $\bar{y}_{2.} - \bar{y}_{1.}$ is equal to $\bar{y}_{3.} - \bar{y}_{2.}$; that is

$$\bar{y}_{3.} - \bar{y}_{2.} = \bar{y}_{2.} - \bar{y}_{1.}$$

or

$$\bar{y}_{3.} - \bar{y}_{2.} - (\bar{y}_{2.} - \bar{y}_{1.}) = 0,$$
$$\bar{y}_{3.} - 2\bar{y}_{2.} + \bar{y}_{1.} = 0.$$

If this relationship among the three means fails to hold, we have a quadratic component of curvature.

Similarly, for $k = 4$, the cubic component, $-\bar{y}_{1.} + 3\bar{y}_{2.} - 3\bar{y}_{3.} + \bar{y}_{4.}$, is equal to the difference between the quadratic component for $\bar{y}_{1.}, \bar{y}_{2.}, \bar{y}_{3.}$ and the quadratic component for $\bar{y}_{2.}, \bar{y}_{3.}, \bar{y}_{4.}$:

$$-\bar{y}_{1.} + 3\bar{y}_{2.} - 3\bar{y}_{3.} + \bar{y}_{4.} = \bar{y}_{2.} - 2\bar{y}_{3.} + \bar{y}_{4.} - (\bar{y}_{1.} - 2\bar{y}_{2.} + \bar{y}_{3.}).$$

PROBLEMS

13.1 Obtain the normal equations in (13.7) from the model in (13.6).

13.2 Obtain $\hat{\boldsymbol{\beta}}$ in (13.12) using $(\mathbf{X}'\mathbf{X})^-$ in (13.11) and $\mathbf{X}'\mathbf{y}$ in (13.7).

13.3 Show that SSE $= \mathbf{y}'[\mathbf{I} - \mathbf{X}(\mathbf{X}'\mathbf{X})^-\mathbf{X}']\mathbf{y}$ in (12.21) is equal to SSE $= \sum_{ij} y_{ij}^2 - \sum_i y_{i.}^2/n$ in (13.13).

13.4 Show that the expressions for SSE in (13.13) and (13.14) are equal.

13.5 **(a)** Show that $H_0: \alpha_1 = \alpha_2 = \cdots = \alpha_k$ in (13.17) is equivalent to $H_0: \alpha_1^* = \alpha_2^* = \cdots = \alpha_k^*$ in (13.18).

(b) Show that $H_0: \alpha_1^* = \alpha_2^* = \cdots = \alpha_k^*$ in (13.18) is equivalent to $H_0: \alpha_1^* = \alpha_2^* = \cdots = \alpha_k^* = 0$ in (13.19).

13.6 Show that $n\sum_{i=1}^k (\bar{y}_{i.} - \bar{y}_{..})^2$ in (13.24) is equal to $\sum_i y_{i.}^2/n - y_{..}^2/kn$ in (13.23).

13.7 Using (13.6) and (13.11), show that $\mathbf{X}(\mathbf{X}'\mathbf{X})^-\mathbf{X}'$ in (13.25) can be written in terms of $\mathbf{J}$ and $\mathbf{O}$ as in (13.26).

13.8 Show that for $\mathbf{C}$ in (13.28), $\mathbf{C}(\mathbf{X}'\mathbf{X})^-\mathbf{C}'$ is given by (13.30).

13.9 Show that $\mathbf{C}(\mathbf{X}'\mathbf{X})^-\mathbf{X}'$ is given by the matrix in (13.32).

13.10 Show that the matrix $(1/4n)\mathbf{A}'\mathbf{J}_3\mathbf{A}$ in (13.33) has the form shown in (13.35).

13.11 Using the model $y_{ij} = \mu^* + \alpha_i^* + \varepsilon_{ij}$ with the assumptions $E(\varepsilon_{ij}) = 0, \text{var}(\varepsilon_{ij}) = \sigma^2, \text{cov}(\varepsilon_{ij}, \varepsilon_{i'j'}) = 0$, and the side condition $\sum_{i=1}^k \alpha_i^* = 0$, obtain the following results used in Table 13.4:

(a) $E(\varepsilon_{ij}^2) = \sigma^2$ for all i, j and $E(\varepsilon_{ij}\varepsilon_{i'j'}) = 0$ for $i,j \neq i',j'$.

(b) $E[\text{SS}(\alpha|\mu)] = (k-1)\sigma^2 + n\sum_{i=1}^k \alpha_i^{*2}$.

(c) $(\text{SSE}) = k(n-1)\sigma^2$.

13.12 Using $\mathbf{C}$ in (13.28), show that $\mathbf{C}'\mathbf{C}$ is given by the matrix in (13.47)

13.13 Show that $\mathbf{C}'\mathbf{J}_3\mathbf{C}$ has the form shown in (13.48).

13.14 Show that if the constraint $\sum_{i=1}^4 \alpha_i = 0$ is not imposed, (13.49) becomes

$$E[\text{SS}(\alpha|\mu)] = 3\sigma^2 + 4\sum_{i=1}^{4}(\alpha_i - \bar{\alpha})^2.$$

13.15 Show that F in (13.52) can be obtained from (13.51).

13.16 Express the sums of squares $(\mathbf{c}_i'\hat{\boldsymbol{\beta}})^2/\mathbf{c}_i'(\mathbf{X}'\mathbf{X})^-\mathbf{c}_i$ and $(\mathbf{c}_j'\hat{\boldsymbol{\beta}})^2/\mathbf{c}_j'(\mathbf{X}'\mathbf{X})^-\mathbf{c}_j$ below (13.54) in Section 13.6.2 as quadratic forms in $\mathbf{y}$, and show that these sums of squares are independent if $\text{cov}(\mathbf{c}_i'\hat{\boldsymbol{\beta}}, \mathbf{c}_j'\hat{\boldsymbol{\beta}}) = 0$ as in (13.53).

13.17 In the proof of Theorem 13.6b, show that $\mathbf{A}_i$ is symmetric and idempotent that $\text{rank}(\mathbf{A}_i) = 1$, and that $\mathbf{A}_i\mathbf{A}_j = \mathbf{O}$.

13.18 **(a)** Show that $\mathbf{J}/kn$ in the first term on the right side of (13.59) is idempotent with one eigenvalue equal to 1 and the others equal to 0.

(b) Show that $\mathbf{j}$ is an eigenvector corresponding to the nonzero eigenvalue of $\mathbf{J}/kn$.

13.19 In Example 13.6b, show that $\mathbf{v}_0'\mathbf{v}_0 = 1$ and $\mathbf{v}_0'\mathbf{v}_1 = 0$.

13.20 Show that $\mathbf{j}'\mathbf{x}_{2\cdot01} = 0$ and $\mathbf{x}_{0\cdot1}'\mathbf{x}_{2\cdot01} = 0$ as in (13.69).

13.21 Show that $\mathbf{x}_{3\cdot012}$ has the form given in (13.70).

13.22 Show that $\mathbf{x}_{3\cdot012}$ is orthogonal to each of $\mathbf{j}$, $\mathbf{x}_{1\cdot0}$, and $\mathbf{x}_{2\cdot01}$, as noted following (13.70).

13.23 Show that $\mathbf{z}_3 = (-1, \ldots, -1, 3, \ldots, 3, -3, \ldots, -3, 1, \ldots, 1)'$ as in (13.74).

13.24 Show that $\beta_0 = \theta_0 - 5\theta_1 + 5\theta_2 - 35\theta_3$, $\beta_1 = 2\theta_1 - 5\theta_2 + (16.7/.3)\theta_3$, $\beta_2 = \theta_2 - 25\theta_3$, and $\beta_3 = \theta_3/.3$, as in (13.75).

13.25 Show that the elements of $\hat{\boldsymbol{\theta}} = (\mathbf{Z}'\mathbf{Z})^{-1}\mathbf{Z}'\mathbf{y}$ are of the form $\mathbf{z}_i'\mathbf{y}/\mathbf{z}_i'\mathbf{z}_i$ as in (13.76).

13.26 Show that $\text{SS}(\theta_k) = \hat{\boldsymbol{\beta}}'\mathbf{X}'\mathbf{y} - \hat{\boldsymbol{\beta}}^{*\prime}\mathbf{X}_1'\mathbf{y}$ as in (13.78).

13.27 If the means $\bar{y}_{1.}, \bar{y}_{2.}, \bar{y}_{3.}$, and $\bar{y}_{4.}$ have the quadratic trend $\bar{y}_{1.} = 1$, $\bar{y}_{2.} = 2$, $\bar{y}_{3.} = 2$, $\bar{y}_{4.} = 1$, show that the linear and cubic contrasts are zero, but the quadratic contrast is not zero.

TABLE 13.6 Blood Sugar Levels (mg/100 g) for 10 Animals from Each of Five Breeds (A–E)

A	B	C	D	E
124	111	117	104	142
116	101	142	128	139
101	130	121	130	133
118	108	123	103	120
118	127	121	121	127
120	129	148	119	149
110	122	141	106	150
127	103	122	107	149
106	122	139	107	120
130	127	125	115	116

13.28 Blood sugar levels (mg/100g) were measured on 10 animals from each of five breeds (Daniel 1974, p. 197). The results are presented in Table 13.6.

(a) Test the hypothesis of equality of means for the five breeds.

(b) Make the following comparisons by means of orthogonal contrasts:

A, B, C, vs. D, E; A, B, vs. C; A vs. B; D vs. E.

13.29 In Table 13.7, we have the amount of insulin released from specimens of pancreatic tissue treated with five concentrations of glucose (Daniel 1974, p. 182).

(a) Test the hypothesis of equality of means for the five glucose concentrations.

(b) Assuming that the levels of glucose concentration are equally spaced, use orthogonal polynomial contrasts to test for linear, quadratic, cubic, and quartic trends.

13.30 A different stimulus was given to each of three groups of 14 animals (Daniel 1974, p. 196). The response times in seconds are given in Table 13.8.

(a) Test the hypothesis of equal mean response times.

(b) Using orthogonal contrasts, make the two comparisons of stimuli: 1 versus 2, 3; and 2 versus 3.

13.31 The tensile strength (kg) was measured for 12 wires from each of nine cables (Hald 1952, p. 434). The results are given in Table 13.9.

(a) Test the hypothesis of equal mean strengths for the nine cables.

(b) The first four cables were made from one type of raw material and the other five from another type. Compare these two types by means of a contrast.

TABLE 13.7 Insulin Released at Five Different Glucose Concentrations (1–5)

1	2	3	4	5
1.53	3.15	3.89	8.18	5.86
1.61	3.96	4.80	5.64	5.46
3.75	3.59	3.69	7.36	5.96
2.89	1.89	5.70	5.33	6.49
3.26	1.45	5.62	8.82	7.81
2.83	3.49	5.79	5.26	9.03
2.86	1.56	4.75	8.75	7.49
2.59	2.44	5.33	7.10	8.98

TABLE 13.8 Response Times (in seconds) to Three Stimuli

Stimulus			Stimulus		
1	2	3	1	2	3
16	6	8	17	6	9
14	7	10	7	8	11
14	7	9	17	6	11
13	8	10	19	4	9
13	4	6	14	9	10
12	8	7	15	5	9
12	9	10	20	5	5

TABLE 13.9 Tensile Strength (kg) of Wires from Nine Cables (1–9)

1	2	3	4	5	6	7	8	9
345	329	340	328	347	341	339	339	342
327	327	330	344	341	340	340	340	346
335	332	325	342	345	335	342	347	347
338	348	328	350	340	336	341	345	348
330	337	338	335	350	339	336	350	355
334	328	332	332	346	340	342	348	351
335	328	335	328	345	342	347	341	333
340	330	340	340	342	345	345	342	347
333	328	335	337	330	346	336	340	348
335	330	329	340	338	347	342	345	341

TABLE 13.10 Scores for Physical Therapy Patients Subjected to Four Treatment Programs (1–14)

1	2	3	4
64	76	58	95
88	70	74	90
72	90	66	80
80	80	60	87
79	75	82	88
71	82	75	85

TABLE 13.11 Weight Gain of Pigs Subjected to Five Treatments (1–5)

1	2	3	4	5
165	168	164	185	201
156	180	156	195	189
159	180	156	195	189
159	180	189	184	173
167	166	138	201	193
170	170	153	165	164
146	161	190	175	160
130	171	160	187	200
151	169	172	177	142
164	179	142	166	184
158	191	155	165	149

13.32 Four groups of physical therapy patients were given different treatments (Daniel 1974, p. 195). The scores measuring treatment effectiveness are given in Table 13.10.

(a) Test the hypothesis of equal mean treatment effects.

(b) Using contrasts, compare treatments 1, 2 versus 3, 4; 1 versus 2; and 3 versus 4.

13.33 Weight gains in pigs subjected to five different treatments are given in Table 13.11 (Crampton and Hopkins 1934).

(a) Test the hypothesis of equal mean treatment effects.

(b) Using contrasts, compare treatments 1, 2, 3 versus 4; 1, 2 versus 3; and 1 versus 2.

14 Two-Way Analysis-of-Variance: Balanced Case

The two-way model without interaction has been illustrated in Section 12.1.2, Example 12.2.2b, and Section 12.8. In this chapter, we consider the two-way ANOVA model with interaction. In Section 14.1 we discuss the model and attendant assumptions. In Section 14.2 we consider estimable functions involving main effects and interactions. In Section 14.3 we discuss estimation of the parameters, including solutions to the normal equations using side conditions and also using a generalized inverse. In Section 14.4 we develop a hypothesis test for the interaction using a full–reduced model, and we obtain tests for main effects using the general linear hypothesis as well as the full–reduced-model approach. In Section 14.5 we derive expected mean squares from the basic definition and also using a general linear hypothesis approach. Throughout this chapter we consider only the balanced two-way model. The unbalanced case is covered in Chapter 15.

14.1 THE TWO-WAY MODEL

The two-way balanced model can be specified as follows:

$$y_{ijk} = \mu + \alpha_i + \beta_j + \gamma_{ij} + \varepsilon_{ijk} \tag{14.1}$$

$$i = 1, 2, \ldots, a, \quad j = 1, 2, \ldots, b, \quad k = 1, 2, \ldots, n.$$

The effect of factor A at the ith level is α_i, and the term β_j is due to the jth level of factor B. The term γ_{ij} represents the interaction AB between the ith level of A and the jth level of B. If an interaction is present, the difference $\alpha_1 - \alpha_2$, for example, is not estimable and the hypothesis H_0: $\alpha_1 = \alpha_2 = \cdots = \alpha_a$ cannot be tested. In Section 14.4, we discuss modifications of this hypothesis that are testable.

There are two experimental situations in which the model in (14.1) may arise. In the first setup, factors A and B represent two types of treatment, for example, various amounts of nitrogen and potassium applied in an agricultural experiment. We apply

Linear Models in Statistics, Second Edition, by Alvin C. Rencher and G. Bruce Schaalje

each of the ab combinations of the levels of A and B to n randomly selected experimental units. In the second situation, the populations exist naturally, for example, gender (males and females) and political preference (Democrats, Republicans, and Independents). A random sample of n observations is obtained from each of the ab populations.

Additional assumptions that form part of the model are the following:

1. $E(\varepsilon_{ijk}) = 0$ for all i, j, k.
2. $\text{var}(\varepsilon_{ijk}) = \sigma^2$ for all i, j, k.
3. $\text{cov}(\varepsilon_{ijk}, \varepsilon_{rst}) = 0$ for $(i, j, k) \neq (r, s, t)$.
4. Another assumption that we sometimes add to the model is that ε_{ijk} is $N(0, \sigma^2)$ for all i, j, k.

From assumption 1, we have $E(y_{ijk}) = \mu_{ij} = \mu + \alpha_i + \beta_j + \gamma_{ij}$, and we can rewrite the model in the form

$$
\begin{aligned}
y_{ijk} &= \mu_{ij} + \varepsilon_{ijk}, \\
&\quad i = 1, 2, \ldots, a, \quad j = 1, 2, \ldots, b, \quad k = 1, 2, \ldots, n,
\end{aligned} \tag{14.2}
$$

where $\mu_{ij} = E(y_{ijk})$ is the mean of a random observation in the (ij)th cell.

In the next section, we consider estimable functions of the parameters α_i, β_j, and γ_{ij}.

14.2 ESTIMABLE FUNCTIONS

In the first part of this section, we use $a = 3$, $b = 2$, and $n = 2$ for expositional purposes. For this special case, the model in (14.1) becomes

$$
y_{ijk} = \mu + \alpha_i + \beta_j + \gamma_{ij} + \varepsilon_{ijk}, \quad i = 1, 2, 3, \quad j = 1, 2, \quad k = 1, 2. \tag{14.3}
$$

The 12 observations in (14.3) can be expressed in matrix form as

$$
\begin{pmatrix} y_{111} \\ y_{112} \\ y_{121} \\ y_{122} \\ y_{211} \\ y_{212} \\ y_{221} \\ y_{222} \\ y_{311} \\ y_{312} \\ y_{321} \\ y_{322} \end{pmatrix}
=
\left(\begin{array}{c|ccc|cc|cccccc}
1 & 1 & 0 & 0 & 1 & 0 & 1 & 0 & 0 & 0 & 0 & 0 \\
1 & 1 & 0 & 0 & 1 & 0 & 1 & 0 & 0 & 0 & 0 & 0 \\
1 & 1 & 0 & 0 & 0 & 1 & 0 & 1 & 0 & 0 & 0 & 0 \\
1 & 1 & 0 & 0 & 0 & 1 & 0 & 1 & 0 & 0 & 0 & 0 \\
1 & 0 & 1 & 0 & 1 & 0 & 0 & 0 & 1 & 0 & 0 & 0 \\
1 & 0 & 1 & 0 & 1 & 0 & 0 & 0 & 1 & 0 & 0 & 0 \\
1 & 0 & 1 & 0 & 0 & 1 & 0 & 0 & 0 & 1 & 0 & 0 \\
1 & 0 & 1 & 0 & 0 & 1 & 0 & 0 & 0 & 1 & 0 & 0 \\
1 & 0 & 0 & 1 & 1 & 0 & 0 & 0 & 0 & 0 & 1 & 0 \\
1 & 0 & 0 & 1 & 1 & 0 & 0 & 0 & 0 & 0 & 1 & 0 \\
1 & 0 & 0 & 1 & 0 & 1 & 0 & 0 & 0 & 0 & 0 & 1 \\
1 & 0 & 0 & 1 & 0 & 1 & 0 & 0 & 0 & 0 & 0 & 1
\end{array}\right)
\begin{pmatrix} \mu \\ \alpha_1 \\ \alpha_2 \\ \alpha_3 \\ \beta_1 \\ \beta_2 \\ \gamma_{11} \\ \gamma_{12} \\ \gamma_{21} \\ \gamma_{22} \\ \gamma_{31} \\ \gamma_{32} \end{pmatrix}
+
\begin{pmatrix} \varepsilon_{111} \\ \varepsilon_{112} \\ \varepsilon_{121} \\ \varepsilon_{122} \\ \varepsilon_{211} \\ \varepsilon_{212} \\ \varepsilon_{221} \\ \varepsilon_{222} \\ \varepsilon_{311} \\ \gamma_{312} \\ \varepsilon_{321} \\ \varepsilon_{322} \end{pmatrix}
\tag{14.4}
$$

or

$$\mathbf{y} = \mathbf{X}\boldsymbol{\beta} + \boldsymbol{\varepsilon},$$

where $\mathbf{y}$ is 12×1, $\mathbf{X}$ is 12×12, and $\boldsymbol{\beta}$ is 12×1. (If we added another replication, so that $n = 3$, then $\mathbf{y}$ would be 18×1, $\mathbf{X}$ would be 18×12, but $\boldsymbol{\beta}$ would remain 12×1.) The matrix $\mathbf{X}'\mathbf{X}$ is given by

$$\mathbf{X}'\mathbf{X} = \left(\begin{array}{c|ccc|cc|cccccc} 12 & 4 & 4 & 4 & 6 & 6 & 2 & 2 & 2 & 2 & 2 & 2 \\ \hline 4 & 4 & 0 & 0 & 2 & 2 & 2 & 2 & 0 & 0 & 0 & 0 \\ 4 & 0 & 4 & 0 & 2 & 2 & 0 & 0 & 2 & 2 & 0 & 0 \\ 4 & 0 & 0 & 4 & 2 & 2 & 0 & 0 & 0 & 0 & 2 & 2 \\ \hline 6 & 2 & 2 & 2 & 6 & 0 & 2 & 0 & 2 & 0 & 2 & 0 \\ 6 & 2 & 2 & 2 & 0 & 6 & 0 & 2 & 0 & 2 & 0 & 2 \\ \hline 2 & 2 & 0 & 0 & 2 & 0 & 2 & 0 & 0 & 0 & 0 & 0 \\ 2 & 2 & 0 & 0 & 0 & 2 & 0 & 2 & 0 & 0 & 0 & 0 \\ 2 & 0 & 2 & 0 & 2 & 0 & 0 & 0 & 2 & 0 & 0 & 0 \\ 2 & 0 & 2 & 0 & 0 & 2 & 0 & 0 & 0 & 2 & 0 & 0 \\ 2 & 0 & 0 & 2 & 2 & 0 & 0 & 0 & 0 & 0 & 2 & 0 \\ 2 & 0 & 0 & 2 & 0 & 2 & 0 & 0 & 0 & 0 & 0 & 2 \end{array}\right). \tag{14.5}$$

The partitioning in $\mathbf{X}'\mathbf{X}$ corresponds to that in $\mathbf{X}$ in (14.4), where there is a column for μ, three columns for the three α's, two columns for the two β's, and six columns for the six γ's.

In both $\mathbf{X}$ and $\mathbf{X}'\mathbf{X}$, the first six columns can be obtained as linear combinations of the last six columns, which are clearly linearly independent. Hence rank($\mathbf{X}$)= rank($\mathbf{X}'\mathbf{X}$) = 6 [in general, rank($\mathbf{X}$) = ab].

Since rank($\mathbf{X}$) = 6, we can find six linearly independent estimable functions of the parameters (see Theorem 12.2c). By Theorem 12.2b, we can obtain these estimable functions from $\mathbf{X}\boldsymbol{\beta}$. Using rows 1, 3, 5, 7, 9, and 11 of $E(\mathbf{y}) = \mathbf{X}\boldsymbol{\beta}$, we obtain $E(y_{ijk}) = \mu_{ij} = \mu + \alpha_i + \beta_j + \gamma_{ij}$ for $i = 1, 2, 3$ and $j = 1, 2$:

$$\begin{aligned} \mu_{11} &= \mu + \alpha_1 + \beta_1 + \gamma_{11}, & \mu_{12} &= \mu + \alpha_1 + \beta_2 + \gamma_{12} \\ \mu_{21} &= \mu + \alpha_2 + \beta_1 + \gamma_{21}, & \mu_{22} &= \mu + \alpha_2 + \beta_2 + \gamma_{22} \\ \mu_{31} &= \mu + \alpha_3 + \beta_1 + \gamma_{31}, & \mu_{32} &= \mu + \alpha_3 + \beta_2 + \gamma_{32}. \end{aligned} \tag{14.6}$$

These can also be obtained from the last six rows of $\mathbf{X}'\mathbf{X}\boldsymbol{\beta}$ (see Theorem 12.2b).

By taking linear combinations of the six functions in (14.6), we obtain the following estimable functions (e.g., $\theta_1 = \mu_{11} - \mu_{21}$ and $\theta_1' = \mu_{12} - \mu_{22}$):

$$\begin{aligned} \mu_{11} &= \mu + \alpha_1 + \beta_1 + \gamma_{11} \\ \theta_1 &= \alpha_1 - \alpha_2 + \gamma_{11} - \gamma_{21} \quad \text{or} \quad \theta_1' = \alpha_1 - \alpha_2 + \gamma_{12} - \gamma_{22} \\ \theta_2 &= \alpha_1 - \alpha_3 + \gamma_{11} - \gamma_{31} \quad \text{or} \quad \theta_2' = \alpha_1 - \alpha_3 + \gamma_{12} - \gamma_{32} \\ \theta_3 &= \beta_1 - \beta_2 + \gamma_{11} - \gamma_{12} \quad \text{or} \quad \theta_3' = \beta_1 - \beta_2 + \gamma_{21} - \gamma_{22} \end{aligned} \tag{14.7}$$

$$\text{or} \quad \theta_3'' = \beta_1 - \beta_2 + \gamma_{31} - \gamma_{32}$$

$$\theta_4 = \gamma_{11} - \gamma_{12} - \gamma_{21} + \gamma_{22}$$

$$\theta_5 = \gamma_{11} - \gamma_{12} - \gamma_{31} + \gamma_{32}.$$

The alternative expressions for θ_4 and θ_5 are of the form

$$\gamma_{ij} - \gamma_{ij'} - \gamma_{i'j} + \gamma_{i'j'}, \quad i, i' = 1, 2, 3, \quad j, j' = 1, 2, \quad i \neq i', \quad j \neq j'. \tag{14.8}$$

[For general a and b, we likewise obtain estimable functions of the form of (14.7) and (14.8).]

In θ_4 and θ_5 in (14.7), we see that there are estimable contrasts in the γ_{ij}'s, but in θ_1, θ_2, and θ_3 (and in the alternative expressions θ_1', θ_2', θ_3', and θ_3'') there are no estimable contrasts in the α's alone or β's alone. (This is also true for the case of general a and b.)

To obtain a single expression involving $\alpha_1 - \alpha_2$ for later use in comparing the α values in a hypothesis test (see Section 14.4.2b), we average θ_1 and θ_1':

$$\begin{aligned} \tfrac{1}{2}(\theta_1 + \theta_1') &= \alpha_1 - \alpha_2 + \tfrac{1}{2}(\gamma_{11} + \gamma_{12}) - \tfrac{1}{2}(\gamma_{21} + \gamma_{22}) \\ &= \alpha_1 - \alpha_2 + \bar{\gamma}_{1.} - \bar{\gamma}_{2.}. \end{aligned} \tag{14.9}$$

For $\alpha_1 - \alpha_3$, we have

$$\begin{aligned} \tfrac{1}{2}(\theta_2 + \theta_2') &= \alpha_1 - \alpha_3 + \tfrac{1}{2}(\gamma_{11} + \gamma_{12}) - \tfrac{1}{2}(\gamma_{31} + \gamma_{32}) \\ &= \alpha_1 - \alpha_3 + \bar{\gamma}_{1.} - \bar{\gamma}_{3.}. \end{aligned} \tag{14.10}$$

Similarly, the average of θ_3, θ_3', and θ_3'' yields

$$\begin{aligned} \tfrac{1}{3}(\theta_3 + \theta_3' + \theta_3'') &= \beta_1 - \beta_2 + \tfrac{1}{3}(\gamma_{11} + \gamma_{21} + \gamma_{31}) - \tfrac{1}{3}(\gamma_{12} + \gamma_{22} + \gamma_{32}) \\ &= \beta_1 - \beta_2 + \bar{\gamma}_{.1} - \bar{\gamma}_{.2}. \end{aligned} \tag{14.11}$$

From (14.1) and assumption 1 in Section 14.1, we have

$$E(y_{ijk}) = E(\mu + \alpha_i + \beta_j + \gamma_{ij} + \varepsilon_{ijk}),$$
$$i = 1, 2, \ldots, a, \quad j = 1, 2, \ldots, b, \quad k = 1, 2, \ldots, n$$

or

$$\mu_{ij} = \mu + \alpha_i + \beta_j + \gamma_{ij} \tag{14.12}$$

[see also (14.2) and (14.6)]. In Section 12.1.2, we demonstrated that for a simple additive (no-interaction) model the side conditions on the α's and β's led to redefined α^*'s and β^*'s that could be expressed as deviations from means, for example, $\alpha_i^* = \bar{\mu}_{i.} - \bar{\mu}_{..}$. We now extend this formulation to an interaction model for μ_{ij}:

$$\begin{aligned} \mu_{ij} &= \bar{\mu}_{..} + (\bar{\mu}_{i.} - \bar{\mu}_{..}) + (\bar{\mu}_{.j} - \bar{\mu}_{..}) + (\mu_{ij} - \bar{\mu}_{i.} - \bar{\mu}_{.j} + \bar{\mu}_{..}) \\ &= \mu^* + \alpha_i^* + \beta_j^* + \gamma_{ij}^*, \end{aligned} \tag{14.13}$$

where

$$\begin{aligned} &\mu^* = \bar{\mu}_{..}, \quad \alpha_i^* = \bar{\mu}_{i.} - \bar{\mu}_{..}, \quad \beta_j^* = \bar{\mu}_{.j} - \bar{\mu}_{..}, \\ &\gamma_{ij}^* = \mu_{ij} - \bar{\mu}_{i.} - \bar{\mu}_{.j} + \bar{\mu}_{..}. \end{aligned} \tag{14.14}$$

With these definitions, it follows that

$$\begin{aligned} &\sum_{i=1}^{a} \alpha_i^* = 0, \quad \sum_{j=1}^{b} \beta_j^* = 0, \\ &\sum_{i=1}^{a} \gamma_{ij}^* = 0 \quad \text{for all} \quad j = 1, 2, \ldots, b, \\ &\sum_{j=1}^{b} \gamma_{ij}^* = 0 \quad \text{for all} \quad i = 1, 2, \ldots, a. \end{aligned} \tag{14.15}$$

Using (14.12), we can write α_i^*, β_j^*, and γ_{ij}^* in (14.14) in terms of the original parameters; for example, α_i^* becomes

$$\begin{aligned} \alpha_i^* &= \bar{\mu}_{i.} - \bar{\mu}_{..} = \frac{1}{b}\sum_{j=1}^{b} \mu_{ij} - \frac{1}{ab}\sum_{ij} \mu_{ij} \\ &= \frac{1}{b}\sum_{j} (\mu + \alpha_i + \beta_j + \gamma_{ij}) - \frac{1}{ab}\sum_{ij} (\mu + \alpha_i + \beta_j + \gamma_{ij}) \\ &= \frac{1}{b}\left(b\mu + b\alpha_i + \sum_{j} \beta_j + \sum_{j} \gamma_{ij} \right) \\ &\quad - \frac{1}{ab}\left(ab\mu + b\sum_{i} \alpha_i + a\sum_{j} \beta_j + \sum_{ij} \gamma_{ij} \right) \\ &= \mu + \alpha_i + \bar{\beta}_{.} + \bar{\gamma}_{i.} - \mu - \bar{\alpha}_{.} - \bar{\beta}_{.} - \bar{\gamma}_{..} \\ &= \alpha_i - \bar{\alpha}_{.} + \bar{\gamma}_{i.} - \bar{\gamma}_{..}. \end{aligned} \tag{14.16}$$

Similarly

$$\beta_j^* = \beta_j - \bar{\beta}_. + \bar{\gamma}_{.j} - \bar{\gamma}_{..}, \tag{14.17}$$

$$\gamma_{ij}^* = \gamma_{ij} - \bar{\gamma}_{i.} - \bar{\gamma}_{.j} + \bar{\gamma}_{..}. \tag{14.18}$$

14.3 ESTIMATORS OF $\boldsymbol{\lambda}'\boldsymbol{\beta}$ AND σ^2

We consider estimation of estimable functions $\boldsymbol{\lambda}'\boldsymbol{\beta}$ in Section 14.3.1 and estimation of σ^2 in Section 14.3.2.

14.3.1 Solving the Normal Equations and Estimating $\boldsymbol{\lambda}'\boldsymbol{\beta}$

We discuss two approaches for solving the normal equations $\mathbf{X}'\mathbf{X}\hat{\boldsymbol{\beta}} = \mathbf{X}'\mathbf{y}$ and for obtaining estimates of an estimable function $\boldsymbol{\lambda}'\boldsymbol{\beta}$.

14.3.1.1 Side Conditions

From $\mathbf{X}$ and $\mathbf{y}$ in (14.4), we obtain $\mathbf{X}'\mathbf{y}$ for the special case $a = 3$, $b = 2$, and $n = 2$:

$$\mathbf{X}'\mathbf{y} = (y_{...}, y_{1..}, y_{2..}, y_{3..}, y_{1..}, y_{.2.}, y_{11.}, y_{12.}, y_{21.}, y_{22.}, y_{31.}, y_{32.})'. \tag{14.19}$$

On the basis of $\mathbf{X}'\mathbf{y}$ in (14.19) and $\mathbf{X}'\mathbf{X}$ in (14.5), we write the normal equations $\mathbf{X}'\mathbf{X}\hat{\boldsymbol{\beta}} = \mathbf{X}'\mathbf{y}$ in terms of general a, b, and n:

$$abn\hat{\mu} + bn\sum_{i=1}^{a}\hat{\alpha}_i + an\sum_{j=1}^{b}\hat{\beta}_j + n\sum_{i=1}^{a}\sum_{j=1}^{b}\hat{\gamma}_{ij} = y_{...},$$

$$bn\hat{\mu} + bn\hat{\alpha}_i + n\sum_{j=1}^{b}\hat{\beta}_j + n\sum_{j=1}^{b}\hat{\gamma}_{ij} = y_{i..}, \quad i = 1, 2, \ldots, a,$$

$$an\hat{\mu} + n\sum_{i=1}^{a}\hat{\alpha}_i + an\hat{\beta}_j + n\sum_{i=1}^{a}\hat{\gamma}_{ij} = y_{.j.}, \quad j = 1, 2, \ldots, b,$$

$$n\hat{\mu} + n\hat{\alpha}_i + n\hat{\beta}_j + n\hat{\gamma}_{ij} = y_{ij.}, \quad i = 1, 2, \ldots, a, \quad j = 1, 2, \ldots, b. \tag{14.20}$$

With the side conditions $\sum_i \hat{\alpha}_i = 0$, $\sum_j \hat{\beta}_j = 0$, $\sum_i \hat{\gamma}_{ij} = 0$, and $\sum_j \hat{\gamma}_{ij} = 0$, the solution of the normal equations in (14.20) is given by

$$
\begin{aligned}
\hat{\mu} &= \frac{y_{...}}{abn} = \bar{y}_{...}, \\
\hat{\alpha}_i &= \frac{y_{i..}}{bn} - \hat{\mu} = \bar{y}_{i..} - \bar{y}_{...}, \\
\hat{\beta}_j &= \frac{y_{.j.}}{an} - \hat{\mu} = \bar{y}_{.j.} - \bar{y}_{...}, \\
\hat{\gamma}_{ij} &= \frac{y_{ij.}}{n} - \frac{y_{i..}}{bn} - \frac{y_{.j.}}{an} + \frac{y_{...}}{abn}, \\
&= \bar{y}_{ij.} - \bar{y}_{i..} - \bar{y}_{.j.} + \bar{y}_{...}.
\end{aligned} \tag{14.21}
$$

These are unbiased estimators of the parameters μ^*, α_i^*, β_j^*, γ_{ij}^* in (14.14), subject to the side conditions in (14.15). If side conditions are not imposed on the parameters, then the estimators in (14.21) are not unbiased estimators of individual parameters, but these estimators can still be used in estimable functions. For example, consider the estimable function $\boldsymbol{\lambda}'\boldsymbol{\beta}$ in (14.9) (for $a = 3$, $b = 2$):

$$\boldsymbol{\lambda}'\boldsymbol{\beta} = \alpha_1 - \alpha_2 + \tfrac{1}{2}(\gamma_{11} + \gamma_{12}) - \tfrac{1}{2}(\gamma_{21} + \gamma_{22}).$$

By Theorem 12.3a and (14.21), the estimator is given by

$$
\begin{aligned}
\boldsymbol{\lambda}'\hat{\boldsymbol{\beta}} &= \hat{\alpha}_1 - \hat{\alpha}_2 + \tfrac{1}{2}(\hat{\gamma}_{11} + \hat{\gamma}_{12}) - \tfrac{1}{2}(\hat{\gamma}_{21} + \hat{\gamma}_{22}) \\
&= \bar{y}_{1..} - \bar{y}_{...} - (\bar{y}_{2..} - \bar{y}_{...}) + \tfrac{1}{2}(\bar{y}_{11.} - \bar{y}_{1..} - \bar{y}_{.1.} + \bar{y}_{...}) \\
&\quad + \tfrac{1}{2}(\bar{y}_{12.} - \bar{y}_{1..} - \bar{y}_{.2.} + \bar{y}_{...}) - \tfrac{1}{2}(\bar{y}_{21.} - \bar{y}_{2..} - \bar{y}_{.1.} + \bar{y}_{...}) \\
&\quad - \tfrac{1}{2}(\bar{y}_{22.} - \bar{y}_{2..} - \bar{y}_{.2.} + \bar{y}_{...}).
\end{aligned}
$$

Since $\bar{y}_{11.} + \bar{y}_{12.} = 2\bar{y}_{1..}$ and $\bar{y}_{21.} + \bar{y}_{22.} = 2\bar{y}_{2..}$, the estimator $\boldsymbol{\lambda}'\hat{\boldsymbol{\beta}} = \hat{\alpha}_1 - \hat{\alpha}_2 + \frac{1}{2}(\hat{\gamma}_{11} + \hat{\gamma}_{12}) - \frac{1}{2}(\hat{\gamma}_{21} + \hat{\gamma}_{22})$ reduces to

$$\boldsymbol{\lambda}'\hat{\boldsymbol{\beta}} = \hat{\alpha}_1 - \hat{\alpha}_2 + \tfrac{1}{2}(\hat{\gamma}_{11} + \hat{\gamma}_{12}) - \tfrac{1}{2}(\hat{\gamma}_{21} + \hat{\gamma}_{22}) = \bar{y}_{1..} - \bar{y}_{2..}. \tag{14.22}$$

This estimator of $\alpha_1 - \alpha_2 + \frac{1}{2}(\gamma_{11} + \gamma_{12}) - \frac{1}{2}(\gamma_{21} + \gamma_{22})$ is the same as the estimator we would have for $\alpha_1^* - \alpha_2^*$, using $\hat{\alpha}_1$ and $\hat{\alpha}_2$ as estimators of α_1^* and α_2^*:

$$\widehat{\alpha_1^* - \alpha_2^*} = \hat{\alpha}_1 - \hat{\alpha}_2 = \bar{y}_{1..} - \bar{y}_{...} - (\bar{y}_{2..} - \bar{y}_{...}) = \bar{y}_{1..} - \bar{y}_{2..}.$$

By Theorem 12.3d, such estimators are BLUE. If we also assume that ε_{ijk} is $N(0, \sigma^2)$, then by Theorem 12.3h, the estimators are minimum variance unbiased estimators.

14.3.1.2 Generalized Inverse

By Corollary 1 to Theorem 2.8b, a generalized inverse of $\mathbf{X}'\mathbf{X}$ in (14.5) is given by

$$(\mathbf{X}'\mathbf{X})^- = \tfrac{1}{2}\begin{pmatrix} \mathbf{O} & \mathbf{O} \\ \mathbf{O} & \mathbf{I}_6 \end{pmatrix}, \tag{14.23}$$

where the $\mathbf{O}$s are 6×6. Then by (12.13) and (14.19), a solution to the normal equations for $a = 3$ and $b = 2$ is given by

$$\begin{aligned} \hat{\boldsymbol{\beta}} &= (\mathbf{X}'\mathbf{X})^-\mathbf{X}'\mathbf{y} \\ &= (0, 0, 0, 0, 0, 0, \bar{y}_{11.}, \bar{y}_{12.}, \bar{y}_{21.}, \bar{y}_{22.}, \bar{y}_{31.}, \bar{y}_{32.})'. \end{aligned} \tag{14.24}$$

The estimators in (14.24) are different from those in (14.21), but they give the same estimators of estimable functions. For example, for $\boldsymbol{\lambda}'\boldsymbol{\beta} = \alpha_1 - \alpha_2 + \frac{1}{2}(\gamma_{11} + \gamma_{12}) - \frac{1}{2}(\gamma_{21} + \gamma_{22})$ in (14.9), we have

$$\begin{aligned} \boldsymbol{\lambda}'\hat{\boldsymbol{\beta}} &= \hat{\alpha}_1 - \hat{\alpha}_2 + \tfrac{1}{2}[\hat{\gamma}_{11} + \hat{\gamma}_{12} - (\hat{\gamma}_{21} + \hat{\gamma}_{22})] \\ &= 0 - 0 + \tfrac{1}{2}[\bar{y}_{11.} + \bar{y}_{12.} - (\bar{y}_{21.} + \bar{y}_{22.})]. \end{aligned}$$

It was noted preceding (14.22) that $\bar{y}_{11.} + \bar{y}_{12.} = 2\bar{y}_{1..}$ and $\bar{y}_{21.} + \bar{y}_{22.} = 2\bar{y}_{2..}$. Thus $\boldsymbol{\lambda}'\hat{\boldsymbol{\beta}}$ becomes

$$\boldsymbol{\lambda}'\hat{\boldsymbol{\beta}} = \tfrac{1}{2}(2\bar{y}_{1..} - 2\bar{y}_{2..}) = \bar{y}_{1..} - \bar{y}_{2..},$$

which is the same estimator as that obtained in (14.22) using $\hat{\boldsymbol{\beta}}$ in (14.21).

14.3.2 An Estimator for σ^2

For the two-way model in (14.1), assumption 2 states that $\text{var}(\varepsilon_{ijk}) = \sigma^2$ for all i, j, k. To estimate σ^2, we use (12.22), $s^2 = \text{SSE}/ab(n-1)$, where abn is the number of rows of $\mathbf{X}$ and ab is the rank of $\mathbf{X}$. By (12.20) and (12.21), we have

$$\text{SSE} = \mathbf{y}'\mathbf{y} - \hat{\boldsymbol{\beta}}'\mathbf{X}'\mathbf{y} = \mathbf{y}'[\mathbf{I} - \mathbf{X}(\mathbf{X}'\mathbf{X})^-\mathbf{X}']\mathbf{y}.$$

With $\hat{\boldsymbol{\beta}}$ from (14.24) and $\mathbf{X}'\mathbf{y}$ from (14.19), SSE can be written as

$$\begin{aligned}\text{SSE} &= \mathbf{y}'\mathbf{y} - \hat{\boldsymbol{\beta}}'\mathbf{X}'\mathbf{y} \\ &= \sum_{i=1}^{a}\sum_{j=1}^{b}\sum_{k=1}^{n} y_{ijk}^2 - \sum_{i=1}^{a}\sum_{j=1}^{b} \bar{y}_{ij.}y_{ij.} \\ &= \sum_{ijk} y_{ijk}^2 - n\sum_{ij}\bar{y}_{ij.}^2. \qquad (14.25)\end{aligned}$$

It can also be shown (see Problem 14.10) that this is equal to

$$\text{SSE} = \sum_{ijk}(y_{ijk} - \bar{y}_{ij.})^2. \qquad (14.26)$$

Thus, s^2 is given by either of the two forms

$$s^2 = \frac{\sum_{ijk}(y_{ijk} - \bar{y}_{ij.})^2}{ab(n-1)} \qquad (14.27)$$

$$= \frac{\sum_{ijk} y_{ijk}^2 - n\sum_{ij}\bar{y}_{ij.}^2}{ab(n-1)}. \qquad (14.28)$$

By Theorem 12.3e, $E(s^2) = \sigma^2$.

14.4 TESTING HYPOTHESES

In this section, we consider tests of hypotheses for the main effects A and B and for the interaction AB. Throughout this section, we assume that $\mathbf{y}$ is $N_{abn}(\mathbf{X}\boldsymbol{\beta}, \sigma^2\mathbf{I})$. For expositional convenience, we sometimes illustrate with $a = 3$ and $b = 2$.

14.4.1 Test for Interaction

In Section 14.4.1.1, we express the interaction hypothesis in terms of estimable parameters, and in Sections 14.4.1.2 and 14.4.1.3, we discuss two approaches to the full–reduced-model test.

14.4.1.1 The Interaction Hypothesis

By (14.8), estimable contrasts in the γ_{ij}'s have the form

$$\gamma_{ij} - \gamma_{ij'} - \gamma_{i'j} + \gamma_{i'j'}, \quad i \neq i', \quad j \neq j'. \qquad (14.29)$$

We now show that the interaction hypothesis can be expressed in terms of these estimable functions.

For the illustrative model in (14.3) with $a = 3$ and $b = 2$, the cell means in (14.12) are given in Figure 14.1. The B effect at the first level of A is $\mu_{11} - \mu_{12}$, the B effect at the second level of A is $\mu_{21} - \mu_{22}$, and the B effect at the third level of A is $\mu_{31} - \mu_{32}$.

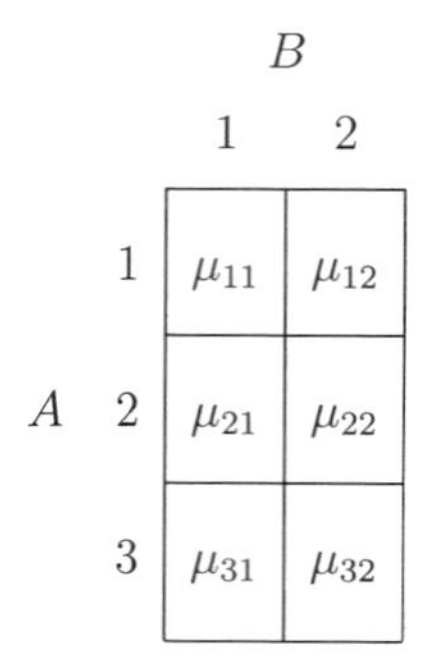

Figure 14.1 Cell means for the model in (14.2) and (14.12).

If these three B effects are equal, we have no interaction. If at least one effect differs from the other two, we have an interaction. The hypothesis of no interaction can therefore be expressed as

$$H_0: \mu_{11} - \mu_{12} = \mu_{21} - \mu_{22} = \mu_{31} - \mu_{32}. \tag{14.30}$$

To show that this hypothesis is testable, we first write the three differences in terms of the γ_{ij}'s by using (14.12). For the first two differences in (14.30), we obtain

$$\begin{aligned}
\mu_{11} - \mu_{12} &= \mu + \alpha_1 + \beta_1 + \gamma_{11} - (\mu + \alpha_1 + \beta_2 + \gamma_{12}) \\
&= \beta_1 - \beta_2 + \gamma_{11} - \gamma_{12}, \\
\mu_{21} - \mu_{22} &= \mu + \alpha_2 + \beta_1 + \gamma_{21} - (\mu + \alpha_2 + \beta_2 + \gamma_{22}) \\
&= \beta_1 - \beta_2 + \gamma_{21} - \gamma_{22}.
\end{aligned}$$

Then the equality $\mu_{11} - \mu_{12} = \mu_{21} - \mu_{22}$ in (14.30) becomes

$$\beta_1 - \beta_2 + \gamma_{11} - \gamma_{12} = \beta_1 - \beta_2 + \gamma_{21} - \gamma_{22}$$

or

$$\gamma_{11} - \gamma_{12} - \gamma_{21} + \gamma_{22} = 0. \tag{14.31}$$

The function $\gamma_{11} - \gamma_{12} - \gamma_{21} + \gamma_{22}$ on the left side of (14.31) is an estimable contrast [see (14.29)]. Similarly, the third difference in (14.30) becomes

$$\mu_{31} - \mu_{32} = \beta_1 - \beta_2 + \gamma_{31} - \gamma_{32},$$

and when this is set equal to $\mu_{21} - \mu_{22} = \beta_1 - \beta_2 + \gamma_{21} - \gamma_{22}$, we obtain

$$\gamma_{21} - \gamma_{22} - \gamma_{31} + \gamma_{32} = 0. \tag{14.32}$$

By (14.29), the function $\gamma_{21} - \gamma_{22} - \gamma_{31} + \gamma_{32}$ on the left side of (14.32) is estimable. Thus the two expressions in (14.31) and (14.32) are equivalent to the interaction hypothesis in (14.30), and the hypothesis is therefore testable.

Since the interaction hypothesis can be expressed in terms of estimable functions of γ_{ij}'s that do not involve α_i's or β_j's, we can proceed with a full–reduced-model approach. On the other hand, by (14.7), the α's and β's are not estimable without the γ's. We therefore have to redefine the main effects in order to get a test in the presence of interaction; see Section 14.4.2.

To get a reduced model from (14.1) or (14.3), we work with $\gamma_{ij}^* = \mu_{ij} - \bar{\mu}_{i.} - \bar{\mu}_{.j} + \bar{\mu}_{..}$ in (14.14), which is estimable [it can be estimated unbiasedly by $\hat{\gamma}_{ij} = \bar{y}_{ij.} - \bar{y}_{i..} - \bar{y}_{.j.} + \bar{y}_{...}$ in (14.21)]. Using (14.13), the model can be expressed in terms of parameters subject to the side conditions in (14.15):

$$y_{ijk} = \mu^* + \alpha_i^* + \beta_j^* + \gamma_{ij}^* + \varepsilon_{ijk}, \tag{14.33}$$

We can get a reduced model from (14.33) by setting $\gamma_{ij}^* = 0$.

In the following theorem, we show that H_0: $\gamma_{ij}^* = 0$ for all i, j is equivalent to the interaction hypothesis expressed as (14.30) or as (14.31) and (14.32). Since all three of these expressions involve $a = 3$ and $b = 2$, we continue with this illustrative special case.

Theorem 14.4a. Consider the model (14.33) for $a = 3$ and $b = 2$. The hypothesis H_0: $\gamma_{ij}^* = 0$, $i = 1, 2, 3$, $j = 1, 2$, is equivalent to (14.30)

$$H_0\colon \mu_{11} - \mu_{12} = \mu_{21} - \mu_{22} = \mu_{31} - \mu_{32}, \tag{14.34}$$

and to the equivalent form

$$H_0\colon \begin{pmatrix} \gamma_{11} - \gamma_{12} - \gamma_{21} + \gamma_{22} \\ \gamma_{21} - \gamma_{22} - \gamma_{31} + \gamma_{32} \end{pmatrix} = \begin{pmatrix} 0 \\ 0 \end{pmatrix} \tag{14.35}$$

obtained from (14.31) and (14.32).

Proof. To establish the equivalence of $\gamma_{ij}^* = 0$ and the first equality in (14.35), we find an expression for each γ_{ij} by setting $\gamma_{ij}^* = 0$. For γ_{12} and γ_{12}^*, for example, we use (14.18) to obtain

$$\gamma_{12}^* = \gamma_{12} - \bar{\gamma}_{1.} - \bar{\gamma}_{.2} + \bar{\gamma}_{..}. \tag{14.36}$$

Then $\gamma_{12}^* = 0$ gives

$$\gamma_{12} = \bar{\gamma}_{1.} + \bar{\gamma}_{.2} - \bar{\gamma}_{..}.$$

Similarly, from (14.18) and the equalities $\gamma_{11}^* = 0, \gamma_{21}^* = 0$, and $\gamma_{22}^* = 0$, we obtain

$$\gamma_{11} = \bar{\gamma}_{1.} + \bar{\gamma}_{.1} - \bar{\gamma}_{..}, \quad \gamma_{21} = \bar{\gamma}_{2.} + \bar{\gamma}_{.1} - \bar{\gamma}_{..}, \quad \gamma_{22} = \bar{\gamma}_{2.} + \bar{\gamma}_{.2} - \bar{\gamma}_{..}.$$

When these are substituted into $\gamma_{11} - \gamma_{12} - \gamma_{21} + \gamma_{22}$, we obtain

$$\begin{aligned}\gamma_{11} - \gamma_{12} - \gamma_{21} + \gamma_{22} &= \bar{\gamma}_{1.} + \bar{\gamma}_{.1} - \bar{\gamma}_{..} - (\bar{\gamma}_{1.} + \bar{\gamma}_{.2} - \bar{\gamma}_{..}) \\ &\quad - (\bar{\gamma}_{2.} + \bar{\gamma}_{.1} - \bar{\gamma}_{..}) + \bar{\gamma}_{2.} + \bar{\gamma}_{.2} - \bar{\gamma}_{..} \\ &= 0,\end{aligned}$$

which is the first equality in (14.35). The second equality in (14.35) is obtained similarly.

To show that the first equality in (14.34) is equivalent to the first equality in (14.35), we substitute $\mu_{ij} = \mu + \alpha_i + \beta_j + \gamma_{ij}$ into $\mu_{11} - \mu_{12} = \mu_{21} - \mu_{22}$:

$$\begin{aligned}0 &= \mu_{11} - \mu_{12} - \mu_{21} + \mu_{22} \\ &= \mu + \alpha_1 + \beta_1 + \gamma_{11} - (\mu + \alpha_1 + \beta_2 + \gamma_{12}) \\ &\quad - (\mu + \alpha_2 + \beta_1 + \gamma_{21}) + \mu + \alpha_2 + \beta_2 + \gamma_{22} \\ &= \gamma_{11} - \gamma_{12} - \gamma_{21} + \gamma_{22}.\end{aligned}$$

Similarly, the second equality in (14.34) is equivalent to the second equality in (14.35). □

In Section 14.4.1.2, we obtain the test for interaction based on the normal equations, and in Section 14.4.1.3, we give the test based on a generalized inverse.

14.4.1.2 Full–Reduced-Model Test Based on the Normal Equations

In this section, we develop the full–reduced-model test for interaction using the normal equations. We express the full model in terms of parameters subject to side conditions, as in (14.33)

$$y_{ijk} = \mu^* + \alpha_i^* + \beta_j^* + \gamma_{ij}^* + \varepsilon_{ijk}, \tag{14.37}$$

where $\mu^* = \bar{\mu}_{..}$, $\alpha_i^* = \bar{\mu}_{i.} - \bar{\mu}_{..}$, $\beta_j^* = \bar{\mu}_{.j} - \bar{\mu}_{..}$, and $\gamma_{ij}^* = \mu_{ij} - \bar{\mu}_{i.} - \bar{\mu}_{.j} + \bar{\mu}_{..}$ are as given in (14.14). The reduced model under H_0: $\gamma_{ij}^* = 0$ for all i and j is

$$y_{ijk} = \mu^* + \alpha_i^* + \beta_j^* + \varepsilon_{ijk}. \tag{14.38}$$

Since we are considering a balanced model, the parameters μ^*, α_i^*, and β_j^* (subject to side conditions) in the reduced model (14.38) are the same as those in the full model (14.37) [in (14.44), the estimates in the two models are also shown to be the same].

Using the notation of Chapter 13, the sum of squares for testing H_0: $\gamma_{ij}^* = 0$ is given by

$$\mathrm{SS}(\gamma|\mu, \alpha, \beta) = \mathrm{SS}(\mu, \alpha, \beta, \gamma) - \mathrm{SS}(\mu, \alpha, \beta). \tag{14.39}$$

The estimators $\hat{\mu}$, $\hat{\alpha}_i$, $\hat{\beta}_j$, $\hat{\gamma}_{ij}$ in (14.21) are unbiased estimators of μ^*, α_i^*, β_j^*, γ_{ij}^*. Extending $\mathbf{X}'\mathbf{y}$ in (14.19) from $a = 3$ and $b = 2$ to general a and b, we obtain

$$\begin{aligned}
\mathrm{SS}(\mu, \alpha, \beta, \gamma) &= \hat{\boldsymbol{\beta}}'\mathbf{X}'\mathbf{y} \\
&= \hat{\mu} y_{...} + \sum_{i=1}^{a} \hat{\alpha}_i y_{i..} + \sum_{j=1}^{b} \hat{\beta}_j y_{.j.} + \sum_{i=1}^{a}\sum_{j=1}^{b} \hat{\gamma}_{ij} y_{ij.} \\
&= \bar{y}_{...} y_{...} + \sum_i (\bar{y}_{i..} - \bar{y}_{...}) y_{i...} + \sum_j (\bar{y}_{.j.} - \bar{y}_{...}) y_{.j.} \\
&\quad + \sum_{ij} (\bar{y}_{ij.} - \bar{y}_{i..} - \bar{y}_{.j.} + \bar{y}_{...}) y_{ij.} \\
&= \frac{y_{...}^2}{abn} + \left(\sum_i \frac{y_{i..}^2}{bn} - \frac{y_{...}^2}{abn} \right) + \left(\sum_j \frac{y_{.j.}^2}{an} - \frac{y_{...}^2}{abn} \right) \\
&\quad + \left(\sum_{ij} \frac{y_{ij.}^2}{n} - \sum_i \frac{y_{i..}^2}{bn} - \sum_j \frac{y_{.j.}^2}{an} + \frac{y_{...}^2}{abn} \right) \qquad (14.40) \\
&= \sum_{ij} \frac{y_{ij.}^2}{n}. \qquad (14.41)
\end{aligned}$$

Note that we would obtain the same result using $\hat{\boldsymbol{\beta}}$ in (14.24) (extended to general a and b).

For the reduced model in (14.38), the $\mathbf{X}_1$ matrix and $\mathbf{X}_1'\mathbf{y}$ vector for $a = 3$ and $b = 2$ consist of the first six columns of $\mathbf{X}$ in (14.4') and the first six elements of $\mathbf{X}'\mathbf{y}$ in (14.19). We thus obtain

$$\mathbf{X}_1'\mathbf{X}_1 = \begin{pmatrix} 12 & 4 & 4 & 4 & 6 & 6 \\ 4 & 4 & 0 & 0 & 2 & 2 \\ 4 & 0 & 4 & 0 & 2 & 2 \\ 4 & 0 & 0 & 4 & 2 & 2 \\ 6 & 2 & 2 & 2 & 6 & 0 \\ 6 & 2 & 2 & 2 & 0 & 6 \end{pmatrix}, \quad \mathbf{X}_1'\mathbf{y} = \begin{pmatrix} y_{...} \\ y_{1..} \\ y_{2..} \\ y_{3..} \\ y_{.1.} \\ y_{.2.} \end{pmatrix}. \tag{14.42}$$

From the pattern in (14.42), we see that for general a and b the normal equations for the reduced model become

$$abn\hat{\mu} + bn\sum_{i=1}^{a}\hat{\alpha}_i + an\sum_{j=1}^{b}\hat{\beta}_j = y_{...},$$

$$bn\hat{\mu} + bn\hat{\alpha}_i + n\sum_{j=1}^{b}\hat{\beta}_j = y_{i..}, \quad i = 1, 2, \ldots, a, \tag{14.43}$$

$$an\hat{\mu} + n\sum_{i=1}^{a}\hat{\alpha}_i + an\hat{\beta}_j = y_{.j.}, \quad j = 1, 2, \ldots, b.$$

Using the side conditions $\sum_i \hat{\alpha}_i = 0$ and $\sum_j \hat{\beta}_j = 0$, we obtain the solutions

$$\hat{\mu} = \frac{y_{...}}{abn} = \bar{y}_{...}, \quad \hat{\alpha}_i = \frac{y_{i..}}{bn} - \hat{\mu} = \bar{y}_{i..} - \bar{y}_{...}, \quad \hat{\beta}_j = \frac{y_{.j.}}{an} - \hat{\mu} = \bar{y}_{.j.} - \bar{y}_{...}. \tag{14.44}$$

These solutions are the same as those for the full model in (14.21), as expected in the case of a balanced model.

The sum of squares for the reduced model is therefore

$$\begin{aligned}\text{SS}(\mu, \alpha, \beta) &= \hat{\boldsymbol{\beta}}_1'\mathbf{X}_1'\mathbf{y} \\ &= \frac{y_{...}^2}{abn} + \left(\sum_i \frac{y_{i..}^2}{bn} - \frac{y_{...}^2}{abn}\right) + \left(\sum_j \frac{y_{.j.}^2}{an} - \frac{y_{...}^2}{abn}\right),\end{aligned}$$

and the difference in (14.39) is

$$\begin{aligned}\text{SS}(\gamma|\mu, \alpha, \beta) &= \text{SS}(\mu, \alpha, \beta, \gamma) - \text{SS}(\mu, \alpha, \beta) \\ &= \sum_{ij}\frac{y_{ij.}^2}{n} - \sum_i \frac{y_{i..}^2}{bn} - \sum_j \frac{y_{.j.}^2}{an} + \frac{y_{...}^2}{abn}.\end{aligned} \tag{14.45}$$

The error sum of squares is given by

$$\begin{aligned}\text{SSE} &= \mathbf{y}'\mathbf{y} - \hat{\boldsymbol{\beta}}'\mathbf{X}'\mathbf{y} \\ &= \sum_{ijk} y_{ijk}^2 - \sum_{ij}\frac{y_{ij.}^2}{n}\end{aligned} \tag{14.46}$$

(see Problem 14.13b). In terms of means rather than totals, (14.45) and (14.46) become

$$\text{SS}(\gamma|\mu, \alpha, \beta) = n\sum_{ij}(\bar{y}_{ij.} - \bar{y}_{i..} - \bar{y}_{.j.} + \bar{y}_{...})^2, \tag{14.47}$$

$$\text{SSE} = \sum_{ijk}(y_{ijk} - \bar{y}_{ij.})^2. \tag{14.48}$$

There are ab parameters involved in the hypothesis H_0: $\gamma_{ij}^* = 0$, $i = 1, 2, \ldots, a$, $j = 1, 2, \ldots, b$. However, the $a + b$ side conditions $\sum_i \gamma_{ij}^* = 0$ for $j = 1, 2, \ldots, b$ and $\sum_j \gamma_{ij}^* = 0$ for $i = 1, 2, \ldots, a$ impose $a - 1 + b - 1$ restrictions. With the additional condition $\sum_{i=1}^a \sum_{j=1}^b \gamma_{ij}^* = 0$, we have a total of $a + b - 2 + 1 = a + b - 1$ restrictions. Therefore the degrees of freedom for $\text{SS}(\gamma|\mu, \alpha, \beta)$ are $ab - (a + b - 1) = (a - 1)(b - 1)$ (see Problem 14.14).

To test H_0: $\gamma_{ij}^* = 0$ for all i, j, we therefore use the test statistic

$$F = \frac{\text{SS}(\gamma|\mu, \alpha, \beta)/(a-1)(b-1)}{\text{SSE}/ab(n-1)}, \tag{14.49}$$

which is distributed as $F[(a-1)(b-1), ab(n-1)]$ if H_0 is true (see Section 12.7.2).

14.4.1.3 Full–Reduced-Model Test Based on a Generalized Inverse

We now consider a matrix development of SSE and $\text{SS}(\gamma|\mu, \alpha, \beta)$ based on a generalized inverse. By (12.21), $\text{SSE} = \mathbf{y}'[\mathbf{I} - \mathbf{X}(\mathbf{X}'\mathbf{X})^-\mathbf{X}']\mathbf{y}$. For our illustrative model with $a = 3$, $b = 2$, and $n = 2$, the matrix $\mathbf{X}'\mathbf{X}$ is given in (14.5) and a generalized inverse $(\mathbf{X}'\mathbf{X})^-$ is provided in (14.23). The 12×12 matrix $\mathbf{X}(\mathbf{X}'\mathbf{X})^-\mathbf{X}'$ is then given by

$$\mathbf{X}(\mathbf{X}'\mathbf{X})^-\mathbf{X}' = \frac{1}{2}\begin{pmatrix} \mathbf{J} & \mathbf{O} & \cdots & \mathbf{O} \\ \mathbf{O} & \mathbf{J} & \cdots & \mathbf{O} \\ \vdots & \vdots & & \vdots \\ \mathbf{O} & \mathbf{O} & \cdots & \mathbf{J} \end{pmatrix} = \frac{1}{2}\begin{pmatrix} \mathbf{jj}' & \mathbf{O} & \cdots & \mathbf{O} \\ \mathbf{O} & \mathbf{jj}' & \cdots & \mathbf{O} \\ \vdots & \vdots & & \vdots \\ \mathbf{O} & \mathbf{O} & \cdots & \mathbf{jj}' \end{pmatrix}, \tag{14.50}$$

where $\mathbf{J}$ and $\mathbf{O}$ are 2×2 and $\mathbf{j}$ is 2×1 (see Problem 14.17). The vector $\mathbf{y}$ in (14.4) can be written as

$$\mathbf{y} = \begin{pmatrix} \mathbf{y}_{11} \\ \mathbf{y}_{12} \\ \mathbf{y}_{21} \\ \mathbf{y}_{22} \\ \mathbf{y}_{31} \\ \mathbf{y}_{32} \end{pmatrix}, \tag{14.51}$$

where $\mathbf{y}_{ij} = \begin{pmatrix} y_{ij1} \\ y_{ij2} \end{pmatrix}$, $i = 1, 2, 3$, $j = 1, 2$. By (12.21), (14.50), and (14.51), SSE becomes

$$\begin{aligned} \text{SSE} &= \mathbf{y}'[\mathbf{I} - \mathbf{X}(\mathbf{X}'\mathbf{X})^-\mathbf{X}']\mathbf{y} = \mathbf{y}'\mathbf{y} - \mathbf{y}'\mathbf{X}(\mathbf{X}'\mathbf{X})^-\mathbf{X}'\mathbf{y} \\ &= \sum_{ijk} y_{ijk}^2 - \tfrac{1}{2}\textstyle\sum_{ij} \mathbf{y}_{ij}'\mathbf{jj}'\mathbf{y}_{ij} = \sum_{ijk} y_{ijk}^2 - \tfrac{1}{2}\sum_{ij} y_{ij.}^2, \end{aligned}$$

which is the same as (14.46) with $n = 2$.

For $\text{SS}(\gamma|\mu, \alpha, \beta)$, we obtain

$$\begin{aligned}
\text{SS}(\gamma|\mu, \alpha, \beta) &= \text{SS}(\mu, \alpha, \beta, \gamma) - \text{SS}(\mu, \alpha, \beta) \\
&= \hat{\boldsymbol{\beta}}'\mathbf{X}'\mathbf{y} - \hat{\boldsymbol{\beta}}_1'\mathbf{X}_1'\mathbf{y} \\
&= \mathbf{y}'[\mathbf{X}(\mathbf{X}'\mathbf{X})^-\mathbf{X}' - \mathbf{X}_1(\mathbf{X}_1'\mathbf{X}_1)^-\mathbf{X}_1']\mathbf{y}, \qquad (14.52)
\end{aligned}$$

where $\mathbf{X}(\mathbf{X}'\mathbf{X})^-\mathbf{X}'$ is as found in (14.50) and $\mathbf{X}_1$ consists of the first six columns of $\mathbf{X}$ in (14.4). The matrix $\mathbf{X}_1'\mathbf{X}_1$ is given in (14.42), and a generalized inverse of $\mathbf{X}_1'\mathbf{X}_1$ is given by

$$(\mathbf{X}_1'\mathbf{X}_1)^- = \frac{1}{12}\begin{pmatrix} -1 & 0 & 0 & 0 & 0 & 0 \\ 0 & 3 & 0 & 0 & 0 & 0 \\ 0 & 0 & 3 & 0 & 0 & 0 \\ 0 & 0 & 0 & 3 & 0 & 0 \\ 0 & 0 & 0 & 0 & 2 & 0 \\ 0 & 0 & 0 & 0 & 0 & 2 \end{pmatrix}. \qquad (14.53)$$

Then

$$\mathbf{X}_1(\mathbf{X}_1'\mathbf{X}_1)^-\mathbf{X}_1' = \frac{1}{12}\begin{pmatrix} 4\mathbf{J} & 2\mathbf{J} & \mathbf{J} & -\mathbf{J} & \mathbf{J} & -\mathbf{J} \\ 2\mathbf{J} & 4\mathbf{J} & -\mathbf{J} & \mathbf{J} & -\mathbf{J} & \mathbf{J} \\ \mathbf{J} & -\mathbf{J} & 4\mathbf{J} & 2\mathbf{J} & \mathbf{J} & -\mathbf{J} \\ -\mathbf{J} & \mathbf{J} & 2\mathbf{J} & 4\mathbf{J} & -\mathbf{J} & \mathbf{J} \\ \mathbf{J} & -\mathbf{J} & \mathbf{J} & -\mathbf{J} & 4\mathbf{J} & 2\mathbf{J} \\ -\mathbf{J} & \mathbf{J} & -\mathbf{J} & \mathbf{J} & 2\mathbf{J} & 4\mathbf{J} \end{pmatrix}, \qquad (14.54)$$

where $\mathbf{J}$ is 2×2. For the difference between (14.50) and (14.54), we obtain

$$\mathbf{X}(\mathbf{X}'\mathbf{X})^-\mathbf{X}' - \mathbf{X}_1(\mathbf{X}_1'\mathbf{X}_1)^-\mathbf{X}_1' = \frac{1}{12}\begin{pmatrix} 2\mathbf{J} & -2\mathbf{J} & -\mathbf{J} & \mathbf{J} & -\mathbf{J} & \mathbf{J} \\ -2\mathbf{J} & 2\mathbf{J} & \mathbf{J} & -\mathbf{J} & \mathbf{J} & -\mathbf{J} \\ -\mathbf{J} & \mathbf{J} & 2\mathbf{J} & -2\mathbf{J} & -\mathbf{J} & \mathbf{J} \\ \mathbf{J} & -\mathbf{J} & -2\mathbf{J} & 2\mathbf{J} & \mathbf{J} & -\mathbf{J} \\ -\mathbf{J} & \mathbf{J} & -\mathbf{J} & \mathbf{J} & 2\mathbf{J} & -2\mathbf{J} \\ \mathbf{J} & -\mathbf{J} & \mathbf{J} & -\mathbf{J} & -2\mathbf{J} & 2\mathbf{J} \end{pmatrix}, \qquad (14.55)$$

where $\mathbf{J}$ is 2×2.

To show that $\text{SS}(\gamma|\mu, \alpha, \beta) = \mathbf{y}'[\mathbf{X}(\mathbf{X}'\mathbf{X})^-\mathbf{X}' - \mathbf{X}_1(\mathbf{X}_1'\mathbf{X}_1)^-\mathbf{X}']\mathbf{y}$ in (14.52) is equal to the formulation of $\text{SS}(\gamma|\mu, \alpha, \beta)$ shown in (14.45), we first write (14.45) in matrix notation:

$$\sum_{i=1}^{3}\sum_{j=1}^{2}\frac{y_{ij.}^2}{2} - \sum_{i=1}^{3}\frac{y_{i..}^2}{4} - \sum_{j=1}^{2}\frac{y_{.j.}^2}{6} + \frac{y_{...}^2}{12} = \mathbf{y}'\left(\tfrac{1}{2}\mathbf{A} - \tfrac{1}{4}\mathbf{B} - \tfrac{1}{6}\mathbf{C} + \tfrac{1}{12}\mathbf{D}\right)\mathbf{y}. \tag{14.56}$$

We now find **A**, **B**, **C**, and **D**. For $\frac{1}{2}\sum_{ij} y_{ij.}^2 = \frac{1}{2}\mathbf{y}'\mathbf{A}\mathbf{y}$, we have by (14.50) and (14.51),

$$\tfrac{1}{2}\sum_{ij} y_{ij.}^2 = \tfrac{1}{2}\sum_{i=1}^{3}\sum_{j=1}^{2}\mathbf{y}_{ij}'\mathbf{j}\mathbf{j}'\mathbf{y}_{ij},$$

where $\mathbf{j}$ is 2×1. This can be written as

$$\begin{aligned} \tfrac{1}{2}\sum_{ij} y_{ij.}^2 &= \tfrac{1}{2}(\mathbf{y}_{11}', \mathbf{y}_{12}', \ldots, \mathbf{y}_{32}')\begin{pmatrix} \mathbf{j}\mathbf{j}' & \mathbf{O} & \cdots & \mathbf{O} \\ \mathbf{O} & \mathbf{j}\mathbf{j}' & \cdots & \mathbf{O} \\ \vdots & \vdots & & \vdots \\ \mathbf{O} & \mathbf{O} & \cdots & \mathbf{j}\mathbf{j}' \end{pmatrix}\begin{pmatrix} \mathbf{y}_{11} \\ \mathbf{y}_{12} \\ \vdots \\ \mathbf{y}_{32} \end{pmatrix} \\ &= \tfrac{1}{2}\mathbf{y}'\mathbf{A}\mathbf{y}, \end{aligned} \tag{14.57}$$

where

$$\mathbf{A} = \begin{pmatrix} \mathbf{J} & \mathbf{O} & \cdots & \mathbf{O} \\ \mathbf{O} & \mathbf{J} & \cdots & \mathbf{O} \\ \vdots & \vdots & & \vdots \\ \mathbf{O} & \mathbf{O} & \cdots & \mathbf{J} \end{pmatrix},$$

and $\mathbf{J}$ is 2×2. Note that by (14.50), we also have $\frac{1}{2}\mathbf{A} = \mathbf{X}(\mathbf{X}'\mathbf{X})^{-}\mathbf{X}'$.

For the second term in (14.56), $\frac{1}{4}\sum_i y_{i..}^2$, we first use (14.51) to write $y_{i..}$ and $y_{i..}^2$ as

$$y_{i..} = \sum_{jk} y_{ijk} = \sum_k y_{i1k} + \sum_k y_{i2k} = \mathbf{y}_{i1}'\mathbf{j} + \mathbf{y}_{i2}'\mathbf{j} = (\mathbf{y}_{i1}', \mathbf{y}_{i2}')\begin{pmatrix} \mathbf{j} \\ \mathbf{j} \end{pmatrix},$$

$$y_{i..}^2 = (\mathbf{y}_{i1}', \mathbf{y}_{i2}')\begin{pmatrix} \mathbf{j} \\ \mathbf{j} \end{pmatrix}(\mathbf{j}', \mathbf{j}')\begin{pmatrix} \mathbf{y}_{i1} \\ \mathbf{y}_{i2} \end{pmatrix} = (\mathbf{y}_{i1}', \mathbf{y}_{i2}')\begin{pmatrix} \mathbf{j}\mathbf{j}' & \mathbf{j}\mathbf{j}' \\ \mathbf{j}\mathbf{j}' & \mathbf{j}\mathbf{j}' \end{pmatrix}\begin{pmatrix} \mathbf{y}_{i1} \\ \mathbf{y}_{i2} \end{pmatrix}.$$

Thus $\frac{1}{4}\sum_{i=1}^{3} y_{i..}^2$ can be written as

$$\frac{1}{4}\sum_{i=1}^{3} y_{i..}^2 = \frac{1}{4}(\mathbf{y}_{11}', \mathbf{y}_{12}', \ldots, \mathbf{y}_{32}') \begin{pmatrix} \mathbf{J} & \mathbf{J} & \mathbf{O} & \mathbf{O} & \mathbf{O} & \mathbf{O} \\ \mathbf{J} & \mathbf{J} & \mathbf{O} & \mathbf{O} & \mathbf{O} & \mathbf{O} \\ \mathbf{O} & \mathbf{O} & \mathbf{J} & \mathbf{J} & \mathbf{O} & \mathbf{O} \\ \mathbf{O} & \mathbf{O} & \mathbf{J} & \mathbf{J} & \mathbf{O} & \mathbf{O} \\ \mathbf{O} & \mathbf{O} & \mathbf{O} & \mathbf{O} & \mathbf{J} & \mathbf{J} \\ \mathbf{O} & \mathbf{O} & \mathbf{O} & \mathbf{O} & \mathbf{J} & \mathbf{J} \end{pmatrix} \begin{pmatrix} \mathbf{y}_{11} \\ \mathbf{y}_{12} \\ \vdots \\ \mathbf{y}_{31} \end{pmatrix} \tag{14.58}$$

$$= \frac{1}{4}\mathbf{y}'\mathbf{B}\mathbf{y}.$$

Similarly, the third term of (14.56), $\frac{1}{6}\sum_{j=1}^{2} y_{.j.}^2$, can be written as

$$\frac{1}{6}\sum_{j=1}^{2} y_{.j.}^2 = \frac{1}{6}\mathbf{y}' \begin{pmatrix} \mathbf{J} & \mathbf{O} & \mathbf{J} & \mathbf{O} & \mathbf{J} & \mathbf{O} \\ \mathbf{O} & \mathbf{J} & \mathbf{O} & \mathbf{J} & \mathbf{O} & \mathbf{J} \\ \mathbf{J} & \mathbf{O} & \mathbf{J} & \mathbf{O} & \mathbf{J} & \mathbf{O} \\ \mathbf{O} & \mathbf{J} & \mathbf{O} & \mathbf{J} & \mathbf{O} & \mathbf{J} \\ \mathbf{J} & \mathbf{O} & \mathbf{J} & \mathbf{O} & \mathbf{J} & \mathbf{O} \\ \mathbf{O} & \mathbf{J} & \mathbf{O} & \mathbf{J} & \mathbf{O} & \mathbf{J} \end{pmatrix} \mathbf{y} = \frac{1}{6}\mathbf{y}'\mathbf{C}\mathbf{y}. \tag{14.59}$$

For the fourth term of (14.56), $y_{...}^2/12$, we have

$$y_{...} = \sum_{ijk} y_{ijk} = \mathbf{y}'\mathbf{j}_{12},$$

$$\frac{1}{12}y_{...}^2 = \frac{1}{12}\mathbf{y}'\mathbf{j}_{12}\mathbf{j}_{12}'\mathbf{y} = \frac{1}{12}\mathbf{y}'\mathbf{J}_{12}\mathbf{y} = \frac{1}{12}\mathbf{y}'\mathbf{D}\mathbf{y}, \tag{14.60}$$

where $\mathbf{j}_{12}$ is 12×1 and $\mathbf{J}_{12}$ is 12×12. To conform with $\mathbf{A}$, $\mathbf{B}$, and $\mathbf{C}$ in (14.57), (14.58), and (14.59), we write $\mathbf{D} = \mathbf{J}_{12}$ as

$$\mathbf{D} = \mathbf{J}_{12} = \begin{pmatrix} \mathbf{J} & \mathbf{J} & \mathbf{J} & \mathbf{J} & \mathbf{J} & \mathbf{J} \\ \mathbf{J} & \mathbf{J} & \mathbf{J} & \mathbf{J} & \mathbf{J} & \mathbf{J} \\ \mathbf{J} & \mathbf{J} & \mathbf{J} & \mathbf{J} & \mathbf{J} & \mathbf{J} \\ \mathbf{J} & \mathbf{J} & \mathbf{J} & \mathbf{J} & \mathbf{J} & \mathbf{J} \\ \mathbf{J} & \mathbf{J} & \mathbf{J} & \mathbf{J} & \mathbf{J} & \mathbf{J} \\ \mathbf{J} & \mathbf{J} & \mathbf{J} & \mathbf{J} & \mathbf{J} & \mathbf{J} \end{pmatrix},$$

where $\mathbf{J}$ is 2×2.

Now, combining (14.57)–(14.60), we obtain the matrix of the quadratic form in (14.56):

$$\frac{1}{2}\mathbf{A} - \frac{1}{4}\mathbf{B} - \frac{1}{6}\mathbf{C} + \frac{1}{12}\mathbf{D} = \frac{1}{12}\begin{pmatrix} 2\mathbf{J} & -2\mathbf{J} & -\mathbf{J} & \mathbf{J} & -\mathbf{J} & \mathbf{J} \\ -2\mathbf{J} & 2\mathbf{J} & \mathbf{J} & -\mathbf{J} & \mathbf{J} & -\mathbf{J} \\ -\mathbf{J} & \mathbf{J} & 2\mathbf{J} & -2\mathbf{J} & -\mathbf{J} & \mathbf{J} \\ \mathbf{J} & -\mathbf{J} & -2\mathbf{J} & 2\mathbf{J} & \mathbf{J} & -\mathbf{J} \\ -\mathbf{J} & \mathbf{J} & -\mathbf{J} & \mathbf{J} & 2\mathbf{J} & -2\mathbf{J} \\ \mathbf{J} & -\mathbf{J} & \mathbf{J} & -\mathbf{J} & -2\mathbf{J} & 2\mathbf{J} \end{pmatrix}, \quad (14.61)$$

which is the same as (14.55). Thus the matrix version of $\text{SS}(\gamma|\mu, \alpha, \beta)$ in (14.52) is equal to $\text{SS}(\gamma|\mu, \alpha, \beta)$ in (14.45):

$$\mathbf{y}'[\mathbf{X}(\mathbf{X}'\mathbf{X})^{-}\mathbf{X}' - \mathbf{X}_1(\mathbf{X}_1'\mathbf{X}_1)^{-}\mathbf{X}_1']\mathbf{y} = \sum_{ij} \frac{y_{ij.}^2}{n} - \sum_i \frac{y_{i..}^2}{bn} - \sum_j \frac{y_{.j.}^2}{an} + \frac{y_{...}^2}{abn}.$$

14.4.2 Tests for Main Effects

In Section 14.4.2.1, we develop a test for main effects using the full–reduced–model approach. In Section 14.4.2.2, a test for main effects is obtained using the general linear hypothesis approach. Throughout much of this section, we use $a = 3$ and $b = 2$, where a is the number of levels of factor A and b is the number of levels of factor B.

14.4.2.1 Full–Reduced-Model Approach

If interaction is present in the two-way model, then by (14.9) and (14.10), we cannot test H_0: $\alpha_1 = \alpha_2 = \alpha_3$ (for $a = 3$) because $\alpha_1 - \alpha_2$ and $\alpha_1 - \alpha_3$ are not estimable. In fact, there are no estimable contrasts in the α's alone or the β's alone (see Problem 14.2). Thus, if there is interaction, the effect of factor A is different for each level of factor B and vice versa.

To examine the main effect of factor A, we consider $\alpha_i^* = \bar{\mu}_{i.} - \bar{\mu}_{..}$, as defined in (14.14). This can be written as

$$\begin{aligned} \alpha_i^* = \bar{\mu}_{i.} - \bar{\mu}_{..} &= \sum_{j=1}^{b} \frac{\mu_{ij}}{b} - \sum_{i=1}^{a}\sum_{j=1}^{b} \frac{\mu_{ij}}{ab} \\ &= \frac{1}{b}\sum_j \left(\mu_{ij} - \sum_i \frac{\mu_{ij}}{a} \right) \\ &= \frac{1}{b}\sum_j (\mu_{ij} - \bar{\mu}_{.j}). \end{aligned} \quad (14.62)$$

The expression in parentheses, $\mu_{ij} - \bar{\mu}_{.j}$, is the effect of the ith level of factor A at the jth level of factor B. Thus in (14.62), $\alpha_i^* = \bar{\mu}_{i.} - \bar{\mu}_{..}$ is expressed as the average effect of the ith level of factor A (averaged over the levels of B). This definition leads to the side condition $\sum_i \alpha_i^* = 0$.

Since the α_i^*'s are estimable [see (14.21) and the comment following], we can use them to express the hypothesis for factor A. For $a = 3$, this becomes

$$H_0\colon \alpha_1^* = \alpha_2^* = \alpha_3^*, \tag{14.63}$$

which is equivalent to

$$H_0\colon \alpha_1^* = \alpha_2^* = \alpha_3^* = 0 \tag{14.64}$$

because $\sum_i \alpha_i^* = 0$.

The hypothesis $H_0\colon \alpha_1^* = \alpha_2^* = \alpha_3^*$ in (14.63) states that there is no effect of factor A when averaged over the levels of B. Using $\alpha_i^* = \bar{\mu}_{i.} - \bar{\mu}_{..}$, we can express $H_0\colon \alpha_1^* = \alpha_2^* = \alpha_3^*$ in terms of means:

$$H_0\colon \bar{\mu}_{1.} - \bar{\mu}_{..} = \bar{\mu}_{2.} - \bar{\mu}_{..} = \bar{\mu}_{3.} - \bar{\mu}_{..},$$

which can be written as

$$H_0\colon \bar{\mu}_{1.} = \bar{\mu}_{2.} = \bar{\mu}_{3.}.$$

The values for the cell means in Figure 14.2 illustrate a situation in which H_0 holds in the presence of interaction.

Because H_0 in (14.63) or (14.64) is based on an average effect, many texts recommend that the interaction AB be tested first, and if it is found to be significant, then the main effects should not be tested. However, with the main effect of A defined as the average effect over the levels of B and similarly for the effect of B, the tests for A and B can be carried out even if AB is significant. Admittedly, interpretation requires more care, and the effect of a factor may change if the number of levels of the other factor is altered. But in many cases useful information can be gained about the main effects in the presence of interaction.

		B 1	2	means
A	1	$\mu_{11} = 5$	$\mu_{12} = 1$	$\bar{\mu}_{1.} = 3$
	2	$\mu_{21} = 4$	$\mu_{22} = 2$	$\bar{\mu}_{2.} = 3$
	3	$\mu_{31} = 3$	$\mu_{32} = 3$	$\bar{\mu}_{3.} = 3$

Figure 14.2 Cell means illustrating $\bar{\mu}_{1.} = \bar{\mu}_{2.} = \bar{\mu}_{3.}$ in the presence of interaction.

Under H_0: $\alpha_1^* = \alpha_2^* = \alpha_3^* = 0$, the full model in (14.33) reduces to

$$y_{ijk} = \mu^* + \beta_j^* + \gamma_{ij}^* + \varepsilon_{ijk}. \tag{14.65}$$

Because of the orthogonality of the balanced model, the estimators of μ^*, β_j^*, and γ_{ij}^* in (14.65) are the same as in the full model. If we use $\hat{\mu}$, $\hat{\beta}_j$, and $\hat{\gamma}_{ij}$ in (14.21) and elements of $\mathbf{X}'\mathbf{y}$ in (14.19) extended to general a, b, and n, we obtain

$$\text{SS}(\mu, \beta, \gamma) = \hat{\mu}y_{...} + \sum_{j=1}^{b} \hat{\beta}_j y_{.j.} + \sum_{i=1}^{a}\sum_{j=1}^{b} \hat{\gamma}_{ij} y_{ij.},$$

which, by (14.40), becomes

$$\begin{aligned}\text{SS}(\mu, \beta, \gamma) &= \frac{y_{...}^2}{abn} + \left(\sum_j \frac{y_{.j.}^2}{an} - \frac{y_{...}^2}{abn}\right) \\ &\quad + \left(\sum_{ij} \frac{y_{ij.}^2}{n} - \sum_i \frac{y_{i..}^2}{bn} - \sum_j \frac{y_{.j.}^2}{an} + \frac{y_{...}^2}{abn}\right).\end{aligned} \tag{14.66}$$

From (14.40) and (14.66), we have

$$\begin{aligned}\text{SS}(\alpha|\mu, \beta, \gamma) &= \text{SS}(\mu, \alpha, \beta, \gamma) - \text{SS}(\mu, \beta, \gamma) \\ &= \sum_{i=1}^{a} \frac{y_{i..}^2}{bn} - \frac{y_{...}^2}{abn}.\end{aligned} \tag{14.67}$$

For the special case of $a = 3$, we see by (14.7) that there are two linearly independent estimable functions involving the three α's. Therefore, $\text{SS}(\alpha|\mu, \beta, \gamma)$ has 2 degrees of freedom. In general, $\text{SS}(\alpha|\mu, \beta, \gamma)$ has $a - 1$ degrees of freedom.

In an analogous manner, for factor B we obtain

$$\begin{aligned}\text{SS}(\beta|\mu, \alpha, \gamma) &= \text{SS}(\mu, \alpha, \beta, \gamma) - \text{SS}(\mu, \alpha, \gamma) \\ &= \sum_{j=1}^{b} \frac{y_{.j.}^2}{an} - \frac{y_{...}^2}{abn},\end{aligned} \tag{14.68}$$

which has $b - 1$ degrees of freedom.

In terms of means, we can express (14.67) and (14.68) as

$$\text{SS}(\alpha|\mu, \beta, \gamma) = bn \sum_{i=1}^{a} (\bar{y}_{i..} - \bar{y}_{...})^2, \tag{14.69}$$

$$\mathrm{SS}(\beta|\mu, \alpha, \gamma) = an \sum_{j=1}^{b} (\bar{y}_{.j.} - \bar{y}_{...})^2. \tag{14.70}$$

It is important to note that the full–reduced-model approach leading to $\mathrm{SS}(\alpha|\mu, \beta, \gamma)$ in (14.67) *cannot* be expressed in terms of matrices in a manner analogous to that in (14.52) for the interaction, namely, $\mathrm{SS}(\gamma|\mu, \alpha, \beta) = \mathbf{y}'[\mathbf{X}(\mathbf{X}'\mathbf{X})^{-}\mathbf{X}' - \mathbf{X}_1(\mathbf{X}_1'\mathbf{X}_1)^{-}\mathbf{X}_1']\mathbf{y}$. The matrix approach is appropriate for the interaction because there are estimable functions of the γ_{ij}'s that do not involve μ or the α_i or β_j terms. In the case of the A main effect, however, we cannot obtain a matrix $\mathbf{X}_1$ by deleting the three columns of $\mathbf{X}$ corresponding to α_1, α_2, and α_3 because contrasts of the form $\alpha_1 - \alpha_2$ are not estimable without involving the γ_{ij}'s [see (14.9) and (14.10)].

If we add the sums of squares for factor A, B, and the interaction in (14.67), (14.68), and (14.45), we obtain $\sum_{ij} y_{ij.}^2/n - y_{...}^2/abn$, which is the overall sum of squares for "treatments," $\mathrm{SS}(\alpha, \beta, \gamma|\mu)$. This can also be seen in (14.40). In the following theorem, the three sums of squares are shown to be independent.

Theorem 14.4b. If $\mathbf{y}$ is $N_{abn}(\mathbf{X}\boldsymbol{\beta}, \sigma^2\mathbf{I})$, then $\mathrm{SS}(\alpha|\mu, \beta, \gamma)$, $\mathrm{SS}(\beta|\mu, \alpha, \gamma)$, and $\mathrm{SS}(\gamma|\mu, \alpha, \beta)$ are independent.

PROOF. This follows from Theorem 5.6c; see Problem 14.23. □

Using (14.45), (14.46), (14.67), and (14.68), we obtain the analysis-of-variance (ANOVA) table given in Table 14.1.

TABLE 14.1 ANOVA Table for a Two-Way Model with Interaction

Source of Variation	df	Sum of Squares
Factor A	$a - 1$	$\sum_i \frac{y_{i..}^2}{bn} - \frac{y_{...}^2}{abn}$
Factor B	$b - 1$	$\sum_j \frac{y_{.j.}^2}{an} - \frac{y_{...}^2}{abn}$
Interaction	$(a - 1)(b - 1)$	$\sum_{ij} \frac{y_{ij.}^2}{n} - \sum_i \frac{y_{i..}^2}{bn} - \sum_j \frac{y_{.j.}^2}{an} + \frac{y_{...}^2}{abn}$
Error	$ab(n - 1)$	$\sum_{ijk} y_{ijk}^2 - \sum_{ij} \frac{y_{ij.}^2}{n}$
Total	$abn - 1$	$\sum_{ijk} y_{ijk}^2 - \frac{y_{...}^2}{abn}$

The test statistic for factor A is

$$F = \frac{\text{SS}(\alpha|\mu, \beta, \gamma)/(a-1)}{\text{SSE}/ab(n-1)}, \tag{14.71}$$

which is distributed as $F[a-1, ab(n-1)]$ if H_0: $\alpha_1^* = \alpha_2^* = \cdots = \alpha_a^* = 0$ is true. For factor B, we use $\text{SS}(\beta|\mu, \alpha, \gamma)$ in (14.68), and the F statistic is given by

$$F = \frac{\text{SS}(\beta|\mu, \alpha, \gamma)/(b-1)}{\text{SSE}/ab(n-1)},$$

which is distributed as $F[b-1, ab(n-1)]$ if H_0: $\beta_1^* = \beta_2^* = \cdots = \beta_b^* = 0$ is true. In Section 14.4.2.2, these F statistics are obtained by the general linear hypothesis approach. The F distributions can thereby be justified by Theorem 12.7c.

Example 14.4. The moisture content of three types of cheese made by two methods was recorded by Marcuse (1949) (format altered). Two cheeses were measured for each type and each method. If *method* is designated as factor A and *type* is factor B, then $a = 2$, $b = 3$, and $n = 2$. The data are given in Table 14.2, and the totals are shown in Table 14.3.

The sum of squares for factor A is given by (14.67) as

$$\begin{aligned} \text{SS}(\alpha|\mu, \beta, \gamma) &= \sum_{i=1}^{2} \frac{y_{i..}^2}{(3)(2)} - \frac{y_{...}^2}{(2)(3)(2)} \\ &= \tfrac{1}{6}[(221.98)^2 + (220.81)^2] - \tfrac{1}{12}(442.79)^2 \\ &= .114075. \end{aligned}$$

TABLE 14.2 Moisture Content of Two Cheeses from Each of Three Different Types Made by Two Methods

	Type of Cheese		
Method	1	2	3
1	39.02	35.74	37.02
	38.79	35.41	36.00
2	38.96	35.58	35.70
	39.01	35.52	36.04

TABLE 14.3 Totals for Data in Table 14.2

		B		
A	1	2	3	Totals
1	$y_{11.} = 77.81$	$y_{12.} = 71.15$	$y_{13.} = 73.02$	$y_{1..} = 221.98$
2	$y_{21.} = 77.97$	$y_{22.} = 71.10$	$y_{23.} = 71.74$	$y_{2..} = 220.81$
Totals	$y_{.1.} = 155.78$	$y_{.2.} = 142.25$	$y_{.3.} = 144.76$	$y_{...} = 442.79$

Similarly, for factor B we use (14.68):

$$\begin{aligned} \text{SS}(\beta|\mu, \alpha, \gamma) &= \sum_{j=1}^{3} \frac{y_{.j.}^2}{(2)(2)} - \frac{y_{...}^2}{12} \\ &= \tfrac{1}{4}[(155.78)^2 + (142.25)^2 + (144.76)^2] - \tfrac{1}{12}(442.79)^2 \\ &= 25.900117. \end{aligned}$$

For error, we use (14.46) to obtain

$$\begin{aligned} \text{SSE} &= \sum_{ijk} y_{ijk}^2 - \tfrac{1}{2}\textstyle\sum_{ij} y_{ij.}^2 \\ &= (39.02)^2 + (38.79)^2 + \cdots + (36.04)^2 - \tfrac{1}{2}[(77.81)^2 + \cdots + (71.74)^2] \\ &= 16{,}365.56070 - 16364.89875 = .661950. \end{aligned}$$

The total sum of squares is given by

$$\text{SST} = \sum_{ijk} y_{ijk}^2 - \frac{y_{...}^2}{12} = 26.978692.$$

The sum of squares for interaction can be found by (14.45) or by subtracting all other terms from the total sum of squares:

$$\begin{aligned} \text{SS}(\gamma|\mu, \alpha, \beta) &= 26.978692 - .114075 - 25.900117 - .661950 \\ &= .302550. \end{aligned}$$

With these sums of squares, we can compute mean squares and F statistics as shown in Table 14.4.

Only the F test for *type* is significant, since $F_{.05,1,6} = 5.99$ and $F_{.05,2,6} = 5.14$. The p value for *type* is .0000155. The p values for *method* and the *interaction* are .3485 and .3233, respectively.

TABLE 14.4 ANOVA for the Cheese Data in Table 14.2

Source of Variation	Sum of Squares	df	Mean Square	F
Method	0.114075	1	0.114075	1.034
Type	25.900117	2	12.950058	117.381
Interaction	0.302550	2	0.151275	1.371
Error	0.661950	6	0.110325	
Total	26.978692	11		

Note that in Table 14.2, the difference between the two replicates in each cell is very small except for the cell with method 1 and type 3. This suggests that the replicates may be repeat measurements rather than true replications; that is, the experimenter may have measured the same piece of cheese twice rather than measuring two different cheeses. □

14.4.2.2 General Linear Hypothesis Approach

We now obtain SS$(\alpha|\mu, \beta, \gamma)$ for $a = 3$ and $b = 2$ by an approach based on the general linear hypothesis. Using $\alpha_i^* = \alpha_i - \bar{\alpha}_. + \bar{\gamma}_{i.} - \bar{\gamma}_{..}$ in (14.16), the hypothesis $H_0\colon \alpha_1^* = \alpha_2^* = \alpha_3^*$ in (14.63) can be expressed as $H_0\colon \alpha_1 + \bar{\gamma}_{1.} = \alpha_2 + \bar{\gamma}_{2.} = \alpha_3 + \bar{\gamma}_{3.}$ or

$$H_0\colon \alpha_1 + \tfrac{1}{2}(\gamma_{11} + \gamma_{12}) = \alpha_2 + \tfrac{1}{2}(\gamma_{21} + \gamma_{22}) = \alpha_3 + \tfrac{1}{2}(\gamma_{31} + \gamma_{32}) \qquad (14.72)$$

[see also (14.9) and (14.10)]. The two equalities in (14.72) can be expressed in the form

$$H_0\colon \begin{pmatrix} \alpha_1 + \frac{1}{2}\gamma_{11} + \frac{1}{2}\gamma_{12} - \alpha_3 - \frac{1}{2}\gamma_{31} - \frac{1}{2}\gamma_{32} \\ \alpha_2 + \frac{1}{2}\gamma_{21} + \frac{1}{2}\gamma_{22} - \alpha_3 - \frac{1}{2}\gamma_{31} - \frac{1}{2}\gamma_{32} \end{pmatrix} = \begin{pmatrix} 0 \\ 0 \end{pmatrix}.$$

Rearranging the order of the parameters to correspond to the order in $\boldsymbol{\beta} = (\mu, \alpha_1, \alpha_2, \alpha_3, \beta_1, \beta_2, \gamma_{11}, \gamma_{12}, \gamma_{21}, \gamma_{22}, \gamma_{31}, \gamma_{32})'$ in (14.4), we have

$$H_0\colon \begin{pmatrix} \alpha_1 - \alpha_3 + \frac{1}{2}\gamma_{11} + \frac{1}{2}\gamma_{12} - \frac{1}{2}\gamma_{31} - \frac{1}{2}\gamma_{32} \\ \alpha_2 - \alpha_3 + \frac{1}{2}\gamma_{21} + \frac{1}{2}\gamma_{22} - \frac{1}{2}\gamma_{31} - \frac{1}{2}\gamma_{32} \end{pmatrix} = \begin{pmatrix} 0 \\ 0 \end{pmatrix}, \qquad (14.73)$$

which can now be written in the form $H_0\colon \mathbf{C}\boldsymbol{\beta} = \mathbf{0}$ with

$$\mathbf{C} = \begin{pmatrix} 0 & 1 & 0 & -1 & 0 & 0 & \frac{1}{2} & \frac{1}{2} & 0 & 0 & -\frac{1}{2} & -\frac{1}{2} \\ 0 & 0 & 1 & -1 & 0 & 0 & 0 & 0 & \frac{1}{2} & \frac{1}{2} & -\frac{1}{2} & -\frac{1}{2} \end{pmatrix}. \tag{14.74}$$

By Theorem 12.7b(iii), the sum of squares corresponding to $H_0\colon \mathbf{C}\boldsymbol{\beta} = \mathbf{0}$ is

$$\text{SSH} = (\mathbf{C}\hat{\boldsymbol{\beta}})'[\mathbf{C}(\mathbf{X}'\mathbf{X})^{-}\mathbf{C}']^{-1}\mathbf{C}\hat{\boldsymbol{\beta}}. \tag{14.75}$$

Substituting $\hat{\boldsymbol{\beta}} = (\mathbf{X}'\mathbf{X})^{-}\mathbf{X}'\mathbf{y}$ from (12.13), SSH in (14.75) becomes

$$\text{SSH} = \mathbf{y}'\mathbf{X}(\mathbf{X}'\mathbf{X})^{-}\mathbf{C}'[\mathbf{C}(\mathbf{X}'\mathbf{X})^{-}\mathbf{C}']^{-1}\mathbf{C}(\mathbf{X}'\mathbf{X})^{-}\mathbf{X}'\mathbf{y} = \mathbf{y}'\mathbf{A}\mathbf{y}. \tag{14.76}$$

Using $\mathbf{C}$ in (14.74), $(\mathbf{X}'\mathbf{X})^{-}$ in (14.23), and $\mathbf{X}$ in (14.4), we obtain

$$\mathbf{C}(\mathbf{X}'\mathbf{X})^{-}\mathbf{X}' = \tfrac{1}{4}\begin{pmatrix} 1 & 1 & 1 & 1 & 0 & 0 & 0 & 0 & -1 & -1 & -1 & -1 \\ 0 & 0 & 0 & 0 & 1 & 1 & 1 & 1 & -1 & -1 & -1 & -1 \end{pmatrix}, \tag{14.77}$$

$$\mathbf{C}(\mathbf{X}'\mathbf{X})^{-}\mathbf{C}' = \tfrac{1}{4}\begin{pmatrix} 2 & 1 \\ 1 & 2 \end{pmatrix}, \qquad [\mathbf{C}(\mathbf{X}'\mathbf{X})^{-}\mathbf{C}']^{-1} = \tfrac{4}{3}\begin{pmatrix} 2 & -1 \\ -1 & 2 \end{pmatrix}. \tag{14.78}$$

Then $\mathbf{A} = \mathbf{X}(\mathbf{X}'\mathbf{X})^{-}\mathbf{C}'[\mathbf{C}(\mathbf{X}'\mathbf{X})^{-}\mathbf{C}']^{-1}\mathbf{C}(\mathbf{X}'\mathbf{X})^{-}\mathbf{X}'$ in (14.76) becomes

$$\mathbf{A} = \tfrac{1}{12}\begin{pmatrix} 2\mathbf{J} & -\mathbf{J} & -\mathbf{J} \\ -\mathbf{J} & 2\mathbf{J} & -\mathbf{J} \\ -\mathbf{J} & -\mathbf{J} & 2\mathbf{J} \end{pmatrix}, \tag{14.79}$$

where $\mathbf{J}$ is 4×4. This can be expressed as

$$\mathbf{A} = \tfrac{1}{12}\begin{pmatrix} 2\mathbf{J} & -\mathbf{J} & -\mathbf{J} \\ -\mathbf{J} & 2\mathbf{J} & -\mathbf{J} \\ -\mathbf{J} & -\mathbf{J} & 2\mathbf{J} \end{pmatrix} = \tfrac{1}{12}\begin{pmatrix} 3\mathbf{J} & \mathbf{O} & \mathbf{O} \\ \mathbf{O} & 3\mathbf{J} & \mathbf{O} \\ \mathbf{O} & \mathbf{O} & 3\mathbf{J} \end{pmatrix} - \tfrac{1}{12}\begin{pmatrix} \mathbf{J} & \mathbf{J} & \mathbf{J} \\ \mathbf{J} & \mathbf{J} & \mathbf{J} \\ \mathbf{J} & \mathbf{J} & \mathbf{J} \end{pmatrix}. \tag{14.80}$$

To evaluate $\mathbf{y}'\mathbf{A}\mathbf{y}$, we redefine $\mathbf{y}$ in (14.51) as

$$\mathbf{y} = \begin{pmatrix} \mathbf{y}_{11} \\ \mathbf{y}_{12} \\ \mathbf{y}_{21} \\ \mathbf{y}_{22} \\ \mathbf{y}_{31} \\ \mathbf{y}_{32} \end{pmatrix} = \begin{pmatrix} \mathbf{y}_1 \\ \mathbf{y}_2 \\ \mathbf{y}_3 \end{pmatrix}, \qquad \text{where } \begin{pmatrix} \mathbf{y}_{i1} \\ \mathbf{y}_{i2} \end{pmatrix} = \mathbf{y}_{i\cdot}. \tag{14.81}$$

Then (14.76) becomes

$$\begin{aligned}
\text{SSH} = \mathbf{y}'\mathbf{A}\mathbf{y} &= \tfrac{1}{12}(\mathbf{y}_1', \mathbf{y}_2', \mathbf{y}_3')\begin{pmatrix} 3\mathbf{J}_4 & \mathbf{O} & \mathbf{O} \\ \mathbf{O} & 3\mathbf{J}_4 & \mathbf{O} \\ \mathbf{O} & \mathbf{O} & 3\mathbf{J}_4 \end{pmatrix}\begin{pmatrix} \mathbf{y}_1 \\ \mathbf{y}_2 \\ \mathbf{y}_3 \end{pmatrix} - \tfrac{1}{12}\mathbf{y}'\mathbf{J}_{12}\mathbf{y} \\
&= \tfrac{3}{12}\sum_{i=1}^{3}\mathbf{y}_i'\mathbf{J}_4\mathbf{y}_i - \tfrac{1}{12}\mathbf{y}'\mathbf{J}_{12}\mathbf{y} \\
&= \frac{1}{4}\sum_i \mathbf{y}_i'\mathbf{j}_4\mathbf{j}_4'\mathbf{y}_i - \tfrac{1}{12}\mathbf{y}'\mathbf{j}_{12}\mathbf{j}_{12}'\mathbf{y} \\
&= \sum_i \frac{y_{i..}^2}{4} - \frac{y_{...}^2}{12},
\end{aligned}$$

which is the same as $\text{SS}(\alpha|\mu, \beta, \gamma)$ in (14.67) with $a = 3$ and $b = n = 2$.

The sum of squares for testing the B main effect can be obtained similarly using a general linear hypothesis approach (see Problem 14.25).

14.5 EXPECTED MEAN SQUARES

We find expected mean squares by direct evaluation of the expected value of sums of squares and also by a matrix method based on the expected value of quadratic forms.

14.5.1 Sums-of-Squares Approach

The expected mean squares for the tests in Table 14.1 are given in Table 14.5. Note that these are expressed in terms of α_i^*, β_j^*, and γ_{ij}^* subject to the side

TABLE 14.5 Expected Mean Squares for a Two-Way ANOVA

Source	Sum of Squares	Mean Square	Expected Mean Square
A	$\text{SS}(\alpha\|\mu, \beta, \gamma)$	$\dfrac{\text{SS}(\alpha\|\mu, \beta, \gamma)}{a-1}$	$\sigma^2 + bn\sum_i \dfrac{\alpha_i^{*2}}{a-1}$
B	$\text{SS}(\beta\|\mu, \alpha, \gamma)$	$\dfrac{\text{SS}(\beta\|\mu, \alpha, \gamma)}{b-1}$	$\sigma^2 + an\sum_j \dfrac{\beta_j^{*2}}{b-1}$
AB	$\text{SS}(\gamma\|\mu, \alpha, \beta)$	$\dfrac{\text{SS}(\gamma\|\mu, \alpha, \beta)}{(a-1)(b-1)}$	$\sigma^2 + n\sum_{ij} \dfrac{\gamma_{ij}^{*2}}{(a-1)(b-1)}$
Error	SSE	$\dfrac{\text{SSE}}{ab(n-1)}$	σ^2

conditions $\sum_i \alpha_i^* = 0$, $\sum_j \beta_j^* = 0$, and $\sum_i \gamma_{ij}^* = \sum_j \gamma_{ij}^* = 0$. These expected mean squares can be derived by inserting the model $y_{ijk} = \mu^* + \alpha_i^* + \beta_j^* + \gamma_{ij}^* + \varepsilon_{ijk}$ in (14.33) into the sums of squares and then finding expected values. We illustrate this approach for the first expected mean square in Table 14.5.

To find the expected value of $\text{SS}(\alpha|\mu, \beta, \gamma) = \sum_i y_{i..}^2/bn - y_{...}^2/abn$ in (14.67), we first note that by using assumption 1 in Section 14.1, we can write assumptions 2 and 3 in the form

$$E(\varepsilon_{ijk}^2) = \sigma^2 \quad \text{for all} \quad i, j, k, \tag{14.82}$$

$$E(\varepsilon_{ijk}\varepsilon_{rst}) = 0 \quad \text{for all} \quad (i,j,k) \neq (r,s,t). \tag{14.83}$$

Using these results, along with assumption 1 and the side conditions in (14.15), we can show that $E(y_{...}^2) = a^2b^2n^2\mu^{*2} + abn\sigma^2$ as follows:

$$\begin{aligned} E(y_{...}^2) &= E\left(\sum_{ijk} y_{ijk}\right)^2 = E\left[\sum_{ijk} (\mu^* + \alpha_i^* + \beta_j^* + \gamma_{ij}^* + \varepsilon_{ijk})\right]^2 \\ &= E\left(abn\mu^* + bn\sum_i \alpha_i^* + an\sum_j \beta_j^* + n\sum_{ij} \gamma_{ij}^* + \sum_{ijk} \varepsilon_{ijk}\right)^2 \\ &= E\left[a^2b^2n^2\mu^{*2} + 2abn\mu^* \sum_{ijk} \varepsilon_{ijk} + \left(\sum_{ijk} \varepsilon_{ijk}\right)^2\right] \\ &= a^2b^2n^2\mu^{*2} + E\left(\sum_{ijk} \varepsilon_{ijk}^2\right) + E\left(\sum_{ijk \neq rst} \varepsilon_{ijk}\varepsilon_{rst}\right) \\ &= a^2b^2n^2\mu^{*2} + abn\sigma^2. \end{aligned}$$

It can likewise be shown that

$$E\left(\sum_{i=1}^{a} y_{i..}^2\right) = ab^2n^2\mu^{*2} + b^2n^2\sum_{i=1}^{a} \alpha_i^{*2} + abn\sigma^2 \tag{14.84}$$

(see Problem 14.27). Thus

$$\begin{aligned} E\left[\frac{\text{SS}(\alpha|\mu, \beta, \gamma)}{a-1}\right] &= \frac{1}{a-1} E\left(\sum_i \frac{y_{i..}^2}{bn} - \frac{y_{...}^2}{abn}\right) \\ &= \frac{1}{a-1}\left[\frac{ab^2n^2\mu^{*2}}{bn} + \frac{b^2n^2\sum_i \alpha_i^{*2}}{bn} + \frac{abn\sigma^2}{bn} - \frac{a^2b^2n^2\mu^{*2}}{abn} - \frac{abn\sigma^2}{abn}\right] \\ &= \frac{1}{a-1}\left[(a-1)\sigma^2 + bn\sum_i \alpha_i^{*2}\right]. \end{aligned}$$

The other expected mean squares in Table 14.5 can be obtained similarly (see Problem 14.28).

14.5.2 Quadratic Form Approach

We now obtain the first expected mean square in Table 14.2 using a matrix approach. We illustrate with $a = 3$, $b = 2$, and $n = 2$. By (14.75), we obtain

$$E[\mathrm{SS}(\alpha|\mu, \beta, \gamma)] = E\{(\mathbf{C}\hat{\boldsymbol{\beta}})'[\mathbf{C}(\mathbf{X}'\mathbf{X})^{-}\mathbf{C}']^{-1}\mathbf{C}\hat{\boldsymbol{\beta}}\}. \tag{14.85}$$

The matrix $\mathbf{C}$ contains estimable functions, and therefore by (12.44) and (12.45), we have $E(\mathbf{C}\hat{\boldsymbol{\beta}}) = \mathbf{C}\boldsymbol{\beta}$ and $\mathrm{cov}(\mathbf{C}\hat{\boldsymbol{\beta}}) = \sigma^2\mathbf{C}(\mathbf{X}'\mathbf{X})^{-}\mathbf{C}'$. If we define $\mathbf{G}$ to be the 2×2 matrix $[\mathbf{C}(\mathbf{X}'\mathbf{X})^{-}\mathbf{C}']^{-1}$, then by Theorem 5.2a, (14.85) becomes

$$\begin{aligned} E[\mathrm{SS}(\alpha|\mu, \beta, \gamma)] &= E[(\mathbf{C}\hat{\boldsymbol{\beta}})'\mathbf{G}(\mathbf{C}\hat{\boldsymbol{\beta}})] \\ &= \mathrm{tr}[\mathbf{G}\ \mathrm{cov}(\mathbf{C}\hat{\boldsymbol{\beta}})] + [E(\mathbf{C}\hat{\boldsymbol{\beta}})]'\mathbf{G}[E(\mathbf{C}\hat{\boldsymbol{\beta}})] \\ &= \mathrm{tr}(\mathbf{G}\sigma^2\mathbf{G}^{-1}) + (\mathbf{C}\boldsymbol{\beta})'\mathbf{G}(\mathbf{C}\boldsymbol{\beta}) \\ &= 2\sigma^2 + \boldsymbol{\beta}'\mathbf{C}'[\mathbf{C}(\mathbf{X}'\mathbf{X})^{-}\mathbf{C}']^{-1}\mathbf{C}\boldsymbol{\beta} \qquad (14.86) \\ &= 2\sigma^2 + \boldsymbol{\beta}'\mathbf{L}\boldsymbol{\beta}, \qquad (14.87) \end{aligned}$$

where $\mathbf{L} = \mathbf{C}'[\mathbf{C}(\mathbf{X}'\mathbf{X})^{-}\mathbf{C}']^{-1}\mathbf{C}$. Using $\mathbf{C}$ in (14.74) and $[\mathbf{C}(\mathbf{X}'\mathbf{X})^{-}\mathbf{C}']^{-1}$ in (14.78), $\mathbf{L}$ becomes

$$\mathbf{L} = \frac{1}{3}\begin{pmatrix} 0 & 0 & 0 & 0 & 0 & 0 & 0 & 0 & 0 & 0 & 0 & 0 \\ 0 & 8 & -4 & -4 & 0 & 0 & 4 & 4 & -2 & -2 & -2 & -2 \\ 0 & -4 & 8 & -4 & 0 & 0 & -2 & -2 & 4 & 4 & -2 & -2 \\ 0 & -4 & -4 & 8 & 0 & 0 & -2 & -2 & -2 & -2 & 4 & 4 \\ 0 & 0 & 0 & 0 & 0 & 0 & 0 & 0 & 0 & 0 & 0 & 0 \\ 0 & 0 & 0 & 0 & 0 & 0 & 0 & 0 & 0 & 0 & 0 & 0 \\ 0 & 4 & -2 & -2 & 0 & 0 & 2 & 2 & -1 & -1 & -1 & -1 \\ 0 & 4 & -2 & -2 & 0 & 0 & 2 & 2 & -1 & -1 & -1 & -1 \\ 0 & -2 & 4 & -2 & 0 & 0 & -1 & -1 & 2 & 2 & -1 & -1 \\ 0 & -2 & 4 & -2 & 0 & 0 & -1 & -1 & 2 & 2 & -1 & -1 \\ 0 & -2 & -2 & 4 & 0 & 0 & -1 & -1 & -1 & -1 & 2 & 2 \\ 0 & -2 & -2 & 4 & 0 & 0 & -1 & -1 & -1 & 1 & 2 & 2 \end{pmatrix}. \tag{14.88}$$

This can be written as the difference

$$
\mathbf{L} = \tfrac{1}{3}\begin{pmatrix}
0 & 0 & 0 & 0 & 0 & 0 & 0 & 0 & 0 & 0 & 0 & 0 \\
0 & 12 & 0 & 0 & 0 & 0 & 6 & 6 & 0 & 0 & 0 & 0 \\
0 & 0 & 12 & 0 & 0 & 0 & 0 & 0 & 6 & 6 & 0 & 0 \\
0 & 0 & 0 & 12 & 0 & 0 & 0 & 0 & 0 & 0 & 6 & 6 \\
0 & 0 & 0 & 0 & 0 & 0 & 0 & 0 & 0 & 0 & 0 & 0 \\
0 & 0 & 0 & 0 & 0 & 0 & 0 & 0 & 0 & 0 & 0 & 0 \\
0 & 6 & 0 & 0 & 0 & 0 & 3 & 3 & 0 & 0 & 0 & 0 \\
0 & 6 & 0 & 0 & 0 & 0 & 3 & 3 & 0 & 0 & 0 & 0 \\
0 & 0 & 6 & 0 & 0 & 0 & 0 & 0 & 3 & 3 & 0 & 0 \\
0 & 0 & 6 & 0 & 0 & 0 & 0 & 0 & 3 & 3 & 0 & 0 \\
0 & 0 & 0 & 6 & 0 & 0 & 0 & 0 & 0 & 0 & 3 & 3 \\
0 & 0 & 0 & 6 & 0 & 0 & 0 & 0 & 0 & 0 & 3 & 3
\end{pmatrix}
- \tfrac{1}{3}\begin{pmatrix}
0 & 0 & 0 & 0 & 0 & 0 & 0 & 0 & 0 & 0 & 0 & 0 \\
0 & 4 & 4 & 4 & 0 & 0 & 2 & 2 & 2 & 2 & 2 & 2 \\
0 & 4 & 4 & 4 & 0 & 0 & 2 & 2 & 2 & 2 & 2 & 2 \\
0 & 4 & 4 & 4 & 0 & 0 & 2 & 2 & 2 & 2 & 2 & 2 \\
0 & 0 & 0 & 0 & 0 & 0 & 0 & 0 & 0 & 0 & 0 & 0 \\
0 & 0 & 0 & 0 & 0 & 0 & 0 & 0 & 0 & 0 & 0 & 0 \\
0 & 2 & 2 & 2 & 0 & 0 & 1 & 1 & 1 & 1 & 1 & 1 \\
0 & 2 & 2 & 2 & 0 & 0 & 1 & 1 & 1 & 1 & 1 & 1 \\
0 & 2 & 2 & 2 & 0 & 0 & 1 & 1 & 1 & 1 & 1 & 1 \\
0 & 2 & 2 & 2 & 0 & 0 & 1 & 1 & 1 & 1 & 1 & 1 \\
0 & 2 & 2 & 2 & 0 & 0 & 1 & 1 & 1 & 1 & 1 & 1 \\
0 & 2 & 2 & 2 & 0 & 0 & 1 & 1 & 1 & 1 & 1 & 1
\end{pmatrix}
$$

$$
= \tfrac{1}{3}\begin{pmatrix} 0 & \mathbf{0}' & \mathbf{0}' & \mathbf{0}' \\ \mathbf{0} & \mathbf{A}_{11} & \mathbf{O} & \mathbf{A}_{12} \\ \mathbf{0} & \mathbf{O} & \mathbf{O} & \mathbf{O} \\ \mathbf{0} & \mathbf{A}_{21} & \mathbf{O} & \mathbf{A}_{22} \end{pmatrix} - \tfrac{1}{3}\begin{pmatrix} 0 & \mathbf{0}' & \mathbf{0}' & \mathbf{0}' \\ \mathbf{0} & \mathbf{B}_{11} & \mathbf{O} & \mathbf{B}_{12} \\ \mathbf{0} & \mathbf{O} & \mathbf{O} & \mathbf{O} \\ \mathbf{0} & \mathbf{B}_{21} & \mathbf{O} & \mathbf{B}_{22} \end{pmatrix}, \qquad (14.89)
$$

where $\mathbf{A}_{11} = 12\mathbf{I}_3$, $\mathbf{B}_{11} = 4\mathbf{j}_3\mathbf{j}_3'$, $\mathbf{B}_{12} = 2\mathbf{j}_3\mathbf{j}_6'$, $\mathbf{B}_{21} = 2\mathbf{j}_6\mathbf{j}_3'$, $\mathbf{B}_{22} = \mathbf{j}_6\mathbf{j}_6'$,

$$
\mathbf{A}_{12} = \begin{pmatrix} 6\mathbf{j}_2' & \mathbf{0}' & \mathbf{0}' \\ \mathbf{0}' & 6\mathbf{j}_2' & \mathbf{0}' \\ \mathbf{0}' & \mathbf{0}' & 6\mathbf{j}_2' \end{pmatrix}, \quad \mathbf{A}_{21} = \begin{pmatrix} 6\mathbf{j}_2 & \mathbf{0} & \mathbf{0} \\ \mathbf{0} & 6\mathbf{j}_2 & \mathbf{0} \\ \mathbf{0} & \mathbf{0} & 6\mathbf{j}_2 \end{pmatrix},
$$

$$\mathbf{A}_{22} = \begin{pmatrix} 3\mathbf{j}_2\mathbf{j}_2' & \mathbf{O} & \mathbf{O} \\ \mathbf{O} & 3\mathbf{j}_2\mathbf{j}_2' & \mathbf{O} \\ \mathbf{O} & \mathbf{O} & 3\mathbf{j}_2\mathbf{j}_2' \end{pmatrix}.$$

If we write $\boldsymbol{\beta}$ in (14.4) in the form

$$\boldsymbol{\beta} = (\mu, \boldsymbol{\alpha}', \beta_1, \beta_2, \boldsymbol{\gamma}')',$$

where $\boldsymbol{\alpha}' = (\alpha_1, \alpha_2, \alpha_3)$ and $\boldsymbol{\gamma}' = (\gamma_{11}, \gamma_{12}, \gamma_{21}, \gamma_{22}, \gamma_{31}, \gamma_{32})$, then $\boldsymbol{\beta}'\mathbf{L}\boldsymbol{\beta}$ in (14.87) becomes

$$\begin{aligned}\boldsymbol{\beta}'\mathbf{L}\boldsymbol{\beta} &= \tfrac{1}{3}\boldsymbol{\alpha}'\mathbf{A}_{11}\boldsymbol{\alpha} + \tfrac{1}{3}\boldsymbol{\alpha}'\mathbf{A}_{12}\boldsymbol{\gamma} + \tfrac{1}{3}\boldsymbol{\gamma}'\mathbf{A}_{21}\boldsymbol{\alpha} + \tfrac{1}{3}\boldsymbol{\gamma}'\mathbf{A}_{22}\boldsymbol{\gamma} - \tfrac{1}{3}\boldsymbol{\alpha}'\mathbf{B}_{11}\boldsymbol{\alpha} \\ &\quad - \tfrac{1}{3}\boldsymbol{\alpha}'\mathbf{B}_{12}\boldsymbol{\gamma} - \tfrac{1}{3}\boldsymbol{\gamma}'\mathbf{B}_{21}\boldsymbol{\alpha} - \tfrac{1}{3}\boldsymbol{\gamma}'\mathbf{B}_{22}\boldsymbol{\gamma}.\end{aligned}$$

Since $\mathbf{A}_{21}' = \mathbf{A}_{12}$ and $\mathbf{B}_{21}' = \mathbf{B}_{12}$, this reduces to

$$\begin{aligned}\boldsymbol{\beta}'\mathbf{L}\boldsymbol{\beta} &= \tfrac{1}{3}\boldsymbol{\alpha}'\mathbf{A}_{11}\boldsymbol{\alpha} + \tfrac{2}{3}\boldsymbol{\alpha}'\mathbf{A}_{12}\boldsymbol{\gamma} + \tfrac{1}{3}\boldsymbol{\gamma}'\mathbf{A}_{22}\boldsymbol{\gamma} - \tfrac{1}{3}\boldsymbol{\alpha}'\mathbf{B}_{11}\boldsymbol{\alpha} \\ &\quad - \tfrac{2}{3}\boldsymbol{\alpha}'\mathbf{B}_{12}\boldsymbol{\gamma} - \tfrac{1}{3}\boldsymbol{\gamma}'\mathbf{B}_{22}\boldsymbol{\gamma}.\end{aligned}$$

If we partition $\boldsymbol{\gamma}$ as $\boldsymbol{\gamma}' = (\boldsymbol{\gamma}_1', \boldsymbol{\gamma}_2', \boldsymbol{\gamma}_3')$, where $\boldsymbol{\gamma}_i' = (\gamma_{i1}, \gamma_{i2})$, then

$$\begin{aligned}\tfrac{2}{3}\boldsymbol{\alpha}'\mathbf{A}_{12}\boldsymbol{\gamma} &= \tfrac{12}{3}\boldsymbol{\alpha}' \begin{pmatrix} \mathbf{j}_2' & \mathbf{0}' & \mathbf{0}' \\ \mathbf{0}' & \mathbf{j}_2' & \mathbf{0}' \\ \mathbf{0}' & \mathbf{0}' & \mathbf{j}_2' \end{pmatrix} \begin{pmatrix} \boldsymbol{\gamma}_1 \\ \boldsymbol{\gamma}_2 \\ \boldsymbol{\gamma}_3 \end{pmatrix} \\ &= 4\boldsymbol{\alpha}' \begin{pmatrix} \mathbf{j}_2'\boldsymbol{\gamma}_1 \\ \mathbf{j}_2'\boldsymbol{\gamma}_2 \\ \mathbf{j}_2'\boldsymbol{\gamma}_3 \end{pmatrix} = 4\sum_{i=1}^{3} \alpha_i \gamma_{i.}.\end{aligned}$$

Now, using the definitions of $\mathbf{A}_{11}$, $\mathbf{A}_{22}$, $\mathbf{B}_{11}$, $\mathbf{B}_{12}$, and $\mathbf{B}_{22}$ following (14.89), we obtain

$$\begin{aligned}\boldsymbol{\beta}'\mathbf{L}\boldsymbol{\beta} &= 4\boldsymbol{\alpha}'\boldsymbol{\alpha} + 4\sum_{i=1}^{3} \alpha_i \gamma_{i.} + \sum_{i=1}^{3} \boldsymbol{\gamma}_i'\mathbf{j}_2\mathbf{j}_2'\boldsymbol{\gamma}_i - \tfrac{4}{3}\boldsymbol{\alpha}'\mathbf{j}_3\mathbf{j}_3'\boldsymbol{\alpha} \\ &\quad - \tfrac{4}{3}\boldsymbol{\alpha}'\mathbf{j}_3\mathbf{j}_6'\boldsymbol{\gamma} - \tfrac{1}{3}\boldsymbol{\gamma}'\mathbf{j}_6\mathbf{j}_6'\boldsymbol{\gamma} \\ &= 4\sum_{i=1}^{3} \alpha_i^2 + 4\sum_{i=1}^{3} \alpha_i \gamma_{i.} + \sum_{i=1}^{3} \gamma_{i.}^2 - \tfrac{4}{3}\alpha_.^2 - \tfrac{4}{3}\alpha_.\gamma_{..} - \tfrac{1}{3}\gamma_{..}^2. \end{aligned} \qquad (14.90)$$

By expressing $\gamma_{i.}$, $\alpha_{.}$, and $\gamma_{..}$ in terms of means, (14.90) can be written in the form

$$\boldsymbol{\beta}'\mathbf{L}\boldsymbol{\beta} = 4\sum_{i=1}^{3}(\alpha_i - \bar{\alpha}_{.} + \bar{\gamma}_{i.} - \bar{\gamma}_{..})^2 = 4\sum_{i=1}^{3}\alpha_i^{*2} \qquad \text{[by (14.16)]}. \tag{14.91}$$

For an alternative approach leading to (14.91), note that since $E(\mathbf{C}\hat{\boldsymbol{\beta}}) = \mathbf{C}\boldsymbol{\beta}$, (14.86) can be written as

$$E[\text{SS}(\alpha|\mu, \beta, \gamma)] = 2\sigma^2 + [E(\mathbf{C}\hat{\boldsymbol{\beta}})]'[\mathbf{C}(\mathbf{X}'\mathbf{X})^{-}\mathbf{C}']^{-1}E(\mathbf{C}\hat{\boldsymbol{\beta}}). \tag{14.92}$$

By (14.75), $\text{SS}(\alpha|\mu, \beta, \gamma) = \text{SSH} = (\mathbf{C}\hat{\boldsymbol{\beta}})'[\mathbf{C}(\mathbf{X}'\mathbf{X})^{-}\mathbf{C}']^{-1}\mathbf{C}\hat{\boldsymbol{\beta}}$. Thus, by (14.92), we can obtain $E[\text{SS}(\alpha|\mu, \beta, \gamma)]$ by replacing $\mathbf{C}\hat{\boldsymbol{\beta}}$ in $\text{SS}(\alpha|\mu, \beta, \gamma)$ with $\mathbf{C}\boldsymbol{\beta}$ and adding $2\sigma^2$. To illustrate, we replace $\bar{y}_{i..}$ and $\bar{y}_{...}$ with $E(\bar{y}_{i..})$ and $E(\bar{y}_{...})$ in $\text{SS}(\alpha|\mu, \beta, \gamma) = 4\sum_{i=1}^{3}(\bar{y}_{i..} - \bar{y}_{...})^2$ in (14.69). We first find $E(\bar{y}_{i..})$:

$$\begin{aligned}
E(\bar{y}_{i..}) &= E\left(\tfrac{1}{4}\sum_{jk} y_{ijk}\right) = \tfrac{1}{4}\sum_{jk} E(y_{ijk}) \\
&= \tfrac{1}{4}\sum_{jk} E(\mu + \alpha_i + \beta_j + \gamma_{ij} + \varepsilon_{ijk}) \\
&= \tfrac{1}{4}\sum_{jk} (\mu + \alpha_i + \beta_j + \gamma_{ij}) \\
&= \tfrac{1}{4}\left(4\mu + 4\alpha_i + 2\sum_j \beta_j + 2\sum_j \gamma_{ij}\right) \\
&= \mu + \alpha_i + \bar{\beta}_{.} + \bar{\gamma}_{i.}.
\end{aligned} \tag{14.93}$$

Similarly

$$E(\bar{y}_{...}) = \mu + \bar{\alpha}_{.} + \bar{\beta}_{.} + \bar{\gamma}_{..}. \tag{14.94}$$

Then,

$$\begin{aligned}
E[\text{SS}(\alpha|\mu, \beta, \gamma)] &= 2\sigma^2 + 4\sum_{i=1}^{3}[E(\bar{y}_{i..}) - E(\bar{y}_{...})]^2 \\
&= 2\sigma^2 + 4\sum_i (\mu + \alpha_i + \bar{\beta}_{.} + \bar{\gamma}_{i.} - \mu - \bar{\alpha}_{.} - \bar{\beta}_{.} + \bar{\gamma}_{..})^2 \\
&= 2\sigma^2 + 4\sum_i (\alpha_i - \bar{\alpha}_{.} + \bar{\gamma}_{i.} - \bar{\gamma}_{..})^2 \\
&= 2\sigma^2 + 4\sum_i \alpha_i^{*2} \qquad \text{[by (14.16)]}.
\end{aligned}$$

PROBLEMS

14.1 Obtain θ_1 and θ_5 in (14.7) from (14.6).

14.2 In a comment following (14.8), it is noted that there are no estimable contrasts in the α's alone or β's alone. Verify this statement.

14.3 Show that $\frac{1}{3}(\theta_3 + \theta_3' + \theta_3'')$ has the value shown in (14.11).

14.4 Verify the following results in (14.15) using the definitions of α_i^*, β_j^*, and γ_{ij}^* in (14.14):

(a) $\sum_i \alpha_i^* = 0$
(b) $\sum_j \beta_j^* = 0$
(c) $\sum_i \gamma_{ij}^* = 0, \quad j = 1, 2, \ldots, b$
(d) $\sum_j \gamma_{ij}^* = 0, \quad i = 1, 2, \ldots, a$

14.5 Verify the following results from (14.15) using the definitions of α_i^*, β_j^*, and γ_{ij}^* in (14.16), (14.17), and (14.18):

(a) $\sum_i \alpha_i^* = 0$
(b) $\sum_j \beta_j^* = 0$
(c) $\sum_i \gamma_{ij}^* = 0, \quad j = 1, 2, \ldots, b$
(d) $\sum_j \gamma_{ij}^* = 0, \quad i = 1, 2, \ldots, a$

14.6 **(a)** Show that $\beta_j^* = \beta_j - \bar{\beta}_. + \bar{\gamma}_{.j} - \bar{\gamma}_{..}$ as in (14.17).
(b) Show that $\gamma_{ij}^* = \gamma_{ij} - \bar{\gamma}_{i.} - \bar{\gamma}_{.j} + \bar{\gamma}_{..}$ as in (14.18).

14.7 Show that $\hat{\alpha}_i$ and $\hat{\gamma}_{ij}$ in (14.21) are unbiased estimators of α_i^* and γ_{ij}^* as noted following (14.21).

14.8 **(a)** Show that $\bar{y}_{11.} + \bar{y}_{12.} = 2\bar{y}_{1..}$ and that $\bar{y}_{21.} + \bar{y}_{22.} = 2\bar{y}_{2..}$, as used to obtain (14.22).
(b) Show that $\hat{\alpha}_1 - \hat{\alpha}_2 + \frac{1}{2}(\hat{\gamma}_{11} + \hat{\gamma}_{12}) - \frac{1}{2}(\hat{\gamma}_{21} + \hat{\gamma}_{22}) = \bar{y}_{1..} - \bar{y}_{2..}$ as in (14.22).

14.9 Show that $(\mathbf{X}'\mathbf{X})^-$ in (14.23) is a generalized inverse of $\mathbf{X}'\mathbf{X}$ in (14.5).

14.10 Show that SSE in (14.26) is equal to SSE in (14.25).

14.11 Show that the second equality in (14.34) is equivalent to the second equality in (14.35); that is, $\mu_{21} - \mu_{22} = \mu_{31} - \mu_{32}$ implies $\gamma_{21} - \gamma_{22} - \gamma_{31} + \gamma_{32} = 0$.

14.12 Show that $\sum_i (\bar{y}_{i..} - \bar{y}_{...}) y_{i..} = \sum_i y_{i..}^2/bn - y_{...}^2/abn$ and that $\sum_{ij} (\bar{y}_{ij.} - \bar{y}_{i..} - \bar{y}_{.j.} + \bar{y}_{...}) y_{ij.} = \sum_{ij} y_{ij.}^2/n - \sum_i y_{i..}^2/bn - \sum_j y_{.j.}^2/an + y_{...}^2/abn$, as in (14.40).

14.13 **(a)** In a comment following (14.41), it was noted that the use of $\hat{\boldsymbol{\beta}}$ from (14.24) would produce the same result as in (14.41), namely, $\hat{\boldsymbol{\beta}}'\mathbf{X}'\mathbf{y} = \sum_{ij} y_{ij.}^2/n$. Verify this.

(b) Show that $\text{SSE} = \sum_{ijk} y_{ijk}^2 - n\sum_{ij} \bar{y}_{ij.}^2$ in (14.25) is equal to $\text{SSE} = \sum_{ijk} y_{ijk}^2 - \sum_{ij} y_{ij.}^2/n$ in (14.46).

14.14 Show that $(a-1)(b-1)$ is the number of independent γ_{ij}^* terms in $H_0\colon \gamma_{ij}^* = 0$ for $i = 1, 2, \ldots, a$ and $j = 1, 2, \ldots, b$, as noted near the end of Section 14.4.1.2.

14.15 Show that $\text{SS}(\gamma|\mu, \alpha, \beta) = n\sum_{ijk} (\bar{y}_{ij.} - \bar{y}_{i..} - \bar{y}_{.j.} + \bar{y}_{...})^2$ in (14.47) is the same as $\text{SS}(\gamma|\mu, \alpha, \beta)$ in (14.45).

14.16 Show that $\text{SSE} = \sum_{ijk} (y_{ijk} - \bar{y}_{ij.})^2$ in (14.48) is equal to $\text{SSE} = \sum_{ijk} y_{ijk}^2 - \sum_{ij} y_{ij.}^2/n$ in (14.46).

14.17 Using $\mathbf{X}'\mathbf{X}$ in (14.5) and $(\mathbf{X}'\mathbf{X})^-$ in (14.23), show that $\mathbf{X}(\mathbf{X}'\mathbf{X})^-\mathbf{X}'$ has the form given in (14.50).

14.18 **(a)** Show that $(\mathbf{X}_1'\mathbf{X}_1)^-$ in (14.53) is a generalized inverse of $\mathbf{X}_1'\mathbf{X}_1$ in (14.42).

(b) Show that $\mathbf{X}_1(\mathbf{X}_1'\mathbf{X}_1)^-\mathbf{X}_1'$ has the form given by (14.54).

14.19 Show that $\frac{1}{6}\sum_{j=1}^2 y_{.j.}^2$ can be written in the matrix form given in (14.59).

14.20 Show that $\frac{1}{2}\mathbf{A} - \frac{1}{4}\mathbf{B} - \frac{1}{6}\mathbf{C} + \frac{1}{12}\mathbf{D}$ has the value shown in (14.61).

14.21 Show that $H_0\colon \alpha_1^* = \alpha_2^* = \alpha_3^*$ in (14.63) is equivalent to $H_0\colon \alpha_1^* = \alpha_2^* = \alpha_3^* = 0$ in (14.64).

14.22 Obtain $\text{SS}(\mu, \alpha, \gamma)$ and show that $\text{SS}(\beta|\mu, \alpha, \gamma) = \sum_{j=1}^b y_{.j.}^2/bn - y_{...}^2/abn$ as in (14.68).

14.23 Prove Theorem 14.4b for the special case $a = 3$, $b = 2$, and $n = 2$.

14.24 **(a)** Using $\mathbf{C}$ in (14.74), $(\mathbf{X}'\mathbf{X})^-$ in (14.23), and $\mathbf{X}$ in (14.4), show that $\mathbf{C}(\mathbf{X}'\mathbf{X})^-\mathbf{X}'$ is the 2×12 matrix given in (14.77).

(b) Using $\mathbf{C}$ in (14.74) and $(\mathbf{X}'\mathbf{X})^-$ in (14.23), show that $\mathbf{C}(\mathbf{X}'\mathbf{X})^-\mathbf{C}'$ is the 2×2 matrix shown in (14.78).

(c) Show that the matrix $\mathbf{A} = \mathbf{X}(\mathbf{X}'\mathbf{X})^-\mathbf{C}'[\mathbf{C}(\mathbf{X}'\mathbf{X})^-\mathbf{C}']^{-1}\mathbf{C}(\mathbf{X}'\mathbf{X})^-\mathbf{X}'$ has the form shown in (14.79).

14.25 For the B main effect, formulate a hypothesis $H_0\colon \mathbf{C}\boldsymbol{\beta} = \mathbf{0}$ and obtain $\text{SS}(\beta|\mu, \alpha, \gamma)$ using SSH in (14.75).

14.26 Using assumptions 1, 2, and 3 in Section 14.1, show that $E(\varepsilon_{ijk}^2) = \sigma^2$ for all i, j, k and $E(\varepsilon_{ijk}\varepsilon_{rst}) = 0$ for $(i, j, k) \neq (r, s, t)$, as in (14.82) and (14.82).

14.27 Show that $E(\sum_{i=1}^a y_{i..}^2) = ab^2n^2\mu^{*2} + b^2n^2\sum_{i=1}^a \alpha_i^{*2} + abn\sigma^2$ as in (14.84).

14.28 **(a)** Show that $E(\sum_{j=1}^b y_{.j.}^2) = a^2bn^2\mu^{*2} + a^2n^2\sum_{j=1}^b \beta_j^{*2} + abn\sigma^2$.

(b) Show that $E(\sum_{ij} y_{ij.}^2) = abn^2\mu^{*2} + bn^2\sum_i \alpha_i^{*2} + an^2\sum_j \beta_j^{*2} + n^2\sum_{ij} \gamma_{ij}^{*2} + abn\sigma^2$.

(c) Show that $E[\text{SS}(\beta|\mu, \alpha, \gamma)/(b-1)] = \sigma^2 + an\sum_j \beta_j^{*2}/(b-1)$.

(d) Show that $E[\text{SS}(\gamma|\mu, \alpha, \beta)/(a-1)(b-1)] = \sigma^2 + n\sum_{ij} \gamma_{ij}^{*2}/(a-1)(b-1)$.

TABLE 14.6 Lactic Acid[a] at Five Successive Time Periods for Fresh and Wilted Alfalfa Silage

	Period				
Condition	1	2	3	4	5
Fresh	13.4	37.5	65.2	60.8	37.7
	16.0	42.7	54.9	57.1	49.2
Wilted	14.4	29.3	36.4	39.1	39.4
	20.0	34.5	39.7	38.7	39.7

[a]In mg/g of silage.

14.29 Using $\mathbf{C}$ in (14.74) and $(\mathbf{X}'\mathbf{X})^-$ in (14.23), show that $\mathbf{L} = \mathbf{C}'[\mathbf{C}(\mathbf{X}'\mathbf{X})^-\mathbf{C}']^{-1}\mathbf{C}$ has the form shown in (14.88).

14.30 Expand $\sum_{i=1}^{3} (\alpha_i - \bar{\alpha}_. + \bar{\gamma}_{i.} - \bar{\gamma}_{..})^2$ in (14.91) to obtain (14.90).

14.31 **(a)** Show that $E(\bar{y}_{...}) = \mu + \bar{\alpha}_. + \bar{\beta}_. + \bar{\gamma}_{..}$ as in (14.94).

(b) Show that $E(\bar{y}_{.j.}) = \mu + \bar{\alpha}_. + \beta_j + \bar{\gamma}_{.j}.$

(c) Show that $E(\bar{y}_{ij.}) = \mu + \alpha_i + \beta_j + \gamma_{ij}.$

14.32 Obtain the following expected values using the method suggested by (14.92) and illustrated at the end of Section 14.5.2. Use the results of Problem 14.31b, c.

(a) $E[\text{SS}(\beta|\mu, \alpha, \gamma)] = \sigma^2 + 6\sum_j \beta_j^{*2}$

(b) $E[\text{SS}(\gamma|\mu, \alpha, \beta)] = 2\sigma^2 + 2\sum_{ij} \gamma_{ij}^{*2}$

TABLE 14.7 Hemoglobin Concentration (g/mL) in Blood of Brown Trout[a]

Rate:	1		2		3		4	
Method:	A	B	A	B	A	B	A	B
	6.7	7.0	9.9	9.9	10.4	9.9	9.3	11.0
	7.8	7.8	8.4	9.6	8.1	9.6	9.3	9.3
	5.5	6.8	10.4	10.2	10.6	10.4	7.8	11.0
	8.4	7.0	9.3	10.4	8.7	10.4	7.8	9.0
	7.0	7.5	10.7	11.3	10.7	11.3	9.3	8.4
	7.8	6.5	11.9	9.1	9.1	10.9	10.2	8.4
	8.6	5.8	7.1	9.0	8.8	8.0	8.7	6.8
	7.4	7.1	6.4	10.6	8.1	10.2	8.6	7.2
	5.8	6.5	8.6	11.7	7.8	6.1	9.3	8.1
	7.0	5.5	10.6	9.6	8.0	10.7	7.2	11.0

[a]After 35 days of treatment at the daily rates of 0, 5, 10, and 15g of sulfamerazine per 100 lb of fish employing two methods for each rate.

14.33 A preservative was added to fresh and wilted alfalfa silage (Snedecor 1948). The lactic acid concentration was measured at five periods after ensiling began. There were two replications. The results are given in Table 14.6. Let factor A be condition (fresh or wilted) and factor B be period. Test for main effects and interactions.

14.34 Gutsell (1951) measured hemoglobin in the blood of brown trout after treatment with four rates of sulfamerazine. Two methods of administering the sulfamerazine were used. Ten fish were measured for each rate and each method. The data are given in Table 14.7. Test for effect of rate and method and interaction.

15 Analysis-of-Variance: The Cell Means Model for Unbalanced Data

15.1 INTRODUCTION

The theory of linear models for ANOVA applications was developed in Chapter 12. Although all the examples used in that and the following chapters have involved balanced data (where the number of observations is equal from one cell to another), the theory also applies to unbalanced data.

Chapters 13 and 14 show that simple and intuitive results are obtained when the theory is applied to balanced ANOVA situations. Intuitive marginal means are informative in analysis of the data [e.g., see (14.69) and (14.70)]. When applied to unbalanced data, however, the general results of Chapter 12 do not simplify to intuitive formulas. Even worse, the intuitive marginal means one is tempted to use can be misleading and sometimes paradoxical. This is especially true for two-way or higher-way data. As an example, consider the unbalanced two-way data in Figure 15.1. The data follow the two-way additive model (Section 12.1.2) with no error

$$y_{ij} = \mu + \alpha_i + \beta_j, \quad i = 1, 2, \quad j = 1, 2,$$

where $\mu = 25$, $\alpha_1 = 0$, $\alpha_2 = -20$, $\beta_1 = 0$, $\beta_2 = 5$. Simple marginal means of the data are given to the right and below the box.

The true effects of factors A and B are, respectively, $\alpha_2 - \alpha_1 = -20$ and $\beta_2 - \beta_1 = 5$. Even for error-free unbalanced data, however, naive estimates of these effects based on the simple marginal means are highly misleading. The effect of factor A appears to be $8.75 - 25.125 = -16.375$, and even more surprisingly the effect of factor B appears to be $15 - 20 = -5$.

Still other complications arise in the analysis of unbalanced data. For example, it was mentioned in Section 14.4.2.1 that many texts discourage testing for main effects in the presence of interactions. But little harm or controversy results from doing so when the data are balanced. The numerators for the main effect F tests are exactly

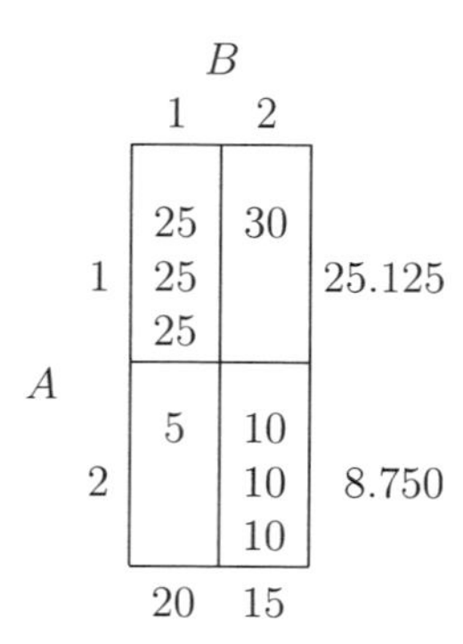

Figure 15.1 Hypotetical error-free data from an unbalanced two-way model.

the same whether the model with or without interactions is being entertained as the full model. Such is not the case for unbalanced data. The numerator sums of squares in these F tests depend greatly on which model is used as the full model, and, obviously, conclusions can be affected. Several types of sums of squares [usually types I, II, and III; see Milliken and Johnson (1984, pp. 138–158)] have been suggested to help clarify this issue.

The issues involved in choosing the appropriate full model for a test are subtle and often confusing. The use of different full models results in different weightings in the sums of squares calculations and expected mean squares. But some of the same weightings also arise for other reasons. For example, the weights might arise because the data are based on "probability proportional to size" (pps) sampling of populations (Cochran 1977, pp. 250–251).

Looking at this complex issue from different points of view has led to completely contradictory conclusions. For example, Milliken and Johnson (1984, p. 158) wrote that "in almost all cases, type III sums of squares will be preferred," whereas Nelder and Lane (1995) saw "no place for types III and IV sums of squares in making inferences from the use of linear models."

Further confusion regarding the analysis of unbalanced data has arisen from the interaction of computing advances with statistical practice. Historically, several different methods for unbalanced data analysis were developed as approximate methods, suitable for the computing resources available at the time. Looking back, however, we simply see a confusing array of alternative methods. Some such methods include weighted squares of means (Yates 1934; Morrison 1983, pp. 407–412), the method of unweighted means (Searle 1971; Winer 1971), the method of fitting constants (Rao 1965, pp. 211–214; Searle 1971, p. 139; Snedecor and Cochran 1967), and various methods of imputing data to make the dataset balanced (Hartley 1956; Healy and Westmacott 1969; Little and Rubin 2002, pp. 28–30).

The overparameterized (non-full rank) model (Sections 12.2, 12.5, 13.1, and 14.1) has some advantages in the analysis of unbalanced data, while the cell means approach (Section 12.1.1) has other advantages. The non-full rank approach builds the structure (additive two-way, full two-way, etc.) of the dataset into the model from the start, but relies on the subtle concepts of estimability, testability, and generalized inverses. The cell means model has the advantages of being a full-rank model, but the structure of the dataset is not an explicit part of the model. Whichever model is used, hard questions about the exact hypotheses of interest have to be faced. Many of the complexities are a matter of statistical practice rather than mathematical statistics.

The most extreme form of imbalance is that in which one or more of the cells have no observations. In this "empty cells" situation, even the cell means model is an overparameterized model. Nonetheless, the cell means approach allows one to deal specifically with nonestimability problems arising from the empty cells. Such an approach is almost impossible using the overparameterized approach.

In the remainder of this chapter we discuss the analysis of unbalanced data using the cell means model. Unbalanced one-way and two-way models are covered in Sections 15.2 and 15.3. In Section 15.4 we discuss the empty-cell situation.

15.2 ONE-WAY MODEL

The non-full-rank and cell means versions of the one-way unbalanced model are

$$\begin{aligned} y_{ij} &= \mu + \alpha_i + \varepsilon_{ij} && (15.1) \\ &= \mu_i + \varepsilon_{ij}, && (15.2) \end{aligned}$$

$$i = 1, 2, \ldots, k, \quad j = 1, 2, \ \ldots, n_i.$$

For making inferences, we assume the ε_{ij}'s are independently distributed as $N(0, \sigma^2)$.

15.2.1 Estimation and Testing

To estimate the μ_i's, we begin by writing the $N = \sum_i n_i$ observations for the model (15.2) in the form

$$\mathbf{y} = \mathbf{W}\boldsymbol{\mu} + \boldsymbol{\varepsilon}, \qquad (15.3)$$

where

$$\mathbf{W} = \begin{pmatrix} 1 & 0 & \cdots & 0 \\ \vdots & \vdots & & \vdots \\ 1 & 0 & \cdots & 0 \\ 0 & 1 & \cdots & 0 \\ \vdots & \vdots & & \vdots \\ 0 & 1 & \cdots & 0 \\ \vdots & \vdots & & \vdots \\ 0 & 0 & & 1 \\ \vdots & \vdots & & \vdots \\ 0 & 0 & \cdots & 1 \end{pmatrix}, \quad \boldsymbol{\mu} = \begin{pmatrix} \mu_1 \\ \mu_2 \\ \vdots \\ \mu_k \end{pmatrix}.$$

The normal equations are given by

$$\mathbf{W}'\mathbf{W}\hat{\boldsymbol{\mu}} = \mathbf{W}'\mathbf{y},$$

where $\mathbf{W}'\mathbf{W} = \text{diag}(n_1, n_2, \ldots, n_k)$ and $\mathbf{W}'\mathbf{y} = (y_{1.}, y_{2.}, \ldots, y_{k.})'$, with $y_{i.} = \sum_{j=1}^{n_i} y_{ij}$. Since the matrix $\mathbf{W}$ is full rank, we have, by (7.6)

$$\hat{\boldsymbol{\mu}} = (\mathbf{W}'\mathbf{W})^{-1}\mathbf{W}'\mathbf{y} \tag{15.4}$$

$$= \bar{\mathbf{y}} = \begin{pmatrix} \bar{y}_{1.} \\ \bar{y}_{2.} \\ \vdots \\ \bar{y}_{k.} \end{pmatrix}, \tag{15.5}$$

where $\bar{y}_{i.} = \sum_{j=1}^{n_i} y_{ij}/n_i$.

To test $H_0: \mu_1 = \mu_2 = \cdots = \mu_k$, we compare the full model in (15.2) and (15.3) with the reduced model $y_{ij} = \mu + \varepsilon_{ij}^*$, where μ is the common value of $\mu_1, \mu_2, \ldots, \mu_k$ under H_0. (We do not use the notation μ^* in the reduced model because there is no μ in the full model $y_{ij} = \mu_i + \varepsilon_{ij}$.) In matrix form, the N observations in the reduced model become $\mathbf{y} = \mu\mathbf{j} + \boldsymbol{\varepsilon}^*$, where $\mathbf{j}$ is $N \times 1$. For the full model, we have $\text{SS}(\mu_1, \mu_2, \ldots, \mu_k) = \hat{\boldsymbol{\mu}}'\mathbf{W}'\mathbf{y}$, and for the reduced model, we have $\text{SS}(\mu) = \hat{\mu}\mathbf{j}'\mathbf{y} = N\bar{y}_{..}^2$, where $N = \sum_i n_i$ and $\bar{y}_{..} = \sum_{ij} y_{ij}/N$. The difference $\text{SS}(\mu_1, \mu_2, \ldots, \mu_k) - \text{SS}(\mu)$ is equal to the regression sum of squares SSR in (8.6), which we denote by SSB for "between" sum of squares

$$\text{SSB} = \hat{\boldsymbol{\mu}}'\mathbf{W}'\mathbf{y} - N\bar{y}_{..}^2 = \sum_{i=1}^{k} \bar{y}_{i.}y_{i.} - N\bar{y}_{..}^2 \tag{15.6}$$

$$= \sum_{i=1}^{k} \frac{y_{i.}^2}{n_i} - \frac{y_{..}^2}{N}, \tag{15.7}$$

where $y_{..} = \sum_{ij} y_{ij}$ and $\bar{y}_{..} = y_{..}/N$. From (15.7), we see that SSB has $k - 1$ degrees of freedom. The error sum of squares is given by (7.24) or (8.6) as

$$\text{SSE} = \mathbf{y}'\mathbf{y} - \hat{\boldsymbol{\mu}}'\mathbf{W}'\mathbf{y}$$

$$= \sum_{i=1}^{k} \sum_{j=1}^{n_i} y_{ij}^2 - \sum_{i=1}^{k} \frac{y_{i.}^2}{n_i}, \tag{15.8}$$

which has $N - k$ degrees of freedom. These sums of squares are summarized in Table 15.1.

TABLE 15.1 One-Way Unbalanced ANOVA

Source	Sum of Squares	df
Between	$\text{SSB} = \sum_i y_{i.}^2/n_i - y_{..}^2/N$	$k-1$
Error	$\text{SSE} = \sum_{ij} y_{ij}^2 - \sum_i y_{i.}^2/n_i$	$N-k$
Total	$\text{SST} = \sum_{ij} y_{ij}^2 - y_{..}^2/N$	$N-1$

The sums of squares SSB and SSE in Table 15.1 can also be written in the form

$$\text{SSB} = \sum_{i=1}^{k} n_i(\bar{y}_{i.} - \bar{y}_{..})^2, \tag{15.9}$$

$$\text{SSE} = \sum_{i=1}^{k} \sum_{j=1}^{n_i} (y_{ij} - \bar{y}_{i.})^2. \tag{15.10}$$

If we assume that the y_{ij}'s are independently distributed as $N(\mu_i, \sigma^2)$, then by Theorem 8.1d, an F statistic for testing H_0: $\mu_1 = \mu_2 = \cdots = \mu_k$ is given by

$$F = \frac{\text{SSB}/(k-1)}{\text{SSE}/(N-k)}. \tag{15.11}$$

If H_0 is true, F is distributed as $F(k-1, N-k)$.

Example 15.2.1. A sample from the output of five filling machines is given in Table 15.2 (Ostle and Mensing 1975, p. 359).

The analysis of variance is given in Table 15.3. The F is calculated by (15.11). There is no significant difference in the average weights filled by the five machines. □

15.2.2 Contrasts

A contrast in the population means is defined as $\delta = c_1\mu_1 + c_2\mu_2 + \cdots + c_k\mu_k$, where $\sum_{i=1}^{k} c_i = 0$. The contrast can be expressed as $\delta = \mathbf{c}'\boldsymbol{\mu}$, where

TABLE 15.2 Net Weight of Cans Filled by Five Machines (A–E)

A	B	C	D	E
11.95	12.18	12.16	12.25	12.10
12.00	12.11	12.15	12.30	12.04
12.25		12.08	12.10	12.02
12.10				12.02

TABLE 15.3 ANOVA for the Fill Data in Table 15.2

Source	df	Sum of Squares	Mean Square	F	p Value
Between	4	.05943	.01486	1.9291	.176
Error	11	.08472	.00770		
Total	15	.14414			

$\text{SST} = \sum_{ij} y_{ij}^2 - y_{..}^2/N$ and $\boldsymbol{\mu} = (\mu_1, \mu_2, \ldots, \mu_k)'$. The best linear unbiased estimator of δ is given by $\hat{\delta} = c_1\bar{y}_{1.} + c_2\bar{y}_{2.} + \cdots + c_k\bar{y}_{k.} = \mathbf{c}'\hat{\boldsymbol{\mu}}$ [see (15.5) and Corollary 1 to Theorem 7.3d]. By (3.42), $\text{var}(\hat{\delta}) = \sigma^2\mathbf{c}'(\mathbf{W}'\mathbf{W})^{-1}\mathbf{c}$, which can be written as $\text{var}(\hat{\delta}) = \sigma^2 \sum_{i=1}^k c_i^2/n_i$, since $\mathbf{W}'\mathbf{W} = \text{diag}(n_1, n_2, \ldots, n_k)$. By (8.38), the F statistic for testing H_0: $\delta = 0$ is

$$F = \frac{(\mathbf{c}'\hat{\boldsymbol{\mu}})'\left[\mathbf{c}'(\mathbf{W}'\mathbf{W})^{-1}\mathbf{c}\right]^{-1}\mathbf{c}'\hat{\boldsymbol{\mu}}}{s^2}, \tag{15.12}$$

$$= \frac{\left(\sum_{i=1}^k c_i\bar{y}_{i.}\right)^2 / \left(\sum_{i=1}^k c_i^2/n_i\right)}{s^2}, \tag{15.13}$$

where $s^2 = \text{SSE}/(N-k)$ with SSE given by (15.8) or (15.10). We refer to the numerator of (15.13) as the sum of squares for the contrast. If H_0 is true, the F statistic in (15.12) or (15.13) is distributed as $F(1, N-k)$, and we reject H_0: $\delta = 0$ if $F \geq F_{\alpha, 1, N-k}$ or if $p \leq \alpha$, where p is the p value.

Two contrasts, say, $\hat{\delta} = \sum_{i=1}^k a_i\bar{y}i.$ and $\hat{\gamma} = \sum_{i=1}^k b_i\bar{y}_{i.}$, are said to be *orthogonal* if $\sum_{i=1}^k a_ib_i = 0$. However, in the case of unbalanced data, two orthogonal contrasts of this type are not independent, as they were in the balanced case (Theorem 13.6a).

Theorem 15.2. If the y_{ij}'s are independently distributed as $N(\mu_i, \sigma^2)$ in the unbalanced model (15.2), then two contrasts $\hat{\delta} = \sum_{i=1}^k a_i\bar{y}_{i.}$ and $\hat{\gamma} = \sum_{i=1}^k b_i\bar{y}_{i.}$ are independent if and only if $\sum_{i=1}^k a_ib_i/n_i = 0$.

Proof. We express the two contrasts in vector notation as $\hat{\delta} = \mathbf{a}'\bar{\mathbf{y}}$ and $\hat{\gamma} = \mathbf{b}'\bar{\mathbf{y}}$, where $\bar{\mathbf{y}} = (\bar{y}_{1.}, \bar{y}_{2.}, \ldots, \bar{y}_{k.})'$. By (7.14), we obtain

$$\text{cov}(\bar{\mathbf{y}}) = \sigma^2(\mathbf{W}'\mathbf{W})^{-1} = \sigma^2 \begin{pmatrix} 1/n_1 & 0 & \ldots & 0 \\ 0 & 1/n_2 & \ldots & 0 \\ \vdots & \vdots & & \vdots \\ 0 & 0 & \ldots & 1/n_k \end{pmatrix} = \sigma^2\mathbf{D}.$$

Then by (3.43), we have

$$\begin{aligned}\text{cov}(\hat{\delta}, \hat{\gamma}) &= \text{cov}(\mathbf{a}'\bar{\mathbf{y}}, \mathbf{b}'\bar{\mathbf{y}}) = \mathbf{a}'\text{cov}(\bar{\mathbf{y}})\mathbf{b} = \sigma^2\mathbf{a}'\mathbf{D}\mathbf{b} \\ &= \sigma^2 \sum_{i=1}^{k} \frac{a_i b_i}{n_i}. \end{aligned} \tag{15.14}$$

Hence, by Theorem 4.4c, $\hat{\delta}$ and $\hat{\gamma}$ are independent if and only if $\sum_i a_i b_i / n_i = 0$. □

We refer to contrasts whose coefficients satisfy $\sum_i a_i b_i / n_i = 0$ as *weighted orthogonal contrasts*. If we define $k - 1$ contrasts of this type, they partition the treatment sum of squares SSB into $k - 1$ independent sums of squares, each with 1 degree of freedom. Unweighted orthogonal contrasts that satisfy only $\sum_i a_i b_i = 0$ are not independent (see Theorem 15.2), and their sums of squares do not add up to the treatment sum of squares (as they do for balanced data; see Theorem 13.6a).

In practice, weighted orthogonal contrasts are often of less interest than unweighted orthogonal contrasts because we may not wish to choose the a_i's and b_i's on the basis of the n_i's in the sample. The n_i's seldom reflect population characteristics that we wish to take into account. However, it is not necessary that the sums of squares be independent in order to proceed with the tests. If we use unweighted orthogonal contrasts with $\sum_i a_i b_i = 0$, the general linear hypothesis test based on (15.12) or (15.13) tests each contrast adjusted for the other contrasts (see Theorem 8.4d).

Example 15.2.2a. Suppose that we wish to compare the means of three treatments and that the coefficients of the orthogonal contrasts $\delta = \mathbf{a}'\boldsymbol{\mu}$ and $\gamma = \mathbf{b}'\boldsymbol{\mu}$ are given by $\mathbf{a}' = (2\ -1\ -1)$ and $\mathbf{b}' = (0\ 1\ -1)$ with corresponding hypotheses

$$H_{01} : \mu_1 = \frac{1}{2}(\mu_2 + \mu_3), \qquad H_{02} : \mu_2 = \mu_3.$$

If the sample sizes for the three treatments are, for example, $n_1 = 10$, $n_2 = 20$, and $n_3 = 5$, then the two estimated contrasts

$$\hat{\delta} = 2\bar{y}_{1.} - \bar{y}_{2.} - \bar{y}_{3.} \quad \text{and} \quad \hat{\gamma} = \bar{y}_{2.} - \bar{y}_{3.}$$

are not independent, and the corresponding sums of squares do not add to the treatment sum of squares.

The following two vectors provide an example of contrasts whose coefficients satisfy $\sum_i a_i b_i / n_i = 0$ for $n_1 = 10$, $n_2 = 20$, and $n_3 = 5$:

$$\mathbf{a}' = (25\ -20\ -5) \quad \text{and} \quad \mathbf{b}' = (0\ 1\ -1). \tag{15.15}$$

However, $\mathbf{a}'$ leads to the comparison

$$H_{03} : 25\mu_1 = 20\mu_2 + 5\mu_3 \quad \text{or} \quad H_{03} : \mu_1 = \frac{4}{5}\mu_2 + \frac{1}{5}\mu_3,$$

which is not the same as the hypothesis $H_{01} : \mu_1 = \frac{1}{2}(\mu_2 + \mu_3)$ that we were initially interested in. □

Example 15.2.2b. We illustrate both weighted and unweighted contrasts for the fill data in Table 15.2. Suppose that we wish to make the following comparisons of the five machines:

A, D	versus	B, C, E
B, E	versus	D
A	versus	D
B	versus	E

Orthogonal (unweighted) contrast coefficients that provide these comparisons are given as rows of the following matrix:

$$\begin{pmatrix} 3 & -2 & -2 & 3 & -2 \\ 0 & 1 & -2 & 0 & 1 \\ 1 & 0 & 0 & -1 & 0 \\ 0 & 1 & 0 & 0 & -1 \end{pmatrix}.$$

We give the sums of squares for these four contrasts and the F values [see (15.13)] in Table 15.4.

Since these are unweighted contrasts, the contrast sums of squares do not add up to the between sum of squares in Table 15.3. None of the p values is less than .05, so we do not reject $H_0: \sum_i c_i\mu_i = 0$ for any of the four contrasts. In fact, the p values should be less than .05/4 for familywise significance (see the Bonferroni approach in Section 8.5.2), since the overall test in Table 15.3 did not reject $H_0 : \mu_1 = \mu_2 \cdots = \mu_5$.

TABLE 15.4 Sums of Squares and F Values for Contrasts for the Fill Data in Table 15.2

Contrast	df	Contrast SS	F	p Value
A, D versus B, C, E	1	.005763	0.75	.406
B, E versus C	1	.002352	0.31	.592
A versus D	1	.034405	4.47	.0582
B versus E	1	.013333	1.73	.215

As an example of two weighted orthogonal contrasts that satisfy $\sum_i a_i b_i / n_i$, we keep the first contrast above and replace the second contrast with (0 2 −6 0 4). Then, for these two contrasts, we have

$$\sum_i \frac{a_i b_i}{n_i} = \frac{3(0)}{4} - \frac{2(2)}{2} - \frac{2(-6)}{3} + \frac{3(0)}{3} - \frac{2(4)}{4} = 0.$$

The sums of squares and F values [using (15.13)] for the two contrasts are as follows:

Contrast	df	Contrast SS	F	p Value
A, D versus B, C, E	1	.005763	.75	.406
B, E versus C	1	.005339	.69	.423

□

15.3 TWO-WAY MODEL

The unbalanced two-way model can be expressed as

$$y_{ijk} = \mu + \alpha_i + \beta_j + \gamma_{ij} + \varepsilon_{ijk} \tag{15.16}$$

$$= \mu_{ij} + \varepsilon_{ijk}, \tag{15.17}$$

$$i = 1, 2, \ldots, a, \quad j = 1, 2, \ldots, b, \quad k = 1, 2, \ldots, n_{ij}.$$

The ε_{ijk}'s are assumed to be independently distributed as $N(0, \sigma^2)$. In this section we consider the case in which all $n_{ij} > 0$.

The cell means model for analyzing unbalanced two-way data was first proposed by Yates (1934). The cell means model has been advocated by Speed (1969), Urquhart et al. (1973), Nelder (1974), Hocking and Speed (1975), Bryce (1975), Bryce et al. (1976, 1980b), Searle (1977), Speed et al. (1978), Searle et al. (1981), Milliken and Johnson (1984, Chapter 11), and Hocking (1985, 1996). Turner (1990) discusses the relationship between (15.16) and (15.17). In our development we follow Bryce et al. (1980b) and Hocking (1985, 1996).

15.3.1 Unconstrained Model

We first consider the *unconstrained model* in which the μ_{ij}'s are unrestricted. To accommodate a no-interaction model, for example, we must place constraints on the μ_{ij}'s. The constrained model is discussed in Section.

To illustrate the cell means model (15.17), we use $a = 2$ and $b = 3$ with the cell counts n_{ij} given in Figure 15.2. This example with $N = \sum_{ij} n_{ij} = 11$ will be referred to throughout the present section and Section 15.3.2.

		B		
		1	2	3
A	1	$n_{11}=2$	$n_{12}=1$	$n_{13}=2$
	2	$n_{21}=1$	$n_{22}=3$	$n_{23}=2$

Figure 15.2 Cell counts for unbalanced data illustration.

For each of the 11 observations in Figure 15.2, the model $y_{ijk} = \mu_{ij} + \varepsilon_{ijk}$ is

$$\begin{aligned}
y_{111} &= \mu_{11} + \varepsilon_{111} \\
y_{112} &= \mu_{11} + \varepsilon_{112} \\
y_{121} &= \mu_{12} + \varepsilon_{121} \\
&\vdots \\
y_{231} &= \mu_{23} + \varepsilon_{231} \\
y_{232} &= \mu_{23} + \varepsilon_{232},
\end{aligned}$$

or in matrix form

$$\mathbf{y} = \mathbf{W}\boldsymbol{\mu} + \boldsymbol{\varepsilon}, \tag{15.18}$$

where

$$\mathbf{y} = \begin{pmatrix} y_{111} \\ y_{112} \\ \vdots \\ y_{232} \end{pmatrix}, \quad \mathbf{W} = \begin{pmatrix} 1 & 0 & 0 & 0 & 0 & 0 \\ 1 & 0 & 0 & 0 & 0 & 0 \\ 0 & 1 & 0 & 0 & 0 & 0 \\ 0 & 0 & 1 & 0 & 0 & 0 \\ 0 & 0 & 1 & 0 & 0 & 0 \\ \vdots & \vdots & \vdots & \vdots & \vdots & \vdots \\ 0 & 0 & 0 & 0 & 0 & 1 \\ 0 & 0 & 0 & 0 & 0 & 1 \end{pmatrix},$$

$$\boldsymbol{\mu} = \begin{pmatrix} \mu_{11} \\ \mu_{12} \\ \mu_{13} \\ \mu_{21} \\ \mu_{22} \\ \mu_{23} \end{pmatrix}, \quad \boldsymbol{\varepsilon} = \begin{pmatrix} \varepsilon_{111} \\ \varepsilon_{112} \\ \vdots \\ \varepsilon_{232} \end{pmatrix}.$$

Each row of $\mathbf{W}$ contains a single 1 that corresponds to the appropriate μ_{ij} in $\boldsymbol{\mu}$. For example, the fourth row gives $y_{131} = (001000)\boldsymbol{\mu} + \varepsilon_{131} = \mu_{13} + \varepsilon_{131}$. In this illustration, $\mathbf{y}$ and $\boldsymbol{\varepsilon}$ are 11×1, and $\mathbf{W}$ is 11×6. In general, $\mathbf{y}$ and $\boldsymbol{\varepsilon}$ are $N \times 1$, and $\mathbf{W}$ is $N \times ab$, where $N = \sum_{ij} n_{ij}$.

Since $\mathbf{W}$ is full-rank, we can use the results in Chapters 7 and 8. The analysis is further simplified because $\mathbf{W}'\mathbf{W} = \text{diag}(n_{11}, n_{12}, n_{13}, n_{21}, n_{22}, n_{23})$. By (7.6), the least-squares estimator of $\boldsymbol{\mu}$ is given by

$$\hat{\boldsymbol{\mu}} = (\mathbf{W}'\mathbf{W})^{-1}\mathbf{W}'\mathbf{y} = \bar{\mathbf{y}}, \tag{15.19}$$

where $\bar{\mathbf{y}} = (\bar{y}_{12.}, \bar{y}_{13.}, \bar{y}_{14.}, \bar{y}_{21.}, \bar{y}_{22.}, \bar{y}_{23.})'$ contains the sample means of the cells, $\bar{y}_{ij.} = \sum_k y_{ijk}/n_{ij}$. By (7.14), the covariance matrix for $\hat{\boldsymbol{\mu}}$ is

$$\begin{aligned} \text{cov}(\hat{\boldsymbol{\mu}}) &= \sigma^2(\mathbf{W}'\mathbf{W})^{-1} = \sigma^2\text{diag}\left(\frac{1}{n_{11}}, \frac{1}{n_{12}}, \dots, \frac{1}{n_{23}}\right) \qquad (15.20) \\ &= \text{diag}\left(\frac{\sigma^2}{n_{11}}, \frac{\sigma^2}{n_{12}}, \dots, \frac{\sigma^2}{n_{23}}\right). \end{aligned}$$

For general a, b, and N, an unbiased estimator of σ^2 [see (7.23)] is given by

$$s^2 = \frac{\text{SSE}}{\nu_E} = \frac{(\mathbf{y} - \mathbf{W}\hat{\boldsymbol{\mu}})'(\mathbf{y} - \mathbf{W}\hat{\boldsymbol{\mu}})}{N - ab}, \tag{15.21}$$

where $\nu_E = \sum_{i=1}^{a}\sum_{j=1}^{b}(n_{ij} - 1) = N - ab$, with $N = \sum_{ij} n_{ij}$. In our illustration with $a = 2$ and $b = 3$, we have $N - ab = 11 - 6 = 5$. Two alternative forms of SSE are

$$\text{SSE} = \mathbf{y}'[\mathbf{I} - \mathbf{W}(\mathbf{W}'\mathbf{W})^{-1}\mathbf{W}']\mathbf{y} \quad \text{[see (7.26)]}, \tag{15.22}$$

$$\text{SSE} = \sum_{i=1}^{a}\sum_{j=1}^{b}\sum_{k=1}^{n_{ij}}(y_{ijk} - \bar{y}_{ij.})^2 \quad \text{[see (14.48)]}. \tag{15.23}$$

Using (15.23), we can express s^2 as the pooled estimator

$$s^2 = \frac{\sum_{i=1}^{a}\sum_{j=1}^{b}(n_{ij} - 1)s_{ij}^2}{N - ab}, \tag{15.24}$$

where s_{ij}^2 is the variance estimator in the (ij)th cell, $s_{ij}^2 = \sum_{k=1}^{n_{ij}}(y_{ijk} - \bar{y}_{ij.})^2/(n_{ij} - 1)$.

The overparameterized model (15.16) includes parameters representing main effects and interactions, but the cell means model (15.17) does not have such parameters. To carry out tests in the cell means model, we use contrasts to express the main effects and the interaction as functions of the μ_{ij}'s in $\boldsymbol{\mu}$. We begin with the main effect of A.

In the vector $\boldsymbol{\mu} = (\mu_{11}, \mu_{12}, \mu_{13}, \mu_{21}, \mu_{22}, \mu_{23})'$, the first three elements correspond to the first level of A and the last three to the second level, as seen in Figure 15.3. Thus, for the main effect of A, we could compare the average of μ_{11},

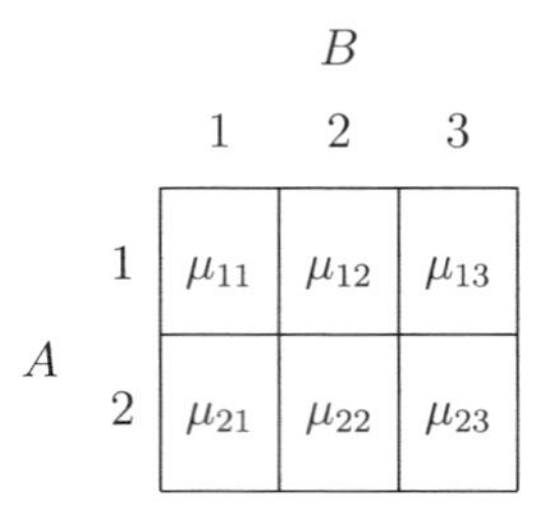

Figure 15.3 Cell means corresponding to Figure 15.1.

μ_{12}, and μ_{13} with the average of μ_{21}, μ_{22}, and μ_{23}. The difference between these averages (sums) can be conveniently expressed by the contrast

$$\begin{aligned}\mathbf{a}'\boldsymbol{\mu} &= \mu_{11} + \mu_{12} + \mu_{13} - \mu_{21} - \mu_{22} - \mu_{23}, \\ &= (1,\ 1,\ 1,\ -1,\ -1,\ -1)\boldsymbol{\mu}.\end{aligned}$$

To compare the two levels of A, we can test the hypothesis H_0: $\mathbf{a}'\boldsymbol{\mu} = 0$, which can be written as $H_0 : (\mu_{11} - \mu_{21}) + (\mu_{12} - \mu_{22}) + (\mu_{13} - \mu_{23}) = 0$. In this form, H_0 states that the effect of A averaged (summed) over the levels of B is 0. This corresponds to a common main effect definition in the presence of interaction; see comments following (14.62). Note that this test is not useful in model selection. It simply tests whether the interaction is "symmetric" such that the effect of A, averaged over the levels of B, is zero.

Factor B has three levels corresponding to the three columns of Figure 15.3. In a comparison of three levels, there are 2 degrees of freedom, which will require two contrasts. Suppose that we wish to compare the first level of B with the other two levels and then compare the second level of B with the third. To do this, we compare the average of the means in the first column of Figure 15.3 with the average in the second and third columns and similarly compare the second and third columns. We can make these comparisons using $H_0 : \mathbf{b}_1'\boldsymbol{\mu} = 0$ and $\mathbf{b}_2'\boldsymbol{\mu} = 0$, where $\mathbf{b}_1'\boldsymbol{\mu}$ and $\mathbf{b}_2'\boldsymbol{\mu}$ are the following two orthogonal contrasts:

$$\begin{aligned}\mathbf{b}_1'\boldsymbol{\mu} &= 2(\mu_{11} + \mu_{21}) - (\mu_{12} + \mu_{22}) - (\mu_{13} + \mu_{23}) \qquad (15.25) \\ &= 2\mu_{11} - \mu_{12} - \mu_{13} + 2\mu_{21} - \mu_{22} - \mu_{23} \\ &= (2, -1, -1, 2, -1, -1)\boldsymbol{\mu}, \\ \mathbf{b}_2'\boldsymbol{\mu} &= (\mu_{12} + \mu_{22}) - (\mu_{13} + \mu_{23}) \qquad (15.26) \\ &= \mu_{12} - \mu_{13} + \mu_{22} - \mu_{23} \\ &= (0, 1, -1, 0, 1, -1)\boldsymbol{\mu}.\end{aligned}$$

We can combine $\mathbf{b}_1'$ and $\mathbf{b}_2'$ into the matrix

$$\mathbf{B} = \begin{pmatrix} \mathbf{b}_1' \\ \mathbf{b}_2' \end{pmatrix} = \begin{pmatrix} 2 & -1 & -1 & 2 & -1 & -1 \\ 0 & 1 & -1 & 0 & 1 & -1 \end{pmatrix}, \qquad (15.27)$$

and the hypothesis becomes H_0: $\mathbf{B}\boldsymbol{\mu} = \mathbf{0}$, which, by (15.25) and (15.26), is equivalent to

$$H_0 : \mu_{11} + \mu_{21} = \mu_{12} + \mu_{22} = \mu_{13} + \mu_{23} \tag{15.28}$$

(see Problem15.9). In this form, H_0 states that the interaction is symmetric such that the three levels of B do not differ when averaged over the two levels of A. Note that other orthogonal or linearly independent contrasts besides those in $\mathbf{b}_1'$ and $\mathbf{b}_2'$ would lead to (15.28) and to the same F statistic in (15.33) below.

By analogy to (14.30), the interaction hypothesis can be written as

$$H_0\colon \mu_{11} - \mu_{21} = \mu_{12} - \mu_{22} = \mu_{13} - \mu_{23},$$

which is a comparison of the "A effects" across the levels of B. If these A effects differ, we have an interaction. We can express the two equalities in H_0 in terms of orthogonal contrasts similar to those in (15.25) and (15.26):

$$\begin{aligned}\mathbf{c}_1'\boldsymbol{\mu} &= 2(\mu_{11} - \mu_{21}) - (\mu_{12} - \mu_{22}) - (\mu_{13} - \mu_{23}) = 0,\\ \mathbf{c}_2'\boldsymbol{\mu} &= (\mu_{12} - \mu_{22}) - (\mu_{13} - \mu_{23}) = 0.\end{aligned}$$

Thus H_0 can be written as $H_0 : \mathbf{C}\boldsymbol{\mu} = \mathbf{0}$, where

$$\mathbf{C} = \begin{pmatrix} \mathbf{c}_1' \\ \mathbf{c}_2' \end{pmatrix} = \begin{pmatrix} 2 & -1 & -1 & -2 & 1 & 1 \\ 0 & 1 & -1 & 0 & -1 & 1 \end{pmatrix}.$$

Note that $\mathbf{c}_1$ can be found by taking products of corresponding elements of $\mathbf{a}$ and $\mathbf{b}_1$ and $\mathbf{c}_2$ can be obtained similarly from $\mathbf{a}$ and $\mathbf{b}_2$, where $\mathbf{a}$, $\mathbf{b}_1$, and $\mathbf{b}_2$ are the coefficient vectors in $\mathbf{a}'\boldsymbol{\mu}$, $\mathbf{b}_1'\boldsymbol{\mu}$ and $\mathbf{b}_2'\boldsymbol{\mu}$. Thus

$$\begin{aligned}\mathbf{c}_1' &= [(1)(2), (1)(-1), (1)(-1), (-1)(2), (-1)(-1), (-1)(-1)]\\ &= (2, -1, -1, -2, 1, 1),\\ \mathbf{c}_2' &= [(1)(0), (1)(1), (1)(-1), (-1)(0), (-1)(1), (-1)(-1)]\\ &= (0, 1, -1, 0, -1, 1).\end{aligned}$$

The elementwise multiplication of these two vectors (the *Hadamard product* — see Section 2.2.4) produces interaction contrasts that are orthogonal to each other and to the main effect contrasts.

We now construct tests for the general linear hypotheses H_0: $\mathbf{a}'\boldsymbol{\mu} = 0$, H_0: $\mathbf{B}\boldsymbol{\mu} = \mathbf{0}$, and H_0: $\mathbf{C}\boldsymbol{\mu} = \mathbf{0}$ for the main effects and interaction. The hypothesis H_0: $\mathbf{a}'\boldsymbol{\mu} = 0$ for the main effect of A, is easily tested using an F statistic similar to (8.38) or (15.12):

$$F = \frac{(\mathbf{a}'\hat{\boldsymbol{\mu}})'[\mathbf{a}'(\mathbf{W}'\mathbf{W})^{-1}\mathbf{a}]^{-1}(\mathbf{a}'\hat{\boldsymbol{\mu}})}{s^2} = \frac{\text{SSA}}{\text{SSE}/\nu_E}, \tag{15.29}$$

where s^2 is given by (15.21) and $\nu_E = N - ab$. [For our illustration, $N - ab = 11 - (2)(3) = 5$.]
If H_0 is true, F in (15.29) is distributed as $F(1, N - a_b)$.

The F statistic in (15.29) can be written as

$$F = \frac{(\mathbf{a}'\hat{\boldsymbol{\mu}})^2}{s^2\mathbf{a}'(\mathbf{W}'\mathbf{W})^{-1}\mathbf{a}} \tag{15.30}$$

$$= \frac{\left(\sum_{ij} a_{ij}\bar{y}_{ij.}\right)^2}{s^2 \sum_{ij} a_{ij}^2/n_{ij}}, \tag{15.31}$$

which is analogous to (15.13). Since $t^2(\nu_E) = F(1, \nu_E)$ (see Problem 5.16), a t statistic for testing H_0: $\mathbf{a}'\boldsymbol{\mu} = 0$ is given by the square root of (15.30)

$$t = \frac{\mathbf{a}'\hat{\boldsymbol{\mu}}}{s\sqrt{\mathbf{a}'(\mathbf{W}'\mathbf{W})^{-1}\mathbf{a}}} = \frac{\mathbf{a}'\hat{\boldsymbol{\mu}} - 0}{\sqrt{\widehat{\text{var}}(\mathbf{a}'\hat{\boldsymbol{\mu}})}}, \tag{15.32}$$

which is distributed as $t(N - ab)$ when H_0 is true. Note that the test based on either of (15.29) or (15.32) is a full–reduced-model test (see Theorem 8.4d) and therefore tests for factor A "above and beyond" (adjusted for) factor B and the interaction.

By Theorem 8.4b, a test statistic for the factor B main effect hypothesis H_0: $\mathbf{B}\boldsymbol{\mu}=\mathbf{0}$ is given by

$$F = \frac{(\mathbf{B}\hat{\boldsymbol{\mu}})'[\mathbf{B}(\mathbf{W}'\mathbf{W})^{-1}\mathbf{B}']^{-1}\mathbf{B}\hat{\boldsymbol{\mu}}/\nu_B}{\text{SSE}/\nu_E} = \frac{\text{SSB}/\nu_B}{\text{SSE}/\nu_E}, \tag{15.33}$$

where $\nu_E = N - ab$ and ν_B is the number of rows of $\mathbf{B}$. (For our illustration, $\nu_E = 5$ and $\nu_B = 2$.) When H_0 is true, F in (15.33) is distributed as $F(\nu_B, \nu_E)$.

A test statistic for the interaction hypothesis H_0: $\mathbf{C}\boldsymbol{\mu} = \mathbf{0}$ is obtained similarly:

$$F = \frac{(\mathbf{C}\hat{\boldsymbol{\mu}})'[\mathbf{C}(\mathbf{W}'\mathbf{W})^{-1}\mathbf{C}']^{-1}\mathbf{C}\hat{\boldsymbol{\mu}}/\nu_{AB}}{\text{SSE}/\nu_E} = \frac{\text{SSAB}/\nu_{AB}}{\text{SSE}/\nu_E}, \tag{15.34}$$

which is distributed as $F(\nu_{AB}, \nu_E)$, where ν_{AB}, the degrees of freedom for interaction, is the number of rows of $\mathbf{C}$. (In our illustration, $\nu_{AB} = 2$.)

Because of the unequal n_{ij}'S, the three sums of squares SSA, SSB, and SSAB do not add to the overall sum of squares for treatments and are not statistically independent, as in the balanced case [see (14.40) and Theorem 14.4b]. Each of SSA, SSB, and SSAB is adjusted for the other effects; that is, the given effect is tested "above and beyond" the others (see Theorem 8.4d).

Example 15.3a. Table 15.5 contains dressing percentages of pigs in a two-way classification (Snedecor and Cochran 1967, p. 480). Let factor A be gender and factor B be breed.

TABLE 15.5 Dressing Percentages (Less 70%) of 75 Swine Classified by Breed and Gender

Breed									
1		2		3		4		5	
Male	Female	Male	Female	Male	Female	Male	Female	Male	Female
13.3	18.2	10.9	14.3	13.6	12.9	11.6	13.8	10.3	12.8
12.6	11.3	3.3	15.3	13.1	14.4	13.2	14.4	10.3	8.4
11.5	14.2	10.5	11.8	4.1		12.6	4.9	10.1	10.6
15.4	15.9	11.6	11.0	10.8		15.2		6.9	13.9
12.7	12.9	15.4	10.9			14.7		13.2	10.0
15.7	15.1	14.4	10.5			12.4		11.0	
13.2		11.6	12.9					12.2	
15.0		14.4	12.5					13.3	
14.3		7.5	13.0					12.9	
16.5		10.8	7.6					9.9	
15.0		10.5	12.9						
13.7		14.5							
		10.9							
		13.0							
		15.9							
		12.8							

We arrange the elements of the vector $\boldsymbol{\mu}$ to correspond to a row of Table 15.5, that is

$$\boldsymbol{\mu} = (\mu_{11}, \mu_{12}, \mu_{21}, \mu_{22}, \ldots, \mu_{52})',$$

where the first subscript represents breed and the second subscript is associated with gender.

The vector $\boldsymbol{\mu}$ is 10×1, the matrix $\mathbf{W}$ is 75×10, the vector $\mathbf{a}$ is 10×1, and the matrices $\mathbf{B}$ and $\mathbf{C}$ are each 4×10. We show $\mathbf{a}$, $\mathbf{B}$, and $\mathbf{C}$:

$$\mathbf{a}' = (1, -1, 1, -1, 1, -1, 1, -1, 1, -1),$$

$$\mathbf{B} = \begin{pmatrix} 3 & 3 & 3 & 3 & -2 & -2 & -2 & -2 & -2 & -2 \\ 1 & 1 & -1 & -1 & 0 & 0 & 0 & 0 & 0 & 0 \\ 0 & 0 & 0 & 0 & 1 & 1 & -2 & -2 & 1 & 1 \\ 0 & 0 & 0 & 0 & 1 & 1 & 0 & 0 & -1 & -1 \end{pmatrix},$$

$$\mathbf{C} = \begin{pmatrix} 3 & -3 & 3 & -3 & -2 & 2 & -2 & 2 & -2 & 2 \\ 1 & -1 & -1 & 1 & 0 & 0 & 0 & 0 & 0 & 0 \\ 0 & 0 & 0 & 0 & 1 & -1 & -2 & 2 & 1 & -1 \\ 0 & 0 & 0 & 0 & 1 & -1 & 0 & 0 & -1 & 1 \end{pmatrix}.$$

TABLE 15.6 ANOVA for Unconstrained Model

Source	df	Sum of Squares	Mean Square	F	p Value
A (gender)	1	1.984	1.984	0.303	.584
B (breed)	4	90.856	22.714	3.473	.0124
AB	4	24.876	6.219	0.951	.440
Error	65	425.089	6.540		
Total	74	552.095			

(Note that other sets of othogonal contrasts could be used in **B**, and the value of F_B below would be the same.) By (15.19), we obtain

$$\hat{\boldsymbol{\mu}} = \bar{\mathbf{y}} = (14.08, 14.60, 11.75, 12.06, 10.40, 13.65, 13.28, 11.03, 11.01, 11.14)'.$$

By (15.22) or (15.23) we obtain $\mathbf{SSE} = 425.08895$, with $\nu_E = 65$. Using (15.29), (15.33), and (15.34), we obtain

$$F_A = .30337, \quad F_B = 3.47318, \quad F_C = .95095.$$

The sums of squares leading to these Fs are given in Table 15.6. Note that the sums of squares for A, B, AB, and error do not add up to the total sum of squares because the data are unbalanced. (These are the type III sums of squares referred to in Section 15.1.) □

15.3.2 Constrained Model

To allow for additivity or other restrictions, constraints on the μ_{ij}'s must be added to the cell means model (15.17) or (15.18). For example, the model

$$y_{ijk} = \mu_{ij} + \varepsilon_{ijk}$$

cannot represent the no-interaction model

$$y_{ijk} = \mu + \alpha_i + \beta_j + \varepsilon_{ijk} \tag{15.35}$$

unless we specify some relationships among the μ_{ij}'s.

In our 2×3 illustration in Section 15.3.1, the two interaction contrasts are expressible as

$$\mathbf{C}\boldsymbol{\mu} = \begin{pmatrix} 2 & -1 & -1 & -2 & 1 & 1 \\ 0 & 1 & -1 & 0 & -1 & 1 \end{pmatrix} \boldsymbol{\mu}.$$

If we wish to use a model without interaction, then $\mathbf{C}\boldsymbol{\mu} = \mathbf{0}$ is not a hypothesis to be tested but an assumption to be included in the statement of the model.

In general, for constraints $\mathbf{G}\boldsymbol{\mu} = \mathbf{0}$, the model can be expressed as

$$\mathbf{y} = \mathbf{W}\boldsymbol{\mu} + \boldsymbol{\varepsilon} \text{ subject to } \mathbf{G}\boldsymbol{\mu} = \mathbf{0}. \tag{15.36}$$

We now consider estimation and testing in this constrained model. [For the case $\mathbf{G}\boldsymbol{\mu} = \mathbf{h}$, where $\mathbf{h} \neq \mathbf{0}$, see Bryce et al. (1980b).]

To incorporate the constraints $\mathbf{G}\boldsymbol{\mu} = \mathbf{0}$ into $\mathbf{y} = \mathbf{W}\boldsymbol{\mu} + \varepsilon$, we can use the Lagrange multiplier method (Section 2.14.3). Alternatively, we can reparameterize the model using the matrix

$$\mathbf{A} = \begin{pmatrix} \mathbf{K} \\ \mathbf{G} \end{pmatrix}, \tag{15.37}$$

where $\mathbf{K}$ specifies parameters of interest in the constrained model. For the no-interaction model (15.35), for example, $\mathbf{G}$ would equal $\mathbf{C}$, the first row of $\mathbf{K}$ could correspond to a multiple of the overall mean, and the remaining rows of $\mathbf{K}$ could include the contrasts for the A and B main effects. Thus, we would have

$$\mathbf{K} = \begin{pmatrix} 1 & 1 & 1 & 1 & 1 & 1 \\ 1 & 1 & 1 & -1 & -1 & -1 \\ 2 & -1 & -1 & 2 & -1 & -1 \\ 0 & 1 & -1 & 0 & 1 & -1 \end{pmatrix},$$

$$\mathbf{G} = \mathbf{C} = \begin{pmatrix} 2 & -1 & -1 & -2 & 1 & 1 \\ 0 & 1 & -1 & 0 & -1 & 1 \end{pmatrix}.$$

The second row of $\mathbf{K}$ is $\mathbf{a}'$ and corresponds to the average effect of A. The third and fourth rows are from $\mathbf{B}$ and represent the average B effect.

If the rows of $\mathbf{G}$ are linearly independent of the rows of $\mathbf{K}$, then the matrix $\mathbf{A}$ in (15.37) is of full rank and has an inverse. This holds true in our example, in which we have $\mathbf{G} = \mathbf{C}$. In our example, in fact, the rows of $\mathbf{G}$ are orthogonal to the rows of $\mathbf{K}$. We can therefore insert $\mathbf{A}^{-1}\mathbf{A} = \mathbf{I}$ into (15.36) to obtain the reparameterized model

$$\begin{aligned} \mathbf{y} &= \mathbf{W}\mathbf{A}^{-1}\mathbf{A}\boldsymbol{\mu} + \boldsymbol{\varepsilon} \quad \text{subject to} \quad \mathbf{G}\boldsymbol{\mu} = \mathbf{0} \\ &= \mathbf{Z}\boldsymbol{\delta} + \boldsymbol{\varepsilon} \quad \text{subject to} \quad \mathbf{G}\boldsymbol{\mu} = \mathbf{0}, \end{aligned} \tag{15.38}$$

where $\mathbf{Z} = \mathbf{W}\,\mathbf{A}^{-1}$ and $\boldsymbol{\delta} = \mathbf{A}\boldsymbol{\mu}$.

In the balanced two-way model, we obtained a no-interaction model by simply inserting $\gamma_{ij}^* = 0$ into $y_{ijk} = \mu + \alpha_i^* + \beta_j^* + \gamma_{ij}^* + \varepsilon_{ij}$ [(see 14.37) and (14.38)]. To analogously incorporate the constraint $\mathbf{G}\boldsymbol{\mu} = \mathbf{0}$ directly into the model in the unbalanced case, we partition $\boldsymbol{\delta}$ into

$$\boldsymbol{\delta} = \mathbf{A}\boldsymbol{\mu} = \begin{pmatrix} \mathbf{K} \\ \mathbf{G} \end{pmatrix}\boldsymbol{\mu} = \begin{pmatrix} \mathbf{K}\boldsymbol{\mu} \\ \mathbf{G}\boldsymbol{\mu} \end{pmatrix} = \begin{pmatrix} \boldsymbol{\delta}_1 \\ \boldsymbol{\delta}_2 \end{pmatrix}.$$

With a corresponding partitioning on the columns of $\mathbf{Z}$, the model can be written as

$$\begin{aligned} \mathbf{y} &= \mathbf{Z}\boldsymbol{\delta} + \boldsymbol{\varepsilon} = (\mathbf{Z}_1, \mathbf{Z}_2)\begin{pmatrix} \boldsymbol{\delta}_1 \\ \boldsymbol{\delta}_2 \end{pmatrix} + \boldsymbol{\varepsilon} \\ &= \mathbf{Z}_1\boldsymbol{\delta}_1 + \mathbf{Z}_2\boldsymbol{\delta}_2 + \boldsymbol{\varepsilon} \quad \text{subject to} \quad \mathbf{G}\boldsymbol{\mu} = \mathbf{0}. \end{aligned} \tag{15.39}$$

Since $\delta_2 = \mathbf{G}\boldsymbol{\mu}$, the constraint $\mathbf{G}\boldsymbol{\mu} = \mathbf{0}$ gives $\boldsymbol{\delta}_2 = \mathbf{0}$ and the constrained model in (15.39) simplifies to

$$\mathbf{y} = \mathbf{Z}_1\boldsymbol{\delta}_1 + \boldsymbol{\varepsilon}. \tag{15.40}$$

An estimator of $\boldsymbol{\delta}_1$ [see (7.6)] is given by

$$\hat{\boldsymbol{\delta}}_1 = (\mathbf{Z}_1'\mathbf{Z}_1)^{-1}\mathbf{Z}_1'\mathbf{y}.$$

To obtain an expression for $\boldsymbol{\mu}$ subject to the constraints, we multiply

$$\mathbf{A}\boldsymbol{\mu} = \begin{pmatrix} \boldsymbol{\delta}_1 \\ \boldsymbol{\delta}_2 \end{pmatrix} = \begin{pmatrix} \boldsymbol{\delta}_1 \\ \mathbf{0} \end{pmatrix}$$

by

$$\mathbf{A}^{-1} = (\mathbf{K}^*, \mathbf{G}^*).$$

If the rows of $\mathbf{G}$ are orthogonal to the rows of $\mathbf{K}$, then

$$(\mathbf{K}^*, \mathbf{G}^*) = [\mathbf{K}'(\mathbf{K}\mathbf{K}')^{-1}, \mathbf{G}'(\mathbf{G}\mathbf{G}')^{-1}] \tag{15.41}$$

(see Problem15.13). If the rows of $\mathbf{G}$ are linearly independent of (but not necessarily orthogonal to) the rows of $\mathbf{K}$, we obtain

$$\mathbf{K}^* = \mathbf{H}_G\mathbf{K}'(\mathbf{K}\mathbf{H}_G\mathbf{K}')^{-1}, \tag{15.42}$$

where

$$\mathbf{H}_G = \mathbf{I} - \mathbf{G}'(\mathbf{G}\mathbf{G}')^{-1}\mathbf{G},$$

and $\mathbf{G}^*$ is similarly defined (see Problem15.14). In any case, we denote the product of $\mathbf{K}^*$ and $\boldsymbol{\delta}_1$ by $\boldsymbol{\mu}_c$:

$$\boldsymbol{\mu}_c = \mathbf{K}^*\boldsymbol{\delta}_1.$$

We estimate $\boldsymbol{\mu}_c$ by

$$\hat{\boldsymbol{\mu}}_c = \mathbf{K}^*\hat{\boldsymbol{\delta}}_1 = \mathbf{K}^*(\mathbf{Z}_1'\mathbf{Z}_1)^{-1}\mathbf{Z}_1'\mathbf{y}, \tag{15.43}$$

which has covariance matrix

$$\text{cov}(\hat{\boldsymbol{\mu}}_c) = \sigma^2\mathbf{K}^*(\mathbf{Z}_1'\mathbf{Z}_1)^{-1}\mathbf{K}^{*\prime}. \tag{15.44}$$

To test for factor B in the constrained model, the hypothesis is H_0: $\mathbf{B}\boldsymbol{\mu}_c = \mathbf{0}$. The covariance matrix of $\mathbf{B}\hat{\boldsymbol{\mu}}_c$ is obtained from (3.44) and (15.44) as

$$\text{cov}(\mathbf{B}\hat{\boldsymbol{\mu}}_c) = \sigma^2\mathbf{B}\mathbf{K}^*(\mathbf{Z}_1'\mathbf{Z}_1)^{-1}\mathbf{K}^{*\prime}\mathbf{B}'.$$

Then, by Theorem 8.4b, the test statistic for H_0: $\mathbf{B}\boldsymbol{\mu}_c = \mathbf{0}$ in the constrained model becomes

$$F = \frac{(\mathbf{B}\hat{\boldsymbol{\mu}}_c)'[\mathbf{B}\mathbf{K}^*(\mathbf{Z}_1'\mathbf{Z}_1)^{-1}\mathbf{K}^{*\prime}\mathbf{B}']^{-1}\mathbf{B}\hat{\boldsymbol{\mu}}_c/\nu_B}{\text{SSE}_c/\nu_{E_c}}, \tag{15.45}$$

where SSE_c (subject to $\mathbf{G}\boldsymbol{\mu} = \mathbf{0}$) is obtained using $\hat{\boldsymbol{\mu}}_c$ [from (15.43)] in (15.21). (In our example, where $\mathbf{G} = \mathbf{C}$ for interaction, SSE_c effectively pools SSE and SSAB from the unconstrained model.) The degrees of freedom ν_{E_c} is obtained as $\nu_{E_c} = \nu_E + \text{rank}(\mathbf{G})$, where $\nu_E = N - ab$ is for the unconstrained model, as defined following (15.21). [In our example, $\text{rank}(\mathbf{G}) = 2$ since there are 2 degrees of freedom for SSAB.] We reject H_0: $\mathbf{B}\boldsymbol{\mu}_c = \mathbf{0}$ if $F \geq F_{\alpha,\nu_B,\nu_{E_c}}$, where F_α is the upper α percentage point of the central F distribution.

For H_0: $\mathbf{a}'\boldsymbol{\mu}_c = 0$, the F statistic becomes

$$F = \frac{(\mathbf{a}'\hat{\boldsymbol{\mu}}_c)'[\mathbf{a}'\mathbf{K}^*(\mathbf{Z}_1'\mathbf{Z}_1)^{-1}\mathbf{K}^{*\prime}\mathbf{a}]^{-1}(\mathbf{a}'\hat{\boldsymbol{\mu}}_c)}{\text{SSE}_c/\nu_{E_c}}, \tag{15.46}$$

which is distributed as $F(1, \nu_{E_c})$ if H_0 is true.

Example 15.3b. For the pigs data in Table 15.5, we test for factors A and B in a no-interaction model, where factor A is gender and factor B is breed. The matrix $\mathbf{G}$ is the same as $\mathbf{C}$ in Example 15.3a. For $\mathbf{K}$ we have

$$\mathbf{K} = \begin{pmatrix} \mathbf{j}' \\ \mathbf{a}' \\ \mathbf{B} \end{pmatrix} = \begin{pmatrix} 1 & 1 & 1 & 1 & 1 & 1 & 1 & 1 & 1 & 1 \\ 1 & -1 & 1 & -1 & 1 & -1 & 1 & -1 & 1 & -1 \\ 3 & 3 & 3 & 3 & -2 & -2 & -2 & -2 & -2 & -2 \\ 1 & 1 & -1 & -1 & 0 & 0 & 0 & 0 & 0 & 0 \\ 0 & 0 & 0 & 0 & 1 & 1 & -2 & -2 & 1 & 1 \\ 0 & 0 & 0 & 0 & 1 & 1 & 0 & 0 & -1 & -1 \end{pmatrix}$$

By (15.43), we obtain

$$\hat{\boldsymbol{\mu}}_c = (14.16, 14.42, 11.77, 12.03, 11.40, 11.65, 12.45, 12.70, 10.97, 11.22).'$$

TABLE 15.7 ANOVA for Constrained Model

Source	df	Sum of Squares	Mean Square	F	p Value
A (gender)	1	1.132	1.132	0.17	.678
B (breed)	4	101.418	25.355	3.89	.00660
Error	69	449.965	6.521		
Total	74	552.0955			

For SSE_c, we use $\hat{\boldsymbol{\mu}}_c$ in place of $\hat{\boldsymbol{\mu}}$ in (15.21) to obtain $\text{SSE}_c = 449.96508$. For ν_{E_c}, we have

$$\nu_{E_c} = \nu_E + \text{rank}(\mathbf{G}) = 65 + 4 = 69.$$

Then by (15.45), we obtain $F_{B_c} = 3.8880003$. The sums of squares leading to F_{B_c} and F_{A_c} are given in Table 15.7. □

15.4 TWO-WAY MODEL WITH EMPTY CELLS

Possibly the greatest advantage of the cell means model in the analysis of unbalanced data is that extreme situations such as empty cells can be dealt with relatively easily. The cell means approach allows one to deal specifically with nonestimability problems arising from the empty cells (as contrasted with nonestimability arising from overparameterization of the model). Much of our discussion here follows that of Bryce et al. (1980a).

Consider the unbalanced two-way model in (15.17), but allow n_{ij} to be equal to 0 for one or more (say m) *isolated* cells; that is, the empty cells do not constitute a whole row or whole column. Assume also that the empty cells are *missing at random* (Little and Rubin 2002, p. 12); that is, the emptiness of the cells is independent of the values that would be observed in those cells.

In the empty cells model, $\mathbf{W}$ is non-full-rank in that it has m columns equal to $\mathbf{0}$. To simplify notation, assume that the columns of $\mathbf{W}$ have been rearranged with the columns of $\mathbf{0}$ occurring last. Hence

$$\mathbf{W} = (\mathbf{W}_1, \mathbf{O}),$$

where $\mathbf{W}_1$ is an $n \times (ab - m)$ matrix and $\mathbf{O}$ is $n \times m$. Correspondingly

$$\boldsymbol{\mu} = \begin{pmatrix} \boldsymbol{\mu}_o \\ \boldsymbol{\mu}_e \end{pmatrix},$$

where $\boldsymbol{\mu}_o$ is the vector of cell means for the occupied cells while $\boldsymbol{\mu}_e$ is the vector of cell means for the empty cells. The model is thus the non-full-rank model

$$\mathbf{y} = (\mathbf{W}_1, \mathbf{O}) \begin{pmatrix} \boldsymbol{\mu}_o \\ \boldsymbol{\mu}_e \end{pmatrix} + \boldsymbol{\varepsilon}. \tag{15.47}$$

The first task in the analysis of two-way data with empty cells is to test for the interaction between the factors A and B. To test for the interaction when there are isolated empty cells, care must be exercised to ensure that a testable hypothesis is being tested (Section 12.6). The full–reduced-model approach [see (8.31)] is useful here. A sensible full model is the unconstrained cell means model in (15.47). Even though $\mathbf{W}$ is not full-rank

$$\mathrm{SSE}_u = \mathbf{y}'[\mathbf{I} - \mathbf{W}(\mathbf{W}'\mathbf{W})^{-}\mathbf{W}']\mathbf{y} \tag{15.48}$$

is invariant to the choice of a generalized inverse (Theorem 12.3e). The reduced model is the additive model, given by

$$\mathbf{y} = \mathbf{W}\mathbf{A}^{-1}\mathbf{A}\boldsymbol{\mu} + \boldsymbol{\varepsilon} \quad \text{subject to} \quad \mathbf{G}\boldsymbol{\mu} = \mathbf{0},$$

where

$$\mathbf{A} = \begin{pmatrix} \mathbf{K} \\ \mathbf{G} \end{pmatrix},$$

in which $\mathbf{K}$ is a matrix specifying the overall mean and linearly independent main effect contrasts for factors A and B, and the rows of $\mathbf{G}$ are linearly independent interaction contrasts (see Section 15.3.2) such that $\mathbf{A}$ is full-rank. We define $\mathbf{Z}_1$ as $\mathbf{W}\mathbf{K}^*$ [see (15.41)]. Because the empty cells are isolated, $\mathbf{Z}_1$ is full-rank even though some of the constraints in $\mathbf{G}\boldsymbol{\mu} = \mathbf{0}$ are nonestimable. The error sum of squares for the additive model is then

$$\mathrm{SSE}_a = \mathbf{y}'[\mathbf{I} - \mathbf{Z}_1(\mathbf{Z}_1'\mathbf{Z}_1)^{-1}\mathbf{Z}_1']\mathbf{y}, \tag{15.49}$$

and the test statistic for the interaction is

$$F = \frac{(\mathrm{SSE}_a - \mathrm{SSE}_u)/[(a-1)(b-1)-m]}{\mathrm{SSE}_u/(n - ab + m)}. \tag{15.50}$$

Equivalently the interaction could be tested by the general linear hypothesis approach in (8.27). However, a maximal set of nonestimable interaction side conditions involving $\boldsymbol{\mu}_e$ must first be imposed on the model. For example, the side conditions could be specified as

$$\mathbf{T}\boldsymbol{\mu} = \mathbf{0}, \tag{15.51}$$

where $\mathbf{T}$ is an $m \times ab$ matrix with rows corresponding to the contrasts $\mu_{ij} - \mu_{i.} - \mu_{.j} + \mu_{..}$ for all m empty cells (Henderson and McAllister 1978). Using (12.37), we obtain

$$\hat{\boldsymbol{\mu}} = (\mathbf{W}'\mathbf{W} + \mathbf{T}'\mathbf{T})^{-1}\mathbf{W}'\mathbf{y} \tag{15.52}$$

and

$$\text{cov}(\hat{\boldsymbol{\mu}}) = \sigma^2(\mathbf{W}'\mathbf{W} + \mathbf{T}'\mathbf{T})^{-1}\mathbf{W}'\mathbf{W}(\mathbf{W}'\mathbf{W} + \mathbf{T}'\mathbf{T})^{-1}. \tag{15.53}$$

The interaction can then be tested using the general linear hypothesis test of H_0: $\mathbf{C}\boldsymbol{\mu} = \mathbf{0}$ where $\mathbf{C}$ is the full matrix of $(a-1)(b-1)$ interaction contrasts. Even though some of the rows of $\mathbf{C}\boldsymbol{\mu}$ are not estimable, the test statistic can be computed using a generalized inverse in the numerator as

$$F = \frac{(\mathbf{C}\hat{\boldsymbol{\mu}})'\{\mathbf{C}[\text{cov}(\hat{\boldsymbol{\mu}})/\sigma^2]\mathbf{C}'\}^{-}(\mathbf{C}\hat{\boldsymbol{\mu}})/[(a-1)(b-1)-m]}{\text{SSE}/(n-ab+m)}. \tag{15.54}$$

The error sum of squares for this model, SSE, turns out to be the same as SSE_u in (15.48). By Theorem 2.8c(v), the numerator of this F statistic is invariant to the choice of a generalized inverse (Problem 15.16).

Both versions of this additivity test involve the unverifiable assumption that the means of the empty cells follow the additive pattern displayed by the means of the occupied cells. If there are relatively few empty cells, this is usually a reasonable assumption.

If the interaction is not significant and is deemed to be negligible, the additive model can be used as in Section 15.3.2 without any modifications. The isolated empty cells present no problems for the use of the additive model.

If the interaction is significant, it may be possible to partially constrain the interaction in an attempt to render all cell means (including those in $\boldsymbol{\mu}_e$) estimable. This is not always possible, because it requires a set of constraints that are both a priori reasonable and such that they render $\boldsymbol{\mu}$ estimable. Nonetheless, it is often advisable to make this attempt because no new theoretical results are needed. The greatest challenges are practical, in that sensible constraints must be used. Many constraints will do the job mathematically, but the results are meaningless unless the constraints are reasonable. Unlike many other methods associated with linear models, the validity of this procedure depends on the parameterization of the model and the specific constraints that are chosen.

We proceed in this attempt by proposing partial interaction constraints

$$\mathbf{G}\boldsymbol{\mu} = \mathbf{0}$$

for the empty cells model in (15.47). We choose $\mathbf{K}$ such that its rows are linearly independent of the rows of $\mathbf{G}$ so that

$$\mathbf{A} = \begin{pmatrix} \mathbf{K} \\ \mathbf{G} \end{pmatrix}$$

is nonsingular. Thus $\mathbf{A}^{-1} = (\mathbf{K}^* \ \mathbf{G}^*)$ as in the comments following (15.41). Suppose that the constraints are realistic, and that they are such that the constrained model is not the additive model; that is, at least a portion of the interaction is unconstrained. Then, if $\mathbf{Z}_1 = \mathbf{W}\mathbf{K}^*$ is full-rank, all the cell means (including $\boldsymbol{\mu}_e$) can be estimated as

$$\hat{\boldsymbol{\mu}} = \mathbf{K}^*(\mathbf{Z}_1'\mathbf{Z}_1)^{-1}\mathbf{Z}_1'\mathbf{y}, \tag{15.55}$$

and cov($\hat{\boldsymbol{\mu}}$) is given by (15.44). Further inferences about linear combinations of the cell means can then be readily carried out. If $\mathbf{Z}_1$ is not full-rank, care must be exercised to ensure that only estimable functions of $\boldsymbol{\mu}$ are estimated and that testable hypotheses involving $\boldsymbol{\mu}$ are tested (see Section 12.2).

A simple way to quickly check whether $\mathbf{Z}_1$ is full-rank (and thus all cell means are estimable) is given in the following theorem.

Theorem 15.4. Consider the constrained empty cells model in (15.47) with m empty cells. Partition $\mathbf{A}$ as

$$\mathbf{A} = \begin{pmatrix} \mathbf{K} \\ \mathbf{G} \end{pmatrix} = \begin{pmatrix} \mathbf{K}_1 & \mathbf{K}_2 \\ \mathbf{G}_1 & \mathbf{G}_2 \end{pmatrix}$$

conformal with the partitioned vector

$$\boldsymbol{\mu} = \begin{pmatrix} \boldsymbol{\mu}_o \\ \boldsymbol{\mu}_e \end{pmatrix}.$$

The elements of $\boldsymbol{\mu}$ are estimable (equivalently $\mathbf{Z}_1$ is full-rank) if and only if rank($\mathbf{G}_2$) $= m$.

Proof. We prove this theorem for the special case in which $\mathbf{G}$ has m rows so that $\mathbf{G}_2$ is $m \times m$. We partition $\mathbf{A}^{-1}$ as

$$\begin{pmatrix} \mathbf{K}_1^* & \mathbf{G}_1^* \\ \mathbf{K}_2^* & \mathbf{G}_2{}^* \end{pmatrix},$$

with submatrices conforming to the partitioning of $\mathbf{A}$. Then

$$\mathbf{Z}_1 = (\mathbf{W}_1, \mathbf{O})\begin{pmatrix} \mathbf{K}_1^* \\ \mathbf{K}_2^* \end{pmatrix} = \mathbf{W}_1\mathbf{K}_1^*.$$

Since $\mathbf{W}_1$ is full-rank and each of its rows consists of one 1 with several 0s, $\mathbf{W}_1\mathbf{K}_1^*$ contains one or more copies of all of the rows of $\mathbf{W}_1$. Thus rank($\mathbf{Z}_1$) = rank($\mathbf{K}_1^*$). Since $\mathbf{A}^{-1}$ is nonsingular, $\mathbf{K}_1^{*-1}$ exists if $\mathbf{K}_1^*$ is full rank. If so, the product

$$\begin{pmatrix} \mathbf{I} & \mathbf{O} \\ -\mathbf{K}_2^*\mathbf{K}_1{}^{*-1} & \mathbf{I} \end{pmatrix}\begin{pmatrix} \mathbf{K}_1^* & \mathbf{G}_1^* \\ \mathbf{K}_2^* & \mathbf{G}_2^* \end{pmatrix} = \begin{pmatrix} \mathbf{K}_1^* & \mathbf{G}_1^* \\ \mathbf{O} & \mathbf{G}_2^* - \mathbf{K}_2^*\mathbf{K}_1^{*-1}\mathbf{G}_1^* \end{pmatrix}$$

is defined and is nonsingular by Theorem 2.4(ii). By Corollary 1 to Theorem 2.9b, $\mathbf{G}_2^* - \mathbf{K}_2^*\mathbf{K}_1^{*-1}\mathbf{G}_1^*$ is also nonsingular. But by equation (2.50), $(\mathbf{G}_2^* - \mathbf{K}_2^*\mathbf{K}_1^{*-1}\mathbf{G}_1^*)^{-1} = \mathbf{G}_2$. Thus, if $\mathbf{A}^{-1}$ is nonsingular, nonsingularity of $\mathbf{K}_1^*$ implies nonsingularity of $\mathbf{G}_2$. Analogous reasoning leads to the converse. Thus $\mathbf{K}_1^*$ is full-rank if and only if $\mathbf{G}_2$ is full-rank. Furthermore, $\mathbf{Z}_1$ is full-rank if and only if rank($\mathbf{G}_2$) $= m$. □

Example 15.4a. For the second-language data of Table 15.8, we test for the interaction of native language and gender. There are two empty cells, and thus $\mathbf{W}$ is a

TABLE 15.8 Comfort in Using English as a Second Language for Students at BYU-Hawaii[a]

Native Language	Gender	
	Male	Female
Samoan	24	28
	3.20	3.38
	0.66	0.68
Tongan	25	39
	3.03	3.10
	0.69	0.61
Hawaiian	4	2
	3.47	3.13
	0.68	0.47
Fijian	1	—
	3.79	—
	—	—
Pacific Islands English	26	49
	3.71	3.13
	0.58	0.73
Maori	3	1
	4.07	3.04
	0.061	—
Mandarin	15	43
	3.33	3.14
	0.74	0.61
Cantonese	—	21
	—	3.00
	—	0.54

[a] Brigham Young University–Hawaii; data classified by gender and native language. *Key to table entries*: number of observations, mean, and standard deviation.

281×16 matrix with two columns of $\mathbf{0}$. For the unconstrained model we use (15.48) to obtain

$$\text{SSE}_u = 113.235.$$

Numbering the cells of Table 15.8 from 1 to 8 for the first column and from 9 to 16 for the second column, we now define

$$\mathbf{A} = \begin{pmatrix} \mathbf{K} \\ \mathbf{G} \end{pmatrix} \tag{15.56}$$

where

$$
\mathrm{K}=\begin{pmatrix}
1 & 1 & 1 & 1 & 1 & 1 & 1 & 1 & 1 & 1 & 1 & 1 & 1 & 1 & 1 & 1 \\
1 & 1 & 1 & 1 & 1 & 1 & 1 & 1 & -1 & -1 & -1 & -1 & -1 & -1 & -1 & -1 \\
1 & -1 & 0 & 0 & 0 & 0 & 0 & 0 & 1 & -1 & 0 & 0 & 0 & 0 & 0 & 0 \\
0 & 0 & 1 & 0 & -1 & 0 & 0 & 0 & 0 & 0 & 1 & 0 & -1 & 0 & 0 & 0 \\
0 & 0 & 0 & 1 & -1 & 0 & 0 & 0 & 0 & 0 & 0 & 1 & -1 & 1 & 0 & 0 \\
0 & 0 & 0 & 0 & -1 & 1 & 0 & 0 & 0 & 0 & 0 & 0 & -1 & -1 & 0 & 0 \\
2 & 2 & -1 & -1 & -1 & -1 & 0 & 0 & 2 & 2 & -1 & -1 & -1 & -1 & 0 & 0 \\
0 & 0 & 0 & 0 & 0 & 0 & 1 & -1 & 0 & 0 & 0 & 0 & 0 & 0 & -1 & 1 \\
1 & 1 & 1 & 1 & 1 & 1 & -3 & -3 & 1 & 1 & 1 & 1 & 1 & 1 & -3 & -3
\end{pmatrix}
$$

and

$$
\mathbf{G}=\begin{pmatrix}
1 & -1 & 0 & 0 & 0 & 0 & 0 & 0 & 1 & -1 & 0 & 0 & 0 & 0 & 0 & 0 \\
0 & 0 & 1 & 0 & -1 & 0 & 0 & 0 & 0 & 0 & -1 & 0 & 1 & 0 & 0 & 0 \\
0 & 0 & 0 & 1 & -1 & 0 & 0 & 0 & 0 & 0 & 0 & -1 & 1 & 0 & 0 & 0 \\
0 & 0 & 0 & 0 & -1 & 1 & 0 & 0 & 0 & 0 & 0 & 0 & 1 & -1 & 0 & 0 \\
2 & 2 & -1 & -1 & -1 & -1 & 0 & 0 & -2 & -2 & 1 & 1 & 1 & 1 & 0 & 0 \\
0 & 0 & 0 & 0 & 0 & 0 & 1 & -1 & 0 & 0 & 0 & 0 & 0 & 0 & -1 & 1 \\
1 & 1 & 1 & 1 & 1 & 1 & -3 & -3 & -1 & -1 & -1 & -1 & -1 & -1 & 3 & 3
\end{pmatrix}
$$

The overall mean and main effect contrasts are specified by **K** while interaction contrasts are specified by **G**. Using (15.49), $\mathrm{SSE}_a = 119.213$. The full–reduced F test for additivity (15.50) yields the test statistic

$$
F = \frac{(119.213 - 113.235)/5}{119.213/267} = 2.82,
$$

which is larger than the critical value of $F_{.05,\,5,\,267} = 2.25$.

As an alternative approach to testing additivity, we impose the nonestimable side conditions $\mu_{8,1} - \mu_{8.} - \mu_{.1} + \mu_{..} = 0$ and $\mu_{4,\,2} - \mu_{4.} - \mu_{.2} + \mu_{..} = 0$ on the model by setting

$$
\mathbf{T}=\begin{pmatrix}
-1 & -1 & -1 & -1 & -1 & -1 & -1 & 7 & 1 & 1 & 1 & 1 & 1 & 1 & 1 & -7 \\
-1 & -1 & -1 & 7 & -1 & -1 & -1 & -1 & 1 & 1 & 1 & -7 & 1 & 1 & 1 & 1
\end{pmatrix}
$$

in (15.51) and

$$\mathbf{C} = \begin{pmatrix} 1 & -1 & 0 & 0 & 0 & 0 & 0 & 0 & -1 & 1 & 0 & 0 & 0 & 0 & 0 & 0 \\ 1 & 0 & -1 & 0 & 0 & 0 & 0 & 0 & -1 & 0 & 1 & 0 & 0 & 0 & 0 & 0 \\ 1 & 0 & 0 & -1 & 0 & 0 & 0 & 0 & -1 & 0 & 0 & 1 & 0 & 0 & 0 & 0 \\ 1 & 0 & 0 & 0 & -1 & 0 & 0 & 0 & -1 & 0 & 0 & 0 & 1 & 0 & 0 & 0 \\ 1 & 0 & 0 & 0 & 0 & -1 & 0 & 0 & -1 & 0 & 0 & 0 & 0 & 1 & 0 & 0 \\ 1 & 0 & 0 & 0 & 0 & 0 & -1 & 0 & -1 & 0 & 0 & 0 & 0 & 0 & 1 & 0 \\ 1 & 0 & 0 & 0 & 0 & 0 & 0 & -1 & -1 & 0 & 0 & 0 & 0 & 0 & 0 & 1 \end{pmatrix}$$

in (15.54). The F statistic for the general linear hypothesis test of additivity (15.54) is again equal to 2.82.

Since the interaction is significant for this dataset, we partially constrain the interaction with contextually sensible estimable constraints in an effort to make all of the cell means estimable. We use $\mathbf{A}$ as defined in (15.56), but repartition it so that

$$\mathbf{K} = \begin{pmatrix} 1 & 1 & 1 & 1 & 1 & 1 & 1 & 1 & 1 & 1 & 1 & 1 & 1 & 1 & 1 & 1 \\ 1 & 1 & 1 & 1 & 1 & 1 & 1 & 1 & -1 & -1 & -1 & -1 & -1 & -1 & -1 & -1 \\ 1 & -1 & 0 & 0 & 0 & 0 & 0 & 0 & 1 & -1 & 0 & 0 & 0 & 0 & 0 & 0 \\ 0 & 0 & 1 & 0 & -1 & 0 & 0 & 0 & 0 & 0 & 1 & 0 & -1 & 0 & 0 & 0 \\ 0 & 0 & 0 & 1 & -1 & 0 & 0 & 0 & 0 & 0 & 0 & 1 & -1 & 0 & 0 & 0 \\ 0 & 0 & 0 & 0 & -1 & 1 & 0 & 0 & 0 & 0 & 0 & 0 & -1 & 1 & 0 & 0 \\ 2 & 2 & -1 & -1 & -1 & -1 & 0 & 0 & 2 & 2 & -1 & -1 & -1 & -1 & 0 & 0 \\ 0 & 0 & 0 & 0 & 0 & 0 & 1 & -1 & 0 & 0 & 0 & 0 & 0 & 0 & 1 & -1 \\ 1 & 1 & 1 & 1 & 1 & 1 & -3 & -3 & 1 & 1 & 1 & 1 & 1 & 1 & -3 & -3 \\ 0 & 0 & 1 & 0 & -1 & 0 & 0 & 0 & 0 & 0 & -1 & 0 & 1 & 0 & 0 & 0 \\ 0 & 0 & 0 & 0 & -1 & 1 & 0 & 0 & 0 & 0 & 0 & 0 & 1 & -1 & 0 & 0 \\ 2 & 2 & -1 & -1 & -1 & -1 & 0 & 0 & -2 & -2 & 1 & 1 & 1 & 1 & 0 & 0 \\ 1 & 1 & 1 & 1 & 1 & 1 & -3 & -3 & -1 & -1 & -1 & -1 & -1 & -1 & 3 & 3 \end{pmatrix}$$

and

$$\mathbf{G} = \begin{pmatrix} 1 & -1 & 0 & 0 & 0 & 0 & 0 & 0 & -1 & 1 & 0 & 0 & 0 & 0 & 0 & 0 \\ 0 & 0 & 0 & 1 & -1 & 0 & 0 & 0 & 0 & 0 & 0 & -1 & 1 & 0 & 0 & 0 \\ 0 & 0 & 0 & 0 & 0 & 0 & 1 & -1 & 0 & 0 & 0 & 0 & 0 & 0 & -1 & 1 \end{pmatrix}.$$

The partial interaction constraints specified by $\mathbf{G}\boldsymbol{\mu} = \mathbf{0}$ seem sensible in that they specify that the male–female difference is the same for Samoan and Tongan speakers, for Fijian and Hawaiian speakers, and for Mandarin and Cantonese speakers. Because the empty cells are the eighth and twelfth cells, we have

$$\mathbf{G}_2 = \begin{pmatrix} 0 & 0 \\ 0 & -1 \\ -1 & 0 \end{pmatrix}$$

which obviously has rank = 2. Thus, by Theorem 15.4, all the cell means are estimable. Using (15.55) to compute the constrained estimates and (15.44) to compute their standard errors, we obtain the results in Table 15.9. □

TABLE 15.9 Estimated Mean Comfort in Using English as Second Language (with Standard Error) for Students at BYU-Hawaii[a]

Native Language	Gender	
	Male	Female
Samoan	3.23 (.11)	3.35 (.11)
Tongan	3.00 (.11)	3.12 (.09)
Hawaiian	3.47 (.33)	3.13 (.46)
Fijian	3.79 (.65)	3.20 (.67)
Pacific Islands English	3.71 (.03)	3.13 (.09)
Maori	4.07 (.38)	3.04 (.65)
Mandarin	3.33 (.17)	3.14 (.10)
Cantonese	3.19 (.24)	3.00 (.14)

[a] On the basis of a constrained empty-cells model.

PROBLEMS

15.1 For the model $\mathbf{y} = \mathbf{W}\boldsymbol{\mu} + \boldsymbol{\varepsilon}$ in (15.2.1), find $\mathbf{W}'\mathbf{W}$ and $\mathbf{W}'\mathbf{y}$ and show that $(\mathbf{W}'\mathbf{W})^{-1}\mathbf{W}'\mathbf{y} = \bar{\mathbf{y}}$ as in (15.5).

15.2 **(a)** Show that for the reduced model $y_{ij} = \mu + \varepsilon^*_{ij}$ in Section 15.3, $\text{SS}(\mu) = N\bar{y}^2_{..}$ as used in (15.6).

(b) Show that $\text{SSB} = \sum_{i=1}^k \bar{y}_{i.}y_{i.} - N\bar{y}^2_{..}$ as in (15.6).

(c) Show that (15.6) is equal to (15.7), that is, $\text{SSB} = \sum_i \bar{y}_{i.}y_{i.} - N\bar{y}^2_{..} = \sum_i y^2_{i.}/n_i - y^2_{..}/N$.

15.3 **(a)** Show that SSB in (15.9) is equal to SSB in (15.7), that is, $\sum_{i=1}^k n_i(\bar{y}_{i.} - \bar{y}_{..})^2 = \sum_{i=1}^k y^2_{i.}/n_i - y^2_{..}/N$.

(b) Show that SSE in (15.10) is equal to SSE in (15.8), that is, $\sum_{i=1}^k \sum_{j=1}^{n_i} (y_{ij} - \bar{y}_{i.})^2 = \sum_{i=1}^k \sum_{j=1}^{n_i} y^2_{ij} - \sum_{i=1}^k y^2_{i.}/n_i$.

15.4 Show that $F = (\Sigma_i c_i \bar{y}_{i.})^2/(s^2 \sum_i c_i^2/n_i)$ in (15.13) follows from (15.12).

15.5 Show that $\mathbf{a}'$ and $\mathbf{b}'$ in (15.15) provide contrast coefficients that satisfy the property $\sum_i a_i b_i/n_i = 0$.

15.6 Show that $\hat{\boldsymbol{\mu}} = \bar{\mathbf{y}}$ as in (15.19).

15.7 Obtain (15.23) from (15.21); that is, show that $(\mathbf{y} - \mathbf{W}\hat{\boldsymbol{\mu}})'(\mathbf{y} - \mathbf{W}\hat{\boldsymbol{\mu}}) = \sum_{i=1}^a \sum_{j=1}^b \sum_{k=1}^{n_{ij}} (y_{ijk} - \bar{y}_{ij.})^2$.

15.8 Obtain (15.24) from (15.23); that is, show that $\sum_{k=1}^{n_{ij}} (y_{ijk} - \bar{y}_{ij.})^2 = (n_{ij} - 1)s_{ij}^2$.

15.9 Show that H_0: $\mathbf{B}\boldsymbol{\mu} = \mathbf{0}$, where $\mathbf{B}$ is given in (15.27), is equivalent to H_0: $\mu_{11} + \mu_{21} = \mu_{12} + \mu_{22} = \mu_{13} + \mu_{23}$ in (15.28).

15.10 Obtain $F = \left(\sum_{ij} a_{ij}\bar{y}_{ij.}\right)^2 / \left(s^2 \sum_{ij} a_{ij}^2/n_{ij}\right)$ in (15.31) from $F = (\mathbf{a}'\hat{\boldsymbol{\mu}})^2 / [s^2\mathbf{a}'(\mathbf{W}'\mathbf{W})^{-1}\mathbf{a}]$ in (15.30).

15.11 Evaluate $\mathbf{a}'(\mathbf{W}'\mathbf{W})^{-1}\mathbf{a}$ in (15.29) or (15.30) for $\mathbf{a}' = (1, 1, 1, -1, -1, -1)$. Use the $\mathbf{W}$ matrix for the 11 observations in the illustration in Section 15.3.1.

15.12 Evaluate $\mathbf{B}(\mathbf{W}'\mathbf{W})^{-1}\mathbf{B}'$ in (15.33) for the matrices $\mathbf{B}$ and $\mathbf{W}$ used in the illustration in Section 15.3.1.

15.13 Show that $\mathbf{A}\mathbf{A}^{-1} = \mathbf{I}$, where $\mathbf{A} = \begin{pmatrix} \mathbf{K} \\ \mathbf{G} \end{pmatrix}$ as in (15.37) and $\mathbf{A}^{-1} = [\mathbf{K}'(\mathbf{K}\mathbf{K}')^{-1}, \mathbf{G}'(\mathbf{G}\mathbf{G}')^{-1}]$ as in (15.41).

15.14 Obtain $\mathbf{G}^*$ analogous to $\mathbf{K}^*$ in (15.42).

15.15 Show that $\text{cov}(\hat{\boldsymbol{\mu}}_c) = \sigma^2\mathbf{K}'(\mathbf{K}\mathbf{K}')^{-1}(\mathbf{Z}_1'\mathbf{Z}_1)^{-1}(\mathbf{K}\mathbf{K}')^{-1}\mathbf{K}$, thus verifying (15.44).

15.16 Show that the numerator of the F statistic in (15.54) is invariant to the choice of a generalized inverse.

15.17 In a feeding trial, chicks were given five protein supplements. Their final weights at 6 weeks are given in Table 15.10 (Snedecor 1948, p. 214).

(a) Calculate the sums of squares in Table 15.1 and the F statistic in (15.11).

(b) Compare the protein supplements using (unweighted) orthogonal contrasts whose coefficients are the rows in the matrix

$$\begin{pmatrix} 3 & -2 & -2 & -2 & 3 \\ 0 & 1 & -2 & 1 & 0 \\ 0 & 1 & 0 & -1 & 0 \\ 1 & 0 & 0 & 0 & -1 \end{pmatrix}.$$

Thus we are making the following comparisons:

L, C	versus	So, Su, M
So, M	versus	Su
So	versus	M
L	versus	C

(c) Replace the second contrast with a weighted contrast whose coefficients satisfy $\sum_i a_i b_i / n_i = 0$ when paired with the first contrast. Find sums of squares and F statistics for these two contrasts.

TABLE 15.10 Final Weights (g) of Chicks at 6 Weeks

Protein Supplement				
Linseed	Soybean	Sunflower	Meat	Casein
309	243	423	325	368
229	230	340	257	390
181	248	392	303	379
141	327	339	315	260
260	329	341	380	404
203	250	226	153	318
148	193	320	263	352
169	271	295	242	359
213	316	334	206	216
257	267	322	344	222
244	199	297	258	283
271	177	318		332
	158			
	248			

15.18 **(a)** Carry out the computations to obtain $\hat{\mu}$, **SSE**, F_{A}, F_{B}, and F_{C} in Example 15.3a.

(b) Carry out the computations to obtain $\hat{\boldsymbol{\mu}}_c$, SSE_c, F_{A_c}, and F_{B_c} in Example 15.3b.

(c) Carry out the tests in parts (a) and (b) using a software package such as SAS GLM.

15.19 Table 15.11 lists weight gains of male rats under three types of feed and two levels of protein.

(a) Let factor A be level of protein and factor B be type of feed. Define a vector **a** corresponding to factor A and matrices **B** and **C** for factor B and interaction AB, respectively, as in Section 15.3.1. Use these to construct general linear hypothesis tests for main effects and interaction as in (15.29), (15.33), and (15.34).

(b) Test the main effects in the no-interaction model (15.35) using the constrained model (15.36). Define **K** and **G** and find $\hat{\boldsymbol{\mu}}_c$ in (15.43), SSE_c, and F for H_0: $\mathbf{a}'\boldsymbol{\mu}_c = 0$ and H_0: $\mathbf{B}\boldsymbol{\mu}_c = \mathbf{0}$ in (15.45).

(c) Carry out the tests in parts (a) and (b) using a software package such as SAS GLM.

15.20 Table 15.12 lists yields when five varieties of plants and four fertilizers were tested. Test for main effects and interaction.

TABLE 15.11 Weight Gains (g) of Rats under Six Diet Combinations

High Protein			Low Protein		
Beef	Cereal	Pork	Beef	Cereal	Pork
73	98	94	90	107	49
102	74	79	76	95	82
118	56	96	90	97	73
104	111	98	64	80	86
81	95	102	86	98	81
107	88	102	51	74	97
100	82		72		106
87	77		90		
	86		95		
	92		78		

Source: Snedecor and Cochran (1967, p. 347).

TABLE 15.12 Yield from Five Varieties of Plants Treated with Four Fertilizers

	Variety				
Fertilizer	1	2	3	4	5
1	57	26	39	23	48
	46	38	—	36	35
	—	20	—	18	—
2	67	44	57	74	61
	72	68	61	47	—
	66	64	—	69	—
3	95	92	91	98	78
	90	89	82	85	89
	89	—	—	—	95
4	92	96	98	99	99
	88	95	93	90	—
	—	—	98	98	—

Source: Ostle and Mensing (1975, p. 368).

16 Analysis-of-Covariance

16.1 INTRODUCTION

In addition to the dependent variable y, there may be one or more quantitative variables that can also be measured on each experimental unit (or subject) in an ANOVA situation. If it appears that these extra variables may affect the outcome of the experiment, they can be included in the model as independent variables (x's) and are then known as *covariates* or *concomitant variables*. *Analysis of covariance* is sometimes described as a blend of ANOVA and regression.

The primary motivation for the use of covariates in an experiment is to gain precision by reducing the error variance. In some situations, analysis of covariance can be used to lessen the effect of factors that the experimenter cannot effectively control, because an attempt to include various levels of a quantitative variable as a full factor may cause the design to become unwieldy. In such cases, the variable can be included as a covariate, with a resulting adjustment to the dependent variable before comparing means of groups. Variables of this type may also occur in experimental situations in which the subjects cannot be randomly assigned to treatments. In such cases, we forfeit the causality implication of a designed experiment, and analysis of covariance is closer in spirit to descriptive model building.

In terms of a one-way model with one covariate, analysis of covariance will be successful if the following three assumptions hold.

1. *The dependent variable is linearly related to the covariate.* If this assumption holds, part of the error in the model is predictable and can be removed to reduce the error variance. This assumption can be checked by testing H_0: $\beta = 0$, where β is the slope from the regression of the dependent variable on the covariate. Since the estimated slope $\hat{\beta}$ will never be exactly 0, analysis of covariance will always give a smaller sum of squares for error than the corresponding ANOVA. However, if $\hat{\beta}$ is close to 0, the small reduction in error sum of squares may not offset the loss of a degree of freedom [see (16.27) and a comment following]. This problem is more likely to arise with multiple covariates, especially if they are highly correlated.

Linear Models in Statistics, Second Edition, by Alvin C. Rencher and G. Bruce Schaalje

2. *The groups (treatments) have the same slope.* In assumption 1 above, a common slope β for all k groups is implied (assuming a one-way model with k groups). We can check this assumption by testing $H_0 : \beta_1 = \beta_2 = \cdots = \beta_k$, where β_i is the slope in the ith group.
3. *The covariate does not affect the differences among the means of the groups (treatments).* If differences among the group means were reduced when the dependent variable is adjusted for the covariate, the test for equality of group means would be less powerful. Assumption 3 can be checked by performing an ANOVA on the covariate.

Covariates can be either fixed constants (values chosen by the researcher) or random variables. The models we consider in this chapter involve fixed covariates, but in practice, the majority are random. However, the estimation and testing procedures are the same in both cases, although the properties of estimators and tests are somewhat different for fixed and random covariates. For example, in the fixed-covariate case, the power of the test depends on the actual values chosen for the covariates, whereas in the random-covariate case, the power of the test depends on the population covariance matrix of the covariates.

As an illustration of the use of analysis of covariance, suppose that we wish to compare three methods of teaching language. Three classes are available, and we assign a class to each of the teaching methods. The students are free to sign up for any one of the three classes and are therefore not randomly assigned. One of the classes may end up with a disproportionate share of the best students, in which case we cannot claim that teaching methods have produced a significant difference in final grades. However, we can use previous grades or other measures of performance as covariates and then compare the students' adjusted scores for the three methods.

We give a general approach to estimation and testing in Section 16.2 and then cover specific balanced models in Sections 16.3–16.5. Unbalanced models are discussed briefly in Section 16.6. We use overparameterized models for the balanced case in Sections 16.2–16.5. and use the cell means model in Section 16.6.

16.2 ESTIMATION AND TESTING

We introduce and illustrate the analysis of covariance model in Section 16.2.1 and discuss estimation and testing for this model in Sections 16.2.2 and 16.2.3.

16.2.1 The Analysis-of-Covariance Model

In general, an analysis of covariance model can be written as

$$\mathbf{y} = \mathbf{Z}\boldsymbol{\alpha} + \mathbf{X}\boldsymbol{\beta} + \boldsymbol{\varepsilon}, \tag{16.1}$$

where $\mathbf{Z}$ contains 0s and 1s, $\boldsymbol{\alpha}$ contains μ and parameters such as α_i, β_i, and γ_{ij} representing factors and interactions (or other effects); $\mathbf{X}$ contains the covariate values; and $\boldsymbol{\beta}$ contains coefficients of the covariates. Thus the covariates appear on the right side of (16.1) as independent variables. Note that $\mathbf{Z}\boldsymbol{\alpha}$ is the same as $\mathbf{X}\boldsymbol{\beta}$ in the ANOVA models in Chapters 12–14, whereas in this chapter, we use $\mathbf{X}\boldsymbol{\beta}$ to represent the covariates in the model.

We now illustrate (16.1) for some of the models that will be considered in this chapter. A one-way (balanced) model with one covariate can be expressed as

$$y_{ij} = \mu + \alpha_i + \beta x_{ij} + \varepsilon_{ij}, \quad i = 1, 2, \ldots, k, \quad j = 1, 2, \ldots, n, \tag{16.2}$$

where α_i is the treatment effect, x_{ij} is a covariate observed on the same sampling unit as y_{ij}, and β is a slope relating x_{ij} to y_{ij}. [If (16.2) is viewed as a regression model, then the parameters $\mu + \alpha_i \quad i = 1, 2, \ldots, k$, serve as regression intercepts for the k groups.] The kn observations for (16.2) can be written in the form $\mathbf{y} = \mathbf{Z}\boldsymbol{\alpha} + \mathbf{X}\boldsymbol{\beta} + \boldsymbol{\varepsilon}$ as in (16.1), where

$$\mathbf{Z} = \begin{pmatrix} 1 & 1 & 0 & \cdots & 0 \\ \vdots & \vdots & \vdots & & \vdots \\ 1 & 1 & 0 & \cdots & 0 \\ 1 & 0 & 1 & \cdots & 0 \\ \vdots & \vdots & \vdots & & \vdots \\ 1 & 0 & 0 & \cdots & 1 \end{pmatrix}, \quad \boldsymbol{\alpha} = \begin{pmatrix} \mu \\ \alpha_1 \\ \vdots \\ \alpha_k \end{pmatrix}, \quad \mathbf{X} = \mathbf{x} = \begin{pmatrix} x_{11} \\ \vdots \\ x_{1n} \\ x_{2n} \\ \vdots \\ x_{kn} \end{pmatrix}, \tag{16.3}$$

and $\boldsymbol{\beta} = \beta$. In this case, $\mathbf{Z}$ is the same as $\mathbf{X}$ in (13.6).

For a one-way (balanced) model with q covariates, the model is

$$y_{ij} = \mu + \alpha_i + \beta_1 x_{ij1} + \cdots + \beta_q x_{ijq} + \varepsilon_{ij}, \quad i = 1, 2, \ldots, k, \quad j = 1, 2, \ldots, n. \tag{16.4}$$

In this case, $\mathbf{Z}$ and $\boldsymbol{\alpha}$ are as given in (16.3), and $\mathbf{X}\boldsymbol{\beta}$ has the form

$$\mathbf{X}\boldsymbol{\beta} = \begin{pmatrix} x_{111} & x_{112} & \cdots & x_{11q} \\ x_{121} & x_{122} & \cdots & x_{12q} \\ \vdots & \vdots & & \vdots \\ x_{kn1} & x_{kn2} & \cdots & x_{knq} \end{pmatrix} \begin{pmatrix} \beta_1 \\ \beta_2 \\ \vdots \\ \beta_q \end{pmatrix}. \tag{16.5}$$

For a two-way model with one covariate

$$y_{ijk} = \mu + \alpha_i + \delta_j + \gamma_{ij} + \beta x_{ijk} + \varepsilon_{ijk}, \tag{16.6}$$

$\mathbf{Z}\boldsymbol{\alpha}$ has the form given in (14.4), and $\mathbf{X}\boldsymbol{\beta}$ is

$$\mathbf{X}\boldsymbol{\beta} = \mathbf{x}\beta = \begin{pmatrix} x_{111} \\ x_{112} \\ \vdots \\ x_{abn} \end{pmatrix} \beta.$$

The two-way model in (16.6) could be extended to include several covariates.

16.2.2 Estimation

We now develop estimators of $\boldsymbol{\alpha}$ and $\boldsymbol{\beta}$ for the general case in (16.1), $\mathbf{y} = \mathbf{Z}\boldsymbol{\alpha} + \mathbf{X}\boldsymbol{\beta} + \boldsymbol{\varepsilon}$. We assume that $\mathbf{Z}$ is less than full rank as in overparameterized ANOVA models and that $\mathbf{X}$ is full-rank as in regression models. We also assume that

$$E(\boldsymbol{\varepsilon}) = \mathbf{0} \quad \text{and} \quad \text{cov}(\boldsymbol{\varepsilon}) = \sigma^2\mathbf{I}.$$

The model can be expressed as

$$\begin{aligned} \mathbf{y} &= \mathbf{Z}\boldsymbol{\alpha} + \mathbf{X}\boldsymbol{\beta} + \boldsymbol{\varepsilon} \\ &= (\mathbf{Z}, \mathbf{X})\begin{pmatrix} \boldsymbol{\alpha} \\ \boldsymbol{\beta} \end{pmatrix} + \boldsymbol{\varepsilon} \\ &= \mathbf{U}\boldsymbol{\theta} + \boldsymbol{\varepsilon}, \end{aligned} \tag{16.7}$$

where $\mathbf{U} = (\mathbf{Z}, \mathbf{X})$ and $\boldsymbol{\theta} = \begin{pmatrix} \boldsymbol{\alpha} \\ \boldsymbol{\beta} \end{pmatrix}$. The normal equations for (16.7) are

$$\mathbf{U}'\mathbf{U}\hat{\boldsymbol{\theta}} = \mathbf{U}'\mathbf{y},$$

which can be written in partitioned form as

$$\begin{pmatrix} \mathbf{Z}' \\ \mathbf{X}' \end{pmatrix}(\mathbf{Z}, \mathbf{X})\begin{pmatrix} \hat{\boldsymbol{\alpha}} \\ \hat{\boldsymbol{\beta}} \end{pmatrix} = \begin{pmatrix} \mathbf{Z}' \\ \mathbf{X}' \end{pmatrix}\mathbf{y},$$

$$\begin{pmatrix} \mathbf{Z}'\mathbf{Z} & \mathbf{Z}'\mathbf{X} \\ \mathbf{X}'\mathbf{Z} & \mathbf{X}'\mathbf{X} \end{pmatrix}\begin{pmatrix} \hat{\boldsymbol{\alpha}} \\ \hat{\boldsymbol{\beta}} \end{pmatrix} = \begin{pmatrix} \mathbf{Z}'\mathbf{y} \\ \mathbf{X}'\mathbf{y} \end{pmatrix}. \tag{16.8}$$

We can express (16.8) as two sets of equations in $\hat{\boldsymbol{\alpha}}$ and $\hat{\boldsymbol{\beta}}$:

$$\mathbf{Z}'\mathbf{Z}\hat{\boldsymbol{\alpha}} + \mathbf{Z}'\mathbf{X}\hat{\boldsymbol{\beta}} = \mathbf{Z}'\mathbf{y}, \tag{16.9}$$

$$\mathbf{X}'\mathbf{Z}\hat{\boldsymbol{\alpha}} + \mathbf{X}'\mathbf{X}\hat{\boldsymbol{\beta}} = \mathbf{X}'\mathbf{y}. \tag{16.10}$$

Using a generalized inverse of $\mathbf{Z}'\mathbf{Z}$, we can solve for $\hat{\boldsymbol{\alpha}}$ in (16.9):

$$\begin{aligned} \hat{\boldsymbol{\alpha}} &= (\mathbf{Z}'\mathbf{Z})^-\mathbf{Z}'\mathbf{y} - (\mathbf{Z}'\mathbf{Z})^-\mathbf{Z}'\mathbf{X}\hat{\boldsymbol{\beta}} \\ &= \hat{\boldsymbol{\alpha}}_0 - (\mathbf{Z}'\mathbf{Z})^-\mathbf{Z}'\mathbf{X}\hat{\boldsymbol{\beta}}, \end{aligned} \tag{16.11}$$

where $\hat{\boldsymbol{\alpha}}_0 = (\mathbf{Z}'\mathbf{Z})^-\mathbf{Z}'\mathbf{y}$ is a solution for the normal equations for the model $\mathbf{y} = \mathbf{Z}\boldsymbol{\alpha} + \boldsymbol{\varepsilon}$ without the covariates [see (12.13)].

To solve for $\hat{\boldsymbol{\beta}}$, we substitute (16.11) into (16.10) to obtain

$$\mathbf{X}'\mathbf{Z}[(\mathbf{Z}'\mathbf{Z})^-\mathbf{Z}'\mathbf{y} - (\mathbf{Z}'\mathbf{Z})^-\mathbf{Z}'\mathbf{X}\hat{\boldsymbol{\beta}}] + \mathbf{X}'\mathbf{X}\hat{\boldsymbol{\beta}} = \mathbf{X}'\mathbf{y}$$

or

$$\mathbf{X}'\mathbf{Z}(\mathbf{Z}'\mathbf{Z})^{-}\mathbf{Z}'\mathbf{y} + \mathbf{X}'[\mathbf{I} - \mathbf{Z}(\mathbf{Z}'\mathbf{Z})^{-}\mathbf{Z}']\mathbf{X}\hat{\boldsymbol{\beta}} = \mathbf{X}'\mathbf{y}. \tag{16.12}$$

Defining

$$\mathbf{P} = \mathbf{Z}(\mathbf{Z}'\mathbf{Z})^{-}\mathbf{Z}', \tag{16.13}$$

we see that (16.12) becomes

$$\mathbf{X}'(\mathbf{I} - \mathbf{P})\mathbf{X}\hat{\boldsymbol{\beta}} = \mathbf{X}'\mathbf{y} - \mathbf{X}'\mathbf{P}\mathbf{y} = \mathbf{X}'(\mathbf{I} - \mathbf{P})\mathbf{y}.$$

Since the elements of $\mathbf{X}$ typically exhibit a pattern unrelated to the 0s and 1s in $\mathbf{Z}$, we can assume that the columns of $\mathbf{X}$ are linearly independent of the columns of $\mathbf{Z}$. Then $\mathbf{X}'(\mathbf{I} - \mathbf{P})\mathbf{X}$ is nonsingular (see Problem 16.1), and a solution for $\hat{\boldsymbol{\beta}}$ is given by

$$\hat{\boldsymbol{\beta}} = [\mathbf{X}'(\mathbf{I} - \mathbf{P})\mathbf{X}]^{-1}\mathbf{X}'(\mathbf{I} - \mathbf{P})\mathbf{y} \tag{16.14}$$

$$= \mathbf{E}_{xx}^{-1}\mathbf{e}_{xy}, \tag{16.15}$$

where

$$\mathbf{E}_{xx} = \mathbf{X}'(\mathbf{I} - \mathbf{P})\mathbf{X} \quad \text{and} \quad \mathbf{e}_{xy} = \mathbf{X}'(\mathbf{I} - \mathbf{P})\mathbf{y}. \tag{16.16}$$

For the analysis-of-covariance model (16.1) or (16.7), we denote SSE as $\text{SSE}_{y\cdot x}$. By (12.20), $\text{SSE}_{y\cdot x}$ can be expressed as

$$\begin{aligned}
\text{SSE}_{y\cdot x} &= \mathbf{y}'\mathbf{y} - \hat{\boldsymbol{\theta}}'\mathbf{U}'\mathbf{y} = \mathbf{y}'\mathbf{y} - (\hat{\boldsymbol{\alpha}}', \hat{\boldsymbol{\beta}}')\begin{pmatrix}\mathbf{Z}'\mathbf{y}\\ \mathbf{X}'\mathbf{y}\end{pmatrix} \\
&= \mathbf{y}'\mathbf{y} - \hat{\boldsymbol{\alpha}}'\mathbf{Z}'\mathbf{y} - \hat{\boldsymbol{\beta}}'\mathbf{X}'\mathbf{y} \\
&= \mathbf{y}'\mathbf{y} - [\hat{\boldsymbol{\alpha}}_0' - \hat{\boldsymbol{\beta}}'\mathbf{X}'\mathbf{Z}(\mathbf{Z}'\mathbf{Z})^{-}]\mathbf{Z}'\mathbf{y} - \hat{\boldsymbol{\beta}}'\mathbf{X}'\mathbf{y} \quad \text{[by (16.11)]} \\
&= \mathbf{y}'\mathbf{y} - \hat{\boldsymbol{\alpha}}_0'\mathbf{Z}'\mathbf{y} - \hat{\boldsymbol{\beta}}'\mathbf{X}'[\mathbf{I} - \mathbf{Z}(\mathbf{Z}'\mathbf{Z})^{-}\mathbf{Z}']\mathbf{y} \\
&= \text{SSE}_y - \hat{\boldsymbol{\beta}}'\mathbf{X}'(\mathbf{I} - \mathbf{P})\mathbf{y},
\end{aligned} \tag{16.17}$$

where $\hat{\boldsymbol{\alpha}}_0$ is as defined in (16.11), $\mathbf{P}$ is defined as in (16.13), and $\text{SSE}_y = \mathbf{y}'\mathbf{y} - \hat{\boldsymbol{\alpha}}_0'\mathbf{Z}'\mathbf{y}$ is the same as the SSE for the ANOVA model $\mathbf{y} = \mathbf{Z}\boldsymbol{\alpha} + \boldsymbol{\varepsilon}$ without the covariates. Using (16.16), we can write (16.17) in the form

$$\text{SSE}_{y\cdot x} = e_{yy} - \mathbf{e}_{xy}'\mathbf{E}_{xx}^{-1}\mathbf{e}_{xy}, \tag{16.18}$$

where

$$e_{yy} = \text{SSE}_y = \mathbf{y}'(\mathbf{I} - \mathbf{P})\mathbf{y}. \tag{16.19}$$

In (16.18), we see the reduction in SSE that was noted in the second paragraph of Section 16.1. The proof that $\mathbf{E}_{xx} = \mathbf{X}'(\mathbf{I} - \mathbf{P})\mathbf{X}$ is nonsingular (see Problem 16.1) can be extended to show that $\mathbf{E}_{xx}$ is positive definite. Therefore, $\mathbf{e}'_{xy}\mathbf{E}_{xx}^{-1}\mathbf{e}_{xy} > 0$, and $\text{SSE}_{y \cdot x} < \text{SSE}_y$.

16.2.3 Testing Hypotheses

In order to test hypotheses, we assume that $\boldsymbol{\varepsilon}$ in (16.1) is distributed as $N_n(\mathbf{0}, \sigma^2\mathbf{I})$, where n is the number of rows of $\mathbf{Z}$ or $\mathbf{X}$. Using the model (16.7), we can express a hypothesis about $\boldsymbol{\alpha}$ in the form $H_0: \mathbf{C}\boldsymbol{\theta} = \mathbf{0}$, where $\mathbf{C} = (\mathbf{C}_1, \mathbf{O})$, so that H_0 becomes

$$H_0: (\mathbf{C}_1, \mathbf{O})\begin{pmatrix} \boldsymbol{\alpha} \\ \boldsymbol{\beta} \end{pmatrix} = \mathbf{0} \quad \text{or} \quad H_0: \mathbf{C}_1\boldsymbol{\alpha} = \mathbf{0}.$$

We can then use a general linear hypothesis test. Alternatively, we can incorporate the hypothesis into the model and use a full–reduced-model approach.

Hypotheses about $\boldsymbol{\beta}$ can also be expressed in the form $H_0: \mathbf{C}\boldsymbol{\theta} = \mathbf{0}$:

$$H_0: \mathbf{C}\boldsymbol{\theta} = (\mathbf{O}, \mathbf{C}_2)\begin{pmatrix} \boldsymbol{\alpha} \\ \boldsymbol{\beta} \end{pmatrix} = \mathbf{0} \quad \text{or} \quad H_0: \mathbf{C}_2\boldsymbol{\beta} = \mathbf{0}.$$

A basic hypothesis of interest is $H_0: \boldsymbol{\beta} = \mathbf{0}$, that is, that the covariate(s) do not belong in the model (16.1). In order to make a general linear hypothesis test of $H_0: \boldsymbol{\beta} = \mathbf{0}$, we need $\text{cov}(\hat{\boldsymbol{\beta}})$, where $\hat{\boldsymbol{\beta}}$ is given by (16.14) as $\hat{\boldsymbol{\beta}} = [\mathbf{X}'(\mathbf{I} - \mathbf{P})\mathbf{X}]^{-1}\mathbf{X}'(\mathbf{I} - \mathbf{P})\mathbf{y}$. Since $\mathbf{I} - \mathbf{P}$ is idempotent (see Theorems 2.13e and 2.13f), $\text{cov}(\hat{\boldsymbol{\beta}})$ can readily be found from (3.44) as

$$\begin{aligned} \text{cov}(\hat{\boldsymbol{\beta}}) &= [\mathbf{X}'(\mathbf{I} - \mathbf{P})\mathbf{X}]^{-1}\mathbf{X}'(\mathbf{I} - \mathbf{P})\sigma^2\mathbf{I}(\mathbf{I} - \mathbf{P})\mathbf{X}[\mathbf{X}'(\mathbf{I} - \mathbf{P})\mathbf{X}]^{-1} \\ &= \sigma^2[\mathbf{X}'(\mathbf{I} - \mathbf{P})\mathbf{X}]^{-1}. \end{aligned} \tag{16.20}$$

Then SSH for testing $H_0: \boldsymbol{\beta} = \mathbf{0}$ is given by Theorem 8.4a(ii) as

$$\text{SSH} = \hat{\boldsymbol{\beta}}'\mathbf{X}'(\mathbf{I} - \mathbf{P})\mathbf{X}\hat{\boldsymbol{\beta}}. \tag{16.21}$$

Using (16.16), we can express this as

$$\text{SSH} = \mathbf{e}'_{xy}\mathbf{E}_{xx}^{-1}\mathbf{e}_{xy}. \tag{16.22}$$

Note that SSH in (16.22) is equal to the reduction in SSE due to the covariates; see (16.17), (16.18), and (16.19).

We now discuss some specific models, beginning with the one-way model in Section 16.3.

16.3 ONE-WAY MODEL WITH ONE COVARIATE

We review the one-way model in Section 16.3.1, consider estimators of parameters in Section 16.3.2, and discuss tests of hypotheses in Section 16.3.3.

16.3.1 The Model

The one-way (balanced) model was introduced in (16.2):

$$y_{ij} = \mu + \alpha_i + \beta x_{ij} + \varepsilon_{ij}, \quad i = 1, 2, \ldots, k, \quad j = 1, 2, \ldots, n. \tag{16.23}$$

All kn observations can be written in the form of (16.1)

$$\mathbf{y} = \mathbf{Z}\boldsymbol{\alpha} + \mathbf{X}\boldsymbol{\beta} + \boldsymbol{\varepsilon} = \mathbf{Z}\boldsymbol{\alpha} + \mathbf{x}\beta + \boldsymbol{\varepsilon},$$

where $\mathbf{Z}$, $\boldsymbol{\alpha}$, and $\mathbf{x}$ are as given in (16.3).

16.3.2 Estimation

By (16.11), (13.11), and (13.12), an estimator of $\boldsymbol{\alpha}$ is obtained as

$$\hat{\boldsymbol{\alpha}} = \hat{\boldsymbol{\alpha}}_0 - (\mathbf{Z}'\mathbf{Z})^- \mathbf{Z}'\mathbf{X}\hat{\boldsymbol{\beta}} = \hat{\boldsymbol{\alpha}}_0 - (\mathbf{Z}'\mathbf{Z})^- \mathbf{Z}'\mathbf{x}\hat{\beta}$$

$$= \begin{pmatrix} 0 \\ \bar{y}_{1.} \\ \bar{y}_{2.} \\ \vdots \\ \bar{y}_{k.} \end{pmatrix} - \begin{pmatrix} 0 \\ \hat{\beta}\bar{x}_{1.} \\ \hat{\beta}\bar{x}_{2.} \\ \vdots \\ \hat{\beta}\bar{x}_{k.} \end{pmatrix} = \begin{pmatrix} 0 \\ \bar{y}_{1.} - \hat{\beta}\bar{x}_{1.} \\ \bar{y}_{2.} - \hat{\beta}\bar{x}_{2.} \\ \vdots \\ \bar{y}_{k.} - \hat{\beta}\bar{x}_{2.} \end{pmatrix} \tag{16.24}$$

(see Problem 16.4). In this case, with a single x, $\mathbf{E}_{xx}$ and $\mathbf{e}_{xy}$ reduce to scalars, along with e_{yy}:

$$\begin{aligned} \mathbf{E}_{xx} &= e_{xx} = \sum_{i=1}^{k} \sum_{j=1}^{n} (x_{ij} - \bar{x}_{i.})^2, \\ \mathbf{e}_{xy} &= e_{xy} = \sum_{ij} (x_{ij} - \bar{x}_{i.})(y_{ij} - \bar{y}_{i.}), \\ e_{yy} &= \sum_{ij} (y_{ij} - \bar{y}_{i.})^2. \end{aligned} \tag{16.25}$$

Now, by (16.15), the estimator of β is

$$\hat{\beta} = \frac{e_{xy}}{e_{xx}} = \frac{\sum_{ij}(x_{ij} - \bar{x}_{i.})(y_{ij} - \bar{y}_{i.})}{\sum_{ij}(x_{ij} - \bar{x}_{i.})^2}. \tag{16.26}$$

By (16.18), (16.19), and the three results in (16.25), $\text{SSE}_{y\cdot x}$ is given by

$$\begin{aligned} \text{SSE}_{y\cdot x} &= e_{yy} - \mathbf{e}'_{xy}\mathbf{E}_{xx}^{-1}\mathbf{e}_{xy} = e_{yy} - \frac{e_{xy}^2}{e_{xx}} \\ &= \sum_{ij}(y_{ij} - \bar{y}_{i.})^2 - \frac{\left[\sum_{ij}(x_{ij} - \bar{x}_{i.})(y_{ij} - \bar{y}_{i.})\right]^2}{\sum_{ij}(x_{ij} - \bar{x}_{i.})^2}, \end{aligned} \tag{16.27}$$

which has $k(n-1) - 1$ degrees of freedom. Note that the degrees of freedom of $\text{SSE}_{y\cdot x}$ are reduced by 1 for estimation of β, since $\text{SSE}_y = e_{yy}$ has $k(n-1)$ degrees of freedom and e_{xy}^2/e_{xx} has 1 degree of freedom. In using analysis of covariance, the researcher expects the reduction from SSE_y to $\text{SSE}_{y\cdot x}$ to at least offset the loss of a degree of freedom.

16.3.3 Testing Hypotheses

For testing hypotheses, we assume that the ε_{ij}'s in (16.23) are independently distributed as $N(0, \sigma^2)$. We begin with a test for equality of treatment effects.

16.3.3.1 Treatments

To test

$$H_{01}: \alpha_1 = \alpha_2 = \cdots = \alpha_k$$

adjusted for the covariate, we use a full–reduced-model approach. The full model is (16.23), and the reduced model (with $\alpha_1 = \alpha_2 = \cdots = \alpha_k = \alpha$) is

$$\begin{aligned} y_{ij} &= \mu + \alpha + \beta x_{ij} + \varepsilon_{ij} \\ &= \mu^* + \beta x_{ij} + \varepsilon_{ij}, \quad i = 1, 2, \ldots, k, \; j = 1, 2, \ldots, n. \end{aligned} \qquad 16.28)$$

This is essentially the same as the simple linear regression model (6.1). By (6.13), SSE for this reduced model (denoted by SSE_{rd}) is given by

$$\text{SSE}_{\text{rd}} = \sum_{i=1}^{k}\sum_{j=1}^{n}(y_{ij} - \bar{y}_{..})^2 - \frac{\left[\sum_{ij}(x_{ij} - \bar{x}_{..})(y_{ij} - \bar{y}_{..})\right]^2}{\sum_{ij}(x_{ij} - \bar{x}_{..})^2}, \tag{16.29}$$

which has $kn - 1 - 1 = kn - 2$ degrees of freedom.

Using a notation adapted from Sections 8.2, 13.4, and 14.4, we express the sum of squares for testing H_{01} as

$$\mathrm{SS}(\alpha|\mu, \beta) = \mathrm{SS}(\mu, \alpha, \beta) - \mathrm{SS}(\mu, \beta).$$

In (16.27), $\mathrm{SSE}_{y\cdot x}$ is for the full model, and in (16.29), $\mathrm{SSE}_{\mathrm{rd}}$ is for the reduced model. They can therefore be written as $\mathrm{SSE}_{y\cdot x} = \mathbf{y}'\mathbf{y} - \mathrm{SS}(\mu, \alpha, \beta)$ and $\mathrm{SSE}_{\mathrm{rd}} = \mathbf{y}'\mathbf{y} - \mathrm{SS}(\mu, \beta)$. Hence

$$\mathrm{SS}(\alpha|\mu, \beta) = \mathrm{SSE}_{\mathrm{rd}} - \mathrm{SSE}_{y\cdot x}, \tag{16.30}$$

which has $kn - 2 - [k(n-1) - 1] = k - 1$ degrees of freedom. The test statistic for $H_{01}\colon \alpha_1 = \alpha_2 = \cdots = \alpha_k$ is therefore given by

$$F = \frac{\mathrm{SS}(\alpha|\mu, \beta)/(k-1)}{\mathrm{SSE}_{y\cdot x}/[k(n-1) - 1]}, \tag{16.31}$$

which is distributed as $F[k-1, k(n-1)-1]$ when H_{01} is true.

By (16.30), we have

$$\mathrm{SSE}_{\mathrm{rd}} = \mathrm{SS}(\alpha|\mu, \beta) + \mathrm{SSE}_{y\cdot x}.$$

Hence, $\mathrm{SSE}_{\mathrm{rd}}$ functions as the "total sum of squares" for the test of treatment effects adjusted for the covariate. We can therefore denote $\mathrm{SSE}_{\mathrm{rd}}$ by $\mathrm{SST}_{y\cdot x}$, so that the expression above becomes

$$\mathrm{SST}_{y\cdot x} = \mathrm{SS}(\alpha|\mu, \beta) + \mathrm{SSE}_{y\cdot x}. \tag{16.32}$$

To complete the analogy with $\mathrm{SSE}_{y\cdot x} = e_{yy} - e_{xy}^2/e_{xx}$ in (16.27), we write (16.29) as

$$\mathrm{SST}_{y\cdot x} = t_{yy} - \frac{t_{xy}^2}{t_{xx}}, \tag{16.33}$$

where

$$\mathrm{SST}_{y\cdot x} = \mathrm{SSE}_{\mathrm{rd}},\ t_{yy} = \sum_{ij}(y_{ij} - \bar{y}_{..})^2,\ t_{xy} = \sum_{ij}(x_{ij} - \bar{x}_{..})(y_{ij} - \bar{y}_{..}),$$
$$t_{xx} = \sum_{ij}(x_{ij} - \bar{x}_{..})^2. \tag{16.34}$$

Note that the procedure used to obtain (16.30) is fundamentally different from that used to obtain $\mathrm{SSE}_{y\cdot x}$ and $\mathrm{SSE}_{\mathrm{rd}}$ in (16.27) and (16.29). The sum of squares $\mathrm{SS}(\alpha|\mu, \beta)$ in (16.30) is obtained as the difference between the sums of squares for full and reduced models, not as an adjustment to $\text{SS}(\alpha|\mu) = n\sum_i (\bar{y}_{i.} - \bar{y}_{..})^2$ in (13.24) analogous to the adjustment used in $\text{SSE}_{y \cdot x}$ and $\text{SST}_{y \cdot x}$ in (16.27) and (16.33). We must use the full–reduced-model approach to compute $\text{SS}(\alpha|\mu, \beta)$, because we do not have the same covariate values for each treatment and the design is therefore unbalanced (even though the n values are equal). If $\text{SS}(\alpha|\mu, \beta)$ were computed in an "adjusted" manner as in (16.27) or (16.33), then $\text{SS}(\alpha|\mu, \beta) + \text{SSE}_{y \cdot x}$ would not equal $\text{SST}_{y \cdot x}$ as in (16.32). In Section 16.4, we will follow a computational scheme similar to that of (16.30) and (16.32) for each term in the two-way (balanced) model.

We display the various sums of squares for testing H_0: $\alpha_1 = \alpha_2 = \cdots = \alpha_k$ in Table 16.1.

16.3.3.2 Slope

We now consider a test for

$$H_{02} : \beta = 0.$$

By (16.22), the general linear hypothesis approach leads to $\text{SSH} = \mathbf{e}'_{xy}\mathbf{E}_{xx}^{-1}\mathbf{e}_{xy}$ for testing H_0: $\boldsymbol{\beta} = \mathbf{0}$. For the case of a single covariate, this reduces to

$$\text{SSH} = \frac{e_{xy}^2}{e_{xx}}, \tag{16.35}$$

where e_{xy} and e_{xx} are as found in (16.25). The F statistic is therefore given by

$$F = \frac{e_{xy}^2/e_{xx}}{\text{SSE}_{y \cdot x}/[k(n-1)-1]}, \tag{16.36}$$

which is distributed as $F[1, k(n-1)-1]$ when H_{02} is true.

16.3.3.3 Homogeneity of Slopes

The tests of H_{01}: $\alpha_1 = \alpha_2 = \cdots = \alpha_k$ and H_{02}: $\beta = 0$ assume a common slope for all k groups. To check this assumption, we can test the hypothesis of equal slopes in the groups

$$H_{03}: \beta_1 = \beta_2 = \cdots = \beta_k, \tag{16.37}$$

where β_i is the slope in the ith group. In effect, H_{03} states that the k regression lines are parallel.

TABLE 16.1 Analysis of Covariance for Testing H_0: $\alpha_1 = \alpha_2 = \cdots = \alpha_k$ in the One-Way Model with One Covariate

Source	SS Adjusted for Covariate	Adjusted df
Treatments	$\text{SS}(\alpha\|\mu, \beta) = \text{SST}_{y \cdot x} - \text{SSE}_{y \cdot x}$	$k-1$
Error	$\text{SSE}_{y \cdot x} = e_{yy} - e_{xy}^2/e_{xx}$	$k(n-1)-1$
Total	$\text{SST}_{y \cdot x} = t_{yy} - t_{xy}^2/t_{xx}$	$kn-2$

The full model allowing for different slopes becomes

$$y_{ij} = \mu + \alpha_i + \beta_i x_{ij} + \varepsilon_{ij}, \quad i = 1, 2, \ldots, k, \; j = 1, 2, \ldots, n. \tag{16.38}$$

The reduced model with a single slope is (16.23). In matrix form, the nk observations in (16.38) can be expressed as $\mathbf{y} = \mathbf{Z}\boldsymbol{\alpha} + \mathbf{X}\boldsymbol{\beta} + \boldsymbol{\varepsilon}$, where $\mathbf{Z}$ and $\boldsymbol{\alpha}$ are as given in (16.3) and

$$\mathbf{X}\boldsymbol{\beta} = \begin{pmatrix} \mathbf{x}_1 & \mathbf{0} & \cdots & \mathbf{0} \\ \mathbf{0} & \mathbf{x}_2 & \cdots & \mathbf{0} \\ \vdots & \vdots & & \vdots \\ \mathbf{0} & \mathbf{0} & \cdots & \mathbf{x}_k \end{pmatrix} \begin{pmatrix} \beta_1 \\ \beta_2 \\ \vdots \\ \beta_k \end{pmatrix}, \tag{16.39}$$

with $\mathbf{x}_i = (x_{i1}, x_{i2}, \ldots, x_{in})'$. By (16.14) and (16.15), we obtain

$$\hat{\boldsymbol{\beta}} = \mathbf{E}_{xx}^{-1}\mathbf{e}_{xy} = [\mathbf{X}'(\mathbf{I}-\mathbf{P})\mathbf{X}]^{-1}\mathbf{X}'(\mathbf{I}-\mathbf{P})\mathbf{y}.$$

To evaluate $\mathbf{E}_{xx}$ and $\mathbf{e}_{xy}$, we first note that by (13.11), (13.25), and (13.26)

$$\begin{aligned} \mathbf{I} - \mathbf{P} &= \mathbf{I} - \mathbf{Z}(\mathbf{Z}'\mathbf{Z})^{-}\mathbf{Z}' \\ &= \begin{pmatrix} \mathbf{I} - \frac{1}{n}\mathbf{J} & \mathbf{O} & \cdots & \mathbf{O} \\ \mathbf{O} & \mathbf{I} - \frac{1}{n}\mathbf{J} & \cdots & \mathbf{O} \\ \vdots & \vdots & & \vdots \\ \mathbf{O} & \mathbf{O} & \cdots & \mathbf{I} - \frac{1}{n}\mathbf{J} \end{pmatrix}, \end{aligned} \tag{16.40}$$

where $\mathbf{I}$ in $\mathbf{I} - \mathbf{P}$ is $kn \times kn$ and $\mathbf{I}$ in $\mathbf{I} - (1/n)\mathbf{J}$ is $n \times n$. Thus

$$\begin{aligned} \mathbf{E}_{xx} = \mathbf{X}'(\mathbf{I}-\mathbf{P})\mathbf{X} &= \begin{pmatrix} \mathbf{x}_1'\left(\mathbf{I} - \frac{1}{n}\mathbf{J}\right)\mathbf{x}_1 & 0 & \cdots & 0 \\ 0 & \mathbf{x}_2'\left(\mathbf{I} - \frac{1}{n}\mathbf{J}\right)\mathbf{x}_2 & \cdots & 0 \\ \vdots & \vdots & & \vdots \\ 0 & 0 & \cdots & \mathbf{x}_k'\left(\mathbf{I} - \frac{1}{n}\mathbf{J}\right)\mathbf{x}_k \end{pmatrix} \\ &= \begin{pmatrix} \sum_j (x_{1j} - \bar{x}_{1.})^2 & 0 & \cdots & 0 \\ 0 & \sum_j (x_{2j} - \bar{x}_{2.})^2 & \cdots & 0 \\ \vdots & \vdots & & \vdots \\ 0 & 0 & \cdots & \sum_j (x_{kj} - \bar{x}_{k.})^2 \end{pmatrix} \end{aligned} \tag{16.41}$$

$$= \begin{pmatrix} e_{xx,1} & 0 & \cdots & 0 \\ 0 & e_{xx,2} & \cdots & 0 \\ \vdots & \vdots & & \vdots \\ 0 & 0 & \cdots & e_{xx,k} \end{pmatrix}, \tag{16.42}$$

where $e_{xx,i} = \sum_j (x_{ij} - \bar{x}_{i.})^2$. To find $\mathbf{e}_{xy}$, we partition $\mathbf{y}$ as $\mathbf{y} = (\mathbf{y}_1', \mathbf{y}_2', \ldots, \mathbf{y}_k')'$, where $\mathbf{y}_i' = (y_{i1}, y_{i2}, \ldots, y_{in})$. Then

$$\mathbf{e}_{xy} = \mathbf{X}'(\mathbf{I} - \mathbf{P})\mathbf{y}$$

$$= \begin{pmatrix} \mathbf{x}_1' & \mathbf{0}' & \cdots & \mathbf{0}' \\ \mathbf{0}' & \mathbf{x}_2' & \cdots & \mathbf{0}' \\ \vdots & \vdots & & \vdots \\ \mathbf{0}' & \mathbf{0}' & \cdots & \mathbf{x}_k' \end{pmatrix} \begin{pmatrix} \mathbf{I} - \frac{1}{n}\mathbf{J} & \mathbf{O} & \cdots & \mathbf{O} \\ \mathbf{O} & \mathbf{I} - \frac{1}{n}\mathbf{J} & \cdots & \mathbf{O} \\ \vdots & \vdots & & \vdots \\ \mathbf{O} & \mathbf{O} & \cdots & \mathbf{I} - \frac{1}{n}\mathbf{J} \end{pmatrix} \begin{pmatrix} \mathbf{y}_1 \\ \mathbf{y}_2 \\ \vdots \\ \mathbf{y}_k \end{pmatrix}$$

$$= \begin{pmatrix} \mathbf{x}_1{}'\left(\mathbf{I} - \frac{1}{n}\mathbf{J}\right)\mathbf{y}_1 \\ \mathbf{x}_2{}'\left(\mathbf{I} - \frac{1}{n}\mathbf{J}\right)\mathbf{y}_2 \\ \vdots \\ \mathbf{x}_k'\left(\mathbf{I} - \frac{1}{n}\mathbf{J}\right)\mathbf{y}_k \end{pmatrix}$$

$$= \begin{pmatrix} \sum_j (x_{1j} - \bar{x}_{1.})(y_{1j} - \bar{y}_{1.}) \\ \sum_j (x_{2j} - \bar{x}_{2.})(y_{2j} - \bar{y}_{2.}) \\ \vdots \\ \sum_j (x_{kj} - \bar{x}_{k.})(y_{kj} - \bar{y}_{k.}) \end{pmatrix} \tag{16.43}$$

$$= \begin{pmatrix} e_{xy,1} \\ e_{xy,2} \\ \vdots \\ e_{xy,k} \end{pmatrix}, \tag{16.44}$$

where $e_{xy,i} = \sum_j (x_{ij} - \bar{x}_{i.})(y_{ij} - \bar{y}_{i.})$. Then, by (16.15), we obtain

$$\hat{\boldsymbol{\beta}} = \mathbf{E}_{xx}^{-1}\mathbf{e}_{xy} = \begin{pmatrix} e_{xy,1}/e_{xx,1} \\ e_{xy,2}/e_{xx,2} \\ \vdots \\ e_{xy,k}/e_{xx,k} \end{pmatrix}. \tag{16.45}$$

By analogy with (16.30), we obtain the sum of squares for the test of H_{03} in (16.37) by subtracting $\text{SSE}_{y \cdot x}$ for the full model from $\text{SSE}_{y \cdot x}$ for the reduced model, that is, $\text{SSE}(R)_{y \cdot x} - \text{SSE}(F)_{y \cdot x}$. For the full model in (16.38), $\text{SSE}(F)_{y \cdot x}$ is given by (16.18), (16.44), and (16.45) as

$$\begin{aligned} \text{SSE}(F)_{y \cdot x} &= e_{yy} - \mathbf{e}_{xy}'\mathbf{E}_{xx}^{-1}\mathbf{e}_{xy} = e_{yy} - \mathbf{e}_{xy}'\hat{\boldsymbol{\beta}} \\ &= e_{yy} - (e_{xy,1}, e_{xy,2}, \ldots, e_{xy,k}) \begin{pmatrix} e_{xy,1}/e_{xx,1} \\ e_{xy,2}/e_{xx,2} \\ \vdots \\ e_{xy,k}/e_{xx,k} \end{pmatrix} \\ &= e_{yy} - \sum_{i=1}^{k} \frac{e_{xy,i}^2}{e_{xx,i}}, \end{aligned} \tag{16.46}$$

which has $k(n-1) - k = k(n-2)$ degrees of freedom. The reduced model in which $H_{03}\colon \beta_1 = \beta_2 = \cdots = \beta_k = \beta$ is true is given by (16.23), for which $\text{SSE}(R)_{y \cdot x}$ is found in (16.27) as

$$\text{SSE}(R)_{y \cdot x} = e_{yy} - \frac{e_{xy}^2}{e_{xx}}, \tag{16.47}$$

which has $k(n-1) - 1$ degrees of freedom. Thus, the sum of squares for testing H_{03} is

$$\text{SSE}(R)_{y \cdot x} - \text{SSE}(F)_{y \cdot x} = \sum_{i=1}^{k} \frac{e_{xy,i}^2}{e_{xx,i}} - \frac{e_{xy}^2}{e_{xx}}, \tag{16.48}$$

which has $k(n-1) - 1 - k(n-2) = k - 1$ degrees of freedom. The test statistic is

$$F = \frac{\left[\sum_{i=1}^{k} e_{xy,i}^2/e_{xx,i} - e_{xy}^2/e_{xx}\right]/(k-1)}{\text{SSE}(F)_{y \cdot x}/k(n-2)}, \tag{16.49}$$

which is distributed as $F[k-1, k(n-2)]$ when $H_{03}\colon \beta_1 = \beta_2 = \cdots = \beta_k$ is true.

TABLE 16.2 Maturation Weight and Initial Weight (mg) of Guppy Fish

Feeding Group					
1		2		3	
y	x	y	x	y	x
49	35	68	33	59	33
61	26	70	35	53	36
55	29	60	28	54	26
69	32	53	29	48	30
51	23	59	32	54	33
38	26	48	23	53	25
64	31	46	26	37	23

If the hypothesis of equal slopes is rejected, the hypothesis of equal treatment effects can still be tested, but interpretation is more difficult. The problem is somewhat analogous to that of interpretation of a main effect in a two-way ANOVA in the presence of interaction. In a sense, the term $\beta_i x_{ij}$ in (16.38) is an interaction. For further discussion of analysis of covariance with heterogeneity of slopes, see Reader (1973) and Hendrix et al. (1982).

Example 16.3. To investigate the effect of diet on maturation weight of guppy fish (*Poecilia reticulata*), three groups of fish were fed different diets. The resulting weights y are given in Table 16.2 (Morrison 1983, p. 475) along with the initial weights x.

We first estimate β, using x as a covariate. By the three results in (16.25), we have

$$e_{xx} = 350.2857, \quad e_{xy} = 412.71429, \quad e_{yy} = 1465.7143.$$

Then by (16.26), we obtain

$$\hat{\beta} = \frac{e_{xy}}{e_{xx}} = \frac{412.7143}{350.2857} = 1.1782.$$

We now test for equality of treatment means adjusted for the covariate, $H_0\colon \alpha_1 = \alpha_2 = \alpha_3$. By (16.27), we have

$$\begin{aligned} \text{SSE}_{y \cdot x} &= e_{yy} - \frac{e_{xy}^2}{e_{xx}} = 1465.7143 - \frac{(412.7143)^2}{350.2857} \\ &= 979.4453 \end{aligned}$$

with 17 degrees of freedom. By (16.29) and (16.33), we have

$$\text{SST}_{y \cdot x} = 1141.4709$$

with 19 degrees of freedom. Thus by (16.30), we have

$$\begin{aligned} \text{SS}(\alpha|\mu, \beta) &= \text{SST}_{y\cdot x} - \text{SSE}_{y\cdot x} = 1141.4709 - 979.4453 \\ &= 162.0256 \end{aligned}$$

with 2 degrees of freedom. The F statistic is given in (16.31) as

$$F = \frac{\text{SS}(\alpha|\mu, \beta)/(k-1)}{\text{SSE}_{y\cdot x}/[k(n-1)-1]} = \frac{162.0256/2}{979.4453/17} = 1.4061.$$

The p value is .272, and we do not reject $H_0\colon \alpha_1 = \alpha_2 = \alpha_3$.

To test $H_0\colon \beta = 0$, we use (16.36):

$$\begin{aligned} F &= \frac{e_{xy}^2/e_{xx}}{\text{SSE}_{y\cdot x}/[k(n-1)-1]} = \frac{(412.7143)^2/350.2857}{979.4453/17} \\ &= 8.4401. \end{aligned}$$

The p-value is .0099, and we reject $H_0 : \beta = 0$.

To test the hypothesis of equal slopes in the groups, $H_0\colon \beta_1 = \beta_2 = \beta_3$, we first estimate β_1, β_2, and β_3 using (16.45):

$$\hat{\beta}_1 = .7903, \quad \hat{\beta}_2 = 1.9851, \quad \hat{\beta}_3 = .8579.$$

Then by (16.46) and (16.47),

$$\text{SSE}(F)_{y\cdot x} = 880.5896, \quad \text{SSE}(R)_{y\cdot x} = 979.4453.$$

The difference $\text{SSE}(R)_{y\cdot x} - \text{SSE}(F)_{y\cdot x}$ is used in the numerator of the F statistic in (16.49):

$$F = \frac{(979.4453 - 880.5896)/2}{880.5896/(3)(5)} = .8420.$$

The p value is .450, and we do not reject $H_0\colon \beta_1 = \beta_2 = \beta_3$. □

16.4 TWO-WAY MODEL WITH ONE COVARIATE

In this section, we discuss the two-way (balanced) fixed-effects model with one covariate. The model was introduced in (16.6) as

$$\begin{aligned} y_{ijk} &= \mu + \alpha_i + \gamma_j + \delta_{ij} + \beta x_{ijk} + \varepsilon_{ijk}, \qquad (16.50) \\ i &= 1, 2, \ldots, a, \; j = 1, 2, \ldots, c, \; k = 1, 2, \ldots, n, \end{aligned}$$

where α_i is the effect of factor A, γ_j is the effect of factor C, δ_{ij} is the AC interaction effect, and x_{ijk} is a covariate measured on the same experimental unit as y_{ijk}.

16.4.1 Tests for Main Effects and Interactions

In order to find $\text{SSE}_{y \cdot x}$, we consider the hypothesis of no overall treatment effect, that is, no A effect, no C effect, and no interaction (see a comment preceding Theorem 14.4b). By analogy to (16.28), the reduced model is

$$y_{ijk} = \mu^* + \beta x_{ijk} + \varepsilon_{ijk}. \tag{16.51}$$

By analogy to (16.29), SSE for the reduced model is given by

$$\begin{aligned} \text{SSE}_{\text{rd}} &= \sum_{i=1}^{a} \sum_{j=1}^{c} \sum_{k=1}^{n} (y_{ijk} - \bar{y}_{...})^2 - \frac{\left[\sum_{ijk} (x_{ijk} - \bar{x}_{...})(y_{ijk} - \bar{y}_{...})\right]^2}{\sum_{ijk} (x_{ijk} - \bar{x}_{...})^2} \\ &= \sum_{ijk} y_{ijk}^2 - \frac{y_{...}^2}{acn} - \frac{\left[\sum_{ijk} (x_{ijk} - \bar{x}_{...})(y_{ijk} - \bar{y}_{...})\right]^2}{\sum_{ijk} (x_{ijk} - \bar{x}_{...})^2}. \end{aligned} \tag{16.52}$$

By analogy to (16.27), SSE for the full model in (16.50) is

$$\begin{aligned} \text{SSE}_{y \cdot x} &= \sum_{ijk} (y_{ijk} - \bar{y}_{ij.})^2 - \frac{\left[\sum_{ijk} (x_{ijk} - \bar{x}_{ij.})(y_{ijk} - \bar{y}_{ij.})\right]^2}{\sum_{ijk} (x_{ijk} - \bar{x}_{ij.})^2} \\ &= \sum_{ijk} y_{ijk}^2 - \sum_{ij} \frac{y_{ij.}^2}{n} - \frac{\left[\sum_{ijk} (x_{ijk} - \bar{x}_{ij.})(y_{ijk} - \bar{y}_{ij.})\right]^2}{\sum_{ijk} (x_{ijk} - \bar{x}_{ij.})^2}, \end{aligned} \tag{16.53}$$

which has $ac(n-1) - 1$ degrees of freedom. Note that the degrees of freedom for $\text{SSE}_{y \cdot x}$ have been reduced by 1 for the covariate adjustment.

Now by analogy to (16.30), the overall sum of squares for treatments is

$$\begin{aligned} \text{SS}(\alpha, \gamma, \delta | \mu, \beta) &= \text{SSE}_{\text{rd}} - \text{SSE}_{y \cdot x} \\ &= \sum_{ij} \frac{y_{ij.}^2}{n} - \frac{y_{...}^2}{acn} + \frac{\left[\sum_{ijk} (x_{ijk} - \bar{x}_{ij.})(y_{ijk} - \bar{y}_{ij.})\right]^2}{\sum_{ijk} (x_{ijk} - \bar{x}_{ij.})^2} \\ &\quad - \frac{\left[\sum_{ijk} (x_{ijk} - \bar{x}_{...})(y_{ijk} - \bar{y}_{...})\right]^2}{\sum_{ijk} (x_{ijk} - \bar{x}_{...})^2}, \end{aligned} \tag{16.54}$$

which has $ac - 1$ degrees of freedom.

Using (14.47), (14.69), and (14.70), we can partition the term $\sum_{ij} y_{ij.}^2/n - y_{...}^2/acn$ in (16.54), representing overall treatment sum of squares, as in (14.40):

$$\begin{aligned}\sum_{ij} \frac{y_{ij.}^2}{n} - \frac{y_{...}^2}{acn} &= cn \sum_i (\bar{y}_{i..} - \bar{y}_{...})^2 + an \sum_j (\bar{y}_{.j.} - \bar{y}_{...})^2 \\ &\quad + n \sum_{ij} (\bar{y}_{ij.} - \bar{y}_{i..} - \bar{y}_{.j.} + \bar{y}_{...})^2 \\ &= \text{SSA}_y + \text{SSC}_y + \text{SSAC}_y. \end{aligned} \tag{16.55}$$

To conform with this notation, we define

$$\text{SSE}_y = \sum_{ijk} (y_{ijk} - \bar{y}_{ij.})^2.$$

We have an analogous partitioning of the overall treatment sum of squares for x:

$$\sum_{ij} \frac{x_{ij.}^2}{n} - \frac{x_{...}^2}{acn} = \text{SSA}_x + \text{SSC}_x + \text{SSAC}_x, \tag{16.56}$$

where, for example

$$\text{SSA}_x = cn \sum_{i=1}^{a} (\bar{x}_{i..} - \bar{x}_{...})^2.$$

We also define

$$\text{SSE}_x = \sum_{ijk} (x_{ijk} - \bar{x}_{ij.})^2.$$

The "overall treatment sum of products" $\sum_{ij} x_{ij.}y_{ij.}/n - x_{...}y_{...}/acn$ can be partitioned in a manner analogous to that in (16.55) and (16.56) (see Problem 16.8):

$$\begin{aligned}\sum_{ij} \frac{x_{ij.}y_{ij.}}{n} - \frac{x_{...}y_{...}}{acn} &= cn \sum_i (\bar{x}_{i..} - \bar{x}_{...})(\bar{y}_{i..} - \bar{y}_{...}) + an \sum_j (\bar{x}_{.j.} - \bar{x}_{...})(\bar{y}_{.j.} - \bar{y}_{...}) \\ &\quad + n \sum_{ij} (\bar{x}_{ij.} - \bar{x}_{i..} - \bar{x}_{.j.} + \bar{x}_{...})(\bar{y}_{ij.} - \bar{y}_{i..} - \bar{y}_{.j.} + \bar{y}_{...}) \\ &= \text{SPA} + \text{SPC} + \text{SPAC}. \end{aligned} \tag{16.57}$$

We also define

$$\text{SPE} = \sum_{ijk} (x_{ijk} - \bar{x}_{ij.})(y_{ijk} - \bar{y}_{ij.}).$$

We can now write $\text{SSE}_{y \cdot x}$ in (16.53) in the simplified form

$$\text{SSE}_{y \cdot x} = \text{SSE}_y - \frac{(\text{SPE})^2}{\text{SSE}_x}.$$

We display these sums of squares and products in Table 16.3.

We now proceed to develop hypothesis tests for factor A, factor C, and the interaction AC. The orthogonality of the balanced design is lost when adjustments are made for the covariate [see comments following (16.34); see also Bingham and Feinberg (1982)]. We therefore obtain a "total" for each term (A, C, or AC) by adding the error SS or SP to the term SS or SP for each of x, y and xy (see the entries for $A + E$, $C + E$, and $AC + E$ in Table 16.3). These totals are analogous to $\text{SST}_{y \cdot x} = \text{SS}(\alpha | \mu, \beta) + \text{SSE}_{y \cdot x}$ in (16.32) for the one-way model. The totals are used to obtain sums of squares adjusted for the covariate in a manner analogous to that employed in the one-way model [see (16.30) or the "treatments" line in Table 16.1]. For example, the adjusted sum of squares $\text{SSA}_{y \cdot x}$ for factor A is obtained as follows:

$$\text{SS}(A + E)_{y \cdot x} = \text{SSA}_y + \text{SSE}_y - \frac{(\text{SPA} + \text{SPE})^2}{\text{SSA}_x + \text{SSE}_x}, \tag{16.58}$$

$$\text{SSE}_{y \cdot x} = \text{SSE}_y - \frac{(\text{SPE})^2}{\text{SSE}_x}, \tag{16.59}$$

$$\text{SSA}_{y \cdot x} = \text{SS}(A + E)_{y \cdot x} - \text{SSE}_{y \cdot x}. \tag{16.60}$$

From inspection of (16.58), (16.59), and (16.60), we see that $\text{SSA}_{y \cdot x}$ has $a - 1$ degrees of freedom. The statistic for testing $H_{01}: \alpha_1 = \alpha_2 = \cdots = \alpha_a$, corresponding to the

TABLE 16.3 Sums of Squares and Products for x and y in a Two-Way Model

	SS and SP Corrected for the Mean		
Source	y	x	xy
A	SSA_y	SSA_x	SPA
C	SSC_y	SSC_x	SPC
AC	SSAC_y	SSAC_x	SPAC
Error	SSE_y	SSE_x	SPE
$A + E$	$\text{SSA}_y + \text{SSE}_y$	$\text{SSA}_x + \text{SSE}_x$	SPA + SPE
$C + E$	$\text{SSC}_y + \text{SSE}_y$	$\text{SSC}_x + \text{SSE}_x$	SPC + SPE
$AC + E$	$\text{SSAC}_y + \text{SSE}_y$	$\text{SSAC}_x + \text{SSE}_x$	SPAC + SPE

TABLE 16.4 Value of Crops y and Size x of Farms in Three Iowa Counties

	County					
	1		2		3	
Landlord–Tenant	y	x	y	x	y	x
Related	6399	160	2490	90	4489	120
	8456	320	5349	154	10026	245
	8453	200	5518	160	5659	160
	4891	160	10417	234	5475	160
	3491	120	4278	120	11382	320
Not related	6944	160	4936	160	5731	160
	6971	160	7376	200	6787	173
	4053	120	6216	160	5814	134
	8767	280	10313	240	9607	239
	6765	160	5124	120	9817	320

Source: Ostle and Mensing (1975, p. 480).

main effect of A, is then given by

$$F = \frac{\text{SSA}_{y\cdot x}/(a-1)}{\text{SSE}_{y\cdot x}/[ac(n-1)-1]}, \tag{16.61}$$

which is distributed as $F[a-1, ac(n-1)-1]$ if H_{01} is true. Tests for factor C and the interaction AC are developed in an analogous fashion.

Example 16.4a. In each of three counties in Iowa, a sample of farms was taken from farms for which landlord and tenant are related and also from farms for which landlord and tenant are not related. Table 16.4 gives the data for $y =$ value of crops produced and $x =$ size of farm.

We first obtain the sums of squares and products listed in Table 16.3, where factor A is relationship status and factor C is county. These are given in Table 16.5, where,

TABLE 16.5 Sums of Squares and Products for x and y

	SS and SP Corrected for the Mean		
Source	y	x	xy
A	2,378,956.8	132.30	17,740.8
C	8,841,441.3	7724.47	249,752.8
AC	1,497,572.6	2040.20	41,440.3
Error	138,805,865	106,870	3,427,608.6
$A+E$	141,184,822	107,002.3	3,445,349.4
$C+E$	147,647,306	114,594.5	3,677,361.4
$AC+E$	140,303,437	108,910.2	3,469,048.9

for example, $\text{SSA}_y = 2378956.8$, $\text{SSA}_y + \text{SSE}_y = 141{,}184{,}822$, and $\text{SPAC} + \text{SPE} = 3{,}469{,}048.9$.

By (16.58), (16.59), and (16.60), we have

$$\text{SS}(A+E)_{y\cdot x} = 30{,}248{,}585, \quad \text{SSE}_{y\cdot x} = 28{,}873{,}230,$$
$$\text{SSA}_{y\cdot x} = 1{,}375{,}355.1.$$

Then by (16.61), we have

$$F = \frac{\text{SSA}_{y\cdot x}/(a-1)}{\text{SSE}_{y\cdot x}/[ac(n-1)-1]}$$
$$= \frac{1{,}375{,}355.1/1}{28{,}873{,}230/23} = \frac{1{,}375{,}355.1}{1{,}255{,}357.8} = 1.0956.$$

The p value is .306, and we do not reject $H_0\colon \alpha_1 = \alpha_2$.

Similarly, for factor C, we have

$$F = \frac{766{,}750.1/2}{1{,}255{,}357.8} = .3054$$

with $p = .740$. For the interaction AC, we obtain

$$F = \frac{932{,}749.5/2}{1{,}255{,}357.8} = .3715$$

with $p = .694$. □

16.4.2 Test for Slope

To test the hypothesis $H_{02}\colon \beta = 0$, the sum of squares due to β is $(\text{SPE})^2/\text{SSE}_x$, and the F statistic is given by

$$F = \frac{(\text{SPE})^2/\text{SSE}_x}{\text{SSE}_{y\cdot x}/[ac(n-1)-1]}, \tag{16.62}$$

which (under H_{02} and also H_{03} below) is distributed as $F[1, ac(n-1)-1]$.

Eample 16.4b. To test $H_0\colon \beta = 0$ for the farms data in Table 16.4, we use SPE and SSE_x from Table 16.5 and $\text{SSE}_{y\cdot x}$ in Example 16.4a. Then by (16.62), we obtain

$$F = \frac{(\text{SPE})^2/\text{SSE}_x}{\text{SSE}_{y\cdot x}/[ac(n-1)-1]}$$
$$= \frac{(3{,}427{,}608.6)^2/106{,}870}{1{,}255{,}357.8} = 87.5708.$$

The p value is 2.63×10^{-9}, and $H_0\colon \beta = 0$ is rejected. □

16.4.3 Test for Homogeneity of Slopes

The test for homogeneity of slopes can be carried out separately for factor A, factor C, and the interaction AC. We describe the test for homogeneity of slopes among the levels of A. The hypothesis is

$$H_{03}: \beta_1 = \beta_2 = \cdots = \beta_a;$$

that is, the regression lines for the a levels of A are parallel. The intercepts, of course, may be different. To obtain a slope estimator $\hat{\beta}_i$ for the ith level of A, we define SSE_x and SPE for the ith level of A:

$$\text{SSE}_{x,i} = \sum_{j=1}^{c} \sum_{k=1}^{n} (x_{ijk} - \bar{x}_{ij.})^2, \quad \text{SPE}_i = \sum_{jk} (x_{ijk} - \bar{x}_{ij.})(y_{ijk} - \bar{y}_{ij.}). \tag{16.63}$$

Then $\hat{\beta}_i$ is obtained as

$$\hat{\beta}_i = \frac{\text{SPE}_i}{\text{SSE}_{x,i}},$$

and the sum of squares due to β_i is $(\text{SPE}_i)^2/\text{SSE}_{x,i}$.

By analogy to (16.46), the sum of squares for the full model in which the β_i's are different is given by

$$\text{SS}(F) = \text{SSE}_y - \sum_{i=1}^{a} \frac{(\text{SPE}_i)^2}{\text{SSE}_{x,i}},$$

and by analogy to (16.47), the sum of squares in the reduced model with a common slope is

$$\text{SS}(R) = \text{SSE}_y - \frac{(\text{SPE})^2}{\text{SSE}_x}.$$

Our test statistic for $H_{03}: \beta_1 = \beta_2 = \cdots = \beta_a$ is then similar to (16.49):

$$\begin{aligned} F &= \frac{[\text{SS}(R) - \text{SS}(F)]/(a-1)}{\text{SS}(F)/[ac(n-1)-1]} \\ &= \frac{\left[\sum_{i=1}^{a} (\text{SPE}_i)^2/\text{SSE}_{x,i} - (\text{SPE})^2/\text{SSE}_x\right]/(a-1)}{[\text{SSE}_y - \sum_{i=1}^{a} (\text{SPE}_i)^2/\text{SSE}_{x,i}]/[ac(n-1)-a]}, \end{aligned} \tag{16.64}$$

which (under H_{03}) is distributed as $F[a-1, ac(n-1)-a]$. The tests for homogeneity of slopes for C and AC are constructed in a similar fashion.

Example 16.4c. To test homogeneity of slopes for factor A, we first find $\hat{\beta}_1$ and $\hat{\beta}_2$ for the two levels of A:

$$\hat{\beta}_1 = \frac{\text{SPE}_1}{\text{SSE}_{x,1}} = \frac{2{,}141{,}839.8}{61{,}359.2} = 34.9066,$$

$$\hat{\beta}_2 = \frac{\text{SPE}_2}{\text{SSE}_{x,2}} = \frac{1{,}285{,}768.8}{45{,}510.8} = 28.2519.$$

Then

$$\text{SS}(F) = \text{SSE}_y - \sum_{i=1}^{2} \frac{(\text{SPE}_i)^2}{\text{SSE}_{x,i}} = 27{,}716{,}088.7,$$

$$\text{SS}(R) = \text{SSE}_y - \frac{(\text{SPE})^2}{\text{SSE}_x} = 28{,}873{,}230.$$

The difference is $\text{SS}(R) - \text{SS}(F) = 1{,}157{,}140.94$. Then by (16.64), we obtain

$$F = \frac{1{,}157{,}140.94/1}{27{,}716{,}088.7/22} = .9185.$$

The p value is .348, and we do not reject $H_0\colon \beta_1 = \beta_2$.

For homogeneity of slopes for factor C, we have

$$\hat{\beta}_1 = 23.2104, \quad \hat{\beta}_2 = 50.0851, \quad \hat{\beta}_3 = 31.6693,$$

$$F = \frac{9{,}506{,}034.16/2}{19{,}367{,}195.5/21} = 5.1537$$

with $p = .0151$. □

16.5 ONE-WAY MODEL WITH MULTIPLE COVARIATES

16.5.1 The Model

In some cases, the researcher has more than one covariate available. Note, however, that each covariate decreases the error degrees of freedom by 1, and therefore the inclusion of too many covariates may lead to loss of power.

For the one-way model with q covariates, we use (16.4):

$$\begin{aligned} y_{ij} &= \mu + \alpha_i + \beta_1 x_{ij1} + \beta_2 x_{ij2} + \cdots + \beta_q x_{ijq} + \varepsilon_{ij} \\ &= \mu + \alpha_i + \boldsymbol{\beta}'\mathbf{x}_{ij} + \varepsilon_{ij}, \qquad (16.65) \\ & i = 1, 2, \ldots, k, \ \ j = 1, 2, \ldots, n, \end{aligned}$$

where $\boldsymbol{\beta}' = (\beta_1, \beta_2, \ldots, \beta_q)$ and $\mathbf{x}_{ij} = (x_{ij1}, x_{ij2}, \ldots, x_{ijq})'$. For this model, we wish to test $H_{01}: \alpha_1 = \alpha_2 = \cdots = \alpha_k$ and $H_{02}: \boldsymbol{\beta} = \mathbf{0}$. We will also extend the model to allow for a different $\boldsymbol{\beta}$ vector in each of the k groups and test equality of these $\boldsymbol{\beta}$ vectors.

The model in (16.65) can be written in matrix notation as

$$\mathbf{y} = \mathbf{Z}\boldsymbol{\alpha} + \mathbf{X}\boldsymbol{\beta} + \boldsymbol{\varepsilon},$$

where $\mathbf{Z}$ and $\boldsymbol{\alpha}$ are given following (16.3) and $\mathbf{X}\boldsymbol{\beta}$ is as given by (16.5):

$$\mathbf{X}\boldsymbol{\beta} = \begin{pmatrix} x_{111} & x_{112} & \cdots & x_{11q} \\ x_{121} & x_{122} & \cdots & x_{12q} \\ \vdots & \vdots & & \vdots \\ x_{kn1} & x_{kn2} & \cdots & x_{knq} \end{pmatrix} \begin{pmatrix} \beta_1 \\ \beta_2 \\ \vdots \\ \beta_q \end{pmatrix}.$$

The vector $\mathbf{y}$ is $kn \times 1$ and the matrix $\mathbf{X}$ is $kn \times q$. We can write $\mathbf{y}$ and $\mathbf{X}\boldsymbol{\beta}$ in partitioned form corresponding to the k groups:

$$\mathbf{y} = \begin{pmatrix} \mathbf{y}_1 \\ \mathbf{y}_2 \\ \vdots \\ \mathbf{y}_k \end{pmatrix}, \quad \mathbf{X}\boldsymbol{\beta} = \begin{pmatrix} \mathbf{X}_1 \\ \mathbf{X}_2 \\ \vdots \\ \mathbf{X}_k \end{pmatrix} \boldsymbol{\beta}, \tag{16.66}$$

where

$$\mathbf{y}_i = \begin{pmatrix} y_{i1} \\ y_{i2} \\ \vdots \\ y_{in} \end{pmatrix} \quad \text{and} \quad \mathbf{X}_i = \begin{pmatrix} x_{i11} & x_{i12} & \cdots & x_{i1q} \\ x_{i21} & x_{i22} & \cdots & x_{i2q} \\ \vdots & \vdots & & \vdots \\ x_{in1} & x_{in2} & \cdots & x_{inq} \end{pmatrix}.$$

16.5.2 Estimation

We first obtain $\mathbf{E}_{xx}$, e_{xy}, and e_{yy} for use in $\hat{\boldsymbol{\beta}}$ and $\text{SSE}_{y \cdot x}$. By (16.16), $\mathbf{E}_{xx}$ can be expressed as

$$\mathbf{E}_{xx} = \mathbf{X}'(\mathbf{I} - \mathbf{P})\mathbf{X}.$$

Using $\mathbf{X}$ partitioned as in (16.66) and $\mathbf{I} - \mathbf{P}$ in the form given in (16.40), $\mathbf{E}_{xx}$ becomes

$$\mathbf{E}_{xx} = \sum_{i=1}^{k} \mathbf{X}_i' \left(\mathbf{I} - \frac{1}{n}\mathbf{J} \right) \mathbf{X}_i \tag{16.67}$$

(see Problem 16.10). Similarly, using $\mathbf{y}$ partitioned as in (16.66), $\mathbf{e}_{xy}$ is given by (16.16) as

$$\mathbf{e}_{xy} = \mathbf{X}'(\mathbf{I} - \mathbf{P})\mathbf{y} = \sum_{i=1}^{k} \mathbf{X}_i'\left(\mathbf{I} - \frac{1}{n}\mathbf{J}\right)\mathbf{y}_i. \tag{16.68}$$

By (16.19) and (16.40), we have

$$e_{yy} = \mathbf{y}'(\mathbf{I} - \mathbf{P})\mathbf{y} = \sum_{i=1}^{k} \mathbf{y}_i'\left(\mathbf{I} - \frac{1}{n}\mathbf{J}\right)\mathbf{y}_i. \tag{16.69}$$

The elements of $\mathbf{E}_{xx}$, $\mathbf{e}_{xy}$, and e_{yy} are extensions of the sums of squares and products found in the three expressions in (16.25).

To examine the elements of the matrix $\mathbf{E}_{xx}$, we first note that $\mathbf{I} - (1/n)\mathbf{J}$ is symmetric and idempotent and therefore $\mathbf{X}_i'[\mathbf{I} - (1/n)\mathbf{J})]\mathbf{X}_i$ in (16.67) can be written as

$$\begin{aligned} \mathbf{X}_i'(\mathbf{I} - (1/n)\mathbf{J})\mathbf{X}_i &= \mathbf{X}_i'(\mathbf{I} - (1/n)\mathbf{J})'(\mathbf{I} - (1/n)\mathbf{J})\mathbf{X}_i \\ &= \mathbf{X}_{ci}'\mathbf{X}_{ci}, \end{aligned} \tag{16.70}$$

where $\mathbf{X}_{ci} = [\mathbf{I} - (1/n)\mathbf{J}]\mathbf{X}_i$ is the centered matrix

$$\mathbf{X}_{ci} = \begin{pmatrix} x_{i11} - \bar{x}_{i.1} & x_{i12} - \bar{x}_{i.2} & \cdots & x_{i1q} - \bar{x}_{i.q} \\ x_{i21} - \bar{x}_{i.1} & x_{i22} - \bar{x}_{i.2} & \cdots & x_{i2q} - \bar{x}_{i.q} \\ \vdots & \vdots & & \vdots \\ x_{in1} - \bar{x}_{i.1} & x_{in2} - \bar{x}_{i.2} & \cdots & x_{inq} - \bar{x}_{i.q} \end{pmatrix} \tag{16.71}$$

[see (7.33) and Problem 7.15], where $\bar{x}_{i.2}$, for example, is the mean of the second column of $\mathbf{X}_i$, that is, $\bar{x}_{i.2} = \sum_{j=1}^{n} x_{ij2}/n$. By Theorem 2.2c(i), the diagonal elements of $\mathbf{X}_{ci}'\mathbf{X}_{ci}$ are

$$\sum_{j=1}^{n} (x_{ijr} - \bar{x}_{i.r})^2, \quad r = 1, 2, \ldots, q, \tag{16.72}$$

and the off-diagonal elements are

$$\sum_{j=1}^{n} (x_{ijr} - \bar{x}_{i.r})(x_{ijs} - \bar{x}_{i.s}), \quad r \neq s. \tag{16.73}$$

By (16.67) and (16.72), the diagonal elements of $\mathbf{E}_{xx}$ are

$$\sum_{i=1}^{k}\sum_{j=1}^{n} (x_{ijr} - \bar{x}_{i.r})^2, \quad r = 1, 2, \ldots, q, \tag{16.74}$$

and by (16.67) and (16.73), the off-diagonal elements are

$$\sum_{i=1}^{k}\sum_{j=1}^{n}(x_{ijr}-\bar{x}_{i.r})(x_{ijs}-\bar{x}_{i.s}), \quad r \neq s. \tag{16.75}$$

These are analogous to $e_{xx} = \sum_{ij}(x_{ij}-\bar{x}_{i.})^2$ in (16.25).

To examine the elements of the vector $\mathbf{e}_{xy}$, we note that by an argument similar to that used to obtain (16.70), $\mathbf{X}_i'[\mathbf{I}-(1/n)\mathbf{J}]\mathbf{y}_i$ in (16.68) can be written as

$$\mathbf{X}_i'[\mathbf{I}-(1/n)\mathbf{J})\mathbf{y}_i = \mathbf{X}_i'[\mathbf{I}-(1/n)\mathbf{J}]'[\mathbf{I}-(1/n)\mathbf{J}]\mathbf{y}_i = \mathbf{X}_{ci}'\mathbf{y}_{ci},$$

where $\mathbf{X}_{ci}$ is as given in (16.71) and

$$\mathbf{y}_{ci} = \begin{pmatrix} y_{i1}-\bar{y}_{i.} \\ y_{i2}-\bar{y}_{i.} \\ \vdots \\ y_{in}-\bar{y}_{i.} \end{pmatrix}$$

with $\bar{y}_{i.} = \sum_{j=1}^{n} y_{ij}/n$. Thus the elements of $\mathbf{X}_{ci}'\mathbf{y}_{ci}$ are of the form

$$\sum_{j=1}^{n}(x_{ijr}-\bar{x}_{i.r})(y_{ij}-\bar{y}_{i.}) \quad r = 1, 2, \ldots, q,$$

and by (16.68), the elements of $\mathbf{e}_{xy}$ are

$$\sum_{i=1}^{k}\sum_{j=1}^{n}(x_{ijr}-\bar{x}_{i.r})(y_{ij}-\bar{y}_{i.}) \quad r = 1, 2, \ldots, q.$$

Similarly, e_{yy} in (16.69) can be written as

$$\begin{aligned} e_{yy} &= \sum_{i=1}^{k}\mathbf{y}_i'\left(\mathbf{I}-\frac{1}{n}\mathbf{J}\right)'\left(\mathbf{I}-\frac{1}{n}\mathbf{J}\right)\mathbf{y}_i = \sum_{i=1}^{k}\mathbf{y}_{ci}'\mathbf{y}_{ci} \\ &= \sum_{i=1}^{k}\sum_{j=1}^{n}(y_{ij}-\bar{y}_{i.})^2. \end{aligned} \tag{16.76}$$

By (16.15), we obtain

$$\hat{\boldsymbol{\beta}} = \mathbf{E}_{xx}^{-1}\mathbf{e}_{xy},$$

where $\mathbf{E}_{xx}$ is as given by (16.67) and $\mathbf{e}_{xy}$ is as given by (16.68). Likewise, by (16.18), we have

$$\text{SSE}_{y\cdot x} = e_{yy} - \mathbf{e}'_{xy}\mathbf{E}_{xx}^{-1}\mathbf{e}_{xy}, \tag{16.77}$$

where e_{yy} is as given in (16.69) or (16.76). The degrees of freedom of $\text{SSE}_{y\cdot x}$ are $k(n-1)-q$.

By (16.11) and (13.12), we obtain

$$\begin{aligned}\hat{\boldsymbol{\alpha}} &= \hat{\boldsymbol{\alpha}}_0 - (\mathbf{Z}'\mathbf{Z})^{-}\mathbf{Z}'\mathbf{X}\hat{\boldsymbol{\beta}} \\ &= \begin{pmatrix} 0 \\ \bar{y}_{1.} \\ \bar{y}_{2.} \\ \vdots \\ \bar{y}_{k.} \end{pmatrix} - \begin{pmatrix} 0 \\ \hat{\boldsymbol{\beta}}'\bar{\mathbf{x}}_{1.} \\ \hat{\boldsymbol{\beta}}'\bar{\mathbf{x}}_{2.} \\ \vdots \\ \hat{\boldsymbol{\beta}}'\bar{\mathbf{x}}_{k.} \end{pmatrix} = \begin{pmatrix} 0 \\ \bar{y}_{1.} - \hat{\boldsymbol{\beta}}'\bar{\mathbf{x}}_{1.} \\ \bar{y}_{2.} - \hat{\boldsymbol{\beta}}'\bar{\mathbf{x}}_{2.} \\ \vdots \\ \bar{y}_{k.} - \hat{\boldsymbol{\beta}}'\bar{\mathbf{x}}_{k.} \end{pmatrix}\end{aligned} \tag{16.78}$$

$$= \begin{pmatrix} \bar{y}_{1.} - (\hat{\beta}_1\bar{x}_{1.1} + \hat{\beta}_2\bar{x}_{1.2} + \cdots + \hat{\beta}_q\bar{x}_{1.q}) \\ \bar{y}_{2.} - (\hat{\beta}_1\bar{x}_{2.1} + \hat{\beta}_2\bar{x}_{2.2} + \cdots + \hat{\beta}_q\bar{x}_{2.q}) \\ \vdots \\ \bar{y}_{k.} - (\hat{\beta}_1\bar{x}_{k.1} + \hat{\beta}_2\bar{x}_{k.2} + \cdots + \hat{\beta}_q\bar{x}_{k.q}) \end{pmatrix}. \tag{16.79}$$

16.5.3 Testing Hypotheses

16.5.3.1 Treatments

To test

$$H_{01}\colon \alpha_1 = \alpha_2 = \cdots = \alpha_k$$

adjusted for the q covariates, we use the full–reduced-model approach as in Section 16.3.3.1. The full model is given by (16.65), and the reduced model (with $\alpha_1 = \alpha_2 = \cdots = \alpha_k = \alpha$) is

$$\begin{aligned} y_{ij} &= \mu + \alpha + \boldsymbol{\beta}'\mathbf{x}_{ij} + \varepsilon_{ij} \\ &= \mu^* + \boldsymbol{\beta}'\mathbf{x}_{ij} + \varepsilon_{ij}, \end{aligned} \tag{16.80}$$

which is essentially the same as the multiple regression model (7.3). By (7.37) and (7.39) and by analogy with (16.33),

$$\text{SSE}_{\text{rd}} = \text{SST}_{y\cdot x} = t_{yy} - t'_{xy}\mathbf{T}_{xx}^{-1}\mathbf{t}_{xy}, \tag{16.81}$$

where t_{yy} is

$$t_{yy} = \sum_{ij} (y_{ij} - \bar{y}_{..})^2,$$

the elements of $\mathbf{t}_{xy}$ are

$$\sum_{ij} (x_{ijr} - \bar{x}_{..r})(y_{ij} - \bar{y}_{..}), \quad r = 1, 2, \ldots, q,$$

and the elements of $\mathbf{T}_{xx}$ are

$$\sum_{ij} (x_{ijr} - \bar{x}_{..r})(x_{ijs} - \bar{x}_{..s}), \quad r = 1, 2, \ldots, q, \quad s = 1, 2, \ldots, q.$$

Thus, by analogy with (16.30), we use (16.81) and (16.77) to obtain

$$\begin{aligned} \mathrm{SS}(\alpha|\mu, \beta) &= \mathrm{SST}_{y\cdot x} - \mathrm{SSE}_{y\cdot x} \\ &= t_{yy} - \mathbf{t}'_{xy}\mathbf{T}_{xx}^{-1}\mathbf{t}_{xy} - e_{yy} + \mathbf{e}'_{xy}\mathbf{E}_{xx}^{-1}\mathbf{e}_{xy} \\ &= \sum_{ij} (y_{ij} - \bar{y}_{..})^2 - \sum_{ij} (y_{ij} - \bar{y}_{i.})^2 - \mathbf{t}'_{xy}\mathbf{T}_{xx}^{-1}\mathbf{t}_{xy} + \mathbf{e}'_{xy}\mathbf{E}_{xx}^{-1}\mathbf{e}_{xy} \\ &= n\sum_{i} (\bar{y}_{i.} - \bar{y}_{..})^2 - \mathbf{t}'_{xy}\mathbf{T}_{xx}^{-1}\mathbf{t}_{xy} + \mathbf{e}'_{xy}\mathbf{E}_{xx}^{-1}\mathbf{e}_{xy}, \end{aligned} \qquad (16.82)$$

which has $k-1$ degrees of freedom (see Problem 16.13). We display these sums of squares and products in Table 16.6.

The test statistic for $H_{01}\colon \alpha_1 = \alpha_2 = \cdots = \alpha_k$ is

$$F = \frac{\mathrm{SS}(\alpha|\mu, \beta)/(k-1)}{\mathrm{SSE}_{y\cdot x}/[k(n-1) - q]}, \qquad (16.83)$$

which (under H_{01}) is distributed as $F[k-1, k(n-1) - q]$.

TABLE 16.6 Analysis-of-Covariance Table for Testing $H_{01}\colon \alpha_1 = \alpha_2 = \cdots = \alpha_k$ in the One-Way Model with q Covariates

Source	SS Adjusted for the Covariate	Adjusted df
Treatments	$\mathrm{SS}(\alpha\|\mu, \beta) = \mathrm{SST}_{y\cdot x} - \mathrm{SSE}_{y\cdot x}$	$k-1$
Error	$\mathrm{SSE}_{y\cdot x} = e_{yy} - \mathbf{e}'_{xy}\mathbf{E}_{xx}^{-1}\mathbf{e}_{xy}$	$k(n-1) - q$
Total	$\mathrm{SST}_{y\cdot x} = t_{yy} - \mathbf{t}'_{xy}\mathbf{T}_{xx}^{-1}\mathbf{t}_{xy}$	$kn - q - 1$

16.5.3.2 *Slope Vector*

To test

$$H_{02}: \boldsymbol{\beta} = \mathbf{0},$$

the sum of squares is given by (16.22) as

$$\text{SSH} = \mathbf{e}'_{xy}\mathbf{E}_{xx}^{-1}\mathbf{e}_{xy},$$

where $\mathbf{E}_{xx}$ is as given by (16.67) and $\mathbf{e}_{xy}$ is the same as in (16.68). The F statistic is then

$$F = \frac{\mathbf{e}'_{xy}\mathbf{E}_{xx}^{-1}\mathbf{e}_{xy}/q}{\text{SSE}_{y\cdot x}/[k(n-1)-q]}, \tag{16.84}$$

which is distributed as $F[q, k(n-1)-q]$ if $H_{02}: \boldsymbol{\beta} = \mathbf{0}$ is true.

16.5.3.3 *Homogeneity of Slope Vectors*

The tests of $H_{01}: \alpha_1 = \alpha_2 = \cdots = \alpha_k$ and $H_{02}: \boldsymbol{\beta} = \mathbf{0}$ assume a common coefficient vector $\boldsymbol{\beta}$ for all k groups. To check this assumption, we can extend the model (16.65) to obtain a full model allowing for different slope vectors:

$$y_{ij} = \mu + \alpha_i + \boldsymbol{\beta}'_i\mathbf{x}_{ij} + \varepsilon_{ij}, \quad i = 1, 2, \ldots, k, \quad j = 1, 2, \ldots, n. \tag{16.85}$$

The reduced model with a single slope vector is given by (16.65). We now develop a test for the hypothesis

$$H_{03}: \boldsymbol{\beta}_1 = \boldsymbol{\beta}_2 = \cdots = \boldsymbol{\beta}_k,$$

that is, that the k regression planes (for the k treatments) are parallel.

By extension of (16.46) and (16.47), we have

$$\text{SSE}(F)_{y\cdot x} = e_{yy} - \sum_{i=1}^{k} \mathbf{e}'_{xy,i}\mathbf{E}_{xx,i}^{-1}\mathbf{e}_{xy,i}, \tag{16.86}$$

$$\text{SSE}(R)_{y\cdot x} = e_{yy} - \mathbf{e}'_{xy}\mathbf{E}_{xx}^{-1}\mathbf{e}_{xy}, \tag{16.87}$$

where

$$\mathbf{E}_{xx,i} = \mathbf{X}'_i[\mathbf{I} - (1/n)\mathbf{J}]\mathbf{X}_i \quad \text{and} \quad \mathbf{e}_{xy,i} = \mathbf{X}'_i[\mathbf{I} - (1/n)\mathbf{J}]\mathbf{y}_i$$

are terms in the summations in (16.67) and (16.68). The degrees of freedom for $\text{SSE}(F)_{y\cdot x}$ and $\text{SSE}(R)_{y\cdot x}$ are $k(n-1) - kq = k(n-q-1)$ and $k(n-1) - q$, respectively. Note that $\text{SSE}(R)_{y\cdot x}$ in (16.87) is the same as $\text{SSE}_{y\cdot x}$ in (16.77). The estimator of $\boldsymbol{\beta}_i$ for the ith group is

$$\hat{\boldsymbol{\beta}}_i = \mathbf{E}_{xx,i}^{-1}\mathbf{e}_{xy,i}. \tag{16.88}$$

By analogy to (16.48), the sum of squares for testing $H_{03}: \boldsymbol{\beta}_1 = \boldsymbol{\beta}_2 = \cdots = \boldsymbol{\beta}_k$ is $\text{SSE}(R)_{y\cdot x} - \text{SSE}(F)_{y\cdot x} = \sum_{i=1}^{k} \mathbf{e}_{xy,i}\mathbf{E}_{xx,i}^{-1}\mathbf{e}_{xy,i} - \mathbf{e}'_{xy}\mathbf{E}_{xx}^{-1}\mathbf{e}_{xy}$, which has $k(n-1) - q - [k(n-1) - kq] = q(k-1)$ degrees of freedom. The test statistic for $H_{03}: \boldsymbol{\beta}_1 = \boldsymbol{\beta}_2 = \cdots = \boldsymbol{\beta}_k$ is

$$F = \frac{[\text{SSE}(R)_{y\cdot x} - \text{SSE}(F)_{y\cdot x}]/q(k-1)}{\text{SSE}(F)_{y\cdot x}/k(n-q-1)}, \tag{16.89}$$

which is distributed as $F[q(k-1), k(n-q-1)]$ if H_{03} is true. Note that if n is not large, $n-q-1$ may be small, and the test will have low power.

Example 16.5. In Table 16.7, we have instructor rating y and two course ratings x_1 and x_2 for five instructors in each of three courses (Morrison 1983, p. 470).

We first find $\hat{\boldsymbol{\beta}}$ and $\text{SSE}_{y\cdot x}$. Using (16.67), (16.68), and (16.69), we obtain

$$\mathbf{E}_{xx} = \begin{pmatrix} 1.0619 & 0.6791 \\ 0.6791 & 1.2363 \end{pmatrix}, \quad \mathbf{e}_{xy} = \begin{pmatrix} 1.0229 \\ 1.9394 \end{pmatrix}, \quad e_{xy} = 3.6036.$$

Then by (16.15), we obtain

$$\hat{\boldsymbol{\beta}} = \mathbf{E}_{xx}^{-1}\mathbf{e}_{xy} = \begin{pmatrix} -0.0617 \\ 1.6026 \end{pmatrix}.$$

By (16.77) and (16.81), we have

$$\text{SSE}_{y\cdot x} = .5585, \qquad \text{SST}_{y\cdot x} = .7840.$$

TABLE 16.7 Instructor Rating y and Two Course Ratings x_1 and x_2 in Three Courses

Course								
1			2			3		
y	x_1	x_2	y	x_1	x_2	y	x_1	x_2
2.14	2.71	2.50	2.77	2.29	2.45	1.11	1.74	1.82
1.34	2.00	1.95	1.23	1.83	1.64	2.41	2.19	2.54
2.50	2.66	2.69	1.37	1.78	1.83	1.74	1.40	2.23
1.40	2.80	2.00	1.52	2.18	2.24	1.15	1.80	1.82
1.90	2.38	2.30	1.81	2.14	2.11	1.66	2.17	2.35

Then by (16.82), we see that

$$\mathrm{SS}(\alpha|\mu, \beta) = \mathrm{SST}_{y\cdot x} - \mathrm{SSE}_{y\cdot x} = .2254.$$

The F statistic for testing $H_0\colon \alpha_1 = \alpha_2 = \alpha_3$ is given by (16.83) as

$$F = \frac{\mathrm{SS}(\alpha|\mu, \beta)/(k-1)}{\mathrm{SSE}_{y\cdot x}/[k(n-1)-q]} = \frac{.2254/2}{.5585/10} = 2.0182, \qquad p = .184.$$

To test $H_{02}\colon \boldsymbol{\beta} = \mathbf{0}$, we use (16.84) to obtain

$$F = \frac{\mathbf{e}_{xy}'\mathbf{E}_{xx}^{-1}\mathbf{e}_{xy}/q}{\mathrm{SSE}_{y\cdot x}/[k(n-1)-q]} = 27.2591, \qquad p = 8.95 \times 10^{-5}.$$

Before testing homogeneity of slope vectors, $H_0\colon \boldsymbol{\beta}_1 = \boldsymbol{\beta}_2 = \boldsymbol{\beta}_3$, we first obtain estimates of $\boldsymbol{\beta}_1$, $\boldsymbol{\beta}_2$, and $\boldsymbol{\beta}_3$ using (16.88):

$$\hat{\boldsymbol{\beta}}_1 = \mathbf{E}_{xx,1}^{-1}\mathbf{e}_{xy,1} = \begin{pmatrix} .4236 & .1900 \\ .1900 & .4039 \end{pmatrix}^{-1} \begin{pmatrix} .2786 \\ .6254 \end{pmatrix} = \begin{pmatrix} -0.0467 \\ 1.5703 \end{pmatrix},$$

$$\hat{\boldsymbol{\beta}}_2 = \begin{pmatrix} .2037 & .2758 \\ .2758 & .4161 \end{pmatrix}^{-1} \begin{pmatrix} .4370 \\ .6649 \end{pmatrix} = \begin{pmatrix} -0.1781 \\ 1.7159 \end{pmatrix},$$

$$\hat{\boldsymbol{\beta}}_3 = \begin{pmatrix} .4346 & .2133 \\ .2133 & .4163 \end{pmatrix}^{-1} \begin{pmatrix} .3073 \\ .6492 \end{pmatrix} = \begin{pmatrix} -0.0779 \\ 1.5993 \end{pmatrix}.$$

Then by (16.86) and (16.87), we obtain

$$\mathrm{SSE}(F)_{y\cdot x} = e_{yy} - \sum_{i=1}^{3} \mathbf{e}_{xy,i}'\mathbf{E}_{xx,i}^{-1}\mathbf{e}_{xy,i} = .55725,$$

$$\mathrm{SSE}(R)_{y\cdot x} = e_{yy} - \mathbf{e}_{xy}'\mathbf{E}_{xx}^{-1}\mathbf{e}_{xy} = .55855.$$

The F statistic for testing $H_0\colon \boldsymbol{\beta}_1 = \boldsymbol{\beta}_2 = \boldsymbol{\beta}_3$ is then given by (16.89) as

$$F = \frac{[\mathrm{SSE}(R)_{y\cdot x} - \mathrm{SSE}(F)_{y\cdot x}]/q(k-1)}{\mathrm{SSE}(F)_{y\cdot x}/k(n-q-1)} = \frac{.0012993/4}{.55725/6} = .003498.$$

□

16.6 ANALYSIS OF COVARIANCE WITH UNBALANCED MODELS

The results in previous sections are for balanced ANOVA models to which covariates have been added. The case in which the ANOVA model is itself unbalanced before the addition of a covariate was treated by Hendrix et al. (1982), who also discussed heterogeneity of slopes. The following approach, based on the cell means model of Chapter 15, was suggested by Bryce (1998).

For an analysis-of-covariance model with a single covariate and a common slope β, we extend the cell means model (15.3) or (15.18) as

$$\mathbf{y} = (\mathbf{W}, \mathbf{x})\begin{pmatrix} \boldsymbol{\mu} \\ \beta \end{pmatrix} + \boldsymbol{\varepsilon} = \mathbf{W}\boldsymbol{\mu} + \beta\mathbf{x} + \boldsymbol{\varepsilon}. \tag{16.90}$$

This model allows for imbalance in the n_{ij}'s as well as the inherent imbalance in analysis of covariance models [see Bingham and Feinberg (1982) and a comment following (16.34)]. The vector $\boldsymbol{\mu}$ contains the means for a one-way model as in (15.2), a two-way model as in (15.17), or some other model. Hypotheses about main effects, interactions, the covariate, or other effects can be tested by using contrasts on $\begin{pmatrix} \boldsymbol{\mu} \\ \beta \end{pmatrix}$ as in Section 15.3.

The hypothesis $H_{02}: \beta = 0$ can be expressed in the form $H_{02}: (0, \ldots, 0, 1)\begin{pmatrix} \boldsymbol{\mu} \\ \beta \end{pmatrix} = 0$. To test H_{02}, we use a statistic analogous to (15.29) or (15.32). To test homogeneity of slopes, $H_{03}: \beta_1 = \beta_2 = \cdots = \beta_k$ for a one-way model (or $H_{03}: \beta_1 = \beta_2 = \cdots = \beta_a$ for the slopes of the a levels of factor A in a two-way model, and so on), we expand the model (16.90) to include the β_i's

$$\mathbf{y} = (\mathbf{W}, \mathbf{W}_x)\begin{pmatrix} \boldsymbol{\mu} \\ \boldsymbol{\beta} \end{pmatrix} + \boldsymbol{\varepsilon} = \mathbf{W}\boldsymbol{\mu} + \mathbf{W}_x\boldsymbol{\beta} + \boldsymbol{\varepsilon}, \tag{16.91}$$

where $\boldsymbol{\beta} = (\beta_1, \beta_2, \ldots, \beta_k)'$ and $\mathbf{W}_x$ has a single value of x_{ij} in each row and all other elements are 0s. (The x_{ij} in $\mathbf{W}_x$ is in the same position as the corresponding 1 in $\mathbf{W}$.) Then $H_{03}: \beta_1 = \beta_2 = \cdots = \beta_k$ can be expressed as $H_{03}: (\mathbf{O}, \mathbf{C})\begin{pmatrix} \boldsymbol{\mu} \\ \boldsymbol{\beta} \end{pmatrix} = \mathbf{C}\boldsymbol{\beta} = \mathbf{0}$, where $\mathbf{C}$ is a $(k-1) \times k$ matrix of rank $k-1$ such that $\mathbf{Cj} = \mathbf{0}$. We can test $H_{03}: \mathbf{C}\boldsymbol{\beta} = \mathbf{0}$ using a statistic analogous to (15.33).

Constraints on the μ's and the β's can be introduced by inserting nonsingular matrices $\mathbf{A}$ and $\mathbf{A}_x$ into (16.91):

$$\mathbf{y} = \mathbf{W}\mathbf{A}^{-1}\mathbf{A}\boldsymbol{\mu} + \mathbf{W}_x\mathbf{A}_x^{-1}\mathbf{A}_x\boldsymbol{\beta} + \boldsymbol{\varepsilon}. \tag{16.92}$$

The matrix $\mathbf{A}$ has the form illustrated in (15.37) for constraints on the μ's. The matrix $\mathbf{A}_x$ provides constraints on the β's. For example, if

$$\mathbf{A}_x = \begin{pmatrix} \mathbf{j}' \\ \mathbf{C} \end{pmatrix},$$

where $\mathbf{C}$ is a $(k-1) \times k$ matrix of rank $k-1$ such that $\mathbf{Cj} = \mathbf{0}$ as above, then the model (16.92) has a common slope. In some cases, the matrices $\mathbf{A}$ and $\mathbf{A}_x$ would be the same.

PROBLEMS

16.1 Show that if the columns of $\mathbf{X}$ are linearly independent of those of $\mathbf{Z}$, then $\mathbf{X}'(\mathbf{I} - \mathbf{P})\mathbf{X}$ is nonsingular, as noted preceding (16.14).

16.2 (a) Show that $\text{SSE}_{y \cdot x} = e_{yy} - \mathbf{e}'_{xy}\mathbf{E}_{xx}^{-1}\mathbf{e}_{xy}$ as in (16.18).

(b) Show that $e_{yy} = \mathbf{y}'(\mathbf{I} - \mathbf{P})\mathbf{y}$ as in (16.19).

16.3 Show that for H_0: $\boldsymbol{\beta} = \mathbf{0}$, we have $\text{SSH} = \hat{\boldsymbol{\beta}}'\mathbf{X}'(\mathbf{I} - \mathbf{P})\mathbf{X}\hat{\boldsymbol{\beta}}$ as in (16.21).

16.4 Show that $\hat{\boldsymbol{\alpha}} = (0,\ \bar{y}_{1.} - \hat{\beta}\bar{x}_{1.}, \ldots, \bar{y}_{k.} - \hat{\beta}\bar{x}_{k.})'$ as in (16.24).

16.5 Show that $e_{xx} = \sum_{ij}(x_{ij} - \bar{x}_{i.})^2$, $e_{xy} = \sum_{ij}(x_{ij} - \bar{x}_{i.})(y_{ij} - \bar{y}_{i.})$, and $e_{yy} = \sum_{ij}(y_{ij} - \bar{y}_{i.})^2$, as in (16.25).

16.6 (a) Show that $\mathbf{E}_{xx}$ has the form shown in (16.41).

(b) Show that $\mathbf{e}_{xy}$ has the form shown in (16.43).

16.7 Show that the sums of products in (16.52) and (16.53) can be written as $\sum_{ijk}(x_{ijk} - \bar{x}_{ij.})(y_{ijk} - \bar{y}_{ij.}) = \sum_{ijk} x_{ijk}y_{ijk} - n\sum_{ij}\bar{x}_{ij.}\bar{y}_{ij.}$ and $\sum_{ijk}(x_{ijk} - \bar{x}_{...})(y_{ijk} - \bar{y}_{...}) = \sum_{ijk} x_{ijk}y_{ijk} - acn\bar{x}_{...}\bar{y}_{...}$.

16.8 Show that the "treatment sum of products" $\sum_{ij} x_{ij.}y_{ij.}/n - x_{...}y_{...}/acn$ can be partitioned into the three sums of products in (16.57).

16.9 (a) Express the sums of squares and test statistic for factor C in a form analogous to those for factor A in (16.58), (16.60), and (16.61).

(b) Express the sums of squares and test statistic for the interaction AC in a form analogous to those for factor A in (16.58), (16.60), and (16.61).

16.10 (a) Show that $\mathbf{E}_{xx} = \sum_{i=1}^{k}\mathbf{X}'_i[\mathbf{I} - (1/n)\mathbf{J}]\mathbf{X}_i$ as in (16.67).

(b) Show that $\mathbf{e}_{xy} = \sum_{i=1}^{k}\mathbf{X}'_i[\mathbf{I} - (1/n)\mathbf{J}]\mathbf{y}_i$ as in (16.68).

(c) Show that $e_{yy} = \sum_{i=1}^{k}\mathbf{y}'_i[\mathbf{I} - (1/n)\mathbf{J}]\mathbf{y}_i$ as in (16.69).

16.11 Show that the elements of $\mathbf{X}'_{ic}\mathbf{X}_{ic}$ are given by (16.72) and (16.73).

16.12 Show that $\hat{\boldsymbol{\alpha}}$ has the form given in (16.78).

16.13 Show that $\sum_{ij}(y_{ij} - \bar{y}_{..})^2 - \sum_{ij}(y_{ij} - \bar{y}_{i.})^2 = n\sum_i(\bar{y}_{i.} - \bar{y}_{..})^2$ as in (16.82).

16.14 In Table 16.8 we have the weight gain y and initial weight x of pigs under four diets (treatments).

(a) Estimate β.

(b) Test H_0: $\alpha_1 = \alpha_2 = \alpha_3 = \alpha_4$ using F in (16.31).

TABLE 16.8 Gain in Weight y and Initial Weight x of Pigs

Treatment							
1		2		3		4	
y	x	y	x	y	x	y	x
165	30	180	24	156	34	201	41
170	27	169	31	189	32	173	32
130	20	171	20	138	35	200	30
156	21	161	26	190	35	193	35
167	33	180	20	160	30	142	28
151	29	170	25	172	29	189	36

Source: Ostle and Malone (1988, p. 445).

(**c**) Test $H_0: \beta = 0$ using F in (16.36).

(**d**) Estimate β_1, β_2, β_3, and β_4 and test homogeneity of slopes $H_0: \beta_1 = \beta_2 = \beta_3 = \beta_4$ using F in (16.49).

16.15 In a study to investigate the effect of income and geographic area of residence on daily calories consumed, three people were chosen at random in each of the 18 income–zone combinations. Their daily caloric intake y and age x are recorded in Table 16.9.

(**a**) Obtain the sums of squares and products listed in Table 16.3, where zone is factor A and income group is factor C.

(**b**) Calculate $\text{SS}(A+E)_{y \cdot x}$, $\text{SSE}_{y \cdot x}$, and $\text{SSA}_{y \cdot x}$ using (16.58), (16.59), and (16.60). For factor A calculate F by (16.61) for $H_0: \alpha_1 = \alpha_2 = \alpha_3$. Similarly, obtain the F statistic for factor C and the interaction.

(**c**) Using SPE, SSE_x, and $\text{SSE}_{y \cdot x}$ from parts (a) and (b), calculate the F statistic to test $H_0: \beta = 0$.

(**d**) Calculate the separate slopes for the three levels of factor A, find SS(F) and SS(R), and test for homogeneity of slopes. Repeat for factor C.

16.16 In a study to investigate differences in ability to distinguish aurally between environmental sounds, 10 male subjects and 10 female subjects were assigned randomly to each of two levels of treatment (experimental and control). The variables were x = pretest score and y = posttest score on auditory discrimination. The data are given in Table 16.10.

We use the posttest score y as the dependent variable and the pretest score x as the covariate. This gives the same result as using the gain score (post–pre) as the dependent variable and the pretest as the covariate (Hendrix et al. 1978).

(**a**) Obtain the sums of squares and products listed in Table 16.3, where treatment is factor A and gender is factor C.

TABLE 16.9 Caloric Intake y and Age x for People Classified by Geographic Zone and Income Group

Income	Zone 1		Zone 2		Zone 3	
Group	y	x	y	x	y	x
1	1911	46	1318	80	1127	74
	1560	66	1541	67	1509	71
	2639	38	1350	73	1756	60
2	1034	50	1559	58	1054	83
	2096	33	1260	74	2238	47
	1356	44	1772	44	1599	71
3	2130	35	2027	32	1479	56
	1878	45	1414	51	1837	40
	1152	59	1526	34	1437	66
4	1297	68	1938	33	2136	31
	2093	43	1551	40	1765	56
	2035	59	1450	39	1056	70
5	2189	33	1183	54	1156	47
	2078	36	1967	36	2660	43
	1905	38	1452	53	1474	50
6	1156	57	2599	35	1015	63
	1809	52	2355	64	2555	34
	1997	44	1932	79	1436	54

Source: Ostle and Mensing (1975, p. 482).

TABLE 16.10 Pretest Score x and Posttest Score y on Auditory Discrimination

Male				Female			
Exp.[a]		Control		Exp.		Control	
x	y	x	y	x	y	x	y
58	71	35	49	64	71	68	70
57	69	31	69	39	71	52	64
63	71	54	69	69	71	53	67
66	70	65	65	56	76	43	63
45	65	54	63	67	71	54	63
51	69	37	55	39	65	35	53
62	69	64	66	32	66	62	65
58	66	69	69	62	70	67	69
52	61	70	69	64	68	51	68
59	63	39	57	66	68	42	61

[a]Experimental.

Source: Hendrix (1967, pp. 154–157).

TABLE 16.11 Initial Age x_1, Initial Weight x_2, and Rate of Gain y of 40 Pigs

Treatment 1			Treatment 2			Treatment 3			Treatment 4		
x_1	x_2	y	x_1	x_2	y	x_1	x_2	y	x_1	x_2	y
78	61	1.40	78	74	1.61	78	80	1.67	77	62	1.40
90	59	1.79	99	75	1.31	83	61	1.41	71	55	1.47
94	76	1.72	80	64	1.12	79	62	1.73	78	62	1.37
71	50	1.47	75	48	1.35	70	47	1.23	70	43	1.15
99	61	1.26	94	62	1.29	85	59	1.49	95	57	1.22
80	54	1.28	91	42	1.24	83	42	1.22	96	51	1.48
83	57	1.34	75	52	1.29	71	47	1.39	71	41	1.31
75	45	1.55	63	43	1.43	66	52	1.39	63	40	1.27
62	41	1.57	62	50	1.29	67	40	1.56	62	45	1.22
67	40	1.26	67	40	1.26	67	40	1.36	67	39	1.36

Source: Snedecor and Cochran (1967, p. 440).

(b) Calculate $\text{SS}(A+E)_{y\cdot x}$, $\text{SSE}_{y\cdot x}$, and $\text{SSA}_{y\cdot x}$ using (16.58), (16.59), and (16.60). For factor A calculate F by (16.61) for $H_0\colon \alpha_1 = \alpha_2$. Similarly, obtain the F statistic for factor C and the interaction.

(c) Using SPE, SSE_x, and $\text{SSE}_{y\cdot x}$ from parts (a) and (b), calculate the F statistic to test $H_0\colon \beta = 0$.

(d) Calculate the separate slopes for the two levels of factor A, find $\text{SS}(F)$ and $\text{SS}(R)$, and test for homogeneity of slopes. Repeat for factor C.

16.17 In an experiment comparing four diets (treatments), the weight gain y (pounds per day) of pigs was recorded along with two covariates, initial age x_1 (days) and initial weight x_2 (pounds). The data are presented in Table 16.11.

(a) Using (16.67), (16.68), and (16.69), find $\mathbf{E}_{xx}$, $\mathbf{e}_{xy}$, and e_{yy}. Find $\hat{\boldsymbol{\beta}}$.

(b) Using (16.77), (16.81), and (16.82), find $\text{SSE}_{y\cdot x}$, $\text{SST}_{y\cdot x}$, and $\text{SS}(\alpha|\mu, \beta)$. Then test $H_0\colon \alpha_1 = \alpha_2 = \alpha_3 = \alpha_4$, adjusted for the covariates, using the F statistic in (16.83).

(c) Test $H_0\colon \boldsymbol{\beta} = \mathbf{0}$ using (16.84).

(d) Find $\hat{\boldsymbol{\beta}}_1$, $\hat{\boldsymbol{\beta}}_2$, $\hat{\boldsymbol{\beta}}_3$, and $\hat{\boldsymbol{\beta}}_4$ using (16.88). Find $\text{SSE}(F)_{y\cdot x}$ and $\text{SSE}(R)_{y\cdot x}$ using (16.86) and (16.87). Test $H_0\colon \boldsymbol{\beta}_1 = \boldsymbol{\beta}_2 = \boldsymbol{\beta}_3 = \boldsymbol{\beta}_4$ using (16.89).

17 Linear Mixed Models

17.1 INTRODUCTION

In Section 7.8 we briefly considered linear models in which the y variables are correlated or have nonconstant variances (or both). We used the model

$$\mathbf{y} = \mathbf{X}\boldsymbol{\beta} + \boldsymbol{\varepsilon}, \quad E(\boldsymbol{\varepsilon}) = \mathbf{0}, \quad \text{cov}(\boldsymbol{\varepsilon}) = \boldsymbol{\Sigma} = \sigma^2\mathbf{V}, \tag{17.1}$$

where $\mathbf{V}$ is a *known* positive definite matrix, and developed estimators for $\boldsymbol{\beta}$ in (7.63) and σ^2 in (7.65). Hypothesis tests and confidence intervals were not given, but they could have been developed by adding the assumption of normality and modifying the approaches of Chapter 8 (see Problems 17.1 and 17.2).

Correlated data are commonly encountered in practice (Brown and Prescott 2006, pp. 1–3; Fitzmaurice et al. 2004, p. xvi; Mclean et al. 1991). We can use the methods of Section 7.8 as a starting point in approaching such data, but those methods are actually of limited practical use because we rarely, if ever, know $\mathbf{V}$. On the other hand, the *structure* of $\mathbf{V}$ is often known and in many cases can be specified up to relatively few unknown parameters. This chapter is an introduction to linear models for correlated y variables where the structure of $\boldsymbol{\Sigma} = \sigma^2\mathbf{V}$ can be specified.

17.2 THE LINEAR MIXED MODEL

Nonindependence of observations may result from serial correlation or clustering of the observations (Diggle et al. 2002). Serial correlation, which will not be discussed further in this chapter, is present when a time- (or space-) varying stochastic process is operating on the units and the units are repeatedly measured over time (or space). Cluster correlation is present when the observations are grouped in various ways. The groupings might be due, for example, to repeated random sampling of subgroups or repeated measuring of the same units. Examples are given in Section 17.3. In many cases the covariance structure of cluster-correlated data can be specified using an extension of the standard linear model (7.4) resembling the partitioned linear model (7.78). If $\mathbf{y}$ is an $n \times 1$ vector of responses, the model is

$$\mathbf{y} = \mathbf{X}\boldsymbol{\beta} + \mathbf{Z}_1\mathbf{a}_1 + \mathbf{Z}_2\mathbf{a}_2 + \cdots + \mathbf{Z}_m\mathbf{a}_m + \boldsymbol{\varepsilon}, \tag{17.2}$$

where $E(\boldsymbol{\varepsilon}) = \mathbf{0}$ and $\text{cov}(\boldsymbol{\varepsilon}) = \sigma^2\mathbf{I}_n$ as usual. Here $\mathbf{X}$ is an $n \times p$ known, possibly non-full-rank matrix of fixed predictors as in Chapters 7, 8, 11, 12, and 16. It could be used to specify a multiple regression model, analysis-of-variance model, or analysis of covariance model. It could be as simple as vector of 1s. As usual, $\boldsymbol{\beta}$ is an $n \times 1$ vector of unknown fixed parameters.

The $\mathbf{Z}_i$'s are known $n \times r_i$ full-rank matrices of fixed predictors, usually used to specify membership in the various clusters or subgroups. The major innovation in this model is that the $\mathbf{a}_i$'s are $r_i \times 1$ vectors of unknown random quantities similar to $\boldsymbol{\varepsilon}$. We assume that $E(\mathbf{a}_i) = \mathbf{0}$ and $\text{cov}(\mathbf{a}_i) = \sigma_i^2\mathbf{I}_{r_i}$ for $i = 1, \ldots, m$. For simplicity we further assume that $\text{cov}(\mathbf{a}_i, \mathbf{a}_j) = \mathbf{O}$ for $i \neq j$, where $\mathbf{O}$ is $r_i \times r_j$, and that $\text{cov}(\mathbf{a}_i, \boldsymbol{\varepsilon}) = \mathbf{O}$ for all i, where $\mathbf{O}$ is $r_i \times n$. These assumptions are often reasonable (McCulloch and Searle 2001, pp. 159–160).

Note that this model is very different from the random-x model of Chapter 10. In Chapter 10 the predictors in $\mathbf{X}$ were random while the parameters in $\boldsymbol{\beta}$ were fixed. Here the opposite scenario applies; predictors in each $\mathbf{Z}_i$ are fixed while the elements of $\mathbf{a}_i$ are random. On the other hand, this model has much in common with the Bayesian linear model of Chapter 11. In fact, if the normality assumption is added, the model can be stated in a form reminiscent of the Bayesian linear model as

$$\begin{aligned} \mathbf{y}|\mathbf{a}_1, \mathbf{a}_2, \ldots, \mathbf{a}_m \quad &\text{is} \quad N_n(\mathbf{X}\boldsymbol{\beta} + \mathbf{Z}_1\mathbf{a}_1 + \mathbf{Z}_2\mathbf{a}_2 + \cdots + \mathbf{Z}_m\mathbf{a}_m, \sigma^2\mathbf{I}_n), \\ \mathbf{a}_i \quad &\text{is} \quad N_{n_i}(\mathbf{0}, \sigma_i^2\mathbf{I}_{r_i}) \text{ for } i = 1, \ldots, m. \end{aligned}$$

The label *linear mixed model* seems appropriate to describe (17.2) because the model involves a mixture of linear functions of fixed parameters in $\boldsymbol{\beta}$ and linear functions of random quantities in the $\mathbf{a}_i$'s. The special case in which $\mathbf{X} = \mathbf{j}$ (so that there is only one fixed parameter) is sometimes referred to as a *random model*. The σ_i^2's (including σ^2) are referred to as *variance components*.

We now investigate $E(\mathbf{y})$ and $\text{cov}(\mathbf{y}) = \boldsymbol{\Sigma}$ under the model in (17.2).

Theorem 17.2. Consider the model $\mathbf{y} = \mathbf{X}\boldsymbol{\beta} + \sum_{i=1}^m \mathbf{Z}_i\mathbf{a}_i + \boldsymbol{\varepsilon}$, where $\mathbf{X}$ is a known $n \times p$ matrix, the $\mathbf{Z}_i$'s are known $n \times r_i$ full-rank matrices, $\boldsymbol{\beta}$ is a $p \times 1$ vector of unknown parameters, $\boldsymbol{\varepsilon}$ is an $n \times 1$ unknown random vector such that $E(\boldsymbol{\varepsilon}) = \mathbf{0}$ and $\text{cov}(\boldsymbol{\varepsilon}) = \sigma^2\mathbf{I}_n$, and the $\mathbf{a}_i'$*s* are $r_i \times 1$ unknown random vectors such that $E(\mathbf{a}_i) = \mathbf{0}$ and $\text{cov}(\mathbf{a}_i) = \sigma_i^2\mathbf{I}_{r_i}$. Furthermore, $\text{cov}(\mathbf{a}_i, \mathbf{a}_j) = \mathbf{O}$ for $i \neq j$, where $\mathbf{O}$ is $r_i \times r_j$, and $\text{cov}(\mathbf{a}_i, \boldsymbol{\varepsilon}) = \mathbf{O}$ for all i, where $\mathbf{O}$ is $r_i \times n$. Then $E(\mathbf{y}) = \mathbf{X}\boldsymbol{\beta}$ and $\text{cov}(\mathbf{y}) = \boldsymbol{\Sigma} = \sum_{i=1}^m \sigma_i^2\mathbf{Z}_i\mathbf{Z}_i' + \sigma^2\mathbf{I}_n$.

PROOF

$$
\begin{aligned}
E(\mathbf{y}) &= E\left(\mathbf{X}\boldsymbol{\beta} + \sum_{i=1}^{m} \mathbf{Z}_i\mathbf{a}_i + \boldsymbol{\varepsilon}\right) \\
&= \mathbf{X}\boldsymbol{\beta} + E\left(\sum_{i=1}^{m} \mathbf{Z}_i\mathbf{a}_i + \boldsymbol{\varepsilon}\right) \\
&= \mathbf{X}\boldsymbol{\beta} + \sum_{i=1}^{m} \mathbf{Z}_i E(\mathbf{a}_i) + E(\boldsymbol{\varepsilon}) \ \ \text{[by (3.21) and (3.38)]} \\
&= \mathbf{X}\boldsymbol{\beta}.
\end{aligned}
$$

$$
\begin{aligned}
\text{cov}(\mathbf{y}) &= \text{cov}\left(\mathbf{X}\boldsymbol{\beta} + \sum_{i=1}^{m} \mathbf{Z}_i\mathbf{a}_i + \boldsymbol{\varepsilon}\right) \\
&= \text{cov}\left(\sum_{i=1}^{m} \mathbf{Z}_i\mathbf{a}_i + \boldsymbol{\varepsilon}\right) \\
&= \sum_{i=1}^{m} \text{cov}(\mathbf{Z}_i\mathbf{a}_i) + \text{cov}(\boldsymbol{\varepsilon}) + \sum_{i \neq j} \text{cov}(\mathbf{Z}_i\mathbf{a}_i, \mathbf{Z}_j\mathbf{a}_j) \\
&\quad + \sum_{i=1}^{m} \text{cov}(\mathbf{Z}_i\mathbf{a}_i, \boldsymbol{\varepsilon}) + \sum_{i=1}^{m} \text{cov}(\boldsymbol{\varepsilon}, \mathbf{Z}_i\mathbf{a}_i) \ \text{[see Problem 3.19]} \\
&= \sum_{i=1}^{m} \mathbf{Z}_i \text{cov}(\mathbf{a}_i)\mathbf{Z}_i' + \text{cov}(\boldsymbol{\varepsilon}) + \sum_{i \neq j} \mathbf{Z}_i \text{cov}(\mathbf{a}_i, \mathbf{a}_j)\mathbf{Z}_j' \\
&\quad + \sum_{i=1}^{m} \mathbf{Z}_i \text{cov}(\mathbf{a}_i, \boldsymbol{\varepsilon}) + \sum_{i=1}^{m} \text{cov}(\boldsymbol{\varepsilon}, \mathbf{a}_i)\mathbf{Z}_i' \ \text{[by Theorem 3.6d and Theorem 3.6e]} \\
&= \sum_{i=1}^{m} \sigma_i^2 \mathbf{Z}_i\mathbf{Z}_i' + \sigma^2 \mathbf{I}_n. \qquad \square
\end{aligned}
$$

Note that the z's only enter into the covariance structure while the x's only determine the mean of $\mathbf{y}$.

17.3 EXAMPLES

We illustrate the broad applicability of the model in (17.2) with several simple examples.

Example 17.3a (Randomized Blocks). An experiment involving three treatments was carried out by randomly assigning the treatments to experimental units within each of four blocks of size 3. We could use the model

$$
y_{ij} = \mu + \tau_i + a_j + \varepsilon_{ij},
$$

where $i = 1, \ldots, 3, j = 1, \ldots, 4$, a_j is $N(0, \sigma_1^2)$, ε_{ij} is $N(0, \sigma^2)$, and $\text{cov}(a_j, \varepsilon_{ij}) = 0$. If we assume that the observations are sorted by blocks and treatments within blocks, we can express this model in the form of (17.2) with

$$m = 1, \mathbf{X} = \begin{pmatrix} \mathbf{j}_3 & \mathbf{I}_3 \\ \mathbf{j}_3 & \mathbf{I}_3 \\ \mathbf{j}_3 & \mathbf{I}_3 \\ \mathbf{j}_3 & \mathbf{I}_3 \end{pmatrix}, \quad \text{and} \quad \mathbf{Z}_1 = \begin{pmatrix} \mathbf{j}_3 & \mathbf{0}_3 & \mathbf{0}_3 & \mathbf{0}_3 \\ \mathbf{0}_3 & \mathbf{j}_3 & \mathbf{0}_3 & \mathbf{0}_3 \\ \mathbf{0}_3 & \mathbf{0}_3 & \mathbf{j}_3 & \mathbf{0}_3 \\ \mathbf{0}_3 & \mathbf{0}_3 & \mathbf{0}_3 & \mathbf{j}_3 \end{pmatrix}.$$

Then

$$\boldsymbol{\sigma} = {\sigma_1}^2 \mathbf{Z}_1 \mathbf{Z}_1' + \sigma^2 \mathbf{I}_{12} = \begin{pmatrix} \boldsymbol{\Sigma}_1 & \mathbf{O} & \mathbf{O} & \mathbf{O} \\ \mathbf{O} & \boldsymbol{\Sigma}_1 & \mathbf{O} & \mathbf{O} \\ \mathbf{O} & \mathbf{O} & \boldsymbol{\Sigma}_1 & \mathbf{O} \\ \mathbf{O} & \mathbf{O} & \mathbf{O} & \boldsymbol{\Sigma}_1 \end{pmatrix},$$

$$\text{where} \quad \boldsymbol{\Sigma}_1 = \begin{pmatrix} \sigma_1^2 + \sigma^2 & \sigma_1^2 & \sigma_1^2 \\ \sigma_1^2 & \sigma_1^2 + \sigma^2 & \sigma_1^2 \\ \sigma_1^2 & \sigma_1^2 & \sigma_1^2 + \sigma^2 \end{pmatrix}.$$

□

Example 17.3b (Subsampling). Five batches were produced using each of two processes. Two samples were obtained and measured from each of the batches. Constraining the process effects to sum to zero, the model is

$$y_{ijk} = \mu + \tau_i + a_{ij} + \varepsilon_{ijk},$$

where $i = 1, 2$; $j = 1, \ldots, 5$; $k = 1, 2$; $\tau_2 = -\tau_1$; a_{ij} is $N(0, \sigma_1^2)$; ε_{ijk} is $N(0, \sigma^2)$; and $\text{cov}(a_{ij}, \varepsilon_{ijk}) = 0$. If the observations are sorted by processes, batches within processes, and samples within batches, we can put this model in the form of (17.2) with

$$m = 1, \mathbf{X} = \begin{pmatrix} \mathbf{j}_{10} & \mathbf{j}_{10} \\ \mathbf{j}_{10} & -\mathbf{j}_{10} \end{pmatrix} \quad \text{and} \quad \mathbf{Z}_1 = \begin{pmatrix} \mathbf{j}_2 & \mathbf{0}_2 & \cdots & \mathbf{0}_2 \\ \mathbf{0}_2 & \mathbf{j}_2 & \cdots & \mathbf{0}_2 \\ \vdots & \vdots & & \vdots \\ \mathbf{0}_2 & \mathbf{0}_2 & \cdots & \mathbf{j}_2 \end{pmatrix}.$$

Hence

$$\boldsymbol{\Sigma} = \sigma_1^2 \mathbf{Z}_1 \mathbf{Z}_1' + \sigma^2 \mathbf{I}_{20} = \begin{pmatrix} \boldsymbol{\Sigma}_1 & \mathbf{O} & \cdots & \mathbf{O} \\ \mathbf{O} & \boldsymbol{\Sigma}_1 & \cdots & \mathbf{O} \\ \vdots & \vdots & & \vdots \\ \mathbf{O} & \mathbf{O} & \cdots & \boldsymbol{\Sigma}_1 \end{pmatrix},$$

$$\text{where} \quad \boldsymbol{\Sigma}_1 = \begin{pmatrix} \sigma_1^2 + \sigma^2 & \sigma_1^2 \\ \sigma_1^2 & \sigma_1^2 + \sigma^2 \end{pmatrix}.$$

□

Example 17.3c (Split-Plot Studies). A 3×2 factorial experiment (with factors A and B, respectively) was carried out using six main units, each of which was subdivided into two subunits. The levels of A were each randomly assigned to two of the main units, and the levels of B were randomly assigned to subunits within main units. An appropriate model is

$$y_{ijk} = \mu + \tau_i + \delta_j + \theta_{ij} + a_{ik} + \varepsilon_{ijk},$$

where $i = 1, \ldots, 3; j = 1, 2; k = 1, 2; a_{ik}$ is $N(0, \sigma_1{}^2)$; ε_{ijk} is $N(0, \sigma^2)$ and $\text{cov}(a_{ik}, \varepsilon_{ijk}) = 0$. If the observations are sorted by levels of A, main units within levels of A, and levels of B within main units, we can express this model in the form of (17.2) with

$$m = 1, \mathbf{X} = \begin{pmatrix} 1 & 1 & 0 & 0 & 1 & 0 & 1 & 0 & 0 & 0 & 0 & 0 \\ 1 & 1 & 0 & 0 & 0 & 1 & 0 & 1 & 0 & 0 & 0 & 0 \\ 1 & 1 & 0 & 0 & 1 & 0 & 1 & 0 & 0 & 0 & 0 & 0 \\ 1 & 1 & 0 & 0 & 0 & 1 & 0 & 1 & 0 & 0 & 0 & 0 \\ 1 & 0 & 1 & 0 & 1 & 0 & 0 & 0 & 1 & 0 & 0 & 0 \\ 1 & 0 & 1 & 0 & 0 & 1 & 0 & 0 & 0 & 1 & 0 & 0 \\ 1 & 0 & 1 & 0 & 1 & 0 & 0 & 0 & 1 & 0 & 0 & 0 \\ 1 & 0 & 1 & 0 & 0 & 1 & 0 & 0 & 0 & 1 & 0 & 0 \\ 1 & 0 & 0 & 1 & 1 & 0 & 0 & 0 & 0 & 0 & 1 & 0 \\ 1 & 0 & 0 & 1 & 0 & 1 & 0 & 0 & 0 & 0 & 0 & 1 \\ 1 & 0 & 0 & 1 & 1 & 0 & 0 & 0 & 0 & 0 & 1 & 0 \\ 1 & 0 & 0 & 1 & 0 & 1 & 0 & 0 & 0 & 0 & 0 & 1 \end{pmatrix}, \quad \text{and}$$

$$\mathbf{Z}_1 = \begin{pmatrix} 1 & 0 & 0 & 0 & 0 & 0 \\ 1 & 0 & 0 & 0 & 0 & 0 \\ 0 & 1 & 0 & 0 & 0 & 0 \\ 0 & 1 & 0 & 0 & 0 & 0 \\ 0 & 0 & 1 & 0 & 0 & 0 \\ 0 & 0 & 1 & 0 & 0 & 0 \\ 0 & 0 & 0 & 1 & 0 & 0 \\ 0 & 0 & 0 & 1 & 0 & 0 \\ 0 & 0 & 0 & 0 & 1 & 0 \\ 0 & 0 & 0 & 0 & 1 & 0 \\ 0 & 0 & 0 & 0 & 0 & 1 \\ 0 & 0 & 0 & 0 & 0 & 1 \end{pmatrix}.$$

Then

$$\boldsymbol{\Sigma} = {\sigma_1}^2\mathbf{Z}_1\mathbf{Z}_1' + \sigma^2\mathbf{I}_{12} = \begin{pmatrix} \boldsymbol{\Sigma}_1 & \mathbf{O} & \cdots & \mathbf{O} \\ \mathbf{O} & \boldsymbol{\Sigma}_1 & \cdots & \mathbf{O} \\ \vdots & \vdots & & \vdots \\ \mathbf{O} & \mathbf{O} & \cdots & \boldsymbol{\Sigma}_1 \end{pmatrix}, \quad \text{where}$$

$$\boldsymbol{\Sigma}_1 = \begin{pmatrix} {\sigma_1}^2 + \sigma^2 & \sigma_1^2 \\ \sigma_1^2 & {\sigma_1}^2 + \sigma^2 \end{pmatrix}.$$

□

Example 17.3d (One-Way Random Effects). A chemical plant produced a large number of batches. Each batch was packaged into a large number of containers. We chose three batches at random, and randomly selected four containers from each batch from which to measure y. The model is

$$y_{ij} = \mu + a_i + \varepsilon_{ij},$$

where $i = 1, \ldots, 3; j = 1, \ldots, 4$; a_j is $N(0, {\sigma_1}^2)$; ε_{ij} is $N(0, \sigma^2)$; and $\text{cov}(a_j, \varepsilon_{ij}) = 0$. If the observations are sorted by batches and containers within batches, we can express this model in the form of (17.2) with

$$m = 1, \mathbf{X} = \mathbf{j}_{12}, \text{ and } \mathbf{Z}_1 = \begin{pmatrix} \mathbf{j}_4 & \mathbf{0}_4 & \mathbf{0}_4 \\ \mathbf{0}_4 & \mathbf{j}_4 & \mathbf{0}_4 \\ \mathbf{0}_4 & \mathbf{0}_4 & \mathbf{j}_4 \end{pmatrix}.$$

Thus

$$\boldsymbol{\Sigma} = {\sigma_1}^2\mathbf{Z}_1\mathbf{Z}_1' + \sigma^2\mathbf{I}_{12} = \begin{pmatrix} \boldsymbol{\Sigma}_1 & \mathbf{O} & \mathbf{O} \\ \mathbf{O} & \boldsymbol{\Sigma}_1 & \mathbf{O} \\ \mathbf{O} & \mathbf{O} & \boldsymbol{\Sigma}_1 \end{pmatrix}, \quad \text{where}$$

$$\boldsymbol{\Sigma}_1 = \begin{pmatrix} {\sigma_1}^2 + \sigma^2 & \sigma_1^2 & \sigma_1^2 & \sigma_1^2 \\ \sigma_1^2 & {\sigma_1}^2 + \sigma^2 & \sigma_1^2 & \sigma_1^2 \\ \sigma_1^2 & \sigma_1^2 & {\sigma_1}^2 + \sigma^2 & \sigma_1^2 \\ \sigma_1^2 & \sigma_1^2 & \sigma_1^2 & {\sigma_1}^2 + \sigma^2 \end{pmatrix}.$$

□

Example 17.3e (Independent Random Coefficients). Three pups from each of four litters of mice were used in an experiment. One pup from each litter was exposed to one of three quantitative levels of a carcinogen. The relationship between weight gain (y) and carcinogen level is a straight line, but slopes and intercepts vary randomly and independently among litters. The three levels of the carcinogen are denoted by $\mathbf{x}$. The model is

$$y_{ij} = \beta_0 + a_i + \beta_1 x_j + b_i x_j + \varepsilon_{ij},$$

where $i = 1, \ldots, 4$; $j = 1, \ldots, 3$; a_i is $N(0, \sigma_1{}^2)$; b_i is $N(0, \sigma_2{}^2)$; ε_{ij} is $N(0, \sigma^2)$, and all the random effects are independent. If the data are sorted by litter and carcinogen levels within litter, we can express this model in the form of (17.2) with

$$m = 2,\ \mathbf{X} = \begin{pmatrix} \mathbf{j}_3 & \mathbf{x} \\ \mathbf{j}_3 & \mathbf{x} \\ \mathbf{j}_3 & \mathbf{x} \\ \mathbf{j}_3 & \mathbf{x} \end{pmatrix},\ \mathbf{Z}_1 = \begin{pmatrix} \mathbf{j}_3 & \mathbf{0}_3 & \mathbf{0}_3 & \mathbf{0}_3 \\ \mathbf{0}_3 & \mathbf{j}_3 & \mathbf{0}_3 & \mathbf{0}_3 \\ \mathbf{0}_3 & \mathbf{0}_3 & \mathbf{j}_3 & \mathbf{0}_3 \\ \mathbf{0}_3 & \mathbf{0}_3 & \mathbf{0}_3 & \mathbf{j}_3 \end{pmatrix}, \quad \text{and}$$

$$\mathbf{Z}_2 = \begin{pmatrix} \mathbf{x} & \mathbf{0}_3 & \mathbf{0}_3 & \mathbf{0}_3 \\ \mathbf{0}_3 & \mathbf{x} & \mathbf{0}_3 & \mathbf{0}_3 \\ \mathbf{0}_3 & \mathbf{0}_3 & \mathbf{x} & \mathbf{0}_3 \\ \mathbf{0}_3 & \mathbf{0}_3 & \mathbf{0}_3 & \mathbf{x} \end{pmatrix}.$$

Then

$$\boldsymbol{\Sigma} = \sigma_1{}^2 \mathbf{Z}_1 \mathbf{Z}_1' + \sigma_2{}^2 \mathbf{Z}_2 \mathbf{Z}_2' + \sigma^2 \mathbf{I}_{12} = \begin{pmatrix} \boldsymbol{\Sigma}_1 & \mathbf{O} & \mathbf{O} & \mathbf{O} \\ \mathbf{O} & \boldsymbol{\Sigma}_1 & \mathbf{O} & \mathbf{O} \\ \mathbf{O} & \mathbf{O} & \boldsymbol{\Sigma}_1 & \mathbf{O} \\ \mathbf{O} & \mathbf{O} & \mathbf{O} & \boldsymbol{\Sigma}_1 \end{pmatrix},$$

where $\boldsymbol{\Sigma}_1 = \sigma_1{}^2 \mathbf{J}_3 + \sigma_2{}^2 \mathbf{x}\mathbf{x}' + \sigma^2 \mathbf{I}_3.$

□

Example 17.3f (Heterogeneous Variances). Four individuals were randomly sampled from each of four groups. The groups had different means and different variances. We assume here that $\sigma^2 = 0$. The model is

$$y_{ij} = \mu_i + \varepsilon_{ij},$$

where $i = 1, \ldots, 4$; $j = 1, \ldots, 4$; ε_{ij} is $N(0, \sigma_i{}^2)$. If the data are sorted by groups and individuals within groups, we can express this model in the form of (17.2) with

$$m = 4,\ \mathbf{X} = \begin{pmatrix} \mathbf{I}_4 \\ \mathbf{I}_4 \\ \mathbf{I}_4 \\ \mathbf{I}_4 \end{pmatrix}, \quad \mathbf{Z}_1 = \begin{pmatrix} \mathbf{I}_4 \\ \mathbf{O}_4 \\ \mathbf{O}_4 \\ \mathbf{O}_4 \end{pmatrix}, \quad \mathbf{Z}_2 = \begin{pmatrix} \mathbf{O}_4 \\ \mathbf{I}_4 \\ \mathbf{O}_4 \\ \mathbf{O}_4 \end{pmatrix}, \quad \mathbf{Z}_3 = \begin{pmatrix} \mathbf{O}_4 \\ \mathbf{O}_4 \\ \mathbf{I}_4 \\ \mathbf{O}_4 \end{pmatrix},$$

$$\text{and} \quad \mathbf{Z}_4 = \begin{pmatrix} \mathbf{O}_4 \\ \mathbf{O}_4 \\ \mathbf{O}_4 \\ \mathbf{I}_4 \end{pmatrix}.$$

Hence

$$\mathbf{\Sigma} = \sigma_1{}^2\mathbf{Z}_1\mathbf{Z}_1' + \sigma_2{}^2\mathbf{Z}_2\mathbf{Z}_2' + \sigma_3{}^2\mathbf{Z}_3\mathbf{Z}_3' + \sigma_4{}^2\mathbf{Z}_4\mathbf{Z}_4' = \begin{pmatrix} \sigma_1{}^2\mathbf{I}_4 & \mathbf{O}_4 & \mathbf{O}_4 & \mathbf{O}_4 \\ \mathbf{O}_4 & \sigma_2{}^2\mathbf{I}_4 & \mathbf{O}_4 & \mathbf{O}_4 \\ \mathbf{O}_4 & \mathbf{O}_4 & \sigma_3{}^2\mathbf{I}_4 & \mathbf{O}_4 \\ \mathbf{O}_4 & \mathbf{O}_4 & \mathbf{O}_4 & \sigma_4{}^2\mathbf{I}_4 \end{pmatrix}.$$

□

These models can be generalized and combined to yield a rich set of models applicable to a broad spectrum of situations (see Problem 17.3). All the examples involved balanced data for convenience of description, but model (17.2) applies equally well to unbalanced situations. Allowing the covariance matrices of the $\mathbf{a}_i$'s and $\boldsymbol{\varepsilon}$ to be nondiagonal (providing for such things as serial correlation) increases the scope of application of these models even more, with only moderate increases in complexity (see Problem 17.4).

17.4 ESTIMATION OF VARIANCE COMPONENTS

After specifying the appropriate model, the next task in using the linear mixed model (17.2) in the analysis of data is to estimate the variance components. Once the variance components have been estimated, $\mathbf{\Sigma}$ can be estimated and the estimate used in the approximate generalized least-squares estimation of $\boldsymbol{\beta}$ and other inferences as suggested by the results of Section 7.8.

Several methods for estimation of the variance components have been proposed (Searle et al. 1992, pp. 168–257). We discuss one of these approaches, that of restricted (or residual) maximum likelihood (REML) (Patterson and Thompson 1971). One reason for our emphasis of REML is that in standard linear models, the usual estimate s^2 in (7.22) is the REML estimate. Also, REML is general; for example, it can be applied regardless of balance. In certain balanced situations the REML estimator has closed form. It is often the best (minimum variance) quadratic unbiased estimator (see Theorem 7.3g).

To develop the REML estimator, we add the normality assumption. Thus the model is

$$\mathbf{y} \text{ is } N_n(\mathbf{X}\boldsymbol{\beta}, \mathbf{\Sigma}), \quad \text{where} \quad \mathbf{\Sigma} = \sum_{i=1}^{m} \sigma_i{}^2\mathbf{Z}_i\mathbf{Z}_i' + \sigma^2\mathbf{I}_n, \tag{17.3}$$

where $\mathbf{X}$ is $n \times p$ of rank $r \leq p$, and $\mathbf{\Sigma}$ is a positive definite $n \times n$ matrix. To simplify the notation, we let $\sigma_0^2 = \sigma^2$ and $\mathbf{Z}_0 = \mathbf{I}_n$ so that (17.3) becomes

$$\mathbf{y} \text{ is } N_n(\mathbf{X}\boldsymbol{\beta}, \mathbf{\Sigma}), \quad \text{where } \mathbf{\Sigma} = \sum_{i=0}^{m} \sigma_i{}^2\mathbf{Z}_i\mathbf{Z}_i'. \tag{17.4}$$

The idea of REML is to carry out maximum likelihood estimation for data $\mathbf{Ky}$ rather than $\mathbf{y}$, where $\mathbf{K}$ is chosen so that the distribution of $\mathbf{Ky}$ involves only the variance components, not $\boldsymbol{\beta}$. In order for this to occur, we seek a matrix $\mathbf{K}$ such that $\mathbf{KX} = \mathbf{O}$. Hence $E(\mathbf{Ky}) = \mathbf{KX} = \mathbf{0}$. For simplicity we require that $\mathbf{K}$ be of full-rank. We also want $\mathbf{Ky}$ to contain as much information as possible about the variance components, so $\mathbf{K}$ must have the maximal number of rows for such a matrix.

Theorem 17.4a. Let $\mathbf{X}$ be as in (17.3). A full-rank matrix $\mathbf{K}$ with maximal number of rows such that $\mathbf{KX} = \mathbf{O}$, is an $(n - r) \times n$ matrix. Furthermore, $\mathbf{K}$ must be of the form $\mathbf{K} = \mathbf{C}(\mathbf{I} - \mathbf{H}) = \mathbf{C}[\mathbf{I} - \mathbf{X}(\mathbf{X}'\mathbf{X})^{-}\mathbf{X}']$ where $\mathbf{C}$ specifies a full-rank transformation of the rows of $\mathbf{I} - \mathbf{H}$.

PROOF. The rows $\mathbf{k}_i'$ of $\mathbf{K}$ must satisfy the equations $\mathbf{k}_i'\mathbf{X} = \mathbf{0}'$ or equivalently $\mathbf{X}'\mathbf{k}_i = \mathbf{0}$. Using Theorem 2.8e, solutions to this system of equations are given by $\mathbf{k}_i = (\mathbf{I} - \mathbf{X}^{-}\mathbf{X})\mathbf{c}$ for all possible $p \times 1$ vectors $\mathbf{c}$. In other words, the solutions include all possible linear combinations of the columns of $\mathbf{I} - \mathbf{X}^{-}\mathbf{X}$.

By Theorem 2.8c(i), $\text{rank}(\mathbf{X}^{-}\mathbf{X}) = \text{rank}(\mathbf{X}) = r$. Also, by Theorem 2.13e, $\mathbf{I} - \mathbf{X}^{-}\mathbf{X}$ is idempotent. Because of this idempotency, $\text{rank}(\mathbf{I} - \mathbf{X}^{-}\mathbf{X}) = \text{tr}(\mathbf{I} - \mathbf{X}^{-}\mathbf{X}) = \text{tr}(\mathbf{I}) - \text{tr}(\mathbf{X}^{-}\mathbf{X}) = n - r$. Hence by the definition of rank (see Section 2.4), there are $n - r$ linearly independent vectors $\mathbf{k}_i$ that satisfy $\mathbf{X}'\mathbf{k}_i = \mathbf{0}$ and thus the maximal number of rows in $\mathbf{K}$ is $n - r$.

Since $\mathbf{k}_i = (\mathbf{I} - \mathbf{X}^{-}\mathbf{X})\mathbf{c}$, $\mathbf{K} = \mathbf{C}(\mathbf{I} - \mathbf{X}^{-}\mathbf{X})$ for some full-rank $(n - r) \times n$ matrix $\mathbf{C}$ that specifies $n - r$ linearly independent linear combinations of the rows of the symmetric matrix $\mathbf{I} - \mathbf{X}^{-}\mathbf{X}$. By Theorem 2.8c(iv)–(v), $\mathbf{K}$ can also be written as $\mathbf{C}(\mathbf{I} - \mathbf{H}) = \mathbf{C}[\mathbf{I} - \mathbf{X}(\mathbf{X}'\mathbf{X})^{-}\mathbf{X}']$. □

There are an infinite number of such $\mathbf{K}$s, and it does not matter which is used. Also, note that $(\mathbf{I} - \mathbf{H})\mathbf{y}$ gives the ordinary residual vector $\hat{\boldsymbol{\varepsilon}}$ in (9.5), so that $\mathbf{Ky} = \mathbf{C}(\mathbf{I} - \mathbf{H})\mathbf{y}$ is a vector of linear combinations of these residuals. Thus the designation *residual maximum likelihood* is appropriate.

The distribution of $\mathbf{Ky}$ for any $\mathbf{K}$ defined as in Theorem 17.4a is given in the following theorem.

Theorem 17.4b. Consider the model in which $\mathbf{y}$ is $N_n(\mathbf{X}\boldsymbol{\beta}, \boldsymbol{\Sigma})$, where $\boldsymbol{\Sigma} = \sum_{i=0}^{m} \sigma_i^2 \mathbf{Z}_i\mathbf{Z}_i'$, and let $\mathbf{K}$ be specified as in Theorem 17.4a. Then

$$\mathbf{Ky} \quad \text{is} \quad N_{n-r}(\mathbf{0}, \mathbf{K\Sigma K}') \quad \text{or} \quad N_{n-r}\left[\mathbf{0}, \mathbf{K}\left(\sum_{i=0}^{m} \sigma_i^2 \mathbf{Z}_i\mathbf{Z}_i'\right)\mathbf{K}'\right]. \tag{17.5}$$

PROOF. Since $\mathbf{KX} = \mathbf{O}$, the theorem follows directly from Theorem 4.4a(ii). □

Thus the distribution of the transformed data $\mathbf{Ky}$ involves only the $m + 1$ variance components as unknown parameters. In order to estimate the variance components, the next step in REML is to maximize the likelihood of $\mathbf{Ky}$ with respect to these variance components. We now develop a set of estimating equations by taking partial derivatives of the log likelihood with respect to the variance components, and setting them to zero.

Theorem 17.4c. Consider the model in which $\mathbf{y}$ is $N_n(\mathbf{X}\boldsymbol{\beta}, \boldsymbol{\Sigma})$, where $\boldsymbol{\Sigma} = \sum_{i=0}^{m} \sigma_i^2 \mathbf{Z}_i \mathbf{Z}_i'$, and let $\mathbf{K}$ be specified as in Theorem 17.4a. Then a set of $m+1$ estimating equations for $\sigma_0^2, \ldots, \sigma_m^2$ is given by

$$\mathrm{tr}[\mathbf{K}'(\mathbf{K}\boldsymbol{\Sigma}\mathbf{K}')^{-1}\mathbf{K}\mathbf{Z}_i\mathbf{Z}_i'] = \mathbf{y}'\mathbf{K}'(\mathbf{K}\boldsymbol{\Sigma}\mathbf{K}')^{-1}\mathbf{K}\mathbf{Z}_i\mathbf{Z}_i'\mathbf{K}'(\mathbf{K}\boldsymbol{\Sigma}\mathbf{K}')^{-1}\mathbf{K}\mathbf{y} \tag{17.6}$$

for $i = 0, \ldots, m$.

Proof. Since $E(\mathbf{K}\mathbf{y}) = \mathbf{0}$, the log likelihood of $\mathbf{K}\mathbf{y}$ is

$$\begin{aligned}
\ln L(\sigma_0^2, \ldots, \sigma_m^2) &= \frac{n-r}{2} \ln(2\pi) - \frac{1}{2} \ln |\mathbf{K}\boldsymbol{\Sigma}\mathbf{K}'| - \frac{1}{2} \mathbf{y}'\mathbf{K}'(\mathbf{K}\boldsymbol{\Sigma}\mathbf{K}')^{-1}\mathbf{K}\mathbf{y} \\
&= \frac{n-r}{2} \ln(2\pi) - \frac{1}{2} \ln \left| \mathbf{K}\left(\sum_{i=0}^{m} \sigma_i^2 \mathbf{Z}_i\mathbf{Z}_i'\right)\mathbf{K}' \right| \\
&\quad - \frac{1}{2}\mathbf{y}'\mathbf{K}'\left[\mathbf{K}\left(\sum_{i=0}^{m} \sigma_i^2 \mathbf{Z}_i\mathbf{Z}_i'\right)\mathbf{K}'\right]^{-1} \mathbf{K}\mathbf{y}
\end{aligned}$$

Using (2.117) and (2.118) to take the partial derivative of $\ln L(\sigma_0^2, \ldots, \sigma_m^2)$ with respect to each of the σ_i^2's, we obtain

$$\begin{aligned}
\frac{\partial}{\partial \sigma_i^2} \ln L(\sigma_0^2, \ldots, \sigma_m^2) &= -\frac{1}{2} \mathrm{tr}\left((\mathbf{K}\boldsymbol{\Sigma}\mathbf{K}')^{-1} \left[\frac{\partial}{\partial \sigma_i^2} (\mathbf{K}\boldsymbol{\Sigma}\mathbf{K}')\right]\right) \\
&\quad + \frac{1}{2}\mathbf{y}'\mathbf{K}'(\mathbf{K}\boldsymbol{\Sigma}\mathbf{K}')^{-1}\left[\frac{\partial}{\partial \sigma_i^2}(\mathbf{K}\boldsymbol{\Sigma}\mathbf{K}')\right](\mathbf{K}\boldsymbol{\Sigma}\mathbf{K}')^{-1}\mathbf{K}\mathbf{y} \\
&= -\frac{1}{2}\mathrm{tr}[(\mathbf{K}\boldsymbol{\Sigma}\mathbf{K}')^{-1}\mathbf{K}\mathbf{Z}_i\mathbf{Z}_i'\mathbf{K}'] \\
&\quad + \frac{1}{2}\mathbf{y}'\mathbf{K}'(\mathbf{K}\boldsymbol{\Sigma}\mathbf{K}')^{-1}\mathbf{K}\mathbf{Z}_i\mathbf{Z}_i'\mathbf{K}'(\mathbf{K}\boldsymbol{\Sigma}\mathbf{K}')^{-1}\mathbf{K}\mathbf{y} \\
&= -\frac{1}{2}\mathrm{tr}[\mathbf{K}'(\mathbf{K}\boldsymbol{\Sigma}\mathbf{K}')^{-1}\mathbf{K}\mathbf{Z}_i\mathbf{Z}_i'] \\
&\quad + \frac{1}{2}\mathbf{y}'\mathbf{K}'(\mathbf{K}\boldsymbol{\Sigma}\mathbf{K}')^{-1}\mathbf{K}\mathbf{Z}_i\mathbf{Z}_i'\mathbf{K}'(\mathbf{K}\boldsymbol{\Sigma}\mathbf{K}')^{-1}\mathbf{K}\mathbf{y}
\end{aligned}$$

Setting these equations to zero, the result follows. □

It is interesting to note that using Theorem 5.2a, the expected value of the quadratic form on the right side of (17.6) is given by the left side of (17.6).

Applying Theorem 17.4c, we obtain $m+1$ equations in $m+1$ unknown σ_i^2's. In some cases these equations can be simplified to yield closed-form estimating equations. In most cases, numerical methods have to be used to solve the equations (McCulloch and Searle 2001, pp. 263–269).

If the solutions to the equations are nonnegative, the solutions are REML estimates of the variance components. If any of the solutions are negative, the log likelihood must be examined to find values of the variance components within the parameter space (i.e., nonnegative values) that maximize the function.

Example 17.4 (One-Way Random Effects). This is an extension of Example 17.3(d). Four containers are randomly selected from each of three batches produced by a chemical plant. Hence

$$\mathbf{X} = \mathbf{j}_{12},\ \mathbf{Z}_0 = \mathbf{I}_{12},\ \mathbf{Z}_1 = \begin{pmatrix} \mathbf{j}_4 & \mathbf{0}_4 & \mathbf{0}_4 \\ \mathbf{0}_4 & \mathbf{j}_4 & \mathbf{0}_4 \\ \mathbf{0}_4 & \mathbf{0}_4 & \mathbf{j}_4 \end{pmatrix} \quad \text{and} \quad \boldsymbol{\Sigma} = \sigma_0^2\mathbf{I}_{12} + \sigma_1^2\mathbf{Z}_1\mathbf{Z}_1'.$$

Then $\mathbf{I} - \mathbf{H} = \mathbf{I}_{12} - \frac{1}{12}\mathbf{J}_{12}$, a suitable $\mathbf{C}$ would be $\mathbf{C} = (\mathbf{I}_{12}, \mathbf{0}_{12})$, and $\mathbf{K} = \mathbf{C}(\mathbf{I} - \mathbf{H})$. Inserting these matrices into (17.6), it can be shown that we obtain the two estimating equations

$$\begin{aligned} 9\sigma_0^2 &= \mathbf{y}'(\mathbf{I}_{12} - \tfrac{1}{4}\mathbf{Z}_1\mathbf{Z}_1')\mathbf{y}, \\ 2(4\sigma_1^2 + \sigma_0^2) &= \mathbf{y}'(\tfrac{1}{4}\mathbf{Z}_1\mathbf{Z}_1' - \tfrac{1}{12}\mathbf{J}_{12})\mathbf{y}. \end{aligned}$$

From these we obtain the closed-form solutions

$$\begin{aligned} \hat{\sigma}_0^2 &= \frac{\mathbf{y}'\left(\mathbf{I}_{12} - \frac{1}{4}\mathbf{Z}_1\mathbf{Z}_1'\right)\mathbf{y}}{9}, \\ \hat{\sigma}_1^2 &= \frac{\mathbf{y}'\left(\frac{1}{4}\mathbf{Z}_1\mathbf{Z}_1' - \frac{1}{12}\mathbf{J}_{12}\right)\mathbf{y}/2 - \hat{\sigma}_0^2}{4}. \end{aligned}$$

If both $\hat{\sigma}_0^2$ and $\hat{\sigma}_1^2$ are positive, they are the REML estimates of σ_0^2 and σ_1^2. Because $(\mathbf{I}_{12} - \frac{1}{4}\mathbf{Z}_1\mathbf{Z}_1')$ is positive definite, $\hat{\sigma}_0^2$ will always be positive. However, $\hat{\sigma}_1^2$ could be negative. In such a case, the REML estimates become

$$\begin{aligned} \hat{\sigma}_0^2 &= \frac{\mathbf{y}'\left(\mathbf{I}_{12} - \frac{1}{12}\mathbf{J}_{12}\right)\mathbf{y}}{11}, \\ \hat{\sigma}_1^2 &= \mathbf{0}. \end{aligned}$$

□

In practice, the equations in (17.6) are seldom used directly to obtain solutions. The usual procedure involves any of a number of iterative methods (Rao 1997 pp. 104–105, McCulloch and Searle 2001, pp. 265–269) To motivate one of these methods, note that the system of $m+1$ equations generated by (17.6) can be written as

$$\mathbf{M}\boldsymbol{\sigma} = \mathbf{q}, \tag{17.7}$$

where $\boldsymbol{\sigma} = (\sigma_0^2 \sigma_1^2 \cdots \sigma_m^2)'$, $\mathbf{M}$ is a nonsingular $(m+1) \times (m+1)$ matrix with (ij)th element $\text{tr}[\mathbf{K}'(\mathbf{K\Sigma K}')^{-1}\mathbf{KZ}_i\mathbf{Z}_i'\mathbf{K}'(\mathbf{K\Sigma K}')^{-1}\mathbf{KZ}_j\mathbf{Z}_j']$, and $\mathbf{q}$ is an $(m+1) \times 1$ vector with ith element $\mathbf{y}'\mathbf{K}'(\mathbf{K\Sigma K}')^{-1}\mathbf{KZ}_i\mathbf{Z}_i'\mathbf{K}'(\mathbf{K\Sigma K}')^{-1}\mathbf{Ky}$ (Problem 17.6). Equation (17.7) is more complicated than it looks because both $\mathbf{M}$ and $\mathbf{q}$ are themselves functions of $\boldsymbol{\sigma}$. Nonetheless, the equation is useful for stepwise improvement of an initial guess $\boldsymbol{\sigma}_{(1)}$. The method proceeds by computing $\mathbf{M}_{(t)}$ and $\mathbf{q}_{(t)}$ using $\boldsymbol{\sigma}_{(t)}$ at step t. Then let $\boldsymbol{\sigma}_{(t+1)} = \mathbf{M}_{(t)}^{-1}\mathbf{q}_{(t)}$. The procedure continues until $\boldsymbol{\sigma}_{(t)}$ converges.

17.5 INFERENCE FOR $\boldsymbol{\beta}$

17.5.1 An Estimator for $\boldsymbol{\beta}$

Estimates of the variance components can be inserted into $\boldsymbol{\Sigma}$ to obtain $\hat{\boldsymbol{\Sigma}} = \sum_{i=0}^{m} \hat{\sigma}_i^2 \mathbf{Z}_i\mathbf{Z}_i'$. A sensible estimator for $\boldsymbol{\beta}$ is then obtained by replacing $\sigma^2\mathbf{V}$ in equation (7.64) by its estimate, $\hat{\boldsymbol{\Sigma}}$. Generalizing the model to accommodate non-full-rank $\mathbf{X}$ matrices, we obtain

$$\hat{\boldsymbol{\beta}} = (\mathbf{X}'\hat{\boldsymbol{\Sigma}}^{-1}\mathbf{X})^{-}\mathbf{X}'\hat{\boldsymbol{\Sigma}}^{-1}\mathbf{y}. \tag{17.8}$$

This estimator, sometimes called the *estimated generalized least-squares* (EGLS) estimator, is a nonlinear function of $\mathbf{y}$ (since $\hat{\boldsymbol{\Sigma}}$ is a nonlinear function of $\mathbf{y}$). Even if $\mathbf{X}$ is full-rank, $\hat{\boldsymbol{\beta}}$ is not in general a (minimum variance) unbiased estimator (MVUE) or normally distributed. However, it is always asymptotically MVUE and normally distributed (Fuller and Battese 1973).

Similarly, a sensible approximate covariance matrix for $\hat{\boldsymbol{\beta}}$ is, by extension of (12.18), as follows:

$$\text{cov}(\hat{\boldsymbol{\beta}}) = (\mathbf{X}'\hat{\boldsymbol{\Sigma}}^{-1}\mathbf{X})^{-}\mathbf{X}'\hat{\boldsymbol{\Sigma}}^{-1}\mathbf{X}(\mathbf{X}'\hat{\boldsymbol{\Sigma}}^{-1}\mathbf{X})^{-}. \tag{17.9}$$

Of course, if $\mathbf{X}$ is full-rank, the expression in (17.9) simplifies to

$$\text{cov}(\hat{\boldsymbol{\beta}}) = (\mathbf{X}'\hat{\boldsymbol{\Sigma}}^{-1}\mathbf{X})^{-1}.$$

17.5.2 Large-Sample Inference for Estimable Functions of $\boldsymbol{\beta}$

Carrying the procedure of replacing $\sigma^2\mathbf{V}$ by its estimate $\hat{\boldsymbol{\Sigma}}$ a bit further, it seems reasonable to extend Theorem 12.7c(ii) and conclude that for a known full-rank $g \times p$ matrix $\mathbf{L}$ whose rows define estimable functions of $\boldsymbol{\beta}$

$$\mathbf{L}\hat{\boldsymbol{\beta}} \text{ is approximately } N_g[\mathbf{L}\boldsymbol{\beta}, \mathbf{L}(\mathbf{X}'\hat{\boldsymbol{\Sigma}}^{-1}\mathbf{X})^{-}\mathbf{L}'] \tag{17.10}$$

and therefore by (5.35)

$$(\mathbf{L}\hat{\boldsymbol{\beta}} - \mathbf{L}\boldsymbol{\beta})'[\mathbf{L}(\mathbf{X}'\hat{\boldsymbol{\Sigma}}^{-1}\mathbf{X})^{-}\mathbf{L}']^{-1}(\mathbf{L}\hat{\boldsymbol{\beta}} - \mathbf{L}\boldsymbol{\beta}) \text{ is approximately } \chi^2(g). \tag{17.11}$$

If so, an approximate general linear hypothesis test for the testable hypothesis $H_0 : \mathbf{L}\boldsymbol{\beta} = \mathbf{t}$ is carried out using the test statistic

$$G = (\mathbf{L}\hat{\boldsymbol{\beta}} - \mathbf{t})'[\mathbf{L}(\mathbf{X}'\hat{\boldsymbol{\Sigma}}^{-1}\mathbf{X})^{-}\mathbf{L}']^{-1}(\mathbf{L}\hat{\boldsymbol{\beta}} - \mathbf{t}). \tag{17.12}$$

If H_0 is true, G is approximately distributed as $\chi^2(g)$. If H_0 is false, G is approximately distributed as $\chi^2(g,\lambda)$ where $\lambda = (\mathbf{L}\boldsymbol{\beta} - \mathbf{t})'[\mathbf{L}(\mathbf{X}'\boldsymbol{\Sigma}^{-1}\mathbf{X})^{-}\mathbf{L}']^{-1}(\mathbf{L}\boldsymbol{\beta} - \mathbf{t})$. The test is carried out by rejecting H_0 if $G \geq \chi^2_{\alpha,\, g}$.

Similarly, an approximate $100(1 - \alpha)\%$ confidence interval for a single estimable function $\mathbf{c}'\boldsymbol{\beta}$ is given by

$$\mathbf{c}'\hat{\boldsymbol{\beta}} \pm z_{\alpha/2}\sqrt{\mathbf{c}'(\mathbf{X}'\hat{\boldsymbol{\Sigma}}^{-1}\mathbf{X})^{-}\mathbf{c}}. \tag{17.13}$$

Approximate joint confidence regions for $\boldsymbol{\beta}$, approximate confidence intervals for individual $\boldsymbol{\beta}_j$'s, and approximate confidence intervals for $E(\mathbf{y})$ can be similarly proposed using (17.10) and (17.11).

17.5.3 Small-Sample Inference for Estimable Functions of $\boldsymbol{\beta}$

The inferences of Section 17.5.2 are not satisfactory for small samples. Exact small-sample inferences based on the t distribution and F distribution are available in rare cases, but are not generally available for mixed models. However, much work has been done on approximate inference for small sample mixed models.

First we discuss the exact small-sample inferences that are available in rare cases, usually involving balanced designs, nonnegative solutions to the REML equations, and certain estimable functions. In order for this to occur, $[\mathbf{L}(\mathbf{X}'\hat{\boldsymbol{\Sigma}}^{-1}\mathbf{X})^{-}\mathbf{L}']^{-1}$ must be of the form $(d/w)\mathbf{Q}$, where w is a central chi-square random variable with d degrees of freedom, and independently $(\mathbf{L}\hat{\boldsymbol{\beta}} - \mathbf{t})'\mathbf{Q}(\mathbf{L}\hat{\boldsymbol{\beta}} - \mathbf{t})$ must be distributed as a (possibly noncentral) chi-square random variable with g degrees of freedom.

Under these conditions, by (5.30), the statistic

$$\frac{(\mathbf{L}\hat{\boldsymbol{\beta}}-\mathbf{t})'\mathbf{Q}(\mathbf{L}\hat{\boldsymbol{\beta}}-\mathbf{t})}{g}\frac{w}{d}=\frac{(\mathbf{L}\hat{\boldsymbol{\beta}}-\mathbf{t})'[\mathbf{L}(\mathbf{X}'\hat{\boldsymbol{\Sigma}}^{-1}\mathbf{X})^{-}\mathbf{L}']^{-1}(\mathbf{L}\hat{\boldsymbol{\beta}}-\mathbf{t})}{g}$$

is F-distributed. We demonstrate this with an example.

Example 17.5 (Balanced Split-Plot Study). Similarly to Example 17.3c, consider a 3×2 balanced factorial experiment carried out using six main units, each of which is subdivided into two subunits. The levels of A are each randomly assigned to two of the main units, and the levels of B are randomly assigned to subunits within main units. We assume that the data are sorted by replicates (with two complete replicates in the study), levels of A, and then levels of B. We use the cell means parameterization as in Section 14.3.1. The means in $\boldsymbol{\beta}$ are sorted by levels of A and then levels of B. Hence

$$\mathbf{X}=\begin{pmatrix}\mathbf{I}_6\\ \mathbf{I}_6\end{pmatrix} \text{ and } \boldsymbol{\Sigma}=\begin{pmatrix}\boldsymbol{\Sigma}_1 & \mathbf{O} & \mathbf{O} & \mathbf{O} & \mathbf{O} & \mathbf{O}\\ \mathbf{O} & \boldsymbol{\Sigma}_1 & \mathbf{O} & \mathbf{O} & \mathbf{O} & \mathbf{O}\\ \mathbf{O} & \mathbf{O} & \boldsymbol{\Sigma}_1 & \mathbf{O} & \mathbf{O} & \mathbf{O}\\ \mathbf{O} & \mathbf{O} & \mathbf{O} & \boldsymbol{\Sigma}_1 & \mathbf{O} & \mathbf{O}\\ \mathbf{O} & \mathbf{O} & \mathbf{O} & \mathbf{O} & \boldsymbol{\Sigma}_1 & \mathbf{O}\\ \mathbf{O} & \mathbf{O} & \mathbf{O} & \mathbf{O} & \mathbf{O} & \boldsymbol{\Sigma}_1\end{pmatrix}, \quad \text{where}$$

$$\boldsymbol{\Sigma}_1=\begin{pmatrix}\sigma_1^2+\sigma^2 & \sigma_1^2\\ \sigma_1^2 & \sigma_1^2+\sigma^2\end{pmatrix}.$$

We test the no-interaction hypothesis $H_0 : \mathbf{L}\boldsymbol{\beta}=\mathbf{0}$, where

$$\mathbf{L}=\begin{pmatrix}1 & -1 & -1 & 1 & 0 & 0\\ 1 & -1 & 0 & 0 & -1 & 1\end{pmatrix}.$$

Assuming that the REML estimating equations yield nonnegative solutions, $\hat{\sigma}^2$ is given by

$$\hat{\sigma}^2=\tfrac{1}{12}\mathbf{y}'\begin{pmatrix}\mathbf{R} & \mathbf{O} & \mathbf{O} & -\mathbf{R} & \mathbf{O} & \mathbf{O}\\ \mathbf{O} & \mathbf{R} & \mathbf{O} & \mathbf{O} & -\mathbf{R} & \mathbf{O}\\ \mathbf{O} & \mathbf{O} & \mathbf{R} & \mathbf{O} & \mathbf{O} & -\mathbf{R}\\ -\mathbf{R} & \mathbf{O} & \mathbf{O} & \mathbf{R} & \mathbf{O} & \mathbf{O}\\ \mathbf{O} & -\mathbf{R} & \mathbf{O} & \mathbf{O} & \mathbf{R} & \mathbf{O}\\ \mathbf{O} & \mathbf{O} & -\mathbf{R} & \mathbf{O} & \mathbf{O} & \mathbf{R}\end{pmatrix}\mathbf{y} \quad \text{where } \mathbf{R}=\begin{pmatrix}1 & -1\\ -1 & 1\end{pmatrix}.$$

Multiplying and simplifying, we obtain

$$\mathbf{X}'\hat{\boldsymbol{\Sigma}}^{-1}\mathbf{X} = 2\begin{pmatrix} \hat{\boldsymbol{\Sigma}}_1^{-1} & \mathbf{O} & \mathbf{O} \\ \mathbf{O} & \hat{\boldsymbol{\Sigma}}_1^{-1} & \mathbf{O} \\ \mathbf{O} & \mathbf{O} & \hat{\boldsymbol{\Sigma}}_1^{-1} \end{pmatrix}.$$

By (2.52), we have

$$(\mathbf{X}'\hat{\boldsymbol{\Sigma}}^{-1}\mathbf{X})^{-1} = \tfrac{1}{2}\begin{pmatrix} \hat{\boldsymbol{\Sigma}}_1^{-1} & \mathbf{O} & \mathbf{O} \\ \mathbf{O} & \hat{\boldsymbol{\Sigma}}_1^{-1} & \mathbf{O} \\ \mathbf{O} & \mathbf{O} & \hat{\boldsymbol{\Sigma}}_1^{-1} \end{pmatrix}.$$

Thus

$$\begin{aligned}
&[\mathbf{L}(\mathbf{X}'\hat{\boldsymbol{\Sigma}}^{-1}\mathbf{X})^{-1}\mathbf{L}']^{-1} \\
&= \left[\tfrac{1}{2}\begin{pmatrix} 1 & -1 & -1 & 1 & 0 & 0 \\ 1 & -1 & 0 & 0 & -1 & 1 \end{pmatrix} \begin{pmatrix} \hat{\boldsymbol{\Sigma}}_1 & \mathbf{O} & \mathbf{O} \\ \mathbf{O} & \hat{\boldsymbol{\Sigma}}_1 & \mathbf{O} \\ \mathbf{O} & \mathbf{O} & \hat{\boldsymbol{\Sigma}}_1 \end{pmatrix} \begin{pmatrix} 1 & 1 \\ -1 & -1 \\ -1 & 0 \\ 1 & 0 \\ 0 & -1 \\ 0 & 1 \end{pmatrix} \right]^{-1} \\
&= -\frac{3}{3\hat{\sigma}^2/\hat{\sigma}^2}\left[\frac{1}{3\sigma^2}\begin{pmatrix} 2 & -1 \\ -1 & 2 \end{pmatrix}\right] \\
&= \frac{3}{w}\mathbf{Q},
\end{aligned}$$

where

$$w = \frac{3\hat{\sigma}^2}{\sigma^2} \quad \text{and} \quad \mathbf{Q} = \frac{1}{3\sigma^2}\begin{pmatrix} 2 & -1 \\ -1 & 2 \end{pmatrix}. \tag{17.14}$$

Also note that in this particular case, the EGLS estimator is equal to the ordinary least-squares estimator for $\boldsymbol{\beta}$ since

$$\begin{aligned}
\hat{\boldsymbol{\beta}} &= (\mathbf{X}'\hat{\boldsymbol{\Sigma}}^{-1}\mathbf{X})^{-1}\mathbf{X}'\hat{\boldsymbol{\Sigma}}^{-1}\mathbf{y} \\
&= \tfrac{1}{2}\begin{pmatrix} \hat{\boldsymbol{\Sigma}}_1 & \mathbf{O} & \mathbf{O} \\ \mathbf{O} & \hat{\boldsymbol{\Sigma}}_1 & \mathbf{O} \\ \mathbf{O} & \mathbf{O} & \hat{\boldsymbol{\Sigma}}_1 \end{pmatrix} \begin{pmatrix} \hat{\boldsymbol{\Sigma}}_1^{-1} & \mathbf{O} & \mathbf{O} & \hat{\boldsymbol{\Sigma}}_1^{-1} & \mathbf{O} & \mathbf{O} \\ \mathbf{O} & \hat{\boldsymbol{\Sigma}}_1^{-1} & \mathbf{O} & \mathbf{O} & \hat{\boldsymbol{\Sigma}}_1^{-1} & \mathbf{O} \\ \mathbf{O} & \mathbf{O} & \hat{\boldsymbol{\Sigma}}_1^{-1} & \mathbf{O} & \mathbf{O} & \hat{\boldsymbol{\Sigma}}_1^{-1} \end{pmatrix}\mathbf{y} \\
&= \tfrac{1}{2}(\,\mathbf{I}_6 \quad \mathbf{I}_6\,)\mathbf{y} \\
&= (\mathbf{X}'\mathbf{X})^{-1}\mathbf{X}'\mathbf{y}.
\end{aligned}$$

Hence

$$(\mathbf{L}\hat{\boldsymbol{\beta}})'\mathbf{Q}(\mathbf{L}\hat{\boldsymbol{\beta}}) = \mathbf{y}'\mathbf{X}(\mathbf{X}'\mathbf{X})^{-1}\mathbf{L}'\mathbf{Q}\mathbf{L}(\mathbf{X}'\mathbf{X})^{-1}\mathbf{X}'\mathbf{y}.$$

It can be shown that $\mathbf{X}(\mathbf{X}'\mathbf{X})^{-1}\mathbf{L}'\mathbf{Q}\mathbf{L}(\mathbf{X}'\mathbf{X})^{-1}\mathbf{X}'\boldsymbol{\Sigma}$ is idempotent, and thus $(\mathbf{L}\hat{\boldsymbol{\beta}})'\mathbf{Q}(\mathbf{L}\hat{\boldsymbol{\beta}})$ is distributed as a chi-square with 2 degrees of freedom. It can similarly be shown that w is a chi-square with 3 degrees of freedom. Furthermore, w and $(\mathbf{L}\hat{\boldsymbol{\beta}})'\mathbf{Q}(\mathbf{L}\hat{\boldsymbol{\beta}})$ are independent chi-squares because of Theorem 5.6b. Thus we can test $H_0 : \mathbf{L}\boldsymbol{\beta} = \mathbf{0}$ using the test statistic $(\mathbf{L}\hat{\boldsymbol{\beta}})'[\mathbf{L}(\mathbf{X}'\hat{\boldsymbol{\Sigma}}^{-1}\mathbf{X})^{-1}\mathbf{L}']^{-1}(\mathbf{L}\hat{\boldsymbol{\beta}})/2$ because its distribution is exactly an F distribution.

If even one observation of this design is missing, exact small-sample inferences are not available for $\mathbf{L}\boldsymbol{\beta}$. Exact inferences are not available even when the design is balanced for estimable functions such as $\mathbf{c}'\boldsymbol{\beta}$ where $\mathbf{c}' = (1 \quad 0 \quad 0 \quad -1 \quad 0 \quad 0)$. □

In most cases, approximate small-sample methods must be used. The exact distribution of

$$t = \frac{\mathbf{c}'\hat{\boldsymbol{\beta}}}{\sqrt{\mathbf{c}'(\mathbf{X}'\hat{\boldsymbol{\Sigma}}^{-1}\mathbf{X})^{-}\mathbf{c}}} \tag{17.15}$$

is unknown in general (McCulloch and Searle 2001, p. 167). However, a satisfactory small-sample test of $H_0 : \mathbf{c}'\boldsymbol{\beta} = \mathbf{0}$ or confidence interval for $\mathbf{c}'\boldsymbol{\beta}$ is available by assuming that t approximately follows a t distribution with unknown degrees of freedom d (Giesbrecht and Burns 1985). To calculate d, we follow the premise of Satterthwaite (1941) to assume, analogously to Theorem 8.4aiii, that

$$\frac{d[\mathbf{c}'(\mathbf{X}'\hat{\boldsymbol{\Sigma}}^{-1}\mathbf{X})^{-}\mathbf{c}]}{\mathbf{c}'(\mathbf{X}'\boldsymbol{\Sigma}^{-1}\mathbf{X})^{-}\mathbf{c}} \tag{17.16}$$

approximately follows the central chi-square distribution. Equating the variance of the expression in (17.16)

$$\operatorname{var}\left[\frac{d[\mathbf{c}'(\mathbf{X}'\hat{\boldsymbol{\Sigma}}^{-1}\mathbf{X})^{-}\mathbf{c}]}{\mathbf{c}'(\mathbf{X}'\boldsymbol{\Sigma}^{-1}\mathbf{X})^{-}\mathbf{c}}\right] = \left[\frac{d}{\mathbf{c}'(\mathbf{X}'\boldsymbol{\Sigma}^{-1}\mathbf{X})^{-}\mathbf{c}}\right]^2 \operatorname{var}[\mathbf{c}'(\mathbf{X}'\hat{\boldsymbol{\Sigma}}^{-1}\mathbf{X})^{-}\mathbf{c}],$$

to the variance of a central chi-square distribution, $2d$ (Theorem 5.3a), we obtain the approximation

$$d \doteq \frac{2[\mathbf{c}'(\mathbf{X}'\hat{\boldsymbol{\Sigma}}^{-1}\mathbf{X})^{-}\mathbf{c}]^2}{\operatorname{var}[\mathbf{c}'(\mathbf{X}'\hat{\boldsymbol{\Sigma}}^{-1}\mathbf{X})^{-}\mathbf{c}]}. \tag{17.17}$$

This approximation cannot be used, of course, unless var$[\mathbf{c}'(\mathbf{X}'\hat{\boldsymbol{\Sigma}}^{-1}\mathbf{X})^{-}\mathbf{c}]$ is known or can be estimated. We obtain an estimate of var$[\mathbf{c}'(\mathbf{X}'\hat{\boldsymbol{\Sigma}}^{-1}\mathbf{X})^{-}\mathbf{c}]$ using the multivariate delta method (Lehmann 1999, p. 315). This method uses the first-order multivariate Taylor series (Harville 1997, p. 288) to approximate the variance of any scalar-valued function of a random vector, say, $f(\boldsymbol{\theta})$. By this method var$[f(\boldsymbol{\theta})]$ is approximated as

$$\text{var}\,[f(\boldsymbol{\theta})] \doteq \left.\frac{\partial f(\boldsymbol{\theta})}{\partial \boldsymbol{\theta}}\right|'_{\boldsymbol{\theta}=\hat{\boldsymbol{\theta}}} \hat{\boldsymbol{\Sigma}}_{\hat{\boldsymbol{\theta}}} \left.\frac{\partial f(\boldsymbol{\theta})}{\partial \boldsymbol{\theta}}\right|_{\boldsymbol{\theta}=\hat{\boldsymbol{\theta}}}$$

where

$$\left.\frac{\partial f(\boldsymbol{\theta})}{\partial \boldsymbol{\theta}}\right|_{\boldsymbol{\theta}=\hat{\boldsymbol{\theta}}}$$

is the vector of partial derivatives of $f(\boldsymbol{\theta})$ with respect to $\boldsymbol{\theta}$ evaluated at $\hat{\boldsymbol{\theta}}$ and $\hat{\boldsymbol{\Sigma}}_{\hat{\boldsymbol{\theta}}}$ denotes an estimate of the covariance matrix of $\hat{\boldsymbol{\theta}}$. In the case of inference for $\mathbf{c}'\boldsymbol{\beta}$ in the mixed linear model (17.4), let $\boldsymbol{\theta} = \boldsymbol{\sigma}$ and $f(\boldsymbol{\sigma}) = [\mathbf{c}'(\mathbf{X}'\Sigma^{-1}\mathbf{X})^{-}\mathbf{c}]$. Then

$$\left.\frac{\partial f(\boldsymbol{\sigma})}{\partial \boldsymbol{\sigma}}\right|_{\boldsymbol{\sigma}=\hat{\boldsymbol{\sigma}}} = -\begin{pmatrix} \mathbf{c}'(\mathbf{X}'\hat{\boldsymbol{\Sigma}}^{-1}\mathbf{X})^{-}\mathbf{X}'\hat{\boldsymbol{\Sigma}}^{-1}\mathbf{Z}_0\mathbf{Z}_0'\hat{\boldsymbol{\Sigma}}^{-1}\mathbf{X}(\mathbf{X}'\hat{\boldsymbol{\Sigma}}^{-1}\mathbf{X})^{-}\mathbf{c} \\ \mathbf{c}'(\mathbf{X}'\hat{\boldsymbol{\Sigma}}^{-1}\mathbf{X})^{-}\mathbf{X}'\hat{\boldsymbol{\Sigma}}^{-1}\mathbf{Z}_1\mathbf{Z}_1'\hat{\boldsymbol{\Sigma}}^{-1}\mathbf{X}(\mathbf{X}'\hat{\boldsymbol{\Sigma}}^{-1}\mathbf{X})^{-}\mathbf{c} \\ \vdots \\ \mathbf{c}'(\mathbf{X}'\hat{\boldsymbol{\Sigma}}^{-1}\mathbf{X})^{-}\mathbf{X}'\hat{\boldsymbol{\Sigma}}^{-1}\mathbf{Z}_m\mathbf{Z}_m'\hat{\boldsymbol{\Sigma}}^{-1}\mathbf{X}(\mathbf{X}'\hat{\boldsymbol{\Sigma}}^{-1}\mathbf{X})^{-}\mathbf{c} \end{pmatrix}.$$

Also $\hat{\boldsymbol{\Sigma}}_{\hat{\boldsymbol{\sigma}}}$, an estimate of the covariance matrix of $\hat{\boldsymbol{\sigma}}$, can be obtained as the inverse of the negative Hessian [the matrix of second derivatives — see Harville (1997, p. 288)] of the restricted log-likelihood function (Theorem 17.4c) evaluated at $\hat{\boldsymbol{\sigma}}$ (Pawitan 2001, pp. 226, 258).

We now generalize this idea obtain the approximate small-sample distribution of

$$F = \frac{(\mathbf{L}\hat{\boldsymbol{\beta}} - \mathbf{L}\boldsymbol{\beta})'[\mathbf{L}(\mathbf{X}'\hat{\boldsymbol{\Sigma}}^{-1}\mathbf{X})^{-}\mathbf{L}']^{-1}(\mathbf{L}\hat{\boldsymbol{\beta}} - \mathbf{L}\boldsymbol{\beta})}{g} \tag{17.18}$$

in order to develop tests for $H_0: \mathbf{L}\boldsymbol{\beta} = \mathbf{t}$ and joint confidence regions for $\mathbf{L}\boldsymbol{\beta}$. We obtain these inferences by assuming that the distribution of F is approximately an F distribution with numerator degrees of freedom g, and unknown denominator degrees of freedom ν (Fai and Cornelius 1996). The method involves the spectral decomposition (see Theorem 2.12b) of $[\mathbf{L}(\mathbf{X}'\hat{\boldsymbol{\Sigma}}^{-1}\mathbf{X})^{-}\mathbf{L}']^{-1}$ to yield

$$\mathbf{P}'[\mathbf{L}(\mathbf{X}'\hat{\boldsymbol{\Sigma}}^{-1}\mathbf{X}^{-}\mathbf{L}']^{-1}\mathbf{P} = \mathbf{D},$$

where $\mathbf{D} = \text{diag}(\lambda_1, \lambda_2, \ldots, \lambda_m)$ is the diagonal matrix of eigenvalues and $\mathbf{P} = (\mathbf{p}_1, \mathbf{p}_2, \ldots, \mathbf{p}_m)$ is the orthogonal matrix of normalized eigenvectors of $[\mathbf{L}(\mathbf{X}'\hat{\boldsymbol{\Sigma}}^{-1}\mathbf{X})^{-}\mathbf{L}']^{-1}$. Using this decomposition, $G = gF$ can be written as

$$G = \sum_{i=1}^{g} \frac{(\mathbf{p}'_i\mathbf{L}\hat{\boldsymbol{\beta}})^2}{\lambda_i} = \sum_{i=1}^{g} t_i^2 \tag{17.19}$$

where the t_i's are approximate independent t-variables with respective degrees of freedom ν_i.

We compute the ν_i values by repeatedly applying equation (17.16). Then we find ν such that $F = g^{-1}G$ is distributed approximately as $F_{g,\nu}$. Since the square of a t-distributed random variable with ν_i degrees of freedom is an F-distributed random variable with 1 and ν_i degrees of freedom:

$$\begin{aligned} E(G) &= E\left(\sum_{i=1}^{g} t_i^2\right) \\ &= \sum_{i=1}^{g} \frac{\nu_i}{\nu_i - 2} \quad \text{[by (5.34)].} \end{aligned}$$

Now, since $E(F) = 1/g\, E(G) = \nu/(\nu - 2)$,

$$\nu = \frac{2E(G)}{E(G) - g} = 2\left(\sum_{i=1}^{g} \frac{\nu_i}{\nu_i - 2}\right) \Bigg/ \left[\left(\sum_{i=1}^{g} \frac{\nu_i}{\nu_i - 2}\right) - g\right]. \tag{17.20}$$

A method due to Kenward and Roger (1997) provides further improvements for small-sample inferences in mixed models.

1. The method adjusts for two sources of bias in $\mathbf{L}(\mathbf{X}'\hat{\boldsymbol{\Sigma}}^{-1}\mathbf{X})^{-}\mathbf{L}'$ as an estimator of the covariance matrix of $\mathbf{L}\hat{\boldsymbol{\beta}}$ in small-sample situations, namely, that $\mathbf{L}(\mathbf{X}'\boldsymbol{\Sigma}^{-1}\mathbf{X})^{-}\mathbf{L}'$ does not account for the variability in $\hat{\boldsymbol{\sigma}}$, and that $\mathbf{L}(\mathbf{X}'\hat{\boldsymbol{\Sigma}}^{-1}\mathbf{X})^{-}\mathbf{L}'$ is a biased estimator of $\mathbf{L}(\mathbf{X}'\boldsymbol{\Sigma}^{-1}\mathbf{X})^{-}\mathbf{L}'$. Kackar and Harville (1984) give an approximation to the first source of bias, and Kenward and Roger (1997) propose an adjustment for the second source of bias. Both adjustments are based on a Taylor series expansion around $\boldsymbol{\sigma}$ (Kenward and Roger 1997, McCulloch and Searle 2001, pp. 164–167). The adjusted approximate covariance matrix of $\mathbf{L}\hat{\boldsymbol{\beta}}$ is

$$\begin{aligned} \hat{\boldsymbol{\Sigma}}^*_{\mathbf{L}\hat{\boldsymbol{\beta}}} = \mathbf{L}[\mathbf{X}'\boldsymbol{\Sigma}^{-1}\mathbf{X})^{-} + 2(\mathbf{X}'\boldsymbol{\Sigma}^{-1}\mathbf{X})^{-}\left\{\sum_{i=0}^{m}\sum_{j=0}^{m} s_{ij}(\mathbf{Q}_{ij} - \mathbf{P}_i\hat{\boldsymbol{\Sigma}}_{\hat{\boldsymbol{\beta}}}\mathbf{P}_j)\right\} \\ \times (\mathbf{X}'\boldsymbol{\Sigma}^{-1}\mathbf{X})^{-}]\mathbf{L}' \end{aligned} \tag{17.21}$$

where s_{ij} is the (i, j)th element of $\hat{\boldsymbol{\Sigma}}_{\hat{\boldsymbol{\sigma}}}$,

$$\mathbf{Q}_{ij} = \mathbf{X}' \frac{\partial \hat{\boldsymbol{\Sigma}}^{-1}}{\partial \sigma_i^2} \hat{\boldsymbol{\Sigma}} \frac{\partial \hat{\boldsymbol{\Sigma}}}{\partial \sigma_i^2} \mathbf{X}, \quad \text{and} \quad \mathbf{P}_i = \mathbf{X}' \frac{\partial \hat{\boldsymbol{\Sigma}}^{-1}}{\partial \sigma_i^2} \mathbf{X}.$$

2. Kenward and Roger (1997) assume that

$$F^* = \delta F_{\mathrm{KR}} = \frac{\delta}{g} (\mathbf{L}\hat{\boldsymbol{\beta}})' \hat{\boldsymbol{\Sigma}}^*_{\mathbf{L}\hat{\boldsymbol{\beta}}} (\mathbf{L}\hat{\boldsymbol{\beta}}) \tag{17.22}$$

is approximately F-distributed with two (rather than one) adjustable constants, a scale factor δ, and the denominator degrees of freedom ν. They use a second-order Taylor series expansion (Harville 1997, p. 289) of $\hat{\boldsymbol{\Sigma}}^{*\,-1}_{\mathbf{L}\hat{\boldsymbol{\beta}}}$ around $\boldsymbol{\sigma}$ and conditional expectation relationships to yield $E(F_{\mathrm{KR}})$ and $\mathrm{var}(F_{\mathrm{KR}})$ approximately. After equating these to the mean (5.29) and variance of the F distribution to solve for δ and ν, they obtain

$$\nu = 4 + \frac{g + 2}{g\gamma - 1}$$

and

$$\delta = \frac{\nu}{E(F_{\mathrm{KR}})(\nu - 2)}$$

where

$$\gamma = \frac{\mathrm{var}(F_{\mathrm{KR}})}{2E(F_{\mathrm{KR}})^2}.$$

These small-sample methods result in confidence coefficients and type I error rates closer to target values than do the large-sample methods. However, they involve many approximations, and it is therefore not surprising that simulation studies have shown that their statistical properties are not universally satisfactory (Schaalje et al. 2002, Gomez et al. 2005, Keselman et al. 1999).

Another approach to small-sample inferences in mixed linear models is the Bayesian approach (Chapter 11). Bayesian linear mixed models are not much harder to specify than Bayesian linear models, and Markov chain Monte Carlo methods can be used to draw samples from exact small-sample posterior distributions (Gilks et al. 1998, pp. 275–320).

17.6 INFERENCE FOR THE $\mathbf{a}_i$

A new kind of estimation problem sometimes arises for the linear mixed model in (17.2)

$$\mathbf{y} = \mathbf{X}\boldsymbol{\beta} + \sum_{i=1}^{m} \mathbf{Z}_i \mathbf{a}_i + \boldsymbol{\varepsilon}, \tag{17.23}$$

namely, the problem of estimation of realized values of the random components (the $\mathbf{a}_i$'s) or linear functions of them. For simplicity, and without loss of generality, we rewrite (17.22) as

$$\mathbf{y} = \mathbf{X}\boldsymbol{\beta} + \mathbf{Z}\mathbf{a} + \boldsymbol{\varepsilon}, \tag{17.24}$$

where $\mathbf{Z} = (\mathbf{Z}_1\mathbf{Z}_2 \ldots \mathbf{Z}_m)$, $\mathbf{a} = (\mathbf{a}_1'\mathbf{a}_2' \ldots \mathbf{a}_m')'$, $\boldsymbol{\varepsilon}$ is $N(\mathbf{0}, \sigma^2\mathbf{I}_n)$, $\mathbf{a}$ is $N(\mathbf{0}, \mathbf{G})$ where

$$G = \begin{pmatrix} \sigma_1^2\mathbf{I}_{n_1} & \mathbf{O} & \cdots & \mathbf{O} \\ \mathbf{O} & \sigma_2^2\mathbf{I}_{n_2} & \cdots & \mathbf{O} \\ \vdots & \vdots & \ddots & \vdots \\ \mathbf{O} & \mathbf{O} & \cdots & \sigma_m^2\mathbf{I}_{n_m} \end{pmatrix},$$

and $\text{cov}(\boldsymbol{\varepsilon}, \mathbf{a}) = \mathbf{0}$. Then the problem can be expressed as that of estimating $\mathbf{a}$ or a linear function $\mathbf{U}\mathbf{a}$. To differentiate this problem from inference for an estimable function of $\boldsymbol{\beta}$, the current problem is often referred to as *prediction of a random effect*.

Prediction of random effects dates back at least to the pioneering work of Henderson (1950) on prediction of the "value" of a genetic line of animals or plants, where the line is viewed as a random selection from a population of such lines. In education the specific effects of randomly chosen schools might be of interest, in medical research the effect of a randomly chosen clinic may be desired, and in agriculture the effect of a specific year on crop yields may be of interest. The phenomenon of *regression to the mean* (Stigler 2000) for repeated measurements is closely related to prediction of random effects.

The general problem is that of predicting $\mathbf{a}$ for a given value of the observation vector $\mathbf{y}$. Note that because of the model in (17.23), $\mathbf{a}$ and $\mathbf{y}$ are jointly multivariate normal, and

$$\begin{aligned} \text{cov}(\mathbf{a}, \mathbf{y}) &= \text{cov}(\mathbf{a}, \mathbf{X}\boldsymbol{\beta} + \mathbf{Z}\mathbf{a} + \boldsymbol{\varepsilon}) \\ &= \text{cov}(\mathbf{a}, \mathbf{Z}\mathbf{a} + \boldsymbol{\varepsilon}) \\ &= \text{cov}(\mathbf{a}, \mathbf{Z}\mathbf{a}) + \text{cov}(\mathbf{a}, \boldsymbol{\varepsilon}) \quad \text{(see Problem 3.19)} \\ &= \mathbf{G}\mathbf{Z}' + \mathbf{O} \\ &= \mathbf{G}\mathbf{Z}'. \end{aligned}$$

By extension of Theorem 10.6 to the case of a random vector $\mathbf{a}$, the predictor based on $\mathbf{y}$ that minimizes the mean squared error is $E(\mathbf{a}|\mathbf{y})$. To be more precise, the vector function $t(\mathbf{y})$ that minimizes $E[\mathbf{a} - t(\mathbf{y})]'[\mathbf{a} - t(\mathbf{y})]$ is given by $t(\mathbf{y}) = E(\mathbf{a}|\mathbf{y})$.

Since $\mathbf{a}$ and $\mathbf{y}$ are jointly multivariate normal, we have, by (4.26)

$$\begin{aligned} E(\mathbf{a}|\mathbf{y}) &= E(\mathbf{a}) + \text{cov}(\mathbf{a}, \mathbf{y})[\text{cov}(\mathbf{y})]^{-1}[\mathbf{y} - E(\mathbf{y})] \\ &= \mathbf{0} + \mathbf{G}\mathbf{Z}'\boldsymbol{\Sigma}^{-1}(\mathbf{y} - \mathbf{X}\boldsymbol{\beta}) \qquad (17.25) \\ &= \mathbf{G}\mathbf{Z}'\boldsymbol{\Sigma}^{-1}(\mathbf{y} - \mathbf{X}\boldsymbol{\beta}). \end{aligned}$$

If $\boldsymbol{\beta}$ and $\boldsymbol{\Sigma}$ were known, this predictor would be a linear function of $\mathbf{y}$. It is therefore sometimes called the *best linear predictor* (BLP) of $\mathbf{a}$. More generally, the BLP of $\mathbf{Ua}$ is

$$E(\mathbf{Ua}|\mathbf{y}) = \mathbf{UGZ}'\boldsymbol{\Sigma}^{-1}(\mathbf{y} - \mathbf{X}\boldsymbol{\beta}). \tag{17.26}$$

Because the BLP is a linear function of $\mathbf{y}$, the covariance matrix of $E(\mathbf{Ua}|\mathbf{y})$ is

$$\text{cov}[E(\mathbf{Ua}|\mathbf{y})] = \mathbf{UGZ}'\boldsymbol{\Sigma}^{-1}\mathbf{ZGU}'. \tag{17.27}$$

Replacing $\boldsymbol{\beta}$ by $\hat{\boldsymbol{\beta}}$ in (17.8), and replacing $\mathbf{G}$ and $\boldsymbol{\Sigma}$ by $\hat{\mathbf{G}}$ and $\hat{\boldsymbol{\Sigma}}$ (based on the REML estimates of the variance components), we obtain

$$\hat{E}(\mathbf{Ua}|\mathbf{y}) = \mathbf{U}\hat{\mathbf{G}}\mathbf{Z}'\hat{\boldsymbol{\Sigma}}^{-1}(\mathbf{y} - \mathbf{X}\hat{\boldsymbol{\beta}}). \tag{17.28}$$

This predictor is neither unbiased nor a linear function of $\mathbf{y}$. Nonetheless, it is an approximately unbiased estimate of a linear predictor, so it is often referred to as the *estimated best linear unbiased predictor* (EBLUP). Ignoring the randomness in $\hat{\mathbf{G}}$ and $\hat{\boldsymbol{\Sigma}}$, we obtain

$$\begin{aligned}
\text{cov}[\hat{E}(\mathbf{Ua}|\mathbf{y})] &\doteq \text{cov}[\mathbf{UGZ}'\boldsymbol{\Sigma}^{-1}(\mathbf{y} - \mathbf{X}\hat{\boldsymbol{\beta}})] \\
&= \text{cov}\{\mathbf{UGZ}'\boldsymbol{\Sigma}^{-1}[\mathbf{I} - \mathbf{X}(\mathbf{X}'\boldsymbol{\Sigma}^{-1}\mathbf{X})^{-}\mathbf{X}'\boldsymbol{\Sigma}^{-1}]\mathbf{y}\} \\
&= \mathbf{UGZ}'\boldsymbol{\Sigma}^{-1}[\mathbf{I} - \mathbf{X}(\mathbf{X}'\boldsymbol{\Sigma}^{-1}\mathbf{X})^{-}\mathbf{X}'\boldsymbol{\Sigma}^{-1}]\boldsymbol{\Sigma}[\mathbf{I} - \boldsymbol{\Sigma}^{-1}\mathbf{X}(\mathbf{X}'\boldsymbol{\Sigma}^{-1}\mathbf{X})^{-}\mathbf{X}'] \\
&\quad \times \boldsymbol{\Sigma}^{-1}\mathbf{ZGU}' \\
&= \mathbf{UGZ}'[\boldsymbol{\Sigma}^{-1} - \hat{\boldsymbol{\Sigma}}^{-1}\mathbf{X}(\mathbf{X}'\boldsymbol{\Sigma}^{-1}\mathbf{X})^{-}\mathbf{X}'\boldsymbol{\Sigma}^{-1}]\mathbf{ZGU}' \\
&\doteq \mathbf{U}\hat{\mathbf{G}}\mathbf{Z}'[\boldsymbol{\Sigma}^{-1} - \hat{\boldsymbol{\Sigma}}^{-1}\mathbf{X}(\mathbf{X}'\hat{\boldsymbol{\Sigma}}^{-1}\mathbf{X})^{-}\mathbf{X}'\hat{\boldsymbol{\Sigma}}^{-1}]\mathbf{Z}\hat{\mathbf{G}}\mathbf{U}'.
\end{aligned} \tag{17.29}$$

Small-sample improvements to (17.28) have been suggested by Kackar and Harville (1984), and approximate degrees of freedom for inferences based on EBLUPs have been investigated by Jeske and Harville (1988).

Example 17.6 (One-Way Random Effects). To illustrate EBLUP, we continue with the one-way random effects model of Examples 17.3d and 17.4 involving four containers randomly selected from each of three batches produced by a chemical plant. In terms of the linear mixed model in (17.23), we obtain

$$X = \mathbf{j}_{12},\ \boldsymbol{\beta} = \mu,\ \mathbf{Z} = \begin{pmatrix} \mathbf{j}_4 & \mathbf{0}_4 & \mathbf{0}_4 \\ \mathbf{0}_4 & \mathbf{j}_4 & \mathbf{0}_4 \\ \mathbf{0}_4 & \mathbf{0}_4 & \mathbf{j}_4 \end{pmatrix}, \quad \mathbf{G} = \sigma_1^2\mathbf{I}_3, \quad \text{and}$$

$$\boldsymbol{\Sigma} = \sigma^2\mathbf{I}_{12} + \sigma_1^2\mathbf{ZZ}' = \begin{pmatrix} \sigma^2\mathbf{I}_4 + \sigma_1^2\mathbf{J}_4 & \mathbf{O}_4 & \mathbf{O}_4 \\ \mathbf{O}_4 & \sigma^2\mathbf{I}_4 + \sigma_1^2\mathbf{J}_4 & \mathbf{O}_4 \\ \mathbf{O}_4 & \mathbf{O}_4 & \sigma^2\mathbf{I}_4 + \sigma_1^2\mathbf{J}_4 \end{pmatrix}.$$

By (2.52) and (2.53),

$$\boldsymbol{\Sigma}^{-1} = \frac{1}{\sigma^2}\begin{pmatrix} \mathbf{I}_4 - \frac{\sigma_1^2}{\sigma^2+4\sigma_1^2}\mathbf{J}_4 & \mathbf{O}_4 & \mathbf{O}_4 \\ \mathbf{O}_4 & \mathbf{I}_4 - \frac{\sigma_1^2}{\sigma^2+4\sigma_1^2}\mathbf{J}_4 & \mathbf{O}_4 \\ \mathbf{O}_4 & \mathbf{O}_4 & \mathbf{I}_4 - \frac{\sigma_1^2}{\sigma^2+4\sigma_1^2}\mathbf{J}_4 \end{pmatrix}.$$

To predict **a**, which in this case is the vector of random effects associated with the three batches, by (17.27) and using the REML estimates of the variance components, we obtain

$$\text{EBLUP}(\mathbf{a}) = \hat{\mathbf{G}}\mathbf{Z}'\hat{\boldsymbol{\Sigma}}^{-1}(\mathbf{y}-\mathbf{X}\hat{\boldsymbol{\beta}}) = \hat{\sigma}_1^2\mathbf{I}_3\begin{pmatrix} \mathbf{j}_4' & \mathbf{0}_4' & \mathbf{0}_4' \\ \mathbf{0}_4' & \mathbf{j}_4' & \mathbf{0}_4' \\ \mathbf{0}_4' & \mathbf{0}_4' & \mathbf{j}_4' \end{pmatrix}\hat{\boldsymbol{\Sigma}}^{-1}(\mathbf{y}-\hat{\mu}\mathbf{j}_{12})$$

$$= \frac{\hat{\sigma}_1^2}{\hat{\sigma}^2}\begin{pmatrix} \mathbf{j}_4' - \frac{4\hat{\sigma}_1^2}{\hat{\sigma}^2+4\hat{\sigma}_1^2}\mathbf{j}_4' & \mathbf{0}_4' & \mathbf{0}_4' \\ \mathbf{0}_4' & \mathbf{j}_4' - \frac{4\hat{\sigma}_1^2}{\hat{\sigma}^2+4\hat{\sigma}_1^2}\mathbf{j}_4' & \mathbf{0}_4' \\ \mathbf{0}_4' & \mathbf{0}_4' & \mathbf{j}_4' - \frac{4\hat{\sigma}_1^2}{\hat{\sigma}^2+4\hat{\sigma}_1^2}\mathbf{j}_4' \end{pmatrix}(\mathbf{y}-\hat{\mu}\mathbf{j}_{12})$$

$$= \frac{\hat{\sigma}_1^2}{\hat{\sigma}^2+4\hat{\sigma}_1^2}\begin{pmatrix} \mathbf{j}_4' & \mathbf{0}_4' & \mathbf{0}_4' \\ \mathbf{0}_4' & \mathbf{j}_4' & \mathbf{0}_4' \\ \mathbf{0}_4' & \mathbf{0}_4' & \mathbf{j}_4' \end{pmatrix}(\mathbf{y}-\hat{\mu}\mathbf{j}_{12})$$

$$= \frac{\hat{\sigma}_1^2}{\hat{\sigma}^2+4\hat{\sigma}_1^2}\begin{pmatrix} y_{1.}-4\hat{\mu} \\ y_{2.}-4\hat{\mu} \\ y_{3.}-4\hat{\mu} \end{pmatrix} = \frac{4\hat{\sigma}_1^2}{\hat{\sigma}^2+4\hat{\sigma}_1^2}\begin{pmatrix} \bar{y}_{1.}-\bar{y}_{..} \\ \bar{y}_{2.}-\bar{y}_{..} \\ \bar{y}_{3.}-\bar{y}_{..} \end{pmatrix}.$$

Thus

$$\text{EBLUP}(a_i) = \frac{4\hat{\sigma}_1^2}{\hat{\sigma}^2+4\hat{\sigma}_1^2}(\bar{y}_{i.}-\bar{y}_{..}). \tag{17.30}$$

If batch had been considered a fixed factor, and the one-way ANOVA model in (13.1) had been used with the constraint $\sum_i \alpha_i = 0$, we showed in (13.9) that

$$\hat{\alpha}_i = (\bar{y}_{i.}-\bar{y}_{..}).$$

Thus $\text{EBLUP}(a_i) = c\hat{\alpha}_i$ where $0 \leq c \leq 1$. For this reason, EBLUPs are sometimes referred to as *shrinkage estimators*.

The approximate covariance matrix of the EBLUPs in (17.29) can be derived using (17.28), and confidence intervals can then be computed or hypothesis tests carried out. □

An extensive development and discussion of EBLUPs is given by Searle et al. (1992, pp. 258–289).

17.7 RESIDUAL DIAGNOSTICS

The assumptions of the linear mixed model in (17.2) and (17.3) are independence, normality, and constant variance of the elements of each of the $\mathbf{a}_i$ vectors, as well as independence, normality, and constant variance of the elements of $\boldsymbol{\varepsilon}$. These assumptions are harder to check than for the standard linear model, and the usefulness of various types of residual plots for mixed model diagnosis is presently not fully understood (Brown and Prescott 1999, p. 77).

As a first step, we can examine each of the EBLUP ($\boldsymbol{a}_i$) vectors as in (17.27) for normality, constant variance and independence (see Section 9.1). This makes sense because, using (4.25) and assuming for simplicity that $\boldsymbol{\Sigma}$ (and therefore $\mathbf{G}$) are known, we have

$$\mathrm{cov}(\mathbf{a}|\mathbf{y}) = \mathbf{G} - \mathbf{G}\mathbf{Z}'\boldsymbol{\Sigma}^{-1}\mathbf{Z}\mathbf{G}.$$

Thus if $\mathbf{U} = (\mathbf{O}\ldots\mathbf{O}\mathbf{I}_{n_i}\mathbf{O}\ldots\mathbf{O})$,

$$\begin{aligned}
\mathrm{cov}(\mathbf{U}\mathbf{a}|\mathbf{y}) &= \mathrm{cov}(\mathbf{a}_i|\mathbf{y}) \\
&= \mathbf{U}\mathbf{G}\mathbf{U}' - \mathbf{U}\mathbf{G}\mathbf{Z}'\boldsymbol{\Sigma}^{-1}\mathbf{Z}\mathbf{G}\mathbf{U}' \\
&= \sigma_i{}^2\mathbf{I}_{n_i} - \sigma_i{}^4\mathbf{Z}_i'\boldsymbol{\Sigma}^{-1}\mathbf{Z}_i \\
&= \sigma_i{}^2(\mathbf{I}_{n_i} - \sigma_i{}^2\mathbf{Z}_i'\boldsymbol{\Sigma}^{-1}\mathbf{Z}_i).
\end{aligned} \tag{17.31}$$

As was the case for the hat matrix in Section 9.1, the off-diagonal elements of the second term in (17.31) are often small in absolute value. Hence the elements of EBLUP($\mathbf{a}_i$) should display normality, constant variance, and approximate independence if the model assumptions are met. It turns out, however, that constant variance and normality of the EBLUP($\mathbf{a}_i$) vectors is a necessary rather than a sufficient condition for the model assumptions to hold. Simulation studies (Verbeke and Molenberghs 2000, pp. 83–87) have shown that EBLUPs tend to reflect the distributional assumptions of the model rather than the actual distribution of random effects in some situations.

The next step is to consider the assumptions of independence, normality, and constant variance for the elements of $\boldsymbol{\varepsilon}$. The simple residual vector $\mathbf{y} - \mathbf{X}\hat{\boldsymbol{\beta}}$ is seldom useful for this purpose because, assuming that $\boldsymbol{\Sigma}$ is known, we have

$$\begin{aligned}
\mathrm{cov}(\mathbf{y} - \mathbf{X}\hat{\boldsymbol{\beta}}) &= \mathrm{cov}\{[\mathbf{I} - \mathbf{X}(\mathbf{X}'\boldsymbol{\Sigma}^{-1}\mathbf{X})^{-}\mathbf{X}'\boldsymbol{\Sigma}^{-1}]\mathbf{y}\} \\
&= [\mathbf{I} - \mathbf{X}(\mathbf{X}'\boldsymbol{\Sigma}^{-1}\mathbf{X})^{-}\mathbf{X}'\boldsymbol{\Sigma}^{-1}]\boldsymbol{\Sigma}[\mathbf{I} - \boldsymbol{\Sigma}^{-1}\mathbf{X}(\mathbf{X}'\boldsymbol{\Sigma}^{-1}\mathbf{X})^{-}\mathbf{X}'],
\end{aligned}$$

which may not exhibit constant variance or independence. However, the vector $\hat{\boldsymbol{\Sigma}}^{-1/2}(\mathbf{y} - \mathbf{X}\hat{\boldsymbol{\beta}})$, where $\hat{\boldsymbol{\Sigma}}^{-1/2}$ is the inverse of the square root matrix of $\hat{\boldsymbol{\Sigma}}$ (2.109), does have the desired properties.

Theorem 17.7. Consider the model in which $\mathbf{y}$ is $N_n(\mathbf{X}\boldsymbol{\beta}, \boldsymbol{\Sigma})$, where $\boldsymbol{\Sigma} = \sigma^2\mathbf{I} + \sum_{i=1}^m \sigma_i^2\mathbf{Z}_i\mathbf{Z}_i'$. Assume that $\boldsymbol{\Sigma}$ is known, and let $\hat{\boldsymbol{\beta}} = (\mathbf{X}'\boldsymbol{\Sigma}^{-1}\mathbf{X})^-\mathbf{X}'\boldsymbol{\Sigma}^{-1}\mathbf{y}$. Then

$$\text{cov}[\boldsymbol{\Sigma}^{-1/2}(\mathbf{y} - \mathbf{X}\hat{\boldsymbol{\beta}})] = \mathbf{I} - \mathbf{H}_* \tag{17.32}$$

where $\mathbf{H}_* = \boldsymbol{\Sigma}^{-1/2}\mathbf{X}(\mathbf{X}'\boldsymbol{\Sigma}^{-1}\mathbf{X})^-\mathbf{X}'\boldsymbol{\Sigma}^{-1/2}$.

PROOF

$$\begin{aligned}
\text{cov}[\boldsymbol{\Sigma}^{-1/2}(\mathbf{y} - \mathbf{X}\hat{\boldsymbol{\beta}})] &= \text{cov}\{\boldsymbol{\Sigma}^{-1/2}[\mathbf{I} - \mathbf{X}(\mathbf{X}'\boldsymbol{\Sigma}^{-1}\mathbf{X})^-\mathbf{X}'\boldsymbol{\Sigma}^{-1}]\mathbf{y}\} \\
&= \boldsymbol{\Sigma}^{-1/2}[\mathbf{I} - \mathbf{X}(\mathbf{X}'\boldsymbol{\Sigma}^{-1}\mathbf{X})^-\mathbf{X}'\boldsymbol{\Sigma}^{-1}] \\
&\quad \times \boldsymbol{\Sigma}[\mathbf{I} - \boldsymbol{\Sigma}^{-1}\mathbf{X}(\mathbf{X}'\boldsymbol{\Sigma}^{-1}\mathbf{X})^-\mathbf{X}']\boldsymbol{\Sigma}^{-1/2} \\
&= \boldsymbol{\Sigma}^{-1/2}\boldsymbol{\Sigma}\boldsymbol{\Sigma}^{-1/2} - \boldsymbol{\Sigma}^{-1/2}\mathbf{X}(\mathbf{X}'\boldsymbol{\Sigma}^{-1}\mathbf{X})^-\mathbf{X}'\boldsymbol{\Sigma}^{-1/2}.
\end{aligned}$$

Now, since $\boldsymbol{\Sigma}^{-1/2} = (\mathbf{C}\mathbf{D}^{1/2}\mathbf{C}')^{-1}$ where $\mathbf{C}$ is orthogonal as in Theorem 2.12d, and $\mathbf{D}^{1/2}$ is a diagonal matrix as in (2.109), we obtain

$$\begin{aligned}
\boldsymbol{\Sigma}^{-1/2}\boldsymbol{\Sigma}\boldsymbol{\Sigma}^{-1/2} &= (\mathbf{C}\mathbf{D}\frac{1}{2}\mathbf{C}')^{-1}\mathbf{C}\mathbf{D}\mathbf{C}'(\mathbf{C}\mathbf{D}\frac{1}{2}\mathbf{C}')^{-1} \\
&= \mathbf{C}\mathbf{D}^{-1/2}\mathbf{C}'\mathbf{C}\mathbf{D}\mathbf{C}'\mathbf{C}\mathbf{D}^{-1/2}\mathbf{C}' \\
&= \mathbf{C}\mathbf{D}^{-1/2}\mathbf{D}\mathbf{D}^{-1/2}\mathbf{C}' \\
&= \mathbf{C}\mathbf{C}' = \mathbf{I}
\end{aligned}$$

and the result follows. □

Thus the vector $\hat{\boldsymbol{\Sigma}}^{-1/2}(\mathbf{y} - \mathbf{X}\hat{\boldsymbol{\beta}})$ can be examined for constant variance, normality and approximate independence to verify the assumptions regarding $\boldsymbol{\varepsilon}$.

A more common approach (Verbeke and Molenberghs 2000, p. 132; Brown and Prescott 1999, p. 77) to verifying the assumptions regarding $\boldsymbol{\varepsilon}$ is to compute and examine $\mathbf{y} - \mathbf{X}\hat{\boldsymbol{\beta}} - \mathbf{Z}\hat{\mathbf{a}}$. To see why this makes sense, assume that $\boldsymbol{\Sigma}$ and $\boldsymbol{\beta}$ are known. Then

$$\begin{aligned}
\text{cov}(\mathbf{y} - \mathbf{X}\boldsymbol{\beta} - \mathbf{Z}\mathbf{a}) &= \text{cov}(\mathbf{y}) - \text{cov}(\mathbf{y}, \mathbf{Z}\mathbf{a}) - \text{cov}(\mathbf{Z}\mathbf{a}, \mathbf{y}) + \text{cov}(\mathbf{Z}\mathbf{a}) \\
&= \boldsymbol{\Sigma} - \mathbf{Z}\mathbf{G}\mathbf{Z}' - \mathbf{Z}\mathbf{G}\mathbf{Z}' + \mathbf{Z}\mathbf{G}\mathbf{Z}' \\
&= \boldsymbol{\Sigma} - \mathbf{Z}\mathbf{G}\mathbf{Z}' \\
&= (\mathbf{Z}\mathbf{G}\mathbf{Z}' + \sigma^2\mathbf{I}) - \mathbf{Z}\mathbf{G}\mathbf{Z}' \\
&= \sigma^2\mathbf{I}.
\end{aligned}$$

PROBLEMS

17.1 Consider the model $\mathbf{y} = \mathbf{X}\boldsymbol{\beta} + \boldsymbol{\varepsilon}$, where $\boldsymbol{\varepsilon}$ is $N_n(\mathbf{0}, \sigma^2\mathbf{V})$, $\mathbf{V}$ is a known positive definite $n \times n$ matrix, and $\mathbf{X}$ is a known $n \times (k+1)$ matrix of rank $k+1$. Also assume that $\mathbf{C}$ is a known $q \times (k+1)$ matrix and $\mathbf{t}$ is a known $q \times 1$ vector such that $\mathbf{C}\boldsymbol{\beta} = \mathbf{t}$ is consistent. Let $\hat{\boldsymbol{\beta}} = (\mathbf{X}'\mathbf{V}^{-1}\mathbf{X})^{-1}\mathbf{X}'\mathbf{V}^{-1}\mathbf{y}$. Find the distribution of

$$F = \frac{(\mathbf{C}\hat{\boldsymbol{\beta}} - \mathbf{t})'[\mathbf{C}(\mathbf{X}'\mathbf{V}^{-1}\mathbf{X})^{-1}\mathbf{C}']^{-1}(\mathbf{C}\hat{\boldsymbol{\beta}} - \mathbf{t})/q}{\mathbf{y}'[\mathbf{V}^{-1} - \mathbf{V}^{-1}(\mathbf{X}'\mathbf{V}^{-1}\mathbf{X})^{-1}\mathbf{X}'\mathbf{V}^{-1}]\mathbf{y}/(n-k-1)}$$

(a) Assuming that $H_0: \mathbf{C}\boldsymbol{\beta} = \mathbf{t}$ is false.

(b) Assuming that $H_0: \mathbf{C}\boldsymbol{\beta} = \mathbf{t}$ is true.

(*Hint*: Consider the model for $\mathbf{P}^{-1}\mathbf{y}$, where $\mathbf{P}$ is a nonsingular matrix such that $\mathbf{PP}' = \mathbf{V}$.)

17.2 For the model described in Problem 17.1, find a $100(1-\alpha)\%$ confidence interval for $\mathbf{a}'\boldsymbol{\beta}$.

17.3 An exercise science experiment was conducted to investigate how ankle roll (y) is affected by the combination of four casting treatments (control, tape cast, air cast, and tape and brace) and two exercise levels (preexercise and postexercise). Each of the 16 subjects used in the experiment was assigned to each of the four casting treatments in random order. Five ankle roll measurements were made preexercise and five measurements were made post exercise for each casting treatment. Thus a total of 40 observations were obtained for each subject. This study can be regarded as a randomized block split-plot study with subsampling. A sensible model is

$$y_{ijkl} = \mu + \tau_i + \delta_j + \theta_{ij} + a_k + b_{ik} + c_{ijk} + \varepsilon_{ijkl},$$

where $i = 1, \ldots, 4$; $j = 1, 2$; $k = 1, \ldots, 16$; $l = 1, \ldots, 5$; a_k is $N(\mathbf{0}, \sigma_1^2)$; $\boldsymbol{b}_{ijk}$ is $N(\mathbf{0}, \sigma_2^2)$; c_{ijk} is $N(\mathbf{0}, \sigma_3^2)$; $\boldsymbol{\varepsilon}_{ijkl}$ is $N(\mathbf{0}, \sigma^2)$, and all of the random effects are independent. If the data are sorted by subject, casting treatment, and exercise level, sketch out the $\mathbf{X}$ and $\mathbf{Z}_i$ matrices for the matrix form of this model as in (17.2).

17.4 **(a)** Consider the model $\mathbf{y} = \mathbf{X}\boldsymbol{\beta} + \sum_{i=1}^{m} \mathbf{Z}_i\mathbf{a}_i + \boldsymbol{\varepsilon}$ where $\mathbf{X}$ is a known $n \times p$ matrix, the $\mathbf{Z}_i$'s are known $n \times r_i$ full-rank matrices, $\boldsymbol{\beta}$ is a $p \times 1$ vector of unknown parameters, $\boldsymbol{\varepsilon}$ is an $n \times 1$ unknown random vector such that $E(\boldsymbol{\varepsilon}) = \mathbf{0}$ and $\text{cov}(\boldsymbol{\varepsilon}) = \mathbf{R} \neq \sigma^2\mathbf{I}_n$, and the $\mathbf{a}_i$'s are $r_i \times 1$ unknown random vectors such that $E(\mathbf{a}_i) = \mathbf{0}$ and $\text{cov}(\mathbf{a}_i) = \mathbf{G}_i \neq \sigma_i^2\mathbf{I}_{r_i}$. As usual, $\text{cov}(\mathbf{a}_i, \mathbf{a}_j) = \mathbf{O}$ for $i \neq j$, where $\mathbf{O}$ is $r_i \times r_j$, and $\text{cov}(\mathbf{a}_i, \boldsymbol{\varepsilon}) = \mathbf{O}$ for all i, where $\mathbf{O}$ is $r_i \times n$. Find $\text{cov}(\mathbf{y})$.

(b) For the model in part (a), let $\mathbf{Z} = (\mathbf{Z}_1\mathbf{Z}_2 \ldots \mathbf{Z}_m)$ and $\mathbf{a} = (\mathbf{a}_1'\mathbf{a}_2' \ldots \mathbf{a}_m')'$ so that the model can be written as $\mathbf{y} = \mathbf{X}\boldsymbol{\beta} + \mathbf{Z}\mathbf{a} + \boldsymbol{\varepsilon}$ and

$$\text{cov}(\mathbf{a}) = \mathbf{G} = \begin{pmatrix} \mathbf{G}_1 & \mathbf{O} & \ldots & \mathbf{O} & \mathbf{O} \\ \mathbf{O} & \mathbf{G}_2 & \ldots & \mathbf{O} & \mathbf{O} \\ \vdots & \vdots & \ddots & \vdots & \vdots \\ \mathbf{O} & \mathbf{O} & \ldots & \mathbf{G}_{m-1} & \mathbf{O} \\ \mathbf{O} & \mathbf{O} & \ldots & \mathbf{O} & \mathbf{G}_m \end{pmatrix}.$$

Express $\text{cov}(\mathbf{y})$ in terms of $\mathbf{Z}$, $\mathbf{G}$, and $\mathbf{R}$.

17.5 Consider the model in which $\mathbf{y}$ is $N_n(\mathbf{X}\boldsymbol{\beta}, \boldsymbol{\Sigma})$, where $\boldsymbol{\Sigma} = \sum_{i=0}^{m} \sigma_i^2 \mathbf{Z}_i\mathbf{Z}_i'$, and let $\mathbf{K}$ be a full-rank matrix of appropriate dimensions as in Theorem 17.4c. Show that for any i,

$$E[\mathbf{y}'\mathbf{K}'(\mathbf{K}\boldsymbol{\Sigma}\mathbf{K}')^{-1}\mathbf{K}\mathbf{Z}_i\mathbf{Z}_i'\mathbf{K}'(\mathbf{K}\boldsymbol{\Sigma}\mathbf{K}')^{-1}\mathbf{K}\mathbf{y}] = \text{tr}[\mathbf{K}'(\mathbf{K}\boldsymbol{\Sigma}\mathbf{K}')^{-1}\mathbf{K}\mathbf{Z}_i\mathbf{Z}_i'].$$

17.6 Show that that the system of $m+1$ equations generated by (17.6) can be written as $\mathbf{M}\boldsymbol{\sigma} = \mathbf{q}$, where $\sigma = (\sigma_0^2\ \sigma_1^2 \ldots \sigma_m^2)'$, $\mathbf{M}$ is an $(m+1) \times (m+1)$ matrix with ijth element $\text{tr}[\mathbf{K}'(\mathbf{K}\boldsymbol{\Sigma}\mathbf{K}')^{-1}\mathbf{K}\mathbf{Z}_i\mathbf{Z}_i'\mathbf{K}'(\mathbf{K}\boldsymbol{\Sigma}\mathbf{K}')^{-1}\mathbf{K}\mathbf{Z}_j\mathbf{Z}_j']$, and $\mathbf{q}$ is an $(m+1) \times 1$ vector with ith element $\mathbf{y}'\mathbf{K}'(\mathbf{K}\boldsymbol{\Sigma}\mathbf{K}')^{-1}\mathbf{K}\mathbf{Z}_i\mathbf{Z}_i'\mathbf{K}'(\mathbf{K}\boldsymbol{\Sigma}\mathbf{K}')^{-1}\mathbf{K}\mathbf{y}$.

17.7 Consider the model in which $\mathbf{y}$ is $N_n(\mathbf{X}\boldsymbol{\beta}, \boldsymbol{\Sigma})$, and let $\mathbf{L}$ be a known full-rank $g \times p$ matrix whose rows define estimable functions of $\boldsymbol{\beta}$.

(a) Show that $\mathbf{L}(\mathbf{X}'\boldsymbol{\Sigma}^{-1}\mathbf{X})^{-}\mathbf{L}'$ is nonsingular.

(b) Show that $(\mathbf{L}\hat{\boldsymbol{\beta}} - \mathbf{L}\boldsymbol{\beta})'[\mathbf{L}(\mathbf{X}'\boldsymbol{\Sigma}^{-1}\mathbf{X})^{-}\mathbf{L}']^{-1}(\mathbf{L}\hat{\boldsymbol{\beta}} - \mathbf{L}\boldsymbol{\beta})$ is $\chi^2(g)$.

17.8 For the model described in Problem 17.7, develop a $100(1-)\%$ confidence interval for $E(\mathbf{y}_0) = \mathbf{x}_0'\boldsymbol{\beta}$.

17.9 Refer to Example 17.5. Show that

$$(\mathbf{X}'\hat{\boldsymbol{\Sigma}}^{-1}\mathbf{X})^{-1} = \tfrac{1}{2}\begin{pmatrix} \hat{\boldsymbol{\Sigma}}_1 & \mathbf{O} & \mathbf{O} \\ \mathbf{O} & \hat{\boldsymbol{\Sigma}}_1 & \mathbf{O} \\ \mathbf{O} & \mathbf{O} & \hat{\boldsymbol{\Sigma}}_1 \end{pmatrix}.$$

17.10 Refer to Example 17.5. Show that the solution to the REML estimating equations is given by

$$\hat{\sigma}^2 = \tfrac{1}{12}\mathbf{y}'\begin{pmatrix} \mathbf{R} & \mathbf{O} & \mathbf{O} & -\mathbf{R} & \mathbf{O} & \mathbf{O} \\ \mathbf{O} & \mathbf{R} & \mathbf{O} & \mathbf{O} & -\mathbf{R} & \mathbf{O} \\ \mathbf{O} & \mathbf{O} & \mathbf{R} & \mathbf{O} & \mathbf{O} & -\mathbf{R} \\ -\mathbf{R} & \mathbf{O} & \mathbf{O} & \mathbf{R} & \mathbf{O} & \mathbf{O} \\ \mathbf{O} & -\mathbf{R} & \mathbf{O} & \mathbf{O} & \mathbf{R} & \mathbf{O} \\ \mathbf{O} & \mathbf{O} & -\mathbf{R} & \mathbf{O} & \mathbf{O} & \mathbf{R} \end{pmatrix}\mathbf{y}, \quad \text{where} \quad \mathbf{R} = \begin{pmatrix} 1 & -1 \\ -1 & 1 \end{pmatrix}.$$

17.11 Refer to Example 17.5. Show that $\mathbf{X}(\mathbf{X}'\mathbf{X})^{-1}\mathbf{L}'\mathbf{Q}\mathbf{L}(\mathbf{X}'\mathbf{X})^{-1}\mathbf{X}'\boldsymbol{\Sigma}$ is idempotent.

17.12 Refer to Example 17.5. Show that w and $(\mathbf{L}\hat{\boldsymbol{\beta}})'\mathbf{Q}(\mathbf{L}\hat{\boldsymbol{\beta}})$ are independent chi-square variables.

17.13 Refer to Example 17.5. Let $c' = (1 \quad 0 \quad 0 \quad -1 \quad 0 \quad 0)$, let w be as in (17.14), and let $d = 3$. Show that if v is such that $v(w/d) = [\mathbf{c}'(\mathbf{X}'\hat{\boldsymbol{\Sigma}}^{-1}\mathbf{X})^{-}\mathbf{c}]^{-1}$ then v is *not* distributed as a central chi-square random variable.

17.14 To motivate Satterthwaite's approximation in expression (17.16), consider the model in which $\mathbf{y}$ is $N_n(\mathbf{X}\boldsymbol{\beta}, \boldsymbol{\Sigma})$, where $\mathbf{X}$ is $n \times p$ of rank k, $\boldsymbol{\Sigma} = \sigma^2\mathbf{I}$ and $\hat{\boldsymbol{\Sigma}} = s^2\mathbf{I}$. If $\mathbf{c}'\boldsymbol{\beta}$ is an estimable function, show that $(n-k)[\mathbf{c}'(\mathbf{X}'\hat{\boldsymbol{\Sigma}}^{-1}\mathbf{X})^{-}\mathbf{c}]/[\mathbf{c}'(\mathbf{X}'\boldsymbol{\Sigma}^{-1}\mathbf{X})^{-}\mathbf{c}]$, is distributed as $\chi^2(n-k)$.

17.15 Given $f(\boldsymbol{\sigma}) = [\mathbf{c}'(\mathbf{X}'\boldsymbol{\Sigma}^{-1}\mathbf{X})^{-}\mathbf{c}]$, where $\boldsymbol{\sigma} = (\sigma_0^2 \sigma_1^2 \cdots \sigma_m^2)'$ and $\boldsymbol{\Sigma} = \sum_{i=0}^{m} \sigma_i{}^2\mathbf{Z}_i\mathbf{Z}_i'$, show that

$$\frac{\partial f(\boldsymbol{\sigma})}{\partial \boldsymbol{\sigma}} = -\begin{pmatrix} \mathbf{c}'(\mathbf{X}'\boldsymbol{\Sigma}^{-1}\mathbf{X})^{-}\mathbf{X}'\boldsymbol{\Sigma}^{-1}\mathbf{Z}_0\mathbf{Z}_0'\boldsymbol{\Sigma}^{-1}\mathbf{X}(\mathbf{X}'\boldsymbol{\Sigma}^{-1}\mathbf{X})^{-}\mathbf{c} \\ \mathbf{c}'(\mathbf{X}'\boldsymbol{\Sigma}^{-1}\mathbf{X})^{-}\mathbf{X}'\boldsymbol{\Sigma}^{-1}\mathbf{Z}_1\mathbf{Z}_1'\boldsymbol{\Sigma}^{-1}\mathbf{X}(\mathbf{X}'\boldsymbol{\Sigma}^{-1}\mathbf{X})^{-}\mathbf{c} \\ \vdots \\ \mathbf{c}'(\mathbf{X}'\boldsymbol{\Sigma}^{-1}\mathbf{X})^{-}\mathbf{X}'\boldsymbol{\Sigma}^{-1}\mathbf{Z}_m\mathbf{Z}_m'\boldsymbol{\Sigma}^{-1}\mathbf{X}(\mathbf{X}'\boldsymbol{\Sigma}^{-1}\mathbf{X})^{-}\mathbf{c} \end{pmatrix}.$$

17.16 Consider the model in which $\mathbf{y}$ is $N_n(\mathbf{X}\boldsymbol{\beta}, \boldsymbol{\Sigma})$, let $\mathbf{L}$ be a known full-rank $g \times p$ matrix whose rows define estimable functions of $\boldsymbol{\beta}$, and let $\hat{\boldsymbol{\Sigma}}$ be the REML estimate of $\boldsymbol{\Sigma}$. As in (17.19), let $\mathbf{D} = \text{diag}(\lambda_1, \lambda_2, \ldots, \lambda_m)$ be the diagonal matrix of eigenvalues and $\mathbf{P} = (\mathbf{p}_1, \mathbf{p}_2, \ldots, \mathbf{p}_m)$ be the orthogonal matrix of normalized eigenvectors of $[\mathbf{L}(\mathbf{X}'\hat{\boldsymbol{\Sigma}}^{-1}\mathbf{X})^{-}\mathbf{L}']^{-1}$.

(a) Show that $(\mathbf{L}\hat{\boldsymbol{\beta}} - \mathbf{L}\boldsymbol{\beta})'[\mathbf{L}(\mathbf{X}'\hat{\boldsymbol{\Sigma}}^{-1}\mathbf{X})^{-}\mathbf{L}']^{-1}(\mathbf{L}\hat{\boldsymbol{\beta}} - \mathbf{L}\boldsymbol{\beta}) = \sum_{i=1}^{g}\left[\mathbf{p}_i'(\mathbf{L}\hat{\boldsymbol{\beta}} - \mathbf{L}\boldsymbol{\beta})\right]^2/\lambda_i$.

(b) Show that $(\mathbf{p}_i'\mathbf{L}\hat{\boldsymbol{\beta}})^2/\lambda_i$ is of the form $\mathbf{c}'\hat{\boldsymbol{\beta}}/\sqrt{\mathbf{c}'(\mathbf{X}'\hat{\boldsymbol{\Sigma}}^{-1}\mathbf{X})^{-}\mathbf{c}}$ as in (17.14).

(c) Show that $\text{cov}(\mathbf{p}_i'\mathbf{L}\hat{\boldsymbol{\beta}}, \mathbf{p}_{i'}'\mathbf{L}\hat{\boldsymbol{\beta}}) = \mathbf{0}$ for $i \neq i'$.

17.17 Consider the model in which $\mathbf{y} = \mathbf{X}\boldsymbol{\beta} + \mathbf{Z}\mathbf{a} + \boldsymbol{\varepsilon}$, where $\boldsymbol{\varepsilon}$ is $N(\mathbf{0}, \boldsymbol{\sigma}^2\mathbf{I}_n)$ and $\mathbf{a}$ is $N(\mathbf{0}, \mathbf{G})$ as in (17.24).

(a) Show that the linear function $\mathbf{B}(\mathbf{y} - \mathbf{X})$ that minimizes $E[\mathbf{a} - \mathbf{B}(\mathbf{y} - \mathbf{X})]'[\mathbf{a} - \mathbf{B}(\mathbf{y} - \mathbf{X})]$ is $\mathbf{G}\mathbf{Z}'\boldsymbol{\Sigma}^{-1}(\mathbf{y} - \mathbf{X}\boldsymbol{\beta})$.

(b) Show that $\mathbf{B} = \mathbf{G}\mathbf{Z}'\boldsymbol{\Sigma}^{-1}(\mathbf{y} - \mathbf{X}\boldsymbol{\beta})$ also "minimizes" $E[\mathbf{a} - \mathbf{B}(\mathbf{y} - \mathbf{X})][\mathbf{a} - \mathbf{B}(\mathbf{y} - \mathbf{X})]'$. By "minimize," we mean that any other choice for $\mathbf{B}$ adds a positive definite matrix to the result.

17.18 Show that $[\mathbf{I} - \mathbf{X}(\mathbf{X}'\boldsymbol{\Sigma}^{-1}\mathbf{X})^{-}\mathbf{X}'\boldsymbol{\Sigma}^{-1}]\boldsymbol{\Sigma}[\mathbf{I} - \mathbf{X}(\mathbf{X}'\boldsymbol{\Sigma}^{-1}\mathbf{X})^{-}\mathbf{X}'\boldsymbol{\Sigma}^{-1}]' = \boldsymbol{\Sigma} - \mathbf{X}(\mathbf{X}'\boldsymbol{\Sigma}^{-1}\mathbf{X})^{-}\mathbf{X}'$ as in (17.29).

17.19 Consider the model described in Problem 17.17.

(a) Show that the best linear predictor of $\mathbf{Ua}$ is $E(\mathbf{Ua}|\mathbf{y}) = \mathbf{UGZ}'\boldsymbol{\Sigma}^{-1}(\mathbf{y} - \mathbf{X}\boldsymbol{\beta})$.

(b) Show that $\text{cov}[E(\mathbf{Ua}|\mathbf{y})] = \mathbf{UGZ}'\boldsymbol{\Sigma}^{-1}\mathbf{ZGU}'$.

(c) Given $\hat{\boldsymbol{\beta}} = (\mathbf{X}'\boldsymbol{\Sigma}^{-1}\mathbf{X})^{-}\mathbf{X}'\boldsymbol{\Sigma}^{-1}\mathbf{y}$, show that

$$\text{cov}[\mathbf{UGZ}'\boldsymbol{\Sigma}^{-1}(\mathbf{y} - \mathbf{X}\hat{\boldsymbol{\beta}})] = \mathbf{UGZ}'[\boldsymbol{\Sigma}^{-1} - \boldsymbol{\Sigma}^{-1}\mathbf{X}(\mathbf{X}'\boldsymbol{\Sigma}^{-1}\mathbf{X})^{-}\mathbf{X}'\boldsymbol{\Sigma}^{-1}]\mathbf{ZGU}'.$$

17.20 Consider the one-way random effects model of Example 17.6. Use (2.52) and (2.53) to derive the expression for $\boldsymbol{\Sigma}^{-1}$.

17.21 Using (17.29), derive the covariance matrix for EBLUP($\mathbf{a}$) where a_i is defined as in (17.30).

17.22 Consider the model described in Problem 17.17. Use (4.27) and assume that $\boldsymbol{\Sigma}$ and $\mathbf{G}$ are known to show that

$$\text{cov}(\mathbf{a}|\mathbf{y}) = \mathbf{G} - \mathbf{GZ}'\boldsymbol{\Sigma}^{-1}\mathbf{ZG}.$$

17.23 Use the model of Example 17.3b (**subsampling**). Find the covariance matrix of the predicted batch effects using (17.31). Comment on the magnitudes of the off-diagonal elements of this matrix.

17.24 Use the model of Example 17.3b. Find the covariance matrix of the transformed residuals $\hat{\Sigma}^{-1/2}(y - x\hat{\beta})$ using (17.32). Comment on the off-diagonal elements of this matrix.

18 Additional Models

In this chapter we briefly discuss some models that are not linear in the parameters or that have an error structure different from that assumed in previous chapters.

18.1 NONLINEAR REGRESSION

A *nonlinear regression model* can be expressed as

$$y_i = f(\mathbf{x}_i, \boldsymbol{\beta}) + \varepsilon_i, \quad i = 1, 2, \ldots, n, \tag{18.1}$$

where $f(\mathbf{x}_i, \boldsymbol{\beta})$ is a nonlinear function of the parameter vector $\boldsymbol{\beta}$. The error term ε_i is sometimes assumed to be distributed as $N(0, \sigma^2)$. An example of a nonlinear model is the exponential model

$$y_i = \beta_0 + \beta_1 e^{\beta_2 x_i} + \varepsilon_i.$$

Estimators of the parameters in (18.1) can be obtained using the method of least squares. We seek the value of $\hat{\boldsymbol{\beta}}$ that minimizes

$$Q(\hat{\boldsymbol{\beta}}) = \sum_{i=1}^{n} [y_i - f(\mathbf{x}_i, \hat{\boldsymbol{\beta}})]^2. \tag{18.2}$$

A simple analytical solution for $\hat{\boldsymbol{\beta}}$ that minimizes (18.2) is not available for nonlinear $f(\mathbf{x}_i, \hat{\boldsymbol{\beta}})$. An iterative approach is therefore used to obtain a solution. In general, the resulting estimators in $\hat{\boldsymbol{\beta}}$ are not unbiased, do not have minimum variance, and are not normally distributed. However, according to large-sample theory, the estimators are almost unbiased, have near-minimum variance, and are approximately normally distributed.

Inferential procedures, including confidence intervals and hypothesis tests, are available for the least-squares estimator $\hat{\boldsymbol{\beta}}$ obtained by minimizing (18.2). Diagnostic procedures are available for checking on the model and on the suitability of the large-sample inferential procedures.

For details of the above procedures, see Gallant (1975), Bates and Watts (1988), Seber and Wild (1989), Ratkowsky (1983, 1990), Kutner et al. (2005, Chapter 13), Hocking (1996, Section 11.2), Fox (1997, Section 14.2), and Ryan (1997, Chapter 13).

18.2 LOGISTIC REGRESSION

In some regression situations, the response variable y has only two possible outcomes, for example, high blood pressure or low blood pressure, developing cancer of the esophagus or not developing it, whether a crime will be solved or not solved, and whether a bee specimen is a "killer" (africanized) bee or a domestic honey bee. In such cases, the outcome y can be coded as 0 or 1 and we wish to predict the outcome (or the probability of the outcome) on the basis of one or more x's.

To illustrate a linear model in which y is binary, consider the model with one x:

$$y_i = \beta_0 + \beta_1 x_i + \varepsilon_i; \quad y_i = 0, 1; \quad i = 1, 2, \ldots, n. \tag{18.3}$$

Since y_i is 0 or 1, the mean $E(y_i)$ for each x_i becomes the proportion of observations at x_i for which $y_i = 1$. This can be expressed as

$$\begin{aligned} E(y_i) &= P(y_i = 1) = p_i, \\ 1 - E(y_i) &= P(y_i = 0) = 1 - p_i. \end{aligned} \tag{18.4}$$

The distribution $P(y_i = 0) = 1 - p_i$ and $P(y_i = 1) = p_i$ in (18.4) is known as the *Bernoulli distribution*. By (18.3) and (18.4), we have

$$E(y_i) = p_i = \beta_0 + \beta_1 x_i. \tag{18.5}$$

For the variance of y_i, we obtain

$$\begin{aligned} \text{var}(y_i) &= E[y_i - E(y_i)]^2 \\ &= p_i(1 - p_i). \end{aligned} \tag{18.6}$$

By (18.5) and (18.6), we obtain

$$\text{var}(y_i) = (\beta_0 + \beta_1 x_i)(1 - \beta_0 - \beta_1 x_i),$$

and the variance of each y_i depends on the value of x_i. Thus the fundamental assumption of constant variance is violated, and the usual least-squares estimators $\hat{\boldsymbol{\beta}}_0$ and $\hat{\boldsymbol{\beta}}_1$ computed as in (6.5) and (6.6) will not be optimal (see Theorem 7.3d).

To obtain optimal estimators of $\boldsymbol{\beta}_0$ and $\boldsymbol{\beta}_1$, we could use generalized least-squares estimators

$$\hat{\boldsymbol{\beta}} = (\mathbf{X}'\mathbf{V}^{-1}\mathbf{X})^{-1}\mathbf{X}'\mathbf{V}^{-1}\mathbf{y}$$

as in Theorem 7.8a, but there is an additional challenge in fitting the linear model (18.5). Since $E(y_i) = p_i$ is a probability, it is limited by $0 \leq p_i \leq 1$. If we fit (18.5) by generalized least squares to obtain

$$\hat{p}_i = \hat{\beta}_0 + \hat{\beta}_1 x_i,$$

then $\hat{p}_i$ may be less than 0 or greater than 1 for some values of x_i. A model for $E(y_i)$ that is bounded between 0 and 1 and reaches 0 and 1 asymptotically (instead of linearly) would be more suitable. A popular choice is the *logistic regression model*.

$$p_i = E(y_i) = \frac{e^{\beta_0 + \beta_1 x_i}}{1 + e^{\beta_0 + \beta_1 x_i}} = \frac{1}{1 + e^{-\beta_0 - \beta_1 x_i}}. \tag{18.7}$$

This model is illustrated in Figure 18.1. The model in (18.7) can be linearized by the simple transformation

$$\ln\left(\frac{p_i}{1 - p_i}\right) = \beta_0 + \beta_1 x_i, \tag{18.8}$$

sometimes called the *logit transformation*.

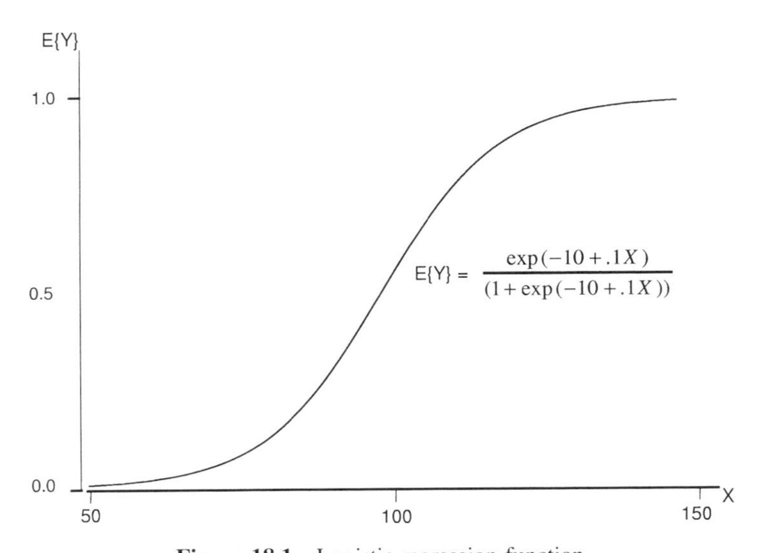

Figure 18.1 Logistic regression function.

The parameters β_0 and β_1 in (18.7) and (18.8) are typically estimated by the method of maximum likelihood (see Section 7.2). For a random sample $y_1, y_2, \ldots, y_n$ from the Bernoulli distribution with $P(y_i = 0) = 1 - p_i$ and $P(y_i = 1) = p_i$, the likelihood function becomes

$$\begin{aligned} L(\beta_0, \beta_1) = f(y_1, y_2, \ldots, y_n; \beta_0, \beta_1) &= \prod_{i=1}^{n} f_i(y_i; \beta_0, \beta_1) \\ &= \prod_{i=1}^{n} p_i^{y_i}(1 - p_i)^{1-y_i}. \end{aligned} \tag{18.9}$$

Taking the logarithm of both sides of (18.9) and using (18.8), we obtain

$$\ln L(\beta_0, \beta_1) = \sum_{i=1}^{n} y_i(\beta_0 + \beta_1 x_i) - \sum_{i=1}^{n} \ln(1 + e^{\beta_0 + \beta_1 x_i}). \tag{18.10}$$

Differentiating (18.10) with respect to β_0 and β_1 and setting the results equal to zero gives

$$\sum_{i=1}^{n} y_i = \sum_{i=1}^{n} \frac{1}{1 + e^{-\hat{\beta}_0 - \hat{\beta}_1 x_i}} \tag{18.11}$$

$$\sum_{i=1}^{n} x_i y_i = \sum_{i=1}^{n} \frac{x_i}{1 + e^{-\hat{\beta}_0 - \hat{\beta}_1 x_i}}. \tag{18.12}$$

These equations can be solved iteratively for $\hat{\boldsymbol{\beta}}_0$ and $\hat{\boldsymbol{\beta}}_1$.

The logistic regression model in (18.7) can be readily extended to include more than one x. Using the notation $\boldsymbol{\beta} = (\beta_0, \beta_1, \ldots, \beta_k)'$ and $\mathbf{x}_i = (1, x_{i1}, x_{i2}, \ldots, x_{ik})'$, the model in (18.7) becomes

$$p_i = E(y_i) = \frac{e^{\mathbf{x}_i'\boldsymbol{\beta}}}{1 + e^{\mathbf{x}_i'\boldsymbol{\beta}}} = \frac{1}{1 + e^{-\mathbf{x}_i'\boldsymbol{\beta}}},$$

and (18.8) takes the form

$$\ln\left(\frac{p_i}{1 - p_i}\right) = \mathbf{x}_i'\boldsymbol{\beta}, \tag{18.13}$$

where $\mathbf{x}_i'\boldsymbol{\beta} = \beta_0 + \beta_1 x_{i1} + \beta_2 x_{i2} + \cdots + \beta_k x_{ik}$. For binary y_i ($y_i = 0, 1; i = 1, 2, \ldots, n$), the mean and variance are given by (18.4) and (18.6). The likelihood function and the value of $\boldsymbol{\beta}$ that maximize it are found in a manner analogous to the approach used to find β_0 and β_1. Confidence intervals, tests of significance, measures of fit, subset selection procedures, diagnostic techniques, and other procedures are available.

Logistic regression has been extended from binary to a *polytomous* logistic regression model in which y has several possible outcomes. These may be ordinal such as large, medium, and small, or categorical such as Republicans, Democrats, and Independents. The analysis differs for the ordinal and categorical cases.

For details of these procedures, see Hosmer and Lemeshow (1989), Hosmer et al. (1989), McCullagh and Nelder (1989), Myers (1990, Section 7.4), Kleinbaum (1994), Stapleton (1995, Section 8.8), Stokes et al. (1995, Chapters 8 and 9), Kutner et al. (2005), Chapter 14, Hocking (1996, Section 11.4), Ryan (1997, Chapter 9), Fox (1997, Chapter 15), Christensen (1997), and McCulloch and Searle (2001, Chapter 5).

18.3 LOGLINEAR MODELS

In the analysis of categorical data, we often use loglinear models. To illustrate a loglinear model for categorical data, consider a two-way contingency table with frequencies (counts) designated as y_{ij} as in Table 18.1, with $y_{i.} = \sum_{j=1}^{s} y_{ij}$ and $y_{.j} = \sum_{i=1}^{r} y_{ij}$. The corresponding cell probabilities p_{ij} are given in Table 18.2, with $p_{i\cdot} = \sum_{j=1}^{s} p_{ij}$ and $p_j = \sum_{i=1}^{r} p_{ij}$.

The hypothesis that A and B are independent can be expressed as $H_0 : p_{ij} = p_{i.}p_{.j}$ for all i, j. Under H_0, the expected frequencies are

$$E(y_{ij}) = np_{i.}p_{.j}.$$

This becomes linear if we take the logarithm of both sides:

$$\ln E(y_{ij}) = \ln n + \ln p_{i.} + \ln p_{.j}.$$

TABLE 18.1 Contingency Table Showing Frequencies y_{ij} (Cell Counts) for an $r \times s$ Classification of Two Categorical Variables A and B

Variable	B_1	B_2	$\dots$	B_s	Total
A_1	y_{11}	y_{12}	$\dots$	y_{1s}	$y_{1\cdot}$
A_2	y_{21}	y_{22}	$\dots$	y_{2s}	$y_{2\cdot}$
$\vdots$	$\vdots$	$\vdots$		$\vdots$	$\vdots$
A_r	y_{r1}	y_{r2}	$\dots$	y_{rs}	$y_{r\cdot}$
Total	$y_{.1}$	$y_{.2}$	$\dots$	$y_{.s}$	$y_{..} = n$

TABLE 18.2 Cell Probabilities for an $r \times s$ Contingency Table

Variable	B_1	B_2	$\cdots$	B_s	Total
A_1	p_{11}	p_{12}	$\cdots$	p_{1s}	$p_{1.}$
A_2	p_{21}	p_{22}	$\cdots$	p_{2s}	$p_{2.}$
$\vdots$	$\vdots$	$\vdots$		$\vdots$	$\vdots$
A_r	p_{r1}	p_{r2}	$\cdots$	p_{rs}	$p_{r.}$
Total	$p_{\cdot 1}$	$p_{\cdot 2}$	$\cdots$	$p_{\cdot s}$	$p_{..} = 1$

To test H_0: $p_{ij} = p_{i.}p_{.j}$, we can use the likelihood ratio test. The likelihood function is given by the multinomial density

$$L(p_{11}, p_{12}, \ldots, p_{rs}) = \frac{n!}{y_{11}!y_{12}! \cdots y_{rs}!} p_{11}^{y_{11}} p_{12}^{y_{12}} \cdots p_{rs}^{y_{rs}}.$$

The unrestricted maximum likelihood estimators of p_{ij} (subject to $\sum_{ij} p_{ij} = 1$) are $\hat{p}_{ij} = y_{ij}/n$, and the estimators under H_0 are $\hat{p}_{ij} = y_{i.}y_{.j}/n^2$ (Christensen 1997, pp. 42–46). The likelihood ratio is then given by

$$\text{LR} = \prod_{i=1}^{r} \prod_{j=1}^{s} \left(\frac{y_{i.}y_{.j}}{ny_{ij}} \right)^{y_{ij}}.$$

The test statistic is

$$-2 \ln \text{LR} = 2 \sum_{ij} y_{ij} \ln \left(\frac{ny_{ij}}{y_{i.}y_{.j}} \right),$$

which is approximately distributed as $\chi^2[(r-1)(s-1)]$.

For further details of loglinear models, see Ku and Kullback (1974), Bishop et al. (1975), Plackett (1981), Read and Cressie (1988), Santner and Duffy (1989), Agresti (1984, 1990) Dobson (1990, Chapter 9), Anderson (1991), and Christensen (1997).

18.4 POISSON REGRESSION

If the response y_i in a regression model is a count, the Poisson regression model may be useful. The Poisson probability distribution is given by

$$f(y) = \frac{\mu^y e^{-\mu}}{y!}, \quad y = 0, 1, 2, \ldots.$$

The *Poisson regression model* is

$$y_i = E(y_i) + \varepsilon_i, \quad i = 1, 2, \ldots, n,$$

where the y_i's are independently distributed as Poisson random variables and $\mu_i = E(y_i)$ is a function of $\mathbf{x}_i'\boldsymbol{\beta} = \beta_0 + \beta_1 x_{i1} + \cdots + \beta_k x_{ik}$. Some commonly used functions of $\mathbf{x}_i'\boldsymbol{\beta}$ are

$$\mu_i = \mathbf{x}_i'\boldsymbol{\beta}, \quad \mu_i = e^{\mathbf{x}_i'\boldsymbol{\beta}}, \quad \mu_i = \ln(\mathbf{x}_i'\boldsymbol{\beta}). \tag{18.14}$$

In each of the three cases in (18.14), the values of μ_i must be positive.

To estimate $\boldsymbol{\beta}$, we can use the method of maximum likelihood. Since y_i has a Poisson distribution, the likelihood function is given by

$$L(\boldsymbol{\beta}) = \prod_{i=1}^{n} f(y_i) = \prod_{i=1}^{n} \frac{\mu_i^{y_i} e^{-\mu_i}}{y_i!},$$

where μ_i is typically one of the three forms in (18.14). Iterative methods can be used to find the value of $\hat{\boldsymbol{\beta}}$ that maximizes $L(\boldsymbol{\beta})$. Confidence intervals, tests of hypotheses, measures of fit, and other procedures are available. For details, see Myers (1990, Section 7.5) Stokes et al. (1995, pp. 471–475), Lindsey (1997), and Kutner et al. (2005, Chapter 14).

18.5 GENERALIZED LINEAR MODELS

Generalized linear models include the classical linear regression and ANOVA models covered in earlier chapters as well as logistic regression in Section 18.2 and some forms of nonlinear regression in Section 18.1. Also included in this broad family of models are loglinear models for categorical data in Section 18.3 and Poisson regression models for count data in Section 18.4. This expansion of traditional linear models was introduced by Wedderburn (1972).

A *generalized linear model* can be briefly characterized by the following three components.

1. Independent random variables $y_1, y_2, \ldots, y_n$ with expected value $E(y_i) = \mu_i$ and density function from the exponential family [described below in (18.15)].
2. A *linear predictor*

$$\mathbf{x}_i'\boldsymbol{\beta} = \beta_0 + \beta_1 x_{i1} + \cdots + \beta_k x_{ik}.$$

3. A *link function* that describes how $E(y_i) = \mu_i$ relates to $\mathbf{x}_i'\boldsymbol{\beta}$:

$$g(\mu_i) = \mathbf{x}_i'\boldsymbol{\beta}.$$

4. The link function $g(\mu_i)$ is often nonlinear.

A density $f(y_i, \theta_i)$ belongs to the *exponential family* of density functions if $f(y_i, \theta_i)$ can be expressed in the form

$$f(y_i, \theta_i) = \exp[y_i\theta_i + b(\theta_i) + c(y_i)]. \tag{18.15}$$

A scale parameter such as σ^2 in the normal distribution can be incorporated into (18.15) by considering it to be known and treating it as part of θ_i. Alternatively, an additional parameter can be inserted into (18.15). The exponential family of density functions provides a unified approach to estimation of the parameters in generalized linear models.

Some common statistical distributions that are members of the exponential family are the binomial, Poisson, normal, and gamma [see (11.7)]. We illustrate three of these in Example 18.5.

Example 18.5. The binomial probability distribution can be written in the form of (18.15) as follows:

$$\begin{aligned} f(y_i, p_i) &= \binom{n_i}{y_i} p_i^{y_i}(1 - p_i)^{n_i - y_i} \\ &= \exp\left[y_i \ln p_i - y_i \ln(1 - p_i) + n_i \ln(1 - p_i) + \ln\binom{n_i}{y_i}\right] \\ &= \exp\left[y_i \ln\left(\frac{p_i}{1 - p_i}\right) + n_i \ln(1 - p_i) + \ln\binom{n_i}{y_i}\right] \\ &= \exp[y_i\theta_i + b(\theta_i) + c(y_i)], \end{aligned} \tag{18.16}$$

where $\theta_i = \ln[p_i/(1 - p_i)]$, $b(\theta_i) = n_i \ln(1 - p_i) = -n_i \ln(1 + e^{\theta_i})$, and $c(y_i) = \ln\binom{n_i}{y_i}$.

The Poisson distribution can be expressed in exponential form as follows:

$$\begin{aligned} f(y_i, \mu_i) &= \frac{\mu_i^{y_i} e^{-\mu_i}}{y_i!} = \exp[y_i \ln \mu_i - \mu_i - \ln(y_i!)] \\ &= \exp[y_i\theta_i + b(\theta_i) + c(y_i)], \end{aligned}$$

where $\theta_i = \ln \mu_i$, $b(\theta_i) = -\mu_i = -e^{\theta_i}$, and $c(y_i) = -\ln(y_i!)$.

The normal distribution $N(\mu_i, \sigma^2)$ can be written in the form of (18.15) as follows:

$$\begin{aligned} f(y_i, \mu_i) &= \frac{1}{(2\pi\sigma^2)^{1/2}} e^{-(y_i-\mu_i)^2/2\sigma^2} \\ &= \frac{1}{(2\pi\sigma^2)^{1/2}} e^{-(y_i^2-2y_i\mu_i+\mu_i^2)/2\sigma^2} \\ &= \exp\left[-\frac{y_i^2}{2\sigma^2} + \frac{y_i\mu_i}{\sigma^2} - \frac{\mu_i^2}{2\sigma^2} - \frac{1}{2}\ln(2\pi\sigma^2)\right] \\ &= \exp[y_i\theta_i + b(\theta_i) + c(y_i)], \end{aligned}$$

where $\theta_i = \mu_i/\sigma^2$, $b(\theta_i) = \sigma^2\theta_i^2/2$, and $c(y_i) = -y_i^2/2\sigma^2 - \frac{1}{2}\ln(2\pi\sigma^2)$. □

To obtain an estimator of β in a generalized linear model, we use the method of maximum likelihood. From (18.15), the likelihood function is given by

$$L(\boldsymbol{\beta}) = \prod_{i=1}^{n} \exp[y_i\theta_i + b(\theta_i) + c(y_i)].$$

The logarithm of the likelihood is

$$\ln L(\boldsymbol{\beta}) = \sum_{i=1}^{n} y_i\theta_i + \sum_{i=1}^{n} b(\theta_i) + \sum_{i=1}^{n} c(y_i). \tag{18.17}$$

For the exponential family in (18.15), it can be shown that

$$E(y_i) = \mu_i = -b'(\theta_i),$$

where $b'(\theta_i)$ is the derivative with respect to θ_i. This relates θ_i to the link function

$$g(\mu_i) = \mathbf{x}_i'\boldsymbol{\beta}.$$

Differentiating (18.17) with respect to each β_i, setting the results equal to zero, and solving the resulting (nonlinear) equations iteratively (iteratively reweighted least squares) gives the estimators $\hat{\beta}_i$. Confidence intervals, tests of hypotheses, measures of fit, subset selection techniques, and other procedures are available. For details, see McCullagh and Nelder (1989), Dobson (1990), Myers (1990, Section 7.6), Hilbe (1994), Lindsey (1997), Christensen (1997, Chapter 9), and McCulloch and Searle (2001, Chapter 5).

PROBLEMS

18.1 For the Bernoulli distribution, $P(y_i = 0) = 1 - p_i$ and $P(y_i = 1) = p_i$ in (18.4), show that $E(y_i) = p_i$ and $\text{var}(y_i) = p_i(1 - p_i)$ as in (18.5) and (18.6).

18.2 Show that $\ln[p_i/(1 - p_i)] = \beta_0 + \beta_1 x_i$ in (18.8) can be obtained from (18.7).

18.3 Verify that $\ln L(\beta_0, \beta_1)$ has the form shown in (18.10), where $L(\beta_0, \beta_1)$ is as given by (18.9).

18.4 Differentiate $\ln L(\beta_0, \beta_1)$ in (18.10) to obtain (18.11) and (18.12).

18.5 Show that $b(\theta_i) = -n \ln(1 + e^{\theta_i})$, as noted following (18.16).

APPENDIX A
Answers and Hints to the Problems

Chapter 2

2.1 Part (i) follows from the commutativity of real numbers, $a_{ij} + b_{ij} = b_{ij} + a_{ij}$. For part (ii), let $\mathbf{C} = \mathbf{A} + \mathbf{B}$. Then, by (2.3), $\mathbf{C}' = (c_{ij})' = (c_{ij}) = (a_{ji} + b_{ji}) = (a_{ji}) + (b_{ji}) = \mathbf{A}' + \mathbf{B}'$.

2.2 **(a)** $\mathbf{A}' = \begin{pmatrix} 7 & 4 \\ -3 & 9 \\ 2 & 5 \end{pmatrix}$.

(b) $(\mathbf{A}')' = \begin{pmatrix} 7 & 4 \\ -3 & 9 \\ 2 & 5 \end{pmatrix}' = \begin{pmatrix} 7 & -3 & 2 \\ 4 & 9 & 5 \end{pmatrix} = \mathbf{A}$.

(c) $\mathbf{A}'\mathbf{A} = \begin{pmatrix} 65 & 15 & 34 \\ 15 & 90 & 39 \\ 34 & 39 & 29 \end{pmatrix}$, $\quad \mathbf{AA}' = \begin{pmatrix} 62 & 11 \\ 11 & 122 \end{pmatrix}$.

2.3 **(a)** $\mathbf{AB} = \begin{pmatrix} 10 & 2 \\ 5 & -6 \end{pmatrix}$, $\quad \mathbf{BA} = \begin{pmatrix} -1 & 13 \\ 5 & 5 \end{pmatrix}$.

(b) $|\mathbf{A}| = 10, \quad |\mathbf{B}| = -7, \quad |\mathbf{AB}| = -70 = (10)(-7)$.

(c) $|\mathbf{BA}| = -70 = |\mathbf{AB}|$.

(d) $(\mathbf{AB})' = \begin{pmatrix} 10 & 5 \\ 2 & -6 \end{pmatrix}$, $\quad \mathbf{B}'\mathbf{A}' = \begin{pmatrix} 10 & 5 \\ 2 & -6 \end{pmatrix}$.

(e) $\text{tr}(\mathbf{AB}) = 4, \quad \text{tr}(\text{BA}) = 4$.

(f) For $\mathbf{AB}, \lambda_1 = 10.6023, \lambda_2 = -6.6023$. For $\mathbf{BA}, \lambda_1 = 10.6023, \lambda_2 = 6.6023$.

Linear Models in Statistics, Second Edition, by Alvin C. Rencher and G. Bruce Schaalje

2.4 **(a)** $\mathbf{A}+\mathbf{B}=\begin{pmatrix} 4 & 1 & 1 \\ 11 & 2 & 9 \end{pmatrix}$, $\mathbf{A}-\mathbf{B}=\begin{pmatrix} -2 & 5 & -9 \\ -1 & -16 & 5 \end{pmatrix}$.

(b) $\mathbf{A}'=\begin{pmatrix} 1 & 5 \\ 3 & -7 \\ -4 & 2 \end{pmatrix}$, $\mathbf{B}'=\begin{pmatrix} 3 & 6 \\ -2 & 9 \\ 5 & 7 \end{pmatrix}$.

(c) $(\mathbf{A}+\mathbf{B})'=\begin{pmatrix} 4 & 11 \\ 1 & 2 \\ 1 & 9 \end{pmatrix}$, $\mathbf{A}'+\mathbf{B}'=\begin{pmatrix} 4 & 11 \\ 1 & 2 \\ 1 & 9 \end{pmatrix}$.

2.5 The (ij)th element of $\mathbf{E}=\mathbf{B}+\mathbf{C}$ is $e_{ij}=b_{ij}+c_{ij}$. The (ij)th element of $\mathbf{AE}$ is $\sum_k a_{ik}e_{kj}=\sum_k a_{ik}(b_{kj}+c_{kj})=\sum_k(a_{ik}b_{kj}+a_{ik}c_{kj})=\sum_k a_{ik}b_{kj}+\sum_k a_{ik}c_{kj}$, which is the (ij)th element of $\mathbf{AB}+\mathbf{AC}$.

2.6 **(a)** $\mathbf{AB}=\begin{pmatrix} 35 & 33 \\ 1 & 37 \end{pmatrix}$, $\mathbf{BA}=\begin{pmatrix} -26 & 19 & -29 \\ 10 & 44 & 0 \\ 56 & -2 & 54 \end{pmatrix}$.

(b) $\mathbf{B}+\mathbf{C}=\begin{pmatrix} -1 & 7 \\ 0 & 8 \\ 8 & 0 \end{pmatrix}$, $\mathbf{AC}\begin{pmatrix} 13 & 47 \\ -23 & -11 \end{pmatrix}$, $\mathbf{A}(\mathbf{B}+\mathbf{C})=\begin{pmatrix} 48 & 80 \\ -22 & 26 \end{pmatrix}$,

$\mathbf{AB}+\mathbf{AC}=\begin{pmatrix} 48 & 80 \\ -22 & 26 \end{pmatrix}$.

(c) $(\mathbf{AB})'=\begin{pmatrix} 35 & 1 \\ 33 & 37 \end{pmatrix}$, $\mathbf{B}'\mathbf{A}'=\begin{pmatrix} 35 & 1 \\ 33 & 37 \end{pmatrix}$.

(d) $\text{tr}(\mathbf{AB})=72$, $\text{tr}(\mathbf{BA})=72$.

(e) $(\mathbf{a}_1'\mathbf{B})=(35 \quad 33)$, $(\mathbf{a}_2'\mathbf{B})=(1 \quad 37)$, $\mathbf{AB}=\begin{pmatrix} 35 & 33 \\ 1 & 37 \end{pmatrix}$.

(f) $(\mathbf{Ab}_1)=\begin{pmatrix} 35 \\ 1 \end{pmatrix}$, $\mathbf{Ab}_2=\begin{pmatrix} 33 \\ 37 \end{pmatrix}$, $\mathbf{AB}=\begin{pmatrix} 35 & 33 \\ 1 & 37 \end{pmatrix}$.

2.7 **(a)** $\mathbf{AB}=\begin{pmatrix} 0 & 0 & 0 \\ 0 & 0 & 0 \\ 0 & 0 & 0 \end{pmatrix}=\mathbf{O}$.

(b) $\mathbf{x}=$ any multiple of $\begin{pmatrix} 1 \\ -1 \\ -1 \end{pmatrix}$.

(c) $\text{rank}(\mathbf{A})=1$, $\text{rank}(\mathbf{B})=1$.

2.8 **(a)** By (2.17), $\mathbf{a}'\mathbf{j}=a_1\cdot 1+a_2\cdot 1+\cdots+a_n\cdot 1=\sum_{i=1}^n a_i$.

(b) If $\mathbf{A}=\begin{pmatrix} \mathbf{a}_1' \\ \mathbf{a}_2' \\ \vdots \\ \mathbf{a}_n' \end{pmatrix}$, then $\mathbf{Aj}=\begin{pmatrix} \mathbf{a}_1'\mathbf{j} \\ \mathbf{a}_2'\mathbf{j} \\ \vdots \\ \mathbf{a}_n'\mathbf{j} \end{pmatrix}=\begin{pmatrix} \sum_j a_{1j} \\ \sum_j a_{2j} \\ \vdots \\ \sum_j a_{nj} \end{pmatrix}$.

2.9 By (2.16), $(\mathbf{ABC})' = [(\mathbf{AB})\mathbf{C}]' = \mathbf{C}'(\mathbf{AB})' = \mathbf{C}'\mathbf{B}'\mathbf{A}'$.

2.10 **(iii)** $(\mathbf{A}'\mathbf{A}') = \mathbf{A}'(\mathbf{A}')' = \mathbf{A}'\mathbf{A}$.

(iv) The ith diagonal element of $\mathbf{A}'\mathbf{A}$ is $\mathbf{a}_i'\mathbf{a}_i$, where $\mathbf{a}_i$ is the ith column of $\mathbf{A}$. Since $\mathbf{a}_i'\mathbf{a}_i = \sum_j a_{ij}^2 = 0$, we have $\mathbf{a}_i = \mathbf{0}$.

2.11 $\mathbf{D}_1\mathbf{A} = \begin{pmatrix} 24 & 9 & 21 \\ 4 & -10 & 6 \end{pmatrix}, \quad \mathbf{AD}_2 = \begin{pmatrix} 40 & 9 & 42 \\ -10 & 5 & -18 \end{pmatrix}.$

2.12 $\mathbf{DA} = \begin{pmatrix} a & 2b & 3c \\ 4a & 5b & 6c \\ 7a & 8b & 9c \end{pmatrix}, \quad \mathbf{DAD} = \begin{pmatrix} a^2 & 2ab & 3ac \\ 4ab & 5b^2 & 6cb \\ 7ac & 8bc & 9c^2 \end{pmatrix}.$

2.13 $\mathbf{y}'\mathbf{Ay} = a_{11}y_1^2 + a_{22}y_2^2 + a_{33}y_3^2 + 2a_{12}y_1y_2 + 2a_{13}y_1y_3 + 2a_{23}y_2y_3$.

2.14 **(a)** $\mathbf{Bx} = \begin{pmatrix} 26 \\ 20 \\ 19 \end{pmatrix}.$

(b) $\mathbf{y}'\mathbf{B} = (40, -16, 29)$.

(c) $\mathbf{x}'\mathbf{Ax} = 108$.

(d) $\mathbf{x}'\mathbf{Cz} = -29$.

(e) $\mathbf{x}'\mathbf{x} = 14$.

(f) $\mathbf{x}'\mathbf{y} = 15$.

(g) $\mathbf{xx}' = \begin{pmatrix} 9 & -3 & 6 \\ -3 & 1 & -2 \\ 6 & -2 & 4 \end{pmatrix}.$

(h) $\mathbf{xy}' = \begin{pmatrix} 9 & 6 & 12 \\ -3 & -2 & -4 \\ 6 & 4 & 8 \end{pmatrix}.$

(i) $\mathbf{B}'\mathbf{B} = \begin{pmatrix} 89 & -11 & 28 \\ -11 & 14 & -21 \\ 28 & -21 & 34 \end{pmatrix}.$

(j) $\mathbf{yz}' = \begin{pmatrix} 6 & 15 \\ 4 & 10 \\ 8 & 20 \end{pmatrix}.$

(k) $\mathbf{zy}' = \begin{pmatrix} 6 & 4 & 8 \\ 15 & 10 & 20 \end{pmatrix}.$

(l) $\sqrt{\mathbf{y}'\mathbf{y}} = \sqrt{29}$.

(m) $\mathbf{C}'\mathbf{C} = \begin{pmatrix} 14 & -7 \\ -7 & 26 \end{pmatrix}.$

2.15 **(a)** $\mathbf{x} + \mathbf{y} = \begin{pmatrix} 6 \\ 1 \\ 6 \end{pmatrix}, \quad \mathbf{x} - \mathbf{y} = \begin{pmatrix} 0 \\ -3 \\ 2 \end{pmatrix}.$

(b) $\text{tr}(\mathbf{A}) = 13, \quad \text{tr}(\mathbf{B}) = 12, \quad \mathbf{A} + \mathbf{B} \begin{pmatrix} 11 & -3 & 6 \\ 6 & 2 & 2 \\ 5 & -1 & 12 \end{pmatrix}, \quad \text{tr}(\mathbf{A} + \mathbf{B}) = 25.$

(c) $\mathbf{AB} = \begin{pmatrix} 29 & -20 & 30 \\ 5 & -3 & 7 \\ 46 & -25 & 44 \end{pmatrix}, \quad \mathbf{BA} = \begin{pmatrix} 41 & -2 & 35 \\ 34 & -6 & 23 \\ 28 & 5 & 35 \end{pmatrix}.$

(d) $\text{tr}(\mathbf{AB}) = 70, \quad \text{tr}(\mathbf{BA}) = 70.$

(e) $|\mathbf{AB}| = -403, \quad |\mathbf{BA}| = -403.$

(f) $(\mathbf{AB})' = \begin{pmatrix} 29 & 5 & 46 \\ -20 & -3 & -25 \\ 30 & 7 & 44 \end{pmatrix}, \quad \mathbf{B}'\mathbf{A}' = \begin{pmatrix} 29 & 5 & 46 \\ -20 & -3 & -25 \\ 30 & 7 & 44 \end{pmatrix}.$

2.16 $\mathbf{Bx} = 3\begin{pmatrix} 6 \\ 7 \\ 2 \end{pmatrix} - 1\begin{pmatrix} -2 \\ 1 \\ -3 \end{pmatrix} + 2\begin{pmatrix} 3 \\ 0 \\ 5 \end{pmatrix} = \begin{pmatrix} 18 \\ 21 \\ 6 \end{pmatrix} + \begin{pmatrix} 2 \\ -1 \\ 3 \end{pmatrix} + \begin{pmatrix} 6 \\ 0 \\ 10 \end{pmatrix} = \begin{pmatrix} 26 \\ 20 \\ 19 \end{pmatrix}.$

2.17 **(a)** $(\mathbf{AB})' = \begin{pmatrix} 27 & 16 \\ -12 & -6 \\ 19 & 11 \end{pmatrix}, \quad \mathbf{B}'\mathbf{A}' = \begin{pmatrix} 27 & 16 \\ -12 & -6 \\ 19 & 11 \end{pmatrix}.$

(b) $\mathbf{AI} = \begin{pmatrix} 2 & 5 \\ 1 & 3 \end{pmatrix}\begin{pmatrix} 1 & 0 \\ 0 & 1 \end{pmatrix} = \begin{pmatrix} 2 & 5 \\ 1 & 3 \end{pmatrix} = \mathbf{A},$

$\mathbf{IB} = \begin{pmatrix} 1 & 0 \\ 0 & 1 \end{pmatrix}\begin{pmatrix} 1 & -6 & 2 \\ 5 & 0 & 3 \end{pmatrix} = \begin{pmatrix} 1 & -6 & 2 \\ 5 & 0 & 3 \end{pmatrix} = \mathbf{B}.$

(c) $|\mathbf{A}| = 1.$

(d) $\mathbf{A}^{-1} = \begin{pmatrix} 3 & -5 \\ -1 & 2 \end{pmatrix}.$

(e) $(\mathbf{A}^{-1})^{-1} = \begin{pmatrix} 3 & -5 \\ -1 & 2 \end{pmatrix}^{-1} = \begin{pmatrix} 2 & 5 \\ 1 & 3 \end{pmatrix} = \mathbf{A}.$

(f) $(\mathbf{A}')^{-1} = \begin{pmatrix} 2 & 1 \\ 5 & 3 \end{pmatrix}^{-1} = \begin{pmatrix} 3 & -1 \\ -5 & 2 \end{pmatrix}, \quad (\mathbf{A}^{-1})' = \begin{pmatrix} 3 & -1 \\ -5 & 2 \end{pmatrix}.$

2.18 **(a)** If $\mathbf{C} = \mathbf{AB}$, then by (2.35), we obtain

$$\begin{aligned} \mathbf{C}_{11} &= \mathbf{A}_{11}\mathbf{B}_{11} + \mathbf{A}_{12}\mathbf{B}_{21} \\ &= \begin{pmatrix} 2 & 1 \\ 3 & 2 \end{pmatrix}\begin{pmatrix} 1 & 1 & 1 \\ 2 & 1 & 1 \end{pmatrix} + \begin{pmatrix} 2 \\ 0 \end{pmatrix}(2 \quad 3 \quad 1) \\ &= \begin{pmatrix} 4 & 3 & 3 \\ 7 & 5 & 5 \end{pmatrix} + \begin{pmatrix} 4 & 6 & 2 \\ 0 & 0 & 0 \end{pmatrix} = \begin{pmatrix} 8 & 9 & 5 \\ 7 & 5 & 5 \end{pmatrix}. \end{aligned}$$

Continuing in this fashion, we obtain

$$\mathbf{AB} = \left(\begin{array}{ccc|c} 8 & 9 & 5 & 6 \\ 7 & 5 & 5 & 4 \\ \hline 3 & 4 & 2 & 2 \end{array}\right).$$

(b) $\mathbf{AB} = \begin{pmatrix} 8 & 9 & 5 & 6 \\ 7 & 5 & 5 & 4 \\ 3 & 4 & 2 & 2 \end{pmatrix}$ when found in the usual way.

2.19 (a) $$\mathbf{AB} = \mathbf{a}_1\mathbf{b}_1' + \mathbf{A}_2\mathbf{B}_2 = \begin{pmatrix} 2 & 2 & 2 & 0 \\ 3 & 3 & 3 & 0 \\ 1 & 1 & 1 & 0 \end{pmatrix} + \begin{pmatrix} 6 & 7 & 3 & 6 \\ 4 & 2 & 2 & 4 \\ 2 & 3 & 1 & 2 \end{pmatrix}$$

$$= \begin{pmatrix} 8 & 9 & 5 & 6 \\ 7 & 5 & 5 & 4 \\ 3 & 4 & 2 & 2 \end{pmatrix}.$$

2.20 $$\mathbf{Ab} = 2\begin{pmatrix} 5 \\ 7 \end{pmatrix} + 4\begin{pmatrix} -2 \\ 3 \end{pmatrix} - 3\begin{pmatrix} 3 \\ 1 \end{pmatrix} = \begin{pmatrix} -7 \\ 23 \end{pmatrix},$$

$\mathbf{Ab} = \begin{pmatrix} -7 \\ 23 \end{pmatrix}$ when found in the usual way.

2.21 By (2.26), $\mathbf{AB} = (\mathbf{Ab}_1, \mathbf{Ab}_2, \ldots, \mathbf{Ab}_p)$. By (2.37) each $\mathbf{Ab}_i$ can be expressed as a linear combination of the columns of $\mathbf{A}$, with coefficients from $\mathbf{b}_i$.

2.22

$$\left[-2\begin{pmatrix} 3 \\ 1 \\ 2 \end{pmatrix} + 3\begin{pmatrix} 0 \\ -1 \\ 1 \end{pmatrix} + \begin{pmatrix} 2 \\ 1 \\ 0 \end{pmatrix}, \; -1\begin{pmatrix} 3 \\ 1 \\ 2 \end{pmatrix} + \begin{pmatrix} 0 \\ -1 \\ -1 \end{pmatrix} - \begin{pmatrix} 2 \\ 1 \\ 0 \end{pmatrix}\right]$$

$$= \left[\begin{pmatrix} -6 \\ -2 \\ -4 \end{pmatrix} + \begin{pmatrix} 0 \\ -3 \\ 3 \end{pmatrix} + \begin{pmatrix} 2 \\ 1 \\ 0 \end{pmatrix}, \; \begin{pmatrix} -3 \\ -1 \\ -2 \end{pmatrix} + \begin{pmatrix} 0 \\ -1 \\ 1 \end{pmatrix} - \begin{pmatrix} 2 \\ 1 \\ 0 \end{pmatrix}\right]$$

$$= \begin{pmatrix} -4 & -5 \\ -4 & -3 \\ -1 & -1 \end{pmatrix} = \mathbf{AB}.$$

2.23 Suppose $\mathbf{a}_i = \mathbf{0}$ in the set of vectors $\mathbf{a}_1, \mathbf{a}_2, \ldots, \mathbf{a}_n$. Then $c_1\mathbf{a}_1 + \cdots + c_i\mathbf{0} + \cdots + c_n\mathbf{a}_n = \mathbf{0}$, where $c_1 = c_2 = \cdots = c_{i-1} = c_{i+1} = \cdots = c_n = 0$ and $c_i \neq 0$. Hence, by (2.40), $\mathbf{a}_1, \mathbf{a}_2, \ldots, \mathbf{a}_n$ are linearly dependent.

2.24 If one of the two matrices, say, $\mathbf{A}$, is nonsingular, multiply $\mathbf{AB} = \mathbf{O}$ by $\mathbf{A}^{-1}$ to obtain $\mathbf{B} = \mathbf{O}$. Otherwise, they are both singular. In fact, as noted following Example 2.3, the columns of $\mathbf{AB}$ are linear combinations of the columns of $\mathbf{A}$, with coefficients from $\mathbf{b}_j$.

$$\begin{aligned} \mathbf{AB} &= (b_{11}\mathbf{a}_1 + \cdots + b_{n1}\mathbf{a}_n, b_{12}\mathbf{a}_1 + \cdots + b_{n2}\mathbf{a}_n, \ldots) \\ &= (\mathbf{0}, \mathbf{0}, \ldots, \mathbf{0}). \end{aligned}$$

Since a linear combination of the columns of $\mathbf{A}$ is $\mathbf{0}$, $\mathbf{A}$ is singular [see (2.40)]. Similarly, by a comment following (2.38), the rows of $\mathbf{AB}$ are linear combinations of the rows of $\mathbf{B}$, and $\mathbf{B}$ is singular.

2.25 $\mathbf{AB} = \begin{pmatrix} 3 & 5 \\ 1 & 4 \end{pmatrix}$, $\mathbf{CB} = \begin{pmatrix} 3 & 5 \\ 1 & 4 \end{pmatrix}$, $\text{rank}(\mathbf{A}) = 2$, $\text{rank}(\mathbf{B}) = 2$, $\text{rank}(\mathbf{C}) = 2$.

2.26 **(a)** $\mathbf{AB} = \begin{pmatrix} 8 & 5 \\ 1 & 1 \end{pmatrix}$, $\mathbf{CB} = \begin{pmatrix} 2c_{11} + c_{13} & c_{11} + 2c_{12} \\ 2c_{21} + c_{23} & c_{21} + 2c_{22} \end{pmatrix}$. $\mathbf{C}$ is not unique.

An example is $\mathbf{C} = \begin{pmatrix} 1 & 2 & 6 \\ -1 & 1 & 3 \end{pmatrix}$.

(b) $\begin{pmatrix} 3 & 1 & 2 \\ 1 & 0 & -1 \end{pmatrix} \begin{pmatrix} x_1 \\ x_2 \\ x_3 \end{pmatrix} = \begin{pmatrix} 0 \\ 0 \end{pmatrix}$ gives two equations in three unknowns with solution vector $x_1 \begin{pmatrix} 1 \\ -5 \\ 1 \end{pmatrix}$, where x_1 is an arbitrary constant. We can't do the same for $\mathbf{B}$ because the columns of $\mathbf{B}$ are linearly independent.

2.27 **(a)** An example is $\mathbf{B} = \begin{pmatrix} 2 & 2 & 3 \\ 1 & 4 & 4 \\ -1 & -1 & 3 \end{pmatrix}$. Although $\mathbf{A}$ and $\mathbf{B}$ can be nonsingular, $\mathbf{A} - \mathbf{B}$ must be singular so that $(\mathbf{A} - \mathbf{B})\mathbf{x} = 0$.

(b) An example is $\mathbf{C} = \begin{pmatrix} -1 & 1 & 1 \\ 1 & -4 & 1 \\ 2 & 1 & -4 \end{pmatrix}$. In the expression $\mathbf{Cx} = \mathbf{0}$, we have a linear combination of the columns of $\mathbf{C}$ that is equal to $\mathbf{0}$, which is the definition of linear dependence. Therefore, $\mathbf{C}$ must be singular.

2.28 $\mathbf{A}'$ is nonsingular by definition because its rows are the columns of $\mathbf{A}$. To show that $(\mathbf{A}')^{-1} = (\mathbf{A}^{-1})'$, transpose both sides of $\mathbf{AA}^{-1} = \mathbf{I}$ to obtain $(\mathbf{AA}^{-1})' = \mathbf{I}'$, $(\mathbf{A}^{-1})'\mathbf{A}' = \mathbf{I}$. Multiply both sides on the right by $(\mathbf{A}')^{-1}$.

2.29 $(\mathbf{AB})^{-1}$ exists by Theorem 2.4(ii). Then

$$\begin{aligned} \mathbf{AB}(\mathbf{AB})^{-1} &= \mathbf{I}, \\ \mathbf{A}^{-1}\mathbf{AB}(\mathbf{AB})^{-1} &= \mathbf{A}^{-1}, \\ \mathbf{B}^{-1}\mathbf{B}(\mathbf{AB})^{-1} &= \mathbf{B}^{-1}\mathbf{A}^{-1}. \end{aligned}$$

2.30 $\mathbf{AB} = \begin{pmatrix} 23 & 1 \\ 13 & 1 \end{pmatrix}$, $\mathbf{B}^{-1} = \frac{1}{10}\begin{pmatrix} 1 & 2 \\ -3 & 4 \end{pmatrix}$, $(\mathbf{AB})^{-1} = \frac{1}{10}\begin{pmatrix} 1 & -1 \\ -13 & 23 \end{pmatrix}$, $\mathbf{B}^{-1}\mathbf{A}^{-1} = \frac{1}{10}\begin{pmatrix} 1 & -1 \\ -13 & 23 \end{pmatrix}$.

2.31 Multiply $\mathbf{A}$ by $\mathbf{A}^{-1}$ in (2.48) to get $\mathbf{I}$.

2.32 Multiply $\mathbf{A}$ by $\mathbf{A}^{-1}$ in (2.49) to get $\mathbf{I}$.

2.33 Muliply $\mathbf{B} + \mathbf{cc}'$ by $(\mathbf{B} + \mathbf{cc}')^{-1}$ in (2.50) to get $\mathbf{I}$.

2.34 Premultiply both sides of the equation by $\mathbf{A} + \mathbf{PBQ}$. The left side obviously equals $\mathbf{I}$. The right side becomes

$$\begin{aligned}
&(\mathbf{A} + \mathbf{PBQ})[\mathbf{A}^{-1} - \mathbf{A}^{-1}\mathbf{PB}(\mathbf{B} + \mathbf{BQA}^{-1}\mathbf{PB})^{-1}\mathbf{BQA}^{-1}] \\
&\quad = \mathbf{AA}^{-1} + \mathbf{PBQA}^{-1} - \mathbf{AA}^{-1}\mathbf{PB}(\mathbf{B} + \mathbf{BQA}^{-1}\mathbf{PB})^{-1}\mathbf{BQA}^{-1} \\
&\qquad - \mathbf{PBQA}^{-1}\mathbf{PB}(\mathbf{B} + \mathbf{BQA}^{-1}\mathbf{PB})^{-1}\mathbf{BQA}^{-1} \\
&\quad = \mathbf{I} + \mathbf{P}[\mathbf{I} - \mathbf{B}(\mathbf{B} + \mathbf{BQA}^{-1}\mathbf{PB})^{-1} - \mathbf{BQA}^{-1}\mathbf{PB}(\mathbf{B} + \mathbf{BQA}^{-1}\mathbf{PB})^{-1}]\mathbf{BQA}^{-1} \\
&\quad = \mathbf{I} + \mathbf{P}[\mathbf{I} - (\mathbf{B} + \mathbf{BQA}^{-1}\mathbf{PB})(\mathbf{B} + \mathbf{BQA}^{-1}\mathbf{PB})^{-1}]\mathbf{BQA}^{-1} \\
&\quad = \mathbf{I} + \mathbf{P}[\mathbf{I} - \mathbf{I}]\mathbf{BQA}^{-1} \\
&\quad = \mathbf{I}.
\end{aligned}$$

2.35 Since $\mathbf{y}'\mathbf{A}'\mathbf{y}$ is a scalar and is therefore equal to its transpose, we have $\mathbf{y}'\mathbf{A}'\mathbf{y} = (\mathbf{y}'\mathbf{A}'\mathbf{y})' = \mathbf{y}'(\mathbf{A})'(\mathbf{y}')' = \mathbf{y}'\mathbf{A}\mathbf{y}$. Then $\frac{1}{2}\mathbf{y}'(\mathbf{A} + \mathbf{A}')\mathbf{y} = \frac{1}{2}\mathbf{y}'\mathbf{A}\mathbf{y} + \frac{1}{2}\mathbf{y}'\mathbf{A}'\mathbf{y} = \frac{1}{2}\mathbf{y}'\mathbf{A}\mathbf{y} + \frac{1}{2}\mathbf{y}'\mathbf{A}\mathbf{y}$.

2.36 Use the proof of part (i) of Theorem 2.6b, substituting ≥ 0 for >0.

2.37 Corollary **1**: $\mathbf{y}'\mathbf{BAB}'\mathbf{y} = (\mathbf{B}'\mathbf{y})'\mathbf{A}(\mathbf{B}'\mathbf{y}) > 0$ if $\mathbf{B}'\mathbf{y} \neq \mathbf{0}$ since $\mathbf{A}$ is positive definite. Then $\mathbf{B}'\mathbf{y} = y_1\mathbf{b}_1 + \cdots + y_k\mathbf{b}_k$, where $\mathbf{b}_i$ is the ith column of $\mathbf{B}'$; that is , $\mathbf{b}_i'$ is the ith row of $\mathbf{B}$. Since the rows of $\mathbf{B}$ are linearly independent, there is no nonzero vector $\mathbf{y}$ such that $\mathbf{B}'\mathbf{y} = \mathbf{0}$.

2.38 We must show that if $\mathbf{A}$ is positive definite, then $\mathbf{A}=\mathbf{P}'\mathbf{P}$, where $\mathbf{P}$ is non singular. By Theorems 2.12d and 2.12f, $\mathbf{A} = \mathbf{CDC}'$, where $\mathbf{C}$ is orthogonal and $\mathbf{D} = \text{diag}(\lambda_1, \lambda_2, \ldots, \lambda_n)$ with all $\lambda_i > 0$. Then $\mathbf{A} = \mathbf{CDC}' = \mathbf{CD}^{1/2}\mathbf{D}^{1/2}\mathbf{C}' = (\mathbf{D}^{1/2}\mathbf{C}')(\mathbf{D}^{1/2}\mathbf{C}') = \mathbf{P}'\mathbf{P}$, where $\mathbf{D}^{1/2} = \text{diag}(\sqrt{\lambda_1}, \sqrt{\lambda_2}, \ldots, \sqrt{\lambda_p})$. Show that $\mathbf{P} = \mathbf{D}^{-1/2}\mathbf{C}'$ is nonsingular.

2.39 This follows by Theorems 2.6c and 2.4(ii).

2.40 **(a)** rank($\mathbf{A}$, $\mathbf{c}$) = rank($\mathbf{A}$) = 3. Solution $x_1 = \frac{7}{6}, x_2 = \frac{-5}{6}, x_3 = \frac{13}{6}$.
(b) rank($\mathbf{A}$) = 2, rank($\mathbf{A}$, $\mathbf{c}$) = 3. No solution.
(c) rank($\mathbf{A}$, $\mathbf{c}$) = rank($\mathbf{A}$) = 2. Solution $x_1 = 7, x_2 + x_3 + x_4 = 1$.

2.41 By definition, $\mathbf{AA}^-\mathbf{A} = \mathbf{A}$. If $\mathbf{A}$ is $n \times m$, then for conformability of multiplication, $\mathbf{A}^-$ must be $m \times n$.

2.42 $\mathbf{AA}_1^- \begin{pmatrix} 1 & 1 & 0 \\ 0 & 1 & 0 \\ 1 & 1 & 0 \end{pmatrix}, \quad \mathbf{AA}_1^-\mathbf{A} = \begin{pmatrix} 2 & 2 & 3 \\ 1 & 0 & 1 \\ 3 & 2 & 4 \end{pmatrix}.$

2.43 $\mathbf{A}_{11}\begin{pmatrix} 2 & 2 \\ 1 & 0 \end{pmatrix}, \quad \mathbf{A}_{11}^{-1} = -\frac{1}{2}\begin{pmatrix} 0 & -2 \\ -1 & 2 \end{pmatrix} = \begin{pmatrix} 0 & 1 \\ \frac{1}{2} & -1 \end{pmatrix}.$

2.44 Let $\mathbf{C}$ be the lower left 2×2 matrix $\mathbf{C} = \begin{pmatrix} 1 & 0 \\ 3 & 2 \end{pmatrix}$. Then $\mathbf{C}^{-1} = \frac{1}{2}\begin{pmatrix} 2 & 0 \\ -3 & 1 \end{pmatrix} = \begin{pmatrix} 1 & 0 \\ -\frac{3}{2} & \frac{1}{2} \end{pmatrix}$ and $(\mathbf{C}^{-1})' = \begin{pmatrix} 1 & -\frac{3}{2} \\ 0 & \frac{1}{2} \end{pmatrix}$.

2.45 **(i)** By Theorem 2.4(i), $\text{rank}(\mathbf{A}^{-}\mathbf{A}) \leq \text{rank}(\mathbf{A})$ and $\text{rank}(\mathbf{A}) = \text{rank}(\mathbf{A}\mathbf{A}^{-}\mathbf{A}) \leq \text{rank}(\mathbf{A}^{-}\mathbf{A})$. Hence $\text{rank}(\mathbf{A}^{-}\mathbf{A}) = \text{rank}(\mathbf{A})$.

(ii) $(\mathbf{A}\mathbf{A}^{-}\mathbf{A})' = \mathbf{A}'(\mathbf{A}^{-})'\mathbf{A}'$

(iii) Let $\mathbf{W} = \mathbf{A}[\mathbf{I} - (\mathbf{A}'\mathbf{A})^{-}\mathbf{A}'\mathbf{A}]$. Show that

$$\begin{aligned} \mathbf{W}'\mathbf{W} &= [\mathbf{I} - (\mathbf{A}'\mathbf{A})^{-}\mathbf{A}'\mathbf{A}][\mathbf{A}'\mathbf{A} - \mathbf{A}'\mathbf{A}(\mathbf{A}'\mathbf{A})^{-}\mathbf{A}'\mathbf{A}] \\ &= [\mathbf{I} - (\mathbf{A}'\mathbf{A})^{-}\mathbf{A}'\mathbf{A}]\mathbf{O} = \mathbf{O}. \end{aligned}$$

Then by Theorem 2.2c(ii), $\mathbf{W} = \mathbf{O}$.

(iv) $\mathbf{A}[(\mathbf{A}'\mathbf{A})^{-}\mathbf{A}']\mathbf{A} = \mathbf{A}(\mathbf{A}'\mathbf{A})^{-}\mathbf{A}'\mathbf{A} = \mathbf{A}$, by part (iii).

(v) (Searle 1982, p. 222) To show that $\mathbf{A}(\mathbf{A}'\mathbf{A})^{-}\mathbf{A}'$ is invariant to the choice of $(\mathbf{A}'\mathbf{A})^{-}$, let $\mathbf{B}$ and $\mathbf{C}$ be two values of $(\mathbf{A}'\mathbf{A})^{-}$. Then by part (iii), $\mathbf{A} = \mathbf{ABA}'\mathbf{A}$ and $\mathbf{A} = \mathbf{ACA}'\mathbf{A}$, so that $\mathbf{ABA}'\mathbf{A} = \mathbf{ACA}'\mathbf{A}$. To demonstrate that this implies $\mathbf{ABA}' = \mathbf{ACA}'$, show that

$$\begin{aligned} (\mathbf{ABA}'\mathbf{A} - \mathbf{ACA}'\mathbf{A})(\mathbf{B}'\mathbf{A}' - \mathbf{C}'\mathbf{A}') &= (\mathbf{ABA}' - \mathbf{ACA}') \\ &\quad \times (\mathbf{ABA}' - \mathbf{ACA}')'. \end{aligned}$$

The left side is $\mathbf{O}$ because $\mathbf{ABA}'\mathbf{A} = \mathbf{ACA}'\mathbf{A}$. The right side is then $\mathbf{O}$, and by Theorem 2.2c(ii), $\mathbf{ABA}' - \mathbf{ACA}' = \mathbf{O}$. To show symmetry, let $\mathbf{S}$ be a symmetric generalized inverse of $\mathbf{A}'\mathbf{A}$ (see Problem 2.46). Then $\mathbf{ASA}'$ is symmetric and $\mathbf{ASA}' = \mathbf{ABA}'$ since $\mathbf{ABA}'$ is invariant to $(\mathbf{A}'\mathbf{A})^{-}$. Thus $\mathbf{ABA}'$ is also symmetric. To show that $\text{rank}[\mathbf{A}(\mathbf{A}'\mathbf{A})^{-}\mathbf{A}'] = r$, use parts (i) and (iv).

2.46 If $\mathbf{A}$ is symmetric and $\mathbf{B}$ is a generalized inverse of $\mathbf{A}$, show that $\mathbf{ABA} = \mathbf{AB}'\mathbf{A}$. Then show that $\frac{1}{2}(\mathbf{B} + \mathbf{B})')$ and $\mathbf{BAB}'$ are symmetric generalized inverses of $\mathbf{A}$.

2.47 **(i)** By Corollary 1 to Theorem 2.8b, we obtain

$$\mathbf{A}^{-} = \begin{pmatrix} 0 & 0 & 0 \\ 0 & \frac{1}{2} & 0 \\ 0 & 0 & \frac{1}{2} \end{pmatrix}.$$

(ii) Using the five-step approach following Theorem 2.8b, with $\mathbf{C} = \begin{pmatrix} 2 & 2 \\ 2 & 0 \end{pmatrix}$ defined as the upper right 2×2 matrix, we obtain $\mathbf{C}^{-1} = \begin{pmatrix} 0 & \frac{1}{2} \\ \frac{1}{2} & -\frac{1}{2} \end{pmatrix}$ and $\mathbf{A}^{-} = \begin{pmatrix} 0 & 0 & 0 \\ 0 & \frac{1}{2} & 0 \\ \frac{1}{2} & -\frac{1}{2} & 0 \end{pmatrix}$.

2.48 **(b)** By definition, $\mathbf{A}\mathbf{A}^{-}\mathbf{A} = \mathbf{A}$. Multiplying on the left by $\mathbf{A}'$ gives $\mathbf{A}'\mathbf{A}\mathbf{A}^{-}\mathbf{A} = \mathbf{A}'\mathbf{A}$. Show that $(\mathbf{A}'\mathbf{A})^{-1}$ exists and multiply on the left by it.

2.49 **(iv)** If $\mathbf{A}$ is positive definite, then by Theorem 2.6d, $\mathbf{A}$ can be expressed as $\mathbf{A} = \mathbf{P}'\mathbf{P}$, where $\mathbf{P}$ is nonsingular. By Theorem 2.9c, we obtain

$$\begin{aligned} |\mathbf{A}| = |\mathbf{P}'\mathbf{P}| &= |\mathbf{P}'||\mathbf{P}| && \text{[by (2.74)]} \\ &= |\mathbf{P}||\mathbf{P}| && \text{[by (2.63)]} \\ &= |\mathbf{P}|^2 > 0 && \text{[by (2.61)]} \end{aligned}$$

(vi)

$$\begin{aligned} &|\mathbf{A}^{-1}\mathbf{A}| = |\mathbf{I}| = 1 \\ &|\mathbf{A}^{-1}||\mathbf{A}| = 1 \qquad \text{[by (2.74)]} \\ &|\mathbf{A}^{-1}| = 1/|\mathbf{A}|. \end{aligned}$$

2.50

$$|\mathbf{A}| = \begin{vmatrix} 2 & 5 \\ 1 & 3 \end{vmatrix} = 1 \neq 0, \quad \text{note that } \mathbf{A} \text{ is nonsingular}$$

$$|\mathbf{A}'| = \begin{vmatrix} 2 & 1 \\ 5 & 3 \end{vmatrix} = 1 = |\mathbf{A}|$$

$$\mathbf{A}^{-1} = \begin{pmatrix} 3 & -5 \\ -1 & 2 \end{pmatrix}, \quad |\mathbf{A}^{-1}| = \begin{vmatrix} 3 & -5 \\ -1 & 2 \end{vmatrix} = 1, \quad \frac{1}{\begin{vmatrix} 2 & 5 \\ 1 & 3 \end{vmatrix}} = 1,$$

2.51 **(a)**

$$10\begin{pmatrix} 2 & 5 \\ 1 & 3 \end{pmatrix} = \begin{pmatrix} 20 & 50 \\ 10 & 30 \end{pmatrix}, \quad \begin{vmatrix} 20 & 50 \\ 10 & 30 \end{vmatrix} = 100,$$

$$10^2\begin{vmatrix} 2 & 5 \\ 1 & 3 \end{vmatrix} = 100(1) = 100$$

(b) $|c\mathbf{A}| = |c\mathbf{I}\mathbf{A}| = |c\mathbf{I}||\mathbf{A}| = c^n|\mathbf{A}|$

2.52 Corollary 4. Let $\mathbf{A}_{11} = \mathbf{B}$, $\mathbf{A}_{22} = 1$, $\mathbf{A}_{21} = \mathbf{c}'$, and $\mathbf{A}_{12} = \mathbf{c}$. Then equate the right sides of (2.68) and (2.69).

2.53

$$\begin{aligned} &|\mathbf{AB}| = |\mathbf{A}||\mathbf{B}| = |\mathbf{B}||\mathbf{A}| = |\mathbf{BA}|, \\ &|\mathbf{A}^2| = |\mathbf{AA}| = |\mathbf{A}||\mathbf{A}| = |\mathbf{A}^2| \end{aligned}$$

2.54 **(a)** $|\mathbf{A}| = 1, \quad |\mathbf{B}| = \begin{vmatrix} 4 & -2 \\ 3 & 1 \end{vmatrix} = 10, \quad \mathbf{AB} = \begin{pmatrix} 23 & 1 \\ 13 & 1 \end{pmatrix}, \quad |\mathbf{AB}| = 10.$

(b) $|\mathbf{A}^2| = 1, \quad \mathbf{A}^2 = \begin{pmatrix} 9 & 25 \\ 5 & 14 \end{pmatrix}, \quad |\mathbf{A}^2| = 1.$

2.55 Define $\mathbf{B} = \begin{pmatrix} \mathbf{A}_{11}^{-1} & \mathbf{O} \\ -\mathbf{A}_{21}\mathbf{A}_{11}^{-1} & \mathbf{I} \end{pmatrix}$. Then

$$\mathbf{BA} = \begin{pmatrix} \mathbf{I} & \mathbf{A}_{11}^{-1}\mathbf{A}_{12} \\ \mathbf{O} & \mathbf{A}_{22} - \mathbf{A}_{21}\mathbf{A}_{11}^{-1}\mathbf{A}_{12} \end{pmatrix}.$$

By Corollary 1 to Theorem 2.9b, $|\mathbf{BA}| = |\mathbf{A}_{22} - \mathbf{A}_{21}\mathbf{A}_{11}^{-1}\mathbf{A}_{12}|$. By Theorem 2.9c. $|\mathbf{BA}| = |\mathbf{B}||\mathbf{A}|$. By Corollary 1 to Theorem 2.9b and (2.64), $|\mathbf{B}| = |\mathbf{A}_{11}^{-1}| = 1/|\mathbf{A}_{11}|$.

2.56 We first show that sine $\mathbf{c}_i'\mathbf{c}_j = 0$ for all $i \neq j$, the columns of $\mathbf{C}$ are linearly independent. Suppose that there exist $a_1, a_2, \ldots, a_p$ such that $a_1\mathbf{c}_1 + a_2\mathbf{c}_2 + \cdots + a_p\mathbf{c}_p = \mathbf{0}$. Multiply by $\mathbf{c}_1'$ to obtain $a_1\mathbf{c}_1'\mathbf{c}_1 + a_2\mathbf{c}_1'\mathbf{c}_2 + \cdots + a_p\mathbf{c}_1'\mathbf{c}_p = \mathbf{c}_1'\mathbf{0} = 0$ or $a_1\mathbf{c}_1'\mathbf{c}_1 = 0$, which implies that $a_1 = 0$. In a similar manner, we can show that $a_2 = a_3 = \cdots = a_p = 0$. Thus the columns of $\mathbf{C}$ are linearly independent and $\mathbf{C}$ is nonsingular. Multiply $\mathbf{C}'\mathbf{C} = \mathbf{I}$ on the left by $\mathbf{C}$ and on the right by $\mathbf{C}^{-1}$.

2.57 **(a)** $$\mathrm{C} = \begin{pmatrix} 1/\sqrt{3} & -1/\sqrt{2} & 1/\sqrt{6} \\ -1/\sqrt{3} & 0 & 2/\sqrt{6} \\ 1/\sqrt{3} & 1/\sqrt{2} & 1/\sqrt{6} \end{pmatrix}.$$

2.58

(i) $|\mathbf{I}| = |\mathbf{C}'\mathbf{C}| = |\mathbf{C}'||\mathbf{C}| = |\mathbf{C}||\mathbf{C}| = |\mathbf{C}|^2$. Thus $|\mathbf{C}|^2 = 1$ and $|\mathbf{C}| = \pm 1$.

(ii) By (2.75), $|\mathbf{C}'\mathbf{AC}| = |\mathbf{ACC}'| = |\mathbf{AI}| = |\mathbf{A}|$.

(iii) Since $\mathbf{c}_i'\mathbf{c}_i = 1$ for all i, we have $\mathbf{c}_i'\mathbf{c}_i = \sum_j c_{ij}^2 = 1$, and the maximum value of any c_{ij}^2 is 1.

2.59

(i) The ith diagonal element of $\mathbf{A} + \mathbf{B}$ is $a_{ii} + b_{ii}$. Hence $\mathrm{tr}(\mathbf{A} + \mathbf{B}) = \sum_i (a_{ii} + b_{ii}) = \sum_i a_{ii} + \sum_i b_{ii} = \mathrm{tr}(\mathbf{A}) + \mathrm{tr}(\mathbf{B})$.

(iv) By Theorem 2.2c(ii), the ith diagonal element of $\mathbf{AA}'$ is $\mathbf{a}_i'\mathbf{a}_i$, where $\mathbf{a}_i'$ is the ith row of $\mathbf{A}$.

(v) By (iii), $\mathrm{tr}(\mathbf{A}'\mathbf{A}) = \sum_i \mathbf{a}_i'\mathbf{a}_i = \sum_i \sum_j a_{ij}^2$, where $\mathbf{a}_i' = (a_{i1}, a_{i2}, \ldots, a_{ip})$.

(vii) By (2.84), $\mathrm{tr}(\mathbf{C}'\mathbf{AC}) = \mathrm{tr}(\mathbf{CC}'\mathbf{A}) = \mathrm{tr}(\mathbf{IA}) = \mathrm{tr}(\mathbf{A})$.

2.60
$$\mathbf{B} = \begin{pmatrix} 2 & 1 \\ 0 & 2 \\ 1 & 0 \end{pmatrix}, \quad \mathbf{B}'\mathbf{B} = \begin{pmatrix} 5 & 2 \\ 2 & 5 \end{pmatrix}, \quad \mathbf{BB}' = \begin{pmatrix} 5 & 2 & 2 \\ 2 & 4 & 0 \\ 2 & 0 & 1 \end{pmatrix},$$

$\mathrm{tr}(\mathbf{B}'\mathbf{B}) = 5 + 5 = 10, \quad \mathrm{tr}(\mathbf{BB}') = 5 + 4 + 1 = 10.$

(iii) Let $\mathbf{b}_i$ be the ith *column* of $\mathbf{B}$. Then

$$\sum_{i=1}^{2} \mathbf{b}_i'\mathbf{b}_i = (2, 0, 1)\begin{pmatrix} 2 \\ 0 \\ 1 \end{pmatrix} + (1, 2, 0)\begin{pmatrix} 1 \\ 2 \\ 0 \end{pmatrix} = 5 + 5 = 10.$$

(iv) Let $\mathbf{b}_i'$ be the ith *row* of $\mathbf{B}$. Then

$$\sum_{i=1}^{3} \mathbf{b}_i'\mathbf{b}_i = (2, 1)\begin{pmatrix} 2 \\ 1 \end{pmatrix} + (0, 2)\begin{pmatrix} 0 \\ 2 \end{pmatrix} + (1, 0)\begin{pmatrix} 1 \\ 0 \end{pmatrix}$$
$$= 5 + 4 + 1 = 10.$$

2.61 $\mathbf{A} = \begin{pmatrix} 3 & 1 & 2 \\ 1 & 0 & -1 \end{pmatrix}$, $\mathbf{A}'\mathbf{A} = \begin{pmatrix} 10 & 3 & 5 \\ 3 & 1 & 2 \\ 5 & 2 & 5 \end{pmatrix}$, $\mathbf{A}\mathbf{A}' = \begin{pmatrix} 14 & 1 \\ 1 & 2 \end{pmatrix}$,

$\mathrm{tr}(\mathbf{A}'\mathbf{A}) = 16$, $\mathrm{tr}(\mathbf{A}\mathbf{A}') = 16$, $\sum_{ij} a_{ij}^2 = 3^2 + 1^2 + 2^2 + 1^2 + 0^2 + (-1)^2 = 16.$

2.62 $(\mathbf{A}^-\mathbf{A})^2 = \mathbf{A}^-\mathbf{A}\mathbf{A}^-\mathbf{A} = \mathbf{A}^-\mathbf{A}$ since $\mathbf{A}\mathbf{A}^-\mathbf{A} = \mathbf{A}$ by definition. Hence $\mathbf{A}^-\mathbf{A}$ is idempotent and $\mathrm{tr}(\mathbf{A}^-\mathbf{A}) = \mathrm{rank}(\mathbf{A}^-\mathbf{A}) = r = \mathrm{rank}(\mathbf{A})$ by Theorem 2.8c(i). Show that $\mathrm{tr}(\mathbf{A}\mathbf{A}^-) = r$ by a similar argument.

2.63 $\mathbf{A} = \begin{pmatrix} 2 & 2 & 3 \\ 1 & 0 & 1 \\ 3 & 2 & 4 \end{pmatrix}$, $\mathbf{A}^- = \begin{pmatrix} 0 & 1 & 0 \\ 0 & -\frac{3}{2} & \frac{1}{2} \\ 0 & 0 & 0 \end{pmatrix}$, $\mathbf{A}^-\mathbf{A} = \begin{pmatrix} 1 & 0 & 1 \\ 0 & 1 & \frac{1}{2} \\ 0 & 0 & 0 \end{pmatrix}$,

$\mathrm{tr}(\mathbf{A}^-\mathbf{A}) = 2$,

$\mathbf{A}\mathbf{A}^- = \begin{pmatrix} 0 & -1 & 1 \\ 0 & 1 & 0 \\ 0 & 0 & 1 \end{pmatrix}$, $\mathrm{tr}(\mathbf{A}\mathbf{A}^-) = 2$, $\mathrm{rank}(\mathbf{A}^-\mathbf{A}) = \mathrm{rank}(\mathbf{A}\mathbf{A}^-) = 2.$

2.64

$$\lambda_2 = 2, (\mathbf{A} - \lambda_2\mathbf{I})\mathbf{x}_2 = \mathbf{0}, \begin{pmatrix} 1-2 & 2 \\ -1 & 4-2 \end{pmatrix}\begin{pmatrix} x_1 \\ x_2 \end{pmatrix} = \begin{pmatrix} 0 \\ 0 \end{pmatrix},$$

$$\begin{aligned} -x_1 + 2x_2 &= 0, \\ -x_1 + 2x_2 &= 0, \\ x_1 &= 2x_2, \\ \mathbf{x}_2 &= \begin{pmatrix} x_1 \\ x_2 \end{pmatrix} = \begin{pmatrix} 2x_2 \\ x_2 \end{pmatrix} = x_2\begin{pmatrix} 2 \\ 1 \end{pmatrix}. \end{aligned}$$

Use $x_2 = 1/\sqrt{5}$ to normalize $\mathbf{x}_2$:

$$\mathbf{x}_2 = \begin{pmatrix} 2/\sqrt{2} \\ 1/\sqrt{5} \end{pmatrix}.$$

2.65 From $\mathbf{A}^2\mathbf{x} = \lambda^2\mathbf{x}$, we obtain $\mathbf{A}\mathbf{A}^2\mathbf{x} = \lambda^2\mathbf{A}\mathbf{x} = \lambda^2\lambda\mathbf{x} = \lambda^3\mathbf{x}$. By induction $\mathbf{A}\mathbf{A}^{k-1}\mathbf{x} = \lambda^{k-1}\mathbf{A}\mathbf{x} = \lambda^{k-1}\lambda\mathbf{x} = \lambda^k\mathbf{x}$.

2.66 By (2.98) and (2.101), $\mathbf{A}^k = \mathbf{C}\mathbf{D}^k\mathbf{C}'$, where $\mathbf{C}$ is an orthogonal matrix containing the normalized eigenvectors of $\mathbf{A}$ and $\mathbf{D}^k = \mathrm{diag}(\lambda_1^k, \lambda_2^k, \ldots, \lambda_p^k)$. If $-1 < \lambda_i < 1$ for all i, then $\mathbf{D}^k \to \mathbf{O}$ and $\mathbf{A}^k \to \mathbf{O}$.

2.67

$$\begin{aligned} (\mathbf{AB} - \lambda\mathbf{I})\mathbf{x} &= \mathbf{0}, \\ (\mathbf{BAB} - \lambda\mathbf{B})\mathbf{x} &= \mathbf{0}, \\ (\mathbf{BA} - \lambda\mathbf{I})\mathbf{Bx} &= \mathbf{0}. \end{aligned}$$

2.68

$$\begin{aligned} 0 &= |\mathbf{P}^{-1}\mathbf{AP} - \lambda\mathbf{I}| = |\mathbf{P}^{-1}\mathbf{AP} - \lambda\mathbf{P}^{-1}\mathbf{P}| \\ &= |\mathbf{P}^{-1}(\mathbf{A} - \lambda\mathbf{I})\mathbf{P}| = |(\mathbf{A} - \lambda\mathbf{I})\mathbf{P}^{-1}\mathbf{P}| \\ &= |\mathbf{A} - \lambda\mathbf{I}|. \end{aligned}$$

Thus $\mathbf{P}^{-1}\mathbf{AP}$ and $\mathbf{A}$ have the same characteristic equation, as in (2.93).

2.69 Writing (2.92) for $\mathbf{x}_i$ and $\mathbf{x}_j$, we have $\mathbf{Ax}_i = \lambda_i\mathbf{x}_i$ and $\mathbf{Ax}_j = \lambda_j\mathbf{x}_j$. Multiplying by $\mathbf{x}_j$ and $\mathbf{x}_i'$ gives

$$\mathbf{x}_j'\mathbf{Ax}_i = \lambda_i\mathbf{x}_j'\mathbf{x}_i, \qquad (1)$$

$$\mathbf{x}_i'\mathbf{Ax}_j = \lambda_j\mathbf{x}_i'\mathbf{x}_j. \qquad (2)$$

Since $\mathbf{A}$ is symmetric, we can transpose (1) to obtain $(\mathbf{x}_j'\mathbf{Ax}_i)' = \lambda_i(\mathbf{x}_j'\mathbf{x}_i)'$ or $\mathbf{x}_i'\mathbf{Ax}_j = \lambda_i\mathbf{x}_i'\mathbf{x}_j$. This has the same left side as (2), and thus $\lambda_i\mathbf{x}_i'\mathbf{x}_j = \lambda_j\mathbf{x}_i'\mathbf{x}_j$ or $(\lambda_i - \lambda_j)\mathbf{x}_i'\mathbf{x}_j = 0$. Since $\lambda_i - \lambda_j \neq 0$, we have $\mathbf{x}_i'\mathbf{x}_j = 0$.

2.70 By (2.101), $\mathbf{A} = \mathbf{CDC}'$. Since $\mathbf{C}$ is orthogonal, we multiply on the left by $\mathbf{C}'$ and on the right by $\mathbf{C}$ to obtain $\mathbf{C}'\mathbf{AC} = \mathbf{C}'\mathbf{CDC}'\mathbf{C} = \mathbf{D}$.

2.71 $\mathbf{C} = \begin{pmatrix} -.5774 & .8165 & 0 \\ .5774 & .4082 & -.7071 \\ .5774 & .4082 & .7071 \end{pmatrix}.$

2.72 **(i)** By Theorem 21.2d, $|\mathbf{A}| = |\mathbf{CDC}'|$. By (2.75), $|\mathbf{CDC}'| = |\mathbf{C}'\mathbf{CD}| = |\mathbf{D}|$. By (2.59), $|\mathbf{D}| = \prod_{i=1}^{n} \lambda_i$.

2.73 **(a)** Eigenvalues of $\mathbf{A}$: 1, 2, -1

Eigenvectors: $\mathbf{x}_1 = \begin{pmatrix} .8018 \\ .5345 \\ .2673 \end{pmatrix}, \quad \mathbf{x}_2 = \begin{pmatrix} .3015 \\ .9045 \\ .3015 \end{pmatrix}, \quad \mathbf{x}_3 = \begin{pmatrix} .7071 \\ 0 \\ .7071 \end{pmatrix}.$

(b) $\text{tr}(\mathbf{A}) = 1 + 2 - 1 = 2, \quad |\mathbf{A}| = (1)(2)(-1) = -2.$

2.74 In the proof of part (i), if $\mathbf{A}$ is positive semidefinite, $\mathbf{x}_i'\mathbf{Ax}_i \geq 0$, while $\mathbf{x}_i'\mathbf{x}_i > 0$. By Corollary 1 to Theorem 2.12d $\mathbf{C}'\mathbf{AC} = \mathbf{D}$, where $\mathbf{D} = \text{diag}(\lambda_1, \lambda_2, \ldots, \lambda_n)$. Since $\mathbf{C}$ is orthogonal and nonsingular, then by Theorem 2.4(ii), the rank of $\mathbf{D}$ is the same as the rank of $\mathbf{A}$. Since $\mathbf{D}$ is diagonal, the rank is the number of nonzero elements on the diagonal, that is, the number of nonzero eigenvalues.

2.75 **(a)** $|\mathbf{A}| = 1$.

(b) The eigenvalues of $\mathbf{A}$ are .2679, 1, and 3.7321, all of which are positive.

2.76 (a) $(\mathbf{A}^{1/2})' = (\mathbf{CD}^{1/2}\mathbf{C}')' = (\mathbf{C}')'(\mathbf{D}^{1/2})'\mathbf{C}' = \mathbf{CD}^{1/2}\mathbf{C}' = \mathbf{A}^{1/2}$.
(b) $(\mathbf{A}^{1/2})^2 = \mathbf{A}^{1/2}\mathbf{A}^{1/2} = \mathbf{CD}^{1/2}\mathbf{C}'\mathbf{CD}^{1/2}\mathbf{C}' = \mathbf{CD}^{1/2}\mathbf{D}^{1/2}\mathbf{C}' = \mathbf{CDC}' = \mathbf{A}$.

2.77 $\lambda_1 = 3,\ \lambda_2 = 1, \mathbf{x}_1 = \begin{pmatrix} \sqrt{2}/2 \\ -\sqrt{2}/2 \end{pmatrix}, \mathbf{x}_2 = \begin{pmatrix} \sqrt{2}/2 \\ \sqrt{2}/2 \end{pmatrix}$.

$$\mathbf{A}^{1/2} = \mathbf{CD}^{1/2}\mathbf{C} = \left(\frac{\sqrt{2}}{2}\right)^2 \begin{pmatrix} 1 & 1 \\ -1 & 1 \end{pmatrix}\begin{pmatrix} \sqrt{3} & 0 \\ 0 & 1 \end{pmatrix}\begin{pmatrix} 1 & -1 \\ 1 & 1 \end{pmatrix}$$

$$= \frac{1}{2}\begin{pmatrix} 1+\sqrt{3} & 1-\sqrt{3} \\ 1-\sqrt{3} & 1+\sqrt{3} \end{pmatrix}.$$

2.78 (i) $(\mathbf{I}-\mathbf{A})^2 = \mathbf{I} - 2\mathbf{A} + \mathbf{A}^2 = \mathbf{I} - 2\mathbf{A} + \mathbf{A} = \mathbf{I} - \mathbf{A}$.
(ii) $\mathbf{A}(\mathbf{I}-\mathbf{A}) = \mathbf{A} - \mathbf{A}^2 = \mathbf{A} - \mathbf{A} = \mathbf{O}$.
(iii) $(\mathbf{P}^{-1}\mathbf{AP})^2 = \mathbf{P}^{-1}\mathbf{APP}^{-1}\mathbf{AP} = \mathbf{P}^{-1}\mathbf{A}^2\mathbf{P} = \mathbf{P}^{-1}\mathbf{AP}$.
(iv) $(\mathbf{C}'\mathbf{AC})^2 = \mathbf{C}'\mathbf{ACC}'\mathbf{AC} = \mathbf{C}'\mathbf{A}^2\mathbf{C} = \mathbf{C}'\mathbf{AC}$,
$(\mathbf{C}'\mathbf{AC})' = \mathbf{C}'\mathbf{A}'(\mathbf{C}')' = \mathbf{C}'\mathbf{AC}$ if $\mathbf{A} = \mathbf{A}'$.

2.79 $(\mathbf{A}^-\mathbf{A})^2 = \mathbf{A}^-\mathbf{AA}^-\mathbf{A} = \mathbf{A}^-\mathbf{A}$, since $\mathbf{AA}^-\mathbf{A} = \mathbf{A}$.
$[\mathbf{A}(\mathbf{A}'\mathbf{A})^-\mathbf{A}']^2 = \mathbf{A}(\mathbf{A}'\mathbf{A})^-\mathbf{A}'\mathbf{A}(\mathbf{A}'\mathbf{A}')^-\mathbf{A}' = \mathbf{A}(\mathbf{A}'\mathbf{A})^-\mathbf{A}'$, since $\mathbf{A} = \mathbf{A}(\mathbf{A}'\mathbf{A})^-\mathbf{A}'\mathbf{A}$ by Theorem 2.8c(iii).

2.80 (**a**) 2, (**e**) 2, (**f**) 1, 1, 0.

2.81 By (2.107), $\text{tr}(\mathbf{A}) = \sum_{i=1}^p \lambda_i$. By case 3 of Section 2.12.2 and (2.107), $\text{tr}(\mathbf{A}^2) = \sum_{i=1}^p \lambda_i^2$. Then $[\text{tr}(\mathbf{A})]^2 = \sum_{i=1}^p \lambda_i^2 + 2\sum_{i\neq j}\lambda_i\lambda_j = \text{tr}(\mathbf{A}^2) + 2\sum_{i\neq j}\lambda_i\lambda_j$.

2.82

$$\frac{\partial \mathbf{H}}{\partial \mathbf{x}} = \frac{\partial \mathbf{B}'(\mathbf{BAB}')^{-1}\mathbf{B}}{\partial \mathbf{x}},$$
$$= \mathbf{B}'\frac{\partial(\mathbf{BAB}')^{-1}}{\partial \mathbf{x}}\mathbf{B},$$
$$= \mathbf{B}'(\mathbf{BAB}')^{-1}\frac{\partial \mathbf{BAB}'}{\partial \mathbf{x}}(\mathbf{BAB}')^{-1}\mathbf{B},$$
$$= -\mathbf{B}'(\mathbf{BAB}')^{-1}\mathbf{B}\frac{\partial \mathbf{A}}{\partial \mathbf{x}}\mathbf{B}'(\mathbf{BAB}')^{-1}\mathbf{B},$$
$$= -\mathbf{H}\frac{\partial \mathbf{A}}{\partial \mathbf{x}}\mathbf{H}.$$

2.83 Let $\mathbf{X} = \begin{pmatrix} a & b \\ b & c \end{pmatrix}$ such that $ac > b^2$. Then,

$$\frac{\partial \ln|\mathbf{X}|}{\partial \mathbf{X}} = \frac{\partial \ln(ac - b^2)}{\partial \mathbf{X}},$$

$$= \begin{pmatrix} \frac{\partial \ln (ac-b^2)}{\partial a} & \frac{\partial \ln (ac-b^2)}{\partial b} \\ \frac{\partial \ln (ac-b^2)}{\partial b} & \frac{\partial \ln (ac-b^2)}{\partial c} \end{pmatrix},$$

$$= \frac{1}{ac - b^2} \begin{pmatrix} c & -2b \\ -2b & a \end{pmatrix},$$

$$= \frac{2}{ac - b^2} \begin{pmatrix} c & -2b \\ -2b & a \end{pmatrix} - \frac{1}{ac - b^2} \begin{pmatrix} c & 0 \\ 0 & a \end{pmatrix},$$

$$= 2\mathbf{X}^{-1} - \text{diag}\mathbf{X}^{-1}.$$

2.84 The constraints can be expressed as $\mathbf{h}(\mathbf{x}) = \mathbf{C}\mathbf{x} - \mathrm{t}$ where $\mathbf{C} = \begin{pmatrix} 1 & 0 \\ 1 & 1 \\ 0 & 1 \end{pmatrix}$ and $t = \begin{pmatrix} 2 \\ 3 \end{pmatrix}$. The Lagrange equations are $2\mathbf{A}\mathbf{x} + \mathbf{C}'\boldsymbol{\lambda} = 0$ and $\mathbf{C}\mathbf{x} = \mathbf{t}$, or

$$\begin{pmatrix} 2\mathbf{A} & \mathbf{C}' \\ \mathbf{C} & \mathbf{O} \end{pmatrix} \begin{pmatrix} \mathbf{x} \\ \boldsymbol{\lambda} \end{pmatrix} = \begin{pmatrix} \mathbf{0} \\ \mathbf{t} \end{pmatrix}.$$

The solution to this system of equations is

$$\begin{pmatrix} \mathbf{x} \\ \lambda \end{pmatrix} = \begin{pmatrix} 2\mathbf{A} & \mathbf{C}' \\ \mathbf{C} & \mathbf{O} \end{pmatrix}^{-1} \begin{pmatrix} \mathbf{0} \\ \mathbf{t} \end{pmatrix}.$$

Subsituting and simplifying using (2.50) we obtain

$$\mathbf{x} = \begin{pmatrix} 1/6 \\ 11/6 \\ 7/6 \end{pmatrix} \text{ and } \boldsymbol{\lambda} = \begin{pmatrix} -1/3 \\ -7 \end{pmatrix}.$$

Chapter 3

3.1 By (3.3) we have

$$E(ay) = \int_{-\infty}^{\infty} ayf(y)\,dy = a\int_{-\infty}^{\infty} yf(y)\,dy = aE(y).$$

3.2

$$\begin{aligned} E(y-\mu)^2 &= E(y^2 - 2\mu y + \mu^2) \\ &= E(y^2) - 2\mu E(y) + \mu^2 \quad \text{[by (3.4) and (3.5)]} \\ &= E(y^2) - 2\mu^2 + \mu^2 = E(y^2) - \mu^2. \end{aligned}$$

3.3

$$\begin{aligned} \text{var}(ay) &= E(ay - a\mu)^2 && \text{[by (3.6)]} \\ &= E[a(y-\mu)]^2 = E[a^2(y-\mu)^2] \\ &= a^2 E(y-\mu)^2 && \text{[by (3.4)].} \end{aligned}$$

3.4 The solution is similar to the answer to Problem 3.2.

3.5

$$\begin{aligned} E(y_i y_j) &= \int_{-\infty}^{\infty}\int_{-\infty}^{\infty} y_i y_j f(y_i y_j)\, dy_i\, dy_j \\ &= \int\int y_i y_j f_i(y_i) f_j(y_j)\, dy_i dy_j \quad \text{[by (3.12)]} \\ &= \int y_j f_j(y_j)\left[\int y_i f_i(y_i)\, dy_i\right] dy_j \\ &= E(y_i)\int y_j f_j(y_j)\, dy_j = E(y_i)E(y_j). \end{aligned}$$

3.6

$$\begin{aligned} \text{cov}(y_i, y_j) &= E(y_i y_j) - \mu_i\mu_j \qquad \text{[by (3.11)]} \\ &= \mu_i\mu_j - \mu_i\mu_j \qquad \text{[by (3.14)].} \end{aligned}$$

3.7 **(a)** Using the quadratic formula to solve for x in $y = 1 + 2x - x^2$ and $y=2x - x^2$, we obtain $x = 1 \pm 1\sqrt{2 - y}$ and $x = 1 \pm \sqrt{1 - y}$, respectively, which become the limits of integration in 3.16 and 3.17.

3.8 **(a)** Area $=\int_1^2 \int_{x-1}^{x} dy\, dx + \int_2^3 \int_{3-x}^{4-x} dy\, dx = 2.$

(b)

$$f_1(x) = \int_{x-1}^{x} \tfrac{1}{2} dy = \tfrac{1}{2}, \quad 1 \leq x \leq 2,$$

$$f_1(x) = \int_{3-x}^{4-x} \tfrac{1}{2} dy = \tfrac{1}{2}, \quad 2 \leq x \leq 3.$$

Hence, $f_1(x) = \frac{1}{2}, \quad 1 \leq x \leq 3.$

$$f_2(y) = \int_1^{y+1} \tfrac{1}{2} dx + \int_{3-y}^{3} \tfrac{1}{2}, \quad 0 \leq dx = y \leq 1,$$

$$f_2(y) = \int_y^{4-y} \tfrac{1}{2} dx = 2 - y, \quad 1 \leq y \leq 2,$$

$$E(x) = \int_1^3 x(\tfrac{1}{2})dx = 2,$$

$$E(y) = \int_0^1 y(y)dy + \int_1^2 y(2 - y)dy = \tfrac{1}{3} + \tfrac{2}{3} = 1,$$

$$E(xy) = \int_1^2 \int_{x-1}^{x} xy(\tfrac{1}{2})dy\, dx + \int_2^3 \int_{3-x}^{4-x} xy\,(\tfrac{1}{2})dy\, dx = 2,$$

$$\sigma_{xy} = 2 - 2(1) = 0.$$

(c)

$$f(y|x) = \frac{f(x, y)}{f_1(x)} = \frac{\frac{1}{2}}{\frac{1}{2}} = 1,$$

$$E(y|x) = \int_{x-1}^{x} y(1)dy = x - \tfrac{1}{2}, \quad 1 \leq x \leq 2,$$

$$E(y|x) = \int_{3-x}^{4-x} y(1)dy = \tfrac{7}{2} - x, \quad 2 \leq x \leq 3.$$

3.9

$$E(\mathbf{x} + \mathbf{y}) = E\begin{pmatrix} x_1 + y_1 \\ x_2 + x_2 \\ \vdots \\ x_p + y_p \end{pmatrix} = \begin{pmatrix} E(x_1 + y_1) \\ E(x_2 + y_2) \\ \vdots \\ E(x_p + y_p) \end{pmatrix}$$

$$= \begin{pmatrix} E(x_1) + E(y_1) \\ E(x_2) + E(y_2) \\ \vdots \\ E(x_p) + E(y_p) \end{pmatrix} = \begin{pmatrix} E(x_1) \\ E(x_2) \\ \vdots \\ E(x_p) \end{pmatrix} + \begin{pmatrix} E(y_1) \\ E(y_2) \\ \vdots \\ E(y_p) \end{pmatrix}$$

$$= E\begin{pmatrix} x_1 \\ x_2 \\ \vdots \\ x_p \end{pmatrix} + E\begin{pmatrix} y_1 \\ y_2 \\ \vdots \\ y_p \end{pmatrix}.$$

3.10 $E[(\mathbf{y} - \boldsymbol{\mu})(\mathbf{y} - \boldsymbol{\mu})'] = E[\mathbf{y}\mathbf{y}' - \mathbf{y}\boldsymbol{\mu}' - \boldsymbol{\mu}\mathbf{y}' + \boldsymbol{\mu}\boldsymbol{\mu}']$

$$= E(\mathbf{y}\mathbf{y}') - E(\mathbf{y})\boldsymbol{\mu}' - \boldsymbol{\mu}E(\mathbf{y}') + E(\boldsymbol{\mu}\boldsymbol{\mu}') \quad \text{[by (3.21) and (3.36)]}$$

$$= E(\mathbf{y}\mathbf{y}') - \boldsymbol{\mu}\boldsymbol{\mu}' - \boldsymbol{\mu}\boldsymbol{\mu}' + \boldsymbol{\mu}\boldsymbol{\mu}'.$$

3.11 Use the square root matrix $\Sigma^{1/2}$ defined in (2.107) to write (3.27) as

$$(\mathbf{y} - \boldsymbol{\mu})'\boldsymbol{\Sigma}^{-1}(\mathbf{y} - \boldsymbol{\mu}) = (\mathbf{y} - \boldsymbol{\mu})'(\boldsymbol{\Sigma}^{1/2}\boldsymbol{\Sigma}^{1/2})^{-1}(\mathbf{y} - \boldsymbol{\mu})$$

$$= [(\boldsymbol{\Sigma}^{1/2})^{-1}(\mathbf{y} - \boldsymbol{\mu})]'[(\boldsymbol{\Sigma}^{1/2})^{-1}(\mathbf{y} - \boldsymbol{\mu})] = \mathbf{z}'\mathbf{z}, \text{ say.}$$

Show that $\text{cov}(\mathbf{z}) = \mathbf{I}$ (see Problem 5.17).

3.13

$$\begin{aligned}
\text{cov}(\mathbf{z}) &= \text{cov}\begin{pmatrix}\mathbf{y}\\ \mathbf{x}\end{pmatrix} = E[(\mathbf{z}-\boldsymbol{\mu}_z)(\mathbf{z}-\boldsymbol{\mu}_z)'] \quad \text{[by (3.24)]}\\
&= E\left[\begin{pmatrix}\mathbf{y}\\ \mathbf{x}\end{pmatrix} - \begin{pmatrix}\boldsymbol{\mu}_y\\ \boldsymbol{\mu}_x\end{pmatrix}\right]\left[\begin{pmatrix}\mathbf{y}\\ \mathbf{X}\end{pmatrix} - \begin{pmatrix}\boldsymbol{\mu}_y\\ \boldsymbol{\mu}_x\end{pmatrix}\right]' \quad \text{[by (3.32)]}\\
&= E\begin{pmatrix}\mathbf{y}-\boldsymbol{\mu}_y\\ \mathbf{x}-\boldsymbol{\mu}_x\end{pmatrix}[(\mathbf{y}-\boldsymbol{\mu}_y)', (\mathbf{x}-\boldsymbol{\mu}_x)']\\
&= E\begin{bmatrix}(\mathbf{y}-\boldsymbol{\mu}_y)(\mathbf{y}-\boldsymbol{\mu}_y)' & (\mathbf{y}-\boldsymbol{\mu}_y)(\mathbf{x}-\boldsymbol{\mu}_x)'\\ (\mathbf{x}-\boldsymbol{\mu}_x)(\mathbf{y}-\boldsymbol{\mu}_y)' & (\mathbf{x}-\boldsymbol{\mu}_x)(\mathbf{x}-\boldsymbol{\mu}_x)'\end{bmatrix}\\
&= \begin{bmatrix}E[(\mathbf{y}-\boldsymbol{\mu}_y)(\mathbf{y}-\boldsymbol{\mu}_y)'] & E[(\mathbf{y}-\boldsymbol{\mu}_y)(\mathbf{x}-\boldsymbol{\mu}_x)']\\ E[(\mathbf{x}-\boldsymbol{\mu}_x)(\mathbf{y}-\boldsymbol{\mu}_y)'] & E[(\mathbf{x}-\boldsymbol{\mu}_x)(\mathbf{x}-\boldsymbol{\mu}_x)']\end{bmatrix}\\
&= \begin{pmatrix}\boldsymbol{\Sigma}_{yy} & \boldsymbol{\Sigma}_{yx}\\ \boldsymbol{\Sigma}_{xy} & \boldsymbol{\Sigma}_{xx}\end{pmatrix}. \quad \text{[by (3.34)]}
\end{aligned}$$

3.14 **(i)** If we write $\mathbf{A}$ in terms of its rows, then

$$\mathbf{Ay} = \begin{pmatrix}\mathbf{a}_1'\\ \mathbf{a}_2'\\ \vdots\\ \mathbf{a}_k'\end{pmatrix}, \quad \mathbf{y} = \begin{pmatrix}\mathbf{a}_1'\mathbf{y}\\ \mathbf{a}_2'\mathbf{y}\\ \vdots\\ \mathbf{a}_k{'}\mathbf{y}\end{pmatrix}$$

Then, by Theorem 3.6a, $E(\mathbf{a}_i'\mathbf{y}) = \mathbf{a}_i'E(\mathbf{y})$, and the result follows by (3.20).

(ii) Write $\mathbf{X}$ in terms of its columns $\mathbf{x}_i$ as $\mathbf{X} = (\mathbf{x}_1, \mathbf{x}_2, \ldots, \mathbf{x}_p)$. Since $\mathbf{Xb}$ is a random vector, we have, by Theorem 3.6a

$$\begin{aligned}
E(\mathbf{a}'\mathbf{Xb}) &= \mathbf{a}'E(\mathbf{Xb})\\
&= \mathbf{a}'E(b_1\mathbf{x}_1 + b_2\mathbf{x}_2 + \cdots + b_p\mathbf{x}_p) && \text{[by (2.37)]}\\
&= \mathbf{a}'[b_1E(\mathbf{x}_1) + b_2E(\mathbf{x}_2) + \cdots + b_pE(\mathbf{x}_p)]\\
&= \mathbf{a}'[E(\mathbf{x}_1), E(\mathbf{x}_2), \ldots, E(\mathbf{x}_p)]\mathbf{b} && \text{[by (2.37)]}\\
&= \mathbf{a}'E(\mathbf{X})\mathbf{b}.
\end{aligned}$$

(iii)

$$E(\mathbf{AXB}) = E\left[\begin{pmatrix}\mathbf{a}_1'\\ \mathbf{a}_2'\\ \vdots\\ \mathbf{a}_k'\end{pmatrix}\mathbf{X}(\mathbf{b}_1, \mathbf{b}_2, \ldots, \mathbf{b}_p)\right]$$

$$= E\begin{pmatrix} \mathbf{a}_1'\mathbf{X}\mathbf{b}_1 & \mathbf{a}_1'\mathbf{X}\mathbf{b}_2 & \cdots & \mathbf{a}_1'\mathbf{X}\mathbf{b}_p \\ \mathbf{a}_2'\mathbf{X}\mathbf{b}_1 & \mathbf{a}_2'\mathbf{X}\mathbf{b}_2 & \cdots & \mathbf{a}_2'\mathbf{X}\mathbf{b}_p \\ \vdots & \vdots & & \vdots \\ \mathbf{a}_k'\mathbf{X}\mathbf{b}_1 & \mathbf{a}_k'\mathbf{X}\mathbf{b}_2 & \cdots & \mathbf{a}_k'\mathbf{X}\mathbf{b}_p \end{pmatrix}$$

$$= \begin{pmatrix} \mathbf{a}_1'E(\mathbf{X})\mathbf{b}_1 & \mathbf{a}_1'E(\mathbf{X})\mathbf{b}_2 & \cdots & \mathbf{a}_1'E(\mathbf{X})\mathbf{b}_p \\ \mathbf{a}_2'E(\mathbf{X})\mathbf{b}_1 & \mathbf{a}_2'E(\mathbf{X})\mathbf{b}_2 & \cdots & \mathbf{a}_2'E(\mathbf{X})\mathbf{b}_p \\ \vdots & \vdots & & \vdots \\ \mathbf{a}_k'E(\mathbf{X})\mathbf{b}_1 & \mathbf{a}_k'E(\mathbf{X})\mathbf{b}_2 & \cdots & \mathbf{a}_k'E(\mathbf{X})\mathbf{b}_p \end{pmatrix}$$

$$= \begin{pmatrix} \mathbf{a}_1' \\ \mathbf{a}_2' \\ \vdots \\ \mathbf{a}_k' \end{pmatrix} E(\mathbf{X})(\mathbf{b}_1, \mathbf{b}_2, \ldots, \mathbf{b}_p) = \mathbf{A}E(\mathbf{X})\mathbf{B}.$$

3.15 By (3.21), $E(\mathbf{Ay} + \mathbf{b}) = E(\mathbf{Ay}) + E(\mathbf{b}) = \mathbf{A}E(\mathbf{y}) + \mathbf{b}$. Show that $E(\mathbf{b}) = \mathbf{b}$. if $\mathbf{b}$ is a constant vector.

3.16 By (3.10) and Theorem 3.6a, we obtain

$$\begin{aligned} \text{cov}(\mathbf{a}'\mathbf{y}, \mathbf{b}'\mathbf{y}) &= E[(\mathbf{a}'\mathbf{y} - \mathbf{a}'\boldsymbol{\mu})(\mathbf{b}'\mathbf{y} - \mathbf{b}'\boldsymbol{\mu})] \\ &= E[(\mathbf{a}'(\mathbf{y} - \boldsymbol{\mu})(\mathbf{y} - \boldsymbol{\mu})'\mathbf{b}] \quad \text{[by (2.18)]} \\ &= \mathbf{a}'E[(\mathbf{y} - \boldsymbol{\mu})(\mathbf{y} - \boldsymbol{\mu})']\mathbf{b} \quad \text{[by Theorem 3.6b (ii)]} \\ &= \mathbf{a}'\boldsymbol{\Sigma}\mathbf{b} \quad \text{[by (3.24)].} \end{aligned}$$

3.17 **(i)** By Theorem 3.6b parts (i) and (iii), we obtain

$$\begin{aligned} \text{cov}(\mathbf{Ay}) &= E[(\mathbf{Ay} - \mathbf{A}\boldsymbol{\mu})(\mathbf{Ay} - \mathbf{A}\boldsymbol{\mu})'] \\ &= E[(\mathbf{A}(\mathbf{y} - \boldsymbol{\mu})(\mathbf{y} - \boldsymbol{\mu})'\mathbf{A}'] \\ &= \mathbf{A}E[(\mathbf{y} - \boldsymbol{\mu})(\mathbf{y} - \boldsymbol{\mu})']\mathbf{A}' \\ &= \mathbf{A}\boldsymbol{\Sigma}\mathbf{A}' \quad \text{[by (3.24)].} \end{aligned}$$

(ii) By (3.34) and Theorem 3.6b(i), $\text{cov}(\mathbf{Ay},\mathbf{By}) = E[(\mathbf{Ay} - \mathbf{A}\boldsymbol{\mu})(\mathbf{By} - \mathbf{B}\boldsymbol{\mu})']$. Show that this is equal to $\mathbf{A}\boldsymbol{\Sigma}\mathbf{B}'$.

3.18 By (3.24) and (3.41), we have

$$\begin{aligned}\text{cov}(\mathbf{Ay}+\mathbf{b}) &= E[\mathbf{Ay}+\mathbf{b}-(\mathbf{A}\boldsymbol{\mu}+\mathbf{b})][\mathbf{Ay}+\mathbf{b}-(\mathbf{A}\boldsymbol{\mu}+\mathbf{b})]' \\ &= E[\mathbf{Ay}-\mathbf{A}\boldsymbol{\mu}][\mathbf{Ay}-\mathbf{A}\boldsymbol{\mu}]'.\end{aligned}$$

Show that this is equal to $\mathbf{A\Sigma A}'$.

3.19 Let $\mathbf{z} = \begin{pmatrix} \mathbf{y} \\ \mathbf{x} \\ \mathbf{v} \\ \mathbf{w} \end{pmatrix}, \mathbf{K} = \begin{pmatrix} \mathbf{A} & \mathbf{O} & \mathbf{O} & \mathbf{O} \\ \mathbf{O} & \mathbf{B} & \mathbf{O} & \mathbf{O} \\ \mathbf{O} & \mathbf{O} & \mathbf{C} & \mathbf{O} \\ \mathbf{O} & \mathbf{O} & \mathbf{O} & \mathbf{D} \end{pmatrix}, \mathbf{L} = \begin{pmatrix} \mathbf{I} & \mathbf{I} & \mathbf{O} & \mathbf{O} \end{pmatrix}$, and $\mathbf{M} = \begin{pmatrix} \mathbf{O} & \mathbf{O} & \mathbf{I} & \mathbf{I} \end{pmatrix}$. Then $\text{cov}(\mathbf{Ay}+\mathbf{Bx}, \mathbf{Cv}+\mathbf{Dw})$

$$\begin{aligned} &= \mathbf{LK}\,\text{cov}(\mathbf{z})\,\mathbf{K}'\mathbf{M}' \\ &= \begin{pmatrix} \mathbf{A} & \mathbf{B} & \mathbf{O} & \mathbf{O} \end{pmatrix} \begin{pmatrix} \boldsymbol{\Sigma}_{yy} & \boldsymbol{\Sigma}_{xy} & \boldsymbol{\Sigma}_{yv} & \boldsymbol{\Sigma} ym \\ \boldsymbol{\Sigma}_{xy} & \boldsymbol{\Sigma}_{xx} & \boldsymbol{\Sigma} xv & \boldsymbol{\Sigma}_{xw} \\ \boldsymbol{\Sigma}_{vy} & \boldsymbol{\Sigma}_{vx} & \boldsymbol{\Sigma}_{vv} & \boldsymbol{\Sigma}_{vw} \\ \boldsymbol{\Sigma}_{wy} & \boldsymbol{\Sigma}_{wx} & \boldsymbol{\Sigma}_{wv} & \boldsymbol{\Sigma}_{ww} \end{pmatrix} \begin{pmatrix} \mathbf{O}' \\ \mathbf{O}' \\ \mathbf{C}' \\ \mathbf{D}' \end{pmatrix} \\ &= \mathbf{A}\boldsymbol{\Sigma}_{yv}\mathbf{C}' + \mathbf{B}\boldsymbol{\Sigma}_{xv}\mathbf{C}' + \mathbf{A}\boldsymbol{\Sigma}_{yw}\mathbf{D}' + \mathbf{B}\boldsymbol{\Sigma}_{xw}\mathbf{D}'.\end{aligned}$$

3.20 **(a)** $E(z) = 8, \quad \text{var}(z) = 2.$

(b) $E(\mathbf{z}) = \begin{pmatrix} 3 \\ -4 \end{pmatrix}, \quad \text{cov}(\mathbf{z}) = \begin{pmatrix} 21 & -14 \\ -14 & 45 \end{pmatrix}.$

3.21 **(a)** $E(\mathbf{w}) = \begin{pmatrix} 6 \\ -10 \\ 6 \end{pmatrix}, \quad \text{cov}(\mathbf{w}) = \begin{pmatrix} 6 & -14 & 18 \\ -14 & 67 & -49 \\ 18 & -49 & 57 \end{pmatrix}.$

(b) $\text{cov}(\mathbf{z}, \mathbf{w}) = \begin{pmatrix} 11 & -25 & 34 \\ -8 & 53 & -31 \end{pmatrix}.$

Chapter 4

4.1 Use (3.2) and (3.8) and integrate directly.

4.2 By (2.67) $|\boldsymbol{\Sigma}^{-1/2}| = |(\boldsymbol{\Sigma}^{1/2})^{-1}| = |\boldsymbol{\Sigma}^{-1/2}|^{-1}$. We now use (2.77) to obtain $|\boldsymbol{\Sigma}| = |\boldsymbol{\Sigma}^{1/2}\boldsymbol{\Sigma}^{1/2}| = |\boldsymbol{\Sigma}^{1/2}||\boldsymbol{\Sigma}^{1/2}| = |\boldsymbol{\Sigma}^{1/2}|^2$, form which it follows that $|\boldsymbol{\Sigma}^{1/2}| = |\boldsymbol{\Sigma}|^{1/2}$.

4.3 Using Theorem 2.14a and the chain rule for differentiation (and assuming that we can interchange integration and differentiation), we obtain

$$\frac{\partial e^{\mathbf{t}'\mathbf{y}}}{\partial \mathbf{t}} = e^{\mathbf{t}'\mathbf{y}} \frac{\partial \mathbf{t}'\mathbf{y}}{\partial \mathbf{t}} = \mathbf{y} e^{\mathbf{t}'\mathbf{y}},$$

$$\frac{\partial M_{\mathbf{y}}(\mathbf{t})}{\partial \mathbf{t}} = \frac{\partial}{\partial \mathbf{t}} \int \cdots \int e^{\mathbf{t}'\mathbf{y}} f(\mathbf{y}) d\mathbf{y} = \int \cdots \int \frac{\partial}{\partial \mathbf{t}} e^{\mathbf{t}'\mathbf{y}} f(\mathbf{y}) d\mathbf{y},$$

$$\frac{\partial M_{\mathbf{y}}(\mathbf{0})}{\partial \mathbf{t}} = \int \cdots \int \mathbf{y} f(\mathbf{y}) d\mathbf{y} = E(\mathbf{y}) \quad [\text{by } (3.2) \text{ and } (3.20)].$$

4.4 $$\frac{\partial^2 e^{\mathbf{t}'\mathbf{y}}}{\partial t_r \partial t_s} = \frac{\partial}{\partial t_r} \left(e^{\mathbf{t}'\mathbf{y}} \frac{\partial \mathbf{t}'\mathbf{y}}{\partial t_s} \right) = \frac{\partial}{\partial t_r} (y_s e^{\mathbf{t}'\mathbf{y}}) = y_r y_s e^{\mathbf{t}'\mathbf{y}}.$$

4.5 Multiply out the third term on the right side in terms of $\mathbf{y} - \boldsymbol{\mu}$ and $\boldsymbol{\Sigma}\mathbf{t}$.

4.6 $M_{\mathbf{y}-\boldsymbol{\mu}}(\mathbf{t}) = E[e^{\mathbf{t}'(\mathbf{y}-\boldsymbol{\mu})}] = E(e^{\mathbf{t}'\mathbf{y}-\mathbf{t}'\boldsymbol{\mu}}) = e^{-\mathbf{t}'\boldsymbol{\mu}} E(e^{\mathbf{t}'\mathbf{y}}) = e^{-\mathbf{t}'\boldsymbol{\mu}} e^{\mathbf{t}'\boldsymbol{\mu}+(1/2)\mathbf{t}'\boldsymbol{\Sigma}\mathbf{t}}.$

4.7 $E(e^{\mathbf{t}'\mathbf{A}\mathbf{y}}) = E(e^{(\mathbf{A}'\mathbf{t})'\mathbf{y}})$. Now use Theorem 4.3 with $\mathbf{A}'\mathbf{t}$ in place of $\mathbf{t}$ to obtain

$$E(e^{\mathbf{t}'\mathbf{A}\mathbf{y}}) = e^{(\mathbf{A}'\mathbf{t})'\boldsymbol{\mu}+(1/2)(\mathbf{A}'\mathbf{t})\boldsymbol{\Sigma}\mathbf{A}'\mathbf{t}} = e^{\mathbf{t}'(\mathbf{A}\boldsymbol{\mu})+(1/2)\mathbf{t}'(\mathbf{A}\boldsymbol{\Sigma}\mathbf{A}')\mathbf{t}}.$$

4.8 Let $K(t) = \ln[M(t)]$. Then $K'(t) = \frac{M'(t)}{M(t)}$ and $K''(t) = \frac{M''(t)}{M(t)} - \left[\frac{M'(t)}{M(t)}\right]^2$. Since $M(0) = 1$, $K''(0) = M''(0) - [M'(0)]^2 = \sigma^2$.

4.9 $\mathbf{C}\boldsymbol{\Sigma}\mathbf{C}' = \mathbf{C}(\sigma^2\mathbf{I})\mathbf{C}' = \sigma^2\mathbf{C}\mathbf{C}' = \sigma^2\mathbf{I}$. Use Theorem 4.4a (ii).

4.10 The moment generating function for $\mathbf{z} = \mathbf{A}\mathbf{y} + \mathbf{b}$ is

$$\begin{aligned} M_{\mathbf{z}}(\mathbf{t}) &= E(e^{\mathbf{t}'\mathbf{z}}) = E(e^{\mathbf{t}'(\mathbf{A}\mathbf{y}+\mathbf{b})}) = E(e^{\mathbf{t}'(\mathbf{A}\mathbf{y}+\mathbf{t}'\mathbf{b})}) = e^{\mathbf{t}'\mathbf{b}} E(e^{\mathbf{t}'\mathbf{A}\mathbf{y}}) \\ &= e^{\mathbf{t}'\mathbf{b}} e^{\mathbf{t}'(\mathbf{A}\boldsymbol{\mu})+\mathbf{t}'(\mathbf{A}\boldsymbol{\Sigma}\mathbf{A}')\mathbf{t}/2} \quad [\text{by } (4.25)] \\ &= e^{\mathbf{t}'(\mathbf{A}\boldsymbol{\mu}+\mathbf{b})+\mathbf{t}'(\mathbf{A}\boldsymbol{\Sigma}\mathbf{A}')\mathbf{t}/2}, \end{aligned}$$

which is the moment generating function for a multivariate normal random vector with mean vector $\mathbf{A}\boldsymbol{\mu} + \mathbf{b}$ and covariance matrix $\mathbf{A}\boldsymbol{\Sigma}\mathbf{A}'$.

4.11 Use (2.35) and (2.36).

4.12 By Theorem 3.6d(ii), $\text{cov}(\mathbf{Ay}, \mathbf{By}) = \mathbf{A\Sigma B}'$.

4.13 Write $g(\mathbf{y}, \mathbf{x})$ in terms of $\begin{pmatrix} \boldsymbol{\mu}_y \\ \boldsymbol{\mu}_x \end{pmatrix}$ and $\boldsymbol{\Sigma} = \begin{pmatrix} \boldsymbol{\Sigma}_{yy} & \boldsymbol{\Sigma}_{yx} \\ \boldsymbol{\Sigma}_{xy} & \boldsymbol{\Sigma}_{xx} \end{pmatrix}$. For $|\boldsymbol{\Sigma}|$ and $\boldsymbol{\Sigma}^{-1}$, see (2.72) and (2.50). After canceling $h(\mathbf{x})$ in (4.28), show that $f(\mathbf{y}|\mathbf{x})$ can be written in the form

$$f(\mathbf{y}|\mathbf{x}) = \frac{1}{(2\pi)^{p/2}|\boldsymbol{\Sigma}_{y\cdot x}|^{1/2}} e^{-(\mathbf{y}-\boldsymbol{\mu}_{y\cdot x})'\boldsymbol{\Sigma}_{y\cdot x}^{-1}(\mathbf{y}-\boldsymbol{\mu}_{y\cdot x})/2},$$

where $\boldsymbol{\mu}_{y\cdot x} = \boldsymbol{\mu}_y + \boldsymbol{\Sigma}_{yx}\boldsymbol{\Sigma}_{xx}^{-1}(\mathbf{x} - \boldsymbol{\mu}_x)$ and $\boldsymbol{\Sigma}_{y\cdot x} = \boldsymbol{\Sigma}_{yy} - \boldsymbol{\Sigma}_{yx}\boldsymbol{\Sigma}_{xx}^{-1}\boldsymbol{\Sigma}_{xy}$.

4.14 $\text{cov}(\mathbf{y} - \mathbf{Bx}, \mathbf{x}) = \text{cov}\left[(\mathbf{I}, -\mathbf{B})\begin{pmatrix} y \\ x \end{pmatrix}, (\mathbf{O}, \mathbf{I})\begin{pmatrix} y \\ x \end{pmatrix}\right]$. Use Theorem 3.6d(ii)

4.16 **(a)** $\begin{pmatrix} y_1 \\ y_3 \end{pmatrix}$ is $N_2\left[\begin{pmatrix} 1 \\ 3 \end{pmatrix}, \begin{pmatrix} 4 & -1 \\ -1 & 5 \end{pmatrix}\right]$.

(b) y_2 is $N(2, 6)$.

(c) z is $N(-4, 79)$.

(d) $\mathbf{z} = \begin{pmatrix} z_1 \\ z_2 \end{pmatrix}$ is $N_2\left[\begin{pmatrix} 2 \\ 9 \end{pmatrix}, \begin{pmatrix} 11 & -6 \\ -6 & 154 \end{pmatrix}\right]$.

(e) $f(y_1, y_2|y_3, y_4) = N_2\left[\begin{pmatrix} 1 + y_3 + \frac{3}{2}y_4 \\ y_3 + \frac{1}{2}y_4 \end{pmatrix}, \begin{pmatrix} 2 & 2 \\ 2 & 4 \end{pmatrix}\right]$.

(f) $$E(y_1, y_3|y_2, y_4) = \begin{pmatrix} 1 \\ 3 \end{pmatrix} + \begin{pmatrix} 2 & 2 \\ 3 & -4 \end{pmatrix}\begin{pmatrix} 6 & -2 \\ -2 & 4 \end{pmatrix}^{-1}\begin{pmatrix} y_2 - 2 \\ y_4 + 2 \end{pmatrix},$$

$$\text{cov}(y_1, y_2|y_2, y_4) = \begin{pmatrix} 4 & -1 \\ -1 & 5 \end{pmatrix} - \begin{pmatrix} 2 & 2 \\ 3 & -4 \end{pmatrix}\begin{pmatrix} 6 & -2 \\ -2 & 4 \end{pmatrix}^{-1}\begin{pmatrix} 2 & 3 \\ 2 & -4 \end{pmatrix}.$$

Thus

$$f(y_1, y_3|y_2, y_4) = N_2\left[\begin{pmatrix} \frac{7}{5} + \frac{3}{2}y_2 + \frac{4}{5}y_4 \\ \frac{4}{5} + \frac{1}{5}y_2 - \frac{9}{10}y_4 \end{pmatrix}, \begin{pmatrix} \frac{6}{5} & \frac{2}{5} \\ \frac{2}{5} & \frac{4}{5} \end{pmatrix}\right].$$

(g) $\rho_{13} = -1/2\sqrt{5}$.

(h) $\rho_{13\cdot 24} = 1/\sqrt{6}$. Note that $\rho_{13\cdot 24}$ is opposite in sign to ρ_{13}.

(i) Using the partitioning

$$\boldsymbol{\mu} = \left(\begin{array}{c} 1 \\ \hline 2 \\ 3 \\ -2 \end{array}\right), \quad \boldsymbol{\Sigma} = \left(\begin{array}{c|ccc} 4 & 2 & -1 & 2 \\ \hline 2 & 6 & 3 & -2 \\ -1 & 3 & 5 & -4 \\ 2 & -2 & -4 & 4 \end{array}\right),$$

we have

$$E(y_1|y_2, y_3, y_4) = 1 + (2 \quad -1 \quad 2)\begin{pmatrix} 6 & 3 & -2 \\ 3 & 5 & -4 \\ -2 & -4 & 4 \end{pmatrix}^{-1}\begin{pmatrix} y_2 - 2 \\ y_3 - 3 \\ y_4 + 2 \end{pmatrix}$$

$$= 1 + (2 \quad -1 \quad 2)\begin{pmatrix} 1/4 & -1/4 & -1/8 \\ -\frac{1}{4} & 5/4 & 9/8 \\ 1/8 & 9/8 & 21/16 \end{pmatrix}\begin{pmatrix} y_2 - 2 \\ y_3 - 3 \\ y_4 + 2 \end{pmatrix}$$

$$= \frac{y_2}{2} + \frac{y_3}{2} + \frac{5y_4}{4} + 1$$

$$\text{var}(y_1|y_2, y_3, y_4) = 4 - (2 \quad -1 \quad 2)\begin{pmatrix} 1/4 & -1/4 & -1/8 \\ -1/4 & 5/4 & 9/8 \\ 1/8 & 9/8 & 21/16 \end{pmatrix}\begin{pmatrix} 2 \\ -1 \\ 2 \end{pmatrix}$$

$$= 4 - 3 = 1.$$

Thus $f(y_1|y_2, y_3, y_4) = N(1 + 1/2y_2 + 1/2y_3 + 5/4y_4, 1)$.

4.17 **(a)** $N(17, 79)$.

(b) $N_2\left[\begin{pmatrix} 6 \\ 0 \end{pmatrix}, \begin{pmatrix} 5 & 4 \\ 4 & 23 \end{pmatrix}\right]$.

(c) $f(y_2|y_1, y_3) = N\left(-\frac{5}{2} + \frac{1}{4}y_1 + \frac{1}{3}y_3, 17/12\right)$.

(d) $f(y_1|y_2, y_3) = N_2\left[\begin{pmatrix} 2 \\ -2 + \frac{1}{3}y_3 \end{pmatrix}, \begin{pmatrix} 4 & 1 \\ 1 & \frac{5}{3} \end{pmatrix}\right]$.

(e) $\rho_{12} = \sqrt{2}/4 = .3536, \rho_{12\cdot 3} = \sqrt{3/20} = .3873$.

4.18 y_1 and y_2 are independent, y_2 and y_3 are independent.

4.19 y_1 and y_2 are independent, (y_1, y_2) and (y_3, y_4) are independent.

4.20 Using the expression in (4.38) for $\boldsymbol{\Sigma}_{yx}$ in terms of its rows $\boldsymbol{\sigma}_{ix}$, show that

$$\boldsymbol{\Sigma}_{yx}\boldsymbol{\Sigma}_{xx}^{-1}\boldsymbol{\Sigma}_{xy} = \begin{pmatrix} \boldsymbol{\sigma}'_{1x}\boldsymbol{\Sigma}_{xx}^{-1}\boldsymbol{\sigma}_{1x} & \boldsymbol{\sigma}'_{1x}\boldsymbol{\Sigma}_{xx}^{-1}\boldsymbol{\sigma}_{2x} & \cdots & \boldsymbol{\sigma}'_{1x}\boldsymbol{\Sigma}_{xx}^{-1}\boldsymbol{\sigma}_{px} \\ \boldsymbol{\sigma}'_{2x}\boldsymbol{\Sigma}_{xx}^{-1}\boldsymbol{\sigma}_{1x} & \boldsymbol{\sigma}'_{2x}\boldsymbol{\Sigma}_{xx}^{-1}\boldsymbol{\sigma}_{2x} & \cdots & \boldsymbol{\sigma}'_{2x}\boldsymbol{\Sigma}_{xx}^{-1}\boldsymbol{\sigma}_{px} \\ \vdots & \vdots & & \vdots \\ \boldsymbol{\sigma}'_{px}\boldsymbol{\Sigma}_{xx}^{-1}\boldsymbol{\sigma}_{1x} & \boldsymbol{\sigma}'_{px}\boldsymbol{\Sigma}_{xx}^{-1}\boldsymbol{\sigma}_{2x} & \cdots & \boldsymbol{\sigma}'_{px}\boldsymbol{\Sigma}_{xx}^{-1}\boldsymbol{\sigma}_{px} \end{pmatrix}.$$

Chapter 5

5.1 $\sum_{i=1}^{n}(y_i-\bar{y})^2=\sum_{i=1}^{n}(y_i^2-2\bar{y}y_i+\bar{y}^2)=\sum_{i=1}^{n}y_i^2-2\bar{y}\sum_i y_i+n\bar{y}^2=\sum_i y_i^2-2n\bar{y}^2+n\bar{y}^2$

5.2 By (2.23)$[(1/n)\mathbf{J}]^2=(1/n^2)\mathbf{jj}'\mathbf{jj}'=(1/n^2)\mathbf{j}(n)\mathbf{j}'=(1/n)\mathbf{jj}'=(1/n)\mathbf{J}$.

5.3 **(a)** By Theorem 5.2b we obtain

$$\begin{aligned}\text{var}(s^2)&=\left[\frac{1}{(n-1)^2}\right]\text{var}\left[\mathbf{y}'\left(\mathbf{I}-\frac{1}{n}\mathbf{J}\right)\mathbf{y}\right]\\&=\left[\frac{1}{(n-1)^2}\right]\left\{2\text{tr}\left[\left(\mathbf{I}-\frac{1}{n}\mathbf{J}\right)\sigma^2\mathbf{I}\right]^2+4\mu^2\sigma^2\mathbf{j}'\left(\mathbf{I}-\frac{1}{n}\mathbf{J}\right)\mathbf{j}\right\}\\&=\left[\frac{1}{(n-1)^2}\right]\left[2\sigma^4\text{tr}\left(\mathbf{I}-\frac{1}{n}\mathbf{J}\right)+4\mu^2\sigma^2(n-n)\right]=\frac{2\sigma^4}{n-1}.\end{aligned}$$

(b)

$$\begin{aligned}\text{var}(s^2)&=\text{var}\left(\frac{\sigma^2 u}{n-1}\right)=\left[\frac{\sigma^4}{(n-1)^2}\right]\text{var}(u)\\&=\left[\frac{\sigma^4}{(n-1)^2}\right](2)(n-1)=\frac{2\sigma^4}{n-1}\end{aligned}$$

5.4 Note that $\boldsymbol{\theta}'\mathbf{V}^{-1}=\boldsymbol{\mu}'\boldsymbol{\Sigma}^{-1}$. Because of symmetry of $\mathbf{V}$ and $\boldsymbol{\Sigma}$, we have $\mathbf{V}^{-1}\boldsymbol{\theta}=\boldsymbol{\Sigma}^{-1}\boldsymbol{\mu}$. Substituting into the expression on the left we obtain

$$\begin{aligned}&|\boldsymbol{\Sigma}|^{-1/2}|\mathbf{I}-2t\mathbf{A}\boldsymbol{\Sigma}|^{-(1/2)}\left|\boldsymbol{\Sigma}^{-1}\right|^{-(1/2)}e^{-[\boldsymbol{\mu}'\boldsymbol{\Sigma}^{-1}\boldsymbol{\mu}-\boldsymbol{\mu}'(\mathbf{I}-2\mathbf{A}\boldsymbol{\Sigma})^{-1}\boldsymbol{\Sigma}^{-1}\boldsymbol{\mu}]/2}\\&=|\mathbf{I}-2t\mathbf{A}\boldsymbol{\Sigma}|^{-(1/2)}e^{-\boldsymbol{\mu}'[\mathbf{I}-(\mathbf{I}-2t\mathbf{A}\boldsymbol{\Sigma})^{-1}]\boldsymbol{\Sigma}^{-1}\boldsymbol{\mu}/2}.\end{aligned}$$

5.5 Expanding the second expression we obtain $e^{-[\boldsymbol{\mu}'\boldsymbol{\Sigma}^{-1}\boldsymbol{\mu}-\boldsymbol{\theta}'\mathbf{V}^{-1}\boldsymbol{\theta}+\mathbf{y}'\mathbf{V}^{-1}\mathbf{y}-2\boldsymbol{\theta}'\mathbf{V}^{-1}\mathbf{y}+\boldsymbol{\theta}'\mathbf{V}^{-1}\boldsymbol{\theta}]/2}$. Substituting $\boldsymbol{\theta}'$ and $\mathbf{V}^{-1}$, simplifying, and noting that $\boldsymbol{\theta}'\mathbf{V}^{-1}=\boldsymbol{\mu}'\boldsymbol{\Sigma}^{-1}$, we obtain the first expression.

5.6 $k'(t)=-\frac{1}{2}\frac{1}{|\mathbf{C}|}\frac{d|\mathbf{C}|}{dt}-\frac{1}{2}\boldsymbol{\mu}'\mathbf{C}^{-1}\frac{d\mathbf{C}}{dt}\mathbf{C}^{-1}\boldsymbol{\Sigma}^{-1}\boldsymbol{\mu}$. Using the chain rule,

$$\begin{aligned}k''(t)=&-\frac{1}{2}\frac{1}{|\mathbf{C}|^2}\left[\frac{d|\mathbf{C}|}{dt}\right]^2-\frac{1}{2}\frac{1}{|\mathbf{C}|}\frac{d^2|\mathbf{C}|}{dt^2}+\frac{1}{2}\boldsymbol{\mu}'\mathbf{C}^{-1}\frac{d\mathbf{C}}{dt}\mathbf{C}^{-1}\frac{d\mathbf{C}}{dt}\mathbf{C}^{-1}\boldsymbol{\Sigma}^{-1}\boldsymbol{\mu}\\&-\frac{1}{2}\boldsymbol{\mu}'\mathbf{C}^{-1}\frac{d^2\mathbf{C}}{dt^2}\mathbf{C}^{-1}\boldsymbol{\Sigma}^{-1}\boldsymbol{\mu}+\frac{1}{2}\boldsymbol{\mu}'\mathbf{C}^{-1}\frac{d\mathbf{C}}{dt}\mathbf{C}^{-1}\frac{d\mathbf{C}}{dt}\mathbf{C}^{-1}\boldsymbol{\Sigma}^{-1}\boldsymbol{\mu}.\end{aligned}$$

5.7

$$\begin{aligned}
\mathbf{y}'\mathbf{A}\mathbf{y} &= (\mathbf{y}-\boldsymbol{\mu}+\boldsymbol{\mu})'\mathbf{A}(\mathbf{y}-\boldsymbol{\mu}+\boldsymbol{\mu}) \\
&= (\mathbf{y}-\boldsymbol{\mu})'\mathbf{A}(\mathbf{y}-\boldsymbol{\mu}) + (\mathbf{y}-\boldsymbol{\mu})'\mathbf{A}\boldsymbol{\mu} + \boldsymbol{\mu}'\mathbf{A}(\mathbf{y}-\boldsymbol{\mu}) + \boldsymbol{\mu}'\mathbf{A}\boldsymbol{\mu} \\
&= (\mathbf{y}-\boldsymbol{\mu})'\mathbf{A}(\mathbf{y}-\boldsymbol{\mu}) + 2(\mathbf{y}-\boldsymbol{\mu})'\mathbf{A}\boldsymbol{\mu} + \boldsymbol{\mu}'\mathbf{A}\boldsymbol{\mu}
\end{aligned}$$

5.8 To show that $E[(\mathbf{y}-\boldsymbol{\mu})(\mathbf{y}-\boldsymbol{\mu})'\,\mathbf{A}(\mathbf{y}-\boldsymbol{\mu})] = \mathbf{0}$, we need to show that all central third moments of the multivariate normal are zero. This can be done by differentiating $M_{\mathbf{y}-\boldsymbol{\mu}}(\mathbf{t})$ from Corollary 1 to Theorem 4.3a. Show that

$$\begin{aligned}
\frac{\partial^3 M_{\mathbf{y}-\boldsymbol{\mu}}(\mathbf{t})}{\partial t_r \partial t_s \partial t_u} = e^{(1/2)\mathbf{t}'\boldsymbol{\Sigma}t}\Bigg[\sigma_{ur}\left(\sum_j t_j\sigma_{sj}\right) + \sigma_{sr}\left(\sum_j t_j\sigma_{uj}\right) \\
+ \left(\sum_j t_j\sigma_{uj}\right)\left(\sum_j t_j\sigma_{sj}\right)\left(\sum_j t_j\sigma_{rj}\right) + \sigma_{us}\left(\sum_j t_j\sigma_{rj}\right)\Bigg]
\end{aligned}$$

Since there is a t_j in every term, $\partial^3 M_{\mathbf{y}-\boldsymbol{\mu}}(\mathbf{t})/\partial t_r\,\partial t_s\,\partial t_u = 0$ for $\mathbf{t}=\mathbf{0}$ and $E[(y_r-\mu_r)(y_s-\mu_s)(y_u-\mu_u)] = 0$ for all r, s, u.
For the second term, we have [by (3.40)]

$$\begin{aligned}
2E[(\mathbf{y}-\boldsymbol{\mu})(\mathbf{y}-\boldsymbol{\mu})'\mathbf{A}\boldsymbol{\mu}] &= 2\{E[(\mathbf{y}-\boldsymbol{\mu})(\mathbf{y}-\boldsymbol{\mu})']\}\mathbf{A}\boldsymbol{\mu} \\
&= 2\boldsymbol{\Sigma}\mathbf{A}\boldsymbol{\mu}.
\end{aligned}$$

For the third term, we have

$$E[(\mathbf{y}-\boldsymbol{\mu})\mathrm{tr}(\mathbf{A}\boldsymbol{\Sigma})] = [E(\mathbf{y}-\boldsymbol{\mu}][\mathrm{tr}(\mathbf{A}\boldsymbol{\Sigma})] = \mathbf{0}[\mathrm{tr}(\mathbf{A}\boldsymbol{\Sigma})] = \mathbf{0}.$$

5.9 By definition

$$\begin{aligned}
\mathrm{cov}(\mathbf{B}\mathbf{y}, \mathbf{y}'\mathbf{A}\mathbf{y}) &= E\{[\mathbf{B}\mathbf{y} - E(\mathbf{B}\mathbf{y})][\mathbf{y}'\mathbf{A}\mathbf{y} - E(\mathbf{y}'\mathbf{A}\mathbf{y})]\} \\
&= E\{[\mathbf{B}(\mathbf{y}-\boldsymbol{\mu})][\mathbf{y}'\mathbf{A}\mathbf{y} - E(\mathbf{y}'\mathbf{A}\mathbf{y})]\} \\
&= \mathbf{B}E\{(\mathbf{y}-\boldsymbol{\mu})][\mathbf{y}'\mathbf{A}\mathbf{y} - E(\mathbf{y}'\mathbf{A}\mathbf{y})]\} \\
&= \mathbf{B}\,\mathrm{cov}(\mathbf{y}, \mathbf{y}'\mathbf{A}\mathbf{y}) = 2\mathbf{B}\boldsymbol{\Sigma}\mathbf{A}\boldsymbol{\mu}.
\end{aligned}$$

5.10 In (3.34), we have $\boldsymbol{\Sigma}_{yx} = E[(\mathbf{y}-\boldsymbol{\mu}_y)(\mathbf{x}-\boldsymbol{\mu}_x)']$. Show that $E(\mathbf{y}\mathbf{x}') = \boldsymbol{\Sigma}_{yx} + \boldsymbol{\mu}_y\boldsymbol{\mu}_x'$. Then

$$\begin{aligned}
E(\mathbf{x}'\mathbf{A}\mathbf{y}) &= E[\mathrm{tr}(\mathbf{x}'\mathbf{A}\mathbf{y})] = E[\mathrm{tr}(\mathbf{A}\mathbf{y}\mathbf{x}')] \\
&= \mathrm{tr}[E(\mathbf{A}\mathbf{y}\mathbf{x}')] = \mathrm{tr}[\mathbf{A}E(\mathbf{y}\mathbf{x}')] \\
&= \mathrm{tr}[\mathbf{A}(\boldsymbol{\Sigma}_{yx} + \boldsymbol{\mu}_y\boldsymbol{\mu}_x')] = \mathrm{tr}(\mathbf{A}\boldsymbol{\Sigma}_{yx} + \mathbf{A}\boldsymbol{\mu}_y\boldsymbol{\mu}_x') \\
&= \mathrm{tr}(\mathbf{A}\boldsymbol{\Sigma}_{yx}) + \mathrm{tr}(\mathbf{A}\boldsymbol{\mu}_y\boldsymbol{\mu}_x') = \mathrm{tr}(\mathbf{A}\boldsymbol{\Sigma}_{yx}) + \mathrm{tr}(\boldsymbol{\mu}_x'\mathbf{A}\boldsymbol{\mu}_y) \\
&= \mathrm{tr}(\mathbf{A}\boldsymbol{\Sigma}_{yx}) + \boldsymbol{\mu}_x'\mathbf{A}\boldsymbol{\mu}_y.
\end{aligned}$$

5.11 (a) $\sum_{i=1}^{n}(x_i-\bar{x})(y_i-\bar{y})=\sum_{i=1}^{n}(x_iy_i-\bar{x}y_i-\bar{y}x_i+\bar{x}\bar{y})=\sum_{i=1}^{n}x_iy_i-\bar{x}\sum_i y_i-\bar{y}\sum_i x_i+n\bar{x}\bar{y}=\sum_i x_iy_i-n\bar{x}\bar{y}-n\bar{x}\bar{y}+n\bar{x}\bar{y}$.

(b) With $\mathbf{x}=(x_1,x_2,\ldots,x_n)'$, $\mathbf{y}=(y_1,y_2,\ldots,y_n)'$, $\bar{x}=(1/n)\mathbf{j}'\mathbf{x}$, and $\bar{y}=(1/n)\mathbf{j}'\mathbf{y}$, we have

$$n\bar{x}\bar{y}=n\left(\frac{1}{n}\right)^2\mathbf{j}'\mathbf{x}\mathbf{j}'\mathbf{y}=\frac{1}{n}\mathbf{x}'\mathbf{j}\mathbf{j}'\mathbf{y}=\mathbf{x}'\left(\frac{1}{n}\mathbf{J}\right)\mathbf{y},$$

$$\sum_{i=1}^{n}x_iy_i-n\bar{x}\bar{y}=\mathbf{x}'\mathbf{y}-\mathbf{x}'\left(\frac{1}{n}\mathbf{J}\right)\mathbf{y}=\mathbf{x}'\left(\mathbf{I}-\frac{1}{n}\mathbf{J}\right)\mathbf{y}.$$

5.12 Apply (5.5), (5.9), and (5.8) with $\boldsymbol{\mu}=0$, $\mathbf{A}=I$, and $\boldsymbol{\Sigma}=\mathbf{I}$. The results follow.

5.13 By (5.9), $\text{var}(\mathbf{y}'\mathbf{A}\mathbf{y})=2\text{tr}(\mathbf{A}\boldsymbol{\Sigma})^2+4\boldsymbol{\mu}'\mathbf{A}\boldsymbol{\Sigma}\mathbf{A}\boldsymbol{\mu}$. In this case, we seek $\text{var}(\mathbf{y}'\mathbf{y})$, where $\mathbf{y}$ is $N_n(\boldsymbol{\mu},\mathbf{I})$. Hence, $\mathbf{A}=\boldsymbol{\Sigma}=\mathbf{I}$, and

$$\text{var}(\mathbf{y}'\mathbf{y})=2\text{tr}(\mathbf{I})^2+4\boldsymbol{\mu}'\boldsymbol{\mu}=2n+8\lambda.$$

Since $\mathbf{I}$ is $n\times n$, $\text{tr}(\mathbf{I})=n$, and by (5.24), $4\boldsymbol{\mu}'\boldsymbol{\mu}=8\lambda$.

5.14

$$\ln M_v(t)=-(n/2)\ln(1-2t)-\lambda[1-(1-2t)^{-1}],$$

$$\frac{d\ln M_v(t)}{dt}=\frac{n}{1-2t}-\lambda[-2(1-2t)^{-2}],$$

$$\frac{d\ln M_v(0)}{dt}=n+2\lambda,$$

$$\frac{d^2\ln M_v(t)}{dt^2}=\frac{2n}{(1-2t)^2}+8\lambda(1-2t)^{-3},$$

$$\frac{d^2\ln M_v(0)}{dt^2}=2n+8\lambda.$$

5.15 Since $v_1,v_2,\ldots,v_k$ are independent, we have

$$\begin{aligned}M_{\Sigma_i v_i}(t)&=E(e^{t\Sigma_i v_i})=E(e^{tv_1}e^{tv_2}\ldots e^{tv_k})\\&=E(e^{tv_1})E(e^{tv_2})\ldots E(e^{tv_k})\\&=\prod_{i=1}^{k}M_{v_i}(t)\prod_{i=1}^{k}\frac{1}{(1-2t)^{n_i/2}}e^{-\lambda_i[1-1/(1-2t)]}\\&=\frac{1}{(1-2t)^{\Sigma_i n_i/2}}e^{-[1-1/(1-2t)]\Sigma_i\lambda_i}.\end{aligned}$$

Thus by (5.25), $\Sigma_i v_i$ is $\chi^2(\Sigma_i n_i,\Sigma_i\lambda_i)$.

5.16 **(a)** $t^2 = x^2/(u/p)$ is $F(1, p)$ since z^2 is χ^2 (1), u is $\chi^2(p)$, and z^2 and u are independent.

(b) $t^2 = y^2/(u/p)$ is $F(1, p, \frac{1}{2}\mu^2)$ since y^2 is $\chi^2(1, \frac{1}{2}\mu^2)$, u is $\chi^2(p)$, and y^2 and u are independent.

5.17 $E[\boldsymbol{\Sigma}^{-1/2}(\mathbf{y}-\boldsymbol{\mu})] = \boldsymbol{\Sigma}^{-1/2}[E(\mathbf{y})-\boldsymbol{\mu}] = \mathbf{0}$. $\text{cov}[\boldsymbol{\Sigma}^{-1/2}(\mathbf{y}-\boldsymbol{\mu})] = \boldsymbol{\Sigma}^{-1/2}\text{cov}(\mathbf{y}-\boldsymbol{\mu})\boldsymbol{\Sigma}^{-1/2} = \boldsymbol{\Sigma}^{-1/2}\boldsymbol{\Sigma}\boldsymbol{\Sigma}^{-1/2} = \boldsymbol{\Sigma}^{-1/2}\boldsymbol{\Sigma}^{1/2}\boldsymbol{\Sigma}^{1/2}\boldsymbol{\Sigma}^{-1/2} = \mathbf{I}$. Then by Theorem 4.4a(ii), $\Sigma^{-1/2}(\mathbf{y}-\boldsymbol{\mu})$ is $N_n(\mathbf{0}, \mathbf{I})$.

5.18 **(a)** In this case, $\boldsymbol{\Sigma} = \sigma^2\mathbf{I}$ and $\mathbf{A}$ is replace by $\mathbf{A}/\sigma^2$. We thus have $(\mathbf{A}/\sigma^2)(\sigma^2\mathbf{I}) = \mathbf{A}$, which is indempotent.

(b) By Theorem 5.5, $\mathbf{y}'(\mathbf{A}/\sigma^2)\mathbf{y}$ is $\chi^2(r, \lambda)$ if $(\mathbf{A}/\sigma^2)\boldsymbol{\Sigma}$ is idempotent. In this case, $\boldsymbol{\Sigma} = \sigma^2\mathbf{I}$, so $(\mathbf{A}/\sigma^2)\boldsymbol{\Sigma} = (\mathbf{A}/\sigma^2)(\sigma^2\mathbf{I}) = \mathbf{A}$. For λ, we have $\lambda = \frac{1}{2}\boldsymbol{\mu}'(\mathbf{A}/\sigma^2)\boldsymbol{\mu} = \boldsymbol{\mu}'\mathbf{A}\boldsymbol{\mu}/2\sigma^2$.

5.19 By Theorem 5.5, $(\mathbf{y}-\boldsymbol{\mu})'\boldsymbol{\Sigma}^{-1}(\mathbf{y}-\boldsymbol{\mu})$ is $\chi^2(n)$ because $\mathbf{A}\boldsymbol{\Sigma} = \boldsymbol{\Sigma}^{-1}\boldsymbol{\Sigma} = \mathbf{I}$ (which is idempotent) and $E(\mathbf{y}-\boldsymbol{\mu}) = \mathbf{0}$. The distribution of $\mathbf{y}'\boldsymbol{\Sigma}^{-1}\mathbf{y}$ is $\chi^2(n, \lambda)$, where $\lambda = \frac{1}{2}\boldsymbol{\mu}'\boldsymbol{\Sigma}^{-1}\boldsymbol{\mu}$.

5.20 All of these are direct applications of Theorem 5.5.

(a) $\lambda = \frac{1}{2}\boldsymbol{\mu}'\mathbf{A}\boldsymbol{\mu} = \frac{1}{2}\mathbf{0}'\mathbf{A}\mathbf{0} = 0$.

(b) $\mathbf{A}\boldsymbol{\Sigma} = (\mathbf{I}/\sigma^2)(\sigma^2\mathbf{I}) = \mathbf{I}$, which is idempotent. $\lambda = \frac{1}{2}\boldsymbol{\mu}'(\mathbf{I}/\sigma^2)\boldsymbol{\mu} = \boldsymbol{\mu}'\boldsymbol{\mu}/2\sigma^2$.

(c) In this case "$\mathbf{A}\boldsymbol{\Sigma}$" becomes $(\mathbf{A}/\sigma^2)(\sigma^2\boldsymbol{\Sigma}) = \mathbf{A}\boldsymbol{\Sigma}$.

5.21 $\mathbf{B}\boldsymbol{\Sigma}\mathbf{A} = \mathbf{B}(\sigma^2\mathbf{I})\mathbf{A} = \sigma^2\mathbf{B}\mathbf{A}$, which is $\mathbf{O}$ if $\mathbf{B}\mathbf{A} = \mathbf{O}$.

5.22 $\mathbf{j}'[\mathbf{I} - (1/n)\mathbf{J}] = \mathbf{j}'[\mathbf{I} - (1/n)\mathbf{j}\mathbf{j}'] = \mathbf{j}' - (1/n)\mathbf{j}'\mathbf{j}\mathbf{j}' = \mathbf{j}' - (1/n)(n)\mathbf{j}' = \mathbf{0}'$.

5.23 $\mathbf{A}\boldsymbol{\Sigma}\mathbf{B} = \mathbf{A}(\sigma^2\mathbf{I})\mathbf{B} = \sigma^2\mathbf{A}\mathbf{B}$, which is $\mathbf{O}$ if $\mathbf{A}\mathbf{B} = \mathbf{O}$.

5.24 **(a)** Use Theorem 4.4a(i). In this case $\mathbf{a} = \mathbf{j}/n$.

(b) $t = \dfrac{z}{\sqrt{u/(n-1)}} = \dfrac{(\bar{y}-\mu)/(\sigma/\sqrt{n})}{\sqrt{[(n-1)s^2/\sigma^2/(n-1)]}}$. Show that $z = (\bar{y}-\mu)/(\sigma/\sqrt{n})$ is $N(0, 1)$.

(c) Let $v = (\bar{y}-\mu_0)/(\sigma/\sqrt{n})$. Then $E(v) = (\mu-\mu_0)/(\sigma/\sqrt{n}) = \delta$, say, and $\text{var}(v) = [1/(\sigma^2/n)]\,\text{var}(\bar{y}) = 1$. Hence v is $N(\delta, 1)$, and by (5.29), we obtain

$$\frac{v}{\sqrt{\dfrac{(n-1)s^2/\sigma^2}{n-1}}} = \frac{\bar{y}-\mu_0}{s/\sqrt{n}} \quad \text{is} \quad t(n-1, \delta).$$

Thus $\delta = (\mu-\mu_0)/(\sigma/\sqrt{n})$.

5.25 By Problem 5.2, $(1/n)\mathbf{J}$ and $\mathbf{I}-(1/n)\mathbf{J}$ are idempotent and $[\mathbf{I}-(1/n)\mathbf{J}]$ $[(1/n)\mathbf{J}]=\mathbf{O}$. By Example 5.5, $\sum_{i=1}^n(y_i-\bar{y})^2/\sigma^2=\mathbf{y}'[\mathbf{I}-(1/n)\mathbf{J}]\mathbf{y}/\sigma^2$ is $\chi^2(n-1)$. Show that $n\bar{y}^2/\sigma^2=\mathbf{y}'[(1/n)\mathbf{J}]\mathbf{y}/\sigma^2$ is $\chi^2(1,\lambda)$, where $\lambda=\frac{1}{2}\boldsymbol{\mu}'\mathbf{A}\boldsymbol{\mu}=n\mu^2/2\sigma^2$. Since $[\mathbf{I}-(1/n)\mathbf{J}(1/n)\mathbf{J}]=\mathbf{O}$, the quadratic forms $\mathbf{y}'[(1/n)\mathbf{J}]\mathbf{y}$ and $\mathbf{y}'[\mathbf{I}-(1/n)\mathbf{J}]\mathbf{y}$ are independent. Thus by (5.26), $n\bar{y}^2/[\sum_{i=1}^n(y_i-\bar{y})^2/(n-1)]$ is $F(1,n-1,\lambda)$, where $\lambda=n\mu^2/2\sigma^2$. If $\mu=0$ (H_0 is true), then $\lambda=0$ and $n\bar{y}^2/[\sum_{i=1}^n(y_i-\bar{y})^2/(n-1)]$ is $F(1,n-1)$.

5.26 **(b)** Since

$$\frac{\sum_{i=1}^n(y_i-\bar{y})^2}{\sigma^2(1-\rho)}=\frac{\mathbf{y}'[\mathbf{I}-(1/n)\mathbf{J}]\mathbf{y}}{\sigma^2(1-\rho)}=\mathbf{y}'\left(\frac{\mathbf{A}}{\sigma^2(1-\rho)}\right)\mathbf{y},$$

we have

$$\frac{\mathbf{A}}{\sigma^2(1-\rho)}\boldsymbol{\Sigma}=\frac{\sigma^2}{\sigma^2(1-\rho)}=[\mathbf{I}-(1/n)\mathbf{J}][(1-\rho)\mathbf{I}+\rho\mathbf{J}].$$

Show that this equals $(\mathbf{I}-\frac{1}{n}\mathbf{J})$, which is idempotent.

5.27 **(a)** $E(\mathbf{y}'\mathbf{A}\mathbf{y})=\text{tr}(\mathbf{A}\boldsymbol{\Sigma})+\boldsymbol{\mu}'\mathbf{A}\boldsymbol{\mu}=-16$.

(b) $\text{var}(\mathbf{y}'\mathbf{A}\mathbf{y})=2\text{tr}(\mathbf{A}\boldsymbol{\Sigma})^2+4\boldsymbol{\mu}'\mathbf{A}\boldsymbol{\Sigma}\mathbf{A}\boldsymbol{\mu}=21{,}138$.

(c) Check to see if $\mathbf{A}\boldsymbol{\Sigma}$ is indepotent.

(d) Check to see if $\mathbf{A}$ is idempotent.

5.28 $\mathbf{A}=\boldsymbol{\Sigma}^{-1}=\text{diag}\left(\frac{1}{2},\frac{1}{4},\frac{1}{3}\right)$, $\frac{1}{2}\boldsymbol{\mu}'\mathbf{A}\boldsymbol{\mu}=2.9167$.

5.29 $\mathbf{A}=\boldsymbol{\Sigma}^{-1}$, $\frac{1}{2}\boldsymbol{\mu}'\mathbf{A}\boldsymbol{\mu}=27$.

5.30 **(a)** Show that $\mathbf{A}$ is idempotent of rank 2, which is equal to tr $(\mathbf{A})$. Therefore, $\mathbf{y}'\mathbf{A}\mathbf{y}/\sigma^2$ is $\chi^2(2,\boldsymbol{\mu}'\mathbf{A}\boldsymbol{\mu}/2\sigma^2)$, where $\frac{1}{2}\boldsymbol{\mu}'\mathbf{A}\boldsymbol{\mu}=\frac{1}{2}(12.6)=6.3$.

(b) $\mathbf{BA}=\begin{pmatrix}0&0&0\\1&0&-1\end{pmatrix}\neq\mathbf{O}$. Hence $\mathbf{y}'\mathbf{A}\mathbf{y}$ and $\mathbf{By}$ are not independent.

(c) $y_1+y_2+y_3=\mathbf{j}'\mathbf{y}$. Show that $\mathbf{j}'\mathbf{A}=\mathbf{0}'$. Hence $\mathbf{y}'\mathbf{A}\mathbf{y}$ and $y_1+y_2+y_3$ are independent.

5.31 **(a)** Show that $\mathbf{B}$ is idempotent of rank 1. Therefore, $\mathbf{y}'\mathbf{B}\mathbf{y}/\sigma^2$ is $\chi^2(1,\boldsymbol{\mu}'\mathbf{B}\boldsymbol{\mu}/2\sigma^2)$. Find $\frac{1}{2}\boldsymbol{\mu}'\mathbf{B}\boldsymbol{\mu}$.

(b) Show that $\mathbf{BA}=\mathbf{O}$. Therefore, $\mathbf{y}'\mathbf{B}\mathbf{y}$ and $\mathbf{y}'\mathbf{A}\mathbf{y}$ are independent.

5.32 **(a)** $\mathbf{A}^2=\mathbf{X}(\mathbf{X}'\mathbf{X})^{-1}\mathbf{X}'\mathbf{X}(\mathbf{X}'\mathbf{X})^{-1}\mathbf{X}'=\mathbf{X}(\mathbf{X}'\mathbf{X})^{-1}\mathbf{X}'=\mathbf{A}$. By Theorem 2.13d rank$(\mathbf{A})=\text{tr}(\mathbf{A})=\text{tr}[\mathbf{X}(\mathbf{X}'\mathbf{X})^{-1}\mathbf{X}']$. By Theorem 2.11(ii), this becomes tr $[\mathbf{X}(\mathbf{X}'\mathbf{X})^{-1}\mathbf{X}'] = \text{tr}(\mathbf{I}_p) = p$. Similarly, rank tr $(\mathbf{I}-\mathbf{A}) = n-p$.

(b) $\text{tr}[\mathbf{A}(\sigma^2\mathbf{I})] = \text{tr}[\sigma^2\mathbf{X}(\mathbf{X}'\mathbf{X})^{-1}\mathbf{X}'] = p\sigma^2$. $\boldsymbol{\mu}'\mathbf{A}\boldsymbol{\mu} = (\mathbf{Xb})'\mathbf{X}(\mathbf{X}'\mathbf{X})^{-1}\mathbf{X}'(\mathbf{Xb}) = \mathbf{b}'\mathbf{X}'\mathbf{Xb}$. Thus $E[\mathbf{y}'\mathbf{Ay}] = p\sigma^2 + \mathbf{b}'\mathbf{X}'\mathbf{Xb}$. Show that tr $[(\mathbf{I}-\mathbf{A})(\sigma^2\mathbf{I})] = \sigma^2(n-p)$ and $\boldsymbol{\mu}'(\mathbf{I}-\mathbf{A})\boldsymbol{\mu} = 0$, if $\boldsymbol{\mu} = \mathbf{Xb}$. Hence $E[\mathbf{y}'(\mathbf{I}-\mathbf{A})\mathbf{y}] = (n-p)\sigma^2$.

(c) $\mathbf{y}'\mathbf{Ay}/\sigma^2$ is $\chi^2(p, \lambda)$, where $\lambda = \boldsymbol{\mu}'\mathbf{A}\boldsymbol{\mu}/2\sigma^2 = \mathbf{b}'\mathbf{X}'\mathbf{Xb}/2\sigma^2$. $\mathbf{y}'(\mathbf{I}-\mathbf{A})\mathbf{y}/\sigma^2$ is $\chi^2(n-p)$.

(d) Show that $\mathbf{X}(\mathbf{X}'\mathbf{X})^{-1}\mathbf{X}'[\mathbf{I} - \mathbf{X}(\mathbf{X}'\mathbf{X})^{-1}\mathbf{X}'] = \mathbf{O}$. Then, by Corollary 1 to Theorem 5.6b, $\mathbf{y}'\mathbf{Ay}$ and $\mathbf{y}'(\mathbf{I}-\mathbf{A})\mathbf{y}$ are independent.

(e) $F(p, n-p, \lambda)$, where $\lambda = b'\mathbf{X}'\mathbf{Xb}/2\sigma^2$.

Chapter 6

6.1 Equations (6.3) and (6.4) can be written as

$$\sum_{i=1}^{n} y_i - n\hat{\beta}_0 - \hat{\beta}_1 \sum_{i=1}^{n} x_i = 0,$$

$$\sum_{i=1}^{n} x_i y_i - \hat{\beta}_0 \sum_{i=1}^{n} x_i - \hat{\beta}_1 \sum_{i=1}^{n} x_i^2 = 0.$$

Solving for $\hat{\beta}_0$ from the first equation gives $\hat{\beta}_0 = \sum_i y_i/n - \hat{\beta}_1 \sum_i x_i/n = \bar{y} - \hat{\beta}_1\bar{x}$. Substituting this into the second equation gives the result for $\hat{\beta}_1$.

6.2 **(a)** Show that $\sum_{i=1}^{n}(x_i - \bar{x})(y_i - \bar{y}) = \sum_{i=1}^{n}(x_i - \bar{x})y_i$. Then $\hat{\beta}_1 = \sum_i (x_i - \bar{x})y_i/c$, where $c = \sum_i (x_i - \bar{x})^2$. Now, using $E(y_i) = \beta_0 + \beta_1 x_i$ from assumption 1 in Section 6.1, we obtain (assuming that the x's are constants)

$$\begin{aligned}
E(\hat{\beta}_1) &= E\left[\sum_{i=1}^{n}(x_i - \bar{x})y_i/c\right] = \sum_i (x_i - \bar{x})E(y_i)/c \\
&= \sum_i (x_i - \bar{x})(\beta_0 + \beta_1 x_i)/c \\
&= \beta_0 \sum_i (x_i - \bar{x})/c + \beta_1 \sum_i (x_i - \bar{x})x_i/c \\
&= 0 + \beta_1 \sum_i (x_i - \bar{x})(x_i - \bar{x})/c = \beta_1 \sum_i (x_i - \bar{x})^2 \Big/ \sum_i (x_i - \bar{x})^2 = \beta_1.
\end{aligned}$$

(b)

$$E(\hat{\beta}_0) = E(\bar{y} - \hat{\beta}_1\bar{x}) = E\left(\sum_{i=1}^{n} y_i/n\right) - [E(\hat{\beta}_1)]\bar{x}$$

$$= \sum_i E(y_i)/n - \beta_1\bar{x} = \sum_i (\beta_0 + \beta_1 x_i)/n - \beta_1\bar{x}$$

$$= \sum_i \beta_0/n + \beta_1 \sum_i x_i/n - \beta_1\bar{x} = n\beta_0/n + \beta_1\bar{x} - \beta_1\bar{x} = \beta_0.$$

6.3 **(a)** Using $\hat{\beta}_1 = \sum_{i=1}^{n}(x_i - \bar{x})y_i/c$, as in the answer to Problem 6.2, and assuming $\text{var}(y_i) = \sigma^2$ and $\text{cov}(y_i, y_j) = 0$, we have

$$\text{var}(\hat{\beta}_1) = \frac{1}{c^2}\sum_{i=1}^{n}(x_i - \bar{x})^2\text{var}(y_i) = \frac{1}{c^2}\sum_{i=1}^{n}(x_i - \bar{x})^2\sigma^2$$

$$= \frac{\sigma^2\sum_{i=1}^{n}(x_i - \bar{x})^2}{\left[\sum_{i=1}^{n}(x_i - \bar{x})^2\right]^2} = \frac{\sigma^2}{\sum_{i=1}^{n}(x_i - \bar{x})^2}.$$

(b) Show that $\hat{\beta}_0$ can be written in the form $\hat{\beta}_0 = \sum_{i=1}^{n} y_i/n - \bar{x}\sum_{i=1}^{n}(x_i - \bar{x})y_i/c$. Then

$$\text{var}(\hat{\beta}_0) = \text{var}\left\{\sum_{i=1}^{n}\left[\frac{1}{n} - \frac{\bar{x}(x_i - \bar{x})}{c}\right]y_i\right\}$$

$$= \sum_{i=1}^{n}\left[\frac{1}{n} - \frac{\bar{x}(x_i - \bar{x})}{c}\right]^2\text{var}(y_i)$$

$$= \sum_{i=1}^{n}\left[\frac{1}{n^2} - \frac{2\bar{x}(x_i - \bar{x})}{nc} + \frac{\bar{x}^2(x_i - \bar{x})^2}{c^2}\right]\sigma^2$$

$$= \sigma^2\left[\frac{n}{n^2} - \frac{2\bar{x}}{nc}\sum_{i=1}^{n}(x_i - \bar{x}) + \frac{\bar{x}^2}{c^2}\sum_{i=1}^{n}(x_i - \bar{x})^2\right]$$

$$= \sigma^2\left[\frac{1}{n} - 0 + \frac{\bar{x}^2\sum_{i=1}^{n}(x_i - \bar{x})^2}{[\sum_{i=1}^{n}(x_i - \bar{x})^2]^2}\right]$$

$$= \sigma^2\left[\frac{1}{n} + \frac{\bar{x}^2}{\sum_{i=1}^{n}(x_i - \bar{x})^2}\right].$$

6.4 Suppose that k of the x_i's are equal to a and the remaining $n-k$ x_i's are equal to b. Then

$$\bar{x} = \frac{ka + (n-k)b}{n},$$

$$\begin{aligned}
\sum_{i=1}^{n}(x_i - \bar{x})^2 &= k\left[a - \frac{ka + (n-k)b}{n}\right]^2 + (n-k)\left[b - \frac{ka + (n-k)b}{n}\right]^2 \\
&= k\left[\frac{n(a-b) - k(a-b)}{n}\right]^2 + (n-k)\left[\frac{k(b-a)}{n}\right]^2 \\
&= \frac{k}{n^2}[(n-k)(a-b)]^2 + \frac{n-k}{n^2}[-k(a-b)]^2 \\
&= \frac{(a-b)^2}{n^2}[k(n-k)^2 + k^2(n-k)] \\
&= \frac{(a-b)^2}{n^2}k(n-k)(n-k+k) = \frac{(a-b)^2 k(n-k)}{n}.
\end{aligned}$$

We then differentiate with respect to k and set the results equal to 0 to solve for k.

$$\frac{\partial \sum_{i=1}^{n}(x_i - \bar{x})^2}{\partial k} = \frac{(a-b)^2}{n}[k(-1) + n - k] = 0,$$

$$k = \frac{n}{2}.$$

6.5

$$\begin{aligned}
\text{SSE} &= \sum_{i=1}^{n}(y_i - \hat{y}_i)^2 = \sum_i (y_i - \hat{\beta}_0 - \hat{\beta}_1 x_i)^2 \\
&= \sum_i (y_i - \bar{y} + \hat{\beta}_1\bar{x} - \hat{\beta}_1 x_i)^2 = \sum_i [y_i - \bar{y} - \hat{\beta}_1(x_i - \bar{x})]^2 \\
&= \sum_i (y_i - \bar{y})^2 - 2\hat{\beta}_1 \sum (y_i - \bar{y})(x_i - \bar{x}) + \hat{\beta}_1^2 \sum_i (x_i - \bar{x})^2.
\end{aligned}$$

Substitute $\hat{\beta}_1$ from (6.5) to obtain the result.

6.6 Show that $\text{SSE} = \sum(y_i - \bar{y})^2 - \hat{\beta}_1^2 \sum(x_i - \bar{x})^2$. Show that $\bar{y} = \beta_0 + \beta_1\bar{x} + \bar{\varepsilon}$, where $\bar{\varepsilon} = \sum_{i=1}^{n} \varepsilon_i/n$. Show that $E\left[\sum_{i=1}^{n}(y_i - \bar{y})^2\right] = E\{\sum_i[\beta_1(x_i - \bar{x}) + \varepsilon_i - \bar{\varepsilon}]^2\} = \beta_1^2 \sum_i (x_i - \bar{x})^2 + (n-1)\sigma^2 + 0$. By (3.8), $E(\hat{\beta}_1^2) = \text{var}(\hat{\beta}_1) + [E(\hat{\beta}_1)]^2 = \sigma^2/\sum_i(x_i - \bar{x})^2 + \beta_1^2$.

6.8 To test $H_0: \beta_1 = c$ versus $H_1: \beta_1 \neq c$, we use the test statistic

$$t = \frac{\hat{\beta}_1 - c}{s/\sqrt{\sum_{i=1}^{n}(x_i - \bar{x})^2}}$$

and reject H_0 if $|t| \geq t_{\alpha/2,n-2}$. Show that t is distributed as $t(n-2, \delta)$, where

$$\delta = \frac{\beta_1 - c}{\sigma/\sqrt{\sum_i (x_i - \bar{x})^2}}$$

6.9 **(a)** To test H_0: $\beta_0 = a$ versus H_1: $\beta_0 \neq a$, we use the test statistic

$$t = \frac{\hat{\beta}_0 - a}{s\sqrt{\dfrac{1}{n} + \dfrac{\bar{x}^2}{\sum_{i=1}^n (x_i - \bar{x})^2}}}$$

and reject H_0 if $|t| > t_{\alpha/2,n-2.}$. Show that t is distributed as $t(n-2, \delta)$, where

$$\delta = \frac{\beta_0 - a}{\sigma\sqrt{\dfrac{1}{n} + \dfrac{\bar{x}^2}{\sum_i (x_i - \bar{x})^2}}}$$

(b) A 100 $(1-\alpha)$% confidence interval for β_0 is given by

$$\hat{\beta} \pm t_{\alpha/2,n-2}s\sqrt{\frac{1}{n} + \frac{\bar{x}^2}{\sum_i (x_i - \bar{x})^2}}.$$

6.10 We add and subtract $\hat{y}_i$ to obtain $\sum_{i=1}^n (y_i - \bar{y})^2 = \sum_{i=1}^n (y_i - \hat{y}_i + \hat{y}_i - \bar{y})^2$. Squaring the right side gives

$$\sum_i (y_i - \bar{y})^2 = \sum_i (y_i - \hat{y}_i)^2 + \sum_i (\hat{y}_i - \bar{y})^2 + 2\sum_i (y_i - \hat{y}_i)(\hat{y}_i - \bar{y}).$$

In the third term on the right side, substitute $\hat{y}_i = \hat{\beta}_0 + \hat{\beta}_1 x_i$ and then $\hat{\beta}_0 = \bar{y} - \hat{\beta}_1\bar{x}$ to obtain

$$\begin{aligned}
\sum_i (y_i - \hat{y}_i)(\hat{y}_i - \bar{y}) &= \sum_i (y_i - \hat{\beta}_0 - \hat{\beta}_1 x_i)(\hat{\beta}_0 + \hat{\beta}_1 x_i - \bar{y}) \\
&= \sum_i (y_i - \bar{y} + \hat{\beta}_1\bar{x} - \hat{\beta}_1 x_i)(\bar{y} - \hat{\beta}_1\bar{x} + \hat{\beta}_1 x_i - \bar{y}) \\
&= \sum_i [(y_i - \bar{y} - \hat{\beta}_1(x_i - \bar{x})][\hat{\beta}_1(x_i - \bar{x})] \\
&= \hat{\beta}_1 \sum_i (y_i - \bar{y})(x_i - \bar{x}) - \hat{\beta}_1^2 \sum_i (x_i - \bar{x})^2.
\end{aligned}$$

This is equal to 0 by (6.5).

6.11

$$\sum_{i=1}^{n}(\hat{y}_i - \bar{y})^2 = \sum_i(\hat{\beta}_0 + \hat{\beta}_1 x_i - \bar{y})^2 = \sum_i(\bar{y} - \hat{\beta}_1\bar{x} + \hat{\beta}_1 x_i - \bar{y})^2$$

$$= \hat{\beta}_1^2 \sum_i (x_i - \bar{x})^2$$

Substituting this into (6.16) and using (6.5) gives the desired result.

6.12 Since $\mathbf{x} - \bar{x}\mathbf{j} = (x_1 - \bar{x}, x_2 - \bar{x}, \ldots, x_n - \bar{x})'$ and $\mathbf{y} - \bar{y}\mathbf{j} = (y_1 - \bar{y}, y_2 - \bar{y}, \ldots, y_n - \bar{y})'$, (6.18) can be written as

$$r = \frac{(\mathbf{x} - \bar{x}\mathbf{j})'(\mathbf{y} - \bar{y}\mathbf{j})}{\sqrt{[(\mathbf{x} - \bar{x}\mathbf{j})'(\mathbf{x} - \bar{x}\mathbf{j})][(\mathbf{y} - \bar{y}\mathbf{j})'(\mathbf{y} - \bar{y}\mathbf{j})]}}.$$

By (2.81), this is the cosine of θ, the angle between the vectors $\mathbf{x} - \bar{x}\mathbf{j}$ and $\mathbf{y} - \bar{y}\mathbf{j}$.

6.13

$$t = \frac{\hat{\beta}_1}{s/\sqrt{\sum_i (x_i - \bar{x}^2}} = \hat{\beta}_1 \frac{\sqrt{\sum_i (x_i - \bar{x})^2}}{\sqrt{\text{SSE}/(n-2)}}$$

$$= \hat{\beta}_1 \frac{\sqrt{\sum_i (x_i - \bar{x})^2}\sqrt{n-2}}{\sqrt{\sum_i (y_i - \bar{y}^2 - \hat{\beta}_1^2 \sum_i (x_i - \bar{x})^2}}$$

$$= \hat{\beta}_1 \frac{\sqrt{\sum_i (x_i - \bar{x})^2}\sqrt{n-2}}{\sqrt{\sum_i (y_i - \bar{y})^2 - \dfrac{\left[\sum_i (x_i - \bar{x})(y_i - \bar{y})\right]^2 \sum_i (x_i - \bar{x})^2}{\left[\sum_i (x_i - \bar{x})^2\right]^2}}}$$

$$= \frac{\sum_i (x_i - \bar{x})(y_i - \bar{y})}{\sum_i (x_i - \bar{x})2} \cdot \frac{\sqrt{\sum_i (x_i - \bar{x})^2}\sqrt{n-2}}{\sqrt{\sum_i (y_i - \bar{y})^2 \left[1 - \dfrac{\left[\sum_i (x_i - \bar{x})(y_i - \bar{y})\right]^2}{\sum_i (y_i - \bar{y})^2 \sum_i (x_i - \bar{x})^2}\right]}}$$

$$= \frac{\sqrt{n-2}\,r}{\sqrt{1 - r^2}}.$$

6.14 **(a)** $\hat{\beta}_0 = 31.752, \quad \hat{\beta}_1 = 11.368.$

(b) $t = 11.109, \quad p = 4.108 \times 10^{-15}.$

(c) $11.368 \pm 2.054, \quad (9.313, 13.422).$

(d) $r^2 = \dfrac{\text{SSR}}{\text{SST}} = \dfrac{6833.7663}{9658.0755} = .7076.$

Chapter 7

7.1 By (2.18), $\hat{\beta}_0 + \hat{\beta}_i x_{i1} + \cdots + \hat{\beta}_k x_{ik} = \mathbf{x}_i' \hat{\boldsymbol{\beta}}$. By (2.20) and (2.27), we obtain

$$\sum_{i=1}^{n} (y_i - \mathbf{x}_i'\hat{\boldsymbol{\beta}})^2 = (y_1 - \mathbf{x}_i'\hat{\boldsymbol{\beta}}, y_2 - \mathbf{x}_2'\hat{\boldsymbol{\beta}}, \ldots, y_n - \mathbf{x}_n'\hat{\boldsymbol{\beta}}) \begin{pmatrix} y_1 - \mathbf{x}_1'\hat{\boldsymbol{\beta}} \\ y_2 - \mathbf{x}_2'\hat{\boldsymbol{\beta}} \\ \vdots \\ y_n - \mathbf{x}_n'\hat{\boldsymbol{\beta}} \end{pmatrix}$$

$$= (\mathbf{y} - \mathbf{X}\hat{\boldsymbol{\beta}})'(\mathbf{y} - \mathbf{X}\hat{\boldsymbol{\beta}}).$$

This can also be seen directly by using estimates in the model $\mathbf{y} = \mathbf{X}\boldsymbol{\beta} + \boldsymbol{\varepsilon}$. Thus $\hat{\boldsymbol{\varepsilon}} = \mathbf{y} - \mathbf{X}\hat{\boldsymbol{\beta}}$, and $\hat{\boldsymbol{\varepsilon}}'\hat{\boldsymbol{\varepsilon}} = (\mathbf{y} - \mathbf{X}\hat{\boldsymbol{\beta}})'(\mathbf{y} - \mathbf{X}\hat{\boldsymbol{\beta}})$.

7.2 Multiply (7.9) using (2.17). Keep $\mathbf{y} - \mathbf{X}\hat{\boldsymbol{\beta}}$ together and $\mathbf{X}\hat{\boldsymbol{\beta}} - \mathbf{X}\mathbf{b}$ together. Factor $\mathbf{X}$ out of $\mathbf{X}\hat{\boldsymbol{\beta}} - \mathbf{X}\mathbf{b}$.

7.3 In (7.12), we obtain

$$\hat{\beta}_1 = \frac{n\sum_i x_i y_i - (\sum_i x_i)(\sum_i y_i)}{n\sum_i x_i^2 - (\sum_i x_i)^2} = \frac{n[\sum_i x_i y_i - (n\bar{x})(n\bar{y})/n]}{n[\sum_i x_i^2 - (n\bar{x})^2/n]}$$
$$= \frac{\sum_i x_i y_i - n\bar{x}\bar{y}}{\sum_i x_i^2 - n\bar{x}^2},$$

which is (6.5). For $\hat{\beta}_0$, we start with (6.6):

$$\hat{\beta}_0 = \bar{y} - \hat{\beta}_1 \bar{x} = \frac{\sum_i y_i}{n} - \left(\frac{\sum_i x_i y_i - n\bar{x}\bar{y}}{\sum_i x_i^2 - n\bar{x}^2} \right) \frac{\sum_i x_i}{n}$$
$$= \frac{(\sum_i y_i)(\sum_i x_i^2 - n\bar{x}^2)}{n(\sum_i x_i^2 - n\bar{x}^2)} - \left(\frac{\sum_i x_i y_i - n\bar{x}\bar{y}}{\sum_i x_i^2 - n\bar{x}^2} \right) \frac{\sum_i x_i}{n}$$
$$= \frac{\sum_i x_i^2 \sum_i y_i - n\sum_i y_i (\sum_i x_i/n)^2 - (\sum_i x_i)(\sum_i x_i y_i) + n(\sum_i x_i/n)(\sum_i y_i/n)(\sum_i x_i)}{n(\sum_i x_i^2 - n\bar{x}^2)}.$$

The second and fourth terms in the numerator add to 0.

7.5 Starting with (6.10), we have

$$\text{var}(\hat{\beta}_0) = \sigma^2\left[\frac{1}{n} + \frac{\bar{x}^2}{\sum_i (x_i - \bar{x})^2}\right] = \sigma^2\left[\frac{\sum_i (x_i - \bar{x})^2 + n\bar{x}^2}{n\sum_i (x_i - \bar{x})^2}\right]$$

$$= \sigma^2\left[\frac{\sum_i x_i^2 - n\bar{x}^2 + n\bar{x}^2}{n\sum_i (x_i - \bar{x})^2}\right].$$

7.6 The two terms missing in (7.17) are

$$[\mathbf{A} - (\mathbf{X}'\mathbf{X})^{-1}\mathbf{X}'][(\mathbf{X}'\mathbf{X})^{-1}\mathbf{X}']' + [(\mathbf{X}'\mathbf{X})^{-1}\mathbf{X}'][\mathbf{A} - (\mathbf{X}'\mathbf{X})^{-1}\mathbf{X}']'.$$

Using $\mathbf{AX} = \mathbf{I}$, the first of these becomes

$$\mathbf{AX}(\mathbf{X}'\mathbf{X})^{-1} - (\mathbf{X}'\mathbf{X})^{-1}\mathbf{X}'\mathbf{X}(\mathbf{X}'\mathbf{X})^{-1} = (\mathbf{X}'\mathbf{X})^{-1} - (\mathbf{X}'\mathbf{X})^{-1} = \mathbf{O}.$$

7.7 **(a)** For the linear estimator $\mathbf{c}'\mathbf{y}$ to be unbiased for all possible $\boldsymbol{\beta}$, we have $E(\mathbf{c}'\mathbf{y}) = \mathbf{c}'\mathbf{X}\boldsymbol{\beta} = \mathbf{a}'\boldsymbol{\beta}$, which requires $\mathbf{c}'\mathbf{X} = \mathbf{a}'$. To express var($\mathbf{c}'\mathbf{y}$) in terms of $\text{var}(\mathbf{a}'\hat{\boldsymbol{\beta}}) = \text{var}[\mathbf{a}'(\mathbf{X}'\mathbf{X})^{-1}\mathbf{X}'\mathbf{y}]$, we write $\text{var}(\mathbf{c}'\mathbf{y}) = \sigma^2\mathbf{c}'\mathbf{c} = \sigma^2[\mathbf{c} - \mathbf{X}(\mathbf{X}'\mathbf{X})^{-1}\mathbf{a} + \mathbf{X}(\mathbf{X}'\mathbf{X})^{-1}\mathbf{a}^{-1}]'[\mathbf{c} - \mathbf{X}(\mathbf{X}'\mathbf{X})^{-1}\mathbf{a} + \mathbf{X}(\mathbf{X}'\mathbf{X})^{-1}\mathbf{a}]$. Show that with $\mathbf{c}'\mathbf{X} = \mathbf{a}'$, this becomes $[\mathbf{c} - \mathbf{X}(\mathbf{X}'\mathbf{X})^{-1}\mathbf{a}]'$ $[\mathbf{c} - \mathbf{X}(\mathbf{X}'\mathbf{X})^{-1}\mathbf{a}] + \mathbf{a}'(\mathbf{X}'\mathbf{X})^{-1}\mathbf{a}$, which is minimized by $\mathbf{c} = \mathbf{X}(\mathbf{X}'\mathbf{X})^{-1}\mathbf{a}$.

(b) To minimize var(**c**'**y**) subject to **c**'**X** = **a**', we differentiate $v = \sigma^2\mathbf{c}'\mathbf{c} - (\mathbf{c}'\mathbf{X} - \mathbf{a}')\boldsymbol{\lambda}$ with respect to $\mathbf{c}$ and $\boldsymbol{\lambda}$ (see Section 2.14.3):

$$\partial v/\partial \lambda = -\mathbf{X}'\mathbf{c} + \mathbf{a} = \mathbf{0} \quad \text{gives } \mathbf{a} = \mathbf{X}'\mathbf{c}.$$

$$\partial v/\partial \mathbf{c} = 2\sigma^2\mathbf{c} - \mathbf{X}\boldsymbol{\lambda} = \mathbf{0} \text{ gives } \mathbf{c} = \mathbf{X}\boldsymbol{\lambda}/2\sigma^2.$$

Substituting $\mathbf{c} = \mathbf{X}\boldsymbol{\lambda}/2\sigma^2$ into $\mathbf{a} = \mathbf{X}'\mathbf{c}$ gives $\mathbf{a} = \mathbf{X}'\mathbf{X}\boldsymbol{\lambda}/2\sigma^2$, or $\boldsymbol{\lambda} = 2\sigma^2(\mathbf{X}'\mathbf{X})^{-1}\mathbf{a}$. Thus $\mathbf{c} = \mathbf{X}\boldsymbol{\lambda}/2\sigma^2 = \mathbf{X}(\mathbf{X}'\mathbf{X})^{-1}\mathbf{a}$.

7.9

$$\hat{\boldsymbol{\beta}}_z = (\mathbf{Z}'\mathbf{Z})^{-1}\mathbf{Z}'\mathbf{y} = (\mathbf{H}'\mathbf{X}'\mathbf{X}\mathbf{H})^{-1}\mathbf{H}'\mathbf{X}'\mathbf{y}$$

$$= \mathbf{H}^{-1}(\mathbf{X}'\mathbf{X})^{-1}(\mathbf{H}')^{-1}\mathbf{H}'\mathbf{X}'\mathbf{y}$$

$$= \mathbf{H}^{-1}(\mathbf{X}'\mathbf{X})^{-1}\mathbf{X}'\mathbf{y} = \mathbf{H}^{-1}\hat{\boldsymbol{\beta}}.$$

For the ith row of $\mathbf{Z} = \mathbf{XH}$, we have $\mathbf{z}_i' = \mathbf{x}_i'\mathbf{H}$, or $\mathbf{z}_i = \mathbf{H}'\mathbf{x}_i$. Thus in general,

$\mathbf{z} = \mathbf{H}'\mathbf{x}$, and

$$\begin{aligned}\hat{y} &= \hat{\boldsymbol{\beta}}_z'\mathbf{z} = (\mathbf{H}^{-1}\hat{\boldsymbol{\beta}})'\mathbf{H}'\mathbf{x} = \hat{\boldsymbol{\beta}}'(\mathbf{H}^{-1})'\mathbf{H}'\mathbf{x} \\ &= \hat{\boldsymbol{\beta}}'(\mathbf{H}')^{-1}\mathbf{H}'\mathbf{x} = \hat{\boldsymbol{\beta}}'\mathbf{x}.\end{aligned}$$

7.10 Since $\hat{\boldsymbol{\beta}}'\mathbf{x} = \mathbf{x}'\hat{\boldsymbol{\beta}}$ is invariant to changes of scale on the x's, $\mathbf{x}_i'\hat{\boldsymbol{\beta}}$ is invariant, where $\mathbf{x}_i'$ is the ith row of $\mathbf{X}$. Therefore, $\mathbf{X}\hat{\boldsymbol{\beta}}$ is invariant, and it follows that $s^2 = (\mathbf{y} - \mathbf{X}\hat{\boldsymbol{\beta}})'(\mathbf{y} - \mathbf{X}\hat{\boldsymbol{\beta}})/(n - k - 1)$ is invariant.

7.11 $(\mathbf{y} - \mathbf{X}\hat{\boldsymbol{\beta}})'(\mathbf{y} - \mathbf{X}\hat{\boldsymbol{\beta}}) = \mathbf{y}'\mathbf{y} - \mathbf{y}'\mathbf{X}\hat{\boldsymbol{\beta}} - \hat{\boldsymbol{\beta}}'\mathbf{X}'\mathbf{y} + \hat{\boldsymbol{\beta}}'\mathbf{X}'\mathbf{X}\hat{\boldsymbol{\beta}}$. Use (7.8).

7.12 By (7.8), $\hat{\boldsymbol{\beta}}'(\mathbf{X}'\mathbf{y}) = \hat{\boldsymbol{\beta}}'(\mathbf{X}'\mathbf{X}\hat{\boldsymbol{\beta}})$. By Theorem 5.2a, $E(\mathbf{y}'\mathbf{y}) = E(\mathbf{y}'\mathbf{I}\mathbf{y}) = \text{tr}(\mathbf{I}\sigma^2\mathbf{I}) + E(\mathbf{y}')\mathbf{I}E(\mathbf{y}) = n\sigma^2 + \boldsymbol{\beta}'\mathbf{X}'\mathbf{X}\boldsymbol{\beta}$. By Theorems 7.3b and 7.3c, $E(\hat{\boldsymbol{\beta}}) = \boldsymbol{\beta}$ and $\text{cov}(\hat{\boldsymbol{\beta}}) = \sigma^2(\mathbf{X}'\mathbf{X})^{-1}$. Thus $E(\hat{\boldsymbol{\beta}}'\mathbf{X}'\mathbf{X}\hat{\boldsymbol{\beta}}) = \text{tr}[(\mathbf{X}'\mathbf{X})\sigma^2(\mathbf{X}'\mathbf{X})^{-1}] + \boldsymbol{\beta}'\mathbf{X}'\mathbf{X}\boldsymbol{\beta}$.

7.13 Let $\mathbf{X}_1$ and $\hat{\beta}_1$ represent a reduced model with $k-1$ x's, and let $\mathbf{X}$ and $\hat{\boldsymbol{\beta}}$ represent the full model with k x's. Then show that SSE for the full model can be expressed as

$$\begin{aligned}\text{SSE}_{k.} &= (\mathbf{y}'\mathbf{y} - \hat{\boldsymbol{\beta}}_1'\mathbf{X}_1'\mathbf{y}) - (\hat{\boldsymbol{\beta}}'\mathbf{X}'\mathbf{y} - \hat{\boldsymbol{\beta}}_1'\mathbf{X}_1'\mathbf{y}) \\ &= \text{SSE}_{k-1} - (\text{ a positive term}).\end{aligned}$$

It is shown in Theorem 8.2d and problem 8.10 that $\hat{\boldsymbol{\beta}}'\mathbf{X}'\mathbf{y} - \hat{\boldsymbol{\beta}}_1'\mathbf{X}_1'\mathbf{y}$ is a positive definite quadratic form.

7.15 First show that $(1/n)\mathbf{j}'\mathbf{X}_1 = \bar{\mathbf{x}}$, where $\bar{\mathbf{x}} = (\bar{x}_1, \bar{x}_2, \ldots, \bar{x}_k)$, which contains the means of the columns of $\mathbf{X}_1$. Then

$$\begin{aligned}\left(\mathbf{I} - \frac{1}{n}\mathbf{J}\right)\mathbf{X}_1 &= \mathbf{X}_1 - \frac{1}{n}\mathbf{J}\mathbf{X}_1 = \mathbf{X}_1 - \frac{1}{n}\mathbf{j}\mathbf{j}'\mathbf{X}_1 = \mathbf{X}_1 - \mathbf{j}\bar{\mathbf{x}}' \\ &= \begin{pmatrix} x_{11} & x_{12} & \cdots & x_{1k} \\ x_{21} & x_{22} & \cdots & x_{2k} \\ \vdots & \vdots & & \vdots \\ x_{n1} & x_{n2} & \cdots & x_{nk} \end{pmatrix} - \begin{pmatrix} \bar{x}_1 & \bar{x}_2 & \cdots & \bar{x}_k \\ \bar{x}_1 & \bar{x}_2 & \cdots & \bar{x}_k \\ \vdots & \vdots & & \vdots \\ \bar{x}_1 & \bar{x}_2 & \cdots & \bar{x}_k \end{pmatrix}.\end{aligned}$$

7.16 By a comment following (2.25), $\mathbf{j}'\mathbf{X}_c$ contains the column sums of $\mathbf{X}_c$. The sum of the second column, for example, is $\sum_{i=1}^n (x_{i2} - \bar{x}_2) = \sum_{i=1}^n x_{i2} - n\bar{x}_2 = n\bar{x}_2 - n\bar{x}_2 = 0$. Alternatively, $\mathbf{j}'\mathbf{X}_c = \mathbf{j}'[\mathbf{I} - (1/n)\mathbf{J}]\mathbf{X}_1 = [\mathbf{j}' - (1/n)\mathbf{j}'\mathbf{j}\mathbf{j}']\mathbf{X}_1 = \mathbf{0}'\mathbf{X}_1 = \mathbf{0}'$ since $\mathbf{j}'\mathbf{j} = n$.

7.17 **(a)** Partition $\mathbf{X}$ and $\hat{\boldsymbol{\beta}}$ as $\mathbf{X}=(\mathbf{j}, \mathbf{X}_1)$ and $\hat{\boldsymbol{\beta}} = \begin{pmatrix} \hat{\beta}_0 \\ \hat{\boldsymbol{\beta}}_1 \end{pmatrix}$. Then show that the normal equations $\mathbf{X}'\mathbf{X}\hat{\boldsymbol{\beta}} = \mathbf{X}'\mathbf{y}$ in (7.8) become

$$\begin{pmatrix} \mathbf{j}'\mathbf{j} & \mathbf{j}'\mathbf{X}_1 \\ \mathbf{X}_1'\mathbf{j} & \mathbf{X}_1'\mathbf{X}_1 \end{pmatrix} \begin{pmatrix} \hat{\beta}_0 \\ \hat{\boldsymbol{\beta}}_1 \end{pmatrix} = \begin{pmatrix} \mathbf{j}'\mathbf{y} \\ \mathbf{X}_1'\mathbf{y} \end{pmatrix},$$

from which we obtain

$$n\hat{\beta}_0 + \mathbf{j}'\mathbf{X}_1\hat{\boldsymbol{\beta}}_1 = n\bar{y} \tag{1}$$

$$\mathbf{X}_1'\mathbf{j}\hat{\beta}_0 + \mathbf{X}_1'\mathbf{X}_1\hat{\boldsymbol{\beta}}_1 = \mathbf{X}_1'\mathbf{y}. \tag{2}$$

Show that (1) becomes $\hat{\beta}_0 + \bar{\mathbf{x}}'\hat{\boldsymbol{\beta}}_1 = \bar{y}$, or $\hat{\alpha} = \bar{y}$. Show that (2) becomes

$$n\bar{\mathbf{x}}\hat{\beta}_0 + \mathbf{X}_1'\mathbf{X}_1\hat{\boldsymbol{\beta}}_1 = \mathbf{X}_1'\mathbf{y}. \tag{3}$$

By (7.33), $\mathbf{X}_c = [\mathbf{I} - (1/n)\mathbf{J}]\mathbf{X}_1$. Show that $\mathbf{X}_c'\mathbf{X}_c = \mathbf{X}_1'\mathbf{X}_1 - (1/n)\mathbf{X}_1'\mathbf{J}\mathbf{X}_1 = \mathbf{X}_1'\mathbf{X}_1' - n\bar{\mathbf{x}}\bar{\mathbf{x}}'$. Similarly, show that $\mathbf{X}_c'\mathbf{y} = \mathbf{X}_1'\mathbf{y} - (1/n)\mathbf{X}_1'\mathbf{J}\mathbf{y} = \mathbf{X}_1'\mathbf{y} - n\bar{\mathbf{x}}\bar{y}$. Now show that the normal equations in (7.34) for the centered model can be written in the form

$$\begin{pmatrix} n & \mathbf{0}' \\ \mathbf{0} & \mathbf{X}_c'\mathbf{X}_c \end{pmatrix} \begin{pmatrix} \hat{\alpha} \\ \hat{\boldsymbol{\beta}}_1 \end{pmatrix} = \begin{pmatrix} n\bar{y} \\ \mathbf{X}_c'\mathbf{y} \end{pmatrix},$$

which becomes

$$n\hat{\alpha} = n\bar{y}, \tag{4}$$
$$\mathbf{X}_c'\mathbf{X}_c\hat{\boldsymbol{\beta}}_1 = \mathbf{X}_c'\mathbf{y}. \tag{5}$$

Thus (4) is the same as (1). Using $\bar{\mathbf{x}}'\hat{\boldsymbol{\beta}}_1 = \hat{\alpha} - \hat{\beta}_0$, show that (5) is the same as (3).

(b) Using (2.50) with $\mathbf{A}_{11} = n$, $\mathbf{A}_{12} = n\bar{\mathbf{x}}'$, $\mathbf{A}_{21} = n\bar{\mathbf{x}}$, $\mathbf{A}_{22} = \mathbf{X}_1'\mathbf{X}_1$, and $\mathbf{X}_c'\mathbf{X}_c = \mathbf{X}_1'\mathbf{X}_1 - n\bar{\mathbf{x}}\bar{\mathbf{x}}'$, show that

$$(\mathbf{X}'\mathbf{X})^{-1} = \begin{pmatrix} n & \mathbf{j}'\mathbf{X}_1 \\ \mathbf{X}_1'\mathbf{j} & \mathbf{X}_1\mathbf{X}_1 \end{pmatrix}^{-1} = \begin{pmatrix} n & n\bar{\mathbf{x}}' \\ n\bar{\mathbf{x}} & \mathbf{X}_1'\mathbf{X}_1 \end{pmatrix}^{-1}$$

$$= \begin{pmatrix} \dfrac{1}{n} + \bar{\mathbf{x}}'(\mathbf{X}_c'\mathbf{X}_c)^{-1}\bar{\mathbf{x}} & -\bar{\mathbf{x}}'(\mathbf{X}_c'\mathbf{X}_c)^{-1} \\ -(\mathbf{X}_c'\mathbf{X}_c)^{-1}\bar{\mathbf{x}} & (\mathbf{X}_c'\mathbf{X}_c)^{-1} \end{pmatrix}$$

and verify by multiplication that $(\mathbf{X}'\mathbf{X})^{-1}\mathbf{X}'\mathbf{X} = \mathbf{I}$. With this partitioned form of $(\mathbf{X}'\mathbf{X})^{-1}$, show that

$$\hat{\boldsymbol{\beta}} = (\mathbf{X}'\mathbf{X})^{-1}\mathbf{X}'\mathbf{y} = [(\mathbf{j}, \mathbf{X}_1)'(\mathbf{j}, \mathbf{X}_1)]^{-1}\begin{pmatrix} \mathbf{j}'\mathbf{y} \\ \mathbf{X}_1'\mathbf{y} \end{pmatrix}$$

$$= \begin{pmatrix} \bar{y} - \bar{\mathbf{x}}'(\mathbf{X}_c'\mathbf{X}_c)^{-1}\mathbf{X}_c'\mathbf{y} \\ (\mathbf{X}_c'\mathbf{X}_c)^{-1}\mathbf{X}_c'\mathbf{y} \end{pmatrix} = \begin{pmatrix} \bar{y} - \hat{\boldsymbol{\beta}}_1'\bar{\mathbf{x}} \\ (\mathbf{X}_c'\mathbf{X}_c)^{-1}\mathbf{X}_c'\mathbf{y} \end{pmatrix},$$

which is the same as (7.37) and (7.38).

7.18 Substitute $x_1 = \bar{x}_1, x_2 = \bar{x}_2, \ldots, x_k = \bar{x}_k$ in $\bar{y} = \bar{y} + \hat{\beta}_1(x_1 - \bar{x}_1) + \cdots + \hat{\beta}_k(x_k - \bar{x}_k)$ to obtain $\hat{y} = \bar{y}$.

7.19

$$\begin{aligned}
\mathbf{y}'\mathbf{y} - \hat{\boldsymbol{\beta}}'\mathbf{X}'\mathbf{y} &= \mathbf{y}'\mathbf{y} - (\hat{\beta}_0, \hat{\boldsymbol{\beta}}_1')(\mathbf{j}, \mathbf{X}_1)'\mathbf{y} \\
&= \mathbf{y}'\mathbf{y} - (\hat{\beta}_0, \hat{\boldsymbol{\beta}}_1')\begin{pmatrix} \mathbf{j}'\mathbf{y} \\ \mathbf{X}_1'\mathbf{y} \end{pmatrix} \\
&= \mathbf{y}'\mathbf{y} - \hat{\beta}_0 n\bar{y} - \hat{\boldsymbol{\beta}}_1'\mathbf{X}_1'\mathbf{y} \\
&= \mathbf{y}'\mathbf{y} - (\bar{y} - \hat{\boldsymbol{\beta}}_1'\bar{\mathbf{x}})n\bar{y} - \hat{\beta}_1'\mathbf{X}_1'\mathbf{y} \\
&= \mathbf{y}'\mathbf{y} - n\bar{y}^2 - \hat{\boldsymbol{\beta}}_1'(\mathbf{X}_1'\mathbf{y} - n\bar{y}\bar{\mathbf{x}}) \\
&= \sum_{i=1}^{n}(y_i - \bar{y})^2 - \hat{\boldsymbol{\beta}}_1'\mathbf{X}_c'\mathbf{y}.
\end{aligned}$$

7.20 **(a)** By Theorem 2.2c(i), $\mathbf{X}_c{}'\mathbf{X}_c$ is obtained as products of columns of $\mathbf{X}_c$. By (7.33), these products are of the form illustrated in the numerators of (7.41) and (7.42).

(b) By (7.43), the numerator of the second element of s_{yx} is $\sum_{i=1}^{n}(x_{i2} - \bar{x}_2)(y_i - \bar{y})$. This can be written as $\sum_i (x_{i2} - \bar{x}_2)y_i - \sum_i (x_{i2} - \bar{x})\bar{y}$, the second term of which vanishes. Note that $\sum_i (x_{i2} - \bar{x}_2)y_i$ is the second element of $\mathbf{X}_c'\mathbf{y}$.

7.21 **(b)** Expand the last term of $\ln L(\boldsymbol{\beta}, \sigma^2)$ in (7.51) to obtain

$$\ln L(\boldsymbol{\beta}, \sigma^2) = -\frac{n}{2}\ln(2\pi) - \frac{n}{2}\ln\sigma^2 - \frac{1}{2\sigma^2}(\mathbf{y}'\mathbf{y} - 2\mathbf{y}'\mathbf{X}\boldsymbol{\beta} + \hat{\boldsymbol{\beta}}'\mathbf{X}'\mathbf{X}\boldsymbol{\beta}).$$

Then

$$\frac{\partial \ln L(\boldsymbol{\beta}, \sigma^2)}{\partial \boldsymbol{\beta}} = -\mathbf{0} - \mathbf{0} - \frac{n}{2\sigma^2}(\mathbf{0} - 2\mathbf{X}'\mathbf{y} + 2\mathbf{X}'\mathbf{X}\boldsymbol{\beta}).$$

Setting this equal to $\mathbf{0}$ gives (7.48)

(c) Use $\ln L(\boldsymbol{\beta}, \sigma^2)$ as in (7.51), to obtain

$$\frac{\partial \ln L(\boldsymbol{\beta}, \sigma^2)}{\partial \sigma^2} = -0 - \frac{n}{2\sigma^2} + \frac{1}{2(\sigma^2)^2}(\mathbf{y} - \mathbf{X}\boldsymbol{\beta})'(\mathbf{y} - \mathbf{X}\boldsymbol{\beta}).$$

Setting this equal to 0 (and substituting $\hat{\boldsymbol{\beta}}$ from $\partial \ln L/\partial \boldsymbol{\beta} = \mathbf{0}$) yields (7.49).

7.22 **(ii)** By (7.26), $\text{SSE} = \mathbf{y}'[\mathbf{I} - \mathbf{X}(\mathbf{X}'\mathbf{X})^{-1}\mathbf{X}']\mathbf{y}$. Show that $\mathbf{I} - \mathbf{X}(\mathbf{X}'\mathbf{X})^{-1}\mathbf{X}'$ is idempotent of rank $n - k - 1$, given that $\mathbf{X}$ is $n \times (k+1)$ of rank $k+1$. Then by Corollary 2 to Theorem 5.5a, SSE/σ^2 is $\chi^2(n-k-1, \lambda)$, where $\lambda = \boldsymbol{\mu}'\mathbf{A}\boldsymbol{\mu}/2\sigma^2 = (\mathbf{X}\boldsymbol{\beta})'[\mathbf{I} - \mathbf{X}(\mathbf{X}'\mathbf{X})^{-1}\mathbf{X}'](\mathbf{X}\boldsymbol{\beta}/2\sigma^2)$. Show that λ=0.

(iii) Show that $(\mathbf{X}'\mathbf{X})^{-1}\mathbf{X}'[\mathbf{I} - \mathbf{X}(\mathbf{X}'\mathbf{X})^{-1}\mathbf{X}'] = \mathbf{O}$. Then by Corollary 1 to Theorem 5.6a, $\hat{\beta} = (\mathbf{X}'\mathbf{X})^{-1}\mathbf{X}'\mathbf{y}$ and $\text{SSE} = \mathbf{y}'[\mathbf{I} - \mathbf{X}(\mathbf{X}'\mathbf{X})^{-1}\mathbf{X}']\mathbf{y}$ are independent.

7.23 The two missing terms in (7.52) are

$$\begin{aligned} &-(\mathbf{y} - \mathbf{X}\hat{\boldsymbol{\beta}})'\mathbf{X}(\hat{\boldsymbol{\beta}} - \boldsymbol{\beta}) - (\hat{\boldsymbol{\beta}} - \boldsymbol{\beta})'\mathbf{X}'(\mathbf{y} - \mathbf{X}\hat{\boldsymbol{\beta}}) \\ &= -(\mathbf{X}'\mathbf{y} - \mathbf{X}'\mathbf{X}\hat{\boldsymbol{\beta}})'(\hat{\boldsymbol{\beta}} - \boldsymbol{\beta}) - (\hat{\boldsymbol{\beta}} - \boldsymbol{\beta})'(\mathbf{X}'\mathbf{y} - \mathbf{X}'\mathbf{X}\hat{\boldsymbol{\beta}}) \\ &= -\mathbf{0}'(\hat{\boldsymbol{\beta}} - \boldsymbol{\beta}) - (\hat{\boldsymbol{\beta}} - \boldsymbol{\beta})'\mathbf{0}. \end{aligned}$$

Note that $\mathbf{X}'\mathbf{y} - \mathbf{X}'\mathbf{X}\hat{\boldsymbol{\beta}} = \mathbf{0}$ by the normal equations $\mathbf{X}'\mathbf{X}\hat{\boldsymbol{\beta}} = \mathbf{X}'\mathbf{y}$ in (7.8).

7.25
$$\hat{\boldsymbol{\beta}}'\mathbf{X}'\mathbf{y} = \begin{pmatrix} \hat{\beta}_0 \\ \hat{\boldsymbol{\beta}}_1 \end{pmatrix}' (\mathbf{j}, \mathbf{X}_1)'\mathbf{y} = (\hat{\beta}_0, \hat{\boldsymbol{\beta}}_1') \begin{pmatrix} \mathbf{j}' \\ \mathbf{X}_1' \end{pmatrix} \mathbf{y} = n\hat{\beta}_0\bar{y} + \hat{\boldsymbol{\beta}}_1'\mathbf{X}_1'\mathbf{y}.$$

With $\hat{\beta}_0 = \bar{y} - \hat{\boldsymbol{\beta}}_1'\bar{\mathbf{x}}$ from (7.38) and $\mathbf{X}_c = [\mathbf{I} - (1/n)\mathbf{J}]\mathbf{X}_1$ from (7.33), this becomes

$$\begin{aligned} \hat{\boldsymbol{\beta}}'\mathbf{X}'\mathbf{y} &= n(\bar{y} - \hat{\boldsymbol{\beta}}_1'\bar{\mathbf{x}})\bar{y} + \hat{\boldsymbol{\beta}}_1'\left(\mathbf{X}_c' + \frac{1}{n}\mathbf{X}_1'\mathbf{J}\right)\mathbf{y} \\ &= n\bar{y}^2 - n(\hat{\boldsymbol{\beta}}_1'\bar{\mathbf{x}})\bar{y} + \hat{\boldsymbol{\beta}}_1'\mathbf{X}_c'\mathbf{y} + \frac{1}{n}\hat{\boldsymbol{\beta}}_1'\mathbf{X}_1'\mathbf{J}\mathbf{y}. \end{aligned}$$

The last term can be written as $(1/n)\hat{\boldsymbol{\beta}}_1'\mathbf{X}_1'\mathbf{J}\mathbf{y} = (1/n)\hat{\boldsymbol{\beta}}_1' \; \mathbf{X}_1'\mathbf{j}\mathbf{j}'\mathbf{y} = (1/n)$ $\hat{\boldsymbol{\beta}}_1' n^2 \bar{\mathbf{x}}\bar{y}$, so that $\hat{\boldsymbol{\beta}}'\mathbf{X}'\mathbf{y} = n\bar{y}^2 + \hat{\boldsymbol{\beta}}_1'\mathbf{X}_c'\mathbf{y}$.

7.26 If $\hat{\beta}_1 = \hat{\beta}_2 = \cdots = \hat{\beta}_k = 0$, then $\hat{\boldsymbol{\beta}}_1 = \mathbf{0}$ and $\hat{\boldsymbol{\beta}}_1'\mathbf{X}_c'\mathbf{X}_c\hat{\boldsymbol{\beta}}_1 = 0$. If $y_i = \hat{y}_i$, $i = 1, 2, \ldots, n$, then $\mathbf{y} = \hat{\mathbf{y}} = \mathbf{X}\hat{\boldsymbol{\beta}}$ and $\hat{\boldsymbol{\beta}}'\mathbf{X}'\mathbf{y} - n\bar{y}^2 = \mathbf{y}'\mathbf{y} - n\bar{y}^2$. Also see formulas below (7.61).

7.27 This follows from the statement following Theorem 7.3f, which notes that an additional x reduces SSE (see Problem 7.13).

7.28 **(a)** A set of full-rank linear transformations on the x's can be represented by $\mathbf{W} = \mathbf{X}\mathbf{H}$, where $\mathbf{H}$ is a nonsingular matrix. Show that $\hat{\boldsymbol{\beta}}_w = (\mathbf{W}'\mathbf{W})^{-1}$ $\mathbf{W}'\mathbf{y} = \mathbf{H}^{-1}(\mathbf{X}'\mathbf{X})^{-1}\mathbf{X}'\mathbf{y} = \mathbf{H}^{-1}\hat{\boldsymbol{\beta}}_x$. Show that $\hat{\boldsymbol{\beta}}_w'\mathbf{W}'\mathbf{y} = \hat{\boldsymbol{\beta}}_x'\mathbf{W}'\mathbf{y}$. Then $R_w^2 = (\hat{\boldsymbol{\beta}}_w'\mathbf{W}'\mathbf{y} - n\bar{y}^2)/(\mathbf{y}'\mathbf{y} - n\bar{y}^2) = (\hat{\boldsymbol{\beta}}_x'\mathbf{X}'\mathbf{y} - n\bar{y}^2)/(\mathbf{y}'\mathbf{y} - n\bar{y}^2) = R_x^2$.

(b) Replacing $\mathbf{y}$ by $\mathbf{z} = c\mathbf{y}$, we have $\bar{z} = (1/n)\mathbf{j}'\mathbf{z} = (1/n)\mathbf{j}'c\mathbf{y} = c\bar{y}$ and $\hat{\boldsymbol{\beta}}_z = (\mathbf{X}'\mathbf{X})^{-1}\mathbf{X}'\mathbf{z} = (\mathbf{X}'\mathbf{X})^{-1}\mathbf{X}'c\mathbf{y} = c\hat{\boldsymbol{\beta}}_y$. Then

$$R_z^2 = \frac{\hat{\boldsymbol{\beta}}_z'\mathbf{X}'\mathbf{z} - n\bar{z}^2}{\mathbf{z}'\mathbf{z} - n\bar{z}^2} = \frac{c\hat{\boldsymbol{\beta}}_y'\mathbf{X}'c\mathbf{y} - n(c\bar{y})^2}{(c\mathbf{y})'(c\mathbf{y}) - n(c\bar{y})^2} = \frac{c^2}{c^2}R_y^2.$$

7.30

$$\sum_{i=1}^{n} \hat{y}i/n = \frac{\mathbf{j}'\hat{\mathbf{y}}}{n} = \frac{\mathbf{j}'\mathbf{X}\hat{\boldsymbol{\beta}}}{n} = \frac{\mathbf{j}'(\mathbf{j}, \mathbf{X}_1)\begin{pmatrix}\hat{\beta}_0 \\ \hat{\boldsymbol{\beta}}_1\end{pmatrix}}{n}$$

$$= n\frac{\hat{\beta}_0}{n} + \frac{\mathbf{j}'\mathbf{X}_1\hat{\boldsymbol{\beta}}_1}{n} = \hat{\beta}_0 + \bar{\mathbf{x}}'\hat{\boldsymbol{\beta}}_1 = \hat{\beta}_0 + (\bar{y} - \hat{\beta}_0),$$

by (7.38)

7.31 By (7.61), we obtain

$$\cos^2\theta = \frac{\hat{\mathbf{y}}'\hat{\mathbf{y}} - \bar{y}\hat{\mathbf{y}}'\mathbf{j} - \bar{y}\mathbf{j}'\hat{\mathbf{y}} + \bar{y}^2\mathbf{j}'\mathbf{j}}{\sum_{i=1}^{n}(y_i - \bar{y})^2} = \frac{\hat{\boldsymbol{\beta}}'\mathbf{X}'\mathbf{X}\hat{\boldsymbol{\beta}} - n\bar{y}^2}{\sum_i (y_i - \bar{y})^2},$$

since $\mathbf{j}'\hat{\mathbf{y}} = n\bar{y}$ by Problem 7.30. By (7.8), $\hat{\boldsymbol{\beta}}'\mathbf{X}'\mathbf{X}\hat{\boldsymbol{\beta}} = \hat{\boldsymbol{\beta}}'\mathbf{X}'\mathbf{y}$.

7.33 **(a)** Using $\hat{\boldsymbol{\beta}} = (\mathbf{X}'\mathbf{V}^{-1}\mathbf{X})^{-1}\mathbf{X}'\mathbf{V}^{-1}\mathbf{y}$, expand $(\mathbf{y} - \mathbf{X}\hat{\boldsymbol{\beta}})'\mathbf{V}^{-1}(\mathbf{y} - \mathbf{X}\hat{\boldsymbol{\beta}})$ to obtain $\mathbf{y}'\mathbf{V}^{-1}\mathbf{y} - \mathbf{y}'\mathbf{V}^{-1}\mathbf{X}(\mathbf{X}'\mathbf{V}^{-1}\mathbf{X})^{-1}\mathbf{X}'\mathbf{V}^{-1}\mathbf{y}$, the second term of which appears twice more with opposite signs.

(b) Use Theorem 5.2a with $\mathbf{A} = \mathbf{V}^{-1} - \mathbf{V}^{-1}\mathbf{X}(\mathbf{X}'\mathbf{V}^{-1}\mathbf{X})^{-1}\mathbf{X}'\mathbf{V}^{-1}$, $\boldsymbol{\Sigma} = \sigma^2\mathbf{V}$, and $\boldsymbol{\mu} = \mathbf{X}\boldsymbol{\beta}$.

7.34 $\ln L(\beta, \sigma^2) = -\frac{n}{2}\ln(2\pi) - \frac{n}{2}\ln\sigma^2 - \frac{1}{2}\ln|\mathbf{V}| - \frac{1}{2\sigma^2}(\mathbf{y} - \mathbf{X}\boldsymbol{\beta})'\mathbf{V}^{-1}(\mathbf{y} - \mathbf{X}\boldsymbol{\beta})$.
Expand the last term to obtain

$$\frac{1}{2\sigma^2}(\mathbf{y}'\mathbf{V}^{-1}\mathbf{y} - \mathbf{y}'\mathbf{V}^{-1}\mathbf{X}\boldsymbol{\beta} - \boldsymbol{\beta}'\mathbf{X}'\mathbf{V}^{-1}\mathbf{y} + \boldsymbol{\beta}'\mathbf{X}'\mathbf{V}^{-1}\mathbf{X}\boldsymbol{\beta}).$$

Differentiate to obtain

$$\frac{\partial \ln L(\boldsymbol{\beta}, \sigma^2)}{\partial \boldsymbol{\beta}} = -\mathbf{0} - \mathbf{0} - \mathbf{0} - \frac{1}{2\sigma^2}(\mathbf{0} - 2\mathbf{X}'\mathbf{V}^{-1}\mathbf{y} + 2\mathbf{X}'\mathbf{V}^{-1}\mathbf{X}\boldsymbol{\beta}),$$

$$\frac{\partial \ln L(\boldsymbol{\beta}, \sigma^2)}{\partial \boldsymbol{\sigma}^2} = -0 - \frac{n}{2\sigma^2} - 0 + \frac{1}{2(\sigma^2)^2}(\mathbf{y} - \mathbf{X}\boldsymbol{\beta})'\mathbf{V}^{-1}(\mathbf{y} - \mathbf{X}\boldsymbol{\beta}).$$

Setting these equal to $\mathbf{0}$ and 0, respectively, gives the results.

7.35 Show that $\mathbf{J}^2 = n\mathbf{J}$. Then multiply $\mathbf{V}$ by $\mathbf{V}^{-1}$ to get $\mathbf{I}$, where $\mathbf{V}$ and $\mathbf{V}^{-1}$ are given by (7.67) and (7.68) respectively.

7.36 **(a)** $\mathbf{j}'\mathbf{V}^{-1}\mathbf{j} = a\mathbf{j}'(\mathbf{I} - b\rho\mathbf{J})\mathbf{j} = a\mathbf{j}'\mathbf{j} - ab\rho\mathbf{j}'\mathbf{j}\mathbf{j}'\mathbf{j} = an - ab\rho^2 n^2 = an(1 - b\rho n)$. Substitute for a and b to show that this is equal to $n/[1 + (n-1)\rho] = bn$. Then $\mathbf{j}'\mathbf{V}^{-1}\mathbf{X}_c = a\mathbf{j}'(\mathbf{I} - b\rho\mathbf{J})\mathbf{X}_c = a\mathbf{j}'\mathbf{X}_c - ab\rho\mathbf{j}'\mathbf{j}\mathbf{j}'\mathbf{X}_c = \mathbf{0}'$ because $\mathbf{j}'\mathbf{X}_c = \mathbf{0}'$. Show that $\mathbf{X}_c'\mathbf{V}^{-1}\mathbf{X}_c = a\mathbf{X}_c'\mathbf{X}_c$.

7.37 $\text{cov}(\hat{\boldsymbol{\beta}}^*) = (\mathbf{X}'\mathbf{X})^{-1}\mathbf{X}'\text{cov}(\mathbf{y})\mathbf{X}(\mathbf{X}'\mathbf{X})^{-1} = \sigma^2(\mathbf{X}'\mathbf{X})^{-1}\mathbf{X}'\mathbf{V}\mathbf{X}(\mathbf{X}'\mathbf{X})^{-1}$.

7.38 **(a)**
$$(\mathbf{X}'\mathbf{V}^{-1}\mathbf{X})^{-1}\mathbf{X}'\mathbf{V}^{-1}\mathbf{y} = \begin{pmatrix} \sum_i \frac{1}{x_i} & n \\ n & \sum_i x_i \end{pmatrix}^{-1} \begin{pmatrix} \sum_i \frac{y_i}{x_i} \\ \sum_i y_i \end{pmatrix}$$
$$= \frac{1}{\sum_i x_i \sum_i \frac{1}{x_i} - n^2} \begin{pmatrix} \sum_i x_i & -n \\ -n & \sum_i \frac{1}{x_i} \end{pmatrix} \begin{pmatrix} \sum_i \frac{y_i}{x_i} \\ \sum_i y_i \end{pmatrix}.$$

7.40 **(a)**
$$\text{var}\left(\frac{\sum_i (x_i - \bar{x})y_i}{\sum_i (x_i - \bar{x})^2}\right) = \frac{1}{[\sum_i (x_i - \bar{x})^2]^2} \sum_i (x_i - \bar{x})^2 \text{var}(y_i)$$
$$= \frac{1}{[\sum_i (x_i - \bar{x})^2]^2} \sum_i (x_i - \bar{x})^2 \sigma^2 x_i.$$

7.42 $\text{cov}(\hat{\boldsymbol{\beta}}_1^*) = E[\hat{\boldsymbol{\beta}}_1^* - E(\hat{\boldsymbol{\beta}}_1^*)][\hat{\boldsymbol{\beta}}_1^* - E(\hat{\boldsymbol{\beta}}_1^*)]'$. Using $E(\hat{\boldsymbol{\beta}}_1^*) = \boldsymbol{\beta}_1 + \mathbf{A}\boldsymbol{\beta}_2$ from (7.80), we have

$$\begin{aligned} \hat{\boldsymbol{\beta}}_1^* - E(\hat{\boldsymbol{\beta}}_1^*) &= \hat{\boldsymbol{\beta}}_1^* - \boldsymbol{\beta}_1 - (\mathbf{X}_1'\mathbf{X}_1)^{-1}\mathbf{X}_1'\mathbf{X}_2\boldsymbol{\beta}_2 \\ &= (\mathbf{X}_1'\mathbf{X}_1)^{-1}\mathbf{X}_1'\mathbf{y} - (\mathbf{X}_1'\mathbf{X}_1)^{-1}\mathbf{X}_1'\mathbf{X}_2\boldsymbol{\beta}_2 - \boldsymbol{\beta}_1 \\ &= (\mathbf{X}_1'\mathbf{X}_1)^{-1}\mathbf{X}_1'(\mathbf{y} - \mathbf{X}_2\boldsymbol{\beta}_2) - \boldsymbol{\beta}_1. \end{aligned}$$

Show that this can be written as $(\mathbf{X}_1'\mathbf{X}_1)^{-1}\mathbf{X}_1'(\mathbf{y} - \mathbf{X}_1\boldsymbol{\beta}_1 - \mathbf{X}_2\boldsymbol{\beta}_2)$, so that

$$\text{cov}(\hat{\boldsymbol{\beta}}_1^*) = E[(\mathbf{X}_1'\mathbf{X}_1)^{-1}\mathbf{X}_1'(\mathbf{y} - \mathbf{X}\boldsymbol{\beta})(\mathbf{y} - \mathbf{X}\boldsymbol{\beta})'\mathbf{X}_1(\mathbf{X}_1'\mathbf{X}_1)^{-1}].$$

7.43 Use Theorem 7.9a and note that

$$\mathbf{x}_0'\boldsymbol{\beta} = (\mathbf{x}_{01}', \mathbf{x}_{02}')\begin{pmatrix} \boldsymbol{\beta}_1 \\ \boldsymbol{\beta}_2 \end{pmatrix} = \mathbf{x}_{01}'\boldsymbol{\beta}_1 + \mathbf{x}_{02}'\boldsymbol{\beta}_2.$$

7.44 Multiply out $(\mathbf{X}_{01} - \mathbf{A}'\mathbf{X}_{01})'\mathbf{G}_{22}(\mathbf{X}_{01} - \mathbf{A}'\mathbf{X}_{01})$, substitute $\mathbf{A} = \mathbf{G}_{11}^{-1}\mathbf{G}_{12}$, and use (2.50).

7.45 $E(\mathbf{X}_{01}'\hat{\boldsymbol{\beta}}_1^*) = \mathbf{x}_{01}'(\boldsymbol{\beta}_1 + \mathbf{A}\boldsymbol{\beta}_2) \neq \mathbf{X}_{01}'\boldsymbol{\beta}_1.$

7.46

$$\begin{aligned} \text{var}(\mathbf{x}_{01}\hat{\boldsymbol{\beta}}_1) &\geq \text{var}(\mathbf{x}_{01}\hat{\boldsymbol{\beta}}_1^*) \\ &= \sigma^2(\mathbf{x}_{01}'\mathbf{G}^{11}\mathbf{x}_{01} - \mathbf{x}_{01}'\mathbf{G}_{11}^{-1}\mathbf{x}_{01}) \\ &= \sigma^2\mathbf{x}_{01}'(\mathbf{G}^{11} - \mathbf{G}_{11}^{-1})\mathbf{x}_{01} \\ &\geq 0 \text{ because } \mathbf{G}^{11} - \mathbf{G}_{11}^{-1} = \mathbf{AB}^{-1}\mathbf{A} \text{ which is positive definite} \\ &\quad [\text{see Theorem 7.9c(ii)}]. \end{aligned}$$

7.47

$$\text{tr}[\mathbf{I} - \mathbf{X}_1(\mathbf{X}_1'\mathbf{X}_1)^{-1}\mathbf{X}_1'] = \text{tr}(\mathbf{I}) - \text{tr}[\mathbf{X}_1'\mathbf{X}_1(\mathbf{X}_1'\mathbf{X}_1)^{-1}] = n - (p+1),$$

$$\boldsymbol{\beta}'\mathbf{X}'[\mathbf{I} - \mathbf{X}_1(\mathbf{X}_1'\mathbf{X}_1)^{-1}\mathbf{X}_1']\mathbf{X}\boldsymbol{\beta} = (\boldsymbol{\beta}_1'\mathbf{X}_1' + \boldsymbol{\beta}_2'\mathbf{X}_2')[\mathbf{I} - \mathbf{X}_1(\mathbf{X}_1'\mathbf{X}_1)^{-1}\mathbf{X}_1'] (\mathbf{X}_1\boldsymbol{\beta}_1 + \mathbf{X}_2\boldsymbol{\beta}_2).$$

Show that three of the resulting four terms vanish, leaving the desired result.

7.48

$$\frac{\partial \sum_{i=1}^n (y_i - \hat{\beta}_1^* x_i)^2}{\partial \hat{\beta}_1^*} = 0,$$

$$2\sum_i (y_i - \hat{\beta}_1^* x_i)(-x_i) = 0.$$

7.49 For the full model $y_i = \beta_0 + \beta_1 x_i + \varepsilon_i$, we have

$$\mathbf{X} = \begin{pmatrix} 1 & x_1 \\ \vdots & \vdots \\ 1 & x_n \end{pmatrix}.$$

For the reduced model $y_i = \beta_1^* x_i + \varepsilon_i^*$, we have $\mathbf{X}_1 = (x_1, x_2, \ldots, x_n)'$. Thus, $\mathbf{X}_2 = (1, 1, \ldots, 1)'$. Then from (7.80), we obtain

$$E(\hat{\boldsymbol{\beta}}_1^*) = \boldsymbol{\beta}_1 + \mathbf{A}\boldsymbol{\beta}_2 = \boldsymbol{\beta}_1 + (\mathbf{X}_1'\mathbf{X}_1)^{-1}\mathbf{X}_1'\mathbf{X}_2\boldsymbol{\beta}_2$$

$$= \beta_1 + \left(\sum_{i=1}^n x_i^2\right)^{-1} \sum_{i=1}^n x_i \cdot \beta_0.$$

7.50 **(a)**

$$\mathbf{X} = \begin{pmatrix} 1 & -3 & 9 & -27 \\ 1 & -2 & 4 & -8 \\ 1 & -1 & 1 & -1 \\ 1 & 0 & 0 & 0 \\ 1 & 1 & 1 & 1 \\ 1 & 2 & 4 & 8 \\ 1 & 3 & 9 & 27 \end{pmatrix}.$$

The first two columns constitute $\mathbf{X}_1$, and the last two columns become $\mathbf{X}_2$. Then by (7.80), we obtain

$$E(\hat{\boldsymbol{\beta}}_1^*) = \boldsymbol{\beta}_1 + (\mathbf{X}_1'\mathbf{X}_1)^{-1}\mathbf{X}_1'\mathbf{X}_2\boldsymbol{\beta}_2.$$

Show that this gives

$$E(\hat{\boldsymbol{\beta}}_1^*) = \begin{pmatrix} \beta_0 \\ \beta_1 \end{pmatrix} + \begin{pmatrix} 7 & 0 \\ 0 & 28 \end{pmatrix}^{-1} \begin{pmatrix} 28 & 0 \\ 0 & 196 \end{pmatrix} \begin{pmatrix} \beta_2 \\ \beta_3 \end{pmatrix}$$

$$= \begin{pmatrix} \beta_0 \\ \beta_1 \end{pmatrix} + \begin{pmatrix} 4 & 0 \\ 0 & 7 \end{pmatrix} \begin{pmatrix} \beta_2 \\ \beta_3 \end{pmatrix},$$

so that $E(\hat{\beta}_0^*) = \beta_0 + 4\beta_2$ and $E(\hat{\beta}_1^*) = \beta_1 + 7\beta_3$.

7.51 $\mathbf{X}_1'\mathbf{X}_{2.1} = \mathbf{X}_1'[\mathbf{X}_2 - \mathbf{X}_1(\mathbf{X}_1'\mathbf{X}_1)^{-1}\mathbf{X}_1'\mathbf{X}_2] = \mathbf{X}_1'\mathbf{X}_2 - \mathbf{X}_1'\mathbf{X}_1(\mathbf{X}_1'\mathbf{X}_1)^{-1}\mathbf{X}_1'\mathbf{X}_2.$

7.52 In the partitioned form, the normal equations $\mathbf{X}'\mathbf{X}\hat{\boldsymbol{\beta}} = \mathbf{X}'\mathbf{y}$ become

$$\begin{pmatrix} \mathbf{X}_1' \\ \mathbf{X}_2' \end{pmatrix} (\mathbf{X}_1, \mathbf{X}_2) \begin{pmatrix} \hat{\boldsymbol{\beta}}_1 \\ \hat{\boldsymbol{\beta}}_2 \end{pmatrix} = \begin{pmatrix} \mathbf{X}_1' \\ \mathbf{X}_2' \end{pmatrix} \mathbf{y},$$

$$\begin{pmatrix} \mathbf{X}_1'\mathbf{X}_1 & \mathbf{X}_1'\mathbf{X}_2 \\ \mathbf{X}_2'\mathbf{X}_1 & \mathbf{X}_2'\mathbf{X}_2 \end{pmatrix} \begin{pmatrix} \hat{\boldsymbol{\beta}}_1 \\ \hat{\boldsymbol{\beta}}_2 \end{pmatrix} = \begin{pmatrix} \mathbf{X}_1'\mathbf{y} \\ \mathbf{X}_2'\mathbf{y} \end{pmatrix},$$

$$\mathbf{X}_1'\mathbf{X}_1\hat{\boldsymbol{\beta}}_1 + \mathbf{X'}_1\mathbf{X}_2\hat{\boldsymbol{\beta}}_2 = \mathbf{X}_1'\mathbf{y}, \qquad (1)$$

$$\mathbf{X}_2'\mathbf{X}_1\hat{\boldsymbol{\beta}}_1 + \mathbf{X}_2'\mathbf{X}_2\hat{\boldsymbol{\beta}}_2 = \mathbf{X}_2'\mathbf{y}. \qquad (2)$$

Solve for $\hat{\boldsymbol{\beta}}_1$ from (1) to obtain $\hat{\boldsymbol{\beta}}_1 = (\mathbf{X}_1'\mathbf{X}_1)^{-1}(\mathbf{X}_1'\mathbf{y} - \mathbf{X}_1'\mathbf{X}_2\hat{\boldsymbol{\beta}}_2)$, and substitute this into (2) to obtain

$$[\mathbf{X}_2'\mathbf{X}_2 - \mathbf{X}_2'\mathbf{X}_1(\mathbf{X}_1'\mathbf{X}_1)^{-1}\mathbf{X}_1'\mathbf{X}_2]\hat{\boldsymbol{\beta}}_2 = \mathbf{X}_2'\mathbf{y} - \mathbf{X}_2'\mathbf{X}_1(\mathbf{X}_1'\mathbf{X}_1)^{-1}\mathbf{X}_1'\mathbf{y}. \quad (3)$$

Multiplying (7.98) by $\mathbf{X}_{2.1}'$, we obtain $\hat{\boldsymbol{\beta}}_2 = (\mathbf{X}_{2.1}'\mathbf{X}_{2.1})^{-1}\mathbf{X}_{2.1}'[\hat{\mathbf{y}} - \hat{\mathbf{y}}(\mathbf{X}_1)]$. Show that this is the same as (3).

7.53 **(a)**

$$\hat{\boldsymbol{\beta}} = \begin{pmatrix} 1.0150 \\ -0.0286 \\ 0.2158 \\ -4.3201 \\ 8.9749 \end{pmatrix}, \quad s^2 = 7.4529.$$

(b)

$$s^2(\mathbf{X}'\mathbf{X})^{-1} = \begin{pmatrix} 3.4645 & .0145 & -.0638 & -1.1620 & 1.0723 \\ 0.0145 & .0082 & -.0019 & -0.1630 & 0.0784 \\ -0.0638 & -.0019 & .0046 & 0.1039 & -0.1250 \\ -1.1620 & -.1630 & .1039 & 8.1280 & -7.2045 \\ 1.0723 & .0784 & -.1250 & -7.2045 & 7.6875 \end{pmatrix}.$$

(c)

$$\hat{\boldsymbol{\beta}}_1 = \mathbf{S}_{xx}^{-1}\mathbf{s}_{yx} = \begin{pmatrix} 380.6684 & 237.6684 & 27.0709 & 25.3549 \\ 237.6684 & 247.5071 & 17.8557 & 18.3362 \\ 27.0709 & 17.8557 & 2.1090 & 1.9909 \\ 25.3549 & 18.3362 & 1.9909 & 1.9369 \end{pmatrix}^{-1}$$

$$\times \begin{pmatrix} 151.0121 \\ 134.0444 \\ 11.8365 \\ 12.0140 \end{pmatrix} = \begin{pmatrix} -0.0286 \\ 0.2158 \\ -4.3201 \\ 8.9749 \end{pmatrix}.$$

$$\hat{\beta}_0 = \bar{y} - \hat{\boldsymbol{\beta}}_1'\bar{\mathbf{x}} = 31.125 - (-0.0286, 0.2158, -4.3201, 8.9749)$$

$$\times \begin{pmatrix} 57.9063 \\ 55.9063 \\ 4.4222 \\ 4.3238 \end{pmatrix} = 1.0150.$$

(d) $R^2 = .9261, \quad R_a^2 = .9151.$

7.54 **(a)**
$$\hat{\boldsymbol{\beta}} = \begin{pmatrix} 332.111 \\ -1.546 \\ -1.425 \\ -2.237 \end{pmatrix}, \quad s^2 = 5.3449.$$

(b)
$$\text{cov}(\hat{\boldsymbol{\beta}}) = s^2(\mathbf{X}'\mathbf{X})^{-1}$$
$$= 5.3449 \begin{pmatrix} 65.37550 & -.33885 & -.31252 & -.02041 \\ -0.33885 & .00184 & .00127 & -.00043 \\ -0.31252 & .00127 & .00408 & -.00176 \\ -0.02041 & -.00043 & -.00176 & .02161 \end{pmatrix}.$$

(c) $R^2 = .9551, \quad R_a^2 = .9462.$

(d)
$$\hat{\boldsymbol{\beta}} = \begin{pmatrix} 964.929 \\ -7.442 \\ -11.508 \\ -2.140 \\ 0.012 \\ 0.033 \\ -0.294 \\ 0.054 \\ 0.038 \\ -0.102 \end{pmatrix}, \quad s^2 = 5.1342.$$

(e) $R^2 = .9741, \quad R_a^2 = .9483.$

7.55 **(a)**
$$\hat{\boldsymbol{\beta}} = \begin{pmatrix} .6628 \\ .7803 \\ .5031 \\ -17.1002 \end{pmatrix} \quad s^2 = 67.9969.$$

(b)
$$\hat{\boldsymbol{\beta}}_1 = \mathbf{S}_{xx}^{-1}\mathbf{s}_{yx} = \begin{pmatrix} 504.2783 & 9.4698 & -1.7936 \\ 9.4698 & 201.9399 & 1.0617 \\ -1.7936 & 1.0617 & 0.0235 \end{pmatrix}^{-1} \begin{pmatrix} 428.9086 \\ 90.8333 \\ -1.2667 \end{pmatrix}$$
$$= \begin{pmatrix} 0.7803 \\ 0.5031 \\ -17.1002 \end{pmatrix}$$
$$\hat{\boldsymbol{\beta}}_0 = \bar{y} - \hat{\boldsymbol{\beta}}_1'\bar{\mathbf{x}} = 41.1553 - (.7803, .5031, -17.1002)$$
$$\times \begin{pmatrix} 42.945 \\ 20.169 \\ 0.185 \end{pmatrix} = .6628.$$

(c) $R^2 = .8667, \quad R_a^2 = .8534.$

Chapter 8

8.1 Substitute $\hat{\boldsymbol{\beta}}_1 = (\mathbf{X}_c'\mathbf{X}_c)^{-1}\mathbf{X}_c'\mathbf{y}$ into SSR $= \hat{\boldsymbol{\beta}}_1'\mathbf{X}_c'\mathbf{y}$.

8.2 (a)
$$\mathbf{H}_c\left(\mathbf{I} - \frac{1}{n}\mathbf{J}\right) = \mathbf{H}_c - \frac{1}{n}\mathbf{H}_c\mathbf{J}$$
$$= \mathbf{X}_c(\mathbf{X}_c'\mathbf{X}_c)^{-1}\mathbf{X}_c' - \frac{1}{n}\mathbf{X}_c(\mathbf{X}_c'\mathbf{X}_c)^{-1}\mathbf{X}_c'\mathbf{j}\mathbf{j}'$$
$$= \mathbf{X}_c(\mathbf{X}_c'\mathbf{X}_c)^{-1}\mathbf{X}_c' - \mathbf{O}$$

since $\mathbf{X}_c'\mathbf{j}\mathbf{j}' = \mathbf{O}\mathbf{j}' = \mathbf{O}$.

(b) Show that $\mathbf{H}_c^2 = \mathbf{H}_c$, where $\mathbf{H}_c = \mathbf{X}_c(\mathbf{X}_c'\mathbf{X}_c)^{-1}\mathbf{X}_c'$. Then, since $\mathbf{H}_c$ is idempotent, rank $(\mathbf{H}_c) = \text{tr}(\mathbf{H}_c)$ by Theorem 2.13d. The centered matrix $\mathbf{X}_c$ is $n \times k$ of rank k [see (7.33)].

(c)
$$\left(\mathbf{I} - \frac{1}{n}\mathbf{J} - \mathbf{H}_c\right)^2 = \left(\mathbf{I} - \frac{1}{n}\mathbf{J}\right)^2 - \left(\mathbf{I} - \frac{1}{n}\mathbf{J}\right)\mathbf{H}_c - \mathbf{H}_c\left(\mathbf{I} - \frac{1}{n}\mathbf{J}\right) + \mathbf{H}_c^2$$
$$= \mathbf{I} - \frac{1}{n}\mathbf{J} - \mathbf{H}_c - \mathbf{H}_c + \mathbf{H}_c.$$

Then rank $\left(\mathbf{I} - \frac{1}{n}\mathbf{J} - \mathbf{H}_c\right) = \text{tr}\left(\mathbf{I} - \frac{1}{n}\mathbf{J} - \mathbf{H}_c\right)$.

(d)
$$\mathbf{H}_c\left(\mathbf{I} - \frac{1}{n}\mathbf{J} - \mathbf{H}_c\right) = \mathbf{H}_c\left(\mathbf{I} - \frac{1}{n}\mathbf{J}\right) - \mathbf{H}_c^2 = \mathbf{H}_c - \mathbf{H}_c = \mathbf{O}.$$

8.3 $\boldsymbol{\mu}'\mathbf{H}_c\boldsymbol{\mu} = \boldsymbol{\beta}'\mathbf{X}'\mathbf{X}_c(\mathbf{X}_c'\mathbf{X}_c)^{-1}\mathbf{X}_c'\mathbf{X}\boldsymbol{\beta}$. By (7.32), we have $\mathbf{X}\boldsymbol{\beta} = \alpha\mathbf{j} + \mathbf{X}_c\boldsymbol{\beta}_1$. Hence $\boldsymbol{\mu}'\mathbf{H}_c\boldsymbol{\mu} = (\alpha\mathbf{j}' + \boldsymbol{\beta}_1'\mathbf{X}_c')\mathbf{X}_c(\mathbf{X}_c'\mathbf{X}_c)^{-1}\mathbf{X}_c'(\alpha\,\mathbf{j} + \mathbf{X}_c\boldsymbol{\beta}_1)$. Three of the resulting four terms vanish because $\mathbf{j}'\mathbf{X}_c = \mathbf{0}'$ (see Problem 7.16).

8.4 By corollary 2 to Theorem 5.5a, SSE$/\sigma^2$ is $\chi^2(n-k-1, \lambda_2)$. Also $\lambda_2 = \boldsymbol{\mu}'[\mathbf{I} - \frac{1}{n}\mathbf{J} - \mathbf{H}_c]\boldsymbol{\mu}/\sigma^2 = (\alpha\mathbf{j}' + \boldsymbol{\beta}_1'\mathbf{X}_c')[\mathbf{I} - \frac{1}{n}\mathbf{J} - \mathbf{H}_c](\alpha\mathbf{j} + \mathbf{X}_c\boldsymbol{\beta}_1)/\sigma^2$. Show that all terms involving either $\mathbf{j}'[\mathbf{I} - \frac{1}{n}\mathbf{J}]$ or $\mathbf{j}'\mathbf{H}_c$ vanish. Show that $\boldsymbol{\beta}_1'\mathbf{X}_c'[\mathbf{I} - \frac{1}{n}\mathbf{J}]\mathbf{X}_c\boldsymbol{\beta}_1 = \boldsymbol{\beta}_1'\mathbf{X}_c'\mathbf{X}_c\boldsymbol{\beta}_1$ and that $\boldsymbol{\beta}_1'\mathbf{X}_c'\mathbf{H}_c\mathbf{X}_c\boldsymbol{\beta}_1 = \boldsymbol{\beta}_1'\mathbf{X}_c'\mathbf{X}_c\beta_1$.

8.6 Most of these results are proved in Problem 5.32, with the adjustment $k + 1 = p$.

8.7 By (8.14), $\mathbf{H}\mathbf{H}_1 = \mathbf{X}(\mathbf{X}'\mathbf{X})^{-1}\mathbf{X}'\mathbf{X}_1(\mathbf{X}_1'\mathbf{X}_1)^{-1}\mathbf{X}_1' = \mathbf{X}_1(\mathbf{X}_1'\mathbf{X}_1)^{-1}\mathbf{X}_1' = \mathbf{H}_1$.

8.9
$$\begin{aligned}\boldsymbol{\mu}'(\mathbf{H}-\mathbf{H}_1)\boldsymbol{\mu} &= \boldsymbol{\beta}'\mathbf{X}'(\mathbf{H}-\mathbf{H}_1)\mathbf{X}\boldsymbol{\beta} \\ &= \boldsymbol{\beta}'\mathbf{X}'\mathbf{X}(\mathbf{X}'\mathbf{X})^{-1}\mathbf{X}'\mathbf{X}\boldsymbol{\beta} - \boldsymbol{\beta}'\mathbf{X}'\mathbf{X}_1(\mathbf{X}_1'\mathbf{X}_1)^{-1}\mathbf{X}_1'\mathbf{X}\boldsymbol{\beta} \\ &= \boldsymbol{\beta}'\mathbf{X}'\mathbf{X}\boldsymbol{\beta} - \boldsymbol{\beta}'\mathbf{X}'\mathbf{X}_1(\mathbf{X}_1'\mathbf{X}_1)^{-1}\mathbf{X}_1'\mathbf{X}\boldsymbol{\beta} \\ &= (\boldsymbol{\beta}_1'\mathbf{X}_1' + \boldsymbol{\beta}_2'\mathbf{X}_2')(\mathbf{X}_1\boldsymbol{\beta}_1 + \mathbf{X}_2\boldsymbol{\beta}_2) \\ &\quad - (\boldsymbol{\beta}_1'\mathbf{X}_1' + \boldsymbol{\beta}_2'\mathbf{X}_2')\mathbf{X}_1(\mathbf{X}_1'\mathbf{X}_1)^{-1}\mathbf{X}_1'(\mathbf{X}_1\boldsymbol{\beta}_1 + \mathbf{X}_2\boldsymbol{\beta}_2).\end{aligned}$$

8.10 Denote the matrix $\mathbf{X}'\mathbf{X}$ by $\mathbf{G}$. Then in partitioned form, we have

$$\begin{aligned}\mathbf{G} = \mathbf{X}'\mathbf{X} = (\mathbf{X}_1, \mathbf{X}_2)'(\mathbf{X}_1, \mathbf{X}_2) &= \begin{pmatrix}\mathbf{X}_1' \\ \mathbf{X}_2'\end{pmatrix}(\mathbf{X}_1, \mathbf{X}_2) \\ &= \begin{pmatrix}\mathbf{X}_1'\mathbf{X}_1 & \mathbf{X}_1'\mathbf{X}_2 \\ \mathbf{X}_2'\mathbf{X}_1 & \mathbf{X}_2'\mathbf{X}_2\end{pmatrix} = \begin{pmatrix}\mathbf{G}_{11} & \mathbf{G}_{12} \\ \mathbf{G}_{21} & \mathbf{G}_{22}\end{pmatrix}.\end{aligned}$$

If we denote the four corresponding blocks of $\mathbf{G}^{-1}$ by $\mathbf{G}^{ij}$, then by (2.48), $\mathbf{G}^{22} = (\mathbf{G}_{22} - \mathbf{G}_{21}\mathbf{G}_{11}^{-1}\mathbf{G}_{12})^{-1}$. By Theorem 2.6e, $\mathbf{G}^{-1}$ is positive definite. By Theorem 2.6f, $\mathbf{G}^{22}$ is positive definite. By Theorem 2.6e, $(\mathbf{G}^{22})^{-1} = \mathbf{G}_{22} - \mathbf{G}_{21}\mathbf{G}_{11}^{-1}\mathbf{G}_{12} = \mathbf{X}_2'\mathbf{X}_2 - \mathbf{X}_2'\mathbf{X}_1(\mathbf{X}_1'\mathbf{X}_1)^{-1}\mathbf{X}_1'\mathbf{X}_2$ is positive definite.

8.11 By Theorem 8.2b(ii), $\text{SS}(\boldsymbol{\beta}_2|\boldsymbol{\beta}_1)/\sigma^2$ is $\chi^2(h, \lambda_1)$. Then $E[\text{SS}(\boldsymbol{\beta}_2|\boldsymbol{\beta}_1)/\sigma^2] = h + 2\lambda_1$ by (5.23).

8.12 $\sigma^2 + \beta_k^2[\mathbf{x}_k'\mathbf{x}_k - \mathbf{x}_k'\mathbf{X}_1(\mathbf{X}_1'\mathbf{X}_1)^{-1}\mathbf{X}_1'\mathbf{x}_k]$.

8.13 For the reduced model $\mathbf{y} = \beta_0^*\mathbf{j} + \boldsymbol{\varepsilon}^*$, we have $\hat{\beta}_0^* = (\mathbf{j}'\mathbf{j})^{-1}\mathbf{j}'\mathbf{y} = (1/n)\sum_{i=1}^n y_i = \bar{y}$ and $\text{SS}(\beta_0^*) = \hat{\beta}_0^*\mathbf{j}'\mathbf{y} = \bar{y}\sum_i y_i = n\bar{y}^2$.

8.14 After multiplying to obtain eight terms, three of the first four terms cancel three of the last four terms. For example, the second of the last four is $\hat{\boldsymbol{\beta}}_1'\mathbf{X}_1'\mathbf{X}_1\mathbf{A}\hat{\boldsymbol{\beta}}_2 = \hat{\boldsymbol{\beta}}_1'\mathbf{X}_1'\mathbf{X}_1(\mathbf{X}_1'\mathbf{X}_1)^{-1}\mathbf{X}_1'\mathbf{X}_2\hat{\boldsymbol{\beta}}_2 = \hat{\boldsymbol{\beta}}_1'\mathbf{X}_1'\mathbf{X}_2\hat{\boldsymbol{\beta}}_2$, which is the same as the second of the first four terms.

8.15 Add and substract $n\bar{y}^2$ in both numerator and denominator of (8.24) and then divide numerator and denominator by $\mathbf{y}'\mathbf{y} - n\bar{y}^2$.

8.16 Express SSH as a quadratic form in $\mathbf{y}$ by substituting $\hat{\boldsymbol{\beta}} = (\mathbf{X}'\mathbf{X})^{-1}\mathbf{X}'\mathbf{y}$. Then use Corollary 1 to Theorem 5.6b.

8.17 This follows from Corollary 1 to Theorem 2.6b.

8.18 By the answer to Problem 7.28, $\hat{\boldsymbol{\beta}}_w'\mathbf{W}'\mathbf{y} = \hat{\boldsymbol{\beta}}'\mathbf{X}'\mathbf{y}$. Thus $\text{SSE} = \mathbf{y}'\mathbf{y} - \hat{\boldsymbol{\beta}}'\mathbf{X}'\mathbf{y}$ is invariant to the full-rank transformation $\mathbf{W} = \mathbf{XH}$. For the numerator of (8.27), we note that $\mathbf{C}$ is transformed the same way as is $\mathbf{X}$, so that $\mathbf{C}_w\hat{\boldsymbol{\beta}}_w = \mathbf{CHH}^{-1}\hat{\boldsymbol{\beta}}_x = \mathbf{C}\hat{\boldsymbol{\beta}}$. Thus the numerator of (8.27) becomes

$$\begin{aligned}(\mathbf{C}_w\hat{\boldsymbol{\beta}}_w)'[\mathbf{C}_w(\mathbf{W}'\mathbf{W})^{-1}\mathbf{C}_w']^{-1}\mathbf{C}_w\hat{\boldsymbol{\beta}}_w &= (\mathbf{C}\hat{\boldsymbol{\beta}})'\{\mathbf{CH}[(\mathbf{XH})'\mathbf{XH}]^{-1}(\mathbf{CH})'\}^{-1}\mathbf{C}\hat{\boldsymbol{\beta}} \\ &= (\mathbf{C}\hat{\boldsymbol{\beta}})'\{\mathbf{CH}[\mathbf{H}'(\mathbf{X}'\mathbf{X})\mathbf{H}]^{-1}\mathbf{H}'\mathbf{C}'\}^{-1}\mathbf{C}\hat{\boldsymbol{\beta}}\end{aligned}$$

$$= (\mathbf{C}\hat{\boldsymbol{\beta}})'[\mathbf{C}\mathbf{H}\mathbf{H}^{-1}(\mathbf{X}'\mathbf{X})^{-1}(\mathbf{H}')^{-1}\mathbf{H}'\mathbf{C}']^{-1}\mathbf{C}\hat{\boldsymbol{\beta}}$$

$$= (\mathbf{C}\hat{\boldsymbol{\beta}})'[\mathbf{C}(\mathbf{X}'\mathbf{X})^{-1}\mathbf{C}']^{-1}\mathbf{C}\hat{\boldsymbol{\beta}}.$$

Show that the transformation $\mathbf{z} = c\mathbf{y}$ also leaves F unchanged.

8.19 **(a)** See Section 2.14.3 $\cdot \dfrac{\partial u}{\partial \boldsymbol{\lambda}} = \mathbf{C}\boldsymbol{\beta}$. Setting this equal to $\mathbf{0}$ gives $\mathbf{C}\hat{\boldsymbol{\beta}}_c = \mathbf{0}$.

(b)
$$u = \mathbf{y}'\mathbf{y} - \mathbf{y}'\mathbf{X}\boldsymbol{\beta} - \boldsymbol{\beta}'\mathbf{X}'\mathbf{y} + \boldsymbol{\beta}'\mathbf{X}'\mathbf{X}\boldsymbol{\beta} + \boldsymbol{\lambda}'\mathbf{C}\boldsymbol{\beta}$$
$$= \mathbf{y}'\mathbf{y} - 2\boldsymbol{\beta}'\mathbf{X}'\mathbf{y} + \boldsymbol{\beta}'\mathbf{X}'\mathbf{X}\boldsymbol{\beta} + \boldsymbol{\lambda}'\mathbf{C}\boldsymbol{\beta},$$
$$= \frac{\partial u}{\partial \boldsymbol{\beta}} = \mathbf{0} - 2\mathbf{X}'\mathbf{y} + 2\mathbf{X}'\mathbf{X}\boldsymbol{\beta} + \mathbf{C}'\boldsymbol{\lambda}.$$

Setting this equal to $\mathbf{0}$ gives

$$\hat{\boldsymbol{\beta}}_c = (\mathbf{X}'\mathbf{X})^{-1}\mathbf{X}'\mathbf{y} - \tfrac{1}{2}(\mathbf{X}'\mathbf{X})^{-1}\mathbf{C}'\boldsymbol{\lambda}$$
$$= \hat{\boldsymbol{\beta}} - \tfrac{1}{2}(\mathbf{X}'\mathbf{X})^{-1}\mathbf{C}'\boldsymbol{\lambda}. \qquad (1)$$

(c)
$$\mathbf{C}\hat{\boldsymbol{\beta}}_c = \mathbf{C}\hat{\boldsymbol{\beta}} - \tfrac{1}{2}\mathbf{C}(\mathbf{X}'\mathbf{X})^{-1}\mathbf{C}'\boldsymbol{\lambda} = \mathbf{0},$$
$$\boldsymbol{\lambda} = 2[\mathbf{C}(\mathbf{X}'\mathbf{X})^{-1}\mathbf{C}']^{-1}\mathbf{C}\hat{\boldsymbol{\beta}}.$$

Substituting this into (1) in part (b) gives the result.

8.20
$$\hat{\boldsymbol{\beta}}_c'\mathbf{X}'\mathbf{X}\hat{\boldsymbol{\beta}}_c = \hat{\boldsymbol{\beta}}_c'\mathbf{X}'\mathbf{X}\{\hat{\boldsymbol{\beta}}' - (\mathbf{X}'\mathbf{X})^{-1}\mathbf{C}'[\mathbf{C}(\mathbf{X}'\mathbf{X})^{-1}\mathbf{C}']^{-1}\mathbf{C}\hat{\boldsymbol{\beta}}\}$$
$$= \hat{\boldsymbol{\beta}}_c'\mathbf{X}'\mathbf{X}\hat{\boldsymbol{\beta}} - \hat{\boldsymbol{\beta}}_c'\mathbf{C}'[\mathbf{C}(\mathbf{X}'\mathbf{X})^{-1}\mathbf{C}']^{-1}\mathbf{C}\hat{\boldsymbol{\beta}}$$
$$= \hat{\boldsymbol{\beta}}_c'\mathbf{X}'\mathbf{y} - \mathbf{0}'[\mathbf{C}(\mathbf{X}'\mathbf{X})^{-1}\mathbf{C}']^{-1}\mathbf{C}\hat{\boldsymbol{\beta}}.$$

Show that $\hat{\boldsymbol{\beta}}_c'\mathbf{C}' = \mathbf{0}'$.

8.21 Substituting $\hat{\boldsymbol{\beta}}_c'$ in (8.30) into SSH $= \hat{\boldsymbol{\beta}}'\mathbf{X}'\mathbf{y} - \hat{\boldsymbol{\beta}}_c'\mathbf{X}'\mathbf{y}$ in (8.31) gives

$$\text{SSH} = \hat{\boldsymbol{\beta}}'\mathbf{X}'\mathbf{y} - \{\hat{\boldsymbol{\beta}}' - \hat{\boldsymbol{\beta}}'\mathbf{C}'[\mathbf{C}(\mathbf{X}'\mathbf{X})^{-1}\mathbf{C}']^{-1}\mathbf{C}(\mathbf{X}'\mathbf{X})^{-1}\}\mathbf{X}'\mathbf{y}$$
$$= \hat{\boldsymbol{\beta}}'\mathbf{X}'\mathbf{y} - \hat{\boldsymbol{\beta}}'\mathbf{X}'\mathbf{y} + \hat{\boldsymbol{\beta}}'\mathbf{C}'[\mathbf{C}(\mathbf{X}'\mathbf{X})^{-1}\mathbf{C}']^{-1}\mathbf{C}\hat{\boldsymbol{\beta}},$$

since $\hat{\boldsymbol{\beta}} = (\mathbf{X}'\mathbf{X})^{-1}\mathbf{X}'\mathbf{y}$.

8.22 In Theorem 8.4e(ii), we have

$$\text{cov}(\hat{\boldsymbol{\beta}}_c) = \text{cov}\{\mathbf{I} - (\mathbf{X}'\mathbf{X})^{-1}\mathbf{C}'[\mathbf{C}(\mathbf{X}'\mathbf{X})^{-1}\mathbf{C}']^{-1}\mathbf{C}\}\hat{\boldsymbol{\beta}}$$
$$= \text{cov}(\mathbf{A}\hat{\boldsymbol{\beta}}) = \mathbf{A}\ \text{cov}(\hat{\boldsymbol{\beta}})\mathbf{A}' = \sigma^2\mathbf{A}(\mathbf{X}'\mathbf{X})^{-1}\mathbf{A}'.$$

Show that $\mathbf{A}(\mathbf{X}'\mathbf{X})^{-1}\mathbf{A}' = (\mathbf{X}'\mathbf{X})^{-1} - (\mathbf{X}'\mathbf{X})^{-1}\mathbf{C}'[\mathbf{C}(\mathbf{X}'\mathbf{X})^{-1}\mathbf{C}']^{-1}\mathbf{C}(\mathbf{X}'\mathbf{X})^{-1}$.

8.23 Replace $\hat{\boldsymbol{\beta}}$ by $(\mathbf{X}'\mathbf{X})^{-1}\mathbf{X}'\mathbf{y}$ in SSH in Theorem 8.4d(ii) to obtain SSH $=$ $[\mathbf{C}(\mathbf{X}'\mathbf{X})^{-1}\mathbf{X}'\mathbf{y}-\mathbf{t}]'[\mathbf{C}(\mathbf{X}'\mathbf{X})^{-1}\mathbf{C}']^{-1}[\mathbf{C}(\mathbf{X}'\mathbf{X})^{-1}\mathbf{X}'\mathbf{y}-\mathbf{t}]$. Show that $\mathbf{C}(\mathbf{X}'\mathbf{X})^{-1}\mathbf{X}'\mathbf{y}-\mathbf{t}=\mathbf{C}(\mathbf{X}'\mathbf{X})^{-1}\mathbf{X}'[\mathbf{y}-\mathbf{X}\mathbf{C}'(\mathbf{C}\mathbf{C}')^{-1}\mathbf{t}]$, so that SSH becomes SSH$=[\mathbf{y}-\mathbf{X}\mathbf{C}'(\mathbf{C}'\mathbf{C})^{-1}\mathbf{t}]'\mathbf{A}[\mathbf{y}-\mathbf{X}\mathbf{C}'(\mathbf{C}\mathbf{C}')^{-1}\mathbf{t}]$, where $\mathbf{A}=\mathbf{X}(\mathbf{X}'\mathbf{X})^{-1}\mathbf{C}'$ $[\mathbf{C}(\mathbf{X}'\mathbf{X})^{-1}\mathbf{C}']^{-1}\mathbf{C}(\mathbf{X}'\mathbf{X})^{-1}\mathbf{X}'$. Show that SSE $=[\mathbf{y}-\mathbf{X}\mathbf{C}'(\mathbf{C}\mathbf{C}')^{-1}\mathbf{t}]'\mathbf{B}[\mathbf{y}-$ $\mathbf{X}\mathbf{C}'(\mathbf{C}\mathbf{C}')^{-1}\mathbf{t}]$, where $\mathbf{B}=\mathbf{I}-\mathbf{X}(\mathbf{X}'\mathbf{X})^{-1}\mathbf{X}'$. Show that $\mathbf{AB}=\mathbf{O}$. Show that $\mathbf{y}-\mathbf{X}\mathbf{C}'(\mathbf{C}\mathbf{C}')^{-1}\mathbf{t}$ is $N_n[\mathbf{X}\boldsymbol{\beta}-\mathbf{X}\mathbf{C}'(\mathbf{C}\mathbf{C}')^{-1}\mathbf{t},\ \sigma^2\mathbf{I}]$. Then by Corollary 1 to Theorem 5.6b, SSH and SSE are independent.

8.24 See Section 2.14.3. Follow the steps in Problems 8.19 using

$$\begin{aligned} u &= (\mathbf{y}-\mathbf{X}\boldsymbol{\beta})'(\mathbf{y}-\mathbf{X}\boldsymbol{\beta})+\boldsymbol{\lambda}'(\mathbf{C}\boldsymbol{\beta}-\mathbf{t}) \\ &= \mathbf{y}'\mathbf{y}-2\boldsymbol{\beta}'\mathbf{X}\mathbf{y}+\boldsymbol{\beta}'\mathbf{X}'\mathbf{X}\boldsymbol{\beta}+\boldsymbol{\lambda}'(\mathbf{C}\boldsymbol{\beta}-\mathbf{t}). \end{aligned}$$

Differentiating with respect to $\boldsymbol{\lambda}$ and $\boldsymbol{\beta}$, we obtain

$$\begin{aligned} \frac{\partial u}{\partial \boldsymbol{\lambda}} &= \mathbf{C}\boldsymbol{\beta}-\mathbf{t}, \\ \frac{\partial u}{\partial \boldsymbol{\beta}} &= \mathbf{0}-2\mathbf{X}'\mathbf{y}+2\mathbf{X}'\mathbf{X}\boldsymbol{\beta}+\mathbf{C}'\boldsymbol{\lambda}. \end{aligned}$$

Setting those equal to $\mathbf{0}$ gives $\mathbf{C}\hat{\boldsymbol{\beta}}_c=\mathbf{t}$ and

$$\hat{\boldsymbol{\beta}}_c=\hat{\boldsymbol{\beta}}-\tfrac{1}{2}(\mathbf{X}'\mathbf{X})^{-1}\mathbf{C}'\boldsymbol{\lambda}. \tag{1}$$

Multiplying (1) by $\mathbf{C}$ and using $\mathbf{C}\hat{\boldsymbol{\beta}}_c=\mathbf{t}$ gives $\lambda=2\ [\mathbf{C}(\mathbf{X}'\mathbf{X})^{-1}$ $\mathbf{C}']^{-1}(\mathbf{C}\hat{\boldsymbol{\beta}}-\mathbf{t})$. Substituting this into (1) gives the result.

8.25 By Theorem 8.4d, we can use the general linear hypothesis test. Use $\mathbf{a}'=$ $(0,\ldots,0,1)$ in place of $\mathbf{C}$ in (8.30) to obtain

$$\begin{aligned} \hat{\boldsymbol{\beta}}_c &= \hat{\boldsymbol{\beta}}-(\mathbf{X}'\mathbf{X})^{-1}\mathbf{a}\left[\mathbf{a}'(\mathbf{X}'\mathbf{X})^{-1}\mathbf{a}\right]^{-1}\mathbf{a}'\hat{\boldsymbol{\beta}} \\ &= \hat{\boldsymbol{\beta}}-\frac{(\mathbf{X}'\mathbf{X})^{-1}\mathbf{a}\mathbf{a}'\hat{\boldsymbol{\beta}}}{g_{kk}}. \end{aligned}$$

By (2.37), $(\mathbf{X}'\mathbf{X})^{-1}\,\mathbf{a}$ is a linear combination of the columns of $(\mathbf{X}'\mathbf{X})^{-1}$. Thus

$$\hat{\boldsymbol{\beta}}_c=\hat{\boldsymbol{\beta}}-\frac{\mathbf{g}_k\hat{\beta}_k}{g_{kk}},$$

where g_{kk} is the kth diagonal element of $(\mathbf{X}'\mathbf{X})^{-1}$ and $\mathbf{g}_k$ is the kth column of

$(\mathbf{X}'\mathbf{X})^{-1}$. Substituting this expression for $\hat{\boldsymbol{\beta}}_c$ into $\hat{\boldsymbol{\beta}}'\mathbf{X}'\mathbf{y} - \hat{\boldsymbol{\beta}}_c'\mathbf{X}'\mathbf{y}$, we obtain

$$\hat{\boldsymbol{\beta}}'\mathbf{X}'\mathbf{y} - \hat{\boldsymbol{\beta}}_c'\mathbf{X}'\mathbf{y} = \hat{\boldsymbol{\beta}}'\mathbf{X}'\mathbf{y} - \left(\hat{\boldsymbol{\beta}}'\mathbf{X}'\mathbf{y} - \frac{\hat{\beta}_k}{g_{kk}}\mathbf{g}_\mathbf{k}'\mathbf{X}'\mathbf{y}\right)$$

$$= \frac{\hat{\beta}_k\hat{\beta}_k}{g_{kk}} = \frac{\hat{\beta}_k^2}{g_{kk}},$$

since $\hat{\mathbf{g}}_k'\ \mathbf{X}'\mathbf{y}$ is the kth element of $\hat{\boldsymbol{\beta}}$.

8.26

$$P\left[-t_{\alpha/2,n-k-1} \leq \frac{\mathbf{a}'\hat{\boldsymbol{\beta}} - \mathbf{a}'\boldsymbol{\beta}}{s\sqrt{\mathbf{a}'(\mathbf{X}'\mathbf{X})^{-1}\mathbf{a}}} \geq t_{\alpha/2,n-k-1}\right] = 1 - \alpha$$

Solve the inequality for $\mathbf{a}'\boldsymbol{\beta}$.

8.27 In the answer to Problem 7.17b, we have

$$(\mathbf{X}'\mathbf{X})^{-1} = \begin{pmatrix} \frac{1}{n} + \bar{\mathbf{x}}_1'(\mathbf{X}_c'\mathbf{X}_c)^{-1}\bar{\mathbf{x}}_1 & -\bar{\mathbf{x}}_1'(\mathbf{X}_c'\mathbf{X}_c)^{-1} \\ -(\mathbf{X}_c'\mathbf{X}_c)^{-1}\bar{\mathbf{x}}_1 & (\mathbf{X}_c'\mathbf{X}_c)^{-1} \end{pmatrix},$$

where $\bar{\mathbf{x}}_1 = (\bar{x}_1, \bar{x}_2, \ldots, \bar{x}_k)'$. Using this form of $(\mathbf{X}'\mathbf{X})^{-1}$, show that

$$\mathbf{X}_0'(\mathbf{X}'\mathbf{X})^{-1}\mathbf{x}_0 = (1,\ \mathbf{x}_{01}')(\mathbf{X}'\mathbf{X})^{-1}\begin{pmatrix} 1 \\ \mathbf{x}_{01} \end{pmatrix}$$

$$= \frac{1}{n} + (\mathbf{x}_{01} - \bar{\mathbf{x}}_1)'(\mathbf{X}_c'\mathbf{X}_c)^{-1}(\mathbf{x}_{01} - \bar{\mathbf{x}}_1).$$

8.28 In this case, $\mathbf{x}_{01} - \bar{\mathbf{x}}_1 = x_0 - \bar{x}$ and

$$\mathbf{X}_c = \begin{pmatrix} x_1 - \bar{x} \\ x_2 - \bar{x} \\ \vdots \\ x_n - \bar{x} \end{pmatrix}.$$

8.29 $E(y_0 - \hat{y}_0) = E(y_0 - \mathbf{x}_0'\hat{\boldsymbol{\beta}}) = \mathbf{x}_0'\boldsymbol{\beta} - \mathbf{x}_0'\boldsymbol{\beta} = 0$. By (8.59), $\text{var}\,(y_0 - \hat{y}_0) = \sigma^2[1 + \mathbf{x}_0'(\mathbf{X}'\mathbf{X})^{-1}\mathbf{x}_0]$. Therefore, $(y_0 - \hat{y}_0)/\sqrt{\sigma^2[1 + \mathbf{x}_0'(\mathbf{X}'\mathbf{X})^{-1}\mathbf{x}_0]}$ is $N(0,1)$ by Theorems 7.6b(i) and 4.4a(i). By Theorem 7.6b(ii), $(n-k-1)s^2/\sigma^2$ is $\chi^2(n-k-1)$. By Theorem 7.6b(iii), $\hat{y}_0$ and s^2 are independent. Use (5.33) to show that $t = (y_0 - \hat{y}_0)/s\sqrt{1 + \mathbf{x}_0'(\mathbf{X}'\mathbf{X})^{-1}\mathbf{x}_0}$ is distributed as $t(n-k-1)$.

8.30 **(a)** Show that $E(\bar{y}_0 - \hat{y}_0) = E(\bar{y}_0 - \mathbf{x}_0'\hat{\boldsymbol{\beta}}) = 0$ and that $\text{var}(\bar{y}_0 - \hat{y}_0) = \sigma^2[1/q + \mathbf{x}_0'(\mathbf{X}'\mathbf{X})^{-1}\mathbf{x}_0]$. For the remaining steps, follow the answer to Problem 8.29.

8.31 Invert (take the reciprocal of) all three numbers of the inequality (which changes the directions of the two inequalities) and multiply by $(n-k-1)s^2$.

8.32 Let $\mathbf{y}_0 = (y_{01}, y_{02}, \ldots, y_{0d})'$ be the vector of d future observations, and let

$$\mathbf{X}_d = \begin{pmatrix} \mathbf{x}_{01}' \\ \vdots \\ \mathbf{x}_{0d}' \end{pmatrix}$$

be the $d \times (k+1)$ matrix of corresponding values $\mathbf{x}_{01}, \mathbf{x}_{02}, \ldots, \mathbf{x}_{0d}$. Show that $\mathbf{y}_0 - \mathbf{X}_d\hat{\boldsymbol{\beta}}$ is $N_d(\mathbf{0}, \sigma^2\mathbf{V}_d)$, where $\mathbf{V}_d = \mathbf{I}_d + \mathbf{X}_d(\mathbf{X}'\mathbf{X})^{-1}\mathbf{X}_d'$ and $\mathbf{X}$ is the $\mathbf{X}$ matrix for the original n observations. Show that

$$\frac{(\mathbf{y}_0 - \mathbf{X}_d\hat{\boldsymbol{\beta}})'\mathbf{V}_d^{-1}(\mathbf{y}_0 - \mathbf{X}_d\hat{\boldsymbol{\beta}})}{ds^2} \quad \text{is} \quad F(d, n-k-1)$$

[for the distribution of the numerator, see (5.27) or Problem 5.12e]. By Theorem 8.5 and (8.71) with $k+1 = d$, we have the simultaneous intervals

$$-s\sqrt{d\mathbf{a}'\mathbf{V}_d^{-1}\mathbf{a}F_{\alpha,d,n-k-1}} \leq \mathbf{a}'(\mathbf{y}_0 - \mathbf{X}_d\hat{\boldsymbol{\beta}}) \leq s\sqrt{d\mathbf{a}'\mathbf{V}_d^{-1}\mathbf{a}F_{\alpha,d,n-k-1}},$$

which hold for all $\mathbf{a}$ with confidence coefficient $1-\alpha$. Setting $\mathbf{a}_1' = (1, 0, \ldots, 0), \ldots, \mathbf{a}_d' = (0, \ldots, 0, 1)$, we obtain

$$\mathbf{x}_{0i}'\hat{\boldsymbol{\beta}} - s\sqrt{d[1 + \mathbf{x}_{0i}'(\mathbf{X}'\mathbf{X})^{-1}\mathbf{x}_{0i}]F_{\alpha,d,n-k-1}}$$
$$\leq y_{0i} \leq \mathbf{x}_{0i}'\hat{\boldsymbol{\beta}} + s\sqrt{d[1 + \mathbf{x}_{0i}'(\mathbf{X}'\mathbf{X})^{-1}\mathbf{x}_{0i}]F_{\alpha,d,n-k-1}}.$$

These intervals hold with confidence coefficient at least $1-\alpha$.

8.33 For (8.77), we have

$$\begin{aligned}
\frac{\partial \ln L(\mathbf{0}, \sigma^2)}{\partial \sigma^2} &= \frac{\partial}{\partial \sigma^2} \ln\left[\frac{1}{(2\pi\sigma^2)^{n/2}} e^{-\mathbf{y}'\mathbf{y}/2\sigma^2}\right] \\
&= \frac{\partial}{\partial \sigma^2}\left[-\frac{n}{2}\ln 2\pi - \frac{n}{2}\ln \sigma^2 - \mathbf{y}'\mathbf{y}/2\sigma^2\right] \\
&= -0 - \frac{n}{2\sigma^2} + \frac{\mathbf{y}'\mathbf{y}}{2(\sigma^2)^2} = 0,
\end{aligned}$$

$$\hat{\sigma}_0^2 = \frac{\mathbf{y}'\mathbf{y}}{n}.$$

For (8.78), we have

$$\begin{aligned}\max_{H_o} L(\beta,\sigma^2) = \max L(\mathbf{0},\sigma^2) &= L(\mathbf{0},\hat{\sigma}_0^2) \\ &= \frac{1}{(2\pi\hat{\sigma}_0^2)^{n/2}} e^{-\mathbf{y}'\mathbf{y}/2\hat{\sigma}_0^2} \\ &= \frac{1}{(2\pi)^{n/2}(\mathbf{y}'\mathbf{y}/n)^{n/2}} e^{-\mathbf{y}'\mathbf{y}/2(\mathbf{y}'\mathbf{y}/n)} \\ &= \frac{n^{n/2}e^{-n/2}}{(2\pi)^{n/2}(\mathbf{y}'\mathbf{y})^{n/2}}\end{aligned}$$

For (8.79), we have

$$\begin{aligned}\left[\frac{(\mathbf{y}-\mathbf{X}\hat{\boldsymbol{\beta}})'(\mathbf{y}-\mathbf{X}\hat{\boldsymbol{\beta}})}{\mathbf{y}'\mathbf{y}}\right]^{n/2} &= \left[\frac{\mathbf{y}'\mathbf{y}-\hat{\beta}'\mathbf{X}'\mathbf{y}}{\mathbf{y}'\mathbf{y}-\hat{\boldsymbol{\beta}}'\mathbf{X}'\mathbf{y}+\hat{\beta}'\mathbf{X}'\mathbf{y}}\right]^{n/2} \\ &= \left[\frac{1}{1+\hat{\boldsymbol{\beta}}'\mathbf{X}'\mathbf{y}/\mathbf{y}'\mathbf{y}-\hat{\boldsymbol{\beta}}'\mathbf{X}'\mathbf{y}}\right]^{n/2}.\end{aligned}$$

8.34 Expanding $(\mathbf{y}-\mathbf{X}\boldsymbol{\beta})'(\mathbf{y}-\mathbf{X}\boldsymbol{\beta})$, we have $v = -(n/2)\ln(2\pi) - (n/2)\ln\sigma^2 - [\mathbf{y}'\mathbf{y} - 2\mathbf{y}'\mathbf{X}\boldsymbol{\beta} + \boldsymbol{\beta}'\mathbf{X}'\mathbf{X}\boldsymbol{\beta}]/2\sigma^2 + \boldsymbol{\lambda}'\mathbf{C}\boldsymbol{\beta}$. Differentiation with respect to $\boldsymbol{\beta}$ gives the result.

8.35 From (8.80), we obtain

$$\hat{\boldsymbol{\beta}}_0 = (\mathbf{X}'\mathbf{X})^{-1}\mathbf{X}'\mathbf{y} + \hat{\sigma}_0^2(\mathbf{X}'\mathbf{X})^{-1}\mathbf{C}'\boldsymbol{\lambda}. \qquad (1)$$

Multiplying $\hat{\boldsymbol{\beta}}_0$ by $\mathbf{C}$ gives $\mathbf{C}\hat{\boldsymbol{\beta}}_0 = \mathbf{C}\hat{\boldsymbol{\beta}} + \hat{\sigma}_0^2\mathbf{C}(\mathbf{X}'\mathbf{X})^{-1}\mathbf{C}'\boldsymbol{\lambda}$. By (8.77), $\mathbf{C}\hat{\boldsymbol{\beta}}_0 = \mathbf{0}$, and we have

$$\boldsymbol{\lambda} = -[\mathbf{C}(\mathbf{X}'\mathbf{X})^{-1}\mathbf{C}']^{-1}\frac{\mathbf{C}\hat{\boldsymbol{\beta}}}{\hat{\sigma}_0^2}.$$

Substituting this into (1) gives

$$\hat{\boldsymbol{\beta}}_0 = \hat{\boldsymbol{\beta}} - (\mathbf{X}'\mathbf{X})^{-1}\mathbf{C}'[\mathbf{C}(\mathbf{X}'\mathbf{X})^{-1}\mathbf{C}']^{-1}\mathbf{C}\hat{\boldsymbol{\beta}},$$

where $\hat{\boldsymbol{\beta}} = (\mathbf{X}'\mathbf{X})^{-1}\mathbf{X}'\mathbf{y}$.

8.36 Substituting (8.83) into (8.84) gives

$$(\mathbf{y}-\mathbf{X}\hat{\boldsymbol{\beta}}_0)'(\mathbf{y}-\mathbf{X}\hat{\boldsymbol{\beta}}_0)=\left\{\mathbf{y}-\mathbf{X}\hat{\boldsymbol{\beta}}+\mathbf{X}(\mathbf{X}'\mathbf{X})^{-1}\mathbf{C}'[\mathbf{C}(\mathbf{X}'\mathbf{X})^{-1}\mathbf{C}']^{-1}\mathbf{C}\hat{\boldsymbol{\beta}}\right\}'$$
$$\times\left\{\mathbf{y}-\mathbf{X}\hat{\boldsymbol{\beta}}+\mathbf{X}(\mathbf{X}'\mathbf{X})^{-1}\mathbf{C}'[\mathbf{C}(\mathbf{X}'\mathbf{X})^{-1}\mathbf{C}']^{-1}\mathbf{C}\hat{\boldsymbol{\beta}}\right\}$$
$$=(\mathbf{y}-\mathbf{X}\hat{\boldsymbol{\beta}})'(\mathbf{y}-\mathbf{X}\hat{\boldsymbol{\beta}})+0+0+(\mathbf{C}\hat{\boldsymbol{\beta}})'[\mathbf{C}(\mathbf{X}'\mathbf{X})^{-1}\mathbf{C}']^{-1}\mathbf{C}\hat{\boldsymbol{\beta}}.$$

Show that the second and third terms vanish and the fourth term is equal to $(\mathbf{C}\hat{\boldsymbol{\beta}})'[\mathbf{C}(\mathbf{X}'\mathbf{X})^{-1}\mathbf{C}']^{-1}\mathbf{C}\hat{\boldsymbol{\beta}}$ as indicated.

8.37 **(a)**

Source	df	SS	MS	F	p Value
Due to β_1	4	2520.2724	630.0681	84.540	7.216×10^{-15}
Error	27	201.2276	7.4529		
Total	31	2721.5000			

(This p value would typically be reported as $p<.0001$). The F value can also be found using (8.23):

$$F=\frac{R^2/k}{(1-R^2)/(n-k-1)}=\frac{.9261/4}{(1-.9261)/27}=84.540.$$

(b) For the reduced model $y_i=\beta_0^*+\beta_2^*x_{i2}+\beta_4^*x_{i4}+\varepsilon_i^*$, we obtain $\hat{\boldsymbol{\beta}}_1^{*\prime}\mathbf{X}_1'\mathbf{y}-n\bar{y}^2=2483.1136$. From the analysis of variance table in part **(a)**, we have $\hat{\boldsymbol{\beta}}'\mathbf{X}'\mathbf{y}-n\bar{y}^2=2520.2724$. The difference is $\hat{\boldsymbol{\beta}}'\mathbf{X}'\mathbf{y}-\hat{\boldsymbol{\beta}}_1^{*\prime}\mathbf{X}_1'\mathbf{y}=37.1588$. By (8.17), we have

$$F=\frac{37.1588/2}{7.4529}=2.4929,$$

with $p=.102$.

(c) The values of $t_j=\hat{\beta}_j/s\sqrt{g_{jj}}$ in (8.39) are given in the following table:

Variable	$\hat{\beta}_j$	$s\sqrt{g_{jj}}$	t_j	p Value
x_1	−0.0286	.0906	−0.316	.755
x_2	0.2158	.0677	3.187	.00362
x_3	−4.3201	2.8510	−1.515	.141
x_4	8.9749	2.7726	3.237	.00319

Comparing each (two-sided) p value to .05, we would reject H_0: $\beta_j=0$ for β_2 and β_4. Comparing each p value to the Bonferroni value of $.05/4=.0125$, we reject H_0 for β_2 and β_4 also.

(d) To test H_0: $\beta_1 = \beta_2 = 12\beta_3 = 12\beta_4$, we write H_0: $\mathbf{C}\boldsymbol{\beta} = \mathbf{0}$ where

$$\mathbf{C} = \begin{pmatrix} 0 & 1 & -1 & 0 & 0 \\ 0 & 0 & 1 & -12 & 0 \\ 0 & 0 & 0 & 1 & -1 \end{pmatrix}.$$

We test H_0 using (8.26). For H_{01}: $\beta_1 = \beta_2$, H_{02}: $\beta_2 = 12\beta_3$, and H_{03}: $\beta_3 = \beta_4$, we test each row of $\mathbf{C}$ separately using (8.37). For H_{04}: $\beta_1 = \beta_2$ and $\beta_3 = \beta_4$, we use the first and third rows of $\mathbf{C}$ and test with (8.26). The results are as follows:

$$H_0 \quad F = \frac{236.3268/3}{201.2276/27} = 10.5698 \quad p = 0.0000899$$

$$H_{01} \quad F = \frac{26.7486}{7.4529} = 3.5890 \quad p = 0.0689$$

$$H_{02} \quad F = \frac{17.2922}{7.4529} = 2.3202 \quad p = 0.139$$

$$H_{03} \quad F = \frac{43.5851}{7.4529} = 5.8481 \quad p = 0.0226$$

$$H_{04} \quad F = \frac{206.2962/2}{7.4529} = 13.8400 \quad p = 0.0000729$$

(e) For $\nu = 27$, we have $t_{.025,27} = 2.0518$ and $t_{.00625,27} = 2.6763$. Using (8.45) and (8.65) and the values in the answer to part (c), we obtain the following lower and upper confidence limits:

$\hat{\beta}_j \pm t_{.025}\, s\sqrt{g_{jj}}$		$\hat{\beta}_j \pm t_{.00625}\, s\sqrt{g_{jj}}$	
−0.2145	0.1573	−0.2711	0.2139
0.0769	0.3548	0.0346	0.3970
−10.1698	1.5297	−11.9500	3.3099
3.2859	14.6639	1.5546	16.3952

8.38 **(a)**

Source	df	SS	MS	F	p Value
Due to β_1	3	13266.8574	4422.2858	65.037	3.112×10^{-13}
Error	30	2039.9062	67.9969		
Total	33	15306.7636			

The F value can also be found using (8.23):

$$F = \frac{R^2}{(1 - R^2)/(n - k - 1)} = \frac{.8667/3}{(1.8667)/30} = 65.037$$

(b) The values of $t_j = \hat{\beta}_j / s\sqrt{g_{jj}}$ in (8.39) are given in the following table:

Variable	$\hat{\beta}_j$	$\sqrt{g_{jj}}$	t_j	p Value
x_1	0.7803	0.0810	9.631	1.09×10^{-10}
x_2	0.5031	0.1251	4.020	0.000361
x_3	−17.1002	13.5954	−1.258	0.218

Comparing each (two-sided) p value to .05, we would reject H_0: $\beta_j = 0$ for β_1 and β_2. Comparing each p value to the Bonferroni value of $.05/3 = .0167$, we reject H_0 for β_1 and β_2 also.

(c) For $\nu = 30$, we have $t_{.025,30} = 2.0423$ and $t_{.00833,30} = 2.5357$. Using (8.47) and (8.67) and the values in the answer to part (b), we obtain the following lower and upper confidence limits:

$\hat{\beta}_j \pm t_{.025}\, s\sqrt{g_{jj}}$		$\hat{\beta}_j \pm t_{.00833}\, s\sqrt{g_{jj}}$	
0.6148	0.9457	0.5748	0.9857
0.2475	0.7587	0.1858	0.8204
−44.8656	10.6652	−51.5745	17.3740

(d) Using (8.52), we have

$$\mathbf{x}_0'\hat{\boldsymbol{\beta}} \pm t_{\alpha/2,n-k-1}s\sqrt{\mathbf{x}_0'(\mathbf{X}'\mathbf{X})^{-1}\mathbf{X}_0}$$

$$18.9103 \pm 2.0423(8.2460)\sqrt{.1615}$$

$$18.9103 \pm 6.7677,$$

$$(12.1426,\ 25.6780)$$

(e) Using (8.61), we have

$$\mathbf{x}_0'\hat{\boldsymbol{\beta}} \pm t_{\alpha/2,n-k-1}s\sqrt{1 + \mathbf{x}_0'(\mathbf{X}'\mathbf{X})^{-1}\mathbf{X}_0}$$

$$18.9103 \pm 2.0423(8.2460)\sqrt{1.1615}$$

$$18.9103 \pm 18.1496,$$

$$(.7609,\ 37.0599)$$

8.39 **(a)** $\mathbf{x}_0'\hat{\boldsymbol{\beta}} \pm t_{.025,15}s\sqrt{\mathbf{x}_0'(\mathbf{X}'\mathbf{X})^{-1}\mathbf{X}_0}$

$$55.2603 \pm (2.1314)(4.0781)\sqrt{.19957}$$

$$55.2603 \pm 3.8849,$$

$$(51.3754,\ 59.1451)$$

(b) $\mathbf{x}_0'\hat{\boldsymbol{\beta}} \pm t_{.025,15}s\sqrt{1 + \mathbf{x}_0'(\mathbf{X}'\mathbf{X})^{-1}\mathbf{X}_0}$

$$55.2603 \pm (2.1314)(4.0781)\sqrt{1.19975}$$
$$55.2603 \pm 9.5205,$$
$$(45.7394, 64.7811)$$

(c) Using $\mathbf{C} = \begin{pmatrix} 0 & 1 & -1 & 0 \\ 0 & 0 & 2 & -1 \end{pmatrix}$, we obtain $\mathbf{C}\hat{\boldsymbol{\beta}} = \begin{pmatrix} .1116 \\ -.4478 \end{pmatrix}$, $\mathbf{C}(\mathbf{X}'\mathbf{X})^{-1}\mathbf{C}' = \begin{pmatrix} .003366 & -.006943 \\ -.006943 & .044974 \end{pmatrix}$, $F = .1577$, $p = .856$.

8.40 **(a)** $\hat{\boldsymbol{\beta}}'\mathbf{X}'\mathbf{y} - n\bar{y}^2 - (\hat{\beta}_1^{*\prime}\mathbf{X}_1'\mathbf{y} - n\hat{y}^2) = 1741.1233 - 1707.1580,$

$$F = \frac{5.6609}{5.1343} = 1.1026, \quad p = .430$$

(b)
$$F = \frac{.9741 - .9551/6}{1 - .9741/9} = 1.1026$$

(c)
$$\begin{aligned}
\beta_0:&\quad 332.1110 \pm 39.8430 \\
&\quad (292.2679, 371.9540), \\
\beta_1:&\quad -1.5460 \pm .21109 \\
&\quad (-1.7571, -1.3349), \\
\beta_2:&\quad -1.4246 \pm .3147 \\
&\quad (1.7393, -1.1098), \\
\beta_3:&\quad -2.2374 \pm .7243 \\
&\quad (2.9617, -1.5130)
\end{aligned}$$

(d)
$$\begin{aligned}
\beta_1:&\quad -1.5460 \pm .2668 \\
&\quad (-1.8127, -1.2792), \\
\beta_2:&\quad -1.4246 \pm .3977 \\
&\quad (-1.8223, -1.0268), \\
\beta_3:&\quad -2.2347 \pm .9154 \\
&\quad (-3.1528, -1.3220)
\end{aligned}$$

(e)
$$20.2547 \pm 2.2024$$
$$(18.0524, 22.4571)$$

(f)
$$20.2547 \pm 5.3975$$
$$(14.8573, 25.6522)$$

Chapter 9

9.1 **(a)** By (9.5), we obtain

$$\begin{aligned} E(\hat{\boldsymbol{\varepsilon}}) &= E[(\mathbf{I}-\mathbf{H})\mathbf{y}] = (\mathbf{I}-\mathbf{H})E(\mathbf{y}) \\ &= [\mathbf{I}-\mathbf{X}(\mathbf{X}'\mathbf{X}')^{-1}\mathbf{X}']\mathbf{X}\boldsymbol{\beta} = \mathbf{X}\boldsymbol{\beta}-\mathbf{X}\boldsymbol{\beta}. \end{aligned}$$

(b) We first note that $\mathbf{I}-\mathbf{H}$ is symmetric and idempotent [see Theorem 2.13e(i)]. Then by Theorem 3.6d(i), we obtain

$$\begin{aligned} \text{cov}(\hat{\boldsymbol{\varepsilon}}) &= \text{cov}[(\mathbf{I}-\mathbf{H})\mathbf{y}] = (\mathbf{I}-\mathbf{H})\sigma^2\mathbf{I}(\mathbf{I}-\mathbf{H})' \\ &= \sigma^2(\mathbf{I}-\mathbf{H})^2 = \sigma^2(\mathbf{I}-\mathbf{H}). \end{aligned}$$

(c) By Theorem 3.6d(ii), we have

$$\begin{aligned} \text{cov}(\hat{\boldsymbol{\varepsilon}}, \boldsymbol{y}) &= \text{cov}[(\mathbf{I}-\mathbf{H})\mathbf{y}, \mathbf{I}\mathbf{y}] \\ &= (\mathbf{I}-\mathbf{H})(\sigma^2\mathbf{I})\mathbf{I} = \sigma^2(\mathbf{I}-\mathbf{H}). \end{aligned}$$

(d)
$$\begin{aligned} \text{cov}(\hat{\boldsymbol{\varepsilon}}, \boldsymbol{y}) &= \text{cov}[(\mathbf{I}-\mathbf{H})\mathbf{y}, \mathbf{H}\mathbf{y}] = (\mathbf{I}-\mathbf{H})(\sigma^2\mathbf{I})\mathbf{H} \\ &= \sigma^2(\mathbf{H}-\mathbf{H}^2) = \sigma^2(\mathbf{H}-\mathbf{H}) \end{aligned}$$

(e) $\bar{\hat{\varepsilon}} = \sum_{i=1}^n \hat{\varepsilon}_i/n = \hat{\boldsymbol{\varepsilon}}'\mathbf{j}/n$. By (9.4) and (9.5), $\hat{\boldsymbol{\varepsilon}}'\mathbf{j} = \mathbf{y}'(\mathbf{I}-\mathbf{H})\mathbf{j} = \mathbf{y}'(\mathbf{j}-\mathbf{j})$.

(f) By (9.5), $\hat{\boldsymbol{\varepsilon}}'\mathbf{y} = \mathbf{y}'(\mathbf{I}-\mathbf{H})\mathbf{y}$.

(g) By (9.2) and (9.5), $\hat{\boldsymbol{\varepsilon}}'\hat{\mathbf{y}} = \mathbf{y}'(\mathbf{I}-\mathbf{H})\mathbf{H}\mathbf{y} = \mathbf{y}'(\mathbf{H}-\mathbf{H}^2)\mathbf{y} = \mathbf{y}'(\mathbf{H}-\mathbf{H})\mathbf{y}$.

(h) By (9.3) and (9.5), $\hat{\boldsymbol{\varepsilon}}'\mathbf{X} = \mathbf{y}'(\mathbf{I}-\mathbf{H})\mathbf{X} = \mathbf{y}'(\mathbf{X}-\mathbf{H}\mathbf{X}) = \mathbf{y}'(\mathbf{X}-\mathbf{X})$.

9.2 **(a)**
$$\frac{d}{dh}(h-h^2) = 1-2h = 0, \quad h = \frac{1}{2}, \quad \frac{1}{2}-\left(\frac{1}{2}\right)^2 = \frac{1}{4}.$$

(b) Let $\mathbf{X}_c = \mathbf{A}$ and $(\mathbf{X}_c'\mathbf{X}_c)^{-1} = \mathbf{B}$. Then

$$\mathbf{ABA}' = \begin{pmatrix} a_1' \\ a_2' \\ \vdots \\ a_n' \end{pmatrix} \mathbf{B}(a_1, a_2, \ldots, a_n)$$

$$= \begin{pmatrix} a_1' \\ a_2' \\ \vdots \\ a_n' \end{pmatrix} (\mathbf{B}\mathbf{a}_1, \mathbf{B}\mathbf{a}_2, \ldots, \mathbf{B}\mathbf{a}_n)$$

$$= \begin{pmatrix} a_1'\mathbf{B}\mathbf{a}_1 & a_1'\mathbf{B}\mathbf{a}_2 & \ldots & a_1'\mathbf{B}\mathbf{a}_n \\ a_2'\mathbf{B}\mathbf{a}_1 & a_2'\mathbf{B}\mathbf{a}_2 & \ldots & a_2'\mathbf{B}\mathbf{a}_n \\ \vdots & \vdots & & \vdots \\ a_n'\mathbf{B}\mathbf{a}_1 & a_n'\mathbf{B}\mathbf{a}_2 & \ldots & a_n'\mathbf{B}\mathbf{a}_n \end{pmatrix}.$$

(c) $\text{tr}(\mathbf{H}) = \text{tr}[\mathbf{X}(\mathbf{X}'\mathbf{X})^{-1}\mathbf{X}'] = \text{tr}[(\mathbf{X}'\mathbf{X})^{-1}\mathbf{X}'\mathbf{X}] = \text{tr}(\mathbf{I}_{k+1}) = k + 1.$

9.3 By Theorem 9.2(iii), $h_{ii} = (1/n) + (\mathbf{x}_{1i} - \bar{\mathbf{x}}_1)'(\mathbf{X}_c'\mathbf{X}_c)^{-1}(\mathbf{x}_{1i} - \bar{\mathbf{x}}_1)$.

By (2.101) and (2.104), this can be written as

$$h_{ii} = \frac{1}{n} + (\mathbf{x}_{1i} - \bar{\mathbf{x}}_1)'\left(\sum_{r=1}^{k} \frac{1}{\lambda_r}\mathbf{a}_r\mathbf{a}_r'\right)(\mathbf{x}_{1i} - \bar{\mathbf{x}}_1)$$

$$= \sum_{r=1}^{k} \frac{1}{\lambda_r}[(\mathbf{x}_{1i} - \bar{\mathbf{x}}_1)'\mathbf{a}_r][\mathbf{a}_r'(\mathbf{x}_{1i} - \bar{\mathbf{x}}_1)]$$

$$= \sum_{r} \frac{1}{\lambda_r}[(\mathbf{x}_{1i} - \bar{\mathbf{x}}_1)'\mathbf{a}_r]^2,$$

where λ_r is the rth eigenvalue of $\mathbf{X}_c'\mathbf{X}_c$ and $\mathbf{a}_r$ is the corresponding (normalized) eigenvector of $\mathbf{X}_c'\mathbf{X}_c$. By (2.81), the consine of the angle θ_{ir} between $\mathbf{x}_{1i} - \bar{\mathbf{x}}_1$

and $\mathbf{a}_r$ is

$$\cos\theta_{ir} = \frac{(\mathbf{x}_{1i} - \bar{\mathbf{x}}_1)'\mathbf{a}_r}{\sqrt{[(\mathbf{x}_{1i} - \bar{\mathbf{x}}_1)'(\mathbf{x}_{1i} - \bar{\mathbf{x}}_1)](\mathbf{a}_r'\mathbf{a}_r)}} = \frac{(\mathbf{x}_{1i} - \bar{\mathbf{x}}_1)'\mathbf{a}_r}{\sqrt{(\mathbf{x}_{1i} - \bar{\mathbf{x}}_1)'(\mathbf{x}_{1i} - \bar{\mathbf{x}}_1)}}$$

since $\mathbf{a}_r'\mathbf{a}_r = 1$. Thus, if we multiply and divide by $(\mathbf{x}_{1i} - \bar{\mathbf{x}}_1)'(\mathbf{x}_{1i} - \bar{\mathbf{x}}_1)$, we can express h_{ii} as

$$\begin{aligned} h_{ii} &= \frac{1}{n} + (\mathbf{x}_{1i} - \bar{\mathbf{x}}_1)'(\mathbf{x}_{1i} - \bar{\mathbf{x}}_1) \sum_{r=1}^{k} \frac{1}{\lambda_r} \frac{[(\mathbf{x}_{1i} - \bar{\mathbf{x}}_1)'\mathbf{a}_r]^2}{(\mathbf{x}_{1i} - \bar{\mathbf{x}}_1)'(\mathbf{x}_{1i} - \bar{\mathbf{x}}_1)} \\ &= \frac{1}{n} + (\mathbf{x}_{1i} - \bar{\mathbf{x}}_1)'(\mathbf{x}_{1i} - \bar{\mathbf{x}}_1) \sum_{r} \frac{1}{\lambda_r} \cos^2\theta_{ir}. \end{aligned}$$

9.4 **(a)** Using (2.51), we obtain

$$\mathbf{H}^* = (\mathbf{X},\mathbf{y}) \begin{pmatrix} \mathbf{X}'\mathbf{X} + \frac{(\mathbf{X}'\mathbf{X})^{-1}\mathbf{X}'\mathbf{y}\mathbf{y}'\mathbf{X}(\mathbf{X}'\mathbf{X})^{-1}}{b} & \frac{-(\mathbf{X}'\mathbf{X})^{-1}\mathbf{X}'\mathbf{y}}{b} \\ \frac{-\mathbf{y}'\mathbf{X}(\mathbf{X}'\mathbf{X})^{-1}}{b} & \frac{1}{b} \end{pmatrix} \begin{pmatrix} \mathbf{X}' \\ \mathbf{y}' \end{pmatrix},$$

where $b = \mathbf{y}'\mathbf{y} - \mathbf{y}'\mathbf{X}(\mathbf{X}'\mathbf{X})^{-1}\mathbf{X}'\mathbf{y}$. Show that $b = \mathbf{y}'(\mathbf{I} - \mathbf{H})\mathbf{y} = \hat{\boldsymbol{\varepsilon}}'\hat{\boldsymbol{\varepsilon}}$. Show that

$$\begin{aligned} \mathbf{H}^* &= \mathbf{X}(\mathbf{X}'\mathbf{X})^{-1}\mathbf{X}' + \frac{\mathbf{X}(\mathbf{X}'\mathbf{X})^{-1}\mathbf{X}'\mathbf{y}\mathbf{y}'\mathbf{X}(\mathbf{X}'\mathbf{X})^{-1}\mathbf{X}'}{b} \\ &\quad - \frac{\mathbf{y}\mathbf{y}'\mathbf{X}(\mathbf{X}'\mathbf{X})^{-1}\mathbf{X}'}{b} - \frac{\mathbf{X}(\mathbf{X}'\mathbf{X})^{-1}\mathbf{X}'\mathbf{y}\mathbf{y}'}{b} + \frac{\mathbf{y}\mathbf{y}'}{b} \\ &= \mathbf{H} + \frac{1}{b}(\mathbf{H}\mathbf{y}\mathbf{y}'\mathbf{H} - \mathbf{y}\mathbf{y}'\mathbf{H} - \mathbf{H}\mathbf{y}\mathbf{y}' + \mathbf{y}\mathbf{y}'). \end{aligned}$$

(b)

$$\begin{aligned} \mathbf{H}^* &= \mathbf{H} + \frac{1}{b}[(\mathbf{H}\mathbf{y}\mathbf{y}' - \mathbf{y}\mathbf{y}')\mathbf{H} + \mathbf{y}\mathbf{y}' - \mathbf{H}\mathbf{y}\mathbf{y}'] \\ &= \mathbf{H} + \frac{1}{b}[(\mathbf{y}\mathbf{y}' - \mathbf{H}\mathbf{y}\mathbf{y}')(\mathbf{I} - \mathbf{H})] \\ &= \mathbf{H} + \frac{1}{b}[(\mathbf{I} - \mathbf{H})\mathbf{y}\mathbf{y}'(\mathbf{I} - \mathbf{H})] \\ &= \mathbf{H} + \frac{\hat{\boldsymbol{\varepsilon}}\hat{\boldsymbol{\varepsilon}}'}{\boldsymbol{\varepsilon}'\boldsymbol{\varepsilon}} \quad \text{[by (9.5)]}. \end{aligned}$$

By (2.21), the diagonal elements of $\hat{\boldsymbol{\varepsilon}}\hat{\boldsymbol{\varepsilon}}'$ are $\hat{\varepsilon}_1^2, \hat{\varepsilon}_2^2, \ldots, \hat{\varepsilon}_n^2$. Therefore, $h_{ii}^* = h_{ii} + \hat{\varepsilon}_i^2/\hat{\boldsymbol{\varepsilon}}'\hat{\boldsymbol{\varepsilon}}$.

(c) Since $\mathbf{H}^*$ is a hat matrix, we have by Theorem 9.2(i), $\frac{1}{n} \leq h_{ii}^* \leq 1$. Therefore, $(1/n) \leq h_{ii} + \hat{\varepsilon}_i^2/\hat{\boldsymbol{\varepsilon}}'\hat{\boldsymbol{\varepsilon}} \leq 1$.

9.5 (a)

$$\mathbf{X}'\mathbf{X} = \begin{pmatrix} \mathbf{x}_1' \\ \mathbf{x}_2' \\ \vdots \\ \mathbf{x}_n' \end{pmatrix}' \begin{pmatrix} \mathbf{x}_1' \\ \mathbf{x}_2' \\ \vdots \\ \mathbf{x}_n' \end{pmatrix} = (\mathbf{x}_1, \mathbf{x}_2, \ldots, \mathbf{x}_n) \begin{pmatrix} \mathbf{x}_1' \\ \mathbf{x}_2' \\ \vdots \\ \mathbf{x}_n' \end{pmatrix}$$

$$= \sum_{j=1}^{n} \mathbf{x}_j\mathbf{x}_j' = \sum_{j \neq i} \mathbf{x}_j\mathbf{x}_j' + \mathbf{x}_i\mathbf{x}_i' = \mathbf{X}_{(i)}'\mathbf{X}_{(i)} + \mathbf{x}_i\mathbf{x}_i'$$

$$\mathbf{X}'\mathbf{y} = (\mathbf{x}_1, \mathbf{x}_2, \ldots, \mathbf{x}_n) \begin{pmatrix} y_1 \\ y_2 \\ \vdots \\ y_n \end{pmatrix} = \sum_{j=1}^{n} \mathbf{x}_j y_j$$

$$= \sum_{j \neq 1} \mathbf{x}_j y_j + \mathbf{x}_i y_i = \mathbf{X}_{(i)}'\mathbf{y}_{(i)} + \mathbf{x}_i y_i.$$

(b)

$$\hat{\boldsymbol{\beta}} = (\mathbf{X}'\mathbf{X})^{-1}\mathbf{X}'\mathbf{y} = (\mathbf{X}'\mathbf{X})^{-1}(\mathbf{X}_{(i)}'\mathbf{y}_{(i)} + \mathbf{x}_i y_i)$$
$$= (\mathbf{X}'\mathbf{X})^{-1}\mathbf{X}_{(i)}'\mathbf{y}_{(i)} + (\mathbf{X}'\mathbf{X})^{-1}\mathbf{x}_i y_i.$$

(c) From $\mathbf{H} = \mathbf{X}(\mathbf{X}'\mathbf{X})^{-1}\mathbf{X}'$, we have $h_{ii} = \mathbf{x}_i'(\mathbf{X}'\mathbf{X})^{-1}\mathbf{x}_i$, where $\mathbf{x}_i'$ is the ith row of $\mathbf{X}$. Then using the result of part (a) and the inverse in the statement of the problem, we obtain

$$\hat{\boldsymbol{\beta}}_{(i)} = (\mathbf{X}_{(i)}'\mathbf{X}_{(i)})^{-1}\mathbf{X}_{(i)}'\mathbf{y}_{(i)}$$
$$= (\mathbf{X}'\mathbf{X} - \mathbf{x}_i'\mathbf{x}_i)^{-1}\mathbf{X}_{(i)}'\mathbf{y}_{(i)}$$
$$= \left[(\mathbf{X}'\mathbf{X})^{-1} + \frac{(\mathbf{X}'\mathbf{X})^{-1}\mathbf{x}_i\mathbf{x}_i'(\mathbf{X}'\mathbf{X})^{-1}}{1 - \mathbf{x}_i'(\mathbf{X}'\mathbf{X})^{-1}\mathbf{x}_i}\right]\mathbf{X}_{(i)}'\mathbf{y}_{(i)}$$
$$= \left[(\mathbf{X}'\mathbf{X})^{-1} + \frac{(\mathbf{X}'\mathbf{X})^{-1}\mathbf{x}_i\mathbf{x}_i'(\mathbf{X}'\mathbf{X})^{-1}}{1 - h_{ii}}\right]\mathbf{X}_{(i)}'\mathbf{y}_{(i)}.$$

(d) From parts (b) and (c), we have

$$\hat{\boldsymbol{\beta}}_{(i)} = (\mathbf{X}'\mathbf{X})^{-1}\mathbf{X}'_{(i)}\mathbf{y}_{(i)} + \frac{(\mathbf{X}'\mathbf{X})^{-1}\mathbf{x}_i\mathbf{x}_i'(\mathbf{X}'\mathbf{X})^{-1}\mathbf{X}'_{(i)}\mathbf{y}_{(i)}}{1-h_{ii}}$$

$$= \hat{\boldsymbol{\beta}} - (\mathbf{X}'\mathbf{X})^{-1}\mathbf{x}_i y_i + \frac{(\mathbf{X}'\mathbf{X})^{-1}\mathbf{x}_i\mathbf{x}_i'[\hat{\boldsymbol{\beta}} - (\mathbf{X}'\mathbf{X})^{-1}\mathbf{x}_i y]}{1-h_{ii}}.$$

With $\mathbf{x}_i'\hat{\boldsymbol{\beta}} = \hat{y}_i$ and $\mathbf{x}_i'(\mathbf{X}'\mathbf{X})^{-1}\mathbf{x}_i = h_{ii}$, we have

$$\hat{\boldsymbol{\beta}}_{(i)} - \hat{\boldsymbol{\beta}} = \frac{-(\mathbf{X}'\mathbf{X})^{-1}\mathbf{x}_i y_i}{1-h_{ii}} + \frac{(\mathbf{X}'\mathbf{X})^{-1}\mathbf{x}_i\hat{y}_i}{1-h_{ii}}$$

$$= \frac{\hat{y}_i - y_i}{1-h_{ii}}(\mathbf{X}'\mathbf{X})^{-1}\mathbf{x}_i = -\frac{\hat{\varepsilon}_i}{1-h_{ii}}(\mathbf{X}'\mathbf{X})^{-1}\mathbf{x}_i.$$

9.6 By (9.27) and (9.29), we obtain

$$\hat{\varepsilon}_{(i)} = y_i - \mathbf{x}_i'\hat{\boldsymbol{\beta}}_{(i)} = y_i - \mathbf{x}_i'\left[\hat{\boldsymbol{\beta}} - \frac{\hat{\varepsilon}_i}{1-h_{ii}}(\mathbf{X}'\mathbf{X})^{-1}\mathbf{x}_i\right]$$

$$= y_i - \mathbf{x}_i'\hat{\boldsymbol{\beta}} + \frac{\hat{\varepsilon}_i}{1-h_{ii}}\mathbf{x}_i'(\mathbf{X}'\mathbf{X})^{-1}\mathbf{x}_i$$

$$= y_i - \hat{y}_i + \frac{\hat{\varepsilon}_i h_{ii}}{1-h_{ii}}$$

$$= \hat{\varepsilon}_i + \frac{\hat{\varepsilon}_i h_{ii}}{1-h_{ii}} = \frac{\hat{\varepsilon}_i}{1-h_{ii}}.$$

9.7 **(a)** Assuming that $\mathbf{X}$ is fixed (constant), we have

$$\operatorname{var}(\hat{\varepsilon}_{(i)}) = \operatorname{var}\left(\frac{\hat{\varepsilon}_i}{1-h_{ii}}\right) = \frac{1}{(1-h_{ii})^2}\operatorname{var}(\hat{\varepsilon}_i) = \frac{\sigma^2(1-h_{ii})}{(1-h_{ii})^2}.$$

9.8 **(a)** $\mathbf{y}'\mathbf{y} = \sum_{j=1}^n y_j^2 = \sum_{j\neq i} y_j^2 + y_i^2 = \mathbf{y}'_{(i)}\mathbf{y}_{(i)} + y_i^2.$

(b)

$$\mathbf{y}'_{(i)}\mathbf{X}_{(i)}\hat{\boldsymbol{\beta}}_{(i)} = \mathbf{y}'\mathbf{X}\hat{\boldsymbol{\beta}} - y_i\mathbf{x}_i'\hat{\boldsymbol{\beta}} - \frac{\hat{\varepsilon}_i}{1-h_{ii}}\mathbf{y}'\mathbf{X}(\mathbf{X}'\mathbf{X})^{-1}\mathbf{x}_i + \frac{\hat{\varepsilon}_i}{1-h_{ii}}y_i\frac{\mathbf{x}_i'(\mathbf{X}'\mathbf{X})^{-1}\mathbf{x}_i}{1-h_{ii}}$$

$$= \mathbf{y}'\mathbf{X}\hat{\boldsymbol{\beta}} - y_i\hat{y}_i - \frac{\hat{\varepsilon}_i}{1-h_{ii}}\hat{\boldsymbol{\beta}}'\mathbf{x}_i + \frac{\hat{\varepsilon}_i}{1-h_{ii}}y_i h_{ii}$$

$$= \mathbf{y}'\mathbf{X}\hat{\boldsymbol{\beta}} - y_i\hat{y}_i - \frac{\hat{\varepsilon}_i\hat{y}_i}{1-h_{ii}} + \frac{\hat{\varepsilon}_i}{1-h_{ii}}y_i h_{ii}.$$

Substituting $\hat{y}_i = y_i - \hat{\varepsilon}_i$, this becomes

$$\begin{aligned}\mathbf{y}'_{(i)}\mathbf{X}_{(i)}\hat{\boldsymbol{\beta}}_{(i)} &= \mathbf{y}'\mathbf{X}\hat{\boldsymbol{\beta}} + \frac{-y_i(y_i - \hat{\varepsilon}_i)(1 - h_{ii}) - \hat{\varepsilon}_i(y_i - \hat{\varepsilon}_i) + \hat{\varepsilon}_i y_i h_{ii}}{1 - h_{ii}} \\ &= \mathbf{y}'\mathbf{X}\hat{\boldsymbol{\beta}} + \frac{-(1 - h_{ii})y_i^2 + \hat{\varepsilon}_i^2}{1 - h_{ii}}.\end{aligned}$$

(c)

$$\begin{aligned}\mathrm{SSE}_{(i)} &= \mathbf{y}'\mathbf{y} - y_i^2 - \left(\mathbf{y}'\mathbf{X}\hat{\boldsymbol{\beta}} - y_i^2 + \frac{\hat{\varepsilon}_i^2}{1 - h_{ii}}\right) \\ &= \mathbf{y}'\mathbf{y} - \mathbf{y}'\mathbf{X}\hat{\boldsymbol{\beta}} - \frac{\hat{\varepsilon}_i^2}{1 - h_{ii}}.\end{aligned}$$

9.9 Substituting (9.29) into (9.35) gives

$$\begin{aligned}D_i &= \frac{\hat{\varepsilon}_i^2}{(1 - h_{ii})^2} \frac{\mathbf{x}'_i(\mathbf{X}'\mathbf{X})^{-1}\mathbf{X}'\mathbf{X}(\mathbf{X}'\mathbf{X})^{-1}\mathbf{x}_i}{(k + 1)s^2} \\ &= \frac{\hat{\varepsilon}_i^2}{(1 - h_{ii})^2} \frac{h_{ii}}{(k + 1)s^2}.\end{aligned}$$

By (9.25), this becomes

$$D_i = \frac{r_i^2}{k + 1} \frac{h_{ii}}{1 - h_{ii}}.$$

9.10 Residuals and Influence Measures for the Gas Vapor Data in Table 7.3[a]

Observations	y_i	$\hat{y}_i$	$\hat{\varepsilon}_i$	h_{ii}	r_i	t_i	D_i
1	29	27.86	1.139	.197	0.466	0.459	.011
2	24	23.76	0.236	.219	0.098	0.096	.001
3	26	25.88	0.120	.179	0.049	0.048	.000
4	22	23.96	−1.961	.289	−0.852	−0.848	.059
5	27	28.42	−1.419	.128	−0.557	−0.550	.009
6	21	21.67	−0.672	.121	−0.262	−0.258	.002
7	33	31.78	1.222	.053	0.460	0.453	.002
8	34	34.22	−0.218	.042	−0.082	−0.080	.000
9	32	31.98	0.017	.055	0.006	0.006	.000
10	34	33.33	0.666	.039	0.249	0.244	.000
11	20	21.54	−1.544	.124	−0.604	−0.597	.010
12	36	32.15	3.846	.040	1.438	1.468	.017
13	34	33.73	0.271	.072	0.103	0.101	.000
14	23	23.98	−0.982	.191	−0.400	−0.394	.008
15	24	19.71	4.287	.418	2.058	2.200	.609
16	32	32.84	−0.841	.060	−0.318	−0.312	.001

Continued

Observations	y_i	$\hat{y}_i$	$\hat{\varepsilon}_i$	h_{ii}	r_i	t_i	D_i
17	40	40.76	−0.762	.285	−0.330	−0.324	.009
18	46	44.39	1.614	.493	0.831	0.826	.134
19	55	52.92	2.083	.243	0.877	0.873	.049
20	52	52.02	−0.018	.224	−0.007	−0.007	.000
21	29	32.38	−3.377	.177	−1.364	1.387	.080
22	22	23.15	−1.155	.169	−0.464	−0.457	.009
23	31	36.59	−5.586	.227	−2.328	−2.555	.319
24	45	47.91	−2.909	.185	−1.180	−1.190	.063
25	37	32.61	4.391	.087	1.683	1.746	.054
26	37	31.89	5.106	.109	1.981	2.103	.096
27	33	30.22	2.775	.124	1.086	1.090	.033
28	27	31.59	−4.593	.102	−1.775	−1.854	.071
29	34	34.40	−0.399	.068	−0.151	−0.149	.000
30	19	19.32	−0.324	.091	−0.124	−0.122	.000
31	16	19.62	−3.623	.102	−1.400	−1.427	.044
32	22	19.39	2.607	.086	0.999	0.999	.019

[a]PRESS=310.443, SSE=201.228.

9.11 Residuals and Influence Measures for the Land Rent data of Table 7.5[a]

Observations	y_i	$\hat{y}_i$	$\hat{\varepsilon}_i$	h_{ii}	r_i	t_i	D_i
1	18.38	17.332	1.048	0.080	0.132	0.130	.000
2	20.00	23.948	−3.948	0.062	−0.494	−0.488	.004
3	11.50	13.855	−2.355	0.141	−0.308	−0.303	.004
4	25.00	26.242	−1.242	0.070	−0.156	−0.154	.000
5	52.50	68.180	−15.680	0.186	−2.107	−2.245	.253
6	82.50	66.431	16.069	0.083	2.035	2.156	.094
7	25.00	32.920	−7.920	0.067	−0.994	−0.994	.018
8	30.67	32.642	−1.792	0.068	−0.248	−0.244	.001
9	12.00	7.715	4.285	0.187	0.576	0.570	.019
10	61.25	57.481	3.769	0.103	0.483	0.476	.007
11	60.00	50.208	9.792	0.058	1.224	1.234	.023
12	57.50	68.846	−11.346	0.100	−1.451	−1.479	.059
13	31.00	31.768	−0.768	0.076	−0.097	−0.095	.000
14	60.00	61.864	−1.864	0.067	−0.234	−0.230	.001
15	72.50	66.773	5.727	0.109	0.736	0.730	.017
16	60.33	66.702	−6.372	0.168	−0.847	−0.843	.036
17	49.75	59.663	−9.913	0.114	−1.278	−1.292	.053
18	8.50	10.790	−2.290	0.192	−0.309	−0.304	.006
19	36.50	24.643	11.857	0.068	1.489	1.522	.040
20	60.00	65.606	−5.606	0.181	−0.751	−0.746	.031
21	16.25	18.016	−1.766	0.505	−0.304	−0.300	.024
22	50.00	47.424	2.576	0.035	0.318	0.313	.001
23	11.50	19.366	−4.866	0.118	−0.628	−0.622	.013

Continued

Observations	y_i	$\hat{y}_i$	$\hat{\varepsilon}_i$	h_{ii}	r_i	t_i	D_i
24	35.00	38.577	−3.577	0.064	−0.448	−0.442	.003
25	75.00	61.694	13.306	0.063	1.667	1.702	.047
26	31.56	35.257	−3.697	0.035	−0.456	−0.450	.002
27	48.50	24.200	6.300	0.063	0.789	0.784	.010
28	77.50	69.889	7.611	0.242	1.060	1.062	.089
29	21.67	22.063	−0.393	0.060	−0.049	−0.048	.000
30	19.75	21.221	−1.471	0.096	−0.188	−0.185	.001
31	56.00	48.174	7.826	0.051	0.974	0.974	.013
32	25.00	41.300	−16.300	0.217	−2.234	−2.406	.346
33	40.00	26.907	16.093	0.214	1.791	1.864	.219
34	56.67	56.585	0.085	0.060	0.011	0.010	.000

[b]PRESS = 2751.18, SSE = 2039.91.

9.12 Residuals and Influence Measures for the Chemical Data with Dependent Variable y_2^a

Observations	y_i	$\hat{y}_i$	$\hat{\varepsilon}_i$	h_{ii}	r_i	t_i	D_i
1	45.9	49.34	−3.442	0.430	−1.118	−1.128	.235
2	53.3	54.51	−1.211	0.310	−0.358	−0.347	.014
3	57.5	53.46	4.039	0.155	1.078	1.084	.053
4	58.8	56.56	2.238	0.139	0.592	0.578	.014
5	60.6	56.04	4.559	0.129	1.198	1.271	.053
6	58.0	59.14	−1.143	0.140	−0.302	−0.293	.004
7	58.6	57.51	1.094	0.228	0.305	0.296	.007
8	52.4	60.61	−8.208	0.186	−2.231	−2.638	.258
9	56.9	56.30	0.598	0.053	0.151	0.146	.000
10	55.4	60.35	−4.947	0.233	−1.385	−1.433	.146
11	46.9	52.26	−5.356	0.240	−1.507	−1.580	.179
12	57.3	57.77	−0.467	0.164	−0.125	−0.121	.001
13	55.0	54.84	0.163	0.146	0.043	0.042	.000
14	58.9	59.40	−0.503	0.245	−0.142	−0.137	.002
15	50.3	53.20	−2.900	0.250	−0.821	−0.812	.056
16	61.1	58.15	2.950	0.258	0.840	0.831	.061
17	62.9	58.15	4.750	0.258	1.352	1.394	.159
18	60.0	56.41	3.592	0.217	0.996	0.955	.069
19	60.6	56.41	4.192	0.217	1.162	1.177	.094

[c]PRESS = 416.039, SSE = 249.462.

Chapter 10

10.1 Since $(\mathbf{v}_i - \bar{\mathbf{v}})' = (y_i - \bar{y}, x_{i1} - \bar{x}_1, \ldots, x_{ik} - \bar{x}_k)$, the element in the (1, 1) position of $(\mathbf{v}_i - \bar{\mathbf{v}})(\mathbf{v}_i - \bar{\mathbf{v}})'$ is $(y_i - \bar{y})^2$. When this is summed over i as in (10.13), we have $\sum_{i=1}^{n} (y_i - \bar{y})^2 = (n-1)s_{yy}$ as in (10.14). Similarly, the (1, 2) element of $(\mathbf{v}_i - \bar{\mathbf{v}})(\mathbf{v}_i - \bar{\mathbf{v}})'$ is $(y_i - \bar{y})(x_{i1} - \bar{x}_1)$, which sums to $(n-1)s_{y1}$, and the (2, 3) element of $(\mathbf{v}_i - \bar{\mathbf{v}})(\mathbf{v}_i - \bar{\mathbf{v}})'$ is $(x_{i1} - \bar{x}_1)(x_{i2} - \bar{x}_2)$, which sums to $(n-1)s_{12}$.

10.2 By a note following Theorem 7.6b, $\hat{\boldsymbol{\mu}}$ and $\mathbf{S}$ are jointly sufficient for $\boldsymbol{\mu}$ and $\boldsymbol{\Sigma}$, if the likelihood function (joint density) in (10.11) factors as $L(\boldsymbol{\mu}, \Sigma) = g(\hat{\mu}, \mathbf{S}, \mu, \Sigma)h(\mathbf{v}_1, \mathbf{v}_2, \ldots, \mathbf{v}_n)$, where $\mathbf{v}_i' = (y_i, \mathbf{x}_i')$, as in the proof of Theorem 10.2a. Noting that a scalar is equal to its trace, we write the exponent in (10.11) in the form

$$\sum_{i=1}^{n}(\mathbf{v}_i - \boldsymbol{\mu})'\boldsymbol{\Sigma}^{-1}(\mathbf{v}_i - \boldsymbol{\mu}) = \sum_{i=1}^{n}\mathrm{tr}(\mathbf{v}_i - \boldsymbol{\mu})'\boldsymbol{\Sigma}^{-1}(\mathbf{v}_i - \boldsymbol{\mu})$$
$$= \mathrm{tr}\left[\boldsymbol{\Sigma}^{-1}\sum_{i=1}^{n}(\mathbf{v}_i - \boldsymbol{\mu})(\mathbf{v}_i - \boldsymbol{\mu})'\right].$$

Adding and subtracting $\bar{\mathbf{v}}$, the sum becomes

$$\sum_{i=1}^{n}(\mathbf{v}_i - \boldsymbol{\mu})(\mathbf{v}_i - \boldsymbol{\mu})' = \sum_{i=1}^{n}(\mathbf{v}_i - \bar{\mathbf{v}} + \bar{\mathbf{v}} - \boldsymbol{\mu})(\mathbf{v}_i - \bar{\mathbf{v}} + \bar{\mathbf{v}} - \boldsymbol{\mu})'$$
$$= \sum_{i=1}^{n}(\mathbf{v}_i - \bar{\mathbf{v}})(\mathbf{v}_i - \bar{\mathbf{v}})' + n(\bar{\mathbf{v}} - \boldsymbol{\mu})(\bar{\mathbf{v}} - \boldsymbol{\mu})'$$
$$= (n-1)\mathbf{S} + n(\bar{\mathbf{v}} - \boldsymbol{\mu})(\bar{\mathbf{v}} - \boldsymbol{\mu})'.$$

Show that the other two terms vanish. Then show that $L(\boldsymbol{\mu}, \boldsymbol{\Sigma})$ can be written as

$$L(\boldsymbol{\mu}, \boldsymbol{\Sigma}) = \frac{1}{(\sqrt{2\pi})^{n(k+1)}|\Sigma|^{n/2}} e^{-[(n-1)\mathrm{tr}(\boldsymbol{\Sigma}^{-1}\mathbf{S}) + n(\bar{\mathbf{v}} - \boldsymbol{\mu})'\boldsymbol{\Sigma}^{-1}(\mathbf{v} - \boldsymbol{\mu})]/2}.$$

10.3

$$\mathrm{DRD} = \begin{pmatrix} s_y & \mathbf{0}' \\ \mathbf{0} & \mathbf{D}_x \end{pmatrix}\begin{pmatrix} 1 & \mathbf{r}_{yx}' \\ \mathbf{r}_{yx} & \mathbf{R}_{xx} \end{pmatrix}\begin{pmatrix} s_y & \mathbf{0}' \\ \mathbf{0} & \mathbf{D}_x \end{pmatrix}$$
$$= \begin{pmatrix} s_y & s_y\mathbf{r}_{yx}' \\ \mathbf{D}_x\mathbf{r}_{yx} & \mathbf{D}_x\mathbf{R}_{xx} \end{pmatrix}\begin{pmatrix} s_y & \mathbf{0}' \\ \mathbf{0} & \mathbf{D}_x \end{pmatrix}$$
$$= \begin{pmatrix} s_y^2 & s_y\mathbf{r}_{yx}'\mathbf{D}_x \\ s_y\mathbf{D}_x\mathbf{r}_{yx} & \mathbf{D}_x\mathbf{R}_{xx}\mathbf{D}_x \end{pmatrix}.$$

10.4 Express y and w in terms of $\begin{pmatrix} y \\ \boldsymbol{x} \end{pmatrix}$ as follows: $\mathrm{y} = (1, 0, \ldots, 0)$ $\begin{pmatrix} y \\ \boldsymbol{x} \end{pmatrix} = \mathbf{a}'\begin{pmatrix} y \\ \boldsymbol{x} \end{pmatrix}$, $w = (0, \sigma_{yx}'\Sigma_{xx}^{-1})\begin{pmatrix} y \\ \boldsymbol{x} \end{pmatrix} +$ constant $= \mathbf{b}'\begin{pmatrix} y \\ \boldsymbol{x} \end{pmatrix} +$ constant. Then use (3.42) and (3.43) with $\boldsymbol{\Sigma}$ partitioned as in (10.3).

10.5 Express w and y as $w = (0, \alpha')\begin{pmatrix} y \\ \boldsymbol{x} \end{pmatrix}$ and $y = (1,0,\ldots,0)\begin{pmatrix} y \\ \boldsymbol{x} \end{pmatrix}$. Then

$$
\begin{aligned}
\text{cov}(y, w) &= \text{cov}\left[(1,0,\ldots,0)\begin{pmatrix} y \\ \boldsymbol{x} \end{pmatrix}, (0, \boldsymbol{\alpha}')\begin{pmatrix} y \\ \boldsymbol{x} \end{pmatrix}\right] \\
&= (1,0,\ldots,0)\begin{pmatrix} \sigma_{yy} & \boldsymbol{\sigma}_{yx}' \\ \boldsymbol{\sigma}_{yx} & \boldsymbol{\Sigma}_{xx} \end{pmatrix}\begin{pmatrix} 0 \\ \boldsymbol{\alpha} \end{pmatrix} \\
&= (\sigma_{yy}, \boldsymbol{\sigma}_{yx}')\begin{pmatrix} 0 \\ \boldsymbol{\alpha} \end{pmatrix} = \boldsymbol{\sigma}_{yx}'\boldsymbol{\alpha},
\end{aligned}
$$

$$
\rho_{yw}^2 = \frac{[\text{cov}(y, w)]^2}{\text{var}(y)\,\text{var}(w)} = \frac{(\boldsymbol{\alpha}'\boldsymbol{\sigma}_{yx})^2}{\sigma_{yy}(\boldsymbol{\alpha}'\boldsymbol{\Sigma}_{xx}\boldsymbol{\alpha})}. \tag{1}
$$

Differentiate ρ_{yw}^2 with respect to $\boldsymbol{\alpha}$ and set the result equal to $\mathbf{0}$ to obtain $\boldsymbol{\alpha} = (\alpha'\boldsymbol{\Sigma}_{xx}\boldsymbol{\alpha}/\boldsymbol{\alpha}'\boldsymbol{\sigma}_{yx})\boldsymbol{\Sigma}_{xx}^{-1}\boldsymbol{\sigma}_{yx}$, with can be substituted into (1) to obtain $\max_\alpha \rho_{yw}^2 = \boldsymbol{\sigma}_{yx}'\boldsymbol{\Sigma}_{xx}^{-1}\boldsymbol{\sigma}_{yx}/\sigma_{yy}$.

10.6 Show that for $\boldsymbol{\Sigma}$ partitioned as in (10.3), (2.75) becomes $|\boldsymbol{\Sigma}| = |\boldsymbol{\Sigma}_{xx}|(\sigma_{yy} - \boldsymbol{\sigma}_{yx}'\boldsymbol{\Sigma}_{xx}^{-1}\boldsymbol{\sigma}_{yx})$. Solve for $\boldsymbol{\sigma}_{yx}'\boldsymbol{\Sigma}_{xx}^{-1}\boldsymbol{\sigma}_{yx}$ and substitute into (10.27).

10.7

$$
\begin{aligned}
\boldsymbol{\sigma}_{uv}' &= \text{cov}(u, \mathbf{v}) = \text{cov}(ay, \mathbf{Bx}) = \text{cov}\left[(a,0,\ldots,0)\begin{pmatrix} y \\ \mathbf{x} \end{pmatrix}, (\mathbf{0}, \mathbf{B})\begin{pmatrix} y \\ \mathbf{x} \end{pmatrix}\right] \\
&= (a,0,\ldots,0)\begin{pmatrix} \sigma_y^2 & \boldsymbol{\sigma}_{yx}' \\ \boldsymbol{\sigma}_{yx} & \boldsymbol{\Sigma}_{xx} \end{pmatrix}\begin{pmatrix} \mathbf{0}' \\ \mathbf{B}' \end{pmatrix} = a\boldsymbol{\sigma}_{yx}'\mathbf{B}', \\
\boldsymbol{\Sigma}_{vv} &= \text{cov}(\mathbf{Bx}) = \mathbf{B}\boldsymbol{\Sigma}_{xx}\mathbf{B}', \\
\sigma_{uu} &= a^2\sigma_{yy}, \\
\rho_{u|v}^2 &= \frac{\boldsymbol{\sigma}_{uv}'\boldsymbol{\Sigma}_{vv}^{-1}\boldsymbol{\sigma}_{uv}}{\sigma_{uu}} = \frac{a\boldsymbol{\sigma}_{yx}'\mathbf{B}'(\mathbf{B}\boldsymbol{\Sigma}_{xx}\mathbf{B}')^{-1}a\mathbf{B}\boldsymbol{\sigma}_{yx}}{a^2\sigma_{yy}} \\
&= \frac{a^2\boldsymbol{\sigma}_{yx}'\mathbf{B}'(\mathbf{B}')^{-1}\boldsymbol{\Sigma}_{xx}^{-1}\mathbf{B}^{-1}\mathbf{B}\boldsymbol{\sigma}_{yx}}{a^2\sigma_{yy}}.
\end{aligned}
$$

10.8

$$\begin{aligned} y - w &= y - \mu_y - \boldsymbol{\sigma}'_{yx}\boldsymbol{\Sigma}_{xx}^{-1}(\mathbf{x} - \boldsymbol{\mu}_x) \\ &= (1, \ -\boldsymbol{\sigma}'_{yx}\boldsymbol{\Sigma}_{xx}^{-1})\begin{pmatrix} y \\ \mathbf{x} \end{pmatrix} + \text{constant} \\ &= \mathbf{a}'\begin{pmatrix} y \\ \mathbf{x} \end{pmatrix} + \text{constant}, \end{aligned}$$

$$\mathbf{x} = (\mathbf{0}, \mathbf{I})\begin{pmatrix} y \\ \mathbf{x} \end{pmatrix} = \mathbf{B}\begin{pmatrix} y \\ \mathbf{x} \end{pmatrix},$$

$$\begin{aligned} \text{cov}(y - w, \mathbf{x}) &= \mathbf{a}'\boldsymbol{\Sigma}\mathbf{B}' = (1, -\boldsymbol{\sigma}'_{yx}\boldsymbol{\Sigma}_{xx}^{-1})\begin{pmatrix} \sigma_y^2 & \boldsymbol{\sigma}'_{yx} \\ \boldsymbol{\sigma}_{yx} & \boldsymbol{\Sigma}_{xx} \end{pmatrix}\begin{pmatrix} \mathbf{0}' \\ \mathbf{I} \end{pmatrix} \\ &= (\sigma_y^2 - \boldsymbol{\sigma}'_{yx}\boldsymbol{\Sigma}_{xx}^{-1}\boldsymbol{\sigma}_{yx}, \boldsymbol{\sigma}'_{yx} - \boldsymbol{\sigma}'_{yx}\boldsymbol{\Sigma}_{xx}^{-1}\boldsymbol{\Sigma}_{xx})\begin{pmatrix} \mathbf{0}' \\ \mathbf{I} \end{pmatrix} \\ &= \mathbf{0}'. \end{aligned}$$

10.9 **(a)** By definition, $r_{y\hat{y}}^2 = [\sum_{i=1}^n (y_i - \bar{y})(\hat{y}_i - \bar{\hat{y}})]^2 / \sum_{i=1}^n (y_i - \bar{y})^2 \sum_{i=1}^n (\hat{y}_i - \bar{\hat{y}})^2$. Show that $\bar{\hat{y}} = \bar{y}$ by using $\mathbf{X}' = \begin{pmatrix} \mathbf{j}' \\ \mathbf{X}_1' \end{pmatrix}$ in $\mathbf{X}'\mathbf{X}\hat{\boldsymbol{\beta}} = \mathbf{X}'\mathbf{y}$ to obtain $\mathbf{j}'\mathbf{X}\hat{\boldsymbol{\beta}} = \mathbf{j}'\mathbf{y}$, from which, $\sum_{i=1}^n \hat{y}_i = \sum_{i=1}^n y_i$. Show that $\sum_{i=1}^n (y_i - \bar{y})(\hat{y}_i - \bar{\hat{y}}) = \sum_i y_i\hat{y}_i - n\bar{y}^2 = \mathbf{y}'\hat{\mathbf{y}} - n\bar{y}^2 = \mathbf{y}'\mathbf{X}\hat{\boldsymbol{\beta}} - n\bar{y}^2$. Show that $\sum_i (\hat{y}_i - \bar{\hat{y}})^2 = \sum_i \hat{y}_i^2 - n\bar{y}^2 = \hat{\mathbf{y}}'\hat{\mathbf{y}} - n\bar{y}^2 = \hat{\boldsymbol{\beta}}\mathbf{X}'\mathbf{X}\hat{\boldsymbol{\beta}} - n\bar{y}^2 = \hat{\boldsymbol{\beta}}'\mathbf{X}'\mathbf{y} - n\bar{y}^2$. Use (7.54).

(b) This follows directly from estimation of (10.25) and the expression following (10.26):

$$r_{y\hat{y}} = \frac{s_{y\hat{y}}}{s_y s_{\hat{y}}} = \frac{\mathbf{s}'_{yx}\mathbf{S}_{xx}^{-1}\mathbf{s}_{yx}}{s_y\sqrt{\mathbf{s}'_{yx}\mathbf{S}_{xx}^{-1}\mathbf{s}_{yx}}}$$

10.10 $r_{y,\mathbf{a}'\mathbf{x}}^2 = (s_{y,\mathbf{a}'\mathbf{x}})^2 / s_y^2 s_{\mathbf{a}'\mathbf{x}}^2$. Express y and $\mathbf{a}'\mathbf{x}$ as $y = (1, 0, \ldots, 0)\begin{pmatrix} y \\ \mathbf{x} \end{pmatrix}$ and $\mathbf{a}'\mathbf{x} = (0, \mathbf{a}')\begin{pmatrix} y \\ \mathbf{x} \end{pmatrix}$. By analogy with (3.40), show that $s_{y,\mathbf{a}'\mathbf{x}} = (1, 0, \ldots, 0)\mathbf{S}\begin{pmatrix} 0 \\ \mathbf{a} \end{pmatrix} = \mathbf{s}'_{yx}\mathbf{a}$, where $\mathbf{S}$ is partitioned as in (10.10). Similarly, by analogy with (3.42) show that $s_{\mathbf{a}'\mathbf{x}}^2 = \mathbf{a}'\mathbf{S}_{xx}\mathbf{a}$. Solve $(\partial/\partial\mathbf{a})(\mathbf{s}'_{yx}\mathbf{a})^2 / s_y^2\mathbf{a}'\mathbf{S}_{xx}\mathbf{a} = \mathbf{0}$ for $\mathbf{a}$ and substitute back into $r_{y,\mathbf{a}'\mathbf{x}}^2$ above to show that $\max_{\mathbf{a}} r_{y,\mathbf{a}'\mathbf{x}}^2 = R^2$.

10.11 Substitute (10.20) and (10.21) into (10.34).

10.12 Adapt (2.51) to obtain

$$\begin{pmatrix} a & \mathbf{b}' \\ \mathbf{b} & \mathbf{C} \end{pmatrix}^{-1} = \begin{pmatrix} 1/d & -\mathbf{b}'\mathbf{C}^{-1}/d \\ -\mathbf{C}^{-1}\mathbf{b}/d & \mathbf{C}^{-1} + \mathbf{C}^{-1}\mathbf{b}\mathbf{b}'\mathbf{C}^{-1}/d \end{pmatrix}, \qquad (1)$$

where $\mathbf{C}$ is symmetric and $d = a - \mathbf{b}'\mathbf{C}^{-1}\mathbf{b}$. Apply (1) to $\mathbf{R}$ partitioned as in (10.18),

$$\mathbf{R} = \begin{pmatrix} 1 & \mathbf{r}'_{yx} \\ \mathbf{r}_{yx} & \mathbf{R}_{xx} \end{pmatrix}.$$

Then

$$\begin{aligned} r^{yy} &= \frac{1}{d} = \frac{1}{a - \mathbf{b}'\mathbf{C}^{-1}\mathbf{b}} \\ &= \frac{1}{1 - \mathbf{r}'_{yx}\mathbf{R}_{xx}^{-1}\mathbf{r}_{yx}} = \frac{1}{1 - R^2}. \end{aligned}$$

10.13 Use (2.71) to show that for $\mathbf{S}$ partitioned as in (10.14), (2.75) can be adapted to the form $|\mathbf{S}| = |\mathbf{S}_{xx}|(s_{yy} - \mathbf{s}'_{yx}\mathbf{S}_{xx}^{-1}\mathbf{s}_{yx})$. Solve for $\mathbf{s}'_{yx}\mathbf{S}_{xx}^{-1}\mathbf{s}_{yx}$ and substitute into (10.34).

10.14 As in Problem 10.4, define $u = ay$ and $\mathbf{v} = \mathbf{B}\mathbf{x}$, so that $\boldsymbol{\sigma}'_{uv} = a\boldsymbol{\sigma}'_{yx}\mathbf{B}'$, $\boldsymbol{\Sigma}_{vv} = \mathbf{B}\boldsymbol{\Sigma}_{xx}\mathbf{B}'$, and $\sigma_{uu} = a^2\sigma_{yy}$. Then by Theorem 10.2c, the maximum likelihood estimators of these are $\mathbf{s}'_{uv} = a\mathbf{s}'_{yx}\mathbf{B}'$, $\mathbf{S}_{vv} = \mathbf{B}\mathbf{S}_{xx}\mathbf{B}'$, and $s_{uu} = a^2 s_{yy}$, respectively. Substitute these into

$$R^2_{u,v} = \frac{\mathbf{s}'_{uv}\mathbf{S}_{vv}^{-1}\mathbf{s}_{uv}}{s_{uu}}.$$

10.15

$$L(\hat{\boldsymbol{\mu}}, \boldsymbol{\Sigma}_0) = \frac{1}{(\sqrt{2\pi})^{n(k+1)}|\boldsymbol{\Sigma}_0|^{n/2}} e^{-\sum_{i=1}^{n}(\mathbf{v}_i - \hat{\boldsymbol{\mu}})'\boldsymbol{\Sigma}_0^{-1}(\mathbf{v}_i - \hat{\boldsymbol{\mu}})/2}.$$

Using $\mathbf{v}_i = \begin{pmatrix} y_i \\ \mathbf{x}_i \end{pmatrix}$, $\hat{\boldsymbol{\mu}}$ in (10.9), amd $\boldsymbol{\Sigma}_0$ in (10.45), show that $L(\hat{\boldsymbol{\mu}}, \boldsymbol{\Sigma}_0)$ becomes

$$\begin{aligned} L(\hat{\boldsymbol{\mu}}, \boldsymbol{\Sigma}_0) &= \frac{1}{(\sqrt{2\pi})^n \sigma_{yy}^{n/2}} e^{-\sum_{i=1}^{n}(y_i - \bar{y})^2/2\sigma_{yy}} \\ &\quad \times \frac{1}{(\sqrt{2\pi})^{kn}|\boldsymbol{\Sigma}_{xx}|^{n/2}} e^{-\sum_{i=1}^{n}(\mathbf{x}_i - \bar{\mathbf{x}})'\boldsymbol{\Sigma}_{xx}^{-1}(\mathbf{x}_i - \bar{\mathbf{x}})/2}. \end{aligned}$$

The first factor is maximized by $\hat{\sigma}_{yy}$ and the second factor by $\hat{\boldsymbol{\Sigma}}_{xx}$. Show that when these are substituted in $L(\hat{\boldsymbol{\mu}}, \boldsymbol{\Sigma}_0)$, the result is given by (10.47).

10.16
$$|\hat{\boldsymbol{\Sigma}}| = \left|\frac{n-1}{n}\mathbf{S}\right| = \left(\frac{n-1}{n}\right)^{k+1}|\mathbf{S}|, \quad |\hat{\boldsymbol{\Sigma}}_{xx}| = \left(\frac{n-1}{n}\right)^{k}|\mathbf{S}_{xx}|.$$

10.17 Multiply by $1/\sqrt{n-3}$, subtract z, multiply by -1 (which reverses the direction of the inequalities), then take tanh (hyperbolic tangent) of all three members.

10.18 **(a)** $\mathbf{V} = \begin{pmatrix} \frac{1}{n_1-3} & 0 & 0 \\ 0 & \frac{1}{n_2-3} & 0 \\ 0 & 0 & \frac{1}{n_3-3} \end{pmatrix}$

(b) Using Theorem 4.4a(ii) and (5.35), $[\mathbf{C}(\mathbf{z}-\boldsymbol{\mu}_z)]'[\mathbf{CVC}']^{-1}[\mathbf{C}(\mathbf{z}-\boldsymbol{\mu}_z)]$ is χ^2 (2).

(c) Calculate $u = \mathbf{z}'\mathbf{C}'[\mathbf{CVC}']^{-1}\mathbf{Cz}$. Reject if $u \geq \chi^2_{2,1-\alpha}$.

10.19 The sample covariance matrix involving y and $\mathbf{w}$ can be expressed in the form

$$\mathbf{S} = \begin{pmatrix} s_y^2 & \mathbf{s}'_{yw} \\ \mathbf{s}_{yw} & \mathbf{S}_{ww} \end{pmatrix},$$

and $\mathbf{s}_{yw}$ and $\mathbf{S}_{ww}$ can be further partitioned as

$$\mathbf{s}_{yw} = \begin{pmatrix} \mathbf{s}_{yx} \\ s_{yz} \end{pmatrix} \quad \text{and} \quad \mathbf{S}_{ww} = \begin{pmatrix} \mathbf{S}_{zx} & \mathbf{s}_{zx} \\ \mathbf{s}'_{zx} & s_z^2 \end{pmatrix}. \tag{1}$$

By (10.34), the squared multiple correlation of y regressed on $\mathbf{w}$ can be written as

$$R^2_{yw} = \frac{\mathbf{s}'_{yw}\mathbf{S}^{-1}_{ww}\mathbf{s}_{yw}}{s_y^2}. \tag{2}$$

Using (2.51) for the inverse of the partitioned matrix $\mathbf{S}_{ww}$ in (1), show that

$$\begin{aligned}
\mathbf{s}'_{yw}\mathbf{S}^{-1}_{ww}\mathbf{s}_{yw} &= \frac{1}{s^2_{z\cdot x}}(s^2_{z\cdot x}\mathbf{s}'_{yx}\mathbf{S}^{-1}_{xx}\mathbf{s}_{yx} + \mathbf{s}'_{yx}\mathbf{S}^{-1}_{xx}\mathbf{s}_{zx}\mathbf{s}'_{zx}\mathbf{S}^{-1}_{xx}\mathbf{s}_{yx} \\
&\quad - s_{yz}\mathbf{s}'_{zx}\mathbf{S}^{-1}_{xx}\mathbf{s}_{yx} - s_{yz}\mathbf{s}'_{yx}\mathbf{S}^{-1}_{xx}\mathbf{s}_{zx} + s^2_{yz}) \\
&= \frac{1}{s^2_{z\cdot x}}\left[s^2_{z\cdot x}s^2_y R^2_{yx} + (\hat{\boldsymbol{\beta}}'_{zx}\mathbf{s}_{yx} - s_{yz})^2\right],
\end{aligned}$$

where $s^2_{z\cdot x} = s^2_z - \mathbf{s}'_{zx}\mathbf{S}^{-1}_{xx}\mathbf{s}_{zx}$ and $\hat{\boldsymbol{\beta}}_{zx} = \mathbf{S}^{-1}_{xx}\mathbf{s}_{zx}$ is the vector of regression coefficients of z regressed on the x's. Then show that (2) becomes

$$R_{yw}^2 = R_{yx}^2 + \frac{(\hat{\boldsymbol{\beta}}_{zx}'\mathbf{s}_{yx} - s_{yz})^2}{s_y^2 s_z^2(1 - R_{zx}^2)}. \tag{3}$$

Simplify (3) to the correlation form shown in (10.58).

10.20 If z is orthogonal to the x's, then $\mathbf{s}_{zx} = \mathbf{0}$. Show that this leads to $\hat{r}_{yx} = 0$ and $R_{zx}^2 = 0$.

10.21 For a linear function $\beta_0 + \boldsymbol{\beta}_1'\mathbf{x}$, the mean squared error is given by $m = E(y - \beta_0 - \boldsymbol{\beta}_1'\mathbf{x})^2$. Adding and subtracting μ_y and $\boldsymbol{\beta}'\boldsymbol{\mu}_x$ leads to

$$m = E[(y - \mu_y) - (\beta_0 - \mu_y + \boldsymbol{\beta}_1'\boldsymbol{\mu}_x) - \boldsymbol{\beta}_1'(\mathbf{x} - \boldsymbol{\mu}_x)]^2.$$

Show that this becomes

$$m = \sigma_y^2 + (\beta_0 - \mu_y + \boldsymbol{\beta}_1'\boldsymbol{\mu}_x)^2 + \boldsymbol{\beta}_1'\boldsymbol{\Sigma}_{xx}\boldsymbol{\beta}_1 - 2\boldsymbol{\beta}_1'\boldsymbol{\sigma}_{yx}.$$

Differentiate m with respect to β_0 and with respect to $\boldsymbol{\beta}_1$ and set the results equal to zero.

10.22 Follow the steps in the answer to Problem 10.19 using $\bar{y}, \bar{\mathbf{x}}, \mathbf{S}_{xx}$, and $\mathbf{s}_{yx}$ in place of μ_y, $\boldsymbol{\mu}_x$, Σ_{xx}, and $\boldsymbol{\sigma}_{yx}$ and using a sample mean in place of expectation.

10.23 From the expression preceding (10.71), we obtain

$$\begin{aligned}\sum_{i=1}^{n} w_{1i}w_{2i} &= \sum_i (y_{1i} - \bar{y}_1)(y_{2i} - \bar{y}_2) - \hat{\beta}_{12}\sum_i (y_{1i} - \bar{y}_1)(y_{3i} - \bar{y}_3) \\ &\quad - \hat{\beta}_{11}\sum_i (y_{3i} - \bar{y}_3)(y_{2i} - \bar{y}_2) + \hat{\beta}_{11}\hat{\beta}_{12}\sum_i (y_{3i} - \bar{y}_3)^2.\end{aligned}$$

Using (10.67) and (10.68), this becomes

$$\begin{aligned}\sum_{i=1}^{n} w_{1i}w_{2i} &= \sum_i (y_{1i} - \bar{y}_1)(y_{2i} - \bar{y}_2) - \hat{\beta}_{12}\hat{\beta}_{11}\sum_i (y_{3i} - \bar{y}_3)^2 \\ &\quad - \hat{\beta}_{11}\hat{\beta}_{12}\sum_i (y_{3i} - \bar{y}_3)^2 + \hat{\beta}_{11}\hat{\beta}_{12}\sum_i (y_{3i} - \bar{y}_3)^2.\end{aligned}$$

10.24 Follow the steps in the answer to Problem 10.23.

10.25 Denote $y_{ki} - \bar{y}_k$ by $y^*_{ki}, k = 1, 2, 3$. Then by (10.76), we obtain

$$r_{w_1 w_2} = \frac{\sum_i y^*_{1i} y^*_{2i} - \hat{\beta}_{11}\hat{\beta}_{12} \sum_i y^{*2}_{3i}}{\sqrt{\left(\sum_i y^{*2}_{1i} - \hat{\beta}^2_{11} \sum_i y^{*2}_{3i}\right)\left(\sum_i y^{*2}_{2i} - \hat{\beta}^2_{12} \sum_i y^{*2}_{3i}\right)}}$$

Substituting for $\hat{\beta}_{11}$ and $\hat{\beta}_{12}$ from (10.67) and (10.68), we have

$$r_{w_1 w_2} = \frac{\sum_i y^*_{1i} y^*_{2i} - \left(\frac{\sum_i y^*_{1i} y^*_{3i}}{\sum_i y^{*2}_{3i}}\right)\left(\frac{\sum_i y^*_{2i} y^*_{3i}}{\sum_i y^{*2}_{3i}}\right)\left(\sum_i y^{*2}_{3i}\right)}{\sqrt{\left[\sum_i y^{*2}_{1i} - \left(\frac{\sum_i y^*_{1i} y^*_{3i}}{\sum_i y^{*2}_{3i}}\right)^2 \left(\sum_i y^{*2}_{3i}\right)\right]\left[\sum_i y^{*2}_{2i} - \left(\frac{\sum_i y^*_{2i} y^*_{3i}}{\sum_i y^{*2}_{3i}}\right)^2 \left(\sum_i y^{*2}_{3i}\right)\right]}}.$$

Dividing numerator and denominator by $\sqrt{\sum_i y^{*2}_{1i} \sum_i y^{*2}_{2i}}$, we obtain

$$r_{w_1 w_2} = \frac{\frac{\sum_i y^*_{1i} y^*_{2i}}{\sqrt{\sum_i y^{*2}_{1i} \sum_i y^{*2}_{2i}}} - \frac{\sum_i y^*_{1i} y^*_{3i}}{\sqrt{\sum_i y^{*2}_{1i} \sum_i y^{*2}_{3i}}} \frac{\sum_i y^*_{2i} y^*_{3i}}{\sqrt{\sum_i y^{*2}_{2i} \sum_i y^{*2}_{3i}}}}{\sqrt{\left[\frac{\sum_i y^{*2}_{1i}}{\sum_i y^{*2}_{1i}} - \frac{(\sum_i y^*_{1i} y^*_{3i})^2}{\sum_i y^{*2}_{1i} \sum_i y^{*2}_{3i}}\right]\left[\frac{\sum_i y^{*2}_{2i}}{\sum_i y^{*2}_{2i}} - \frac{(\sum_i y^*_{2i} y^*_{3i})^2}{\sum_i y^{*2}_{2i} \sum_i y^{*2}_{3i}}\right]}}$$

$$= \frac{r_{12} - r_{13} r_{23}}{\sqrt{(1 - r^2_{13})(1 - r^2_{23})}}.$$

10.26 By (10.78),

$$\begin{aligned}
\sum_{i=1}^{n} [\mathbf{y}_i - \hat{\mathbf{y}}_i(x)] &= \sum_{i=1}^{n} [\mathbf{y}_i - \bar{\mathbf{y}} - \mathbf{S}_{yx}\mathbf{S}^{-1}_{xx}(\mathbf{x}_i - \bar{\mathbf{x}})] \\
&= \sum_i (\mathbf{y}_i \bar{\mathbf{y}}) - \mathbf{S}_{yx}\mathbf{S}^{-1}_{xx} \sum_i (\mathbf{x}_i - \bar{\mathbf{x}}) \\
&= \mathbf{0} - \mathbf{S}_{yz}\mathbf{S}^{-1}_{xx}\mathbf{0}.
\end{aligned}$$

10.27 By definition, the partitioned S can be written as

$$\mathbf{S} = \begin{pmatrix} \mathbf{S}_{yy} & \mathbf{S}_{yx} \\ \mathbf{S}_{xy} & \mathbf{S}_{xx} \end{pmatrix} = \frac{1}{n-1}\sum_{i=1}^{n}\left[\begin{pmatrix} \mathbf{y}_i \\ \mathbf{x}_i \end{pmatrix} - \begin{pmatrix} \bar{\mathbf{y}} \\ \bar{\mathbf{x}} \end{pmatrix}\right]\left[\begin{pmatrix} \mathbf{y}_i \\ \mathbf{x}_i \end{pmatrix} - \begin{pmatrix} \bar{\mathbf{y}} \\ \bar{\mathbf{x}} \end{pmatrix}\right]$$

$$= \frac{1}{n-1}\sum_i \begin{pmatrix} \mathbf{y}_i - \bar{\mathbf{y}} \\ \mathbf{x}_i - \bar{\mathbf{x}} \end{pmatrix}\begin{pmatrix} \mathbf{y}_i - \bar{\mathbf{y}} \\ \mathbf{x}_i - \bar{\mathbf{x}} \end{pmatrix}$$

$$= \frac{1}{n-1}\sum_i \begin{pmatrix} y_i - \bar{y} \\ x_i - \bar{x} \end{pmatrix}[(y_i - \bar{y})\prime,(x_i - \bar{x})\prime]$$

$$= \frac{1}{n-1}\sum_i \begin{bmatrix} (\mathbf{y}_i - \bar{\mathbf{y}})(\mathbf{y}_i - \bar{\mathbf{y}})' & (\mathbf{y}_i - \bar{\mathbf{y}})(\mathbf{x}_i - \bar{\mathbf{x}})' \\ (\mathbf{x}_i - \bar{\mathbf{x}})(\mathbf{y}_i - \bar{\mathbf{y}})' & (\mathbf{x}_i - \bar{\mathbf{x}})(\mathbf{x}_i - \bar{\mathbf{x}})' \end{bmatrix}.$$

10.28

(a) $\mathbf{S}_{xx} = \begin{pmatrix} 271.9298 & 10.0830 & 1.4011 \\ 10.0830 & 1.4954 & 0.0514 \\ 1.4011 & 0.0514 & 0.0074 \end{pmatrix}, \quad \mathbf{s}_{yx} = \begin{pmatrix} .2204 \\ .0220 \\ .0017 \end{pmatrix},$

$\hat{\boldsymbol{\beta}}_1 = \begin{pmatrix} -0.0212 \\ 0.0143 \\ 4.1781 \end{pmatrix}, \quad \hat{\boldsymbol{\beta}}_0 = .2659, \quad s^2 = .004978$

(b) $\mathbf{R}_{xx} = \begin{pmatrix} 1.000 & 0.500 & 0.990 \\ 0.500 & 1.000 & 0.490 \\ 0.990 & 0.490 & 1.000 \end{pmatrix}, \quad r_{yx} = \begin{pmatrix} .151 \\ .203 \\ .228 \end{pmatrix},$

$\hat{\boldsymbol{\beta}}_1^* = \begin{pmatrix} -3.960 \\ 0.198 \\ 4.052 \end{pmatrix}$

(c) $R^2 = .3639$

(d) $F = \dfrac{R^2/3}{(1 - R^2)/15} = 2.86045, \quad p = .072$

10.29 **(a)** $x_2: r_1 = .5966$, $r_2 = .0721$ $z_1 = .6878$, $z_2 = .722$, $v = 2.0642$, limits for ρ_1 are .2721 and .7992, limits for ρ_2 are $-.3325$ and .4543.

(b) $x_3: r_1 = .7012$, $r_2 = .8209$, $z_1 = .8697$, $z_2 = 1.1594$, $v = -.9716$, limits fo ρ_1 are .4309 and .8561, limits for ρ_2 are .6301 and .9182.

(c) $x_4: r_1 = .0400$, $r_2 = .0714$, $z_1 = .04002$, $z_2 = .07154$, $v = -.1057$, limits for ρ_1 are $-.3528$ and .4208, limits for ρ_2 are $-.3331$ and .4537.

(d) $x_5: r_1 = .3391$, $r_2 = .2683$, $z_1 = .3531$, $z_2 = .2751$, $v = .2617$, limits for ρ_1 are $-.0555$ and .6421, limits for ρ_2 are $-.1418$ and .5999.

10.30

z	$\hat{r}_{yz}$	r_{yz}	R^2_{zx}	$R^2_{yw}-R^2_{yx}$	F	p Value
x_1	.227	.0228	.9808	.3010	7.099	.018
x_2	.055	.0413	.2513	.0292	.689	.417
x_3	.149	.0518	.9806	.3193	7.531	.015

10.31

(a) To find $r_{y1\cdot 23}$, the sample covariance matrix is partitioned as

$$\mathbf{S} = \left(\begin{array}{cc|cc} .0078 & .2204 & .0220 & .0017 \\ .2204 & 271.9298 & 10.0830 & 1.4011 \\ \hline .0220 & 10.0830 & 1.4954 & .0514 \\ .0017 & 1.4011 & .0514 & .0074 \end{array}\right) = \begin{pmatrix} \mathbf{S}_{yy} & \mathbf{S}_{yx} \\ \mathbf{S}_{xy} & \mathbf{S}_{xx} \end{pmatrix},$$

where $\mathbf{y} = (y, x_1)'$ and $\mathbf{x} = (x_2, x_3)'$. From $\mathbf{S}_{yy}$, $\mathbf{S}_{yx}$, $\mathbf{S}_{xy}$, and $\mathbf{S}_{xx}$, we obtain

$$\mathbf{D}_s = \begin{pmatrix} .0856 & 0 \\ 0 & 2.2846 \end{pmatrix}, \quad \mathbf{R}_{y\cdot x} = \begin{pmatrix} 1.000 & -.567 \\ -.567 & 1.000 \end{pmatrix}.$$

Thus $r_{y1\cdot 23} = -.567$, as compared to $r_{y1} = .151$.

(b) For $r_{y2\cdot 13}$, we have $\mathbf{y} = (y, x_2)'$ and $\mathbf{x} = (x_1, x_3)'$,

$$\mathbf{S}_{yy} = \begin{pmatrix} .0078 & 0.0220 \\ .0220 & 1.4954 \end{pmatrix}, \quad \mathbf{D}_s = \begin{pmatrix} 0 & 1.0581 \\ .0722 & 0 \end{pmatrix},$$

$$\mathbf{R}_{y\cdot x} = \begin{pmatrix} 1.000 & 0.2097 \\ 0.2097 & 1.000 \end{pmatrix}.$$

(c) For $\mathbf{R}_{y.x}$ corresponding to $\mathbf{y} = (y,\ x_1,\ x_2)'$ and $\mathbf{x} = x_3$, we have

$$\mathbf{S}_{yy} = \begin{pmatrix} .0078 & .2204 & .0220 \\ .2204 & 271.9298 & 10.0830 \\ .0220 & 10.0830 & 1.4954 \end{pmatrix},$$

$$\mathbf{D}_s = \begin{pmatrix} .0861 & 0 & 0 \\ 0 & 2.3015 & 0 \\ 0 & 0 & 1.0660 \end{pmatrix}$$

$$\mathbf{R}_{y\cdot x} = \begin{pmatrix} 1.000 & -0.546 & 0.108 \\ -0.546 & 1.000 & 0.121 \\ 1.108 & 0.121 & 1.000 \end{pmatrix}$$

Chapter 11

11.1 $(\boldsymbol{\beta}-\boldsymbol{\phi})'\mathbf{V}^{-1}(\boldsymbol{\beta}-\boldsymbol{\phi})+(\mathbf{y}-\mathbf{X}\boldsymbol{\beta})'(\mathbf{y}-\mathbf{X}\boldsymbol{\beta})+\delta_*$

$$
\begin{aligned}
&= \boldsymbol{\beta}'\mathbf{V}^{-1}\boldsymbol{\beta}-2\boldsymbol{\beta}'\mathbf{V}^{-1}\boldsymbol{\phi}+\boldsymbol{\phi}'\mathbf{V}^{-1}\boldsymbol{\phi}+\mathbf{y}'\mathbf{y}-2\boldsymbol{\beta}'\mathbf{X}'\mathbf{y}+\boldsymbol{\beta}\mathbf{X}'\mathbf{X}\boldsymbol{\beta}+\delta_*\\
&= \boldsymbol{\beta}'(\mathbf{V}^{-1}+\mathbf{X}'\mathbf{X})\boldsymbol{\beta}-2\boldsymbol{\beta}'(\mathbf{V}^{-1}+\mathbf{X}'\mathbf{X})(\mathbf{V}^{-1}+\mathbf{X}'\mathbf{X})^{-1}(\mathbf{V}^{-1}\boldsymbol{\phi}+\mathbf{X}'\mathbf{y})\\
&\quad+(\mathbf{V}^{-1}\boldsymbol{\phi}+\mathbf{X}'\mathbf{y})'(\mathbf{V}^{-1}+\mathbf{X}'\mathbf{X})^{-1}(\mathbf{V}^{-1}+\mathbf{X}'\mathbf{X})(\mathbf{V}^{-1}+\mathbf{X}'\mathbf{X})^{-1}\\
&\quad\times(\mathbf{V}^{-1}\boldsymbol{\phi}+\mathbf{X}'\mathbf{y})-(\mathbf{V}^{-1}\boldsymbol{\phi}+\mathbf{X}'\mathbf{y})'(\mathbf{V}^{-1}+\mathbf{X}'\mathbf{X})^{-1}(\mathbf{V}^{-1}+\mathbf{X}'\mathbf{X})\\
&\quad(\mathbf{V}^{-1}+\mathbf{X}'\mathbf{X})^{-1}(\mathbf{V}^{-1}\boldsymbol{\phi}+\mathbf{X}'\mathbf{y})+\boldsymbol{\phi}'V^{-1}\boldsymbol{\phi}+\mathbf{y}'\mathbf{y}+\delta_*\\
&= \boldsymbol{\beta}'\mathbf{V}_*^{-1}\boldsymbol{\beta}-2\boldsymbol{\beta}'\mathbf{V}_*^{-1}\boldsymbol{\phi}_*+\boldsymbol{\phi}_*'\mathbf{V}_*^{-1}\boldsymbol{\phi}_*-\boldsymbol{\phi}_*'\mathbf{V}_*^{-1}\boldsymbol{\phi}_*\\
&\qquad+\boldsymbol{\phi}'V^{-1}\boldsymbol{\phi}+\mathbf{y}'\mathbf{y}+\delta_*\\
&= (\boldsymbol{\beta}-\boldsymbol{\phi}_*)'\mathbf{V}_*^{-1}(\boldsymbol{\beta}-\boldsymbol{\phi}_*)+\delta_{**}.
\end{aligned}
$$

11.2

$$
\begin{aligned}
\int_0^\infty t^a e^{-bt}\,dt &= b^{-a}\int_0^\infty (bt)^a e^{-(bt)}\,dt\\
&= b^{-a}\int_0^\infty (bt)^a e^{-(bt)}\,d(bt)\\
&= b^{-(a+1)}\int_0^\infty s^a e^{-s}\,ds \text{ (letting } s=bt)\\
&= b^{-(a+1)}\Gamma(a+1) \text{ [by definition of } \Gamma(a+1)].
\end{aligned}
$$

11.3 **(a)** Use (2.54) with $\mathbf{A}=\mathbf{I}$, $\mathbf{P}=\mathbf{XV}$, $\mathbf{B}=\mathbf{V}^{-1}$, and $\mathbf{Q}=\mathbf{VX}'$.

(b) $(\mathbf{I}+\mathbf{XVX}')^{-1}\mathbf{X}-\mathbf{X}(\mathbf{X}'\mathbf{X}+\mathbf{V}^{-1})^{-1}\mathbf{V}$

$$
\begin{aligned}
&= [\mathbf{I}-\mathbf{X}(\mathbf{X}'\mathbf{X}+\mathbf{V}^{-1})^{-1}\mathbf{X}']\mathbf{X}-\mathbf{X}(\mathbf{X}'\mathbf{X}+\mathbf{V}^{-1})^{-1}\mathbf{V}^{-1} \quad \text{(Problem 11.3a)}\\
&= \mathbf{X}-\mathbf{X}(\mathbf{X}'\mathbf{X}+\mathbf{V}^{-1})^{-1}\mathbf{X}'\mathbf{X}-\mathbf{X}(\mathbf{X}'\mathbf{X}+\mathbf{V}^{-1})^{-1}\mathbf{V}\\
&= \mathbf{X}-\mathbf{X}(\mathbf{X}'\mathbf{X}+\mathbf{V}^{-1})^{-1}(\mathbf{X}'\mathbf{X}+\mathbf{V}^{-1})\\
&= \mathbf{X}-\mathbf{X}\\
&= \mathbf{O}.
\end{aligned}
$$

(c) $\mathbf{V}^{-1} - \mathbf{V}^{-1}(\mathbf{X}'\mathbf{X} + \mathbf{V}^{-1})^{-1}\mathbf{V}^{-1}$

$$
\begin{aligned}
&= [\mathbf{V} + (\mathbf{X}'\mathbf{X})^{-1}(\mathbf{X}'\mathbf{X})(\mathbf{X}'\mathbf{X})^{-1}]^{-1} && \text{[use (2.54) in reverse]} \\
&= [(\mathbf{X}'\mathbf{X})^{-1} + \mathbf{V}]^{-1} && \text{(simplify)} \\
&= \mathbf{X}'\mathbf{X} - \mathbf{X}'\mathbf{X}(\mathbf{X}'\mathbf{X} + \mathbf{V}^{-1})^{-1}\mathbf{X}'\mathbf{X} && \text{[use (2.54)]} \\
&= \mathbf{X}'[\mathbf{I} - \mathbf{X}(\mathbf{X}'\mathbf{X} + \mathbf{V}^{-1})^{-1}\mathbf{X}']\mathbf{X} && \text{(factor)} \\
&= \mathbf{X}'(\mathbf{I} + \mathbf{X}\mathbf{V}\mathbf{X}')^{-1}\mathbf{X} && \text{(Problem 11.3a).}
\end{aligned}
$$

11.4 $\mathbf{y}'\mathbf{y} + \boldsymbol{\phi}'\mathbf{V}^{-1}\boldsymbol{\phi} - \boldsymbol{\phi}_*'\mathbf{V}_*^{-1}\boldsymbol{\phi}_*$

$$
\begin{aligned}
&= (\mathbf{y} - \mathbf{X}\boldsymbol{\phi})'(\mathbf{I} + \mathbf{X}\mathbf{V}\mathbf{X}')^{-1}(\mathbf{y} - \mathbf{X}\boldsymbol{\phi}) - (\mathbf{y} - \mathbf{X}\boldsymbol{\phi})' \\
&\quad \times (\mathbf{I} + \mathbf{X}\mathbf{V}\mathbf{X}')^{-1}(\mathbf{y} - \mathbf{X}\boldsymbol{\phi}) + \mathbf{y}'\mathbf{y} + \boldsymbol{\phi}'\mathbf{V}^{-1}\boldsymbol{\phi} - \boldsymbol{\phi}_*'\mathbf{V}_*^{-1}\boldsymbol{\phi}_* \\
&= (\mathbf{y} - \mathbf{X}\boldsymbol{\phi})'(\mathbf{I} + \mathbf{X}\mathbf{V}\mathbf{X}')^{-1}(\mathbf{y} - \mathbf{X}\boldsymbol{\phi}) - \mathbf{y}'(\mathbf{I} + \mathbf{X}\mathbf{V}\mathbf{X}')^{-1}\mathbf{y} + 2\mathbf{y}'(\mathbf{I} + \mathbf{X}\mathbf{V}\mathbf{X}')^{-1} \\
&\quad \times \mathbf{X}\boldsymbol{\phi} - \boldsymbol{\phi}'\mathbf{X}'(\mathbf{I} + \mathbf{X}\mathbf{V}\mathbf{X}')^{-1}\mathbf{X}\boldsymbol{\phi} + \mathbf{y}'\mathbf{y} + \boldsymbol{\phi}'\mathbf{V}^{-1}\boldsymbol{\phi} - (\mathbf{X}'\mathbf{y} + \mathbf{V}^{-1}\boldsymbol{\phi})' \\
&\quad \times (\mathbf{X}'\mathbf{X} + \mathbf{V}^{-1})^{-1}(\mathbf{X}'\mathbf{X} + \mathbf{V}^{-1})(\mathbf{X}'\mathbf{X} + \mathbf{V}^{-1})^{-1}(\mathbf{X}'\mathbf{y} + \mathbf{V}^{-1}\boldsymbol{\phi}) \\
&= (\mathbf{y} - \mathbf{X}\boldsymbol{\phi})'(\mathbf{I} + \mathbf{X}\mathbf{V}\mathbf{X}')^{-1}(\mathbf{y} - \mathbf{X}\boldsymbol{\phi}) + \mathbf{y}'[\mathbf{I} - (\mathbf{I} + \mathbf{X}\mathbf{V}\mathbf{X}')^{-1} \\
&\quad - \mathbf{X}(\mathbf{X}'\mathbf{X} + \mathbf{V}^{-1})^{-1}\mathbf{X}']\mathbf{y} + 2\mathbf{y}'[(\mathbf{I} + \mathbf{X}\mathbf{V}\mathbf{X}')^{-1}\mathbf{X} - \mathbf{X}(\mathbf{X}'\mathbf{X} + \mathbf{V}^{-1})^{-1}\mathbf{V}^{-1}]\boldsymbol{\phi} \\
&\quad + \boldsymbol{\phi}'[\mathbf{V}^{-1} - \mathbf{X}'(\mathbf{I} + \mathbf{X}\mathbf{V}\mathbf{X}')^{-1}\mathbf{X} - \mathbf{V}^{-1}(\mathbf{X}'\mathbf{X} + \mathbf{V}^{-1})^{-1}\mathbf{V}^{-1}]\boldsymbol{\phi} \\
&= (\mathbf{y} - \mathbf{X}\boldsymbol{\phi})'(\mathbf{I} + \mathbf{X}\mathbf{V}\mathbf{X}')^{-1}(\mathbf{y} - \mathbf{X}\boldsymbol{\phi}) + \mathbf{y}'\mathbf{O}\mathbf{y} + 2\mathbf{y}'\mathbf{O}\boldsymbol{\phi} + \boldsymbol{\phi}'\mathbf{O}\boldsymbol{\phi} \\
&\quad \text{(see Problems 11.3a, b, and c)} \\
&= (\mathbf{y} - \mathbf{X}\boldsymbol{\phi})'(\mathbf{I} + \mathbf{X}\mathbf{V}\mathbf{X}')^{-1}(\mathbf{y} - \mathbf{X}\boldsymbol{\phi}).
\end{aligned}
$$

11.5 The prior density for $\boldsymbol{\beta}$ is $p_1(\boldsymbol{\beta}) = c_1$. Since the prior density for $\ln(\tau^{-1})$ is uniform, the prior density for τ is $p_2(\tau) = c_2\tau^{-1}$. The likelihood for $\mathbf{y}|\boldsymbol{\beta}$, τ is the multivariate normal density with mean $\mathbf{X}\boldsymbol{\beta}$ and covariance matrix $\tau^{-1}\mathbf{I}$. Using Bayes theorem in (11.4), the joint posterior density is

$$
\begin{aligned}
g(\boldsymbol{\beta}, \tau|\mathbf{y}) &= c_4c_1c_2\tau^{-1}c_3\tau^{n/2}e^{-\tau(\mathbf{y}-\mathbf{X}\boldsymbol{\beta})'(\mathbf{y}-\mathbf{X}\boldsymbol{\beta})/2} \\
&= c_5\tau^{(n-2)/2}e^{-\tau(\mathbf{y}-\mathbf{X}\boldsymbol{\beta})'(\mathbf{y}-\mathbf{X}\boldsymbol{\beta})/2}.
\end{aligned}
$$

The marginal posterior density of $\boldsymbol{\beta}|\boldsymbol{y}$ is

$$u(\boldsymbol{\beta}|\mathbf{y}) = c_5 \int_0^\infty \tau^{(n-2)/2} e^{-\tau(\mathbf{y}-\mathbf{X}\boldsymbol{\beta})'(\mathbf{y}-\mathbf{X}\boldsymbol{\beta})/2} d\tau$$

$$= c_5 \boldsymbol{\Gamma}(n/2)[(\mathbf{y}-\mathbf{X}\boldsymbol{\beta})'(\mathbf{y}-\mathbf{X}\boldsymbol{\beta})/2]^{-n/2} \quad \text{(Problem 11.2)}$$

$$= c_6[(n-k-1)s^2 + (\boldsymbol{\beta}-\hat{\boldsymbol{\beta}})'(\mathbf{X}'\mathbf{X})(\boldsymbol{\beta}-\hat{\boldsymbol{\beta}})]^{-n/2}$$

(proof to Theorem 7.6c)

$$= c_7[1 + (\boldsymbol{\beta}-\hat{\boldsymbol{\beta}})'(\mathbf{X}'\mathbf{X})(\boldsymbol{\beta}-\hat{\boldsymbol{\beta}})/(n-k-1)s^2]^{-n/2}$$

$$= c_7[1 + (\boldsymbol{\beta}-\hat{\boldsymbol{\beta}})'[s^2(\mathbf{X}'\mathbf{X})^{-1}]^{-1}(\boldsymbol{\beta}-\hat{\boldsymbol{\beta}})/(n-k-1)]^{-[(n-k-1)+(k+1)]/2}$$

which is the density function of the multivariate t-distribution (Gelman, et al. 2004, pp. 576–577) with parameters $(n-k-1, \hat{\boldsymbol{\beta}}, s^2(\mathbf{X}'\mathbf{X})^{-1})$.

11.6 Using (7.64), the generalized least squares estimate of $\boldsymbol{\beta}$ for the augmented data is

$$\left[\begin{pmatrix}\mathbf{X}\\ \mathbf{I}\end{pmatrix}'\begin{pmatrix}\mathbf{I} & \mathbf{O}\\ \mathbf{O} & \mathbf{V}\end{pmatrix}^{-1}\begin{pmatrix}\mathbf{X}\\ \mathbf{I}\end{pmatrix}\right]^{-1}\begin{pmatrix}\mathbf{X}\\ \mathbf{I}\end{pmatrix}'\begin{pmatrix}\mathbf{I} & \mathbf{O}\\ \mathbf{O} & \mathbf{V}\end{pmatrix}^{-1}\begin{pmatrix}\mathbf{y}\\ \boldsymbol{\phi}\end{pmatrix}$$

$$= \left[(\mathbf{X}' \quad \mathbf{I}')\begin{pmatrix}\mathbf{I} & \mathbf{O}\\ \mathbf{O} & \mathbf{V}\end{pmatrix}^{-1}\begin{pmatrix}\mathbf{X}\\ \mathbf{Y}\end{pmatrix}\right]^{-1}(\mathbf{X}' \quad \mathbf{I}')\begin{pmatrix}\mathbf{I} & \mathbf{O}\\ \mathbf{O} & \mathbf{V}\end{pmatrix}^{-1}\begin{pmatrix}\mathbf{y}\\ \boldsymbol{\phi}\end{pmatrix}$$

$$= (\mathbf{X}'\mathbf{X} + \mathbf{V}^{-1})^{-1}(\mathbf{X}'\mathbf{y} + \mathbf{V}^{-1}\boldsymbol{\phi}).$$

11.7

$$E(\tau) = \int_0^\infty \tau \frac{\delta^\alpha}{\Gamma(\alpha)} \tau^{\alpha-1} e^{-\delta\tau} d\tau$$

$$= \frac{\delta^\alpha}{\Gamma(\alpha)} \int_0^\infty \tau^\alpha e^{-\delta\tau} d\tau$$

$$= \frac{\delta^\alpha}{\Gamma(\alpha)} \delta^{-(\alpha+1)} \Gamma(\alpha+1) \quad \text{[using prob. 11.2]}$$

$$= \frac{\alpha}{\delta}.$$

$$\text{var}(\tau) = \int_0^\infty \tau^2 \frac{\delta^\alpha}{\Gamma(\alpha)} \tau^{\alpha-1} e^{-\delta\tau} d\tau - \left(\frac{\alpha}{\delta}\right)^2$$

$$= \frac{\delta^\alpha}{\Gamma(\alpha)} \int_0^\infty \tau^{\alpha+1} e^{-\delta\tau} d\tau - \left(\frac{\alpha}{\delta}\right)^2$$

$$= \frac{\delta^\alpha}{\Gamma(\alpha)} \delta^{-(\alpha+2)} \Gamma(\alpha+2) - \left(\frac{\alpha}{\delta}\right)^2$$

$$= \frac{\delta^\alpha}{\Gamma(\alpha)} \delta^{-(\alpha+2)} \Gamma(\alpha+2) - \left(\frac{\alpha}{\delta}\right)^2$$

$$= \frac{(\alpha+1)(\alpha)}{\delta^2} - \left(\frac{\alpha}{\delta}\right)^2$$

$$= \frac{\alpha}{\delta^2}.$$

11.8 The density function of $\tau|\mathbf{y}$ is given in (11.14). Using the change-of-variable technique, the marginal posterior density of $\sigma^2|\mathbf{y}$ is

$$\mathbf{w}(\sigma^2|\mathbf{y}) = c_5(\sigma^{-2})^{(\alpha+n/2)-1} e^{-[(-\boldsymbol{\phi}_*'\mathbf{V}_*^{-1}\boldsymbol{\phi}_* + \boldsymbol{\phi}'\mathbf{V}^{-1}\boldsymbol{\phi} + \mathbf{y}'\mathbf{y} + 2\delta)/2](\sigma^{-2})} (\sigma^2)^{-2}$$

$$= c_6(\sigma^2)^{-(\alpha+n/2)-1} e^{-[(-\boldsymbol{\phi}_*'\mathbf{V}_*^{-1}\boldsymbol{\phi}_* + \boldsymbol{\phi}'\mathbf{V}^{-1}\boldsymbol{\phi} + \mathbf{y}'\mathbf{y} + 2\delta)/2]/\sigma^{-2}}.$$

11.9 (a) This is the model of Section 11.2.1, with $k = 1$, $\boldsymbol{\phi} = \begin{pmatrix} 0 \\ 0 \end{pmatrix}$, and $\mathbf{V} = \begin{pmatrix} \sigma_0^2 & 0 \\ 0 & \sigma_1^2 \end{pmatrix}$.

Using Theorem 11.2b and the expression in (11.18), $\beta_1|\tau$ is t-distributed with parameters $n + 2\alpha$, $\boldsymbol{\phi}_{*2}$, and w_{22}^* where

$$\boldsymbol{\phi}_{*2} = \frac{(\sigma_0^{-2} + n)\sum_i x_i y_i - \sum_i y_i \sum_i x_i}{(\sigma_0^{-2} + n)(\sigma_1^{-2} + \sum_i x_i^2) - (\sum_i x_i)^2}$$

and

$$w_{*22} = \frac{\mathbf{y}'(\mathbf{I} - \mathbf{XVX}')^{-1}\mathbf{y} + 2\delta}{n + 2\alpha} \frac{(\sigma_0^{-2} + n)}{(\sigma_0^{-2} + n)(\sigma_0^{-2} + \sum_i x_i^2) - (\sum_i x_i)^2}.$$

(b) A point estimate is given by ϕ_{*2} and a $(1-\omega)\times 100\%$ confidence interval is given by $\boldsymbol{\phi}_{*2} \pm t_{\omega/2,n+2\alpha}\omega_{*22}$.

11.10 (a) The joint prior density is

$$\begin{aligned} p(\boldsymbol{\beta}, \tau) &= p_1(\boldsymbol{\beta}|\tau)p_2(\tau) \\ &= c_1 e^{-(\boldsymbol{\beta}-\boldsymbol{\phi})'\mathbf{V}^{-1}(\boldsymbol{\beta}-\boldsymbol{\phi})/2}\tau^{\alpha-1}e^{-\delta\tau} \\ &= c_1\tau^{\alpha-1}e^{-(\boldsymbol{\beta}-\boldsymbol{\phi})'\mathbf{V}^{-1}(\boldsymbol{\beta}-\boldsymbol{\phi})/2-\delta\tau.} \end{aligned}$$

Using (11.4), the joint posterior density is

$$\begin{aligned} g(\boldsymbol{\beta}, \tau|\mathbf{y}) &= cp(\boldsymbol{\beta}, \tau)\, L(\boldsymbol{\beta}, \tau|\mathbf{y}) \\ &= c_2\tau^{\alpha-1}e^{-(\boldsymbol{\beta}-\boldsymbol{\phi})'\mathbf{V}^{-1}(\boldsymbol{\beta}-\boldsymbol{\phi})/2-\delta\tau}\tau^{n/2}e^{-\tau(\mathbf{y}-\mathbf{X}\boldsymbol{\beta})'(\mathbf{y}-\mathbf{X}\boldsymbol{\beta})/2} \\ &= c_2\tau^{n/2+\alpha-1}e^{-[(\boldsymbol{\beta}-\boldsymbol{\phi})'\mathbf{V}^{-1}(\boldsymbol{\beta}-\boldsymbol{\phi})+\tau(\mathbf{y}-\mathbf{X}\boldsymbol{\beta})'(\mathbf{y}-\mathbf{X}\boldsymbol{\beta})]/2-\delta\tau}. \end{aligned}$$

(b) Picking the terms out of the joint density $g(\boldsymbol{\beta}, \tau|\mathbf{y})$ that involve $\boldsymbol{\beta}$, and considering everything else to be part of the normalizing constant, the conditional posterior density of $\boldsymbol{\beta}|\tau$, $\mathbf{y}$ is

$$\begin{aligned} \varphi(\boldsymbol{\beta}|\tau, \mathbf{y}) &= c_3\, e^{-[(\boldsymbol{\beta}-\boldsymbol{\phi})'\mathbf{V}^{-1}(\boldsymbol{\beta}-\boldsymbol{\phi})+\tau(\mathbf{y}-\mathbf{X}\boldsymbol{\beta})'(\mathbf{y}-\mathbf{X}\boldsymbol{\beta})]/2} \\ &= c_3\, e^{-\tau(\tau^{-1}\boldsymbol{\beta}'\mathbf{V}^{-1}\boldsymbol{\beta}-2\tau^{-1}\boldsymbol{\beta}'\mathbf{V}^{-1}\boldsymbol{\phi}+\tau^{-1}\boldsymbol{\phi}'\mathbf{V}^{-1}\boldsymbol{\phi}+\mathbf{y}'\mathbf{y}-2\boldsymbol{\beta}'\mathbf{X}'\mathbf{y}+\boldsymbol{\beta}'\mathbf{X}'\mathbf{X}\boldsymbol{\beta})/2} \\ &= c_4\, e^{-\tau[\boldsymbol{\beta}'(\mathbf{X}'\mathbf{X}+\tau^{-1}\mathbf{V}^{-1})\boldsymbol{\beta}-2\boldsymbol{\beta}'(\mathbf{X}'\mathbf{y}+\tau^{-1}\mathbf{V}^{-1}\boldsymbol{\phi})]/2} \\ &= c_4\, e^{-\tau[\boldsymbol{\beta}'(\mathbf{X}'\mathbf{X}+\tau^{-1}\mathbf{V}^{-1})\boldsymbol{\beta}-2\boldsymbol{\beta}'(\mathbf{X}'\mathbf{X}+\tau^{-1}\mathbf{V}^{-1})(\mathbf{X}'\mathbf{X}+\tau^{-1}\mathbf{V}^{-1})^{-1}(\mathbf{X}'\mathbf{y}+\tau^{-1}\mathbf{V}^{-1}\boldsymbol{\phi})]/2} \\ &= c_5\, e^{-\tau(\boldsymbol{\beta}'\mathbf{V}_n^{-1}\boldsymbol{\beta}-2\boldsymbol{\beta}'\mathbf{V}_n^{-1}\boldsymbol{\phi}_n+\boldsymbol{\phi}_n'\mathbf{V}_n^{-1}\boldsymbol{\phi}_n)/2} \end{aligned}$$

where $\mathbf{V}_n = (\mathbf{X}'\mathbf{X} + \tau^{-1}\mathbf{V}^{-1})^{-1}$ and $\boldsymbol{\phi}_n = \mathbf{V}_n(\mathbf{X}'\mathbf{y} + \tau^{-1}\mathbf{V}^{-1}$

$$= c_5\, e^{-\tau[(\boldsymbol{\beta}-\boldsymbol{\phi}_n)'\mathbf{V}_n^{-1}(\boldsymbol{\beta}-\boldsymbol{\phi}_n)]/2}.$$

Hence $\boldsymbol{\beta}|\tau, \mathbf{y}$ is $N_{k+1}(\boldsymbol{\phi}_n, \tau^{-1}\mathbf{V}_n)$.

(c) Picking the terms out of the joint density $g(\boldsymbol{\beta}, \tau|\mathbf{y})$ that involve τ, and considering everything else to be part of the normalizing constant, the conditional posterior density of $\tau|\boldsymbol{\beta}, \mathbf{y}$ is

$$\psi(\tau|\boldsymbol{\beta}, \mathbf{y}) = c_6\, \tau^{n/2+\alpha-1} e^{-\tau(\mathbf{y}-\mathbf{X}\boldsymbol{\beta})'(\mathbf{y}-\mathbf{X}\boldsymbol{\beta})/2+\delta\tau}$$

$$= c_6 \tau^{n/2+\alpha-1} e^{-[(\mathbf{y}-\mathbf{X}\boldsymbol{\beta})'(\mathbf{y}-\mathbf{X}\boldsymbol{\beta})/2+\delta]\tau}.$$

Hence $\tau|\boldsymbol{\beta}, \mathbf{y}$ is $Gamma[n/2 + \alpha, (\mathbf{y} - \mathbf{X}\boldsymbol{\beta})'(\mathbf{y} - \mathbf{X}\boldsymbol{\beta})/2 + \delta]$.

(d)

- Specify $1/s^2$ from (7.23) as a starting value τ_0.
- For $i = 1$ to M:
  calculate $\mathbf{V}_{n,i-1} = (\mathbf{X}'\mathbf{X} + \tau_{i-1}^{-1}\mathbf{V}^{-1})^{-1}$,
  calculate $\boldsymbol{\phi}_{n,i-1} = \mathbf{V}_{n,i-1}(\mathbf{X}'\mathbf{y} + \tau_{i-1}^{-1}\mathbf{V}^{-1}\boldsymbol{\phi})$,
  draw $\boldsymbol{\beta}_i$ from $N_{k+1}(\boldsymbol{\phi}_{n,i-1}, \tau_{i-1}^{-1}\mathbf{V}_{n,i-1})$,
  draw τ_i from $Gamma[n/2 + \alpha, (\mathbf{y} - \mathbf{X}\boldsymbol{\beta}_i)'(\mathbf{y} - \mathbf{X}\boldsymbol{\beta}_i)/2 + \delta]$,
  calculate τ_i^{-1}
- Consider all draws $(\boldsymbol{\beta}_i, \tau_i^{-1})$ to be from the joint posterior distribution.

11.11 **(a)** Bayesian estimates of $\boldsymbol{\beta}_1$, $\boldsymbol{\beta}_2$, and $\boldsymbol{\beta}_3$ are 0.7820, 0.5007, −16.6443. Lower 95% confidence limits are 0.6281, 0.2627, −42.2511. Upper 95% confidence limits are 0.9358, 0.7386, 8.9625.

(b) Answers will vary. We obtained Bayesian estimates of 0.7817, 0.4990, −16.5490, lower 95% confidence limits of 0.6332, 0.2627, −42.6158, and upper 95% confidence limits of 0.9358, 0.7358, 9.5144.

(c) Answers will vary. We obtained a Bayesian prediction of 18.9113, with lower and upper 95% limits of 2.3505 and 35.7526.

(d) Answers will vary. We obtained Bayesian estimates of 0.8170, 0.4399, −6.1753, lower 95% confidence limits of 0.6831, 0.2142, −21.4675, and upper 95% confidence limits of 0.9523, 0.6567, 9.7605.

11.12 Use Problem 11.2 with $t = \tau$, $a = (\alpha_{**} + k + 2)/2$, and $b = [(\boldsymbol{\beta} - \boldsymbol{\phi}_*)'\mathbf{V}_*^{-1}(\boldsymbol{\beta} - \boldsymbol{\phi}_*) + (y_0 - \mathbf{x}_0'\boldsymbol{\beta})^2 + \delta_{**}]/2$.

Chapter 12

12.1

$$\bar{\mu}_{1.} + \bar{\mu}_{2.} = \frac{\mu_{11} + \mu_{12}}{2} + \frac{\mu_{21} + \mu_{22}}{2} = \frac{\mu_{11} + \mu_{12} + \mu_{21} + \mu_{22}}{2}$$

$$= 2\left(\frac{\mu_{11} + \mu_{12} + \mu_{21} + \mu_{22}}{4}\right) = 2\bar{\mu}_{..}$$

12.2 The deficiency in the rank of $\mathbf{X}$ does not affect the differentiation of $\hat{\varepsilon}'\hat{\varepsilon}$ in (12.10). Thus

$$\frac{\partial \hat{\boldsymbol{\varepsilon}}'\hat{\boldsymbol{\varepsilon}}}{\partial \hat{\boldsymbol{\beta}}} = \mathbf{0} - 2\mathbf{X}'\mathbf{y} + 2\mathbf{X}'\mathbf{X}\hat{\boldsymbol{\beta}} = \mathbf{0},$$

which yields (12.11).

12.3 For Theorem 2.7a, the coefficient matrix is $\mathbf{A} = \mathbf{X}'\mathbf{X}$, and the augmented matrix is $\mathbf{B} = (\mathbf{X}'\mathbf{X}, \mathbf{X}'\mathbf{y})$. We can write $\mathbf{B}$ as $\mathbf{X}'(\mathbf{X}, \mathbf{y})$. which leads to $\text{rank}(\mathbf{B}) \leq \text{rank}(\mathbf{X}') = \text{rank}(\mathbf{A})$. On the other hand, $\text{rank}(\mathbf{B}) \geq \text{rank}(\mathbf{A})$ because augmenting a matrix by a column vector cannot decrease the column rank. Hence $\text{rank}(\mathbf{B}) = \text{rank}(\mathbf{A})$; that is, $\text{rank}(\mathbf{X}'\mathbf{X},\mathbf{X}'\mathbf{y}) = \text{rank}(\mathbf{X}'\mathbf{X})$, and the system is consistent.

12.4 **(a)** We can obtain $\boldsymbol{\lambda}'\mathbf{X}'\mathbf{X}(\mathbf{X}'\mathbf{X})^- = \boldsymbol{\lambda}'$ from the expression $\mathbf{X}(\mathbf{X}'\mathbf{X})^-\mathbf{X}'\mathbf{X} = \mathbf{X}$ given in Theorem 2.8c(iii). Since $\boldsymbol{\lambda}' = \mathbf{a}'\mathbf{X}$, multiplying by $\mathbf{a}'$ gives the result; that is, $\mathbf{a}'\mathbf{X}(\mathbf{X}'\mathbf{X})^-\mathbf{X}'\mathbf{X} = \mathbf{a}'\mathbf{X}$ implies $\boldsymbol{\lambda}'(\mathbf{X}'\mathbf{X})^-\mathbf{X}'\mathbf{X} = \boldsymbol{\lambda}'$.

(b) The condition $\mathbf{X}'\mathbf{X}(\mathbf{X}'\mathbf{X})^-\boldsymbol{\lambda} = \boldsymbol{\lambda}$ follows from Theorem 2.8f, which states that $\mathbf{Ax} = \mathbf{c}$ has a solution if and only if $\mathbf{AA}^-\mathbf{c} = \mathbf{c}$ for any generalized inverse of $\mathbf{A}$. Thus, $\mathbf{X}'\mathbf{Xr} = \boldsymbol{\lambda}$, has a solution if and only if $\mathbf{X}'\mathbf{X}(\mathbf{X}'\mathbf{X})^-\boldsymbol{\lambda} = \boldsymbol{\lambda}$.

12.5 **(a)** $\mathbf{a}'\mathbf{X} = (0, 0, 0, 1, 0, 0)\mathbf{X} = (1, 0, 1)$. $\mathbf{X}'\mathbf{Xr} = \lambda$, where $\mathbf{r} = (0, 0, \frac{1}{3})'$. Show that $\mathbf{X}'\mathbf{X}(\mathbf{X}'\mathbf{X})^-\boldsymbol{\lambda} = (1, 0, 1)'$. These values of $\mathbf{a}$ and $\mathbf{r}$ are illustrative. Many others are possible.

(b) We attempt to find a vector $\mathbf{a}$ such that $\mathbf{a}'\mathbf{X} = \boldsymbol{\lambda}' = (0, 1, 1)$. Since $\mathbf{X}$ has only two distinct rows, $\mathbf{a}'\mathbf{X}$ is of the form $a_1(1, 1, 0) + a_2(1, 0, 1) = (a_1 + a_2, a_1, a_2)$ which must equal $(0, 1, 1)$. This gives $a_1 + a_2 = 0$, $a_1 = 1$, and $a_2 = 1$, which is clearly impossible. By Theorem 2.8f, the system of equations $\mathbf{X}'\mathbf{Xr} = \boldsymbol{\lambda}$ has a solution if and only if $\mathbf{X}'\mathbf{X}(\mathbf{X}'\mathbf{X})^-\boldsymbol{\lambda} = \boldsymbol{\lambda}$. This is also condition (iii) of Theorem 11.2b. We find that

$$\mathbf{X}'\mathbf{X}(\mathbf{X}'\mathbf{X})^-\boldsymbol{\lambda} = \begin{pmatrix} 0 & 1 & 1 \\ 0 & 1 & 0 \\ 0 & 0 & 1 \end{pmatrix} \begin{pmatrix} 0 \\ 1 \\ 1 \end{pmatrix} = \begin{pmatrix} 2 \\ 1 \\ 1 \end{pmatrix},$$

which is not equal to $\boldsymbol{\lambda}$.

12.6 Multiply the two sets of normal equations by $\mathbf{r}'$, where $\mathbf{r}'\mathbf{X}'\mathbf{X} = \boldsymbol{\lambda}'$:

$$\mathbf{r}'\mathbf{X}'\mathbf{X}\hat{\boldsymbol{\beta}}_1 = \mathbf{r}'\mathbf{X}'\mathbf{y}$$
$$\mathbf{r}'\mathbf{X}'\mathbf{X}\hat{\boldsymbol{\beta}}_2 = \mathbf{r}'\mathbf{X}'\mathbf{y}.$$

Since the right sides are equal, we obtain $\mathbf{r}'\mathbf{X}'\mathbf{X}\hat{\boldsymbol{\beta}}_1 = \mathbf{r}'\mathbf{X}'\mathbf{X}\hat{\boldsymbol{\beta}}_2$, or $\lambda'\hat{\boldsymbol{\beta}}_1 = \lambda'\hat{\boldsymbol{\beta}}_2$.

12.7 In the answer to Problem 11.5a, a solution to $\mathbf{X}'\mathbf{X}\mathbf{r} = \lambda$ is given as $\mathbf{r} = (0, 0, \frac{1}{3})'$. Thus

$$\mathbf{r}'\mathbf{X}'\mathbf{y} = (0, 0, \tfrac{1}{3})\begin{pmatrix} y_{..} \\ y_{1.} \\ y_{2.} \end{pmatrix} = \frac{y_{2.}}{3} = \bar{y}_{2.}.$$

For $\boldsymbol{\lambda}'\hat{\boldsymbol{\beta}}$, we use

$$\hat{\boldsymbol{\beta}} = \begin{pmatrix} \hat{\mu} \\ \bar{y}_{1.} - \hat{\mu} \\ \bar{y}_{2.} - \hat{\mu} \end{pmatrix}$$

from Example 12.3.1. Then

$$\boldsymbol{\lambda}'\hat{\boldsymbol{\beta}} = (1, 0, 1)\begin{pmatrix} \hat{\mu} \\ \bar{y}_{1.} - \hat{\mu} \\ \bar{y}_{2.} - \hat{\mu} \end{pmatrix} = \hat{\mu} + \bar{y}_{2.} - \hat{\mu} = \bar{y}_{2.}.$$

12.8 (a) $\mathbf{X}'\mathbf{X}\mathbf{r} = \boldsymbol{\lambda}$ is given by

$$\begin{pmatrix} 6 & 3 & 3 \\ 3 & 3 & 0 \\ 3 & 0 & 3 \end{pmatrix}\begin{pmatrix} r_1 \\ r_2 \\ r_3 \end{pmatrix} = \begin{pmatrix} 1 \\ 1 \\ 0 \end{pmatrix}, \quad \text{or}$$

$$6r_1 + 3r_2 + 3r_3 = 1$$
$$3r_1 + 3r_2 = 1$$
$$3r_1 + 3r_3 = 0$$

Using the last two equations, we obtain

$$r_1 = -r_3$$
$$r_2 = r_3 + \tfrac{1}{3},$$

or

$$r = \begin{pmatrix} r_1 \\ r_2 \\ r_3 \end{pmatrix} = \begin{pmatrix} -r_3 \\ r_3 + \frac{1}{3} \\ r_3 \end{pmatrix} = r_3 \begin{pmatrix} -1 \\ 1 \\ 1 \end{pmatrix} + \begin{pmatrix} 0 \\ \frac{1}{3} \\ 0 \end{pmatrix},$$

where r_3 is an arbitrary constant that we can denote by c.

(b) The BLUE, $\mathbf{r}'\mathbf{X}'\mathbf{y}$, is given by

$$\begin{aligned} \mathbf{r}'\mathbf{X}'\mathbf{y} &= (-c, c + \tfrac{1}{3}, c) \begin{pmatrix} y_{..} \\ y_{1.} \\ y_{2.} \end{pmatrix} \\ &= -cy_{..} + cy_{1.} + \tfrac{1}{3}y_{1.} + cy_{2.} \\ &= -c(y_{1.} + y_{2.}) + cy_{1.} + \tfrac{1}{3}y_{1.} + cy_{2.} = \tfrac{1}{3}y_{1.}. \end{aligned}$$

Show that $y_{..} = y_{1.} + y_{2.}$

12.9 (a) From Example 12.2.2(b), we have

$$\boldsymbol{\beta} = \begin{pmatrix} \mu \\ \alpha_1 \\ \alpha_2 \\ \beta_1 \\ \beta_2 \end{pmatrix}, \quad \mu + \alpha_1 + \beta_1 = (1, 1, 0, 1, 0)\boldsymbol{\beta} = \boldsymbol{\lambda}_1'\boldsymbol{\beta}.$$

Show that

$$\mathbf{X}'\mathbf{X} = \begin{pmatrix} 4 & 2 & 2 & 2 & 2 \\ 2 & 2 & 0 & 1 & 1 \\ 2 & 0 & 2 & 1 & 1 \\ 2 & 1 & 1 & 2 & 0 \\ 2 & 1 & 1 & 0 & 2 \end{pmatrix}, \quad \mathbf{X}'\mathbf{y} = \begin{pmatrix} y_{..} \\ y_{1.} \\ y_{2.} \\ y_{.1} \\ y_{.2} \end{pmatrix}.$$

The value $\mathbf{r}' = (0, \frac{1}{2}, 0, \frac{1}{4}, -\frac{1}{4})$ gives $\mathbf{r}'\mathbf{X}'\mathbf{X} = \boldsymbol{\lambda}_1'$. Then

$$\mathbf{r}'\mathbf{X}'\mathbf{y} = \frac{y_{1.}}{2} + \frac{y_{.1}}{4} - \frac{y_{.2}}{4}.$$

For $\boldsymbol{\lambda}_2'\boldsymbol{\beta} = \boldsymbol{\beta}_1 - \boldsymbol{\beta}_2 = (0, 0, 0, 1, -1)\boldsymbol{\beta}$, a convenient value for $\mathbf{r}$ is $\mathbf{r}' = (0, 0, 0, \frac{1}{2}, -\frac{1}{2})$, which gives $\mathbf{r}'\mathbf{X}'\mathbf{y} = \frac{1}{2}y_{.1} - \frac{1}{2}y_{.2} = \bar{y}_{.1} - \bar{y}_{.2}$. The function $\boldsymbol{\lambda}_3'\boldsymbol{\beta} = \alpha_1 - \alpha_2 = (0, 1, -1, 0, 0)\boldsymbol{\beta}$ can be obtained using $\mathbf{r}' = (0, \frac{1}{2}, -\frac{1}{2}, 0, 0)$, which leads to $\mathbf{r}'\mathbf{X}'\mathbf{y} = \frac{1}{2}y_{1.} - \frac{1}{2}y_{2.} = \bar{y}_{1.} - \bar{y}_{2.}$

(b)

$$\begin{aligned}E(\mathbf{r}_1'\mathbf{X}'\mathbf{y}) &= E\left(\frac{y_{1.}}{2}+\frac{y_{.1}}{4}-\frac{y_{.2}}{4}\right)\\ &= \tfrac{1}{4}E[2(y_{11}+y_{12})+(y_{11}+y_{21})-(y_{12}+y_{22})]\\ &= \tfrac{1}{4}E(3y_{11}+y_{12}+y_{21}-y_{22})\\ &= \tfrac{1}{4}[3(\mu+\alpha_1+\beta_1)+\mu+\alpha_1\\ &\quad+\beta_2+\mu+\alpha_2+\beta_1-(\mu+\alpha_2+\beta_2)]\\ &= \tfrac{1}{4}(4\mu+4\alpha_1+4\beta_1)=\mu+\alpha_1+\beta_1=\boldsymbol{\lambda}_1'\boldsymbol{\beta}.\end{aligned}$$

12.10 **(a)** The function $\boldsymbol{\lambda}'\boldsymbol{\beta} = (0, c_1, c_2, \ldots, c_k)\boldsymbol{\beta} = \sum_{i=1}^k c_i\tau_i$ is estimable if there exists a vector a such that $\boldsymbol{\lambda}' = \mathbf{a}'\mathbf{X}$. The k distinct rows of $\mathbf{X}$ are of the form $\mathbf{x}_i' = (1, 0, \ldots, 0, 1, 0, \ldots, 0)$, so that

$$\boldsymbol{\lambda}' = \mathbf{a}'\mathbf{X} = \sum_{i=1}^{k} a_i\mathbf{x}_i' = \left(\sum_i a_i, a_1, a_2, \ldots, a_k\right).$$

Equating this to $\boldsymbol{\lambda}' = (0, c_1, c_2, \ldots, c_k)$, we obtain $\sum_{i=1}^k a_i = 0$, $a_i = c_i$ $i = 1, 2, \ldots, k$. Thus $\sum_i c_i = 0$.

(b) Any estimable function can be found as $\mathbf{a}'\mathbf{X}\boldsymbol{\beta}$, which gives

$$\begin{aligned}\mathbf{a}'\mathbf{X}\boldsymbol{\beta} = \mathbf{a}'E(\mathbf{y}) &= \sum_{i=1}^{k}\sum_{j=1}^{n} a_{ij}E(y_{ij})\\ &= \sum_i\sum_j a_{ij}(\mu+\tau_i) = \sum_i\left[(\mu+\tau_i)\sum_j a_{ij}\right]\\ &= \sum_i(\mu+\tau_i)a_{i.} = \mu\sum_i a_{i.} + \sum_i a_{i.}\tau_i\\ &= \mu\sum_i c_i + \sum_i c_i\tau_i,\end{aligned}$$

where $c_i = a_{i.} = \sum_j a_{ij}$. Thus $\sum_i c_i\tau_i$ is estimable if and only if $\sum_i c_i = 0$.

12.11 In Example 12.2.2(a), part (ii), we have

$$\mathbf{X}'\mathbf{X} = \begin{pmatrix} 6 & 3 & 3\\ 3 & 3 & 0\\ 3 & 0 & 3\end{pmatrix}.$$

Then for $\boldsymbol{\lambda} = (0, 1, -1)'$, $\mathbf{X}'\mathbf{X}\mathbf{r} = \boldsymbol{\lambda}$ becomes

$$\begin{aligned} 6r_1 + 3r_2 + 3r_3 &= 0 \\ 3r_1 + 3r_2 &= 1 \\ 3r_1 + 3r_3 &= -1. \end{aligned}$$

Show that all solutions are given by

$$\mathbf{r} = c\begin{pmatrix} 1 \\ -1 \\ -1 \end{pmatrix} + \tfrac{1}{3}\begin{pmatrix} 0 \\ 1 \\ -1 \end{pmatrix},$$

where c is arbitrary. Show that $\mathbf{r}'\mathbf{X}'\mathbf{y} = \bar{y}_{1.} - \bar{y}_{2.}$ for all values of c.

12.12 Use to Corollary 2 Theorem 3.6d(ii) to obtain $\text{cov}(\mathbf{r}_1'\mathbf{X}'\mathbf{y}, \mathbf{r}_2'\mathbf{X}'\mathbf{y}) = \mathbf{r}_1'\mathbf{X}'\text{cov}(\mathbf{y})\mathbf{X}\mathbf{r}_2$ and $\text{cov}(\boldsymbol{\lambda}_1'\hat{\boldsymbol{\beta}}, \boldsymbol{\lambda}_2'\hat{\boldsymbol{\beta}}) = \boldsymbol{\lambda}_1'\text{cov}(\hat{\boldsymbol{\beta}})\boldsymbol{\lambda}_2$.

12.13 **(a)** $(y - \mathbf{X}\hat{\boldsymbol{\beta}})'(\mathbf{y} - \mathbf{X}\hat{\boldsymbol{\beta}}) = \mathbf{y}'\mathbf{y} - \mathbf{y}'\mathbf{X}\hat{\boldsymbol{\beta}} - \hat{\boldsymbol{\beta}}'\mathbf{X}'\mathbf{y} + \hat{\boldsymbol{\beta}}'\mathbf{X}'\mathbf{X}\hat{\boldsymbol{\beta}}$. Since $\mathbf{y}'\mathbf{X}\hat{\boldsymbol{\beta}}$ is a scalar, it is equal to its transpose $\hat{\boldsymbol{\beta}}'\mathbf{X}'\mathbf{y}$. The last term, $\hat{\boldsymbol{\beta}}\mathbf{X}'\mathbf{X}\hat{\boldsymbol{\beta}}$, becomes $\hat{\boldsymbol{\beta}}'\mathbf{X}'\mathbf{y}$ because $\mathbf{X}'\mathbf{X}\hat{\boldsymbol{\beta}} = \mathbf{X}'\mathbf{y}$.

(b) Using $\hat{\boldsymbol{\beta}} = (\mathbf{X}'\mathbf{X})^{-}\mathbf{X}'\mathbf{y}$, we have

$$\begin{aligned} \mathbf{y}'\mathbf{y} - \hat{\boldsymbol{\beta}}'\mathbf{X}'\mathbf{y} &= \mathbf{y}'\mathbf{y} - \mathbf{y}'\mathbf{X}[(\mathbf{X}'\mathbf{X})^{-}]'\mathbf{X}'\mathbf{y} \\ &= \mathbf{y}'\mathbf{y} - \mathbf{y}'\mathbf{X}(\mathbf{X}'\mathbf{X})^{-}\mathbf{X}'\mathbf{y} \end{aligned}$$

by Theorem 2.8c(ii).

12.14 $\boldsymbol{\beta}'\mathbf{X}'[\mathbf{I} - \mathbf{X}(\mathbf{X}'\mathbf{X})^{-}\mathbf{X}']\mathbf{X}\boldsymbol{\beta} = \boldsymbol{\beta}'\mathbf{X}'\mathbf{X}\boldsymbol{\beta} - \boldsymbol{\beta}'\mathbf{X}'\mathbf{X}(\mathbf{X}'\mathbf{X})^{-}\mathbf{X}'\mathbf{X}\boldsymbol{\beta}$.
By (2.58), $\mathbf{X}'\mathbf{X}(\mathbf{X}'\mathbf{X})^{-}\mathbf{X}'\mathbf{X} = \mathbf{X}'\mathbf{X}$.

12.15 Follow the steps in the answer to Problem 7.21. Is there any step that must be altered because $\mathbf{X}$ is not full-rank?

12.16 **(a)** Since $\hat{\boldsymbol{\beta}} = (\mathbf{X}'\mathbf{X})^{-}\mathbf{X}'\mathbf{y}$ is a linear function of $\mathbf{y}$ for a particular choice of $(\mathbf{X}'\mathbf{X})^{-}$, we can use Theorem 4.4a(ii) directly.

(b) Show that $\mathbf{I} - \mathbf{X}(\mathbf{X}'\mathbf{X})^{-}\mathbf{X}'$ is idempotent. Then use Corollary 2 to Theorem 5.5.

(c) Show that $(\mathbf{X}'\mathbf{X})^{-}\mathbf{X}'[\mathbf{I} - \mathbf{X}(\mathbf{X}'\mathbf{X})^{-}\mathbf{X}'] = \mathbf{O}$, and then invoke Corollary 1 to Theorem 5.6a.

12.17 Since $\boldsymbol{\gamma} = \mathbf{U}\boldsymbol{\beta}$ and $\mathbf{X}\boldsymbol{\beta} = \mathbf{Z}\boldsymbol{\gamma}$, we have $\boldsymbol{\lambda}'\boldsymbol{\beta} = \mathbf{a}'\mathbf{X}\boldsymbol{\beta} = \mathbf{a}'\mathbf{Z}\boldsymbol{\gamma} = \mathbf{b}'\boldsymbol{\gamma}$. Thus $\widehat{\boldsymbol{\lambda}'\boldsymbol{\beta}} = \widehat{\mathbf{b}'\boldsymbol{\gamma}} = \mathbf{b}'\hat{\boldsymbol{\gamma}}$. Similarly, with $\boldsymbol{\delta} = \mathbf{V}\boldsymbol{\beta}$ and $\mathbf{X}\boldsymbol{\beta} = \mathbf{W}\boldsymbol{\delta}$, we have $\boldsymbol{\lambda}'\boldsymbol{\beta} = \mathbf{a}'\mathbf{X}\boldsymbol{\beta} = \mathbf{a}'\mathbf{W}\boldsymbol{\delta} = \mathbf{c}'\boldsymbol{\delta}$ and $\widehat{\boldsymbol{\lambda}'\boldsymbol{\beta}} = \widehat{\mathbf{c}'\boldsymbol{\delta}} = \mathbf{c}'\hat{\boldsymbol{\delta}}$.

12.18

$$\mathbf{XU}' = \begin{pmatrix} 2 & 1 \\ 2 & 1 \\ 1 & 2 \\ 1 & 2 \end{pmatrix}, \quad \mathbf{UU}' = \begin{pmatrix} 2 & 1 \\ 1 & 2 \end{pmatrix}.$$

12.19

$$\mathbf{Z} = \begin{pmatrix} 1 & 0 \\ 1 & 0 \\ 1 & -1 \\ 1 & -1 \end{pmatrix}, \quad \mathbf{U} = \begin{pmatrix} 1 & 1 & 0 \\ 0 & 1 & -1 \end{pmatrix}.$$

12.20 The normal equations are given by

$$\begin{aligned} 4\hat{\mu} + 2\hat{\tau}_1 + 2\hat{\tau}_2 &= y_{..} \\ 2\hat{\mu} + 2\hat{\tau}_1 &= y_{1.} \\ 2\hat{\mu} + 2\hat{\tau}_2 &= y_{2.} \end{aligned}$$

Substituting $\hat{\boldsymbol{\beta}}$ in (12.39) into the first of these, for example, gives

$$\begin{aligned} 4\bar{y}_{..} + 2(\bar{y}_{1.} - \bar{y}_{..}) + 2(\bar{y}_{2.} - \bar{y}_{..}) &= y_{..} \\ \frac{4y_{..}}{4} + 2\left(\frac{y_{1.}}{2} - \frac{y_{..}}{4}\right) + 2\left(\frac{y_{2.}}{2} - \frac{y_{..}}{4}\right) &= y_{..} \\ y_{..} + y_{1.} + y_{2.} - y_{..} &= y_{..} \\ y_{..} &= y_{..} \end{aligned}$$

12.21 $\alpha_1 - \alpha_2 = 0$ gives $\alpha_1 = \alpha_2$. Substituting this into $\alpha_1 + \alpha_2 - 2\alpha_3 = 0$ gives $2\alpha_2 - 2\alpha_3 = 0$ or $\alpha_2 = \alpha_3$.

12.22 Express SSH as a quadratic form in $\mathbf{y}$ by substituting $\hat{\boldsymbol{\beta}} = (\mathbf{X}'\mathbf{X})^{-}\mathbf{X}'\mathbf{y}$. Show that SSH is independent of SSE in (12.21) by use of Corollary 1 to Theorem 5.6b. Use either $\mathbf{C}(\mathbf{X}'\mathbf{X})^{-}\mathbf{X}'\mathbf{X} = \mathbf{C}$ or $\mathbf{X}'\mathbf{X}(\mathbf{X}'\mathbf{X})^{-}\mathbf{X}' = \mathbf{X}'$.

12.23 The first normal equation, for example, is $6\hat{\mu} + 2\hat{\alpha}_1 + 2\hat{\alpha}_2 + 2\hat{\alpha}_3 + 3\hat{\boldsymbol{\beta}}_1 + 3\hat{\boldsymbol{\beta}}_2 = y_{..}$, which simplifies to $6\hat{\mu} = y_{..}$ when we use the two side conditions.

12.24

$$\mathbf{X}_2 = \begin{pmatrix} 1 & 1 & 0 \\ 1 & 1 & 0 \\ 1 & 1 & 0 \\ 1 & 0 & 1 \\ 1 & 0 & 1 \\ 1 & 0 & 1 \end{pmatrix}, \quad \mathbf{X}_2'\mathbf{X}_2 = \begin{pmatrix} 6 & 3 & 3 \\ 3 & 3 & 0 \\ 3 & 0 & 3 \end{pmatrix}, \quad \boldsymbol{\beta}_2 = \begin{pmatrix} \mu \\ \boldsymbol{\beta}_1 \\ \boldsymbol{\beta}_2 \end{pmatrix},$$

$$\mathbf{X}_2'\mathbf{y} = \begin{pmatrix} y_{..} \\ y_{.1} \\ y_{.2} \end{pmatrix}.$$

Then $\mathbf{X}_2'\mathbf{X}_2\hat{\boldsymbol{\beta}}_2 = \mathbf{X}_2'\mathbf{y}$ gives the result in (12.51).

12.25 (a)

$$\mathbf{X} = \begin{pmatrix} 1 & 1 & 0 & 0 \\ 1 & 1 & 0 & 0 \\ 1 & 1 & 0 & 0 \\ 1 & 0 & 1 & 0 \\ 1 & 0 & 1 & 0 \\ 1 & 0 & 1 & 0 \\ 1 & 0 & 0 & 1 \\ 1 & 0 & 0 & 1 \\ 1 & 0 & 0 & 1 \end{pmatrix}, \quad \mathbf{X}'\mathbf{X} = \begin{pmatrix} 9 & 3 & 3 & 3 \\ 3 & 3 & 0 & 0 \\ 3 & 0 & 3 & 0 \\ 3 & 0 & 0 & 3 \end{pmatrix},$$

$$\mathbf{X}'\mathbf{y} = \begin{pmatrix} y_{..} \\ y_{1.} \\ y_{2.} \\ y_{3.} \end{pmatrix}.$$

The normal equations are given by

$$\begin{pmatrix} 9 & 3 & 3 & 3 \\ 3 & 3 & 0 & 0 \\ 3 & 0 & 3 & 0 \\ 3 & 0 & 3 & 0 \end{pmatrix} \begin{pmatrix} \hat{\mu} \\ \hat{\tau}_1 \\ \hat{\tau}_2 \\ \hat{\tau}_3 \end{pmatrix} = \begin{pmatrix} y_{..} \\ y_{1.} \\ y_{2.} \\ y_{3.} \end{pmatrix}, \quad \text{or}$$

$$9\hat{\mu} + 3\hat{\tau}_1 + 3\hat{\tau}_2 + 3\hat{\tau}_3 = y_{..}$$
$$3\hat{\mu} + 3\hat{\tau}_i = y_{i.} \quad i = 1, 2, 3$$

(b) Three possible sets of linearly independent estimable functions are

$$\{\mu + \tau_1, \quad \mu + \tau_2, \quad \mu + \tau_3\}$$
$$\{3\mu + \tau_1 + \tau_2 + \tau_3, \quad \tau_1 - \tau_2, \quad \tau_2 - \tau_3\}$$
$$\{\mu + \tau_1, \quad \tau_1 - \tau_2, \quad \tau_2 - \tau_3\}.$$

(c) The side condition $\hat{\tau}_1 + \hat{\tau}_2 + \hat{\tau}_3 = 0$ gives

$$\hat{\mu} = \frac{y_{..}}{9} = \bar{y}_{..}$$
$$\hat{\tau}_i = \tfrac{1}{3}y_{i.} - \tfrac{1}{9}y_{..} = \bar{y}_{i.} - \bar{y}_{..} \quad i = 1, 2, 3.$$

(d) The hypothesis $H_0\colon \tau_1 = \tau_2 = \tau_3$ is equivalent to $H_0\colon \tau_1 - \tau_2 = 0$ and $\tau_1 - \tau_3 = 0$; hence H_0 is testable:

$$\begin{aligned}
\text{SS}(\mu, \tau) &= \hat{\boldsymbol{\beta}}'\mathbf{X}'\mathbf{y} = \hat{\mu}y_{..} + \sum_{i=1}^{3} \hat{\tau}_i y_{i.} \\
&= \bar{y}_{..}y_{..} + \sum_{i=1}^{3}\left(\frac{y_{i.}}{3} - \frac{y_{..}}{9}\right)y_{i.} \\
&= \frac{y_{..}^2}{9} + \sum_{i=1}^{3}\frac{y_{i.}^2}{3} - \frac{y_{..}^2}{9} = \sum_{i=1}^{3}\frac{y_{i.}^2}{3}.
\end{aligned}$$

The reduced model is $y_{ij} + \mu + \varepsilon_{ij}$, the $\mathbf{X}_2$ matrix reduces to a single column of 1's, and the normal equations become

$$9\hat{\mu} = y_{..}$$
$$\hat{\mu} = \frac{y_{..}}{9} = \bar{y}_{..}.$$

Hence

$$\text{SS}(\mu) = \hat{\boldsymbol{\beta}}_2'\mathbf{X}_2'\mathbf{y} = \bar{y}_{..}y_{..} = \frac{y_{..}^2}{9}.$$

(e)

Analysis of Variance for $H_0: \tau_1 = \tau_2 = \tau_3$

Sum of Squares	df	F Statistic
$\text{SS}(\tau\vert\mu) = \sum_{i=1}^{3}\frac{y_{i.}^2}{3} - \frac{y_{..}^2}{9}$	2	$\frac{\text{SS}(\tau\vert\mu)/2}{\text{SSE}/6}$
$\text{SSE} = \sum_{ij} y_{ij}^2 - \sum_i \frac{y_{i.}^2}{3}$	6	
$\text{SST} = \sum_{ij} y_{ij}^2 - \frac{y_{..}^2}{9}$	8	

12.26 (a) The normal equations $\mathbf{X}'\mathbf{X}\hat{\boldsymbol{\beta}} = \mathbf{X}'\mathbf{y}$ are given by

$$\begin{pmatrix} 12 & 6 & 6 & 6 & 6 & 3 & 3 & 3 & 3 \\ 6 & 6 & 0 & 3 & 3 & 3 & 3 & 0 & 0 \\ 6 & 0 & 6 & 3 & 3 & 0 & 0 & 3 & 3 \\ 6 & 3 & 3 & 6 & 0 & 3 & 0 & 3 & 0 \\ 6 & 3 & 3 & 0 & 6 & 0 & 3 & 0 & 3 \\ 3 & 3 & 0 & 3 & 0 & 3 & 0 & 0 & 0 \\ 3 & 0 & 3 & 3 & 0 & 0 & 0 & 3 & 0 \\ 3 & 0 & 3 & 0 & 3 & 0 & 0 & 0 & 3 \end{pmatrix}, \quad \text{or} \quad \begin{pmatrix} \hat{\mu} \\ \hat{\alpha}_1 \\ \hat{\alpha}_2 \\ \hat{\boldsymbol{\beta}}_1 \\ \hat{\boldsymbol{\beta}}_2 \\ \hat{\gamma}_{11} \\ \hat{\gamma}_{12} \\ \hat{\gamma}_{21} \\ \hat{\gamma}_{22} \end{pmatrix} = \begin{pmatrix} y_{...} \\ y_{1..} \\ y_{2..} \\ y_{.1.} \\ y_{.2.} \\ y_{11.} \\ y_{12.} \\ y_{21.} \\ y_{22.} \end{pmatrix}, \quad \text{or}$$

$$12\hat{\mu} + 6\sum_{i=1}^{2}\hat{\alpha}_i + 6\sum_{j=1}^{2}\hat{\boldsymbol{\beta}}_j + 3\sum_{ij}\hat{\gamma}_{ij} = y_{..}$$

$$6\hat{\mu} + 6\hat{\alpha}_i + 3\sum_{j=1}^{2}\hat{\boldsymbol{\beta}}_j + 3\sum_{j=1}^{2}\hat{\gamma}_{ij} = y_{i..} \quad i = 1, 2$$

$$6\hat{\mu} + 3\sum_{i=1}^{2}\hat{\alpha}_i + 6\hat{\boldsymbol{\beta}}_j + 3\sum_{i=2}^{2}\hat{\gamma}_{ij} = y_{.j.} \quad j = 1, 2$$

$$3\hat{\mu} + 3\hat{\alpha}_i + 3\hat{\boldsymbol{\beta}}_j + 3\hat{\gamma}_{ij} = y_{ij.} \quad i = 1, 2 \quad j = 1, 2$$

(b) The rank of $\mathbf{X}'\mathbf{X}$ is 4. From the last four rows, which are linearly independent, we obtain

$$\begin{aligned} &\mu + \alpha_1 + \beta_1 + \gamma_{11} \\ &\mu + \alpha_1 + \beta_2 + \gamma_{12} \\ &\mu + \alpha_2 + \beta_1 + \gamma_{21} \\ &\mu + \alpha_2 + \beta_2 + \gamma_{22} \end{aligned}$$

or

$$\begin{aligned} &\mu + \alpha_1 + \beta_1 + \gamma_{11} \\ &\alpha_1 - \alpha_2 + \gamma_{11} - \gamma_{21} \quad (\text{or } \alpha_1 - \alpha_2 + \gamma_{12} - \gamma_{22}) \\ &\beta_1 - \beta_2 + \gamma_{11} - \gamma_{12} \quad (\text{or } \beta_1 - \beta_2 + \gamma_{21} - \gamma_{22}) \\ &\gamma_{11} - \gamma_{12} - \gamma_{21} + \gamma_{22}. \end{aligned}$$

12.27 **(a)** The normal equations are

$$\begin{pmatrix} 8 & 4 & 4 & 4 & 4 & 4 & 4 \\ 4 & 4 & 0 & 2 & 2 & 2 & 2 \\ 4 & 0 & 4 & 2 & 2 & 2 & 2 \\ 4 & 2 & 2 & 4 & 0 & 2 & 2 \\ 4 & 2 & 2 & 0 & 4 & 2 & 2 \\ 4 & 2 & 2 & 2 & 2 & 4 & 0 \\ 4 & 2 & 2 & 2 & 2 & 0 & 4 \end{pmatrix} \begin{pmatrix} \hat{\mu} \\ \hat{\alpha}_1 \\ \hat{\alpha}_2 \\ \hat{\beta}_1 \\ \hat{\beta}_2 \\ \hat{\gamma}_1 \\ \hat{\gamma}_2 \end{pmatrix} = \begin{pmatrix} y_{...} \\ y_{1..} \\ y_{2..} \\ y_{.1.} \\ y_{.2.} \\ y_{..1} \\ y_{..2} \end{pmatrix}$$

(b)

$$\mu + \alpha_1 + \beta_1 + \gamma_1$$
$$\alpha_1 - \alpha_2$$
$$\beta_1 - \beta_2$$
$$\gamma_1 - \gamma_2$$

(c) Using the side conditions $\hat{\alpha}_1 + \hat{\alpha}_2 = 0$, $\hat{\beta}_1 + \hat{\beta}_2 = 0$, $\hat{\gamma}_1 + \hat{\gamma}_2 = 0$, we obtain $\hat{\mu} = \bar{y}_{...}$, $\hat{\alpha}_i = \bar{y}_{i..} - \bar{y}_{...}$, $\hat{\beta}_j = \bar{y}_{.j.} - \bar{y}_{...}$, $\hat{\gamma}_k = \bar{y}_{..k} - \bar{y}_{...}$.

(d)

$$\begin{aligned} \text{SS}(\mu, \alpha, \beta, \gamma) &= \hat{\boldsymbol{\beta}}'\mathbf{X}'\mathbf{y} = \bar{y}_{...}y_{...} + \sum_i (\bar{y}_{i..} - \bar{y}_{...})y_{i..} \\ &\quad + \sum_j (\bar{y}_{.j.} - \bar{y}_{...})y_{.j.} + \sum_k (\bar{y}_{..k} - \bar{y}_{...})y_{..k} \\ &= \frac{y_{...}^2}{8} + \sum_i \frac{y_{i..}^2}{4} - \frac{y_{...}^2}{8} + \sum_j \frac{y_{.j.}^2}{4} - \frac{y_{...}^2}{8} + \sum_k \frac{y_{..k}^2}{4} - \frac{y_{...}^2}{8} \\ &= \text{SS}(\mu) + \text{SS}(\alpha) + \text{SS}(\beta) + \text{SS}(\gamma). \end{aligned}$$

Using this same notation, the reduced normal equations under $H_0 : \alpha_1 = \alpha_2$ become $\text{SS}(\mu, \beta, \gamma) + \text{SS}(\mu) + \text{SS}(\beta) + \text{SS}(\gamma)$.

(e)

Analysis of Variance for H_0: $\tau_1 = \tau_2 = \tau_3$

Source	df	Sum of Squares	F
$\text{SS}(\alpha\|\mu, \beta, \gamma)$	1	$\text{SS}(\mu, \alpha, \beta, \gamma) - \text{SS}(\mu, \beta, \gamma) = \text{SS}(\alpha)$	$\dfrac{\text{SS}(\alpha\|\mu, \beta, \gamma)}{\text{SSE}/4}$
Error	4	$\text{SSE} = \sum_{ijk} y_{ijk}^2 - \text{SSE}(\mu, \alpha, \beta, \gamma)$	

12.28

$$\mathbf{X}'\mathbf{X} = \begin{pmatrix} 8 & 4 & 4 & 4 & 4 & 2 & 2 & 2 & 2 \\ 4 & 4 & 0 & 2 & 2 & 2 & 2 & 0 & 0 \\ 4 & 0 & 4 & 2 & 2 & 0 & 0 & 2 & 2 \\ 4 & 0 & 2 & 4 & 0 & 2 & 0 & 2 & 0 \\ 4 & 2 & 2 & 0 & 4 & 0 & 2 & 0 & 2 \\ 2 & 2 & 0 & 2 & 0 & 2 & 0 & 0 & 0 \\ 2 & 2 & 0 & 0 & 2 & 0 & 2 & 0 & 0 \\ 2 & 0 & 2 & 2 & 0 & 0 & 0 & 2 & 0 \\ 2 & 0 & 2 & 0 & 2 & 0 & 0 & 0 & 2 \end{pmatrix}.$$

Chapter 13

13.1

$$\mathbf{X}'\mathbf{X} = \begin{pmatrix} k\mathbf{j}'\mathbf{j} & \mathbf{j}'\mathbf{j} & \mathbf{j}'\mathbf{j} & \dots & \mathbf{j}'\mathbf{j} \\ \mathbf{j}'\mathbf{j} & \mathbf{j}'\mathbf{j} & 0 & \dots & 0 \\ \mathbf{j}'\mathbf{j} & 0 & \mathbf{j}'\mathbf{j} & \dots & 0 \\ \vdots & \vdots & \vdots & & \vdots \\ \mathbf{j}'\mathbf{j} & 0 & 0 & \dots & \mathbf{j}'\mathbf{j} \end{pmatrix} = \begin{pmatrix} kn & n & n & \dots & n \\ n & n & 0 & \dots & 0 \\ n & 0 & n & \dots & 0 \\ \vdots & \vdots & \vdots & & \vdots \\ n & 0 & 0 & \dots & n \end{pmatrix},$$

$$\mathbf{X}'\mathbf{y} = \begin{pmatrix} \sum_i \mathbf{j}'\mathbf{y}_i \\ \mathbf{j}'\mathbf{y}_1 \\ \mathbf{j}'\mathbf{y}_2 \\ \vdots \\ \mathbf{j}'\mathbf{y}_k \end{pmatrix} = \begin{pmatrix} \sum_i y_{i.} \\ y_{1.} \\ y_{2.} \\ \vdots \\ y_{k.} \end{pmatrix} = \begin{pmatrix} y_{..} \\ y_{1.} \\ y_{2.} \\ \vdots \\ y_{k.} \end{pmatrix}.$$

13.2

$$\hat{\boldsymbol{\beta}} = (\mathbf{X}'\mathbf{X})^{-}\mathbf{X}'\mathbf{y} = \begin{pmatrix} 0 & 0 & \dots & 0 \\ 0 & 1/n & \dots & 0 \\ \vdots & \vdots & & \vdots \\ 0 & 0 & \dots & 1/n \end{pmatrix} \begin{pmatrix} y_{..} \\ y_{1.} \\ \vdots \\ y_{k.} \end{pmatrix} = \begin{pmatrix} 0 \\ y_{1.}/n \\ \vdots \\ y_{k.}/n \end{pmatrix} = \begin{pmatrix} 0 \\ \bar{y}_{1.} \\ \vdots \\ \bar{y}_{k.} \end{pmatrix}.$$

13.3

$$\begin{aligned} \sum_{ij} (y_{ij} - \bar{y}_{i.})^2 &= \sum_{i=1}^{k} \sum_{j=1}^{n} (y_{ij}^2 - 2y_{ij}\bar{y}_{i.} + \bar{y}_{i.}^2) \\ &= \sum_{ij} y_{ij}^2 - 2\sum_i \left(\frac{y_{i.}}{n} \sum_j y_{ij} \right) + n\sum_i \frac{y_{i.}^2}{n^2} \\ &= \sum_{ij} y_{ij}^2 - 2\sum_i \frac{y_{i.}^2}{n} + \sum_i \frac{y_{i.}^2}{n}. \end{aligned}$$

13.4

$$\mathbf{X}(\mathbf{X}'\mathbf{X})^-\mathbf{X}' = \begin{pmatrix} \mathbf{j} & \mathbf{j} & \mathbf{0} & \dots & \mathbf{0} \\ \mathbf{j} & \mathbf{0} & \mathbf{j} & \dots & \mathbf{0} \\ \vdots & \vdots & \vdots & & \vdots \\ \mathbf{j} & \mathbf{0} & \mathbf{0} & \dots & \mathbf{j} \end{pmatrix} \begin{pmatrix} 0 & 0 & 0 & \dots & 0 \\ 0 & 1/n & 0 & \dots & 0 \\ 0 & 0 & 1/n & \dots & 0 \\ \vdots & \vdots & \vdots & & \vdots \\ 0 & 0 & 0 & \dots & 1/n \end{pmatrix}$$

$$\times \begin{pmatrix} \mathbf{j}' & \mathbf{j}' & \mathbf{j}' & \dots & \mathbf{j}' \\ \mathbf{j}' & \mathbf{0}' & \mathbf{0}' & \dots & \mathbf{0}' \\ \mathbf{0}' & \mathbf{j}' & \mathbf{0}' & \dots & \mathbf{0}' \\ \vdots & \vdots & \vdots & & \vdots \\ \mathbf{0}' & \mathbf{0}' & \mathbf{0}' & \dots & \mathbf{j}' \end{pmatrix}$$

$$= \begin{pmatrix} \mathbf{0} & \frac{1}{n}\mathbf{j} & \mathbf{0} & \dots & \mathbf{0} \\ \mathbf{0} & \mathbf{0} & \frac{1}{n}\mathbf{j} & \dots & \mathbf{0} \\ \vdots & \vdots & \vdots & & \vdots \\ \mathbf{0} & \mathbf{0} & \mathbf{0} & \vdots & \frac{1}{n}\mathbf{j} \end{pmatrix} \begin{pmatrix} \mathbf{j}' & \mathbf{j}' & \mathbf{j}' & \dots & \mathbf{j}' \\ \mathbf{j}' & \mathbf{0}' & \mathbf{0}' & \dots & \mathbf{0}' \\ \mathbf{0}' & \mathbf{j}' & \mathbf{0}' & \dots & \mathbf{0}' \\ \vdots & \vdots & \vdots & & \vdots \\ \mathbf{0}' & \mathbf{0}' & \mathbf{0}' & \dots & \mathbf{j}' \end{pmatrix}$$

$$= \frac{1}{n} \begin{pmatrix} \mathbf{jj}' & \mathbf{O} & \dots & \mathbf{O} \\ \mathbf{O} & \mathbf{jj}' & \dots & \mathbf{O} \\ \vdots & \vdots & & \vdots \\ \mathbf{O} & \mathbf{O} & \dots & \mathbf{jj}' \end{pmatrix},$$

$$\mathbf{y}'[\mathbf{I} - \mathbf{X}(\mathbf{X}'\mathbf{X})^-\mathbf{X}']\mathbf{y} = \mathbf{y}'\mathbf{y} - \mathbf{y}'\mathbf{X}(\mathbf{X}'\mathbf{X})^-\mathbf{X}'\mathbf{y}$$

$$= \sum_{ij} y_{ij}^2 - \frac{1}{n}(\mathbf{y}_1', \mathbf{y}_2', \dots, \mathbf{y}_k') \begin{pmatrix} \mathbf{jj}' & \mathbf{O} & \dots & \mathbf{O} \\ \mathbf{O} & \mathbf{jj}' & \dots & \mathbf{O} \\ \vdots & \vdots & & \vdots \\ \mathbf{O} & \mathbf{O} & \dots & \mathbf{jj}' \end{pmatrix} \begin{pmatrix} \mathbf{y}_1 \\ \mathbf{y}_2 \\ \vdots \\ \mathbf{y}_k \end{pmatrix}$$

$$= \sum_{ij} y_{ij}^2 - \frac{1}{n}\sum_{i=1}^{k} \mathbf{y}_i'\mathbf{jj}'\mathbf{y}_i = \sum_{ij} y_{ij}^2 - \frac{1}{n}\sum_i y_{i.}^2$$

13.5 **(a)** With $\alpha_i^* = \mu_i - \bar{\mu}.$ in (13.5), H_0: $\alpha_i^* = \alpha_2^* = \dots = \alpha_k^*$ in (13.18) becomes $H_0: \mu_1 - \bar{\mu}. = \mu_2 - \bar{\mu}. = \dots = \mu_k - \bar{\mu}.$ or $H_0: \mu_1 = \mu_2 = \cdots \mu_k$, which is equivalent to (13.7).

(b) Denote by α^* the common value of α_i^* in $H_0: \alpha_1^* = \alpha_2^* = \dots = \alpha_k^*$ in (13.18). Then $\sum_i \alpha_i^* = 0$ give $\alpha^* = 0$, since $\sum_{i=1}^k \alpha^* = \alpha_i^* = 0$, $i = 1, 2, \dots, k$. Thus, $\alpha_i^* = 0$, $i = 1, 2, \dots, k$.

13.6

$$\begin{aligned}
n\sum_{i=1}^{k}(\bar{y}_{i.}-\bar{y}_{..})^2 &= n\sum_i(\bar{y}_{i.}^2-2\bar{y}_{i.}\bar{y}_{..}+\bar{y}_{..}^2)\\
&= n\sum_i\bar{y}_{i.}^2-2n\bar{y}_{..}\sum_i\bar{y}_{i.}+kn\bar{y}_{..}^2\\
&= n\sum_i\left(\frac{y_{i.}}{n}\right)^2-2n\frac{y_{..}}{kn}\sum_i\frac{y_{i.}}{n}+kn\left(\frac{y_{..}}{kn}\right)^2\\
&= \frac{1}{n}\sum_i y_{i.}^2-2\frac{y_{..}}{k}\frac{y_{..}}{n}+\frac{y_{..}^2}{kn}.
\end{aligned}$$

13.7 See the first part of the answer to Problem 13.3.

13.9 Using $\mathbf{X}$ in (13.6), we have

$$\begin{aligned}
\mathbf{C}(\mathbf{X}'\mathbf{X})^{-}\mathbf{X}' &= \begin{pmatrix} 0 & 1 & -1 & 0 & 0\\ 0 & 1 & 0 & -1 & 0\\ 0 & 1 & 0 & 0 & -1\end{pmatrix}\begin{pmatrix} 0 & 0 & 0 & 0 & 0\\ 0 & 1 & 0 & 0 & 0\\ 0 & 0 & 1 & 0 & 0\\ 0 & 0 & 0 & 1 & 0\\ 0 & 0 & 0 & 0 & 1\end{pmatrix}\\
&\quad\times\begin{pmatrix} \mathbf{j}_n & \mathbf{j}_n & \mathbf{0} & \mathbf{0} & \mathbf{0}\\ \mathbf{j}_n & \mathbf{0} & \mathbf{j}_n & \mathbf{0} & \mathbf{0}\\ \mathbf{j}_n & \mathbf{0} & \mathbf{0} & \mathbf{0} & \mathbf{j}_n\end{pmatrix}'\\
&= \begin{pmatrix} 0 & 1 & -1 & 0 & 0\\ 0 & 1 & 0 & -1 & 0\\ 0 & 1 & 0 & 0 & -1\end{pmatrix}\begin{pmatrix} \mathbf{j}_n' & \mathbf{j}_n' & \mathbf{j}_n' & \mathbf{j}_n'\\ \mathbf{j}_n' & \mathbf{0}' & \mathbf{0}' & \mathbf{0}'\\ \mathbf{0}' & \mathbf{j}_n' & \mathbf{0}' & \mathbf{0}'\\ \mathbf{0}' & \mathbf{0}' & \mathbf{j}_n' & \mathbf{0}'\\ \mathbf{0}' & \mathbf{0}' & \mathbf{0}' & \mathbf{j}_n'\end{pmatrix}\\
&= \begin{pmatrix} \mathbf{j}_n' & -\mathbf{j}_n' & \mathbf{0}' & \mathbf{0}'\\ \mathbf{j}_n' & \mathbf{0}' & -\mathbf{j}_n' & \mathbf{0}'\\ \mathbf{j}_n & \mathbf{0} & \mathbf{0}' & -\mathbf{j}_n'\end{pmatrix}.
\end{aligned}$$

13.10 By (2.37), we obtain

$$\mathbf{Aj}_3 = \begin{pmatrix} \mathbf{j}_n & \mathbf{j}_n & \mathbf{j}_n \\ -\mathbf{j}_n & \mathbf{0} & \mathbf{0} \\ \mathbf{0} & -\mathbf{j}_n & \mathbf{0} \\ \mathbf{0} & \mathbf{0} & -\mathbf{j}_n \end{pmatrix} \begin{pmatrix} 1 \\ 1 \\ 1 \end{pmatrix}$$

$$= \begin{pmatrix} \mathbf{j}_n \\ -\mathbf{j}_n \\ \mathbf{0} \\ \mathbf{0} \end{pmatrix} + \begin{pmatrix} \mathbf{j}_n \\ \mathbf{0} \\ -\mathbf{j}_n \\ \mathbf{0} \end{pmatrix} + \begin{pmatrix} \mathbf{j}_n \\ \mathbf{0} \\ \mathbf{0} \\ -\mathbf{j}_n \end{pmatrix} = \begin{pmatrix} 3\mathbf{j}_n \\ -\mathbf{j}_n \\ -\mathbf{j}_n \\ -\mathbf{j}_n \end{pmatrix},$$

$$\mathbf{AJ}_3\mathbf{A}' = \mathbf{Aj}_3\mathbf{j}_3'\mathbf{A}' = \begin{pmatrix} 3\mathbf{j}_n \\ -\mathbf{j}_n' \\ -\mathbf{j}_n' \\ -\mathbf{j}_n' \end{pmatrix} (3\mathbf{j}_n, -\mathbf{j}_n, -\mathbf{j}_n, -\mathbf{j}_n)$$

$$= \begin{pmatrix} 9\mathbf{J}_n & -3\mathbf{J}_n & -3\mathbf{J}_n & -3\mathbf{J}_n \\ -3\mathbf{J}_n & \mathbf{J}_n & \mathbf{J}_n & \mathbf{J}_n \\ -3\mathbf{J}_n & \mathbf{J}_n & \mathbf{J}_n & \mathbf{J}_n \\ -3\mathbf{J}_n & \mathbf{J}_n & \mathbf{J}_n & \mathbf{J}_n \end{pmatrix}.$$

13.11 (a)

$$\begin{aligned} E(\varepsilon_{ij})^2 &= E(\varepsilon_{ij} - 0)^2 = E[\varepsilon_{ij} - E(\varepsilon_{ij})]^2 = \text{var}(\varepsilon_{ij}) = \sigma^2, \\ E(\varepsilon_{ij}\varepsilon_{i'j'}) &= E[\varepsilon_{ij} - 0)(\varepsilon_{i'j'} - 0)] \\ &= E[\varepsilon_{ij} - E(\varepsilon_{ij})][\varepsilon_{i'j'} - E(\varepsilon_{i'j'})] \\ &= \text{cov}(\varepsilon_{ij}, \varepsilon_{i'j'}) = 0. \end{aligned}$$

(b)

$$\begin{aligned} E(y_{..}^2) &= E\left(\sum_{ij}^{n} y_{ij}\right)^2 = E\left[\sum_{i=1}^{n}\sum_{j=1}^{n}(\mu^* + \alpha_i^* + \varepsilon_{ij})\right]^2 \\ &= E\left[kn\mu^* + n\sum_i \alpha_i^* + \sum_{ij}\varepsilon_{ij}\right]^2 \\ &= E\left[k^2n^2\mu^{*2} + \left(\sum_{ij}\varepsilon_{ij}\right)^2 + 2kn\mu^*\sum_{ij}\varepsilon_{ij}\right] \\ &= E\left(k^2n^2\mu^{*2} + \sum_{ij}\varepsilon_{ij}^2 + \sum_{ij \neq lm}\varepsilon_{ij}\varepsilon_{lm} + 2kn\mu^*\sum_{ij}\varepsilon_{ij}\right) \\ &= k^2n^2\mu^{*2} + kn\sigma^2, \end{aligned}$$

$$
\begin{aligned}
E\left(\sum_{i=1}^{k} y_{i.}^2\right) &= E\left[\sum_{i=1}^{k}\left(\sum_{j=1}^{n} y_{ij}\right)^2\right] \\
&= E\left\{\sum_{i}\left[\sum_{j}(\mu^* + \alpha_i^* + \varepsilon_{ij})\right]^2\right\} \\
&= E\left[\sum_{i}\left(n\mu^* + n\alpha_i^* + \sum_{j}\varepsilon_{ij}\right)^2\right] \\
&= E\left\{\sum_{i}\left[n^2\mu^{*2} + n^2\alpha_i^{*2} + \left(\sum_{j}\varepsilon_{ij}\right)^2 + 2n^2\mu^*\alpha_i^*\right.\right. \\
&\quad \left.\left. + 2n\mu^*\sum_{j}\varepsilon_{ij} + 2n\sum_{j}\alpha_i^*\varepsilon_{ij}\right]\right\} \\
&= E\left[kn^2\mu^{*2} + n^2\sum_{i}\alpha_i^{*2} + \sum_{i}\left(\sum_{j}\varepsilon_{ij}^2 + \sum_{j\neq 1}\varepsilon_{ij}\varepsilon_{il}\right)\right. \\
&\quad \left. + 2n^2\mu^*\sum_{i}\alpha_i^* + 2n\mu^*\sum_{ij}\varepsilon_{ij} + 2n\sum_{i}\sum_{j}\alpha_i^*\varepsilon_{ij}\right] \\
&= kn^2\mu^{*2} + n^2\sum_{i}\alpha_i^{*2} + kn\sigma^2,
\end{aligned}
$$

$$
\begin{aligned}
E[\text{SS}(\alpha|\mu)] &= E\left(\frac{1}{n}\sum_{i} y_{i.}^2 - \frac{1}{kn}y_{..}^2\right) \\
&= \frac{1}{n}\left(kn^2\mu^{*2} + n^2\sum_{i}\alpha_i^{*2} + kn\sigma^2\right) - \frac{1}{kn}(k^2n^2\mu^{*2} + kn\sigma^2) \\
&= kn\mu^{*2} + n\sum_{i}\alpha_i^{*2} + k\sigma^2 - kn\mu^{*2} - \sigma^2 \\
&= (k-1)\sigma^2 + n\sum_{i}\alpha_i^{*2}.
\end{aligned}
$$

(c)

$$
\begin{aligned}
E\left(\sum_{i=1}^{k}\sum_{j=1}^{n} y_{ij}^2\right) &= E\left[\sum_{ij}(\mu^* + \alpha_i^* + \varepsilon_{ij})^2\right] \\
&= E\left[\sum_{ij}(\mu^{*2} + \alpha_i^{*2} + \varepsilon_{ij}^2 + 2\mu^*\alpha_i^* + 2\mu^*\varepsilon_{ij} + 2\alpha_i^*\varepsilon_{ij})\right]
\end{aligned}
$$

$$= E\left[kn\mu^{*2} + n\sum_i \alpha_i^{*2} + \sum_{ij} \varepsilon_{ij}^2 + 2n\mu^* \sum_i \alpha_i^* \right.$$

$$\left. +2\mu^* \sum_{ij} \varepsilon_{ij} + 2\sum_{ij} \alpha_i^* \varepsilon_{ij}\right]$$

$$= kn\mu^{*2} + n\sum_i \alpha_i^{*2} + kn\sigma^2,$$

$$E(\text{SSE}) = E\left(\sum_{ij} y_{ij}^2 - \frac{1}{n}\sum_i y_{i.}^2\right)$$

$$= kn\mu^{*2} + n\sum_i \alpha_i^{*2} + kn\sigma^2 - kn\mu^{*2} - n\sum_i \alpha_i^{*2} - k\sigma^2$$

$$= k(n-1)\sigma^2.$$

13.13 By (2.37),

$$\mathbf{C}'\mathbf{j}_3 = \begin{pmatrix} 0 & 0 & 0 \\ 1 & 1 & 1 \\ -1 & 0 & 0 \\ 0 & -1 & 0 \\ 0 & 0 & -1 \end{pmatrix} \begin{pmatrix} 1 \\ 1 \\ 1 \end{pmatrix} = \begin{pmatrix} 0 \\ 1 \\ -1 \\ 0 \\ 0 \end{pmatrix} + \begin{pmatrix} 0 \\ 1 \\ 0 \\ -1 \\ 0 \end{pmatrix} + \begin{pmatrix} 0 \\ 1 \\ 0 \\ 0 \\ -1 \end{pmatrix} + \begin{pmatrix} 0 \\ 3 \\ -1 \\ -1 \\ -1 \end{pmatrix}.$$

Thus

$$\mathbf{C}'\mathbf{J}_3 = \mathbf{C}'(\mathbf{j}_3, \mathbf{j}_3, \mathbf{j}_3) = (\mathbf{C}'\mathbf{j}_3, \mathbf{C}'\mathbf{j}_3, \mathbf{C}'\mathbf{j}_3) = \begin{pmatrix} 0 & 0 & 0 \\ 3 & 3 & 3 \\ -1 & -1 & -1 \\ -1 & -1 & -1 \\ -1 & -1 & -1 \end{pmatrix},$$

$$\mathbf{C}'\mathbf{J}_3\mathbf{C} = \begin{pmatrix} 0 & 0 & 0 \\ 3 & 3 & 3 \\ -1 & -1 & -1 \\ -1 & -1 & -1 \\ -1 & -1 & -1 \end{pmatrix} \begin{pmatrix} 0 & 1 & -1 & 0 & 0 \\ 0 & 1 & 0 & -1 & 0 \\ 0 & 1 & 0 & 0 & -1 \end{pmatrix}$$

$$= \begin{pmatrix} 0 & 0 & 0 & 0 & 0 \\ 0 & 9 & -3 & -3 & -3 \\ 0 & -3 & 1 & 1 & 1 \\ 0 & -3 & 1 & 1 & 1 \\ 0 & -3 & 1 & 1 & 1 \end{pmatrix}.$$

13.15 Using $(\mathbf{X}'\mathbf{X})^-$ in (13.11) and $\hat{\boldsymbol{\beta}}$ in (13.12), we obtain

$$\mathbf{c}'\hat{\beta} = (0, c_1, c_2, \ldots, c_k)\begin{pmatrix} 0 \\ \bar{y}_{1.} \\ \bar{y}_{2.} \\ \vdots \\ \bar{y}_{k.} \end{pmatrix} = \sum_{i=1}^{k} c_i \bar{y}_{1.},$$

$$\mathbf{c}'(\mathbf{X}'\mathbf{X})^-\mathbf{c} = (0, c_1, c_2, \ldots, c_k)\begin{pmatrix} 0 & 0 & \ldots & 0 \\ 0 & 1/n & \ldots & 0 \\ \vdots & \vdots & & \vdots \\ 0 & 0 & \ldots & 1/n \end{pmatrix}\begin{pmatrix} 0 \\ c_1 \\ c_2 \\ \vdots \\ c_k \end{pmatrix} = \sum_{i=1}^{k} \frac{c_i^2}{n}.$$

13.16 Using $\hat{\boldsymbol{\beta}} = (\mathbf{X}'\mathbf{X})^-\mathbf{X}'\mathbf{y}$, the sum of squares for the contrast $\mathbf{c}_i'\hat{\boldsymbol{\beta}}$ can be expressed as

$$\frac{(\mathbf{c}_i'\hat{\boldsymbol{\beta}})^2}{\mathbf{c}_i'(\mathbf{X}'\mathbf{X})^-\mathbf{c}_i} = \frac{\hat{\boldsymbol{\beta}}'\mathbf{c}_i\mathbf{c}_i'\hat{\boldsymbol{\beta}}}{\mathbf{c}_i'(\mathbf{X}'\mathbf{X})^-\mathbf{c}_i} = \frac{\mathbf{y}'\mathbf{X}(\mathbf{X}'\mathbf{X})^-\mathbf{c}_i\mathbf{c}_i'(\mathbf{X}'\mathbf{X})^-\mathbf{X}'\mathbf{y}}{\mathbf{c}_i'(\mathbf{X}'\mathbf{X})^-\mathbf{c}_i},$$

with a similar expression for the sum of squares for $\mathbf{c}_j'\hat{\boldsymbol{\beta}}$. By Corollary 1 to Theorem 5.6b, these two quadratic forms are independent if

$$\mathbf{X}(\mathbf{X}'\mathbf{X})^-\mathbf{c}_i\mathbf{c}_i'(\mathbf{X}'\mathbf{X})^-\mathbf{X}'\mathbf{X}(\mathbf{X}'\mathbf{X})^-\mathbf{c}_j\mathbf{c}_j'(\mathbf{X}'\mathbf{X})^-\mathbf{X}' = \mathbf{O}.$$

This holds if $\mathbf{c}_i'(\mathbf{X}'\mathbf{X})^-\mathbf{X}'\mathbf{X}(\mathbf{X}'\mathbf{X})^-\mathbf{c}_j = 0$, which reduces to $\mathbf{c}_i'(\mathbf{X}'\mathbf{X})^-\mathbf{c}_j = 0$, since $\mathbf{c}_i\boldsymbol{\beta}$ is an estimable function and therefore by Theorem 11.2b(iii), we have $\mathbf{c}_i'(\mathbf{X}'\mathbf{X})^-\mathbf{X}'\mathbf{X} = \mathbf{c}_i'$. Now by Theorem 12.3c, we obtain

$$\text{cov}(\mathbf{c}_i'\hat{\boldsymbol{\beta}}, \mathbf{c}_j'\hat{\boldsymbol{\beta}}) = \sigma^2\mathbf{c}_i'(\mathbf{X}'\mathbf{X})^-\mathbf{c}_j.$$

13.17 $(\mathbf{A}_i)' = (\mathbf{v}_i\mathbf{v}_i')' = (\mathbf{v}_i')'\mathbf{v}_i' = \mathbf{v}_i\mathbf{v}_i' = \mathbf{A}_i$.

$(\mathbf{A}_i)^2 = \mathbf{v}_i\mathbf{v}_i'\mathbf{v}_i\mathbf{v}_i' = \mathbf{v}_i\mathbf{v}_i' = \mathbf{A}_i$ since $\mathbf{v}_i'\mathbf{v}_i = 1$.

By Theorem 2.4(iii), $\text{rank}(\mathbf{A}_i) = \text{rank}(\mathbf{v}_i\mathbf{v}_i') = \text{rank}(\mathbf{v}_i) = 1$.
$\mathbf{A}_i\mathbf{A}_j = \mathbf{v}_i\ \mathbf{v}_i'\mathbf{v}_j\ \mathbf{v}_j' = \mathbf{O}$ because $\mathbf{v}_i'\mathbf{v}_j = 0$ by Theorem 2.12c(ii).

13.18 (a)
$$\left(\frac{\mathbf{J}}{abn}\right)^2 = \frac{(\mathbf{jj}')^2}{(abn)^2} = \frac{\mathbf{jj}'\,\mathbf{jj}'}{(abn)^2} = \frac{\mathbf{j}(abn)\,\mathbf{j}'}{(abn)^2} = \frac{\mathbf{J}}{abn}.$$

(b)
$$\left(\frac{\mathbf{J}}{abn}\right)\mathbf{x}_1 = \lambda_1\mathbf{x}_1 = \mathbf{x}_1, \quad \text{since } \lambda_1 = 1$$

$$\frac{\mathbf{jj}'\mathbf{x}_1}{abn} = \mathbf{x}_1.$$

Clearly $\mathbf{x}_1 = \mathbf{j}$ is a solution, since $\mathbf{j}'\mathbf{j} = abn$.

13.19
$$\mathbf{v}_0'\mathbf{v}_0 = \frac{\mathbf{j}_{4n}'\,\mathbf{j}_{4n}}{4n} = \frac{4n}{4n} = 1,$$

$$\mathbf{v}_0'\mathbf{v}_1 = \frac{1}{\sqrt{4n}\sqrt{2n}}(\mathbf{j}_n', \mathbf{j}_n', \mathbf{j}_n', \mathbf{j}_n')\begin{pmatrix} \mathbf{j}_n \\ -\mathbf{j}_n \\ \mathbf{0} \\ \mathbf{0} \end{pmatrix} = 0.$$

13.20
$$\begin{aligned} \mathbf{j}'\mathbf{x}_{2.01} &= \mathbf{j}'\mathbf{x}_2 - \frac{\mathbf{j}'\mathbf{x}_2}{\mathbf{j}'\mathbf{j}}\,\mathbf{j}'\mathbf{j} - \frac{\mathbf{x}_{1.0}'\mathbf{x}_2}{\mathbf{x}_{1.0}'\mathbf{x}_{1.0}}\,\mathbf{j}'\mathbf{x}_{1.0} \\ &= \mathbf{j}'\mathbf{x}_2 - \mathbf{j}'\mathbf{x}_2 - 0 \qquad [\text{by } (13.66)], \\ \mathbf{x}_{1.0}'\mathbf{x}_{2.01} &= \mathbf{x}_{1.0}'\mathbf{x}_2 - \frac{\mathbf{j}'\mathbf{x}_2}{\mathbf{j}'\mathbf{j}}\mathbf{x}_{1.0}'\mathbf{j} - \frac{\mathbf{x}_{1.0}'\mathbf{x}_2}{\mathbf{x}_{1.0}'\mathbf{x}_{1.0}}\mathbf{x}_{1.0}'\mathbf{x}_{1.0} \\ &= \mathbf{x}_{1.0}'\mathbf{x}_2 - 0 - \mathbf{x}_{1.0}'\mathbf{x}_2 \qquad [\text{by } (13.66)]. \end{aligned}$$

13.21 By (7.97), we have $\mathbf{x}_{3.012} = \mathbf{x}_3 - \mathbf{Z}_1(\mathbf{Z}_1'\mathbf{Z}_1)^{-1}\mathbf{Z}_1'\mathbf{x}_3$, where $\mathbf{Z}_1 = (\mathbf{j}, \mathbf{x}_{1.0}, \mathbf{x}_{2.01})$. Thus

$$\begin{aligned} \mathbf{x}_{3.012} &= \mathbf{x}_3 - (\mathbf{j}, \mathbf{x}_{1.0}, \mathbf{x}_{2.01})\begin{pmatrix} \mathbf{j}'\mathbf{j} & 0 & 0 \\ 0 & \mathbf{x}_{1.0}'\mathbf{x}_{1.0} & 0 \\ 0 & 0 & \mathbf{x}_{2.01}'\mathbf{x}_{2.01} \end{pmatrix}^{-1}\begin{pmatrix} \mathbf{j}'\mathbf{x}_3 \\ \mathbf{x}_{1.0}'\mathbf{x}_3 \\ \mathbf{x}_{2.01}'\mathbf{x}_3 \end{pmatrix} \\ &= \mathbf{x}_3 - \frac{\mathbf{j}'\mathbf{x}_3}{\mathbf{j}'\mathbf{j}}\,\mathbf{j} - \frac{\mathbf{x}_{1.0}'\mathbf{x}_3}{\mathbf{x}_{1.0}'\mathbf{x}_{1.0}}\mathbf{x}_{1.0} - \frac{\mathbf{x}_{2.01}'\mathbf{x}_3}{\mathbf{x}_{2.01}'\mathbf{x}_{2.01}}\mathbf{x}_{2.01}. \end{aligned}$$

13.22 Using (13.66) and (13.69), we have

$$\begin{aligned} \mathbf{j}'\mathbf{x}_{3.012} &= \mathbf{j}'\mathbf{x}_3 - \frac{\mathbf{j}'\mathbf{x}_3}{\mathbf{j}'\mathbf{j}}\,\mathbf{j}'\mathbf{j} - \frac{\mathbf{x}_{1.0}'\mathbf{x}_3}{\mathbf{x}_{1.0}'\mathbf{x}_{1.0}}\,\mathbf{j}'\mathbf{x}_{1.0} - \frac{\mathbf{x}_{2.01}'\mathbf{x}_3}{\mathbf{x}_{2.01}'\mathbf{x}_{2.01}}\,\mathbf{j}'\mathbf{x}_{2.01} \\ &= \mathbf{j}'\mathbf{x}_3 - \mathbf{j}'\mathbf{x}_3 - 0 - 0, \\ \mathbf{x}_{1.0}'\mathbf{x}_{3.012} &= \mathbf{x}_{1.0}'\mathbf{x}_3 - \frac{\mathbf{j}'\mathbf{x}_3}{\mathbf{j}'\mathbf{j}}\mathbf{x}_{1.0}'\mathbf{j} - \frac{\mathbf{x}_{1.0}'\mathbf{x}_3}{\mathbf{x}_{1.0}'\mathbf{x}_{1.0}}\mathbf{x}_{1.0}'\mathbf{x}_{1.0} - \frac{\mathbf{x}_{2.01}'\mathbf{x}_3}{\mathbf{x}_{2.01}'\mathbf{x}_{2.01}}\mathbf{x}_{1.0}'\mathbf{x}_{2.01}. \end{aligned}$$

13.23 Show that the coefficients in (13.70) are given by

$$\frac{\mathbf{j}'x_3}{\mathbf{j}'\mathbf{j}} = \frac{100n}{4n} = 25,$$

$$\frac{\mathbf{x}_{1.0}'\mathbf{x}_3}{\mathbf{x}_{1.0}'\mathbf{x}_{1.0}} = \frac{208}{20} = 10.4,$$

$$\frac{\mathbf{x}_{2.01}'\mathbf{x}_3}{\mathbf{x}_{2.01}'\mathbf{x}_{2.01}} = \frac{30}{4} = 7.5.$$

Then by (13.70).

$$\begin{aligned}\mathbf{z}_3 &= \mathbf{x}_3 - 25\,\mathbf{j} - 10.4\mathbf{x}_{1.0} - 7.5\mathbf{x}_{2.01}\\ &= (-.3, \ldots, -.3, .9, \ldots, .9, -.9, \ldots, -.9, .3, \ldots, .3)',\end{aligned}$$

which we divide by .3 to obtain

$$\mathbf{z}_3 = (-1, \ldots -1, 3, \ldots 3, -3, \ldots, -3, 1, \ldots, 1)'.$$

13.24

$$\begin{aligned}\mathbf{z}_0 &= \mathbf{x}_0 = \mathbf{j},\\ \mathbf{z}_1 &= \mathbf{x}_{1.0} = 2(\mathbf{x}_1 - 2.5\mathbf{j}),\\ \mathbf{z}_2 &= \mathbf{x}_{2.01} = \mathbf{x}_2 - 7.5\mathbf{j} - 2.5\mathbf{x}_{1.0} = \mathbf{x}_2 - 7.5\mathbf{j} - 2.5[2(\mathbf{x}_1 - 2.5\mathbf{j})]\\ &= \mathbf{x}_2 + 5\mathbf{j} - 5\mathbf{x}_1,\\ \mathbf{z}_3 &= \frac{\mathbf{x}_3 - 25\mathbf{j} - 10.4\mathbf{x}_{1.0} - 7.5\mathbf{x}_{2.01}}{.3}\\ &= \frac{\mathbf{x}_3 - 25\mathbf{j} - 10.4(2\mathbf{x}_1 - 5\mathbf{j}) - 7.5(\mathbf{x}_2 + 5\mathbf{j} - 5\mathbf{x}_1)}{.3}\\ &= \frac{\mathbf{x}_3}{.3} - 35\mathbf{j} + \left(\frac{16.7}{.3}\right)\mathbf{x}_1 - 25\mathbf{x}_2.\end{aligned}$$

Then $\mathbf{X}\boldsymbol{\beta} = \mathbf{Z}\boldsymbol{\theta}$ can be written as

$$\begin{aligned}\beta_0\mathbf{j} + \beta_1\mathbf{x}_1 + \beta_2\mathbf{x}_2 + \beta_3\mathbf{x}_3 &= \theta_0\mathbf{j} + \theta_1\mathbf{z}_1 + \theta_2\mathbf{z}_2 + \theta_3\mathbf{z}_3\\ &= \theta_0\mathbf{j} + \theta_1(2\mathbf{x}_1 - 5\mathbf{j}) + \theta_2(\mathbf{x}_2 + 5\mathbf{j} - 5\mathbf{x}_1)\\ &\quad + \theta_3\left[\left(\frac{\mathbf{x}_3}{.3}\right) - 35\mathbf{j} + \left(\frac{16.7}{.3}\right)\mathbf{x}_1 - 25\mathbf{x}_2\right]\\ &= (\theta_0 - 5\theta_1 + 5\theta_2 - 35\theta_3)\mathbf{j}\\ &\quad + \left[2\theta_1 - 5\theta_2 + \left(\frac{16.7}{.3}\right)\theta_3\right]\mathbf{x}_1\\ &\quad + (\theta_2 - 25\theta_3)\mathbf{x}_2 + \left(\frac{\theta_3}{.3}\right)\mathbf{x}_3.\end{aligned}$$

Thus

$$\beta_0 = \theta - 5\theta_1 + 5\theta_2 - 35\theta_3$$

$$\beta_1 = 2\theta_1 - 5\theta_2 + \left(\frac{16.7}{.3}\right)\theta_3$$

$$\beta_2 = \theta_2 - 25\theta_3$$

$$\beta_3 = \frac{\theta_3}{.3}.$$

13.25

$$\mathbf{Z}'\mathbf{y} = (\mathbf{j}, \mathbf{z}_1, \mathbf{z}_2, \mathbf{z}_3)'\mathbf{y} = \begin{pmatrix} \mathbf{j}' \\ \mathbf{z}_1' \\ \mathbf{z}_2' \\ \mathbf{z}_3' \end{pmatrix} \mathbf{y} = \begin{pmatrix} \mathbf{j}'\mathbf{y} \\ \mathbf{z}_1'\mathbf{y} \\ \mathbf{z}_2'\mathbf{y} \\ \mathbf{z}_3'\mathbf{y} \end{pmatrix},$$

$$\hat{\boldsymbol{\theta}} = (\mathbf{Z}'\mathbf{Z})^{-1}\mathbf{Z}'\mathbf{y} = \begin{pmatrix} \mathbf{j}'\mathbf{j} & 0 & 0 & 0 \\ 0 & \mathbf{z}_1'\mathbf{z}_1 & 0 & 0 \\ 0 & 0 & \mathbf{z}_2'\mathbf{z}_2 & 0 \\ 0 & 0 & 0 & \mathbf{z}_3'\mathbf{z}_3 \end{pmatrix}^{-1} \begin{pmatrix} \mathbf{j}'\mathbf{y} \\ \mathbf{z}_1'\mathbf{y} \\ \mathbf{z}_2'\mathbf{y} \\ \mathbf{z}_3'\mathbf{y} \end{pmatrix}$$

$$= \begin{pmatrix} \mathbf{j}'\mathbf{y}/\mathbf{j}'\mathbf{j} \\ \mathbf{z}_1'\mathbf{y}/\mathbf{z}_1'\mathbf{z}_1 \\ \mathbf{z}_2'\mathbf{y}/\mathbf{z}_2'\mathbf{z}_2 \\ \mathbf{z}_3'\mathbf{y}/\mathbf{z}_3'\mathbf{z}_3 \end{pmatrix}.$$

13.26 Since the columns of $\mathbf{Z}$ are linear transformations of the columns of $\mathbf{X}$ [see (13.65), (13.68), and (13.70)], we can write $\mathbf{Z} = \mathbf{XH}$ and $\mathbf{Z}_1 = \mathbf{X}_1\mathbf{H}_1$, where $\mathbf{H}$ and $\mathbf{H}_1$ are nonsingular. Thus

$$\begin{aligned} \hat{\boldsymbol{\beta}}'\mathbf{X}'\mathbf{y} - \hat{\boldsymbol{\beta}}^{*\prime}\mathbf{X}_1'\mathbf{y} &= \mathbf{y}'\mathbf{X}(\mathbf{X}'\mathbf{X})^{-1}\mathbf{X}'\mathbf{y} - \mathbf{y}'\mathbf{X}_1(\mathbf{X}_1'\mathbf{X}_1)^{-1}\mathbf{X}_1'\mathbf{y} \\ &= \mathbf{y}'\mathbf{ZH}^{-1}[(\mathbf{ZH}^{-1})'(\mathbf{ZH}^{-1})]^{-1}(\mathbf{ZH}^{-1})'\mathbf{y} \\ &\quad - \mathbf{y}'\mathbf{Z}_1\mathbf{H}_1^{-1}[(\mathbf{Z}_1\mathbf{H}_1^{-1})'(\mathbf{Z}_1\mathbf{H}_1^{-1})]^{-1}(\mathbf{Z}_1\mathbf{H}_1^{-1})'\mathbf{y}. \end{aligned}$$

Show that this reduces to $(\mathbf{z}_k'\mathbf{y})^2/\mathbf{z}_k'\mathbf{z}_k$.

13.27

Linear: $-3(1) - (2) + 2 + 3(1) = 0$

Quadratic: $1 - 2 - 2 + 1 = -2$

Cubic: $-1 + 3(2) - 3(2) + 1 = 0$

13.28 The orthogonal contrasts that can be used in $H_0\colon \sum_{i=1}^{k} c_i\mu_i = 0$ in part (b) are

$$2\mu_1 + 2\mu_2 + 2\mu_3 - 3\mu_4 - 3\mu_5 = 0$$
$$2\mu_1 - \mu_2 - \mu_3 = 0$$
$$\mu_2 - \mu_3 = 0$$
$$\mu_4 - \mu_5 = 0.$$

The results for parts (a) and (b) are given in the following ANOVA table.

Source	df	Sum of Squares	Mean Square	F	p Value
Breed	4	4,276.1327	1069.0332	8.47	.000033
Contrasts					
A, B, C vs. D, E	1	211.7289	211.7289	1.68	.202
A, B, vs. C	1	370.6669	370.6669	2.94	.0933
A vs. B	1	708.0500	708.0500	5.61	.0221
D vs. E	1	2,885.4545	2885.4545	22.86	.0000182
Error	46	5,806.4556	126.2273		
Total	50	10,082.5882			

13.29 The orthogonal polynomial contrast coefficients are the rows of the following matrix see Table (13.5):

$$\begin{pmatrix} -2 & -1 & 0 & 1 & 2 \\ 2 & -1 & -2 & -1 & 2 \\ -1 & 2 & 0 & -2 & 1 \\ 1 & -4 & 6 & -4 & 1 \end{pmatrix}.$$

The results for parts (a) and (b) are given in the following ANOVA table.

Source	df	Sum of Squares	Mean Square	F	p Value
Glucose	4	154.9210	38.7303	29.77	7.902×10^{-11}
Contrasts					
Linear	1	140.1587	140.1587	107.74	3.168×10^{-12}
Quadratic	1	0.0065	0.0065	0.006	.944
Cubic	1	14.7319	14.7319	11.32	.002
Quartic	1	0.0241	0.0241	0.021	.893
Error	35	45.5322	1.3009		
Total	39	200.4532			

The means for the five glucose concentrations are 2.66, 2.69, 4.94, 7.09, and 7.10. From the Fs we see that there is a large linear effect and a small cubic effect.

13.30 The contrast coefficients are given in the following matrix:

$$\begin{pmatrix} 2 & -1 & -1 \\ 0 & 1 & -1 \end{pmatrix}.$$

The results for parts (a) and (b) are given in the following ANOVA table.

Source	df	Sum of Squares	Mean Square	F	p Value
Stimulus	2	561.5714	280.7857	67.81	2.018×10^{-13}
Contrasts					
1 vs. 2, 3	1	525.0000	252.0000	126.78	8.005×10^{-14}
2 vs. 3	1	36.5714	36.5714	8.83	.00505
Error	39	161.5000	4.1410		
Total	41	723.0714			

13.31 For contrast coefficients comparing the two types of raw materials, we can use those in the vector $(5, 5, 5, 5, -4, -4, -4, -4, -4)$. The results for parts (a) and (b) are in the following ANOVA table.

Source	df	Sum of Squares	Mean Square	F	p Value
Cable	8	1924.2963	240.5370	9.07	2.831×10^{-9}
Contrast	1	1543.6463	1543.6463	58.18	1.493×10^{-11}
Error	99	2626.9167	26.5345		
Total	107	4551.2130			

13.32 Contrast coefficients are given in the following matrix:

$$\begin{pmatrix} 1 & 1 & -1 & -1 \\ 1 & -1 & 0 & 0 \\ 0 & 0 & 1 & -1 \end{pmatrix}.$$

The results for parts (a) and (b) are given in the following ANOVA table.

Source	df	Sum of Squares	Mean Square	F	p Value
Treatments	3	1045.4583	348.8461	6.03	.0043
Contrasts					
1, 2 vs. 3, 4	1	7.0417	7.0417	0.12	.731
1 vs. 2	1	30.0833	30.0833	0.52	.479
3 vs. 4	1	1008.3333	1008.3333	17.44	.0005
Error	20	1156.5000	57.8250		
Total	23	2201.9583			

13.33 Contrast coefficients are given in the following matrix:

$$\begin{pmatrix} 1 & 1 & 1 & -3 \\ 1 & 1 & -2 & 0 \\ 1 & -1 & 0 & 0 \end{pmatrix}.$$

The results for parts (a) and (b) are given in the following ANOVA table.

Source	df	Sum of Squares	Mean Square	F	p Value
Treatments	3	3462.500	1154.167	6.71	.00103
Contrasts					
1, 2, 3 vs. 4	1	1968.300	1968.300	11.44	.00175
1, 2 vs. 3	1	66.150	66.150	.385	.539
1 vs. 2	1	1428.050	1428.050	8.30	.0066
Error	36	6193.400	172.039		
Total	39	9655.900			

Chapter 14

14.1 $\theta_1 = \mu_{11} - \mu_{21} = \mu + \alpha_1 + \beta_1 + \gamma_{11} - (\mu + \alpha_2 + \beta_1 + \gamma_{21})$

$\theta_5 = \mu_{11} - \mu_{12} - \mu_{31} + \mu_{32}$

14.2 By Theorem 12.2b, all estimable functions can be obtained from $\mu_{ij} = \mu + \alpha_i + \beta_j + \gamma_{ij}$. To obtain an estimable contrast of the form $\sum_i c_i\alpha_i$, where $\sum_i c_i = 0$, we consider

$$\begin{aligned}\sum_{i=1}^{a} c_i \mu_{ij} &= \sum_{i=1}^{a} c_i \mu_i + \sum_i c_i \alpha_i + \sum_i c_i \beta_j + \sum_i c_i \gamma_{ij} \\ &= \sum_i c_i \alpha_i + \sum_i c_i \gamma_{ij}.\end{aligned}$$

Thus an estimable function of the α's also involves the γ's.

14.3
$$\begin{aligned}\tfrac{1}{3}(\theta_3 + \theta_3' + \theta_3'') &= \tfrac{1}{3}(\beta_1 - \beta_2 + \gamma_{11} - \gamma_{12} + \beta_1 - \beta_2 + \gamma_{21} - \gamma_{22} \\ &\quad + \beta_1 - \beta_2 + \gamma_{31} - \gamma_{32}) \\ &= \tfrac{1}{3}(3\beta_1 - 3\beta_2 + \gamma_{11} + \gamma_{21} + \gamma_{31} - \gamma_{12} - \gamma_{22} - \gamma_{32}).\end{aligned}$$

14.4 (a)
$$\begin{aligned}\sum_{i=1}^{a} \alpha_i^* &= \sum_i (\bar{\mu}_{i.} - \bar{\mu}_{..}) = \sum_i \bar{\mu}_{i.} - a\bar{\mu}_{..} \\ &= \sum_i \frac{\mu_{i.}}{b} - \frac{a\mu_{..}}{ab} = \frac{\mu_{..}}{b} - \frac{\mu_{..}}{b}\end{aligned}$$

(c)
$$\begin{aligned}\sum_{i=1}^{a} \gamma_{ij}^* &= \sum_i (\mu_{ij} - \bar{\mu}_{i.} - \bar{\mu}_{.j} + \bar{\mu}_{..}) \\ &= \sum_i \mu_{ij} - \sum_i \bar{\mu}_{i.} - a\bar{\mu}_{.j} + a\bar{\mu}_{..} \\ &= \mu_{.j} - \sum_i \frac{\mu_{i.}}{b} - \frac{a\mu_{.j}}{a} + \frac{a\mu_{..}}{ab} \\ &= \mu_{.j} - \frac{\mu_{..}}{b} - \mu_{.j} + \frac{\mu_{..}}{b}.\end{aligned}$$

14.5 (a)
$$\begin{aligned}\sum_{i=1}^{a} \alpha_i^* &= \sum_i (\alpha_i - \bar{\alpha}_. + \bar{\gamma}_{i.} - \bar{\gamma}_{..}) \\ &= \sum_i (\alpha_i - \bar{\alpha}_.) + \sum_i (\bar{\gamma}_{i.} - \bar{\gamma}_{..}) \\ &= \alpha_. - a\frac{\alpha_.}{a} + \sum_i \left(\frac{\gamma_{i.}}{b} - \frac{\gamma_{..}}{ab}\right) \\ &= \alpha_. - \alpha_. + \frac{\gamma_{..}}{b} - a\frac{\gamma_{..}}{ab}.\end{aligned}$$

(c) $$\sum_{i=1}^{a} \gamma_{ij}^* = \sum_i (\gamma_{ij} - \bar{\gamma}_{i.} - \bar{\gamma}_{.j} + \bar{\gamma}_{..}) = \sum_i \left(\gamma_{ij} - \frac{\gamma_{i.}}{b} - \frac{\gamma_{.j}}{a} + \frac{\gamma_{..}}{ab}\right)$$

$$= \gamma_{.j} - \frac{\gamma_{..}}{b} - a\frac{\gamma_{.j}}{a} + a\frac{\gamma_{..}}{ab}.$$

14.6 **(b)**

$$\begin{aligned}
\gamma_{ij}^{*} &= \mu_{ij} - \bar{\mu}_{i.} - \bar{\mu}_{.j} + \bar{\mu}_{..} \\
&= \mu_{ij} - \frac{1}{b}\sum_{j=1}^{b}\mu_{ij} - \frac{1}{a}\sum_{i=1}^{a}\mu_{ij} + \frac{1}{ab}\sum_{i=1}^{a}\sum_{j=1}^{b}\mu_{ij} \\
&= \mu + \alpha_i + \beta_j + \gamma_{ij} - \frac{1}{b}\sum_{j=1}^{b}(\mu + \alpha_i + \beta_j + \gamma_{ij}) \\
&\quad - \frac{1}{a}\sum_{i=1}^{a}(\mu + \alpha_i + \beta_j + \gamma_{ij}) + \frac{1}{ab}\sum_{ij}(\mu + \alpha_i + \beta_j + \gamma_{ij}) \\
&= \mu + \alpha_i + \beta_j + \gamma_{ij} - \mu - \alpha_i - \frac{1}{b}\sum_{j}\beta_j - \frac{1}{b}\sum_{j}\gamma_{ij} \\
&\quad - \mu - \frac{1}{a}\sum_{i}\alpha_i - \beta_j - \frac{1}{a}\sum_{i}\gamma_{ij} + \mu + \frac{1}{a}\sum_{i}\alpha_i \\
&\quad + \frac{1}{b}\sum_{j}\beta_j + \frac{1}{ab}\sum_{ij}\gamma_{ij} \\
&= \gamma_{ij} - \bar{\gamma}_{i.} - \bar{\gamma}_{.j} + \bar{\gamma}_{..}.
\end{aligned}$$

14.7

$$\begin{aligned}
E(\hat{\alpha}_i) &= E(\bar{y}_{i..} - \bar{y}_{...}) = E\left(\frac{y_{i..}}{bn}\right) - E\left(\frac{y_{...}}{abn}\right) \\
&= E\left(\frac{\sum_{jk} y_{ijk}}{bn}\right) - E\left(\frac{\sum_{ijk} y_{ijk}}{abn}\right) \\
&= \frac{\sum_{jk} E(y_{ijk})}{bn} - \frac{\sum_{ijk} E(y_{ijk})}{abn} \\
&= \frac{\sum_{jk}(\mu^{*} + \alpha_i^{*} + \beta_j^{*} + \gamma_{ij}^{*})}{bn} - \frac{\sum_{ijk}(\mu^{*} + \alpha_i^{*} + \beta_j^{*} + \gamma_{ij}^{*})}{abn} \\
&= \frac{bn\mu^{*} + bn\alpha_i^{*} + n\sum_j \beta_j^{*} + n\sum_j \gamma_{ij}^{*}}{bn} \\
&\quad - \frac{abn\mu^{*} + bn\sum_i \alpha_i^{*} + an\sum_j \beta_j^{*} + n\sum_{ij}\gamma_{ij}^{*}}{abn} \\
&= \mu^{*} + \alpha_i^{*} - \mu^{*}.
\end{aligned}$$

14.8 **(a)** For $b = 2$ and $n = 2$, we have

$$\begin{aligned}
\bar{y}_{11.} + \bar{y}_{12.} &= \frac{1}{2}\sum_{k} y_{11k} + \frac{1}{2}\sum_{k} y_{12k} = \frac{1}{2}\sum_{jk} y_{1jk} \\
&= \frac{1}{2}\left(4\sum_{jk}\frac{y_{1jk}}{4}\right) = 2\bar{y}_{1..}
\end{aligned}$$

14.9 Write $\mathbf{X}'\mathbf{X}$ as $\mathbf{X}'\mathbf{X} = \begin{pmatrix} \mathbf{A}_{11} & \mathbf{A}_{12} \\ \mathbf{A}_{21} & 2\mathbf{I} \end{pmatrix}$. Then

$$\mathbf{X}'\mathbf{X}(\mathbf{X}'\mathbf{X})^- = \begin{pmatrix} \mathbf{A}_{11} & \mathbf{A}_{12} \\ \mathbf{A}_{21} & 2\mathbf{I} \end{pmatrix}\begin{pmatrix} \mathbf{O} & \mathbf{O} \\ \mathbf{O} & \frac{1}{2}\mathbf{I} \end{pmatrix} = \begin{pmatrix} \mathbf{O} & \frac{1}{2}\mathbf{A}_{12} \\ \mathbf{O} & \mathbf{I} \end{pmatrix},$$

$$\mathbf{X}'\mathbf{X}(\mathbf{X}'\mathbf{X})^-\mathbf{X}'\mathbf{X} = \begin{pmatrix} \frac{1}{2}\mathbf{A}_{12}\mathbf{A}_{21} & \mathbf{A}_{12} \\ \mathbf{A}_{21} & 2\mathbf{I} \end{pmatrix}.$$

Show that $\frac{1}{2}\mathbf{A}_{12}\mathbf{A}_{21} = \mathbf{A}_{11}$; that is, show that

$$\frac{1}{2}\begin{pmatrix} 2 & 2 & 2 & 2 & 2 & 2 \\ 2 & 2 & 0 & 0 & 0 & 0 \\ 0 & 0 & 2 & 2 & 0 & 0 \\ 0 & 0 & 0 & 0 & 2 & 2 \\ 2 & 0 & 2 & 0 & 2 & 0 \\ 0 & 2 & 0 & 2 & 0 & 2 \end{pmatrix}\begin{pmatrix} 2 & 2 & 0 & 0 & 2 & 0 \\ 2 & 2 & 0 & 0 & 0 & 2 \\ 2 & 0 & 2 & 0 & 2 & 0 \\ 2 & 0 & 2 & 0 & 0 & 2 \\ 2 & 0 & 0 & 2 & 2 & 0 \\ 2 & 0 & 0 & 2 & 0 & 2 \end{pmatrix}$$
$$= \begin{pmatrix} 12 & 4 & 4 & 4 & 6 & 6 \\ 4 & 4 & 0 & 0 & 2 & 2 \\ 4 & 0 & 4 & 0 & 2 & 2 \\ 4 & 0 & 0 & 4 & 2 & 2 \\ 6 & 2 & 2 & 2 & 6 & 0 \\ 6 & 2 & 2 & 2 & 0 & 6 \end{pmatrix}.$$

14.10

$$\begin{aligned} \sum_{ijk}(y_{ijk} - \bar{y}_{ij.})^2 &= \sum_{ijk}(y_{ijk}^2 - 2y_{ijk}\bar{y}_{ij.} + \bar{y}_{ij.}^2) \\ &= \sum_{ijk} y_{ijk}^2 - 2\sum_{ij}\bar{y}_{ij.}\sum_k y_{ijk} + n\sum_{ij}\bar{y}_{ij.}^2 \\ &= \sum_{ijk} y_{ijk}^2 - 2\sum_{ij}\bar{y}_{ij.}n\bar{y}_{ij.} + n\sum_{ij}\bar{y}_{ij.}^2. \end{aligned}$$

14.11 From $\mu_{21} - \mu_{22} = \mu_{31} - \mu_{32}$, we have

$$\begin{aligned} 0 &= \mu_{21} - \mu_{22} - \mu_{31} + \mu_{32} \\ &= \mu + \alpha_2 + \beta_1 + \gamma_{21} - (\mu + \alpha_2 + \beta_2 + \gamma_{22}) \\ &\quad - (\mu + \alpha_3 + \beta_1 + \gamma_{31}) + \mu + \alpha_3 + \beta_2 + \gamma_{32} \\ &= \gamma_{21} - \gamma_{22} - \gamma_{31} + \gamma_{32}. \end{aligned}$$

14.12

$$\sum_i (\bar{y}_{i..} - \bar{y}_{...})y_{i..} = \sum_i \left(\frac{y_{i..}}{bn} - \frac{y_{...}}{abn}\right) y_{i..} = \sum_i \frac{y_{i..}^2}{bn} - \frac{y_{...}}{abn}\sum_i y_{i..}$$

$$= \sum_i \frac{y_{i..}^2}{bn} - \frac{y_{...}^2}{abn},$$

$$\sum_{ij} (\bar{y}_{ij.} - \bar{y}_{i..} - \bar{y}_{.j.} + \bar{y}_{...})y_{ij.} = \sum_{ij}\left(\frac{y_{ij.}}{n} - \frac{y_{i..}}{bn} - \frac{y_{.j.}}{an} + \frac{y_{...}}{abn}\right) y_{ij.}$$

$$= \sum_{ij} \frac{y_{ij.}^2}{n} - \sum_i \left(y_{i..} \sum_j \frac{y_{ij.}}{bn}\right)$$

$$- \sum_j \left(y_{.j.} \sum_i \frac{y_{ij.}}{an}\right) + y_{...} \sum_{ij} \frac{y_{ij.}}{abn}$$

$$= \sum_{ij} \frac{y_{ij.}^2}{n} - \sum_i \frac{y_{i..}^2}{bn} - \sum_j \frac{y_{.j.}^2}{an} + \frac{y_{...}^2}{abn}.$$

14.13 **(a)** Using $\hat{\boldsymbol{\beta}}$ from (14.24) and $\mathbf{X}'\mathbf{y}$ from (14.19) (both extended to general a and b), we obtain

$$\hat{\boldsymbol{\beta}}'\mathbf{X}'\mathbf{y} = \sum_{ij} \bar{y}_{ij.} y_{ij.} = \sum_{ij}\left(\frac{y_{ij.}}{n}\right) y_{ij.} = \sum_{ij} \frac{y_{ij.}^2}{n}.$$

(b)

$$n\sum_{ij} \bar{y}_{ij.}^2 = n\sum_{ij}\left(\frac{y_{ij.}}{n}\right)^2 = n\sum_{ij}\frac{y_{ij.}^2}{n^2}$$

14.14 In the following array, we see that the γ_{ij}^*'s in the margins can all be obtained from the remaining $(a-1)(b-1)$ γ_{ij}^*'s by using side conditions:

$$\begin{array}{cccc|c}
\gamma_{11}^* & \gamma_{12}^* & \cdots & \gamma_{1,b-1}^* & \gamma_{1b}^* \\
\gamma_{21}^* & \gamma_{22}^* & \cdots & \gamma_{2,b-1}^* & \gamma_{2b}^* \\
\vdots & \vdots & & \vdots & \vdots \\
\gamma_{a-1,1}^* & \gamma_{a-1,2}^* & \cdots & \gamma_{a-1,b-1}^* & \gamma_{a-1,b}^* \\
\hline
\gamma_{a1}^* & \gamma_{a2}^* & \cdots & \gamma_{a,b-1}^* & \gamma_{ab}^*
\end{array}$$

14.15 By (5.1), $\sum_{i=1}^n (y_i - \bar{y})^2 = \sum_{i=1}^n y_i^2 - n\bar{y}^2$. Then

$$
\begin{aligned}
n\sum_{ij} &(\bar{y}_{ij.} - \bar{y}_{i..} - \bar{y}_{.j.} + \bar{y}_{...})^2 \\
&= n\sum_{ij}[(\bar{y}_{ij.} - \bar{y}_{i..}) - (\bar{y}_{.j.} - \bar{y}_{...})]^2 \\
&= n\sum_{ij}(\bar{y}_{ij.} - \bar{y}_{i..})^2 + an\sum_{j}(\bar{y}_{.j.} - \bar{y}_{...})^2 - 2n\sum_{ij}(\bar{y}_{ij.} - \bar{y}_{i..})(\bar{y}_{.j.} - \bar{y}_{...}) \\
&= n\sum_{ij}\bar{y}_{ij.}^2 - bn\sum_{i}\bar{y}_{i..}^2 + an\sum_{j}\bar{y}_{.j.}^2 - abn\bar{y}_{...}^2 \\
&\quad - 2n\sum_{j}\left[\left(\frac{y_{.j.}}{an} - \frac{y_{...}}{abn}\right)\sum_{i}\left(\frac{y_{ij.}}{n} - \frac{y_{i..}}{bn}\right)\right] \\
&= n\sum_{ij}\frac{y_{ij.}^2}{n^2} - bn\sum_{i}\frac{y_{i..}^2}{b^2n^2} + an\sum_{j}\frac{y_{.j.}^2}{a^2n^2} - abn\frac{y_{...}^2}{a^2b^2n^2} \\
&\quad - 2n\sum_{j}\left[\left(\frac{y_{.j.}}{an} - \frac{y_{...}}{abn}\right)\left(\frac{y_{.j.}}{n} - \frac{y_{...}}{bn}\right)\right] \\
&= \sum_{ij}\frac{y_{ij}^2}{n} - \sum_{i}\frac{y_{i..}^2}{bn} + \sum_{j}\frac{y_{.j.}^2}{an} - \frac{y_{...}^2}{abn} - \frac{2}{an}\sum_{j}\left(y_{.j.}^2 - \frac{2y_{...}y_{.j.}}{b} + \frac{y_{...}^2}{b^2}\right) \\
&= \sum_{ij}\frac{y_{ij}^2}{n} - \sum_{i}\frac{y_{i..}^2}{bn} + \sum_{j}\frac{y_{.j.}^2}{an} - \frac{y_{...}^2}{abn} - \frac{2}{an}\sum_{j}y_{.j.}^2 + \frac{2y_{...}^2}{abn} \\
&= \sum_{ij}\frac{y_{ij}^2}{n} - \sum_{i}\frac{y_{i..}^2}{bn} - \sum_{j}\frac{y_{.j.}^2}{an} + \frac{y_{...}^2}{abn}.
\end{aligned}
$$

14.16 By (5.1), we obtain

$$
\begin{aligned}
\text{SSE} = \sum_{ijk}(y_{ijk} - \bar{y}_{ij.})^2 &= \sum_{ij}\sum_{k}(y_{ijk} - \bar{y}_{ij.})^2 \\
&= \sum_{ij}\left(\sum_{k}y_{ijk}^2 - n\bar{y}_{ij.}^2\right) \\
&= \sum_{ij}\left(\sum_{k}y_{ijk}^2 - \frac{y_{ij.}^2}{n}\right) \\
&= \sum_{ijk}y_{ijk}^2 - \sum_{ij}\frac{y_{ij.}^2}{n}.
\end{aligned}
$$

14.17 Partitioning $\mathbf{X}$ into $\mathbf{X} = (\mathbf{X}_1, \mathbf{X}_2)$, where $\mathbf{X}_1$ contains the first six columns and $\mathbf{X}_2$ constitutes the last six columns, we have

$$\mathbf{X}(\mathbf{X}'\mathbf{X})^{-}\mathbf{X}' = \tfrac{1}{2}(\mathbf{X}_1, \mathbf{X}_2)\begin{pmatrix} \mathbf{O} & \mathbf{O} \\ \mathbf{O} & \mathbf{I} \end{pmatrix}\begin{pmatrix} \mathbf{X}_1' \\ \mathbf{X}_2' \end{pmatrix}$$
$$= \tfrac{1}{2}(\mathbf{O}, \mathbf{X}_2)\begin{pmatrix} \mathbf{X}_1' \\ \mathbf{X}_2' \end{pmatrix} = \tfrac{1}{2}\mathbf{X}_2\mathbf{X}_2'.$$

We can express $\mathbf{X}_2$ as

$$\mathbf{X}_2 = \begin{pmatrix} \mathbf{j} & \mathbf{0} & \dots & \mathbf{0} \\ \mathbf{0} & \mathbf{j} & \dots & \mathbf{0} \\ \vdots & \vdots & & \vdots \\ \mathbf{0} & \mathbf{0} & \dots & \mathbf{j} \end{pmatrix},$$

where $\mathbf{j}$ and $\mathbf{0}$ are 2×1. Hence $\frac{1}{2}\mathbf{X}_2\mathbf{X}_2'$ assumes the form given in (14.50).

14.18 **(a)**

$$\mathbf{X}_1'\mathbf{X}_1(\mathbf{X}_1'\mathbf{X}_1)^{-} = \begin{pmatrix} -1 & 1 & 1 & 1 & 1 & 1 \\ -\frac{1}{3} & 1 & 0 & 0 & \frac{1}{3} & \frac{1}{3} \\ -\frac{1}{3} & 0 & 1 & 0 & \frac{1}{3} & \frac{1}{3} \\ -\frac{1}{3} & 0 & 0 & 1 & \frac{1}{3} & \frac{1}{3} \\ -\frac{1}{2} & \frac{1}{2} & \frac{1}{2} & \frac{1}{2} & 1 & 0 \\ -\frac{1}{2} & \frac{1}{2} & \frac{1}{2} & \frac{1}{2} & 0 & 1 \end{pmatrix}$$

Multiply by $\mathbf{X}_1'\mathbf{X}_1$ on the right to show that $\mathbf{X}_1'\mathbf{X}_1(\mathbf{X}_1'\mathbf{X}_1)^{-}\mathbf{X}_1'\mathbf{X}_1 = \mathbf{X}_1'\mathbf{X}_1$.

(b)

$$\mathbf{X}_1(\mathbf{X}_1'\mathbf{X}_1)^{-} = \frac{1}{12}\begin{pmatrix} -1 & 3 & 0 & 0 & 2 & 0 \\ -1 & 3 & 0 & 0 & 2 & 0 \\ -1 & 3 & 0 & 0 & 0 & 2 \\ -1 & 3 & 0 & 0 & 0 & 2 \\ -1 & 0 & 3 & 0 & 2 & 0 \\ -1 & 0 & 3 & 0 & 2 & 0 \\ -1 & 0 & 3 & 0 & 0 & 2 \\ -1 & 0 & 3 & 0 & 0 & 2 \\ -1 & 0 & 0 & 3 & 2 & 0 \\ -1 & 0 & 0 & 3 & 2 & 0 \\ -1 & 0 & 0 & 3 & 0 & 2 \\ -1 & 0 & 0 & 3 & 0 & 2 \end{pmatrix}$$

Multiply on the right side by $\mathbf{X}_1'$ to obtain $\mathbf{X}_1(\mathbf{X}_1'\mathbf{X}_1)^{-}\mathbf{X}_1'$ in (14.54).

14.19 We first consider $y_{.1.}$ and $y_{.1.}^2$:

$$\begin{aligned} y_{.1.} &= \sum_{ik} y_{i1k} = \sum_k y_{11k} + \sum_k y_{21k} + \sum_k y_{31k} \\ &= \mathbf{y}_{11}'\mathbf{j} + \mathbf{y}_{21}'\mathbf{j} + \mathbf{y}_{31}'\mathbf{j} \\ &= (\mathbf{y}_{11}', \mathbf{y}_{12}', \mathbf{y}_{21}', \mathbf{y}_{22}', \mathbf{y}_{31}', \mathbf{y}_{32}') \begin{pmatrix} \mathbf{j} \\ \mathbf{0} \\ \mathbf{j} \\ \mathbf{0} \\ \mathbf{j} \\ \mathbf{0} \end{pmatrix}, \end{aligned}$$

$$y_{.1.}^2 = \mathbf{y}' \begin{pmatrix} \mathbf{j} \\ \mathbf{O} \\ \mathbf{j} \\ \mathbf{O} \\ \mathbf{j} \\ \mathbf{O} \end{pmatrix} (\mathbf{j}', \mathbf{O}', \mathbf{j}', \mathbf{O}', \mathbf{j}', \mathbf{O}')\mathbf{y} = \mathbf{y}' \begin{pmatrix} \mathbf{J} & \mathbf{O} & \mathbf{J} & \mathbf{O} & \mathbf{J} & \mathbf{O} \\ \mathbf{O} & \mathbf{O} & \mathbf{O} & \mathbf{O} & \mathbf{O} & \mathbf{O} \\ \mathbf{J} & \mathbf{O} & \mathbf{J} & \mathbf{O} & \mathbf{J} & \mathbf{O} \\ \mathbf{O} & \mathbf{O} & \mathbf{O} & \mathbf{O} & \mathbf{O} & \mathbf{O} \\ \mathbf{J} & \mathbf{O} & \mathbf{J} & \mathbf{O} & \mathbf{J} & \mathbf{O} \\ \mathbf{O} & \mathbf{O} & \mathbf{O} & \mathbf{O} & \mathbf{O} & \mathbf{O} \end{pmatrix} \mathbf{y}.$$

Similarly

$$y_{.2.}^2 = \mathbf{y}' \begin{pmatrix} \mathbf{O} \\ \mathbf{j} \\ \mathbf{O} \\ \mathbf{j} \\ \mathbf{O} \\ \mathbf{j} \end{pmatrix} (\mathbf{O}', \mathbf{j}', \mathbf{O}', \mathbf{j}', \mathbf{O}', \mathbf{j}')\mathbf{y} = \mathbf{y}' \begin{pmatrix} \mathbf{O} & \mathbf{O} & \mathbf{O} & \mathbf{O} & \mathbf{O} & \mathbf{O} \\ \mathbf{O} & \mathbf{J} & \mathbf{O} & \mathbf{J} & \mathbf{O} & \mathbf{J} \\ \mathbf{O} & \mathbf{O} & \mathbf{O} & \mathbf{O} & \mathbf{O} & \mathbf{O} \\ \mathbf{O} & \mathbf{J} & \mathbf{O} & \mathbf{J} & \mathbf{O} & \mathbf{J} \\ \mathbf{O} & \mathbf{O} & \mathbf{O} & \mathbf{O} & \mathbf{O} & \mathbf{O} \\ \mathbf{O} & \mathbf{J} & \mathbf{O} & \mathbf{J} & \mathbf{O} & \mathbf{J} \end{pmatrix} \mathbf{y}.$$

If we denote the above matrices as $\mathbf{C}_1$ and $\mathbf{C}_2$, we have

$$\tfrac{1}{6}\sum_{j=1}^{2} y_{.j.}^2 = \tfrac{1}{6}y_{.1.}^2 + \tfrac{1}{6}y_{.2.}^2 = \tfrac{1}{6}\mathbf{y}'\mathbf{C}_1\mathbf{y} + \tfrac{1}{6}\mathbf{y}'\mathbf{C}_2\mathbf{y} = \tfrac{1}{6}\mathbf{y}'(\mathbf{C}_1 + \mathbf{C}_2)\mathbf{y}.$$

Then

$$\mathbf{C} = \mathbf{C}_1 + \mathbf{C}_2 = \begin{pmatrix} \mathbf{J} & \mathbf{O} & \mathbf{J} & \mathbf{O} & \mathbf{J} & \mathbf{O} \\ \mathbf{O} & \mathbf{J} & \mathbf{O} & \mathbf{J} & \mathbf{O} & \mathbf{J} \\ \mathbf{J} & \mathbf{O} & \mathbf{J} & \mathbf{O} & \mathbf{J} & \mathbf{O} \\ \mathbf{O} & \mathbf{J} & \mathbf{O} & \mathbf{J} & \mathbf{O} & \mathbf{J} \\ \mathbf{J} & \mathbf{O} & \mathbf{J} & \mathbf{O} & \mathbf{J} & \mathbf{O} \\ \mathbf{O} & \mathbf{J} & \mathbf{O} & \mathbf{J} & \mathbf{O} & \mathbf{J} \end{pmatrix} \mathbf{y}.$$

14.20 We show the result of $\frac{1}{2}\mathbf{A} - \frac{1}{4}\mathbf{B} - \frac{1}{6}\mathbf{C} + \frac{1}{12}\mathbf{D}$ for the first two "rows":

$$\frac{1}{2}\begin{pmatrix} \mathbf{J} & \mathbf{O} & \mathbf{O} & \mathbf{O} & \mathbf{O} & \mathbf{O} \\ \mathbf{O} & \mathbf{J} & \mathbf{O} & \mathbf{O} & \mathbf{O} & \mathbf{O} \end{pmatrix} - \frac{1}{4}\begin{pmatrix} \mathbf{J} & \mathbf{J} & \mathbf{O} & \mathbf{O} & \mathbf{O} & \mathbf{O} \\ \mathbf{J} & \mathbf{J} & \mathbf{O} & \mathbf{O} & \mathbf{O} & \mathbf{O} \end{pmatrix}$$

$$-\frac{1}{6}\begin{pmatrix} \mathbf{J} & \mathbf{O} & \mathbf{J} & \mathbf{O} & \mathbf{J} & \mathbf{O} \\ \mathbf{O} & \mathbf{J} & \mathbf{O} & \mathbf{J} & \mathbf{O} & \mathbf{J} \end{pmatrix} + \frac{1}{12}\begin{pmatrix} \mathbf{J} & \mathbf{J} & \mathbf{J} & \mathbf{J} & \mathbf{J} & \mathbf{J} \\ \mathbf{J} & \mathbf{J} & \mathbf{J} & \mathbf{J} & \mathbf{J} & \mathbf{J} \end{pmatrix}$$

$$= \frac{1}{12}\begin{pmatrix} 2\mathbf{J} & -2\mathbf{J} & -\mathbf{J} & \mathbf{J} & -\mathbf{J} & \mathbf{J} \\ -2\mathbf{J} & 2\mathbf{J} & \mathbf{J} & -\mathbf{J} & \mathbf{J} & -\mathbf{J} \end{pmatrix}$$

since

$$\begin{aligned} \tfrac{6}{12}\mathbf{J} - \tfrac{3}{12}\mathbf{J} - \tfrac{2}{12}\mathbf{J} + \tfrac{1}{12}\mathbf{J} &= \tfrac{2}{12}\mathbf{J}, \\ \mathbf{O} - \tfrac{3}{12}\mathbf{J} - \mathbf{O} + \tfrac{1}{12}\mathbf{J} &= -\tfrac{2}{12}\mathbf{J}, \\ \mathbf{O} - \mathbf{O} - \tfrac{2}{12}\mathbf{J} + \tfrac{1}{12}\mathbf{J} &= -\tfrac{1}{12}\mathbf{J}, \end{aligned}$$

and so on.

14.21 If $\alpha_1^* = \alpha_2^* = \alpha_3^* = \alpha^*$, say, then $\sum_{i=1}^3 \alpha_i^* = 0$ implies $0 = \sum_{i=1}^3 \alpha_i^* = \sum_{i=1}^3 \alpha^* = 3\alpha^*$, or $\alpha^* = 0$.

14.22

$$\begin{aligned} \mathrm{SS}(\mu, \alpha, \gamma) &= \hat{\mu} y_{\ldots} + \sum_{i=1}^{a} \hat{\alpha}_i y_{i..} + \sum_{i=1}^{a}\sum_{j=1}^{b} \hat{\gamma}_{ij} y_{ij.}. \\ &= \frac{y_{\ldots}^2}{abn} + \left(\sum_i \frac{y_{i..}^2}{bn} - \frac{y_{\ldots}^2}{abn}\right) + \left(\sum_{ij} \frac{y_{ij.}^2}{n} - \sum_i \frac{y_{i..}^2}{bn} - \sum_j \frac{y_{.j.}^2}{an} + \frac{y_{\ldots}^2}{abn}\right). \end{aligned}$$

14.23 As noted preceding Theorem 14.4b, $\mathrm{SS}(\alpha, \beta, \gamma|\mu) = \mathrm{SS}(\alpha|\mu, \beta, \gamma) + \mathrm{SS}(\beta|\mu, \alpha, \gamma) + \mathrm{SS}(\gamma|\mu, \alpha, \beta)$, where $\mathrm{SS}(\alpha, \beta, \gamma|\mu) = \sum_{ij} y_{ij.}^2/n - y_{\ldots}^2/abn$. For $a = 3$, $b = 2$, and $n = 2$, we have by (14.57) and (14.60), $\mathrm{SS}(\alpha, \beta, \gamma|\mu) = \mathbf{y}'(\frac{1}{2}\mathbf{A} - \frac{1}{12}\mathbf{D})\mathbf{y}$, where

$$\mathbf{A} = \begin{pmatrix} \mathbf{J} & \mathbf{O} & \mathbf{O} & \mathbf{O} & \mathbf{O} & \mathbf{O} \\ \mathbf{O} & \mathbf{J} & \mathbf{O} & \mathbf{O} & \mathbf{O} & \mathbf{O} \\ \mathbf{O} & \mathbf{O} & \mathbf{J} & \mathbf{O} & \mathbf{O} & \mathbf{O} \\ \mathbf{O} & \mathbf{O} & \mathbf{O} & \mathbf{J} & \mathbf{O} & \mathbf{O} \\ \mathbf{O} & \mathbf{O} & \mathbf{O} & \mathbf{O} & \mathbf{J} & \mathbf{O} \\ \mathbf{O} & \mathbf{O} & \mathbf{O} & \mathbf{O} & \mathbf{O} & \mathbf{J} \end{pmatrix}, \quad \mathbf{D} = \mathbf{J}_{12} = \begin{pmatrix} \mathbf{J} & \mathbf{J} & \mathbf{J} & \mathbf{J} & \mathbf{J} & \mathbf{J} \\ \mathbf{J} & \mathbf{J} & \mathbf{J} & \mathbf{J} & \mathbf{J} & \mathbf{J} \\ \mathbf{J} & \mathbf{J} & \mathbf{J} & \mathbf{J} & \mathbf{J} & \mathbf{J} \\ \mathbf{J} & \mathbf{J} & \mathbf{J} & \mathbf{J} & \mathbf{J} & \mathbf{J} \\ \mathbf{J} & \mathbf{J} & \mathbf{J} & \mathbf{J} & \mathbf{J} & \mathbf{J} \\ \mathbf{J} & \mathbf{J} & \mathbf{J} & \mathbf{J} & \mathbf{J} & \mathbf{J} \end{pmatrix},$$

and $\mathbf{J}$ and $\mathbf{O}$ are 2×2. Show that $\frac{1}{2}\mathbf{A} - \frac{1}{2}\mathbf{D}$ is idempotent, so that condition (c) of Theorem 5.6c is satisfied. To show that condition (d) holds, note that the degrees of freedom of $\sum_{ij} y_{ij.}^2/n - y_{...}^2/abn$ are $ab - 1$, which is easily shown to equal $(a-1)+(b-1)+(a-1)(b-1)$.

14.25 For $b = 2$, the sum of squares has only 1 degree of freedom and $\mathbf{C}$ has only one row. From (14.11), we obtain H_0: $\beta_1^* - \beta_2^* = 0$ or H_0: $\beta_1 - \beta_2 + \frac{1}{3}\gamma_{11} + \frac{1}{3}\gamma_{21} + \frac{1}{3}\gamma_{31} - \frac{1}{3}\gamma_{12} - \frac{1}{3}\gamma_{22} - \frac{1}{3}\gamma_{32} = 0$. Thus $\mathbf{C} = \mathbf{c}' = (0, 0, 0, 0, 1, -1, \frac{1}{3}, -\frac{1}{3}, \frac{1}{3}, -\frac{1}{3}, \frac{1}{3}, -\frac{1}{3})$, $\mathbf{c}'(\mathbf{X}'\mathbf{X})^-\mathbf{c} = 1/3$, $[\mathbf{c}'(\mathbf{X}\mathbf{X})^-\mathbf{c}]^{-1} = 3$, and

$$\mathbf{X}(\mathbf{X}'\mathbf{X})^-\mathbf{c}[\mathbf{c}'(\mathbf{X}\mathbf{X})^-\mathbf{c}]^{-1}\mathbf{c}'(\mathbf{X}'\mathbf{X})^-\mathbf{X}' = \frac{1}{12}\begin{pmatrix} \mathbf{J} & -\mathbf{J} & \mathbf{J} & -\mathbf{J} & \mathbf{J} & -\mathbf{J} \\ -\mathbf{J} & \mathbf{J} & -\mathbf{J} & \mathbf{J} & -\mathbf{J} & \mathbf{J} \\ \mathbf{J} & -\mathbf{J} & \mathbf{J} & -\mathbf{J} & \mathbf{J} & -\mathbf{J} \\ -\mathbf{J} & \mathbf{J} & -\mathbf{J} & \mathbf{J} & -\mathbf{J} & \mathbf{J} \\ \mathbf{J} & -\mathbf{J} & \mathbf{J} & -\mathbf{J} & \mathbf{J} & -\mathbf{J} \\ -\mathbf{J} & \mathbf{J} & -\mathbf{J} & \mathbf{J} & -\mathbf{J} & \mathbf{J} \end{pmatrix}.$$

where $\mathbf{J}$ is 2×2. This can be expressed as

$$\frac{1}{12}\begin{pmatrix} 2\mathbf{J} & \mathbf{O} & 2\mathbf{J} & \mathbf{O} & 2\mathbf{J} & \mathbf{O} \\ \mathbf{O} & 2\mathbf{J} & \mathbf{O} & 2\mathbf{J} & \mathbf{O} & 2\mathbf{J} \\ 2\mathbf{J} & \mathbf{O} & 2\mathbf{J} & \mathbf{O} & 2\mathbf{J} & \mathbf{O} \\ \mathbf{O} & 2\mathbf{J} & \mathbf{O} & \mathbf{J} & \mathbf{O} & 2\mathbf{J} \\ 2\mathbf{J} & \mathbf{O} & 2\mathbf{J} & \mathbf{O} & 2\mathbf{J} & \mathbf{O} \\ \mathbf{O} & 2\mathbf{J} & \mathbf{O} & 2\mathbf{J} & \mathbf{O} & 2\mathbf{J} \end{pmatrix} - \frac{1}{12}\mathbf{J}_{12} = \frac{1}{12}\mathbf{B} - \frac{1}{12}\mathbf{J}_{12}.$$

Since $\frac{1}{12}\mathbf{B}$ is the same as $\frac{1}{6}\mathbf{C}$ in (14.59), the result is obtained.

14.26

$$E(\varepsilon_{ijk}^2) = E(\varepsilon_{ijk} - 0)^2 = E[\varepsilon_{ijk} - E(\varepsilon_{ijk})]^2 = \text{var}(\varepsilon_{ijk}) = \sigma^2,$$

$$\begin{aligned} E(\varepsilon_{ijk}\varepsilon_{lmn}) &= E[(\varepsilon_{ijk} - 0)(\varepsilon_{lmn} - 0)] = E\{[\varepsilon_{ijk} - E(\varepsilon_{ijk})][\varepsilon_{lmn} - E(\varepsilon_{lmn})]\} \\ &= \text{cov}(\varepsilon_{ijk}, \varepsilon_{lmn}) = 0. \end{aligned}$$

14.27

$$E\left(\sum_i y_{i..}^2\right) = E\left[\sum_i\left(\sum_{jk} y_{ijk}\right)^2\right]$$

$$= E\left\{\sum_i\left[\sum_{jk}(\mu^* + \alpha_i^* + \beta_j^* + \gamma_{ij}^* + \varepsilon_{ijk})\right]^2\right\}$$

$$= E\left\{\sum_i\left[bn\mu^* + bn\alpha_i^* + n\sum_j \beta_j^* + n\sum_j \gamma_{ij}^* + \sum_{jk}\varepsilon_{ijk}\right]^2\right\}$$

$$= E\left\{\sum_i\left[b^2n^2\mu^{*2} + b^2n^2\alpha_i^{*2} + \left(\sum_{jk}\varepsilon_{ijk}\right)^2 + 2b^2n^2\mu^*\alpha_i^*\right.\right.$$

$$\left.\left. + 2bn\mu^*\sum_{jk}\varepsilon_{ijk} + 2bn\alpha_i^*\sum_{jk}\varepsilon_{ijk}\right]\right\}$$

$$= E\left\{ab^2n^2\mu^{*2} + b^2n^2\sum_i \alpha_i^{*2} + \sum_{ijk}\varepsilon_{ijk}^2 + \sum_i\left(\sum_{jk\neq lm}\varepsilon_{ijk}\varepsilon_{ilm}\right)\right.$$

$$\left. + 2b^2n^2\mu^*\sum_i \alpha_i^* + 2bn\mu^*\sum_{ijk}\varepsilon_{ijk} + 2bn\sum_i\left(\alpha_i^*\sum_{jk}\varepsilon_{ijk}\right)\right\}$$

$$= ab^2n^2\mu^{*2} + b^2n^2\sum_i \alpha_i^{*2} + abn\sigma^2.$$

14.30 Using $\alpha_{.} = 3\bar{\alpha}_{.}$, $\gamma_{i.} = 2\bar{\gamma}_{i.}$, and $\gamma_{..} = 6\bar{\gamma}_{..}$, (14.90) becomes $\boldsymbol{\beta}'\mathbf{H}\boldsymbol{\beta} = 4\sum_i \alpha_i^2 + 8\sum_i \alpha_i\bar{\gamma}_i + 4\sum_i \bar{\gamma}_{i.} - 12\bar{\alpha}_{.}^2 - 24\bar{\alpha}_{.}\,\bar{\gamma}_{..} - 12\bar{\gamma}_{..}^2$. Show that the 10 terms of $4\sum_i(\alpha_i - \bar{\alpha}. + \bar{\gamma}_{i.} - \bar{\gamma}_{..})^2$ in (14.91) collapse to the same expression for $\boldsymbol{\beta}'\mathbf{H}\boldsymbol{\beta}$ involving 6 terms.

14.31 **(c)**

$$E(\bar{y}_{ij.}) = E\left(\tfrac{1}{2}\sum_{k=1}^{n} y_{ijk}\right) = \tfrac{1}{2}\sum_k E(y_{ijk})$$

$$= \tfrac{1}{2}\sum_k(\mu + \alpha_i + \beta_j + \gamma_{ij}) = \tfrac{1}{2}(2\mu + 2\alpha_i + 2\beta_j + 2\gamma_{ij}).$$

14.32 **(b)** By (14.47), (14.93), (14.94) and Problem 14.31(b, c), we have

$$E[\mathrm{SS}(\gamma|\mu, \alpha, \beta)] = 2\sigma^2 + 2\sum_{ij}[E(\bar{y}_{ij.}) - E(\bar{y}_{i..}) - E(\bar{y}_{.j.}) + E(\bar{y}_{...})]^2$$

$$= 2\sigma^2 + 2\sum_{ij}[\mu + \alpha_i + \beta_j + \gamma_{ij} - \mu - \alpha_i - \bar{\beta}_. - \bar{\gamma}_{i.}$$

$$- \mu - \bar{\alpha}_. - \beta_j - \bar{\gamma}_{.j} + \mu + \bar{\alpha}_. + \bar{\beta}_. + \bar{\gamma}_{..}]^2$$

$$= 2\sigma^2 + 2\sum_{ij}(\gamma_{ij} - \bar{\gamma}_{i.} - \bar{\gamma}_{.j} + \bar{\gamma}_{..})^2.$$

14.33

Analysis of Variance for the Lactic Acid Data in Table 13.5

Source	Sum of Squares	df	Mean Square	F
A	533.5445	1	533.5445	30.028
B	2974.0180	4	746.5045	41.844
AB	441.1580	4	110.2895	6.207
Error	177.6850	10	17.7685	
Total	4126.4055	19		

The p values for these three F's are .0003, .000003, and .009.

14.34

Analysis of Variance for the Hemoglobin Data in Table 13.7

Source	Sum of Squares	df	Mean Square	F
Rate	90.560375	3	30.186792	19.469
Method	2.415125	1	2.415125	1.380
Interaction	4.872375	3	1.624125	1.558
Error	111.637000	72	1.550514	
Total	209.484875	79		

The p value for the first F is 2.404×10^{-9}. The other two p values are .2161 and .3769.

Chapter 15

15.1

$$\mathbf{W}'\mathbf{W} = \begin{pmatrix} n_1 & 0 & \ldots & 0 \\ 0 & n_2 & \ldots & 0 \\ \vdots & \vdots & & \vdots \\ 0 & 0 & \ldots & n_k \end{pmatrix}, \mathbf{W}'\mathbf{y} = \begin{pmatrix} y_{1\cdot} \\ y_{2\cdot} \\ \vdots \\ y_{k\cdot} \end{pmatrix},$$

$$(\mathbf{W}'\mathbf{W})^{-1}\mathbf{W}'\mathbf{y} = \begin{pmatrix} y_{1\cdot}/n_1 \\ y_{2\cdot}/n_2 \\ \vdots \\ y_{k\cdot}/n_k \end{pmatrix} = \begin{pmatrix} \bar{y}_{1\cdot} \\ \bar{y}_{2\cdot} \\ \vdots \\ \bar{y}_k \end{pmatrix}.$$

15.2 **(a)** The reduced model $y_{ij} = \mu + \varepsilon_{ij}^*$ can be written in matrix form as $\mathbf{y} = \mu\mathbf{j} + \varepsilon^*$, from which we have $\hat{\mu} = \bar{y}_{..}$ and $\hat{\mu}\mathbf{j}'\mathbf{y} = \bar{y}_{..}y_{..} = y_{..}^2/N = N\bar{y}_{..}^2$.

(b) $\text{SSB} = \hat{\boldsymbol{\mu}}'\mathbf{W}'\mathbf{y} - N\bar{y}_{..}^2 = (\bar{y}_{1.}, \bar{y}_{2.}, \ldots, \bar{y}_{k.})\begin{pmatrix} y_{1.} \\ \vdots \\ y_k \end{pmatrix} - N\bar{y}_{..}^{-2}$.

(c) $\sum_{i=1}^k \bar{y}_{i.}y_{i.} - N\bar{y}_{..}^2 = \sum_i \dfrac{y_{i.}}{n_i}y_{i.} - N\left(\dfrac{y_{..}}{N}\right)^2$.

15.3 **(a)**

$$\begin{aligned}\sum_{i=1}^k n_i(\bar{y}_{i.} - \bar{y}_{..})^2 &= \sum_i (n_i\bar{y}_{i.}^2 - 2n_i\bar{y}_{i.}\bar{y}_{..} + n_i\bar{y}_{..}^2) \\ &= \sum_i n_i\bar{y}_{i.}^2 - 2\bar{y}_{..}\sum_i n_i\bar{y}_{i.} + \bar{y}_{..}^2\sum_i n_i \\ &= \sum_i n_i\left(\frac{y_{i.}}{n_i}\right)^2 - \frac{2y_{..}}{N}\sum_i \frac{n_iy_{i.}}{n_i} + N\left(\frac{y_{..}}{N}\right)^2 \\ &= \sum_i \frac{y_{i.}^2}{n_i} - \frac{2y_{..}^2}{N} + \frac{y_{..}^2}{N}.\end{aligned}$$

(b)

$$\begin{aligned}\sum_{i=1}^k\sum_{j=1}^{n_i}(y_{ij} - \bar{y}_{i.})^2 &= \sum_i\sum_j (y_{ij}^2 - 2y_{ij}\bar{y}_{i.} + \bar{y}_{i.}^2) \\ &= \sum_{ij} y_{ij}^2 - 2\sum_i\left(\frac{y_{i.}}{n_i}\sum_j y_{ij}\right) + \sum_i\sum_j\left(\frac{y_{i.}}{n_i}\right)^2 \\ &= \sum_{ij} y_{ij}^2 - 2\sum_i \frac{y_{i.}^2}{n_i} + \sum_i n_i\frac{y_{i.}^2}{n_i^2} \\ &= \sum_{ij} y_{ij}^2 - \sum_i \frac{y_{i.}^2}{n_i}.\end{aligned}$$

15.4 $(c'\hat{\boldsymbol{\mu}})'[c'(\mathbf{W}'\mathbf{W})^{-1}c]^{-1}c'\hat{\boldsymbol{\mu}}/s^2 = \left(\sum_i c_i\bar{y}_{i.}\right)\left(\sum_i c_i^2/n_i\right)^{-1}\sum_i c_i\bar{y}_{i.}/s^2.$

15.5 $\sum_i \frac{a_i b_i}{n_i} = \frac{(25)(0)}{10} + \frac{(-20)(1)}{20} + \frac{(-5)(-1)}{5} = -1 + 1 = 0.$

15.6
$$(\mathbf{W}'\mathbf{W})^{-1}\mathbf{W}'\mathbf{y} = \begin{pmatrix} n_1 1 & 0 & \dots & 0 \\ 0 & n_{12} & \dots & 0 \\ \vdots & \vdots & & \vdots \\ 0 & 0 & \dots & n_{23} \end{pmatrix}^{-1} \begin{pmatrix} y_{11.} \\ y_{12.} \\ \vdots \\ y_{23.} \end{pmatrix} = \begin{pmatrix} \bar{y}_{11.} \\ \bar{y}_{12.} \\ \vdots \\ \bar{y}_{23.} \end{pmatrix}.$$

15.7 $\mathbf{W}\hat{\boldsymbol{\mu}} = (\bar{y}_{11.}, \bar{y}_{11.}, \bar{y}_{12.}, \bar{y}_{13.}, \bar{y}_{13.}, \dots, \bar{y}_{23.})'$. Note that $\hat{\boldsymbol{\mu}}$ is 6×1 and $\mathbf{W}\hat{\boldsymbol{\mu}}$ is 11×1.

15.8 This follows by definition; see, for example, (7.41).

15.9 Since $\mathbf{B}\boldsymbol{\mu} = \mathbf{0}$, that is $\mathbf{b}_1'\boldsymbol{\mu} = 0$, we can equate $\mathbf{b}_1'\boldsymbol{\mu}$ and $\mathbf{b}_2'\boldsymbol{\mu}$ to obtain $2\mu_{11} - \mu_{12} - \mu_{13} + 2\mu_{21} - \mu_{22} - \mu_{23} = \mu_{12} - \mu_{13} + \mu_{22} - \mu_{23}$, which reduces to $2\mu_{11} + 2\mu_{21} = 2\mu_{12} + 2\mu_{22}$. We can obtain $\mu_{12} + \mu_{22} = \mu_{13} + \mu_{23}$ from $\mathbf{b}_2'\boldsymbol{\mu} = 0$.

15.10 $\mathbf{a}'\hat{\boldsymbol{\mu}} = \sum_{ij} a_{ij}\bar{y}_{ij.}$ and $\mathbf{a}'(\mathbf{W}'\mathbf{W})^{-1}\mathbf{a} = \sum_{ij} a_{ij}^2/n_{ij.}$

15.11 $\mathbf{a}'(\mathbf{W}'\mathbf{W})^{-1}\mathbf{a} = 3.833.$

15.12 $\mathbf{B}(\mathbf{W}'\mathbf{W})^{-1}\mathbf{B}' = \frac{1}{3}\begin{pmatrix} 25 & -1 \\ -1 & 7 \end{pmatrix}.$

15.13 Show that $\mathbf{KG}' = \mathbf{O}$.

15.14 By (3.42), $\text{cov}(\hat{\boldsymbol{\mu}}_c) = \mathbf{K}'(\mathbf{KK}')^{-1}\text{cov}(\hat{\boldsymbol{\delta}}_c)(\mathbf{KK}')^{-1}\mathbf{K}$.

15.15 (a)

Analysis of Variance for the Weight Data of Table 14.6

Source	df	Sum of Squares	Mean Square	F	p Value
Protein	4	111,762.28	27940.57	8.36	.0000169
Error	56	181,256.71	3236.73		
Total	60	293,018.98			

(b)

***F* Tests for Unweighted Contrasts**

Contrasts	df	Contrast SS	F	p Value
L, C vs. So, Su, M	1	2,473.61	0.76	.386
So, M vs Su	1	36,261.93	11.20	.00147
So vs. M	1	5,563.22	1.72	.195
L vs. C	1	65,940.17	20.37	.0000332

(c)

***F* Tests for Two Weighted Contrasts**

Contrasts	df	Contrast SS	F	p Value
L, C vs. So, Su, M	1	2,473.61	0.76	.386
So, M vs. Su	1	31,673.79	9.79	.00278

15.17 **(a)** $\boldsymbol{\mu} = (\mu_{11}, \mu_{12}, \mu_{13}, \mu_{21}, \mu_{22}, \mu_{23})'$, $\mathbf{W}$ is 47×6, $\mathbf{a}$ is 6×1, $\mathbf{B}$ and $\mathbf{C}$ are each 2×6.

$$\mathbf{a} = (1,\ 1,\ 1,\ -1,\ -1)$$

$$\mathbf{B} = \begin{pmatrix} 1 & -2 & 1 & 1 & -2 & 1 \\ 1 & 0 & -1 & 1 & 0 & -1 \end{pmatrix}$$

$$\mathbf{C} = \begin{pmatrix} 1 & -2 & 1 & -1 & 2 & -1 \\ 1 & 0 & -1 & -1 & -1 & 1 \end{pmatrix}$$

$$\hat{\boldsymbol{\mu}} = \bar{\mathbf{y}} = (96.50,\ 85.90,\ 95.17,\ 79.20, 91.83,\ 82.00)'$$

$$\text{SSE} = 8436.1667, \quad \nu_E = 41$$

$$F_A = 3.65104, \quad F_B = .022053, \quad F_C = 2.90567.$$

(b) $\mathbf{G}$ is the same as $\mathbf{C}$ in part (a)

$$\mathbf{K} = \begin{pmatrix} \mathbf{j}' \\ \mathbf{a}' \\ \mathbf{B} \end{pmatrix} = \begin{pmatrix} 1 & 1 & 1 & 1 & 1 & 1 \\ 1 & 1 & 1 & -1 & -1 & -1 \\ 1 & -2 & 1 & 1 & -2 & 1 \\ 1 & 0 & -1 & 1 & 0 & -1 \end{pmatrix}$$

$$\hat{\boldsymbol{\mu}}_c = (91.61, 91.31, 92.66, 83.11, 82.81, 84.15)'$$

$$\text{SSE}_c = 9631.9072, \quad \nu_{E_c} = 43$$

$$F_{A_c} = 3.687, \quad F_{B_c} = .03083.$$

(c)

Analysis of Variance for Unconstrained Model

Source	df	Sum of Squares	Mean Square	F	p Value
Level	1	751.238	751.238	3.65	.0630
Type	2	9.075	4.538	.02	.978
Level × type	2	1195.7406	597.870	2.91	.0661
Error	41	8436.167	205.760		
Total	46	10474.851			

Analysis of Variance for Constrained Model

Source	df	Sum of Squares	Mean Square	F	p Value
Level	1	826.544	826.544	3.69	.0614
Type	2	13.810	6.905	0.03	.970
Error	43	9,631.907	223.998		
Total	46	10,474.851			

15.18 **Analysis of Variance for Data in Table 15.12**

Source	df	Sum of Squares	Mean Square	F	p Value
Fertilizer	3	20,979.042	9663.014	111.52	5.773×10^{-15}
Variety	4	306.621	76.655	1.22	0.325
Fertilizer × variety	12	997.589	83.132	1.33	0.263
Error	26	1,630.333	62.705		
Total	45	28,486.370			

Chapter 16

16.1 Verify that $\mathbf{I} - \mathbf{P}$ is symmetric and idempotent. Then $\mathbf{X}'(\mathbf{I} - \mathbf{P})\mathbf{X} = \mathbf{X}'(\mathbf{I} - \mathbf{P})'(\mathbf{I} - \mathbf{P})\mathbf{X}$, and by Theorem 2.4(iii), rank$[\mathbf{X}'(\mathbf{I} - \mathbf{P})\mathbf{X}] =$ rank $[(\mathbf{I} - \mathbf{P})\mathbf{X}]$. The matrix $(\mathbf{I} - \mathbf{P})\mathbf{X}$ is $n \times q$. To show that rank $[(\mathbf{I} - \mathbf{P})\mathbf{X} = \mathbf{q}$, we demonstrate that the columns are linearly independent. By the definition of linear independence in (2.40), the columns of $(\mathbf{I} - \mathbf{P})\mathbf{X}$ are lineraly independent if $(\mathbf{X} - \mathbf{PX})\mathbf{a} = \mathbf{0}$ implies $\mathbf{a} = \mathbf{0}$. Suppose that there is a vector $\mathbf{a} \neq \mathbf{0}$ such that $(\mathbf{X} - \mathbf{PX})\mathbf{a} = \mathbf{0}$. Then

$$\mathbf{Xa} = \mathbf{PXa} = \mathbf{Z}(\mathbf{Z}'\mathbf{Z})^{-}\mathbf{Z}'\mathbf{Xa}.$$

By Theorem 2.8c(iii), a solution to this is $\mathbf{Xa} = \mathbf{z}_i$, where $\mathbf{z}_i$ is the ith column of $\mathbf{Z}$. But this is impossible since the columns of $\mathbf{Z}$ are linearly independent of those of $\mathbf{X}$. We therefore have $\mathbf{Xa} = \mathbf{0}$, which implies that $\mathbf{a} = \mathbf{0}$, since $\mathbf{X}$ is full-rank. This contradicts the possibility that $\mathbf{a} \neq \mathbf{0}$.

16.2 **(a)** By (16.11) and (16.14) to (16.17), we obtain

$$\begin{aligned} \text{SSE}_{y.x} &= \mathbf{y}'\mathbf{y} - \mathbf{y}'\mathbf{Z}(\mathbf{Z}'\mathbf{Z})^{-}\mathbf{Z}'\mathbf{y} - \hat{\boldsymbol{\beta}}'\mathbf{X}'(\mathbf{I} - \mathbf{P})\mathbf{y} \\ &= \mathbf{y}'(\mathbf{I} - \mathbf{P})\mathbf{y} - \mathbf{e}_{xy}'\mathbf{E}_{xx}^{-1}\mathbf{e}_{xy} = e_{yy} - \mathbf{e}_{xy}'\mathbf{E}_{xx}^{-1}\mathbf{e}_{xy.} \end{aligned}$$

(b) By (12.21), $\text{SSE}_y = \mathbf{y}'[\mathbf{I} - \mathbf{Z}(\mathbf{Z}'\mathbf{Z})^{-}\mathbf{Z}']\mathbf{y}$, which, by (16.13), equals $\mathbf{y}'(\mathbf{I} - \mathbf{P})\mathbf{y}$.

16.3 By Theorem 8.4a(ii), SSH $= (\mathbf{C}\hat{\boldsymbol{\beta}})'\mathbf{A}(\mathbf{C}\hat{\boldsymbol{\beta}})$, where $\mathbf{A} = [\mathrm{cov}(\mathbf{C}\hat{\boldsymbol{\beta}})]^{-1}/\sigma^2$. In this case, $\mathbf{C} = \mathbf{I}$ and by (15.19) $\mathrm{cov}(\hat{\boldsymbol{\beta}}) = \sigma^2[\mathbf{X}(\mathbf{I} - \mathbf{P})\mathbf{X}']^{-1}$.

16.4 By (13.12), a solution $\hat{\boldsymbol{\alpha}}_0$ is given by $\hat{\boldsymbol{\alpha}}_0 = (0, \bar{y}_{1.}, \ldots, \bar{y}_{k.})'$. By analogy to (13.7), $\mathbf{Z}'\mathbf{x} = (x_{..}, x_{1.}, \ldots, x_{k.})'$, and by (13.11), a generalized inverse is $(\mathbf{Z}'\mathbf{Z})^{-} = \mathrm{diag}(0, 1/n, \ldots, 1/n)$. Then $(\mathbf{Z}'\mathbf{Z})^{-}\mathbf{Z}'\mathbf{x} = (0, \bar{x}_{1.}, \ldots, \bar{x}_{k.})'$.

16.5 By (16.13) and (16.16), $e_{xx} = x'(\mathbf{I} - \mathbf{P})\mathbf{x} = \mathbf{x}'\mathbf{x} - \mathbf{x}'\mathbf{Z}(\mathbf{Z}'\mathbf{Z})^{-}\mathbf{Z}'\mathbf{x}$. From the answer to Problem 16.4, $\mathbf{x}'\mathbf{Z} = (x_{..}, x_{1.}, \ldots, x_{k.})$ and $(\mathbf{Z}'\mathbf{Z})^{-}\mathbf{Z}'\mathbf{x} = (0, \bar{x}_{1.}, \ldots, \bar{x}_{k.})'$. Thus $\mathbf{x}'\mathbf{Z}(\mathbf{Z}'\mathbf{Z})^{-}\mathbf{Z}'\mathbf{x} = \sum_{i=1}^{k} x_{i.}\bar{x}_{i.}$ $= n\sum_i \bar{x}_{i.}^2$, and $e_{xx} = \sum_{ij} x_{ij}^2 - n\sum_i \bar{x}_{i.}^2$. Show that e_{xx} can be written as $e_{xx} = \sum_{ij}(x_{ij} - \bar{x}_{i.})^2$. The quantities e_{xy} and e_{yy} can be found in an analogous manner.

16.6 **(a)** By (16.39) and (16.40), we have

$$\mathbf{X}'(\mathbf{I} - \mathbf{P}) = \begin{pmatrix} \mathbf{x}_1' & \mathbf{0}' & \cdots & \mathbf{0}' \\ \mathbf{0}' & \mathbf{x}_2' & \cdots & \mathbf{0}' \\ \vdots & \vdots & & \vdots \\ \mathbf{0}' & \mathbf{0}' & \cdots & \mathbf{x}_k' \end{pmatrix} \begin{pmatrix} \mathbf{I} - \frac{1}{n}\mathbf{J} & \mathbf{O} & \cdots & \mathbf{O} \\ \mathbf{O} & \mathbf{I} - \frac{1}{n}\mathbf{J} & \cdots & \mathbf{O} \\ \vdots & \vdots & & \vdots \\ \mathbf{O} & \mathbf{O} & \cdots & \mathbf{I} - \frac{1}{n}\mathbf{J} \end{pmatrix}$$

$$= \begin{pmatrix} \mathbf{x}_1'(\mathbf{I} - \frac{1}{n}\mathbf{J}) & \mathbf{0}' & \cdots & \mathbf{0}' \\ \mathbf{0}' & \mathbf{x}_2'(\mathbf{I} - \frac{1}{n}\mathbf{J}) & \cdots & \mathbf{0}' \\ \vdots & \vdots & & \vdots \\ \mathbf{0}' & \mathbf{0}' & \cdots & \mathbf{x}_k'(\mathbf{I} - \frac{1}{n}\mathbf{J}) \end{pmatrix},$$

$$\mathbf{X}'(\mathbf{I} - \mathbf{P})\mathbf{X} = \begin{pmatrix} \mathbf{x}_1'(\mathbf{I} - \frac{1}{n}\mathbf{J})\mathbf{x}_1 & 0 & \cdots & 0 \\ 0 & \mathbf{x}_2'(\mathbf{I} - \frac{1}{n}\mathbf{J})\mathbf{x}_2 & \cdots & 0 \\ \vdots & \vdots & & \vdots \\ 0 & 0 & \cdots & \mathbf{x}_k'(\mathbf{I} - \frac{1}{n}\mathbf{J})\mathbf{x}_k \end{pmatrix}.$$

To show that this equal to (16.41), we have, for example

$$\mathbf{x}_2'\left(\mathbf{I}-\frac{1}{n}\mathbf{J}\right)\mathbf{x}_2=\mathbf{x}_2'\mathbf{x}_2-\frac{\mathbf{x}_2\mathbf{j}\mathbf{j}'\mathbf{x}_2}{n}=\sum_j x_{2j}^2-\frac{x_{2.}^2}{n}.$$

Show that this equals $\sum_j (x_{2j}-\bar{x}_{2.})^2$.

(b) Using $\mathbf{X}'(\mathbf{I}-\mathbf{P})$ from part (a), we have

$$\mathbf{X}'(\mathbf{I}-\mathbf{P})\mathbf{y}=\begin{pmatrix}\mathbf{x}_1'\left(\mathbf{I}-\frac{1}{n}\mathbf{J}\right) & \mathbf{0}' & \cdots & \mathbf{0}' \\ \mathbf{0}' & \mathbf{x}_2'\left(\mathbf{I}-\frac{1}{n}\mathbf{J}\right) & \cdots & \mathbf{0}' \\ \vdots & \vdots & & \vdots \\ \mathbf{0}' & \mathbf{0}' & \cdots & \mathbf{x}_k'\left(\mathbf{I}-\frac{1}{n}\mathbf{J}\right)\end{pmatrix}\begin{pmatrix}\mathbf{y}_1\\ \mathbf{y}_2\\ \vdots\\ \mathbf{y}_k\end{pmatrix}$$

$$=\begin{pmatrix}\mathbf{x}_1'\left(\mathbf{I}-\frac{1}{n}\mathbf{J}\right)\mathbf{y}_1\\ \mathbf{x}_2'(\mathbf{I}-\frac{1}{n}\mathbf{J})\mathbf{y}_2\\ \vdots\\ \mathbf{x}_k'\left(\mathbf{I}-\frac{1}{n}\mathbf{J}\right)\mathbf{y}_k\end{pmatrix}.$$

The elements of this vector are, for example

$$\mathbf{x}_2'\left(\mathbf{I}-\frac{1}{n}\mathbf{J}\right)\mathbf{y}_2=\mathbf{x}_2'\mathbf{y}_2-\frac{\mathbf{x}_2'\mathbf{j}\mathbf{j}'\mathbf{y}_2}{n}=\sum_j x_{2j}y_{2j}-\frac{\bar{x}_{2.}\bar{y}_{2.}}{n.}$$

Show that this equals $\sum_j (x_{2j}-\bar{x}_{2.})(y_{2j}-\bar{y}_{2.})$.

16.7

$$\sum_{ijk}(x_{ijk}-\bar{x}_{ij.})(y_{ijk}-\bar{y}_{ij.}) = \sum_{ijk}x_{ijk}y_{ijk} - \sum_{ijk}x_{ijk}\bar{y}_{ij.} - \sum_{ijk}\bar{x}_{ij.}y_{ijk} + n\sum_{ij}\bar{x}_{ij.}\bar{y}_{ij.}$$

$$= \sum_{ijk}x_{ijk}y_{ijk} - \sum_{ij}\left(\bar{y}_{ij.}\sum_{k}x_{ijk}\right) - \sum_{ij}\left(\bar{x}_{ij.}\sum_{k}y_{ijk}\right) + n\sum_{ij}\bar{x}_{ij.}\bar{y}_{ij.}$$

$$= \sum_{ijk}x_{ijk}y_{ijk} - n\sum_{ij}\bar{x}_{ij.}\bar{y}_{ij.} - n\sum_{ij}\bar{x}_{ij.}\bar{y}_{ij.} + n\sum_{ij}\bar{x}_{ij.}\bar{y}_{ij.}.$$

16.8 In (14.40) and (14.41), we have

$$\sum_{ij}\frac{y_{ij.}^2}{n} - \frac{y_{...}^2}{abn} = \left(\sum_i\frac{y_{i..}^2}{bn} - \frac{y_{...}^2}{abn}\right) + \left(\sum_j\frac{y_{.j.}^2}{an} - \frac{y_{...}^2}{abn}\right) + \left(\sum_{ij}\frac{y_{ij.}^2}{n} - \sum_i\frac{y_{i..}^2}{bn} - \sum_j\frac{y_{.j.}^2}{an} + \frac{y_{...}^2}{abn}\right).$$

By an analogous identity, we have (note that b is replaced by c)

$$\sum_{ij}\frac{x_{ij.}y_{ij.}}{n} - \frac{x_{...}y_{...}}{acn} = \left(\sum_i\frac{x_{i..}y_{i..}}{cn} - \frac{x_{...}y_{...}}{acn}\right) + \left(\sum_j\frac{x_{.j.}y_{.j.}}{an} - \frac{x_{...}y_{...}}{acn}\right) + \left(\sum_{ij}\frac{x_{ij.}y_{ij.}}{n} - \sum_i\frac{x_{i..}y_{i..}}{cn} - \sum_j\frac{x_{.j.}y_{.j.}}{an} + \frac{x_{...}y_{...}}{acn}\right).$$

Show that the right side is equal to

$$cn\sum_i(\bar{x}_{i..}-\bar{x}_{...})(\bar{y}_{i..}-\bar{y}_{...}) + an\sum_j(\bar{x}_{.j.}-\bar{x}_{...})(\bar{y}_{.j.}-\bar{y}_{...}) + n\sum_{ij}(\bar{x}_{ij.}-\bar{x}_{i..}-\bar{x}_{.j.}+\bar{x}_{...})(\bar{y}_{ij.}-\bar{y}_{i..}-\bar{y}_{.j.}+\bar{y}_{...})$$

$$= \text{SPA} + \text{SPC} + \text{SPAC}.$$

16.9 **(a)**

$$\text{SS}(C+E)_{y.x} = \text{SSC}_{\hat{y}} + \text{SSE}_y - \frac{(\text{SPC}+\text{SPE})^2}{\text{SSC}_x + \text{SSE}_x}$$

$$\text{SSC}_{y.x} = \text{SS}(C+E)_{y.x} - \text{SSE}_{y.x}$$

$$F = \frac{\text{SSAC}_{y.x}/(c-1)}{\text{SSE}_{y.x}/[ac(n-1)-1]}.$$

The F statistic is distributed as $F[c-1,\ ac(n-1)-1]$ if H_0 is true.

(b)

$$\text{SS}(AC+E)_{y.x} = \text{SSAC}_y + \text{SSE}_y - \frac{(\text{SPAC}+\text{SPE})^2}{\text{SSAC}_x + \text{SSE}_x}$$

$$\text{SSAC}_{y.x.} = \text{SS}(AC+E)_{y.x} - \text{SSE}_{y.x.}$$

$$F = \frac{\text{SSAC}_{y.x.}/(a-1)(c-1)}{\text{SSE}_{y.x}/[ac(n-1)-1]}.$$

The F statistic is distributed as $F[(a-1)\,(c-1),\ ac\,(n-1)-1]$ if H_0 is true.

16.10 (a) $\mathbf{E}_{xx} = \mathbf{X}'(\mathbf{I}-\mathbf{P})\mathbf{X}$

$$= (\mathbf{X}_1', \mathbf{X}_2', \ldots, \mathbf{X}_k')\begin{pmatrix} \mathbf{I}-\frac{1}{n}\mathbf{J} & \mathbf{O} & \ldots & \mathbf{O} \\ \mathbf{O} & \mathbf{I}-\frac{1}{n}\mathbf{J} & \ldots & \mathbf{O} \\ \vdots & \vdots & & \vdots \\ \mathbf{O} & \mathbf{O} & \ldots & \mathbf{I}-\frac{1}{n}\mathbf{J} \end{pmatrix}\begin{pmatrix} \mathbf{X}_1 \\ \mathbf{X}_2 \\ \vdots \\ \mathbf{X}_k \end{pmatrix}$$

$$= \left[\mathbf{X}_1'\left(\mathbf{I}-\frac{1}{n}\mathbf{J}\right), \mathbf{X}_2'\left(\mathbf{I}-\frac{1}{n}\mathbf{J}\right), \ldots, \mathbf{X}_k'\left(\mathbf{I}-\frac{1}{n}\mathbf{J}\right)\right]\begin{pmatrix} \mathbf{X}_1 \\ \mathbf{X}_2 \\ \vdots \\ \mathbf{X}_k \end{pmatrix}$$

$$= \sum_{i=l}^{k} \mathbf{X}_i'\left(\mathbf{I}-\frac{1}{n}\mathbf{J}\right)\mathbf{X}_i$$

16.11 By Theorem 2.2c(i), the diagonal elements of $\mathbf{X}_{ci}'\mathbf{X}_{ci}$ are products of columns of $\mathbf{X}_{ci}$. Thus, for example, the seond diagonal element of $\mathbf{X}_{ci}'\mathbf{X}_{ci}$ in (16.70) is

$$(x_{i12}-\bar{x}_{i.2},\ \ldots,\ x_{in2}-\bar{x}_{i.2})\begin{pmatrix} x_{i12}-\bar{x}_{i.2} \\ \vdots \\ x_{in2}-\bar{x}_{i.2} \end{pmatrix} = \sum_{j=1}^{n}(x_{ij2}-\bar{x}_{i.2})^2.$$

Similarly, the (1,2) element of $\mathbf{X}_{ci}{}'\mathbf{X}_{ci}$ is

$$(x_{i11} - \bar{x}_{i.1}, \ \ldots, x_{in1} - \bar{x}_{i.1}) \begin{pmatrix} x_{i12} - \bar{x}_{i.2} \\ \vdots \\ x_{in2} - \bar{x}_{i.2} \end{pmatrix} = \sum_{j=1}^{n} (x_{ij1} - \bar{x}_{i.1})(x_{ij2} - \bar{x}_{i.2}).$$

16.12

$$(\mathbf{Z}'\mathbf{Z})^{-}\mathbf{Z}'\mathbf{X}\hat{\boldsymbol{\beta}} = \begin{pmatrix} 0 & 0 & \ldots & 0 \\ 0 & \frac{1}{n} & \ldots & 0 \\ \vdots & \vdots & & \vdots \\ 0 & 0 & \ldots & \frac{1}{n} \end{pmatrix} \begin{pmatrix} \mathbf{j}' & \mathbf{j}' & \ldots & \mathbf{j}' \\ \mathbf{j}' & \mathbf{0}' & \ldots & \mathbf{0}' \\ \mathbf{0}' & \mathbf{j}' & \ldots & \mathbf{0}' \\ \vdots & \vdots & & \vdots \\ \mathbf{0}' & \mathbf{0}' & \ldots & \mathbf{j}' \end{pmatrix} \begin{pmatrix} \mathbf{X}_1 \\ \mathbf{X}_2 \\ \vdots \\ \mathbf{X}_k \end{pmatrix} \hat{\boldsymbol{\beta}}$$

$$= \frac{1}{n} \begin{pmatrix} \mathbf{0}' & \mathbf{0}' & \ldots & \mathbf{0}' \\ \mathbf{j} & \mathbf{0}' & \ldots & \mathbf{0}' \\ \mathbf{0}' & \mathbf{j}' & \ldots & \mathbf{0}' \\ \vdots & \vdots & & \vdots \\ \mathbf{0}' & \mathbf{0}' & \ldots & \mathbf{j}' \end{pmatrix} \begin{pmatrix} \mathbf{X}_1 \\ \mathbf{X}_2 \\ \vdots \\ \mathbf{X}_k \end{pmatrix} \hat{\boldsymbol{\beta}}$$

$$= \frac{1}{n} \begin{pmatrix} \mathbf{j}'\mathbf{X}_1 \\ \mathbf{j}'\mathbf{X}_2 \\ \vdots \\ \mathbf{j}'\mathbf{X}_k \end{pmatrix} \hat{\boldsymbol{\beta}} = \begin{pmatrix} \bar{\mathbf{x}}_1' \\ \bar{\mathbf{x}}_2' \\ \vdots \\ \bar{\mathbf{x}}_k' \end{pmatrix} \hat{\boldsymbol{\beta}} = \begin{pmatrix} \bar{\mathbf{x}}_1'\hat{\boldsymbol{\beta}} \\ \bar{\mathbf{x}}_2'\hat{\boldsymbol{\beta}} \\ \vdots \\ \bar{\mathbf{x}}_k'\hat{\boldsymbol{\beta}} \end{pmatrix}.$$

16.13

$$\sum_{ij} (y_{ij} - \bar{y}_{..})^2 - \sum_{ij} (y_{ij} - \bar{y}_{i.})^2$$

$$= \sum_{ij} y_{ij}^2 - 2\bar{y}_{..} \sum_{ij} y_{ij} + kn\bar{y}_{..}^2$$

$$- \left[\sum_{ij} y_{ij}^2 - 2 \sum_{i} \left(\bar{y}_{i.} \sum_{j} y_{ij} \right) + n \sum_{i} \bar{y}_{i.}^2 \right]$$

$$= \sum_{ij} y_{ij}^2 - kn\bar{y}_{..}^2 - \sum_{ij} y_{ij}^2 + 2n \sum_{i} \bar{y}_{i.}^2 - n \sum_{i} \bar{y}_{i.}^2$$

$$= n \sum_{i} \bar{y}_{i.}^2 - kn\bar{y}_{..}^2 = n \sum_{i} (\bar{y}_{i.} - \bar{y}_{..})^2.$$

16.14

(a) $e_{xx} = 358.1667$, $e_{xy} = 488.5000$, $e_{yy} = 5937.8333$, $\hat{\beta} = 1.3639$.

(b) $\text{SSE}_{y \cdot x} = 5271.5730$ with 19df, $\text{SST}_{y \cdot x} = 6651.1917$ with 22 df, $\text{SS}(\alpha|\mu, \beta) = 1379.6188$ with 3df, $F = 1.6575$, $p = .210$.

(c) $F = 2.4014, \quad p = .138.$

(d) $\hat{\beta}_1 = 1.9950, \quad \hat{\beta}_2 = -.9878, \quad \hat{\beta}_3 = -1.2687, \quad \hat{\beta}_4 = 3.1646,$

$$F = \frac{(5271.5730 - 4178.2698)/3}{4178.2698/16} = 1.3955, \quad p = .280.$$

16.15 (a)

Sum of Squares and Products for *x* and *y*

	SS and SP Corrected for the Mean		
Source	y	x	xy
A	268,043.37	811.11	14,627.22
C	588,510.81	2485.11	1,468.89
$-AC$	1,789,999.1	2411.11	−2,736.222
Error	7,717,172.7	5632.67	−168,409.7
$A + C$	7,985,216	6443.78	−183,036.9
$C + E$	8,305,683.5	8117.78	− 182,678.6
$AC + E$	950,7171.7	8043.78	− 171,145.9

(b)

$$\text{SS}(A+E)_{y\cdot x} = 2,786014, \text{SSE}_{y\cdot x} = 2,681,934.9, \text{SSA}_{y.x} = 104,079.06.$$

$$\text{For factor } A, F = \frac{104,079.06/2}{2,681,934.92/35} = .6791, p = .514.$$

$$\text{For factor } C, F = \frac{1,512,838.46/5}{2,681,934.92/35} = 3.9486, p = .0061.$$

$$\text{For interaction } AC, F = \frac{3,183,799.17/10}{2,681,934.92/35} = 4.1549, p = .000783.$$

(c)
$$F = \frac{(\text{SPE})^2/\text{SSE}_x}{\text{SSE}_{y\cdot x}/[ac(n-1)-1]} = \frac{(-168,409.7)^2/5632.67}{2,681,934.92/35}$$
$$= 65.7113, p = 1.516 \times 10^{-9}.$$

(d) For factor A

$$\hat{\beta}_1 = \frac{\text{SPE}_1}{\text{SSE}_{x,1}} = \frac{-50,869.67}{1,274.67} = -39.9082,$$

$$\hat{\beta}_2 = \frac{\text{SPE}_2}{\text{SSE}_{x,2}} = \frac{-37,796.33}{1,987.33} = -19.0187,$$

$$\hat{\beta}_3 = \frac{\text{SPE}_3}{\text{SSE}_{x,3}} = \frac{-79,743.67}{2,370.67} = -33.6377,$$

$$\text{SS}(F) = \text{SSE}_y - \sum_{i=1}^{3} \frac{(\text{SPE}_i)^2}{\text{SSE}_{x,i}} = 2,285,831.3,$$

$$\text{SS}(R) = \text{SSE}_y - \frac{(\text{SPE})^2}{\text{SSE}_2} = 2,681,934.9.$$

By (16.64), we obtain

$$F = \frac{(2,681,934.9 - 2,285,831.3)/2}{2,285,831.3/33} = 2.8592, p = .0716.$$

For factor C

$$\hat{\beta}_1 = -32.8195, \quad \hat{\beta}_2 = -30.3492, \quad \hat{\beta}_3 = -26.2928, \quad \hat{\beta}_4 = -27.8251,$$

$$\hat{\beta}_5 = -53.1667, \quad \hat{\beta}_6 = -28.2191,$$

$$F = \frac{156,728.91/5}{2,525,206.01/30} = .3724, \quad p = .864.$$

16.16

(a)

Sums of Squares and Products for *x* and *y*

	SS and SP Corrected for the Mean		
Source	y	x	xy
A	235.225	176.4	203.70
C	30.625	0.400	−3.50
AC	3.025	12.10	−6.05
Error	867.500	5170.2	1253.10
$A+E$	1102.725	5346.6	1456.80
$C+E$	898.125	5170.6	1249.60
$AC+E$	870.525	5182.3	1247.05

(b) $\text{SS}(A+E)_{y\cdot x} = 705.7875$, $\text{SSE}_{y\cdot x} = 563.7865$, $\text{SSA}_{y\cdot x} = 142.0010$.

$$\text{For factor } A,\ F = \frac{142.0010/1}{563.7865/35} = 8.8155, \quad p = .0054.$$

$$\text{For factor } C,\ F = \frac{32.3426/1}{563.7865/35} = 2.0078, \quad p = .165.$$

$$\text{For interaction } AC,\ F = \frac{6.6529/1}{563.7865/35} = .4130, \quad p = .525.$$

(c)

$$F = \frac{(\text{SPE})^2/\text{SSE}_x}{\text{SSE}_{y\cdot x}/[ac(n-1)-1]} = \frac{(1253.10)^2/5170.200}{563.7865/35}$$

$$= 18.8546, \quad p = .000115.$$

(d) For factor A

$$\hat{\beta}_1 = \frac{\text{SPE}_1}{\text{SSE}_{x,1}} = \frac{996.10}{3109.7} = .3203,$$

$$\hat{\beta}_2 = \frac{\text{SPE}_2}{\text{SSE}_{x,2}} = \frac{257.00}{2060.50} = .1247,$$

$$\text{SS}(F) = \text{SSE}_y - \sum_{i=1}^{2} \frac{(\text{SPE}_i)^2}{\text{SSE}_{x,i}} = 516.3741,$$

$$\text{SS}(R) = \text{SSE}_y - \frac{(\text{SPE})^2}{\text{SSE}_x} = 563.7865.$$

By (16.64),

$$F = \frac{(563.7865 - 516.3741)/1}{516.3741/34} = 3.1218, \quad p = .0862.$$

For factor C, $\hat{\beta}_1 = .2034$, $\hat{\beta}_2 = .2870$,
$F = 8.9930/1/554.7935/34 = .5511, \quad p = .463.$

16.17 **(a)**

$$\mathbf{E}_{xx} = \begin{pmatrix} 4548.2 & 2877.4 \\ 2877.4 & 4876.9 \end{pmatrix}, \quad \mathbf{e}_{xy}\begin{pmatrix} 5.623 \\ 26.219 \end{pmatrix}$$

$$e_{yy} = .8452, \quad \hat{\boldsymbol{\beta}} = \begin{pmatrix} -.003454 \\ .007414 \end{pmatrix}.$$

(b)

$$\text{SSE}_{y \cdot x} = .67026, \quad \text{SST}_{y \cdot x} = .84150,$$

$$\text{SS}(\alpha|\mu, \beta) = \text{SST}_{y \cdot x} - \text{SSE}_{y \cdot x} = .17124,$$

$$F = \frac{.17124/3}{.67026/34} = 2.8955, \quad p = .0493.$$

(c) To test $H_0 : \boldsymbol{\beta} = \mathbf{0}$, we use (16.84):

$$F = \frac{\mathbf{e}'_{xy}\mathbf{E}_{xx}^{-1}\mathbf{e}_{xy}/q}{\text{SSE}_{y \cdot x}/[k(n-1) - q]} = 4.4378, \quad p = .0194.$$

(d) $$\hat{\boldsymbol{\beta}}_1 = \begin{pmatrix} 1268.9 & 983.4 \\ 983.4 & 1076.4 \end{pmatrix}^{-1} \begin{pmatrix} 2.984 \\ 5.694 \end{pmatrix} = \begin{pmatrix} -.00599 \\ .01076 \end{pmatrix},$$

$$\hat{\boldsymbol{\beta}}_2 = \begin{pmatrix} 1488.4 & 836.0 \\ 836.0 & 1512.0 \end{pmatrix}^{-1} \begin{pmatrix} -2.636 \\ 3.95 \end{pmatrix} = \begin{pmatrix} -.004697 \\ .005209 \end{pmatrix},$$

$$\hat{\boldsymbol{\beta}}_3 = \begin{pmatrix} 502.9 & 513.0 \\ 513.0 & 1552.0 \end{pmatrix}^{-1} \begin{pmatrix} 1.735 \\ 12.94 \end{pmatrix} = \begin{pmatrix} -.00763 \\ .01086 \end{pmatrix},$$

$$\hat{\boldsymbol{\beta}}_4 = \begin{pmatrix} 1288.0 & 545.0 \\ 545.0 & 736.5 \end{pmatrix}^{-1} \begin{pmatrix} 3.540 \\ 3.635 \end{pmatrix} = \begin{pmatrix} .000961 \\ .004224 \end{pmatrix}$$

$$\text{SSE}(F)_{y\cdot x} = e_{yy} - \sum_{i=1}^{4} \mathbf{e}'_{xy,i}\mathbf{E}^{-1}_{xx,\,i}\mathbf{e}_{xy,\,i} = .62284,$$

$$\text{SSE}(R)_{y\cdot x} = e_{yy} - \mathbf{e}'_{xy}\mathbf{E}^{-1}_{xx}\mathbf{e}_{xy} = .67026.$$

By (16.89)

$$F = \frac{[\text{SSE}(R)_{y\cdot x} - \text{SSE}(F)_{y\cdot x}]/q(k-1)}{\text{SSE}(F)_{y\cdot x}/k(n-q-1)}$$

$$= \frac{.047425/6}{.62284/28} = .3553, \quad p = .901.$$

Chapter 17

17.1 If $\mathbf{V} = \mathbf{PP}'$, let $\mathbf{v} = \mathbf{P}^{-1}\,\mathbf{y}$ and $\mathbf{W} = \mathbf{P}^{-1}\,\mathbf{X}$. Then $\mathbf{v}$ is $N(\mathbf{W}\boldsymbol{\beta}, \sigma^2\mathbf{I})$, $\hat{\boldsymbol{\beta}} = (\mathbf{W}'\mathbf{W})^{-1}\,\mathbf{W}'\mathbf{y}$, and

$$F = \frac{(\mathbf{C}\hat{\boldsymbol{\beta}} - t)'[\mathbf{C}(\mathbf{W}'\mathbf{W})^{-1}\mathbf{C}'](\mathbf{C}\hat{\boldsymbol{\beta}} - t)/q}{\mathbf{v}'(\mathbf{I} - \mathbf{W}(\mathbf{W}'\mathbf{W})^{-1}\mathbf{W}')\mathbf{v}/(n-k-1)}.$$

(a) By Theorem 8.4g(ii), F is $F(q, n-k-1)$.

(b) By Theorem 8.4g(i), F is $F(q, n-k-1, (\mathbf{C}\hat{\boldsymbol{\beta}} - t)'[\mathbf{C}(\mathbf{W}'\mathbf{W})^{-1}\mathbf{C}']^{-1}(\mathbf{C} - t)/2)$.

17.2 As in Problem 17.1 and using (8.49), the confidence interval is

$$\mathbf{a}'\hat{\boldsymbol{\beta}} \pm t_{\alpha/2,\,n-k-1}\sqrt{\mathbf{a}'(\mathbf{W}'\mathbf{W})^{-1}\mathbf{a}}$$

$$\text{or } \mathbf{a}'\hat{\boldsymbol{\beta}} \pm t_{\alpha/2,\,n-k-1}\sqrt{\mathbf{a}'(\mathbf{X}'(\mathbf{X}'\mathbf{V}^{-1}\mathbf{X})^{-1}\mathbf{a}}$$

where s is given by (7.67)

17.3 Let

$$\mathbf{X}_0 = \begin{pmatrix} \mathbf{j}_5 & \mathbf{j}_5 & \mathbf{0} & \mathbf{0} & \mathbf{0} & \mathbf{j}_5 & \mathbf{0} & \mathbf{j}_5 & \mathbf{0} & \mathbf{0} & \mathbf{0} & \mathbf{0} & \mathbf{0} & \mathbf{0} & \mathbf{0} \\ \mathbf{j}_5 & \mathbf{j}_5 & \mathbf{0} & \mathbf{0} & \mathbf{0} & \mathbf{0} & \mathbf{j}_5 & \mathbf{0} & \mathbf{j}_5 & \mathbf{0} & \mathbf{0} & \mathbf{0} & \mathbf{0} & \mathbf{0} & \mathbf{0} \\ \mathbf{j}_5 & \mathbf{0} & \mathbf{j}_5 & \mathbf{0} & \mathbf{0} & \mathbf{j}_5 & \mathbf{0} & \mathbf{0} & \mathbf{0} & \mathbf{j}_5 & \mathbf{0} & \mathbf{0} & \mathbf{0} & \mathbf{0} & \mathbf{0} \\ \mathbf{j}_5 & \mathbf{0} & \mathbf{j}_5 & \mathbf{0} & \mathbf{0} & 0 & \mathbf{j}_5 & \mathbf{0} & \mathbf{0} & \mathbf{0} & \mathbf{j}_5 & \mathbf{0} & \mathbf{0} & \mathbf{0} & \mathbf{0} \\ \mathbf{j}_5 & \mathbf{0} & 0 & \mathbf{j}_5 & \mathbf{0} & \mathbf{j}_5 & \mathbf{0} & \mathbf{0} & \mathbf{0} & \mathbf{0} & \mathbf{0} & \mathbf{j}_5 & \mathbf{0} & \mathbf{0} & \mathbf{0} \\ \mathbf{j}_5 & \mathbf{0} & 0 & \mathbf{j}_5 & \mathbf{0} & \mathbf{0} & \mathbf{j}_5 & \mathbf{0} & \mathbf{0} & \mathbf{0} & \mathbf{0} & \mathbf{0} & \mathbf{j}_5 & \mathbf{0} & \mathbf{0} \\ \mathbf{j}_5 & \mathbf{0} & 0 & 0 & \mathbf{j}_5 & \mathbf{j}_5 & \mathbf{0} & \mathbf{0} & \mathbf{0} & \mathbf{0} & \mathbf{0} & \mathbf{0} & \mathbf{0} & \mathbf{j}_5 & \mathbf{0} \\ \mathbf{j}_5 & \mathbf{0} & 0 & 0 & \mathbf{j}_5 & \mathbf{0} & \mathbf{j}_5 & \mathbf{0} & \mathbf{0} & \mathbf{0} & \mathbf{0} & \mathbf{0} & \mathbf{0} & \mathbf{0} & \mathbf{j}_5 \end{pmatrix}.$$

Then

$$\mathbf{X} = \begin{pmatrix} \mathbf{X}_0 \\ \mathbf{X}_0 \\ \vdots \\ \mathbf{X}_0 \end{pmatrix}, \quad \mathbf{Z}_1 = \begin{pmatrix} \mathbf{j}_{40} & \mathbf{0} & \cdots & \mathbf{0} \\ \mathbf{0} & \mathbf{j}_{40} & \cdots & \mathbf{0} \\ \vdots & \vdots & & \vdots \\ \mathbf{0} & \mathbf{0} & \cdots & \mathbf{j}_{40} \end{pmatrix},$$

$$\mathbf{Z}_2 = \begin{pmatrix} \mathbf{j}_{10} & \mathbf{0} & \cdots & \mathbf{0} \\ \mathbf{0} & \mathbf{j}_{10} & \cdots & \mathbf{0} \\ \vdots & \vdots & & \vdots \\ \mathbf{0} & \mathbf{0} & \cdots & \mathbf{j}_{10} \end{pmatrix}, \text{ and } \mathbf{Z}_3 = \begin{pmatrix} \mathbf{j}_5 & \mathbf{0} & \cdots & \mathbf{0} \\ \mathbf{0} & \mathbf{j}_5 & \cdots & \mathbf{0} \\ \vdots & \vdots & & \vdots \\ \mathbf{0} & \mathbf{0} & \cdots & \mathbf{j}_5 \end{pmatrix}.$$

17.4 (a)

$$\text{cov}(\mathbf{y}) = \text{cov}\left(\sum_{i=1}^{m} \mathbf{Z}_i\mathbf{a}_i + \boldsymbol{\varepsilon}\right) = \sum_{i=1}^{m} \mathbf{Z}_i\mathbf{G}_i\mathbf{Z}_i' + \mathbf{R}.$$

(b)

$$\text{cov}(\mathbf{y}) = \mathbf{ZGZ}' + \mathbf{R}.$$

17.5 Using (5.4),

$$\begin{aligned} E[&\mathbf{y}\mathbf{K}'(\mathbf{K}\boldsymbol{\Sigma}\mathbf{K}')^{-1}\mathbf{K}\mathbf{Z}_i\mathbf{Z}_i'\mathbf{K}'(\mathbf{K}\boldsymbol{\Sigma}\mathbf{K}')^{-1}\mathbf{K}\mathbf{y}] \\ &= \text{tr}[\mathbf{K}'(\mathbf{K}\boldsymbol{\Sigma}\mathbf{K}')^{-1}\mathbf{K}\mathbf{Z}_i\mathbf{Z}_i'\mathbf{K}'(\mathbf{K}\boldsymbol{\Sigma}\mathbf{K}')^{-1}\mathbf{K}\boldsymbol{\Sigma}] \\ &\quad + \boldsymbol{\beta}'\mathbf{X}'\mathbf{K}'(\mathbf{K}\boldsymbol{\Sigma}\mathbf{K}')^{-1}\mathbf{K}\mathbf{Z}_i\mathbf{Z}_i'\mathbf{K}'(\mathbf{K}\boldsymbol{\Sigma}\mathbf{K}')^{-1}\mathbf{K}\mathbf{X}\boldsymbol{\beta} \\ &= \text{tr}[\mathbf{K}\boldsymbol{\Sigma}\mathbf{K}'(\mathbf{K}\boldsymbol{\Sigma}\mathbf{K}')^{-1}\mathbf{K}\mathbf{Z}_i\mathbf{Z}_i'\mathbf{K}'(\mathbf{K}\boldsymbol{\Sigma}\mathbf{K}')^{-1}] + 0 \qquad (\text{since } \mathbf{KX} = \mathbf{O}) \\ &= \text{tr}[\mathbf{K}\mathbf{Z}_i\mathbf{Z}_i'\mathbf{K}'(\mathbf{K}\boldsymbol{\Sigma}\mathbf{K}')^{-1}] \\ &= \text{tr}[\mathbf{K}'(\mathbf{K}\boldsymbol{\Sigma}\mathbf{K}')^{-1}\mathbf{K}\mathbf{Z}_i\mathbf{Z}_i']. \end{aligned}$$

17.6 $\mathbf{q}$ is obvious;

$$\begin{aligned}\text{tr}[\mathbf{K}'(\mathbf{K}\boldsymbol{\Sigma}\mathbf{K}')^{-1}\mathbf{K}\mathbf{Z}_i\mathbf{Z}_i'] &= \text{tr}[(\mathbf{K}\boldsymbol{\Sigma}\mathbf{K}')^{-1}\mathbf{K}\mathbf{Z}_i\mathbf{Z}_i'\mathbf{K}'] \\ &= \text{tr}[(\mathbf{K}\boldsymbol{\Sigma}\mathbf{K}')^{-1}\mathbf{K}\mathbf{Z}_i\mathbf{Z}_i'\mathbf{K}'(\mathbf{K}\boldsymbol{\Sigma}\mathbf{K}')^{-1}\mathbf{K}\boldsymbol{\Sigma}\mathbf{K}'] \\ &= \text{tr}\left[\mathbf{K}'(\mathbf{K}\boldsymbol{\Sigma}\mathbf{K}')^{-1}\mathbf{K}\mathbf{Z}_i\mathbf{Z}_i'\mathbf{K}'(\mathbf{K}\boldsymbol{\Sigma}\mathbf{K}')^{-1}\mathbf{K}\sum_{i=0}^{m}\mathbf{Z}_i\mathbf{Z}_i'\sigma_i^2\right] \\ &= \mathbf{m}_1'\boldsymbol{\sigma}, \quad \text{where the } j\text{th element of } \mathbf{m}_1 \text{ is} \\ &\quad\ \text{tr}[\mathbf{K}'(\mathbf{K}\boldsymbol{\Sigma}\mathbf{K}')^{-1}\mathbf{K}\mathbf{Z}_i\mathbf{Z}_i'\mathbf{K}'(\mathbf{K}\boldsymbol{\Sigma}\mathbf{K}')^{-1}\mathbf{K}\mathbf{Z}_j\mathbf{Z}_j'].\end{aligned}$$

17.7 **(a)** If $\boldsymbol{\Sigma} = \mathbf{PP}'$, let $\mathbf{v} = \mathbf{P}^{-1}\mathbf{y}$ and $\mathbf{W} = \mathbf{P}^{-1}\mathbf{X}$. Then $\mathbf{v}$ is $N(\mathbf{W}\boldsymbol{\beta}, \mathbf{I})$ and $\mathbf{L}(\mathbf{X}'\boldsymbol{\Sigma}^{-1}\mathbf{X})^{-}\mathbf{L}' = \mathbf{L}(\mathbf{W}'\mathbf{W})^{-}\mathbf{L}'$. Since $\mathbf{L}$ is estimable, $\mathbf{L} = \mathbf{AX} = \mathbf{APP}'\mathbf{X} = \mathbf{BW}$. Hence the rows of $\mathbf{L}$ also define estimable functions of $\mathbf{v}$. Thus by Theorem 12.7b, $\mathbf{L}(\mathbf{X}'\boldsymbol{\Sigma}^{-1}\mathbf{X})^{-}\mathbf{L}'$ is nonsingular.

(b) Since $\mathbf{L}\hat{\boldsymbol{\beta}} - \mathbf{L}\boldsymbol{\beta}$ is $N[\mathbf{0}, \mathbf{L}(\mathbf{W}'\mathbf{W})^{-}\mathbf{L}']$, note that $[\mathbf{L}(\mathbf{W}'\mathbf{W})^{-}\mathbf{L}']^{-1}\mathbf{L}(\mathbf{W}'\mathbf{W})^{-}\mathbf{L}' = \mathbf{I}$, which is idempotent of rank g.

17.8 $\mathbf{x}_0'\hat{\boldsymbol{\beta}}$ is estimable and is $N[\mathbf{x}_0'\boldsymbol{\beta}, \mathbf{x}_0'(\mathbf{W}'\mathbf{W})^{-}\mathbf{x}_0]$. Hence

$$\frac{\mathbf{x}_0'\hat{\boldsymbol{\beta}} - \mathbf{x}_0'\boldsymbol{\beta}}{\sqrt{\mathbf{x}_0'(\mathbf{W}'\mathbf{W})^{-}\mathbf{x}_0}} \quad \text{is} \quad N(0, 1).$$

Thus a $100(1 - \alpha)\%$ confidence interval for $\mathbf{x}_0'\hat{\boldsymbol{\beta}}$ is

$$\mathbf{x}_0'\hat{\boldsymbol{\beta}} \pm z_{\alpha/2}\sqrt{\mathbf{x}_0'(\mathbf{W}'\mathbf{W})^{-}\mathbf{x}_0} \quad \text{or}$$

$$\mathbf{x}_0'\hat{\boldsymbol{\beta}} \pm z_{\alpha/2}\sqrt{\mathbf{x}_0'(\mathbf{X}'\boldsymbol{\Sigma}^{-1}\mathbf{X})^{-}\mathbf{x}_0}.$$

17.9

$$(\mathbf{X}'\hat{\boldsymbol{\Sigma}}^{-1}\mathbf{X})^{-1} = \left[(\mathbf{I}_6 \quad \mathbf{I}_6) \begin{pmatrix} \hat{\boldsymbol{\Sigma}}_1^{-1} & \mathbf{O} & \cdots & \mathbf{O} \\ \mathbf{O} & \hat{\boldsymbol{\Sigma}}_1^{-1} & \cdots & \mathbf{O} \\ \vdots & \vdots & & \vdots \\ \mathbf{O} & \mathbf{O} & \cdots & \hat{\boldsymbol{\Sigma}}_1^{-1} \end{pmatrix} \begin{pmatrix} \mathbf{I}_6 \\ \mathbf{I}_6 \end{pmatrix} \right]^{-1}$$

$$= \left[2 \begin{pmatrix} \hat{\boldsymbol{\Sigma}}_1^{-1} & \mathbf{O} & \mathbf{O} \\ \mathbf{O} & \hat{\boldsymbol{\Sigma}}_1^{-1} & \mathbf{O} \\ \mathbf{O} & \mathbf{O} & \hat{\boldsymbol{\Sigma}}_1^{-1} \end{pmatrix} \right]^{-1}$$

$$= \frac{1}{2} \begin{pmatrix} \hat{\boldsymbol{\Sigma}}_1 & \mathbf{O} & \mathbf{O} \\ \mathbf{O} & \hat{\boldsymbol{\Sigma}}_1 & \mathbf{O} \\ \mathbf{O} & \mathbf{O} & \hat{\boldsymbol{\Sigma}}_1 \end{pmatrix}.$$

17.10 Let $\mathbf{C} = (\mathbf{I}_6, \ \mathbf{O})$ and $\mathbf{K} = \mathbf{C}(\mathbf{I} - \mathbf{H}) = \frac{1}{2}(\mathbf{I}_6, \ -\mathbf{I}_6)$. Then $\mathbf{K}'(\mathbf{K}\boldsymbol{\Sigma}\mathbf{K}')^{-1}\ \mathbf{K}$ $= \frac{1}{2}\begin{pmatrix} \mathbf{T} & -\mathbf{T} \\ -\mathbf{T} & \mathbf{T} \end{pmatrix}$ where $\mathbf{T} = \begin{pmatrix} \hat{\boldsymbol{\Sigma}}_1^{-1} & \mathbf{O} & \mathbf{O} \\ \mathbf{O} & \hat{\boldsymbol{\Sigma}}_1^{-1} & \mathbf{O} \\ \mathbf{O} & \mathbf{O} & \hat{\boldsymbol{\Sigma}}_1^{-1} \end{pmatrix}$. Since $\mathbf{Z}_0 = \mathbf{I}_{12}$ and

$\mathbf{Z}_1 = \begin{pmatrix} \mathbf{j}_2 & \mathbf{0} & \cdots & \mathbf{0} \\ \mathbf{0} & \mathbf{j}_2 & \cdots & \mathbf{0} \\ \vdots & \vdots & & \vdots \\ \mathbf{0} & \mathbf{0} & \cdots & \mathbf{j}_2 \end{pmatrix}$, the REML equations become

$$3\ \mathrm{tr}(\boldsymbol{\Sigma}^{-1}) = \mathbf{y}'[\mathbf{K}'(\mathbf{K}\boldsymbol{\Sigma}\mathbf{K}')]^{-1}\mathbf{K}\mathbf{K}'(\mathbf{K}\boldsymbol{\Sigma}\mathbf{K}')\mathbf{K}]\mathbf{y}, \quad \text{and}$$
$$3\ \mathrm{tr}(\boldsymbol{\Sigma}^{-1}\mathbf{J}_2) = \mathbf{y}'[\mathbf{K}'(\mathbf{K}\boldsymbol{\Sigma}\mathbf{K}')^{-1}\mathbf{K}\mathbf{J}_2\mathbf{K}'(\mathbf{K}\boldsymbol{\Sigma}\mathbf{K}')\mathbf{K}]\mathbf{y}.$$

Noting that $\hat{\boldsymbol{\Sigma}}^{-1} = \frac{1}{\hat{\sigma}^2(\hat{\sigma}^2 + 2\hat{\sigma}_1^2)}\begin{pmatrix} \hat{\sigma}^2 + \hat{\sigma}_1^2 & -\hat{\sigma}_1^2 \\ -\hat{\sigma}_1^2 & \hat{\sigma}^2 + \hat{\sigma}_1^2 \end{pmatrix}$, the REML equations can be written as

$$6\frac{\hat{\sigma}^2+\hat{\sigma}_1^2}{\hat{\sigma}^2(\hat{\sigma}^2+2\hat{\sigma}_1^2)}=\frac{1}{2}\mathbf{y}'\begin{pmatrix}\mathbf{T}^2 & -\mathbf{T}^2\\ -\mathbf{T}^2 & \mathbf{T}^2\end{pmatrix}\mathbf{y}, \quad \text{and}$$

$$\frac{6}{\hat{\sigma}^2(\hat{\sigma}^2+2\hat{\sigma}_1^2)}=\frac{1}{2}\frac{u}{(\hat{\sigma}^2+2\hat{\sigma}_1^2)^2}$$

where $u=\mathbf{y}'\begin{pmatrix}\mathbf{U} & -\mathbf{U}\\ -\mathbf{U} & \mathbf{U}\end{pmatrix}\mathbf{y}$ and $\mathbf{U}=\begin{pmatrix}\mathbf{J}_2 & \mathbf{O} & \mathbf{O}\\ \mathbf{O} & \mathbf{J}_2 & \mathbf{O}\\ \mathbf{O} & \mathbf{O} & \mathbf{J}_2\end{pmatrix}$. The second REML equation can be rearranged $\hat{\sigma}_1^2=\frac{u}{24}-\frac{\hat{\sigma}^2}{2}$. Substituting this expression into the first REML equation and then simplifying, we obtain $\frac{u^2\hat{\sigma}^2}{24}+\frac{u\hat{\sigma}^4}{2}=\frac{u\hat{\sigma}^4}{2}+\frac{u^2}{288}\mathbf{y}'\mathbf{P}\mathbf{y}$ where $\mathbf{P}=$

$$\begin{pmatrix}\mathbf{R} & \mathbf{O} & \mathbf{O} & -\mathbf{R} & \mathbf{O} & \mathbf{O}\\ \mathbf{O} & \mathbf{R} & \mathbf{O} & \mathbf{O} & -\mathbf{R} & \mathbf{O}\\ \mathbf{O} & \mathbf{O} & \mathbf{R} & \mathbf{O} & \mathbf{O} & -\mathbf{R}\\ -\mathbf{R} & \mathbf{O} & \mathbf{O} & \mathbf{R} & \mathbf{O} & \mathbf{O}\\ \mathbf{O} & -\mathbf{R} & \mathbf{O} & \mathbf{O} & \mathbf{R} & \mathbf{O}\\ \mathbf{O} & \mathbf{O} & -\mathbf{R} & \mathbf{O} & \mathbf{O} & \mathbf{R}\end{pmatrix}, \text{ which simplifies to } \hat{\sigma}^2=\frac{1}{12}\mathbf{y}'\mathbf{P}\mathbf{y}.$$

17.11

$$\mathbf{X}(\mathbf{X}'\mathbf{X})^{-1}\mathbf{L}'\mathbf{Q}\mathbf{L}(\mathbf{X}'\mathbf{X})^{-1}\mathbf{X}'\boldsymbol{\Sigma}$$

$$=\frac{1}{2}\begin{pmatrix}\mathbf{I}_6\\ \mathbf{I}_6\end{pmatrix}\frac{1}{3\sigma^2}\begin{pmatrix}2 & -1\\ -1 & 2\end{pmatrix}\frac{1}{2}(\mathbf{I}_6 \quad \mathbf{I}_6)\boldsymbol{\Sigma}$$

$$=\frac{1}{12\sigma^2}\begin{pmatrix}\mathbf{W} & \mathbf{W}\\ \mathbf{W} & \mathbf{W}\end{pmatrix}\begin{pmatrix}\boldsymbol{\Sigma}_1 & \mathbf{O} & \cdots & \mathbf{O}\\ \mathbf{O} & \boldsymbol{\Sigma}_1 & \cdots & \mathbf{O}\\ \vdots & \vdots & & \vdots\\ \mathbf{O} & \mathbf{O} & \cdots & \boldsymbol{\Sigma}_1\end{pmatrix}$$

$$\text{where } \begin{pmatrix}2 & -2 & -1 & 1 & -1 & 1\\ -2 & 2 & 1 & -1 & 1 & -1\\ -1 & 1 & 2 & -2 & -1 & 1\\ 1 & -1 & -2 & 2 & 1 & -1\\ -1 & 1 & -1 & 1 & 2 & -2\\ 1 & -1 & 1 & -1 & -2 & 2\end{pmatrix}$$

$$=\frac{1}{12}\begin{pmatrix}\mathbf{W} & \mathbf{W}\\ \mathbf{W} & \mathbf{W}\end{pmatrix}.$$

Now,

$$\frac{1}{12}\begin{pmatrix}\mathbf{W} & \mathbf{W}\\ \mathbf{W} & \mathbf{W}\end{pmatrix}\frac{1}{12}\begin{pmatrix}\mathbf{W} & \mathbf{W}\\ \mathbf{W} & \mathbf{W}\end{pmatrix}$$
$$=\frac{1}{144}\begin{pmatrix}2\mathbf{W}^2 & 2\mathbf{W}^2\\ 2\mathbf{W}^2 & 2\mathbf{W}^2\end{pmatrix}$$
$$=\frac{1}{12}\begin{pmatrix}\mathbf{W} & \mathbf{W}\\ \mathbf{W} & \mathbf{W}\end{pmatrix}.$$

17.12

$$\frac{1}{12\sigma^2}\begin{pmatrix}\mathbf{W} & \mathbf{W}\\ \mathbf{W} & \mathbf{W}\end{pmatrix}\boldsymbol{\Sigma}\frac{1}{2}\mathbf{P}$$
$$=\frac{1}{24}\begin{pmatrix}\mathbf{W} & \mathbf{W}\\ \mathbf{W} & \mathbf{W}\end{pmatrix}\begin{pmatrix}\mathbf{R} & \mathbf{O} & \mathbf{O} & -\mathbf{R} & \mathbf{O} & \mathbf{O}\\ \mathbf{O} & \mathbf{R} & \mathbf{O} & \mathbf{O} & -\mathbf{R} & \mathbf{O}\\ \mathbf{O} & \mathbf{O} & \mathbf{R} & \mathbf{O} & \mathbf{O} & -\mathbf{R}\end{pmatrix}$$
$$=\mathbf{O}.$$

17.13

$$[\mathbf{c}'(\mathbf{X}'\hat{\boldsymbol{\Sigma}}^{-1}\mathbf{X})^{-1}\mathbf{c}]^{-1}=\hat{\sigma}^2+2\hat{\sigma}_1^2=\frac{(\hat{\sigma}^2+2\hat{\sigma}_1^2)/3\sigma^2}{w/d}.$$

Using results from the solution to Problem 17.10, $v=\frac{1}{3\sigma^2}(\frac{1}{12}\mathbf{y}'\mathbf{P}\mathbf{y}+\frac{1}{12}u-\hat{\sigma}^2)=\frac{1}{36}\mathbf{y}'\mathbf{D}\mathbf{y}$ where $\mathbf{D}$ is a nonzero square matrix not involving σ^2. Hence $(\frac{1}{36\sigma^2}\mathbf{D})^2\neq\frac{1}{36\sigma^2}\mathbf{D}$.

17.14

$$\frac{(n-k)[\mathbf{c}'(\mathbf{X}'\hat{\boldsymbol{\Sigma}}^{-1}\mathbf{X})^{-}\mathbf{c}]}{[\mathbf{c}'(\mathbf{X}'\boldsymbol{\Sigma}^{-1}\mathbf{X})^{-}\mathbf{c}]}=\frac{(n-k)[\mathbf{c}'(\mathbf{X}'\mathbf{X})^{-}\mathbf{c}]s^2}{[\mathbf{c}'(\mathbf{X}'\mathbf{X})^{-}\mathbf{c}]\sigma^2}=\frac{(n-k)s^2}{\sigma^2}.$$

17.15

$$\frac{\partial \mathbf{f}(\boldsymbol{\sigma})}{\partial\sigma_i^2}=\frac{\partial}{\partial\sigma_i^2}[\mathbf{c}'(\mathbf{X}'\boldsymbol{\Sigma}^{-1}\mathbf{X})^{-}\mathbf{c}]$$
$$=-\mathbf{c}'(\mathbf{X}'\boldsymbol{\Sigma}^{-1}\mathbf{X})-\left(\frac{\partial}{\partial\sigma_i^2}(\mathbf{X}'\boldsymbol{\Sigma}^{-1}\mathbf{X})\right)(\mathbf{X}'\boldsymbol{\Sigma}^{-1}\mathbf{X})^{-}\mathbf{c}$$

[by an extension of (2.117)]

$$=\mathbf{c}'(\mathbf{X}'\boldsymbol{\Sigma}^{-1}\mathbf{X})^{-}\mathbf{X}'\boldsymbol{\Sigma}^{-1}\left(\frac{\partial}{\partial\sigma_i^2}\boldsymbol{\Sigma}\right)\boldsymbol{\Sigma}^{-1}\mathbf{X}(\mathbf{X}'\boldsymbol{\Sigma}^{-1}\mathbf{X})^{-}\mathbf{c}$$
$$=\mathbf{c}'(\mathbf{X}'\boldsymbol{\Sigma}^{-1}\mathbf{X})^{-}\mathbf{X}'\boldsymbol{\Sigma}^{-1}\left(\frac{\partial}{\partial\sigma_i^2}\sum_{j=0}^{m}\sigma_j^2\mathbf{Z}_j\mathbf{Z}_j'\right)\boldsymbol{\Sigma}^{-1}\mathbf{X}(\mathbf{X}'\boldsymbol{\Sigma}^{-1}\mathbf{X})^{-}\mathbf{c}$$
$$=\mathbf{c}'(\mathbf{X}'\boldsymbol{\Sigma}^{-1}\mathbf{X})^{-}\mathbf{X}'\boldsymbol{\Sigma}^{-1}\mathbf{Z}_i\mathbf{Z}_i'\boldsymbol{\Sigma}^{-1}\mathbf{X}(\mathbf{X}'\boldsymbol{\Sigma}^{-1}\mathbf{X})^{-}\mathbf{c}.$$

17.16 **(a)**

$$
\begin{aligned}
&(\mathbf{L}\hat{\boldsymbol{\beta}} - \mathbf{L}\boldsymbol{\beta})'[\mathbf{L}(\mathbf{X}'\hat{\boldsymbol{\Sigma}}^{-1}\mathbf{X}^{-}]^{-1}(\mathbf{L}\hat{\boldsymbol{\beta}} - \mathbf{L}\boldsymbol{\beta}) \\
&= (\mathbf{L}\hat{\boldsymbol{\beta}} - \mathbf{L}\boldsymbol{\beta})'\mathbf{P}\mathbf{D}^{-1}\mathbf{P}'(\mathbf{L}\hat{\boldsymbol{\beta}} - \mathbf{L}\boldsymbol{\beta}) \\
&= (\mathbf{L}\hat{\boldsymbol{\beta}} - \mathbf{L}\boldsymbol{\beta})' \sum_{i=1}^{g} \frac{\mathbf{p}_i\mathbf{p}_i'}{\lambda_i} (\mathbf{L}\hat{\boldsymbol{\beta}} - \mathbf{L}\boldsymbol{\beta}) \\
&= \sum_{i=1}^{g} \frac{[\mathbf{p}_i'(\mathbf{L}\hat{\boldsymbol{\beta}} - \mathbf{L}\boldsymbol{\beta})]^2}{\lambda_i}.
\end{aligned}
$$

(b) note that $\text{var}[\mathbf{p}_i'(\mathbf{L}\hat{\boldsymbol{\beta}} - \mathbf{L}\boldsymbol{\beta})] = \text{var}(\mathbf{p}_i'\mathbf{L}\hat{\boldsymbol{\beta}})$

$$
\begin{aligned}
&= \mathbf{p}_i'\mathbf{L}(\mathbf{X}'\hat{\boldsymbol{\Sigma}}^{-1}\mathbf{X})^{-}\mathbf{L}'\mathbf{p}_i \\
&= \mathbf{p}_i'\mathbf{L}(\mathbf{X}'\hat{\boldsymbol{\Sigma}}^{-1}\mathbf{X})^{-}\mathbf{L}'\mathbf{p}_i \\
&= [\mathbf{P}'\mathbf{L}(\mathbf{X}'\hat{\boldsymbol{\Sigma}}^{-1}\mathbf{X})^{-}\mathbf{L}'\mathbf{P}]_{ii} \\
&= [\mathbf{P}'\mathbf{P}\mathbf{D}\mathbf{P}'\mathbf{P}]_{ii} = \mathbf{D}_{ii} = \lambda_i.
\end{aligned}
$$

(c)

$$
\begin{aligned}
\text{cov}(\mathbf{p}_i'\mathbf{L}\hat{\boldsymbol{\beta}}, \mathbf{p}_{i'}', \mathbf{L}\hat{\boldsymbol{\beta}}) &= \mathbf{p}_i'\mathbf{L}(\mathbf{X}'\hat{\boldsymbol{\Sigma}}^{-1}\mathbf{X})^{-}\mathbf{L}'\mathbf{p}_i' \\
&= \mathbf{p}_i'\mathbf{P}\mathbf{D}\mathbf{P}'\mathbf{p}_{i'} \\
&= \lambda_i\mathbf{p}_i'\mathbf{p}_i\mathbf{p}_{i'}'\mathbf{p}_{i'} = \lambda_i\mathbf{p}_i'\mathbf{O}\mathbf{p}_{i'} = 0.
\end{aligned}
$$

17.17 **(a)** We use Theorems 5.2a and 5.2e to obtain

$$
\begin{aligned}
&E[\mathbf{a} - \mathbf{B}(\mathbf{y} - \mathbf{X}\boldsymbol{\beta})]'[\mathbf{a} - \mathbf{B}(\mathbf{y} - \mathbf{X}\boldsymbol{\beta})] \\
&\quad = E(\mathbf{a}'\mathbf{a}) - E[\mathbf{a}'\mathbf{B}(\mathbf{y} - \mathbf{X}\boldsymbol{\beta}) - E(\mathbf{y} - \mathbf{X}\boldsymbol{\beta})'\mathbf{B}'\mathbf{a}] \\
&\qquad + E[(\mathbf{y} - \mathbf{X}\boldsymbol{\beta})'\mathbf{B}'\mathbf{B}(\mathbf{y} - \mathbf{X}\boldsymbol{\beta})] \\
&\quad = \text{tr}(\mathbf{V}) - \text{tr}[\mathbf{B}\,\text{cov}(\mathbf{y}, \mathbf{a})] - \text{tr}[\mathbf{B}'\text{cov}(\mathbf{a}, \mathbf{y}] + \text{tr}(\mathbf{B}'\mathbf{B}\boldsymbol{\Sigma}) \\
&\quad = \text{tr}(\mathbf{V}) - \text{tr}(\mathbf{B}\mathbf{Z}\mathbf{V})' - \text{tr}(\mathbf{B}'\mathbf{V}\mathbf{Z}') + \text{tr}(\mathbf{B}\boldsymbol{\Sigma}\mathbf{B}') \\
&\quad = \text{tr}(\mathbf{V}) + \text{tr}(\mathbf{B}\mathbf{Z}\mathbf{V}' - \mathbf{B}'\mathbf{V}\mathbf{Z}' + \mathbf{B}\boldsymbol{\Sigma}\mathbf{B}') \\
&\quad = \text{tr}(\mathbf{V}) + \text{tr}[\mathbf{B} - \mathbf{V}\mathbf{Z}'\boldsymbol{\Sigma}^{-1})\boldsymbol{\Sigma}(\mathbf{B} - \mathbf{V}\mathbf{Z}'\boldsymbol{\Sigma}^{-1})' - \mathbf{V}\mathbf{Z}'\boldsymbol{\Sigma}^{-1}\mathbf{Z}\mathbf{V}'].
\end{aligned}
$$

(b) Since $E(\mathbf{a})=0$, $E(\mathbf{y} - \mathbf{X}\boldsymbol{\beta})=0$, and $\text{cov}(\mathbf{a}, \mathbf{y})=\mathbf{V}\mathbf{Z}'$, we have

$$
\begin{aligned}
&E[\mathbf{a} - \mathbf{B}(\mathbf{y} - \mathbf{X}\boldsymbol{\beta})][\mathbf{a} - \mathbf{B}(\mathbf{y} - \mathbf{B}\boldsymbol{\beta})]' \\
&\quad = E(\mathbf{a}\mathbf{a}') - E[\mathbf{a}(\mathbf{y} - \mathbf{X}\boldsymbol{\beta})'\mathbf{B}'] - E[\mathbf{B}(\mathbf{y} - \mathbf{X}\boldsymbol{\beta})\mathbf{a}'] \\
&\qquad + E[\mathbf{B}(\mathbf{y} - \mathbf{X}\boldsymbol{\beta})(\mathbf{y} - \mathbf{X}\boldsymbol{\beta})'\mathbf{B}'] \\
&\quad = \text{cov}(\mathbf{a}) - [\text{cov}(\mathbf{a}, \mathbf{y})]\mathbf{B}' - \mathbf{B}\,\text{cov}(\mathbf{y}, \mathbf{a}) + \mathbf{B}\,\text{cov}(\mathbf{y})\mathbf{B}' \\
&\quad = \mathbf{V} - \mathbf{V}\mathbf{Z}'\mathbf{B}' - \mathbf{B}\mathbf{Z}\mathbf{V}' + \mathbf{B}\boldsymbol{\Sigma}\mathbf{B}' \\
&\quad = \mathbf{V} + (\mathbf{B} - \mathbf{V}\mathbf{Z}'\boldsymbol{\Sigma}^{-1})\boldsymbol{\Sigma}(\mathbf{B} - \mathbf{V}\mathbf{Z}'\boldsymbol{\Sigma}^{-1})' - \mathbf{V}\mathbf{Z}'\boldsymbol{\Sigma}^{-1}\mathbf{Z}\mathbf{V}'.
\end{aligned}
$$

The first and third terms do not involve $\mathbf{B}$, and the second term is "minimized" by $\mathbf{B}=\mathbf{VZ}'\boldsymbol{\Sigma}^{-1}$. By "minimize," we mean that any other choice for $\mathbf{B}$ adds a positive definite matrix to the result. This holds because $\boldsymbol{\Sigma}$ is positive definite.

17.18

$$
\begin{aligned}
&[\mathbf{I}-\mathbf{X}(\mathbf{X}'\boldsymbol{\Sigma}^{-1}\mathbf{X})^{-}\mathbf{X}'\boldsymbol{\Sigma}^{-1}]\boldsymbol{\Sigma}[\mathbf{I}-\mathbf{X}(\mathbf{X}'\boldsymbol{\Sigma}^{-1}\mathbf{X})^{-}\mathbf{X}'\boldsymbol{\Sigma}^{-1}]' \\
&=[\boldsymbol{\Sigma}-\mathbf{X}(\mathbf{X}'\boldsymbol{\Sigma}^{-1}\mathbf{X})^{-}\mathbf{X}'][\mathbf{I}-\boldsymbol{\Sigma}^{-1}\mathbf{X}(\mathbf{X}'\boldsymbol{\Sigma}^{-1}\mathbf{X})^{-}\mathbf{X}'] \\
&=\boldsymbol{\Sigma}-\mathbf{X}(\mathbf{X}'\boldsymbol{\Sigma}^{-1}\mathbf{X})^{-}\mathbf{X}'-\mathbf{X}(\mathbf{X}'\boldsymbol{\Sigma}^{-1}\mathbf{X})^{-}\mathbf{X}' \\
&\quad+\mathbf{X}(\mathbf{X}'\boldsymbol{\Sigma}^{-1}\mathbf{X})^{-}\mathbf{X}'\boldsymbol{\Sigma}^{-1}\mathbf{X}(\mathbf{X}'\boldsymbol{\Sigma}^{-1}\mathbf{X})^{-}\mathbf{X}' \\
&=\boldsymbol{\Sigma}-\mathbf{X}(\mathbf{X}'\boldsymbol{\Sigma}^{-1}\mathbf{X})^{-}\mathbf{X}'.
\end{aligned}
$$

17.19 **(a)** Using Problem 17.17, the BLP OF $\mathbf{Ua}$ is $E(\mathbf{Ua}|\mathbf{y})=\mathbf{U}E(\mathbf{a}|\mathbf{y})=\mathbf{UGZ}'\boldsymbol{\Sigma}^{-1}(\mathbf{y}-\mathbf{X}\boldsymbol{\beta})$.

(b) $\text{cov}[\mathbf{UGZ}'\boldsymbol{\Sigma}^{-1}(\mathbf{y}-\mathbf{X}\boldsymbol{\beta})]=\mathbf{UGZ}'\boldsymbol{\Sigma}^{-1}\boldsymbol{\Sigma}\boldsymbol{\Sigma}^{-1}\mathbf{ZGU}'=\mathbf{UGZ}'\boldsymbol{\Sigma}^{-1}\mathbf{ZGU}'$.

(c)

$$
\begin{aligned}
&\text{cov}[\mathbf{UGZ}'\boldsymbol{\Sigma}^{-1}(\mathbf{y}-\mathbf{X}\hat{\boldsymbol{\beta}})] \\
&=\text{cov}\{\mathbf{UGZ}'\boldsymbol{\Sigma}^{-1}[\mathbf{I}-\mathbf{X}(\mathbf{X}'\boldsymbol{\Sigma}^{-1}\mathbf{X})^{-}\mathbf{X}'\boldsymbol{\Sigma}^{-1}]\mathbf{y}\} \\
&=\mathbf{UGZ}'\boldsymbol{\Sigma}^{-1}[\mathbf{I}-\mathbf{X}(\mathbf{X}'\boldsymbol{\Sigma}^{-1}\mathbf{X})^{-}\mathbf{X}'\boldsymbol{\Sigma}^{-1}] \\
&\quad\times\boldsymbol{\Sigma}[\mathbf{I}-\mathbf{X}(\mathbf{X}'\boldsymbol{\Sigma}^{-1}\mathbf{X})^{-}\mathbf{X}'\boldsymbol{\Sigma}^{-1}]\boldsymbol{\Sigma}^{-1}\mathbf{ZGU}' \\
&=\mathbf{UGZ}'\boldsymbol{\Sigma}^{-1}[\boldsymbol{\Sigma}-\mathbf{X}(\mathbf{X}'\boldsymbol{\Sigma}^{-1}\mathbf{X})^{-}\mathbf{X}']\boldsymbol{\Sigma}^{-1}\mathbf{ZGU}' \\
&\qquad\text{[using Problem 17.18]} \\
&=\mathbf{UGZ}'[\boldsymbol{\Sigma}^{-1}-\boldsymbol{\Sigma}^{-1}\mathbf{X}(\mathbf{X}'\boldsymbol{\Sigma}^{-1}\mathbf{X})^{-}\mathbf{X}'\boldsymbol{\Sigma}^{-1}]\mathbf{ZGU}'.
\end{aligned}
$$

17.20 $\boldsymbol{\Sigma}^{-1} = (\sigma^2\mathbf{I}_{12} + \sigma_1^2\mathbf{ZZ}')^{-1}$

$$= \begin{pmatrix} \sigma^2\mathbf{I}_4 + \sigma_1^2\mathbf{j}_4\mathbf{j}_4' & \mathbf{O} & \cdots & \mathbf{O} \\ \mathbf{O} & \sigma^2\mathbf{I}_4 + \sigma_1^2\mathbf{j}_4\mathbf{j}_4' & \cdots & \mathbf{O} \\ \vdots & \vdots & & \vdots \\ \mathbf{O} & \mathbf{O} & \cdots & \sigma^2\mathbf{I}_4 + \sigma_1^2\mathbf{j}_4\mathbf{j}_4' \end{pmatrix}^{-1}$$

$$= \begin{pmatrix} (\sigma^2\mathbf{I}_4 + \sigma_1^2\mathbf{j}_4\mathbf{j}_4')^{-1} & \mathbf{O} & \cdots & \mathbf{O} \\ \mathbf{O} & (\sigma^2\mathbf{I}_4 + \sigma_1^2\mathbf{j}_4\mathbf{j}_4')^{-1} & \cdots & \mathbf{O} \\ \vdots & \vdots & & \vdots \\ \mathbf{O} & \mathbf{O} & \cdots & (\sigma^2\mathbf{I}_4 + \sigma_1^2\mathbf{j}_4\mathbf{j}_4')^{-1} \end{pmatrix}$$

[by(2.52)]

$$= \begin{pmatrix} \frac{1}{\sigma^2}\left(\mathbf{I}_4 - \frac{\sigma^2}{\sigma^2+4\sigma_1^2}\mathbf{J}_4\right) & \mathbf{O} & \cdots & \mathbf{O} \\ \mathbf{O} & \frac{1}{\sigma^2}\left(\mathbf{I}_4 - \frac{\sigma^2}{\sigma^2+4\sigma_1^2}\mathbf{J}_4\right) & \cdots & \mathbf{O} \\ \vdots & \vdots & & \vdots \\ \mathbf{O} & \mathbf{O} & \cdots & \frac{1}{\sigma^2}\left(\mathbf{I}_4 - \frac{\sigma^2}{\sigma^2+4\sigma_1^2}\mathbf{J}_4\right) \end{pmatrix}.$$

[by(2.53)]

17.21

cov[EBLUP(**a**)]

$$= \hat{\mathbf{G}}\mathbf{Z}'[\hat{\boldsymbol{\Sigma}}^{-1} - \hat{\boldsymbol{\Sigma}}^{-1}\mathbf{X}(\mathbf{X}'\hat{\boldsymbol{\Sigma}}^{-1}\mathbf{X})^{-}\mathbf{X}'\hat{\boldsymbol{\Sigma}}^{-1}]\mathbf{Z}\hat{\mathbf{G}}$$

$$= \hat{\sigma}_1^4 \begin{pmatrix} \mathbf{j}_4' & \mathbf{0}' & \mathbf{0}' \\ \mathbf{0}' & \mathbf{j}_4' & \mathbf{0}' \\ \mathbf{0}' & \mathbf{0}' & \mathbf{j}_4' \end{pmatrix} [\hat{\boldsymbol{\Sigma}}^{-1} - \hat{\boldsymbol{\Sigma}}^{-1}\mathbf{j}_{12}(\mathbf{j}_{12}'\hat{\boldsymbol{\Sigma}}^{-1}\mathbf{j}_{12})^{-1}\mathbf{j}_{12}'\hat{\boldsymbol{\Sigma}}^{-1}] \begin{pmatrix} \mathbf{j}_4 & \mathbf{0} & \mathbf{0} \\ \mathbf{0} & \mathbf{j}_4 & \mathbf{0} \\ \mathbf{0} & \mathbf{0} & \mathbf{j}_4 \end{pmatrix}$$

$$= \hat{\sigma}_1^4 \begin{pmatrix} \mathbf{j}_4' & \mathbf{0}' & \mathbf{0}' \\ \mathbf{0}' & \mathbf{j}_4' & \mathbf{0}' \\ \mathbf{0}' & \mathbf{0}' & \mathbf{j}_4' \end{pmatrix} \left[\hat{\boldsymbol{\Sigma}}^{-1} - \frac{12}{\hat{\sigma}^2 + 4\hat{\sigma}_1^2}\mathbf{J}_{12}\right] \begin{pmatrix} \mathbf{j}_4 & \mathbf{0} & \mathbf{0} \\ \mathbf{0} & \mathbf{j}_4 & \mathbf{0} \\ \mathbf{0} & \mathbf{0} & \mathbf{j}_4 \end{pmatrix}$$

$$= \frac{\hat{\sigma}_1^4}{3(\hat{\sigma}^2 + 4\hat{\sigma}_1^2)} \begin{pmatrix} 8 & -4 & -4 \\ -4 & 8 & -4 \\ -4 & -4 & 8 \end{pmatrix}.$$

17.22 $\text{cov}(\mathbf{a}, \mathbf{y}) = \text{cov}(\mathbf{a}, \mathbf{X}\boldsymbol{\beta} + \mathbf{Za} + \boldsymbol{\varepsilon}) = \text{cov}(\mathbf{a}, \mathbf{Za}) = \mathbf{GZ}'$. Hence, $\text{cov}(\mathbf{a}|\mathbf{y}) = \boldsymbol{\Sigma}_{aa} - \boldsymbol{\Sigma}_{ay}\boldsymbol{\Sigma}_{yy}^{-1}\boldsymbol{\Sigma}_{ya} = \mathbf{G} - \mathbf{GZ}'\boldsymbol{\Sigma}^{-1}\mathbf{ZG}$.

17.23 $\text{cov}(\mathbf{a}|\mathbf{y}) = \sigma_1^2(\mathbf{I}_{10} - \sigma_1^2\mathbf{Z}'\boldsymbol{\Sigma}^{-1}\mathbf{Z}) = \sigma_1^2\mathbf{I}_{10} - \frac{2\sigma_1^4}{\sigma^2+2\sigma_1^2}\mathbf{I}_{10}$. Note that the off-diagonal elements are 0's.

17.24 Since $\boldsymbol{\Sigma}^{-1/2}$ is symmetric, let $\boldsymbol{\Sigma}^{-1/2} = \begin{pmatrix} a & b \\ b & a \end{pmatrix}$. Then

$$
\begin{aligned}
\mathbf{I}_{20} - \mathbf{H}_* &= \mathbf{I}_{20} - \boldsymbol{\Sigma}^{-1/2}\mathbf{X}(\mathbf{X}'\boldsymbol{\Sigma}^{-1}\mathbf{X})^{-1}\mathbf{X}'\boldsymbol{\Sigma}^{-1/2} \\
&= \mathbf{I}_{20} - \begin{pmatrix} a & a \\ a & a \\ \vdots & \vdots \\ a & -a \\ a & -a \end{pmatrix} \begin{pmatrix} \frac{1}{20a^2} & 0 \\ 0 & \frac{1}{20a^2} \end{pmatrix} \begin{pmatrix} a & a & \cdots & a & a \\ a & a & \cdots & -a & -a \end{pmatrix} \\
&= \mathbf{I}_{20} - \begin{pmatrix} 0.1\mathbf{J}_{10} & \mathbf{O} \\ \mathbf{O} & 0.1\mathbf{J}_{10} \end{pmatrix}
\end{aligned}
$$

The off-diagonal elements are either 0 or -0.1 (corresponding to correlations of either 0 or -0.11).

Chapter 18

18.1

$$
\begin{aligned}
E(y_i) &= (0)P(y_i = 0) + (1)P(y_i = 1) = 1.p_i = p_i, \\
\text{var}(y_i) &= E[y_i - E(y_i)]^2 \\
&= (0 - p_i)^2P(y_i = 0) + (1 - p_i)^2P(y_i = 1) \\
&= p_i^2(1 - p_i) + (1 - p_i)^2p_i \\
&= p_i(1 - p_i)[p_i + (1 - p_i)].
\end{aligned}
$$

18.2 Let $\theta_i = \beta_0 + \beta_1x_i$. Then (18.7) becomes $p_i = e^{\theta_i}/(1 + e^{\theta_i})$. From this we obtain

$$
\begin{aligned}
1 - p_i &= 1 - \frac{e^{\theta_i}}{1 + e^{\theta_i}} = \frac{1 + e^{\theta_i} - e^{\theta_i}}{1 + e^{\theta_i}} = \frac{1}{1 + e^{\theta_i}}, \\
\frac{p_i}{1 - p_i} &= \frac{e^{\theta_i}}{1 + e^{\theta_i}} \Big/ \frac{1}{1 + e^{\theta_i}} = \frac{e^{\theta_i}(1 + e^{\theta_i})}{1 + e^{\theta_i}} = e^{\theta_i}, \\
\ln\left(\frac{p_i}{1 - p_i}\right) &= \theta_i.
\end{aligned}
$$

18.3

$$
\begin{aligned}
\ln L(\beta_0, \beta_1) &= \ln\left[\prod_{i=1}^{n} p_i^{y_i}(1-p_i)^{1-y_i}\right] \\
&= \sum_{i=1}^{n} [y_i \ln p_i + (1-y_i)\ln(1-p_i)] \\
&= \sum_i y_i[\ln p_i - \ln(1-p_i)] + \sum_i \ln(1-p_i) \\
&= \sum_i y_i \ln\left(\frac{p_i}{1-p_i}\right) + \sum_i \ln(1-p_i).
\end{aligned}
$$

By Problem 17.2, this becomes

$$
\ln L(\beta_0, \beta_1) = \sum_i y_i(\beta_0 + \beta_1 x_i) + \sum_i \ln(1-p_i).
$$

To show that $\ln(1-p_i) = -\ln(1+e^{\beta_0+\beta_1 x_i})$, let $\theta_i = \ln[p_i/(1-p_i)]$. Then

$$
e^{\theta_i} = \frac{p_i}{1-p_i}.
$$

Solve this no obtain $p_i = e^{\theta_i}/(1+e^{\theta_i})$. Then show that $1 - p_i = 1/(1+e^{\theta_i})$ and that $\ln(1-p_i) = -\ln(1+e^{\theta_i}) = -\ln(1+e^{\beta_0+\beta_1 x_i})$.

18.4

$$
\frac{\partial \ln L(\beta_0, \beta_1)}{\partial \beta_0} = \sum_{i=1}^{n} y_i - \sum_{i=1}^{n} \frac{e^{\beta_0+\beta_1 x_i}}{1+e^{\beta_0+\beta_1 x_i}}.
$$

$$
\frac{\partial \ln L(\beta_0, \beta_1)}{\partial \beta_1} = \sum_{i=1}^{n} x_i y_i - \sum_{i=1}^{n} \frac{e^{\beta_0+\beta_1 x_i}}{1+e^{\beta_0+\beta_1 x_i}}.
$$

18.5 $b(\theta_i) = n_i \ln(1-p_i) = -n_i \ln(1+e^{\theta_i})$, as shown in the answer to Problem 17.3.

References

Agresti, A. (1984). *Analysis of Ordinal Categorical Data*. New York: Wiley.

Agresti, A. (1990). *Categorical Data Analysis*. New York: Wiley.

Anderson, E. B. (1991). *The Statistical Analysis of Categorical Data* (2nd ed.). New York: Springer-Verlag.

Anderson, T. W. (1984). *Introduction to Multivariate Statistical Analysis* (2nd ed.). New York: Wiley.

Andrews, D. F. (1974). A robust method for multiple linear regression. *Technometrics 16*, 523–531.

Andrews, D. F. and A. M. Herzberg (1985). *Data*. New York: Springer-Verlag.

Bailey, B. J. R. (1977). Tables of the Bonferroni *t*-statistic. *Journal of the American Statistical Association 72*, 469–479.

Bates, D. M. and D. G. Watts (1988). *Nonlinear Regression and Its Applications*. New York: Wiley.

Beckman, R. J. and R. D. Cook (1983). Outliers (with comments). *Technometrics 25*, 119–163.

Belsley, D. A., E. Kuh, and R. E. Welsch (1980). *Regression Diagnostics: Identifying Data and Sources of Collinearity*. New York: Wiley.

Benjamini, Y. and Y. Hochberg (1995). Controlling the false discovery rate: A practical and powerful approach to multiple testing. *Journal of the Royal Statistical Society, Series B: Methodological 57*, 289–300.

Benjamini, Y. and D. Yekutieli (2001). The control of the false discovery rate in multiple testing under dependency. *The Annals of Statistics 29*(4), 1165–1188.

Benjamini, Y. and D. Yekutieli (2005). False discovery rate-adjusted multiple confidence intervals for selected parameters. *Journal of the American Statistical Association 100*(469), 71–81.

Bingham, C. and S. E. Feinberg (1982). Textbook analysis of covariance—is it correct? *Biometrics 38*, 747–753.

Birch, J. B. (1980). Some convergence properties of iterated reweighted least squares in the location model. *Communications in Statistics B9*(4), 359–369.

Bishop, Y., S. Fienberg, and P. Holland (1975). *Discrete Multivariate Analysis: Theory and Practice*. Cambridge, MA: Massachusetts Institute of Technology Press.

Linear Models in Statistics, Second Edition, by Alvin C. Rencher and G. Bruce Schaalje

Bloomfield, P. (2000). *Fourier Analysis of Time Series: An Introduction*. New York: Wiley.

Bonferroni, C. E. (1936). Il calcolo delle assicurazioni su gruppi di teste. Studii in Onore del Profesor S. O. Carboni. Roma.

Box, G. E. P. and P. V. Youle (1955). The exploration and exploitation of response surfaces: An example of the link between the fitted surface and the basic mechanism of the system. *Biometrics 11*, 287–323.

Broadbent, K. L. (1993). *A Comparison of Six Bonferroni Procedures*. Master's thesis, Department of Statistics, Brigham Young University.

Brown, H. and R. Prescott (1999). *Applied Mixed Models in Medicine*. New York: Wiley & Sons.

Brown, H. and R. Prescott (2006). *Applied Mixed Models in Medicine* (2nd ed.). Hoboken, NJ: Wiley.

Bryce, G. R. (1975). The one-way model. *The American Statistician 29*, 69–70.

Bryce, G. R. (1998). Personal communication.

Bryce, G. R., M. W. Carter, and M. W. Reader (1976). Nonsingular and singular transformations in the fixed model. Annual Meeting of the American Statistical Association, Boston, Aug. 1976.

Bryce, G. R., M. W. Carter, and D. T. Scott (1980a). *Recovery of Estimability in Fixed Models with Missing Cells*. Technical Report SD-022-R, Department of Statistics, Brigham Young University.

Bryce, G. R., D. T. Scott, and M. W. Carter (1980b). Estimation and hypothesis testing in linear models—a reparameterization approach. *Communications in Statistics—Series A, Theory and Methods 9*, 131–150.

Casella, G. and E. I. George (1992). Explaining the Gibbs sampler. *The American Statistician 46*, 167–174.

Chatterjee, S. and A. S. Hadi (1988). *Sensitivity Analysis in Linear Regression*. New York: Wiley.

Christensen, R. (1996). *Plane Answers to Complex Questions: The Theory of Linear Models* (2nd ed.). New York: Springer-Verlag.

Christensen, R. (1997). *Log-Linear Models and Logistic Regression* (2nd ed.). New York: Springer-Verlag.

Cochran, W. G. (1934). The distribution of quadratic forms in a normal system with applications to the analysis of variance. *Proceedings, Cambridge Philosophical Society 30*, 178–191.

Cochran, W. G. (1977). *Sampling Techniques*. New York: Wiley.

Cook, R. D. (1977). Detection of influential observations in linear regression. *Technometrics 19*, 15–18.

Cook, R. D. and S. Weisberg (1982). *Residuals and Influence in Regression*. New York: Chapman & Hall.

Crampton, E. W. and J. W. Hopkins (1934). The use of the method of partial regression in the analysis of comparative feeding trial data. Part II. *J. Nutrition 8*, 329–339.

Daniel, W. W. (1974). *Biostatistics: A Foundation for Analysis in the Health Sciences*. New York: Wiley.

Devlin, S. J., R. Gnanadesikan, and J. R. Kettenring (1975). Robust estimation and outlier detection with correlation coefficients. *Biometrika 62*, 531–546.

Diggle, P., P. Heagerty, K.-Y. Liang, and S. L. Zeger (2002). *Analysis of Longitudinal Data*. Oxford University Press.

Dobson, A. J. (1990). *An Introduction to Generalized Linear Models*. New York: Chapman & Hall.

Draper, N. R. and H. Smith (1981). *Applied Regression Analysis* (2nd ed.). New York: Wiley.

Draper, N. R. and H. Smith (1998). *Applied Regression Analysis*. New York: Wiley.

Driscoll, M. F. (1999). An improved result relating quadratic forms and chi-square distributions. *The American Statistician 53*, 273–275.

Eubank, R. L. and R. L. Eubank (1999). *Nonparametric Regression and Spline Smoothing*. New York: Marcel Dekker.

Evans, M. and T. Swartz (2000). *Approximating Integrals via Monte Carlo and Deterministic Methods*. Oxford University Press.

Ezekiel, M. (1930). *Methods of Correlation Analysis*. New York: Wiley.

Fai, A. H.-T. and P. L. Cornelius (1996). Approximate F-tests of multiple degree of freedom hypotheses in generalized least squares analyses of unbalanced split-plot experiments. *Journal of Statistical Computation and Simulation 54*, 363–378.

Fisher, R. A. (1921). On the probable error of a coefficient of correlation deduced from a small sample. *Metron 1*, 1–32.

Fitzmaurice, G. M., N. M. Laird, and J. H. Ware (2004). *Applied Longitudinal Analysis*. Hoboken, NJ: Wiley.

Flury, B. W. (1989). Understanding partial statistics and redundancy of variables in regression and discriminant analysis. *The American Statistician 43*(1), 27–31.

Fox, J. (1997). *Applied Regresion Analysis, Linear Models, and Related Methods*. Thousand Oaks, CA: SAGE Publications.

Freund, R. J. and P. D. Minton (1979). *Regression Methods: A Tool for Data Analysis*. New York: Marcel Dekker.

Fuller, W. A. and G. E. Battese (1973). Transformations for estimation of linear models with nested-error structure. *Journal of the American Statistical Association 68*, 626–632.

Gallant, A. R. (1975). Nonlinear regression. *The American Statistician 29*, 73–81.

Gelman, A., J. B. Carlin, H. S. Stern, and D. B. Rubin (2004). *Bayesian Data Analysis* (2nd ed.). Chapman & Hall/CRC.

Ghosh, B. K. (1973). Some monotonicity theorems for chi-square. F and t distributions with applications. *Journal of the Royal Statistical Society 35*, 480–492.

Giesbrecht, F. G. and J. C. Burns (1985). Two-stage analysis based on a mixed model: Large-sample asymptotic theory and small-sample simulation results. *Biometrics 41*, 477–486.

Gilks, W. R. E., S. E. Richardson, and D. J. E. Spiegelhalter (1998). *Markov Chain Monte Carlo in Practice*. London: Chapman & Hall.

Gomez, E., G. Schaalje, and G. Fellingham (2005). Performance of the kenward-roger method when the covariance structure is selected using aic and bic. *Communications in Statistics: Simulation and Computation 34*(2), 377–392.

Graybill, F. A. (1954). On quadratic estimates of variance components. *Annals of Mathematical Statistics 25*(2), 367–372.

Graybill, F. A. (1969). *Introduction to Matrices with Applications in Statistics*. Belmont, CA: Wadsworth Publishing Company.

Graybill, F. A. (1976). *Theory and Application of the Linear Model*. North Scituate, MA: Duxbury Press.

Graybill, F. A. and H. K. Iyer (1994). *Regression Analysis: Concepts and Applications*. North Scituate, MA: Duxbury Press.

Graybill, F. A. and A. W. Wortham (1956). A note on uniformly best, unbiased estimators for variance components. *Journal of the American Statistical Association 51*, 266–268.

Gutsell, J. S. (1951). The effect of sulfamerazine on the erythrocyte and hemoglobin content of trout blood. *Biometrics* 7(2), 171–179.

Guttman, I. (1982). *Linear Models: An Introduction*. New York: Wiley.

Hald, A. (1952). *Statistical Theory with Engineering Applications*. New York: Wiley.

Hamilton, D. (1987). Sometimes $R^2 > r^2_{yx1} + r^2_{yx2}$: Correlated variables are not always redundant. *The American Statistician 41*(2), 129–132.

Hampel, F. R. (1974). The influence curve and its role in robust estimation. *Journal of the American Statistical Association 69*, 383–393.

Hartley, H. O. (1956). Programming analysis of variance for general purpose computers. *Biometrics 12*, 110–122.

Harville, D. A. (1997). *Matrix Algebra from a Statistician's Perspective*. New York: Springer-Verlag.

Healy, M. J. R. and M. Westmacott (1969). Missing values in experiments analysed on automatic computers. *Applied Statistics 5*, 203–206.

Helland, I. S. (1987). On the interpretation and use of R^2 in regression analysis. *Biometrics 43*, 61–69.

Henderson, C. R. (1950). Estimation of genetic parameters. *Annals of Mathetmatical Statistics 21*, 309–310.

Henderson, C. R. and A. J. McAllister (1978). The missing subclass problem in two-way fixed models. *Journal of Animal Science 46*, 1125–1137.

Hendrix, L. J. (1967, Aug.). *Auditory Discrimination Differences between Culturally Deprived and Nondeprived Preschool Children*. PhD thesis, Brigham Young University.

Hendrix, L. J., M. W. Carter, and J. Hintze (1978). A comparison of five statistical methods for analyzing pretest-post designs. *Journal of Experimental Education 47*, 96–102.

Hendrix, L. J., M. W. Carter, and D. T. Scott (1982). Covariance analysis with heterogeneity of slopes in fixed models. *Biometrics 38*, 641–650.

Hilbe, J. M. (1994). Generalized linear models. *The American Statistician 48*, 255–265.

Hoaglin, D. C. and R. E. Welsch (1978). The hat matrix in regression and ANOVA. *The American Statistician 32*, 17–22.

Hochberg, Y. (1988). A sharper Bonferroni procedure for multiple tests of significance. *Biometrika 75*, 800–802.

Hocking, R. R. (1976). The analysis and selection of variables in linear regression. *Biometrics 32*, 1–51.

Hocking, R. R. (1985). *The Analysis of Linear Models*. Monterey, CA: Brooks/Cole.

Hocking, R. R. (1996). *Methods and Applications of Linear Models*. New York: Wiley.

Hocking, R. R. (2003). *Methods and Applications of Linear Models* (2nd ed.). New York: Wiley.

Hocking, R. R. and F. M. Speed (1975). A full rank analysis of some linear model problems. *Journal of the American Statistical Association 70*, 706–712.

Hoerl, A. E. and R. W. Kennard (1970). Ridge regression: Biased estimation for nonorthogonal problems. *Technometrics 12*, 55–67.

Hogg, R. V. and A. T. Craig (1995). *Introduction to Mathematical Statistics* (5th ed.). Englewood Cliffs, NJ: Prentice-Hall.

Holland, B. (1991). On the application of three modified Bonferroni procedures to pairwise multiple comparisons in balanced repeated measures designs. *Computational Statistics Quarterly 3*, 219–231.

Holland, B. and M. D. Copenhaver (1987). An improved sequentially rejective Bonferroni test procedure. *Biometrics 43*, 417–423.

Holm, S. (1979). A simple sequentially rejective multiple test procedure. *Scandinavian Journal of Statistics 6*, 65–70.

Hommel, G. (1988). A stagewise rejective multiple test procedure based on a modified Bonferroni test. *Biometrika 75*, 383–386.

Hosmer, D., B. Jovanovic, and S. Lemeshow (1989). Best subsets logistic regression. *Biometrics 45*, 1265–1270.

Hosmer, D. W. and S. Lemeshow (1989). *Applied Logistic Regression*. New York: Wiley.

Huber, P. J. (1973). Robust regression: Asymptotics, conjectures, and Monte Carlo. *Annals of Statistics 1*, 799–821.

Hummel, T. J. and J. Sligo (1971). Empirical comparison of univariate and multivariate analysis of variance procedures. *Psychological Bulletin 76*, 49–57.

Jammalamadaka, S. R. and D. Sengupta (2003). *Linear Models an Integrated Approach*. Singapore: World Scientific Publications.

Jeske, D. R. and D. A. Harville (1988). Prediction-interval procedures and (fixed-effects) confidence-interval procedures for mixed linear models. *Communications in Statistics: Theory and Methods 17*, 1053–1087.

Jørgensen, B. (1993). *The Theory of Linear Models*. New York: Chapman & Hall.

Kackar, R. N. and D. A. Harville (1984). Approximations for standard errors of estimators of fixed and random effects in mixed linear models. *Journal of the American Statistical Association 79*, 853–862.

Kendall, M. G. and A. Stuart (1969). *The Advanced Theory of Statistics* (3rd ed.), Vol. 1. New York: Hafner.

Kenward, M. G. and J. H. Roger (1997). Small sample inference for fixed effects from restricted maximum likelihood. *Biometrics 53*, 983–997.

Keselman, H. J., R. K. Kowalchuk, J. Algina, and R. D. Wolfinger (1999). The analysis of repeated measurements: A comparison of mixed-model Satterthwaite F tests and a nonpooled adjusted degrees of freedom multivariate test. *Communications in Statistics: Theory and Methods* 28, 2967–2999.

Kleinbaum, D. G. (1994). *Logistic Regression*. New York: Springer-Verlag.

Krasker, W. S. and R. Welsch (1982). Efficient bounded-influence regression estimation. *Journal of the American Statistical Association 77*, 595–604.

Kshirsagar, A. M. (1983). *A Course in Linear Models*. New York: Marcel Dekker.

Ku, H. H. and S. Kullback (1974). Loglinear models in contingency table analysis. *The American Statistician 28*, 115–122.

Kutner, M. H., C. J. Nachtsheim, J. Neter, and W. Li (2005). *Applied Linear Statistical Models* (5th ed.). New York: McGraw-Hill/Irwin.

Lehmann, E. L. (1999). *Elements of Large-Sample Theory*. New York: Springer-Verlag.

Lindley, D. V. and A. F. M. Smith (1972). Bayes estimates for the linear model (with discussion). *Journal of the Royal Statistical Society, Series B: Methodological 34*, 1–41.

Lindsey, J. K. (1997). *Applying Generalized Linear Models*. New York: Springer-Verlag.

Little, R. J. A. and D. B. Rubin (2002). *Statistical Analysis with Missing Data*. Hoboken, NJ: Wiley.

Mahalanobis, P. C. (1936). On the generalized distance in statistics. *Proceedings of the National Institute of Science of India 12*, 49–55.

Mahalanobis, P. C. (1964). Professor Ronald Aylmer Fisher. *Biometrics 20*, 238–250.

Marcuse, S. (1949). Optimum allocation and variance components in nested sampling with an application to chemical analysis. *Biometrics 5*(3), 189–206.

McCullagh, P. and J. A. Nelder (1989). *Generalized Linear Models* (2nd ed.). New York: Chapman & Hall.

McCulloch, C. E. and S. R. Searle (2001). *Generalized, Linear, and Mixed Models*. New York: Wiley.

Mclean, R. A., W. L. Sanders, and W. W. Stroup (1991). A unifed approach to mixed linear models. *American Statistician 45*, 54–64.

Mendenhall, W. and T. Sincich (1996). *A Second Course in Statistics: Regression Analysis*. Englewood Cliffs, NJ: Prentice-Hall.

Milliken, G. A. and D. E. Johnson (1984). *Analysis of Messy Data,* Vol. 1: *Designed Experiments*. New York: Van Nostrand-Reinhold.

Montgomery, D. C. and E. A. Peck (1992). *Introduction to Linear Regresion Analysis* (2nd ed.). New York: Wiley.

Morrison, D. F. (1983). *Applied Linear Statistical Methods*. Englewood Cliffs, NJ: Prentice-Hall.

Mosteller, F. and J. W. Tukey (1977). *Data Analysis and Regression*. Reading, MA: Addison-Wesley.

Muller, K. E. and M. C. Mok (1997). The distribution of Cook's *D* statistic. *Communications in Statistics: Theory and Methods 26*, 525–546.

Myers, R. H. (1990). *Classical and Modern Regression with Applications* (2nd ed.). Boston: Duxbury Press.

Myers, R. H. and J. S. Milton (1991). *A First Course in the Theory of Linear Statistical Models*. Boston: PWS-Kent.

Nelder, J. A. (1974). Letter to the editor. *Journal of the Royal Statistical Society, Series C 23*, 232.

Nelder, J. A. and P. W. Lane (1995). The computer analysis of factorial experiments: In memoriam—Frank Yates. *The American Statistician 49*, 382–385.

Nelder, J. A. and R. W. M. Wedderburn (1972). Generalized linear models. *Journal of the Royal Statistical Society, Series A 135*, 370–384.

Ogden, R. T. (1997). *Essential Wavelets for Statistical Applications and Data Analysis*. Birkhauser.

Ostle, B. and L. C. Malone (1988). *Statistics in Research: Basic Concepts and Techniques for Research Workers* (4th ed.). Ames: Iowa State University Press.

Ostle, B. and R. W. Mensing (1975). *Statistics in Research* (3rd ed.). Ames: Iowa State University Press.

Patterson, H. D. and R. Thompson (1971). Recovery of inter-block information when block sizes are unequal. *Biometrika 58*, 545–554.

Pawitan, Y. (2001). *In All Likelihood: Statistical Modelling and Inference Using Likelihood.* Oxford University Press.

Pearson, E. S., R. L. Plackett, and G. A. Barnard (1990). *Student: A Statistical Biography of William Sealy Gossett*. New York: Oxford University Press.

Plackett, R. L. (1981). *The Analysis of Categorical Data* (2nd ed.). London: Griffin.

Rao, C. R. (1965). *Linear Statistical Inference and Its Applications*. New York: Wiley.

Rao, P. S. R. S. (1997). *Variance Components Estimation*. London: Chapman & Hall.

Ratkowsky, D. A. (1983). *Nonlinear Regression Modelling: A Unified Approach*. New York: Marcel Dekker.

Ratkowsky, D. A. (1990). *Handbook of Nonlinear Regression Models*. New York: Marcel Dekker.

Read, T. R. C. and N. A. C. Cressie (1988). *Goodness-of-Fit Statistics for Discrete Multivariate Data*. New York: Springer-Verlag.

Reader, M. W. (1973). *The Analysis of Covariance with a Single Linear Covariate Having Heterogeneous Slopes*. Master's thesis, Department of Statistics, Brigham Young University.

Rencher, A. C. (1993). The contribution of individual variables to Hotelling's T^2, Wilks' Λ and R^2. *Biometrics 49*, 217–225.

Rencher, A. C. (1995). *Methods of Multivariate Analysis*. New York: Wiley.

Rencher, A. C. (1998). *Multivariate Statistical Inference and Applications*. New York: Wiley.

Rencher, A. C. (2002). *Multivariate Statistical Inference and Applications*. Hoboken, NJ: Wiley.

Rencher, A. C. and D. T. Scott (1990). Assessing the contribution of individual variables following rejection of a multivariate hypothesis. *Communications in Statistics—Series B, Simulation and Computation 19*, 535–553.

Rom, D. M. (1990). A sequentially rejective test procedure based on a modified Bonferroni inequality. *Biometrika 77*, 663–665.

Ross, S. M. (2006). *Introduction to Probability Models* (9th ed.). San Diego, CA: Academic Press.

Royston, J. P. (1983). Some techniques for assessing multivariate normality based on the Shapiro-Wilk W. *Applied Statistics 32*, 121–133.

Ryan, T. P. (1997). *Modern Regression Methods*. New York: Wiley.

Santner, T. J. and D. E. Duffy (1989). *The Statistical Analysis of Discrete Data*. New York: Springer-Verlag.

Satterthwaite, F. E. (1941). Synthesis of variances. *Psychometrika 6*, 309–316.

Saville, D. J. (1990). Multiple comparison procedures: The practical solution (C/R: 91V45 p165–168). *The American Statistician 44*, 174–180.

Schaalje, G. B., J. B. McBride, and G. W. Fellingham (2002). Adequacy of approximations to distributions of test statistics in complex mixed linear models. *Journal of Agricultural, Biological, and Environmental Statistics 7*(4), 512–524.

Scheffé, H. (1953). A method of judging all contrasts in the analysis of variance. *Biometrika 40*, 87–104.

Scheffé, H. (1959). *The Analysis of Variance*. New York: Wiley.

Schott, J. R. (1997). *Matrix Analysis for Statistics*. New York: Wiley.

Schwarz, G. (1978). Estimating the dimension of a model. *Annals of Statistics 6*, 461–464.

Searle, S. R. (1971). *Linear Models*. New York: Wiley.

Searle, S. R. (1977). *Analysis of Variance of Unbalanced Data from 3-Way and Higher-Order Classifications*. Technical Report BU-606-M, Cornell University, Biometrics Units.

Searle, S. R. (1982). *Matrix Algebra Useful for Statistics*. New York: Wiley.

Searle, S. R., G. Casella, and C. E. McCulloch (1992). *Variance Components*. New York: Wiley.

Searle, S. R., F. M. Speed, and H. V. Henderson (1981). Some computational and model equivalencies in analysis of variance of unequal-subclass-numbers data. *The American Statistician 35*, 16–33.

Seber, G. A. F. (1977). *Linear Regression Analysis*. New York: Wiley.

Seber, G. A. F. and A. J. Lee (2003). *Linear Regression Analysis* (2nd ed.). Hoboken, NJ: Wiley.

Seber, G. A. F. and C. J. Wild (1989). *Nonlinear Regression*. New York: Wiley.

Sen, A. and M. Srivastava (1990). *Regression Analaysis: Theory, Methods, and Applications*. New York: Springer-Verlag.

Shaffer, J. P. (1986). Modified sequentially rejective multiple test procedures. *Journal of the American Statistical Association 81*, 826–831.

Silverman, B. W. (1999). *Density Estimation for Statistics and Data Analysis*. London: Chapman & Hall.

Simes, R. J. (1986). An improved Bonferroni procedure for multiple tests of significance. *Biometrika 73*, 751–754.

Snedecor, G. W. (1948). Answer to query. *Biometrics 4*(2), 132–134.

Snedecor, G. W. and W. G. Cochran (1967). *Statistical Methods* (6th ed.). Ames: Iowa State University Press.

Snee, R. D. (1977). Validation of regression models: Methods and examples. *Technometrics 19*, 415–428.

Speed, F. M. (1969). *A New Approach to the Analysis of Linear Models*. Technical report, National Aeronautics and Space Administration, Houston, TX; a NASA Technical memo, NASA TM X-58030.

Speed, F. M., R. R. Hocking, and O. P. Hackney (1978). Methods of analysis of linear models with unbalanced data. *Journal of the American Statistical Association 73*, 105–112.

Spiegelhalter, D. J., N. G. Best, B. P. Carlin, and A. van der Linde (2002). Bayesian measures of model complexity and fit (pkg: P583-639). *Journal of the Royal Statistical Society, Series B: Statistical Methodology 64*(4), 583–616.

Stapleton, J. H. (1995). *Linear Statistical Models*. New York: Wiley.

Stigler, S. M. (2000). The problematic unity of biometrics. *Biometrics 56*(3), 653–658.

Stokes, M. E., C. S. Davis, and G. G. Koch (1995). *Categorical Data Analysis Using the SAS System*. Cary, NC: SAS Institute.

Theil, H. and C. Chung (1988). Information-theoretic measures of fit for univariate and multivariate linear regressions. *The American Statistician 42*, 249–252.

Tiku, M. L. (1967). Tables of the power of the *F*-test. *Journal of the American Statistical Association 62*, 525–539.

Turner, D. L. (1990). An easy way to tell what you are testing in analysis of variance. *Communications in Statistics—Series A, Theory and Methods 19*, 4807–4832.

Urquhart, N. S., D. L. Weeks, and C. R. Henderson (1973). Estimation associated with linear models: A revisitation. *Communications in Statistics 1*, 303–330.

Verbeke, G. and G. Molenberghs (2000). *Linear Mixed Models for Longitudinal Data*. Springer-Verlag.

Wald, A. (1943). Tests of statistical hypotheses concerning several parameters when the number of observations is large. *Transactions of the American Mathematical Society 54*, 426–483.

Wang, S. G. and S. C. Chow (1994). *Advanced Linear Models: Theory and Applications*. New York: Marcel Dekker.

Weisberg, S. (1985). *Applied Linear Regression*. New York: Wiley.

Welsch, R. E. (1975). Confidence regions for robust regression. *Paper presented at Annual Meeting of the American Statistical Association*, Washington, DC.

Winer, B. J. (1971). *Statistical Principles in Experimental Design* (2nd ed.). New York: McGraw-Hill.

Working, H. and H. Hotelling (1929). Application of the theory of error to the interpretation of trends. *Journal of the American Statistical Association, Suppl. (Proceedings) 24*, 73–85.

Yates, F. (1934). The analysis of multiple classifications with unequal numbers in the different classes. *Journal of the American Statistical Association 29*, 52–66.

Index

Linear Models in Statistics, Second Edition, by Alvin C. Rencher and G. Bruce Schaalje